TUNNELLING AND GROUND CONDITIONS

PROCEEDINGS OF THE INTERNATIONAL CONGRESS ON TUNNELLING AND GROUND CONDITIONS / CAIRO / EGYPT / 3-7 APRIL 1994

Tunnelling and Ground Conditions

Edited by

M.E.ABDEL SALAM

National Authority for Tunnels & Egyptian Tunnelling Society, Cairo, Egypt

A.A.BALKEMA / ROTTERDAM / BROOKFIELD / 1994

The texts of the various papers in this volume were set individually by typists under the supervision of each of the authors concerned.

Published by
A.A.Balkema, P.O.Box 1675, 3000 BR Rotterdam, Netherlands
A.A.Balkema Publishers, Old Post Road, Brookfield, VT 05036, USA

ISBN 90 5410 363 9

Printed in the Netherlands

Tunnelling and Ground Conditions, Abdel Salam (ed.) © 1994 Balkema, Rotterdam, ISBN 90 5410 363 9

Table of contents

6 *Repair and management of underground projects*

7 *General tunnelling topics*

Tunnelling and Ground Conditions, Abdel Salam (ed.) © 1994 Balkema, Rotterdam, ISBN 90 5410 363 9

Foreword

Since the beginning of human history, man, by his own initiative, has carried out underground works due to his needs for shelter, security and comfort. During the earliest days of civilization Ancient Egyptians developed the use of underground structures for religious purposes. Their discovered tombs with their entrance tunnels still exist as evidence and, perhaps those tombs still to be discovered will reveal even more advanced underground features.

Human need for underground space has continued from those early days up to present times with many varied demands for many different activities. This underground space offers a stable and safe place for these many activities. Those countries which have suitable geological conditions for underground structures should take advantage of this natural resource which should be utilized fully in the same way as any other resource.

The key to the functional use and economic benefit of underground development will always depend, in the first place, on the availability of suitable ground conditions followed by the technological knowledge and its economic and technical characteristics.

This may be the main reason, when Cairo was chosen, by the 18th General Assembly of the International Tunnelling Association (ITA) in April 1991 in London, as the venue for its 1994 Congress, we chose the title of 'Tunnelling and Ground Conditions' as the main theme of the Cairo Congress to illustrate the challenges which face the construction of tunnels and other underground structures from the various ground and hydrological conditions likely to be encountered. This challenge also comes from the presence of water which is also related to the ground conditions.

The contents of these proceedings show that the scope of this Congress 'Tunnelling and Ground Conditions' gives a wide range of studies for a variety of underground conditions ranging from hard rock to plastic clay with or without the presence of water. These proceedings also record the state of the art in tunnelling and underground works for the choice of construction methods, management, financing and risks. Fascinating case studies from projects from all over the world illustrate all this and we were proud to receive such excellent papers and we are grateful to all the authors.

I appreciate the work carried out by the Scientific Committee who really contributed an enormous effort in revising the abstracts and papers for inclusion in these proceedings and I offer them my sincere thanks.

Many thanks also go to the Organizing Committee for the preparation of all the activities associated with this Congress and to the sponsors and all those who offered us their assistance in many ways.

Finally, it is for the participants and delegates to this Congress to evaluate and judge for themselves the results of these efforts.

M. E. Abdel Salam
Executive Chairman

Tunnelling and Ground Conditions, Abdel Salam (ed.) © 1994 Balkema, Rotterdam, ISBN 90 5410 363 9

Préface

Depuis la nuit des temps, l'homme a, de lui-même, entrepris des travaux souterrains pour satisfaire ses besoins en habitat, sécurité et confort. Durant les premiers temps de la civilisation, les anciens Egyptiens ont développé l'utilisation des structures souterraines à des fins religieuses. La découverte de leurs tombes avec leurs accès en tunnels en est toujours la preuve existante, et peut être d'autres tombes encore à découvrir nous révèleront des caractéristiques souterraines plus développées.

Les besoins de l'homme en espace souterrain ont continué, depuis ces premiers temps jusqu'à nos jours, avec des demandes variées pour de nombreuses activités. Cet espace souterrain offre en effet un endroit stable et sûr pour ces différentes activités. Les pays ayant des conditions géologiques propices aux structures souterraines doivent exploiter cette ressource au même titre que toute autre.

La clé pour l'utilisation fonctionnelle et le profit économique du développement souterrain dépendront toujours en premier lieu de la disponibilité des conditions géologiques favorables suivies par la connaissance technologique et ses caractéristiques économiques et techniques.

Ceci serait peut être la raison essentielle pour laquelle Le Caire a été choisi, par la 18 ième Assemblée Générale de l'ITA tenue à Londres en Avril 1991, pour accueillir son Congrès de 1994. Le titre 'Tunnelling and Ground Conditions' a été choisi comme thème principal du Congrès du Caire pour illustrer les challenges affrontés lors de la construction de tunnels ou d'autres structures souterraines, dûs aux différentes conditions géologiques et hydrologiques qui pourraient être rencontrées. Ce challenge vient également de la présence de l'eau qui dépend aussi des conditions du sous-sol.

Le contenu des communications montre que la portée de ce Congrès 'Tunnelling and Ground Conditions' offre un champs d'études très étendu, des conditions du sous-sol allant de la roche dure à l'argile plastique, avec ou sans le présence d'eau. Ces communications traitent également des règles de l'art relatives aux travaux souterrains concernant le choix des méthodes de construction, de gestion, de financement et des risques. Des exemples passionnants de projets dans le monde entier illustrent ces propos et nous sommes fiers de recevoir des communications d'une telle qualité dont nous remercions tous les auteurs.

J'apprécie le travail réalisé par le Comité Scientific qui a vraiement consenti un énorme effort dans la révision de toutes les communications en vue de les intégrer dans ce manuel et je leur adresse mes sincères remerciements.

Je remercie également le Comité Organisateur pour la préparation des activités connexes au Congrès, et tous les répondants qui nous ont offert leurs aides de différentes manières.

Finalement, il revient aux participants et délégués de ce Congrès d'évaluer et de juger par eux-mêmes le résultat de ces efforts.

M.E. Abdel Salam
Président

Tunnelling and Ground Conditions, Abdel Salam (ed.) © 1994 Balkema, Rotterdam, ISBN 90 5410 363 9

مقدمـــة

منذ بداية التاريخ -قام الانسان بمبادرة أوحت بها سليقته بتنفيذ منشآت تحت الارض استخدامها فى توفير احتياجاته التى تتطلبها معيشته وحياته من ايواء أو أمن أو راحة -وفى بداية عصور الحضارة المسجلة تاريخيا ، قام قدماء المصريون بتطوير المنشآت تحت سطح الارض واستخدموها فى تحقيق متطلبات أخرى تتمشى مع عقيدتهم الدينية · ولعل المقابر والانفاق الموصلة الى تلك المقابر والتى تم اكتشافها تعطينا خير دليل على ذلك · بل ومن المرجح أيضا ان المقابر التى لم يتم اكتشافها بعد سوف توضح اعمالا أكثر تطورا فى اقامة المنشآت تحت سطح الارض ·

ولقد استمرت حاجة الانسان الى المنشآت تحت سطح الارض تتزايد منذ تلك العصور وحتى عصرنا الحديث مع تنوع فى المتطلبات والانشطة وزيادة فى حجمها · ذلك ان المنشآت تحت سطح الارض توفر المكان الآمن والمستقر للكثير من الانشطة · ويجب على تلك الدول التى حباها الله بخواص جيلوجية جيدة وتربة صالحة لتنفيذ الاعمال تحت سطح الارض ان تدرك ان ذلك موردا ضمن مواردها الطبيعية وتنظر اليه بنفس النظرة التى تنظر بها الى الموارد الطبيعية الاخرى التى تمتلكها ·

والعامل الرئيسى المؤثر فى الاستفادة من الفراغ المتاح تحت سطح الارض استفادة اقتصادية للانشطة المختلفة - ظل وسيظل دائما هو خواص التربة وطبيعة تكوينها الجيلوجى ويأتى ذلك فى المقام الأول وتليه الموضوعات الفنية الخاصة لاستيعاب والالمام بالتكنولوجيا الحديثة وأسلوب الاستخدام الاقتصادى والامثل لها ·

ولقد كان ذلك هو السبب الرئيسى لاختيارنا موضوع " **تأثير النواحى الجيوتكنيكية على الأنفاق** " عندما اختيرت القاهرة - بواسطة الجمعية العمومية للاتحاد الدولى للانفاق فى دورة انعقادها الثامنة عشرة فى لندن فى ابريل ١٩٩١ لتكون مكانا لانعقاد المؤتمر السنوى للاتحاد عام ١٩٩٤ · ان اختيارنا ذلك الموضوع ليكون مادة البحث فى المؤتمر كان بغرض ان يوضح التحديات والمخاطر المحتمل ان تواجهها صناعة الانفاق والمنشآت تحت الارض طبقا لما يعترضها من خصائص هيدروجيلوجية للانواع المختلفة من التربة · وهذه التحديات قد تأتى أيضا من تواجد المياة الجوفية أو عدم تواجدها ·

والمحتويات الى يضمها مجلد أبحاث المؤتمر توضح ان اختيارنا لمادة البحث هذه اعطانا مجالا واسعا للحصول على دراسات متنوعة للانواع المتعددة من التربة ابتداء من التربة الرخوة الى التربة الصخرية سواء فى وجود أو عدم

وجود المياة الجوفية - وتسجل تلك الابحاث ما وصلت اليه التكنولوجيا الحديثة من وسائل فى تنفيذ مشروعات الانفاق أو المنشآت تحتسطح الارض لاختيار أسلوب الانشاء أو الادارة أو التمويل ٠

وكلها أبحاث ممتازة ومثيرة للاهتمام وصلتنا من جميع أنحاء العالم ٠ ونحن فخورون لتلقينا هذه الابحاث وفى نفس الوقت نشعر بالامتنان والشكر لمؤلفيها ٠

كما أعرب عن امتنانى وشكرى للجنة العلمية التى ساهمت بمجهود وافر سواء فى مراجعة ملخصات الابحاث فى البداية ثم مراجعة وتنقيح تلك الابحاث عند ورودها فى شكلها النهائى ٠

وأتقدم أيضا بخالص الشكر والعرفان للجنة التنظيمية للمؤتمر التى قامت بالاعداد والتجهيز لكافة الانشطة وبذلت فى سبيل ذلك مجهودا كبيرا كما أشكر أيضا وزارة النقل المصرية وكافة الهيئات الاخرى التى قدمت لنا مساعدتها بطرق مختلفة لانجاح هذا المؤتمر ٠

وفى النهاية فان الامر متروك لكل المشاركين وأعضاء الوفود فى هذا المؤتمر ليحكموا بأنفسهم ويقيموا نتائج تلك الجهود ٠

مهندس / محمد الحسينى عبد السلام

الرئيس التنفيذى للمؤتمر

Tunnelling and Ground Conditions, Abdel Salam (ed.) © 1994 Balkema, Rotterdam, ISBN 90 5410 363 9

Organization

Papers presented at the International Congress 'Tunnelling and Ground Conditions' organized by the Egyptian Ministry of Transport, The National Authority for Tunnels and the Egyptian Tunnelling Society.

CONGRESS HONORARY CHAIRMAN
Eng. S. Metwalli, Minister of Transport, Communications and Civil Aviation

CONGRESS EXECUTIVE CHAIRMAN
Eng. M.E. Abdel Salam, NAT Chairman & Chairman of Egyptian Tunnelling Society

SCIENTIFIC COMMITTEE
Prof. A. Moharram (Egypt), Chairman
Prof. E. Broch (Norway)
Prof. H. Duddeck (Germany)
Prof. A. Glerum (Netherlands)
Prof. A. El Ramly (Egypt)
Prof. A. Hosny (Egypt)
Prof. A. Abou Eleid (Egypt)
Prof. F. El Nahhas (Egypt)

TECHNICAL SECRETARIAT
Eng. Sh. Elbedaiwy

ORGANIZING COMMITTEE
Dr Y. Khattab, Chairman
Eng. A. Madkour
Dr A. Hassan
Eng. M. Mahgoub
Eng. Sh. Elbedaiwy
Mr I. Elbialy
Eng. A. El Salamawy
Eng. H. El Sharkawy
Eng. A. Ghanem
Mr H. Khidr
Mr H. Affifi

GENERAL SECRETARIAT
Mr H. Khidr
Mr A. Attwa

SPONSORS
Ministry of Transport
National Authority for Tunnels
Egyptian Tunnelling Society

1 Recent case histories of tunnels in different ground conditions

Tunnelling and Ground Conditions, Abdel Salam (ed.) © 1994 Balkema, Rotterdam, ISBN 90 5410 363 9

Avise tunnel revisited

A. Da Vià
Torno, Milan, Italy

A. Mahtab, G. Carrieri & P. Grasso
Geodata SpA, Turin, Italy

S. Pelizza
Politecnico of Turin, Italy

ABSTRACT: A comparison between the choice of "a pilot bore plus enlargement and a full face bore" using TBM was discussed previously by the authors for the case of the 3 km-long, twin-bore Avise tunnel. The Avise tunnel is now revisited for information on the performance of the TBM reaming machine from 4.5 m pilot hole to 11.4 m final (right) bore compared with the drill and blast excavation of part of the left bore and drill and blast enlargement of the remainder of the left bore.

1. INTRODUCTION

1.1 Project Description

The Avise Tunnel is one of the 10 tunnels along the section of the Italian motorway A5 that connects the town of Aosta (100 km N of Turin) to the French border through the Monte Bianco Tunnel. The twin-bore Avise tunnel is 3 km long and lies under a cover of 40 to 330m. The two bores are separated by 10m, each has a diameter of 11.4 and will house 2 traffic lanes.

The construction of Avise tunnel was begun in late 1989 and is expected to be completed in early 1994. The owner of the motorway is RAV of Rome; the Contractor is a consortium of Torno (Milan) and Fioroni (Perugia), assisted by Geodata as engineering consultant, Murer Italia as tunnel boring contractor, and Consonda as consolidation contractor.

The tunnel provides an interesting opportunity to compare the following situations (see Fig. 1.1) which are discussed in Sec. 2:

(1) pilot bore plus TBM enlargement (TBE) vs. drill and blast excavation (DBExc) of the entire section (without a pilot bore),
(2) TBE vs. drill and blast enlargement (DBEnl) after a pilot bore, and
(3) the expected vs. actual performance of the TBE.

The third situation, the performance of TBE, is a follow-up on the previous article about Avise tunnel (Da Viá et al., 1992) and has provided the title of this paper.

1.2 Geologic Setting

The Avise tunnel traverses the S. Bernardo formation, mainly consituted by intensely fractured and faulted gneiss and micaschist with intercalations of chlorite and phyllite. Except for the portal sections, which are constituted by detritus or moraines under a cover upto 40m, the 2.7 km long section through rock is under a cover of 180-330m.

The principal source of the geological information is the 4.5m diameter pilot bore. This information was coupled with some seismic measurements in the field and mechanical property tests in the laboratory. For example, the geomechanical properties of the gneiss were obtained from one series of tests as follows:
Density: 2.7 kg/cm^3
Unconfined compressive strength C_o: 80MPa
Brazilian tensile strength: 9 MPa
Youngs modulus: 10-17 GPa
Comp. wave velocity: 4452 m/s

A longitudinal section through the tunnel is given in Fig. 1.2. The most significant and, by experience, the most critical geologic feature encountered during excavation was the Vertosan fault. The fault is located about half way through the tunnel and is associated with a 60m wide fault zone. The negative influence of the Vertosan fault on rock quality and advance rate (by TBM or drill and blast excavation) is discussed in Sec.2.

The RMR and classification system of Bieniawski (1979) is used for comparing the advance rate between the various excavation schemes. For convenience,

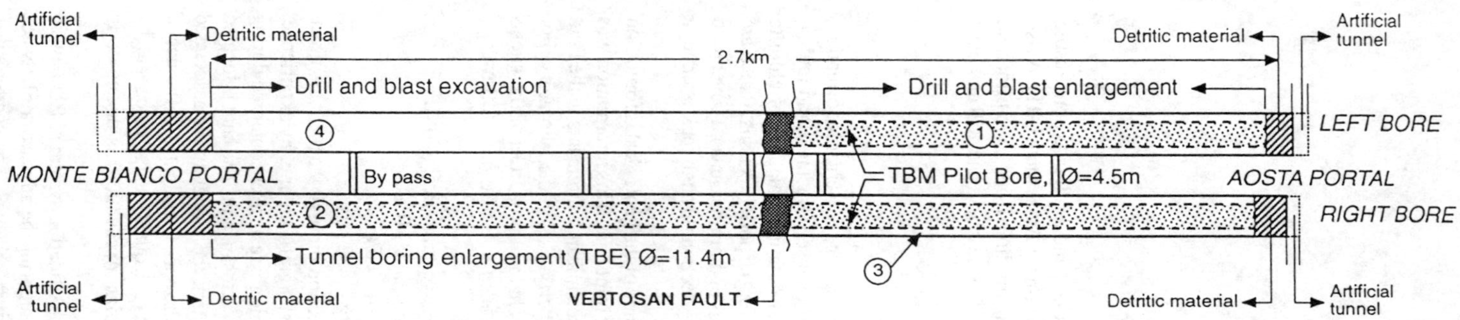

Fig. 1.1 Plan view of 2 bores of Avise Tunnel showin the excavation scheme:
1 Left pilot bore, 1.5km, from Aosta (AO) toward Monte Bianco (M.B.); 2 Right pilot bore, 2.7km AO toward M.B.; 3 TBE, 2.7km M.B. to AO; 4 D&B from M.B. to AO

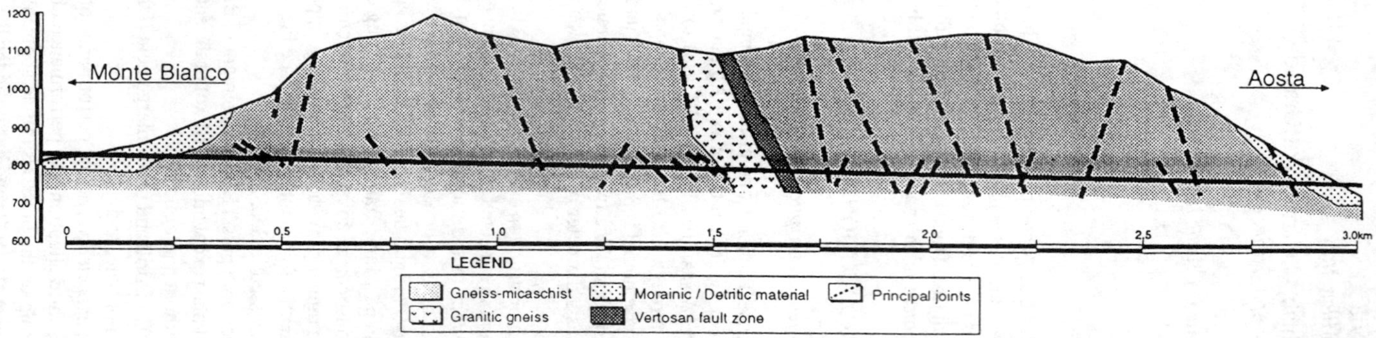

Fig. 1.2 Longitudinal geologic section along Avise tunnel

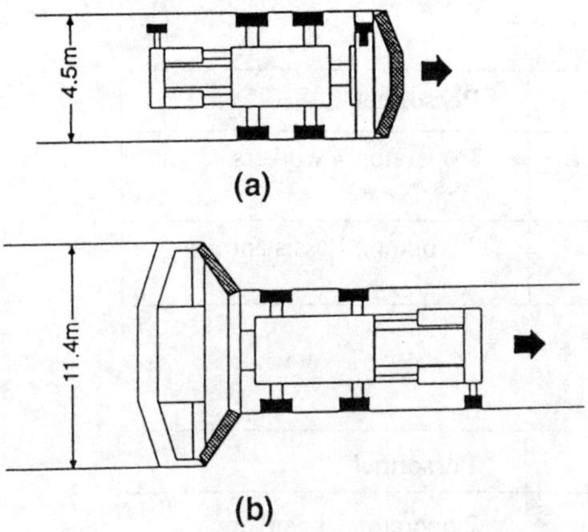

Fig. 2.1 Excavation by TBM
(a) Pilot bore
(b) Enlargement by reaming machine (TBE)

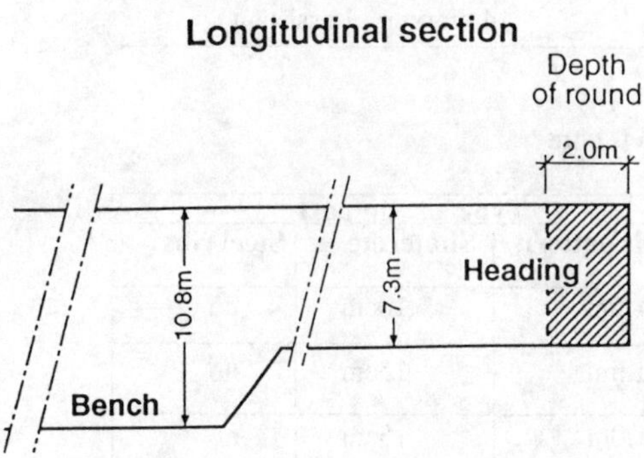

2.2 Sequence by Drill & Blast Excavation

Bieniawski's classes IV and V were further subdivided into classes IVa, IVb and Va, Vb. It is noted that other classification schemes, such as the Q-system of Barton et al. (1973), may be equally applicable.

2. EXCAVATION METHODOLOGY

The originally planned scheme was to make a pilot hole of diameter 4.5m with a TBM at the center of the right and left bores (11.4m diameter) of Avise tunnel. The pilot holes were to be subsequently enlarged by a reaming machine. However, due to the problems associated with the Vertosan fault, the methodology actually used evolved differently for the two bores: the TBM pilot bore and TBM enlargement (TBE) was used for the right bore (Fig. 2.1) whereas drill-and-blast excavation (DBExc) was used for the (Monte Bianco) half of the left bore (Fig. 2.2) and TBM pilot bore plus drill-and-blast enlargement (DBEnl) was used for the other half (adjacent to Aosta portal). A cross-section of the two bores is shown in Fig. 2.3.

2.1 Right Bore

2.1.1 Pilot bore

A 4.5m diameter TBM (WIRTH TBS III 450) was used to bore the pilot hole in the center of the right bore proceeding from Aosta (AO) portal toward Monte Bianco (M.B.) portal. A major portion of the rock (70%) was given class III while the rest of the rock horizon was associated with the subdivided class intervals IVa (12.4%), IVb (16.0%), and V (1.6%). Depending on the number of shifts, the advance rate generally ranged between 5 and 45m/day, see Fig. 2.4.

When required, the pilot hole was reinforced with 1.5m long, 22m diameter, fiber-glass bolts. The bolted section comprised 266m out of a total length of 2676m, with an average frequency of 1.2 bolts/m. Another 300m of the pilot bore in very poor rock was supported by steel ribs with a frequency of 3.7 steel arches/m. The excavation cycle for the pilot hole is given in Table 2.1.

2.1.2 TBM enlargement of pilot hole (TBE)

The right bore of the tunnel was enlarged from the 4.5m diameter pilot hole to the 11.4m diameter full bore for the entire (2,676m) rock mass horizon. The enlargement was made using a WIRTH TBS III machine. Generally, the entire length of the bore is supported with:

- resin-grouted fiber-glass bolts, 4.5m long, 25mm diameter, spaced at 1.8m
- a 150mmx1500mm, 5mm diameter wire mesh bolted to the periphery of the bore; and
- a steel-fiber reinforced, 10cm thick layer of shotcrete.

The rock mass was divided into 3 support classes, or sections: A, B, and C, with slightly different pre-reinforcement design used for each section. In some portions of the bore, both steel bolts and steel arches were required for support. The final lining was provided by 30cm thick concrete, that was cast in place. The TBE cycle is shown in Table 2.2.

2.2 Left Bore

The left bore of the tunnel had been explored by a 4.5m diameter pilot hole for about half of its length (see Fig. 1.1) from AO portal to the Vertosan fault. This section

Table 2.1 Right bore, TBM pilot hole excavation

Work phase	Equipment used	Personnel
Excavation and support	1 wirth TBS III 4.50m diameter, drilling machine	1 operator, 4 workers
Mucking	2 electric locomotives with 5 wagons	3 workers, 1 assistant

Table 2.2 Right bore, TBE cycle

Work phase	Equipment used	Personnel
Excavation and support	1 wirth TBE IV 450/1140 H	2 operators, 8 workers
Install service lines in invert and complete	1 roller	1 operator, 5 workers
Mucking	2 electric locomotives with 5 wagons	2 locomotive operators 4 workers, 1 assistant

Table 2.3 Type of support used in the DBExc section of left bore

Rock class (Bieniawski)	% of excavated length	Type of support		
		Bolt pattern	Shotcrete	Steel ribs
III	54.3	1.5m x 1.5m	10cm	no
IVa	29.1	1.0m x 1.5m	15cm	no
IVb	13.1	1.0m x 1.0m	15cm	no
Va	3.5	-	20cm	yes

Table 2.4 Left bore, drill & blast excavation cycle for the heading

Work phase	Equipment used	Duration (h)	Personnel
Drill & blast	1 jumbo drill (3 arms)	2	2 operators, 4 workers 1 assistant
Mucking	1 shovel, 4 dump trucks	2	1 operator, 4 drivers
Scaling	1 hammer	1	1 operator, 1 assistent
Shotcrete	1 shotcrete machine 2 or 3 cement trucks	1	1 operator, 2 or 3 drivers 1 assistant
Support	1 jumbo drill	2	2 operators 4 workers 1 assistant

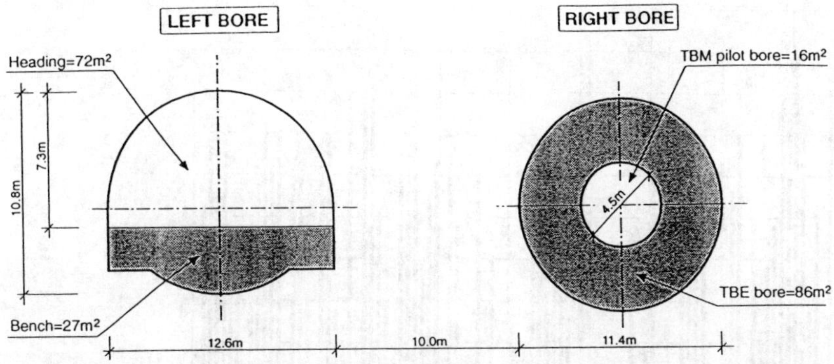

2.3 Cross-section of Avise tunnel showing the drill-and-blast excavation (Left Bore) and TBE (Right Bore)

was enlarged using drill and blast excavation (DBEnl) advancing both from the Vertosan fault and AO portal. The remaining section was excavated by drill and blast (DBExc) proceeding from M.B. Portal to Vertosan fault.

2.2.1. Drill & blast excavation (DBExc)

Drill and blast excavation (DBExc) was used to excavate 1343m of the left bore at an average advance rate of 5.4m/day. Controlled excavation was made to avoid any damage to the rock around the right bore.

Two-step excavation was used (with heading and benching, see Fig. 2.2), limiting the advance per blasting round to 2.0m for controlling the peak particle velocity. The support provided was a function of the rock class as shown in Table 2.3. The bolts are 24mm diameter, 4.5m, injection-grouted steel or friction bolts. The final lining of cast concrete varied in thickness from 50cm to 60cm. The DBExc cycle is given in Table 2.4.

2.2.2 Pilot hole and drill & blast enlargement (DBEnl)

As shown in Fig. 1.1, the 4.5m diameter pilot hole was made at the center of the left bore from AO portal to the Vertosan fault for about 1340m. The rock mass classification and support requirements were similar to those observed in the parallel pilot hole in the right bore. The DBEnl was made from two faces (Fig. 1.1) with support provided in a manner similar to that given in Table 2.3.

3. COMPARATIVE ANALYSIS

3.1 TBE versus DBExc

As mentioned earlier, the approximately 1340m section of the tunnel between the M.B. portal and Vertosan fault was excavated usign TBE (in the right bore) and DBExc (in the left bore). The daily advance rates for the two operations are given in Figs. 3.1 and 3.2, respectively. The advance as function of rock class is given in Table 3.1 for this (M.B.) section of the tunnel.

The TBE advance from M.B. portal to Vertosan fault (1336m) was made in 181 days using 1 shift for 62 days, 2 shifts for 20 days, and 3 shifts for 99 days. The average advance of 7.38m/d was influenced by organizational factors (machine operation, services, personnel). Almost 50% of the total period was used for bringing the job site operations to an optimum level. Note, however, that the average rate achieved here falls within the range of experience of large TBM's (Kovari et al., 1993).

The DBExc advance in the 1343m of the M.B. section reached an average of 5.4 m/d over a period of 249 days. The lower advance (compared to TBE) reflects the effect of controlled blasting to minimize the damage to the already excavated right bore, that was separated by a 10m-wide pillar. It can be assumed that, in the absence of controlled blasting, the advance rate will increase, but with the probable increase of overbreak of the tunnel section by a % that is significantly more than the present 3.5 to 4%.

3.2 TBE versus DBEnl

The comparison between the enlargement of the pilot hole by the reaming machine (TBE) and by the conventional means (DBEnl) refers to the AO-section of the tunnel, from Vertosan fault to AO portal.

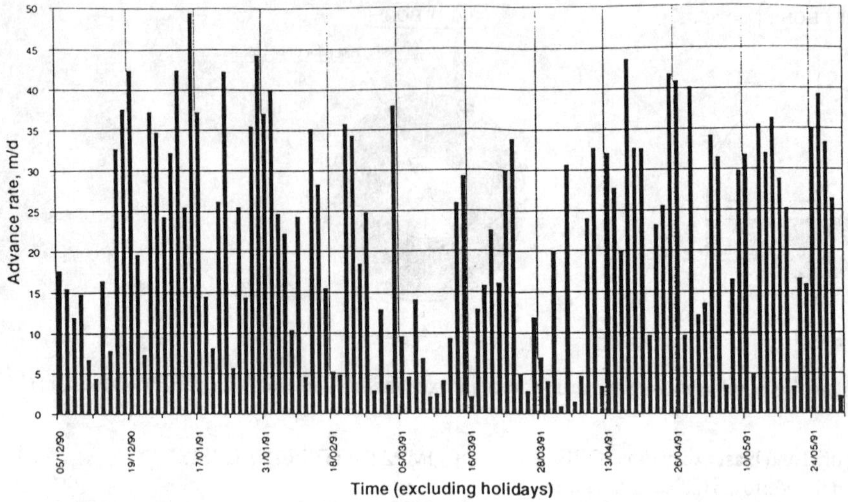

Fig. 2.4 Daily advance of the 4.5m diameter TBM in the pilot hole for the entire length of the right bore

Table 3.1 Rock class and advance rates in the section of Avise tunnel from M.B. portal to Vertosan fault using DBExc (left bore) and TBE (right bore).

Rock Class	DBExc, Left Bore		TBE, right bore	
	% of excav. length	Advance rate m/d	% of Excav. length	Advance rate m/shift
III	54.3	5.9	85.7	3.9
IVa	29.1	5.4	7.0	3.2
IVb	13.1	3.8	7.3	1.6
Va	3.5	2.7	0	-

Table 3.2 Reduction in advance rate as a function of decreasing rock quality, assuming 100% advance for class III.

Rock Class	Reduction % in advance rate	
	DBEnl	TBE
III	0	0
IVa	9	19
IVb	30	62
Va	55	88

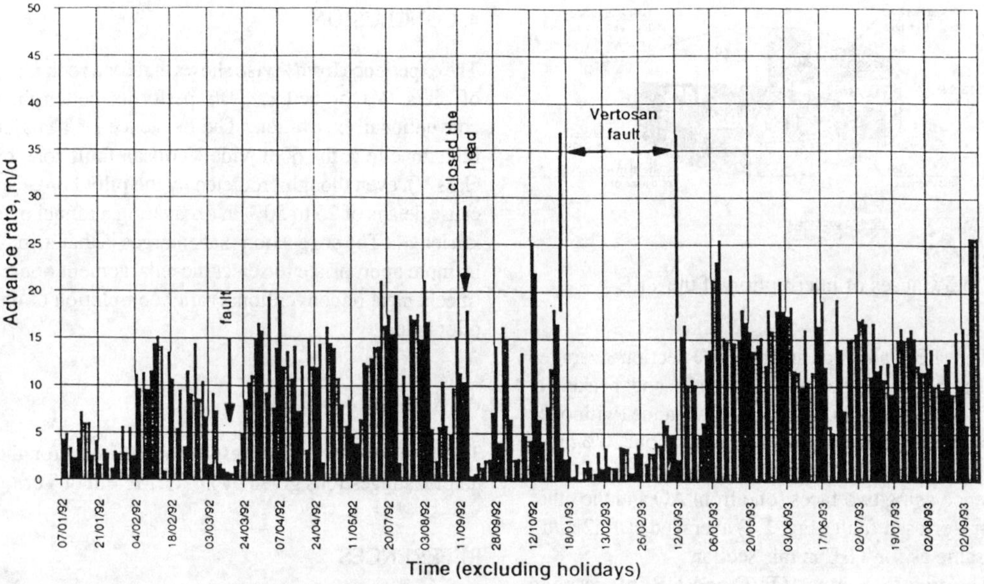

Fig. 3.1 Daily advance of TBE, for the entire length of the right bore

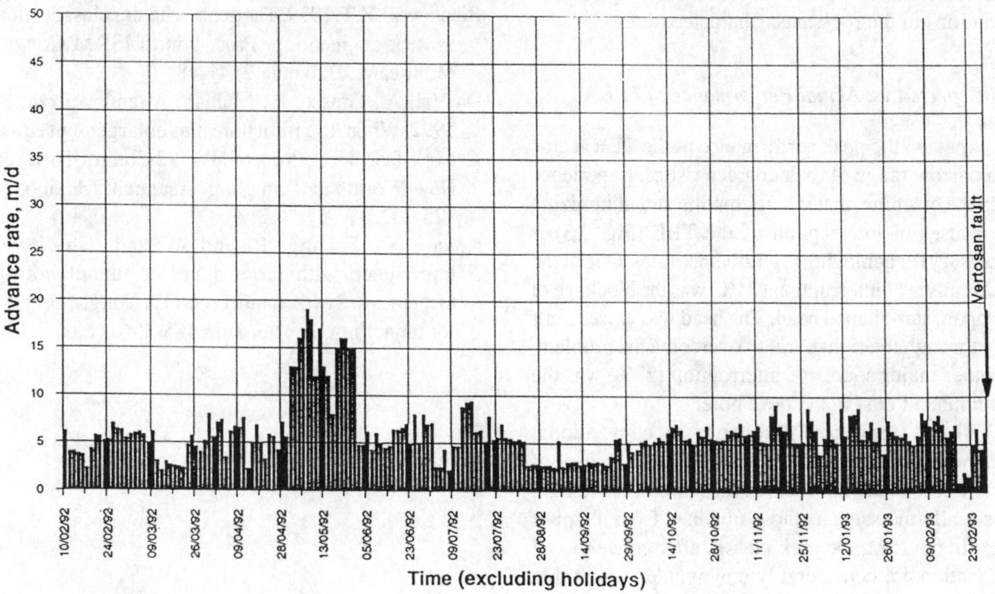

Fig. 3.2 Daily advance of the D&B, heading excavation in left bore from M.B. Portal to Vertosan fault.

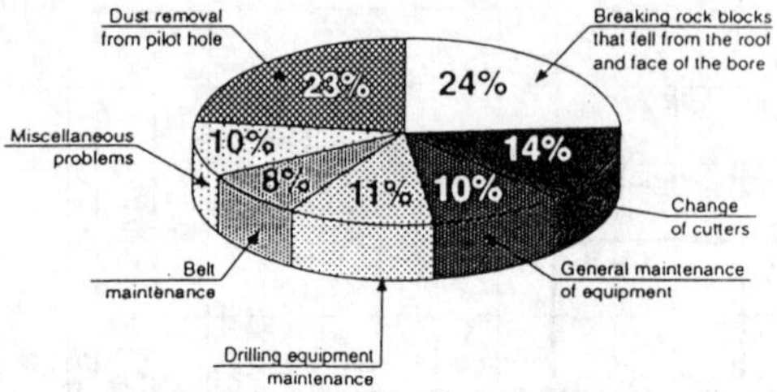

Fig. 3.3 Causes of interruption of the TBE

The TBE advance in the AO-section averaged 13.2m/d, over a period of 85 days and peaked at 25.50m/d (Fig. 3.1). The DBEnl advance (with once face from AO) averaged at 7.7 m/d - about 30% more than the advance of DBExc. However, the DBEnl advance using two faces (one from AO and the other from Vertosan fault - Fig. 1.1) averaged at 13.2 m/d - the same as for TBE in this section.

The two techniques (TBE and DBEnl) respond differently to the rock mass characteristics. The production data (Table 3.2) demonstrate that conventional excavation responds better to marked variations in the rock mass characteristics.

3.3 Expected vs. Actual Performance of TBE

As expected the peak performance of the TBE accurs in a narrow range of rock characteristics. Experience with the machine, that is, anticipating and minimizing the causes of interruption of the TBE (Fig. 3.3) is necessary for achieving high advance rates. One of the main causes of interruption (24%) was the blockage of the open, star-shaped head. The head was closed, half way through the right bore, to overcome the problem. Another major source of interruption (23%) was the accumulated dust in the pilot bore.

TBE is significantly superior to conventional excavation with regard to preserving the quality of rock. For istance, the rock classes after TBE remain essentially the same as those obtained from the pilot hole. In contrast, the rock classes after conventional excavation are considerably downgraded (see Table 2.4).

TBE requires a stable pilot hole and good anchoring capacity of the ground. In poor ground (rock class V), the grippers may not be properly anchored. As a result, the pistons, instead of pulling the head forward, can pull the ground into the pilot hole

4. CONCLUSION

The experience from Avise shows that for a rock mass of class IVa or better, TBE performs better than conventional excavation. The existence of a major fault zone (e.g. the 60m wide Vertosan fault zone of class V), even though predicted by the pilot bore, can cause delays of 25 to 50% in excavating a tunnel of 3 km length. The suggestion that emerges is that it may be more appropriate to select the enlargement option (mechanical or conventional) after completion of the pilot hole.

AKNOWLEDGMENT

The authors wish to express their gratitude for the helpful suggestions given by Dr. G. Metelli of Torno.

REFERENCES

Barton, N., Lien, R. and Lunde, J. 1974. Engineering classification of rock masses for the design of tunnel support. Rock Mechanics 6: 183-236.

Bieniawski, Z.T. 1979. The geomechanics classfication in rock engineering. Proc. Fourth ISRM Congr., Montreaux, Balkema 2: 41-48.

Da Viá, A., Grasso, P., Mahtab, A. and Pelizza, S. 1992. When do a pilot bore plus enlargement equal a full-face bore? Proc. of ITA '92 Congr., Towards New Worlds in Tunneling, Acapuco, Mexico, 1: 125-132.

Kovari, K., Fechtig, R. and Amstad, Ch. 1993. Experience with large diameter tunnel boring machines in Switzerland. Proc. ITA Congr., Options for tunnelling, Amsterdam: 485-496.

Tunnelling and Ground Conditions, Abdel Salam (ed.) © 1994 Balkema, Rotterdam, ISBN 90 5410 363 9

Some geotechnical aspects of shield tunnelling in the Cairo area

Fathalla El-Nahhas
Ain Shams University, Faculty of Engineering, Abbassia, Cairo, Egypt

ABSTRACT : The variety of tunnelling activities within the Greater Cairo area during the last decade have provided an excellent opportunity to compile factual data on quantitative performance of different shield technologies which have been used for the first time under Cairo geotechnical conditions.

This paper describes the geotechnical constrains identified during tunnelling under Cairo using compressed air technology, bentonite slurry machines and earth pressure ballanced TBMs. In-situ measurements compiled during construction of shield tunnels using these different technologies were utilized to present practical evaluation and guidlines for future use of similar systems within the Cairo area or other regions having similar geological conditions

1 INTRODUCTION

Proposals for the rehabilitation program of the Greater Cairo wastewater infrastructure and the new network of its mass transportation system included construction of tens of kilometers of soft ground tunnels to be completed during the 1980s and 1990s. All these tunnels are located under significant head of groundwater of more than 10 meters, and through different ground conditions including compact to dense sand, soft to very stiff cohesive soils, and highly jointed weak to moderately strong limestone interbedded with layers of hard silty clay.

Most of the relatively new buildings in Cairo have reinforced concrete skeletons built on either shallow or pile foundations. Majority of older structures are wall bearing type having up to five stories. Both types of buildings exist very close to the proposed alignment of the new tunnels. Hence, control of ground movement during tunnelling has become an essential element in the design and construction of these tunnels.

Different types of shield machines have been used during the last 10 years for constructing the first phase of these tunnels which represent the spain and branch tunnels of the Greater Cairo Wastewater Project. The employed techniques include, open face shields with compressed air, bentonite slurry shields, and earth pressure balanced (EPB) machines. The final choice of the specific shield method for each tunnel depended mainly on the size of the tunnel, condition of existing buildings within the tunnel effective zone as well as the ground conditions along the tunnel route.

Based on the updated available field

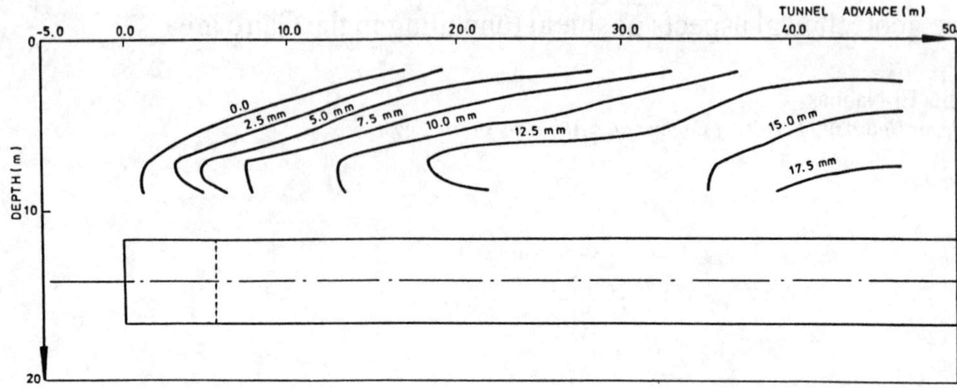

Figure 1 : Contour map of ground subsidence above compressed air tunnel

measurements compiled during construction of these tunnels, this paper proposes practical geotechnical assessment which can be used as guidelines for geotechnical design and construction of future similar tunnels.

2 COMPRESSED AIR TUNNELLING

Several open face shields equipped with backhoes were used to excavate some of the smaller diameter tunnels of up to 5.15 m. Most of those tunnels were excavated through dense sand layers containing pockets and/or layers of clay and silt. Stabilization of tunnel face and control of groundwater flow into the tunnel during excavation were achieved using compressed air of up to 22 psi (150 kPa) within the whole tunnel. Geotechical performance of these tunnels based on field measurements compiled at instrumented test sections is documented by El-Nahhas et al (1990) and Shalaby (1990).

One of these open face shields was also used to construct a 500-m section of one of those tunnels through jointed, weathered, weak to moderately strong limestone interbedded with layers of hard calcareous silty clay. This required the use of drill-and-blast technique through the open face shield combined with compressed air (13 psi - 90 kPa) to control water flow into the tunnel face.

Tunnelling under Cairo using the compressed air technique was successful in minimizing the ground loss . A typical contour map of measured subsidence above one of these tunnels is shown on Figure 1. It is interesting to note that only very small settlement occurred at the tunnel face and around the shield. Final maximum settlement at ground surface after removing compressed air was in the order of 15 to 25 mm for a 5.15-m diameter tunnels. Only one significant case of blow-out had occurred during construction.

However, it is evident that compressed air tunnelling results in a severe rising of groundwater table within a large area surrounding the tunnel which can cause considereable flooding of basements with significant damage or at least some inconveniences.

The reported average rate of advance for the 5.15 m diameter compressed air tunnel was about 7 meters per day when advancing through soil and working three 8-hour shifts. This rate of advance reduced to less than 2 meters per day when the drill-and-blast procedure through the limestone formation was necessary.

3 TUNNELLING USING BENTONITE SLURRY SHIELDS

The majority of executed shield tunnelling in Cairo (13.5 km) was constructed using five bentonite slurry machines having two different designs. Excavated diameters were either 6.1 or 5.15 m driven entirely through alluvial layers of compact to dense sand and soft to very stiff clay under 12 to 14 meters of groundwater. These tunnels formed a main portion of the spine sewer of the east bank section of the Greater Cairo Wastewater Project. More details concerning construction procedure and settlement measurements of these tunnels are given by El-Nahhas (1991 & 1993).

Typical maximum settlements at ground surface above these tunnels were in the order of 20 to 65 mm. Field measurements confirmed that most of this settlement developed at and behind the tail of the slurry shields (Figure 2). Larger settlement and excessive ground losses occurred at a number of locations. In general, control of ground loss was adequate keeping in mind that almost all of these tunnels were advanced under relatively wide streets.

Design of the cutterhead of the smaller bentonite slurry machines had to be modified to allow advancing through sand containing boulders (of up to 1 m size) and cobbles (of up to 200 mm size). Their compressive strength ranged between 19 to 500 MPa. These boulders and cobbles were unexpectedly encountered during sinking one of the access shafts for the spine tunnel. Data complied from large diameter boreholes excavated specially in the area (using piling machines) confirmed the intersection of this feature with the tunnel cross-section along a stretch of about 325-m long as shown on Figure 3.

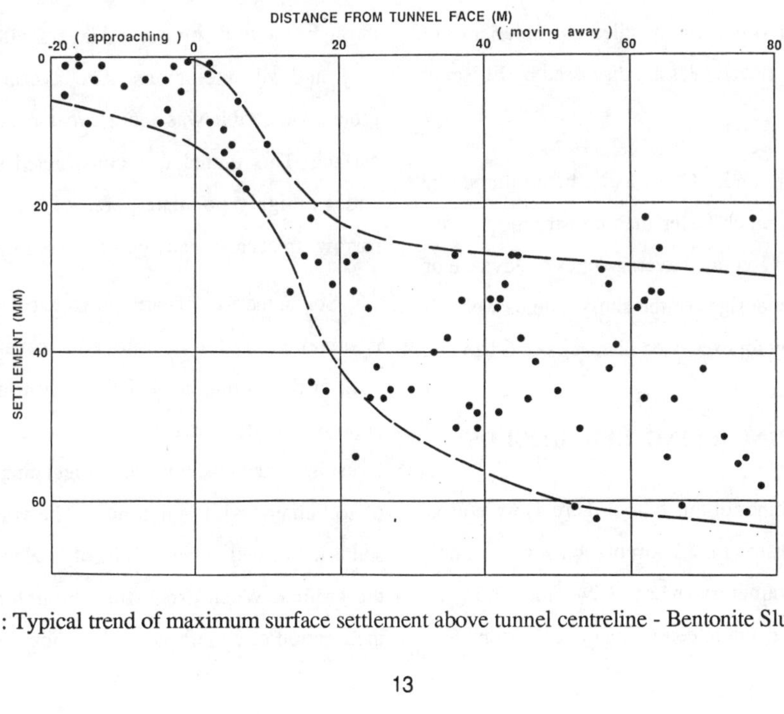

Figure 2 : Typical trend of maximum surface settlement above tunnel centreline - Bentonite Slurry Shield.

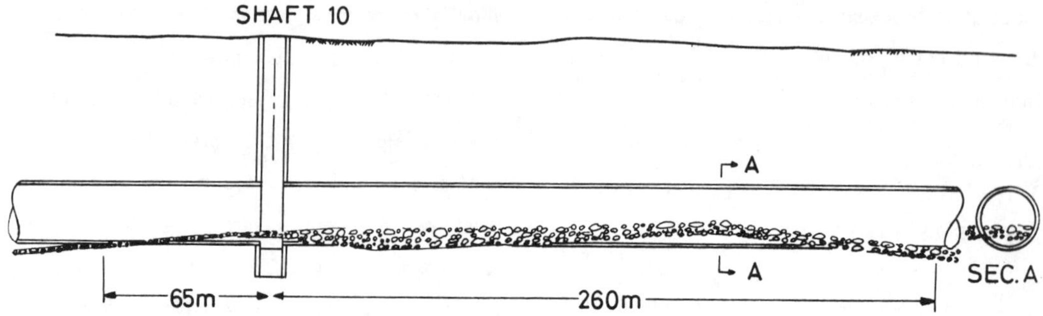

Figure 3 : Boulders and cobbles encountered at the route of slurry shields - Greater Cairo Wastewater Project

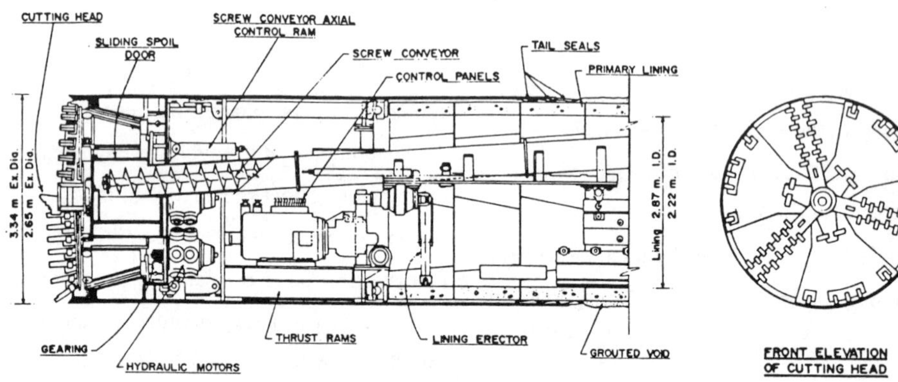

Figure 4 : Configuration of Earth Pressure Ballanced (EPB) machine (Lovat Tunnel Equipment)

The geological origin and detailed characteristics of these boulders and cobbles are discussed by El-Ramli (1992).

Tunnelling under Cairo using bentonite slurry machines was much faster than construction under compressed air. Reported average rates of advance of the two differet designs of the slurry shields were 12 m and 24 m/day for excavated diameters of 6.1 m

4 TUNNELLING USING EPB SHIELDS

Two machines of this type (Figure 4) were used in Cairo to construct a 6.2 km branch sewer tunnel. Excavated diameters were 3.24 and 2.54 m advancing at a depth to centreline of 12 to 16 meters through fine to medium sand layers with occasional clay and silt inclusions. The average depth of groundwater table was about 2 meters below ground surface. This tunnel was constructed successfully under a highly populated area of Cairo with very narrow streets and many weak and old structures.

Soil at the face of earth pressure balanced shields is supported by the pressurized excavated material itself at the cutting head. This pressure is generated ,against in-situ horizontal pressures, by forward thrusting of the machine while regulating the amount of soil entering into the tunnel. The volume of this soil is kept equal to the volume of shield advance into the ground. When excavating through dense sand, the rotation of cutterhead will require a huge torque capacity of the machine particularly for relatively large diameters unless a lubricating slurry is injected into the cutter chamber. On the other hand, excavated soil using the EBP shields is transported through the tunnel in large quantities on mine cars which is much faster and simpler than the hydraulic transportation utilized in the slurry shields.

The reported rates of advance of these machines in Cairo tunnels compared favourably with those achieved by the larger bentonite slurry machines (Flint, 1992). Field measurements showed that the typical maximum settlements at the ground surface were in the order of 20 to 40 mm. Few larger settlements of up to 90 mm were recorded at some locations.

5 ASSESSMENT AND DISCUSSION

Trial of different types of shield tunnelling through Cairo ground conditions provided factual data for evaluating the advantages and limitations of each machine as pointed out in the previous sections. The employed three types of shields were outstanding in controlling the ground loss ahead and around the machines. However, it is evident from the field measurements that most of the soil deformations took place behind the machines tail. For all these tunnels, grouting around the precast concrete lining was performed several rings back from the shield through holes in the segments. The injected slurry contained cement, bentonite and limestone powder grouted at pressure in the order of about 200 kPa to volume of about 150 % of theoretical volume.

It is apparent that in order to attain better control of ground deformations, particularly for tunnels with larger diameters, a more sophisticated grouting system should be utilized. Recent generation of shields have been built with grouting tubes through the tail seal in order to allow simultaneous grouting with shield advance eliminating development of any gap behind the precast lining (Babendererde, 1991, Fukuchi, 1991 and Eisenstein & Ezzeldine, 1992).

On the other hand, the advantages of slurry shields and EBP shields have been combined in newly developed machines (Robbins, 1992).

6 CONCLUSIONS

Geotechnical evaluation of the performance of different shield methods employed for construction of tunnels under Cairo showed the advantages and limitations of each type. The field measurements confirm that better control of ground deformation associated with future tunnels, particularly of larger diameters, will require utilization of simultaneous grouting with shield advance directly behind the machine tail.

REFERENCES

Babendererde, S. 1991. Tunnelling machines in soft ground - a comparison of slurry and EBP shield systems. Tunnelling and Underground Space Technology, Vol. 6, No. 2, pp. 169-174.

Eisenstein, Z. and Ezzeldine, O. 1992. Ground control by shields with positive soil support. In Proc. of International Symposium on Current Experiences in Tunnelling, Cairo, pp. 75-90.

El-Nahhas, F. 1991. Ground settlement above urban tunnels constructed using bentonite slurry machines. In Proc. of an International Symposium on Tunnelling in Congested Cities, Cairo, pp. 61-74.

El-Nahhas, F. 1993. Tunnelling under Cairo using bentonite slurry machines. In Proc. of International Congress on Options for Tunnelling, Amsterdam, pp. 351-361.

El-Nahhas, F., El-Kadi, F. and Shalaby, A.G. 1990. Field measurements during construction of a compressed air tunnel in Cairo. In Proc. of International Congress on Tunnels and Underground Works Today and Future, pp. 259-266.

El-Ramli, A.H. 1992. Geology and geotechnics in some areas in Egypt. In Proc. of International Symposium on Current Experiences in Tunnelling, Cairo, pp. 93-117.

Flint, G.R. 1992. Boulac spine sewers for the Greater Cairo Wastewater Project. In Proc. of International Symposium on Current Experiences in Tunnelling, Cairo, pp. 299-325.

Fukuchi, G. 1991. The present and future of machanized tunnel works in soft ground. Tunnelling and Underground Space Technology, Vol. 6, No.2, pp. 175-183.

Robbins, R.J. 1992. Machine excavation of large diameter tunnels in rock and mixed ground below the water table. In Proc. of International Symposium on Current Experiences in Tunnelling, Cairo, pp. 213-223.

Shalaby, A.G. 1990. Behaviour of tunnels in some Egyptian soils. Ph.D. Thesis, Ain Shams University, Cairo, Egypt.

Tunnelling and Ground Conditions, Abdel Salam (ed.) © 1994 Balkema, Rotterdam, ISBN 90 5410 363 9

Connection between two tunnels under a busy crossing in a big city

Yutaka Horibe, Hazime Yagi, Seizi Okayasu & Takashi Ukawa
NTT Corporation, Tokyo, Japan

ABSTRACT: It grows harder and harder to execute excavation in the center of cities because the traffic volume has remarkably increased, and too many facilities for traffic, communication,energy and waste disposal are buried under the road. In addition,shortage of land and uprising of land price due to overcrowded urban areas has made it difficult to proceed with such construction. And therefore, it is getting harder to find a small amount of site space for a shaft for use in shielding. For that reason, each new turned and another existing one are connected underground without using a shaft to be excavated from the road surface. NTT has been successful in connecting two tunnels under a busy crossing on an artery in the vicinity of important buildings in the center.

1 Introduction

Recent construction projects in urban areas are limited in scope and duration of surface work due to heavy road traffic and/or conflict with underground utilities which are laid densely. In particular, for construction work near existing facilities, an underground tunnel excavated by the shield machine is increasingly used to avoid adverse effects.

Yet, the skyrocketed land price and shortage of suitable land in urban areas has made it difficult to secure a site to excavate a vertical shaft for the shield machine. Instead, the method to join a new tunnel with an existing tunnel underground has been increasingly used.

The project has successfully joined a new tunnel with a trunk utility tunnel beneath an intersection with heavy traffic along one of major arterial highways in Tokyo.

This paper reports stress fluctuation occurred at the opening of the existing segments in the process of the underground joining operation, and the physical effects of approaching and reaching of a shield machine to the existing segments, on the basis of field data collected.

2 Outline of Tunnel Joining Work

In the project, a shield machine was advanced from an existing shaft and advanced approximately 30m to excavate a tunnel which joins an existing utility tunnel

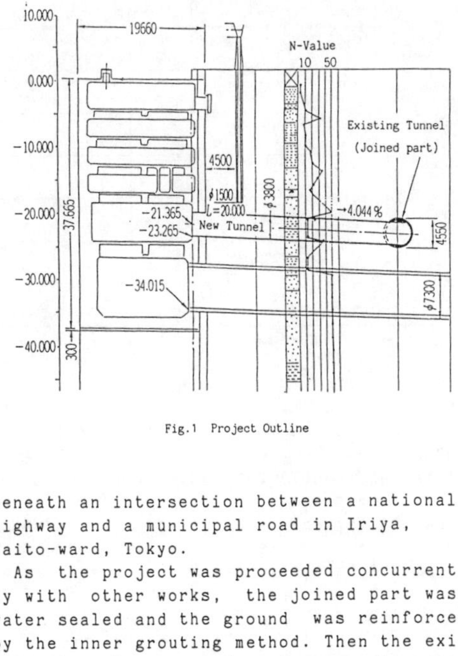

Fig.1 Project Outline

beneath an intersection between a national highway and a municipal road in Iriya, Taito-ward, Tokyo.

As the project was proceeded concurrently with other works, the joined part was water sealed and the ground was reinforced by the inner grouting method. Then the existing tunnel was reinforced, segments were cut out, and an opening was reinforced with mortar. Finally, the shield machine was advanced for joining. The project site and the outline of construction are shown Figures 1.

Construction method: Earth pressure type shield machine

Segment: (New shield machine) Outer diam-

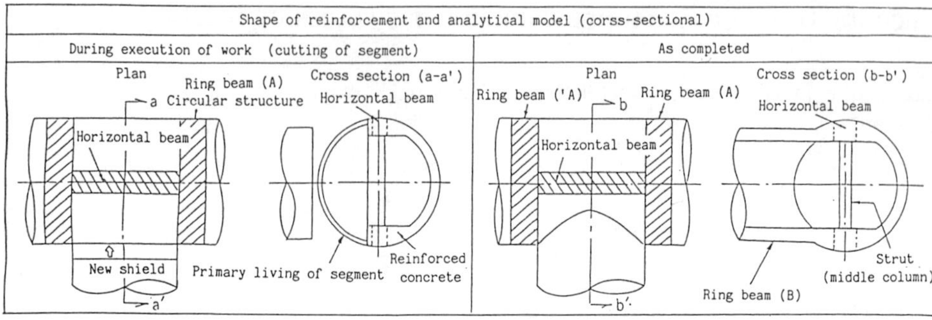

Fig.2 Reinforcement Structure

Table 1 Geological Structure

Period	Stratum name	Nature of the soil	Depth (G.L-m)	Thickness (m)	N Value (N)
This world	———	Fill up soil	0 ↓ 2.4	2.4	2
Alluvial strata	Yurakucho	Fine sand Fine sand (includ silt)	↓ 7.8	5.4	2-33 (15)
Diluvial formation	Shield upper level -22.0m	Silt	↓ 14.7	9.3	7-22 (15)
		Medium sand Fine sand	↓ 21.7	7.0	25-50up (37)
	Tokyo	Silt	↓ 25.7	4.0	11,12
	-26.0m Shield Lower level	Fine sand	↓ 27.9	2.2	30
		Silt	↓ 30.8	2.9	13
	Tokyo gravel	Pebble	↓ 33.1	2.3	50up
	Edogawa	Fine sand	↓ 43.9	10.8	50up
		Hard silt	↓ 45.9	2.0	50up
		Fine sand	↓ 48.4up	2.5up	50up

——— Shield position

eter —— 3,800mm
Steel SM members (main girder 125 x 14mm, 900mm wide)
(Existing tunnel) Outer diameter
—— 4,550mm
Steel SS members (main girder 150 x 22mm, 900mm wide)
Depth: 22.5m on average

3 General Geological Features

The project site is located in the western margin of alluvial lowland, called the Tokyo lowland, around which a burial upper wave-cut plateau, called the Asakusa plateau, is developed. Major strata include the Yurakucho stratum, the Tokyo stratum, the Tokyo gravel stratum, and the Edogawa stratum. Alluvial strata ended in the Yurakucho stratum, around 8m deep, relatively shallow in the alluvial lowland.

In and around the underground joining point is hard silt stratum showing the N value of 11 - 12 within the Tokyo stratum, with sand included occasionally.

Pore-water pressure is 1.9kgf/cm² at around the center of the shield machine.

4 Design

4-1 Design Outline

As shown in Fig.2, the underground joining part consists of segmental rings, ring beams on both sides (A), and upper and lower horizontal beams, and an interior pillar of the segmental ring section.

The underground joining part is designed on the basis of structural calculation:

(1) The lining is designed as a composite structure to overlay primary lining (segments) and secondary lining (reinforced concrete).

(2) Design earth pressure for the segmental section is assumed to be "loose earth pressure" for temporary work (before joining, short-term allowable stress) and "total earth pressure" for permanent structure (after completion, long-term allowable stress), with static water pressure.

(3) Earth pressure in a longitudinal direction of the existing tunnel is calculated as the state immediately before the joining of the shield machine with the existing tunnel. (model without earth pressure on the shield side)

(4) The effect of the jack's propulsion force when the shield machine approaches the existing tunnel is not considered.

(5) The effect of chemical injection is neglected because it proceeds in the construction sequence.

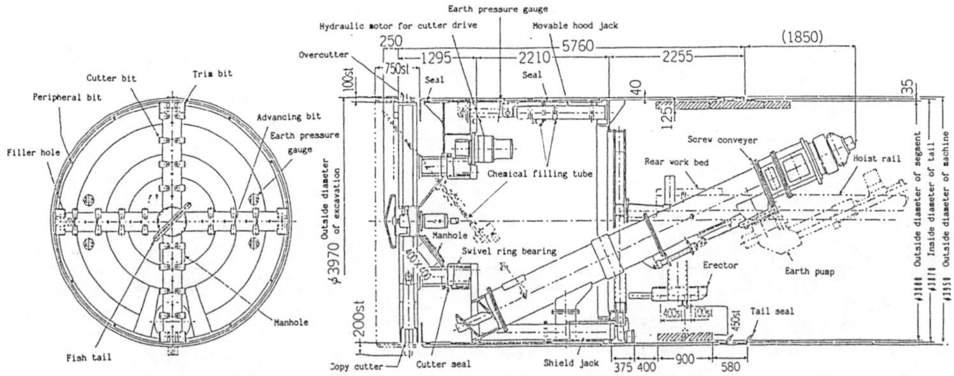

Fig.3 Enclosed, Slurry Type Shield Tunnel Boring Machine

4-2 Outline of Shield Machine

Considering that the shield machine advances in the deep part of the ground subject to high water pressure, near a major structure (Hibiya line of Tokyo Metropolitan Subway), and the excavated ground contains a sand stratum with a small uniformity coefficient, the earth pressure type shield machine has been selected.

Fig.3 shows the shield machine using hydraulic cutter drive. In addition, an extrusion type movable hood is used to prevent earth and ground water from entering at the time of underground joining.

5. Auxiliary Work Method

The auxiliary work was carried out to secure the safe and effective underground joining work.

Since the project site is located beneath the intersection with heavy traffic and is surrounded by underground structures, the inner grouting method was used to improve the ground from the existing tunnel. The scope of ground improvement by using the inner grouting method is shown in Fig.5.

6. Construction Sequence

The work flow is shown in Fig.4 and the construction schedule in Table 2.

The grouting work was conducted at both ends of the project, the start side and the existing tunnel side, due to relatively a short construction period required for interface with other works.

7. Measurement Plan

Field measurement was take to secure saf-

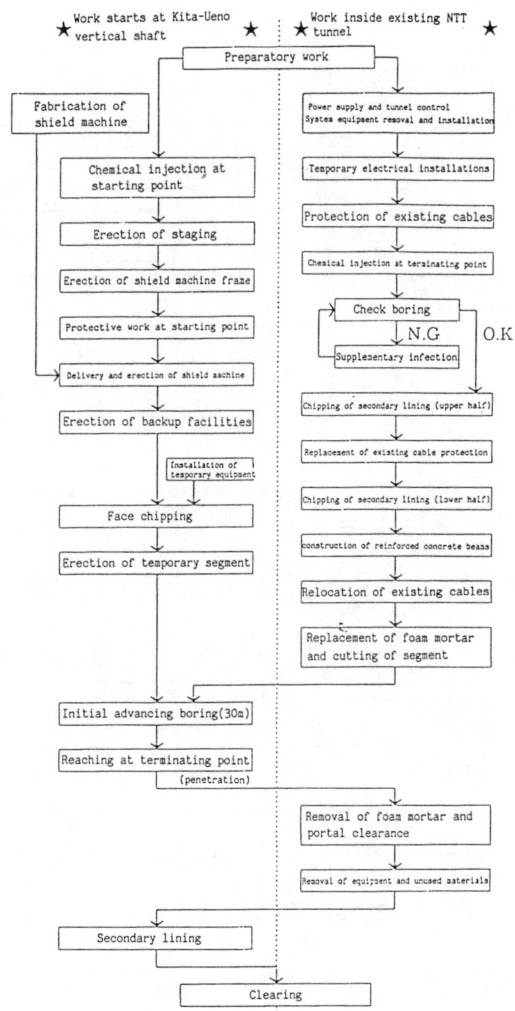

Fig.4 General Work Flow

Table 2 Construction Schedule

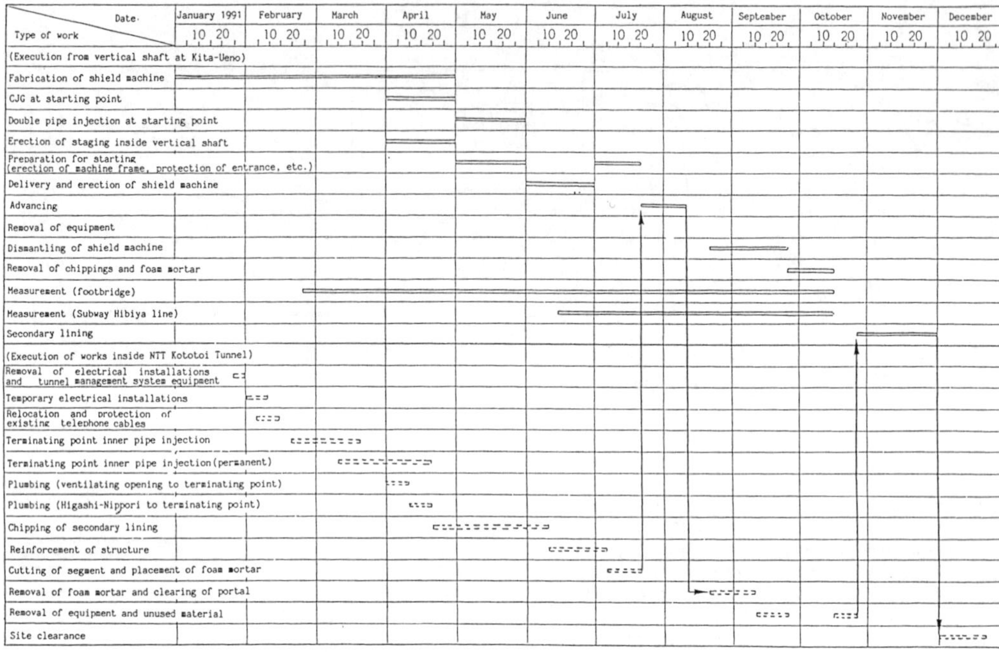

Execution from existing tunnel

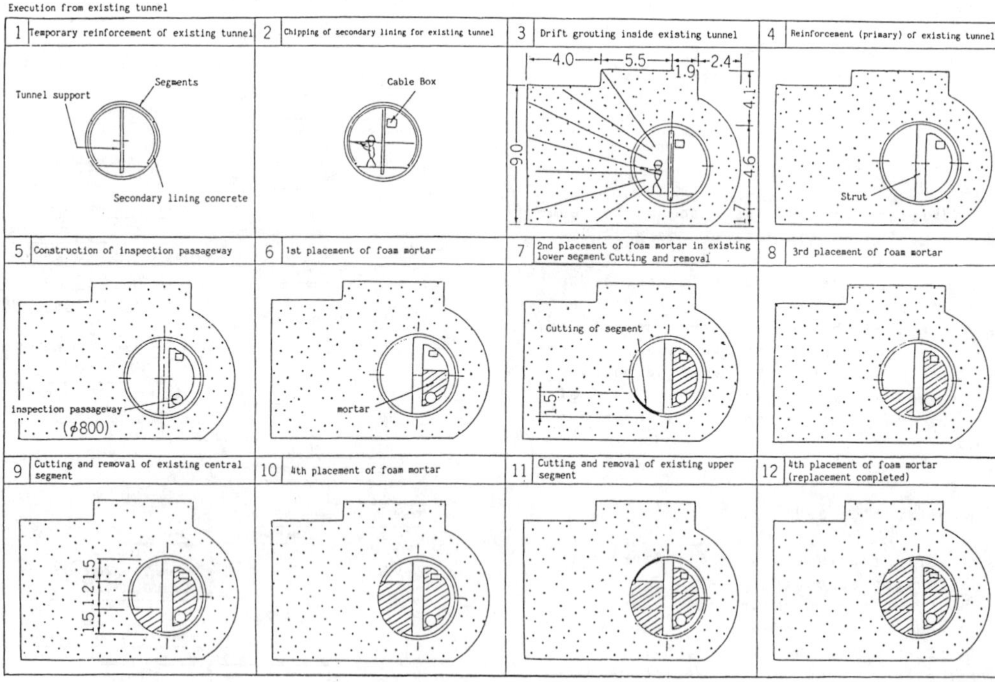

Fig.5 Construction Sequence

ety of construction in the process of underground joining and feasibility of structural design. For the benefit of construction sequence, measuring instruments were installed after ground improvement.

7-1 Items of Measurement

Stress fluctuations in main beams for segments, reinforcement for segmental beams, and reinforcement for auxiliary interior

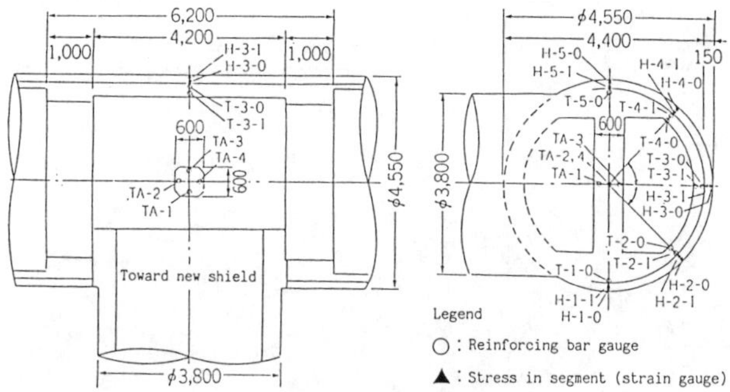

Fig.6 Arrangement of Measuring Instruments

Table 3 Items of Measurement

Items	Number of Instrument	Remarks
Stress fluctuations for existing segme-nt	8	Measure stress fluctuations for inside and outside reinforcing bar at 5 points of opening part of existing segment. (except the part where no inside reinforcing bar)
Stress fluctuations for existing segme-nt	10	Measure stress fluctuations for existing segment by strain gauges at same points of reinforcing bar gauges.
Stress fluctuations for existing segme-nt	4	Measure stress fluctuations for reinforci-ng bar at center pillar by reinforcing bar gauges.
Earth pressure		Measure earth pressure for Shield Machine by earth pressure gauges.

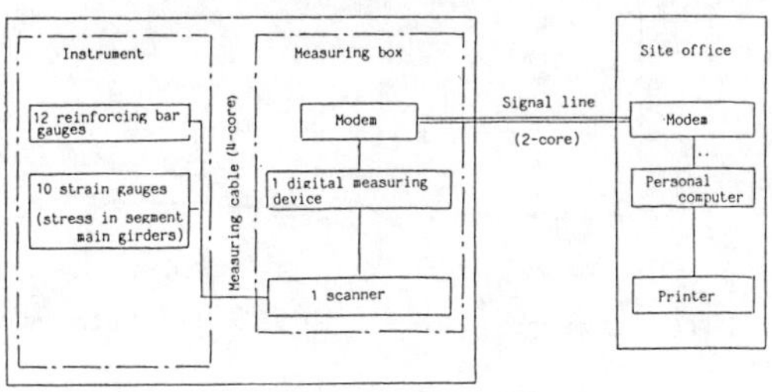

Fig.7 Measurement System Block Diagram

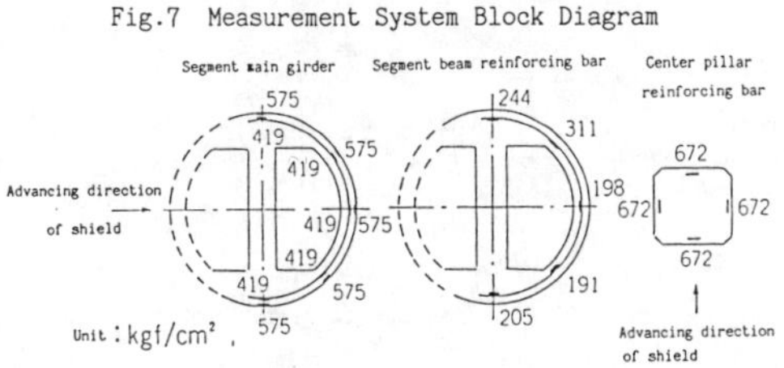

Fig.8 Control Values

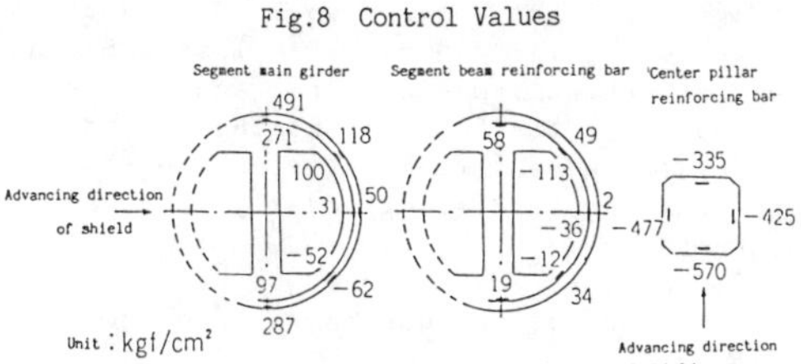

Amount of variation = Variation 1 week after opening of segment variation before opening of segment (Aug.19)

Fig.9 Variation in segment opening work (1 week after opening)

pillars due to the opening of the existing segment, and the approaching and arriving of the new shield machine. Items of measurement and arrangement of measuring of measuring instruments are shown Table 3 and Fig.6.

7-2 Method of Measurement

Measurement was taken continuously for the effects of underground joining work and the advancement of the shield machine on the existing tunnel, which were compared with control values to verify work safety. For this purpose, digital recorder linked to a personal computer was used. Fig.7 shows a block diagram of the measuring system.

The measurement was taken for 4 months from the installation of instrumens to the finishing of the tunnel face.

7-3 Control Values for Measurement

The existing segments to which the new tunnel would be joined were a few years old after completion, thus it was assumed that they were subject to stress due to earth and water pressures. Thus, for the purpose of calculation, loose earth pressure was assumed.

Assuming the calculated stress occurs in the existing segments, control values were set at minimum level on the basis of judgment that stress produced by making the opening would be within a safety range if would remain within a difference between the allowable value (1,400kgf/cm²) and the calculated value.

Control value for measuring instruments installed inside 419kfg/cm²

Control value for measuring instruments installed outside 575kfg/cm²

For reinforcements for segmental beams and auxiliary interior pillars, design values were assumed to be control values. Fig.8 shows control values.

8. Results of Measurement

8-1 Stress Fluctuation Due to the Opening of Existing Segments

The opening of the existing tunnel was carried out for 4 days after the completion of the reinforcement work. As described in the permanent work, the work was conducted in the following sequence:

Day 1 Placing of foam mortar in the service path section

Table 4 Measurement Results (During Opening Work)

	Instruments	August 19 before opening	August 20 passageway mortar	August 21 lower opening	August 22 middle opening	August 23 upper opening	August 30 after 1 week	Sept. 6 after 2 weeks
Segment main girder	H-1-I	24	24	63	59	54	121	159
	H-1-0	21	25	44	68	104	308	383
	H-2-I	14	17	13	−3	−14	−38	−43
	H-2-0	19	22	8	−14	−30	−43	−17
	H-3-I	3	5	39	49	58	34	32
	H-3-0	10	13	49	69	77	60	63
	H-4-I	−2	−2	5	32	75	98	99
	H-4-0	−4	−4	3	40	89	114	117
	H-5-I	2	1	−20	7	164	273	270
	H-5-0	2	1	13	159	452	493	497
Segment beam reinforcing bar	T-1-0	2	2	56	41	29	21	29
	T-2-I	7	14	−28	−37	−38	−5	20
	T-2-0	2	7	33	25	25	36	46
	T-3-I	7	5	−76	−110	−99	−29	−4
	T-3-0	31	36	12	−16	−23	33	88
	T-4-I	14	15	17	−10	−6	−99	−51
	T-4-0	10	12	20	35	60	59	81
	T-5-0	16	17	22	38	85	74	101
Pillar reinforcing bar	TA-1	−8	−5	−58	−148	−283	−578	−534
	TA-2	0	2	−55	−149	−255	−477	−423
	TA-3	1	−3	−102	−214	−242	−336	−307
	TA-4	2	1	−56	−143	−232	−427	−357

Unit: kgf/cm^2 Mark: (+) Tension), (−) Compression

Table 5 Mixture Design ($1m^3$)

Cement	Potter's clay	Water	Foamer	Silicic soda
225kg	225kg	354kg	2.2kg	85.4kg

* Strength of foam mortar is more than $20kgf/cm^2$

Day 2 Cutting of lower segments and placing of foam mortar
Day 3 Cutting of central segments and placing of foam mortar
Day 4 Cutting of upper segments and placing of foam mortar

Measured values during each work are summarized in TAble 4. Stress fluctuation one week after the opening work is shown in Fig.9. Also, design strength of foam mortar is shown in Table 5.

Chronological fluctuation during the segment opening work is summarized as follows:

(1) As the opening work proceeded from the lower section to the central and upper sections, stress fluctuation of the segments moved accordingly.

(2) In the interior pillar, compressive force occurred concurrently with the opening of the segments and increased with the progress of the opening work. Stress on the interior pillar did not stabilize immediately after the opening work and increased gradually after then. It reached a maximum level around one week after completion of the opening work and gradually decreased thereafter.

(3) Large stress occurred in upper and lower sides of the opening, and the interior pillar.

(4) Large stress occurred at all 4 measuring points of the interior pillar, and

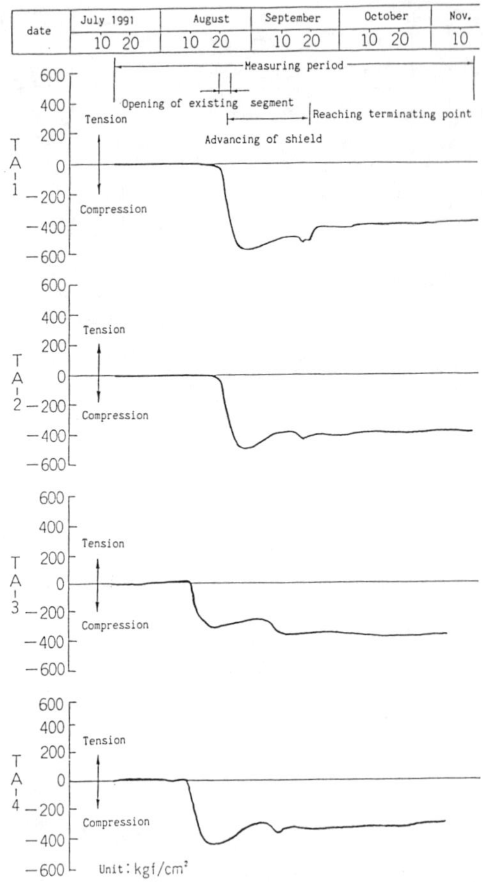

Fig.10 Chronogical Stress Fluctuation in Reinforcement

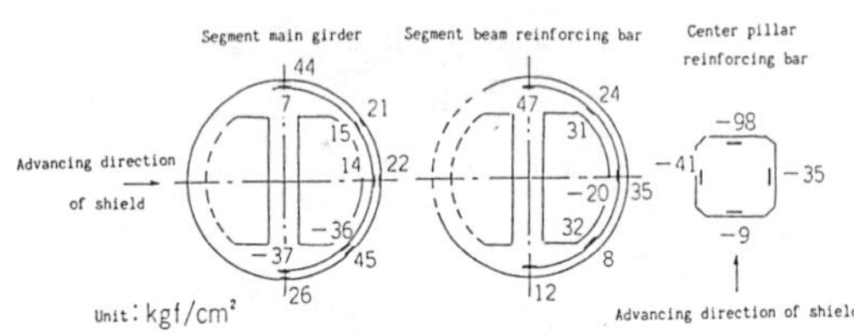

Fig.11 Stress Fluctuation Due to Arrival of the Shield

Table 6 Measurement Results (During Advance of Shield Machine)

	Instruments	September 14 5 days before	September 17 2 days before	September 19 arrival day	September 21 2 days after	September 24 5 days after
Segment main girder	H-1-I	163.15	166.38	129.23	137.31	129.23
	H-1-0	458.78	479.78	484.62	482.20	466.85
	H-2-I	−72.69	−80.77	−90.46	−98.54	−108.23
	H-2-0	14.54	24.23	32.31	43.62	59.77
	H-3-I	34.64	39.97	44.42	47.97	48.86
	H-3-0	86.20	77.29	82.75	87.29	90.02
	H-4-I	98.14	103.44	112.28	113.17	112.28
	H-4-0	121.12	128.20	137.04	139.69	142.34
	H-5-I	272.06	277.38	275.61	280.93	280.93
	H-5-0	503.80	517.69	540.84	546.40	547.33
Segment beam reinforcing bar	T-1-0	37.60	41.93	44.10	48.44	49.89
	T-2-I	38.11	46.74	48.17	66.15	69.74
	T-2-0	51.98	58.48	57.04	58.48	59.93
	T-3-I	6.48	6.48	−5.04	−13.68	−12.96
	T-3-0	133.76	150.38	147.49	157.61	168.46
	T-4-I	−19.41	−5.75	−4.31	5.75	11.50
	T-4-0	101.52	112.32	115.92	120.96	126.00
	T-5-0	135.17	154.59	168.97	179.75	182.63
Pillar reinforcing bar	TA-1	−488.20	−503.3	−497.55	−449.38	−430.68
	TA-2	−378.65	−388.06	−419.20	−410.51	−403.27
	TA-3	−285.59	−291.37	−352.82	−378.13	−383.19
	TA-4	−301.08	−303.07	−336.07	−326.59	−322.22

Unit : kgf/cm^2 Mark : (+) Tension , (−) Compression

external load due to the opening work concentrated in the interior pillar.

8-2 Stress Fluctuation Due to Advance and Arrival of the Shield Machine

The shield machine started to advance on August 23 and reached the existing tunnel around one month later, on September 19.

Daily changes in measured data on the interior pillar are shown in Fig.10.

Due to the advance of the shield machine, slight fluctuation occurred 5 days before the arrival (September 14). Fluctuation continued for a few days after the shield machine reached the existing tunnel and became stable thereafter.

Fluctuation measured by each instrument due to the advance of the shield machine is shown in Table 6, and that due to the arrival of the shield machine in Fig.11.

Stress fluctuation due to the approaching shield machine was within the range less than at each measuring point, smaller than that at the time 100kfg/cm² of opening the segments.

Table 7 Chamber Earth Pressure

Date	Chamber Earth Pressure	Position of Shield Machine
September 16	0.7 kgf/cm²	
September 17	0.6 kgf/cm²	Near the existing segment
September 18	0.2 kgf/cm²	
September 19	0.2 kgf/cm²	Joining position

Table 8 Comparison of Underground Jointing Work on Existing Tunnel Side

Name of construction	Construction method	Results of measurment	
A (this time)	Before arrival of Shield Machine, existing tunnel was reinforced and segment was opened.	Segment main girder	491
		Segment beam	113
		Center pillar	570
B (other Work)	After arrival of Shield Machine, segment was opened.	Segment main girder	104
		Reinforced beam	36
		Supporter	149

Unit:kgf/cm²

9. Result and Conclusion

9-1 Fluctuation Due to Opening of Segments

9-1-1 Segments and Segmental Beams

Stress due to the opening of main beams of segments mostly occurred at upper and lower ends.

The measured stress was approximately 85% of the control value of 575kgf/cm². On the other hand, stress occurred in reinforcement for segmental beams was less than 35% of the control value (311kgf/cm²).

Larger stress occurred on the segment side probably because stress concentrated on the segments with smaller rigidity and in contact with ground load.

9-1-2 Interior Pillarin Segmental Section

Due to the opening of the segments, significant compressive force occurred at each of instruments installed in the interior pillar. The largest value was observed on the side of the arrival point of the new shield machine. (instrument No.TA-1), with compressive force (-570kgf/cm²) being 85% of the design value.

The fluctuation occurred because external loads (earth and water pressures) which were countered by the circular lining concentrated on the interior pillar in the form axial force in a compressive direction due to the opening of the segments.

9-2 Fluctuation upon Tunnel Joining

Stress fluctuation in the interior pillar due to the approaching and joining of the shield machine was less than 100kgf/cm², one-fifth that during the opening of the segments.

Chamber earth pressure due to the approaching and joining of the shield machine is summarized in Table 7, estimated to be 5% - 25% of design pressures of 3.0 - 3.5kgf/cm².

Judging from stress fluctuation due to the approaching and joining of the shield machine, the effect of the new shield machine on the existing tunnel can be controlled through sufficient safety management (earth pressure management).

9-3 Comparison with Similar Works

The results of measurement on stress fluctuation during the underground joining of the existing tunnel with the new shield machine at other projects are compared in Table 8.

While varying with soil conditions and construction methods, work A where the new shield machine was joined after the opening of the segments showed stress on the segments and reinforcements, around 3 times as much as that for work B where the opening of the segments was done while joining the new shield machine to the old tunnel.

This is probably because work B opened up the segments while joining the new shield machine to the existing tunnel, so that load around the existing tunnel was transmitted to the new shield, resulting in stress dispersion three dimensionally.

10. Conclusion

The opening of the existing segments during the underground joining work sometimes causes large stress on major beams of segments.

In the project, segments were reinforced before the shield machine reached the existing tunnel, then the opening of the segments was made. As a result, effective stress sharing occurred by building the auxiliary interior pillar.

Also, it was confirmed that the effect of the approach and arrival of the new shield machine on the existing lining could be controlled through appropriate earth pressure control.

Technical issues to be dealt with in the future include stress sharing between segments and reinforced concrete.

For this purpose, further data on various types of underground joining should be collected for more efficient design and construction.

Finally, we thank those who have extended cooperation during the project.

Reference

Yagy,Takigasaki,Ukawa and Okayasu 1992. The stress fluctuation of the opening of the existing segments during the underground joining work. Journal of The Japan Society of Civil Engineers.

Tunnelling and Ground Conditions, Abdel Salam (ed.) © 1994 Balkema, Rotterdam, ISBN 90 5410 363 9

Urban double deck circular section tunnels for private vehicles

M. Marec
Centre d'Etudes des Tunnels, France

ABSTRACT: Since 1987 Mssrs. G.T.M., then COFIROUTE, motorway manager, which count G.T.M. among their shareholders, have proposed to build shield driven tunnels with circular section. Such tunnels can accept two superposed road levels, each of them with one traffic direction. The tunnels will be reserved to light vehicles of maximal 2 m in height. The accesses to the surface will use branches of very flexible lay-out, which will allow to collect and distribute traffic to the various streets in the open. This system will allow making an underground infrastructure of motorway character at a relatively acceptable cost. All aspects - geometrical design, ventilation, safety, access of rescue staffs, civil engineering arrangements - have been elaborated and accepted by the French Authorities. Two projects are being started in the Paris area : one of 10 km underground (A86 West), the other of 14 km (RSP).

French contractors have proposed for several years the construction of urban motorway tunnels, shield-driven and reserved to private vehicles with a maximal height of 2 m. Thus, on two levels, it is then possible to have twice three vehicle lanes, and therefore reach an exceptional capacity with regard to the excavated cross section. This was made for the first time in 1987 for the LASER Project proposed by Mssrs. Grands Travaux de Marseille (GTM) for Paris.

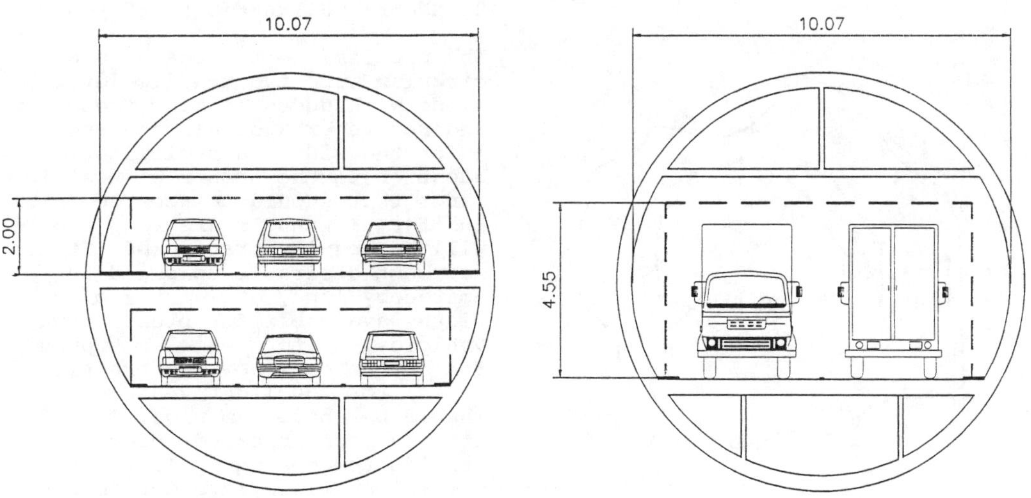

Fig.1 6 lines of light vehicles use the space required by 2 lorries

A tunnel reserved to private vehicles is likely to show another advantage : an easier integration into the urban context than for a tunnel authorizing the heavy vehicle transit - going down underground can be possible with gradients - upgrades and downgrades - over 10%; upgrades inside the tunnel can exceed 5% without any inconvenience. On the other hand an obvious drawback of the method is to keep the heavy vehicles on surface streets.

Underground passages with reduced sizes already existed in France and foreign countries to cross heavily trafficked intersections in cities. They have been codified in a guide published by SETRA in 1971. However, their small length generally under 100 m - allowed to design and build them without a too acute analysis of safety issues (operational conditions, evacuation of users in case of sinister, access of rescue vehicles) and environmental issues inside the tunnel (how would the users behave in tunnels with ceiling at about 2.5 m above the road and several kilometers, even several tens kilometers, in length ?).

Relevant studies have therefore increased in France. A special Inter-Departmental Commission on Safety has been set up in October 1991, and produced a report in June 1992 after 30 working meetings. By this time two first projects are

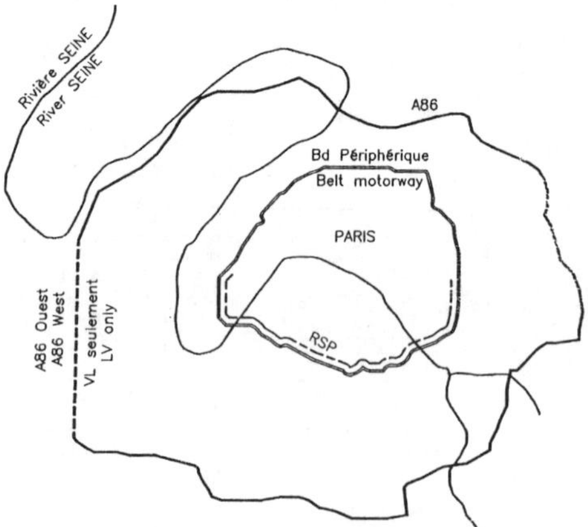

Fig.2 Location of the A86 West and RSP Projects in the Paris area

well advanced : completion of the A86 circle motorway on 10 km, 20 km west of Paris; and doubling of the Paris belt motorway on 14 km, south and east of Paris (RSP Project). These two cases plan toll tunnels.

1 SELECTION OF THE VEHICLE GAUGE TO BE ADMITTED IN THE TUNNEL

The distribution of heights and widths of vehicles registered in France has been examined for the year 1990. Lorries and coaches were not considered, since too high, and the same for vans; private vehicles could be accepted. The height of light vans and all-purpose vehicles required a closer examination. The Commission first hesitated between maximal gauges ranking from 1.95 m and 2.05 m, and finally recommended 2.00 m. This concerned 93% of the registered vehicles, i.e. 80% of the traffic (because lorries travel much more than private cars), including practically all all-purpose vehicles and the greatest number of light vans. Keeping a restricted gauge also allowed to keep a tunnel with a reasonable diameter at a rather low cost.

Bicycles, motor cycles and pedestrians were excluded.

The population of accepted vehicles was therefore quite homogeneous, thus in favour of safety.

2 NUMBER OF TRAFFIC LANES

The problem was to locate in a circular tube two traffic levels, along with ducts to blow fresh air and remove exhaust air and smoke.

For the width, a preliminary question is the number of traffic lanes that should be accepted. Insofar as tunnels are long and will be heavily trafficked, it seemed most desirable to build an emergency lane, at least 2 m in width, available for broken down vehicles, or for vehicles stopped for any reason. Now the traffic lanes, for the light vehicles, should be about 3 m wide. Therefore it very soon appeared adequate to provide the emergency lane with a sufficient width in order to use it as a traffic lane and decelerating or accelerating lane at the entran-

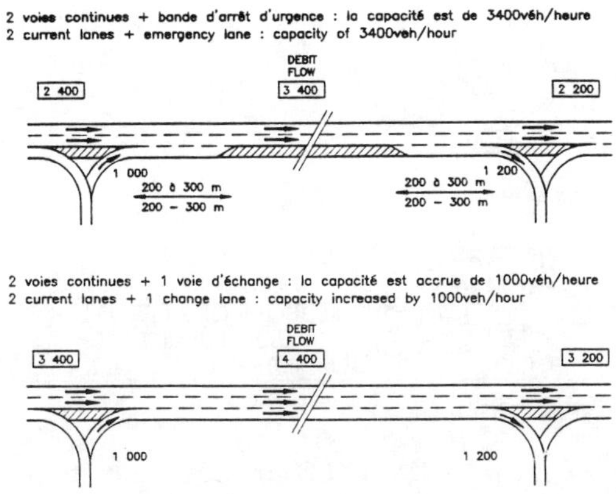

Fig.3 Comparison of capacities of a 2-lane tunnel and 3-lane tunnel

ce or exit point of an interchange. The tunnel, thus, has been designed to allow 3 traffic lanes, without any emergency lane.

If traffic is possible on all 3 lanes, the tunnel capacity is increased - with respect to 2 lanes - by about 30-50%, depending on the interchange conditions.

3 INNER DIMENSIONS OF THE TUNNEL

They depend on the usable width, also on the available height under the ceiling.
Concerning the usable width, the basis generally is the lane number, lane width, shoulder width. Possible sidewalks are added. But in this case, it appeared that no sidewalks should be planned, because they would have led to too large diameters, thus making the project too expensive to be built. We will see later how safety conditions have been defined to dispense with them. Therefore, the vehicles were likely to drive close to the walls - which are curved. It was then essential to check that the usable width ensured a sufficient distance to the walls. Finally a minimal distance between the tunnel walls has been defined, at 1 m above the road level. This is the sum of the minimal values for the distance to the wall (1.20 m), distance between vehicles (1 m), and vehicle width (3 vehicles in file of 1.80 m, 1.70 m and 1.70 m). This gave 9.60 m.

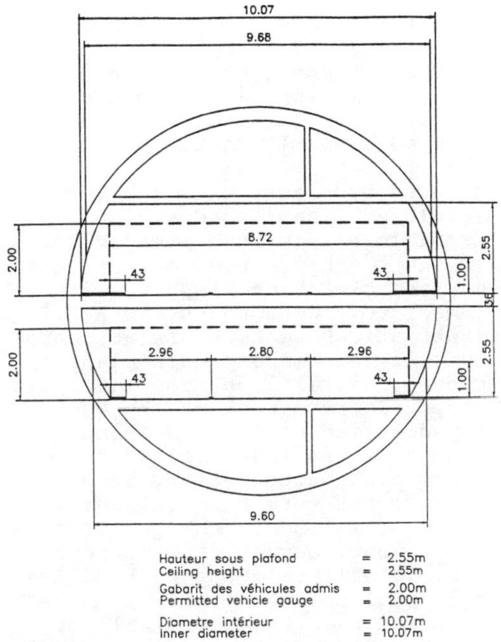

Fig.4 Inner dimensions

Concerning the height, it was necessary to plan a sufficient distance between the vehicle top (2 m) and the ceiling bottom. It was fixed at 55 cm, consisting of 40 cm for the lighting and signalling equipment, and 15 cm for the safety and signalling margins. Sufficient areas had also to be planned for the technical ventilation and smoke removal ducts.

The city of Nice already has tunnels with a clearance of 2.60 m, up to 180 m in length, well accepted by the users. We also are experienced in car-parks with only 2 m in clearance. All this showed that 2.55 m should be quite acceptable for the user, while inciting him to drive at a moderate speed, as it was whished.

The combination of all these minimal sizes - width and height - led to the same inner diameter of about 10 m (finally fixed at 10.07 m), which is worth noticing ! This result expresses in itself the high value of the considered concept.

4 SPECIAL OPERATIONAL CONDITIONS

In these tunnels the only and very rare pedestrians are safety employees or users going out of their stopped cars. It should be noticed that the tunnel maintenance will be ensured by night, when the tunnel is closed. Then a pedestrian is possible only with a stopped car. It has therefore been planned that any vehicle which stops be immediately detected by an A.D.I. (Automatic Detection of Incidents) system. The corresponding lane is closed to traffic by red lit cross-stays. Therefore it can be used by a pedestrian to go to a phone recess (every 200 m) or push an alarm push-button (every 50 m).

It is not sure that all automobilists will observe the red traffic light, but, anyway, the stopped vehicle will oblige them to drive off the lane.

In addition the users will be recommended not to go out of their cars. A visual announcement will be made (in front of push-buttons) stating that the rescue staffs are coming.

These arrangements will prevent an investment (the sidewalk), very expensive in this case since it leads to increase the diameter and which would be almost never used.

The inner sizes of the tunnel allow the operation both with 2 lanes and emergency lane, and with 3 lanes. The running conditions, therefore, will be closely analyzed when the tunnel is opened to traffic.

5 ASSESSMENT OF THE OBTAINED SAFETY LEVEL

We can argue only from hypotheses and compare to the accident statistics in existing motorway (or unidirectional) tunnels. The main differences that should be considered are : absence of heavy vehicles, absence of motor cyclists, most limited speed, absence of inclemencies, on the other part absence of an emergency lane, width of lanes, interchange facilities, ceiling at 2.55 m.

The absence of heavy vehicles would allow a homogeneous, therefore steady traffic flow; the absence of motor cyclists is

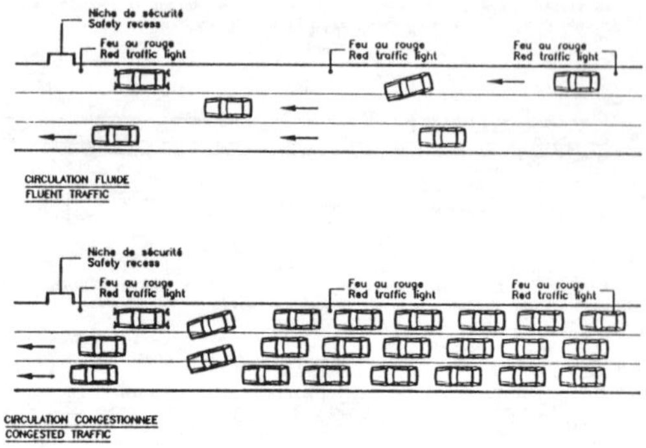

Fig.5 Vehicle stopped in a tunnel

	A86 Ouest–West Véhicules légers light vehicles	Autoroutes Motorways	Routes Roads
Accidents corporels légers Light injuries	12.5	9.3	38
Accidents corporels graves Severe injuries	1.3	1.8	8.5
Accidents mortels Fatalities	0.4	0.6	2.0

Fig.6 Safety ratios compared for motorways, roads and tunnel

favourable, also the absence of inclemencies. But the decisive factor is speed.

The decision to install an A.D.I. system made it adequate its use to check speed, and come to the automatic and systematic fining of speed excesses. The Safety Commission recommended 70 km/h as authorized speed. Any excess beyond 80 km/h will be fined automatically. The quasi-absolute control of speed is the essential condition for the preventive safety : these tunnels will be driven through calmly. The users who cannot accept it will always have the possibility to stay on open streets.

The average lane width is about 2.90 m, which is comfortable for private cars driving at a moderate speed. The interchange facilities have been specially investigated (see below).

Finally it has been estimated that safety will be higher than in ordinary motorway tunnels.

6 INTERCHANGE FACILITIES

The lateral accesses to the main tube will all be on the right for one level, and on the left for the other. As a matter of fact the access lanes will reach back the surface through tubes called branches, connected to only one side of the tunnel for cost reasons.

Traffic will enter the tunnel and go out on the third lane, and the traffic flow will use only two lanes between an entrance and an exit. Solid obstacles will prohibit driving on the neutralized part of the third lane. This will allow the entering vehicles to drive safe in spite of the lack of lateral visibility.

To manage the exit or entrance of a lane it should be built an opening or 'window' in the tunnel wall, where the shield-built wall will be lined with prefabricated segments. For certain grounds encountered in the Paris area such windows can reasonably be opened only for maximal spans of 25-30 m. Otherwise a true underground station - or box - must be built, crossed by the tunnel. This restrictive condition was retained when determining the maximal permitted exit speed, i.e. 50 km/h.

7 VENTILATION

The smoke emissions are reduced, as no lorries are admitted. Therefore the two features determining the design of the ventilation systems are the CO carbon monoxide emissions and the consequences of possible fire accidents. The following general arrangement has been selected : fresh air is aspired through vertical shafts located about every 3,200 m and distributed on 2x1,600 m all along the tunnel. The exhaust air is also removed through shafts located every 3,200 m, alternately with the previous ones; exhaust air is collected all along the tunnel. This is then a transverse ventilation system.

Fire accidents will concern only light vehicles. We can foresee, on 10 km, about 3 beginning fires yearly, 2 of them being put out by means of the portable extinguishers installed in the tunnel and one by

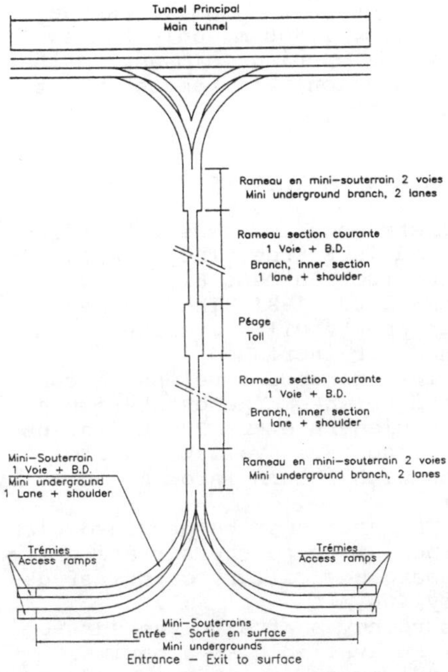

Fig.7 Scheme of a branch

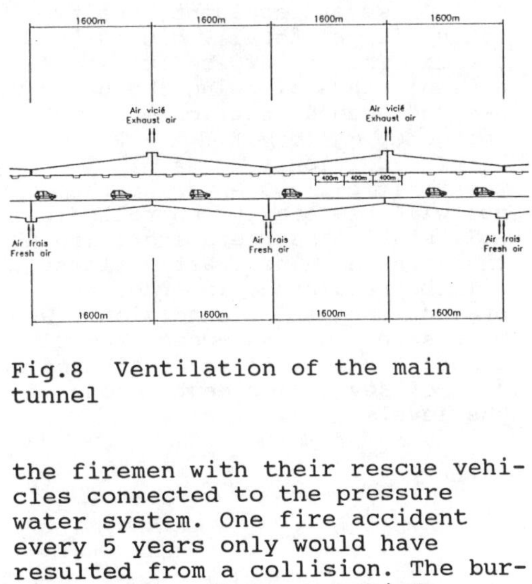

Fig.8 Ventilation of the main tunnel

the firemen with their rescue vehicles connected to the pressure water system. One fire accident every 5 years only would have resulted from a collision. The burning of more than one vehicle, hence after a collision, will occur only every 10 to 20 years. As a precaution the smoke removal equipment will be planned to aspire smoke produced by three vehicles burning at the same time. Extraction will occur through exhaust air ducts and shafts. The exhaust openings will be located every 400 m.

On the length of a smoke removal duct, i.e. 1,600 m, only the two downwards openings closest to the fire location will remain open to aspire smoke.

8 SAFETY FACILITIES

The reference is the safety rules applied in France. They are fixed by Instruction Memo 81-109 of December 29, 1981. The planned arrangements will be either maintained, or increased.

- They foresee a telephone connected to an emergency station and extinguishers every 200 m. The user will also have available alarm boxes with return information every 50 m.
- The user must be provided with escape issues to the surface or to an emergency gallery or to refuges every 400 m :

Emergency shafts to the surface will be available at a maximal interdistance of 1,250 m. Refuges will also be available every 400 m in front of staircases connecting the two traffic levels. Lastly, in case of severe accident, the two tubes will be totally closed to traffic (see below, rescue staff access), thus allowing the users to use the second traffic level as emergency gallery.

- In addition :

the vehicle flows entering the tunnel will be totally controlled, via toll station barriers installed at the accesses. The traffic flows will be regulated, in order to prevent congestion conditions. We have seen that the access to both tunnel levels will be prohibited in case of severe accident in one of the levels.

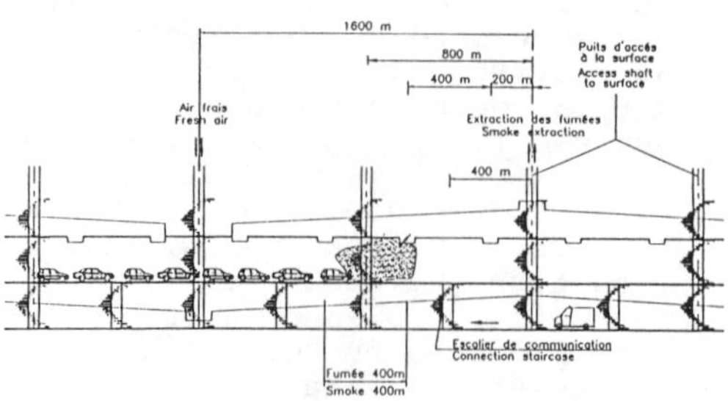

Fig.9 Intervention of the rescue staffs in case of severe accident

The whole tunnel will be permanently controlled by a video system of television monitors connected to the central command room.

Let us deal now with the access conditions for the rescue staffs:

The rescue staffs (firemen, police, operator's services) will have equipped vehicles of reduced sizes, located at the necessary access points. They will be able to reach the sinister location either through the concerned level, or through the other level, or through the emergency shafts from the surface. In every case they will therefore be at less than 200 m by foot from the accident or fire location, even if it happens in the other traffic level than the access level.

An automatic fire detection system covering the whole network will be installed. A water network to fight against fire will supply hydrants located in the safety recesses every 200 m.

9 CONCLUSIONS

All the outstandingly acute reflections made around this concept confirmed its interest. It will allow limiting the investments (it must be said that they remain relatively expensive - about 70 millions ECU per km). It proves adaptable to the urban environmental constraints. The access points to the surface can be distributed to the various streets thanks to the flexible access branches. In addition the connection ramps with the surface network are very short (about 30 m).

It remains to have it known and appreciated by the users. Henceforth it should become possible before the year 2000.

Tunnelling and Ground Conditions, Abdel Salam (ed.) © 1994 Balkema, Rotterdam, ISBN 90 5410 363 9

Highway tunnel being driven through a volcanic zone

Masao Matsuyama
Kajima Corporation, Tokyo, Japan

ABSTRACT: The construction of Abo Tunnel through a volcanic zone which lies across the central region of Japan involves difficult problems including high temperature rock and poisonous volcanic gases. There being no past examples of construction under conditions such as these, an investigation tunnel 4340m in length was excavated commencing in 1980 to investigate the geological conditions and study methods for constructing the main tunnel. Excavation was advanced fighting against the severe working conditions of a thermal zone in which the temperature of the ground reached a maximum of 75°C and the danger of sudden emission of poisonous volcanic gas until the investigation tunnel was holed through in 1991. From these investigations it was decided that construction of the main tunnel was possible and excavation was commenced in that same year. Based upon the experience achieved in the investigation tunnel, various measures were developed including those regarding ventilation equipment, volcanic gas, excavation methods and the lining concrete in the thermal zone. Construction of the main tunnel employing these measures is now progressing smoothly.

1 FOREWORD

The most prominent feature of the central region of Japan is a range of mountains with peaks exceeding 3000m in elevation which divides the country into east and west. Because of their beauty and forcefulness, these mountains are called the "Japanese Alps." However, these Japanese Alps cause the Hokuriku and Hida districts, which lie on their western side, to be cut off from Tokyo. Thus, the construction of Abo Tunnel as the only tunnel to penetrate the Japanese Alps has been looked forward to for many long years.

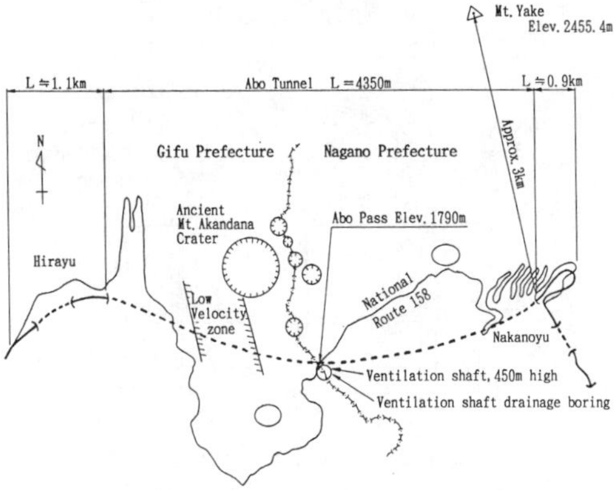

Figure 1. General plan.

Abo Tunnel was planned to be constructed under Abo Pass which is located on National Route 158 at the boundary between Gifu and Nagano Prefectures. The tunnel has a total length of 4350m. (Figure 1.)

National Route 158, a main highway passing through the city of Takayama on the west side and extending to the city of Matsumoto on the east side, is an important route which connects the Hokuriku district to the Kanto district surrounding Tokyo by the shortest distance. However, because the section of this highway over Abo Pass, which has an elevation of 1,790m, is marked by steeply sloping topography and weak ground characteristic of volcanic zones, vehicles using the highway are constantly exposed to danger, and the highway is frequently closed to traffic because of rainfall. Further, because the highway is also closed to traffic for a period of six months each winter due to heavy snow, it is not in a condition to function satisfactorily as a main highway.

Because of this, investigations were commenced in 1967 by the Ministry of Construction to make this section of highway safely passable throughout the year, and in 1978 it was decided to launch a project to construct a new 6.3km long section of highway including this tunnel.

However, because the tunnel would be constructed through a volcanic zone, the following problems were forecast during construction:

1) The work will be affected due to excavating close to an ancient crater located near the middle of the tunnel.

2) On the Nagano Prefecture end, the portal is only a short distance of 3km from an active volcano, Mt. Yake, and there is a thermal zone in which the ground has a high temperature and poisonous volcanic gases exist.

3) On the Gifu Prefecture end, there is a low velocity zone which consists of deposits of volcanic ejecta in which the ground water is under the high pressure of 22 kg/cm^2 and seepage occurs in large quantities.

Because a tunnel such as this had never been constructed before, it was decided to excavate an investigation tunnel before constructing the main tunnel in order to investigate the geological conditions and study methods for constructing the main tunnel.

Excavation of the investigation tunnel, which was commenced in 1980, was interrupted several times by events such as massive seepage on the Gifu Prefecture end, but the tunnel was finally holed through on July 31, 1991 at the point 2260m from the Nagano Prefecture portal (2080m from Hirayu). As a result, it was judged that construction of the main tunnel was possible, and excavation of the main tunnel was commenced from the Gifu Prefecture end in October 1989 and from the Nagano Prefecture end in February 1991.

This paper presents a report on the execution of the Nakanoyu contract section of the investigation tunnel at the Nagano Prefecture end of Abo Tunnel and the main tunnel construction which was planned and is being executed on the basis of the experience obtained there.

2 OUTLINE OF WORK

2.1 General outline

The official title of the project reported here is Abo Tunnel Nakanoyu Contract Section on the Chubu Judan Expressway. The owner is the Ministry of Construction Chubu Regional Construction Bureau and the contractor is a joint venture of Kajima Corporation and Taisei Corporation. The project is located at Nakanoyu, Azumi Town, Minami Azumi County, Nagano Prefecture.

The investigation tunnel, which had an excavation cross section of $25m^2$ and an excavated length of 2260m (4340m in total), was constructed during the period from August 23, 1980 to July 31, 1991. The main tunnel, which has a standard excavation cross section of $100m^2$ ($120m^2$ at emergency parking zones) and an excavation length of 2150m (4350m in total), was commenced on February 5, 1991 and is being excavated by upper heading and short bench method using the NATM system of support. (Figure 2.)

2.2 Topography and geology

Abo Tunnel is located in the center of the Japan Alps Norikura volcanic zone which includes numerous peaks exceeding 2000m and 3000m marked by steeply sloping topography. Mt. Akandana which is located

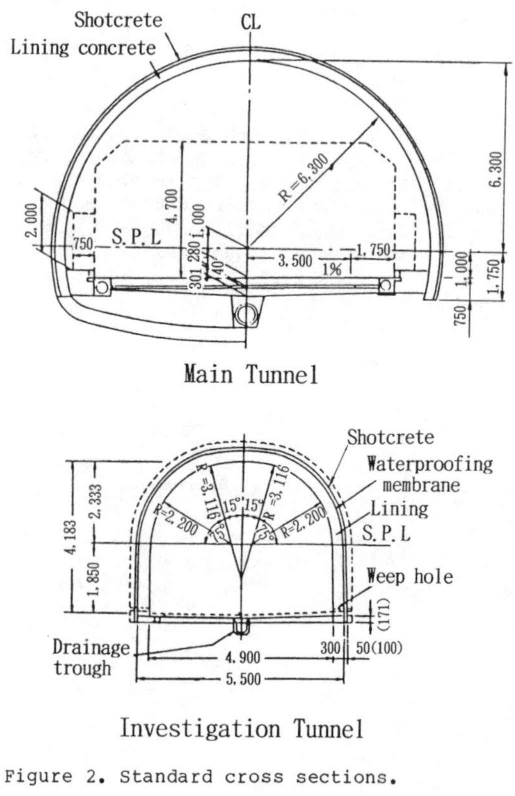

Figure 2. Standard cross sections.

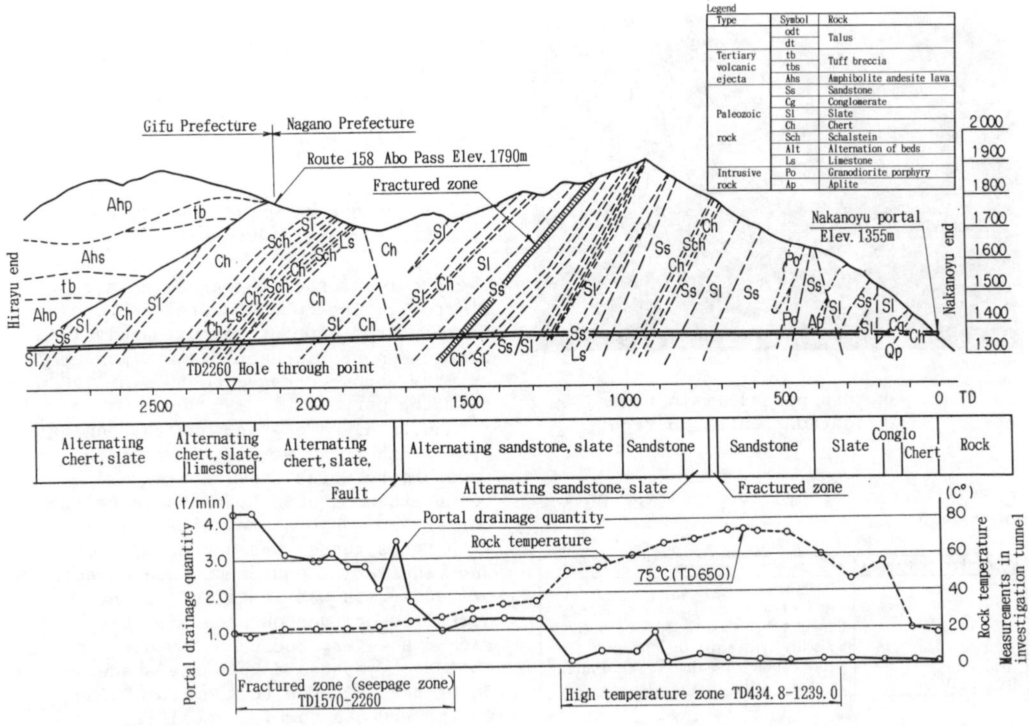

Figure 3. Geological profile and rock temperature record.

near the center of the tunnel (Figure 1.) has ceased volcanic activity, but Mt. Yake which is located approximately 3km from the Nakanoyu portal continues to emit white smoke even at the present day.

The geological profile along the line of tunnel is shown in Figure 3. The geology of the Nakanoyu contract section consists of alternating beds of paleozoic sedimentary rock such as sandstone, slate and chert. In the first 1000m long section of tunnel this bedrock has been intruded by granite. The rock is cracky, has undergone thermal metamorphism and as a whole is relatively soft. The intruded rock is hard but cracky, and forms a passage for heat energy. Because of this, the section from 300m to 1200m from the portal forms the "Nakanoyu high temperature zone" in which the rock temperature exceeds 50°C. The highest temperature exceeds 75°C, and sudden emissions of poisonous volcanic gases, such as hydrogen sulfide and methane, and hot water were foreseen. Beyond 1200m the rock temperature gradually decreases, and becomes approximately 20°C at 1800m and beyond.

Beyond 1575m the geology continues to consist of alternating beds of sandstone, slate and chert but there are fault fractured zones accompanied by seepage water under the high pressure of 20 kgf/cm^2; especially a large scale fractured zone 25m wide between 1610m and 1635m.

3 HIGH TEMPERATURE SECTION OF INVESTIGATION TUNNEL

From the geological investigations which were commenced in 1967, the Nakanoyu high temperature zone was found in which the rock temperature reaches as high as 160°C, and sudden emissions of poisonous volcanic gases, such as hydrogen sulfide and methane, and hot water were foreseen.

3.1 Problems in construction

The existence of the high temperature zone presented the following problems in construction.

(1) Securement of working environment within the tunnel.

(2) Possibility of sudden emissions of volcanic gases.

Figure 4. Nakanoyu portal; main tunnel in center, investigation tunnel on right.

Table 1. Effect of hydrogen sulfide on the human body.

Concentration (ppm)	Effect
3	Strong odor
10	Eye irritation
20 to 30	Strong irritation of mucus membranes
200 to 300	Subacute poisoning in 1 hour
600	Fatal acute poisoning in 30 minutes
1000 to 1500	Instant death

(3) Possibility of hot water emissions.
(4) Durability of secondary lining concrete against hot spring water.
(5) Safety of explosives under high temperatures.

Because these problems would be difficult to solve merely by performing geological investigations by boring, it was decided to excavate an investigation tunnel for the purpose of investigating the possibility of constructing the main tunnel and to study construction methods.

In the high temperature zone the rock temperature increases in the direction of Mt. Yake and also in the direction of lower elevation. However, from the results of horizontal boring performed at elevation 1360m with the direction at the tunnel portal swung toward the southwest, it was found that within the high temperature zone in which the rock temperature exceeded 50°C, the maximum temperature was on the order of 75°C and the quantity of volcanic gases was small.

As shown in Figure 4, the investigation tunnel was located on the Mt. Yake side of the main tunnel at a distance of 30m between center lines, at almost the same elevation so that it can be utilized as an emergency tunnel after the main tunnel is put into service, and with the portal located close to the position of the horizontal boring.

The investigation tunnel had an excavation cross section of $25m^2$ and an internal cross section of $18m^2$ within the lining. Excavation was performed by NATM and a double track railway system was employed for transportation.

3.2 Securement of working environment in high temperature zone

The record of rock temperature is shown in Figure 3. The rock temperature rose sharply from a point approximately 100m from the portal, reached a maximum of 75°C at 650m, and decreased to 50°C at 1200m from the portal.

Industrial safety and health regulations provide that the working environment in underground works shall not exceed a temperature of 38°C. At Abo Tunnel the target air temperature to be maintained within the tunnel was set at 30°C or less, and ventilation equipment with a capacity of $2000m^3$/min was installed at the portal to deliver outside air to the face. However, because this outside air could not penetrate into the space within 5m of the face due to the heat being emitted by the rock, a local fan was provided at a point 20m from the face to deliver air directly into the face. In this way it was finally possible to reduce the air temperature.

Also, because the heat given off by the muck after blasting causes the working environment to become bad during the mucking operation, spraying with water at a temperature of 16° to 18°C (200 liter/min) was tried, but this had the opposite effect of causing the humidity in the tunnel to rise. Finally, the air temperature within the tunnel was kept generally within the range of 28° to 30°C by delivering the greatest amount of air possible to the face.

3.3 Measures against volcanic gases

From the results of the geological investigations, it was known that hydrogen sulfide and methane existed in the volcanic gases in the Nakanoyu contract section.

Hydrogen sulfide is a poisonous gas which has a specific gravity of 1.55 compared to air, and which affects the human body as shown in Table 1.

Methane gas is a flammable gas which has a specific gravity of 0.559 compared to air, which can be ignited by even an almost invisible spark when present in a concentration of 4.8 to 15%, and which has its greatest explosive force when present in a concentration on the order of 9.5%.

In order to secure worker safety in the excavation, the following measures were taken.

1) After each 30m advance of the excavation, advance boring was performed to 40m ahead of the face, the concentration of gas within the bore hole was measured and the next excavation was started only after confirming safety.

2) Concentration of gas, rock temperature and air temperature in the tunnel were measured twice at each round of blasting, after the completion of face drilling and after mucking.

3) At the face the miners foreman guarded against gas by carrying a pocket type gas warning device.

4) Fixed sensors to measure the concentration of gas within the tunnel were installed at 200m intervals and portable type sensors were also provided at the face to measure during the drilling operation. The data from these sensors was constantly gathered and recorded at the central observation room. An "automatic gas warning system" was provided to automatically sound an emergency warning siren and broadcast an evacuation order when the concentration of poisonous gases reached a certain level (hydrogen sulfide 25ppm, methane gas 1%, oxygen concentration less than 18%). At the same time, signals at the portal and inside the tunnel change to "red" to stop entering.

5) An emergency evacuation train (battery type, explosion proofed, 16 passenger) was stationed near the face to carry personnel out of the tunnel when any emergency situation occurred such as a gas emission.

6) Personnel training and evacuation practice were performed periodically.

7) To protect personnel from the dangers of high temperature and volcanic gases, a completely automatic hydraulic jumbo which can perform the drilling operation by computer control was newly developed and introduced into the work as a step in advancing the automation of face operations and elimination of personnel.

In the excavation of the investigation tunnel, the only volcanic gases detected was a very small quantity of hydrogen sulfide (32ppm at TD625m) detected in an advance boring hole. No volcanic gases were detected at all in the working area and no emergency situation occurred. It can thus be said that the tunnel route which was determined through careful study based on the geological investigations had been determined appropriately.

3.4 Hot water emissions

No hot water emissions occurred. There was only a small quantity of seepage water in the high temperature zone, and no problem occurred to obstruct work progress.

3.5 Secondary lining concrete in high temperature zone

There were two problems concerning the secondary lining concrete in the high temperature zone; durability against hot spring water, and deterioration of concrete strength due to high temperature.

1) Durability against hot spring water
The hot spring water which occurs in the Nakanoyu contract section is simple spring water of pH6.6 having only a small salinity content. However, it was imagined that the salts contained in the spring water might react with the concrete hydration product calcium hydroxide and cause spalling and dissolution on the concrete surface.

Therefore, the Abo Tunnel Construction Methods Investigation Committee of the Ministry of Construction investigated the durability of concrete specimens made with three types of cement, ordinary portland cement, moderate heat cement plus fly ash, and blast furnace cement Type B, by soaking these specimens in hot spring water for a period of five years. The result of this test was that, although the specimens which were made using ordinary portland cement and moderate heat cement plus fly ash were eroded to the extent that the coarse aggregates became exposed, those which were made using blast furnace cement Type B suffered almost no erosion at all, their surfaces being only blackened a little. This is believed to be due to the calcium hydroxide of the blast furnace cement Type B being used up in neutralization of slag, leaving only a very small quantity with the result that almost no erosion occurs.

From these tests it was decided to use blast furnace cement Type B, which possesses this superior durability, in the secondary lining concrete.

2) Effect of high temperature on concrete strength
To investigate the effect of high temperature on concrete strength, the Abo Tunnel Construction Methods Investigation

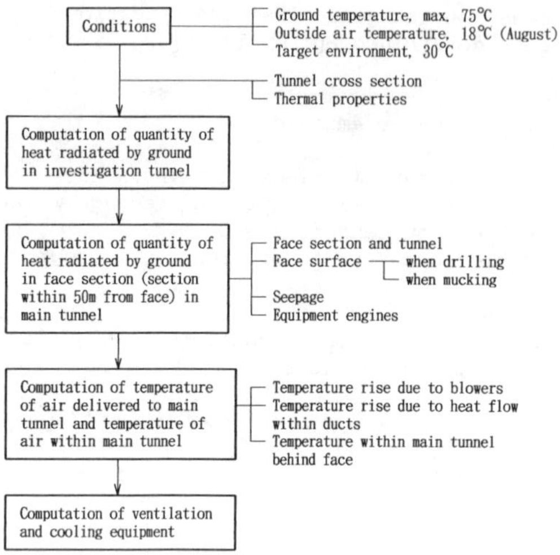

Figure 5. Flow of computation of ventilation equipment.

Committee of the Ministry of Construction subjected concrete specimens made with four types of cement, ordinary portland cement, moderate heat cement, blast furnace cement Type B and Type C, to high temperature curing at a temperature of 80°C.

The result of this test was that, when cured at high temperature, hydration occurs only in the very early stage and continues to only a very small extent afterwards, with the result that the increase in strength with increasing age which occurs in normal curing can not be expected. However, it was found that when blast furnace cement Type B was used, the pozzolan reaction of the admixture and latent hydraulicity of the cement were activated by being cured at high temperature, thus preventing reduction of concrete strength to a certain extent.

For this reason also, blast furnace cement Type B was judged to be suitable.

3.6 Safety of explosives under high temperature

Explosives used under high temperature conditions must satisfy certain conditions including:

(1) Will not become fluid at high temperature.

(2) Will not ignite and explode due to decomposition into its components.

(3) Explosive performance will not be altered.

The temperature range within which general explosives can be used with stability has an upper limit at 70°C. Therefore, under temperature conditions of 65°C and over, heat resistant detonators which use lead azide as the detonating agent were used in combination with an emulsion explosive Chita-Mite (manufactured by NOF Corporation). These have high stability up to 120°C, and the blasting effect was excellent.

4 MAIN TUNNEL EXCAVATION IN HIGH TEMPERATURE SECTION

The plan for excavation of the main tunnel was prepared after reviewing the problems in construction which emerged from the experience in executing the investigation tunnel.

4.1 Plan of ventilation equipment

The study of ventilation equipment was performed as shown in Figure 5 based on the experience in the investigation tunnel.

Computed from the empirical data on ventilation in the investigation tunnel, the maximum quantity of heat radiation at the face would occur during the mucking operation following blasting and would be 571 Mcal/hr. In order to maintain the air temperature within the tunnel at 30°C or lower at this time, the quantity of fresh air delivered to the face would have to be 3000m^3/min, the total quantity of ventilation air flowing in the tunnels as a whole would have to be 4000m^3/min, and, in addition, equipment would be required to cool the air to be delivered and to provide local air chilling at the face.

However, the expense of such cooling equipment would be excessive, and further, it would be difficult to secure cooling water in large quantities during the hard winter months. Therefore, excavation was performed in the high temperature section using ventilation equipment only, but supplemented by the following measures.

1) The equipment and vehicles used in the mucking operation, when the quantity of heat radiation is at its maximum, were equipped with enclosed operator's cabins with air conditioning.

2) That portion of the high temperature section in which the rock temperature was at its highest was excavated during the winter months.

Because ventilation was required in the great quantity of 3000 m^3/min in the high temperature section, the "tunnel passageway type" of ventilation was employed in which fresh air is delivered to the face by utilizing the already holed

Stage	Face position	Mucking passage	Ventilation equipment	Materials passage
1st stage	Past 350	330	0	220
2nd stage	Past 660	600	330	220
3rd stage	Past 1005	975	600	220
4th stage	Past 1375	1345	975	220

Figures are distance from investigation tunnel portal.

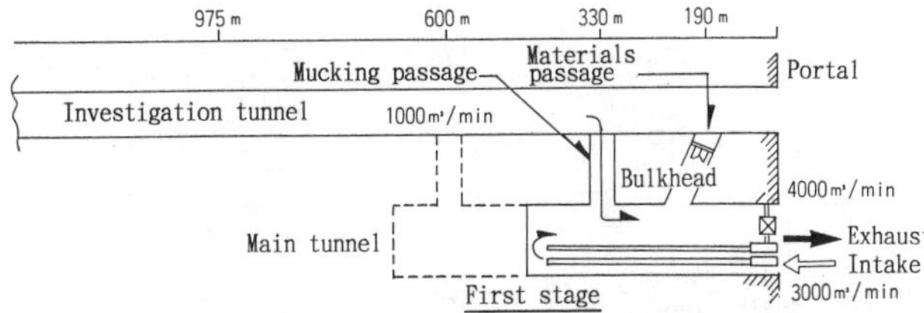

Item	Specification	Number	Type
A blower	4000 m /min 80 mm Aq	1	90kw, one stage, variable vane
B, C blowers	1500 m /min 350 mm Aq	2	55kw, two stage, inverter control
Duct	1500 diameter nonflammable vinyl fabric		

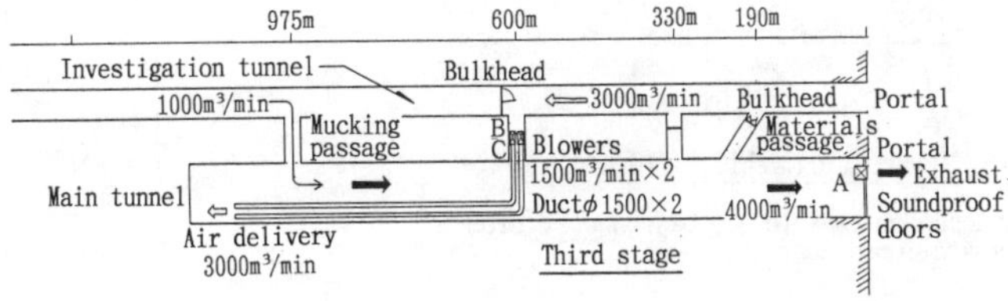

Figure 6. Ventilation equipment.

through investigation tunnel as the air supply passage and installing the ventilation equipment in the connecting passages which connect the main tunnel to the investigation tunnel at intervals of 375m. Because the connecting passages would be excavated successively as the tunnel face advances, it was planned to move the ventilation equipment successively as the connecting passages are completed. Consequently, the maximum distance between the ventilation equipment and the face is 750m or less, thus making large quantity ventilation possible. The arrangement of the ventilation equipment is shown in Figure 6.

Also, in order to perform the ventilation efficiently, a control system was employed by which the flow of ventilation air can be varied automatically in accordance with the concentration of dust and the temperature in the tunnel indicated by dust and temperature sensors located at the face. The variation of air flow is performed by variable vane pitch control of the exhaust blower (A) and inverter control of the intake-delivery blowers (B, C) in an attempt to reduce power costs.

In the main tunnel excavation the rock temperature reached a maximum of 72°C in the vicinity of TD650m where the temperature had reached 75°C in the investigation tunnel, but through the use of this system of ventilation, the temperature of the working environment within the tunnel was approximately 28°C. Because the tunnel is large in cross section, the flow of air is good with a flow velocity of 2 m/sec at the face, thus helping to reduce the concentration of dust generated during operations.

However, due to the great amount of heat radiation during mucking the engines on the loading equipment sometimes overheated, but the operation was not hindered because the operators' cabins were air conditioned.

One problem which was not forecast was the great amount of dust which arose from the travelled way within the tunnel due to the drying effect of the high rock temperature. This problem was solved by deliberately spraying water before the start of mucking.

4.2 Measures against volcanic gases

The measures which were taken against volcanic gases in the main tunnel were on the same order as in the investigation tunnel. Communicators and sensor devices were located at 200m intervals by branching off from the cables laid in the investigation tunnel and leading into the main tunnel by way of the connecting passages. The automatic gas warning system is shown in Figure 7.

4.3 Temperature cracking of lining concrete

In general there are numerous examples of cracking which has developed in the secondary lining concrete of tunnels due to temperature stresses resulting from the heat of hydration and strong restraint on the concrete by the underlying rock or shotcrete.

In a high temperature section of 70°C such as in the Nakanoyu contract section, it would be more difficult for the heat of hydration of the cement to escape than in the general case, and the temperature difference between the face of the lining

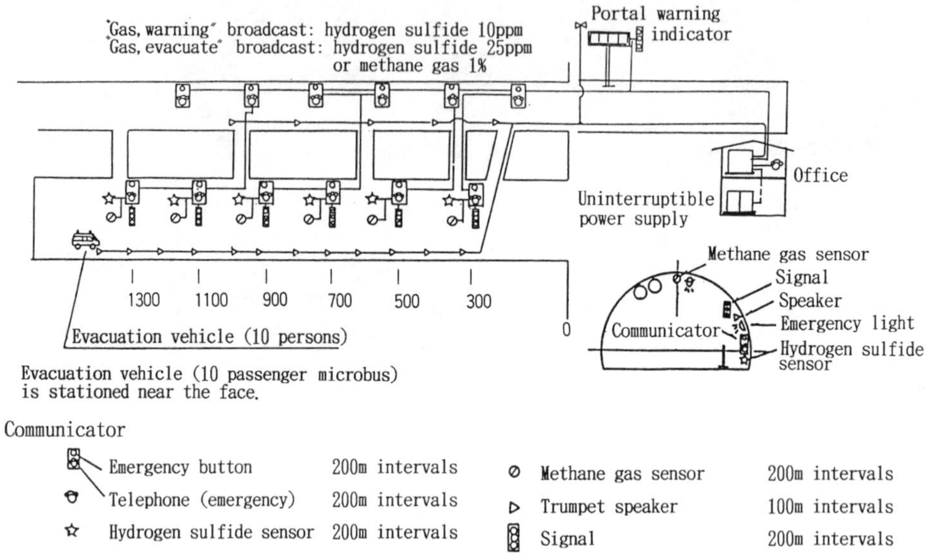

Figure 7. Automatic gas warning system.

concrete and the rock would become larger because the face which is in contact with the air inside the tunnel would be cooled rapidly by the ventilation to a temperature near 30°C. Thus, it is believed that a large temperature difference would develop within the concrete, causing the development of cracking at a relatively early age and windening of these cracks due to causes such as the drying shrinkage which follows.

In the lining concrete placed in the high temperature section of the investigation tunnel, cracking developed in both the directions parallel to and perpendicular to the tunnel axis.

Therefore, the Abo Tunnel Construction Methods Investigation Committee of the Ministry of Construction performed a temperature analysis by finite element method, a temperature stress analysis by two dimensional finite element method on the cracking which develops in the direction parallel to the tunnel axis, and a three dimensional temperature stress analysis by compensation plane method on the cracking which develops in the direction perpendicular to the tunnel axis.

From these analyses it was found that

1) The possibility that cracking will develop in the direction parallel to the tunnel axis is small.

2) The probability that cracking will develop in the direction perpendicular to the tunnel axis is high, 88%.

3) If a 5cm thickness of rigid polyurethane foam is placed as an insulation material between the lining concrete and the shotcrete as shown in Figure 8, the temperature difference between the rock side and air side of the concrete can be reduced to 5 to 10°C, and the probability that cracking will develop in the direction perpendicular to the tunnel axis is reduced to 5%.

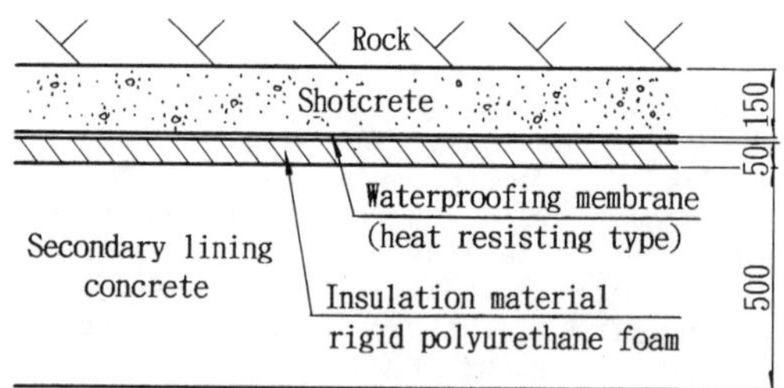

Figure 8. Insulation material.

Consequently, because it was believed that there was a great possibility that temperature cracking would develop in the lining concrete in the high temperature section, the following measures were taken:

1) Apply a 5cm thickness of rigid polyurethane foam (Table 2.) as an insulation material on the waterproofing membrane behind the lining concrete, and increase the excavation cross section by this amount.

2) Because construction joints are effective in the control of cracking in

the direction perpendicular to the tunnel axis, their spacing was reduced from the usual 10.5m to 6m.

3) Use blast furnace cement to suppress the temperature rise in concrete.

In adopting the insulation material, a trial experiment was performed within the tunnel using a water tank filled with water at a temperature of 80°C, measuring the record of temperature change from before through after concrete placing. The results of this site experiment were generally the same as the results of the temperature stress analysis described above, thus confirming the performance of the insulation material.

4.4 Excavation method

The main tunnel is being excavated by upper heading and short bench method using the NATM system of support. The face is drilled by two 3-boom hydraulic jumbos. (Figure 9.)

Unlike the investigation tunnel, there is no large-scale spoil site available at the Nakanoyu end of the main tunnel. Because of this, all of the muck from the excavation of the main tunnel has to be loaded into muck cars and hauled by battery locomotive through the 4.34km long investigation tunnel to Hirayu at the opposite end portal.

In the high temperature section, it is desirable to load the muck into the muck cars without delay in order to prevent deterioration of the work environment due to the great heat radiation from the muck after blasting.

Thus, the following method of handling muck was adopted after studying various methods.

1) Muck is loaded directly from the face by large capacity load-haul-dump carriers (3 units, $6m^3$ capacity, EIMCO 928) which dump the muck directly into the muck cars at the intersection of the investigation tunnel and the connecting passage nearest to the face.

2) The mucking operation is completed at once by loading the muck into fifty $6m^3$ capacity muck cars, the number corresponding to one round of blasting.

3) While other operations are being performed at the face, the mucking team hauls the loaded muck cars, 3 cars at a time, by battery locomotive to Hirayu. One round trip takes 80 minutes.

5 CONCLUSION

At this project site which is located at an elevation of 1360m, the outside air temperature falls to below -20°C in the winter, 50°C below the air temperature inside the tunnel. Also, the national highway is closed to traffic for a period of six months including winter. Thus, the work environment is extremely severe. However, backed by the fervent expectations of the local people, we continue our assiduous endeavors to complete this project by the time of the Winter Olympic Games which are scheduled to be held in Nagano in 1998.

Table 2. Rigid polyurethane foam.

Material

	Component	Name	Contents
Spray Material	A fluid	Premix resin	Polyol Foaming agent Cell stabilizer Catalyst
	B fluid	Polyiso-cyanate	Crude MD-1

Performance

Heat conductivity	0.022 kcal/mhr°C or less
Incombustibility	Class 3 (JIS-A1321)
Compressive strength	1.5 kgf/cm^2 or more
Density	30 kg/m^3 or more
Water absorption	3.0 $g/100cm^2$ or less

Figure 9. Drilling face in upper heading of main tunnel by two 3-boom jumbos.

Tunnelling and Ground Conditions, Abdel Salam (ed.) © 1994 Balkema, Rotterdam, ISBN 90 5410 363 9

Construction of large sized sewers by slurry machines in Hong Kong's bouldery soils

Ian McFeat-Smith & Pulasthi S. Herath
Charles Haswell & Partners (Far East) Ltd, Hong Kong

ABSTRACT : This paper describes the problems that exist in Hong Kong with regard to the construction of Hong Kong's first sewer tunnels constructed by pipejacked slurry machines. Particular reference is made to limitations imposed by the local geological conditions which include the occurrence of hard igneous boulders in soft soils. The paper presents case histories of two projects which involved the successful excavation of large sized sewers in these extreme conditions.

1 INTRODUCTION

As with most major cities, the roads of Hong Kong are underlain by an extensive network of utilities. Many of these are in an advanced state of decay due to corrosion, internal growth, loading from traffic or adjacent works - leading to leakages and ultimately collapse. As a result, substantial trench works are undertaken for refurbishment, creating much disturbance in the main urban areas. This situation is aggravated by Hong Kong's very dense traffic, estimated in 1989 at 236 vehicles per km of road, and the rapid growth of both the economy and population since the Second World War. In the urban areas, for example, the average population density now exceeds 100,000 persons per km^2. This growth necessitates continuous installation of new and larger capacity utilities thus intensifying the general disruption created.

Construction work, including sheet piling and trench excavation has traditionally taken place with little control on noise, emissions and safety. The effect of such works in the overcrowded urban areas is significant and often results in substantial impact and traffic delays with associated loss of travel time. In the past, the population has accepted such inconvenience as a way of life. However, over the last few years, a growing respect for environmental control has taken place, accelerated to some extent by powers given to the Environmental Protection Department. At the present time Hong Kong faces yet another period of rapid expansion of infrastructure works with, for example, upgrading of all major utilities; and installation of a new network of drainage and sewerage facilities associated with a series of Sewerage Master Plans. A clash between environmental control and the need for such expansion seems inevitable unless new, less disruptive, methods of installation can be found to avoid needless excavation of trenches.

One of the principal problems in Hong Kong is that, excepting water supply and sewerage, most utilities in Hong Kong are privately owned, hence it can be difficult to achieve the necessary co-ordination for the building and maintenance of multi purpose conduits. Government has recently established a Utility Technical Liaison Co-ordination Committee which hopefully has the powers needed to effect such measures and ensure forward planning.

Clearly the real solution to Hong Kong's problem of utility placement, if the full impact of trench excavation is to be avoided, is trenchless technology, as discussed in the following sections of this paper.

2 TUNNELLING CONDITIONS

The majority of Hong Kong's urban areas are located on a narrow strip of land bordering a steep rocky coastline. Previous work by McFeat-Smith et al (1989) illustrates how this platform has been developed by cut and fill techniques and gives typical geological conditions. The predominant rock type of the former shoreline in the urban areas is granite, whilst fill material for extensive reclamations often consists of completely weathered granite with boulders. The fill overlies a typical sequence of marine deposits, alluvium, colluvium, a zone of completely to highly weathered granite and an erratically weathered rockhead respectively.

The size of boulders that can be encountered in the local soils varies considerably, normally from cobbles in the marine deposits, up to very large boulders in both landslip colluvium deposits and in weathered granitic and volcanic soils due to corestone formation. Recent fill materials have been screened for boulders up to 300 mm in size whilst many of the older fill materials were less closely controlled.

Hong Kong can be described as being one of the main tunnelling centres in the world with over 80 kms of tunnels being built at shallow depths in the urban areas for projects like the Mass Transit Railway (McFeat-Smith, et al, 1985).

For rock tunnelling, excavation has traditionally been carried out by drill and blast. However the current lack of experienced miners, rapidly increasing labour costs, and the need to achieve higher rates of advance with reduced ground disturbance and environmental impact led in 1990 to the first installations of two hard rock tunnel boring machines (TBMs). These have been required for the second Hong Kong Electric cable tunnel and the Tolo Effluent Export Scheme. These two have been successful and herald a fundamental change in local attitudes towards mechanisation of underground works, which hitherto have been highly conservative.

For the 25 kms of tunnels built in soft ground and mixed face conditions for the MTR, excavation has been largely by hand held tools or by shield, sometimes with mechanical diggers and normally used in conjunction with compressed air or ground treatment. However fully mechanised slurry shield systems for these early works were not favoured due largely to the presence of the hard granite boulders and a highly variable rockhead.

3 TRENCHLESS TECHNIQUES FOR HONG KONG'S EXTREME GROUND CONDITIONS

The history of mini-tunnelling and micro-tunnelling in Hong Kong has been extremely short to-date. Pipejacking, in its traditional form, has infrequently been used for short railway and road crossings. The first major example of trenchless technology for new works occurred in mid-1989 on the Fanling Trunk Sewer project for the Territory Development Department. Details of this important case history are presented later in this paper.

Many other applications are now being planned, most notably those for sewerage projects like the East Kowloon Sewerage Improvement and Pollution Control Scheme, involving both installation of new sewers by slurry machine and refurbishment of existing pipelines. This project provides us with a second case study as discussed later in the paper.

The degree of risk of a drive in the local soils encountering rock obstructions varies with proximity to rockhead, the size and in particularly the length of tunnel, the soil strata and the local topography. In some areas only a few cobbles will be experienced whilst others will be less fortunate. Comprehensive site investigations will reduce the risk of a tunnel unexpectially encountering many large boulders (as well as other utilities). The drilling of long horizontal boreholes can minimise this risk, but will not eliminate it. Thus it is clear that any method to be adopted in Hong Kong must have some capacity to deal with cobbles and at least occasional large boulders.

For mini-tunnelling, techniques that allow access to the face for manual excavation, including the traditional timber-headings used in the UK, eliminate the risk of failure of a drive due to the presence boulders. Other remote control mini-tunnelling rigs are available that can be used in conjunction with a primary lining system. These allow the rig to be withdrawn and the obstructions to be dealt with manually. Another recent system allows the central core of the rig and cutterhead to be withdrawn and serviced. However most methods do not provide for this option hence could involve abandoning the tunnel.

The more popular techniques now being used elsewhere for mini-tunnelling, and particularly micro-tunnelling, employ remote control methods of excavation in conjunction with pipejacking. Some of these techniques include heads fitted with rock roller or disc cutters which have the capability to cut occasional hard boulders. For long drives in bouldery soils the rig should allow, replacement of worn cutters without removal of the rig. In order to cut through zones of large competent boulders, or rockhead, such rigs must also be used in conjunction with specially designed pipes that can sustain the high thrusts required or be fitted with gripper pads.

The option of using directional drilling and reaming out to large diameters offers potential. This method has recently been used to construct hard rock tunnels for sewers in Scandinavia. Thus micro-tunnelling in granite rocks is practical at present. For larger sizes, thrust, torque and directional control are of concern although this appears simply to be a question of applying existing technology rather than one of feasibility.

For directional drilling in soils bentonite flush is used to power the mud motor drill bit, minimise settlement, and avoid collapse of the hole. It is also understood that, using recent innovations, techniques are available for drilling and reaming through occasional hard boulders. To achieve this suitable standby equipment must be available on site.

Thus it is evident that the further application of available techniques now offer low risk, lower cost solutions for Hong Kong's extreme geological conditions.

4 CASE HISTORY: FANLING SOUTH TRUNK SEWER

This contract called for the construction of 370 metres of 1350 mm internal diameter sewer located 6-14 metres below ground level. The alignment passed beneath new roads, footpaths, a cycle track, a pedestrian subway, the embankment of the recently constructed Fanling Bypass and many existing services.

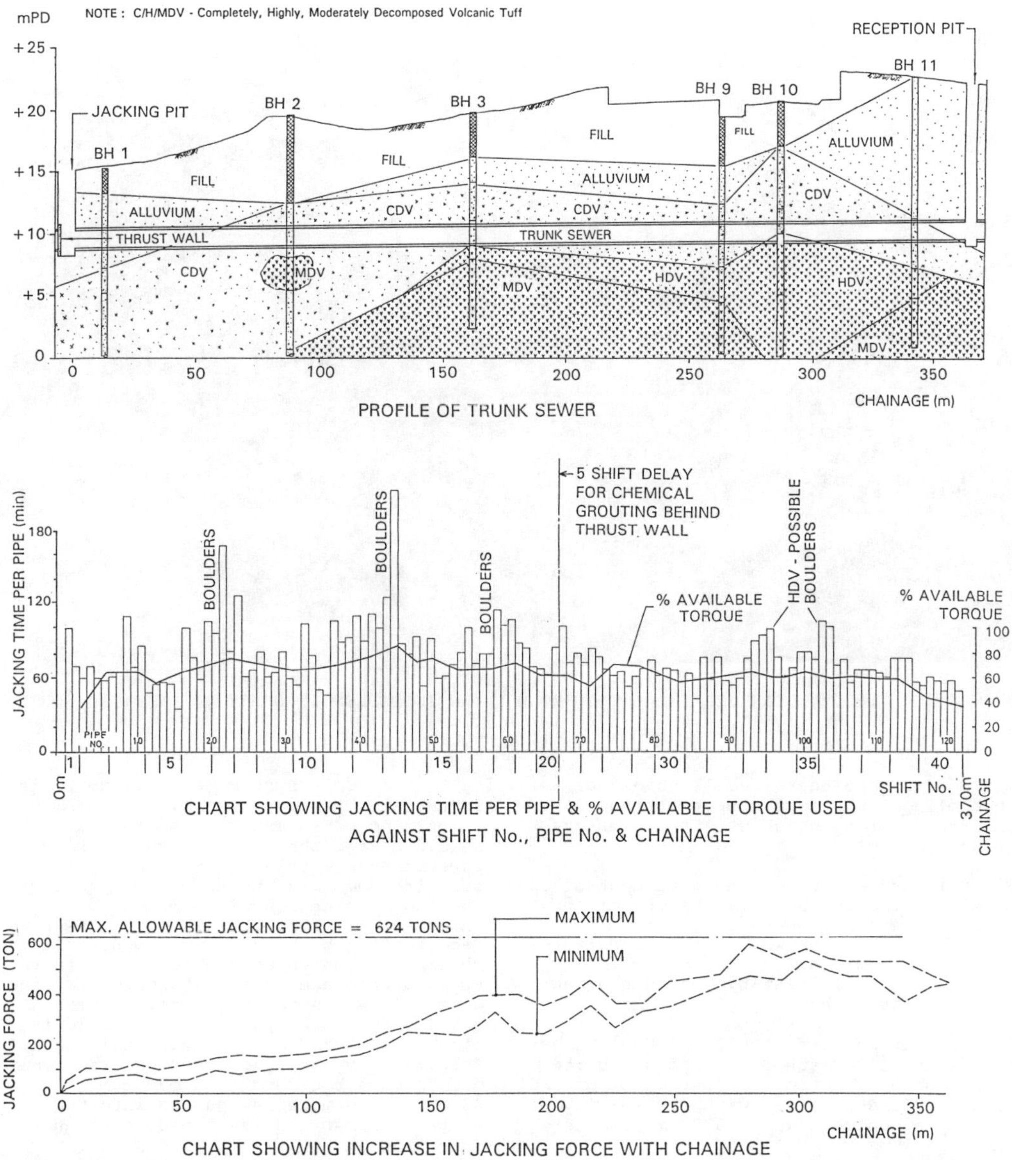

FIGURE 1

Figure 1 shows the longitudinal section which illustrates that the geological sequence of the area consisted of fill and alluvium deposits overlying moderately to completely decomposed volcanic rock (mainly silt). The water head above the tunnel soffit ranged up to 7 m.

The contract stipulated that the maximum ground movement allowable was 25 mm and that the works would be suspended if movement exceeded 10 mm until remedial proposals had been approved by the Engineer.

Given these difficult tunnelling conditions and the probability of encountering boulders Charles Haswell & Partners (Far East) Ltd was appointed by the successful tenderer, Chun Wo Construction Ltd, to determine the most appropriate method of construction within extremely tight budget and programme constraints. As trench excavation was not permitted even on revised alignments the options considered were:

A. Conventional soft ground tunnelling

FIGURE 2 1620MM DIAMETER ISEKI UNCLEMOLE BEING INSTALLED IN SHAFT FOR FANLING SOUTH TRUNK SEWER

methods using either a small shield or forepoling with close support to the face and face boards to achieve the necessary ground control.

B. Soft ground tunnelling with control of water inflows and hence stability of the soils achieved by means of:
1. Dewatering by well point lowering
2. Compressed air
3. Ground treatment by the tube-a-manchetts method

C. Excavation using the Iseki Unclemole mini-tunnelling method incorporating the slurry shield system for control of ground water and a cutterhead with the ability to cope with boulders up to 500 mm diameter.

From this review it was determined that options A and B would take 6-9 months to construct, which was unacceptable for the project programme and option A in particular would require the use of skilled miners, who were in short supply at that time, to minimise the risk of face collapse. It was also considered that option B.1. would have resulted in unacceptable settlements whilst options B.2. and B.3. would be too expensive.

Thus option C was recommended as the only acceptable solution in spite of its relatively high cost, the possible time constraints to import the plant from Japan and some uncertainty about the maximum size of boulders to be encountered.

The Unclemole system (see Figure 2) is essentially a slurry tunnelling machine comprising of an articulated shield. The shield houses the cutter head which is powered by electric motors, the hydraulic steering rams and an instrument panel showing machine functions and guidance information. This is relayed to the surface monitor by a TV Camera mounted in the shield. As shown on Figure 2, the four spoke cutter head has 21 tungsten carbide bits and is mounted on a cone shaped rotor. The fixed part of the shield is also conical and as the machine is jacked forward, a gyratory crusher progressively grinds down boulders to a maximum size of 40 mm. This material passes into the slurry chamber and is pumped to the surface through the slurry pipe lines.
In order to change worn cutters and if necessary, to rescue the TBM from boulder obstructions, it was planned that intervening manholes could be used for this purpose.

Three intermediate jacking stations were installed. These consisted of a steel skin and spreader ring containing 10 number 50 ton hydraulic jacks with a 300 mm stroke. The steel skin fitted over the specially constructed trailing pipes.

The mucking system consisted of charge pumps, discharge pumps, pit by-pass unit, electromagnetic flow meter, slurry tank, settling tank and slurry pipeline.

Bentonite was mixed with water at the slurry tank and pumped to the cutter head through the pit by-pass unit installed in the jacking pit. Spoil excavated by the shield was pumped through the discharge pipes into the settling tanks on the surface. The spoil was settled out and the slurry recycled. The electromagnetic flow meter was used to measure the slurry flow and ensure that the flow rate was higher than the critical sedimentation flow velocity. The operator was able to monitor and control this on the surface. The slurry was also used to maintain a pressure balance with the ground water by means of a valve in the jacking pit. Injection of bentonite lubricant to minimise ground friction was carried out at the shield, interjacks and through concrete jacking pipe grouts holes at 5 m intervals. In total 3300 litres of bentonite lubricant was used representing 1.6 tonnes of Poly Glort.

The spun concrete jacking pipes were manufactured under licence by Doran (HK) Ltd and designed and tested to BS 5911 and Japanese Industrial Standards. An embedded steel collar formed the pipe socket and each pipe had two grout holes for lifting and injection of cement grout around the pipes on completion of pipe jacking.

The reinforced pipes were 3 m long, 125 mm thick weighed 4.5 tonnes and manufactured from 50 N/mm^2 concrete. Each pipe was designed to withstand a jacking force of 624 ton. Purpose designed rubber seals prevented the ground, groundwater and bentonite lubricant entering the pipeline at the joints between pipes, interjacks and the shield.

A 7.5 x 6 m x 7 m deep sheet pile jacking pit was excavated at chainage 0 m complete with reinforced concrete thrust wall. The jacking equipment consisted of four 200 ton hydraulic main jacks with a stroke of 800 mm. These were installed complete with pressure plate, spacer blocks, thrust ring and guide rails which provided the initial directional control of the shield and jacking pipes. A 5 m diameter receiption pit was constructed at chainage 370 m by the hand dug caisson method.

Jacking commenced on 24.7.89 and was completed on 2.9.89. Apart from a delay of 5 days required to chemical grout behind the thrust wall which moved when the jacking pressure reached 400 ton, there were no interuptions to progress. The average production per day was 3.5 pipes (10.4 m/day). The jacking progress per day, percentage torque, jacking time per pipe and jacking forces are all shown on Figure 1.

Evidence that the shield passed through boulders was found on three occasions in the form of crushed rock in the spoil. This coincided with the longer jacking times for pipes 21, 44-45 and 59-61.

The three interjacks installed in place of pipes 15, 47 and 86 were not used as the actual jacking pressure stayed within the capacity of both the main jacks and the concrete pipes, the silty nature of the volcanic soil providing an ideal jacking medium.

Before construction commenced calculations where carried out to predict the settlement caused by pipe jacking. In order to stay within the 10 mm limit, the maximum face loss allowed was 2% (0.4 m^3/m). The worst actual case monitored gave maximum settlements of 3 mm at 10 m from the tunnel centreline and 1 mm heave above this point. The average settlement recorded was 0.5 mm.

5 CASE HISTORY: EAST KOWLOON SEWERAGE IMPROVEMENTS STAGE I

This project was designed to alleviate the sewerage problems due to a rapid increase of the urban population in the Eastern region of Kowloon peninsular. The project involved both open cut excavations and trenchless techniques for the installation of new pipe sewers. A 104m long section of 1950mm i.d. pipeline was installed at a depth of 9-11m below ground level by pipe jacking in order to avoid any disturbance to the surrounding urban areas. This was carried out beneath a major highway where opencut construction was not a feasible alternative. One of the main objectives of the minitunnelling work was to avoid clashes with existing utilities including cables, ducts and pipes which wholly occupied the pedestrian footways and about 1/3 of the carriageway. These were located at a depths ranging from 1m to 3m from the road surface.

Charles Haswell & Partners (Far East) Ltd in association with Donohue / JRP were responsible for the design and supervision of the scheme on behalf of the Drainage Services Department of Hong Kong. Aoki Corporation was the Contractor.

As illustrated by Figure 3, the geological formation of the area consisted of fill materials, namely clay and silty clay with small to large sized boulders (up several to metres) at the proposed level of the pipe jacking route. A coarse grained alluvium layer was found below this. Ground water was located 3m below the ground level.

The stipulated allowable ground settlement was 5mm and, in the event that this was exceeded, boring was to be ceased and remedial measures undertaken in consultation with the Engineer.

A Komatsu Semi Slurry Shield (see Figure 3) was chosen for tunnel boring as it demonstrated a satisfactory track record in soil conditions similar to those anticipated on this contract. It is capable of crushing boulders up to 400mm diameter. In the event that larger boulders were encountered these were to be removed by hand means. Access to the face is gained through the machine chamber and removal of boulders would be affected using handbreakers to reduce them to a size that could be removed by hand or crushed by the boring machine.

In view of the soil types expected a high water Absorbent Polymer was used as an admixture. A specially designed jack system called Molemeister was employed.

PROFILE OF SEWER ALIGNMENT

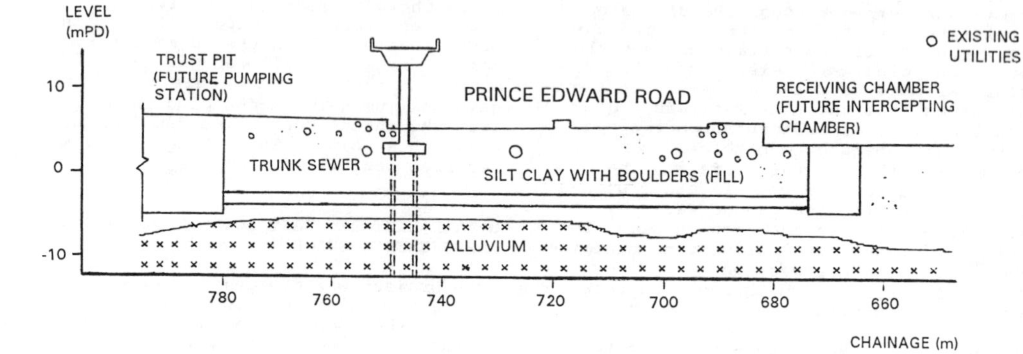

THRUST Vs CHAINAGE

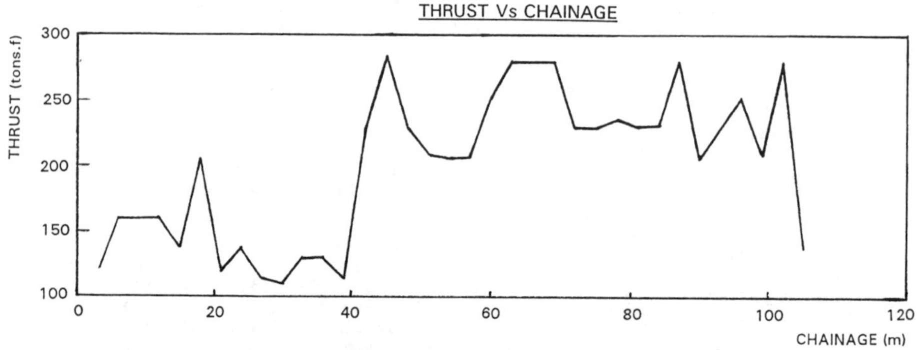

KOMATSU SLURRY SHIELD

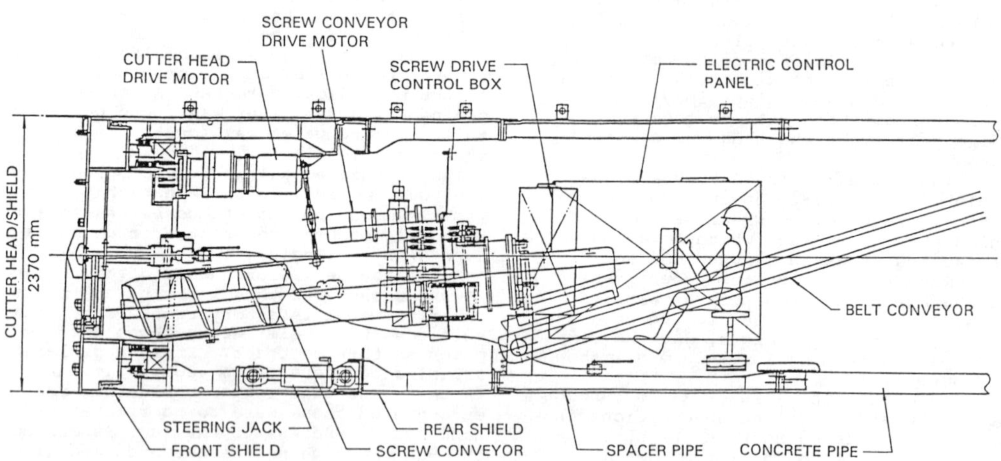

EAST KOWLOON SEWERAGE IMPROVEMENT & POLLUTION CONTROL - STAGE 1

FIGURE 3

It has a stroke of 3000mm and is capable of handling 3000mm long jacking pipes. The type of precast concrete pipes used for this project was identical to those used for the Fanling South Trunk Sewer discussed previously, although larger in diameter.

The steel piled jacking shaft was sunk whilst ground treatment by chemical injection was provided around the sheet piles to seal off the ingress of groundwater and to prevent the collapse of the ground around the entrance ring at the commencement of the pipe jacking operation. A concrete base slab was placed in the bottom of the jacking shaft and sump pumps were installed to drain the groundwater. It was necessary to provide ground treatment by chemical injection on the outside of the sheet piles of both the jacking and receiving pits to stabilize the soil against collapse. A rubber seal was fitted to the entrance ring in order to prevent ingress of ground water and mud during pipe jacking operations.

Once the jacking equipment was in place the counter flow unit was placed in the first concrete pipe and connected to slurry lines using flexible slurry hoses to allow lubrication of the periphery. A test run was carried out both on the jacking equipment and slurry pumps to ensure the whole system was ready for operation.

During driving the operator controlled the jacking speed, cutter head tarque, earth pressure and the inclination of the shield. The proposed maximum jacking speed was 100mm per minute but the actual production rate averaged 5m per day and good control of line and level was achieved.

The number of boulders encountered was less than the expected number calculated from probability calculations. The maximum boulder size was 300mm and in practice these did not have a significant effect on the performance of the machine.

Lubricant grouting was carried out through the grout holes in the 3rd pipe behind the machine using 'Subelu' lubricant. These holes were also used for grouting the annulus around the pipeline upon completion. While the pipejacking operation was in progress the lubricant was pumped continuously to the periphery of the concrete pipes to reduce the jacking friction.

On completion of jacking operation it was necessary to grout the outer periphery of the jacking pipes for the full tunnel length. A grout mix consisting of cement, fine sand, bentonite, fly ash, dispersing agent and water was pumped through the grout holes which were sealed after completion of grouting.

Settlement monitoring points had been set up at 5m intervals along the centre line of the jacking pipe and points offset by 5m and 10m to either side. Additional monitoring points had been placed on all adjacent structures. The settlements recorded were significantly less than the allowable 5mm at all settlement monitoring points.

6 CONCLUSIONS

The benefits of adopting trenchless technology for Hong Kong's urban environment are substantial and becoming increasingly evident to Government. Utility organisations tend to consider only the installation cost and thus select the lowest cost option unless statutory requirements prevent this. In view of the very high costs associated with trench works to the community such controls must be applied.

The success of both installations is evident, particularly with regard to the ability of the machines to cope with boulders derived from hard igneous rock, and to the avoidance of settlement. The size of the boulders encountered on the Fanling South Trunk Sewer project could not be determined however it is postulated that these ranged up to the maximum size specified by the manufacturer, ie approximately 1/3 of the area of the face, or 500 mm in this case.

The degree of risk accepted by the Contractor in this case was reduced by his undertaking additional ground investigations and his retaining the option of using the intervening manholes to rescue of the TBM in the event of an impassible boulder being encountered.

Such contingency plans are essential unless the methods of construction to be adopted in such extreme ground conditions are capable of excavating through large boulders. A number of options have been investigated and presented in this paper. It is evident that these rely more on the application and adaptation of proven methods rather than the development of new ones. As with all tunnelling, selection of the most appropriate techniques depends upon access, layout, geological conditions and accurate identification and consideration of all the other constraints. Clearly each project is unique hence the availability of specialist advice is a prerequisit for ensuring success.

7 ACKNOWLEDGEMENTS

The writers gratefully acknowledge permission given by Chun Wo Construction and Engineering Co Ltd and the Drainage Services Department of The Hong Kong Government to publish this paper.

8 REFERENCES

McFeat-Smith I., Workman D., Burnett A.D. & Chau E.P.Y. 'Geology of Hong Kong'. Bulletin of the Association of Engineering Geologists, Vol. XXVI, No. 1 1989, pp. 23-107.

McFeat-Smith I., Turner V.D. and Brace-girdle D. Tunnelling Conditions in Hong Kong. Hong Kong Engineer, Vol. 13, No. 6, June 1985.

Tunnelling and Ground Conditions, Abdel Salam (ed.) © 1994 Balkema, Rotterdam, ISBN 90 5410 363 9

Underground openings for public use – Some results from the research programme for the Gjøvik Olympic Mountain Hall

Øystein Meland
SINTEF, Trondheim, Norway

Einar Broch
Norwegian Institute of Technology, Trondheim, Norway

ABSTRACT: In may 1993 the Gjøvik Olympic Mountain Hall in Norway was officially opened by His Majesty King Harald V. This world's largest underground cavern for public use will have seating for 5500 people for icehockey games during the Winter Olympic Games 1994.

In connection with planning and construction of the cavern a comprehensive research programme was initiated. The programme includes a wide variety of studies within engineering geology and rock mechanics as well as studies within energy consumption, fire protection and other safety aspects related to public use of the underground. The programme aims at establishing desing criteria, codes of practise and regulations for public underground openings. The paper gives a description of the programme and also presents some of the more important findings.

1 INTRODUCTION

Utilizing the underground has a long tradition in Norway. 200 hydroelectric powerhouses are already put underground, and more than 3.500 km of tunnels are excavated for the hydropower industry. 750 railway tunnels and 850 road tunnels are vital parts of the country's transportation system. Important infrastructure elements to be found underground are water and sewage treatment plants, tunnels for drinking water as well as for waste water, caverns for storing of drinking water, oil and other hydrocarbons, car parks etc (Broch, 1989).

Twenty years ago the first underground sports-hall was put into use, and in 1975 the first underground swimming pool was opened to the public in Gjøvik. Next to the swimming pool the regional telecommunication centre was safely put into caverns excavated in the bedrock close to the town centre, (Broch and Rygh, 1976).

When the organizers of the 1994 Winter Olympic Games decided that an ice hockey rink should be built in Gjøvik, it was only natural that an underground solution came up as an alternative. Gjøvik already had long and positive experience from the operation of its swimming pool, and there was plenty of easy accessible rock volume.

In January 1991 the Gjøvik City Council made the historic decision to build the world's largest rock cavern for public use: THE GJØVIK OLYMPIC MOUNTAIN HALL.

The main cavern has a span of 61 m (by far the largest in the world), a length of 91 m and a maximum height of 25 m. During ice hockey games the total seating capacity is 5.800. The cavern is located in jointed gneiss of average RQD = 60-70 % and with Q-values varying from 1 to 30. It has a rock cover of only 25 to 50 m, thus posing challenging geotechnical design problems. Figure 1 shows an early plan of the pre-existing caverns and the new ice-hockey cavern.

Construction work started in April 1991 and after only 8 months a total of 140.000 m^3 of rock was blasted out of the mountain, three months faster than originally planned. Figure 2 shows a picture taken during excavation.

The construction work was finished in April 1993 which gives a total construction period of exactly two years. The Gjøvik Olympic Mountain Hall, as it is officially named, was opened on the 6th of May 1993.

A comprehensive research and development programme has been integrated as an important part of the planning and design as well as the construction of this record-breaking underground structure. The research is guided by a steering committee consisting of representatives of all companies and organizations involved, and is carried out by the leading research institutes in Norway.

The programme is carried under the following headlines:

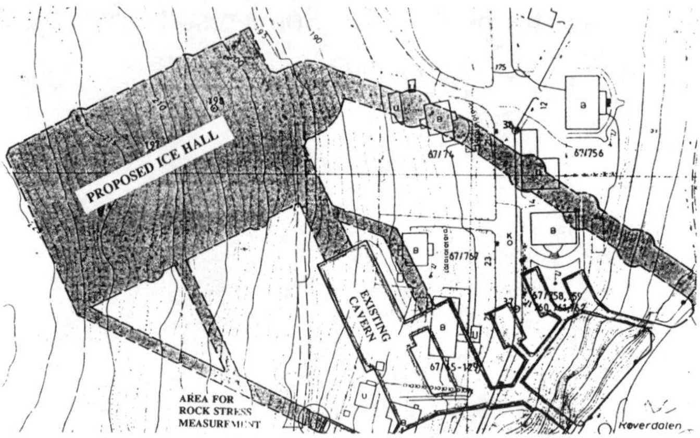

Figure 1. Early plan view of the Gjøvik caverns (NOTEBY AS 1990)

Figure 2. Picture during excavation

* Ventilation, energy fire safety
* Rock engineering and engineering geology
* Environment, safety, documentation
* Codes and regulations

An additional activity within the programme is the international marketing of Norwegian rock engineering and underground technology in general, and the publishing of results from all the research projects related to the design, construction and operation of the Gjøvik Olympic Mountain Hall. A number of scientific and technical papers have already been published in international journals and presented at international meetings like this congress here in Cairo.

The most important part of this publishing effort will take place in Gjøvik later this year (1994). The international audience is invited to take part in a symposium devoted to "Underground Openings for Public Use" from 14 to 17 June. The participants will, of course, be invited to study the Gjøvik Olympic Mountain Hall "*in situ*", as well as to visit other interesting underground openings in the Oslo area. The symposium is sponsored by a number of organizations, among them the International Tunnelling Association.

The conference themes are defined as:

* Rock engineering
* Heating, ventilation and air conditioning in rock caverns
* Fire safety and escape strategies
* Psychological Impacts
* Case histories
* Special session on the Gjøvik Olympic Mountain Hall

Two months before the submission dead line some 40 abstracts from 15 countries had been received.

The succeeding chapters of this paper will firstly (chapter 2) contain a brief description of the total research programme with key words for the different sub-tasks. While the work on some of these are almost completed by summer 1993, work on other projects is still in an early stage. The intention is to have all projects completed and properly reported by the end of 1994.

In chapters 3 and 4 selected topics from the sub-projects "Rock engineering and engineering geology" and "Ventilation, energy, fire safety", respectively, will be presented. These are the projects in which work for some tasks have been completed, and which contain results that should be of particular interest to this audience.

2 RESEARCH PROGRAMME "ROCK CAVERN STADIUM" Organization and financial basis

The research program is organized with a project manager and a steering committee with representatives from the contractors, the financial contributors, and the participating research organizations. A project committee and the subtask leaders coordinate the activities and the relations between the five different subtasks. To ensure that results from recent research and development were included in the ongoing construction and design, the contractors received information such as stress measurements, geological mapping and rock-mass classification, crosshole seismic tomography, georadar, core index testing, numerical modelling, vibration monitoring and results from stress and deformation monitoring.

The program has an estimated budget of NOK 25 mill (approximately USD 3.5 mill) and is carried out in collaboration with The Foundation for Scientific and Industrial Research at the Norwegian Institute of Technology, The Norwegian Geotechnical Institute and The Eastern Norway Research Center.

The research program is split into five subtasks, each with a different number of activities.

Subtask 1: Energy consumption, HVAC installations, Fire Safety design
Activity 1 Quality control of energy and climatic units
Activity 2 Development, verification and documentation of simulation models related to heat accumulation and energy consumption
Activity 3 Optimal energy facilities for rock cavern stadiums
Activity 4 Fire safety in rock cavern stadium

Subtask 2: Engineering geology and Rock mechanics
Activity 5 Instrumentation and rock mechanics
Activity 6 Evaluation of rock reinforcement methods
Activity 7 Follow-up of geoengineering field methods
Activity 8 Geophysical measurements
Activity 9 Numerical analyses
Activity 10 Vibration control and blasting technology
Activity 15 PC-based system for dynamic measurements
Activity 16 Development of CW radar for foundation studies

Subtask 3: Environmental aspects
Activity 11 Experience of safety and risk

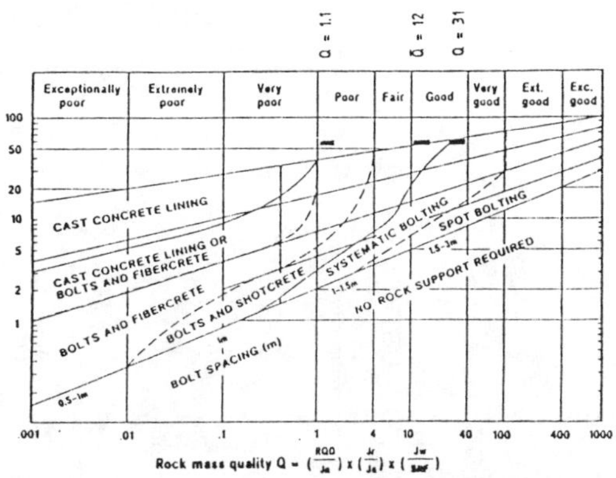

Figure 3 The 61 m span places the Olympic Cavern at the top of the reinforcement data base. (GRIMSTAD et. al., 1986)

conditions in rock stadiums
Activity 13 Information neighbourhood
Activity 14 The public, psychology and the design of facility

Subtask 4: Guidelines, codes and regulations
Activity 12 Revision of regulations for public facilities in rock caverns

Subtask 5: International marketing of products and technology related to preinvestigations, design, construction and use of underground caverns
Activity 17 International Marketing

3 ROCK ENGINEERING AND ENGINEERING GEOLOGY

The different tasks within this sub-task have been carried out in a cooperation between the research organization SINTEF at the Norwegian Institute of Technology, the Norwegian Geotechnical Institute (NGI) and NOTEBY (consulting engineers). The latter company already had valuable experience from its earlier involvement in the design and construction of the underground swimming pool and telecommunication centre.

3.1 Rock Mass Classification

Engineering geological mapping for rock mass classification was in the initial phase done in the already existing caverns, see Fig 1. For the second phase of the pre-investigations four diamond cored holes of 50 to 70 m length were drilled.

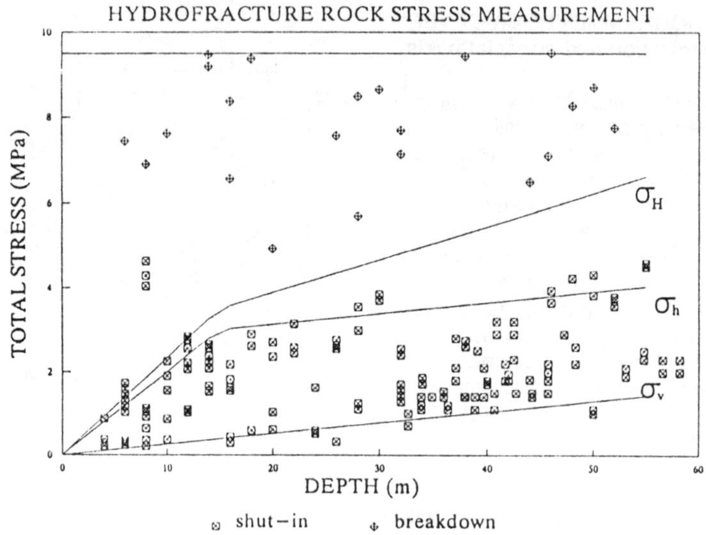

Figure 4 Results of rock stress measurements using hydraulic facturing and joint jacking-shut-in (BARTON et al, 1992).

Based on the combined cavern mapping and core logging the following typical rock qualities were expected in the actual cavern.

1. Typical best quality

$$Q = \frac{90}{6} \times \frac{2}{1} \times \frac{1}{1} = 30$$

2. Typical poorest quality

$$Q = \frac{30}{9} \times \frac{1.5}{3} \times \frac{0.66}{1} = 1.1$$

3. Weighted average:

$$Q = \frac{73}{6.6} \times \frac{2.2}{1.8} \times \frac{0.9}{1.0} = 12$$

Figure 3 shows the planned 61 m span Ice Hockey cavern plotted on the Q-system rock support diagram. The exceptionally large span and the high safety requirement (ESR=1.0 to 0.8) places the cavern right at the top or even above the present data base. The need for careful numerical analysis to support the empirically derived reinforcement prediction was therefore evident. During the construction phase key geological parameters including the Q-system data for rock mass classification and for subsequent jointed rock modelling studies were recorded. Mean Q-values observed in the cavern pilot tunnel of 80 m^2 cross-section were 7.4 (typical range 4-27), and in the cavern arch as a whole 9.4. The typical range observed here was 1 to 30 as predicted.

The jointing in the cavern is irregular, rough walled and with quite large variations in dip and strike.

3.2 Rock Stress Measurements

Preliminary measurements of rock stress using overcoring were performed in a 9 m deep hole drilled from an existing cavern. These measurements indicated a surprisingly high horizontal major principal stress of about 4 MPa with an NE-SW orientation, approximately normal to the cavern axis. However, these first measurements needed to be verified by additional stress measurements since they were carried out close to an existing cavern.

Subsequent hydraulic fracturing and hydraulic jacking measurements confirmed the generally high horizontal stress levels, but suggested a N-S principal stress orientation.

The major horizontal stress was estimated to be 3.5 MPa, 5 metres above the arch of the planned cavern, and oriented N 174° E (approximately N-S). The intermediate stress estimated from joint jacking tests was estimated to be 2.0 MPa, 5 metres above the arch, and oriented N 084° E (approximately E-W). The vertical stress was calculated to be approximately 1.0 MPa at this same location, some 40 m beneath the surface.

Shortly before cavern construction commenced in April, 1991, a further set of hydraulic fracturing stress measurements were made in the upper 30 m of rock. The combined data sets are shown in Figure 4. The high stress to within about 10 m of the surface is a very positive aspect of the site for ensuring the stability of large span excavations. Key properties favourable to large cavern construction have been the high horizontal stresses and the relative roughness or waviness of foliation planes and of the more continuous joints. The rock reinforcement with rock-bolts and shotcrete has taken care of less favourable features such as the less favourable joint orientations and mean RQDs of only 65 to 70% in the disturbed rock exposed by the blasting.

3.3 Numerical modelling

Numerical modelling has been done at different stages during the planning, design and construction of the cavern. At an early stage simple finite element models and boundary element models were indicating arch deformations in the range of 2 to 20 mm. Maximum horisontal stresses were mostly in the range 3 to 5 MPa. For these studies rock mass moduli of 10, 30 and 51 GPa were utilized.

UDEC-BB modelling with a jointed model was done at three stages as new information about the stress situation and the rock mass properties were obtained. The second phase of UDEC modelling was performed following the core drilling, stress measurement and cross-hole seismic tomography. The excavation procedure was still simplified, since no final designs were yet developed for the cavern. Figure 5 shows the fracture pattern and the boundary stress conditions assumed in the modelling. This modelling verified the stability of the cavern and gave maximum deformations of 5 mm.

A diagram of the bolting pattern and excavation sequence that was modelled in the final phase is shown in Figure 6. The numbers in the cavern area refer to the excavation steps. The rock mass in the cavern arch and the walls were numerically reinforced by untensioned fully grouted rock bolts and anchors after each numerical excavation step. The UDEC-BB results for each of the excavated steps are given in Table 1.

A new series of FEM modelling was conducted subsequent to cavern excavation. The aim of this was to compare results from FEM- and UDEC analysis. A 2D continuous, non-linear finite element program was used for the modelling. The study included four different combinations of *in-situ* stresses and Mohr-Coulomb parameters. Real ground surface profile was modelled due to the limited overburden, and a five-stage excavation sequence including the rock support was followed (Figure 6).

The mesh adopted for the analysis, shown in Figure 7 represents the central cross section of the cavern. Figure 3 also shows the two different horizontal stress levels used in the FEM calculations. Based on the stress measurements, these are believed to represent the upper and lower bound. Rock bolts are taken into account in each excavation stage by means of incorporating corresponding bolt elements.

Table 2 summarizes input parameters and predicted maximum displacement of cavern crown in each excavation stage for the four combinations of stress levels land Mohr-Coulomb parameters. The calculations confirm the general stability of the cavern although a yielding zone extending more than 15 m above the cavern crown has been identified. The measured crown displacement is 8 mm.

Details about the numerical modelling and a

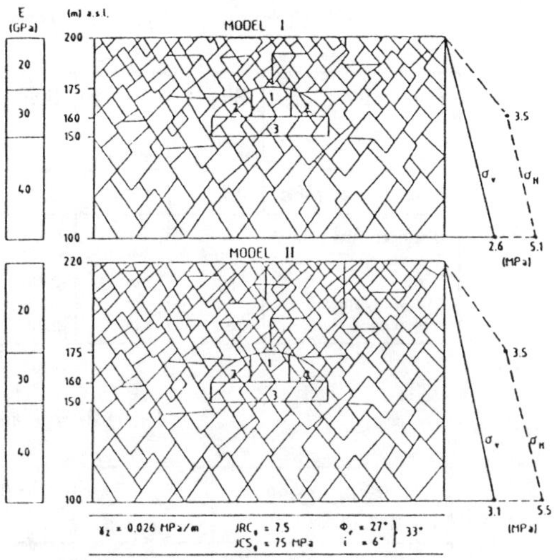

Figure 5 Phase II assumptions concerning boundary stress conditions, deformation modulus (E) and joint properties. Model I =n 25 m overburden, Model II = 45 m overburden (BARTON et al, 1992).

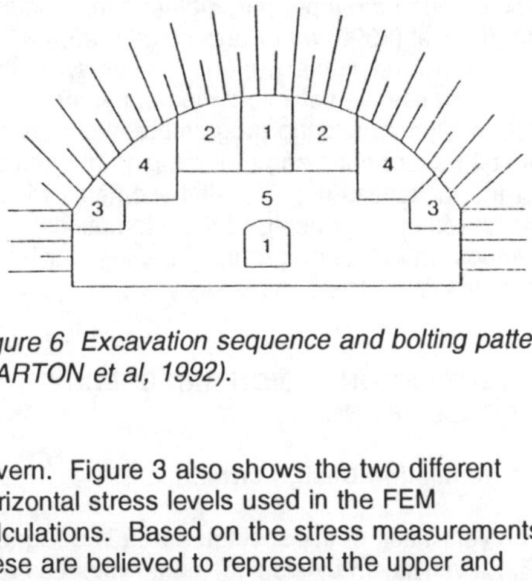

Figure 6 Excavation sequence and bolting pattern (BARTON et al, 1992).

Table 1 Summary of Gjøvik Ice Hockey Cavern run (with Postal Service caverns)

Parameter	Step 1	Step 2	Step 3	Step 4	Step 5	Excav. of 1st cavern	Excav. of 2nd cavern	Excav. of 3rd cavern
Max. Stress MPa	9.29	11.49	9.91	8.39	8.37	8.56	8.71	8.83
Displacements (mm) * maximum * wall * Crown (vertical component)	 1.85 - 0.50	 1.80 - 1.08	 2.63 - 2.62	 6.99 1.33 4.05	 8.16 3.78 4.33	 8.28 3.88 4.39	 8.43 3.92 4.87	 8.65 3.97 7.01

Table 2 Results from FEM-modelling

Boundary stress level	Mohr-Coulomb parameters		Maximum crown deformation (mm)
	< (MPa)	ϕ	
Low	0.3	39.5	3.9
High	0.3	39.5	14.5
Low	0.3	45.0	3.5
High	0.3	45.0	8,7

discussion of the input parameters can be found in Barton et.al (1992) and Barton et al (1993). These papers also describes the drill core analysis, the use of cross-hole seismic tomography, the deformation monitoring programme, the monitoring of bolt performance and the mapping programme for the rock mass quality. All these items will be discussed in great depth at the international symposium on "Underground Openings for Public Use" in Gjøvik 14-17 June this year.

4 VENTILATION DESIGN AND ENERGY CONSUMPTION

4.1 Ventilation Design criteria

The ventilation of the spectator area is designed with a ventilation-by-displacement system as an alternative to ventilation by complete mixing. Applying this method of ventilation, implies that air is supplied at low levels and extracted near the ceiling utilizing the buoyancy forces to create a thermal stratification. In this design, air is let into the spectator area through the openings from the surrounding hall as well as from devices located at the upper level of the arena. Extraction devices are located beneath the ceiling. Figure 8.

The heating, ventilating and air-conditioning installations for underground facilities are not much different from those of conventional buildings. However, during the design some topics should be especially addressed. These topics are mainly related to humidity, climate during different and changing types of activity and smoke control in case of fire.

Compared with ordinary insulation materials the thermodynamic properties of rocks have much higher heat accumulation capacity, heat transfer capacity and density. As the walls in this type of constructions are "infinitely" thick, the insulation theoretically becomes very good. Assuming a situation where the cavern has a constant air temperature about 10 K above the constant rock temperature, a temperature gradient of 0,1 K/m is reached at a depth of about 50 m. In the deeper part, the temperature change becomes very slow, and the time constant is in the order of some hundred days. This is only correct when the rock is dry. Often water is penetrating and highly influencing the thermodynamic properties.

Short period changes in the room temperature, for instance from day to night, will only cause changes in the rock temperature in a thin layer. Using the thermal properties for a granite type of rock, the thickness of this layer can be estimated to be in the range of 15 cm. Most often the inside ceiling is covered by a suspended ceiling and/or insulation of some kind, which may have an influence. When designing the air-conditioning system, these parameters should always be taken into account before deciding the capacities.

As the rock mass is completely airtight, the only infiltration of outdoor air will be through the doors in the access tunnels. Normally these will be closed. Thus the infiltration parameter will have only a minor influence on the total energy consumption for underground facilities like this.

If water diffusion through the shotcrete layer covering sidewalls could not be avoided, a minimum ventilation airflow should always be present to avoid damages and growth of fungi inside cracks and joints which for sensitive people could cause allergenic reactions.

In areas where radon is a problem this should be taken into account when calculating the airflow capacities of the ventilation system. In any other situations the ventilation airflow rate for underground installations due to air quality should be calculated as for any other type of buildings.

In situations where the ventilation air inlet is through excavated shafts without any type of a watertight lining, water will tend to penetrate. In the winter time with low air temperatures , ice can build up and restrict or close the air inlet opening. This can be taken care of by preheating the air before entering the shaft. Shafts and tunnels should also be supplied with a drainage system for water to avoid this problem and to better control the climate inside the installation.

To calculate the energy consumption and decide upon the design criteria for the ventilation system, different computer codes with various complexity can be used. For the actual design at Gjøvik Olympic Cavern Arena the computer codes ROCKHALL[1], FRES[2] and KAMELEON[3] have been applied. At the time when this paper is written, extensive measurements are carried out to verify the results from using this codes.

Fire safety design

Initially underground spaces for public use like this rock cavern are more dangerous in case of fire because they are closed spaces with few and long escape routes.

With daylight buildings a lot of research has focused on fire safety through the proper design of escape routes according to codes and guidelines describing the size of the area and net width. Basically these criteria can be applied to rock caverns for public use as well. In practice, geotechnical aspects make it necessary to design solutions which provide adequate safety without excessive cost.

To achieve the same standard of safety as in daylight buildings, underground openings for public use like this rock cavern, can be designed with inside "safe areas". No combustible or hazardous materials should ever be allowed to be used or stored in this "safe areas". With a properly designed smoke ventilation system, it is possible to

[1] *Computer code for calculation of heat exchange between surrounding surface inside a rock cavern ref [Thorbergsen, 1993]*

[2] *Computer code for calculation of energy flows and energy transfer in building constructions ref [Rømen, 1993]*

[3] *The CFD-code in use is KAMELEON developed at SINTEF Applied Thermodynamics.*

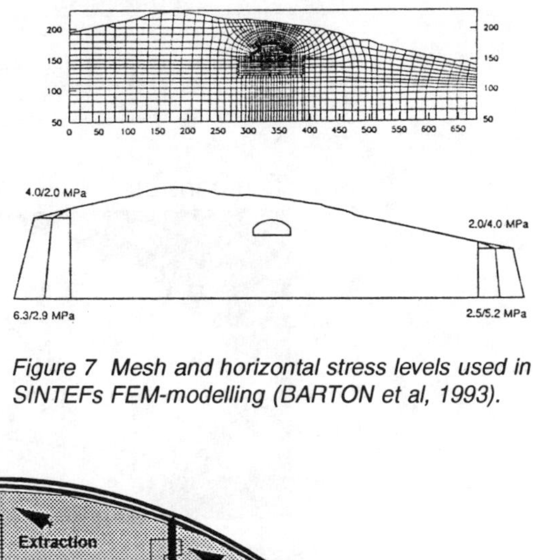

Figure 7 Mesh and horizontal stress levels used in SINTEFs FEM-modelling (BARTON et al, 1993).

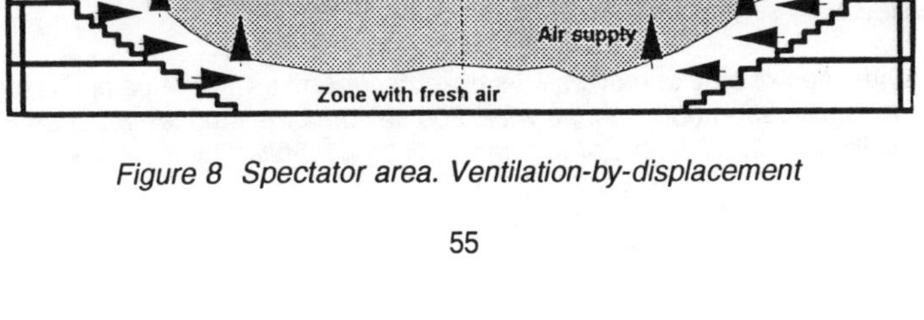

Figure 8 Spectator area. Ventilation-by-displacement

maintain these areas without smoke when there is a fire. However, attention should always be paid to the refinement of the operating and maintenance procedures for these life safety systems.
The arena and the spectator area are normally occupied areas. In case of fire, this is assumed to start in one of this areas. In the surrounding area

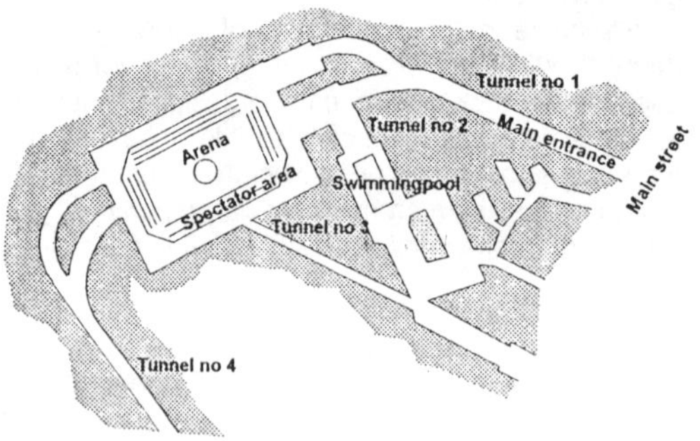

Figure 9 Gjøvik Olympic Mountain Hall.

including the tunnels 1-4, the extent of combustible and hazardous materials is highly restricted. The smoke control system is designed to keep the surrounding areas free from smoke by extracting air from the spectator area/arena and supplying air from outside through the tunnels .
With the actual design, in case of fire in the spectator area, smoke will be extracted by using an extraction system with a designed capacity of about 80 m^3/s. The airflow temperature through this extraction system should never exceed 70 degrees Celsius. The system will be activated from smoke detectors located beneath the ceiling. From that time on, the air supply system will be stopped and air will be supplied from the outside through the main tunnel (no.1) as well as through the two emergency tunnels (no.2 and no.3). The strategy with the mechanical smoke ventilation system implies that evacuation from the spectator area always will be in a direction against the inflowing air. The air flow pattern is verified by using computerized fluid dynamic calculations as shown in Fig. 10.

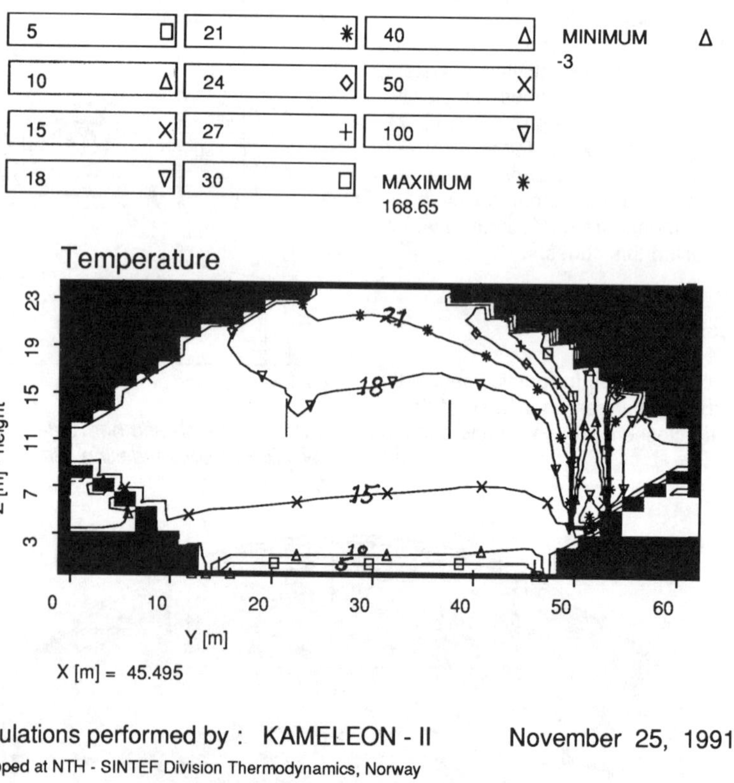

Figure 10 From simulation of smoke movement/smoke control by using the computer code KAMELEON, A fire is assumed to start in the public arena. After 860 seconds the smoke produced forms an upper layer beneath the ceiling. The thickness of this layer will be controlled by the smoke control system.

The fire development i.e. heat output in kW versus time, for these simulations is based on full scale fire tests [9]. Really high efforts have been spent through this research programme by using alternative computer codes with various complexity to study the smoke movement from alternative fire scenarios and the efficiency of the smoke control system.

ACKNOWLEDGEMENTS

This work has been supported by the research programme "Rock Cavern Stadium", which is funded by, The Royal Norwegian Council for Scientific and Industrial Research, Gjøvik municipality as well as the following companies: Statkraft, Teledirektoratet, Noteby A/S, Dyno Industrier A/S, Veidekke A/S, Selmer A/S, Fortifikasjon A/S, The Foundation of Scientific and Industrial Research at The Norwegian Institute of Technology (SINTEF), The Norwegian Geotechnical Institute (NGI) and The Eastern Norway Research Center.

REFERENCES

Broch,E. and RYGH, J.A. 1976. Permanent underground openings in Norway - design approach and some examples. Underground Space. Vol 1.

Broch,E. 1989. Use of the underground in Norway. Proc.Infl.Conf. Storage of Gases in Rock Caverns. Balkema, Rotterdam pp 3-13.

Barton, N. et al 1993. A review of rock mechanical and engineering geological research at Gjøvik Olympic Cavern. Proc. Seoul Symposium on Underground structures 8-9 Oct. (in print)

Barton, N. et al 1992. Comparison of prediction and performance for a 62 m span sports hall in jointed gneiss. Proc. Conf. di Meccanica e Ingegneria delle Rocce. Toriona 24-27 Nov. pp 17.1-17.15.

Laksâ, B. et al 1991. Kameleon II, SINTEF STF15 F91048, Trondheim, Norway.

Mathisen, H.M. et al 1992. Design of Smoke Controlin Large Rock Premises, International Symposium on Room Air Convection and Ventilation Effectiveness, Tokyo, June.

Mathisen, H.M. 1993. Planning, design and construction of the cavern arena: Heating, Ventilation and Air Conditioning. Proc. Seoul Sumposium on Underground structures, 8-9 Oct.

Meland, Ø. 1990. Personsikkerhet ved brann i ishall, SINTEF, Norway. Report no STF15 F90048.

Pauls, J. 1983. The Movement of People in Buildings and Design Solutions for Means of Egress", Ninth Australian Conference on Fire, Sydney, September.

Rømen, B.H. 1993. FRES - Version 2.0 Flexible Room Climate and Energy Simulator. Users Manual, SINTEF, Norway. Report no STF15 A93071.

Thorbergsen, E. 1993. RochHall, a program for simulating public halls in rock. Draft report.

Tunnelling and Ground Conditions, Abdel Salam (ed.) © 1994 Balkema, Rotterdam, ISBN 90 5410 363 9

Excavating tunnels of large cross-section with a thin overburden beneath densely populated areas (Konan Tunnel on Ring Road No. 2 in Yokohama City)

Eiki Mizuno
Primary Road Division, Road & Highway Bureau, Yokohama City Government, Japan

Itsuo Hirano & Kuniaki Kataoka
Taisei Corporation, Tokyo, Japan

ABSTRACT: The Konan Tunnel, under construction in a densely populated area of Yokohama City, is a twin tunnel designed to carry three lanes and a sidewalk in each direction. It is 530 m long. Its specifications are a 16.2 m width, a separation of 8.7 m between the tunnels, a cross-sectional area of 143 m^2, and an overburden of 5 to 24 m. The geology of the ground through which it is being excavated consist of mainly diluvial sand and silt. This paper reports on how this tunnel was planned and designed to prevent ground subsidence under very constraining conditions. The construction method, auxiliary excavation methods, and measurements of the bedrock are also explained.

1 Profile of the Konan Tunnel

Yokohama City lies on a narrow coastal plain that has developed around the port and on the hilly hinterland surrounding it. As a result of this geography, the city's main roads have developed mainly as a radial network. With recent rises in population density, traffic has become much heavier, and serious congestion now occurs in the city center as well as in many other places throughout the city. This congestion adversely affects the local economy and the day-to-day activities of citizens. Construction of a Ring Road No. 2 has been regarded as an essential part of measures to overcome some of these problems.

The Ring Road No. 2, the planned solution to these congestion problems, is a main trunk route with a standard width of 42 m and a length of 24.5 km. The first top-class road within Yokohama City, the Ring Road No. 2 will be grade-separated from all other roads.

The Konan Tunnel is south of the city and is the only tunnel along the route of Ring Road No. 2. It has two separate parallel roadways, one for each direction, and is 530 m long. The northernmost of the two east-west tunnels is called the inner tunnel, since it is closer to the city center, while the more southerly one is called the outer tunnel. Construction of these twin tunnels faces the following forbidding problems:

(1) Each tunnel is to have a width of 16.2 m and a height of 10.5 m. The total excavated cross section is 143 m^2, to allow for three lanes and a sidewalk. The two tunnels are to be separated by only 8.7 m from each other.

(2) The tunnel is to have a very thin overburden of 5 to 24 m along its whole length, and there is a densely packed residential area on the ground above the tunnel.

(3) The tunnel must penetrate a geology consisting mainly of diluvial sandy and silty layers. In addition, the alluvial clay layer reaches the top edge of the tunnel at the center of the tunnel where the overburden is thinnest. The groundwater level in the diluvial sandy layer is higher and extends from around the tunnel top to the tunnel spring line.

2 OUTLINE OF THE KONAN TUNNEL CONSTRUCTION PLAN

(1) Design speed of highway: 60 km/h
(2) Roadway width: 2.50 (sidewalk) + 0.50 + 3 @ 3.25 (roadway) + 0.25 + 0.75 (inspection walk) = 13.75 m
(3) Tunnel length: 531.132 m (inner tunnel) and 528.848 m (outer tunnel)
(4) Traffic: Three lane, one-way flow
(5) Planned traffic volume: 43,150 vehicles per day
(6) Longitudinal gradient(i): 1.20%
(7) Paving: Deflection-resistant pavement
(8) Interior finish: Absorbing finishing plates
(9) Design volumes
- Tunnel excavation: 148,134 m^3
- Shotcrete: 78,041 m^2
- Rockbolts: 97,968 m
- Steel supports: 3,302.7 t
- Concrete lining: 27,639 m^3

3 GEOGRAPHY AND GEOLOGY

The main geographic feature of the area in which the tunnel is being built is hills of 40 to 60 m in height

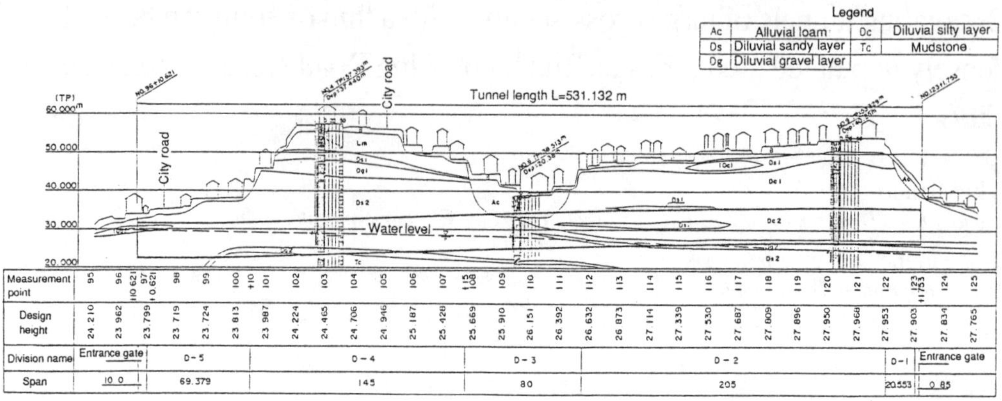

Fig. 1 Geological profile (inner tunnel)

above sea level. However, the original topographical features have almost completely disappeared as a result of land development for residential purposes and the subsequent construction of houses.

The geological classification is diluvium of the Quarternary period. As for the actual geology of the ground to be tunneled, it can be divided into two according to features. There is low land comprising an old river bed at the middle of the tunnel, and alluvium is dominant here. The eastern half is mainly composed of a yellowish-gray fine sandy layer (D_{S2}), while the western half is a dark gray to dark brown silty layer (D_{c2}).

The fine sandy layer is characterized by a uniformity of particle size, with uniformity coefficient of 2 to 4. More specifically, 10% of particles are 0.1 mm and below and 60% are 0.3 mm and below. Since the fine particle content is almost 20%, quicksand tends to be prominent. This layer ranges from 19 to 50 in N-value and has a permeability coefficient of the order of 10^{-3} cm/s, indicating good water permeation.

The silt layer is assumed to be a deposit left in a marsh, including a large amount of organic matter, dotted with fine sand. It has an N-value range of 17 to 50 and has an average uniaxial compressive strength of around 1 N/mm^2. The geological profile is shown in Figure 1.

4 DESIGN OF THE CROSS SECTION

The tunnel's cross section was designed to ensure sufficient space to suit the architectural constraints imposed by the sidewalk and roadway. Since the road requires a width of 13.7 m, unusually large for this type of tunnel, the arch section is designed to consist of concentric sections, while the invert is of two concentric sections. The tunnel is also designed with a flatness (height/width) of more than 0.6. The dimensions of the cross section are shown.

	Inner tunnel	Outer tunnel
Inner sectional area	107.7 m^2	101.6 m^2
Radius of arch section	8.05 m	8.05 m
Radius of side walls	5.75 m	5.75 m
Spring height level	1.35 m	1.10 m
Inner volume flatness	0.632	0.613
Excavation flatness	0.671	0.655

5 EXCAVATION METHOD AND DETERMINATION OF SECTION DIVISIONS FOR TUNNEL EXCAVATION

Construction of the Konan Tunnel under such a densely populated area requires an excavation method that minimizes subsidence of the surface. The method chosen is the New Austrian Tunneling Method (NATM), since it ensures safety and accuracy while having minimum impact on the bedrock, and it features the combined use of absolute surveys and measurements. As for the excavation, the center diaphragm method and the side drift method were compared as regards their suitability for excavation of a large cross section. This comparison basically leads to choice of the side drift method, since the side drifts effectively give an understanding of the geological features of the ground and effect water level drops.

The entire cross section is first divided into several smaller sections for piece-by-piece excavation. In order to determine how to divide the section and where to dig first, several factors were analyzed using the finite element method (FEM). As an example of the techniques used in this analysis, the size of the heading was varied to find the optimum size as regards excavation workability. The deformation of the bedrock excavation which would occur at each division was also studied. According to the results of this analysis, highly rigid timbering would be needed to restrict deformation and prevent subsidence of the ground surface. In an optimum sectional division, the width of a drift should be almost equal to the width of excavation at the upper half section in the center, and the inner diameter of the drift should be larger than this. To ensure stability of the face, the bench cut method was chosen for the separate excavation of upper and lower halves of both the drift and center drift. Figure 2 shows the sectional division selected for this work.

This division makes drifts larger than the normal side-drift method, and requires two diaphragms. We decided to call this excavation technique the "double-diaphragm method."

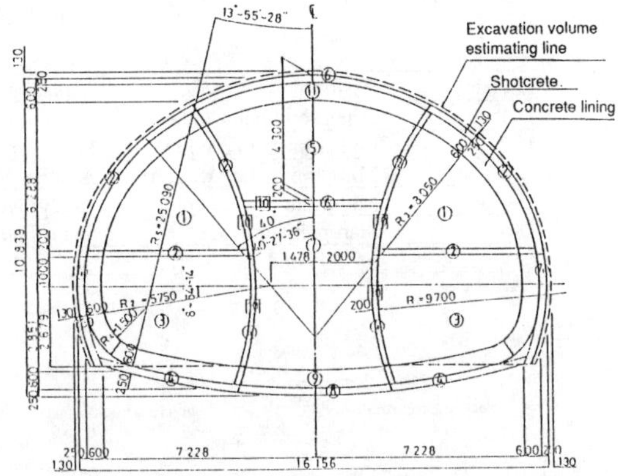

	Item	Remarks
1	Excavation of upper half of side drift	
2	Shotcreting of upper half of side drift	Temporary invert included
3	Excavation of lower half of side drift	
4	Shotcreting of lower half of side drift	
5	Excavation of upper half of center drift	
6	Shotcreting of upper half of center drift	Temporary invert included
7	Excavation of lower half of center drift Shotcreting of the invert	
8	Shotcreting of the invert	
9	Invert concrete	
10	Removal of middle walls	
11	Arch	Temporary invert included

Fig. 2 Section division

6. GROUND CLASSIFICATION FOR SUPPORT AND TIMBERING

Timbering pattern is classified by the geological features of a silty layer, an alluvial clay layer, or a sandy layer. It is further subclassified according to the geographical features whether the overburden is more or less than 1D (D = tunnel width), as shown in Table 1

The support structure is designed to be more rigid than normal to prevent deformation of the ground. The exterior arch is designed as steel timbering of H-200 with a 25 cm shotcrete covering. For the diaphragms, H-150 steel timbering with a 20 cm shotcrete covering is adopted and for the invert, H-150 steel timbering with 25 cm shotcrete. All rockbolts require the insertion type to ensure adequate anchorage; they are 4 m in the outward direction, 6 m into the area between the two bores to reinforce the pillars, and 3 m on the center drift side. For the upper half cross sections of the side drift and center drift, struts of H-150 with a 20 cm shotcrete covering are to be used as a temporary invert.

Injection forepoling to a distance of 3 m is adopted for the arch section to prevent the top of the tunnel collapsing and to ensure the safety of the face. Figure 3 shows a typical pattern of timbering.

As regards the lining concrete of the secondary concreting, structural calculations are attempted to allow for safety even after possible load changes in the future caused by environmental alterations in the urban area. As a result, a double-reinforcement concrete structure with a thickness of 60 cm is adopted for the whole length of the tunnel.

7 TUNNEL CONSTRUCTION METHOD

Our FEM analysis concludes that the invert concrete should be placed with the two side-drift diaphragms left in place in order to constrain excavation-related displacement of the bedrock. In accordance with this conclusion, the tunneling work is being done using the following procedure:

(1) Excavation and timbering of the upper half of the side drift (including temporary invert construction). The inner drift is to be completed first.

(2) Excavation and timbering of the lower half of the side drift

(3) Excavation and timbering of the upper half of the center drift (including temporary invert construction)

(4) Excavation and timbering of the lower half of the center drift

(5) Placement of invert concrete

(6) Removal of the two diaphragms

(7) Placement of waterproofing sheet, reinforcement, and final concrete lining on arch section

Excavation is to take place on a downward gradient and in one direction from the work yard,within the main line at the end point.

Construction machinery consists of cutter loaders for upper half excavations and 0.2 m^3 back hoes for the lower half. Debris is transported using 4-ton dump trucks. Shotcreting, steel timbering, and rockbolt placement are all to be done manually, since the excavation cross section is too small for the work to be done by machines other than cutter loaders.

Table 1 Section division

Division	Position	Typical geology	Overburden	Inner extension	Outer extension
D-1	Tunnel end-point entrance	Silt layer	<1D	20.553 m	22.206 m
D-2	End point to mid-point	Silt layer	>1D	205.0 m	190.0 m
D-3	Mid-point valley section	Alluvial clay layer	<1D	80.0 m	85.0 m
D-4	Mid-point to start point	Sandy layer	>1D	145.0 m	180.0 m
D-5	Start point entrance	Sandy layer	<1D	69.379 m	40.442 m
	Total			519.932 m	517.648 m

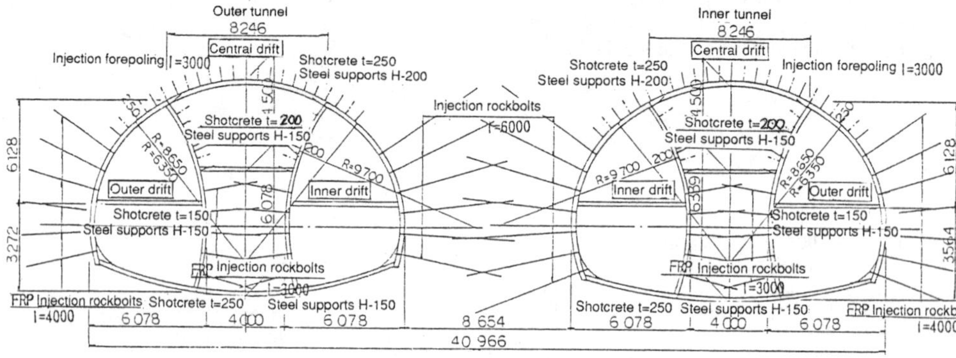

Fig. 3 Support structure diagrams

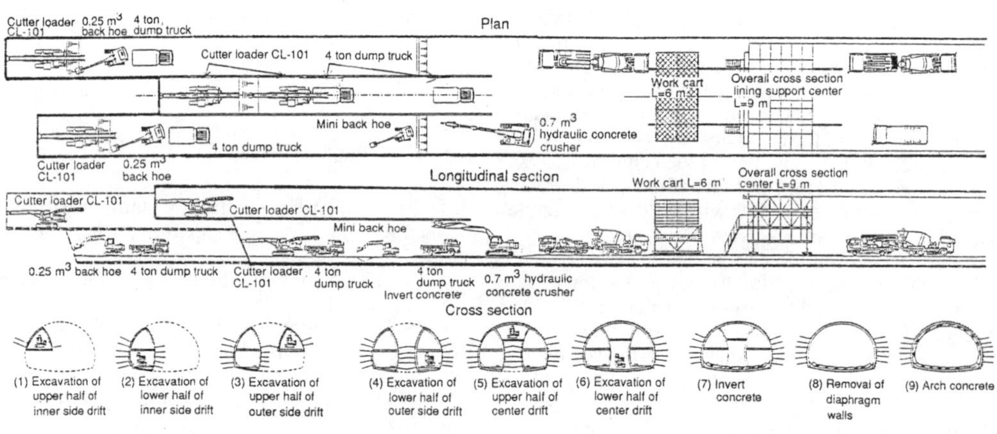

Fig.4 Tunneling work procedure and construction process

Figure 4 shows the tunneling work procedure and construction process diagram.

8 AUXILIARY TUNNELING METHODS

It is very important when excavating a tunnel under difficult conditions, such as in the case of the Konan Tunnel, to utilize proper auxiliary excavation methods. In this project, the auxiliary excavation methods play an important role as support structure.

The auxiliary methods chosen in this case are basically aimed at forepoling, thus ensuring face stability and preventing collapse of the tunnel top. These methods included an embankment stabilizing method for the tunnel entrances, a ground improvement method for soft ground (the alluvial clay layer) where the overburden is thin, a pipe roofing method where the top of the tunnel entrance is just under the road at the far end of the tunnel, and a groundwater lowering method for the sandy layer.

Table 2 lists the auxiliary methods and where they are used. Auxiliary Method A is a technique carried out simultaneously with ordinary tunnel excavation, without special large equipment. Auxiliary Method B is a type of relatively heavy work which is done prior to actual tunnel excavation. An outline of Method B is illustrated below.

Table 2: List of auxiliary construction methods

Div.	Supporting system A	Supporting system B
D-1	Mortar injction forepoling	Vertical bolting method
D-2	Mortar injection forepoling	-----
D-3	Mortar injection forepoling	Rodin jet piling method (RJP)
D-4	Urethane injection forepoling	Drift dewatering well-point method
D-5	-----	Pipe roofing and deep-well method

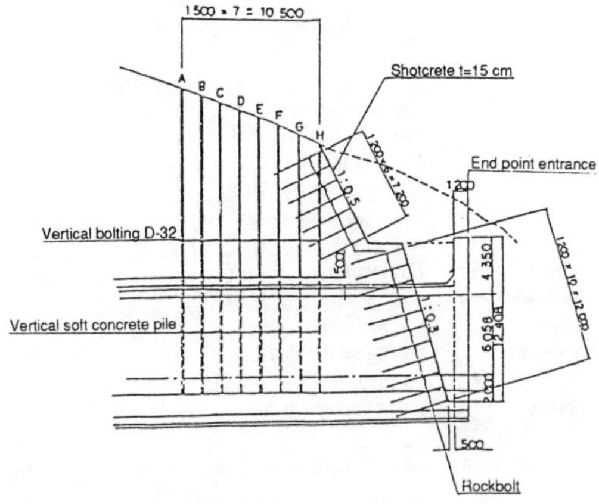

Fig.5 Typical diagram of vertical bolting work

(1) Vertical bolting

Since there is a possibility of embankment collapse at the end point entrance during tunnel excavation, vertical bolting is to be carried out from the surface to add strength to the ground. The work procedure is to construct a stage on the embankment at the entrance, to bore a hole 100 mm in diameter using a boring machine, to place an injection hose into the hole, and insert a reinforcing bar of ø32 mm. Mortar for injection is mixed so as to maintain a strength of σ_{28} = 19.6 N/mm^2. A total of 270 bars, each about 18.9 m long, are used at a pitch of 1.5 m x 1.5 m. Figure 5 shows a typical cross section.

(2) Ground improvement

The RJP method, one of jet-grouting, is chosen to enhance the strength of the soft alluvial clay layer and to reduce the construction period. With this method, a large cross section is improved by destroying the structure of the ground using kinetic energy—an extremely high-pressure jet is used—and mixing the broken soil particles with a hardening agent. Figure 6 shows a schematic (conceptual) drawing of ground improvement using the RJP method. The improved strength of the clay layer is more than 1.5 N/mm^2. The volume of soil to be improved is 18,500 m^3.

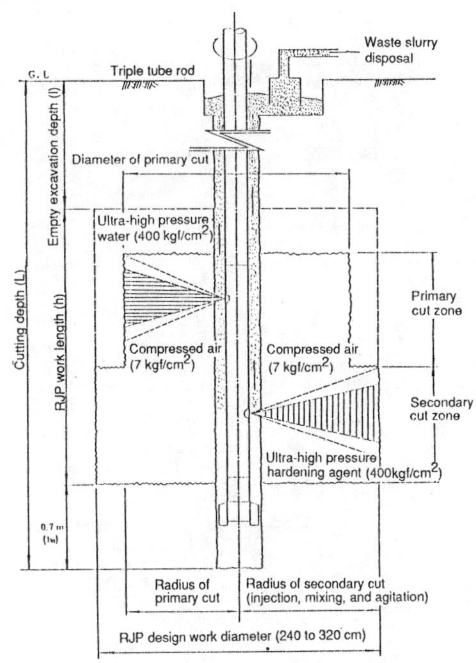

Fig.6 Schematic diagram of RJP method

9 MEASUREMENT AND MEASUREMENT CONTROL

When excavating a tunnel with a thin overburden, it is important to ensure stability of the tunnel itself and to obtain the earliest possible indications of ground subsidence. A measurement is planned not only to measure the displacement of the tunnel itself but also to predict the behavior of the bedrock ahead of the face. A cross section through a typical measurement sensor is shown in Fig. 7.

The method of underground advance subsidence measurement shown in the figure is based on an electrical means of detecting subsidence over time. The measurement range is ±50 mm with an accuracy of ±1 mm. The procedure is to bore a horizontal hole 275 mm in diameter at a distance of 2.5 m above the tunnel top for 100 to 130 m long and to place measuring equipment every 5 to 10 m within it. The readings can be scaled to give an absolute measure of subsidence around the top of the tunnel, helping with an understanding of how widely the bedrock ahead of the face will be affected. It can be used as a numerical index for construction management procedures.

Measurement control standards are determined by

Table 3 Measurement standard values for tunnel top edge subsidence

		Normal Level (mm)	Alert Level I (mm)	Alert Level II (mm)
Inner drift top-edge subsidence	Excavation of lower half of inner side drift	<10	10-15	>15
	Excavation of lower half of outer side drift	<15	15-20	>20
	Excavation of lower half of center drift	<20	20-25	>25
Outer side drift top-edge subsidence	Excavation of lower half of outer side drift	<10	10-15	>15
	Excavation of lower half of center drift	<15	15-20	>20
Central drift top-edge subsidence	Excavation of lower half of center drift	<20	20-25	>25
Measurement management system		• Ordinary work	• Increase in measurement frequency: more than once a day at least • Automatic measurements: every four hours • Understanding of cause and study of what action to take	• Increase in measurement frequency: at least twice a day • *Addition of cross sections to be measured • Understanding of causes
Construction management system		• Ordinary work	• Enhanced inspection of supports • Inspection with supervisor • Survey and observation of the bedrock • Selection of measures • Implementation of measures	• Termination of excavation • Enhanced inspection of inner and outer side drifts • Selection and implementation of measures • Change in support pattern

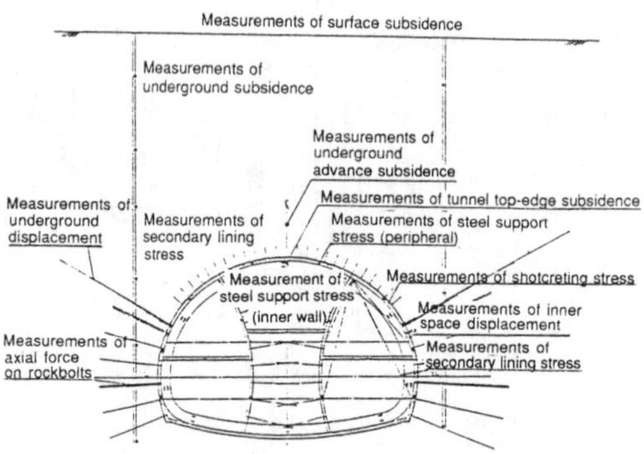

Fig.7 Arrangement of measuring equipment

taking as standards the subsidence at the top edge of the tunnel for each excavation cross section and the construction procedure. Standard values of deformation of the ground surface, especially non-uniform subsidence and deformation angles related to structural impediments, were selected. The standard tolerable deformation for the foundations of a Japanese wooden house is $\delta = (1 \text{ to } 2) \times 10^{-3}$ rad. In this case, $\delta = 2 \times 10^{-3}$ rad was adopted. The standard values for measurement management were selected by comparing this limit deformation angle with the FEM analysis results. The standard values for measurement management for top-edge subsidence of the tunnel shown in Table 3.

10 MEASUREMENT RESULTS

(1) Ground surface subsidence and advance underground subsidence

The average subsidence of the ground surface by the end of center drift excavations was 8 mm. A maximum of 18 mm was measured just above the inner tunnel. These values agree quite well with the results of FEM analysis used to simulate advance subsidence as excavation of the inner drift went ahead. No damage to the houses standing on the ground was seen.

As shown in Fig. 8, the measurement system of underground advance subsidence showed a subsidence of about 36 mm of the ground where the maximum subsidence occurred. This reveals that about 50% of all the bedrock subsidence near the top edge of the tunnel appeared on the surface. The overburden in this area is about 20 m.

(2) Subsidence of tunnel top edge

The maximum subsidence of the tunnel top edge

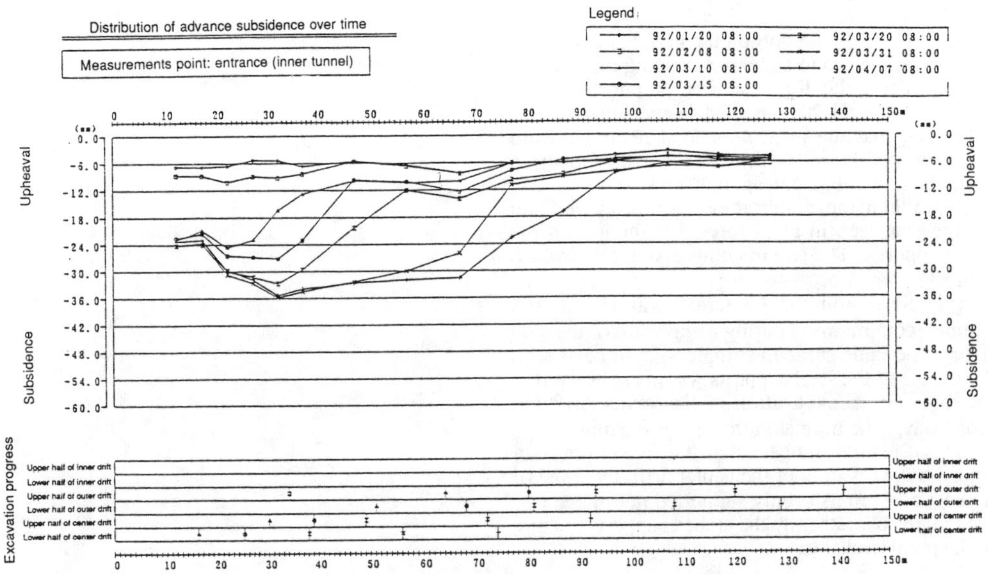

Fig.8 Measurements of underground advance subsidence

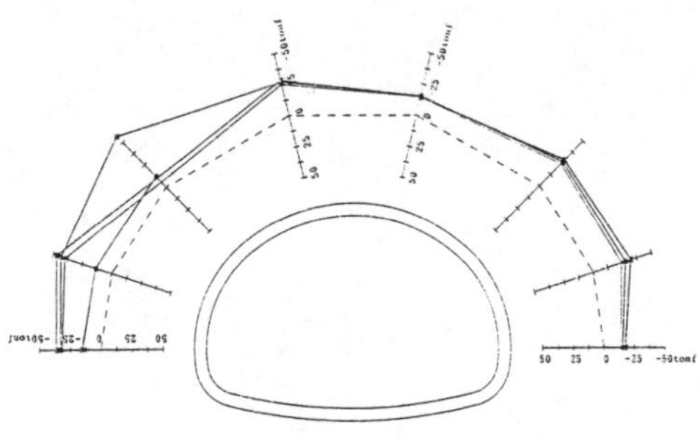

Fig.9 Axial force distribution on steel supports

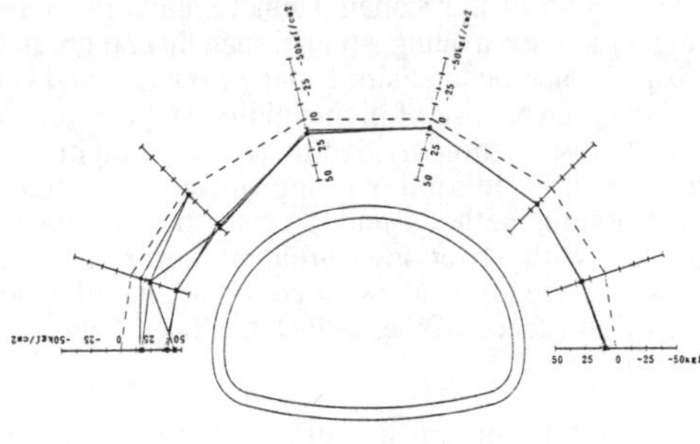

Fig.10 Stress distribution in shotcrete

occurred where, in the previously excavated inner side drift, the center drift was excavated. As a whole, the tunnel top-edge subsidence was 12 mm. The following trends were seen in top-edge subsidence:

(i) For the inner side drifts excavated in advance, the top-edge subsidence was measured at 5 to 8 mm. This is less than the management normal level of 10 mm.

(ii) Subsidence of the inner side-drift top edge during excavation of the outer side drifts in the second phase was almost zero. This is presumably because support rigidity is higher for the size of the side-drift cross section. The projected values for the outer side drift itself were almost the same as those for the inner side drift

(iii) Subsidence of the center-drift top edge during center drift excavation was greater than that for both side drifts; it measured around 10 mm. Excavation of the center drift aggravated subsidence at the top edge of both side drifts. After center drift excavation, additional subsidence of about 5 mm was measured at the side drifts.

(3) Axial force on steel support and shotcrete stress

Measurements of the shotcrete and the steel supports, which form a major part of the timbering, yielded the results shown in Figs. 9 and10, which give the axial force distribution on the steel supports and the shotcrete stress distribution, respectively.

(i) Steel supports on the peripheral side (H-200) were loaded with a compressive force of 15 N/mm^2 at the time of side-drift excavation, but this force rapidly increased to 46 N/mm^2 as excavation of the center drift began.

(ii) Steel supports on the diaphragm side (H-150) saw no increase in compressive force even during excavation of the center drift. The force remained at 15 N/mm^2, the value measured when the side drift was excavated.

(iii) The steel support on the peripheral side (H-200) was subjected to almost no bending moment.

(iv) The axial force on steel supports on the peripheral side (H-200) showed an even distribution, indicating that the maximum load resulting from excavation of the center drift was equally supported by all the timbering on the peripheral side.

(v) When the diaphragms were removed, no change was seen in axial force distribution on the steel supports. The rise in compressive force was 8 N/mm^2.

(vi) Stress on the shotcrete was 0.5 to 0.8 N/mm^2 (compressive) during excavation of the side drifts. The same surge in compressive force as seen in the case of the steel supports was also observed in the shotcrete as excavation of the center drift got under way. The ultimate force was 2.0 N/mm^2.

These results indicate that the maximum load caused by excavation of the center drift, the issue of greatest concern, is evenly supported not by the two diaphragms but rather by the arched support structure on the peripheral side.

11 CONCLUSION

Construction of the Konan Tunnel entails the very tough task of excavating a tunnel such that no ground ©3subsidence occurs, since many houses stand on the land above. Use of high-rigidity supports and a special cross section divided by two diaphragms, as well as introduction of appropriate auxiliary construction methods and precise measurements coupled with strict measurement control, have allowed excavation work to continue smoothly to date. Subsidence of the surface has been held to a minimum.

The Konan Tunnel consists of twin bores with cross sections which are large by both Japanese and world standards. The work is expected to provide information useful in future projects involving similar large-cross-section tunnels.

Tunnelling and Ground Conditions, Abdel Salam (ed.) © 1994 Balkema, Rotterdam, ISBN 90 5410 363 9

The evaluation of effects by motorway tunnel excavation on a nearby railway tunnel in operation and the supervising system to maintain the tunnel stability

Masuo Nishimura, *Osaka Prefectural Highway Corporation, Japan*

Michiyoshi Wadabayashi, *Construction Section, Construction Dept., Kinki Nippon Railway Co., Ltd, Japan*

Kazunori Watanabe, *The Second Hanna Toll Road, Obayashi Corporation, Japan*

Akihisa Daikoku, *Technical Administration Department, Nihon Shinko Co., Ltd, Japan*

Yoshinobu Iai, *Civil Engineering Technical Department, Obayashi Corporation, Japan*

ABSTRACT: This paper deals with an evaluation of effects by motorway tunnel construction on a nearby railway tunnel and the supervising system and protective measures to maintain the tunnel stability. The evaluation was conducted based on surveys of the status and the measurements of the existing tunnel taken prior to the new tunnel excavation for a period of about one year, and FEM analyses reflecting a back analysis of the data acquired when the existing tunnel had been excavated. For supervising purposes, monitoring items and standard and critical values for the monitoring were established based on the above evaluation. Thereafter protective measures were selected by reassessing the excavation effects on the existing tunnel through FEM analysis, using the data acquired during an early phase of the new tunnel excavation so that monitoring items would measure lower than the deremined critical values.

1 INTRODUCTION

The construction plan is to excavate two two-lane motorway tunnels named Hanna Tunnel to connect Osaka and Nara - westbound one for Osaka and eastbound one for Nara. The new tunnel is 5,600 m long each with a cross-section of 11.4 m wide and 9.7 m high. The tunnels are now being excavated by NATM from Osaka side entrance. For 600 m from the Osaka side entrance, the tunnels run through soft soils and overburden depth is as shallow as 20m to 50m. On the ground surface of the entrance area there are many private houses and public structures. In addition, the Ikoma Tunnel, a railway tunnel in operation (Width: 9.1m, Height:7.3m), exists in the close neighborhood. The tunnels are being driven under such severe conditions. These conditions required estimation of any adverse effects by the excavation, a study of protective

Table 1 Outline of construciton

Name of construction	Hanna tunnel
Place of construction	Second Hanna toll Road between Osaka and Nara
Road classification	Type 1, class 3
Design speed	60 km/hr
Total length of tunnel	5,600 m
Width of tunnel	11.4 m (for 2 lanes) 2 tunnels (one for Osaka, the other for Nara)
Inside cross-section of tunnel	75.0m^2
Longitudial section incline	25.62‰

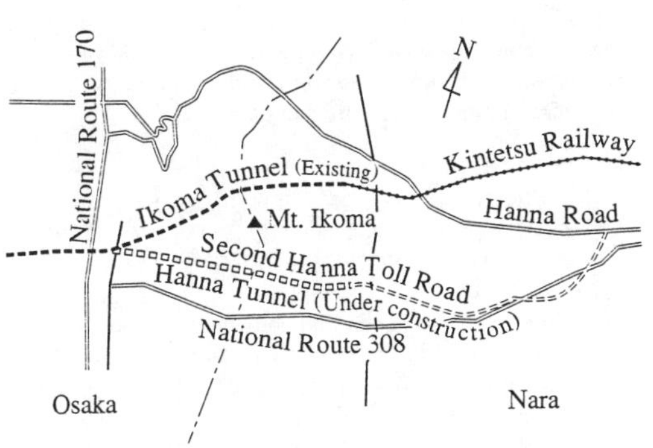

Fig. 1 Location

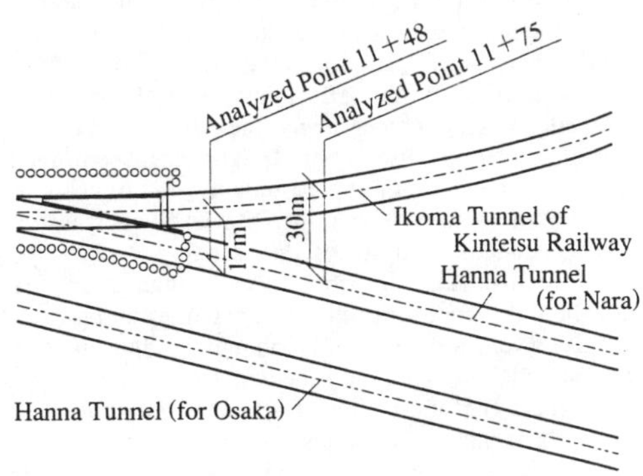

Fig. 2 Plan

Table 2 Physical characteristics values of soil mechanism

Ground physical characteristics	Alluvium	First layer (Diluvium)	Second layer (Diluvium)	Third layer (Diluvium)	Fourth layer (Diluvium)
Unconfined compressive strength (kgf/cm^2)	0.2~1.5	0.5~3.5	1.0~7.0	3.0~10.0	0.5~20.0
Deformation modulus (kgf/cm^2)	2~60	20~100	30~200	100~700	200~1,500
Failure strain (%)	3~6	2~5	1~3	1~3	1~2
Cohesion (kgf/cm^2)	0.1~0.8	0.2~1.5	0.5~3.0	2.5~7.0	7.5~11.0

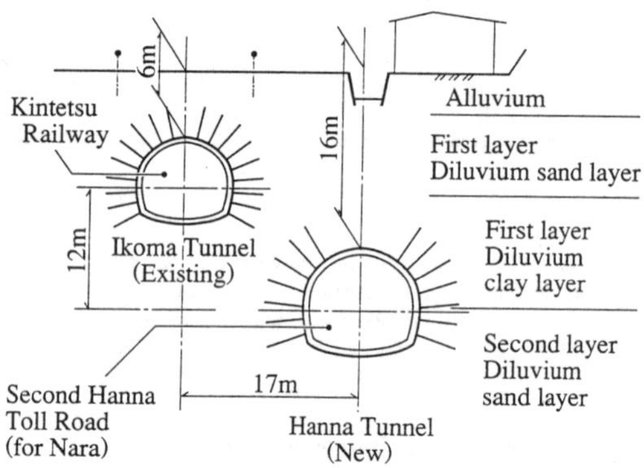

Fig. 3 Cross section (Analyzed Point 11+48)

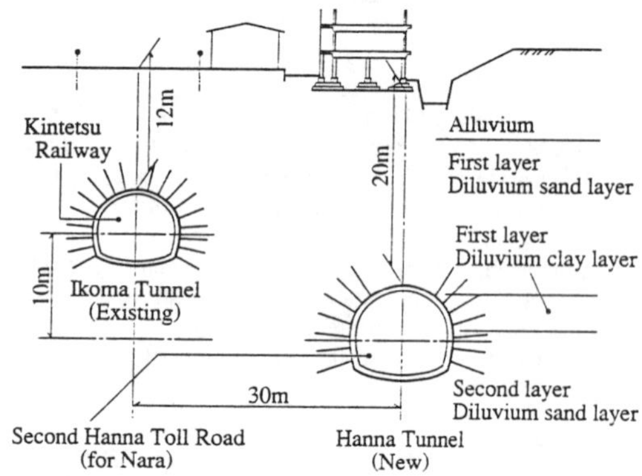

Fig. 4 Cross section (Analyzed Point 11+75)

measures and supervising system to be adopted, and establishment of standard values for monitoring. Table 1 shows the outline of the construction and Fig. 1 shows the location.

The conditions are severe, the distance between the centers of the existing tunnel and the new one being mere 17 m horizontally and 12 m vertically at the extreme proximity, as shown in Fig. 2, 3 and 4.

2 TOPOGRAPHIC AND GEOLOGIC OUTLINE OF THE GROUND

The location is the borderline area between the plain and the mountain. Geologically, as shown in Fig. 3 and 4, the ground consists of Quarternary Alluvium and Diluvium. As shown in Fig. 5, the Alluvium and the Diluvium are elevated, marking a steep angle due to the effect of the Ikoma Fault at the west side of Mt. Ikoma, and continue to the fan at the foot of the mountain. As it nears the mountainous area of Ikoma, the Diluvial layers transform into alternation of strata of thin sandy and clay layers whose total thickness expands to more than 20 m. The Alluvium also transforms into laminated ground of sandy and clay layers at the foot of the mountain, thus obscuring the distinction between the Alluvium and the sand layer of Diluvium. The alternation of strata at the foot of the mountain is assumed to be fan sediment.

It had been reported, at the time of the excavation of the existing tunnel, that the uniform coefficient (U_c) of the Diluvial sandy layer ranged between 1.8 and 3.0, and that the excavation work faced difficulties due to instability of the face, and that in the Diluvial clay layers the influence of confined groundwater caused occasional face collapse. Table 2 summarizes the results of soil tests conducted in the area involved.

3 PRE-CONSTRUCTION STATUS SURVEY OF THE EXISTING TUNNEL

Protection of the existing tunnel required the establishment of monitoring standard values. To that end, pre-construction status surveys, such as measurement of side wall radial deformation, were conducted on the existing tunnel for a period of about one year.

The stress status of the secondary concrete lining was measured by stress relief technique. Table 3 shows the results of these surveys and uniaxial compression tests of secondary lining, using core samples.

4 PREDICTION OF THE EFFECTS OF CONSTRUCTION ON THE EXISTING TUNNEL AND THE RANGE OF EFFECTS

To evaluate adverse effects and deformation tendency to be caused by the new tunnel construction on the existing one, two-dementional FEM analysis was conducted on the new tunnel together with the existing one. In this analysis, each layer was treated as isotropic and homogeneous continuum and as a linear elastic body. The local safety factor of the ground (F) was set at 1.5 following equation (1); Mohr and Coulomb's yield criterion (Fig. 6)

Table 3 Pre-construction status survey of existing tunnel

	Monitoring items	Measuring point	Measuring instrument	Measurement result	Notes
Tunnel interior	Crown settlement	Crown	Level	Max =3.0mm Average=0.6mm	Seasonal fluctuation ±2.3mm
	Side wall settlement	S.L-50cm	Settlement gauge	Max =1.4mm Average=0.2mm	Seasonal fluctuation ±0.5mm
	Side wall radial deformation	S.L	Convergence measure	Max =−2.0mm Average=−1.0mm	Seasonal fluctuation ±0.9mm
	Stress measurement on concrete lining	Tangential direction	Plane strain gauge	Max. compressive stress σc =7.0kgf/cm² Max. tensile stress $\sigma \tau$=17.0kgf/cm²	Uniaxial compressive strength of lining concrete 6c=389kgf/cm²
	Existing cracks of concrete	Cracks	Contact strain gauge	Max =−0.6mm	Seasonal fluctuation ±0.2mm
	Deformation measurement at concrete lining joint	Joint	Joint gauge	Max =−0.4mm Average=−0.1mm	

Table 4 Estimated values of ground physical characteristics

Layers		Standard penetration test N value	Internal friction angle (Degree)	Cohesion (kg/cm²)	Elastic modulus (tf/m²)	Poisson's ratio	Unit weight (t/m³)
Alluvium		5	0	0.3	600	0.45	2.0
First layer	(Diluvium sand layer)	14	30	0.0	2,100	0.35	2.0
	(Diluvium clay layer)	26	0	1.5	3,600	0.40	2.0
Second layer	(Diluvium sand layer)	107	45	0.0	6,000	0.35	2.0

$$F = \frac{a}{r} = \frac{2 \cdot c \cdot \cos\phi + (\sigma_1 + \sigma_3)\sin\phi}{(\sigma_1 - \sigma_3)} \qquad (1)$$

c : cohesion (kgf/cm²)
τ : shear stress (kgf/cm²)
ϕ : angle of internal friction (°)
σ : normal stress (kgf/cm²)
$\sigma 1$: major principal stress (kgf/cm²)
$\sigma 3$: minor principal stress (kgf/cm²)

Table 4 shows estimated values of ground physical characteristics. To obtain these values the results of lateral loading tests (LLT) in the bore holes near the tunnel entrance and triaxial compression tests were used as references. As for the sandy layers, the angle of internal friction was obtained using equation $\phi = \sqrt{20N} + 15$ (°), based on Ohsaki's equation; the deformation modulus was obtained using $E = 24 + 5.3$ N (kfg/cm²), from the graph by Schuitze, Menzenbach et al., which shows the relationship between soil deformation modulus (Es) and N value. For clay layers, uniaxial compressive strength was obtained from q_u = N/8 (kfg/cm²), the cohesion from C = q_u/2 (kfg/cm²), both equations being based on the equation by Terzaghi and Peck; the deformation modulus was calculated from E = 210 C (kfg/cm²) based on Takenaka's equation.

The lateral pressure coefficinet (Ko) was obtained through a back-analysis method as shown in Fig. 7, using

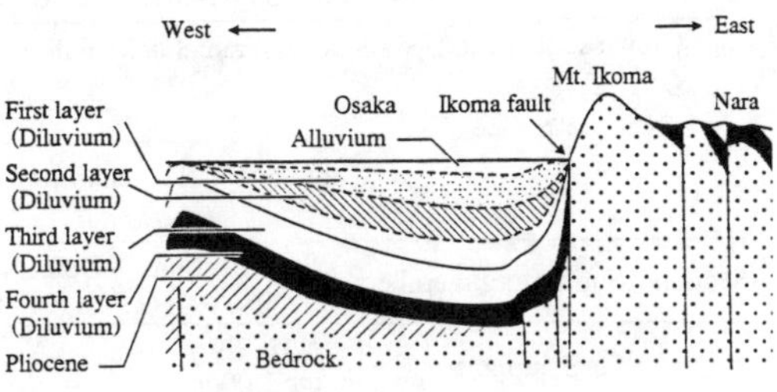

Fig. 5 Topographic/geologic features (Osaka & Nara), cross section

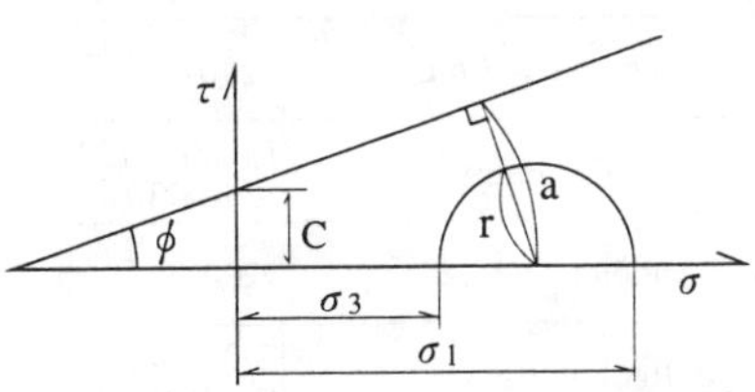

Fig. 6 Mohr and Coulomb's yield criterion

measurement results of crown settlement and sidewall radial deformation in Table 5 which were obtained during the excavation of the existing tunnel. Physical characteristics values of supporting members are shown in Table 6.

Based on the analysis results, it was predicted that the maximum displacement of the existing tunnel would be 15 mm horizontally and 14 mm vertically to the direction of the new tunnel as shown Fig. 8. Collation of analysis results on stress of the primary support and the secondary concrete lining indicated that the primary support would have no serious problem. However it showed that the secondary concrete lining at a point 11 + 48 suffered bending tensile stress of 32.4 kgf/cm^2 at maximum, which exceeded an allowable bending tensile stress of 14.3 kgf/cm^2. (Table 7).

Since the second concrete lining of the existing tunnel is reinforced with steel fiber reinforced concrete(SFRC), the allowable bending stress was determined based on the relationship between compressive and bending strength by steel fiber content ratio (Fig. 9).

Steel fiber content ratio V_f = 0.5% (40 kg/m^3);
Uniaxial compressive strength σ_c = 389 kgf/cm^2 (survey result);
Allowable compressive stress $\sigma_{ca} = \sigma_c/4$ = 389/4 = 97.3 kfg/cm^2;
Allowable bending tensile stress $\sigma_{ba} = \sigma_{bk}/4$ = 57/4 = 14.3 kfg/cm^2

As the results in Table 7 indicated, the maximum stress decreased as the distance between the tunnel entrance and an analyzed point got bigger. Therefore the range subject to the construction effect was estimated by drawing a line of the maximum bending tensile stresses of the second lining to come to intersect the allowable stress as indicated in Fig. 10. The result indicated the estimated range to be approximately 35m close to the tunnel entrance between points 11 + 40 and 11 + 75.

5 MONITORING ITEMS AND MONITORING STANDARD VALUES

Monitoring items were selected to make the best use of the analysis results in Fig. 8, and it was decided that the monitorng would be conducted through observation of the main items for regular monitoring and subordinate items to complement the main items.

5.1 *Establishment of critical values of main monitoring items*

a Absolute displacement

Critical values for crown settlement and side wall displacement was determined by equation (2):

$$C_v = (\sigma_{ba}/\sigma_b)\ S \qquad (2)$$

Table 5 Analysis Quantities

Analyzed point	Overburden above crown	U_0	U_{90}	U_{90}/U_0
11+75m	12m	11.27mm	10.81mm	0.96

U_0 : Crown settlement, U_{90} : Side wall radial deformation

Table 7 Maximum stress of secondary concrete lining

Analyzed section	Bending tensile stress	Bending compressive stress
11+48	−32.4 kg f/cm^2	31.3 kg f/cm^2
11+75	−14.1 kg f/cm^2	13.1 kg f/cm^2

Table 6 Physical charateristics values of supporting members

Section	Supporting members	Specifications (mm)	Deformation modulus (tf/m^2)	Cross section (m^2)	Geometrical moment of inertia (m^4)
Ikoma tunnel (Existing)	Steel support	MU−29 ctc 1,000	2.1×10^7	0.0037	5.81×10^{-6}
	Shotcrete	t=200	1.0×10^6 (1.0×10^5)*	0.2	6.67×10^{-5} **
	Rock bolts	TD21×3,000 ctc 1,000	2.1×10^7	3.69×10^{-4}	1.08×10^{-8}
	Secondary lining	t=250	2.4×10^6	0.25	1.30×10^{-3}
Hanna tunnel (New)	Steel support	H−150×150 ctc 1,000	2.1×10^7	4.01×10^{-3}	1.64×10^{-5}
	Shotcrete	t=200	1.0×10^6 (1.0×10^5)*	0.2	6.67×10^{-5} **
	Rock bolts	TD24×4,000 ctc 1,000	2.1×10^7	4.46×10^{-4}	1.58×10^{-8}

* The deformation modulus of the shotcrete is assumed to be E=1.0×10^5 (t/m^2) at the early stage, considering hardening time.

** When shotcrete had tensile cracks due to bending moment, analysis was done with bending moment bearing ratio reduced by assuming the geometrical moment of inertia of shotcrete to be 1/10, since in such a case the steel support members bear the bending moment.

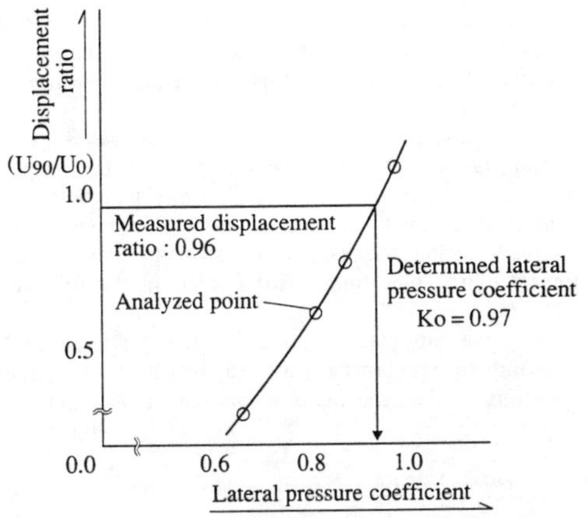

Fig. 7 Maximum stress of secondary concrete lining

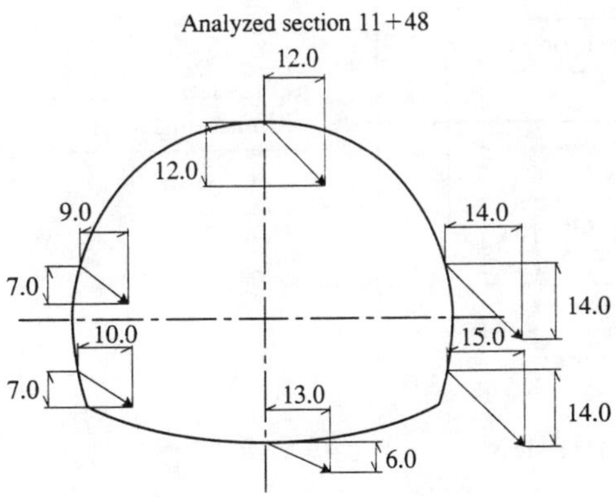

Fig. 8 Behavior of existing tunnel according to analysis

where Cv = critical value; σ_{ba} =allowable bending tensile stress; σ_b = maximum bending tensile stress;
S = displacement.

b Relative displacement

Critical value of relative settlement in longitudial direction was determined to be 7 mm/10 m in accordance with the Manual for the Track Maintenance.

Critical value of side wall radial deformation was set at the value obtained by equation (2), using the difference in displacement of right and left side walls.

c Crack width of secondary concrete lining

Greater importance was placed on cracks in the longitudinal direction than those in the traverse direction. The critical value of crack width was set at 3 mm.

5.2 *Establishment of critical values of sobordinate monitoring items*

a Absolute displacement

Ground level settlement is considered to be in accordance

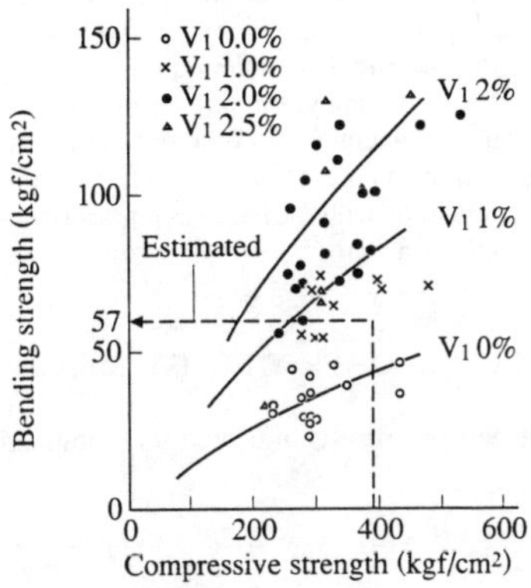

Fig. 9 Relationship between compressive and bending strength

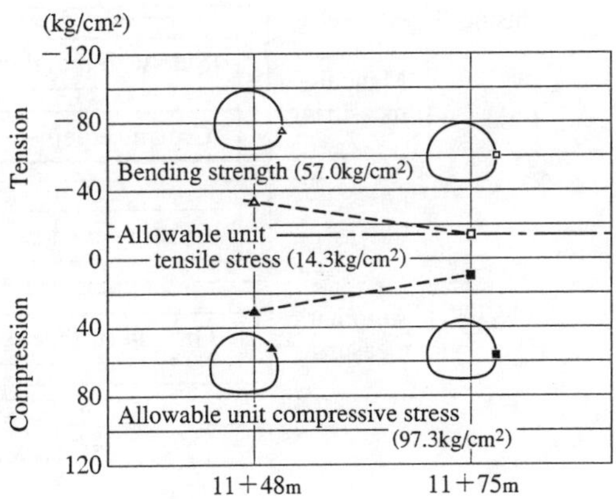

Fig. 10 Estimated range of effect of maximum stress

with crown settlement, since the overburden is as shallow as 20 m. Underground displacement is monitored with an inclinometer set at the midpoint between the existing tunnel and the new tunnel. Therefore critical value of underground settlement was deterined by equation (2), using the analysis value at the place of inclinometer.

b Relative displacement

Critical value of relative settlement in traverse direction was determined by equation (2), using the difference in settlement of right side wall and left side wall.

c Secondary concrete lining plane strain

Critical value of secondary lining plane strain was set according to equation (3)

$$\varepsilon = \sigma_a / E_c \qquad (3)$$

The critical values described above are summarized in Table 8.

As shown in Table 9, three categories of standard values and three levels of attention were identified to classify monitored values and measures to be taken.

As for monitoring method, automatic monitoring and manual monitoring are being conducted and measurements are being supervised by a monitoring room at site. (Fig. 11) Following the monitoring system in Fig. 12, monitored value are to be analyzed and studied for further construction management. When a monitored value reaches to Attention Level III in Table 9, a meeting for investigation is to be called to discuss countermeasures and future construction plan.

6 SELECTION OF PROTECTIVE MEASURES

The FEM Analysis results indicated that monitoring items would measure more than 50% of the determined critical values due to the new tunnel construction. Therefore it was judged necessary to adopt protective measures to control displacement.

In order to select appropriate protective measures for the existing tunnel, physical characteristics of the ground including the effect of tunnel supporting members were revaluated through a back-analysis of measurements obtained during the excavation of main and working tunnels of the new tunnel (for Osaka) in the following manner:

The apparent elastic modulus (E') was calculated according to equations (4) and (5) which are simplified equations to describe radial displacement of a perfectly

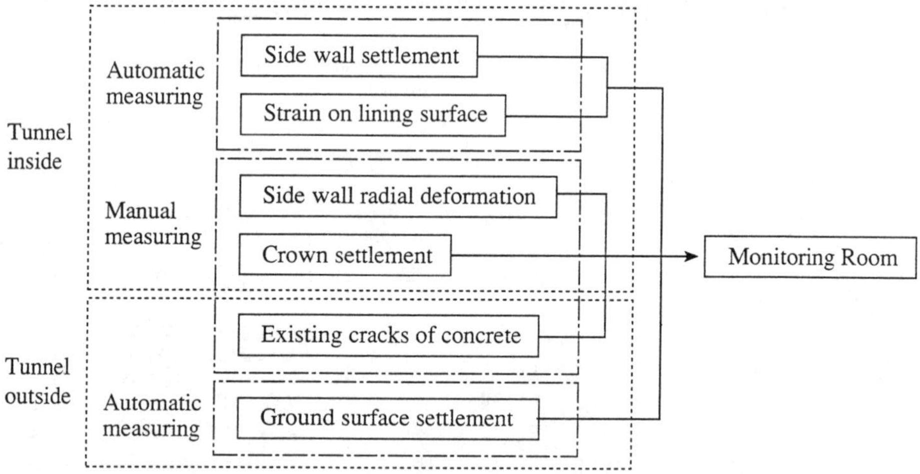

Fig. 11 Monitoring system

Table 8 Critical values for monitoring

Monitoring items			Critical values for monitoring
Main monitoring	Absolute displacement	Crown, sidewall	7 mm
	Relative displacement	Longitudinal direction	7 mm/10m
		Side wall radial deformation	3 mm
	Secondary lining crack width		3 mm
Subordinate monitoring	Absolute displacement	Ground level	7 mm
		Underground	9 mm
	Relative displacement	Traverse direction	4 mm
	Secondary lining plane strain	Conpression	81×10^{-5}
		Tension	12×10^{-5}

Table 9 Standard values

Attention level	Monitored value	Measures
I	50% of Cv※ > Mv※※	<Safe> No special measure to be taken
II	50% of Cv ≦ Mv but 70% of Cv > Mv	<Attention required> Existing tunnel to be under strict guard Monitoring to be reflected in subsequent execution steps
III	70% of Cv ≦ Mv	<Emergency> Construction works to be suspended and counter measures to be taken

※ Cv : Critical value
※※ Mv : Monitored value

round tunnel, where Ua is a side wall displacement obtained from side wall radial deformation and Ut is a wall displacement obtained from crown settlement. The lateral pressure coefficient (Ko) was then calculated from equation (6).

$$E_a' = \frac{(1+\nu)}{U_a} \cdot \gamma H \cdot a\{2(1-\nu)K_o-(1-2\nu)\} \quad (4)$$

$$E_t' = \frac{(1+\nu)}{U_t} \cdot \gamma H \cdot a\{2(1-\nu)-(1-2\nu)K_o\} \quad (5)$$

$$K_o' = \frac{2(1-\nu)U_a+(1-2\nu)U_t}{2(1-\nu)U_t+(1-2\nu)U_a} \quad (6)$$

where
Ea' : apparent eleatic modulus on sidewall
Et' : apparent eleatic modulus at crown
Ko : lateral pressure coefficient ($\sigma x/\sigma y$)
H : overburden
a : tunnel radius
ν : Poisson ratio
r : weight of unit volume
Ua : side wall displacement
Ut : crown settlement

Because of displacement ahead of the tunnel face and other reasons, values of Ua and Ut were 2.5 times of the measured values.

Based on *The Evaluation of NATM Support in City NATM* (Okabe et al. 1991), it was assumed that the supporting effect would improve the physical characteristic values of the ground by 1.5 to 2.0 times apparently. Taking this into consideration, the elastic modulus (Er) gained through the back-analysis was estimated to be 5,000 (tf/m^2) which is approximately 1.5 times greater (weighted mean), and the lateral pressure coefficient (Ko) to be 0.65. By feeding back these results on the figures shown in Table 4 to update, FEM analysis in consideration of protective measures was done.

As shown in Fig. 13, the range to apply protective measures was determined to be 47 m from the tunnel entrance based on the estimated range of effect of maximum stress (Fig. 10), surrounding environments and FEM analysis.

It was decided to apply column jet to the extreme

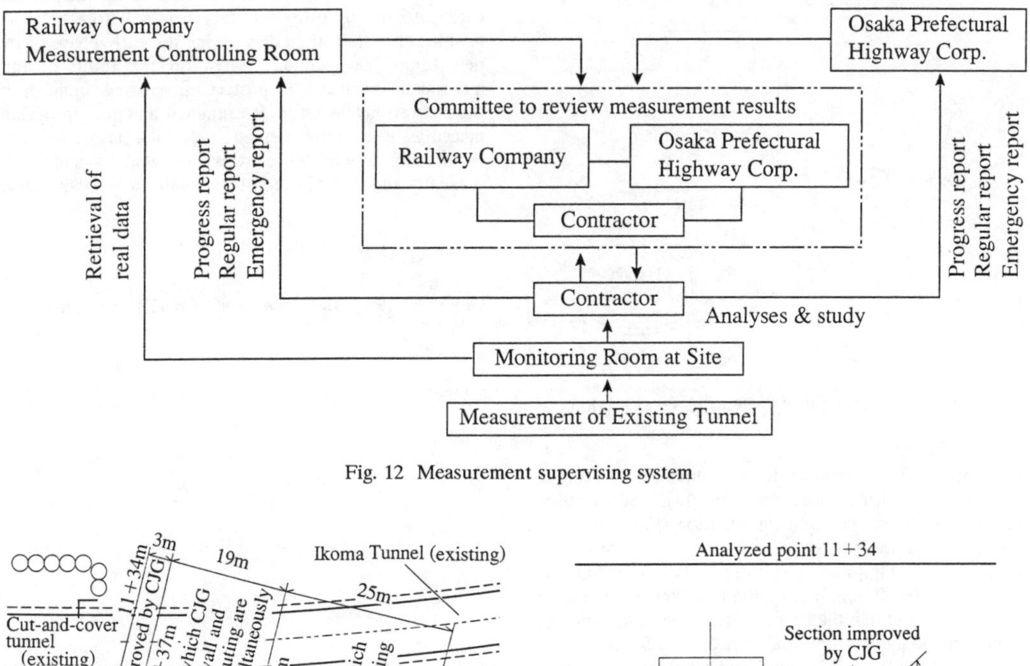

Fig. 12 Measurement supervising system

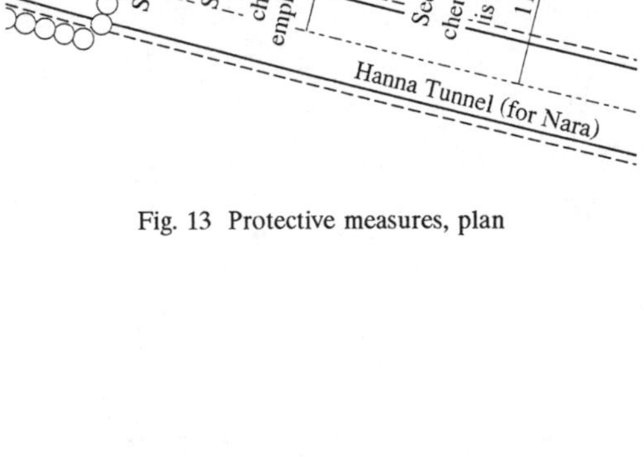

Fig. 13 Protective measures, plan

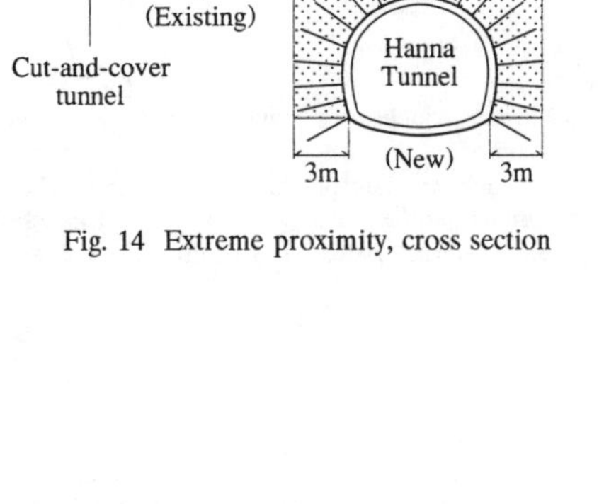

Fig. 14 Extreme proximity, cross section

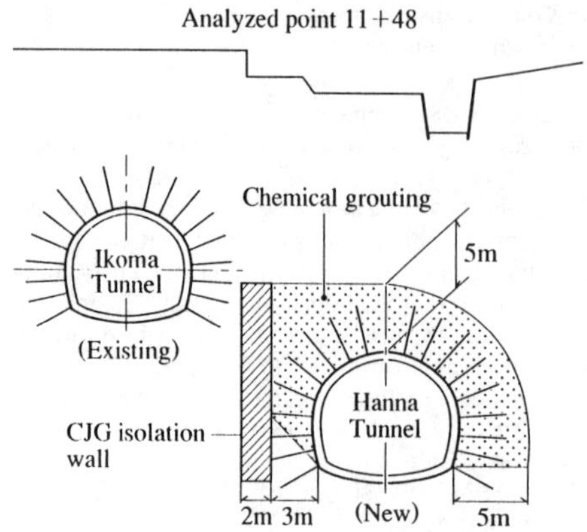

Fig. 15 Range requiring measures, cross section

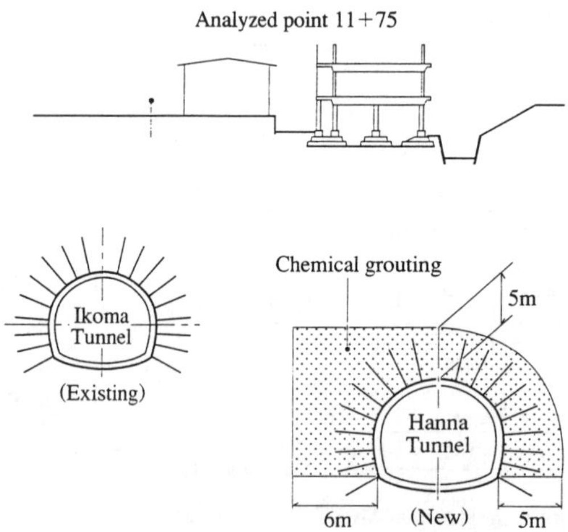

Fig. 16 Range requiring attention, cross section

proximity (Fig. 14), column jet and chemical grouting to the range requiring measures (Fig. 15), and chemical grouting to the range requiring attention (Fig.16).

In the FEM analysis it was assumed that the protective measures would improve the elastic modulus of column jet (C.J.G.) to 20,000 (tf/m^2), cohesion to 4.0kgf/cm^2, Poisson's ratio to 0.25 and that the chemical grouting would improve cohesion and angle of internal friction by 1.5 times.

The results of the FEM analysis indicated that monitoring items would measure less than 50% of the determined critical value after the above measures were taken.

7 CONCLUSION

The following conclusions have been found from the discussion above.

1. The back analysis of the data acquired when the existing tunnel had been excavated was useful in assessing the physical characteristics of the ground, and actually resulted in the determination of the very reliable physical characteristics values.
2. Based on the FEM analysis, the effect of the excavation was predicted to be such that the stress at the existing tunnel's secondary lining would increase and that the existing tunnel would suffer deformation. Also, it was assessed that in the extreme proximity the stresss would exceed the allowable stress in a certain position.
3. The deformation tendency of the existing tunnel, determined through analysis, was used as valuable information in establishing monitoring items and standard values for the monitoring.
4. To ensure the safe operation of the trains, the monitoring items were selected carefully and separated into main items and subordinate ones and the standard values for monitoring each item were determined, based on the FEM analysis results and the deformation studies/pre-excavation (of the new tunnel) measurements of the existing tunnel. In addition, the automatic measuring and monitoring system; a network that unites the Highway Corporation, the Railway Company, and the Contractor, was established.
5. After the new tunnel excavation was started, the excavation effect on the existing tunnel was re-assessed through FEM analysis, which evaluated the ground physical characteristics by using the data acquired from the back analysis and the soil studies during the early phase of the new tunnel excavation. Based on the result of that assessment, measures to protect the existing tunnel from the excavation effect were re-examined, and the appropriate measures were implemented under the measuring and monitoring system. The excavation work is under way presently and it is expected that it will progress smoothly.

REFERENCE

Okabe Y. 1991. *The Evaluation of NATM Support in City NATM*

Tunnelling and Ground Conditions, Abdel Salam (ed.) © 1994 Balkema, Rotterdam, ISBN 90 5410 363 9

Geological optimization of underground facilities of the Superconducting Super Collider, Texas, USA

Donald P. Richards
Parsons Brinckerhoff International, Cairo, Egypt

Jeffrey S. Robbins
PB/MK Team, Dallas, Tex., USA

ABSTRACT: The Superconducting Super Collider (SSC) currently under construction in Dallas, Texas, U.S.A., involves an 87 km (54 mile) ring of circular tunnel as well as approximately 26 km (16 miles) of secondary tunnels. All tunnels are located in sedimentary strata of cretaceous age, but which is variable in nature and quality. Anticipated difficulties in construction prompted the need to re-evaluate the collider position in terraspace in order to optimize the alignment by maximizing the favorable geology and still satisfying the operational boundary conditions. This positioning adjustment not only solved several construction problems, but also was estimated to provide a construction cost savings of approximately $40 million.

1 INTRODUCTION

The Superconducting Super Collider is an underground high energy physics experimental facility which when complete will produce two 20 TeV counter rotating proton beams. The facility is composed of approximately 26 km (16 miles) of accelerator tunnels in which the protons undergo initial acceleration before being transferred to the main collider tunnel which is an approximately circular ring 87 km (54 miles) in circumference.

In addition to the basic tunnels, auxiliary facilities include 47 shafts for magnet installation, electrical/mechanical connection to the surface and personnel access, connecting adits between tunnels and shafts, and equipment niches located every 450 m (1476 ft.) along the tunnel.

The tunnel itself is simply a container for the technical systems. The tunnel will house the superconducting magnets which will guide and propel the high energy protons around the ring at near the speed of light. Physics experiments where the proton beams collide are located in underground detector halls which have been discussed in detail by Schmidt and Moeller (1994).

Preliminary siting of the underground facilities located the tunnel complex in a sedimentary deposit of cretaceous age composed of a sequence of chalk, marl and shale. The Site Specific Conceptual Design (SSCL, 1990) based conceptual tunnel design and layout on preliminary geological information which included general interpretations of contact zones between stratigraphic horizons and estimated location and displacement of faults. However, when detailed geological information became available, it became increasingly evident that the detector halls would be founded in the poorer geological materials and that a considerable part of the tunnel excavation would require substantial support due to relatively poor geological conditions. The study being reported herein was then initiated in an attempt to optimize the geology, which would minimize construction problems and reduce construction risk in both tunnel and detector hall excavations.

2 GEOLOGY

The SSC project is located in the gently rolling landscape of Ellis County, Texas, in the south-central U.S.A., and is centered about 55 km (35 miles) south of the central business district of the city of Dallas, near Waxahachie, Texas as illustrated in Figure 1.

The project area lies within the northwestern margin of the West Gulf Coastal Plain which is bounded on the east by the Mississippi River and on the west by the Edwards and Balcones escarpments. Project facilities will be constructed in the geologic units which are Upper Cretaceous sedimentary rocks belonging to the Gulf Series.

Rock units dip gently to the east-southeast and include the younger Ozan Formation, known as the lower Taylor Marl (TM), the Austin Chalk limestone (AC), and the older Eagle Ford Formation shale (EFS). A general geologic profile is shown in Figure 2. Contacts between rock units are

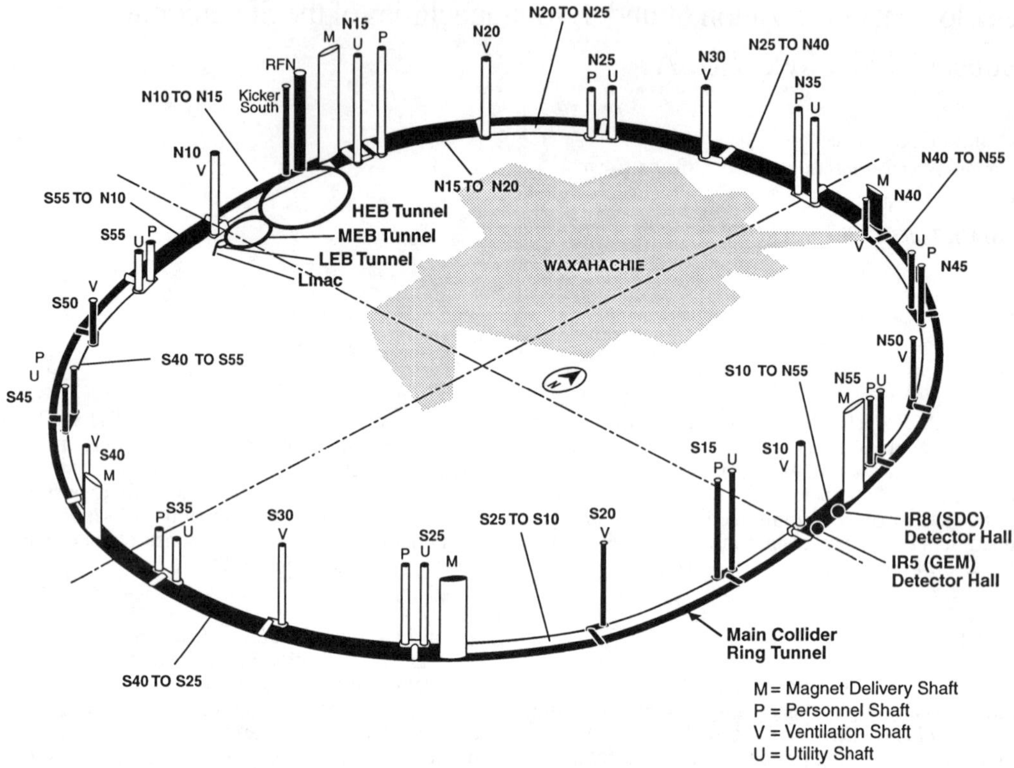

Figure 1. Project Location

unconformable. Both the TM and the AC outcrop within the project area and are the source of a relatively thin residual soils created by chemical and mechanical weathering which is common throughout the site. Some recent Quaternary clays, sands and gravels, and silty to clayey sands are located within the project area and can be found in thickness up to 15.2 m (50 ft.). General geological descriptions of these formations have been summarized by Lundin et al (1993) and are described below.

2.1 Taylor marl

The lowest unit of the Ozan Formation, known as the Taylor Marl (TM), is generally a green-gray to blue-gray, fine grained, laminated, calcareous claystone with interbedded thin limestone beds and is massive when fresh. The TM normally contains about 60-70 percent illite and montmorillonite clays and an average calcium carbonate content of about 25 percent. Thin beds of calcareous bentonite are present but are not prevalent.

The relatively high percentage of clay minerals in the TM, about two-thirds of which is montmorillonite, makes it moisture sensitive and subject to slaking and swelling with variations in moisture content. Excavated surfaces in the TM are required to be covered with shotcrete within six hours of exposure. This formation presents average tunneling conditions on this project.

2.2 Austin chalk

The Austin Chalk Limestone (AC) is generally a light to medium gray, fine to very fine grained, slightly to moderately fossiliferous limestone. The AC is composed of an average 85 percent calcium carbonate and is interbedded with thin beds of calcareous claystone (shaley beds), argillaceous beds and occasional thin beds of bentonite. The hardness of the AC ranges from very soft, where weathered, to moderately hard for fresh rock.

The occasional beds of bentonite, argillaceous and shaley materials may be susceptible to slabbing in the crown of excavated openings due to exposure to air and moisture variations. When encountered in the invert of excavations, these beds, when given access to water, sometimes become soft which requires attention during excavation. This formation presents the best tunneling conditions on the project.

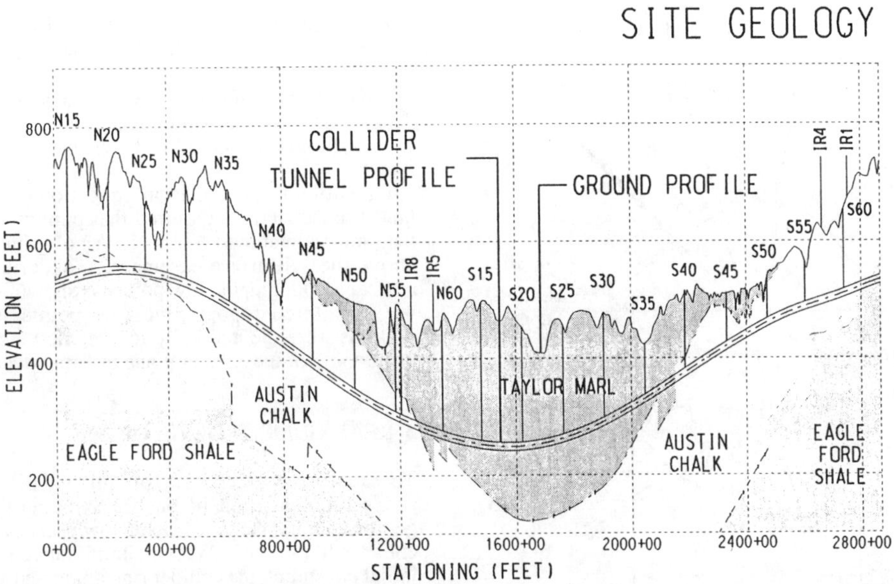

Figure 2. Collider Geologic Profile

2.3 Eagle Ford Shale

The Eagle Ford Formation is divided into two units in the Dallas-Fort Worth area. Only the upper unit, the South Bosque Formation shale, is encountered in the project area and is referred to in this paper as the Eagle Ford Shale (EFS). The EFS is a dark gray to black, very fined grained, calcareous to noncalcareous marine shale with occasional thin limestone beds and calcareous concretions. The EFS is a montmorillonitic shale with high shrink-swell potential.

Lying uncomformably under the AC, the upper 3 m (10 ft.) of the EFS generally has a slightly increased moisture content and a corresponding decrease in strength. This zone also tends to exhibit an increased frequency of discontinuities which may result in lower rock quality in this area. In general, higher RQD values in the EFS may give a misleading impression of the basic competence of the material. Other properties and characteristics, such as compressive strength, may be a better indicator of the competence of the EFS.

The weak nature of the EFS soil-like rock, coupled with excavation induced stress redistribution, creates a tendency for the materials strength to be exceeded resulting in stress slabbing, ravelling, yielding or squeezing of the ground. The stress regime and the shrink-swell characteristics of the material require that excavated surfaces be protected immediately upon exposure to minimize adverse ground behavior. This formation presents the worst tunneling conditions on this project.

3 TUNNEL DESIGN CONCEPTS

The basic tunnel design concepts for the SSC have been addressed in detail by Clark, 1993.

There are two basic types of tunnels being designed and constructed for this project, reinforced precast concrete and unlined (Figure 3). The fully lined tunnels are constructed in the softer rocks (TM and EFS). The lining consists of precast 23 cm (9 in.) thick 55, 160 kPa (8,000 psi) concrete segments. Each liner ring is nominally 1.5 m (5 ft.) wide and is made up of four segments. The finish diameter of these tunnels is a minimum of 4.3 m (14 ft.); therefore, the minimal bored diameter is 4.7 m (15.5 ft.).

The unlined tunnel is constructed in the Austin Chalk. Although not lined, any unstable areas are reinforced with shotcrete and rockbolts. The basic design calls for systematic rockbolts in the crown with a utility support channel in between which acts as a minestrap. A nominal 15 cm (6 in.) thick fiber reinforced shotcrete is utilized to stabilize the tunnel in areas deemed necessary. All unlined tunnels and shotcrete are sprayed with a special rock sealing treatment to prevent deterioration of the rock due to air or water, to improve light reflection, and to reduce dust.

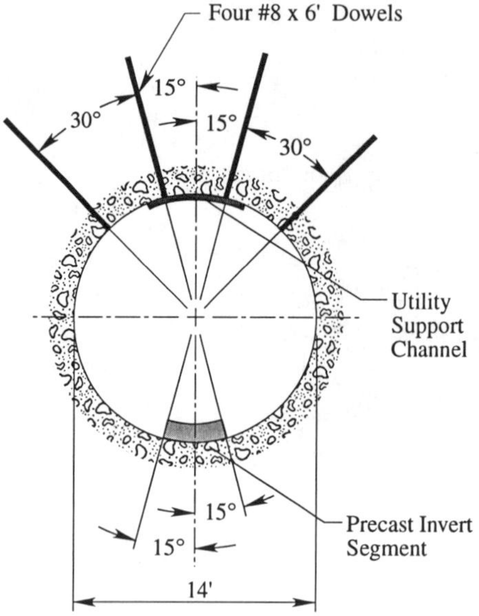

Typical Unlined Collider
Tunnel Cross Section-Chalk

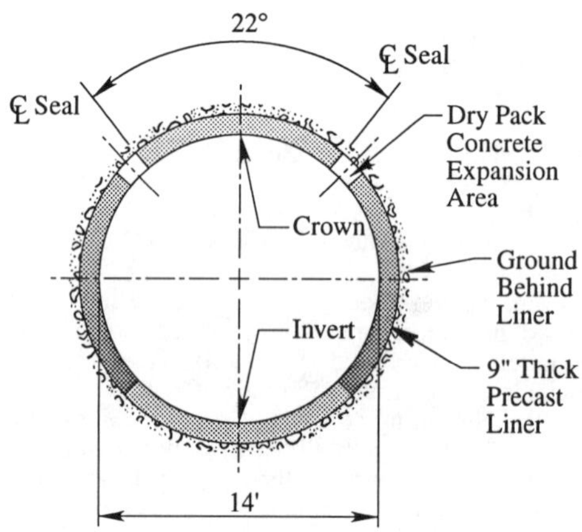

Typical Concrete Reinforced
Collider Tunnel Cross Section
-Shale and Marl

Figure 3.

4 CONSTRAINTS

The underground layout of the main collider ring had several design constraints for operational and safety reasons SSC Site Characterization Document (SSCC,1989). These are noted below:

- Radiation exposure limit of 10 mrem/year at the boundary of areas controlled by the project. (This included the subsurface easement surrounding the tunnel.)
- Minimum 9.1 m (30 ft.) of cover over tunnel, even for land controlled by the project.
- Governmental requirement for an additional 4.6 m (15 ft.) of ground cover in areas where the surface is not controlled by the project.

In addition to the radiation protection constraints, land use restrictions required that public or private land maintain at least 4.6 m (15 ft.) of ground cover above the subsurface easement. This criteria was critical at topographic low points as shown in Figure 4. The three topographic low points in creek bottoms were the main physical restrictions in final positioning of the collider plane in terraspace.

5 REALIGNMENT DEVELOPMENT

To satisfy the criteria for optimizing the collider ring's final position in the site's topography and geologic strata, the topographic low points were first considered as noted. With the minimum cover as an initial constraint, the collider ring alignment was then reoriented by a series of iterative manipulations in both elevation and dip as described in the following paragraphs.

In order to redefine the orientation of the conceptual design footprint, the following criteria were established:

- Maintain the projection to the surface (the footprint) of the subsurface easement.
- Maintain the lattice as a true plane.
- Maintain an average dip of the collider plane of less than 0.20 degrees.
- Maintain a minimum of 9.1 m (30 ft.) of cover above the crown of the tunnel at Chambers Creek Tributary and Waxahachie Creeks and 13.7 m (45 ft.) of cover at Red Oak Creek.
- Minimize the requirements for additional land acquisition at the surface.
- Maximize lattice footprint elevations in the west, in the vicinity of the experiment halls, within these criteria.

In keeping with the above referenced criteria, the lattice footprint plane was raised to obtain a minimum clearance (from the crown of the tunnel to the bottom of the creek) of 9.1 m (30 ft.) at Waxahachie Creek (see Figure 5 for reference) and then tilted at this point (the intersection of Waxahachie Creek and the ring) in the direction of Chambers Creek Tributary to obtain an initial clearance of approximately 10 m (33 ft.) at that location.

The intersection of this newly derived plane and the local geodetic plane at the initial lattice origin was

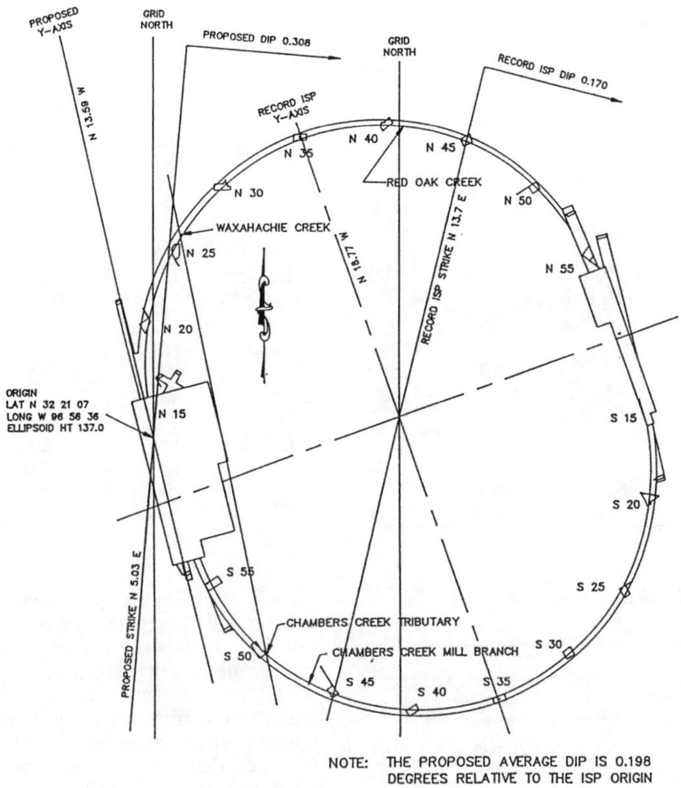

Figure 4. Axis Orientation of Recommended Realignment

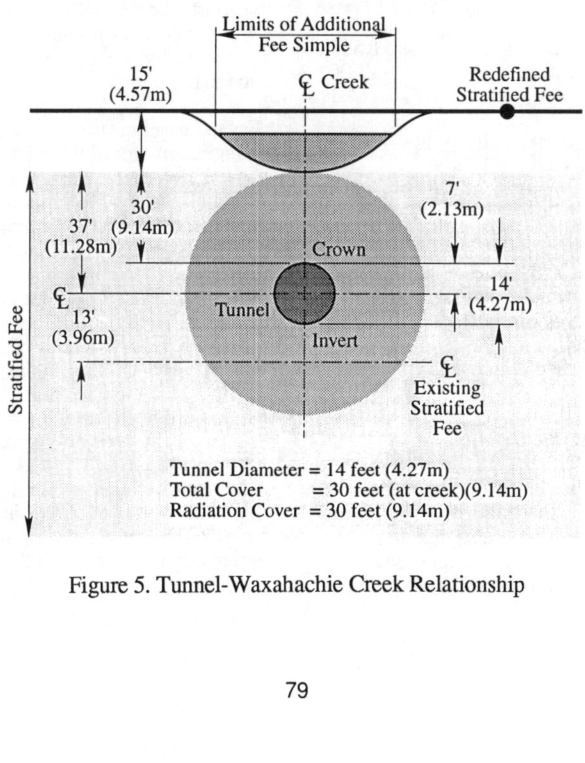

Figure 5. Tunnel-Waxahachie Creek Relationship

Table 1. Collider tunnel geology

Geology	Units	Baseline	Realign-ment	Difference
Austin Chalk	l.m.	43,260	48,990	+5,730
	l.f.	141,900	160,700	+18,800
	%	49.7	56.2	+6.5
Eagle Ford Shale or Taylor Marl	l.m.	39,020	34,540	-4,480
	l.f.	128,000	113,300	-14,700
	%	44.8	39.7	-5.1
Mixed Face	l.m.	4,850	3,600	-1,250
	l.f.	15,900	11,800	-4,100
	%	5.5	4.1	-1.4
TOTAL	l.m.	87,130	87,130	0
	l.f.	285,800	285,000	0
	%	100.0	100.0	0

Table 2. HEB tunnel geology

Geology	Units	Baseline	Realign-ment	Difference
Austin Chalk	l.m.	9,020	10,790	+1,770
	l.f.	29,600	35,400	+5,800
	%	83.6	100.0	+16.4
Eagle Ford Shale or Taylor Marl	l.m.	240		-240
	l.f.	800	0	-800
	%	2.3	0	-2.3
Mixed Face	l.m.	1,530		-1,530
	l.f.	5,000	0	-5,000
	%	14.1	0	-14.1
TOTAL	l.m.	10,790	10,790	
	l.f.	35,400	35,400	-
	%	100.0	100.0	-

computed yielding a new strike line. The plane was subsequently tilted about this new strike line to obtain a maximum effective dip of 0.198 degrees at the origin (i.e., at the center of the ring along the original strike azimuth). (See Figure 4.)

6 GEOLOGICAL IMPACTS

Based on the existing geological data base, the geological intercepts of the tunnel through the various formations were estimated for both the baseline layout and the proposed realignment. The net differences that would affect construction costs are shown in Table 1.

Table 3. Transfer tunnel geology

Geology	Units	Baseline	Realign-ment	Difference
Austin Chalk	l.m.		140	+140
	l.f.	0	450	+450
	%	0	20.9	+20.9
Eagle Ford Shale or Taylor Marl	l.m.	650	300	-350
	l.f.	2,150	1,000	-1,150
	%	100.0	46.5	-53.5
Mixed Face	l.m.		210	+210
	l.f.	0	700	+700
	%	0	32.6	+32.6
TOTAL	l.m.	650	650	
	l.f.	2,150	2,150	-
	%	100.0	100.0	-

It was a functional requirement that the high energy booster (HEB) tunnel ring move unilaterally with the collider main tunnel ring for any realignment/reorientation. Accordingly, the geological intercept differences between the baseline and the recommended realignment for the HEB tunnel were estimated as shown in Table 2. It should be noted that the proposed realignment raised the HEB entirely out of shale and mixed face conditions moving it entirely into chalk.

The transfer tunnels between the HEB ring and the main collider ring require an enlargement of both the HEB and collider tunnels at the point of intersection, connected by a short length of tunnel between the two enlargements.

For the baseline alignment, the entire transfer tunnel arrangement would be in Eagle Ford Shale. For the revised alignment, the transfer tunnels penetrate up into chalk as the HEB tunnel rises into chalk. The estimated differences between the baseline alignment and the recommended alignment are given in Table 3.

With the revised collider ring alignment, the west side experiment halls are raised in elevation approximately 5.8 and 5.5 m (19 and 18 ft.) respectively at the beamline/tunnel centerline position. On the east side, the experiment halls are lowered approximately 7.6 and 7.3 m (25 and 24 ft.) respectively at the beamline/tunnel centerline.

The results of this realignment effectively raise the foundation level out of the shale and into the chalk on the west, and lower the foundation level out of the marl and into the chalk on the east.

Table 4. Conceptual cost summary

Feature	Analyzed cost change
Collider Tunnels	-$ 11,890 k
Collider Shafts	-$ 210 k
HEB Tunnels	-$ 4,570 k
HEB Shafts	-$ 980 k
Niches and Transfer Tunnels	-$ 1,950 k
Experiment Halls	-$ 17,000 k
Conventional M & E	-$ 0 k
AE/CM Costs	-$ 4,000 k
Subtotal	-$ 40,660 k
Additional Fee Simple & Stratified Fee	TBD*
Net Savings	TBD*

7 COST IMPACTS

To evaluate the construction cost implications of revising the tunnel alignment, several simplifying assumptions/boundary conditions were necessary due to the preliminary nature of the design process at the time of the study. The simplifying boundary conditions are noted below:

- All tunnel construction costs assume a 4.3 m (14 ft.) finished inside diameter tunnel.
- In evaluating the construction cost differentials, only the changes in conventional construction were considered. No adjustment was considered for shortening or lengthening the technical electrical and mechanical systems.
- At this early stage in the design process, many of the design concepts were preliminary and represented the best information available at the time of the study.
- The tunnel excavation costs assumed unit costs in different rock types. No adjustment was made for contract specific conditions in which a segment of tunnel may be driven in chalk using a machine designed for shale or marl at a consequent higher unit cost than that for a tunnel fully in chalk.
- Considering the decreased efficiency and the resulting increased cost of mixed face tunneling (marl over chalk or chalk over shale) an adjustment was made for the cost differentials resulting from addition or deletion of mixed faced tunnels.
- TBM excavation costs were used for assessment of the HEB tunnels since only about 4 percent of the length will be non-TBM.
- Cost differentials for excavation and support of adits and alcoves were not considered due to estimated minimal changes in geology.

The results of the cost analysis are presented in Table 4.

8 CONCLUSIONS

By adopting the change in elevation of the main collider ring, substantial improvements were made in operational stability of the beamline by minimizing the amount of shale, marl, and mixed face tunnel conditions in the collider and HEB tunnels while maximizing the amount of chalk tunnel which is the most desirable and the least costly tunneling medium. Similarly, the west side experiment halls were raised sufficiently to minimize or eliminate the exposure of shale in the bottom of the excavation, substantially improving foundation conditions and reducing construction risk due to excavation instability in the shale. Other components of construction experienced similar improvements.

The change in elevation produced an estimated construction cost savings of approximately $41 million, without accounting for additional real estate and subsurface easement costs.

Although the adoption of the changes required acquisition of additional land and subsurface easements, the associated costs of this additional procurement were small relative to the total estimated construction cost savings.

The change in elevation satisfied all radiation protection requirements. Minimal changes in the environmental basis used to evaluate environmental impacts resulted from this change.

REFERENCES/BIBLIOGRAPHY

Clark, G., 1993. "Analysis and Design of SSC Underground Structures", Proc. RETC, Boston, AIME.

Lundin, T.K., J.E. Shively & F. Fong, 1993. "Underground Facilities for the Superconducting Super Collider", Proc. RETC, Boston, AIME.

Matthes, G.H. & J.H. Albino, 1993. "Superconducting Super Collider Shaft Excavation", Proc. RETC, Boston, AIME.

PB/MK Team, 1991. Collider Tunnel Elevation Study Report, Document No. CPB-000955.
Prince, R., J. Bhore, T. Martin, & R. Richards, 1993. "Superconducting Super Collider Construction of Tunnels and Shafts", Proc. RETC, Boston, AIME.
Robbins, J.R. & T.J. Greening, 1992. "Geodetic Design of the Superconducting Super Collider", Proc. ASCE Conference, Dallas.
Schmidt B. & D.T. Moeller, 1994. "Tieback Anchors Support Record Setting Walls for Super Collider Experiment Halls in Texas", International Congress on Tunneling and Ground Conditions, Cairo.
Superconducting Super Collider Laboratory, 1989. Footprint Characterization Document, Report No. SSCL-SR-1041.
Superconducting Super Collider Laboratory, 1990. Site Specific Conceptual Design Report, Report No. SSCL-SR-1056.

Tunnelling and Ground Conditions, Abdel Salam (ed.) © 1994 Balkema, Rotterdam, ISBN 90 5410 363 9

Influence forecast and field measurement of a tunnel excavation crossing right above existing tunnels

A. Saitoh, K. Gomi & T. Shiraishi
East Japan Railway Company, Japan

ABSTRACT: In comparatively weak alluvial layered ground using a pipe roof construction, a box shape tunnel (height 9m, width 14m) crossing directly above existing in-use twin railway shield tunnels (external diameters 7.1m) was excavated (not by open cut method). An influence forecast analysis was carried out before starting construction, and this forecast is compared with the on-site observed measurements. From these results the following points were understood.
(1) For circular tunnels the change in behaviour of existing tunnels depends on the excavation of the tunnel crossing above; when there is a rise in level by several millimeters due to the excavation locally above, it will be seen that the dynamic influence on the tunnel lining will be especially small.
(2) It was confirmed that the above can be forecast by using comparatively simple two-dimensional elastic FEM analysis.
(3) It was confirmed that the behaviour of a pipe roof during excavation can be expressed according to successive analysis with a beam spring model.

1 INTRODUCTION

In recent years, in urban areas, accompanying the high level of use of ground space, there has been a sharp increase in construction work carried out very close to existing structures. In particular, for conditions where space underground must be used, for example, to construct railway lines in urban areas, there has been an increase, by necessity, in those constructions which cross close to existing tunnels. Indeed it would take too long to give all the recent examples of this kind of construction in Japan.

In cases where new tunnel construction is carried out close to existing structures in weak unconsolidated ground, it is considered especially important to forecast the influence on the existing tunnel before starting work, to take measurements at the time of construction, and to confirm structural safety by comparing calculated and measured values.

Therefore the authors carried out a comparative investigation of the influence forecast analysis before starting construction work on a box shape tunnel (whole body), not excavated by open cut but with a pipe roof construction method. The work was part of the extension construction for the Keihin Kyuko Haneda-Kuko Line, and measurements were taken on the section which crosses directly above the JR Tokaido Freight Branch Line shield tunnels (Haneda Tunnel). It was confirmed that it is possible to carry out an influence forecast to a reliable level using a comparatively simple two-dimensional analysis with FEM and a beam spring model, if the ground values are established adequately. In this report, the results of the comparison of the influence forecast analysis with the on-site measured data for the behaviour of the pipe roof and, in particular, for the behaviour of the lining in the existing tunnel at the time of excavation of the new tunnel, are discussed.

2 SUMMARY OF CONSTRUCTION FOR CROSSING TUNNEL

2.1 *Relationship between Geology and Position*

Figure 1 shows a simplified geological profile in the vicinity of the area concerned. Generally the ground being filled-in Tokyo Bay, fine grain alluvial layers principally with N values (Standard Penetration Test Values) of 5 to 20 have accumulated to a depth of 15m. The new tunnel was excavated in these fine grain alluvial layers but it includes material with a coefficient of uniformity of 4 and 11% fine grain particles, so when excavating the self supportability of the ground was considered to be a problem (Japan National Railways Construction Bureau, 1977). Beneath this there are 5-8m thick layers of Upper Diluvial clays, and below this Upper Diluvial sandstones with N value of above 30 have accumulated. The Upper Diluvial clays are comparatively well compacted with a uniaxial compressive strength of 2-7kgf/cm^2. The Upper Diluvial layers are not continually connected with the inclined muddy sands of the Lower Diluvial (Kazusa) layers. This non-continuous face is steeply inclined downwards from the existing tunnel Down Line side towards the Up Line side. The level of consolidation of the Lower Diluvial layers is somewhat high compared to that of the Upper Diluvial layers. With the 16m overburden over the existing tunnels, the larger half of the tunnel cross-sections are within the Upper Diluvial clay layers.

Figure 2 shows the relative positions of all the tunnels. From this diagram, the pipe roof, for the new tunnel, crosses directly above the existing tunnel at a height of 1m. In addition, the angle of crossing is about 80°. The pipe roof (width 13.8m, height about 9m) was constructed as a box shape surrounding the whole body, and within this, having excavated, the new tunnel was constructed.

2.2 Structure of Existing Tunnels

The existing tunnels were constructed using an open shield method on single track parallel circular tunnels over about 20 years. As Figure 3 shows, the external diameters are 7.1m, with 10m between the two tunnels. The lining was built of box shape segments, and there was no secondary lining. Regarding the tunnel soundness, the segments connected with bolts showed some corrosion but the condition of the whole body was good. Before starting construction a shear strength test was carried out on these bolts and it was confirmed that they were resistant enough.

2.3 Outline of New Tunnel Construction

The separation was extremely small, the existing tunnels were built only with segments, and also the earth and sand ground was unconsolidated, so various investigations were carried out at the

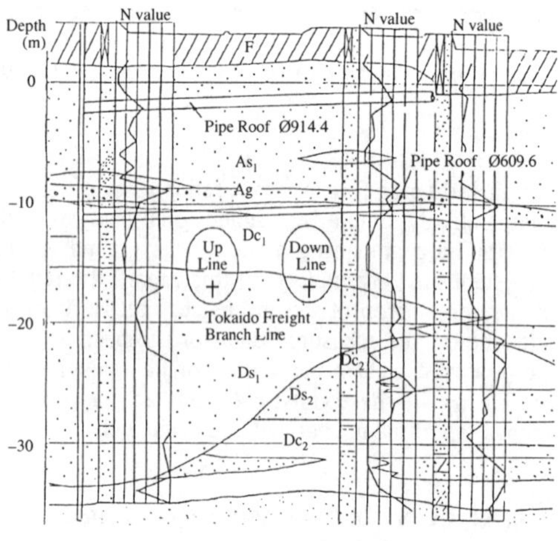

Figure 1. Geological Profile (section showing crossing above existing tunnels)

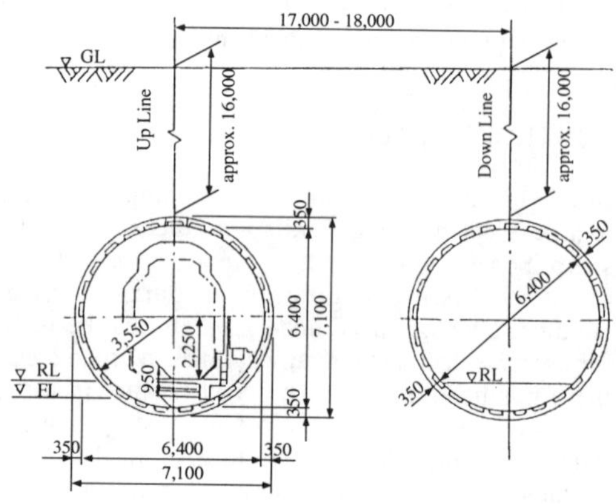

Figure 3. Structure of the Existing Tunnels

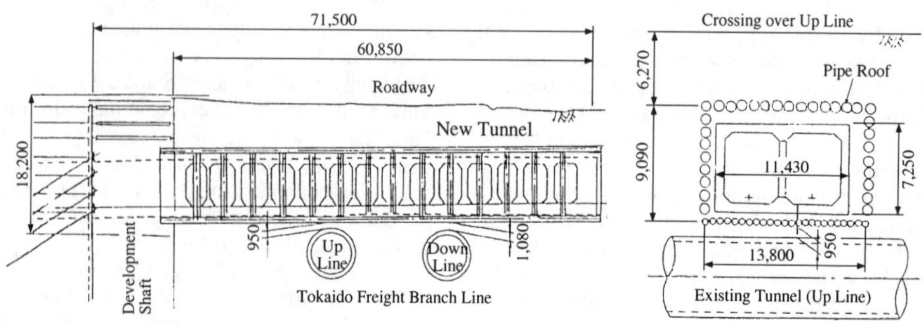

Figure 2. Relative Positions at the Tunnel Crossing

Table 1. Ground Values used in FEM Analysis

STEP 1				STEP 2				Notes
	E	ν	γ		E'	ν	γ	
1	3 500	0.35	1.90	1'	10 500	0.35	1.90	E = 3E'
2	5 000	0.45	1.80	2'	15 000	0.45	1.80	
3	9 000	0.35	1.90	3'	27 000	0.35	1.90	
4	10 000	0.35	1.90	4'	30 000	0.35	1.90	
5	1.17×10^6	0.167	1.61		-	-	-	
6	-	-	-	6'	1 885	0.30	0.665	

E, E': Modulus of Deformation (tf/m^2) ν: Poisson's Ratio γ: Unit Weight (tf/m^3)

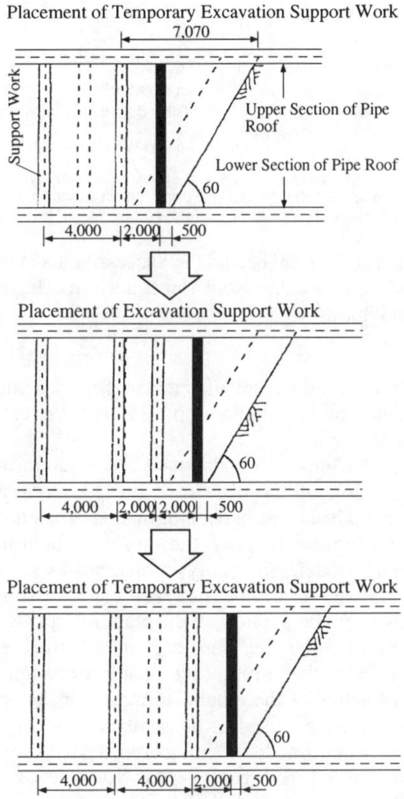

Figure 4. Excavation Sequence for the New Tunnel

planning stage regarding the method of construction for the tunnel (Fukuyama & Gomi, 1992). An investigation into cut and cover excavation was done, but from elastic FEM analysis, the stress relief at the time of excavation giving a rise in the existing tunnel level (about 21mm) was incompatible with the standard maintenance of rail track, and it was forecast that the stress would also exceed permitted stress values. For these reasons and to reduce the influence of the upper tunnel excavation as far as

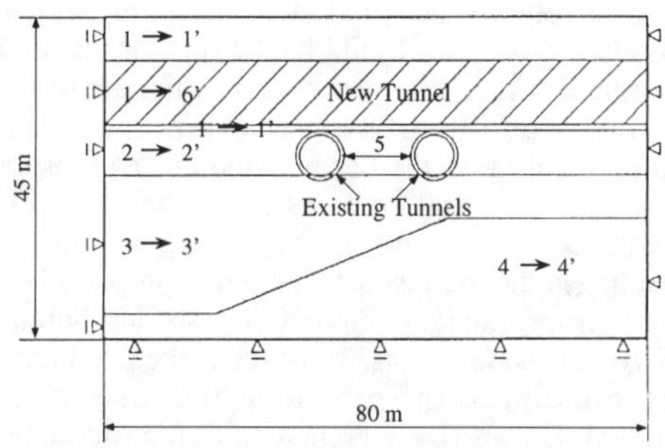

Figure 5. FEM Analysis Model

possible, a complete box pipe roof construction was employed instead of cut and cover.

The sequence of construction is summarized as follows. Namely, from the base of the development shaft, with the purpose of strengthening the ground (within the pipe roof) and stopping water ingress (around the pipe roof), injection grouting was carried out from the shaft in a horizontal direction (Hisanaga & Yoshinaga. 1981). Next, after driving the box pipe roof (pipes in top section and sides: external diameter 914.4mm, bottom pipes 609.6mm) in from the development shaft, support work was built as the inside of the pipe roof was excavated, and the tunnel body was built. The length of the pipe roof was set at about 60m (see Figure 2), having considered the area of influence on the existing tunnels. Having studied the work and machines, a full face excavation method was chosen. To maintain the independence of the cutting face it was decided to install horizontal rock bolts (self perforating IBO anchors ø42.2mm) from the shaft, in addition to the above mentioned injection grouting. In addition, as shown on Figure 4, while keeping the inclination of the cutting face at 60°, it was decided to limit the excavation step advance to 2m.

3 INFLUENCE FORECAST ANALYSIS

In order to understand the influence of the excavation on the existing tunnels, two-dimensional elastic FEM analysis was carried out initially. Figure 5 shows a typical analysis model, Table 1 gives the ground values used in the analysis. Having carried out dead load analysis on the existing state of the tunnels (Step 1), analysis was carried out again after the excavation inside the pipe roof (Step 2). In order to carry out a precise forecast, it was necessary to analyse having considered each step of the excavation advance. On this analysis, step analysis was not carried out, the principal object being to forecast roughly the final state behavior, which it was also considered would be the time of maximum change in the existing tunnels. Concerning the ground Modulus of Deformation used in the analysis, based on the Secant Modulus of Elasticity taken from uniaxial compression tests on the clay, and the N value of the sandy ground, a value was established having also considered appropriately the results from triaxial compression tests and borehole horizontal loading tests. In general, the Modulus of Deformation of the ground at the time of load removal, E', was large compared to the Modulus of Deformation at the time of loading, E. In Step 2 a value E' = 3E was used.

According to the analysis cross-section the rigidity of the tunnel, having input the values when there were changes in the rigidity, becomes equal to the sum of the bending rigidity of the whole pipe roof. The support work was not thought to contribute to the bending rigidity of the tunnel and was not modeled. Furthermore for the rigidity of the shield segments, the effective ratio of bending rigidity was considered to be h = 0.5 (Japan National Railways Construction Bureau, 1977).

The crown of the existing tunnels showed a final rise in level of 6mm, while the upper section of the pipe roof directly above the existing tunnels showed a settlement of 70mm. The area of influence of the excavation on the existing tunnels, according to the results of the two-dimensional elastic FEM analysis carried out directly along the section of the new tunnel, was 54m. The whole body of the existing tunnels will rise, and for the mode change, the lateral pressure will be relatively large so there will be a tendency towards a reduction in the horizontal direction but, the level of displacement will be slight. Regarding the lining stress, at a right angle to the tunnel axis, the stress will decrease and the bending moments will be relatively small, so the whole body will be in a state of compression, with the result that it will be far less than the permitted stress levels. Regarding the tunnel axial stress too, depending on the results estimated from the changes due from FEM analysis of the axial cross-section, it is considered that the values will be safe enough.

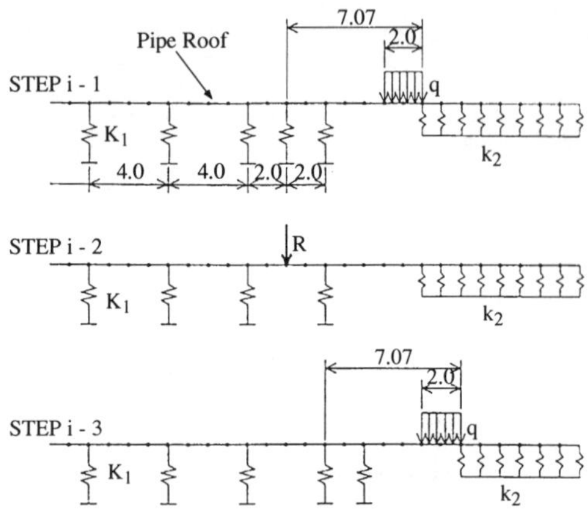

K_1 : Spring Constant for the Support Work (3,340tf/m)
k_2 : Spring Constant for the Ground (750tf/m^2)
q : Increase in the Perpendicular Ground Stress accompanying the Cutting Face Advance
R : Reaction accompanying the Support Work Removal

Note: at Step i increased forces in the beam cross-sections at each point are ΔMi, ΔSi while the displacement at this time is ΔU; at Step i M, S, U are $M=\Sigma\Delta Mi$, $S=\Sigma\Delta Si$, $U=\Sigma\Delta Ui$.

Figure 6. Concept behind the Successive Excavation Analysis of the Pipe Roof Upper Section using a Beam Spring Model

From the above results it was forecast that the influence on the shield from the tunnel excavation would be small.

Next, in order to understand the behavior of the top section of the pipe roof during excavation, elastic analysis was carried out using a beam spring model. Figure 6 shows the concepts behind this analysis model. The analysis method, as shown in this figure, involves successively addition and removal of the support work material springs and loading of the increase in load, in accordance with the shifts in the cutting face position changing the area of action of the ground springs. For the ground spring constants used in the analysis, the Modulus of Deformation, E = 350kgf/cm^2, was shown in documentary records (Japan Society of Civil Engineers, 1986), and the calculations for the Horizontal Coefficient of Subsoil Reaction for tunnels were referred to. Also the excavation of the working face was carried out at an angle of 60° and rock bolts were installed horizontally, so it was estimated that the looseness of the ground in the vicinity of the working face would be small, and that there would be no reduction in the ground spring constants.

Some of the results of the successive analysis are shown on Figure 7. From this figure the maximum value for the pipe roof settlement is 31mm. The maximum value for the stress is 1,400kgf/cm^2, but the maximum stress in the neighbourhood of the cutting

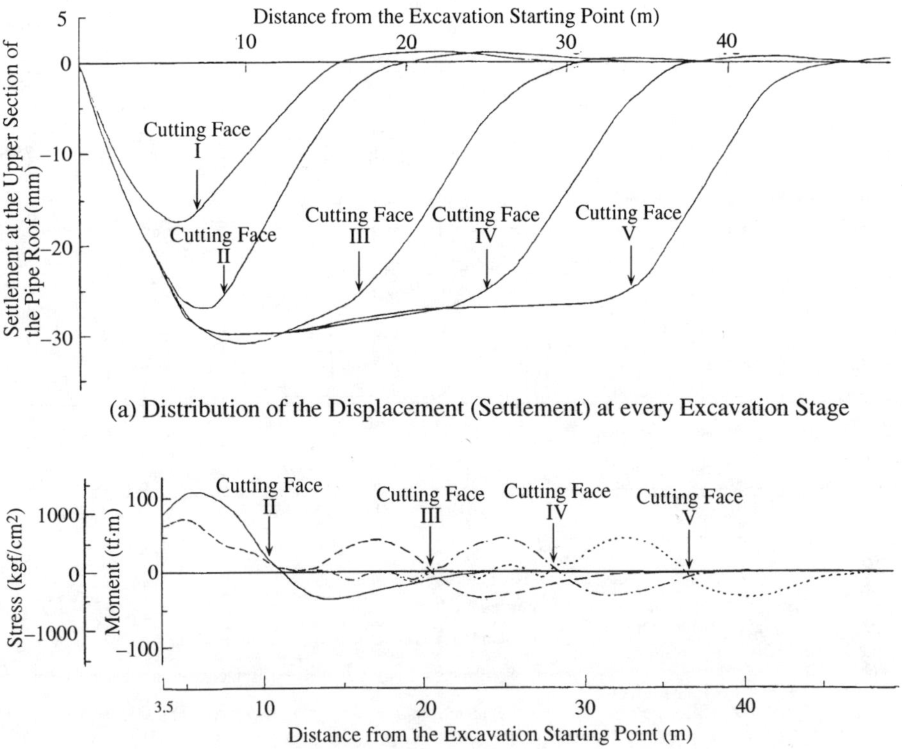

(a) Distribution of the Displacement (Settlement) at every Excavation Stage

Figure 7. Results of the Analysis with the Beam Spring Model

face on and after Stage III is almost uniform, and about 4m after the excavation has passed it is about 600kgf/cm^2. These values are well within the permitted stress values (1,710kgf/cm^2) for the pipe roof.

4 COMPARISON WITH OBSERVED DATA

4.1 Details of Measurements

Based on the forecast analysis results, the measurements below were detailed carefully during the excavation of the tunnel.

(1) Existing Tunnels: Rise in Level, Displacement inside the Tunnels, Segment Joint Displacement.
(2) New Tunnel (Pipe Roof): Settlement of Upper Section of Pipe Roof, Stress in Upper and Lower Sections of Pipe Roof.

The position of the measurement equipment for the existing tunnels is shown on Figure 8.

4.2 Behaviour of Existing Tunnels

1) Rise in Level

Figure 9 shows the rise in level in the existing tunnels accompanying the excavation. From the distribution along the tunnel axis of the cumulative rise in level between the start and finish of excavation, Figure 9(a), the maximum value for both tunnels was at Point ⑤ (beneath the right side of the new tunnel cross-section), where the rises in level were 4.9 and 5.1mm. These values are almost equal to the values forecast by the FEM analysis (a rise in level of 6mm). Also, in Figure 9(b) the change over time for Point ⑤ is shown. Studying this graph, it is possible to see the following conditions:

(1) From the start of excavation, advance displacement accompanying the approach of the cutting face steadily occurred.
(2) At the point when the cutting face passed directly above the tunnels there was a sharp rise in level.
(3) After this until completion of the excavation the preceding displacements rose at almost the same rate.
(4) The maximum value was seen at the point when the excavation was completed.

After the completion of the excavation the settlement dropped away, due to the increased loading as the new tunnel body was constructed.

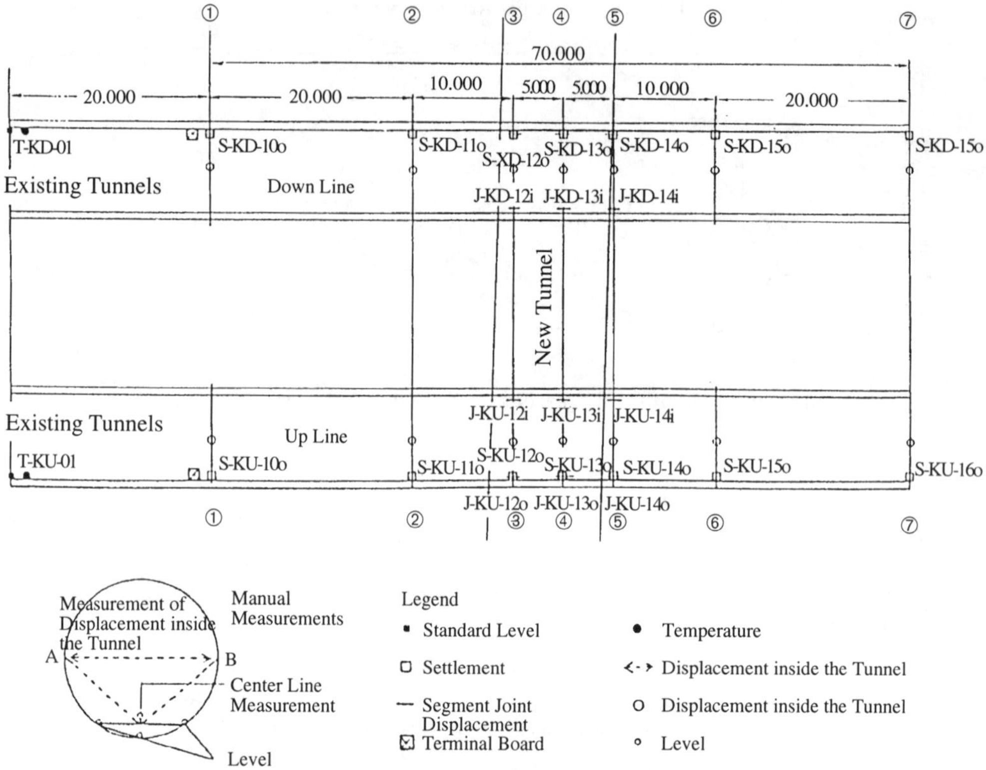

Figure 8. Position of the Measurement Equipment in the Existing Tunnels

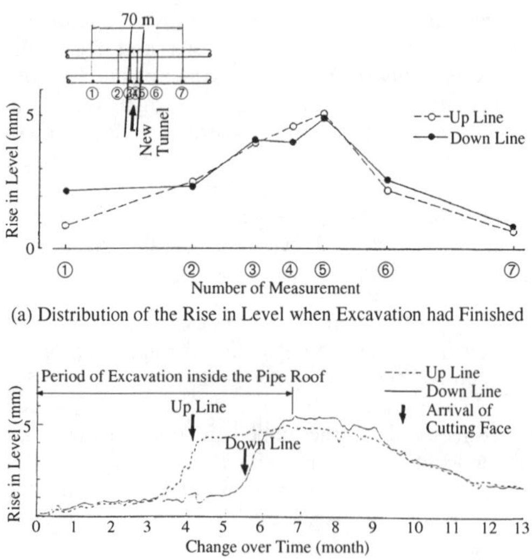

(a) Distribution of the Rise in Level when Excavation had Finished

(b) Change of Rise in Level over Time

Figure 9. Rise in Level in the Existing Tunnels

2) Cross-Section Displacement inside the Tunnels
From the results of measurements of the cross-section displacement inside the tunnels, a tendency towards a small reduction horizontally (less than 0.5mm) in the Down Line tunnel was confirmed, but no perceptible change in the behaviour of the Up Line tunnel was observed. For the change of mode, results were obtained agreeing generally with the tendencies in the forecast analysis.

3) Segment Joint Displacement
From the results of the measured displacement both along the axis and circumferentially on the joints of the key segment in the arch part, no increase or decrease was seen due to the excavation. The cross-section displacement inside the tunnel was minute, and in agreement with this the change along the axis due to the rise in level was also small.

4.3 Behaviour of New Tunnel

1) Displacement of the Upper Section of the Pipe Roof
Figure 10(a) shows the distribution of the

displacement in the upper section of the pipe roof at every stage of the excavation. From this the following points can be made.

(1) The settlement of the pipe roof increases with the cutting face position accompanying the advance of the cutting face, but after the cutting face has passed by, support work is successively installed and so settlement displacement ceases.
(2) The settlement increases linearly until about Measurement Point 5, and this is the ceiling level.
(3) The settlement reaches a maximum at Measurement Point 6 (56mm); Measurement Points 7 and 10 are directly above the existing tunnels, and the settlement at these points is 5 and 15mm respectively less than the maximum. This is considered to be due to the action of the loading on the existing tunnels, after support work is placed on the lower section of the pipe roof.

Comparing the observed results in Figure 10(a) with the forecast analysis results according to the beam spring model in Figure 7, it can be seen that the whole body settlement condition due to analysis was well expressed. For the maximum settlement value, the beam spring model forecast a value (31mm), less than the actual value. The forecast accuracy was not especially high, but on the whole it is thought that the forecast has a practical relationship with the actual observed values. However it is necessary to increase the accuracy of the forecast.

2) Stress in the Upper and Lower Sections of the Pipe Roof

Figure 10(b) shows the distribution of the stress in the upper and lower sections of the pipe roof at every stage of the excavation. The stress distribution varies successively corresponding with both the position of the cutting face and support work, and shows almost the same wave forms. Due to the erection of support work a reaction force occurred and at Measurement Points 6 and 7 the stress reverses. On the upper and lower sections, the direction of the ground reactions are opposed so it is reasonable that the modes of occurrence of the stress are also opposed. Also the upper section receives the pressure due to the overburden directly so the stress will be comparatively higher than for the lower section. The stress values for each stage of the excavation are high both in the vicinity of the cutting face position and in front of the cutting face itself. The stress in the vicinity of the cutting face position in both the upper and lower sections is almost uniform, and for the upper section it is 750kgf/cm^2.

On comparing these results with those from the beam spring model analysis (Figure 7), the stress distributions may be seen to be almost the same. Stress values of a similar degree were obtained but the actual observed values were approximately twenty percent larger.

From the results in (1) and (2) above, it may be

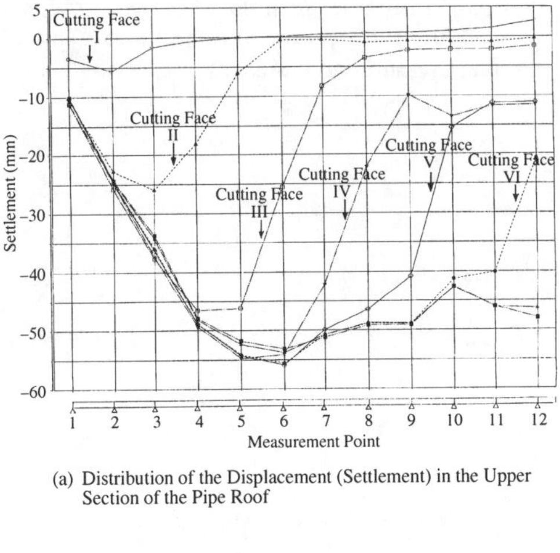

(a) Distribution of the Displacement (Settlement) in the Upper Section of the Pipe Roof

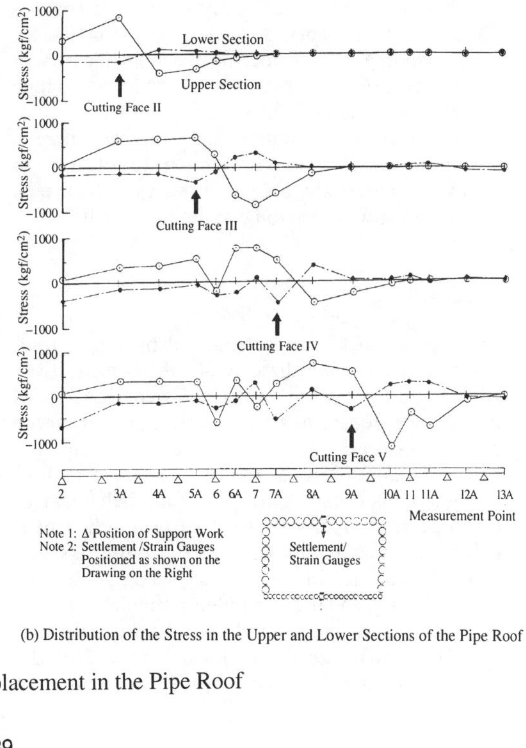

(b) Distribution of the Stress in the Upper and Lower Sections of the Pipe Roof

Figure 10. Stress and Displacement in the Pipe Roof

seen that the successive analysis model here can simulate the behaviour of the pipe roof. To increase the accuracy of the forecast, concerning the modeling for the analysis, it will be necessary henceforth to investigate considering methods such as the use of a non-linear ground spring.

5 OVERVIEW

From the forecast analysis and the on-site measurements discussed above, the facts are summarized below.

(1) The possibility of forecasting the change in behaviour of the existing tunnels when there is an excavation above for a tunnel crossing, using two-dimensional elastic FEM analysis, was confirmed by on-site measurements. In particular, on this kind of construction, in cases where the strain in the ground originating from the rise in level is small, a highly reliable forecast is considered possible.

(2) You can accurately forecast the rise in level, depending on setting the Modulus of Deformation of the ground at the time of loading as three times that of the Modulus of Deformation at the time of load removal. Hereafter, regarding the method for setting the Modulus of Deformation, it will be necessary to carry out investigations of the change over depth.

(3) The behaviour of the pipe roof accompanying the excavation, as expressed by successive analysis with a beam spring model, was confirmed. Depending on considering the non-linearity of the ground spring, behaviour forecast with a high level of accuracy is possible.

(4) For shield type circular tunnels, where there is several millimeters rise in level due to excavation locally above, the stress influence on the lining was seen to be extremely small.

6 CONCLUSION

For a box type tunnel, surrounded by a pipe roof, constructed crossing directly above in-use railway shield tunnels, a comparative investigation of influence forecast analysis and on-site observed results was carried out on the behaviour of the existing tunnel lining at the time of excavation of the new tunnel, having looked at the behaviour of the pipe roof. From these results much useful information was obtained. Nevertheless the investigations carried out on this give only a limited understanding, and many problems remain.

In this report, the example project was taken up directly to increase the benefits, but one of the problems in the calculation work is forecasting at a high level of accuracy employing a simple to use method. The tunnel crossing is a three dimensional problem, so to carry out an exact influence forecast analysis it is necessary to use a three dimensional analysis. Nevertheless to carry out a three dimensional analysis much time and labour is needed, so the development of a reliable enough two dimensional method is necessary. From this point of view, the authors, due to the comparatively simple influence forecast analysis method used here, want to research in depth how to obtain a forecast with a high level of accuracy. For these reasons, it is considered important to collect a large amount of highly reliable on-site observed data, and to evaluate appropriately the results of the forecast analysis, so as to discover the points of variance of the forecast with the actual observations and then to carry out a simulation analysis.

Also, hereafter it is necessary during the construction stage, to actively introduce information on the work being carried out, while revising forecast values based on the results of observations.

REFERENCES

Fukuyama & Gomi 1992. Extension Construction for the Keihin Kyuko Haneda-Kuko Line - Pipe Roof crossing directly above Shield Tunnels, *Tunnels and Underground*, Vol. 23, No. 8,

Hisanaga & Yoshinaga 1981. Displacement of Surrounding Ground accompanying Chemical Grouting Works, *Foundation Works*, Vol. 19, No. 3,

Japan National Railways Construction Bureau, Structure Planning Office 1977. *Guide to Design Construction of Shield Tunnels* (Proposed)

Japan Society of Civil Engineers 1986. *Japan National Railways Structure Design Standards Commentary* - Structure Foundations, Tunnel Ground Pressure Structures -

2 Urban tunnels and tunnelling through water bearing deposits

Tunnelling and Ground Conditions, Abdel Salam (ed.) © 1994 Balkema, Rotterdam, ISBN 90 5410 363 9

Design and construction of an urban tunnel, in muddy soils subject to water pressure, in Bilbao, Spain

Roberto Anero
Agromán Empresa Constructora, Madrid, Spain

José R. Madinaveitia
Ingeniería para el Metro de Bilbao 'Imebisa', Spain

ABSTRACT: A double-track tunnel running throught muddy soils has been built under Iruña Street for line nº 1 of the Bilbao subway. The rails elevation is located at an average 20 m. depth, below the surface grade. A sizeable water pressure is present (18 m. of water column at the rails level). A cut and cover system has been applied and reinforced concrete retainer walls were built. An improved soil slab, located at 13 m. below grade, was built applying the Jet Grouting, type III process so that a second level of rigid props is provided, before the actual excavation of the existing soil, in between the retainer walls, was carried out.

1 INTRODUCTION

Bilbao, as a city was founded in the Midle age, to benefit from the most excellent characteristics of the mouth of the river Nervion to build a harbour, which is nowadays one of the most important Spanish ports.

The town's population has extended over each bank of the river Nervion, which are connected by several highway's bridges. Comunication inside each bank is quite good, in fact there is a railway train for each one. These two railway trains run almost parallel to the river.

The Basque Government has designed the underground railway's net to increase remarkably the integration of people who live in both river's banks, and of visitors who reach the city from other towns.

The first subway's line, crosses twice the river below it's bed; and the second line is supposed to cross it once, in the same way as well. Both lines have two tracks.

Several firms tendered for the subfluvial's sections of the first subway line's contract, and AGROMAN E.C.S.A. was the one who won the legal bid which included the construction and the project in detail. These two subfluvial's sections cross the river Nervion with the purpose of connecting Deusto with Olaveaga and Ripa with Arenal. Inside the subfluvial section between Deusto and Olaveaga there is a part of the tunnel built using a "cut and cover" system all along Iruña street which is the subject of this paper. (Figures 1 and 2).

Fig.1 Tunnel in Iruña Street

2 GEOLOGY AND GEOTECHNICS

Geologically, Bilbao's Area is placed upon materials dated from Cretaceus age, which consist of limestone and marl, locally known as "cayuela", whose compression's resistance is between 60 and 300 kg/cm^2.

Upon this mass of rocks, cuaternary materials of many kinds, according to different areas: sands, gravels, muds and clays..., are deposited.

The cuaternary deposits in this section, between Deusto and Olabeaga, are mostly clayly and silty muds very soft with a high content of water, coming from 27 to 55%, the relation water's weight to dry

Fig. 2 View of buildings from the trench

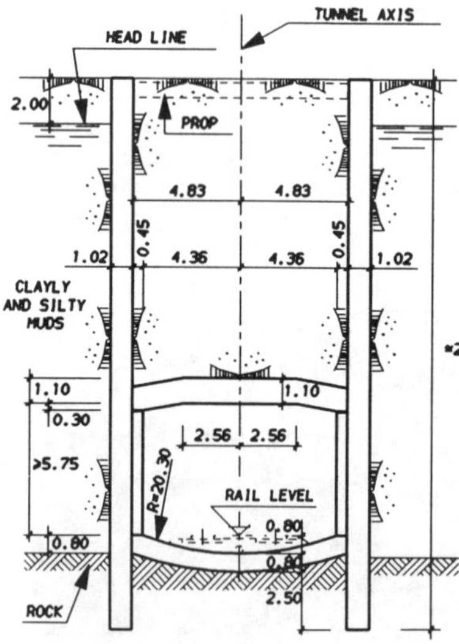

Fig. 3 Typical cross section

material's weight. The average of their compression's stress is 0,28 kg/cm^2, and its natural slope is between 8:1 and 10:1. Its granulometric composition averages:

. coarse sand 1%
. medium sand 1%
. fine sand 13%
. muds 60%
. clays 25%

3 THE SOLUTION TAKEN

The part of the tunnel that had to be built, with a "cut and cover" system along the Iruña street, showed the usual problems of tunnels placed in urban zones: there were cellars, buildings foundations at the surface, and also there were piles etc. ...

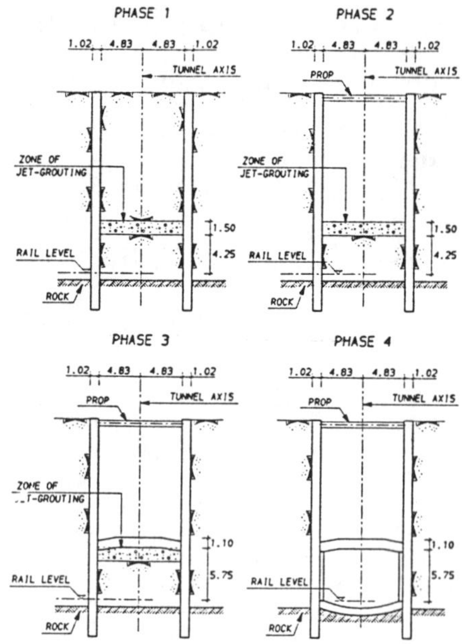

Fig. 4 Construction phases

After several studies were made, a solution was taken and it solved all the problems. The solution was to use two parallel diaphragms of reinforced concrete, introducing them in hard rock and fixed by props between them (Figure 3).
The great height of diaphragms, more than 24 m., the great pressure from the outside water and the poor-quality muds, made impossible to use the one meter thickness standard diaphragms. This problem was found a solution by introducing into the soil, before excavating, a passive prop, made up of a 1,50 m. thickness slab of ground ameliorated with cement injected according to the jet-grouting type III technique, that has worked perfectly well.
The tunnel's vault has been built over this slab of ground ameliorated with cement injected, as we said before, using the jet-grouting type III technique.
All the diaphragms are embeded in their lowest part in the hard rock. This was made only by using a special machinery known as "hidrofesa", which has excavated rocks whose stress to simple compression, reaches the amount of 400 kg/cm^2.

4 CONSTRUCTION'S PHASES

The tunnel's construction as "cut and cover" system was carried out in four parts (figure 4):

1st. Casting the diaphragms and the slab of ground ameliorated with cement injected according to J.G.III.

2nd. Placing, in the upper part, metallic props and excavating between the diaphragms until, the slab of ground ameliorated with cement, is found.

3th. Constructing of the tunnel's vault over the slab of ground ameliorated with cement.

4th. Excavating under the vault and supporting of the vault over the invert.

Once the whole vault is supported, the hole between the diaphragms is fulfilled, until the original surface is replaced.

5 SLOPING WAY TO ACCESS

With the purpose of performing the excavations under the tunnel's vault and the concrete setting of the invert and the supporting walls for the vault, and also in order to have an easy way of access for the materials and for the machinery, which is needed to settle down the track's infrastructure and superstructure of most part of Bilbao Underground's line 1, it has been built a sloping way so as to allow lorries pass trough, and it has a 20% slope (figure 5).

The tunnel's vault has been modified in the sloping way to access, so as to leave a hole of 14,25 x 5,20 m., next to one row of the diaphragms. These diaphragms in the side of the hole have been anchored in the rock.

The structure of this vault's part is supposed to be prepared for restoring the definitive vault's shape, when the sloping way to access is not going to be used any more.

6 QUALITY CONTROL OF GROUND AMELIORATED WITH JG-TYPE III

Previous to the project in detail, several trials were made in the area of works, so as to optimize the results of the ground ameliorated with the system of Jet-grouting type III. The trials were fundamentally about the performance of fifteen columns of ground ameliorated with J.G.III, varying in each one the system's essential characteristics: Grouting pressure, water pressure, air pressure, drill pipes elevating time, cement dosage, grouting injection's cement dosage per m^3.

In each column of ground ameliorated according to the technique jet-grouting type III, three vertical drillholes with drill core recovering were taken: one of

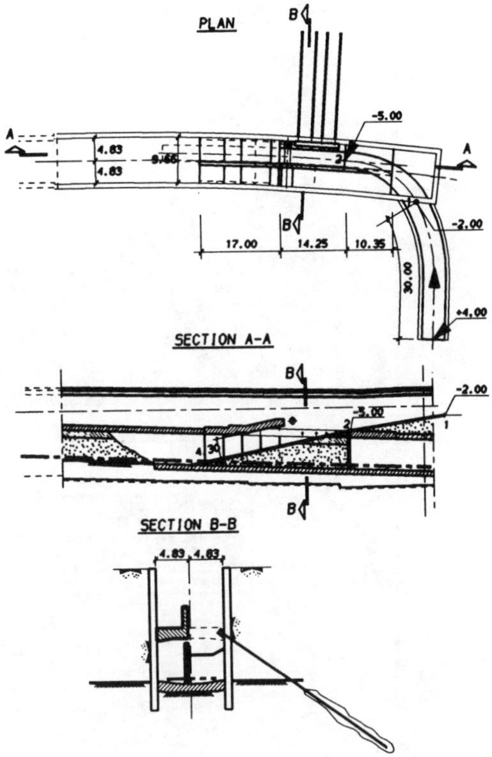

Fig. 5 Slope way to access

them in the centre, and the other two at different distances from the centre. Several pieces of these drill cores were tested with the purpose of obtaining their mechanic functioning. Finally during the injections's process the following values were taken:

Water pressure: 450 kg/cm^2
Air pressure: 7 kg/cm^2
Cement pressure: 90 kg/cm^2
Injection's time: 20 sec/4 cm.
Dosage: 960 kg. of cement/m^3 of grouting

The distance chosen between continuous columns of treatment was 1,50 m. The minimum mechanic parameters required to the ground ameliorated with cement injected according to the J.G.III technique were:

Young modulus = 10.000 kg/cm^2
Compression's stress = 40 kg/cm^2

7 INSTRUMENTATION AND CONTROL DURING CONSTRUCTION

Before the beginning of ground's excavations between the diaphragms recognizing drillholes with drill core extraction

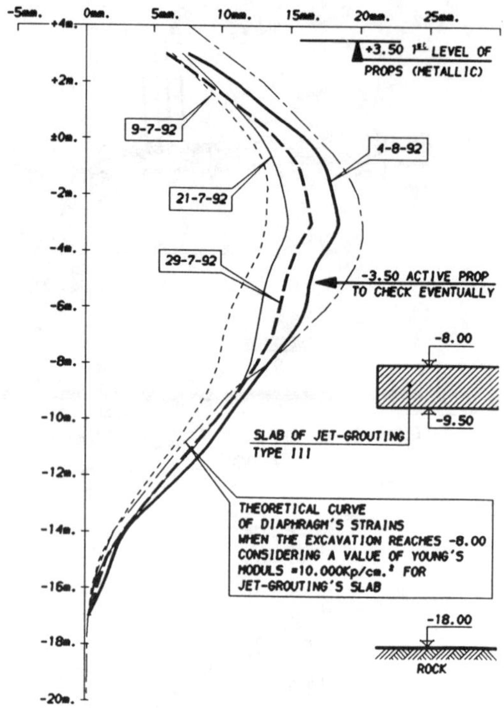

Fig.6 Horizontal strains controlled by inclinometer inside the diaphragm.

were made to evaluate again the characteristics of the slab of ground ameliorated with J.G.III. From the drill core, test tubes were obtained, which were tested at simple compression, obtaining their stress and deformability, so it was proved that its mechanic characteristics exceeded the minimum ones required in the project. It was chosen the most adverse zone to carry out a quite intensive control.

The disposed instrumentation consisted of:

. Pressure cells placed in upper props, complemented with strain gauges.
. Inclinometers placed in diaphragms.
. Piezometers settled down on the ground between the diaphragms and the outer zone of them.
. Convergence measurements.

In this intensive control zone, the security of the performance process was increased, placing intermediate active props, in order to check up without any risk, in a wide zone, for the concrete's quality of the diaphragms and of the slab of ground ameliorated with J.G.III. Once it was proved that the structural elements worked as it was planned in the project, the rest of the works were carried out in accordance with the phases defined in the project. In figure number 6 is a diagram of horizontal strains controlled by an inclinometer inside the diaphragm.

8 CONCLUSIONS

It is very important to remark that there was a great accordance between the mesures obtained from the instrumentation and the values obtained from the project's calculations. This fact was due to many real trials made before the calculations started. This way of acting is absolutely necessary to do not make mistakes about the mechanic characteristcs of the ground ameliorated with cement injected according to the Jet-grouting type III technique.

Tunnelling and Ground Conditions, Abdel Salam (ed.) © 1994 Balkema, Rotterdam, ISBN 90 5410 363 9

Surface settlements in urban tunnelling works: Design approach for Toulon underground motorway crossing

J.-P. Durand
DDE 83, France

M. Deffayet, C. Jassionnesse & J.-L. Reith
Centre d'Etudes des Tunnels CETu, France

ABSTRACT

Surface settlements limitation is one of the main target in the designing of tunnels in soft ground under urban area. The first step is the determination of the acceptable limits for the different types of buildings and foundations. Tunnel design must then select and adapt construction methods to comply with these limits. Application of this problem on the Toulon underground motorway crossing is described with contractual implications.

I- DESCRIPTION OF TOULON TUNNEL

The motorway crossing of Toulon is designed to link motorway A50 arriving from the west (Marseille-Toulon) and motorway A57 arriving from the east (Nice). During a first phase, only the North tunnel of this structure will be completed, to be assigned to traffic running in the east-west direction. At the end of the first quarter of 1994, the west and east access structures will be under construction ; they include the access ramps and the cut-and-cover sections. Works on the central section, which consists of a tunnel 1,857 m long, will begin in the second quarter of 1994. The main access shaft, to be used for a major part of these works, was completed at the end of 1993.

Prior studies were very detailed and included two exploratory galleries, one having a length of 278 m.
They allowed the characterization of three types of ground :
- surface ground of the quaternary which, when they affect the tunnel (generally in the arch) will be treated beforehand by grouting, from the surface of underground ;
- the autochthonous upper Permian, on the one hand, and the quaternary filling of a paleothalweg followed by Triassic scale (both being treated previously by grouting), on the other hand, which will be excavated using a conventional method ;
- the other Permians, which affect about two-thirds of the tunnel length.

The latter type of ground, composed of sandstone and tectonized argilaceous pelites, was considered to be very difficult and led to the recommendation of a special execution method.

Further, owing to the significant heterogeneity of the ground (requiring here the use of blasting and there leading to instability of the working face) and the small length of the tunnel, it was not possible to use confinement shield.

II - DEFINITION OF PERMISSIBLE SETTLEMENT THRESHOLDS

In order to set a clear objective with regard to settlement limitations, the assessment of execution quality and the evaluation of the overall safety factor of the structure under construction within urban surroundings, the designer must define permissible settlement thresholds. To meet these three objectives and take into account the financial, technological and human constraints of the project, the thresholds must be as realistic as possible.

In a project within an urban site, the determination of permissible settlements is a procedure quite comparable to geological studies. In the present case, the procedure followed may be broken down into five main steps :

1 - The first step consists in gathering information on the buildings and the sensitive networks (water, gas, etc), the construction techniques and the history of existing structures. Along with this, an assessment of the present condition of outstanding buildings was conducted.

2 - The second step is the grouping of this information in the form of a typological classification of existing structures and net works according to :
* the position of the structure in distance, altitude

and orientation in relation to the centreline of the project,
* the geometry of the existing structures -length, width, elevation-,
* the design of the structures and the materiels used,
* type of fondations,
* present condition and history of existing structures.

This led to a brief classification of existing structures in three categories :
*structures very sensitive to settlements, owing to their position or their design (gas pipelines and connection of cast-iron piping, for example),
*structures moderately sensitive to differential settlements if they remain within the limits allowed during this design,
*structures able to whithstand significant deformations owing to their rugged design, such as recent bridges, plastic piping or buildings supported by a general foundation.

3 - The third step is an attempt to model the existing structures, affected by the land movements induced by the digging of the tunnel. The objective is to quantify the influence of the mechanical and geometrical parameters issued from the typological classification.

For this purpose, simplifying assumptions, often encountered in the literature, were chosen :
* the buildings, generally located on the street edge parallel to the centreline of the project and at a small distance from it, are loaded like thick beams under compound bendind according to the "hogging mode", (Fig. 1).
* distribution of settlements (settlement crater Fig.2) at the level of the foundations approximated by a normal probability law (Peck R.B., 1969) - Gaussian curve- whose shape depends of the depth of the tunnel (O'Reilly et al., 1982),
* the half-length of the beam representing a building is the distance from the maximum curvature of the crater to the centreline of the tunnel (Wahls H.E., 1981). On the other hand, the height of the beam is determined according to the real proportions of the building,
* the damage evaluation variable is the deformation of extension in the beam (Boscardin M.D. et al, 1989),

A parametric study on the elastic solution based upon these assumptions (Boscardin M.D. et al, 1989) shows that the important parameters are :
* the ratio between the elastic moduli of the beam, in extension and in distorsion, whose value depends on the nature of the links between the bearing structures of the building;
* the ratio between the height and length of the beam (lenght representative of the system) directly related to the geometry of the building and to its orientation in relation to the tunnel centreline ;
* possibly, the value of the resulting extension of

Transverse surface settlement profile
and building response for the hogging mode

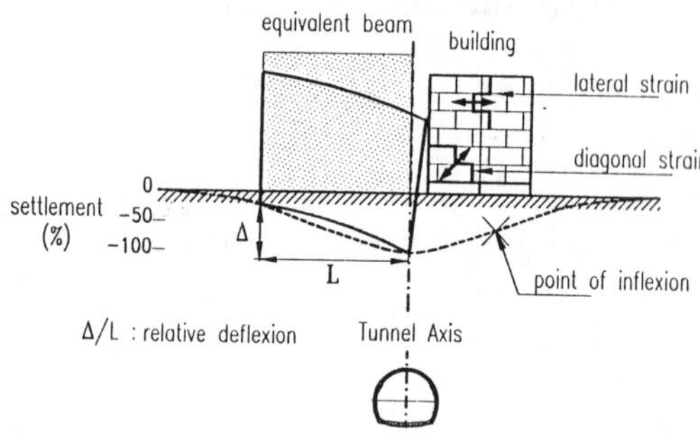

Figure 1

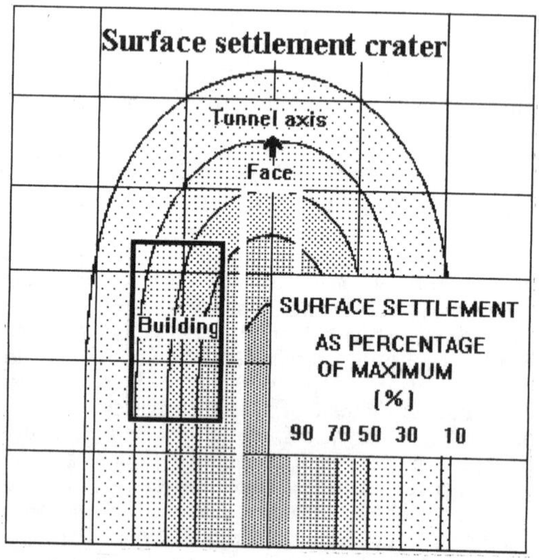

Figure 2

former damage to the excavation of the tunnel.

Based upon data provided by the literature, values were given to these different parameters according to the typological classification of the buildings. At the end of this phase, the extension deformation in the modelled building is directly parameterized by the class of the building, the extension of the crater and the intensity of the settlement (value on centreline).

4 - The fourth step consists in choosing the limit value of the extension as a function of the intensity of the permissible damage in the masonry walls (Boscardin M.D. et al, 1989) present in most of the buildings. A classical degradation of the damages in three degrees - aesthetic, functional and structural - was applied according to the value of the extension. As the basis choice of the design is not to tolerate any damage or disturbance leading to

plastering or prior renforcement works, the limit value chosen remains well short of the critical value of functional damage appearance.

The value of the permissible differential settlement is deduced from formulas developed in step 3 by a countdown calculation. At the end of this step, the permissible settlement is directly parameterized by the class of the building and the extension of the settlement crater, itself dependent on the depth of the tunnel, and to a smaller degree, on geotechnical conditions. The modelling used for the buildings was not applied to the networks, owing to the complexity of their spatial distribution. For significant pipelines, the permissible settlement values adopted are dependent only on the materials composing them, classified according to their mechanical fragility.

Similarly, factors not always identified with certainty, such as nature of the foundations, are taken into account by the application of a more or less high safety margin on the permissible settlements.

5 - The last step is the application of the values determined in the preceding step to clearly identifiable sections of the route (block of houses, for example) assuming uniform and rather unfavourable geotechnical conditions.
It is evident that the most sensitive structure (building or network) of a section determines the values adopted for the permissible settlement on the tunnel centreline and for the differential settlement.
Finally, the values adopted for the differential settlement are between 1/1333 and 1/750 and the settlement on the centreline between 15 mm and 35 mm, depending on the sections.

III - TAKING SETTLEMENT LIMITATION INTO ACCOUNT WHEN CHOOSING A METHOD

The choice of a tunnel construction method allowing compliance with permissible settlement thresholds is based upon the taking into account of many constraints.

It is evident that the nature of the surrounding ground is the main factor determining the feasibility of the method. In the case of Toulon, il was moreover the decisive factor in selecting the layout and the longitudinal profile.

The gauge of the structure, which is an essential imposed constraint, is also a determining factor. Il is to be noted that, in the case of Toulon, the section to be built is sizeable, about 100 m^2.
The hydrogeology, the nature and the extent of the covering ground dictate the main options in the selection of the method.

However, an approach based upon the knowledge accumulated during many projects throughout the word enables the possible choices to be identified rapidly : based upon permissible settlements and the characteristics of the structure -gauge, depth - and by means of conservative hypotheses concerning the distribution of settlements, it is possible to isolate the value of the permissible relative ground loss, which is the ratio between the volume of the permissible settlement crater and the volume of the tunnel, for a unit lenght of heading :
Z : depth (15 m < Z < 35 m)
L : half width of the crater
W : maximum settlement (on the centreline) (15 mm < W < 35 mm)
V : permissible relative ground loss.

For a first approximation, it is assumed that L is not very different from Z, then it comes 0,25 < V (%) < 1,25

These relatively low value require the use of very safe methods, either by extremely careful phasing or by the setup of powerful and effective facilities.
Geological reconnaissance completed has helped considerably in determining the possible choices.
The structural context of Toulon, making the suboil of the city a multiple geological contact zone, two exploratory galeries were completed to determine the nature and position of different contacts.

The observations made during the digging of one of the reconnaissance galleries revealed some outstanding phenomena :
- at very precise points, surface settlements reached high values (up to 10 cm) as a result of accidentally poorly controlled convergences not highly dampened by the covering ground ;
- the measured convergences, in particular of the side walls, were significant ;
- the general stability of the structure finally required quite heavy resources ;
- the deformatibility of the ground, measured by plate bearing tests, was important, at least in the pelite formations of the Permian.

Based upon these observations, interpretation hypotheses were made in accordance with a back analysis study of the parameters and a simple analytical approach :
- the significant difference between the deformability modulus measured in the tunnel and that measured during the pressure meter tests in boreholes (much higher) can be explained only by the significant damage to the ground during deconfinement in connection with excavation ;
- the considerable horizontal displacement are probably the consequence of the local tectonics, creating a quasi-isotropic stress states, and even a major horizontal stress ;
- finally, to conserve the good mechanical properties of the ground, stress relaxation on the extrados of the tunnel should not exceed 50 % of inital stress, the other half to be taken up by the support.

Allowing for these different assumptions, designers were able to define the basic characteristics of the method allowing maximum limitation of the deconfinement and loss of ground :
- the setup, ahead of the working face, of a relatively rigid pre-support capable of maintaining a high level of stress in the ground,
- the use of a working face reinforcement capable of applying an effective confinement,
- the fast closing of the structure near the face.

IV-DESCRIPTION OF CONSIDERED METHOD

A pre-support method is one in which all or part of the support of a tunnel section is placed before beginning its excavation. Studies carried out in the case of Toulon had revealed the importance of the support remaining beyond the face of each advance. Further, it was of course necessary to remain realistic with regard to deadlines and costs. Based upon these requirements, two pre-support methods were considered :
- the mechanical precutting method with pre-vault, which we shall call PM,
- the "umbrella roof" method, which we shall call VP.

In the PM method, the pre-support requires : (Fig.3)
- the opening on the periphery of the excavation of a trench dug mechanically by a cut-off type tool
- the filling of this trench with concrete setting fast enough to allow the start of the excavation about four hours after its placement.

The concrete filling the trench constitutes a pre-vault which, after excavation, is reinforced by an additional support with bolts, cross-linked arches and shotcrete. The equipement specific to the use of this method is the precutting machine on which is mounted a rack giving the support of the sawing tool its transverse movement.
In the Toulon project, the lenght of the trench is 4 m and its thickness is 22 cm.

In the VP method, pre-support requires : (Fig. 4)
- the provision, on the periphery of the future excavation, of a serie of sub-horizontal micropiles reinforced with metallic tubes ;
- as the excavation advances, the supporting of these micropiles by heavy steel arches and shotcrete.
A forced and immediate contact must be obtained between the arches and the tubes, and this can be achieved by using polypropylene ducting with a cement mortar filling under pressure.

The material specific to the use of this method is the machine for inserting the micropiles, for which placement accuracy is extremely important. In the Toulon project, the length of the tubes is 12m and a new serie of tubes is placed when only 3,5 cm of the preceding one remains.

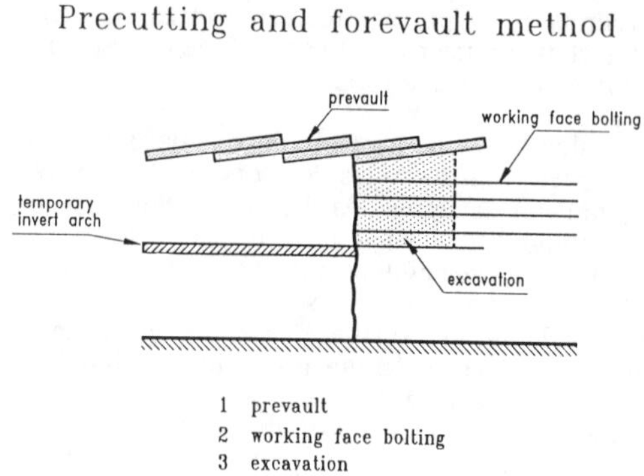

Figure 3

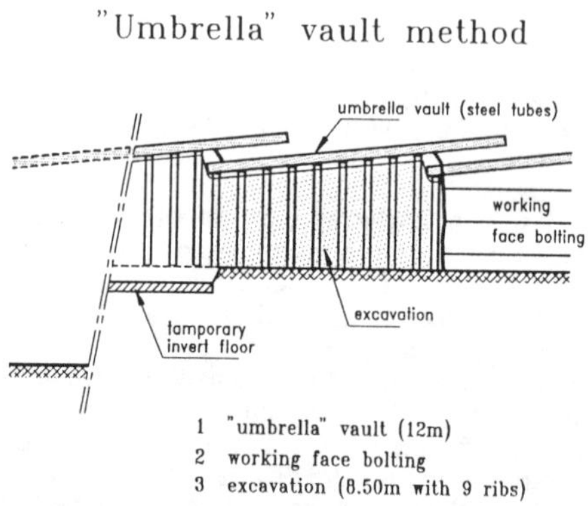

Figure 4

Whether with the PM or VP method, pre-support is not sufficient to ensure the stability of the face. Calculations have shown that this stability required a confinement pressure of 1.1 bar. In both methods, this confinement pressure is provided by shotcrete on the face and the use of long, horizontal anchor bolts. These bolts are in reinforced fibre-glass resin so that they may be easily destroyed while advancing. In the upper half-section, for bolts with a service strength of 25 tonnes, the required useful density is one bolt for 2,3 m².

The two methods described above are those which are recommended in the tender documents. To date, calculations carried out in connection with the VP method have not been completed, but those conducted as part of the first method (PM) are summarized in the following paragraph.

V - ESTIMATION OF SETTLEMENTS IN A PARTICULAR SECTION

One of the most difficult sections of the structure is located in the Red Permian zone. This formation, consisting of red pelites including sandstone and

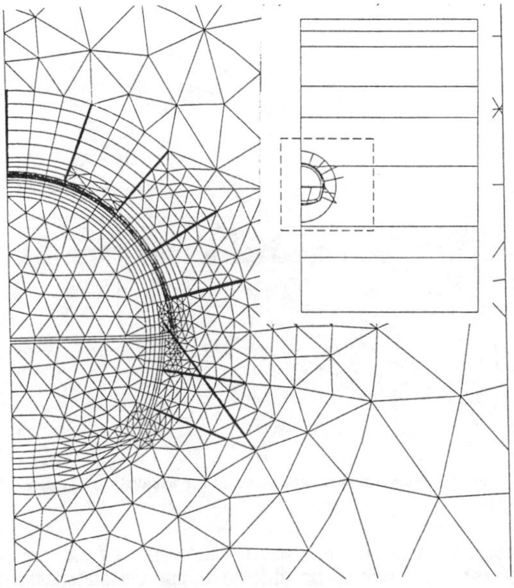

Figure 5 - General structure for the 2D approach

schist elements, is covered with a thickness of 25 m of colluvia.

The exploratory gallery had revealed the strong deformability of the ground and allowed the correlation of geotechnical properties.

Several finite element calculations were carried out with the CESAR-LCPC program in order to take into account all the excavation phases, to justify the supporting systems used and to provide indications on possible settlements. The excavation method considers the completion of the upper half-section, and then the excavation of the lower half section in halves, the final lining being placed in the last phase.

1 - Two-dimensional modelling of the structure - calculation phases:

The mesh used is shown in figure 5. Reinforcements by means of bolts, as well as arches or micropiles are introduced as linear elements (beams or bars).
There are nine excavation and supporting phases, from the pre-support of the upper half-section, its excavation until the long-term study of the lining ; only the most interesting phases from the settlement viewpoint are discussed in the present document.
Prior explorations demonstrated the significant difference between the properties of the ground in place and its behaviour when deconfinement takes place. This factor, of primary importance in this project, was taken into account by means of assumptions regarding the evolution of the deformability modulus of the ground during the phases. After several calculations, it is thus possible to propose a range of results encompassing the movements and settlements.
The deformability modulus of Red Permian varies from 300 MPa ahead of the face, to 50 MPa in the long term after excavation. Classical elastoplastic calculations have also been carried out in order to determine the range of settlements to be considered.

2 - Main results - Problems encountered :

* In phase 1, the pre-support trench is made over a length of 4 m and a thickness of 22 cm. The movements of this trench vary significantly depending on how it is made.
Thus, completing the trench entirely before concreting leads to a movement of 2 cm for a deconfinement of 20 %. Completion with successively concreted panels limits to 0,7 cm the reduction in the thickness of the trench during this phase. The corresponding surface settlement is 1,3 mm. In the latter calculation, the modulus of the concrete increases gradually as the other panels are excavated.

* In phase 2, the pre-vault is assumed to be active but, as a safety measure, its early age concrete modulus is taken equal to 4000 MPa. The excavation will be carried out under the pre-vault.
The estimation of the forces to be taken up and of the deconfinement rate λ to be considered is a delicate point. In fact, the usual convergence-confinement method defines a variation in the confinement rate independent of whether or not a support is present. While this poses few problems in the case of a conventional excavation, the reasoning is wrong when a pre-support is used effectively very early.
Kielbassa et al (1991) proposed an analytical method for taking into account this effect of the support ; Bernaud et al (1992) adapted the assumptions of the convergence-confinement method.
In the present case, we have considered the following principes :
- in the absence of a pre-support, the face is crushed (displacement U_1) and exerts over its periphery a given contact pressure (P_1) ; the Sakuraï formulation (1978) being used to set the variatio of the deconfinement rate.
- with the pre-vault, the face is less crushed owing to the smaller radial displacements (U_2) and then exerts only a pressure P_2 lower than P_1.
Assuming the same rigidity of the face, at the same distance from the face, we have :

$$\frac{P_1}{U_1} = \frac{P_2}{U_2} \quad \text{then} \quad \frac{1-\lambda_1}{U_1} = \frac{1-\lambda_2}{U_2} \qquad (1)$$

This formulation makes it possible to deduce the correction to be made on the deconfinement rate as a result of the presence of the pre-crown. A force $\lambda_2\sigma_o$ is then applied on the periphery of the exca-

vation. σ_o designates the initial stress condition (not necessarily isotropic).
It will be noted that this approach requires us to carry out a first calculation without pre-vault to determine U1 as a function of λ_2 and U_2 used verify the relation (1).

During this phase, by grouping the different cases examined -modulus of 300 MPa in the best case and 100 MPa in the worst case - the displacement of the crown remains between 0,9 and 1,3 cm, whereas the surface settlements range from 0,2 to 0,5 cm.
The forces in the pre-support do not exceed 5 MPa.

* In phase 3, the additional support is placed in the upper half-section and the face advances. The reinforcement of the support includes arches and shotcrete bearing on a concrete beam. A temporary floor is also provided near the face.
The forces exerted on the periphery of the excavation are calculated by means of a deconfinement of the ground will be clearly lower (about 0,5).
The displacement at the crown will range from 2 to 2,2 cm and the corresponding surface settlement from 0,8 to 1 cm.
The overall sinking of the base of the pre-vault is 0,5 cm, while the lateral micropiles take up globally almost 25 tonnes with maximum normal forces near 40 tonnes locally.
It should be noted that the maximum settlement is obtained during this phase. In fact, during subsequent phases, considering the good stability of the support, and taking into account a high coefficient Ko (initial earth pressure ratio), the horizontal convergence tends to ovalize the section and to recompress the crown. This same phenomenon had already been observed in the measurements carried out in the exploratory galery. Also, the settlement crater will become larger, decreasing proportionately the extent of the differential settlements.

* During subsequent phases, the maximul settlement no longer increases, and the crater becomes larger. Horizontal convergences of 3 to 4 cm are noted essentially, and an increase of almost 15 tonnes in the loading of the micropiles.
Summarizing, it should be noted that the different calculations carried out, with variable assumptions concerning the changes in the mechanical properties of the ground, yield a maximum settlement in the centreline not exceeding 1 cm. This figure presupposes good stabilization of the supports in the upper half-section, hence requiring the placement of micropiles.

3 - Face stability - Axisymmetric approach

The two-dimensional approach does not allow determination of the displacements likely to occur on the face. It is however certain that the volume lost at the face will continue to increase the surface settlements.
This particular point was the subject of an axisymmetrical calculation, itself made up of several phases in order to be able to adopt the stationary solution. Figure 6 shows the geometry adopted.

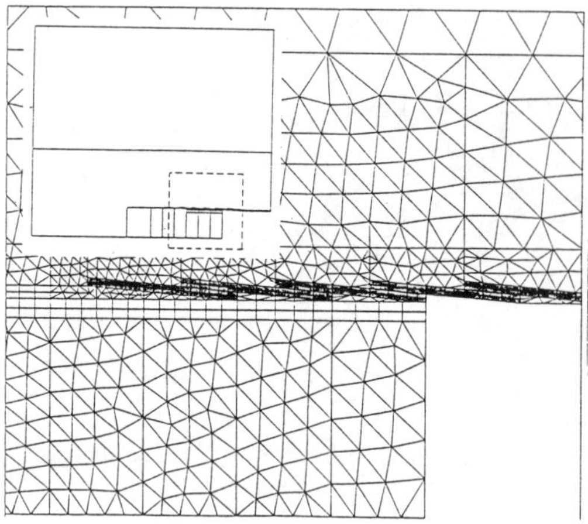

Figure 6 - Axisymmetric approach of the excavation advance.

The calculation preparation process was the following :
a) Determination, by empirical formulations, of the stabilization pressure to be applied to the face. The developments of Terzaghi (1951) and Feder (1979) were used for the calculation of the veertical pressures to be taken up, as well as the geometry of the instability zone. The presence of the pre-vault is also considered.

b) Taking into account a greater cohesion of the ground in the area of the bolt-strengthened face. This bolting is supposed to offer the required confinement determined in a). This equivalent cohesion is obtained by translating the Mohr Coulomb criterion line so that the reinforced ground has a compressive strength identical to that of the non-reinforced ground but confined by the stabilization pressure.
c) Application of axisymmetric calculation simulating six advance steps in order to eliminate the side effects and to get a stationary solution. The ground is considered to be elastoplastic and the cohesion ahead of the face is reinforced on almost 6 metres.

After these steps, the volume lost at the face is estimated at 0,3 to 0,4 %, which is added to the 0,8 % determined from the lateral convergences. The total of 1,1 to 1,2 % remains lower than the threshold set for the considered sections. Let us note that the evaluation of the volume lost at the face and in the cross-sections is rather difficult and that the approach presented above is a rough approximation.

VI - PROVISIONAL STUDY OF SETTLEMENTS OVER THE ENTIRE TUNNEL LAY OUT

The purpose of these calculations is to confirm the overall feasibility of the structure within the urban environment and especially to determine, section by section, the safety margin offered by the execution method.
To achieve these goals, it is thus necessary to take into account in the calculation the geometry of the structure, the longitudinal profile and the geology on the homogeneous sections. It is also necessary to be able to parameterize the results by the "offer" of support furnished by the different adaptations of the execution method.

The offer of support determines :
- the final stress relaxation level in the ground (the deconfinement rate),
- the evolution of the mechanical properties of the ground (deformability modulus).

The depth of the structure determines the initial stress level (earth weight) and the damping of displacements. The mechanical properties of all the geological formations encountered between the tunnel and the surface are taken into account by a homogenization calculation.
In addition, consolidation settlements generated by the partial drainage of the ground during excavation are calculated and taken into account in the estimation of the total settlement value.
In fact, it is probable that this method overestimates the ground loss in the tunnel, but underestimates the transmission to the surface.
Finally, the distribution of surface settlements forms the subject of a calculation with the classical assumptions already described in paragraph II. The result of this approach is the value, section by section, of the surface settlement on the centreline of the tunnel and the predictable differential settlement.
The results obtained by this method on the particular sections described in paragraph V are comparable to the results of the finite element calculations, giving surface settlements of the order of 10 mm.
The values estimated for the differential settlement are between 1/5000 and 1/500 and the settlement on the centreline between 5 mm and 35 mm depending on the sections. In general, the safety margin offered by the execution method is about 50% of the permissible settlements and it should be noted that the maximum values correspond to the tunnel driving zone at a shallow depth in the alluvial formations partially grouted or drained by pumping.

Taking into account the large inaccuracy of this kind of analysis, the safety margin about 50 % should not allow a significant reduction of support resources without measurements to confirm real low settlement values during the work.

VII - PROJECT MANAGEMENT BY SETTLEMENT CONTROL

During the completion of the structure, the continuous control of settlements will make it possible to determine the effectiveness of the measures taken ant the need for possible adaptations in order to limit them. In addition, in order to prevent any damage to existing structures, the control should make it possible to anticipate harmful evolution of ground movements.
For this reason, the designers considered it necessary to control intensively the settlements ahead of and in the vicinity of the face. The values measured in this zone are one of the criteria for halting the work in the event of that an alert threshold is exceeded, set at half the permissible settlement threshold.

Accordingly, the measurement points located less than 5 radii from the vertical of the face are used for daily measurements.
Measurements carried out behind this zone are less frequent and their purpose is to allow the evaluation of the extend of deferred movements as well as the evaluation of the safety factor really offered bu the methods used. Likewise, measurements on the cross-sections will make it possible to determine the real value of the differential settlements.
These measurements will be the subject of an interpretation that may lead to the adaptation of the execution method and to a modification of the alert threshold. In order to complete this interpretation, the various players will be able to use a computerized measurement transmission and acquisition system with which they will have access to the measurements of different kings carried ou on the site (convergence measurements, extensometry, etc;).

The settlements will be controlled by high-precision altimetric measurements and, for this purpose, 650 levelling marks have been placed on the buildings along the alignment. During 1992, they were used for a premliminary series of measurements which led to the determination of the extent of natural altimetric variations resulting, in particular, from fluctuations in the ground water levels during the seasons.

VIII - CONCLUSIONS

The urban environment is of great importance in the design of the project, the choice of methods and the inspection of the works.

Determining the permissible settlement values makes it possible to check that they can be obtained with the planned methods. Permanent checking of the values actually obtained on the site allows the following of the work, i.e. the adaptation of supporting devices.

REFERENCES

- BERNAUD D., ROUSSET G. (1992) : "la nouvelle méthode implicite pour l'étude du dimensionnement des tunnels"
Revue française de géotechnique n° 60, pp 5 - 26

-BOSCARDIN M.D CORDIN E.J.(1989): :"Building response to excavation-induced settlement"
Journal of geotechnical engineering division , ASCE Vol. 115, No.GT1, pp 1 -21

- FEDER G. (1979) "Einfluβ von Bauverfahren, Anisotropie und Ausbruchsform aud die Konvergenz und deu Stützmittelbedarf trefliegender Hohlraumbauten"
Bundesministerium für Bauten und Technik.Straβenforschung-Heft 124

- KIELBASSA S., DUDDECK H. (1991) "stress - strain fields at the tunnelling face - three dimensional analysis for two - dimensional technical approach"
Rock mechanics and rock engineering 24, pp 115 - 132

- O' REILLY M.P., NEW B.M. (1982) "Settlements above tunnels in the United Kingdom - their magnitude and predictions", Proc. Tunneling' 82, ed. M.P. Jones, IMM.. London.

- PECK R.B. (1969) "deep excavations and tunneling in soft ground", state of the art report, 7th inf. conf. on soil mechanics and foundation engineering, Mexico city, pp 225 - 290

- SAKURAI S. (1978) : "Approximate time-dependent analysis of tunnel support structure considering progress of tunnel face", Int. j. for num and analyt. meth. in geom.
Vol 2, pp 159 - 175

- TERZAGHI K. (1951) : "mécanique théorique des sols" Edition Dunod - Paris

- WAHLS H.E. (1981) "tolerable settlement of buildings", Journal of geotechnical engineering division, ASCE, vol 17, No GT 11, pp 1489 - 1504.

Tunnelling and Ground Conditions, Abdel Salam (ed.) © 1994 Balkema, Rotterdam, ISBN 90 5410 363 9

Modelling braced excavation for subway stations

F. El-Nahhas & A. A. Ahmed
Ain Shams University, Faculty of Engineering, Abbassia, Cairo, Egypt

M. A. El-Gammal & A. H. Abdel-Rahman
National Research Center, Dokky, Giza, Egypt

ABSTRACT: Implementation of subway networks are universally accepted as an effective solution to overcome or completely eliminate some environmental problems associated with highly populated congested cities.

Design of braced excavations for subway stations has been conventionally carried out using empirical methods and/or statical analysis of the walls with their bracing system utilizing oversimplified assumptions regarding the surrounding soil mass. Both methods have proven to be unrealistic and insufficient. In particular, both methods fail to take into account the effect of proposed construction sequence and the associated changes in the characteristics of surrounding ground.

The present study provides a numerical modelling for evaluating the interaction between braced reinforced concrete diaphragm walls and the surrounding ground. This modelling avoids the main shortcomings of commonly used design methods. The results of the present study are compared with field measurements complied during the construction of Cairo Metro (line No. 1). Furthermore, the same modelling procedure is used to predict the expected behaviour of the new subway stations which are presently under construction for line No. 2 of Cairo Metro.

1 INTRODUCTION:

Realistic modelling of deep excavations of subway stations is usually considered as one of the most complicated soil-structure interaction problems. Modelling such problems must fulfill the sophisticated details of excavation and the associated alternation in the in-situ stresses field. Figure 1 shows a typical configuration of a subway station of the Regional Line of Cairo Metro. Numerical modelling of such station must take into account the characteristics of the ground continuum, diaphragm walls, injected ground zones, and bracinq system. Also the compatibility and equilibrium conditions at the interface between soil and bracing system should be idealized in the numerical model.

In this study, the soil-braced excavation system and the construction procedure are formulized and idealized using nonlinear finite element techniques. The ground continuum, reinforced concrete wall, bracing system, and the interface medium are simulated using the appropriate finite elements.

The ground continuum constitutive behaviour is updated in the analysis employing incremental nonlinear technique utilizing modified Duncan model (Duncan et al. 1984).

2 IDEALIZATION OF THE PROBLEM

Finite element method is considered

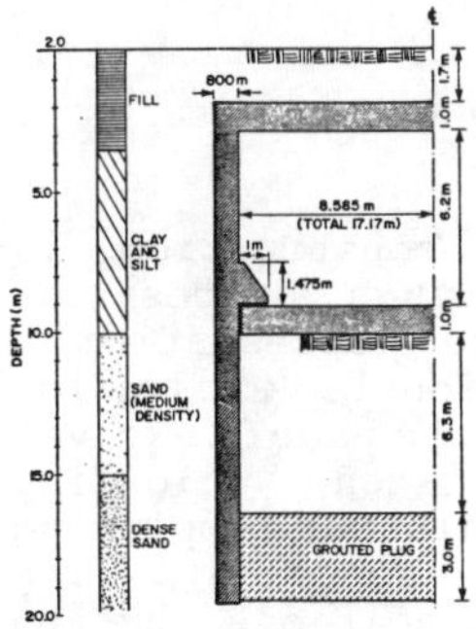

Fig.1 Typical configuration of a subway station of the regional line of Cairo Metro

as a very powerful technique in simulating the soil-structure interaction problems. It is capable for providing all the necessary finite elements needed for the realistic idealization of the present work. It also simulates materials of great complexities such as nonlinearity and time dependency.

In this research, the eight noded isoparametric finite elements are utilized in the numerical model to simulate both ground continuum and reinforced concrete diaphragm walls. Spring type interface finite elements are inserted between the ground elements and diaphragm wall elements to ensure the compatibility conditions at the interface surface between them, as well as the associated shear stresses and strains along the interface surface. This type of finite elements is used to link adjacent nodes and characterized by two stiffness components, known as normal and shear stiffness (K_n & K_s). These components can be determined after performing a series of direct shear tests to evaluate the element parameters (Dessouki, 1985). The nonlinear behaviour of the interface surface between ground continuum and diaphragm wall can be formulized and programed employing special technique of iterations (Ahmed, 1991).

The constitutive relation of the reinforced concrete diaphragm walls and the grouted plug are assigned utilizing the classical linear elastic model. This model is recommended to be used with structural elements characterized with relatively high stiffness exposed to minimum deformations.

2.1 Modelling of construction procedure

The procedure of construction is simulated also through process of iteration. Each one representing a stage of construction or excavation stage. For every stage of excavation, the associated unloading conditions are evaluated employing stress reversal approach. The external loading conditions due to excavation are then applied to soil-wall system to set up the final state of stresses developed in ground continuum. These stresses are then integrated and added to the external loading conditions associated with next stage of excavation to evaluate net nodal forces due sequential excavation. This procedure is continued and used as input data for proceeding stage of construction until the final stage of construction is reached.

2.2 Ground nonlinear idealization

The ground continuum constitutive behaviour is treated using the modified Duncan model (Duncan et al., 1984). This model is considered as a foreword step on application of hyperbolic nonlinear approach which was initiated by Duncan and Chang (1970) for tangent deformation modulus "E_t".

Similarly, Wong and Duncan (1974) proposed nonlinear tangential Poisson's ratio "ν_t" to allow nonlinear volumetric strain. Duncan (1980) also expressed the relationship between the values of bulk modulus (B) and the confining pressure by the following relation:

$$B = K_b\,P_a(\sigma_3 / P_a)^m \qquad (1)$$

where K_b is the bulk modulus number and m is bulk modulus exponent. Duncan et al. (1984) modified the previous model in order to overcome two main problems. They distinguished between primary loading and unloading-reloading based on stress state (SS) as:

$$SS=(1-\sin\phi)(\sigma_1 - \sigma_3)/(2c\cos\phi+2\sigma_3\sin\phi)\sqrt[4]{(\sigma_3/P_a)} \qquad (2)$$

where:

c, ϕ cohesion and angle of shearing resistance of soil,

P_a the atmospheric pressure, and

σ_1, σ_3 the major and minor principle stresses.

The modified model then assumes primary loading moduli E_t are appropriate when current stress state is greater than or equal to the maximum previous stress state ($SS \geq SS_{max\ past}$), and otherwise assumes unloading-reloading moduli (E_{ur}). Based on this criteria, the critical stress level (SL_{crit}) which primary loading confining stress (σ_3), will be given as:

$SL_{crit} = SS_{max\ past} / (\sqrt[4]{(\sigma_3/P_a)}$ (3)

This model also assumes a transition zone between the unloading-reloading and the primary loading as illustrated on Figure 2. This can be used as follows:

a) knowing the stress state ($SS_{max\ past}$) which defines the boundary between primary loading and unloading-reloading behaviour, the stress level (SL_{crit}) at which primary loading begins for a given confining stress can be evaluated using Eq. 3.
b) when $SL \geq SL_{crit}$, the primary loading modulus (E_t) is used.
c) when $SL \leq 0.75\ SL_{crit}$, the unloading-reloading modulus (E_{ur}) is used.
d) when $SL_{crit} > SL > 0.75\ SL_{crit}$ linear interpolation is used.

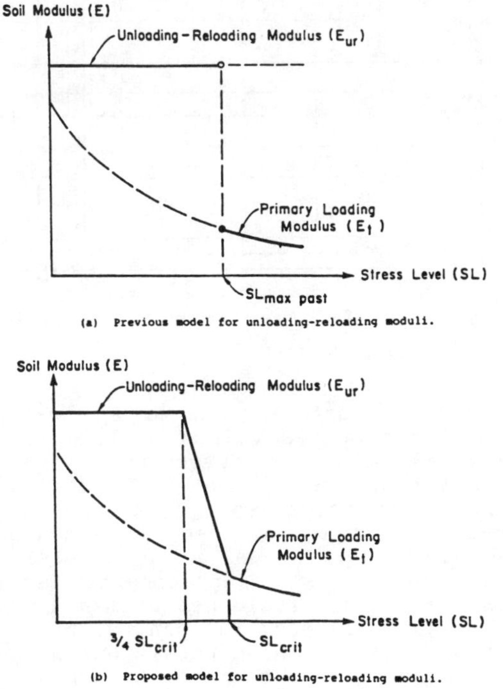

Fig.2 An illustration for the modified Duncan model (1984)

3 CASE HISTORY

The cut-and-cover tunnel of Cairo Metro Regional Line was extensively instrumented during the construction period by El-Nahhas et al. (1988) and Shalaby (1989). It is employed here to evaluate the proposed numerical model. The tunnel configuration and the soil profile at this location are shown on Figure 3. Table 1 contains summary of the soil nonlinear parameters and concrete properties.

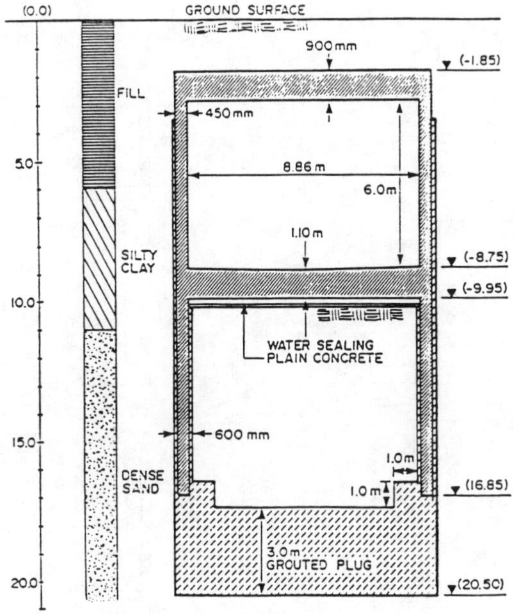

Fig.3 The tunnel configuration and the soil profile at its location

Table 1 Hyperbolic model parameters (After Shalaby, 1989)

Material \ Parameter	K	K_u	n	R	C kPa	$\phi°$	K_b	m
Dry fill	200	450	0.79	0.69	60	33	120	0.37
Wet fill	225	450	0.79	0.69	60	33	120	0.37
Dry silty clay	225	450	0.45	0.7	60	32	120	0.37
Wet silty clay	225	450	0.45	0.7	60	32	140	0.37
Medium density sand	250	450	0.4	0.7	-	25	140	0.2
Dense sand	300	500	0.4	0.7	-	30	150	0.2

CONCRETE PROPERTIES

Unit Weight kN/m^3 23.750
Compressive Strength MPa 40.00
Modulus of Elasticity MPa 31250.0

Generally, the cut-and-cover tunnel was constructed using the diaphragm walls. Figure 4 illustrates the general construction stages for the tunnel test section as simulated in the finite element analysis. These stages are summarized as:

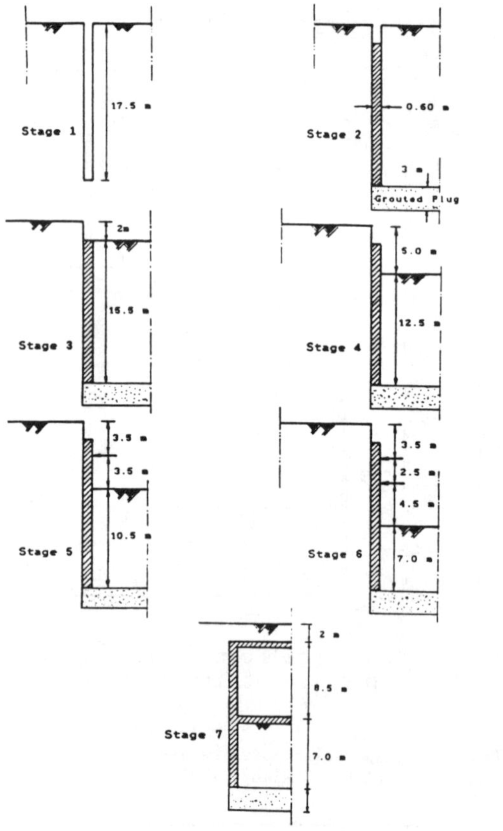

Fig.4 The general construction stages for the tunnel test section as simulated in the finite element analysis

Stage 1: Excavation of slurry trench
Stage 2: Installation of the precast concrete units
Stage 3: Excavation of the inner soil to -2.0m
Stage 4: Excavation of another 3.0m deeper
Stage 5: Installation of the upper bracing, then excavation of another 2.0m deeper.
Stage 6: Installation of the lower bracing, then excavation of another 3.5m deeper.
Stage 7: Pouring the raft and slab and removing all temporary bracings.

Figures 5 illustrates the finite element mesh employed for analysis. Figure 6 indicates the propagation of wall horizontal movement during the different stages of construction. Figure 7 shows the settlement troughs associated with the construction stages. Results of detailed parametric study for this case history is given by Abdel-Rahman (1993).

A comparison between the measured

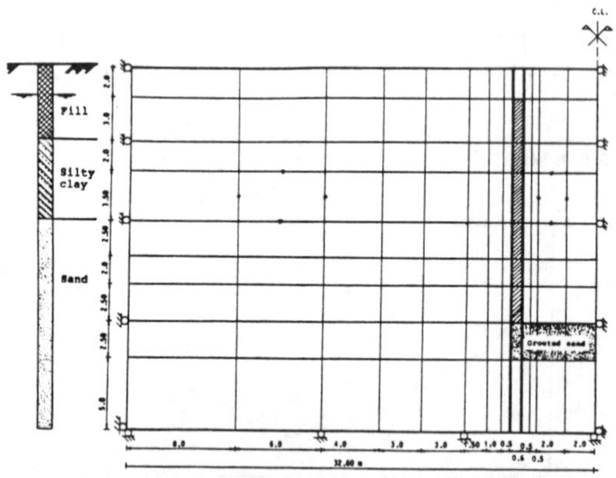

Fig.5 The finite element mesh employed in the analysis

horizontal deformations of the wall and the computed values outlined on Figures 8&9, for stage No. 6 and final stage respectively. Figure 10 shows a comparison between the measured and computed subsidence adjacent to the wall. In general, the calculated displacements are similar to the field measurements of this case history. The general trend of behaviour is also consistent with the observations given by Clough and O'Rourke (1990).

4 UNDERGROUND STATIONS OF CAIRO METRO - LINE 2

Five subway stations are presently under construction for phase 1 of Line 2 for Cairo Metro. It was

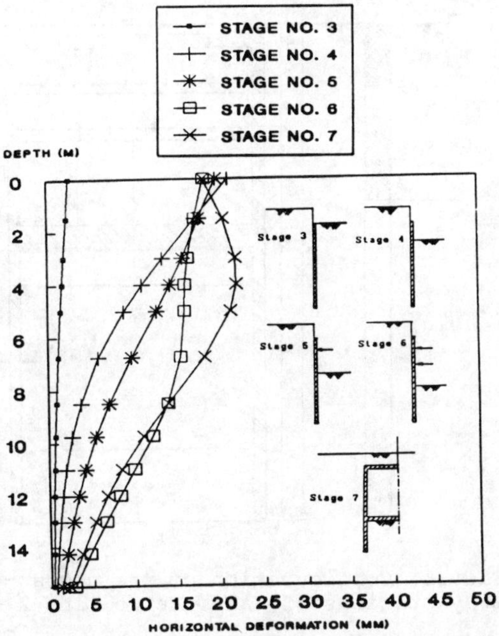

Fig.6 Propagation of the wall horizontal movement during the construction stages

decided to apply the numerical modelling described above for predicting the behaviour of one of these stations. It is noteworthy that an in-situ monitoring program is planned to measure the actual performance of the diaphragm walls, surrounding soil and existing near-by buildings. It will be interesting to compare the results of type A - prediction given here for different stages with the field measurements.

Configuration of the new subway station is shown on Figure 11 and the proposed stages of construction are given on Figure 12. The predicted lateral deflections of the diaphragm wall at different stages of construction are illustrated on Figure 13 and the associated settlement troughs adjacent to the subway station are shown on Figure 14.

5 CONCLUSION

Although modelling all the boundary conditions controlling the interaction characteristics between the ground and the diaphragm walls seems to be impossible, the proposed numerical modelling of the present work has proven good results. The

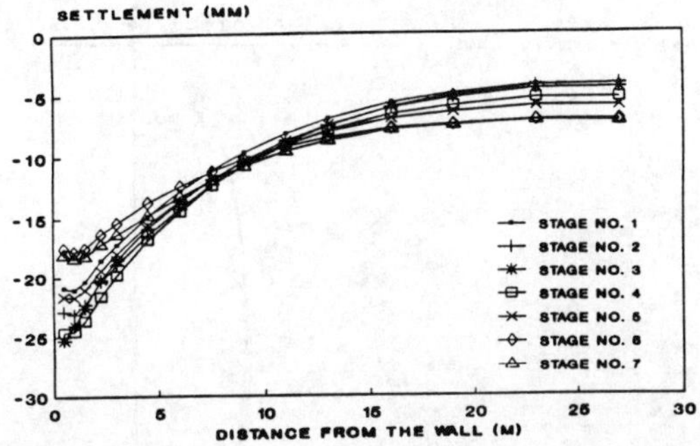

Fig.7 The settlement troughs associated with the construction stages

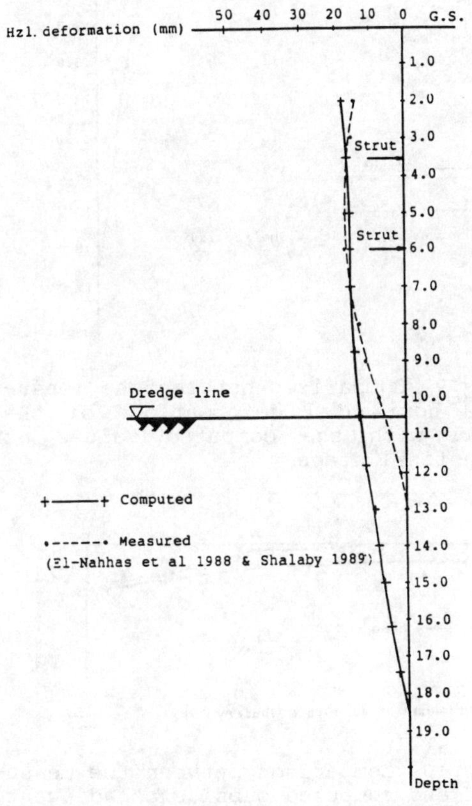

Fig.8 Comparison between the measured horizontal deformations of the wall with the computed values at stage No.6

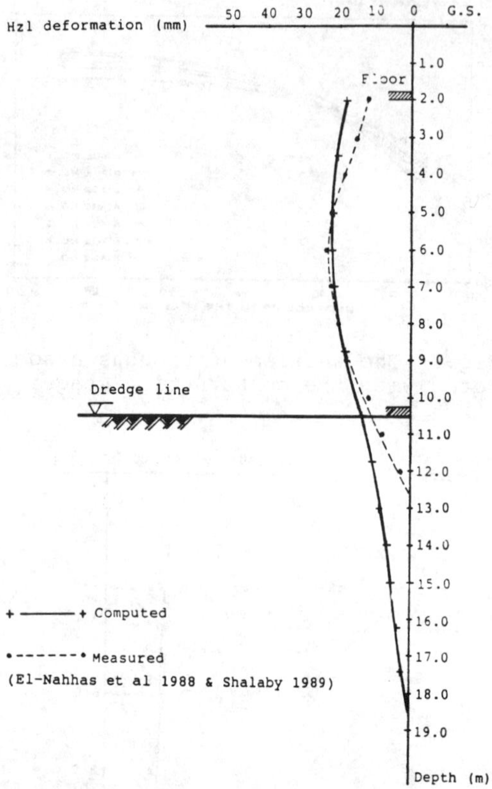

Fig.9 Comparison between the measured horizontal deformations of the wall with the computed values at the final stage

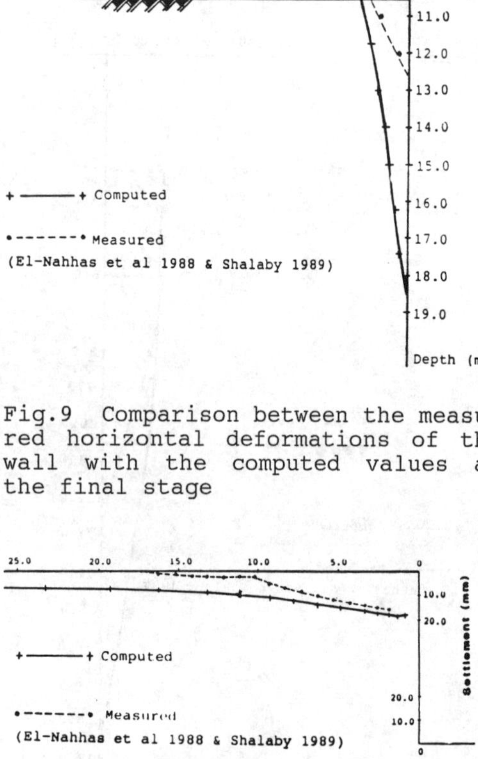

Fig.10 Comparison between the measured and computed subsidence adjacent to the wall

comparison between results of this model and the field measurements compiled during construction of Cairo Underground Tunnel (Line No. 1) indicated the ability to simulate such complicated soil-structure interaction problem. The results also confirmed, the need for realis-

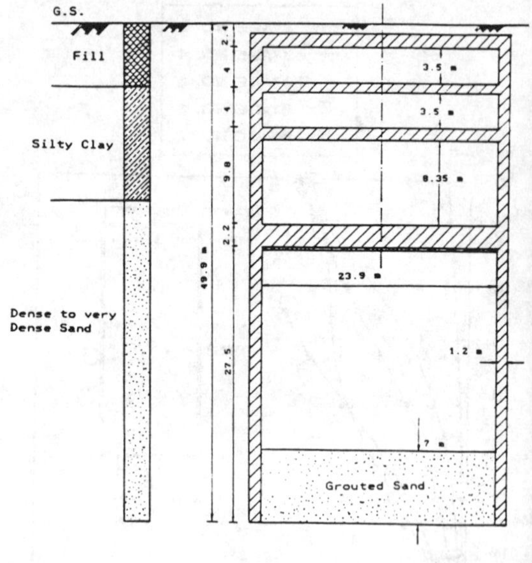

Fig.11 Configuration of the new subway stations of Cairo Metro-Line 2

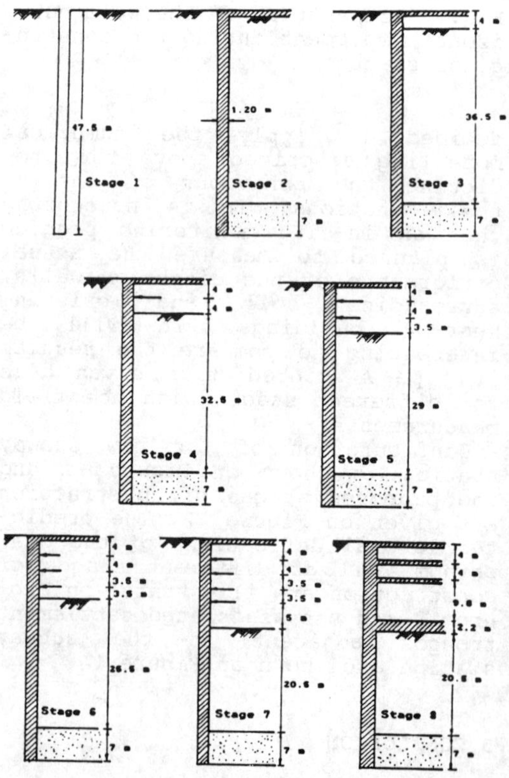

Fig.12 The proposed construction stages of the new stations

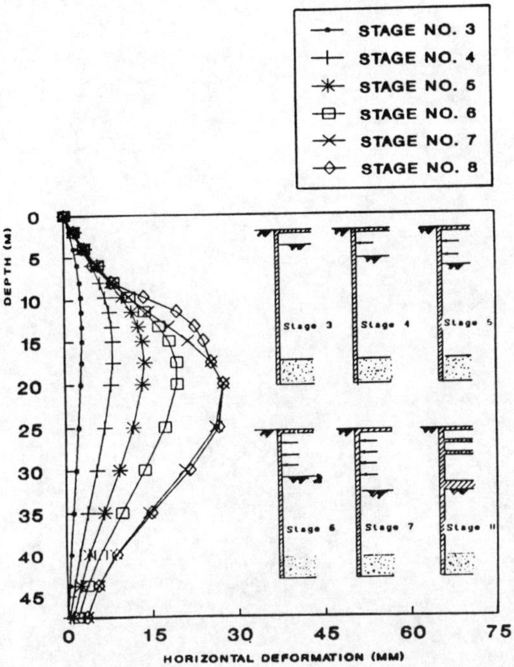

Fig.13 The predicted lateral deflections of the wall during the construction stages

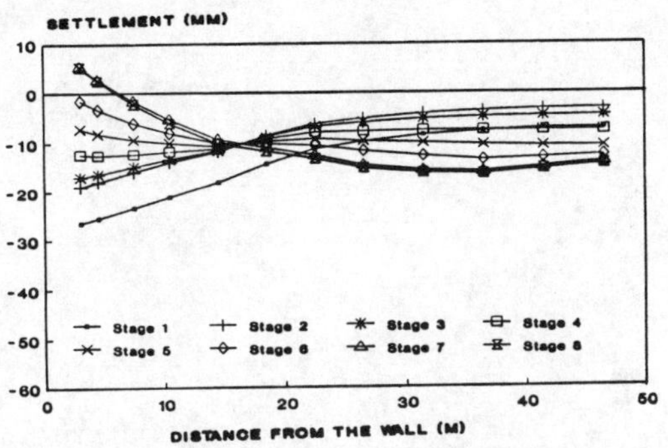

Fig.14 The predicted settlement troughs adjacent to the subway station during the construction stages

tic establishment of construction procedure in the numerical model because it is considered as the main factor controlling the ground-diaphragm wall interaction characteristics. Prediction of behaviour of the new subway station of Line 2 of Cairo Metro which are presently under construction are offered.

REFERENCES

Abdel-Rahman,A.H. 1993. Numerical Modelling of Concrete Diaphragm Walls. M. Sc. thesis, Ain Shams University, Egypt.

Ahmed, A.A. (1990). Interaction of Tunnel Lining and Ground. Ph. D. thesis, Ain Shams Univ., Egypt.

Clough,G.W. and O'Rourke,T.D. 1990. Construction induced movements of insitu walls. In Proc. ASCE special publication No. 25, pp. 439-470.

Dessouki,A.K. 1985. Stability of soil-steel structures. Ph. D. thesis, Univ. of Windsor,Canada.

Duncan,J.M. 1980. Hyperbolic stress-strain relationships. Proc. of Workshop on Limit Equilibrium, Plasticity and generalized stress-strain in Geotech. Eng., McGill University, Canada.

Duncan,J.M. and Chang,C.Y. 1970. Nonlinear analysis of stress and strain in soils. J. of the Soil Mech. and Found. Div., ASCE, Vol. 96, No.SM5, September.

Duncan,J.M.;Seed,R.B.;Wong,K.S.; and Ozawa,Y. 1984. FEADAM84, A computer program for finite element analysis of dams, Virginia Polytechnic Inst. and State Univ., Dept. of Civil Engineering, U.S.A.

El-Nahhas,F.;Eisenstein,Z.;Shalaby,A Abdel Salam,M.; and El-Bedaiwy, S. 1988. Geotechnical performance of diaphragm walls during construction of Cairo Metro Tunnel. In Proc. of Int. Cong. on Tunnels and Water, Madrid, Spain, Vol. 1, pp 259-266.

Shalaby,A.M. 1989. Geotechnical behaviour of diaphragm walls. Ph. D. thesis, University of Alberta, Canada.

Wong,K.S. and Duncan,J.M. 1974. Hyperbolic stress-strain parameters for nonlinear finite element analysis of stresses and movements in earth masses. Report No. TE 74-3, University of California, Berkeley, California.

Tunnelling and Ground Conditions, Abdel Salam (ed.) © 1994 Balkema, Rotterdam, ISBN 90 5410 363 9

Excavation through an old masonry wall in the 'Principe Pio' Station, Madrid

S. Estefania
Intecsa, Spain

A. Fernández
Dragados y Construcciones, Spain

ABSTRACT: This paper describes the construction methods of a junction tunnel of about 20 m long between a shaft placed in the parking of the station and the inside of the building. Two parts can be distinguished: recent fills with a low degree of cohesion and an old masonry wall 6 m thick. Actually the works are successfully finished.

1. INTRODUCTION

The "Pasillo Verde Ferroviario" in Madrid is an urban promotion owned by a Consortium made up by the City Council and RENFE (Spanish Railways Company). The aim of the work consists in covering the existing link between the stations of Principe Pio and Atocha, respectively in the North and South sides of the town. In such way a strip of ground placed in the city center is recovered for buildings, gardens and leisure zones. The first lot, between the Principe Pio's Station and the Imperial's Station consists in two bored single tunnels, 1.200 m in length, connecting with a cut and cover double rail section.

The bored tunnels are excavated with an EPBS (Earth Pressure Balanced Shield). The machine diameter is 7,40 m. For shield extraction a shaft of 20 m in diameter and 18 m in depth has been excavated in the parking of the Principe Pio's Station.

From the shaft to the connection with the rails inside the station a single rail tunnel of 16 m has been constructed. This tunnel crosses a zone of fills and a masonry wall 6 m thick. This wall withstands the metallic structure of the station built in the 1870's. (See fig. 1).

The construction methods were choosen in order to avoid settlements in the metallic structure.

The part of tunnel in fills was excavated under a vault of "jet grouting" columns between the diaphragm walls of the shaft and the masonry wall.

The wall crossing was performed by means of an arch for withstanding the loads of the wall. This arch consists in rings of concrete coffered with corrugated iron sheets.

In order to avoid unstabilities when excavating the stross, bolts of 32 mm in diameter and 4 m in length were placed in both lower sides of tunnel vault. Those bolts were anchored in the whole length with cement mortar.

2. TUNNEL IN FILLS

2.1. DESCRIPTION

This part of tunnel between the shaft and the masonry wall is 11 m long. The cross section of the tunnel is shown in the fig. 2.

Due to the low geotechnical properties of fills, with a cohesion value between 0,5 and $1 T/m^2$, a vault of "jet grouting" was considered necessary for protecting the excavation.

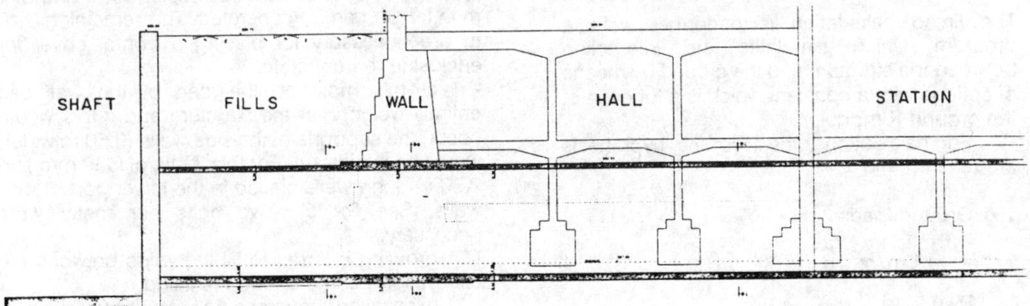

FIG 1. LONGITUDINAL PROFILE

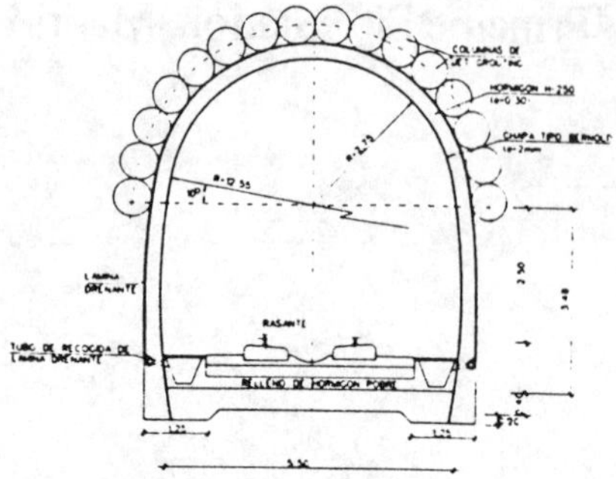

FIG 2. TUNNEL SECTION IN FILLS

For this, drills were performed from the shaft (See fig. 3). The length of the grouting columns reaches the masonry wall. The estimated diameter was about 0,70 m, and the grouting pressure 400 kg/cm^2.

The average cement consumption is 400 kg/m of column. Under the "jet grouting" vault, excavation was performed in two steps without any particular problem, even without temporary support.

The lining with the dimensions of the fig. 2 was in unreinforced concrete coffered by iron sheets. The iron sheets are mountered on steel sets which remain in place until the concrete hardening. Then, they are removed and re-utilised.

FIG 3. TUNNEL PORTAL IN SHAFT (See "jet grouting" drills)

2.2. CALCULATIONS

The lining calculation is performed with a program which simulates the interaction between the structure and the ground by means of springs with a constant which is a function of the ground K-modulus.

For calculations, the following hypothesis are taken:

1. Ground loads

$p_v = 14\ Tm/m^2$; $p_h = 7\ T/m^2$

2. Own load

3. Sum 1+2

Concerning the values of K modulus.

$K_H = K_V = 10.000\ T/m^3$ for the springs in contact with the "jet grouting" vault.

$K_H = K_V = 2000\ T/m^3$ for the other springs.

3. TUNNEL IN MASONRY WALL

3.1. DESCRIPTION

As we can see in the fig.1 the masonry wall has a stairs shape and the thickness at the bench level is about 4,50 m. The tunnel invert is lower than the wall fondation.

The wall is built by blocks jointed by mortar. The main problem to solve consists in the fact that the metallic structure of the station hall lies down the upper part of the wall and, consequently, no settlements can be allowed.

From the beginning, it was clear that a good solution should be the construction of an structure able to withstand the wall loads but, evidently, the dimensions of wall don't allow the construction in a single step.

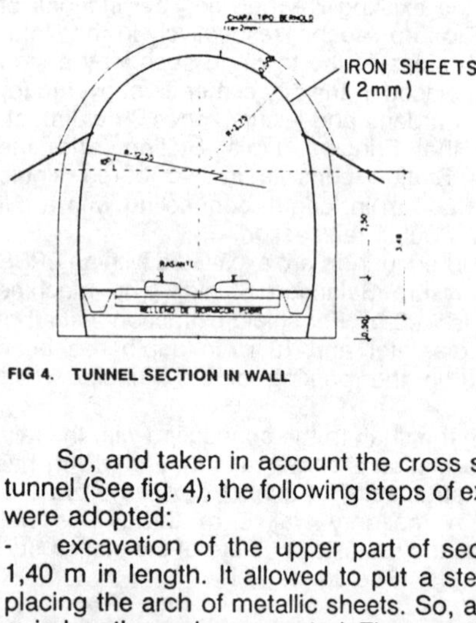

FIG 4. TUNNEL SECTION IN WALL

So, and taken in account the cross section of tunnel (See fig. 4), the following steps of excavation were adopted:

- excavation of the upper part of section with 1,40 m in length. It allowed to put a steel set for placing the arch of metallic sheets. So, a ring of 1 m in length can be concreted. The remaining 0,40 m are necessary for placing a frontal cover for enclosing the concrete.
- at this moment, the load of the wall can entirely weigh over the structure and if this would occur, the supports of the side-walls (0,50 m wide), would be insuffisant. For this, bolts of Ø32 mm and 2,50 m long were placed in the lower part of side walls (See fig. 5). Two bolts per meter were necessary.
- grouting in order to fill the voids between the excavation perimeter and the lining.
- excavating the stross from shaft to station.

FIG 5. BOLTS IN SIDE WALLS

First, the central part of the stross. The side walls remain liyind down the unexcavated bench level. After this, the remaining sides were excavated by alternated parts of 2,50 m. (fig. 6)

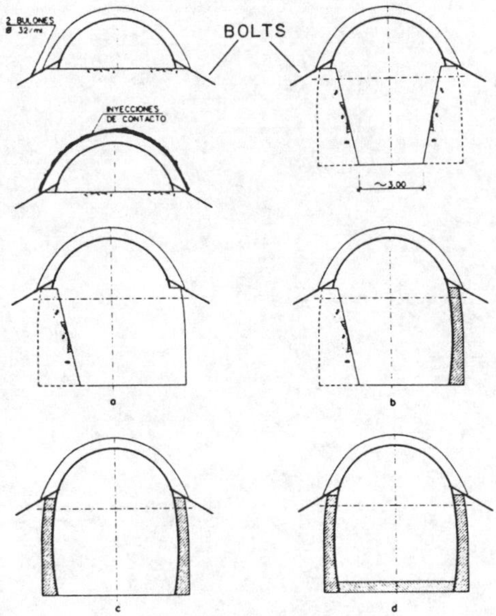

FIG 6. STEPS FOR STROSS EXCAVATION

3.2. CALCULATIONS

3.2.1. Tunnel lining

The program employed is the same described thereafter. For the tunnel lining, two calculations were performed, only for the bench section.

The following loads were applied:

1. Load of wall and of the superstructure as P_v
2. Horizontal load consisting in a 10% of the vertical.
3. Own weight of tunnel lining assuming an unreinforced concrete with γ = 2,4 T/m^3 and 0,50 m thick.
4. Sum of 1+2+3.

3.2.2.-Bolts

Concerning the bolts, the shearing force from the calculation results in 3.2.1 was taken with the bolts drilled with an angle of 30º from the horizontal.

Consequently, 2 bolts were needed per m of tunnel.

4. CONCLUSIONS

The part of single tunnel between the shaft and the station building with a length of 16 m has been constructed towards recent fills and an old masonry wall. In the fills, a vault of "jet grouting" columns allowed to excavate the tunnel without problems. The unreinforced concrete lining was coffered with iron corrugated sheets which remain as a part of the lining.

For the wall crossing, the excavation was made in two parts, upper and lower. The lining was built by the same method described and the arch stability is reinforced by bolts Ø32 and 2,50 m length drilled in the side wall basis. Two bolts per meter of side wall were necessary. The works are now finished (See fig. 7) and no settlements were registered in the structure of the station.

FIG 7. FINISHED TUNNEL

5. ACKNOWLEDGEMENTS

The authors are grateful to the "Consorcio del Pasillo Verde", INECO and RENFE for their contribution to this paper.

Tunnelling and Ground Conditions, Abdel Salam (ed.) © 1994 Balkema, Rotterdam, ISBN 90 5410 363 9

Ameria tunnel junction using the freezing technique

G.R. Flint
AMBRIC, Cairo, Egypt

ABSTRACT: The paper "Bentonite Tunnelling for the Greater Cairo Wastewater Project" (Flint and Foreman 1991) described the use of bentonite tunnelling machines for driving the new large diameter trunk sewers under Cairo. This paper describes the methods envisaged at the design stage for junctioning the downstream end of the new sewer system with the Distribution Chamber (DC) constructed as part of the new Ameria pumping station complex. This required the construction of a machine reception chamber, but the tunnel machine arrived before the chamber could be constructed, which created major problems and constructional changes. Eventually freezing techniques were adopted to enable a short tunnel to be driven out from the DC to the tunnel machine and an insitu R.C. junction to be made in "free air". The internal parts of the abandoned tunnel machine were cut up and removed piecemeal, whilst the skin was left in place to form the back shutter to the internal insitu R.C. lining. Thereafter the secondary lining from the end of the existing tunnelled sewer into the DC was constructed as originally envisaged.

1 INTRODUCTION

This paper must be read in conjunction with the paper on "Bentonite Tunnelling for the Greater Cairo Wastewater Project" by the author and W. Foreman [(1)] presented to the International Symposium on Tunnelling in Cairo 1991. That paper outlined the history of Cairo's sewerage system leading to major resewering works of which the new Ameria pumping station complex and the main collector tunnel system form essential parts. The contracts for these works have been completed, and the new system is now in service.

Figure 1 shows the relevant works, and the demarcation between contracts No.1 and No.3. Christiani-Misr Concrete (CMC)* carried out contract No.1 of which the circular Distribution Chamber (DC) formed a major part, and Cairo Wastewater Consortium (CWC)** carried out contract No.3 comprising the main collector tunnel (MCT) of which the downstream end junctioned with the DC. Contract No.1 required a reception chamber to be constructed integral with the DC to receive the tunnel machine.

The ground at Ameria comprises fine, medium and coarse sands with some gravel, and occasional bands of clay. The water-table is generally approximately 5m below ground level, although this can vary as it is influenced to some extent by a well-field approximately 700m away. The combination of mainly fine granular soil and high ground-water creates extremely onerous excavation conditions, especially with respect to tunnelling operations at the large diameters under consideration.

The DC was sunk to a depth of approximately 31m using the wet caisson technique, and the MCT was driven at a depth of approximately 22m to invert using a Bade and Theelin designed bentonite tunnelling machine. This machine incorporated a cutting head of the planet wheel type

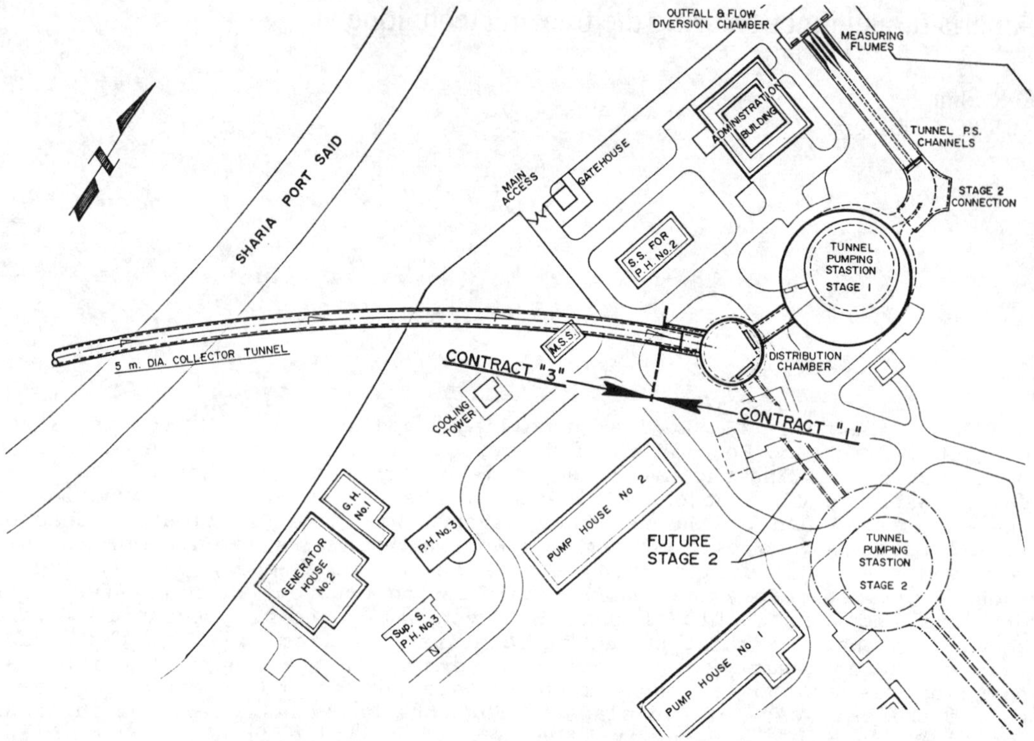

FIGURE 1
LAYOUT OF AMERIA PUMPING STATION, DISTRIBUTION CHAMBER AND INCOMING MAIN COLLECTOR TUNNEL

with spokes and adjustable vanes for regulating the face excavation. These techniques are described more fully in other papers.[(1)(2)]

Progress on these two inter-dependent contracts got seriously out of phase, and the tunnelling machine reached its contract limit before the reception chamber could be constructed.

This paper describes the problems that developed with the junctioning of the tunnel to the DC and also the various, and initially unsuccessful, attempts to achieve a revised form of junction, until freezing techniques were adopted from within the DC. Freezing provided the solution to the ground stabilization and ground-water problems, and enabled a short tunnel to be driven from the DC to the abandoned machine.

2 ORIGINAL RECEPTION CHAMBER

The form of junction between the tunnel and the DC envisaged at the design stage required that the Ameria pumping station contractor construct an oversize reception chamber integral with the DC to receive the tunnelling machine, and to enable it to be dismantled, or removed as one unit. The 12m long reception chamber would have an internal diameter of 7.1m with its horizontal axis 245mm lower than the centre of the tunnel to accommodate the transition from a circular to the horse-shoe shaped sewer required for hydraulic purposes. The reception chamber would be filled with weak mortar to allow easy penetration by the tunnelling machine, and also to facilitate a water-tight seal with the primary tunnel lining once the machine was safely inside the chamber.

Figure 2 shows the details of the segmental reception chamber and the relevant part of the DC as put out to tender. Figure 3 shows cross sections of the circular tunnel within the oversize reception chamber, and also the transition to the horse-shoe shape on entry into the DC.

The tender method for constructing the reception chamber was to consolidate a block of ground adjacent to the DC, and to progressively construct the chamber out from the DC within the protection of the consolidated ground. During the contract the contractor decided to use jet-grouting techniques from the surface for ground consolidation.

This technique had been used on one of the other tunnel contracts, but without complete success even when supplemented by compressed air, and on one occasion a massive collapse had occurred. In the event the jet-grouting technique for constructing the reception chamber was not put to the test at Ameria, being overtaken by events.

The reception chamber was required to be ready by February 1987, but the sinking of the DC ran into severe difficulties, and it was not sunk and bottomed out until September 1987. Conversely the tunnel machine whilst having some difficulties of its own, [such as tail-seal problems initially[1] and other delays along the way[3]], had managed to make up for lost time and arrived approximately on schedule in October 1987. This left insufficient time for constructing the reception chamber as originally envisaged.

This unforeseen development also resulted in a classic engineering (and contractual) dilemma in that on the one hand the presence of the tunnelling machine prevented, for all practical purposes, construction of the reception chamber as originally envisaged (and also made it partially superfluous), yet on the other hand the carcass of the machine could not be removed because it constituted the only means of sealing the end of the tunnel against the adverse ground conditions. Therefore, a modified method of dealing with the machine and junctioning the tunnel with the DC had to be devised.

The initial revised form of junction is shown in Figure 2, although this was further modified because of potential difficulties in trying to insert the heavy segments and build the rings within the limited working space between the extrados of the machine and the ground. In the event the collar around the front of the machine was changed from segments to insitu R.C.

3 REVISED JUNCTIONING ATTEMPTS

Initially the two contractors agreed to consolidate (by jet-grouting) a 1m thick layer of ground immediately adjacent to the DC, and to drive the machine through the remaining 3m into the preformed mortar filled eye in the DC wall. This would entail strutting the back of the eye to resist the high thrust forces that would develop as the cutting head ground its way into the mortar and buried the front of the shield to the depth required to achieve a watertight seal.

Although the jet-grouting was done apparently successfully, when it came to restart the tunnelling machine the cutting head remained immobile, and after repeated attempts to lubricate the head with bentonite and unlock it the gearbox failed. During this period bentonite forced its way up to the surface, disturbing the full depth of ground in the process. Further attempts to continue tunnelling with the bentonite machine in its normal mode had, therefore, to be abandoned.

All the internal mechanical equipment and non-essential structural members at the rear of the machine were stripped or burnt out, and removed from within the tunnel leaving only the bulkhead and cutting head to be removed later. Figure 4 shows the carcass of the abandoned machine at this stage.

Application of compressed air and driving the machine as a normal hand-shield were considered. However, the combination of the potential fire risk associated with burning out the bulkhead at the relatively high air pressures required to resist the head of water at the face, and the fact that the ground ahead of the machine had already been ruptured during the attempts to free the cutting head ruled this out as a viable and safe method of proceeding.

The only viable technique left capable of dealing with the extremely difficult situation that had developed, and retrieving the obstructing machine was freezing, and the decision to adopt this solution was taken in June 1988.

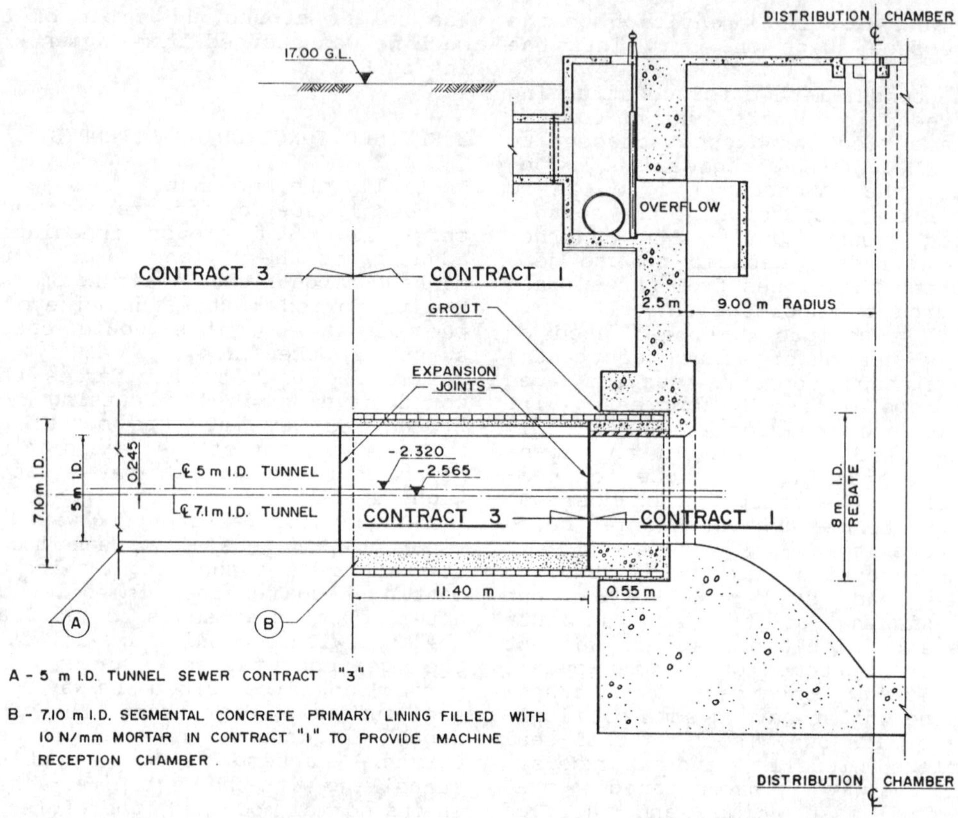

ORIGINAL RECEPTION CHAMBER AND JUNCTION DETAILS

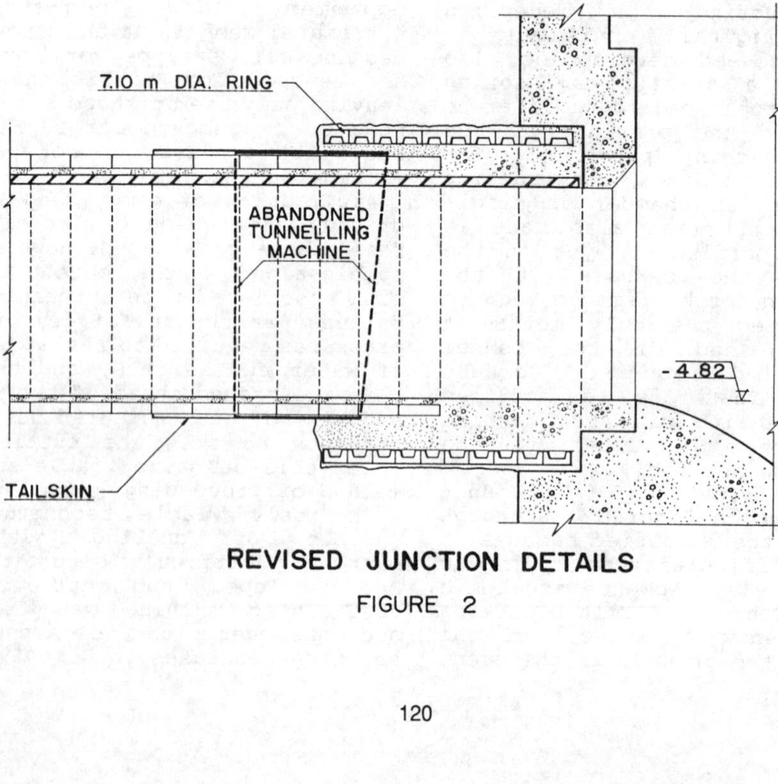

REVISED JUNCTION DETAILS

FIGURE 2

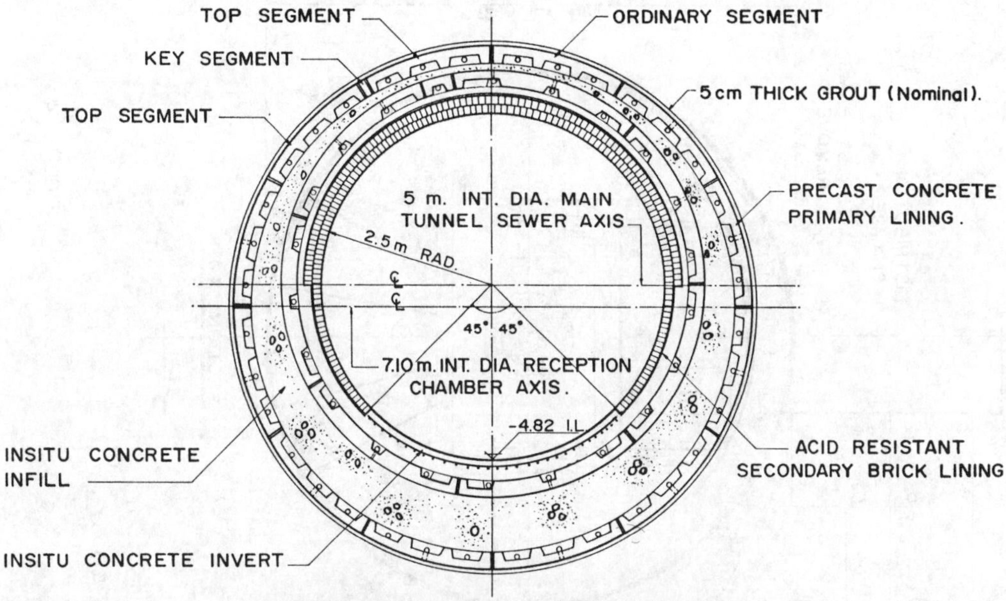

SECTION SHOWING MAIN TUNNEL WITHIN RECEPTION CHAMBER

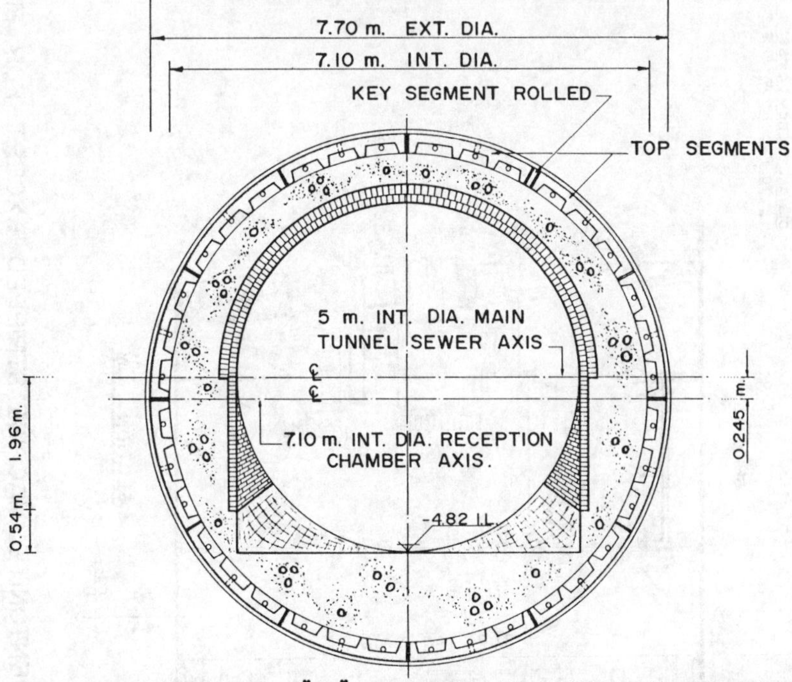

SECTION SHOWING "D" SHAPED TRANSITION ON ENTRY TO DISTRIBUTION CHAMBER

FIGURE 3

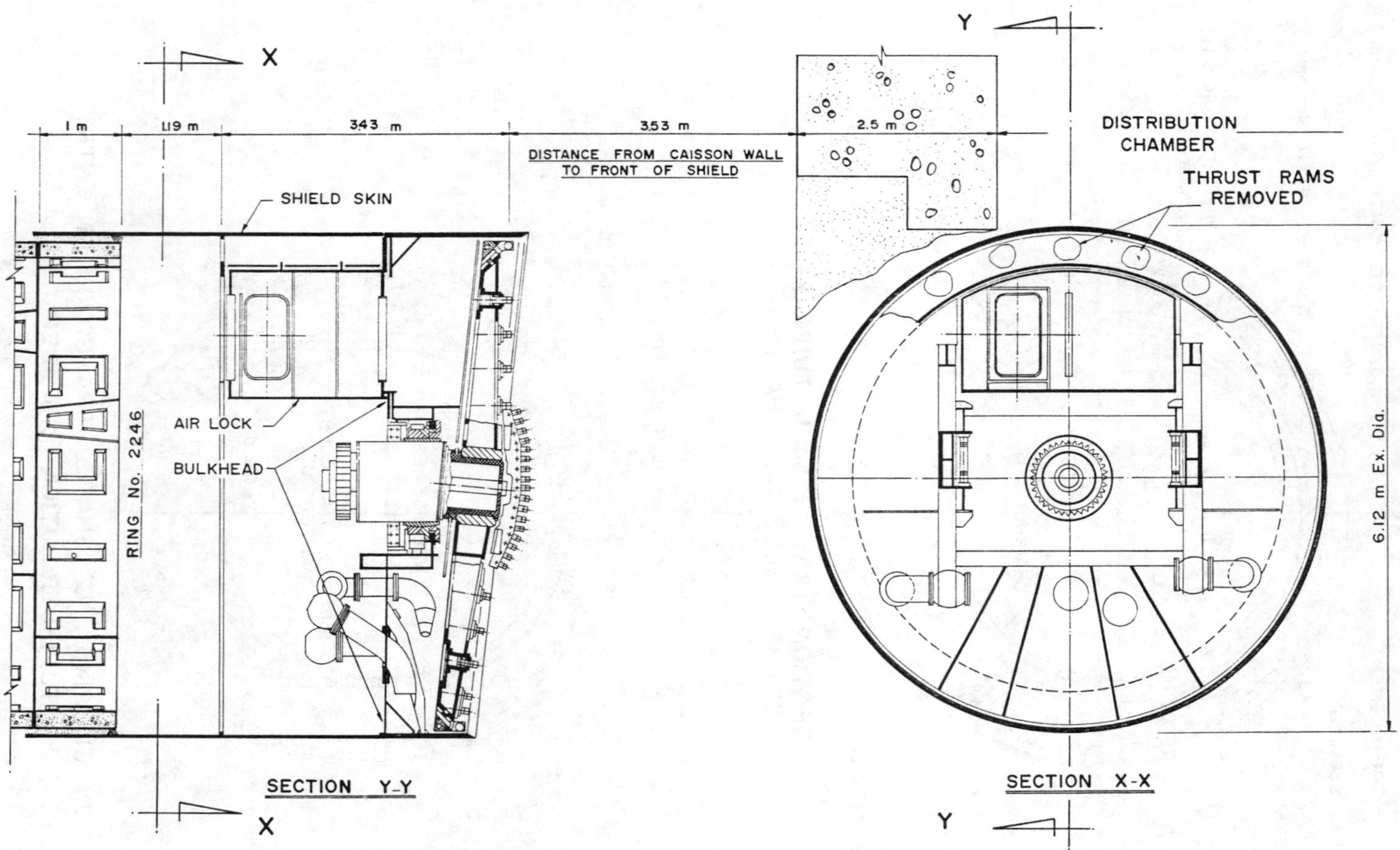

BENTONITE MACHINE STRIPPED EXCEPT FOR BULKHEAD AND CUTTINGHEAD

FIGURE 4

4 FREEZING

The freezing operation was designed to produce a cylinder of frozen ground leading out from the DC to envelop the forward part of the abandoned tunnelling machine, and to enable the driving of a short tunnel within the frozen ground to expose the machine.

The freezing was carried out by British Drilling and Freezing Ltd. (BDF) as sub-contractors to CMC.

The large size and slightly irregular shape of the frozen ground required to ensure adequate working space around the machine, and the 7.1m I.D. rings, especially at the invert, involved an extensive and complicated freeze tube pattern, as indicated in Figure 5. The freeze tubes around the preformed tunnel eye were cored through stuffing boxes grouted into the DC wall to ensure no inflow of ground during the drilling process. Generally, the freeze tubes converged towards the machine, although some of those under the machine were horizontal and others were inclined downwards to ensure adequate frozen ground in case some of the former became exposed and had to be abandoned or diverted during excavation. The system incorporated a series of observation holes (with thermocouples at 2m intervals) located strategically around the periphery of the freeze tubes to monitor the freeze growth and to enable the thickness of frozen ground to be estimated. Thermocouples were also placed within the tail-skin of the machine to monitor and estimate the extent of frozen ground beyond the bulkhead and around the last few tunnel rings. By October 1989 the freezing circuit had been commissioned and was functioning as designed in preparation for breaking through the DC wall.

Unfortunately on 25th October a collapse occurred whilst driving the tunnel connecting the DC to the Tunnel Pumping Station (TPS), which was also being carried out using the freezing technique[4]. As the same refrigeration plant was supplying the coolant for both tunnels, the main tunnel circuit was switched off so that all freezing capacity could be directed to the interconnecting tunnel system, in an endeavor to retrieve the situation. This resulted in the resumption of the freezing and the breakout for the MCT being postponed for about one year.

During this interval, a reappraisal of the calculated thickness of frozen ground above the leading edge of the machine likely to be produced by the horizontal freeze system resulted in the installation of a supplementary vertical freeze system to reinforce the frozen canopy above the soffit. This was similar in some respects to the procedure adopted for the recovery of the freeze following the interconnecting tunnel collapse[4].

Insulation curtains were also erected in the tunnel at the rear of the machine to minimize heat transfer to the freeze zone.

Following the successful completion of the interconnecting tunnel in October 1990, all freezing capacity was directed to the main tunnel freeze circuit, and by late December 1990 break out commenced.

5 FREEZE TUNNEL TO ABANDONED BENTONITE MACHINE

With all blind tunnel junctions there is some concern regarding the precise position of the tunnel, even with modern sophisticated surveying techniques. In this case there were added complications of machine alignment because the tunnel had approached the DC on a relatively sharp curve, and the eye in the caisson had been constructed prior to finalization of the precise tunnel route. Breakout started with a central timbered heading to locate the boss of the cutting head, and to establish its precise position relative to the tunnel eye. The heading was insulated to prevent subsequent freeze losses.

Once the machine had been confirmed in its calculated position, the DC eye was broken out in three stepped operations working downwards from the top. All the mortar was removed back to the structural concrete, and the first 0.61m long ring concreted and sealed into the rebated eye. Hydrotite was used around the periphery of the ring to facilitate the sealing process. Subsequent rings were built (and grouted) one at a time progressively out from the eye with just sufficient excavation taken out

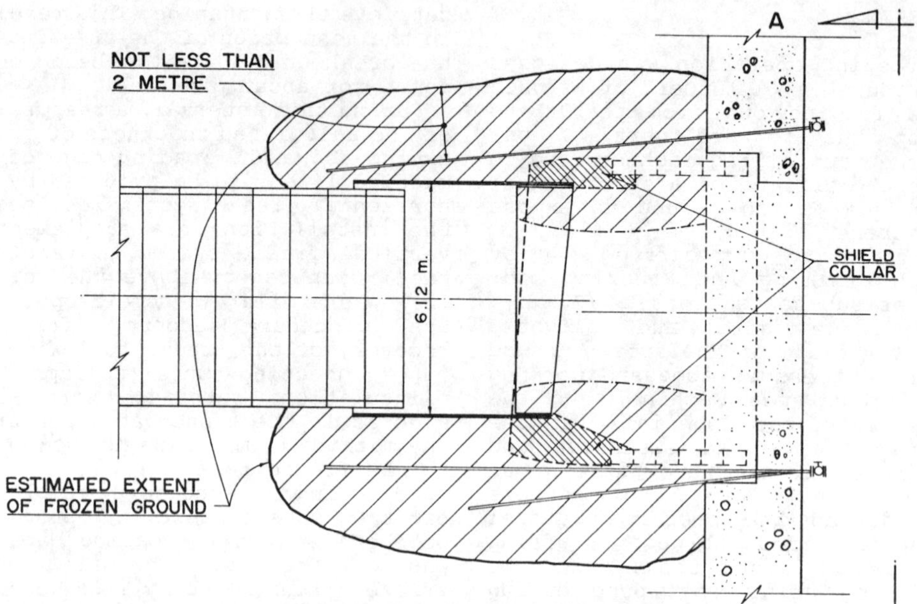

DIAGRAMMATIC SECTION B-B

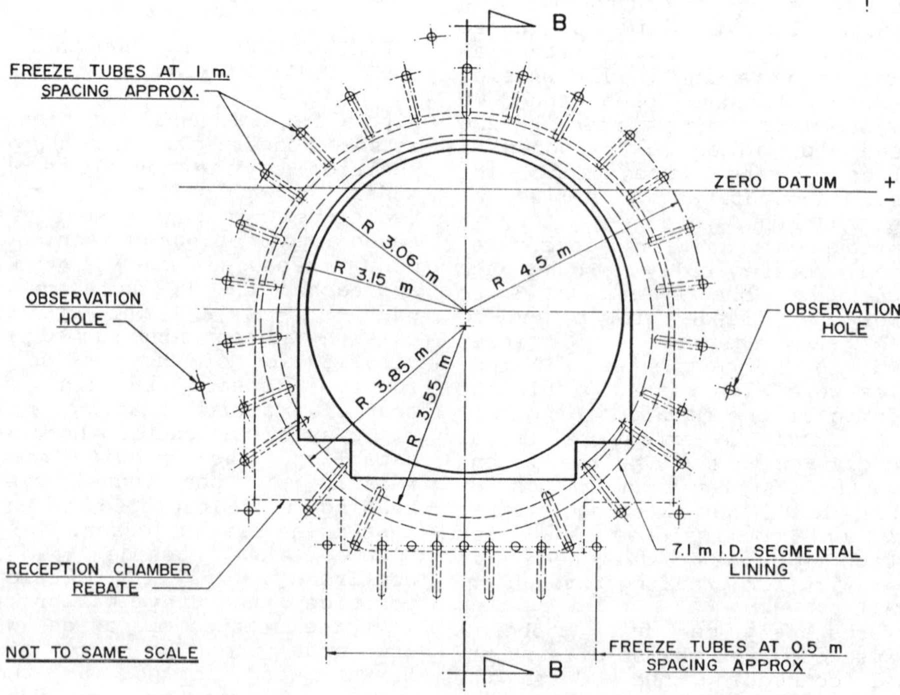

ELEVATION A-A

CONFIGURATION OF FREEZE TUBES AND FROZEN GROUND

FIGURE 5

FIGURE 6

for the erection of one ring (i.e. approximately 0.65m). Excavation was done in three benches progressively from the soffit, with each bench being insulated and supported as the tunnel progressed forward. Experience gained on the interconnecting tunnel proved invaluable, and throughout the whole operation the status of the frozen ground was monitored hourly (by the thermocouples) for any adverse trends before proceeding to the next tunnelling stage.

At all times great care was taken to minimize the area of unsupported ground, which was lagged immediately on exposure.

Special care was required whilst excavating for the invert segments because of the close proximity and vulnerability of the freeze tubes, as shown on Figure 5. On exposure these were immediately lagged, and also diverted where they interfered with the permanent works.

Construction of the insitu R.C. collar joining the machine to the 7.1m ID rings commenced after the sixth ring had been built. The collar was designed to overlap and bond with the shield by 1m, and to be integral with the last ring through 25mm bars locked into the bolt holes. This constituted the most difficult operation of the whole exercise because it necessitated exposing greater areas of frozen ground, and for longer periods, than had been the case hitherto. Sufficient working space was required ahead of the last ring and around the front of the shield to enable installation of reinforcement cages, the welding of shear connectors, concrete to be placed and vibrated, and grout tubes to be fixed into position. Throughout these operations exposed ground remained unsupported, although lagged.

The most critical conditions existed above the soffit of the shield where the freeze tubes were closest and access was limited for fixing the reinforcement and placing the overhead concrete. (See Figure 5).

The R.C. collar was cast in six segments, starting with the invert section which supported the shield, and ending with the closing section at the soffit. Each segment was constructed from start to finish in one operation. Excavation was done in pockets just sufficient to accommodate one section at a time, and all exposed ground was immediately lagged and stop-ends insulated.

Excavation for adjacent sections was not permitted until the freeze monitoring data had confirmed that any localized reduction in freeze caused by the previous concreting operations had recovered, and the freeze temperatures returned to "status quo".

The upper sections of the R.C. collar were shuttered with liner plates, using the shield as a template. Hydrotite was fixed around the periphery of the shield and to the leading face of the last ring for sealing purposes. On completion the R.C. collar was back-grouted through the tubes installed with the reinforcement.

Once the tunnel junction was secured, the cutting head and bulk-head were exposed and cut out piecemeal as shown in Figure 6, and the inside of the shield thoroughly cleaned in preparation for the construction of the insitu R.C. primary lining. This part of the primary lining was approximately 6.5m long, and stretched from the tunnel ring in the tail-skin to the start of the transition zone in the 7.1m ID rings. This was cast in two pours of two lifts. Here again the insitu lining was mechanically fixed to the last tunnel ring, using 25mm lacer bars through the bolt holes.

Thereafter the freezing system was switched off, the transition was formed and the secondary lining of acid resistant bricks installed to complete the tunnelled sewer into the DC.

6 CONCLUSION

The adoption of freezing techniques at Ameria solved a very difficult tunnelling problem which had threatened the successful completion of the new trunk sewer.

The advantage of freezing is that it solidifies the ground and solves the ground-water problem in one process. Nevertheless, it is a sophisticated and costly technique requiring a relatively long period of time to establish the necessary thickness of frozen ground, and once established it requires expert operation and control to ensure 100% reliability.

ACKNOWLEDGEMENT

The author wishes to thank Engineer Mahmoud Abdel Halim, Chairman of Cairo Wastewater Organization for the Greater Cairo Wastewater Project for his approval to publish his paper.

REFERENCES

1) Flint, G.R. and Foreman, W. 1991. Bentonite Tunnelling for the Greater Cairo Wastewater Project. International Tunnelling Symposium, Cairo, Egypt.
2) Grimes, J.F. et al. 1993. Conveyance System: Pumping Stations and Culverts. Proc. Instn. Civ. Engrs. Civ. Engng, Greater Cairo Wastewater Project. UK.
3) Flint, G.R. et al. 1993. Collection System: Tunnels. Proc. Instn Civ. Engrs. Civ. Engng, Greater Cairo Wastewater Project. UK.
4) Agib Wassif A. 1991. Interconnecting Tunnel, Ameria Pumping Station. International Tunnelling Symposium, Cairo, Egypt.

*CMC is a consortium of Christiani and Nielsen Ltd. (UK), and Misr Concrete (Egypt).

**CWC is a consortium of Tarmac Overseas Ltd. (UK), Balfour Beatty Construction Ltd. (UK), Cementation International Ltd. (UK) and The Arab Contractors (Egypt).

Tunnelling and Ground Conditions, Abdel Salam (ed.) © *1994 Balkema, Rotterdam, ISBN 90 5410 363 9*

Ground upheaval at an underground station due to rising groundwater level and subsequent restoration work

T. Ishibashi, H. Ikeda & N. Kamata
East Japan Railway Company, Japan

ABSTRACT: An abnormal rise in the groundwater level due to a spell of heavy rain resulted in ground upheaval over a 100 m section at an underground station operated by the East Japan Railway Co. in the early morning of 12th October, 1991. The underground station is located in a cutting protected by U-shaped retaining walls, which are approximately 12 m high and weigh approximately 120 tf/m. The railway line passes through tunnels on either side of the station.

In the restoration work, the elevation of the station structure was reduced by approximately 50 cm by lowering the groundwater level. The space underneath the retaining walls was filled with low-strength non-segregating underwater mortar to ensure the stability of the lowered structure. As the station structure itself was undamaged, it was decided that it should be retained in use. In sections where sufficient lowering could not be achieved, the slabs were demolished and the ground was excavated for installation of new U-shaped slabs. As a measure against buoyancy, the structure was fixed down with ground anchors to provide it with an adequate resistance even if the water level were to rise as far as the ground surface level in future.

1 INTRODUCTION

During the night of 11th October, 1991, damage was caused by the rain due to Typhoon No. 21 at Shin-Kodaira Station on the Musashino Line, which is located in a cutting (U-shaped retaining walls) with tunnels on either side. Rainfall had been more or less continuous up to this point since 6th October, and the precipitation during this period was 227 mm. Prior to this, a cumulative rainfall of 724 mm had been observed between 1st August and 30th September, which was more than twice the average rainfall for these two months in the Musashino Area. As a result of this rain, the groundwater level showed a sharp rise (to approx. 2.5 m below the ground level), and the uplift (buoyancy) due to the groundwater caused upheaval (max. 1.3 m) of the U-shaped retaining walls over a 100 m section in the station. The upheaval, in turn, resulted in a maximum opening of 70 cm at the tops of the expansion joints in the retaining walls. Large quantities of groundwater and earth flowed into the station through these openings, inundating the railway tracks. A major disruption of the passenger and freight traffic resulted during the subsequent two months until the restoration work was completed (Figures1, 2 and 3).

The report below is concerned with the design and execution of the restoration work that was necessitated by this disaster.

2 SHIN-KODAIRA STATION

The Musashino Line is railway line which was constructed with the aims of improving the means of transportation by rail to central Tokyo and the connection between the satellite cities, as well as of promoting the development of the nearby areas. It was planned as a new circular line along a 20 to 30 km radius from the centre of Tokyo to supplement the existing circular line (Yamanote Line) and runs from Nishi-Funabashi to Fuchu-Honmachi. The construction work, implemented by the Japan Railway Construction Public Corporation, was begun in November 1966, and took eleven years to complete before the opening of the last section in March 1976.

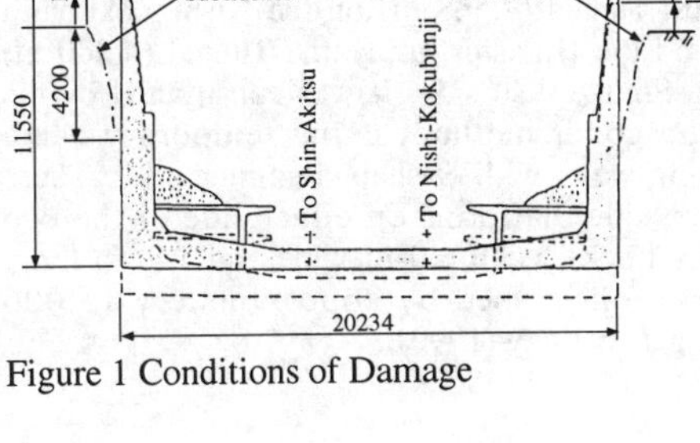

Figure 1 Conditions of Damage

Figure 2 Conditions of Damage

Figure 3 Conditions of Damage

Shin-Kodaira Station, opened on 1st April, 1973, is located in a cutting sandwiched between the Kodaira Tunnel (2,563 m) on the Nishi-Kokubunji side and the Higashi-Murayama Tunnel (4,380 m) on the Shin-Akitsu side. It was constructed by the cut and cover method, using temporary earth retaining walls with H-shaped soldier piles. There are two side platforms on either side of the two railway tracks, each 6.0 m wide and 140 m long. The station is used by approximately 15,000 passengers per day (1990).

3 GEOLOGICAL CONDITIONS

The Musashino Plateau, the diluvial plateau on which Shin-Kodaira Station is located, is known as the largest plateau in Japan. Around Shin-Kodaira Station, the Musashino gravel layer has a depth of 20 m, below which are found alternating layers of clay containing shells and of sandy gravel. The Musashino gravel layer is covered with around 5 m of Kanto loam soil. Groundwater is found under rather different conditions in the Musashino gravel layer and the sand layers further down. The Musashino gravel layer has a high void ratio and permeability and contains large quantities of groundwater. The groundwater in this area flows southwestwards towards the edge of the Musashino Plateau, crossing below the station (Figure 4).

4 RESTORATION WORK (OUTLINE)

4-1 Original Design

The U-shaped retaining walls that underwent upheaval were 11 to 12 m high and were approximately 30 m wide in the section supporting the bridge with the station building and 20 m wide in the ordinary section. The ordinary section was divided into eight blocks with expansion joints at intervals of 10 to 15 m. As the original calculation documents were not available, calculations were made back from such factors as the weight of the structure (W = 117.21 t/m) for the original U-shaped retaining walls approximately 20 m in width and 12 m in height, and it was estimated that the weight of structure would balance with the buoyancy when the groundwater level was 6 m below

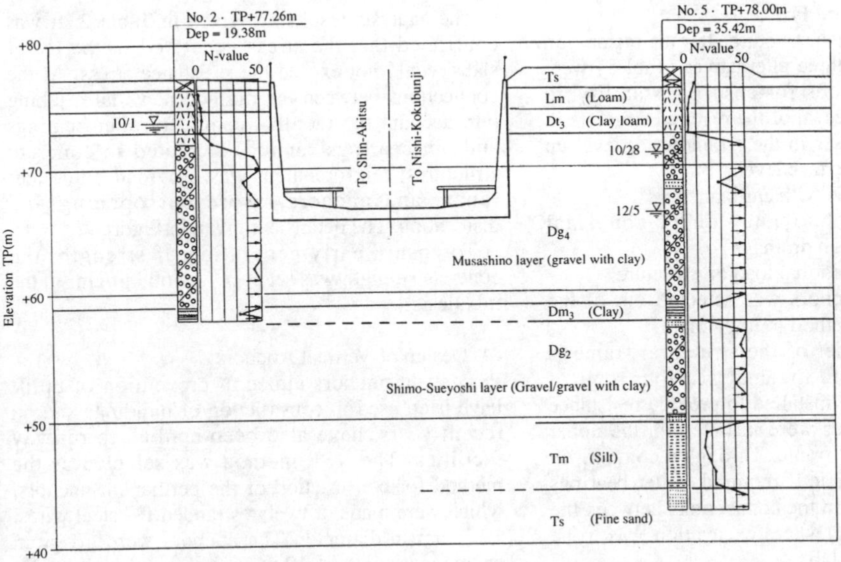

Figure 4 Geological Profile

the bottom of the slabs (5 m below the ground surface). A slight margin would have to be allowed in this figure in view of such factors as the friction between the side wall concrete and the soil.

4-2 Restoration Design

Although there was little damage to the U-shaped retaining walls themselves, the gradient of the railway line was altered from 8 per mil to 21 to 35 per mil by the upheaval. While the structure was lowered as far as possible, other measures had to be taken where sufficient lowering could not be achieved.

(1) Investigations on Restoration Method
The alternatives for the restoration methods included reconstruction, lowering and rail-level alteration. Since an early reopening of the Musashino Line was desired in view of the importance of the line as a major freight and commuter route, the lowering plan was adopted, and it was decided that where adequate lowering could not be achieved the track sections of the lower slabs should be demolished and removed for construction of new shallow U-shaped slabs. As a measure against future rises in the groundwater level, the structure was to be fixed down with ground anchors to raise its stability.

(2) Lowering of Structural Frame
Reduction of the buoyancy through lowering of the water level was chosen as the method for lowering the structural frame, as this was judged to be the quickest and most efficient method. To lower the water level, a total of 26 deep wells were created, seven inside the U-shaped retaining walls (*l* = 7 m, ø = 650 mm, lifting pipes: 100 A, high-lift underwater pumps: 15 kW) and nineteen on the periphery of the retaining walls (*l* = 20 m, ø = 700 mm, lifting pipes: 100 A, high-lift underwater pumps: 15 kW). Pumping began on 22nd October and continued until 30th November, after the completion of the new slabs. There had been no past cases, however, where complete lowering of the structural frame was achieved after an upheaval as large as 1.3 m and there were concerns as to whether adequate lowering of the water level could be achieved through use of deep wells. (Figure5)

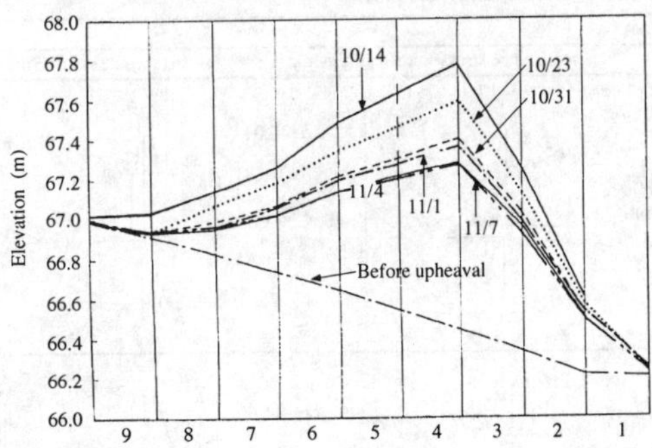

Figure 5 Elevation of Lower Slabs (Slab Crests): Measurement Results

(3) Design of Structural Frame

For the final form of the structure after the restoration work, the following three alternatives (Table1) were considered as methods for ensuring stability, to prevent repeated upheaval of the restored structure in the event of future rises in the groundwater level up as far as the ground surface level.

1) Fixing to position with earth anchors
2) Prevention of buoyancy under abnormal conditions through drainage
3) Provision of additional load on structure

The use of earth anchors was selected out of the three as the easiest method to implement.

The analysis model of the structural frame is shown in Figure 6. In the analysis, four rows of vertical anchors were installed to provide resistance to buoyancy and struts were installed on the upper parts of the structural frame. Through a comparison of the construction periods required, roller bearings were selected for use in the connections between the old slabs and sections where excavation was to be conducted below the slabs.

Table 1 Comparison of Measures against Buoyancy

	Measure	Remarks	Evaluation
Alternative 1	Vertical anchors 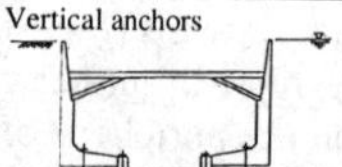	• Need for construction of reliable vertical anchor • Short execution period	○
Alternative 2	Water level lowering under abnormal conditions 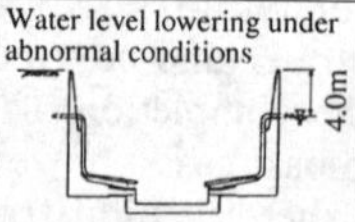	• Maintenance of drainage equipment required to retain its durability • Piping required within the station structure	△
Alternative 3	Additional load 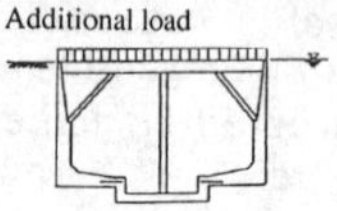	• Reinforcement of side walls required for placement of additional load on U-shaped retaining walls • Stable appearance • Long execution period	△

Table 2 Structural Stress

Block	2		3		4	5	6	7
	Resistance stress	Stress generated	Resistance stress	Stress generated				
Moment (t·m)	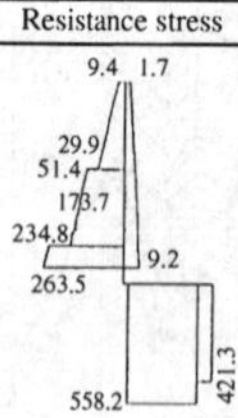	a = 6.01 b = −6.81 c = 72.85 d = 76.65	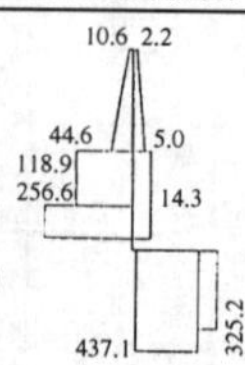	a = 6.48 b = −14.14 c = 49.51 d = 52.32	6.94 −12.90 49.66 52.00	7.91 −11.57 51.06 53.35	3.51 −14.34 78.86 82.69	3.82 −14.34 81.05 84.38
Shear force (t)	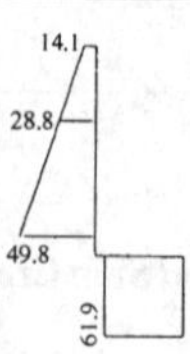	a = −11.46 b = 25.88 c = −35.46	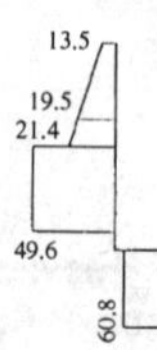	a = −11.46 b = 25.88 c = −35.46	−15.87 23.66 −14.54	−15.87 22.75 −21.36	−11.22 29.63 −37.29	−11.63 29.29 −37.89

The analysis results are given in Table 2. It was confirmed that the stress generated in the lower slabs would not exceed the resistance stress. At the connections between the old and new slabs, taking into account the fact that complete roller bearings and pin bearings cannot be created in concrete structures, reinforcement was provided with steel rebars and anchors to prevent opening and dislocation. The details are given in Figure 7.

To ensure early generation of strength, the concrete strength was set at σ_{28} = 360 kgf/cm^2 in the mix design.

(4) Design of Vertical Anchors

Permanent anchors aimed at prevention of uplift have been used in construction of buildings and, in recent years, have also been applied to railway facilities. The VSL method was selected as the method for construction of the permanent anchors, which were made of twelve stranded PC steel wires, 12.7 mm in diameter. The anchors were given an anchorage length of 10 m in the gravel layer and a free length of 5 m. The design anchoring force (per anchor) was 134 tf for the side wall anchors (pitch: 4.0 m) and 92 tf for the anchors for new slabs in the middle (pitch: 2.0 m). The details of the anchor tendons are given in Figure 8.

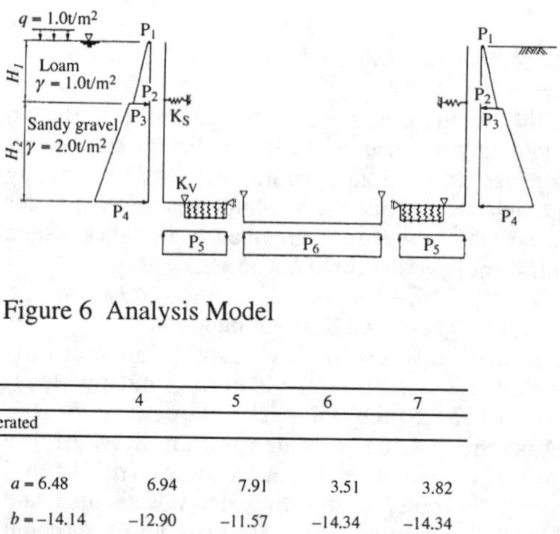

Figure 6 Analysis Model

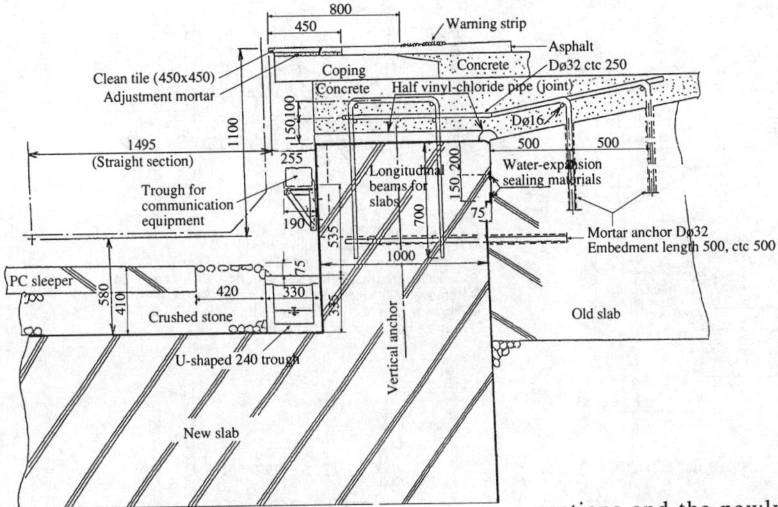

Figure 7 Connection (Details)

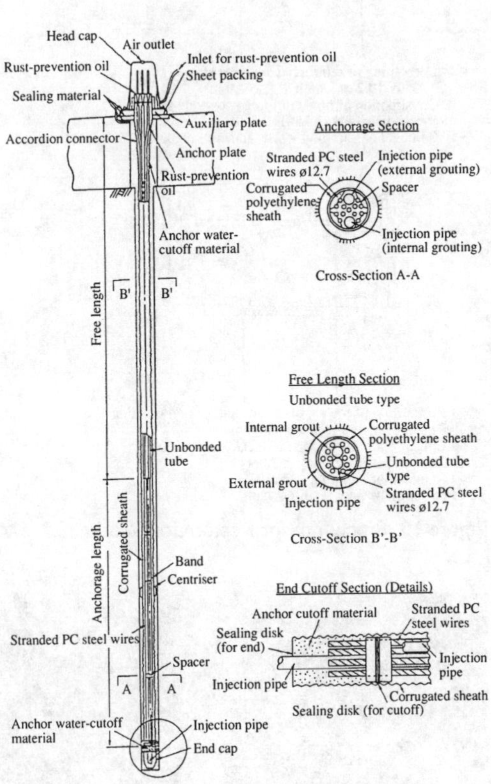

Figure 8 Anchor Tendon

(5) Design of Struts

After the completion of the restoration work, the U-shaped retaining walls would comprise a composite structure consisting of the original L-shaped sections and the newly-constructed U-shaped sections. It was decided that struts should be provided in the L-shaped sections which were liable to destabilisation through overturning at times of rises in the water level.

The struts were to be approximately 19 m long. As a measure relating to the buckling stress intensity, angle braces with hinge metals at both ends were installed at points approximately 4.4 m away from the two ends of the struts to reduce the buckling length. The struts were to be connected to each other with longitudinal beams to prevent sideways movement due to vibration and to ensure structural stability.

The members used for the struts and angle braces were H 400 x 400 x 13 x 21 H-shaped SS 400 steels galvanised by hot-dipping.

4-3 Execution

The cross-section after restoration and the restoration procedure are shown in Figures 9 and 10, respectively.

(1) Lowering of Groundwater Level

Deep wells were used to lower the groundwater level. In this method, the permeability coefficient provides the basis for the design. Valid values for the permeability coefficients were not available at the initial stage of the design, but a value of $k = 7.69 \times 10^{-3}$ (m/sec.) was obtained by a deep well pumping test and used in the planning of the water level lowering. The required reduction in the water level was calculated by Theis' method and used for the determination of the number and pumping capacity required of the deep wells. Nineteen deep wells were created on the periphery of the U-shaped retaining walls and seven inside the retaining walls to lower the water level. During the water level

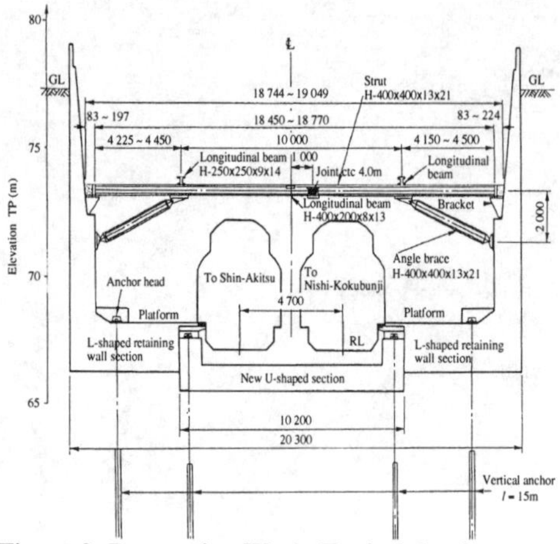

Figure 9 Restoration Work: Typical Section

lowering, inflow of earth was expected in the parts below the ends of the old slabs, which had been raised by the buoyancy of water. It was found that old slabs would be ruptured by the combined action of the weights of the old slabs themselves and the lateral earth pressure from the side walls with the ends of the old slabs providing the fulcra as the buoyancy was reduced. For this reason, temporary supports were constructed inside the U-shaped retaining walls to provide reinforcement (Figure 11).

The permeability coefficient calculated after the lowering of the water level from the reduction in the water level and the number of days required was k = 7.0 to 8.0 x 10^{-3} (m/sec.), indicating that the value obtained through measurement in the deep well was close to the actual value.

(2) Concrete Grouting underneath Slabs

The progress of the water level lowering and lowering of the structural frame was monitored and concrete was injected into the space underneath the slabs once a drop was observed in the rate of lowering. Utmost care was taken in the determination of the timing of the concrete grouting underneath the slabs, as this had a major effect on the subsequent full-scale restoration work. The grouting was implemented through tremie pipes installed on the slabs. The uniaxial compressive strength of the concrete was set at σ = 80 kg/cm^2, bearing in mind the fact that a part of the concrete would need to be removed at a later stage. The water level was still above the level of the slabs at the time of the concrete grouting, which as a result took place under pressurised conditions.

To ensure satisfactory filling performance and resistance to segregation under water, non-segregating underwater concrete was used in the grouting, with the mix proportions determined through tests, which are given in Table 3.

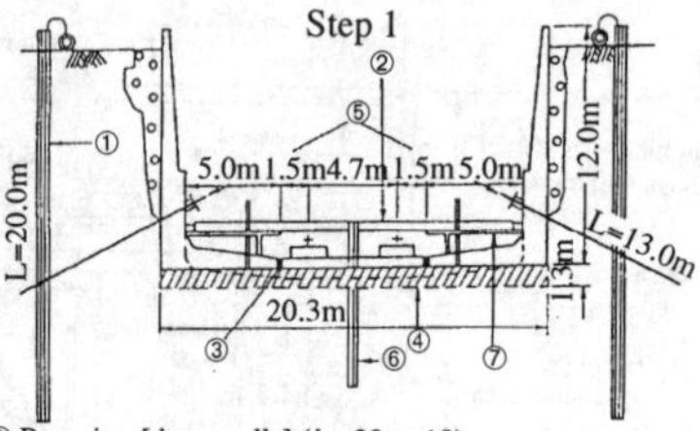

① Pumping [deep wells] (l = 20, x 19)
② Temporary supports (Blocks 2 to 8)
③ Coring of lower slabs for drainage (ø 50 mm)
④ Concrete grouting below slabs (coring ø 150 mm, x 24)
⑤ Removable horizontal anchors (ctc = 2.5 m, x 70)
⑥ Pumping at centre of slabs (l= 5 m, x 7)
⑦ Removal of supports and platforms

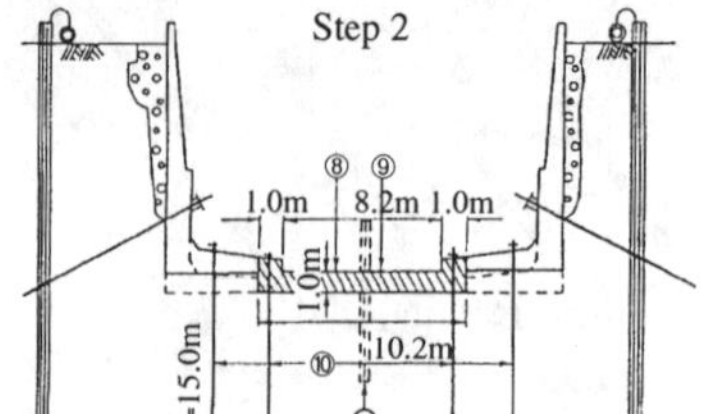

⑧ Demolition of reinforced concrete slabs (width: 10.2 m, depth: 0.8 to 1.0 m)
⑨ Construction of new reinforced concrete slabs
⑩ Vertical anchors (x 150)
⑪ Removal of pumping works at centre of slabs

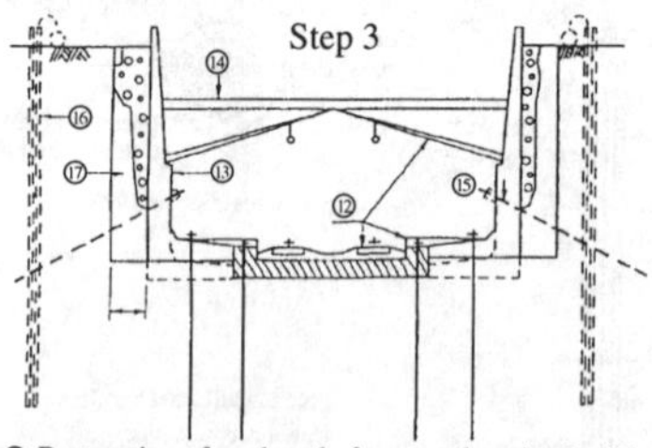

⑫ Restoration of tracks, platforms and station building
⑬ Repair of retaining walls and tunnels
⑭ Construction of supports (Blocks 2 to 7)
⑮ Removal of horizontal anchors
⑯ Removal of pumping works
⑰ Grouting behind retaining walls

Figure 10 Procedure for Restoration Work

Figure 11 Temporary Supports

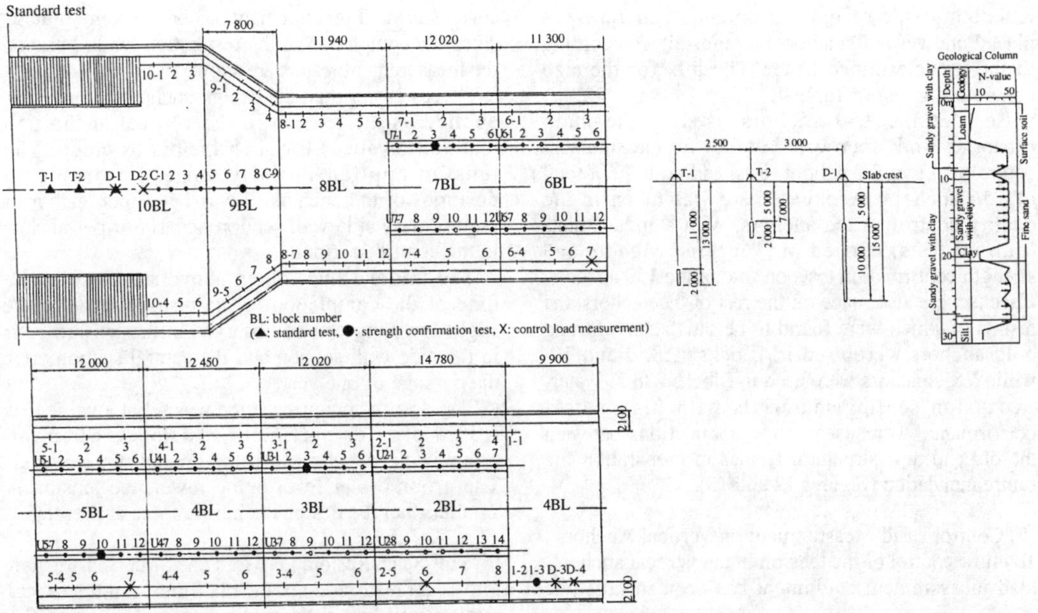

Figure 12 Vertical Anchor Test Sites

Figure 13 Restoration Work

(3) Removable Horizontal Anchors and Demolition of Lower Slabs

Before demolishing the lower slabs, seventy removable horizontal anchors 13.0 m in length (ø = 12.7 mm, nine stranded PC steel wires) were first installed at 2.5 m pitches to stabilise the L-shaped section. The design anchoring force was 100 t.

The lower slabs were demolished after the removal of the obstructions such as the temporary supports and platforms. As the old slabs below the platforms were to be retained as they were, they were isolated from the demolition section by concrete coring along the boundary. Concrete coring was selected as it was possible to use a large number of coring machines at one time, allowing a reduction

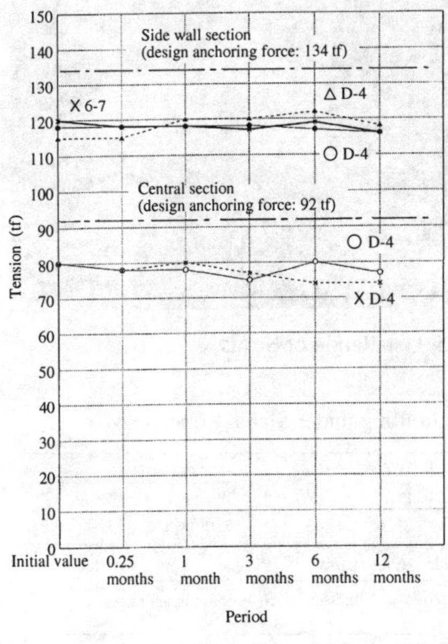

Figure 14 Vertical Anchor Load: Measurement Results

in the time required. After isolation, the concrete in the demolition section was crushed with giant breakers and removed from the site.

(4) Slab Concrete and Vertical Anchors

After creating a shallow sump on the Nishi-

Kokubunji side of the station, slab concrete was placed and vertical anchors were installed according to the predetermined order. The mix for the slab concrete is given in Table 4.

Besides the 150 anchors used in the main anchoring work, four were installed for measurement control and two for pullout tests, adding up to a total of 156 anchors. Adequate care was taken in the quality control of the anchors, with standard tests (pullout tests) carried out on three anchors and strength confirmation tests on five of the 150 anchors. Checks were also made on the rest of the anchors, the results of which were found to be satisfactory. Strong hold anchors were used in Blocks 1, 8, 9 and 10, while VSL anchors were used in Blocks 2 to 7. Water-expansion sealing materials with high cutoff performance were used at the connections between the old and new structural frames in preparation for future inundation (Figures 12 and 13).

(5) Control Load Measurement on Vertical Anchors

To allow control of the tension in the vertical anchors, load measurement equipment has been installed at seven sites and measurement has been taken on these since the completion of the restoration work. Parts of the measurement results are shown in Figure 14.

Forces larger than the design anchoring force was applied as the tension in the vertical anchors to make allowances for such factors as creep. The tension applied initially varies according to positions of the anchors. It was 144 tf per anchor in the ordinary side wall section and 106 tf per anchor in the central section.

The initial values given above are those at the time of the completion of the restoration work. A slight lowering can be observed in the tension both in the side wall section and the central section after the passage of one year.

The design anchoring force was set at a value that would provide sufficient resistance when the groundwater level reached the ground surface level. The groundwater level being lower, the tension is smaller than the design anchoring force at present.

(6) Full-Scale Restoration Work and Strut Construction

Full-scale restoration of the platforms, rail tracks and station building was implemented along with the restoration of the power and communication equipment. At the same time, liquid containing suspended water glass was injected into the ground behind the U-shaped retaining walls to fill the gaps left by the outflow of earth and for cutoff of water flowing towards the station structure. Struts were also installed at 4.0 m pitches as reinforcements for the retaining walls in preparation for future abnormal rises in the groundwater level, thus completing the series of restoration work (Figure 15).

Figure 15 Installation of Struts

Table 3 Grouting under Slabs: Concrete Mix

Cement	W/C (%)	S/a (%)	Unit contents W	C	S	G
Ordinary	83.3	45	250	300	720	902

Chemical admixtures (kg/m³) Non-segregation agent	Fluidiser	Air-entraining agent	Slump flow (cm)
3.0	10	0.6	60 x 60

Table 4 Slabs: Concrete Mix

Nominal strength (kg/cm²)	Cement	W/C (%)	S/a (%)	Slump (cm)	Unit contents(kg/m³) W	C	S	G	Chemical admixture (kg/m³)
360	Ordinary	42.1	42.0	8±2.5	157	373	749	1,060	3.79

5 CONCLUSION

The damage due to ground upheaval caused by Typhoon No. 21 led one to recognise anew the dangers of the forces of nature acting in the form of buoyancy. The total damage on the Musashino Line, a major artillery for commuter and freight traffic, is estimated at ¥ 8,600 million. The line was reopened after two months of restoration work implemented day and night, with the cooperation of the agencies concerned and the local residents.

The authors would like to take this opportunity to express their gratitude to all those who helped in one way or another in the restoration work, and to declare the resolve on the part of the staff of JR East to prevent the repetition of such a disaster.

Tunnelling and Ground Conditions, Abdel Salam (ed.) © 1994 Balkema, Rotterdam, ISBN 90 5410 363 9

Construction of large cross-sectional diversion tunnels through confined water-bearing sandy layers in urban areas: Katabiragawa River Diversion Channel Project (Lot No. 2, Yokohama)

Toshinobu Ishii & Saburo Ishii
Shimizu Corporation, Japan

Masahiro Kobayashi
Kanagawa Prefectural Government, Japan

ABSTRACT: Katabiragawa divesion tunnel is a large sectional urban tunnel of 111 m² in excavation area, iconstructed in complex geological conditions of Kanagawa prefecture in Japan. The ground consist of alternating layer of silt and water confined water-bearing sandy soil. Most of the tunnel was constructed under the residental areas in Yokohama city and the tunnel is located to under pass through the Shinkansen, conventional railway and several maijor highways.

The Side drift NATM(New Austrian Tunnelling Method) was selected for tunnelling to prevent surface settlement and to execute water cut-off by drainage of the heading.

Subsidual tunnelling methods were succsessfully applied for stabilize the cutting face and the inner tunnel displacements kept within 20 mm, though the soil was very sensitive for teh seepage water. This tunnel was completed in November 1992 and the results areas follows.

1 OUTLINE OF THE SCHEME

Katabiragawa is a river under direct supervision of the Governor of Kanagawa, and the flow extends 17 km in length with a drainage area of 58 km².

The river became the cause of a major disaster in 1958 with No. 22nd Typhoon and found a rise in flood reports with urbanization. River improvement works have been undertaken from 1959 to prevent flooding, however, the recent rapid progress of urbanization has made it difficult to further expand the width of the river, and thus, has lead way to construct a tunnel-type diversion channel instead.

As indicated in Figure 1, the project planed for a diversion channel at the middle course of Katabiragawa. It is designed to divert 350 m³/sec. of water flood, to be discharged into Yokohama Port via the tunnel with total length of 5320 m and the cross section of the tunnel amounts about 111m² in excavation area.

This large-sectional tunnel faced complex construction conditions since it was excavated in a sand consist of alternating layers of consolidated silt and sandy water bearing bed.

Most of the tunnel was constructed under the residential areas, however also extended below grounds for the Shinkansen super express, conventional railways and major highways.

The tunnel works was started in 1981, however,

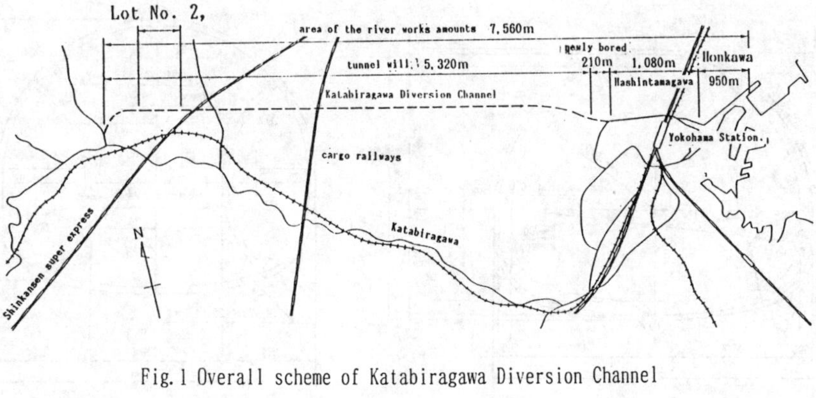

Fig. 1 Overall scheme of Katabiragawa Diversion Channel

the Nishiya Lot No. 2 construction Lod (435 m of tunnel) was started in April 1991 toward the downstream after completion of the uppermost stream in the Kawashika area (boring length 580 m).

The Side drift NATM(New Austrian Tunnelling Method) was selected to excavate the tunnel for preventation of the surface settlement.

The sandy layer had minimum cohesion and was so sensitive as easy to collapse the cutting face due to the slightest seepage of water.
For the first time, horizontal drain boring from the side wall of the heading was planned. However, in areas of sand dominant, boring works had offen be prevented by overflows of sand with water, and more happened to colapse of the cutting face due to inefective pre-drainage. Considering of these in site condition, chemical grouting was adapted successfully.

Tunnel construction in the urban areas, it is common probrem to keep ground surface stable and so it becom most important to keep stability of the cutting tunnel face. The tunnel was completed successfully in November 1992 and the surface settlement and the inner tunnel displacement was kept under 5 mm and under 20 mm, respectively, by adapting thoes subsidual metohods in excavation.

2 OUTLINE OF CONSTRUCTION

Owner: Yokohama Flood Control Office, Kanagawa Prefecture
Construction period: March, 1991∼June, 1993
Outline of construction
- Constructive length of tunnel: L = 435 m
- Excavation netod: Side Lot NATM
- Amount of soil excaxated: 50,126 m^3
- Amount of °oncrete: 10,165 m^3
- Support works: ctc 0.9 m
  - (Heading) H-125x125(inner), H-100x100(outer)
  - (Centre crown) Mu-29
  - (Invert) H-125x125
- Subsidual works:
  - Chemical Grouting: 6,430kl, Drain Boring: 568m

Figure 2 indicates the standard sectional diagram, and Figure 3, the division of heading section and Figure 4 indicates the actual procedures.

3 GEOLOGY

The geology of Nishiya Lot No. 2 has a Hoshikawa foundational group on the Kazusa layer that is covered with a uonconformed Sagami layer.

The tunnel is located around 40∼50 m below ground level in between the alternating consolidated silt and alluvial loose sandy layer of the upper general layer group of the tertriaera. Most of the sandy layers are exist in lens and block like distributions that are irregular.
The consolidated silt marked a single axis compression strength of qu=25∼73 kgf/cm², and posed no problems to construct a tunnel. The angle of internal frictiona was some large at ϕ=40°, howeverlacked cohesive compornents. The uniformity coefficient was small (Uc=2∼7), and there is minimal mixture of fine particles with a diameter under 74 μm. Uniform sands are generally considered to easy to flow down with minimum seepage of grouud water by the characteristics as follows: Uc<5 and/or F_r < 10%, wher Fr mean the composition ratio of fine (<74 μm) particles of soil contents.

The welling water pressure was some large at 0.3∼2.5 kgf/cm² and coefficient of permeability, k= 10^{-2}∼10^{-4}cm/sec, so it is recognized as to welling water may easily cause fluidization. Then the major ploblems that had to be cleared for successful completion of this tunnel was to take effective measures against underground water and stabilization of the cutting face sand layer.

Figure 5 indicates the londitudinal section of the geology, Figure 6 compares the accumulation curve of the past sandy layers particle diameters that fluidized (Ikuta & Uramoto Tunnel) compared to distribution range of particles in sandy layers of this construction area.

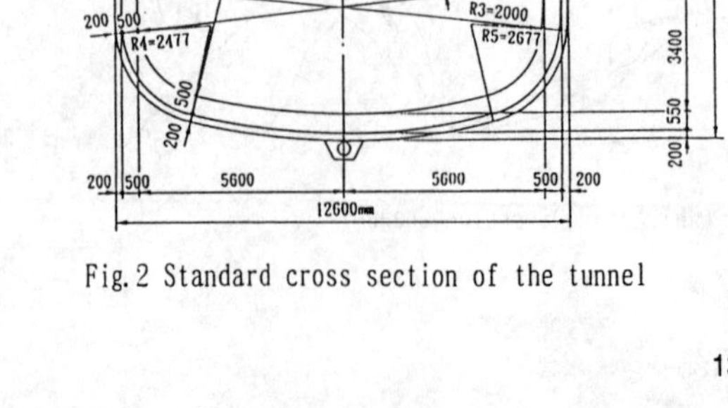

Fig. 2 Standard cross section of the tunnel

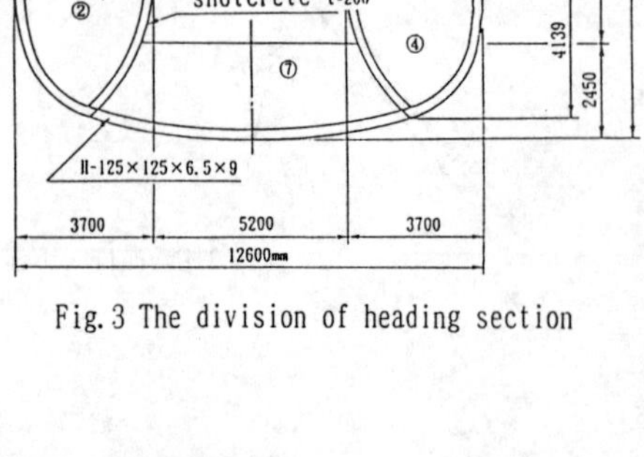

Fig. 3 The division of heading section

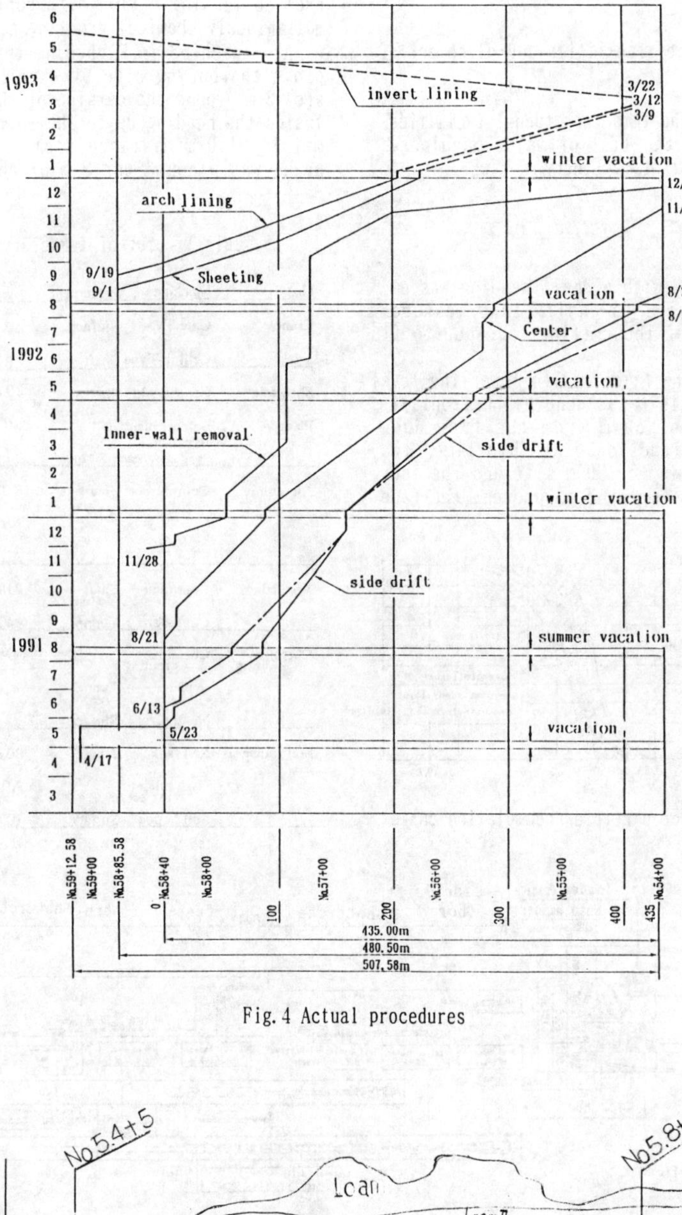

Fig. 4 Actual procedures

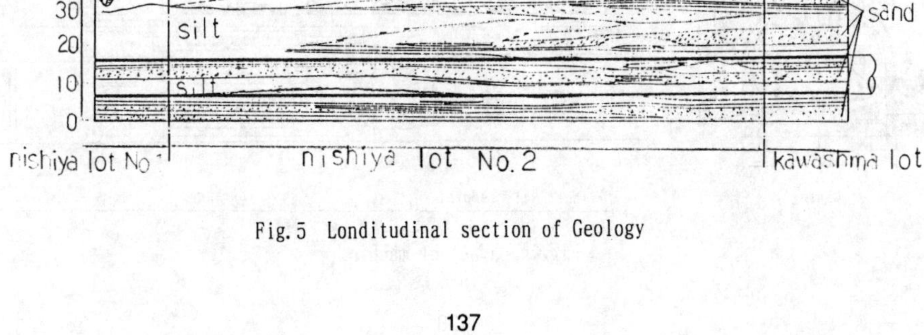

Fig. 5 Londitudinal section of Geology

4 WORKS

4-1 Temporary tunnel facilities and placement of machinery

Table 1 indicates the temporary tunnel facilities, Table 2 indicates the list of major machinery. and Figure 7, layout of machinery.

4-2 Boring

The side drift NATM with a small heading was employed for this site to easily take measures against collapse of the cutting face due to fluidization.

The first measure taken was to lower the level of ground level. This was done by horizontally boring drainage of 50 m from the face upon starting the construction. At this time, two piles were driven in the 50 m site and the four piling measure was employed to prevent the face from collapsing. Side dift boring was done symultaniously chemical grouting.

As indicated in Figure 8, the plan was to improve the boring site by chemical grouting the area 3 m around the outside of the tunnel and 2 m inside the heading up to the 85 m point. The remaining 140 m distance from the 85 m to 225 m point was planned for 2 m of chemical grouting

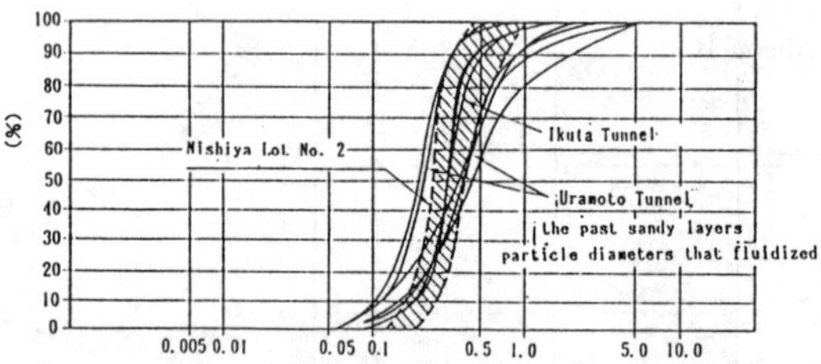

Fig. 6 Comparison of particle accumulation curves

Table 1 List of temporary facilities

Use	Machine	Placement	Model/Specifications	Amt.
Crane	Gate-Type	Surface	2.8 t	1
	Ceiling-Type	Shaft	2.8 t	1
Shotcrete	Cement silo	Shaft	30 t	1
Plant	Batcher plant		MCP-500P-B 25m^3/h	1
	Aggregate/ soil tank each 20 m^3		2	
Air supply	Compressor	Shaft	KST75A	3
			Air discharge: 12 m^3/min.	
			Air pressure: 7 kgf/cm^3	
Ventilation	Contrafan	Shaft	ø1000	1
	Contrafan	Tunnel	ø400	3
Chemical grouting	Chemical grouting plant	Tunnel		1
	Neutralizer	Surface		1
Soil transport	Soil tank	Tunnel	80 m^3	2
	Hydraulic bucket		3.0 m^3	1
	Soil hopper	Surface	100 m^3	2

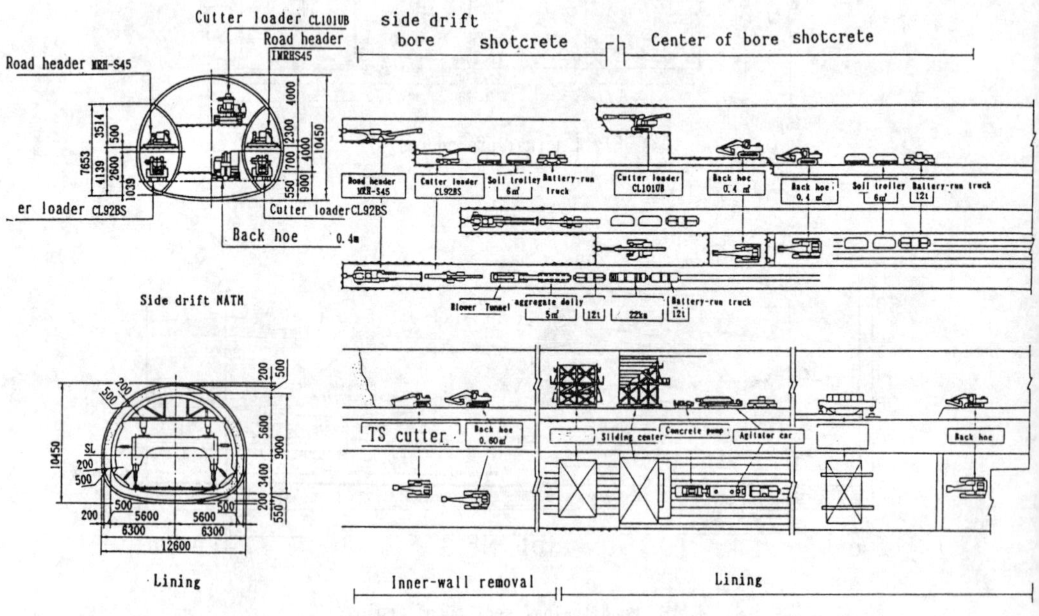

Fig. 7 Layout of machinery

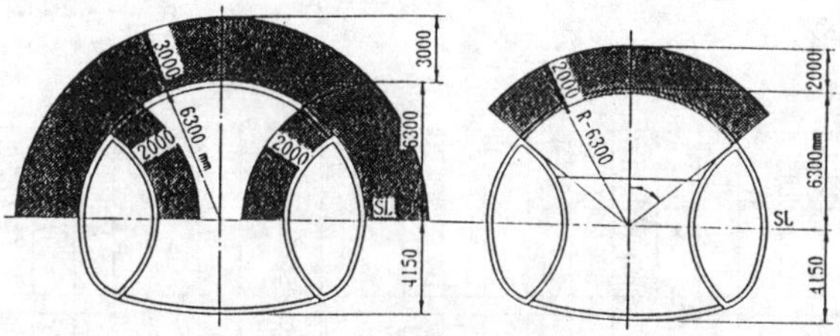

85 m section   140 m section

Fig. 8 Area Improved By Chemical Grouting

Table 2 List of major machinery

Use	Machine	Placement	Model/ Specifications	Amt.
Boring Machine	Road header	L/R Side Lot	MRH-S45-14	2
	Cutter loader		CL92BS	2
	Cutter loader	Upper middle	CL101-VB	1
	Back hoe	Mid&lo middle	E120B	1
Soil transport	Battery-run truck	Tunnel	Toshiba 12t battery -run machine	5
	Soil trolley	Tunnel	8 m^3	5
Shotcrete	Blower	Tunnel	Ariba 280 (dry type)	2
	Aggregate truck		4.5 m^3	2
Chemical grouting	Midget drill	L/R Side Lot	HC-42B	8
	Boring machine	Upper middle	D2-G	4
Drain boring	Boring machine	L/R Side Lot	TOP-MB (unit type)	2
	Boring pump		NES-100	2
Inner-wall removal	Back hoe	Tunnel	E200B	1
	TS cutter		TS500RCL	1
Secondary works	Reinforcing frame	Tunnel		1
	Sliding center			1
	Concrete pump		60 m^3/h	1
	Agitator car		6 m^3	2
	Sheeting dolly			1

Table 3. Grouting Rate

Ground layer	Instant Agent	Loose Agent	Total
Consolidated Silt	1%	3%	4%
Sandy Layer	6.75%	20.25%	27%

around the outside of the mid-point of the crown portion

Drainage boring was capable of draining 15～20 l/min. of water per hole, and was a successful measure in lowering the underground water level to allow for smooth excavation.

However, after hitting the 85 m heading and chemical grouting point, we experienced an inflow of spring water, causing the sandy layer to fluidize and the face to collapse. The second at tempt of drainage boring also resulted in spring water and sand gushing out of the ϕ116 mm water drainage hole.

Jamming took place at the 20 m bore site which made boring impossible to continue, thus, hoced to halt the excavation of side drift.

Additional investigation was performed via boring to study the geology from the 85 m point in hope to reconsider the boring method. As a result, we found that the overall geology consisted of extremely complex alternating layers of consolidated silt and sandy layers between the 85 m and 260 m point. The sandy layer was especially dominant in this section and was found in the Side drift and the upper half of the middle portion. What's more, it was found that consolidated silt was widely spread in the lower half after the 260 m point, in which the consolidated silt and sandy layer was found to be in a fairly clear pattern.

Thus, we performed chemical grouting up to the 260 m point and adapted drainage boring from the 260 m point.

Furthermore, chemical grouting was performed 2 m outside the arch in the central portion from the 80 m section to the downstream side since the soi l layer extremely thinned out (around 15 m), with the surface concentrated with residents and stores, and the Tokaido Shinkansen running nearby.

Figure 9 is a detailed vertical section of the geology based on an observation of the face during boring works.

The double tube compound chemical grouting method was chosen for the following reasons:

(1) Long gel time grouting with good permeability to even out the fine sandy layer.
(2) Ideal packing around the rod as necessary.
(3) Instant adhesion for high compaction by vessel grouting along the border of the consolidated silt and sandy layers.
(4) Instant measures against spring water during boring.
(5) Good work performance in narrow areas.

Grouting from the heading was performed by installing a frame on the upper half of the heading.

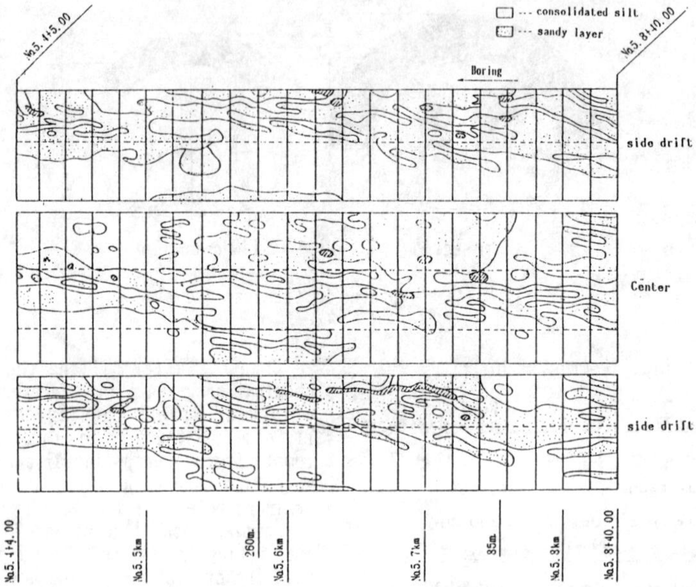

Figure 9. Vertical section of geology based on sketch of face

Table 4. Shotcrete mixture

W/C (%)	C (kg)	Aggregate (m3)	Rough Aggregate (m3)	Instant Agent (kg)	Dust Suppresser
45	360	0.8	0.47	19.8	3.6 kg

Dimension of Max. Rough Aggregate	Cement Specs.	
under 13 mm	N	σ_{28}= 210 kgf/cm3

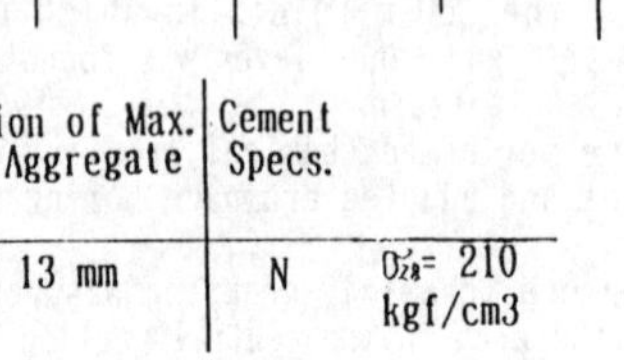

Figure 10. tunnel shape

Boring and grouting was alternated by day and night in 1.8 m boring increments. (half of the heading bore proceeded at 1.8 m, with a supporting width of 0.9 m; 2 base). Table 3 indicates the grouting rate.

The drainage boring executed from the 260 m point was performed 30 m from the leading heading and 50 m boring lengths were drifted each on the left and right. There was a 15～20 l/min. drainage from each boring to allow for the works to successfully proceed.

4-3 Shotcrete

The dry shotcrete method was employed for this project. The materials dry mixed in the Batcher plant (25 m3/h) in the shaft was transported via the aggregate dolly (4.5 m3) and shotcreting was performed using the blower (Ariba 280). Manual shotcreting was employed since each face was narrow.

Table 4 indicates the shotcrete mixture.

4-4 Lining

The general procedures entailed in lining can largely be categorized into (1) invert lining covered with a concrete lining, and (2) arch concreting finished with invert concrete.

However, the diversion tunnel, for the meantime will adopt a shape as indicated in Figure 10, in which the arch bases are circular, unlike general horseshoe tunnels. Therefore, it was thought that construction of the latter may accompany difficulties. After studying the stability of the young aging arch concrete and the concrete coating of other areas, arch concrete, advantageous in procedural and facility aspects, was employed.

The secondary concrete linlng was structured with ferroconcrete as indicated in Figure 11 with due consideration to the water flow and the inside and outside water pressure of the temporary tank in the tunnel. What's more, waterproof sheets were placed behind the linig.

5. Dynamic characteristic of grouting for soil improvement

Chemical grouting was performed on this tunnel to cut-off the water in the area designated for grouting on all sandy layers and areas to protect (and stabilize) the arch in the central portion. The dynamic characteristics of the improved soil becomes a major issue in both cases; in the former, the stability of the surrounding soil against ground and water pressure and in the latter, the face stability becomes the issue. For this reason, the sandy layer improved by chemical grouting was block sampled for dynamic testing (unconfined and trcaxial compression test) on the interior to grasp the compaction strength of this construction area. The results are as follows:

5-1 Physical Characteristics of Sandy Layer

Table 5 indicates the basic physical characteristics of the sandy layer improved by chemical grouting. N was over 50 and indicated a tight compacted sandy layer, however the uniformity coefficient was Uc=6.8, which indicates a sandy layer with "poor particle distribution" according to the Japanese Unified soil classification format.

5-2 Results of compaction test

The single axis compression compaction test was performed on 7 samples. As a result, it was found that the maximum compaction was 11.75 kgf/cm^2, with a minimum compaction of 9.65 kgf/cm^2, and a mean compaction of 10.63 kgf/cm^2(σ=0.818 kgf/cm^2). The results indicate that the compaction strength surpasses the improvement goal of 6~8 kgf/cm^2, with a small dispersion. What's more, the triple compression test gave us a Mohr's circle and a standard collapse line as indicated in Figure 12. The viscosity was fairly high at c=3.93 kgf/cm^2, and the inside friction angle was found to be ϕ=35.1 °

6. Conclusion

The Nishiya Lot No. 2 section of the Katabiragawa Diversion Channel Project was noted as a large section tunnel (111 m3 sectional bore area) in an aquifier alluvial loose sandy mountain in the urban area. However, the adverse effects to the peripheral mountains were successfully minimized by employing the Side Lot NATM, special method (chemical grouting, drainage boring) and appropriate analysis of the conditions to complete boring the tunnel. The project is currently in the secondary concreting stage that is carefully being undertaken and is favorably progressing without changes in the tunnel shape by removing the side lot wall, as initially feared.

This is an intermediate report of the tunnel boring works performed in this project, in which the following conclusions were drawn:

(1) It was found that Side Lot NATM is useful when used in complex alternating acquifier alluvial loose sandy layers both in terms of water drainage and cut-off (grouting). It was also noted that this method allowed for immediate measures against trouble such as face collapse and was valid in that it had minimal effect on the peripheral mountain area.

(2) Since Side Lot NATM entails multiple boring, the joints of the supporting works often becomes the weakspot and thus requires careful works when removing the inner wall. However, upon this project we experienced practically no changes in the tunnel upon removing the inner wall.

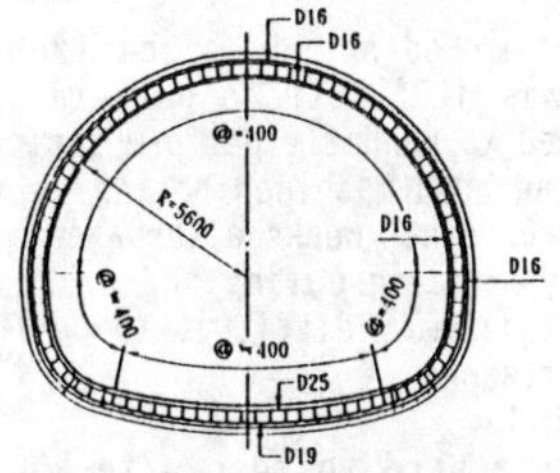

Figure 11. Section of concrete lining

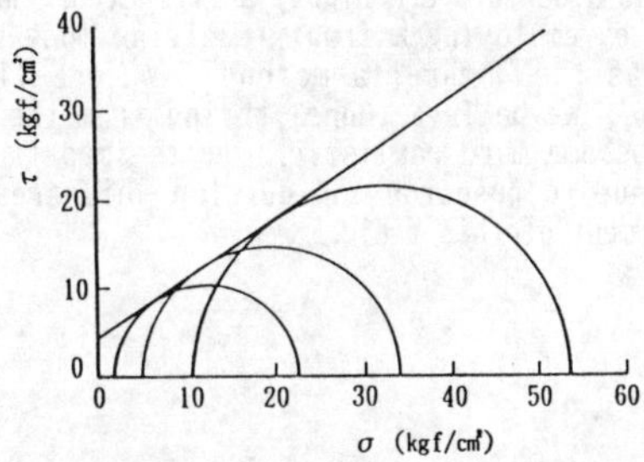

Fig. 12 Mohr's circle and standard collapse line

Table 5 Characteristics of sandy layer improved by chemical grouting

Weight	r t	2.08 tf/m3
Dry Unit Volumetric Weight	r d	1.71 tf/m3
Pore Ratio	e	0.58
Moisture Ratio	w	20.2%
Particle Composition	Sand	84.0%
	Silt	12.0%
	Clay	4.0%
Uniformity Coefficient	UC	6.8

(3) Due to limited space, mechanized Side Lot heading was difficult to perform. Thus, we were forced to manually perform shotcreting and four piling upon machined boring in this area. What's more, other works alternated with boring such as chemical grouting and drainage boring was also extremely difficult to perform due to the same reasons.

There is a need to mechanize and eliminate waste, when considering recent construction labor conditions and productability. Thus, developments for multi-function machines for small sections such as side lots are highly anticipated. What's more, by employing a front receiving long blade such as the "umbrella method" or "pre-lining method", we believe tunnel boring from the back will become more realistic. It is important to continue to research and develop this area for betterment of this field.

Tunnelling and Ground Conditions, Abdel Salam (ed.) © 1994 Balkema, Rotterdam, ISBN 90 5410 363 9

Construction of triple face tunnel by urban NATM (Yokohama City Subway No. 3: Azamino Station Works)

Iwao Iwamura, Souichirou Usuda & Jun Mitsuo
Tokyu Construction Co., Ltd, Tokyo, Japan

ABSTRACT: This report is on the case of Azamino station of Yokohama City Subway No. 3, where NATM method with triple face was applied under the condition of earth covering of 10 meters for 64.6 meters of the line.
This paper also refers to the design concept of NATM in the urban area in the future, in relation to the comparison of calculated values and measured values.

1 INTRODUCTION

Yokohama City Subway No.3,which was opened in March 1993, has been an important line connecting Yokohama City Centre with residential areas.

The works discussed here were carried out as part of the Azamino station terminus construction; but as surface works included the bus terminal, and the preparation of ticketing and commercial-use structures in the station area, the underground construction at Azamino station was completed using the New Austrian Tunnelling Method (NATM) to construct a Triple Face Tunnel for the final 64.6m.

Regarding the method of construction, initially cut and cover was planned but the tunnels would have had to be placed under a road junction, so almost all of the work would have involved regulation of the traffic and night time working, and local residents and existing station users would not have been able to avoid this traffic obstacle. Also, above the planned tunnel layout, there were 2.5m square box culvert sewers and other underground utilities which it would be necessary to move or protect. In addition, there was concern that the influence of a large excavation at a depth of about 20m could affect neighboring structures.

To resolve the problems above, an investigation of the possibility of applying NATM in this urban area was carried out. With regards to NATM as a construction method, a single tunnel construction would have had an especially large cross-sectional area and there may have been difficulties over the possibility of groundsoil damage, with earth cover limited to a few meters. Thus an investigation was carried out both for a single tunnel and for a triple face tunnel construction.

As Table 1 shows, the result of this comparative investigation was that the triple face NATM tunnel was adopted for this project, as it was concluded both that the effects on the neighboring area would be smallest and that cost wise it would be better.

2 GROUND CONDITIONS

As for the ground conditions, the Kazusa Formation Ouzenji layers were formed between the Tertiary Period, Pliocene Epoch, and the beginning of the Diluvial Age, and as Figure 1 shows are principally layers of mudstone and sand. The thickness of the layers varies from several centimeters to as much as 5 meters but the inclination of each layer are almost level and they distribute continuously. The mudstone (Tm) is homogeneous and shows an N value (Standard Penetration Test value) of above 50 except the shallow weathered layers.

The sand layers (Ts) also show an N value of above 50, and have a very dense, tight formation but are for the most part unbonded.

The water table is between 1.0m and 4.0m below ground

Table 1 Comparison of Construction Methods

	Cut and Cover	NATM	
		Large Single Cross-Section Tunnel (approx 120-170 m²)	Triple Tube Tunnel (approx 80-117 m²)
Traffic Obstacles	Many	Few	Few
Effect on Neighboring Buildings	Large	Large	Some
Construction Time	△	○	○
Costs	△	○	○
Cutting Face Stability	—	△	○
Project Execution	Many underground conduits to be moved and many restrictions on working	Working away from the shaft portal will be tight and restrictions on working many. Also with either face independence or confined water it is important to have supplementary construction methods.	
Examples of Construction	Many	None	Few
Overall Appraisal	△	○	◎
Notes		It is important to use automatic measuring equipment in order to make real-time appraisals and judgements on tunnel stability and effects on the neighboring area.	

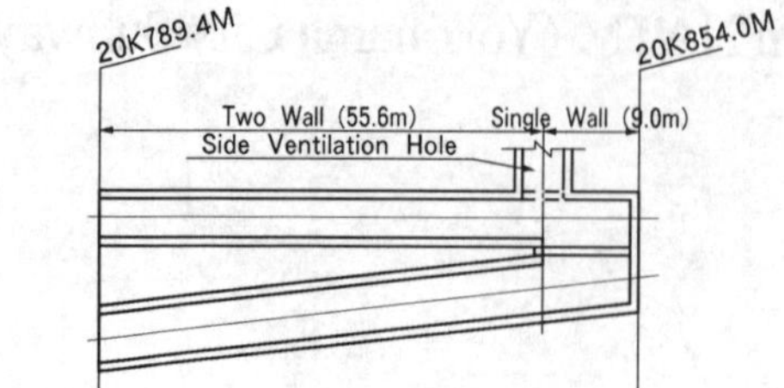

(1) Plan

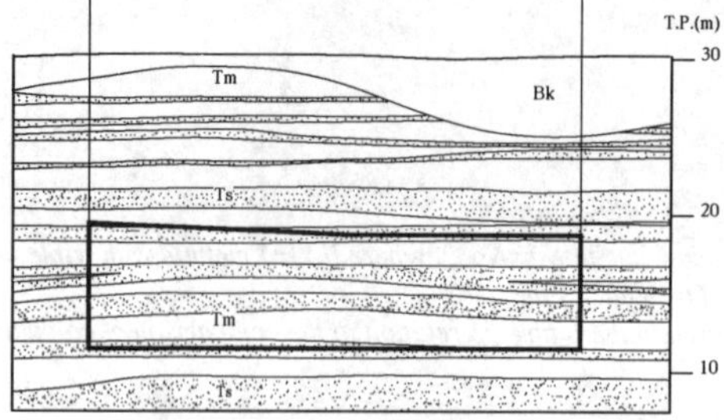

(2) Vertical Geological Section

Figure 1 Project Plan and Vertical Section showing Geology

level, and within the layers of sand the ground water pressure is usually 5 − 7 tf/m².

3 DESIGN

3.1 Tunnel Construction and Support Pattern

Figure 1 shows reduced cross-sections of the tunnel construction facing the terminus end; from the station part for 55.6m there are two props in the tunnel and for a 9.0m section there is a single wall.

As the excavated cross-section was 80.5 − 117.2 m², including the large cross-section of the triple faces, the standard support pattern were determined as follows, referring the results of the Elastoplastic Finite Element Method (FEM) analysis and framework calculations which are discussed later:

Shotcrete thickness	t=250mm
Steel support size	H-150(1m intervals)
concrete lining thickness (side tunnels)	t=500mm (all reinforced concrete)

Figure 2 shows two vertical cross-sections of the tunnels.

3.2 Supplementary construction methods

On this project, there were natural alternating layers of hardpan and unbonded saturated sand where NATM was to be employed, so it was decided to give extra consideration to supplementary methods of construction in the following cases:

① Dealing with underground water.
② Cutting Face stability.

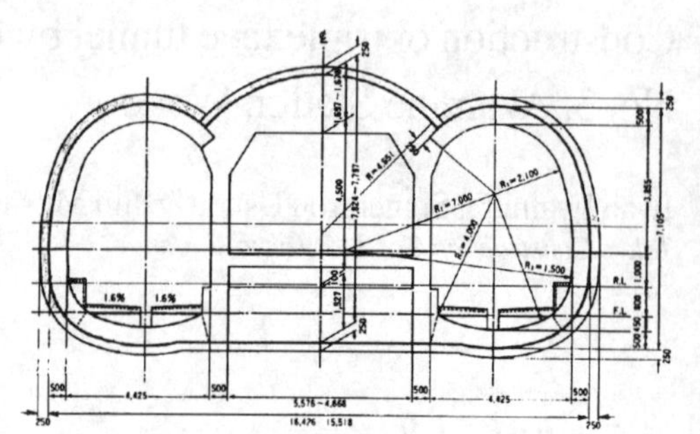

(1) Two Wall Section

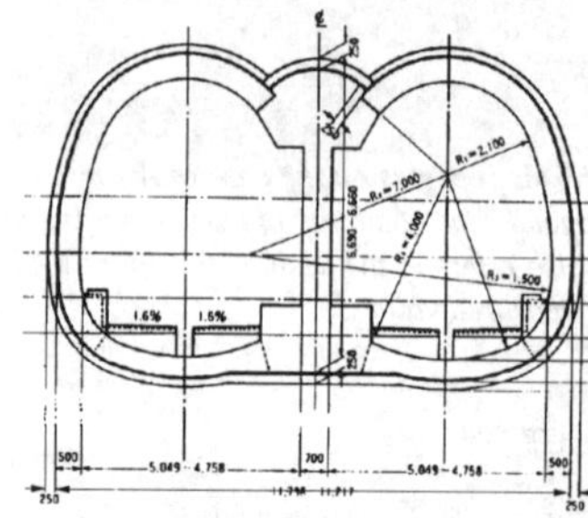

(2) Single Wall Section

Figure 2 Tunnel Cross-Sections

③ Neighborhood environmental measures for control of surface settlement.

Tunnel safety, cost and construction time have a large influence so it was important to hold these points in mind when choosing appropriate supplementary construction methods. For this work, with reference to previous experiences, it was decided to employ the following methods.

(1) For underground water

Where there is much underground water in a tunnel, there can be problems with quicksand occuring at the working face, shotcrete not fixing well, an increase in the level of surface settlement, and a low rate of work efficiency. Either deep wells or the well point method are used to eliminate this underground seepage, or the natural ground is grouted to increase the ground strength and stop the water flow. On this tunnel project it was thought that both methods should be applied.

First, it was judged necessary to stop the water flow, due to the existence of a 3 − 4m sand layer in the crown vicinity along the whole of the tunnel. To prevent the occurrence of quicksand and with the purpose of reducing surface settlement, chemical grouting to stop this water flow was carried out in this layer. Grouting of the whole sand base led to the practical application of the double packer sleeved pipe method. Figure 3 shows the extent of the grouting.

Next, drainage was deemed to be necessary in the sand layers situated beneath, rather than in, the sand layer just mentioned. Due to drainage in these layers, face freedom would increase, shotcrete quality would improve, and progress would be faster, so six deep wells were placed on the outside of the tunnel, as shown on Figure 4.

(2) Cutting Face stability measures

As regards general face stability, due to the grouting undertaken above it was thought that this would be enough protection, but cutting face stability measures for the initial excavation from the shaft are discussed below.

Working away from the main shaft, in the vicinity of the entrance, a loosened zone can occur depending on the shaft excavation, and the tunnel at this part can become very unstable. Therefore, to protect against this, a very effective arch piperoof construction was installed starting from the shaft portal, thereby increasing face stability and controlling settlement (Figure 5).

(3) Neighborhood environmental measures

Regarding the above mentioned measures, it was felt that they could keep the effects of surface settlement and effects on structures in the vicinity to a minimum, and so it was judged that other measures thought of before work started were not especially necessary. Depending on the conditions measured during the construction, additional measures could be applied if necessary.

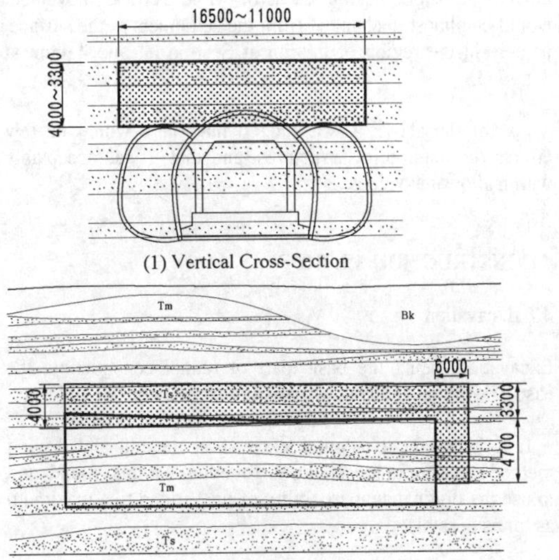

(1) Vertical Cross-Section

(2) Section Lengthwise

Figure 3 Extent of Grouting Works

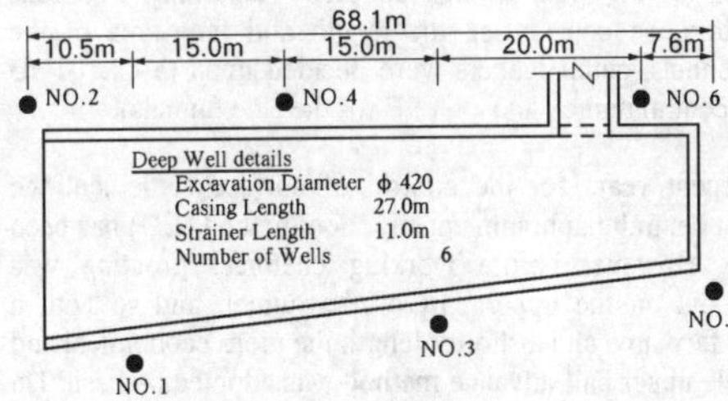

Figure 4 Plan of Deep Well Positions

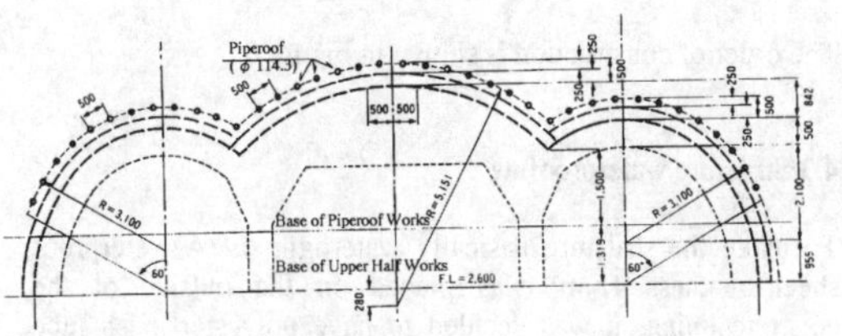

Figure 5 Piperoof Works

3.3 Examination of the effects on the neighborhood by use of FE analysis

Generally, when investigating tunnel design and construction method, standard design and construction methods are prepared according to ground conditions. However, it being essential to take special care over the design and choice of construction method under the special conditions of having carried out examinations of the shallow cover tunnel, an urban tunnel in the vicinity of structures or a varying shape tunnel.

Especially, for special shape tunnels such as Azamino Station, there were many cases where design and method of construction were decided with reference to numerical analysis such as FEM. Below, according to elastoplastic FE analysis considering the order of excavation, the results of investigations into the influence of tunnel stability, ground surface and underground structures in the area are discussed.

(1) Investigation of cross-section and analysis model

For the calculations, it was decided to take largest area cross-sections from both the two wall part and the single wall part (Section A and Section B respectively). Figure 6 shows collectively the analysis values for the Section A analysis model.

(2) Analysis results

Surface and underground settlement are shown on Figure 7.

According to the results of the analysis, ground settlement, on the surface directly above the tunnel, would be 8.7mm at Section A and 4.6mm at Section B. Depending on the surface settlement the angle of inclination of the ground, would be less than 2/10000 at Section A and less than 1/10000 at Section B.

The Japan Tunneling Association's "Investigation Research Report on NATM Monitoring and Measurement" (March 1983), gives permissible ground inclinations as:

Reinforced concrete structures	2/1000
Concrete block structures	1/1000
Wooden structures	5/1000

Therefore depending on the excavation of the tunnel it was judged that the effects on neighboring buildings would be almost nil.

As for the displacement of underground conduits, horizontal

movement would be almost zero, while vertical movement would be almost unchanged from the settlement at the surface, namely in the region of 8.7mm at Section A, and 4.6mm at Section B.

As for the above it was judged that there would be few effects on neighboring structures, and that it was acceptably within allowable values.

4 CONSTRUCTION METHOD

4.1 Excavation

Excavation, being the basic part of tunnel construction, the ease of working and cost will affect a project considerably.

It is important to adopt a excavation method and driving method so as not to loosen the surrounding natural ground, and to use the original support ability of the ground to a maximum as far as possible.

Thus for driving method, so as not to disturb the surrounding natural ground and to enable smooth operation, machine driving method was adopted. Regarding free face excavators, on considering the length and face area of the triple tunnels, cutter loaders were decided upon (a CL101UB for the central tunnel, and a CL9E for the side tunnels).

In recent years for the control of surface settlement the efficient central diaphragm construction method (CD) has been applied. However before working chemical grouting was carried out on the upper part of the tunnel, and so with a smaller face area and a shorter length the more economical and workable upper half advance method was adopted. A basic 1m length excavation cycle was decided upon.

4.2 Order of construction

The order of construction is shown in Figure 8.

4.3 Structure waterproofing

To make the structure basically watertight, EVA waterproof sheet(thickness 1mm) was placed on the outside of the concrete lining. It was decided to have polyester mesh tube drainage in the side tunnels in the longitudinal direction, with a line of four vertical pipes on each side, total eight. The horizontal pipework resumes at each construction joint on the concrete lining; these water collection pipes leading to the lowest point where there is an internal water pump leading up the shaft.

For the waterproofing works the greatest problems were the areas where the central tunnel connects with the side tunnels, and between the shaft and concrete tunnel lining.

First, as measures against leakage where the central tunnel connects with the side tunnels, in the long term the water drains via vertical drainage pipes in the central wall to the lowest part, and from the side tunnels towards the shaft (Figure 9(1)).

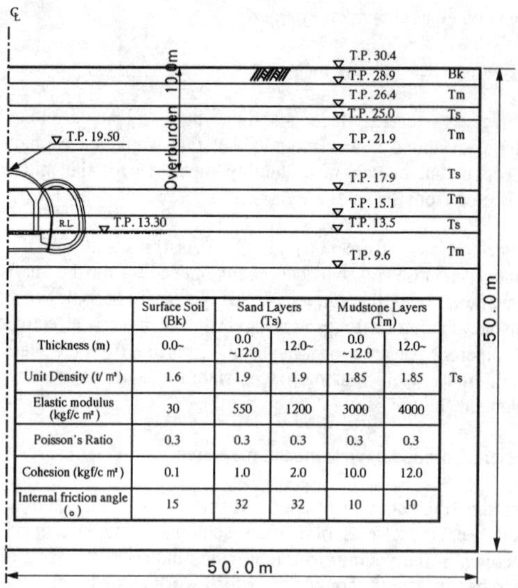

	Surface Soil (Bk)	Sand Layers (Ts)		Mudstone Layers (Tm)	
Thickness (m)	0.0~	0.0 ~12.0	12.0~	0.0 ~12.0	12.0~
Unit Density (t/m³)	1.6	1.9	1.9	1.85	1.85
Elastic modulus (kgf/cm²)	30	550	1200	3000	4000
Poisson's Ratio	0.3	0.3	0.3	0.3	0.3
Cohesion (kgf/cm²)	0.1	1.0	2.0	10.0	12.0
Internal friction angle (°)	15	32	32	10	10

Figure 6 Analysis Model (Cross-Section A, 20k789.4m)

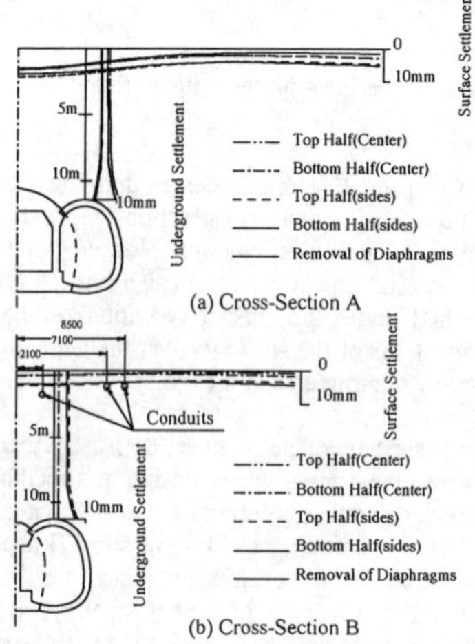

Figure 7 FEM Analysis Results (Surface and Underground Settlement)

On the other hand, between the body of the shaft and the tunnel concrete lining, a 70x70mm U-type rubber gutter was used, and on the rest of the joint a sealing material (thickness 20mm) was employed (Figure 9(2)).

5 MONITORING and MEASURING

5.1 Measurements

On this project, various measurements were carried out for the following purposes:

① Control of the behaviour of the ground area around the tunnels.

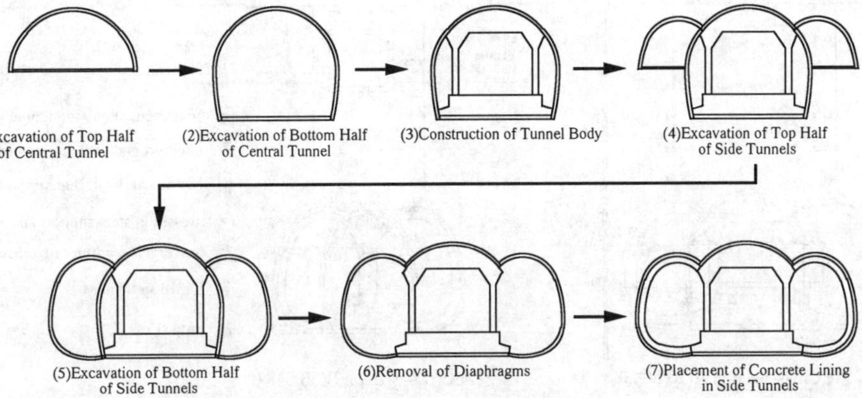

Figure 8 Order of Construction

② Confirmation of the safety of both surface and underground structures.
③ Appraisal of the safety of the tunnel itself (verification of design and construction).

With regards to the cross-sectional measurements, the first measurement cross-section (cross-section a) was established 10m from the entrance to the shaft, while the second (cross-section b) was 50m from the shaft entrance in the single wall section. For the excavated cross-section convergence and the crown settlement, measuring instruments were placed at 10m intervals respectively, for measurement of the longitudinal distribution of the settlement along the tunnel crown, they were placed at 5m intervals at points 1m above the tunnel crown. The instrumentation arrangement is shown in Figure 10.

Moreover, to feed the measurements back into the construction work, a real-time automatic measuring system was employed which could register any changes immediately. The monitored data acquisition flowchart is shown in Figure 11.

5.2 Control requirements

So that measurements were fed back into construction work as soon as possible, it was essential to carry out control based on the appropriate control criteria for permitted values of tunnel stability and behaviour of neighboring structures, and to have appropriate countermeasures decided beforehand.

Techniques for the control criteria preparation were not established and there were many cases where these values were determined with reference to previous results in similar cases, FEM analyses, and framework calculations.

The control criteria levels on this project are shown on Figure 12. The three stages undertaken to establish the control standard levels are given below:

Control Level 1 Results of FEM analysis
Control Level 2 80% of Control Level 3
Control Level 3 Either permitted values from public agencies

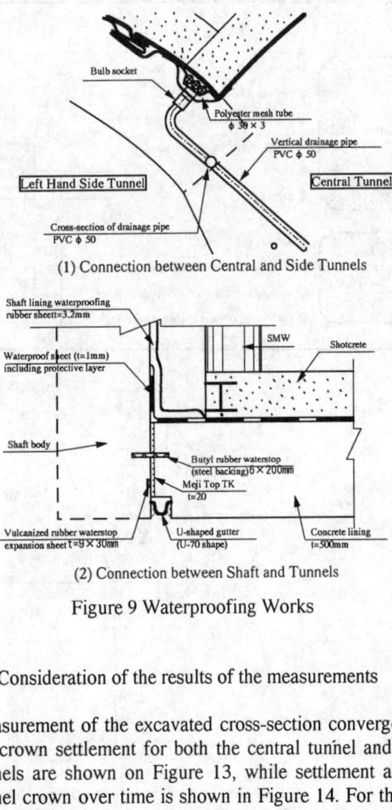

(1) Connection between Central and Side Tunnels

(2) Connection between Shaft and Tunnels

Figure 9 Waterproofing Works

5.3 Consideration of the results of the measurements

Measurement of the excavated cross-section convergence and the crown settlement for both the central tunnel and the side tunnels are shown on Figure 13, while settlement above the tunnel crown over time is shown in Figure 14. For the central tunnel excavated cross-section displacement, the lateral earth pressure at the side walls exceeds the vertical earth pressure. Accordingly, during the central tunnel excavation there was almost no vertical displacement. However, during the excavation of the side tunnels, in reducing the lateral pressure, the cross-sections at a and b showed 7.0mm vertical displacement. There was almost no surface settlement during all stages of construction.

Figure 15 shows the steel support stress.

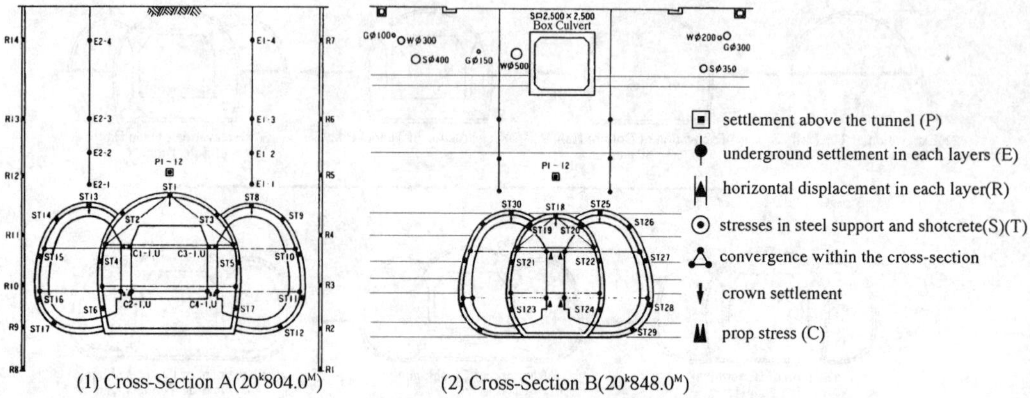

Figure 10 Cross-Sections Showing Measurements

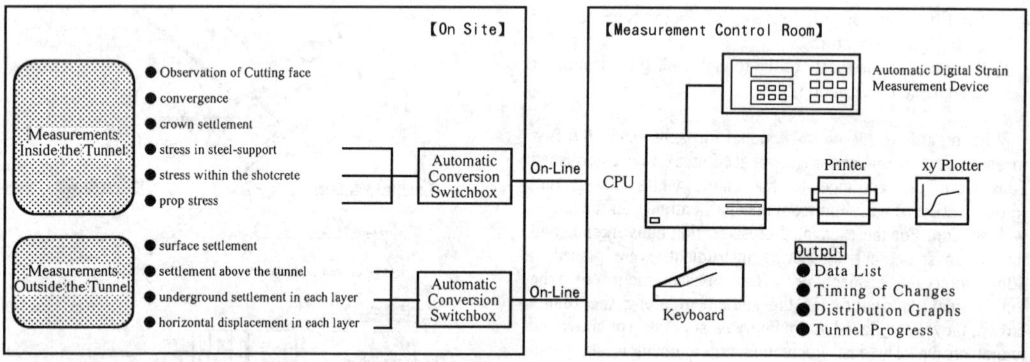

Figure 11 Measurement System

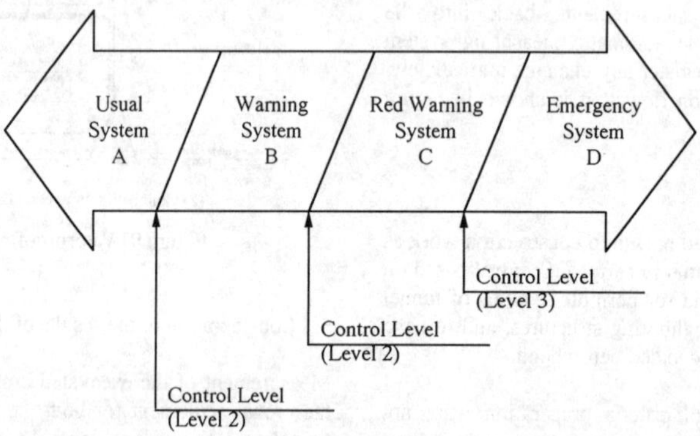

A: Usual System · · Fixed time measurement,standard observation of cutting face

B: Warning System · · Increased frequency of measurement, check of the field, increased caution by employees

C: Red Warning System · · Strengthening of system, estimate of final displacement , execution of countermeasures

D: Emergency System · · Stopping all face excavation, analysis of cause of displacement and any further inclination ,re-examine supporing pattern and countermeasures

Figure 12 Appraisal of Measured Data and Related Safety Management

For the steel support stress, at the time of the excavation of the central tunnel, as previously mentioned for the corresponding excavated cross-section displacement, there was a large compressive stress at the tunnel crown, but it became very small on the side walls. Also, at the time of the excavation of the side tunnels the stress in the steelwork supports in the side tunnels was a maximum of 650kgf/c m², less than the allowable stress.

Directly above the whole length of the tunnel there was no surface settlement, therefore there were minimal effects on the underground conduits, and on neighboring structures; this can be thought to be the result of the chemical grouting directly above the tunnel.

5.4 Comparison of calculated and measured values

In a comparison of the preliminary analysis and measured values, the actual lateral stress exceeded the lateral stress calculated from the FEM analysis; therefore both the excavated cross-section displacement and the steel support stress tended to the opposite.

The causes behind these developments are thought to be due to the following reasons.

① Due to the ground improvements directly above the tunnel the vertical ground stress is thought to have been reduced, but in the analysis improvements were not considered.

② The problem in appraising Poisson's ratio of the ground or the coefficient of lateral pressure due to the overconsolidation in the hardpan.

To estimate the magnitude of correlation between the coefficient of lateral pressure and the analysis results, FEM was carried out on only the central tunnel using decided upon parameters for the coefficient of lateral pressure. These results are shown in Figure 16.

On the preliminary analysis the coefficient of lateral pressure, k, was taken as 0.43 (Poisson's ratio ν =0.3) but, on Figure 16(1), when taking k=1.2, the stress at the crown was consistent with the measured results. As the ground improvements were not considered in the analysis, the calculated lateral stress were extremely larger compared with the measured results. For the excavated cross-section convergence and the crown settlement, when taking k=2.0, the results were consistent with the measured results. Therefore in this kind of sand/mudstone layered ground, for the preliminary analysis it is appropriate to use values of k of between 1.2 and 2.0. In order to evaluate the coefficient of lateral pressure precisely, the variation in the deformation modulus and the rate of excavation (set at 33% in the preliminary analysis) must be considered in future projects.

6 CONCLUSION

An example of the application of a triple face NATM tunnel in an urban area, is explained in this report, which focuses on the construction method and results of measurements.

In order to work on the shallow overburden and under the condition in alternation sand and mudstone including confined ground water in layers, this project was designed to place importance on the control of surface settlement and the cutting

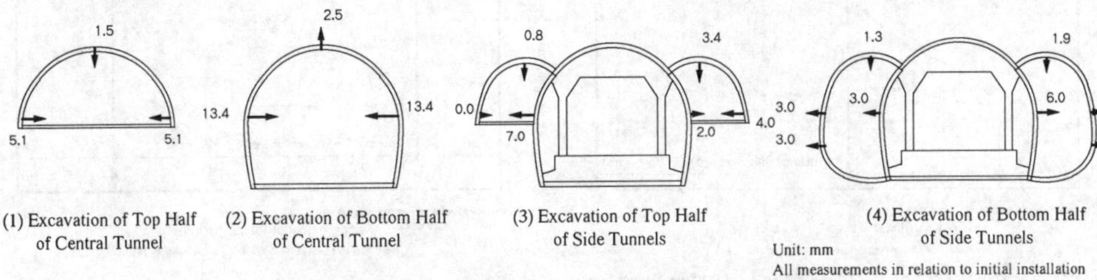

Figure 13 Measured Crown Settlement and Displacement

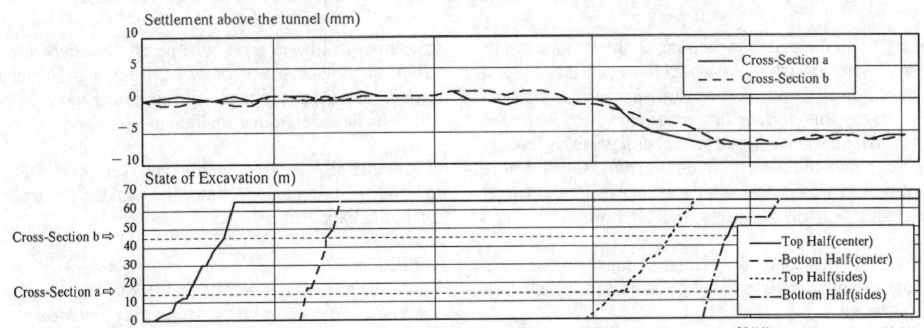

Figure 14 Displacement directly above the Tunnel in Relation to Time

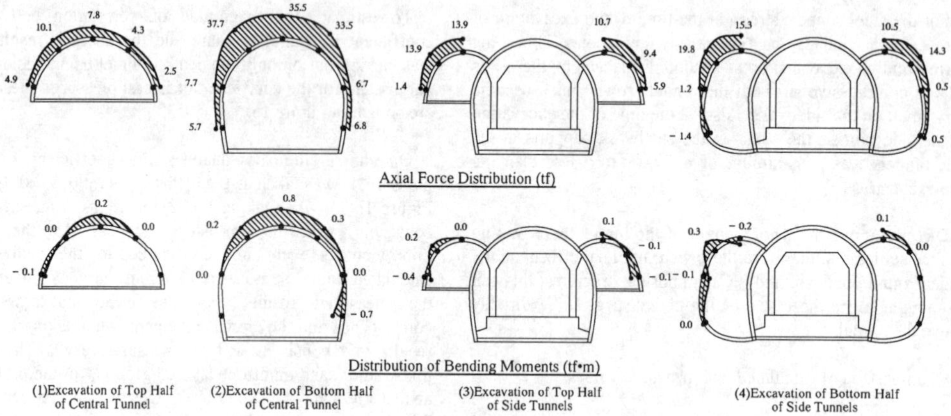

Table Showing Total Stress

Excavation Stage \ Area	Central Tunnel		Side Tunnels	
	Crown	Side	Crown	Side
Top Half(Center)	286	168	–	–
Bottom Half(Center)	1250	192	–	–
Top Half(Sides)	–	–	438	218
Bottom Half(Sides)	–	–	630	76

Unit: kgf/c m²

Figure 15 Measured Stresses in Steel Support

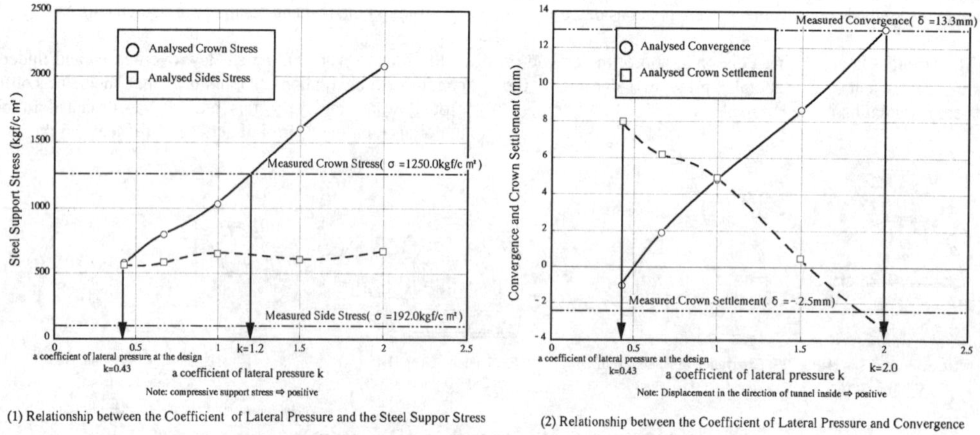

Figure 16 Results of FEM Analysis with Specified Parameters for the Coefficient of Lateral Pressure

face stability from the outset. According to the various results from the supplementary construction methods and measurements carried out, the construction was completed without damaging the surrounding area. However, while the origin of the overconsolidation in the upper layers was thought to be the magnitude of the lateral pressure, the estimate before the tunnel project was in fact wrong as shown by the lateral stress and the large stress on the top of the steelwork.

Therefore, for this kind of overconsolidated ground it is important to grasp the initial ground pressure and include it in the rational design.

In the future in Japan there is expected to be a further increase in tunnelling projects in urban areas, and accompanying this, the projects will get closer to neighboring structures, will be in areas where soil conditions are poor or will be large in cross-section and triple bored. The tendency is towards important projects with more and more complicated and difficult underground conditions .

A tunnel project, such as this one, can offer valuable data and other information with the kinds of underground conditions experienced.

REFERENCE

The Japan Tunneling Association 1983.3. *Investigation Research Report on NATM Monitoring and Measurement (in Japanese)*、p88

Tunnelling and Ground Conditions, Abdel Salam (ed.) © 1994 Balkema, Rotterdam, ISBN 90 5410 363 9

Effect of slurry shield tunneling in soft alluvial clay on an adjacent underground subway structure

Shozo Katano & Tamotsu Ogawa
The Tokyo Electric Power Co., Inc., Japan

ABSTRACT: There is an increasing amount of tunneling work being carried out under difficult conditions—such as in soft ground or at great depth—in metropolitan areas. In this case, the project was to build a tunnel to carry a power transmission cable with an outside diameter of 5.5 m through soft cohesive soil. The slurry-pressure shield method was adopted. The design and construction were successfully accomplished without undue influence on the operation of an existing subway operating nearby in a tunnel of outside diameter 6.9 m. Underground deformation caused by the shield tunneling work was measured during the advance at a point short of the cable tunnel's closest approach to the subway. Based on this measurement, the effects of the work on the subway tunnel at the closest approach were predicted analytically, allowing a safe distance to be maintained from the subway tunnel. Tunneling of the section of the cable tunnel closest to the subway was carried out while continually measuring the settlement of the subway tunnel. Using this technique, the work was completed safely with a measured settlement of 1.8 mm; the prediction had been 6.9 mm. Through this study, we have developed a powerful tool for predicting the effects of tunneling work in soft cohesive soil on adjacent structures.

1 PROJECT OVERVIEW

Table 1 outlines the project and Figure 1 shows a cross-sectional view of the completed cable tunnel at its closest approach to the subway tunnel.

Table 1. Outline of project.

Project name	: New transmission line installation project under Harumi Avenue
Tunnel dimensions	: Length : 1,342 m Inner diameter : ø4,700 mm, 4,950 mm Depth : 12.0 m - 45.0 m
Primary lining	: RC Segment ø5,550 mm Width : 1,200 mm Thickness : 300 mm, 425 mm Steel segment (R = 35 portion) ø5,550 mm Width : 300 mm Thickness : 374 mm
Plane alignment	: Minimum radius curve of R = 35 m
Tunnel inclination	: -4.2 to 4.9%
Ventilation shaft	: 4 places
Owner	: The Tokyo Electric Power Co., Inc.
Contractor	: Taisei, Mitsui, Toda, Dainihon Doboku, Shiraishi, Wakachiku Joint Venture Enterprise

2. CONSTRUCTION PLAN

(1) Topographical and geological profiles

This cable tunnel was to run below reclaimed land at the southern edge of Tokyo. Along its path there were deposits of thick alluvial soil down to a depth of about 50 m. The aim was to drive the cable tunnel through a soft cohesive soil layer with an N-value

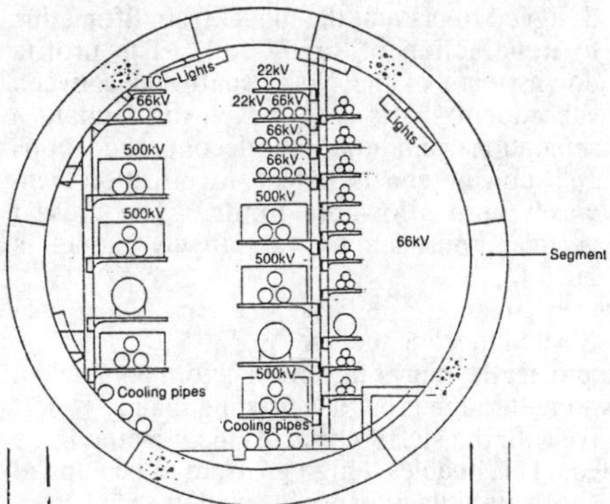

Fig. 1 Standard section of shield tunnel (segment thickness = 425 mm)

Table 2. Physical and mechanical properties

Soils		Strata classifica tion	Symbol	soils	Layer thick-ness (m)	N-value	Unit weight	Cohe-sion c (kgf.cm^2)	Internal friction angle ø (deg)	Modulus of de-forma-tion E (tf/m^2)
Quater-nary	Allu-vium	Filler earth	B	Sandy soil	1-6	0-15	-	-	-	-
		Alluvial layer	A_{S1}	Sandy soil	3-6	0-4	1.45-1.65	0.2-0.3	1-5	770
			A_{C1}	Cohesive soil	3-7	2-15	1.83-2.04	0.1-0.4	32-38	640
			A_{S2}	Sandy soil	23-35	0-4	1.55-1.77	0.2-0.8	1-8	2100
			A_{C2}	Cohesive soil	1-4	4-40	1.90-1.99	0.2-0.9	32-35	2700
			A_{S3}	Sandy soil	6-28	8-15	1.62-1.89	0.5-1.6	-	2400
			A_{C3}	Cohesive soil	-	40-5	-	-	-	-
	Dilu-vium	Diluvial layer	D_S	Sandy soil	-	50 or more	-	-	-	-

ranging from 0 to 4. Table 2 lists the physical and mechanical properties of this soil.

(2) General layout and longitudinal profile

a. Ground plan

As shown in Figure 2, the cable tunnel was to run under Harumi Avenue. After crossing the Shinonome Canal, it was to detour to the west around the 65 m-deep piles of the Shinonome Bridge foundations. The subway and cable tunnels cross over each other at different depths.

b. Longitudinal profile

Figure 3 shows the longitudinal profile of the cable tunnel. Between the No. 1 and No. 2 temporary vertical shafts, the profile was designed to prevent the tunnel from lifting due to liquefaction of sandy soil. The profile downstream of the No. 2 shaft and between subsequent shafts was chosen to maintain a separation from adjacent telecommunications and subway tunnels so that any effect on them was within allowable limits. The closest vertical approach to the subway tunnel is 20.5 m.

(3) Shield tunneling method

To protect the tunnel face from deformation while excavating through ground consisting mainly of soft cohesive soil, the slurry pressure shield method was adopted. This enables a high pressure to be applied to the face with little pressure fluctuation.

(4) Shield tunnel segments

Concrete segments of 1.2 m in width were used.

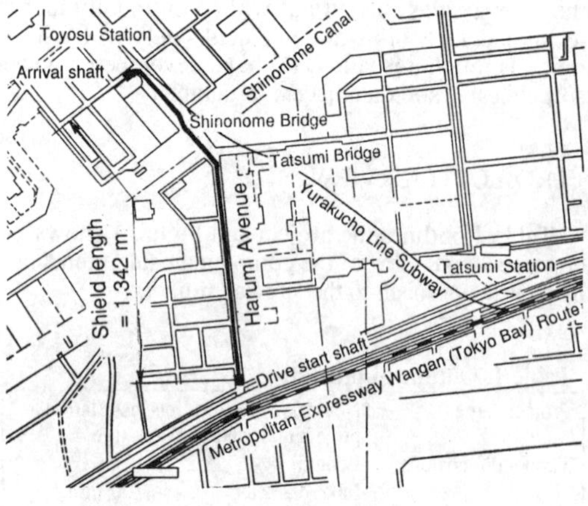

Fig. 2 General layout of the tunnel

For the curved section of tunnel with a radius of 35 m, 0.3 m-wide steel segments were used. Wedge-type key segments were adopted to improve rigidity as well as to ensure water tightness, allowing segments to be inserted longitudinally from within the shield machine.

(5) Backfiller injection

Since the tunnel was driven through ground consisting mainly of soft cohesive soil, backfiller was injected from within the shield machine as soon as tail voids were formed. A plastic backfiller was adopted to ensure flowability and to allow efficient filling of tail voids.

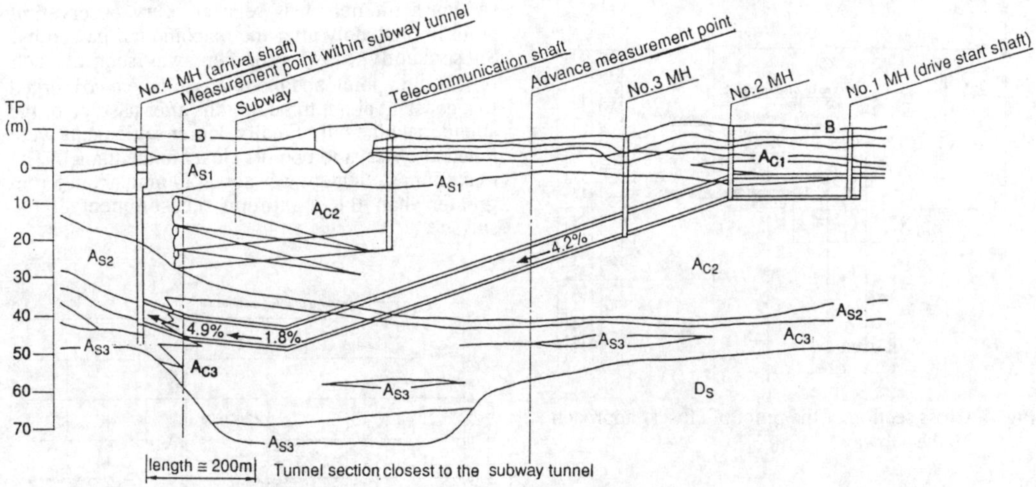

Fig. 3 Longitudinal profile of the tunnel

3 SHIELD TUNNEL CONTROL METHODS

(1) Tunnel control plan

To keep the tunnel face stable, and to minimize ground deformation during tunneling through the soft cohesive soil, it was necessary to maintain control over the following two items in particular:

a. Slurry pressure

To minimize ground deformation when driving through soft ground adjacent to the telecommunications and subway tunnels, the slurry pressure was controlled using the criterion below.

Slurry pressure = underground water pressure + 0.2 kg f/cm^2

b. Quantity of backfiller

The quantity of backfiller was carefully controlled. In addition, to minimize the effects of high backfilling pressure on surrounding ground, the injection pressure was controlled using pressure sensors fitted around the backfiller discharge nozzles within the shield machine. Injection was suspended when the injection pressure rose to the slurry pressure plus 0.2 kg/cm^2. The injection ratio was set at 160% after referring to construction records of other shield tunnels built near the project site.

(2) Control value verification during initial tunneling work

The effectiveness of the predetermined control values (slurry pressure and injection ratio) was checked by measuring surface settlement during initial tunneling work. Neither settlement nor uplift was measured as the shield machine approached, so the predetermined slurry pressure was adopted. After the shield machine had passed underneath, however, a surface settlement of 5 to 6 mm was measured. Although tests were also conducted at injection ratios of 180 and 200%, no significant suppression of this settlement was observed. Consequently, for reasons of economy, the planned injection ratio of 160% was retained.

4 PREDICTING EFFECTS ON SUBWAY TUNNEL

The effects of the tunneling work on significant adjacent structures, and the subway tunnel in particular, were predicted using the methods outlined below.

i) Measurement of ground deformation caused by tunneling work before approaching the section of closest approach to the subway tunnel (that is, advance measurements).
ii) Definition of an excavation stress release ratio satisfying the measurements using the Finite Element Method (FEM) which takes into account the ground consolidation proposed by Mori and Akagi.[1)] Here, the excavation stress release ratio is expressed as the ratio of release stress acting on the tunnel face to the initial ground stress before excavation.
iii) Implementation of an FEM ground deformation analysis at the point closest to the subway tunnel based on the excavation release ratio and the consolidation shrinkage coefficient.

Figure 4 is a cross section illustrating the location of the cable tunnel and other tunnels.

(1) Advance measurements

The advance measurements were made at the point shown in Figure 3. Figure 5 shows the cross section at this advance measurement point, indicating the layout of measuring instruments for vertical and horizontal displacements, and for pore water pressure.

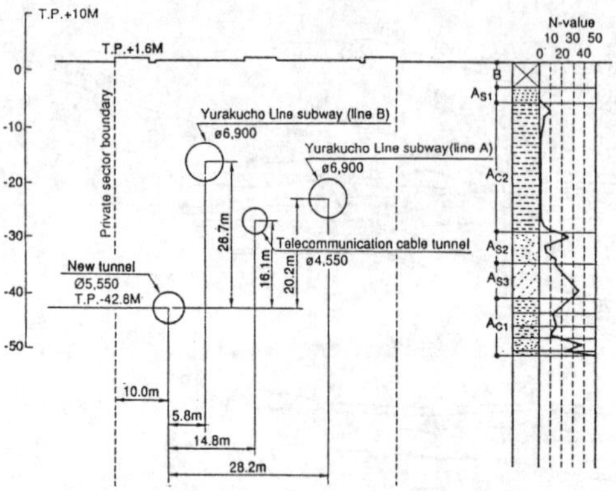

Fig. 4 Cross section of the point of closest approach to the subway

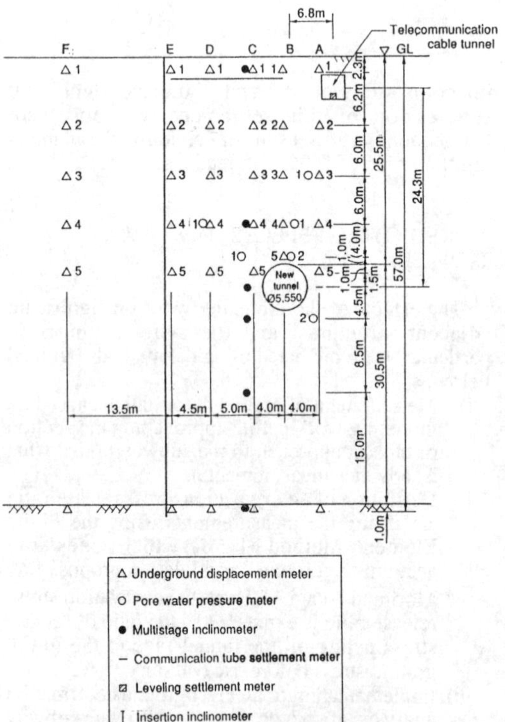

Fig. 5 Layout of measuring instruments on measurement cross section

(2) Results of advance measurements

Figure 6 shows the change in underground vertical displacement along traverse A with time. As is clear from the figure, an uplift was observed at the same time when the shield machine passed the underground displacement measuring point A5 closest to the cable tunnel. This was probably due to the effects of backfiller injection. In addition, a sudden settlement was seen at every observation point immediately after the machine tail passed by. Subsequently, slow settlement was seen at each observation point, and the total settlement converged to a constant about three months after passage of the shield machine tail. Finally, larger settlements were noted at observation points closer to the tunnel. The maximum settlement was about 20 mm, about 5 mm greater than the maximum measurement at the surface.

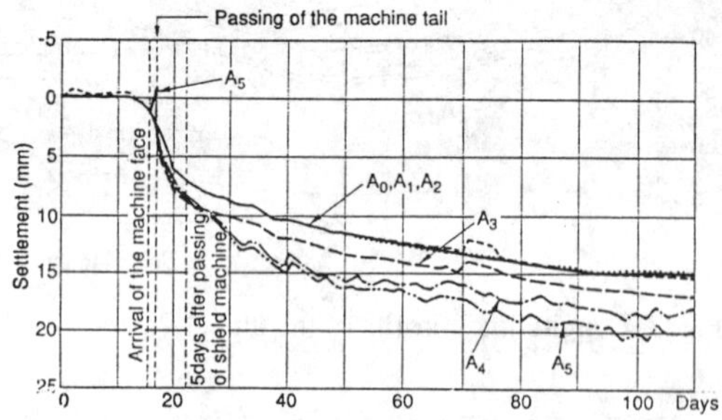

Fig.6 Results of measurement changes in underground displacement (alignment A)

Figure 7 shows the changes in both vertical and horizontal underground displacement before and after passage of the shield machine. Although no significant vertical displacement was observed as the machine passed, as shown in Figure 7 (a), large vertical displacements along traverse lines A and B close to the cable tunnel were observed later, as shown in Figure 7 (b). Three months after passage of the machine, when settlement can be considered to have nearly converged, a large horizontal settlement was observed, as shown in Figure 7 (c). However, this influence did not extend horizontally as far as the traverse line F, or to a distance approximately equal to the soil cover over the cable tunnel. As is clear from the measured lateral displacements shown in these figures, ground adjacent to the cable tunnel was forced outward as the machine arrived and then pulled back towards the tunnel after the shield tail passed by.

Figure 8 shows the change in pore water pressure near the cable tunnel with time. The pressure at the pore water pressure measuring point B2 directly above the tunnel rose nearly as high as the backfiller pressure when the machine was passing, dropped immediately after the machine passed, and then rose again. The pore water pressure reached a maximum again about five days after passage of the machine; it returned to its previous value about three months after the machine passed. These advance measurements confirm that immediate settlement occurs during the period up to five days after the machine passes while excessive water pressure is generated, and that consolidation settlement due to dispersion of excessive pore water pressure begins at five days.

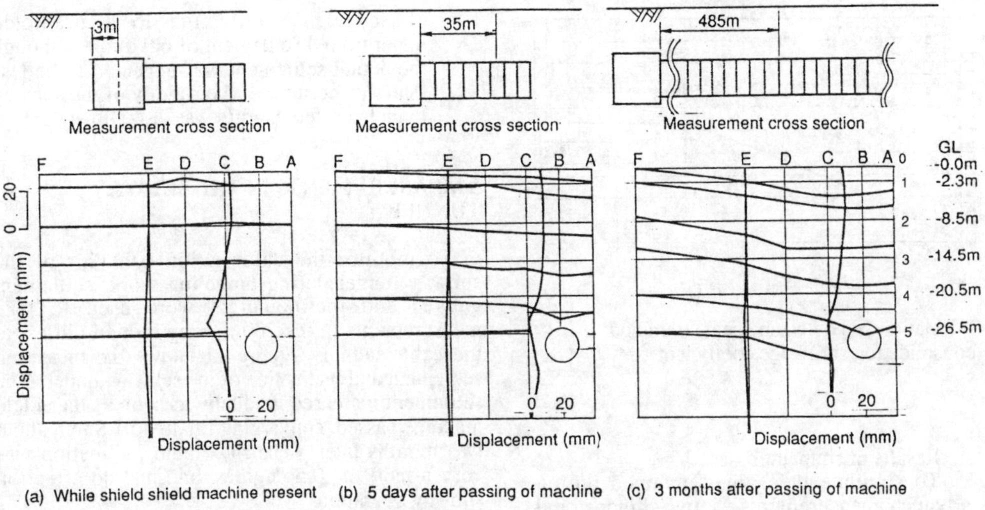

Fig. 7 Results of measurement of ground deformation

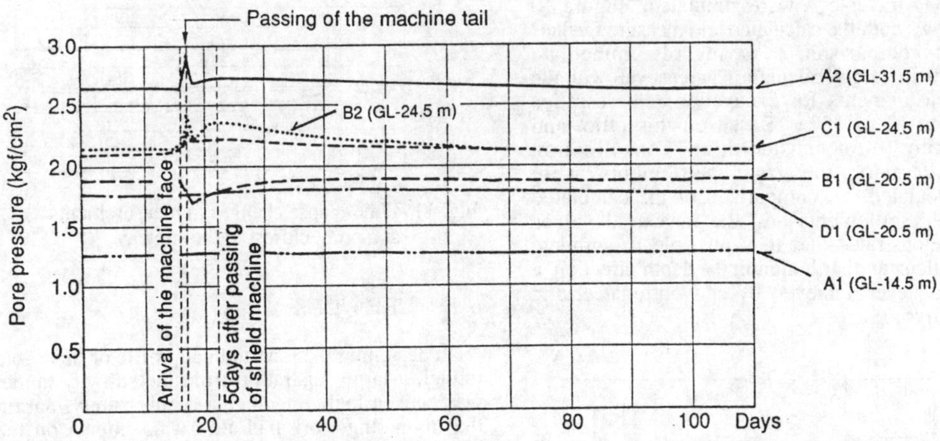

Fig.8 Results of measurement changes in pore pressure

(3) Simulation of effects

a. Analytical method

The analysis was undertaken using an FEM technique designed to take account of the consolidation settlement mentioned above. The actual analytical procedure is described below.

i) The stress release load is applied radially on the tunnel wall in a two-dimensional elastic analysis, yielding the immediate displacements resulting from ground settlement into the tail voids and the resulting shear strains.
ii) The calculated shear strains are converted into consolidation displacements using the relationship between shear strain and consolidation shrinkage coefficient.
iii) The final displacements are obtained by adding the immediate displacements to the consolidation displacements.

b. Assumed physical properties

The physical properties input into the calculation of immediate displacements were taken from the physical and mechanical characteristics listed in Table 2. The relationship between shear strain and consolidation shrinkage coefficient, needed for calculation of the consolidation displacements, was obtained through triaxial tests in a laboratory using undisturbed samples taken from the project site. Figure 9 gives an outline of these tests and the results.

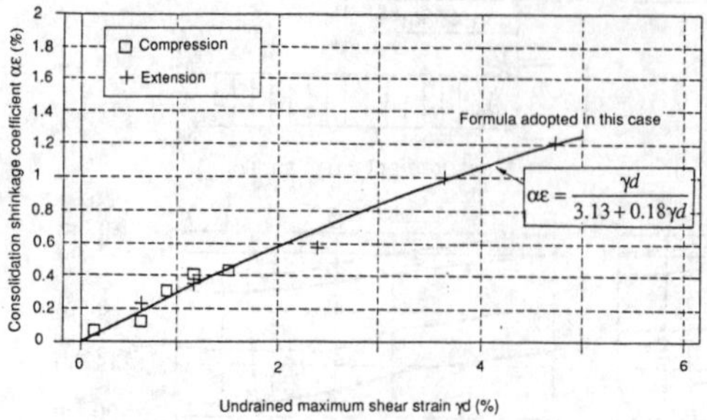

Fig. 9 Relationship between shear strain and consolidation shrinkage coefficient

c. Results of simulation

To simulate the values obtained during advance measurements, a two-dimensional elastic analysis was carried out with excavation stress release ratio as a parameter. Measurements of underground displacement along traverse A were simulated. Figure 10 shows both the calculated and measured values for comparison. As regards immediate settlement, the calculation agrees well with the measurements for excavation release ratios from 12 to 15%. Based on this ratio, and taking into consideration consolidation settlement, the final settlements were calculated. A comparison of the calculated final settlements and the measured values demonstrates that it is possible to simulate settlement distribution in the depth direction, a feat never achieved by conventional elastic analysis.

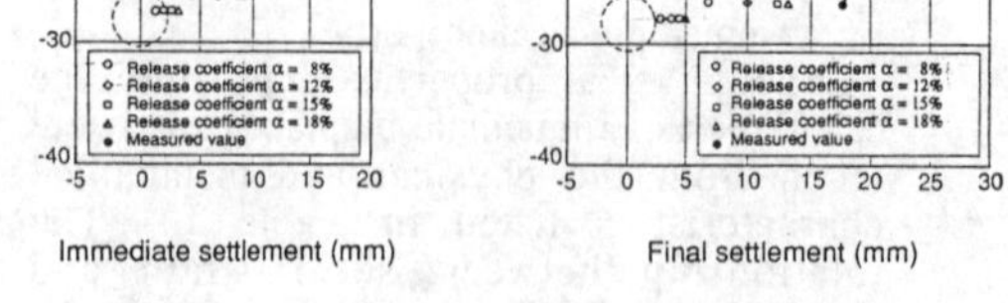

Fig. 10 Results of numerical analyses (alignment A)

Based on the results given above, the design excavation stress release ratio was set at 15%. An analysis of the influence of tunneling on the ground at the closest approach to the subway tunnel based on this design excavation stress release ratio led to a predicted underground settlement of 6.9 mm. Although the actual settlement was expected to be less than this because of the rigidity of the subway tunnel, a safety margin was also allowed.

5 MEASUREMENTS IN THE SUBWAY TUNNEL

To monitor the deformation behavior of the subway tunnel during tunneling work, settlement gauges and inclinometers were used to take measurements in the 200 m-long section closest to the cable tunnel. Figure 11 shows the measured settlement and inclination of the subway tunnel. The settlement increased gradually as soon as the shield machine passed, converging finally to 1.8 mm about two months later. The maximum inclination was very minor, or -0.5 degrees, and had no effect on subway operations.

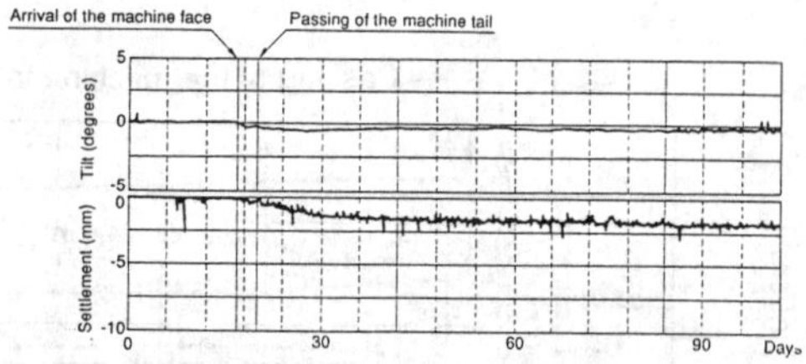

Fig. 11 Time-elapse changes in measurements at a point closest to the subway

6 FUTURE WORK

In designing the longitudinal profile of this cable tunnel, a large separation from the subway tunnel was chosen including such a secure safety margin that tunneling work had almost no impact on the subway tunnel. That is why the tunnel is located so deep in the ground. For similar projects in the future, therefore, greater economy would be assured by choosing more appropriate separations.

REFERENCE

Akira Mori and Kanichi Akagi, Consolidation of Normally Consolidated Clay Based on Disturbance Caused by Undrained Shear Deformation, Proceedings of The Japan Society of Civil Engineers, No. 335, July 1983 (in Japanese)

Tunnelling and Ground Conditions, Abdel Salam (ed.) © 1994 Balkema, Rotterdam, ISBN 90 5410 363 9

Observational construction of a large sectional minimum interval twin road tunnel in urban alluvial loose sand soil

Masakuni Kobayashi
Tokyo Metropolitan Government, Japan

Takashi Kawata, Yoshinori Suganuma, Hidekuni Takasaki & Masayuki Ohtsuka
Shimizu Corporation, Japan

ABSTRACT: Most of Japanese cities are located on soft grounds with high water level. For this reason, construction of large sectional tunnels in urban areas accompany many difficulties, in addition to their topographical and geological conditions. Limitations on the planned site, environmental preservation, prevention of surface settlement, and scenic harmony are also major issues in safely constructing large sectional tunnels in soft grounds.

OYAMA-DAIRI Tunnel is a typical twin urban road tunnel constructed in alluvial loose sandy soil. The CD method was applied to construct the twin tunnel, without any supplementary methods by effectively utilizing the measurement data. This report indicates the progress of deformation and stress in the ground mass, by means of careful measurements and analysis, especially on the mutual effects upon excavation of both of the twin tunnel at the Machida construction site.

1 OUTLINE OF THE SCHEME

The OYAMA-DAIRI tunnel, as shown in Fig. 1, is a large sectional twin tunnel with a 389 m extent, that is located at the west tip of Tokyo Tama New Town and connects Hachioji City and Machida City. The cut and cover method was initially planned for this site, however, was changed to a tunnel in stead, to preserve the historical relics and greenery within the planned area. Both twin and spectacle-shaped tunnels were considered upon the planning stage for the double lane 2-way road tunnel with a sidewalk. However, the twin type tunnel construction was chosen in terms of dynamic stability, economic feasibility and workability.

Fig. 1 Location of OYAMA-DAIRI Tunnel

Generally, the clearance required for two tunnels without affecting each other is expressed as 2-d (d=excavation width of tunnel), and when the excavation width is less than 1-d, it can not be treated as a single tunnel since the ground between both tunnels becomes unstable by excavation. (Fig. 2)

Since two tunnels had to be constructed within the planned width of 30 m, the absolute clearance for OYAMA-DAIRI tunnel was confined to 5.5 m, and the large sectional twin tunnel was forced to be constructed within minimum interval in unconfined ground conditions. (Fig. 3)

2 OUTLINE OF WORKS

2.1 Topographical and geological conditions

The OYAMA-DAIRI tunnel was constructed in a hilly area rising 170～180 m in altitude, covered with

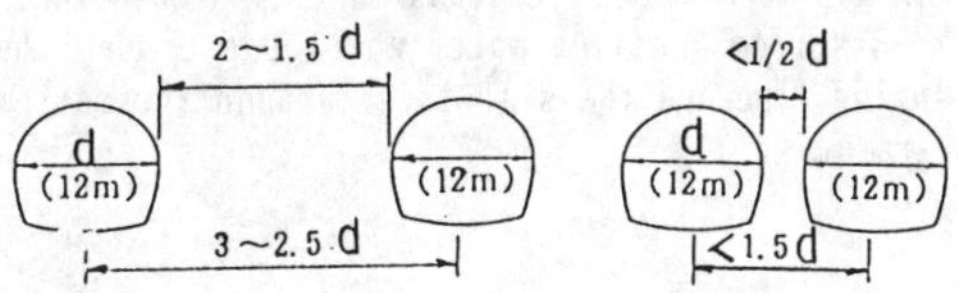

Fig. 2 Interval of the twin tunnel

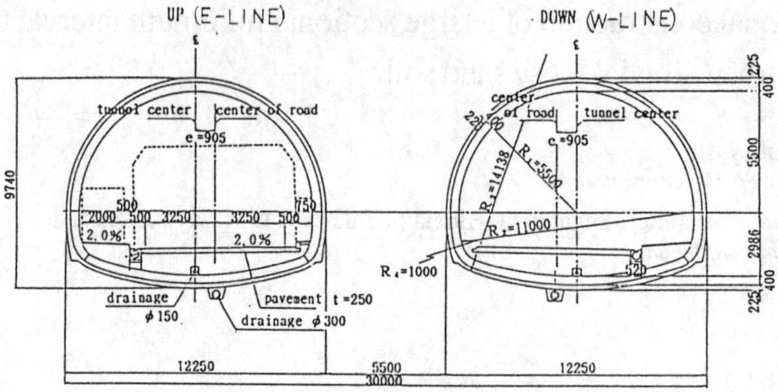

Fig.3 Cross section of the OYAMA-DAIRI Tunnel

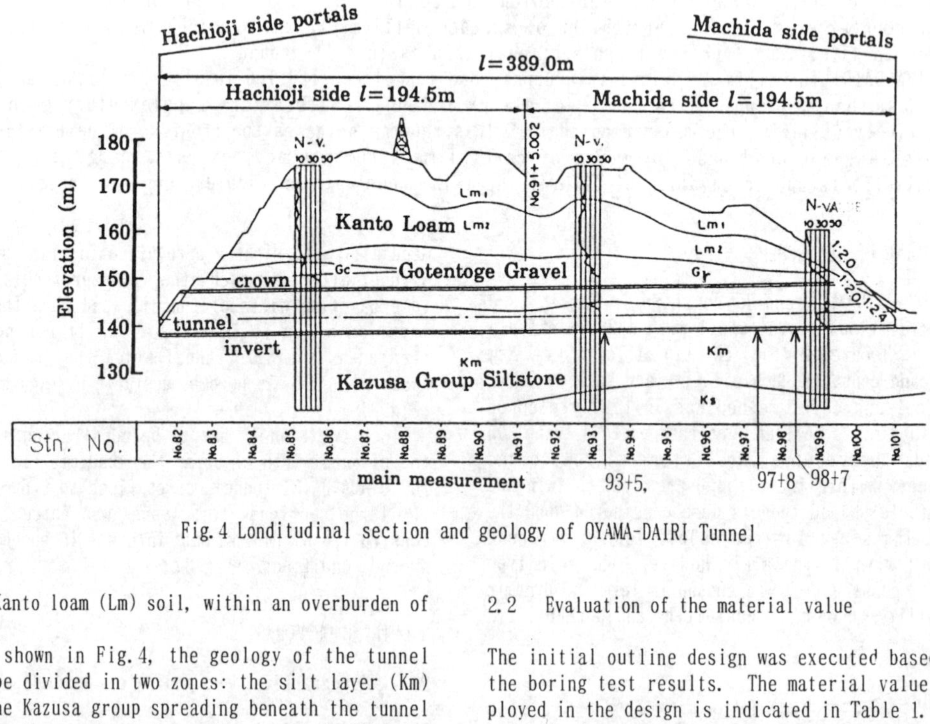

Fig.4 Londitudinal section and geology of OYAMA-DAIRI Tunnel

the Kanto loam (Lm) soil, within an overburden of 26 m.

As shown in Fig.4, the geology of the tunnel can be divided in two zones: the silt layer (Km) of the Kazusa group spreading beneath the tunnel spring line, and the Goten Toge gravel (Gr) in the upper half. The silt layer is not confined and the average uniaxial compression strength (qu) was between 3 ~12kgf/cm². The gravel layer composed with particles between several 10 ~100 mm, was much weathered, thus marking a qu between 2~3 kgf/cm² Welling water was found around the spring line and the silt layer around the bottom stratum.

2.2 Evaluation of the material value

The initial outline design was executed based on the boring test results. The material value employed in the design is indicated in Table 1.

The horizontal loading test was performed in the bored holes since triaxial sampling was not gained in the gravel stratum.

That's more, undisturbed samples of the gravel and silt layers were obtained with the cutting face upon excavation, for re-evaluation of the initial value. A triaxial test was performed for the block samples that were cut off at the site and shaped by freezing. These material values were compared with measurements at the site and then applied for re-designing and analyzation. The material value obtained here is also showed in Table 1.

Table 1. Material Value of Geology

ITEMS	Kantoh Loam	Gotentohge Gravel		Silt of Kazusa Group
		UPPER	LOWER	
UNIT WEIGHT (tf/m³)	1. 7 0	1. 8 0	1. 8 0	1. 9 0
MODULUS OF DEFORMATION (kgf/cm²)	1 0 0	4 0 0 (560)	4 0 0	1 0 0 0
COHHESION (kgf/cm²)	0. 5	2. 0 0 (1. 0)	2. 0 0	4. 0 (1. 3 8)
ANGLE OF INT. FRICTION (deg.)	1 0	2 0 (3 6. 1)	3 0	1 0 (2 6. 7)

( ) show the values obtained by block sampling at the site

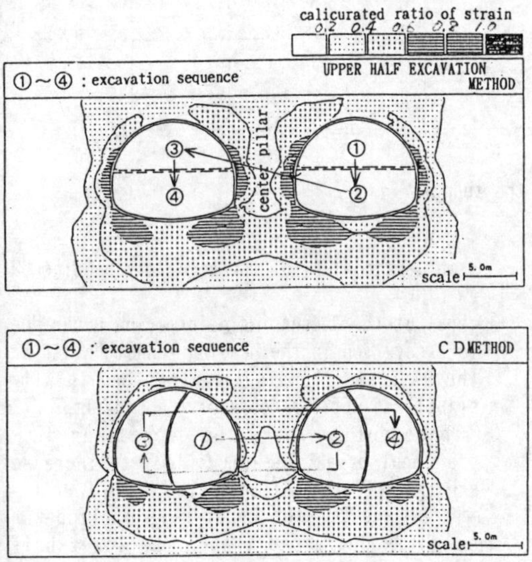

Fig. 5 Comparision of the stability of the twin tunnel by excavation method

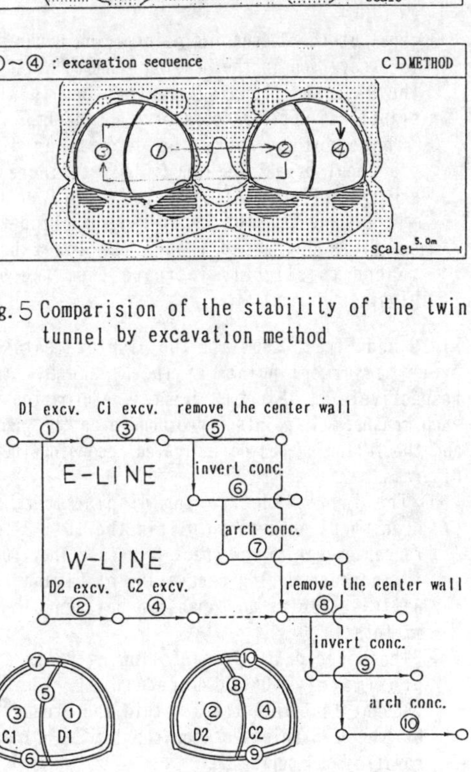

Fig. 6. Work execution procedure

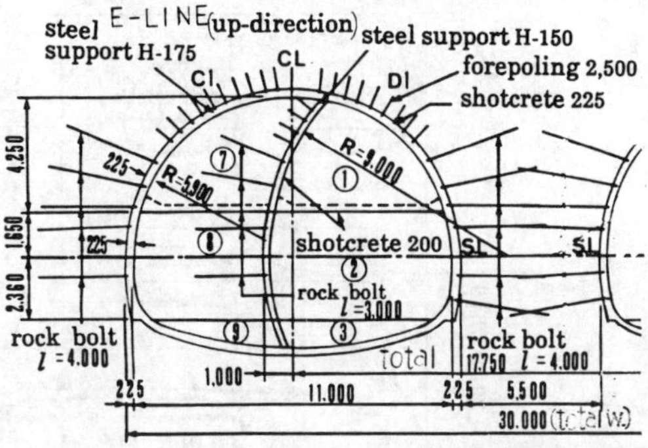

Fig. 7. Standard support structure and excavation sequence

2.3 Outline of execution and construction

The excavation procedures becomes the major topic in constructing with minimum interval twin tunnels since mutual effect upon excavation and insurance of stable grounds and support works in the center pillar of both tunnels is essential. Fig. 5 compares the loosening zone by choice of excavating method for the upper half method and the CD method that vertically divides the excavation portion.

CD (center diaphragm) method, as we call it, leaves a temporary center wall (diaphragm). This is extremely advantageous in suppressing settlement of the ground surface. Based on its form, we will call the heading portion D and the following portion C. As a result of analyzing the design, D drift was designed at the center of both the up and down directional tunnel. It was determined that alternating excavation would be optimal and as shown in Fig. 6, resulting in D1(upward=E-LINE)→D2(downward=W-LINE)→C1(E-LINE)→C2(W-LINE) procedures. The standard support patterns are shown in Fig. 7.

To prevent any effect on both tunnels the face clearance between D and C was decided as between 50 ～100 m, and each portion was excavated by a 2 ～14 meter short bench cutting. Thus, the invert was excavated within a 5 m distance during excavation of the lower half was performed by shotcreting and closing the ground to stabilize it.

The construction was separated into two lots: HACHIOJI and MACHIDA. This report would be mainly menshoned about the results from HACHIOJI site.

3 FIELD MEASUREMENTS

The ground displacements and steel support stress

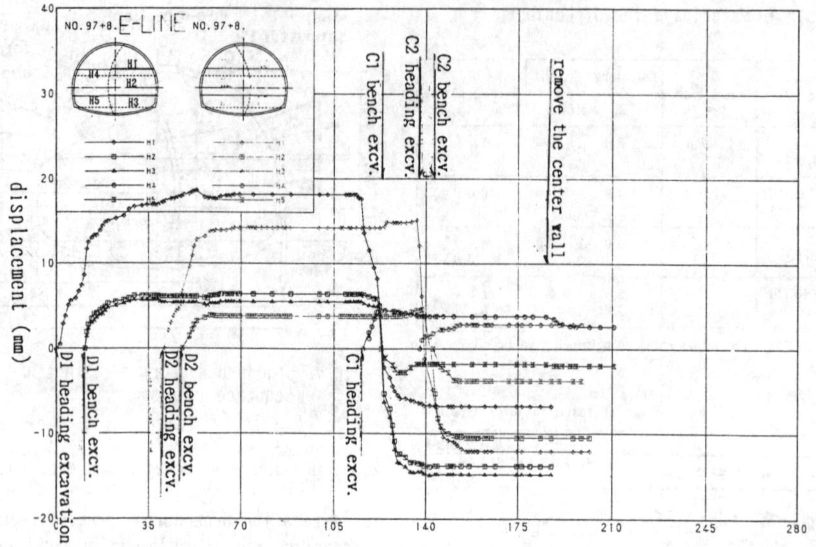

Fig. 8 All displacements of the tunnels in No. 87+5

in the twin tunnel CD excavation method is exposed to a 9-step sequence including the removal of the center wall. Therefore, it is especially important to accurately monitor the mutual effects of alternating excavation of double headings, as to maintain the ground deformation within the range of elastic limit and also it was to reflect the support pattern. We made a timely and precise evaluation employing the NATM automatic measurement system for this reason.

Upon purchasing the tunnel excavation, the displacement monitoring points were established in approximately 10m intervals in the tunnel. Main measuring sections added at points No. 98+7, 97+8 and 93+5, with extenso-meters and stress meters to measure the displacements of the ground and stresses in the steel support and shotcreat lining.

Surface settlement was monitored at points No. 99+0 and 97+0 from the ground the tunnel. Section No. 97+0 was monitored of the ground deformation in the vertical direction according to depth, using multi-point extensometer.

3.1 Tunnel Measurements

1) Convergence

Fig. 8 shows, for an exsample, procecc of the entire horizontal displacement (H_1 ~H_5) of the tunnel at the section of No. 87+5 according to excavation procedue. The results were mentionod below according the diagrams.

① Displacement of H_1 in the advanced heading D1 (E-LINE) was the largest, and more than the same sectional h_1 of the secondary heading D2 (W-LINE).

② Most of the decreasing deformation occur due to excavation in the heading tunnel D. After the D drift, the displacements are totally stabilized untill access of the C drift.

③ Those points where the decreasing displacement break out immediately to increase are the influence of the neighboring C drift.

④ In C1 and C2, the upper half portion was slightly degreased though the lower half tended to slightly increase from the beginning.

Fig. 9 and Fig. 10 show all the displacements on every measurment points by the H_1 and h_1 line respectively in D1 and D2 drift. Deformation of each measurement points corresponded on the whole and the following can be indicated according these diagrams.

① The initial decreasing displacement of heading D1 amounts to within the 16 ~21 mm range, except in few special cases. They turn to increase during excavation of C1 and finally settle at an expansion between 0 ~10 mm totally.

② Thoes decrease of the following D2 is approximately 80% compared to D1. It is thought that the stress within the ground in D2 had been released when the leading D1 excavation was performed.

③ The deformation in the portal portion at the final step of C drift tends to be small. The ground movement indicate the rebounds due to

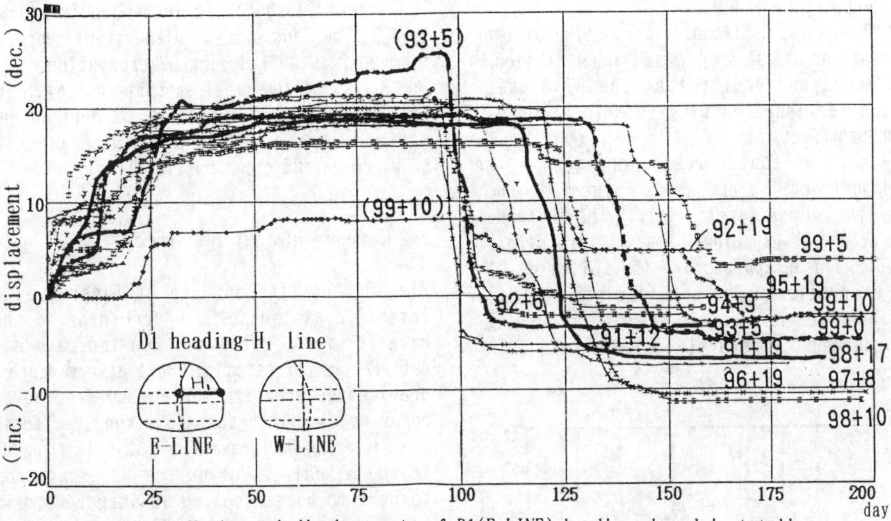

Fig. 9. Horizontal displacements of D1(E-LINE) heading showed in totally

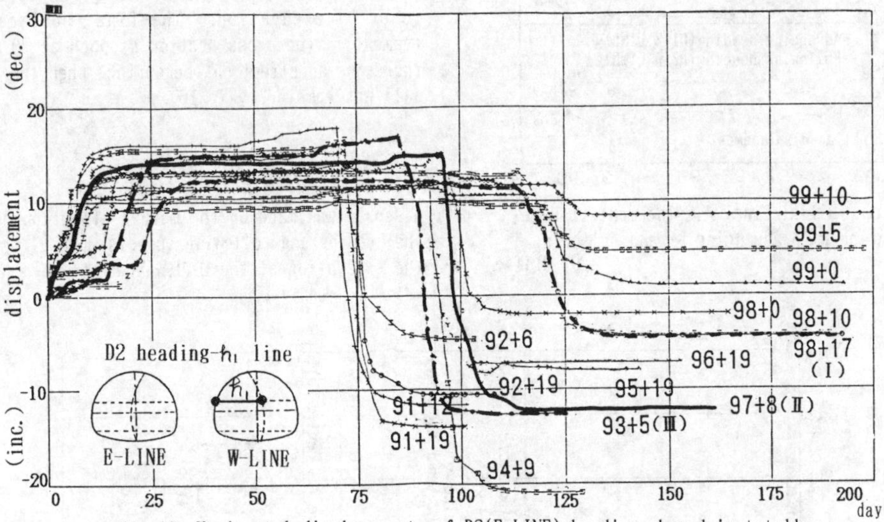

Fig. 10. Horizontal displacements of D2(E-LINE) heading showed in totally

deweight when C portion was excavated.

Those mentiond above can easily understand by Fig. 11, whitch shows the initial decrease and the final deformation along with the tunnel length.

The influence of overburden is clear in potal area.

The point No. 94+10 where indicates large displacements corresponds to the place where mutch ammount of welling water was recognized at the site. The settlement of crouwn was reported large around this area and so the section should be tend to enlarge the tunnel section.

2) Measurement of the Main Monitoring Sections
The following items were measured to ideally reflect in the support pattern.

① Extensometer for monitoring of the ground deformation
② Axial tensions in the rockbolts
③ Stress inside shotcrete
④ Axial compression of steel support
⑤ Stress inside lining concrete

Tipical results are showed in Fig. 12~Fig. 13

Fig. 12 shows the history of the stress in the shotcrete in No. 93+5.

Fig. 13 shows the history of the axial stress in

the steel support in No. 93+5.

Fig. 14 showes the sectional distribution of the displacements or stresses on tunnel inner section, steel support and scotcret at the 97+8 main monitoring section according to main steps of excavaion sequences.

It was shown that stress increases in the steel support and the amount is largest at the joints of the S3 and S4 side wall. The increase was less around the shoulder, however, the stress increased again as approaching the following C1 portion excavation. The shoulder stress (S1) sharply increased immediately before the cutting face approached, temporarily dropped, and then rose again.

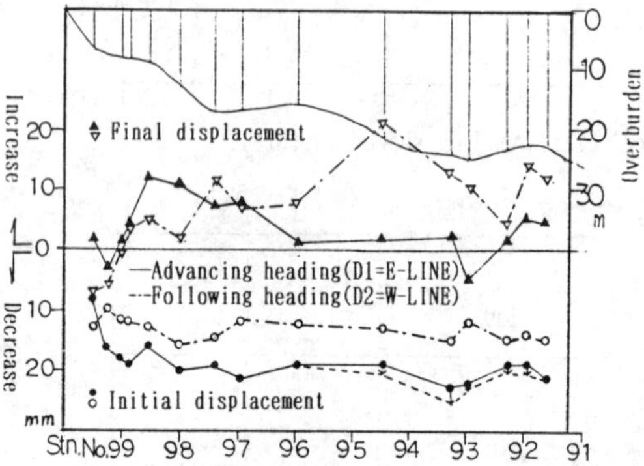

Fig. 11. Initial/Final displacement of H_1/h_1 (relation with advancing D1 and following D2 drift)

Stress in shotcrete increased from 37kgf/cm² to 43kgf/cm² and stress in the steel support rose from 20tf to 30tf during D2 excavation.

Stress of the steel support rose again in parallel with excavation of the C2 portion and the maximum stress of the shotcrete amounted to 65kgf/cm² at D2 crown portion finaly.

3.2 Measurements outside tunnel

Fig. 15 shows the advance of inner ground deformation at the No. 90+0 according to the excavatio steps. This diagram indicats all the patterrn is correspondent well at eny depth. The previous deformation had not be seen before D1 upper heading to reach, and strongly affected by the following D2 excavation. This is thought that to the mutual effect during D2 excavation and thought to have resulted in larger deformation since the upper and lower D2 excavation was performed when the ground strength of the pillar portion already declined during the upper and lower D1 excavation. The final surface settlement in the cross section stopped at 20 mm and there was no effect on the surface when the center wall was removed. (Fig. 16)

4. Summary

A report was made on the mutual affection on both the heading and following tunnels upon alternating the excavation of the OYAMA-DAIRI tunnel. The re

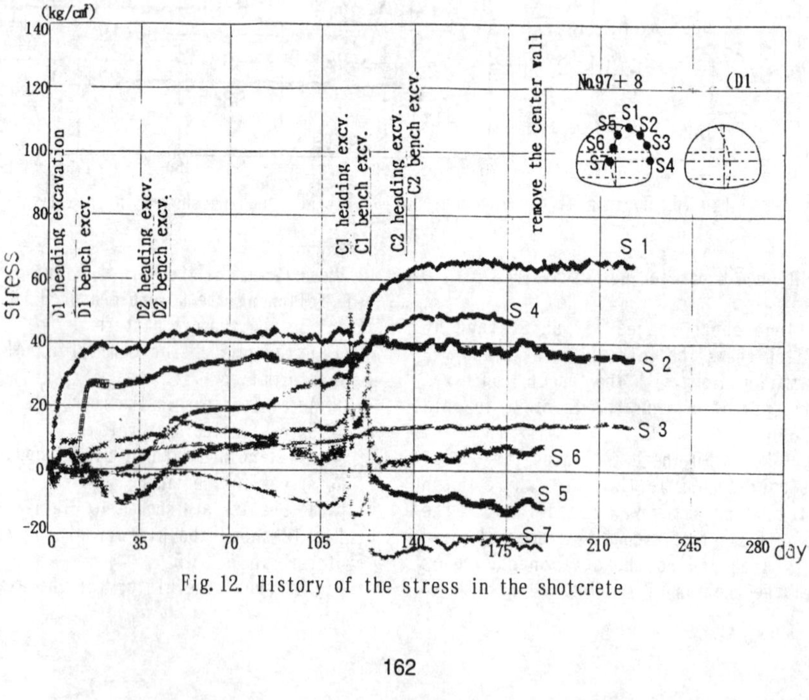

Fig. 12. History of the stress in the shotcrete

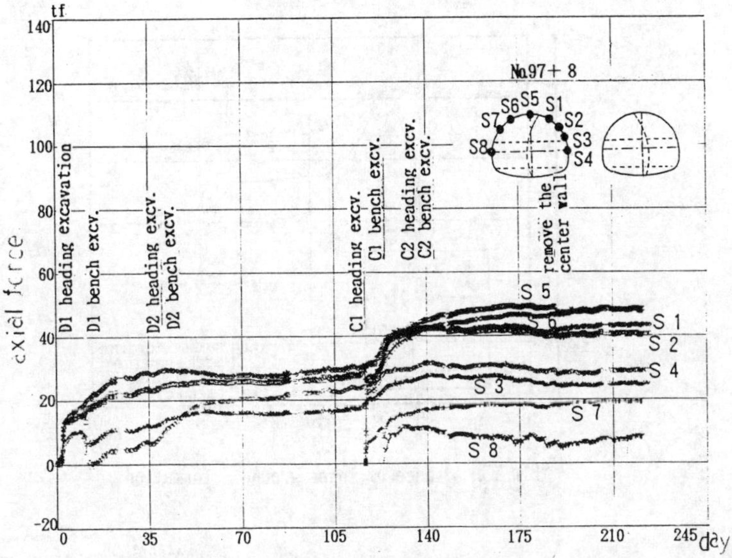

Fig. 13. History of the axial stress in the steel support

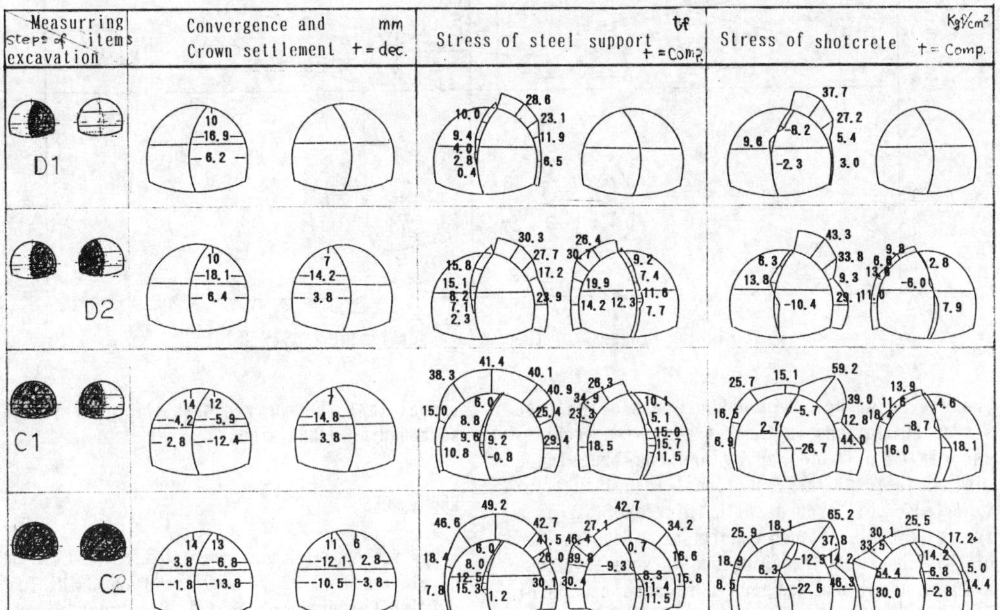

Fig. 14. Distribution of the displacements and stresses according excavation steps

sults can be summarized as follows.

1) The deformation was small in D of the heading tunnel following works on section D of the following tunnel. However, the increase in support strength on the center pillar side was significant. The increase and stress around the spring line where the remaining ground width is the smallest indicated a large value and was strongly affected upon D2 excavation.

2) The increased stress in the steel support at the sandy ground is affected by the location of the gravel stratum and its degree of consolidation.

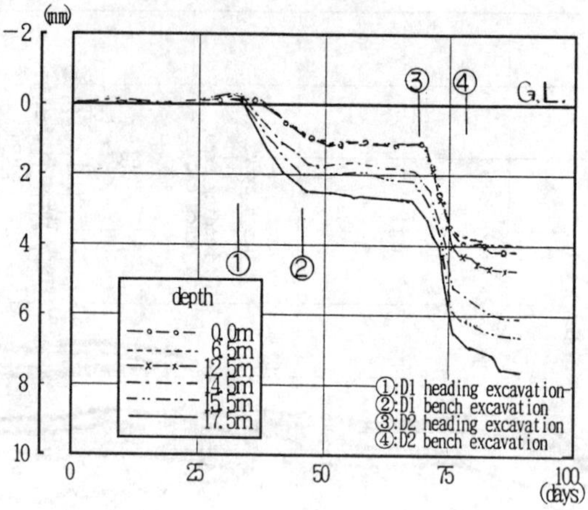

Fig. 15. Advance of inner ground deformation

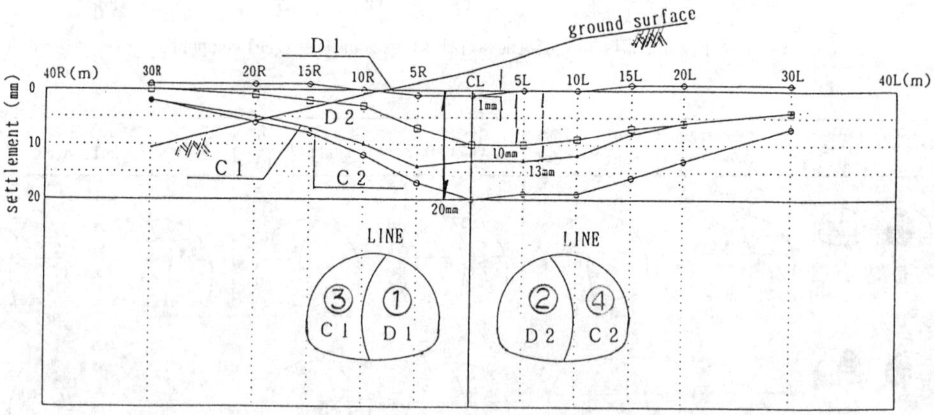

Fig. 16. Progress of the surface settlement (No.97＋0)

Tunnels constructed in urban areas often face strict limitations in their plans. The problems associated with designing and constructing a minimal interval large sectional tunnel in alluvial loose soil grounds with thin covering, is a typical example of such conditions. Construction of this tunnel by the CD method is sure to become a major asset and impetus for construction of future tunnels under the same conditions.

We would like to pay our respect for those involved in this project for their excellent work, and would like to take this opportunity to note that the OYAMA-DAIRI tunnel received the Technical Award from the Civil Engineering Society of Japan in fiscal 1990 for its excellence ensured from the designing to execution stage, ideally monitored under consolidated and sophisticated technology and for the excellent design of the tunnel that harmonizes with the scenery of the surrounding urban area.

REFERENCES

Tokyo Metropolitan government Office of TAMA urban developement, 1992. 3 REPORT of OYAMA-DAIRI Tunnel.
Shimizu Corporation, 1991. 3. Construction report of Minami-Osawa tunnel.
M. Ohtsuka, K. Watanabe, M. Kobayashi, 1992. 5. Annual Sympogium proceedings;Japan Society of Engineering Geology. Tokyo.
K. Watanabe, M. Ohtsuka, H. Kumasaka, M. Kobayashi, 1992. Results of field measurements and analyses of Oyamadairi Tunnel.

Tunnelling and Ground Conditions, Abdel Salam (ed.) © 1994 Balkema, Rotterdam, ISBN 90 5410 363 9

Environmental measures by means of groundwater control considering effects after construction of subway structures

Itsuki Koshiishi, Akiyoshi Nakamura & Takayuki Suzuki
East Japan Railway Company, Tohoku Construction Office, Japan

Yoshio Takagi
East Japan Railway Company, Construction Department, Japan

ABSTRACT: Sendai City, the capital of the Tohoku District in Japan, is working on placing the Senseki Line underground, and the part that is to be placed under the streets are being constructed by the "Cut and Cover Method." For the construction at the gravel-bed section where the groundwater level is high, the authors have adopted SMW (Soil Mixing Wall), that is an impervious earth retaining wall, and planned to penetrate the embedment into bed rock.
This wall cuts off most of the aquifer, and as a result, it is feared that plant roots will die due to a dam-up phenomena on the upper stream side and that wells and plants will dry up on the down straem side in the future.

In order to deal with this problem, the authors have conceived to control the groundwater flow by means of removing parts of the retaining wall and to backfill with permeable materials. In carrying out these measures, influence prediction was made using the finite element analysis in advance, and on the basis of the result of the analysis, a laboratory test on permeability of the backfill material (which is also reported here) has been carried out.

1 GEOLOGY

Around the planned route are distributed, respectively, the surface layer and loam (about 2 m in thickness), the river terrace of the Quaternary period underneath and the tuff group of the Tertiary period still underneath.

The terrace is a gravel layer mixed with round stones of about 200 to 300 mm in diameter, and is divided into Nakamachi Terrace distributed around an altitude of about 30 m and Shimomachi Terrace distributed around an altitude of about 20 m. Their formation thicknesses are 2 to 5 m and 10 to 16 m, respectively.

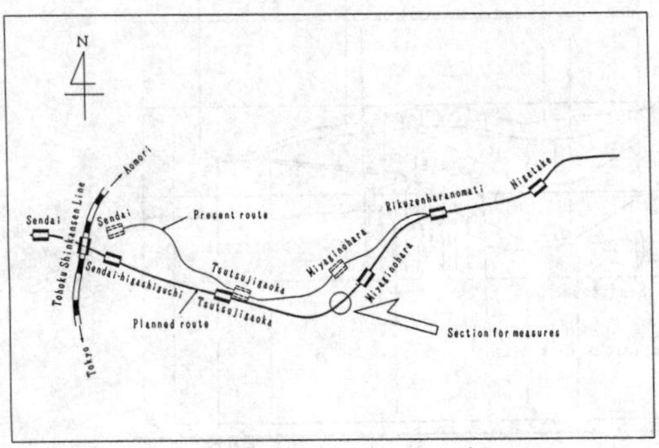

Figure 1. Location map.

Further, both the terraces have a borderline consisting of a comparatively clearly defined terrace cliff (their upper and lower altitudes are 33 to 36 m and about 16 m, respectively) and the section for measures is located right below them.

The lower New Tertiary formation consists of tuff, siltstone, mudstone, sandstone and lignite, and is in an unconformity contact with the upper terrace formation.

The groundwater flow around the section for measures is judged to be predominant in the direction of [northwest → southeast] in view of the groundwater contour line, and shows good agreement with the landform.

Namely, the groundwater altitudes are greatly different from each other (the upper altitude is about 26 m and the lower altitude is about 8 m) at the borderline, and a "groundwater cliff," which has a hydraulic gradient of about 10/200 is judged to have been formed near the terrace cliff.

2 NECESSITY OF MEASURES

Since the section for measures runs in parallel with the "Groundwater Contour Line" shown in Figure 2, a possibility is feared that underground structures (culvert for the planned route for the subway and culvert for telephone wires), which are to be constructed as the work progresses, will possibly obstruct the groundwater flow.

When a structure is constructed so as to cut off the most part of the aquifer, changes in the groundwater level (groundwater level rise on the upper-stream side and groundwater level drop on the lower-stream side) will probably occur, and supposing that the aquifer is completely cut off, the condition shown in Figure 4 is foreseen.

In order to prevent "changes in groundwater levels," it is necessary to secure most of the present groundwater flow also after the completion of the construction work, and for this means of permeation, several methods have been proposed.

For the present section for measures, it has been decided, after considering ground conditions, etc., to adopt the method of using highly permeable material for backfill and to cause groundwater to flow down through this part.

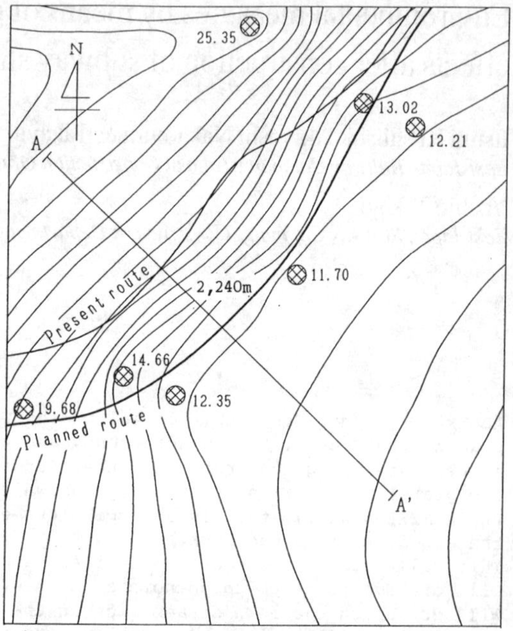

Figure 3. Groundwater contour line around the section for measures (cited from "Report on Groundwater Research concerning JR Senseki Line between Sendai and Nigatake (Mar, 1992)" by East Japan Railway Company.

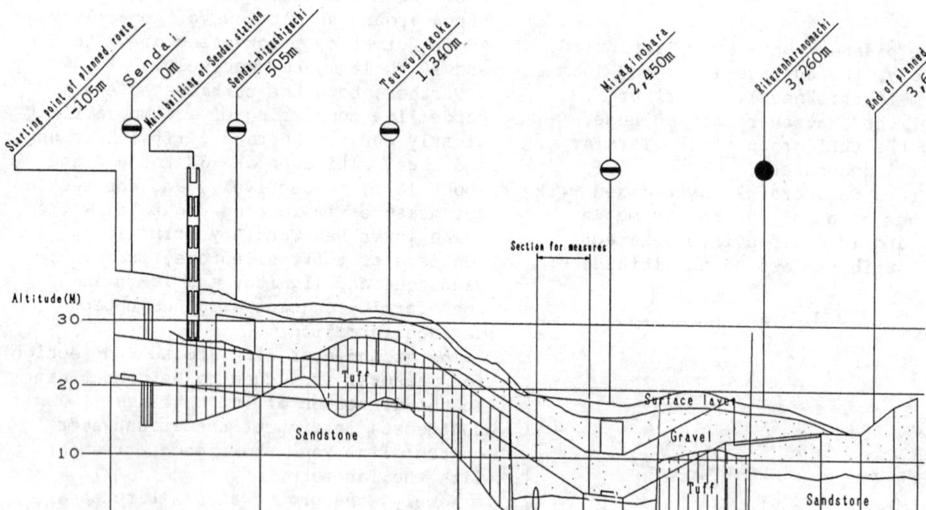

Figure 2. Vertical cross section of planned route.

In such a case, however, it is necessary to grasp beforehand the permeability (coefficient of permeability) of necessary backfilling materials in order to restore the current groundwater flow, and it has been decided to examine this matter in the present study.

3 WAY TO EXAMINE

First, from the section for measures, a representative cross section is selected which should become the object of the analysis, and changes in groundwater levels after the completion of the construction have been analyzed by the simulation of two-dimensional seepage flow.

Further, on the basis of the result of the analysis, the authors have examined appropriate permeability of materials for backfill and carried out a permeability test of soil materials for backfill with the aim of obtaining a reference material for determining the appropriate backfilling material.

In the following, the sequence of the analysis and examination is summarized.

1) Setting-up of Initial Condition and Structuring of Model of Present Condition
In order to permit reproducing the pre-construction groundwater condition, the initial condition is set up and the present condition model (Case 1) is structured.

2) Simulation of Each Case
Three kinds of backfilling materials having different coefficients of permeability are foreseen, and changes in groundwater levels in the respective cases have been simulated (Cases 2 to 4).

3) Examination of Simulation Results
First, "Changes in Groundwater Levels" is examined.

Next, examination is given to "wetting of soil" and "flow of soil particles" which are liable to be caused depending upon changes in groundwater levels.

4) Selection of Soil Materials of Backfill
Permeability tests are given to general-use crushed stones for road construction (C-40 and C-20) and mountain sand as backfill materials.

4 SIMULATION

4.1 Theoretical Expression and Using program

Simulation has been carried out using the finite element analysis of the saturated-unsaturated seepage flow.

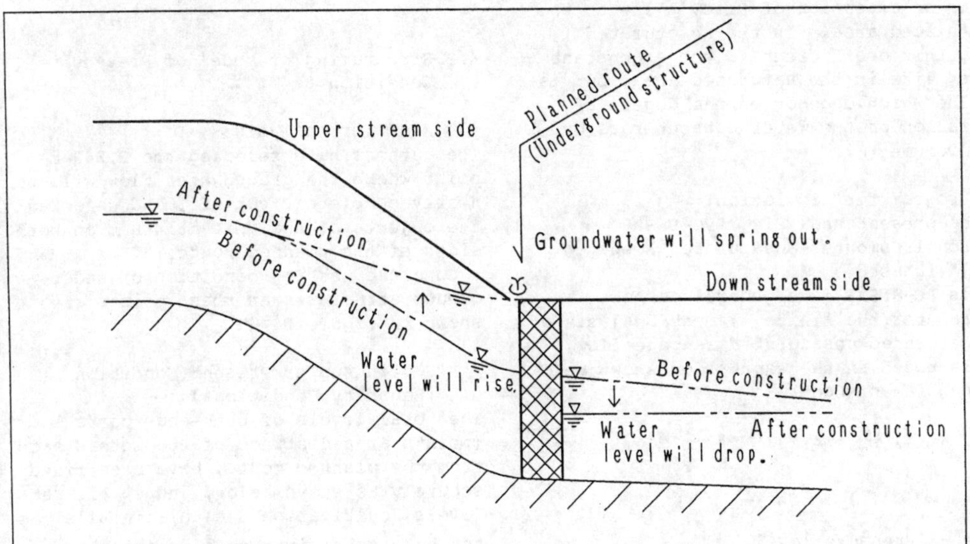

Figure 4. Dam-up phenomena.

<<Control equation in Saturated-Unsaturated Area>>

<Equation of motion → Darcy's expression>

$$v = k \frac{\delta h}{\delta z} \quad \cdots\cdots\cdots\cdots\cdots\cdots \quad (1)$$

where v: Darcy's velocity
k: Coefficient of permeability
h: Total level

<Equation of continuance → Law of conservation of mass>

$$\frac{\delta V}{\delta z} + \frac{\delta \theta}{\delta t} = 0 \quad \cdots\cdots\cdots\cdots\cdots \quad (2)$$

where t: Time
θ: Water content rate by unit volume ($\theta = Vw/V$)
The above indicates the water volume (Vw) contained in the void of unit volume (V), and is equal to the void rate in the saturated condition (n).

From eqs. (1) and (2), the following eq. (3) is derived.

$$\frac{\delta}{\delta z}\left\{ k \frac{\delta h}{\delta z} \right\} = \frac{\delta \theta}{\delta t} \quad \cdots\cdots\cdots \quad (3)$$

Coefficient of permeability (k) of eq. (3) is an extension of Darcy's law to the unsaturated area. In the unsaturated area, this coefficient is not a constant value, like in the saturated area, but is a value which depends on the degree of saturation or the water content rate by unit volume (θ).

<<Program for Simulation>>
In the present analysis, "PC-UNSAF (Personal Computer-Unsaturated Seepage Flow)" is used.
This PC-UNSAF is a general-purpose program for the finite element analysis of the saturated unsaturated seepage flow, and is based on the expression shown below:

<Equation of Control>

$$\frac{\delta}{\delta z}\left\{ k(\psi) \frac{\delta \psi}{\delta z} + k(\psi) \right\} + \frac{\delta}{\delta z}\left\{ k(\psi) \frac{\delta \psi}{\delta z} \right\} = \left\{ C(\psi) + \alpha Ss \right\} \frac{\delta \psi}{\delta t}$$

where ψ: Pressure level
k (ψ): Coefficient of permeability (= Krks)
C (ψ): Water content ratio
α: Parameter (Unsaturated → 0, saturated → 1)
Ss: Coefficient of stock ratio
Kr: Relative coefficient of permeability
Ks: Coefficient of saturated permeability

<Calculations of Relative Coefficient of Permeability (Kr) and Water Content>

At the time of calculating the water content rate of soil and relative coefficient of permeability (Kr) in the unsaturated zone, the following Irmay's expression has been used.

$$Kr = Se^{m}$$

Where m is a constant and Se is expressed by the following equation:

$$Se = \frac{(\theta - \theta_0)}{(\theta_s - \theta_0)}$$

where θ: Water content of soil
θ_0: Minimum water content of soil
θ_s: Water content of saturated soil

Further, the relation between the water content rate of soil and the pressure level is divided into two kinds, namely, a water content rate of 0 to 50% and 50 to 100% which have been approximated to linear equation respectively.
Water content rates of soil have been calculated from pressure levels using these linear equation.

4.2 Structuring of Model of Present Condition

<<Mesh for Analysis>>
The authors have selected the 2,240-m point where the groundwater flow will be nearly completely cut off, and selected the objects of analysis at 400 m on both sides of the planned route.
Concrete setting condition of underground structures at these points are shown in Figure 5.

<<Setting-up of Present Condition (Boundary Conditions)>>
The total levels of both-edge parts (points at a distance of 400 meters each from the planned route) have been fixed securely as shown below, and in all water levels, cultivation due to rainfalls has not been taken into consideration.

Upstream side: G.H. + 26.0 m
Downstream side : G.H. + 8.0 m

Void rate has been set up at 0.2.
Coefficients of permeability of the

current ground have been set up as follows, after the cross section has been divided into three parts:

Section of upper terrace
→ 1.7×10^{-2} cm/sec

Section of terrace cliff
→ 1.7×10^{-3} cm/sec

Section of lower terrace
→ 1.7×10^{-2} cm/sec
(incl. planned route)

When estimated on the basis of Darcy's expression, the coefficient of permeability near the terrace cliff seems to be relatively smaller than those of the sections on both sides.

4.3 Cases of Simulation

At the time of examination, three kinds of coefficients of permeability of backfill have been set up, and it has been decided to make calculations concerning changes in groundwater levels in the respective cases.

Table 1 summarizes four cases including the model of the current condition.

Table 1. Cases of simulation.

Case	Coefficient of permeability (=k) about soil materials for backfill		Comparison of permeability
1	–	(Before construction)	(Present condition)
2	I	1.7×10^{-2} cm/sec	Equal to (k) of ground
3	II	1.7×10^{-1} cm/sec	10 times larger than (k) of ground
4	III	8.5×10^{-3} cm/sec	Half of (k) of ground

5 EXAMINATION

5.1 Changes in Groundwater Levels

Table 2 shows changes in groundwater levels in Cases 1 to 4.

<Case 1> Before Construction Work
It is judged that this Case has deftly reproduced the groundwater level at the present condition.

<Case 2> After Construction Work:
Soil materials for backfill have the same

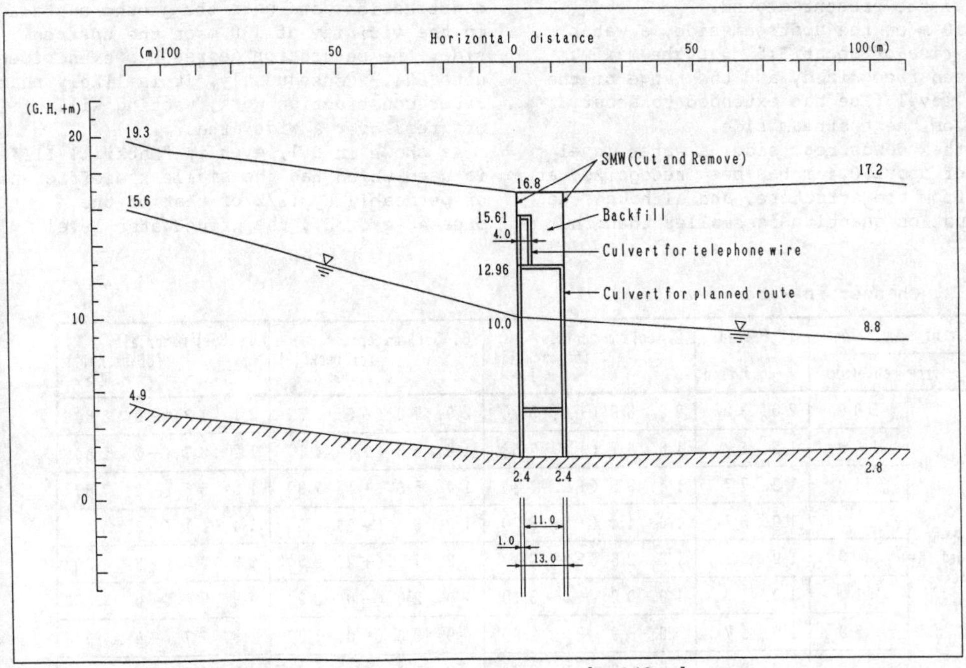

Figure 5. Condition of underground at structures [2,240 m].

permeability as that of the present ground.

At the position of 10 m on the upstream side, a water level rise of about 2.9 m at the maximum has been recognized, and the range of this water level rise has extended to about 300 m on the upstream side.

At 10 m from the structure on the downstream side, a water level drop of about 0.3 m has been recognized, and although the fluctuation quantity is smaller than that on the upstream side, the range affected by the water level drop has similarly extended to about 300 m.

<Case 3> After Construction Work: The backfill has permeability 10 times as large as that of the present ground.

At 10 m on the upstream side, a water level rise of about 2.4 m at the maximum has been recognized, and the range of this water level rise has extended to about 300 m on the upstream side.

At 10 m from the structure on the downstream side, a water level drop of about 0.2 m has been recognized, and although the fluctuation quantity is smaller than that on the upstream side, the range affected by the water level drop has similarly extended to about 300 m.

<Case 4> After Construction Work: The backfill has permeability one half of that of the present ground.

At 10 m on the upstream side, a water level rise of about 3.5 m at the maximum has been recognized, and the range of the water level rise has extended to about 300 m on the upstream side.

On the downstream side, a water level drop of about 0.4 m has been recognized at 10 m from the structure, and although the fluctuation quantity is smaller than that on the upstream side, the range to be affected by the water level drop is extended similarly to about 300 m.

Upon comparisons of various cases, it is judged that the better the permeability of the backfill, the smaller the fluctuation of the groundwater level, and conversely, the worse the permeability of the back-fill, the greater the fluctuations of the groundwater levels.

In any case, the groundwater level will not rise above the structure, and flows away from the upstream to the downstream through the unsaturated area.

5.2 Wetting of Soil

When "Backfill II" (having the greatest permeability) is used, velocity of water in the unsaturated soil changes within the range of 1.2 to 1.5 x 10^{-5} m/sec, and the degree of saturation at this time will change to about 50%.

When a comparison is made between the saturation degree before construction and that after construction, it has been found that near the ground surface in the vicinity of the structure, the degree of saturation, which is about 10% before construction, has risen to 35 to 50% after construction, and near the ground surface in the vicinity of 100 m on the upstream side, the saturation degree has even risen upto 60%. Consequently, it is likely that after construction work, wetting will progress over a wide area.

As shown in 5.1, even if "Backfill III" is used which has the smallest coefficient of permeability (1/2 of that of the present ground), the groundwater level

Table 2. Changes in water levels.

Case of simulation / Distance from route (m)		1. Present condition	2. Soil materials for backfill < I >	3. Soil materials for backfill < II >	4. Soil materials for backfill < III >
Up-stream side ⇧	200	20.32	20.89 (+0.57)	20.70 (+0.38)	21.18 (+0.86)
	100	15.12	16.87 (+1.75)	16.53 (+1.41)	17.37 (+2.25)
	50	12.77	14.95 (+2.18)	14.56 (+1.79)	15.55 (+2.78)
	10	10.57	13.44 (+2.87)	12.97 (+2.40)	14.11 (+3.54)
Position of planned route ⇨	0	10.52	13.26 (+2.74)	12.77 (+2.25)	13.95 (+3.43)
⇩ Down-stream side	10	10.40	10.09 (−0.31)	10.20 (−0.20)	9.96 (−0.44)
	50	10.17	9.89 (−0.28)	9.98 (−0.19)	9.77 (−0.40)
	90	9.94	9.68 (−0.26)	9.77 (−0.17)	9.56 (−0.38)

[UNIT] Water level: G.H.+m<( )···Figure in bracket is a difference in a Water level (m)>

will not rise above the top of the structure, after completion of the construction work.

However, as the coefficient of permeability of the backfill becomes smaller, the rate of the water-level rise quantity will become larger; hence the rise quantity of the saturation degree near the ground surface will increase by that amount, and it is highly liable that wetting will progress further.

Consequently, it is judged desirable that the coefficient of permeability of the backfill will be as high as possible (more than 10 times the coefficient of permeability of the present ground).

5.3 Flow of Soil Particles

When the velocity of groundwater increases as a result of construction work, it is considered possible that surrounding soil particles will flow as a result of this increase. The authors have examined the possibility of the soil particle flow on the basis of the groundwater velocity obtained by simulation. In general, it is considered that there is a significant relation between the limited velocity of the flow of soil particles and their coefficients of permeability.

Namely, when the velocity of water flow exceeds the coefficient of permeability of water, soil particles receive significant resistance; hence it is considered that the flow of soil particles will occur.

When water flows sideways as in the present case, a soil particle flow will not occur, as long as resistance on the bottom surface due to dead weight of soil particles is larger than the pushing force of water.

This problem is studied in the field of river engineering as "tractive current force" and hence examination has been carried out in conformity with this concept.

When the tractive current force in the river increases and exceeds a certain limit, sand particles on the riverbed begin to move. This "increase in tractive current force" is generated mainly by an increase in flow velocity, and the velocity at this time is called the "limited velocity of tractive current."

In the case of soil particle flow in the aquifer, a nearly similar concept also will apply as the condition of the riverbed, when the water flows from the aquifer to the backfill layer.

Consequently, when the limited velocity of flow is to be examined in the present study, it is necessary to compare the foreseen groundwater velocity and limited velocity of the tractive current.

In the case of the present study, the maximum value of predicted flow velocity of groundwater is 1×10^{-5} m/sec. This value is smaller than that of the limited velocity of flow and, therefore, it is judged that there is no problem as far as the flow of soil particles is concerned.

5.4 Selection of Material

In order to select backfill materials having a coefficient of permeability of more than 10 times (on the order of 10^{-1} cm/sec) that of the present ground, the following generally-used soil materials for back-fill have been put to a laboratory permeability test:

Tested materials
- C-40 (Crusher run with a diameter of 0 to 40 mm)
- C-20 (Crusher run with a diameter of 0 to 20 mm)
- Mountain sand

Test method has conformed to the standards set up by JSSMFE, and samples have been prepared by changing them into four stages of those in minimum density condition and those made by tamping efforts (mountain sand is in two stages).

As a result of the test, a coefficient of permeability has been confirmed on the order of 10^{-1} cm with the compactive effort upto about 0.5 Ec in respect of C-40 and C-20 (refer to Figure 6).

Here the term "Ec" means the compacting effort ($Ec \doteqdot 5.6$ cm · kgf/cm^3) defined by Proctor.

6 SUMMARY

<Flow of Analysis>

Examination has been made on an appropriate soil material for backfill which can limit to the minimum the "changes in groundwater flow" caused by construction of underground structures.

The authors have selected a typical cross section in the section for measures, assumed three kinds of soil materials for backfill having different values for the coefficient of permeability and analyzed changes in groundwater levels.

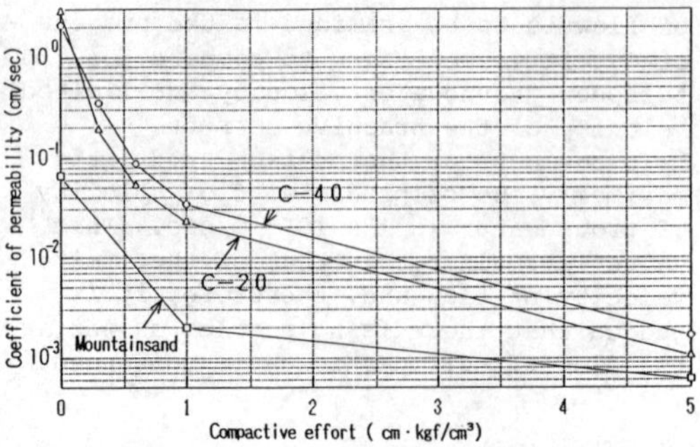

Figure 6. Relation between coefficient of permeability and compactive effort.

Backfill I : Permeability coefficient $k = 1.7 \times 10^{-2}$ cm/sec (same as present ground)
Backfill II : Permeability coefficient $k = 1.7 \times 10^{-1}$ cm/sec (10 times of present ground)
Backfill III: Permeability coefficient $k = 8.5 \times 10^{-2}$ cm/sec (1/2 of present ground)

<Changes in Water Level>
When the permeability of the backfill is improved, the result has been obtained in which changes in groundwater levels have become smaller.

Even when the backfill is given permeability as large as about 10 times that for the present ground, a certain degree of changes in groundwater levels will occur.

<Wetting of Soil>
In any case, the groundwater level will not rise as high as above structures (the vicinity of the ground surface will not be saturated).

However, when the coefficient of permeability of the backfill is small, naturally the rate of the water-level rise becomes larger, and the degree of saturation at the unsaturated zone will become higher than that in the preconstruction period, showing a strong possibility of wetting the soil.

Consequently, as an appropriate coefficient of permeability of backfill, it is desirable to secure a coefficient of permeability which is more than 10 times that of the present ground.

<Flow of Soil Particles>
The predicted velocity of groundwater is 1×10^{-5} m/sec even at the maximum, and is judged to be below the llimited velocity of groundwater flow. Therefore, no problem will occur regarding the flow of soil particles caused by the groundwater flow.

<Selection of Backfill Materials>
When a permeability test is carried out for general materials suitable for backfill, the required permeability (on the order of 10^{-1} cm/sec) has been confirmed in crushed stone for road construction (C-40 and C-20).

The authors have decided on C-40 for the most suitable soil material for backfill, taking also into consideration marketability and economy.

REFERENCES

1) Report on Groundwater Research concerning JR Senseki Line between Sendai and Nigatake (Mar, 1992 and Mar, 1993)" by East Japan Railway Company.

2) "Report on Permeability Test about Soil Materials for Backfill (Feb, 1993)" by East Japan Railway Company.

Tunnelling and Ground Conditions, Abdel Salam (ed.) © 1994 Balkema, Rotterdam, ISBN 90 5410 363 9

Field measurement controlling method of CD-NATM in urban tunnel

Masayuki Okada & Yasuhiro Hamazuka
Hokkaido Engineering Consultants Co., Ltd, Japan

Shigetoshi Okano
Mitsui Construction Co., Ltd, Japan

ABSTRACT: The Tsukisamu Tunnel is a 638m-long typical urban NATM, which runs through thin overburden under residential areas and the main street. Therefore, we adopted the CD-NATM. When this tunnel was constructed, the measuring control system was improved to determine necessary additional countermeasures. Overburden under a national road reduced to less than 1 D. This resulted in increased surface settlement. Therefore, ground around this tunnel was reinforced with polyurethane injection-type rock bolts. The reinforing effect of the rock bolts was evaluated by measuring the axial stress andbending moment. The measurement data and data on Japan's other similar tunnels were compared. The comparison shows that the CD-NATM hasa large controlling effect on surface settlement and is effective in the construction of urban tunnels.

1. INTRODUCTION

Many urban tunnels in Japan are constructed primarily in uncemented, thin overburden of Quaternary diluvium. Furthermore, urban tunnels are constructed beneath towns. Therefore, we must pay special attention to surrounding natural and man-made structures to prevent occurrences such as ground settlement and damage to buildings on the surface. In other words, various restrictions are imposed on tunnel construction.

The Tsukisamu Tunnel of the Sapporo Subway Toho Line is a 638m-long typical urban NATM, which runs through thin overburden (16.5-9.0m) under residential areas and National Route 36. (Fig. 1)

The geology in the tunnel area consists of weak, uncemented volcanic ashes, ashy silts and gravel, in this order from top to bottom. When the Tsukisamu Tunnel was constructed, we faced unstable ground and severe environmental restrictions. Therefore, we adopted the CD-NATM (Center Dividing Wall Type NATM with three-step short bench cut).

Measuring control of ground settlement, deformation and other critical factors is particularly important for such a construction site. When this tunnel was constructed, the measuring control system was improved to determine necessary additional measures. As a result, we minimized adverse effect on the ground and completed excavation safely.

This paper aims to examine the results of the above measurements in sections with thin overburden and the effect of rock bolts for reinforcement, and to inspect the effectiveness of the CD-NATM based on the results of the measurement.

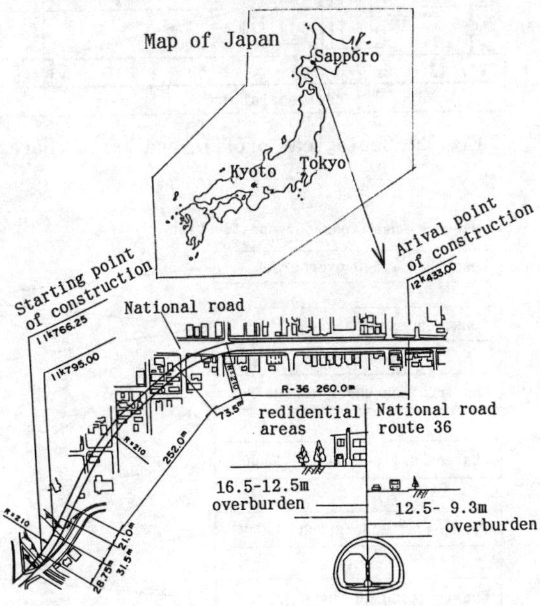

Fig. 1 Plan of the tunnel ruoteLocation of the Tsukisamu tunnel

2. OUTLINE OF THE TSUKISAMU TUNNEL

2.1 *Design*

The Tsukisamu tunnel was constructed in ground with thin, sandy overburden.(Fig. 2) When this tunnel was designed, we faced the technical problems as shown in Fig. 3. Therefore, we had to adopt methods other than the "empirical design method," in which the standard supporting patterns are established to construct ordinary mountain tunnels.

In September 1988, test construction was carried out and basic design items were examined by the FEM analysis. The results are summarized as follows (Fig. 2).

1. Prevention of displacement by high stiffness support members
2. Watertight (Full-water-proof type) tunnel
3. Adoption of the CD-NATM (Center Dividing Wall Type NATM)
4. Auxiliary method by polyurethane injection-type fore-piling

Figure 4 shows a typical support pattern for this tunnel.

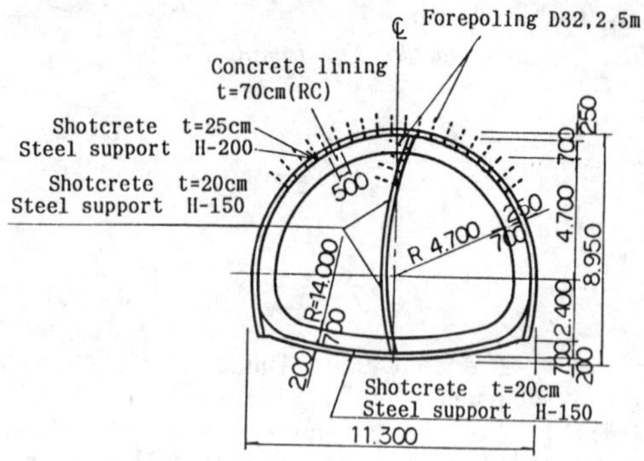

Fig. 4 Support pattern (R-36)

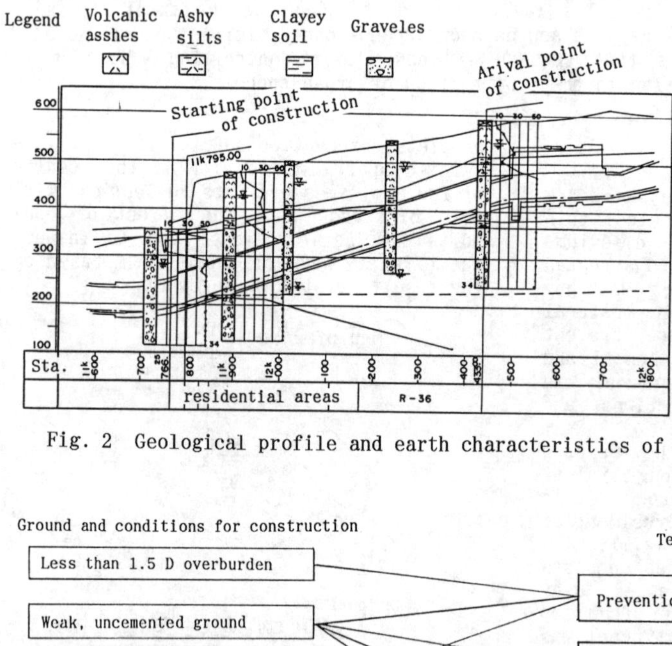

Results of soil test

Earth characteristics	Volcanic ashes	Ashy silts	Graveles
Water content(%)	55	40	20
Unit weight(tf/m3)	1.5	1.85	2.1
Void ratio	1.75	0.98	0.57
Coefficient of deformation(kgf/cm2)	200	240	1,000
Angle of internal friction(deg.)	33	16	37
Cohision(kgf/cm2)	3.0	3.5	0

Fig. 2 Geological profile and earth characteristics of the Tsukisamu tunnel

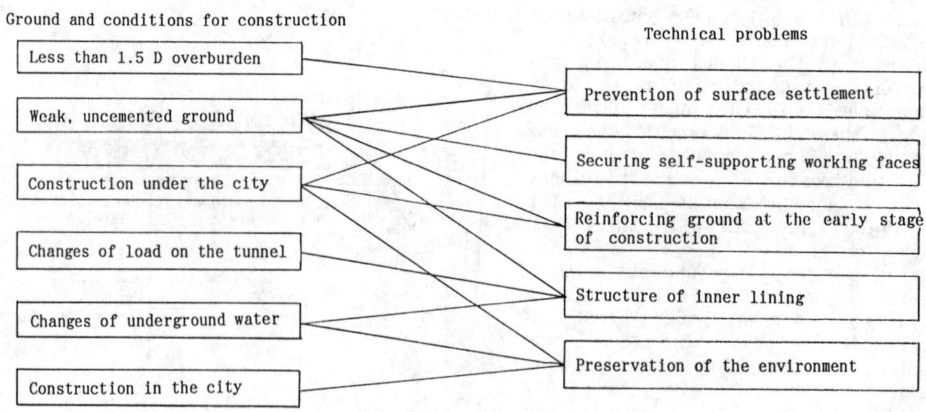

Fig. 3 Technical problems for design

2.2 *Construction*

The principal considerations for the construction of this tunnel were to:

1. prevent surface settlement and minimize adverse effects on the ground;
2. secure the stability of the tunnel by stabilizing working faces and sealing them as soon as possible;
3. prevent noise, vibration and dust, thus preserving the living environment;
4. and prevent adverse effects on buildings and underground facilities around the tunnel.

The order of construction is shown in Fig. 5. After considering the results of a test construction, the three-step short bench cut method (6-divisional excavation) was adopted for division of heading. One round excavation covers 0.9m. Excavation was carried out with a backhoe with rock breaker.

As for the CD-NATM, it is important to remove dividing walls at the right time. While this tunnel was being constructed, dividing walls were removed only after confirming that all cross sections had been sealed and the axial force of the dividing walls had been released. According to the results of measurement, the following removal standards were established.

1. Dividing walls will be removed in the areas of which all cross sections have been sealed for three days or more.
2. Dividing walls that are located 5m or more from the area in which all cross sections have been sealed will be removed.
3. Ten or fewer dividing walls will be removed in a day.

3. MEASUREMENT CONTROL

3.1 *Outline*

Measurement was carried out for this construction primarily to:

1. secure the safety of construction;
2. know the effect of construction on ground surface and important buildings;
3. and confirm the effectiveness of the design to contribute to the aforementioned two items.

Generally, when tunnels are excavated, ground displacement occurs before the working face is excavated. Construction in these cases in particular requires displacement control in accordance with total displacement (including ground displacement before the working face is excavated), because ground displacement before the working face is excavated is larger in uncemented ground. Therefore, we conducted measurement inside and outside the tunnel and on the ground. Figure 6 shows a typical cross section for measurement.

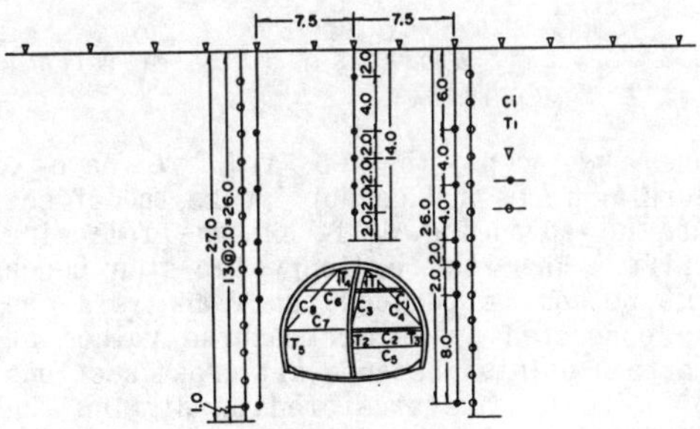

Fig. 6 Typical cross section for measurement

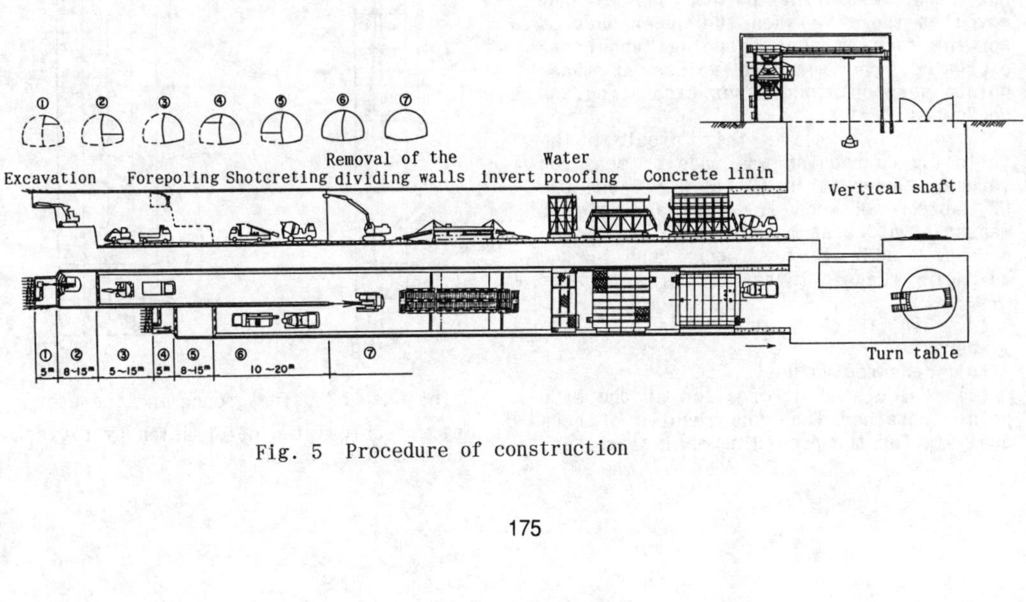

Fig. 5 Procedure of construction

We adopted the Direct Strain Control Method as a control method for measured values. Critical strain was calculated by using the results of pressure meter tests in the borehole, because uniaxial compression tests could not be conducted on samples of gravel which was to be excavated. As for the tunnel section of a national highway, the geology changed and silt layer subsided during excavation, so samples were collected from the silt layer of the working face. A uniaxial compression test was carried out on the samples to conduct more accurate measurement control by establishing a new critical strain of the silt layer.

Table 1 shows control criteria values for the measurement of displacement in the tunnel.

Table 1 Control criteria

Control level	settlement (mm)	Convergence (mm)	Critical strain (%)
Level I	-22	-44	-0.4
Level II	22-33	44-66	0.4-0.6
Level III	33-	66-	0.6-

3.2 *Back analysis for divisional excavation*

When we adopt the CD-NATM, we have 6 working faces and cannot ignore the effect of the advancing drift on the following drift. When the ordinary two-step bench cut method is adopted, back analysis can be conducted by using measured values at certain points, because all cross sections of a tunnel are excavated by digging the upper part of working faces. As for the CD-NATM, we cannot use measured values for analysis because values measured at measuring points T1 and C1 reflect total deformation after the start of excavation, but other measuring points had already moved at the time when the upper part of working face of the advancing tunnel was excavated, and measured values at those points were obtained after excavation was conducted.

In order to solve this problem, the following correction was made on measured values. Deformation (Z) at T2-T5 and C2-C8, which was used for back analysis of measured values at measuring points in the second - and other - steps and in the advancing tunnel, is as follows:

$$\{Z\} = \{D\} + \{d\}$$

where

$\{D\}$: Measured values

$\{d\}$: Calculated deformation at the same points obtained from the results of back analysis for the preceding analyzing step

Analysis was conducted by regarding $\{d\}$ as initial displacement before the working face was excavated. In order to confirm analytical accuracy, cross sections for measurement were established under residential areas shortly after excavation started, and under a national highway, under which the geological structure was changing. As shown in Fig. 6, extensometers were installed at surface settlement measuring points to observe deformation mode and deformation.

Back analysis produced fully satisfactory results by using the aforementioned revised deformation. The deformation mode of ground in particular produced highly accurate results by using a three-directional extensometer.(Fig.7)

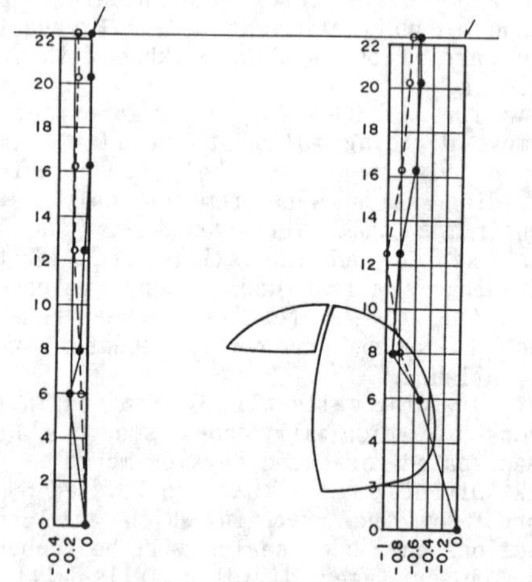

(a) Case of using ground settlement values

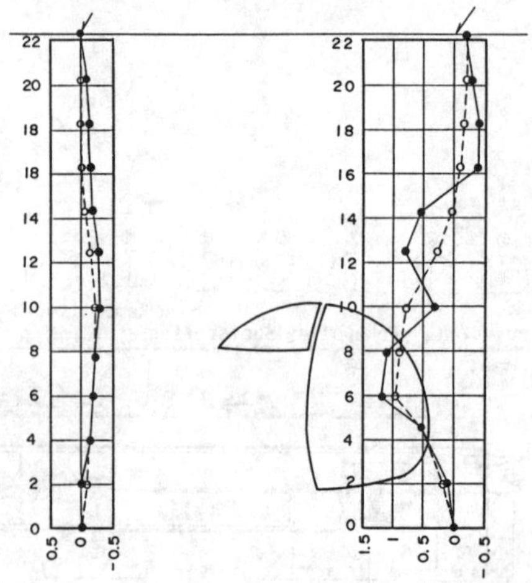

(b) Case of using ground inclinometer

Fig. 7 Calculated deformation by revised and measured values

4. THE RESULTS OF MEASUREMENT UNDER A NATIONAL HIGHWAY

4.1 *Outline of the results of measurement*

Overburden under a national highway decreased to less than 1 D. Because the earth's surface tends to subside, we were worried that surface settlement might reach the limit of the control criteria.

An outline of the results of measurement under a national highway follows.

1. Figure 8 is a typical graph showings changes in convergence. The maximum convergence (about 30mm) remained within Control Level I (44mm), reaching about 70% of the level. Therefore, ground stability was secured around the tunnel.
2. Measured values of convergence and of crown and surface settlement were larger in the overburden under a national highway than in the thick overburden under residential areas (16.5-12.5m). Surface settlement showed such a tendency particularly clearly.
3. Figure 9 shows the iongitudinal distribution of surface settlement and overburden. After overburden was less than 1 D, and the silt layer (which was considered to contribute to the prevention of surface settlement) was excavated, advancing settlement increased and final settlement also tended to do so.
4. It was confirmed that both the ground surface and the tunnel crown subsided, and it was revealed that maximum surface settlement has the following relationship. [Maximum surface settlement] = [Advancing surface settlement] + [Crown settlement in the tunnel]
5. The following equation was obtained to estimate surface settlement (*G8*) by conducting multiple regressive analysis of deformation factors of surface settlement.

$$G_s = -1.1\,H + 0.04\,C_1 + 0.12\,C_2 + 20.9$$

where

G_s: Surface settlement (m)

C_1 and C_2: Convergence of the upper and middle steps of the advancing drift$H$: Overburden (m)

This equation was used to forecast final settlement at the early stage of construction and to examin measures as early as possible against surface settlement.

4.2 *The effect of rock bolts for reinforcement*

Self-boring and polyurethane injection-type rock bolts (hereafter called "rock bolts for reinforcement") were chosen to keep surface settlement within the control

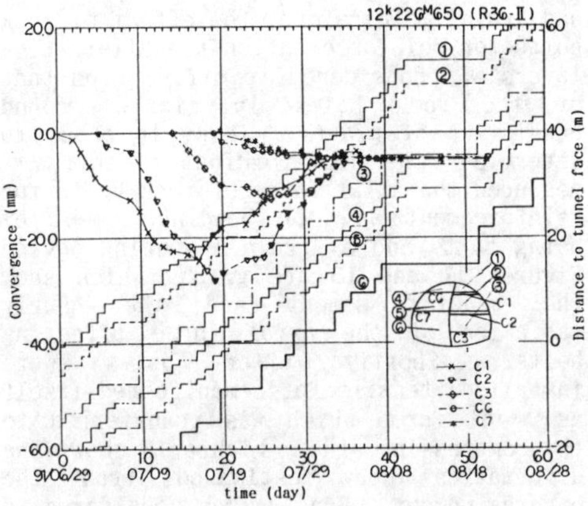

Fig. 8 Changes in convergence

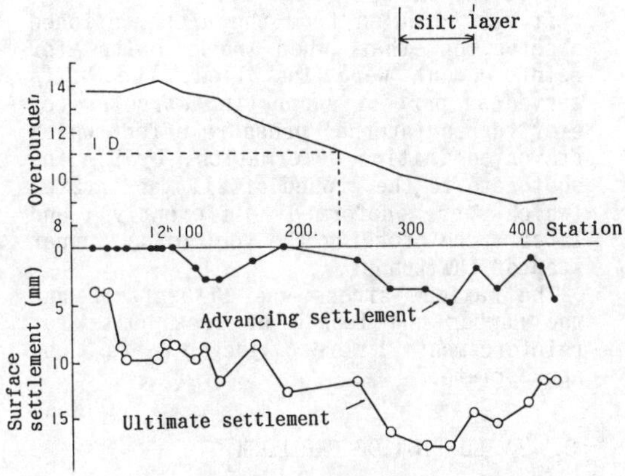

Fig. 9 Longitudinal distribution of surface settlement and overburden

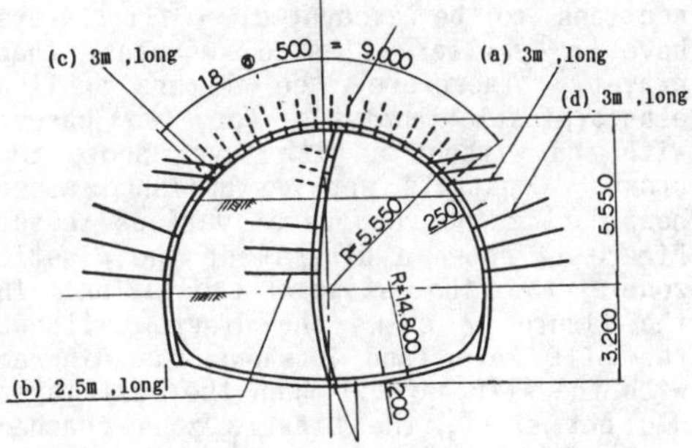

Fig.10 Positions of rock bolts for reinforcement

value. Figure 10 shows positions at which rock bolts for reinforcement were installed.

Between 12.290km and 12.355km, silt layer subsided and the working face consisted of alternating layers of silt and gravel. Therefore, the effect of rock bolts for reinforcement on the alternating layers was considered to differ from that of the rock bolts in simple ground consisting of gravel. Thus, in order to determine the reinforcing effect, we measured the axial force of rock bolts for reinforcement and the bending moment by using bolts and a strain measuring device Figures 11 and 12 are graphs which show the bending moment and the hourly variation of the axial force affecting bolts. Shortly after bolts were installed, tensile force manifested itself as axial force which was transferred to the compressing side. Figure 13 shows the deformation mode estimated from the bending moment. Bending was considered to occur against the deformation of the tunnel, as shown in Fig. 13.

It was revealed from the aforementioned observation that when rock bolts for reinforcement were installed, the bolts served as part of supporting structure by exerting an internal pressure effect which prevented initial deformation, by binding shotcrete to the ground of silt and gravel (which were deformed differently) and later by reinforcing the foot of the upper step of the tunnel.

The maximum stress was 1710kgf/cm2 and the number and length of rock bolts for reinforcement were judged to be appropriate.

5. EVALUATION OF CAP ROCK

When this tunnel was constructed, the increase and decrease of overburden of silt layer and surface settlement were closely correlated, as shown in the cross sections to be excavated. Silt layers have a smaller elastic modulus than gravel. Therefore, the writers applied elasto-plastic analysis for two cases: with and without a silt layer above the crown. Constants were established after considering the results of various tests. Figure 14 shows a diagram of the plastic zone at the final stage of excavation. In the figure, A shows the diagram without the silt layer and B shows the diagram with the silt layer. When the silt layer did not exist, the plastic zone reached the upper volcanic ash layer. On the other hand, when silt layer existed, the plastic zone was distributed only in the gravel because the silt layer prevented the spreading of the plastic zone. In this case, surface settlement was limited to about 60% of that without the silt layer. The surface settlement almost equaled the measured value.

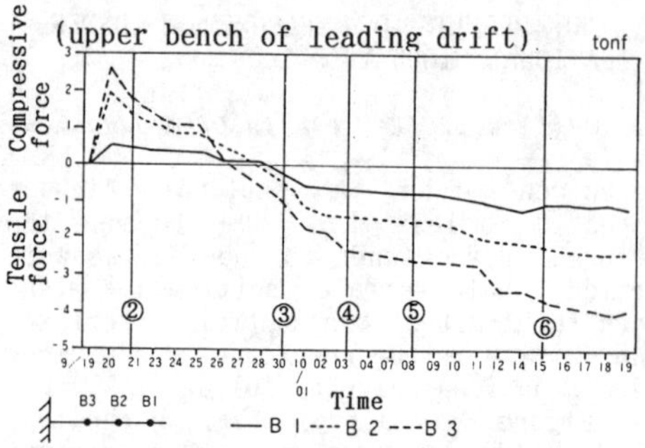

Fig.11 Changes in axial force of rock bolt for reinforcement

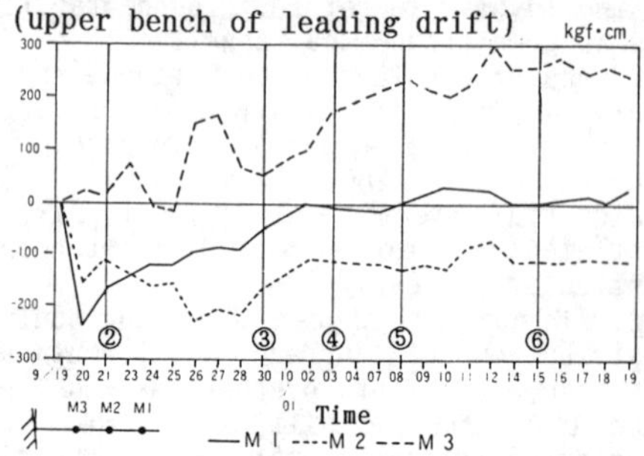

Fig.12 Changes in bending moment of rock bolts for reinforcement

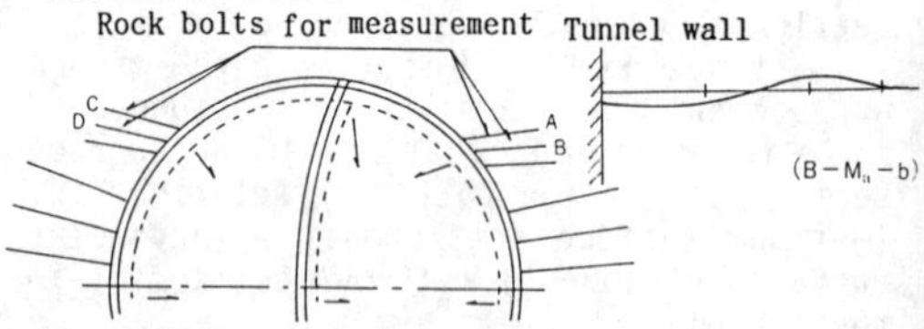

Fig.13 Estimated deformation mode of this tunnel

In other words, the silt layer serves to reduce plastic deformation that develops during excavation, because the silt layer has a large cohesion and does not readily permit intrusion of the plastic zone. When the silt layer is absent, plastic deformation develops in the volcanic ash layer above the tunnel and the surface settlement increases. These phenomena reflect the cap rock effect of the silt layer.

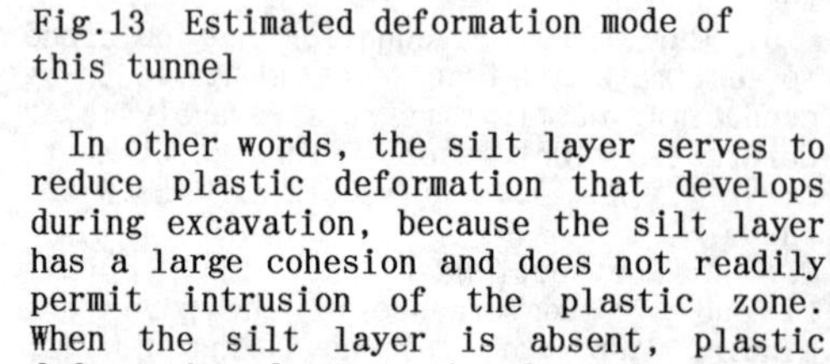

A Without silt layer B With silt layer

Fig. 14 Diagram of plastic zone

6. CHARACTERISTICS OF BEHAVIOR OF THE CD-NATM

Generally, the CD-NATM (Center Dividing Wall Type NATM) has the following characteristics.

1. The section of heading is smaller and the working face becomes stable by conducting divisional excavation on both sides.
2. In this method, excavation is carried out on vertically long cross sections. Therefore, when overburden is thin and vertical earth pressure is large, this method is effective in preventing deformation (settlement).
3. Dividing walls serve as pillars and greatly support the ground until arches are established.
4. In accordance with ground conditions, excavation methods can easily be changed from this method to the Dividing Wall Method only for the upper half or the CRD Method.

We would like to examine the aforementioned benefits of the CD-NATM by using the results of measurement in the Tsukisamu Tunnel.

6.1 *Surface settlement control effect*

We gathered data on surface settlement in similar tunnels from books published in Japan (for example,Yokoyama 1987), and we examined the surface settlement tendency in the CD-NATM. Figure 15 shows the relationship between overburden (H/D) and surface settlement. The figure compares data in the CD-NATM and that in the Short Bench Cut Method (with the Ring Cut Method), which is used for the construction of ordinary mountain tunnels.

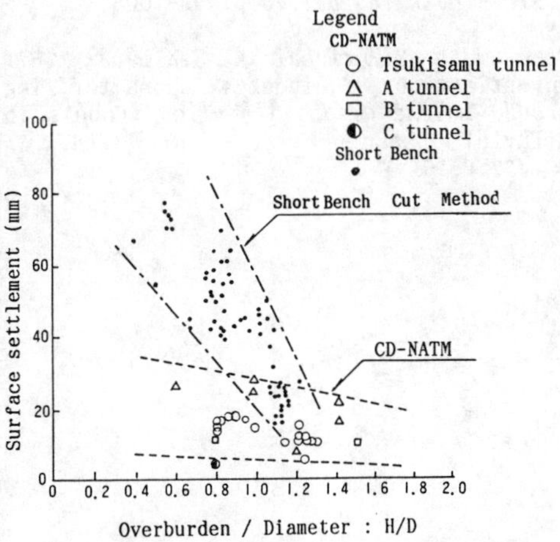

Fig.15 Rerationship between overburden and surface settlement

It can be seen from the figure that surface settlement tends to increase with the decrease of overburden in both the Short Bench Cut Method and the CD-NATM. However, we can say that the increase of surface settlement with the decrease of overburden in the Short Bench Cut Method tends to be larger than in the CD-NATM, though we cannot necessarily reach a final conclusion due to differences in ground conditions and other factors.

It is known from the aforementioned facts that the CD-NATM has a high controlling effect on surface settlement and is effective in the construction of tunnels under severe ground restrictions.

6.2 *Effect of dividing walls*

In the CD-NATM, the dividing walls are typically the source of support. Figure 16 shows changes in shotcrete stress of dividing walls. Stress at Point A (an arch part of the advancing drift) and that at Point B (a dividing wall part of the drift) increased almost equally, and peaked when the upper step of the following drift was being excavated. Later, when the middle step of the following drift was being excavated, stress at Point A increased further, but that at the dividing wall part was released and decreased to near zero. On the other hand, stress at Point C (an arch part of the following drift) increased, and stress at both Points A and C reduced to zero when all cross sections were sealed.

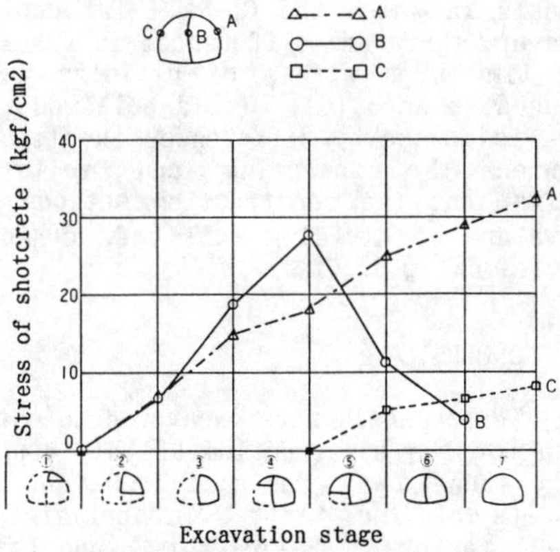

Fig.16 Changes in shotcrete stress of dividing wall

Figure 16 shows changes in the axial force of the steel support. Behavior of the steel support resembles that of shotcrete, and thus, both steel support and shotcrete work harmoniously in

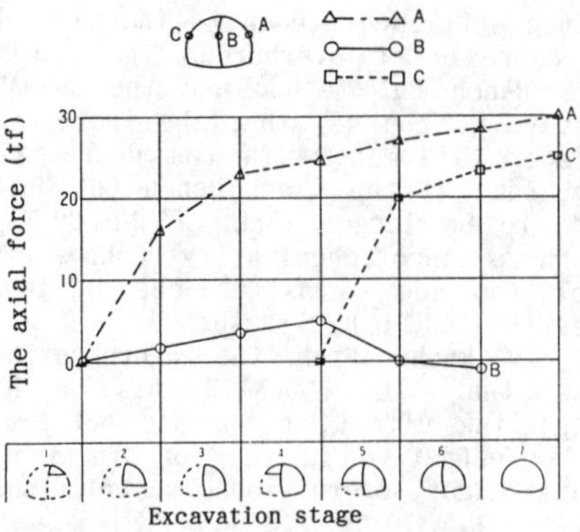

Fig.17 Changes in the axial force of the steel support

combination. It is known from the aforementioned results that while the advancing drift was excavated, dividing walls directly supported the ground along with the outer steel support, but that as the following drift, which was behind the advancing drift, was excavated, dividing walls came to serve as pillars. Furthermore, as the support effect of the following drift came into play, stress at dividing walls was redistributed to Points A and C (arch parts). From this redistribution process of dividing wall members over time, it is known that the CD-NATM is effective in the control of vertical deformation until arches are established.

Such dividing wall behavior can been seen in other similar construction of tunnels in which the CD-NATM is adopted. However, there are differences in terms of the time when stress of dividing walls reduced to zero (6). It is believed that this is caused by differences in distance between the advancing and following drifts, and the construction methods and curvature of dividing walls of advancing and following drifts.

7. CONCLUSION

1. Test tunnels were excavated to obtain data for the construction of this tunnel. This tunnel was designed based on the results of the test, FEM analysis and other factors. While this tunnel was being constructed, efforts were made to minimize surface settlement and preserve the nearby environment.

2. The Direct Strain Control Method, which was established by Dr. Sakurai, was adopted as a measuring control method. In this method, advancing settlement can be included as a factor by improving the conventional back analysis method.

3. Overburden under a national highway reduced to less than 1 D. Ground to be excavated was made up of complex layers of gravel and silt. This resulted in increased surface settlement. Therefore, ground around this tunnel was reinforced with polyurethane injection-type rock bolts. The reinforcing effect of the rock bolts was evaluated by measuring the axial stress and bending moment.

4. While this tunnel was being constructed, the silt layer was considered to serve as a "cap rock," contributing to the prevention of surface settlement. In order to prove this, FEM elasto-plastic analysis was carried out, and the results of the analysis were compared with measured values. As a result, it was proved from an analytical viewpoint that silt layers serve as a cap rock.

5. The results of measurement in this tunnel and data on Japan's other similar tunnels were compared. The comparison shows that the CD-NATM has a large controlling effect on surface settlement and is effective in the construction of tunnels in city areas in which severe restrictions are imposed.

ACKNOWLEDGMENT

We are grateful to the precious guide and aid of Sapporo Bureau of Traffic and the parties concerned.

REFERENCES

Soma,H.,M.Okada & S.Okana 1992. CD-NATM in uncemented ground beneath towns. *Proc. of J.S.C.E.Hokkaido,vol.48*:p.999-1004.

Yokoyama,a.,Y.Horiuchi & K.Kimura 1987. Investigation of indexes characterizing ground movements above shallow tunnels in diluvial sand. *Proc. of J.S.C.E. NO.457*:P.161-170.

Tunnelling and Ground Conditions, Abdel Salam (ed.) © 1994 Balkema, Rotterdam, ISBN 90 5410 363 9

El Salam syphon under Suez Canal

Ahmed Mazen
North Sinai Development Organisation, MPWWR

Rodney Craig
HALCROW

ABSTRACT: The Ministry of Public Works and Water Resources of the Arab Republic of Egypt has a comprehensive Scheme to develop 400,000 feddans in Northern Sinai on the east side of the Suez Canal. The irrigation for this development will require up to 160 cubic metres per second of water to be transferred from the existing El Salam Canal on the west bank to the new canal system on the east bank. This water will be taken in syphon under the Suez Canal.

Four tunnels of 5.1 metre internal diameter, 775 metres in length, will be constructed below the Suez Canal with a clearance to the future deepening and widening of 9m. The tunnels will be constructed with a closed face machine and lined with precast segments and a secondary lining. The construction will commence in early 1994 and will be completed in three years.

1 EL SALAM CANAL PROJECT

Agriculture is an important sector of the Egyptian Economy. The rapid population growth in Egypt has made land reclamation even more important. One of the largest reclamation projects is El Salam Canal Project which was first conceived in the late 1970s. Because of the low rainfall all crop water has to be provided by irrigation. The project comprises three main elements.

- El Salam Canal, west of the Suez Canal
- Syphon under Suez Canal
- El Sheikh Gaber Canal, in Northern Sinai

Construction of El Salam Canal to the west commenced in 1989 and was completed to within 500m of the Suez Canal in 1991.

The Ministry of Public Works and Water Resources (MPWWR) of the Arab Republic of Egypt is developing a total area of 600,000 feddans of which 200,000 feddans is on the west of the Suez Canal and 400,000 feddans in Northern Sinai on the east side of the Suez Canal. See Figure 1. Irrigation water for the development will be taken from the Damietta branch of the River Nile to the south of Damietta City. It is also supplied by water from El-Sirw drain and Bahr Hadous drain. This additional intake is controlled to ensure that the salinity does not exceed 800 parts per million. The total water requirement is 4450 million cubic metres per year with a planned supply from the River Nile of 2110 million cubic metres. The water will pass in tunnel below the Suez Canal and connect into the El Sheikh Gaber Canal, to be constructed in Northern Sinai with feeder canals along the whole development. The project area is bounded by the Mediterranean Sea and Bardawil Lake in the north, the Suez Canal in the west and adjoins sand dune zones to the southern and eastern edges. The land slopes gently from the south to the north from 30m to sea level. The El Sheikh Gaber Canal will extend to El Arish City, a total length of 160km. The water which will be lifted at 7 pump stations, will be used for crop cultivation, livestock rearing and fish farming.

The infrastructure will be constructed as the development proceeds. New settlements will be developed on the basis of 10,000 feddan units with five villages per unit, with one village providing the

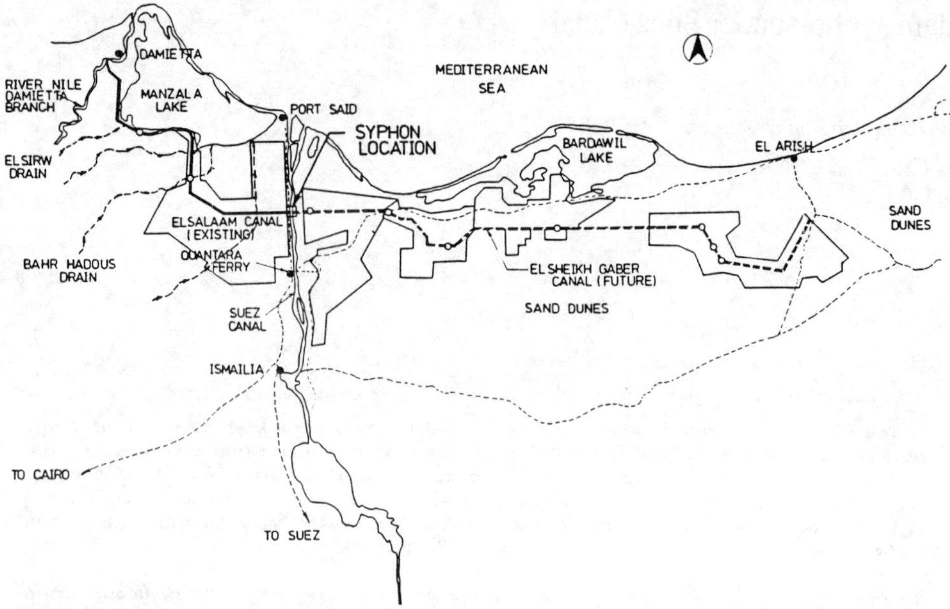

NORTH SINAI DEVELOPMENT PROJECT AREA

Fig.1

community facilities and adminstration. The project is being funded by the Kuwait Fund for Arab Economic Development.

2 HYDRAULIC DESIGN

The hydraulic design formed an important part of the design of the syphon. The maximum design flow is 160 cubic metres per second but there is a wide variation in the monthly design flows. (See table 1). The minimum flow of 42 cubic metres per second is in October and the peak flows in June and July. The peak velocity in the open El Salam canal will be 0.7 metres per second. For the tunnel syphon the minimum velocity will be 1 metre per second and the maximum velocity 2 metres per second. The design head loss in the syphon is 0.9 metres.

During the conceptual design phase of the project two, three and four tunnels schemes were considered for both a segmental lined tunnel and for a cast in-situ lining inside a precast segmental lining. A range of roughness heights was used from 0.3 to 3.0mm.

The initial hydraulic design showed that all three schemes were feasible and that the decision on the number of tunnels should be made on the basis of tunnelling aspects, programme and costs, operations and maintenance considerations. In all cases the desirable flow velocity within the tunnel was at least twice the flow velocity in the upstream canal, thus minimising sediment deposits in the syphon. Overall the four tunnel scheme was marginally the best scheme for hydraulic performance.

Following the design of the scheme an hydraulic model of the syphon has been built, at the MPWWR's Research Laboratory outside Cairo, which has been used to confirm the hydraulic design.

TABLE 1

Monthly Flows

	m^3/s		m^3/s
January	64	July	160
February	84	August	148
March	82	September	117
April	86	October	42
May	82	November	94
June	160	December	108

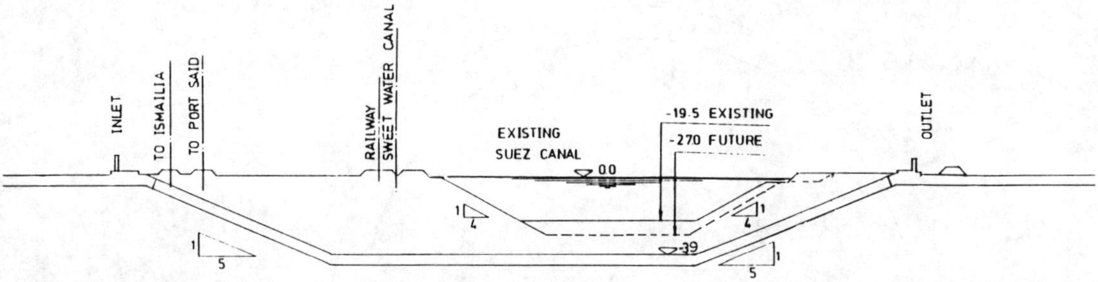

VERTICAL ALIGNMENT

Fig 2

3 CHOICE OF NUMBER OF TUNNELS

The analysis of the number of tunnels showed that two, three or four tunnels were all suitable for the syphon. The four tunnel scheme provided an improved flexibility of operation, particularly during the low flow months, and for maintenance. There was very little difference in the construction costs and the operations and maintenance costs. The four tunnel scheme could be constructed within the overall project programme of three years and provided more flexibility for phased operation.

Taking into account all these factors the four tunnel option was chosen as the most appropriate scheme.

4 TUNNEL ALIGNMENT

Figure 2 shows the vertical alignment for the syphon for the proposed four tunnel scheme. The two inclined parts of the syphon are at gradients of 1 in 5 with a 1 in 200 grade under the Suez Canal to a low point at the east side of the Suez Canal.

The vertical curves have a radius of 500 metres. The syphon on the west bank has a portal at the inlet structure, passes beneath the Port Said to Ismailia dual carriageway and railway, the Sweet Water Canal and the Suez Canal Authority road, before passing below the Suez Canal and then rises to the outlet structure. The vertical alignment allows for the future widening, on the east side, and deepening of the Suez Canal, with a minimum clearance of 9m to the deepened canal. In discussion with the Suez Canal Authority it has been agreed that no dredging works will be carried out within the canal during the construction of the syphon tunnels.

For the horizontal alignment the tunnels are at 12 metre centres at the inlet and outlet structures. They then diverge to a constant 15 metres centres for the majority of the horizontal alignment. The horizontal curves have a 500 metres radius.

The strata along the route of the syphon have been investigated on both sides of the Suez Canal in two site investigation programmes. No boreholes have, however, been sunk within the Suez Canal. The strata along the route are predominantly waterbearing cohesive and non-cohesive soils. These are:

- grey silty clays
- yellowish grey sands

The water table is close to the surface with piezometer heads of 1 metre above sea level, giving a maximum water head of some 45 metres above invert level of the syphon.

5 TUNNEL LINING

The tunnel will be constructed through water bearing ground conditions and in an aggressive environment. The groundwater is saline. During the design phase particular attention has been paid to the durability of the tunnel lining and the long term life of the tunnel. Two alternative linings were considered in detail a single skin tunnel with a precast segmental lining, and a double skin tunnel, with a precast segmental lining, and a secondary cast in situ lining with a waterproofing membrane between.

The double skin lining was chosen on the

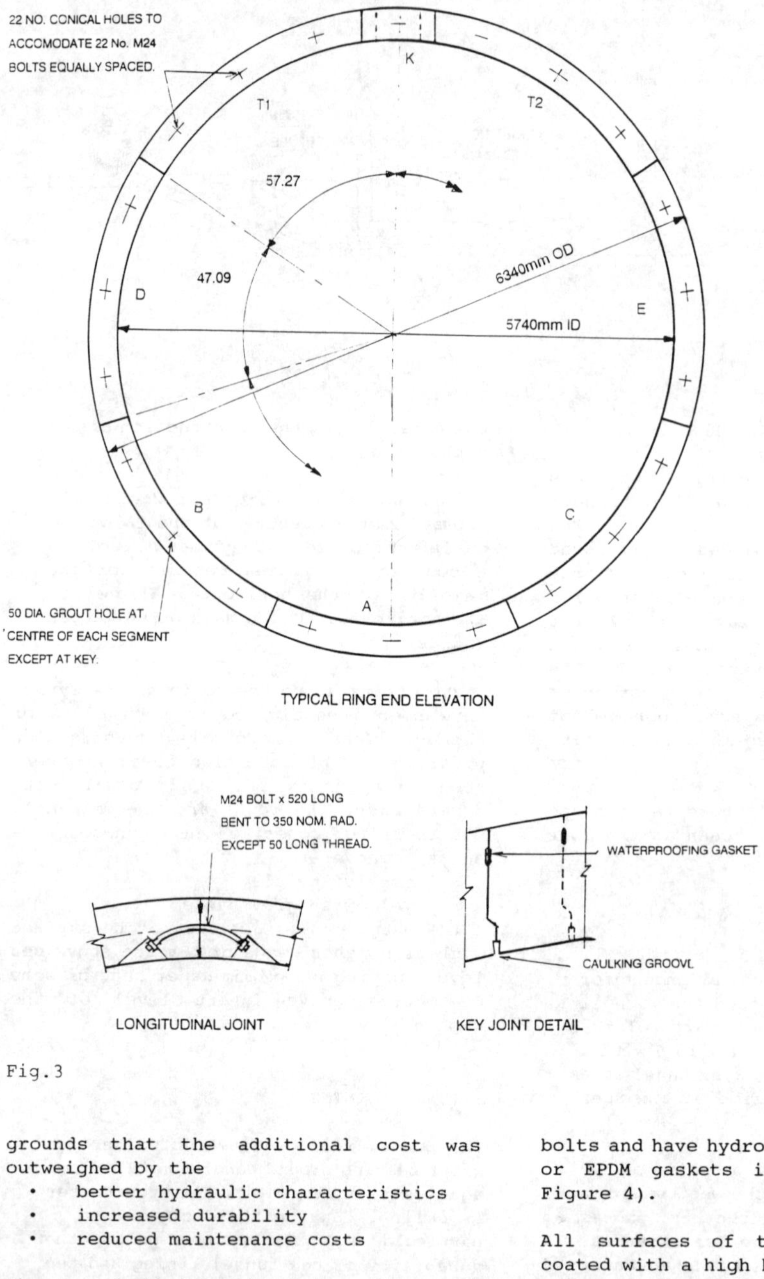

Fig.3

grounds that the additional cost was outweighed by the

- better hydraulic characteristics
- increased durability
- reduced maintenance costs

The tunnel lining scheme (see Figure 3) provides for a 5.74 metre internal diameter precast concrete segmental lining consisting of seven segments and a wedged shaped key. The lining is 300mm thick and the rings have a nominal 1200mm width with a 20mm taper with left and right hand tapered rings.

The segments are connected with curved bolts and have hydrophillic/EPDM composite or EPDM gaskets in the joints. (See Figure 4).

All surfaces of the segments will be coated with a high build solvent coal tar epoxy coating. The segments are of grade 45/20 concrete with 45mm minimum cover to the reinforcement. The cement type has been selected in accordance with the recommendation of the British BRE digest 363 (July 1991), taking into account the local conditions and classified as Class 5.

With the particular emphasis on the

durability, the cement specified is a high slag blastfurnace cement of 30% Portland Cement to British Standard 12 and 70% ground granular blastfurnace slag to BS 6699.

A laminated PVC waterproof membrane will be fixed to the primary lining and an unreinforced concrete secondary lining cast inside of 320mm thickness to give 5.1m internal diameter tunnels. The secondary lining will be of grade 40/20 concrete.

The precast tunnel lining has been designed for 75% of the full ground loadings with full water pressure, and the cast in situ lining for the full ground loadings and water pressure. The design has been checked for a seismic acceleration of 0.15g and for the subsequent deepening and widening of the Suez Canal.

6 METHOD OF CONSTRUCTION

The four tunnels will be driven through waterbearing non-cohesive and cohesive soils with a water pressure above the crown of about 4 bars. The tunnels will be constructed at 1 in 5 up and down grades and pass below the Suez Canal. These are very challenging criteria and the Contractor will have to pay particular attention to them in choosing his method of construction.

The depth of the tunnel and the high water table mean that the tunnels cannot be constructed by conventional compressed air methods. Grouting and freezing methods were considered but were discounted on programme and, costs and for health and safety reasons. The tunnels must therefore be constructed using a tunnel boring machine with a closed face. The alternatives are

- Earth Pressure Balance Machine (EPBM)
- Slurry or mix shield TBM

The choice of TBM will be the decision of the contractors under the conditions specified. The three year programme allows for the four tunnels to be constructed using one machine.

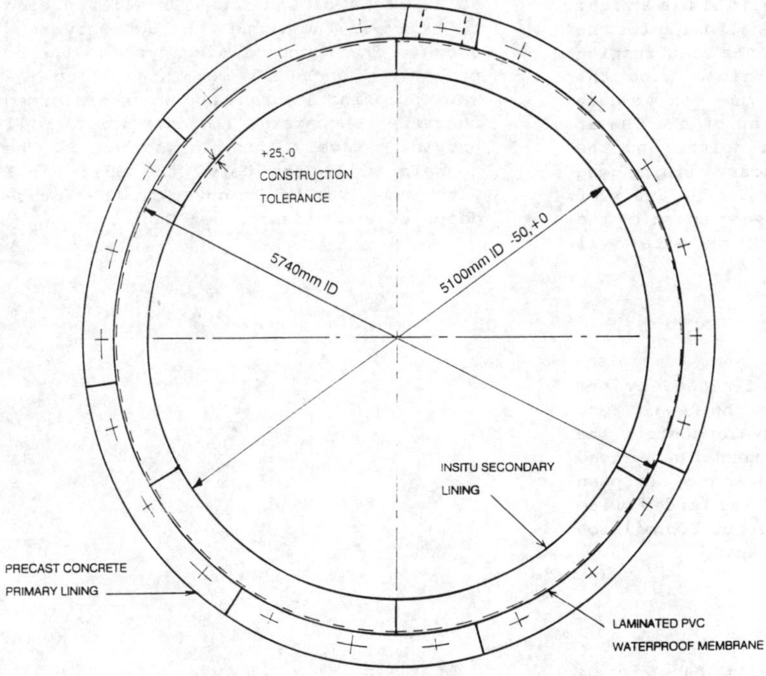

TYPICAL CROSS SECTION
OF SYPHON TUNNEL

Fig 4

The specification requires that the unreinforced concrete secondary lining shall not be cast until the adjacent tunnel has been excavated and lined. This will allow any minor deformation of the tunnel caused by the relaxation of the ground during excavation of the adjacent tunnel to occur prior to casting the secondary lining. The movement of the precast tunnel linings will be monitored during the excavation of the adjacent tunnel.

The main working site proposed is on the east bank of the Suez Canal where a very large area is available. However, access is difficult by road and use will be made of a ferry across the Suez Canal. The working site on the west bank for the inlet structure is too limited for a large site set up and precasting yard. The choice of driving the tunnels from the east bank or from the west bank or alternating will be the choice of the contractors.

7 PROGRAMME

The programme specified in the Tender Document is three years. This is a tight but realistic programme allowing for the manufacture of the TBM, the construction of the four tunnel drives with the removal of the TBM at the end of each drive and the setting up of the TBM at the beginning of each drive, and the casting of the secondary lining and completion of the works. CMC/Besix of Italy/Belgium have been awarded the Contract and the work on site will commence in January 1994.

8 COSTS

The project is being financed by the MPWWR with support from the Kuwait Fund for Arab Economic Development. The CMC/Besix contractors' tender was priced in Egyptian Pounds (LE) and in European Currency Units (ECUs). The tender price will be equivalent to about 200 million LE (60 million US Dollars).

9 DEWATERING

The syphon tunnels will have to be dewatered for inspection and maintenance although they will normally be kept full when not in use. Closure of the tunnels at the inlet and at the outlet will be with gates and stoplogs.

Several methods of dewatering of the tunnels have been considered. Dewatering shafts have not been included in the final scheme partly on account of the potential difficulty of their construction through the waterbearing ground and for forming the junctions between the shafts and the tunnels and partly on account of the potential waterpaths at the junction and their effect on the overall durability of the scheme.

The chosen scheme will be to lower dewatering equipment down the inclined adits at the outlet. The contract for the dewatering equipment and the electrical and mechanical equipment is based on a performance specification which will be developed during the contract.

10 CONCLUSION

The four tunnel syphon under the Suez Canal, of 775m in length, will be constructed during the three years of 1994 to 1996. A closed face TBM will be used to excavate the tunnel through clays and sands. The tunnel lining system will be precast concrete segments with a waterproofing membrane and an unreinforced concrete secondary lining, which will provide a very durable system. The tunnels will pass below the Suez Canal with a minimum of 9m cover to the future deepened and widened Suez Canal.

Tunnelling and Ground Conditions, Abdel Salam (ed.) © 1994 Balkema, Rotterdam, ISBN 90 5410 363 9

Construction of a very deep subway station using cut-and-cover method under high groundwater pressure

J. Ono
Teito Rapid Transit Authority, Japan

K. Murata
Pacific Consultants, Japan

ABSTRACT : There are two typical methods for the construction of underground subway stations. These are the cut-and-cover and shield tunneling methods. The latter is thought to be economically advantageous when the depth of excavation exceeds 20 to 25 m, and it also ensures greater work safety. On the other hand, there are some situations where the cut-and-cover method has to be adopted even for very deep excavation because of the necessity of passing close to existing underground structures or of joining up with other railways.

This report concerns the construction of a subway station 40 m deep, 18 m wide and 251 m long. The station crosses existing structures such as a multi service tunnel, sewerage tunnel, the foundations of a huge surface structure called Air Dome, and the like, and also crosses other railways and makes connections with their stations. This station is situated in an aquiferous sand layer with high groundwater pressure. The characteristics of its design are as follows:

1) Since the construction site is in downtown Tokyo, a highly rigid RC diaphragm wall was adopted to minimize the influence of excavation upon neighboring buildings and elevated railway structures.

2) A deep well system was applied to lower the groundwater level and then to stabilize the bottom surface of the excavation.

Additionally, large scale steel supports and reinforced concrete supports that were used as structural slabs were adopted to enhance the safety of excavation.

3) As a very deep construction was carried out, the RC diaphragm walls were used as the earth-retaining wall. To reduce the construction cost significantly, an up-to-date method where side walls consisting of diaphragm walls and cast-in-place reinforced concrete was adopted to become a monolithic structure.

1 INTRODUCTION

The first Tokyo subway started operation for the 2.2 km section between Ueno and Asakusa in December, 1927. Since that time new subway lines have been constructed in rapid succession. Presently, the Teito Rapid Transit Authority (TRTA) operates 162.2 km of underground railway lines on eight routes.

The excavation depth of the subway (depth from the ground surface to the rail surface) was 8.5 m on average for the routes constructed up to 1962. In view of grade-separated crossings between subway lines and the relationships with other urban facilities, new lines tend to be constructed at greater depths. Actually, the route constructed in 1990 was about 2.5 times deeper, that is, at an average depth of about 21.3 m.

Because of the concentration of the population and industries in Tokyo, the aboveground space has become saturated with buildings and other urban facilities. In this situation, spaces for urban public facilities have to be sought under the ground, so the new subway line routes are inevitably being constructed at increasing depths.

This paper introduces large-scale very-deep excavation (depth of 40 m, width of 18 m, and length of 251 m) through an aquiferous sand layer with high groundwater pressure. This current excavation at Korakuen Station on the Nanboku Line became necessary in view of the relationship with other public facilities.

2 OUTLINE OF CONSTRUCTION OF THE NANBOKU LINE

As shown in Fig. 1, the Nanboku Line is a 21.4 km long route running through the Tokyo area approximately from south to north. The 6.8 km northern section between Komagome and Akabane-Iwabuchi has been operating since November, 1991 and the 14.6 km section between Meguro and Komagome is currently under construction.

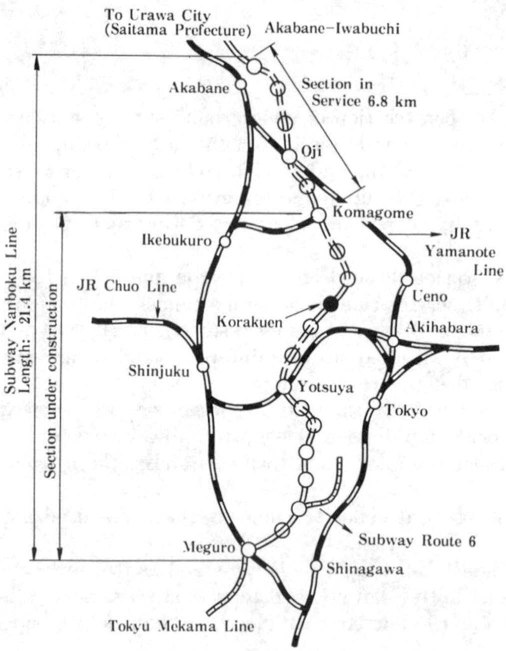

Fig. 1 Route of Tokyo Subway Nanboku Line

The Nanboku Line will have the right of through trackage with the Tokyu Mekama Line at Meguro Station, thereby providing a way of connecting downtown Tokyo directly with Kanagawa. It is also planned to extend the route from Akabane-Iwabuchi through the center of Kawaguchi to the east of Urawa. The areas which are within a 10 - 20 km radius from downtown Tokyo, but not currently readily accessible will have greater access to the inner city area.

When the entire section is opened, the Nanboku Line will become the main artery for railway transport in a north-south direction.

3 OUTLINE OF KORAKUEN STATION ON THE NANBOKU LINE

3.1 *Outline of geology*

The geomorphology of downtown Tokyo is divided into upland areas, including the Musashino plateau (Toshima and Hongo uplands) and the Yodobashi upland, and Tokyo lowland along the coast.

Korakuen Station on the Nanboku Line is located in the valley between the Toshima and Hongo uplands (both belonging to the Musashino upland). The geological conditions include an 11.4 m top layer of Aluvium over a layer of Diluvium. The former is classified into humus, and cohesive soil. The Diluvium is classified into sandy soil of the Tokyo formation with clayey soil of the Tokyo formation found in-between, the Tokyo gravel bed, and sandy soil of the Edogawa formation. The ground water is located on both sides of the clay soil of the Tokyo formation and found in two aquiferous sand layers; one is the upper sandy soil deposit of the Tokyo formation and the other is the Tokyo gravel bed and sandy soil of the Edogawa formation. The amount of ground-water is extremely large.

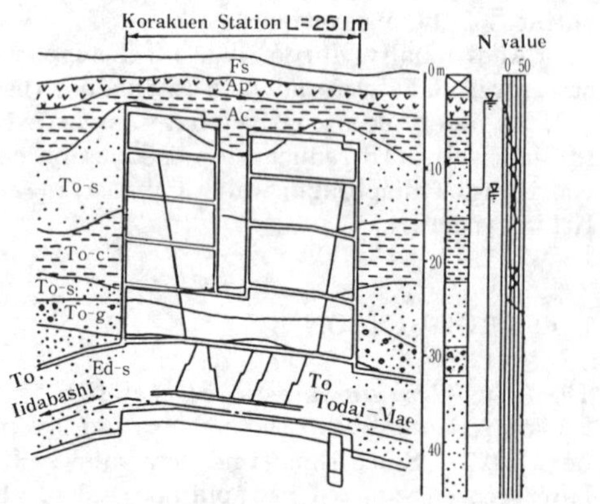

Fig. 2 Geological profile

3.2 *Outline of the structure*

Korakuen Station was planned to be located under Senkawa Street, as shown in Fig. 3, thereby enabling transfer to the Marunouchi Line, the Mita Line, and Route 12 currently under planning, connection to the Air Dome, and the Bunkyo ward office, and the commercial facilities in line with the redevelopment plan.

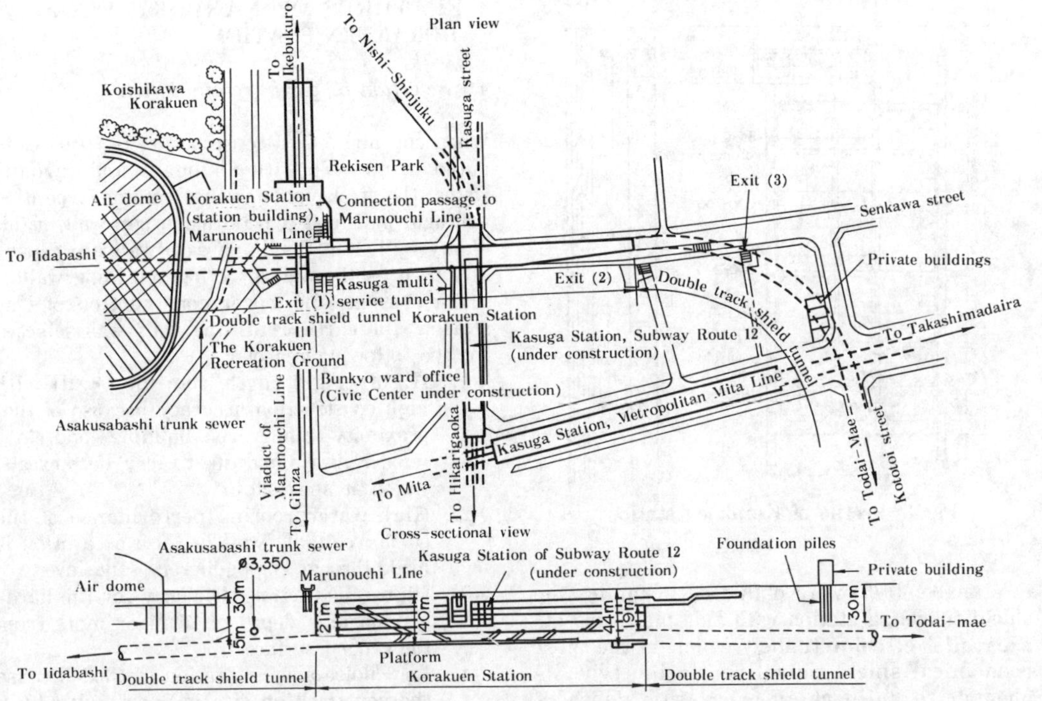

Fig. 3 Plan view and cross-sectional view of Korakuen Station

The depth of the station was 40 m on average in order to maintain sufficient distance from the foundation piles of the Air Dome, the Asakusabashi trunk sewer, the Kasuga multi service tunnel, Route 12, and existing buildings within the 1-chome area of Koishikawa.

In order to accomodate a train of eight cars the length of the station was determined to be 170 m. The platform width was determined to be as large as possible, that is, 9.0 m in order to cope with high concentrations of passengers when events are held in the neighboring Air Dome and to allow passengers to transter to other subway lines. This width also brings the station structure into conformity with the road width of Senkawa Street. As a result, the length of the station section to be constructed by the cut-and-cover method became 251 m and the width 18 m.

The station has a standard sectional shape with five stories, three spans and six stories, three spans. The first and fifth basements are for concourses, the sixth basement for the platform, and the second to fourth basements for station facilities. As the station itself is located at a considerable depth, four staircases and escalators are provided to connect the platform in the sixth basement to the concourse in the fifth basement while seven escalators are provided from the fifth to the first basement, thereby enabling rapid transit of passengers. In addition, an elevator to connect the platform in the sixth basement directly to the concourse in the first basement is provided for physically handicapped passengers.

3.3 *Construction method of Korakuen Station*

Conventionally, in Japan subway stations have been constructed by cut-and-cover methods. As the grade-separated crossings increase under the ground and the new lines have to pass below other underground facilities, the shield tunneling stations have come to be constructed in deep locations.

In comparing the economies between stations being constructed by the shield tunneling and cut-and-cover methods under standard construction conditions, the results shown in Fig. 5

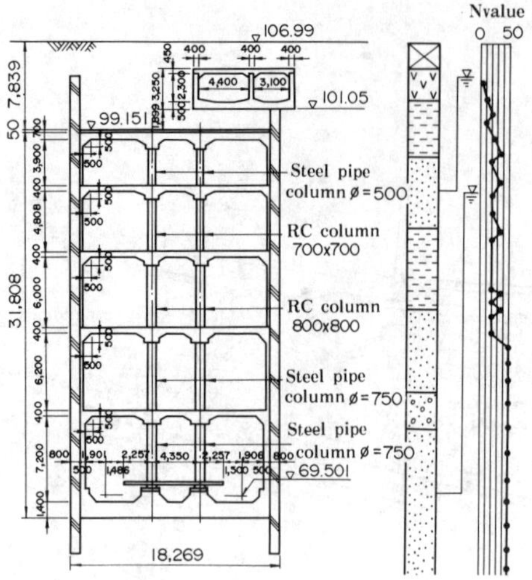

Fig.4 Profile of Korakuen station

(Watanabe, 1973) were obtained. That is, the shield tunneling station with side platforms of a parallel circular tunnel, which is the most economical shield tunneling station type, is cheaper to build at an excavation depth of about 20 m or more.

For Korakuen Station, however, cut–and–cover constructing station was chosen because of a small distance of only a few meters from the Kasuga multi service tunnel and the necessity of connecting with the route 12 subway line and the Bunkyo ward office (See Fig. 3).

4 FEATURES OF NANBOKU LINE KORAKUEN STATION

4.1 *Selection of earth–retaining wall*

When cut–and–cover construction is to be executed in an area with abundant underground water, the steel pipe sheeting, column–type diaphragm wall, soil cement diaphragm wall, rigid slurry wall, and RC diaphragm wall are generally used as the waterproofing retaining walls.

For the cut–and–cover construction of this station, the earth–retaining wall was selected for the following reasons:

1. Highly rigid earth–retaining wall with high construction accuracy because of the proximity to high–rise buildings and elevated railways and due to very–deep excavation at about 40 m
2. High waterproofing performance of the earth–retaining wall because excavation is made through the aquiferous sand layer
3. High construction efficiency for the hard ground at a depth of 25 m or more from the ground surface
4. Low–noise and –vibration method because the construction site is surrounded by a built–up area

The result of a study of the type of earth–retaining walls is summarized as follows.

1. The steel pipe sheeting is not economical because the space around the steel pipes must be filled with slurry to improve the waterproofing performance.

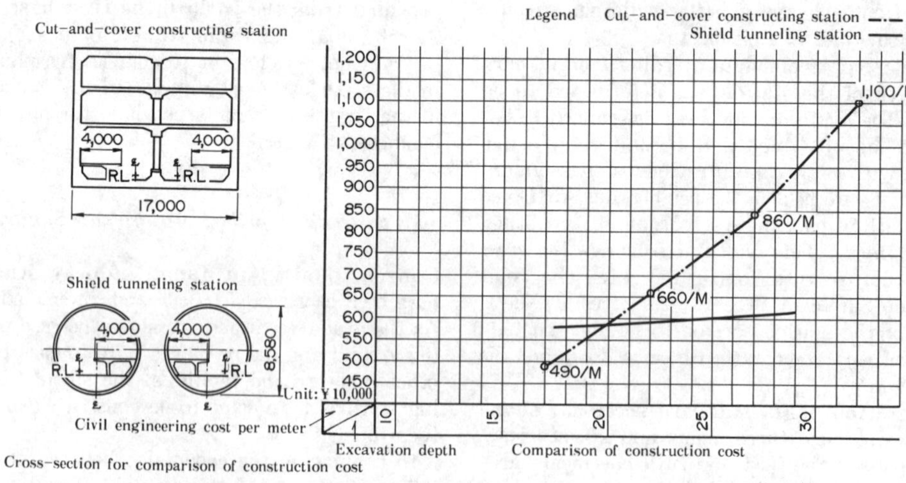

Fig.5 Cost comparison between stations being constructed by the shield tunneling and cut–and–cover methods

2. With the column type diaphragm wall, excavation of the hard ground is impossible.
3. With the soil cement diaphragm wall, the construction efficiency is lowered considerably because of the necessity of excavating the hard ground for 20 m or more.
4. The rigid slurry wall is not appropriate for this section and cannot be used as a part of the permanent wall because of safety concerns and the extreme depth of this excavation.
5. The RC diaphragm wall causes less excavation problems at extreme depths and can be used as a part of the permanent wall, ensuring economic feasibility.

On the basis of the above, the RC diaphragm wall was decided upon. At the intersection of Kasuga Street where work can only be carried out in the nighttime and there is a multi service tunnel buried under the ground, the column-type diaphragm wall was used from above the ground and the steel diaphragm wall for the section below the road.

4.2 *Countermeasures against confined high grandwater pressure*

(1) Stabilization of excavation bottom surface

The ground around the excavation bottom surface at the site being constructed by the cut-and-cover method is an aquiferous sand layer with high confined groundwater pressure (sandy soil of the Edogawa formation). Therefore, measures against boiling are necessary to stabilize the excavation bottom surface.

There are two methods available; one is to increase the depth of embedment for the earth-retaining wall and the other is to lower the groundwater level. Lowering the groundwater-level was selected for the following reasons:

1. Where the depth of embedment of the earth-retaining wall is increased, the length of embedment must be extended, significantly to about 20 m because of the lack of an impermeable layer below the excavation bottom surface.

2. Where the groundwater level is lowered, settlement problems may arise in the surrounding ground and in the draw-off capacity of pumped groundwater. But, it is known that the surrounding ground will not settle because both the aquiferous sandy soil and the impervious cohesive soil above the sandy soil belong to the diluvial ground. Besides, there will be no problem in terms of draw-off of the pumped groundwater because the Senkawa trunk sewer is running by the work site.

Because of the necessity of lowering the groundwater level by about 30 m at this site, the deep well method was employed, with deep wells drilled at about 30 m intervals as shown in Fig. 6.

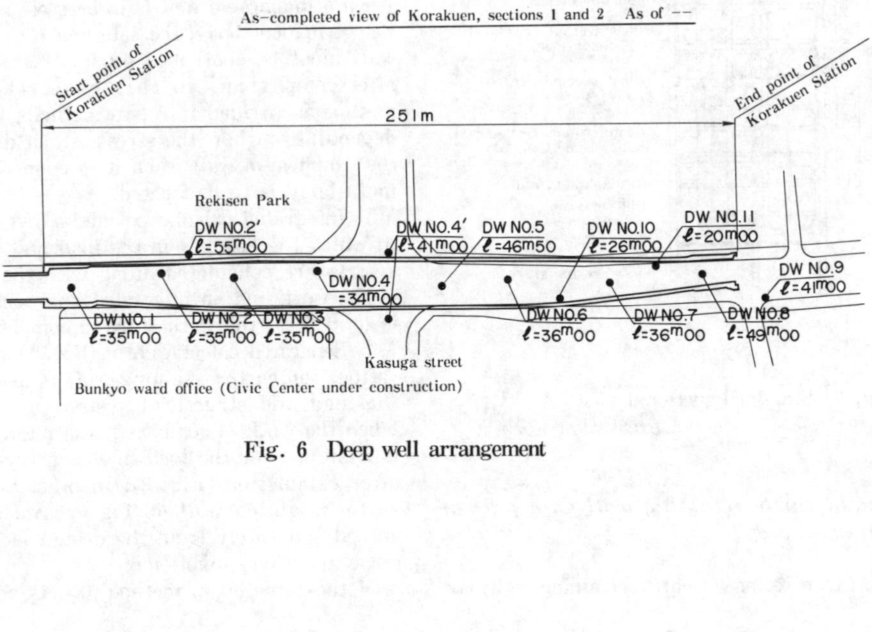

Fig. 6 Deep well arrangement

(2) Enhancement of the safety of excavation work

This cut-and-cover construction involves excavation at an extreme depth through aquiferous sandy soil with high groundwater pressure, resulting in high water pressure acting on the earth-retaining wall. It is usual that significant deformation of the earth-retaining wall under such high groundwater pressure causes loosening of the ground behind the wall, resulting in its substantial settlement or unexpectedly high lateral pressure on the earth-retaining wall.

This work must therefore minimize the deformation of the earth-retaining wall to enhance the safety of excavation work.

As a countermeasure against the above problem, the RC diaphragm walls which offer the highest rigidity as an earth-retaining wall were used. In addition, highly rigid supports, which are large scale steel supports of H-400 x 408 and reinforced concrete supports being used as structural slabs, were also used to suppress the deformation of the earth-retaining wall. (See Fig. 7)

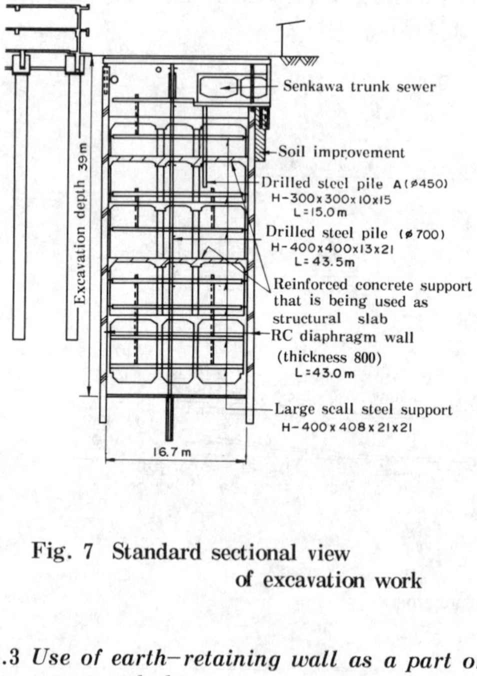

Fig. 7 Standard sectional view of excavation work

4.3 *Use of earth-retaining wall as a part of permanent body*

Though expensive as an earth-retaining wall, the RC diaphragm walls are such that they can be incorporated into the permanent body. For this station, the RC diaphragm wall which was originally the earth-retaining wall was used as a part of the permanent body, thereby achieving the widest possible platform and reducing the construction cost.

There are various design concepts for the use of the earth-retaining wall as a part of the permanent body. The following concept was adopted for this station:

(1) Method for use as a part of the permanent body

When the RC diaphragm wall was used as a part of the permanent body initially, either the independent or overlap wall method was usually adopted. Recently, however, the employment of a unified wall has increased for economic reasons as various test and measurement results confirmed the feasibility of combining the diaphragm wall and inner wall into a composite wall.

The TRTA also adopted the unified wall method for this station because the measurement result of back-filling in other work sites confirmed that the diaphragm wall and inner wall could behave as a unified wall.

(2) Permanent body structure design method

When a diaphragm wall is to be used as a part of the permanent body, the safety of the diaphragm wall must be confirmed during excavation and after completion. In this case, the design method is divided into two methods as follows depending on how the stress which develops in the diaphragm wall when it is connected to the main body is to be treated:

① Integrated calculation method: A method in which the changes in the load and structural system are considered along with the sequence of the work and an integrated analytical method is used from the work start to completion.

② Separated calculation method: During excavation, the earth-retaining wall is designed for the load and structural system at that time. When the work is completed, the main structure is designed for the load and structural system after completion (Fig. 8). In other words, the earth-retaining wall during excavation is designed separately from the design of the main structure after completion.

Of these methods, method ① has advantages

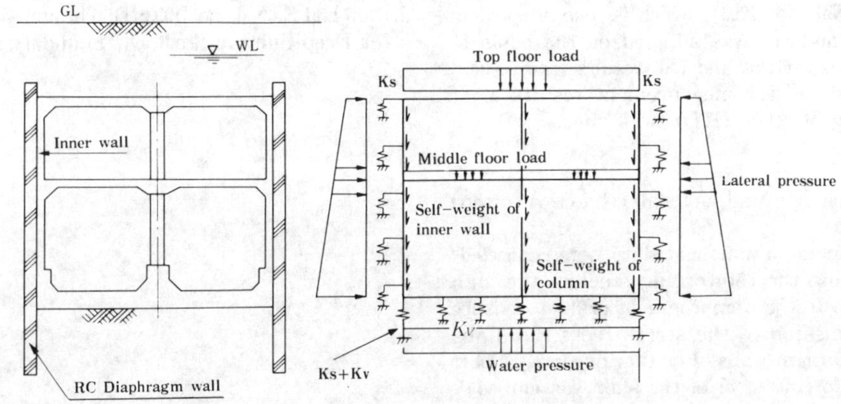

Fig. 8 Structural drawing and model

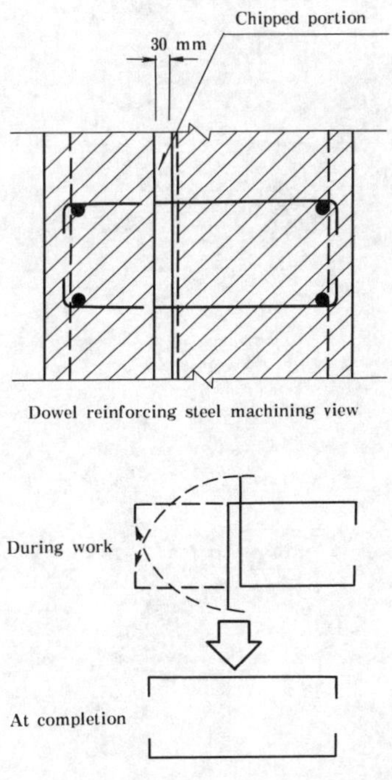

Fig. 9 Inner wall joint method

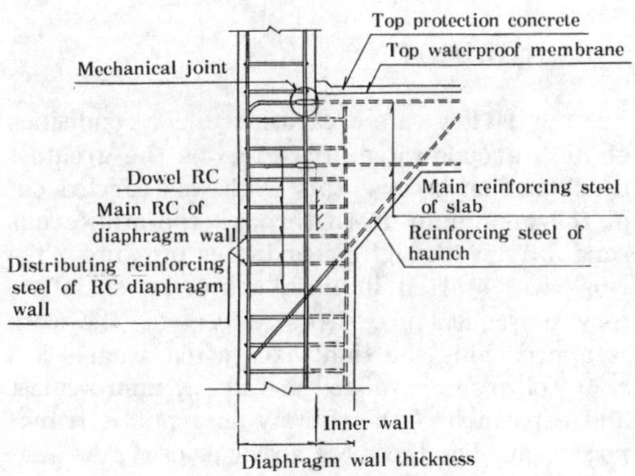

Fig. 10 Standard reinforcing bar arrangement for the connection between the diaphragm wall and slab

because an acceptable design is possible through the consolidation of disparate member stresses, but suffers because the calculation becomes extremely complicated. Besides, the stress developed in the earth-retaining wall during work does not have much influence on the ultimate strength when sectional failure occurs (Kinki Branch of Japan Construction Consultants Association, 1979). In view of these considerations, method ② is applicable in most cases.

As a consequence, design method ② was adopted for this station. Note here that the stress of the side wall is kept below 90 % of the allowable stress because the high stress developed during work may possibly remain in certain sections.

(3) Integration of diaphragm wall with inner wall

Integration of the diaphragm wall with the inner wall was made by chipping out the dowel reinforcing steel embedded in the diaphragm wall for anchorage with the inner wall (Fig. 9). To determine the dowel reinforcing steel amount, a method based on the shear-friction hypothesis

(Ono and Nakada, 1993), which is also adopted in the ACI standard, was adopted on the basis of various measurement and test results from other enterprises and the measurement results after back-filling at other TRTA work sites.

(4) RC diaphragm wall and slab connection method

The RC diaphragm walls and slabs were connected by anchoring the reinforcing steel of the slab and haunch to the diaphragm in order to ensure the transmission of the stress from the slabs to the diaphragm walls. For this purpose, a part of the reinforcing steel of the slab was embedded in the diaphragm wall beforehand, which was then bent to rise during construction of the slab reinforcement.

5 CONCLUSIONS

For the TRTA, safe excavation under conditions of high groundwater pressure was the greatest consideration because this work was carried out at the maximum depth through the aquiferous sand-layer with high groundwater pressure. The work was started in 1988, achieving satisfactory subsequent progress. At present, the work is approaching the final excavation stage.

In Tokyo's overcrowded downtown, improvement and expansion of the subway as a public transport means has become a social priority. As various public facilities are constructed under the ground, the depth of the subways will increase further in the future.

The TRTA intends to utilize the experience obtained during this work and various measurement data for construction in the future while making efforts to open the entire Nanboku Line as early as possible. Further improvement and expansion of the subway service is also what we strive for.

References

K. Watanabe, 1973: Study on the structural design of Subway Shield Tunneling Stations, Teito Rapid Transit Authorities, Construction Headquarters, P22 – 24.

Kinki Branch of Japan Construction Consultants Association, 1979: Ultimate Strength of Double-casted RC Walls, P65 – 68.

J. Ono and S. Nakano, 1993: Diaphragm Walls used for Deep Subway Stations, Foundation, P73 – 79.

Tunnelling and Ground Conditions, Abdel Salam (ed.) © 1994 Balkema, Rotterdam, ISBN 90 5410 363 9

Compressed air tunneling through saturated porous media

Ashour Ghareeb Abd El-Maksoud Shalaby
Structural Department, Faculty of Engineering, Ain Shams University, Egypt

ABSTRACT: The use of compressed air for tunneling through saturated porous media is one of the common tunneling techniques specially when large size boulders are expected at the tunnel face. Upward flow of ground water, associated with compressed air tunneling, occurs and the stability of the surface and subsurface utilities is influenced.

In this paper, factors controlling the properties of upward ground water flow are numerically investigated. Moreover, the effect of that upward flow and consequently the upward seepage force on the mobilization of ground movements is studied.

I. INTRODUCTION:

Ground movements associated with tunnelling at different levels have a great effect on surface and subsurface utilities. Many research works were performed to investigate sources of ground loss which causes ground movements. Ground control is necessary to avoid much ground loss specially in the case of tunneling through porous deposits below ground water level. Different techniques as grouting, freezing, lowering of ground water level and the use of compressed air are the available methods used for controlling ground movements. Among these techniques, lowering of ground water level and the use of compressed air technique are the common and economic for tunneling. In some cases, the use of lowering of ground water leads to land subsidence which leads to a damage effect on utilities. The compressed air tunneling technique is widely used for tunneling through difficult soil conditions.

This paper presents a numerical solution to study the effect of using the compressed air on the feature of the flow of ground water around tunnel section, the variation of the porewater pressure and consequenly the created changes in the effective stresses. Isotropic and anisotropic porous media are considered.

2. MATHEMATICAL MODELLING

The problem of flow through porous media under the condition of steady state is mathematically expressed by Laplace equation which takes the form:

$$kx\,(\delta^2h/\delta x^2) + ky\,(\delta^2h/\delta y^2) + kz\,(\delta^2h/\delta z^2) = 0$$

where: kx, ky, and kz are coefficients of permeability in x, y and z directions respectively and h is the potential head.

In the case of two dimensional flow, Laplace equation is reduced to the form:

$$kx\,(\delta^2h/\delta x^2) + ky\,(\delta^2h/\delta y^2) = 0$$

Alternatively, this problem is numerically solved using finite element model (Zienkiewiez, 1986 & Hassan, 1989 and Attabi et al., 1991 & Hassan et al., 1992). The model considered here is applicable for two dimensional flow through homogeneous and non-homongenous as well as isotropic and anisotropic porous media. The problem considered here is a tunnel of 4m diameter and 12m depth of crown below the surface which is excavated through porous media. The water level is assumed at depth 2m below the surface as shown on Figure 1. To prevent water flow into the tunnel section, an air pressure 18 t/m^2 is applied which is equivalent to 34m and 30m potential heads at crown and invert respectively assuming reference level is 28m below ground surface. One half of the section is considered and divided into 268 triangular finite elements and 157 nodes as shown on Figure 2. The following boundary conditions are considered:

AB is an equipotential line at which the excess potential ceases, and CD and EF are flow lines.

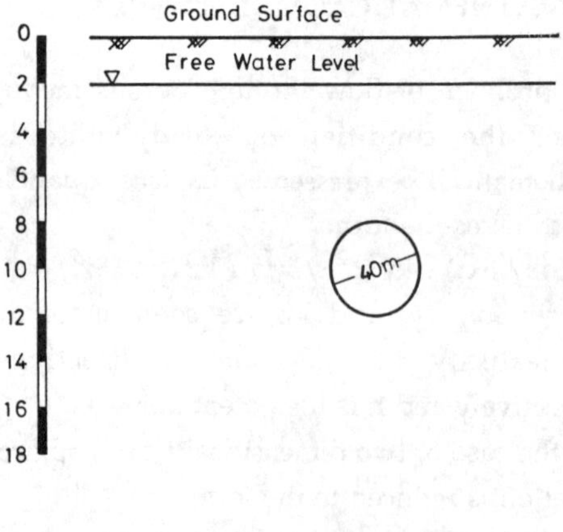

Fig. 1 Tunnel configuration

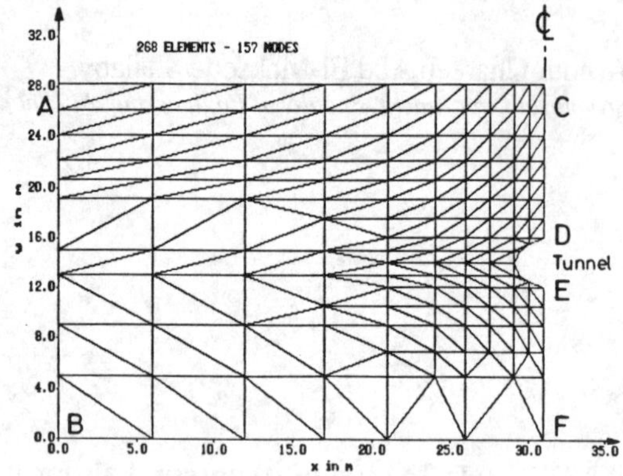

Fig. 2 Finite element mech and tunnel configuration

3. RESULTS AND DISCUSSION

To investigate the effect of anisotropy ratio (kr) which is defined as the ratio between permeability in vertical direction to that of horizontal direction on the flow pattern, porewater pressure and consequently effecitve pressure due to the effect of air pressure, different values of 0, 1, 0.5, 1.0, 5.0 and 10.0 are considered. Figure 3 shows the flow pattern in case of isotropic media (kr = 1). Figure 4 and 5 shows samples of the computer output in the case of anisotropic media having kr = 0.1 and 10.

Investigation of these figures show how the flow pattern is affected by the anisotropy ratio. As the value of anisotropy ratio increases, the possibility of flow in the vertical direction increases. At the same time and although the value of air pressure is higher than the porewater pressure at the invert by about 29%, flow of water into the tunnel section is occurred at higher values of anisotropy ratio.

Figure 6 shows how the overflow on the surface is influenced by anisotropy ratio. Examination of this figure shows that the phnomenon of overflow on the surface increases with the increase of anisotropy ratio. Overflow on the surface may affect the stability of surface grains and consequently the stability of surface utilities. On the other hand, overflow phenomenon causes difficulties to the surface use of the ground surface.

Figure 7 shows the effect of the anisotropy ratio on the distribution of porewater pressure on a vertical plane passing through tunnel center line. It is clear that the porewater pressure increases with the anisotropy ratio. The excess water pressure due to compressed air tends to be maximum around the tunnel and decreasing with different rates above and below crown and invert respectively.

The differences between initial porewater pressure (hydrostatic pressure) and the final porewater pressure represent the seepage forces which have an effect on the soil grains. Seepage forces improve the stability of soil grains around tunnel because of its direction is upward above the crown and downward below th invert. The upward and downward seepage forces caused a decrease and increase of effective stresses respectively. These changes in the effective stress values cause less ground loss and consequently ground movements. Figure 8 illustrates the variation of porewater pressure on horizontal plane passing through the spring line. It is obvious that the anisotropy ratio has a small effect on the value of porewater pressure. On the other hand, the porewater pressure value is mainly affected by the air pressure value.

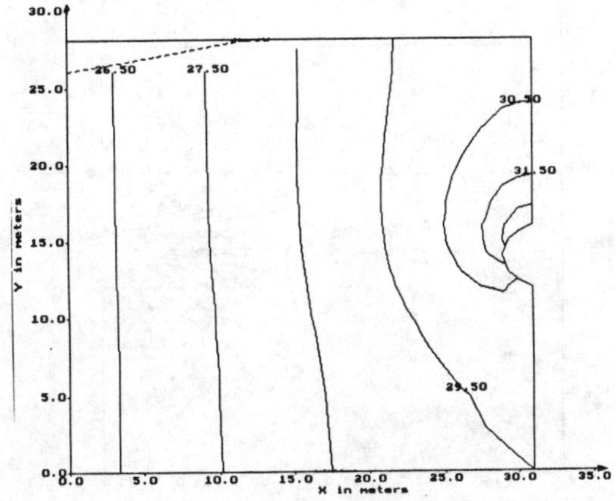

Fig. 3.a Potntial head distribution
(kr = 1.0)

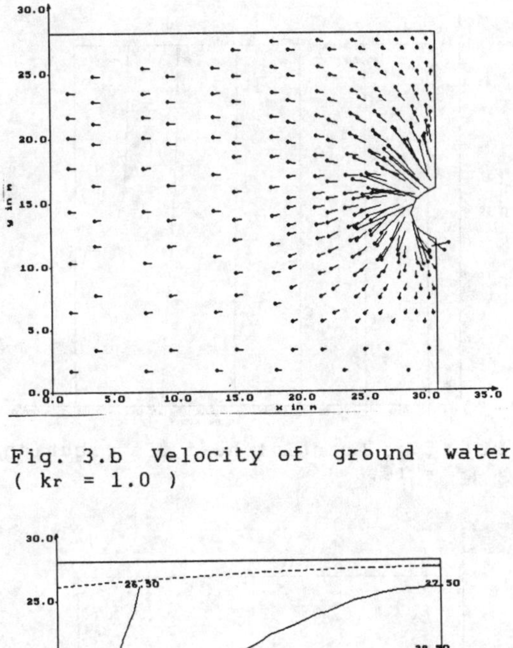

Fig. 3.b Velocity of ground water
(kr = 1.0)

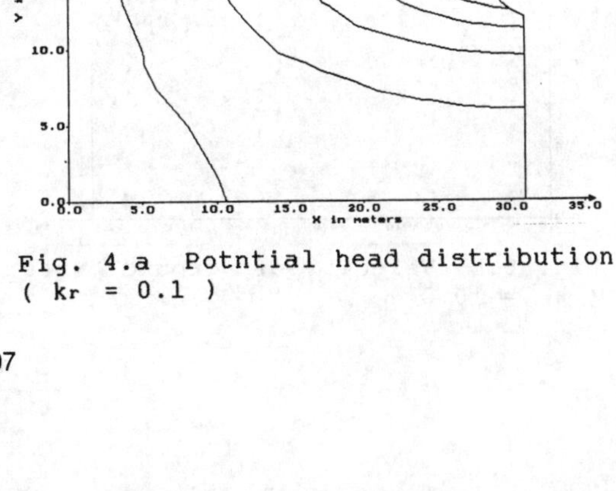

Fig. 4.a Potntial head distribution
(kr = 0.1)

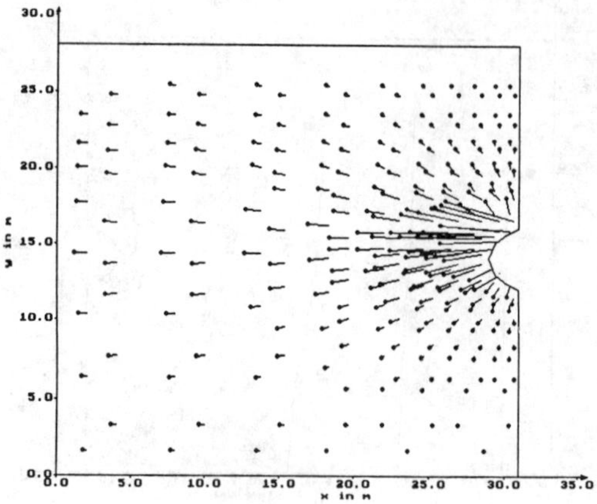

Fig. 4.b Velocity of ground water (kr = 0.1)

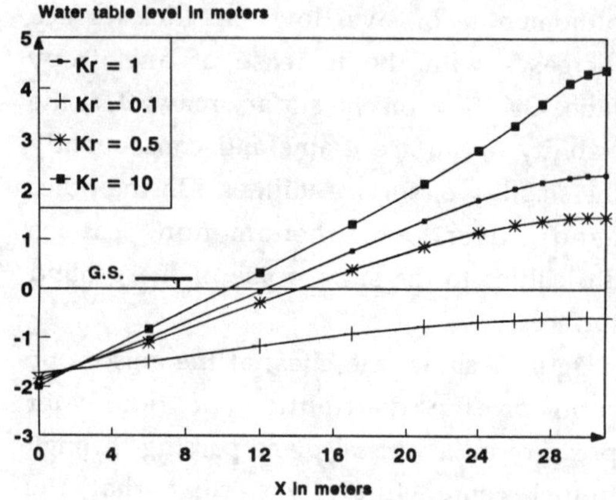

Fig. 6 Effect of anisotropy ratio on surface overflow

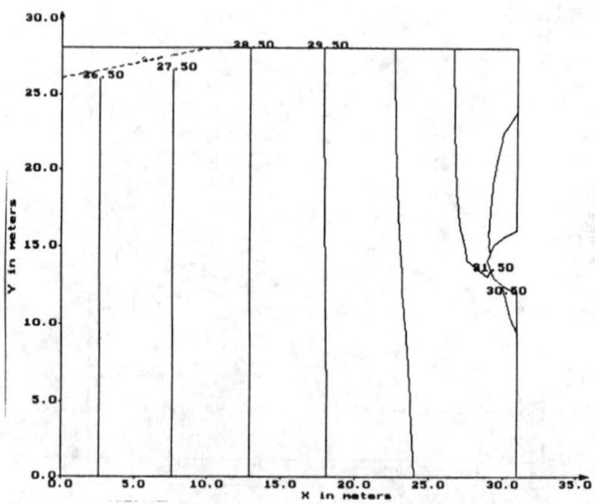

Fig. 5.a Potntial head distribution (kr = 10.0)

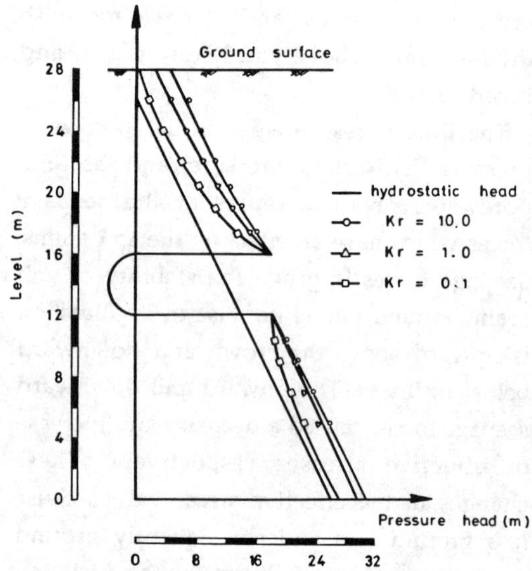

Fig. 7 Effect of anisotropy ratio on porewater pressure at different levels

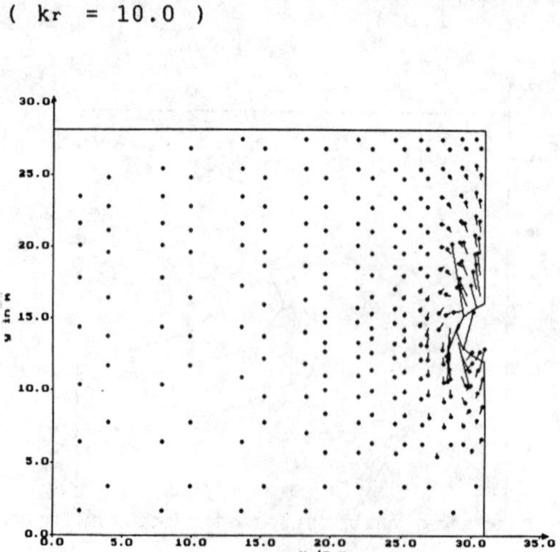

Fig. 5.b Velocity of ground water (kr = 10.0)

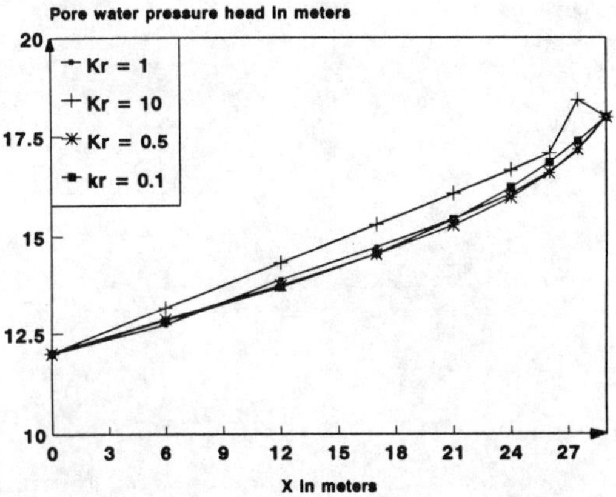

Fig. 8 Pressure head on horizontal plane passing through tunnel centerline

4. CONCLUSIONS

Based on the previous analysis, the following conclusions can be drawn:

1. Anisotropy ratio has a great effect on water flow pattern associated with compressed air tunnelling. Flow of water tends to occur vertically at higher values of anisotropy ratio.
2. Effective stress was found to be greatly affected by the anisotropy ratio specially on vertical planes passing through the tunnel centerline.
3. Effective stress on horizontal plane passing through tunnel centerline is not practically affected by anisotropy ratio.
4. Loss of ground associated with waterflow due to compressed air, and consequently ground movements, decreased as the anisotropy ratio increased. Vice versa, stability of surface soil grain and surface utilities is affected by the increase of anisotropy ratio.
5. Flow of water into tunnel section around invert occurred at high values of anisotropy ratio although air pressure increased by about 29% than water pressure.

5. REFERENCES:

Attabi, et al., 1991. Evaluation of uplift pressure distribution on raft foundation. C.E. Conf., Cairo, March 5-6.

Hassan, A.A., 1989. Hydrology of wadis in arid regions with application to quality models. Ph.D. thesis, Ain Shams University, Cairo, Egypt.

Hassan, A.A., Shalaby, A.G. and Ahamed, A.A., 1992. Coffer-dam-flow hydraulics interaction in porous medium. Soil mechanics & foundation, Vol. 7 pp. 45-49.

Kinzelbach, W., 1986. Groundwater modelling. Elsever, New York.

Zienkiewicz, O.C., 1971. The finite element method in engineering science, McGraw-Hill, London.

Tunnelling and Ground Conditions, Abdel Salam (ed.) © 1994 Balkema, Rotterdam, ISBN 90 5410 363 9

Behaviors of ground and adjacent underground piping during shield tunnelling

T. Yoshida
Nippon Koei Co., Ltd, Japan

M. Kusabuka
Department of Civil Engineering, Hosei University, Japan

ABSTRACT: Analyzing the in-situ measurements from neighboring constructions, it was confirmed that the most serious influences are ground heaving caused by additional earth pressure due to excess thrusting force, and friction between shield skinplate and surrounding subsoils as a shield passes. The existing underground pipes and tunnels are influenced more by the additional earth pressure than by subsidence due to stress release. It was considerd that the displacement and deformation of the existing underground piping varies according to the section area ratio as well as the standoff distance ratio.

1 INTRODUCTION

Recently, due to the concentration of the population in urban areas in Japan, a number of shield tunnels, designed as part of lifeline facilities, have been constructed in these areas. Such tunnels are inevitably constructed very close to existing underground pipes and tunnels, because underground space is limited. Naturally this situation has caused many problems, such as damage to existing underground piping induced by settlement associated with shield tunnelling. However, countermeasures are not always used rationally, because the behavior of the ground and the underground piping during the passing of the shield were not known in detail. In order to clarify this, the authors conducted field investigations at several tunnelling sites neighboring existing undergrond piping in Tokyo. From the investigations, some significant information was obtained with regard to the items below.

1. Ground behavior (particularly ground heaving)
2. Behavior of underground piping due to ground displacement.
3. Factors affecting the behavior of the ground and underground piping.

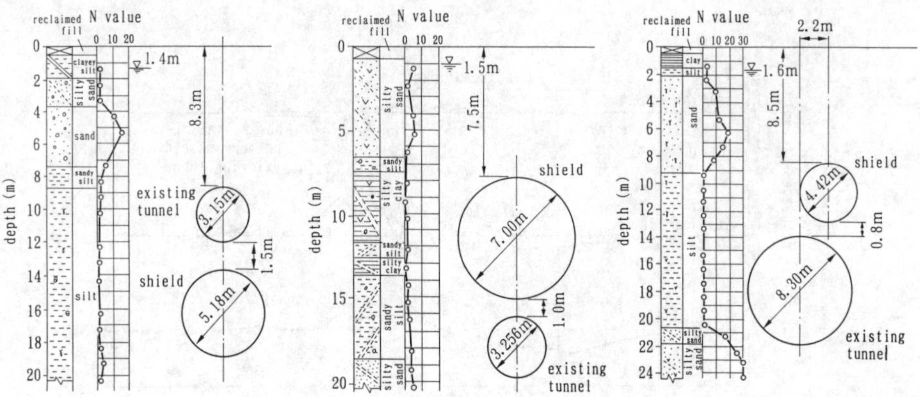

(a) site A (b) site B (c) site C

Fig. 1 Cross sections of investigation sites

2 SITE CONDITIONS AND MEASUREMENTS TAKEN

For the above mentioned purposes, the authors chose three shield construction sites in soft ground, in the eastern area of Tokyo, each with different site conditions. Cross sections for the three sites are shown in Fig. 1. The shield diameter (D) ranged from appoximately 4 to 7m. The underground pipes were shield tunnels (diameter ≈3-8m), and the distance between the shield and the existing tunnels ranged from 0.8 to 1.5m. At site A, the shield was driven below the existing tunnel, however for sites B and C, the shields were driven above the existing tunnels. The subsoil of the three sites was composed of layers of alluvial silt. The shield machines used were blind types.

Measurements included the vertical and horizontal displacement of the ground, earth pressure, overall displacement and sectional deformation of the existing underground tunnels. Investigation equipment and typical layout, site B are shown in Fig. 2.

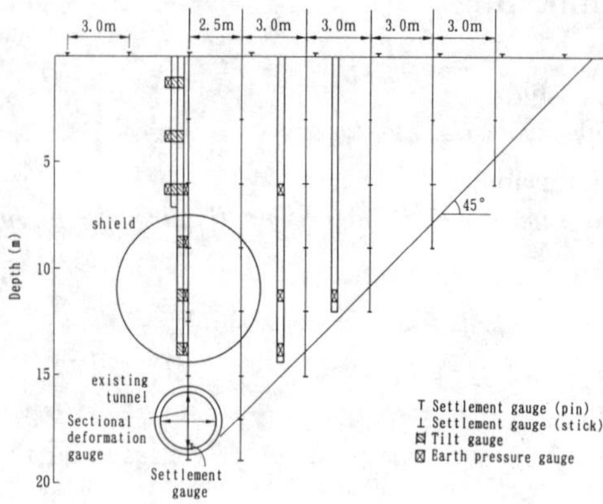

Fig. 2 Installation of gauges (site B)

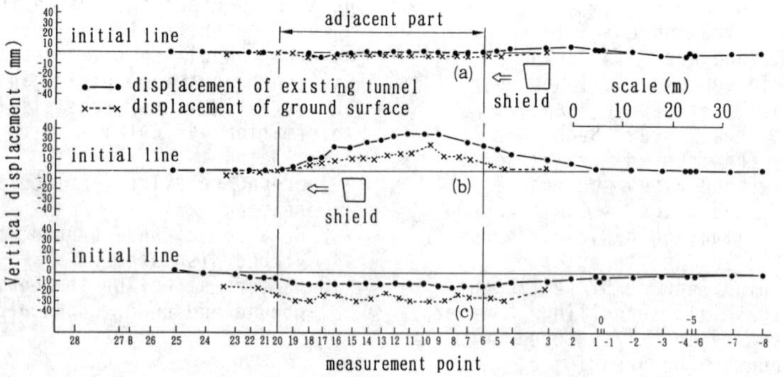

Fig. 3 Vertical displacement of existing tunnel and ground (site A)

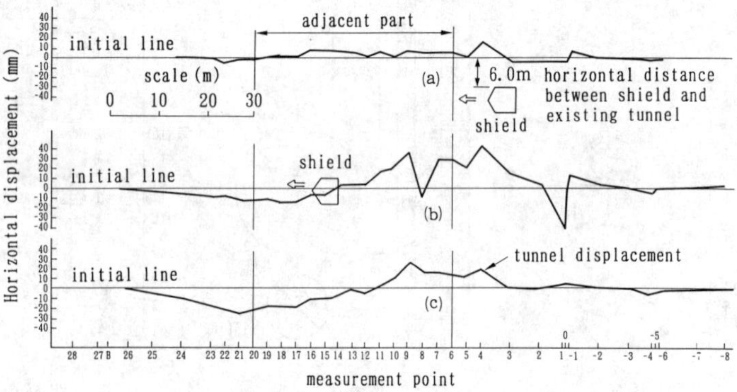

Fig. 4 Horizontal desplacement of existing tunnel (site A)

3 RESULTS OF INVESTIGATIONS

3.1 Site A

The vertical displacement of the ground and existing tunnel at site A is shown in Fig. 3. The tunnel was sinking at the final steady state due to stress release at the yield of the tail void. However, during the shield passes the tunnel rose together with ground heaving. The tunnel showed a convex shape with the peak at the shield position, and this peak moved with the shield as it progressed forward. Thus, it was considered that the heaving was caused by the additional earth pressure due to the excess thrusting force, and the friction between the shield skinplate and the surrounding subsoil. The horizontal displacement, as shown in Fig. 4, was different from the vertical one. The distribution shows that the tunnel deflected to the left and right along the center line due to the action of the passive earth pressure. The horizontal displacement did not recover up to the initial center line of the tunnel at the final steady state.

Fig. 5 shows time-displacement curves for the ground between the existing tunnel and shield. Three settlement gauges were installed at different depths. The curves indicate displacement relative to the existing tunnel. In the figure, the displacement becomes larger closer to the shield. This indicates that the soil moved around the tunnel surface in a plastic flow state, as i Fig. 6. This was because the ground was soft and the existing tunnel rigid. Due to this ground movement the additional earth pressure influencing the existing tunnel is assumed to have decreased.

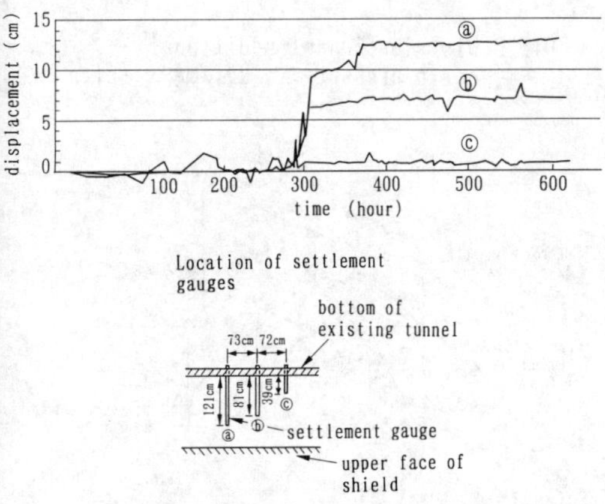

Fig. 5 Time-displacement curve for gruond between existing tunnel and shield (site A)

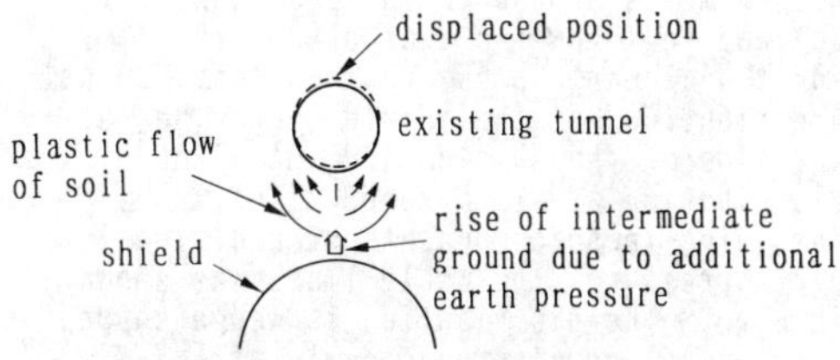

Fig. 6 The movement of ground between existing tunnel and shield

3.2 Site B

The Vertical displacement of the existing tunnel at site B was as shown in Fig. 7. Although the shield passed above the existing tunnel, the existing tunnel was deformed downwards, and this tendency remained at the final steady state. This indicates that the additional earth pressure caused by the shield thrusting was acting not only in an upwards but also downwards. This suggests that neighboring underground piping are influenced more by additional earth pressure than by subsidence. In order to protect the existing underground piping, it is therefore very important to understand correctly in which area of the ground the additional earth pressure due to the shield thrusting force occurs, and how the magnitude varies according to the distance from the shield.

Fig. 8 shows the distribution of the

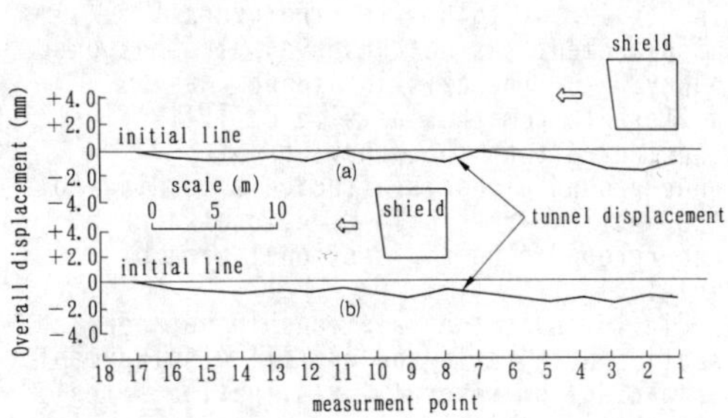

Fig. 7 Vertical displacement of existing tunnel (site B)

additional earth pressure obtained by measurements around the shield tunnel. (a), (b) and (c) show the contours of the additional earth pressure in a longitudinal distribution, at the shield center and at the side of the shield, and the plane distribution at 11.5m respectively. Combining these components of additional earth pressure, the whole image was shown in Fig. 9. The distribution is wedge shaped with almost equal expansion in all directions, the center of which is in the shield machine. The additional earth pressure behind the shield tail was not observed. The maximum additional earth pressure occured at the bottom (toe) of the cutting face and at the side of shield.

From figures 8 and 9, it can be seen that the additional earth pressure, as well as the ground rise, varied in accordance with the distance from the shield, and that it decreased greatly beyond one shield diameter from the shield. This suggests, that the influence on underground piping differs considerably depending upon whether the distance from the shield to the existing underground piping is longer or shorter than 1 D. At the front of the cutting face the soil mass is taken in through openings, and therefore additional earth pressure trends to decrease closer to the cutting face.

3.3 Site C

Fig. 10 shows the time-strain curve measured by concrete gauges installed on the secondary lining of the existing tunnel. Tension occured on the concrete surfaces when the shield cutting face reached the measurement point. Fig. 11 shows the time-sectional deformation curve. Vertical deformation (contraction) and horizontal deformation (extension) were observed at the same time. From this, it may be seen, that the existing tunnel deformed as shown in Fig. 12. At this site, the overall displacement was not observed, the response behavior of the existing tunnel was different from that at site B. It is considered that the behaviors of underground pipes is affected according to the section area ratio (sectional area of underground pipe a / sectional area of shield A).

Fig. 13 shows the relationship between the section area ratio and overall displacement obtained from several investigation sites neighboring existing underground piping. From this, it is clear that displacement is large if the section area ratio is low, and that displacement does not occure if the section area ratio is more than 1.0. This suggests that the influence on existing

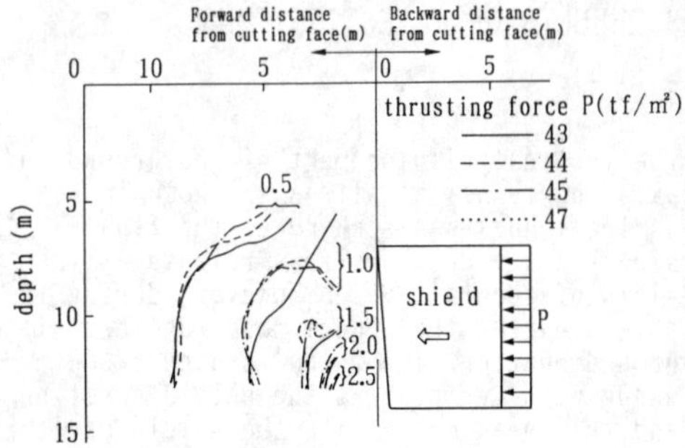

(a) longitudinal distribution at shield center

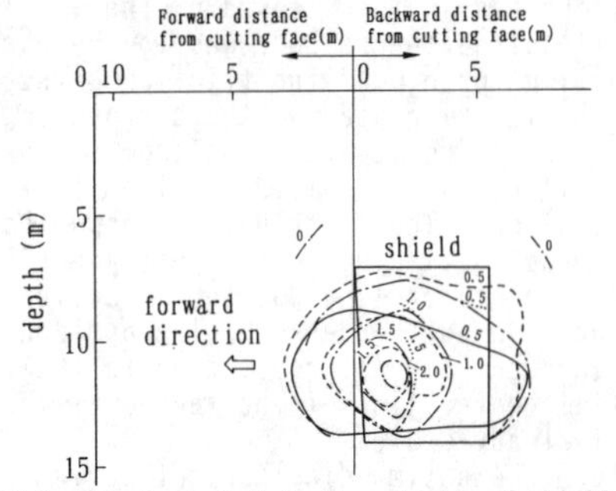

(b) longitudinal distribution at side of shield

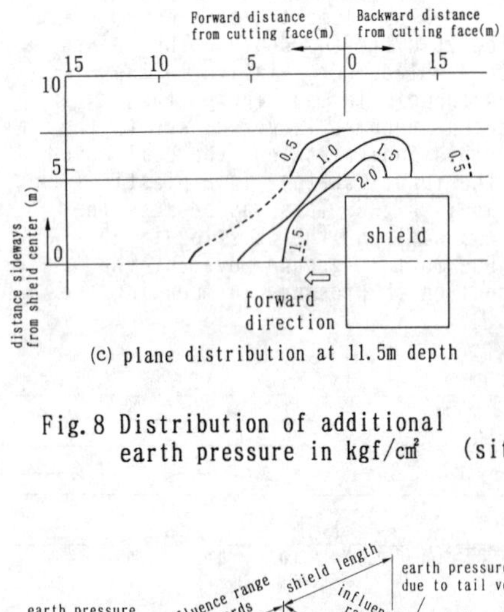

(c) plane distribution at 11.5m depth

Fig. 8 Distribution of additional earth pressure in kgf/cm² (site B)

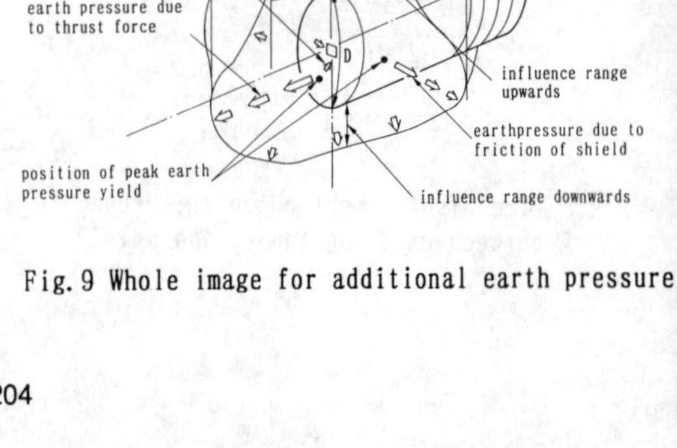

Fig. 9 Whole image for additional earth pressure

underground piping depends upon whether the section area ratio is larger or smaller than 1.0.

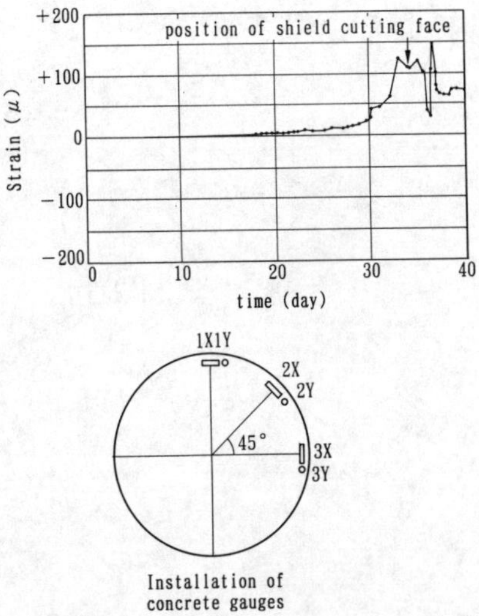

Fig. 10 Strain in secondary lining (1X) (site C)

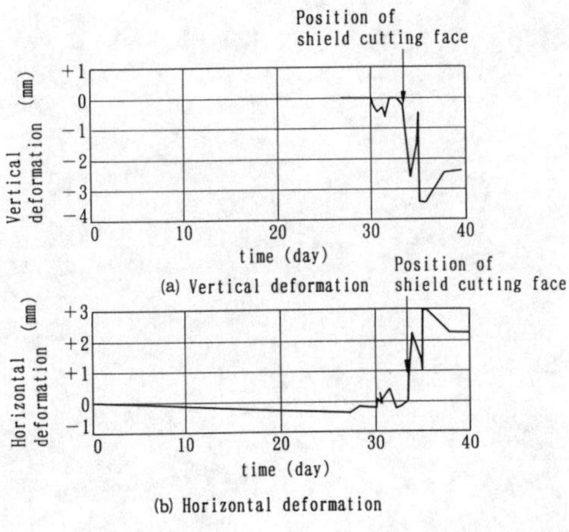

Fig. 11 Sectional deformation for the existing tunnel (site C)

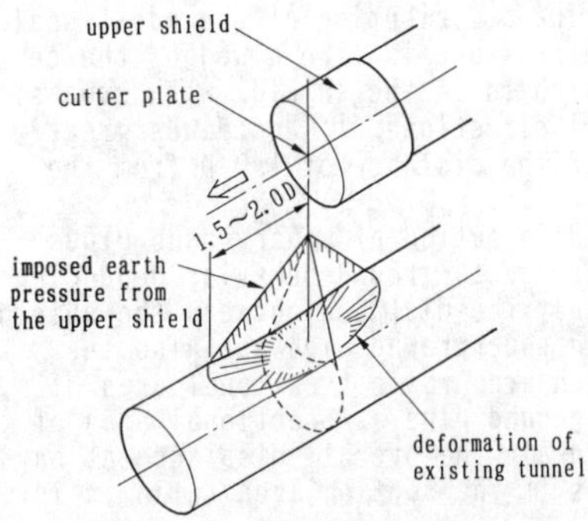

Fig. 12 Imposed earth pressure and deformation of existing tunnel

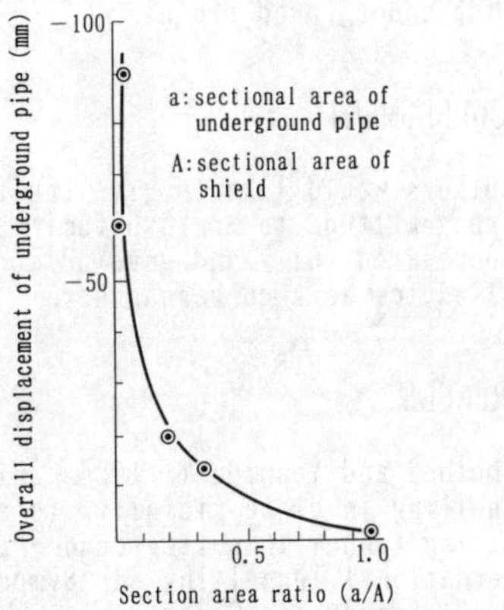

Fig. 13 Relationship between section area ratio and overall displacement of underground pipe

4 CONCLUSION

The following points were clarified by the field investigations.

1. As a shield passes, additional earth pressure occurs temporarily around the shield due to the excess thrusting force and the friction between the shield skin plate and the surrounding subsoil. Existing underground piping are influenced more by this additional earth pressure than by subsidence due to stress release.

2. Vertical and horizontal displacement behavior in existing underground piping is different; horizontal displacement does not recover to it's initial position.

3. Soil between an existing tunnel and a shield moves around the existing tunnel in a plastic flow state, and therefore the displacement of the existing tunnel is restained.

4. The distribution of the additional earth pressure is like a wedge, the center of which is in the shield, with expansion in all directions. It decreases greatly beyond the distance of 1.0 D from the shield.

5. Deformation of underground pipes varies due to ground behavior depending on not only the distance between the shield and an underground pipe but also the section area ratio (sectional area of underground pipe a / sectional area of shield A). The overall displacement hardly occurs if the section area ratio is more than 1.0.

6. When shield tunnelling close to the existing underground piping, construction in real time measurement system is both essential and effective in protecting existing underground piping.

ACKNOWLEDGMENT

The authors would like to give their sincere gratitude to Mr. Yoshifumi Takahashi who cooperated fully and gave us much useful advice as a co-researcher.

REFERENCES

Hashimoto. S and Yoshida. T. 1984. Shield tunnelling in close vicinity, to an existing tunnel in soft ground. Proc. of International Tunnelling 84' Symposium : 493-501. Venezuela, Caracas.

Hashimoto. S. 1984. Results of shield tunnelling in vertical colse to existing shield tunnels. Proc. of JSCE No 352 : 1-22 (in Japanese)

3 Underground structures in rock formations

Tunnelling and Ground Conditions, Abdel Salam (ed.) © *1994 Balkema, Rotterdam, ISBN 90 5410 363 9*

Excavating through swelling ground

Teiichi Asai
Kumagai Gumi Co., Ltd, Kanazawa, Japan

ABSTRACT: In excavating a tunnel to shorten a narrow and winding existing road into a straight-line route,ground of much poorer geological condition than originally assumed was encountered. It was found as a result of investigation that this ground corresponded to one of the most swelling in Japan. To grasp the behavior of the natural ground,the final displacement was predicted by analyses at the stage when the upper-half cross section had been excavated employing various methods, and further excavation was performed upon providing measures to deal with the amount of deformation. With steps taken for improvements in performing construction, it was succeeded in holding the displacement quantity below the predicted value.

1 INTRODUCTION

Nakaya Tunnel is a highway tunnel located at a peninsula which juts out into the Sea of Japan side at the middle of the Japanese island arc. This highway is a trunk road supporting industry, economy, and tourism on the peninsula, but traffic during the tourist season and in the wintertime is greatly hindered by section after section crossing over rugged mountainland where grades are steep and the road is narrow and winding.

Thus, it was decided to construct the 2-1ane Nakaya Tunnel of length 1,260 m, width 9.75 m, and excavation crosssectional area 83 to 87 m^2.

2 OUTLINE OF GEOLOGY

The topography in the vicinity of Nakaya Tunnel is that of a mountainland of 200 to 300 m with many ups and downs where erosion has progressed to a considerable degree. The geology consists of Neogine Miocene sedimentary rocks and volcanic rocks. The rock characters at the tunnel location are mainly basalt at the southwest side overlain by tuff and alternations of mudstone and sandstone, and alternations of tuffaceous sandstone,mudstone, and conglomerate at the northeast side, it being a monoclinics structure of approximately 20 deg. from the southwest to northeast. The geological profile is shown in Fig.1.

3 OUTLINE OF CONSTRUCTION

Construction of the tunnel was commenced simultaneously from both portals by NATM (short bench cut construction method). The standard section of the tunnel and the typical standard timbering pattern are shown in Fig.2. The outline of construction execution is given below.

1) Portalling In
Since portal sections were where landslides and slope collapses had occurred, soil cement of approximately 1,200 m^3 was placed as counterweight fill.

2) Excavation and Mucking
For excavation, one hydraulic breaker and one backhoe (0.4 m^3 loading capacity class) were used as auxiliary equipment. Muck produced in excavation was hauled outside to a temporary stockyard by dump truck ($11t$).

3) Timbering
For shotcreting, shotcrete equipment of the wet-mix system (pump type) equipped with a ceramic nozzle was used aiming for reduction in dust and rebound. For drilling holes for longer rock bolts (6, $9m$), special coupling sleeves (outside diameter 41 mm, with partition wall) were used aiming to shorten drilling time. The construction sequence is shown in Fig.3.

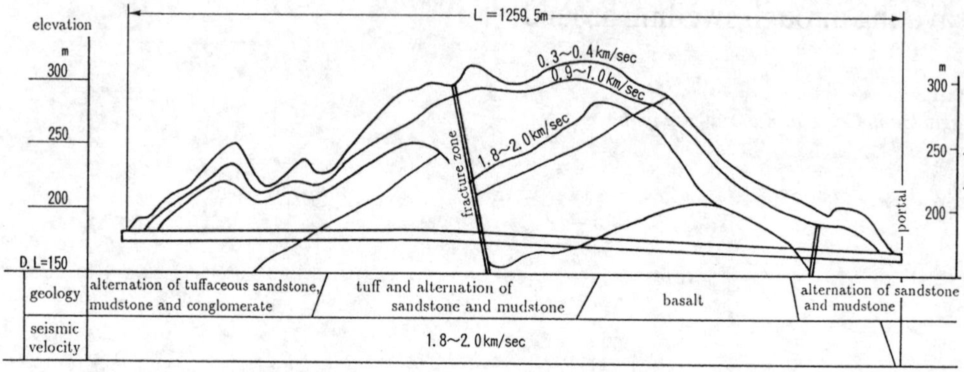

Fig.1 Geological profile

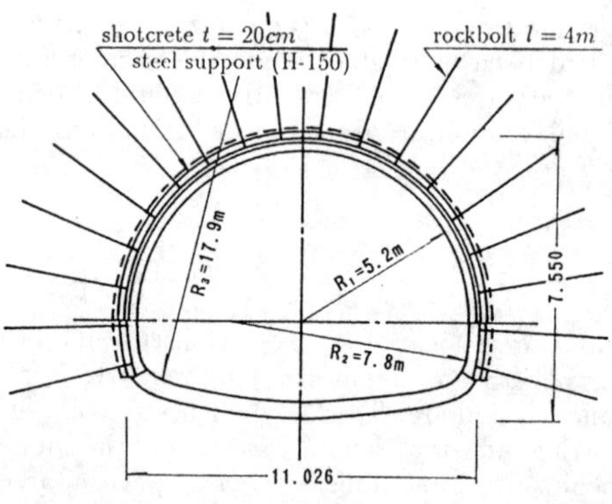

Fig.2 Standerd section and typical timbering pattern

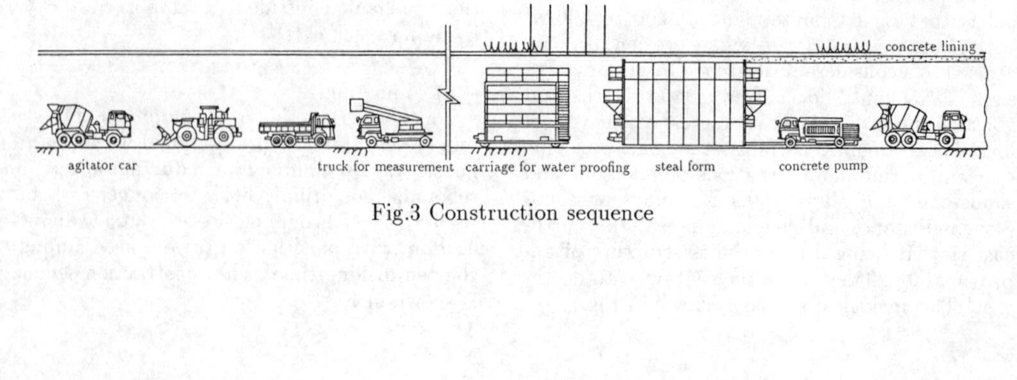

Fig.3 Construction sequence

4 CONDITION OF DEFORMATION OCCURRENCES

Excavation form the southwest portal side was through comparatively hard sandstone-mudstone alternations ane basalt layers, and this progressed smoothly, but from around 340 *m* from the portal,a soft clay layer thought to be strongly weathered fine tuff which had not been anticipated in the original geological survey was encountered. This clay layer began to appear from the left shoulder looking toward the face, and displacement became correspondingly large. As the proportion of this layer increased,the convergence increased and symptoms of heading appeared.

As a measure to cope with this situation, what was first done was to increase shotcrete placement and drive in additional rock bolts, after which, larger deformation allowances were sequentially provided and temporary closure was achieved by shotcrete at the top heading invert.

After a continuation of approximately 30 to 40 *m* of the strongly weathered layer of fine tuff,as an altered basalt layer of boulder form overlying the tuff appeared,the face changed into ground which was non-selfstanding and of collapsing nature. This was coped with by auxiliary measures such as forepoling and splaying on cutting face for face stabilization, but at 411 *m*, where the boulder-form altered basalt layer came to make up the greater part of the face, falling of the face crown occurred. As emergency measures, placing of shotcrete on cutting face and air milk injection into voids from feeler holes at the collapsing crown portion were carried out aiming to prevent enlargement of the loosened zone.

5 CAUSES OF OCCURRENCES OF DEFORMATION AND REMEDIAL MEASURES

The geology in this vicinity, as shown in Fig.4, differed from what had been imagined initially, and the condition was that of a thick fine tuff layer into which there was a basalt layer intruded in lens form. Consequently, the fine tuff and basalt layers had rubbed against each other for extreme deterioration, so that the fine tuff layer had become argillaceous, while the basalt layer was in the form of boulders with, moreover, the surroundings having become gravel-bearing soil due to autobrecciation. The reason for the fine tuff layer underlying the intermediate basalt layer having been overlooked was that basalt had a higher elastic wave velocity and the fine tuff of lower velocity was not detected.

The cause of collapse was considered to be the fact that the consolidation of the rock mass was extremely poor and loosened to a degree that it could not be resisted by forepoling, at the boundary of two layers where the basalt layer, hard but in the form of boulders and brecciated, overlaid a soft fine tuff layer. As a result of contemplating various auxiliary construction methods as measures to deal with the situation, the forepoling with grouting method, which was the most promising strength-wise from among ground reinforcing methods and, moreover,could be incorporated in the excavation cycle, was adopted.

Meanwhile, the fine tuff layer which had become argillaceous was considered to be the cause of the abnormal convergence, and as a result of carrying out tests of swelling properties concerning the

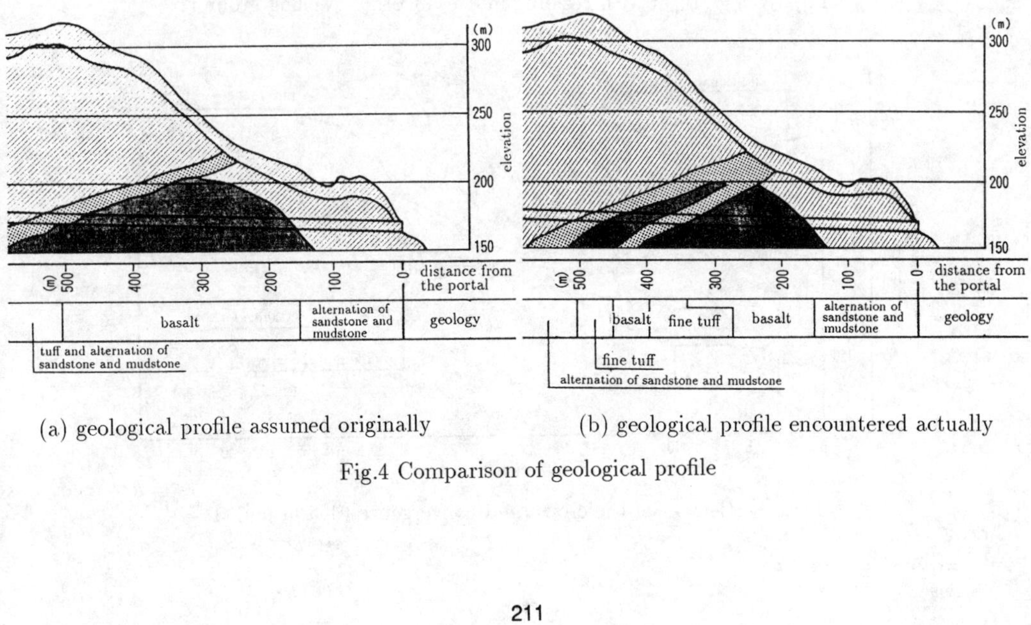

(a) geological profile assumed originally (b) geological profile encountered actually

Fig.4 Comparison of geological profile

fine tuff, cation exchange capacity was found to be very large and, in addition, distinct montmorillonite peaks were observed in X-ray diffraction results. Fig.5 gives comparisons with swelling indices at other swelling ground tunnels in Japan, and it can be seen that there was a swelling capability comparable to the others. In view of the prominent slickensides on the fine tuff, the causes of deformation are thought to have been the fact that the ratio between unconfined compressive strength q_u and earth cover pressure γh was very low ($q_u/\gamma h = 1.1$), and plastic deformation of the natural ground occurred in a composite manner because of relief of residual stress due to folding action and redistribution of stress occurring with excavation.

The results of measuring settlement of the crown and convergence at the 356-m point are shown in Fig.6. Photo 1 and Photo 2 show the deformation of plates of rockbolts and the buckling of steel support, respectively.

Regarding deformation, remedial measures were provided in accordance with the conditions of occurrence, the measures being such as additional placement of shotcrete, additional rock bolts, construction of top heading temporary invert, besides which, there was also an excavation pattern securing a deformation allowance of a maximum 300mm.

At the swelling section, the top heading cross section was being maintained through a closed cross section by means of a temporary invert, but failure occurred due to heaving.

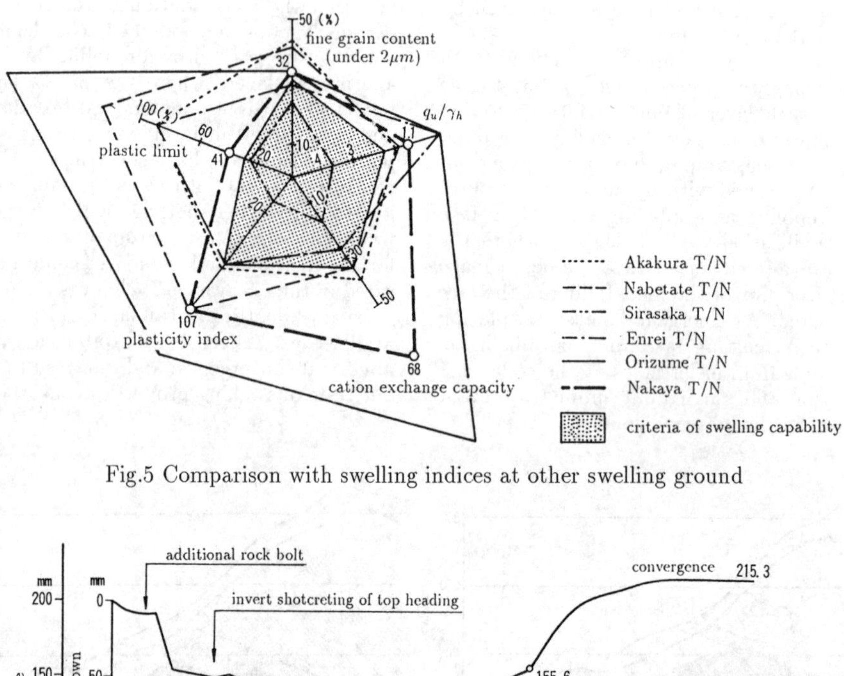

Fig.5 Comparison with swelling indices at other swelling ground

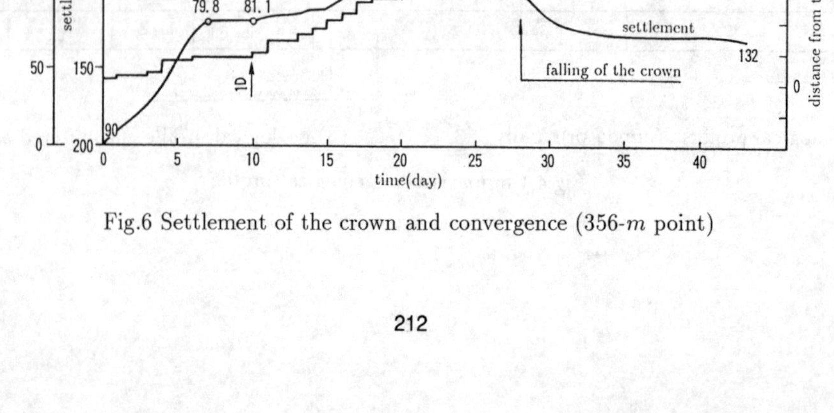

Fig.6 Settlement of the crown and convergence (356-m point)

Photo 1 Deformation of plates of rockbolts

Photo 2 Buckling of steel support

6 CONSTRUCTION OF BOTTOM SECTION IN SWELLING GROUND

Collapsing ground was passed through by grouted forepoling and multiple heading excavation of the face, and at that time (460 m from the portal) excavation of the top heading face was temporarily suspended. The reasons for this were that it was confirmed through horizontal boring investigations that the same fine tuff as that which had previously produced deformation lay ahead beyond this point, and because it was feared that the previous deformation would encroach greatly on the lining gap when excavating the bottom section. In carrying out excavation, a prediction was made of the convergence displacement after excavation of the bottom section (hereinafter referred to as "final displacement")

6.1 Prediction of Final Displacement

There are the follwing three methods of predicting displacement in a tunnel:

(1) Statistical method based on measurement data

(2) Method based on records from sites of similar ground

(3) Method employing numerical analysis

Of these,(1)relates initial stage displacement to final displacement by statistical processing of the measurement data accumulated. However, since past data on the swelling ground encountered here did not exist, it was decided to use relationships according to past literature for the rations of initial displacement velocity and initial (1D) displacement with final displacement.

(1) Linear regressions of final displacement (maximum displacement) and initial displacement velocity, initial displacement (1D displacement) resulted in

(Maximum Displacement) = 8.2 × (Initial Displacement Velocity)

[Scope of Estimation according to 70% Confidence Region:(Maximum Displacement)≈ 3 to 20 × (Initial Displacement Velocity)]

(Final Displacement) = 2.1 to 2.9 × (1D Displacement)

On calculation from the average value of measurement data on the three cross sections given in Table 1 using the above equations, it will be 152.5 mm for initial displacement velocity (18.6 mm/day), and 142.2 to 196.3 mm for 1D displacement (67.7 mm), and it may be seen that the top heading displacement(191.3 mm) is roughly equal.

Table 1 Displacement at various stages

	distance from the portal			average
	359m	365m	373m	
initial displacement velocity (mm/day)	36.0	9.0	10.9	18.6
initial(1D) displacement (mm)	65.7	81.1	56.4	67.7
top heading displacement (mm)	192.7	215.3	165.9	191.3

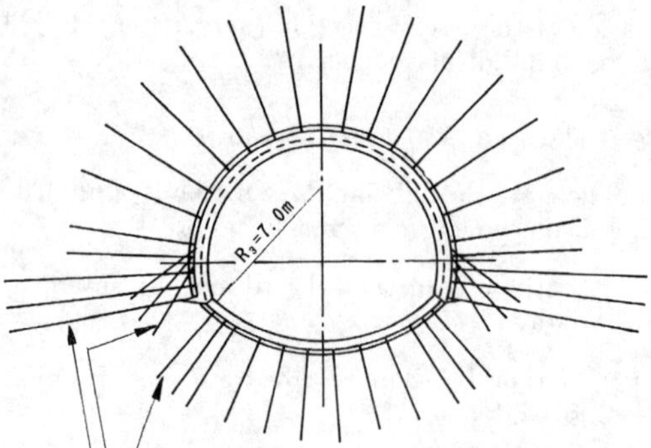

Fig.7 Shape of invert and timbering pattern of bottom-section

Table 2 Result of FEM analysis

mesured value (mm)	calculated displacement		
373m-point	top heading (mm)	final (mm)	magnification
156	145	281	1.9

Meanwhile, when the upper limit of the 70% confidence region for the initial displacement velocity is taken and 20 is used as the regression coefficient, it will be 18.6 $mm \times 20 = 372\ mm$, and the final displacement will be approximately 1.9 times the average value 191.3 mm of top heading displacement.

Therefore, this value was used as the predicted value upon considering scatter in the statistical data, and final displacement was taken to be approximately 1.9 times the top heading displacement.

(2) For records at similar natural ground, measurement data of tunnels where swelling properties had been exhibited according to existing literature and other sources were used as the basis, and the ratio between top-heading convergence displacement or top-heading displacement immeditely before starting to be affected by bottom-section excavation and the final displacement (hereinafter referred to as "displacement magnification") was investigated (Yoshida 1991). According to the results, the displacement magnification was from 1.5 to 3.0, averaging 2.0.

(3) As the method of employing numerical analysis the physical property values of the ground (coefficient of deformation and lateral pressure coefficient) were estimated using FEM and performing back analysis based on top-heading displacements, and displacement magnification was determined by calculating final displacement simulating bottom-section excavation employing these physical property values. The analyses were performed classifying ground strata based on the geological conditions of the face at the 373-m point as the representative cross section, and with the physical property values of the surrounding basalt layer fixed, the physical property values of fine tuff were trial-calculated in sequence. The results are as given in Table 2, and the displacement magnification was 1.9.

Further, giving consideration to the fact that the ground strata appearing at the fact vary, calculations of several cases were performed by a method other than the above changing geological and other conditions. It was learned that even though the physical property values of fine tuff varied depending on the conditions set up, the displacement magnification was between 1.9 and 2.1, and not very much different from the results in Table 2.

Therefore, it was judged that even if the distribution of fine tuff at the face were to change,the displacement magnification obtained in the end would remain almost unchanged, and the displacement magnification of the final displacement over the top-heading displacement was taken to be 1.9.

From (1) to (3), the displacement magnification of the final displacement after bottom-section excavation over the top-heading displacement was taken to be 1.9.

6.2 Result of Construction

Replacing of timbering excavation was done from the portal side toward the face in the sequence of top-heading replacing of timbering, slipway removal, bottom-section excavation, invert shotcrete, and slipway installation with the length of one bench 5 to 6 m for early closure (Fig.7 and Photo 3). It was possible for work to be executed with convergence less than expected as shown in Table 3 and Fig.8 , and held to about 1.2 to 1.3 times the top-heading displacement.

7 FEEDBACK TO SWELLING GROUND AHEAD

Since it was predicted that swelling ground of a fine tuff layer would exist at the face for approximately 90 m beyond 460 m from the portal, a study was made of the timbering pattern for this section based on the construction experience up to that point and on the results of measurements.

(1) According to the results of measuring rock bolt axial forces, axial forces were as much as 18

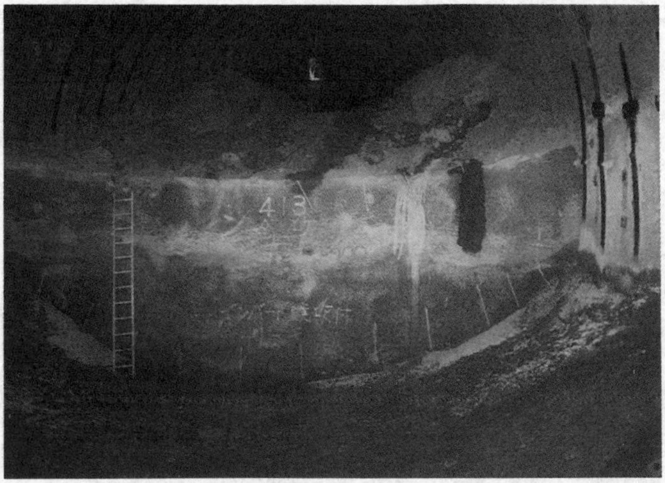

Photo 3 Excavation of bottom-section

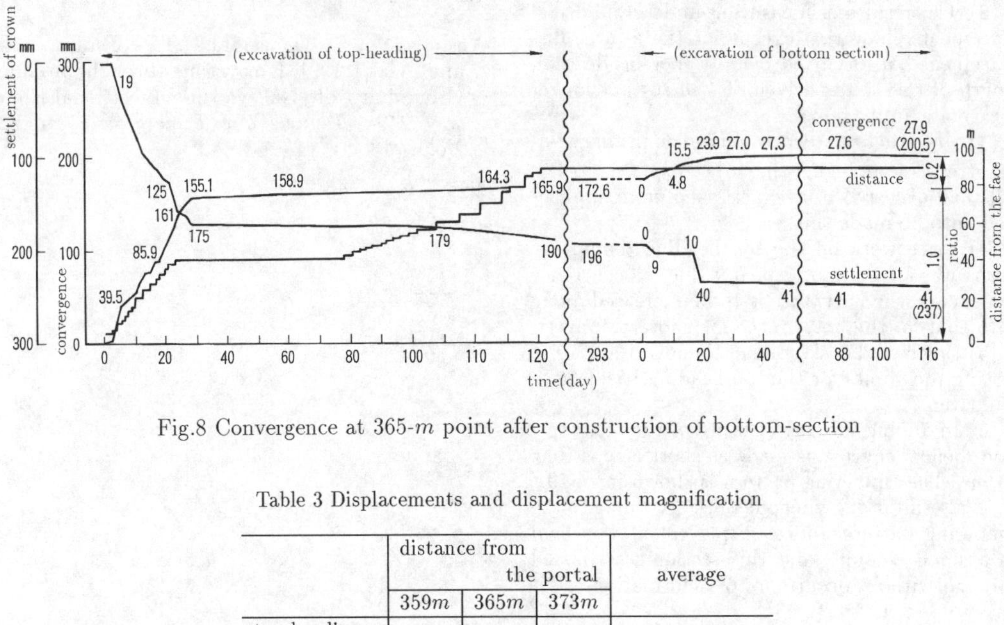

Fig.8 Convergence at 365-m point after construction of bottom-section

Table 3 Displacements and displacement magnification

	distance from the portal			average
	359m	365m	373m	
top heading displacement (mm)	192.7	215.3	165.9	191.3
final displacement (mm)	250.9	257.1	200.5	236.2
displacement magnification	1.30	1.19	1.21	1.23

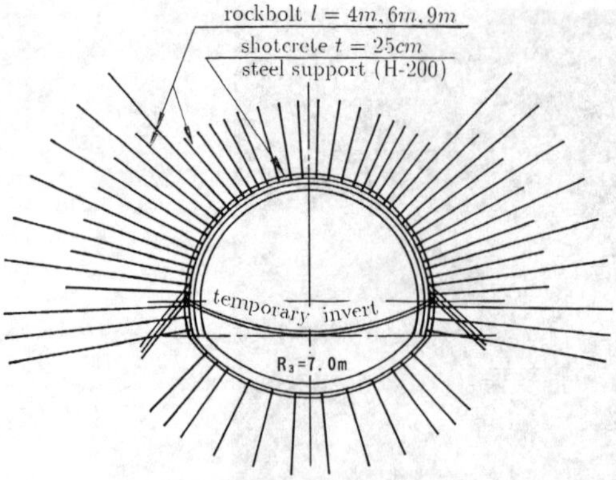

Fig.9 Timbering pattern
($460m$ ~$550m$ from the portal)

tf with part of the bolts broken, and therefore, driving intervals are to be made smaller.

(2) The results of measuring underground displacements show that,except for the crown, displacement had occurred from deeper inside than length of rock bolts driven of 4 m so that longer rock bolts are to be used.

(3) Since there were locations when primary lining stresses were higher than $\sigma = 70kgf/cm^2$, and falls had occurred from cracks, shotcrete application is to be made thicker.

(4) Since work on the top heading temporary invert was forced to be repeated due to occurrence of heaving, the curvature is to be increased along with that for the bottom-section shotcrete invert.

(5) Downward rock bolts of the invert and reinforcing rock bolts for the top heading base are to be driven.

(6) In aiming for early closure of the cross section, bench length was made as short as possible: 15 m . The timbering pattern is shown in Fig.9.

As a result of carrying out work according to the timbering pattern above, displacement was held to a small amount, and this section was passed through without occurrence of deformation.

8 CONCLUSION

Although Nakaya tunnel was constructed by NATM with real time construction control system, it is not an easy matter for the characteristics of natural ground with swelling properties to be accurately grasped beforehand. Even when the characteristics of the swelling property has been grasped, the fact that it dose not directly lead to grasping deformation is thought to make execution of work at swelling ground difficult.

Ground of worse conditions than initially anticipated was encountered at this tunnel, so that at the beginning there was much concern over how to deal with the situation. Based on phenomena which occurred during construction, it was tried predicting the final displacement behavior. Measures to deal with the situation were studied based on the prediction results obtained and it was succeeded in getting through the swelling ground without mishap.

It would be gratifying if this report were to serve as a reference for future tunnel work in swelling ground.

As this report is prepared, the authors wish to express their gratitude to the persons concerned in Ishikawa Prefecture, the owner, and the Public Works Research Institute of the Ministry of Construction, and the various persons who provided considerable assistance in carrying out the construction.

REFERENCES

Yoshida, K., Yamashita, T., Tsurui, H., and Asai,T.1991 Excavating through Swelling Ground at Noto-Nakaya tunnel, National Highway 249- ,*Tunnels and Underground*, vol.22, No.9:7-13

Tunnelling and Ground Conditions, Abdel Salam (ed.) © 1994 Balkema, Rotterdam, ISBN 90 5410 363 9

Excavation problems of an underground water reservoir in a jointed rock mass

F. Cafiso & D. Viviano
Ingegnere Libero Professionista, Palermo, Italy

N. Nocilla
Dipartimento di Ingegneria Strutturale e Geotecnica, Università di Palermo, Italy

ABSTRACT: In order to obtain an underground water reservoir in a "medium" strenght dolomitic rock mass, an excavation of total volume 150.000 m³ was realized. Due to the overburden less than 150 m, the application of semiempirical methods and inclined hemisphere proiections lead to the adoption of systematic supports, while in situ observations confirmed the possibility of adopting local support. During the stand up time, which turned out to be greater than 5 years, the excavations were maintained without lining. A back-analysis, by means of the distinct element method, was carried out. The results proved the real behaviour to be influenced by the dilatant discontinuities and the block interlockings producing a stress distribution favorable to the global stability. The "over break", occurred during the excavation, agree with the theoretical results.

1. INTRODUCTION

The meaningful geomechanical features, regarding the artificial tunnels stand-up, have been examined within a research on the prevision and evolution of rock mass instability phenomena. Both not supported tunnels, built in Sicily during the immediate pre-war period, and under execution tunnels, excavated in fractured rock mass formed of low to medium-high strength carbonatic rock, have been taken into consideration as sampling and experimental evidence (Viviano D., 1991; Cafiso F. et al., 1993).
This paper reports the surveys and analysis relatives to an underground water reservoir meant for the water supply of Palermo.
The reservoir, excavated inside M. Grifone, includes an access tunnel, a head tunnel and four parallel tanks, connected in pairs (Fig. 1).
The tunnel width is between 11,70 m and 12,20 m; the tunnel height is between 9.20 m and 15.10 m. The total length of the tunnels system is 1.540 m; the total volume is 150.000 m³.
The overburden, equal to about 30 m in the access and head tunnel, reaches 150 m on the tanks.
The excavation has been executed full face and, on the basis of the experience made analysing the behaviour of the tunnels realized in carbonatic rock mass belonging to the same geological complex, the support has been realized 5 years after the beginning of the excavations; temporary supports as bolts, wire mesh and shotcrete were necessary only in limited parts.
This has made possible a detailed mesostructural survey as well as a control of the tunnel behaviour, later analysed by means of the distinct element method.

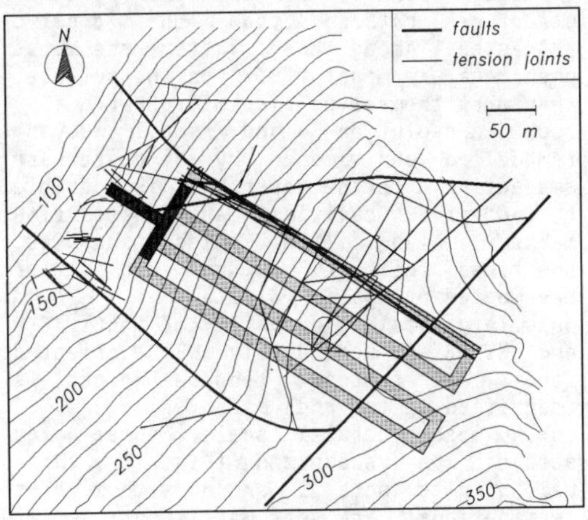

FIG.1 – Lay-out of the M. Grifone underground water reservoir and megastructural pattern

It has been observed that it is possible to have a more reliable prevision of the real behaviour of an underground excavation taking into account the termination and dilatation of joints.

2. MORPHOLOGICAL AND GEOLOGICAL FEATURES - MEGA AND MESOSTRUCTURE

The reservoir is situated on the North-West slope of M. Grifone, one of the mountains, that closes the Piana di Palermo in the south-west, formed of resedimented dolomitic breccias and intact bad stratified dolomitic calcarenites.
The slope is characterized by gradients decreasing from the botton upwards; the medium inclination of the slope is approximately 45° from the base until halfway the total difference in level and about 30° until the top of the ridge.
Grand rock prisms, like "horst", stand in the steeper zone; they are delimited by subvertical faces parallel to the strike or to the maximum inclination of slope.
All the above mentioned features point out the morphological evolution to be more strongly conditioned by the tectonic structure rather than the erosive processes. Among these latter, the karst phenomena are meaningful: on the surface, they show themselves like micro Karren, rock dissolution and small caves, remodelled and widened by the Quaternary sea action; in the depth, it is possible to observe cavities and dissolution marks along the discontinuities, sometimes full of residual red silt or recemented by spatic calcite.
On a large scale, a system of mostly E-W and NW-SE and secondarily NE-SW oriented faults and tectonic tension joints is identified in the rock mass.
On a mesostructural scale, three joint sets can be recognized (Fig. 2); those belonging to sets 1 and 3, with *x* joint terminations, are very persistent; set 2, with a reduced frequency, group discontinuities, with *d* joint terminations that, in NW-SW direction, stop on set 1 joints and, in N-S direction, on set 3 joints.
Set 1 orientation agrees with the most frequently recognized megastructural tectonic tension joints; this is confirmed by the fact that the *xx* joint termination predominates in this set (Fig. 2). Also set 3 joints, which among the great extension tectonic tension joints are characterised by a low percentage, often show *x* joint terminations (Fig. 2).
The fracture intercept (λ), measured along scanlines, is small.
In cross sections of the underground tank (α=90°, β=210°), the minimum fracture intercept shows itself S-Ewards, with a plunge of 35° and is equal to 1,91 disc/m; the maximum fracture intercept, which occurs northwards, is 3.16 disc/m. On such planes, the anisotropy value is, therefore, A_n = 1.65.

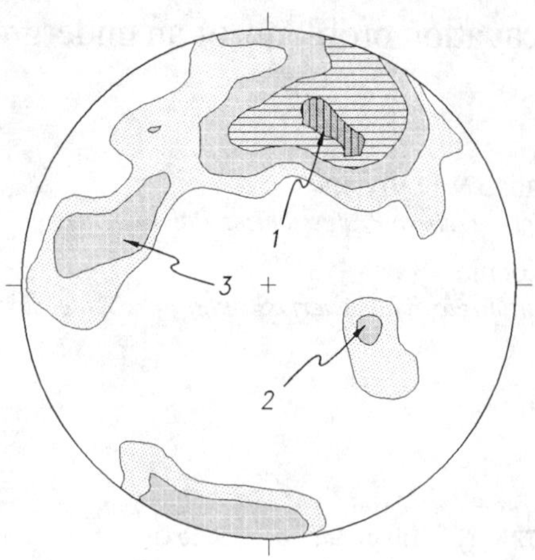

Set n°	α (°)	β (°)	%	prevalent ending
1	60	200	8	xx
2	36	292	5	dd
3	55	107	5	xx

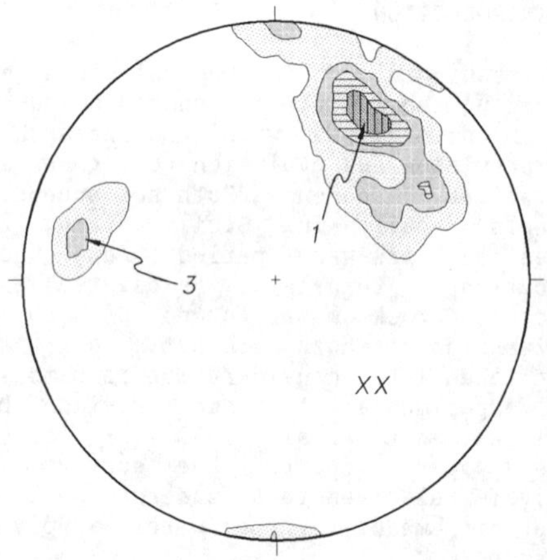

FIG.2–Contour plots from lower hemisphere equal-area projection of joints recogniwed in the scanline survey (N=632)

The RQD values, measured on cores pulled out from deep boreholes carried out by "longyear" system, are very high: so that more than 80% of the samples consists of "good" or "excellent" rock. The lowest values regard the most superficial zones or the cores with discontinuities filled with residual red silt. The cores seem limited only by discontinuities not coring system induced: at a depth of about 100 m, the value of fracture intercept, in vertical direction, is less then 0,5 disc/m.
Boreholes results lead to believe that the discontinuities fracture intercepts, measured along the scanlines and illustrated in Fig. 3, represent the highest values, because the meteoric agents on the surface and the excavations can have determined the re-opening of cemented discontinuities.

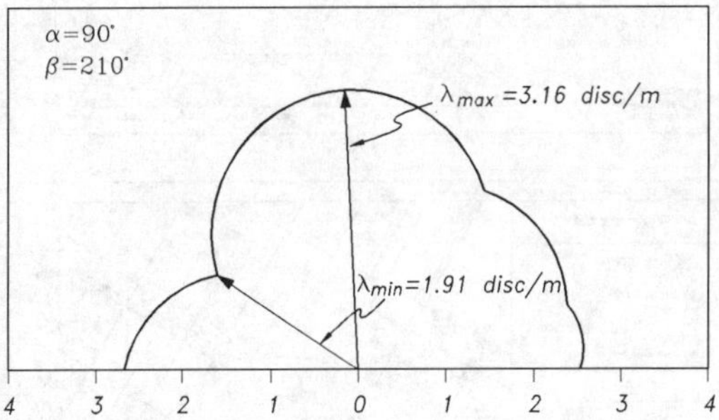

FIG.3–Polar distribution of joint linear frequency in cross sections of the underground tank

3. GEOTECHNICAL CHARACTERIZATION

The unit weight of the intact rock varies between 26,0 kN/m³ and 28,5 kN/m³, depending on the diffusion of the karst phenomena that have caused dissolutions in the material; in 63% of the cases the unit weight is between 27,0 kN/m³ and 27,5 kN/m³.
The uniaxial compressive strength varies between 30 MPa and 130 MPa, 69% of the obtained values is included in the interval 80 MPa÷110 MPa (Fig. 4); the lowest values, however about ten times greater than the lithostatic tension, have been obtained testing samples intersected by discontinuities. In Miller's diagram, the experimental points mostly fall in the zone limited by the N=200 and N=500 lines (Fig. 4): intact rock strength and deformability are medium.
In the samples subject to laboratory direct shear tests, the first order asperities showed an about 1 cm base width

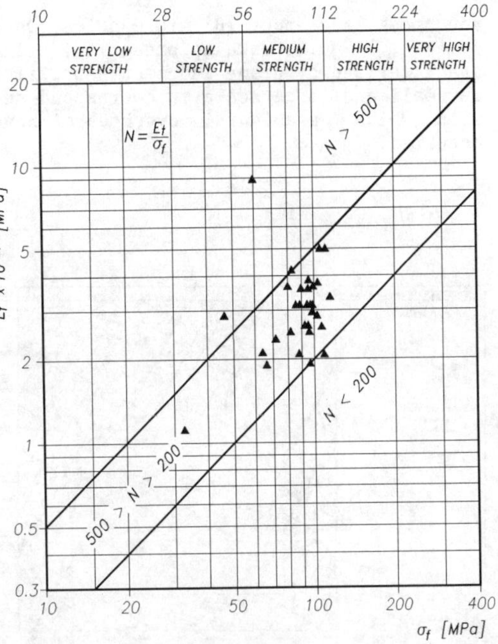

FIG.4–M. Grifone dolomite – Classification for intact rock substance

and dip angles between 10° and 20°; the discontinuities show a dilatant behaviour with 15°÷16° angles both in correspondence with the peak shear strenght, defined by the equation $\tau = \sigma \cdot tg54°$, and after displacements comparable with the asperity semi-wideness ($\tau = \sigma \cdot tg43°$) (Fig. 5).

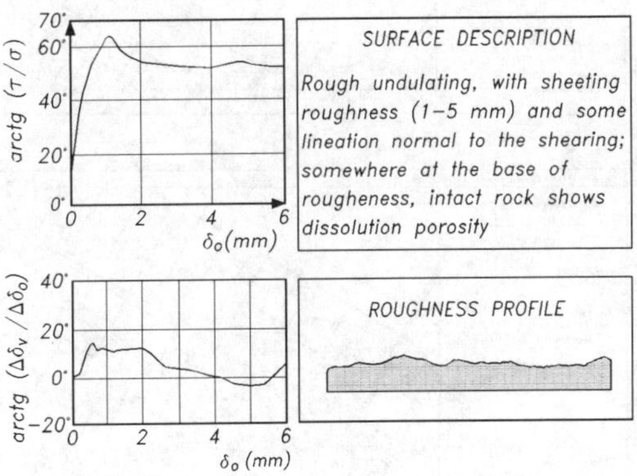

FIG.5–Variation of mobilized friction and dilation angles with shear displacement

Such dilatation values have also been considered representative of the large scale discontinuities, generally characterized by about 50 cm wide roughness and about 10° dip angles. Only some set 1 joint make an exception, being characterized by plane and smooth profiles on small and large scale in correspondence with which meaningful overbreaks have occurred.

4. A BACK ANALYSIS BY MEANS OF THE DISTINCT ELEMENT METHOD

In the analysis of the tunnel behaviour by means of the distinct element method a simplified rock mass mesostructure, with two groups of pervasive discontinuities and *xx* joint termination, has been assumed. In the model, on the basis of the

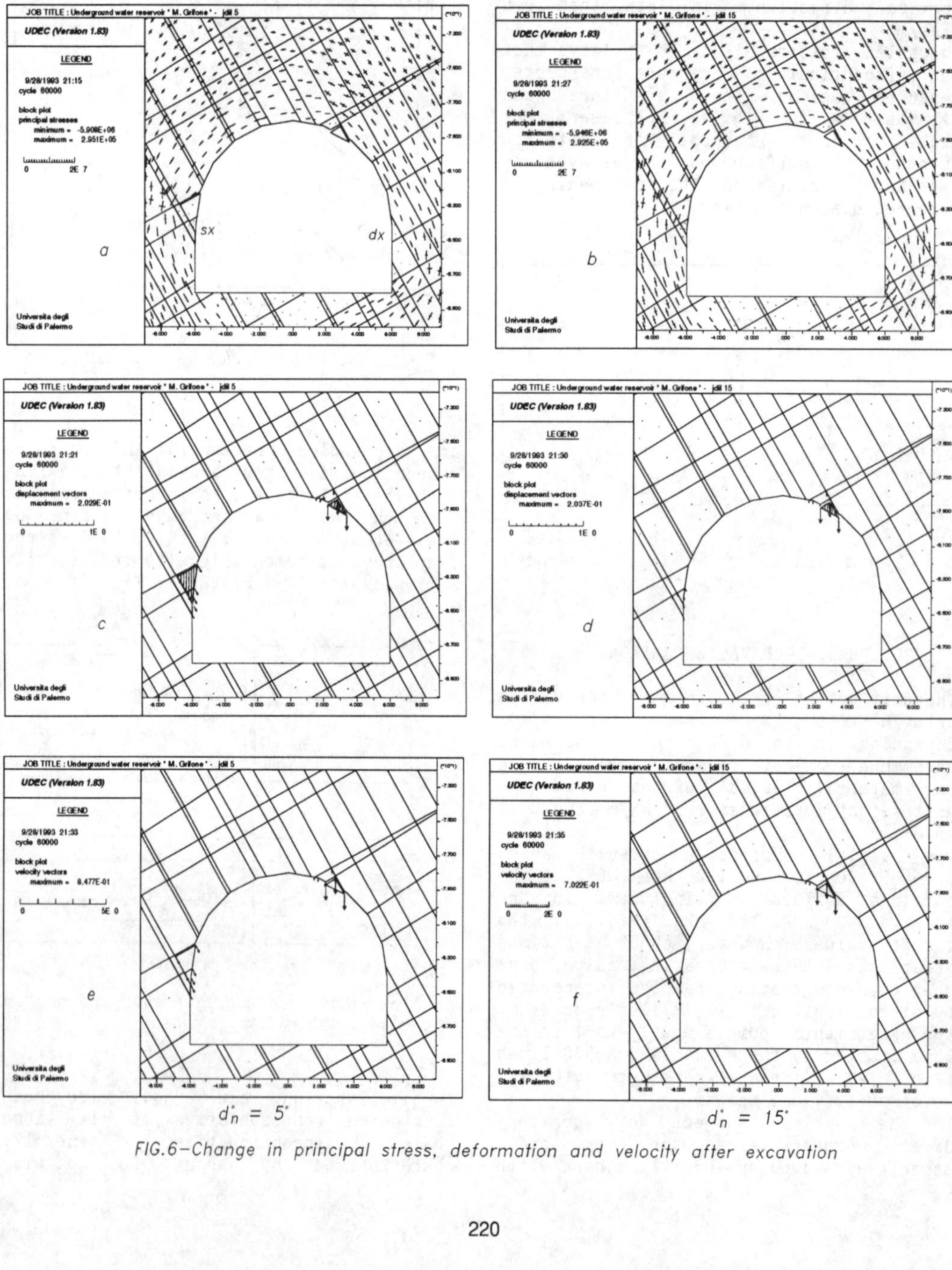

FIG.6–Change in principal stress, deformation and velocity after excavation

observation on the cores (λ = 0,5 disc/m) and of the fracture intercept relating to the *xx* joint termination, varying spacing values have been assigned model, to sets 1 and 3 joints, respectively, between 1.5 m ÷ 2.1 m and 3.0 m ÷ 4.0 m. A 5° random fluctuation about dip direction modal values has also been considered.
The wide gaping tension joints filled with residual silt, both HW-SEwards and NE-SWwards (Fig. 1), have induced to evaluate the horizontal stress as a percentage of the vertical one.
Before the excavation, due to the oriented structure influence, the major principal stress is subvertical with an about 5° shifting the same direction of set 1 trace.
The first important result, which emphasites the dilatation role, derives from the principal stress redistribution as a consequence of the excavation.
When the dilatation of joints is negligible ($d°_n$ = 5° - Fig. 6a), the major principal stresses tend to place towards a direction concentric to the tunnel profile, next to the sidewalls and the roof.
If $d°_n$ = 15° the arch effect in the roof lasts; but near the sidewalls the dilatation creates an about subvertical major principal stress shape (Fig. 6b).
The increase of stresses normal to discontinuities, due to dilatation, determines, in most roof and sidewalls blocks, at the most millimetric displacements.
The pre-existing principal stresses vanish only in correspondence with the darkened blocks (Figg. 6c,d); moreover, in the roof, the oriented structure is such to make a gravity fall possible both if $d°_n$ = 5° and $d°_n$ = 15°.
In corrispondence with the left sidewall, the darkened blocks slide along set 1 joints, with displacements that can mobilize the final shear strenght, only when reduced or null dilatation values are supposed (Fig. 6c). As a matter of fact, only in this case velocity vectors different from zero and parallel to set 1 joints have been associated to these blocks (Fig. 6e), with the progressive opening of set 3 joints, that delimit them at the top.
On the opposite sidewall, both if $d°_n$ = 5° and $d°_n$ = 15° the displacements of about one millimeter mobilize the peak shear strenght, which is sufficient to ensure also the local stability (Figg. 6c,d); in fact, velocity vectors are null and the contact of set 1 or 3 joints lasts along the walls (Figg. 6e,f).

5. CONCLUSION

A research on the stability of underground excavation has been carried out using the distinct element method. The excavation has been executed full face and in medium strength and deformability dolomitic rock. It has been considered a numerical model of a simplified mesastructure deriving only from the *xx* joint termination, which are considered pervasive and characterized by dialtation angles equal to 5° and 15°, respectively in relation to plane and undulating discontinuities.
The application of the method shows that the block displacement due to sliding can occur only locally, depending on the dilatation value and provides with coherent results with in situ observations. During works, in fact, over break inside the roof and the sidewalls have occurred with a geometric shape agreeing with the back-analysis results; on the sidewalls, besides, the over breaks were back delimited by plane and smooth discontinuities.
The research so far developed has enabled to note that a reliable prevision on the behaviour of an underground excavation at small depth can be achieved considering, in the model, the termination and dilatation of joints.

Reference

Barton, N. - Lien, R. - Lunde, J. 1973. Engineering Classification of Rock Masses for the Design of Tunnel Support - Rock Mechanics, Vol. VI No. 4, pp. 183-236.

Barton, N. 1976. Recent Experience with the Q-system for Tunnel Support - Proceeding, Symposium Exploration for Rock Engineering, ed. Z.T. Bieniawski, A.A. Balkema, Rotterdam, Vol I, pp. 107-114.

Barton, N. 1977. Estimating the Shear Strength of Complex Discontinuities - Discussion - A.G.I. Associazione Geotecnica Italiana. The Geotechinics of Structurally Complex Formation, Int. Symp., Vol II, pp. 226-232.

Barton, N. e Choubey 1977. The Shear Strength of Rock Joints in Theory and Practice - Rock Mechanics, Springer Verlag, Vol. X, p. 154.

Barton, N. - By, T.L. - Chryssanthakis, P. - Tunbridge, L. - Kristiansen, J. - Loset, F. - Bhasin, R.K. - Westerdahl,

H. - Vik, G. 1992. Comparison of prediction and performance for a 62 m span sports hall in jointed gneiss. Quarto Ciclo di Conferenze di Meccanica e Ingengeria delle Rocce - Politecnico di Torino.

Bieniawski, Z.T. 1976. Rock Mass Classification in Rock Engineering - Proc. Symp. on Exploration for Rock Engineering, ed. Z.T. Bieniawski, A.A. Balkema Press, Rotterdam, Vol. II, pp. 51-58.

Bieniawski, Z.T. 1984. Geomechanics Classification of rock masses and its application in tunneling - Proc. 3rd Int. Congress Rock Mechanics, International Society for Rock Mechanics, Denver, Vol. IIA, pp. 27-32.

Bieniawski, Z.T. 1984. Rock Mechanics Design in Mining and Tunneling - A.A. Balkema Press, Rotterdam.

Cafiso, F. E AL. 1993. Valutazione della stabilità di un serbatoio in sotterraneo nei monti di Palermo - XVIII Convegno Nazionale di Geotecnica, Rimini.

Catalano, R. - Abate, B. - Renda, P. 1979. Carta geologica dei monti di Palermo.

Cundall, P.A. 1971. A computer in blocky rock systems. Proceedings of the Symposium of the International Society for Rock Mechanics, Nancy, 4-6 October 1971, paper II, p. 8.

Cundall, P.A. 1974. Rational design of tunnel supports: a computer model for rock mass behavior using interactive graphics for the input and output of geometrical data. Technical Report MRD - 2 - 74. Missouri River Division, Corp of Engineers - Omaha, Nebraska 6810.

Cundall, P.A. 1987. Distinct element models of rock and soil structure. Analytical and Computational Methods in Engineering Rock Mechanics, a cura di E.T. Brown, Allen and Unwin, Londra.

Viviano, D. 1990/91. Proporzionamento del prerivestimento di cavità artificiali in ammassi rocciosi a struttura orientata - Università di Palermo.

Acknowledgements

The M. Grifone underground water reservoir has been executed by FERROCEMENTO S.p.A. - ROME, to which Authors are greatly indebted on account of the kind collaboration and the paper authorization.

Tunnelling and Ground Conditions, Abdel Salam (ed.) © 1994 Balkema, Rotterdam, ISBN 90 5410 363 9

Voussoir beam response of bedded limestone roofs in Greek underground mining excavations

J.N. Economopoulos, A.I. Sofianos & N.J. Koronakis
Laboratory of Mining Technology, National Technical University of Athens (NTUA), Greece

ABSTRACT: When an underground opening is driven into well bedded rock formations, the lower layers of the immediate roof, will almost always be intersected by pre-existing or induced cracks, thus allowing a discontinuous beam arching action to take place. The stability of such roofs is evaluated through a compiled computer code which simulates them as voussoir beams. Characteristic correlations between important parameters such as the modulus of deformation of the rock mass, the span of the opening, the strata layer thickness and the modes of failure involved are demonstrated. A typical chart for the theoretical prediction of stable and unstable voussoir beam roofs is checked against numerous observed failures in Greek bauxite mines.

1 INTRODUCTION

Roof failures and consequent falls have been always considered to be the major and most important source of fatalities and serious injuries in underground mining and civil engineering works. Although safer excavation procedures, modern production methods, sophisticated techniques and efficient training have significantly contributed to the general goal of reducing such serious accidents, it is also essential that engineers, foremen and even underground workers increase their understanding and theoretical knowledge of the basic mechanics of roof failure.

The introduction of complicated mining equipment and the lack of shallow deposits have necessitated mining in larger working spans and at increasing depths. The main aim of modern mining activities is to establish operations characterized by optimum ore recovery, fast production rates, improved safety and minimum support costs. Modern standards in the construction of tunnels and large storage caverns exhibit the general requirement of reliable structural design. Ground control for underground workings in bedded formations has become increasingly complex and requires a rather more rational scientific approach.

The engineering significance of "bedded rock" is defined as applying to those rocks in which the bed thickness is small compared to the roof span and the bond between the layers is weak, so that bed separation and slip may occur (Adler and Sun, 1976). These rocks are usually involved in the construction of underground openings within layered formations i.e. limestones, sandstones, shales, conglomerates etc., as well as in mining of rather flat dipping, bedded deposits of coal, potash, trona, bauxite, uranium, copper, iron ore etc.

Outcrops of well bedded rock formations are quite often in the Greek mainland and numerous underground workings are completely or partially driven into such rocks. Almost all Greek bauxite and nickel mines as well as a great part of numerous civil tunnels are surrounded by clearly stratified rock formations.

When an underground opening is driven into bedded rock formations, the main effects due to the overburden loads can be estimated as follows : At remote distances, subsidence may occur on surface. At intermediate distances, a pressure dome or arch structure is developed, and at close-in distances, the immediate roof will deflect downward in beam or plate action (Adler and Sun, 1976).

In the past, ground control design was based mainly on observations, practice and experience and sometimes a rock mass classification system was used, especially for the estimation of support requirements of tunnels. Ground control was based on concepts of continuous elastic solid mechanics. Linear elastic continuous beam theory assumes that the rock of the immediate roof behaves as a series of continuous beams or plates loaded by their own weight. The roof span was then designed for a specified allowable tensile stress of the rock. Nevertheless, observations showed that roof strata, although cracked transversely, resist bending moments which are larger than evaluated by linear elastic analysis. Such beams are formed in the roof of excavations driven within well bedded rock. There, the rock mass is separated at the bedding planes due to sagging and tensile cracks occur at the exposed surface of the roof. It is necessary for the engineer, before starting any theoretical analyses, to go back to the rock and find out what actually happens. This approach may not be very fascinating to him because it comprises dirty experimental work and rather dangerous observations, but that's what is really needed.

The existing openings contain the real conditions on a full scale, just asking to be observed, measured and analyzed. These beds cannot be simulated adequately by continuous, linear, elastic beams. The lower strata which will almost always be intersected by pre-existing or induced cracks, will allow a discontinuous beam arching action to take place. No one discontinuous beam model may simulate accurately roof response but it will approach the discrete phase comprehensively. A simple such model is the voussoir beam.

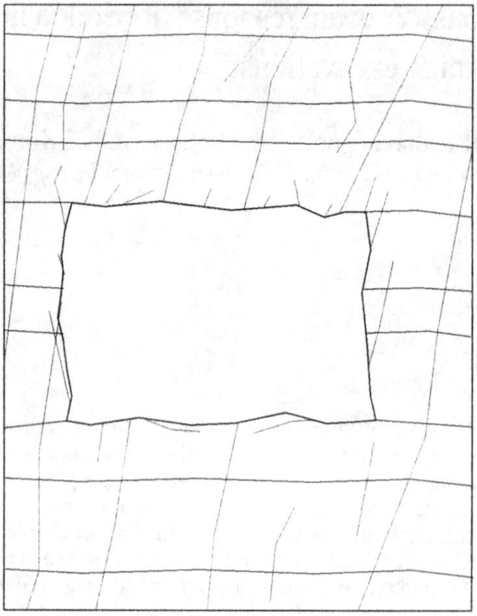

Figure 1. Prototype underground excavation in layered jointed rock.

2 VOUSSOIR BEAM MECHANICS

Voussoir beam theory came on stream in order to throw a new light on some of the many factors involved in roof control. It recognizes that in a confined situation the ultimate strength of a beam is larger than its evaluated elastic strength and that pre-existing cross fractures may not allow tensile stresses. Frictional resistance of the joint surfaces should provide adequate shear strength for the stability of the beams. Thus, the beam which consists of a no-tension material, is assumed to carry its own weight by arching. The term "arching" does not denote exclusively the formation of an arch or a dome-shaped opening. It mainly refers to the process by which fractured rocks may become partially self-supporting by the development of a compression zone above an opening, which transfers vertical load to the abutments on each side (Adler and Sun, 1976).

A voussoir beam is a beam composed of individual blocks (voussoirs). This beam may be considered as a limiting condition of a voussoir arch of zero curvature. Such arches are used in structures such as masonry bridges. In Fig. 1 the in situ conditions of an opening driven into bedded rock formations are illustrated. An idealized opening with a roof beam, crossed by vertical joints which separate the voussoirs, is shown in Fig.2. These joints may carry negligible tension and have a low shear resistance. This renders them inappropriate for ordinary beam analysis. Vertical loading of such horizontal (or subhorizontal) beams will develop cracks underneath at midspan and over the abutments, which in turn will cause vertical movement at midspan and rotation at the abutments.

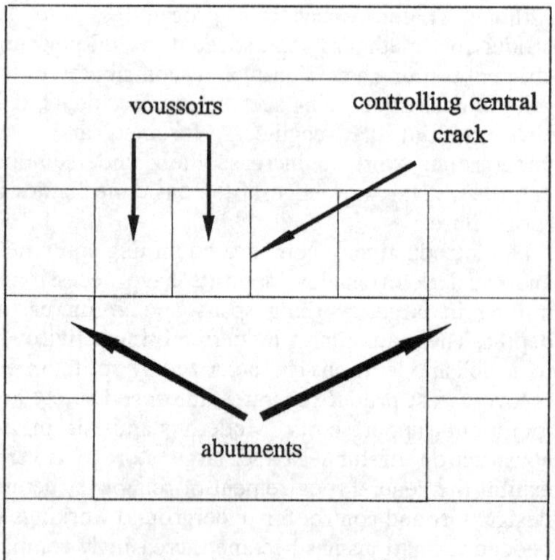

Figure 2. Idealized opening with Voussoir beam roof.

This, will squeeze the lower section of the beam at the abutments and the upper section at midspan, which will develop progressively an arch shaped thrust line (Fig.3). Movement will continue until the formed arch will be capable of withstanding the acting load.

γ : is the unit weight of the rock,

s : is the clear distance between the abutments,

n : is the ratio of the thickness of the arch, at midspan or at the abutments, to the thickness t of the beam,

z_f : is the moment arm of the thrust line after movement,

z_o : is the moment arm of the thrust line before movement.

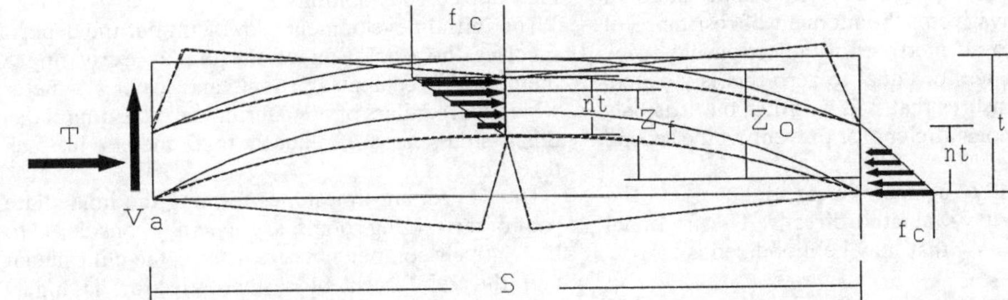

Figure 3. Voussoir beam model.

Three modes of failure may be considered. The first assumes that the developed lateral stress will be greater than the compressive strength of the rock mass. The second assumes that buckling of the beam occurs. The third assumes that shear failure occurs at the abutments. Stability analysis examines the safety against each mode of failure.

Evans (1940-41) and Beer and Meek (1982) developed models to simulate the behaviour of such beams formed in bedded formations. The assumptions made were :

a. Rock is an elastic no-tension material.

b. A thrust zone will be formed with an arch shaped thrust line.

c. The thickness of the formed thrust zone is the same at the midspan and at the abutments.

d. The distribution of normal stress at any cross section is triangular.

e. The arch is loaded by the weight of the beam only.

In order to attain equilibrium of the arch the following equations have to be satisfied :

$$f_c = 0.25 \cdot \gamma \cdot s^2 / (n \cdot z_f) \qquad (1)$$

$$z_o = t \cdot (1 - n \cdot 2/3) \qquad (2)$$

$$0 < n < 1 \qquad (2a)$$

$$1/3 < z_o/t < 1 \qquad (2b)$$

f_c : is the maximum stress at the abutment or the midspan,

The final moment arm z_f may be evaluated by considering the elastic squeezing of the beam due to the vertical movement at midspan and the lateral constraint at the abutments. In order to evaluate this squeezing, a certain distribution of stress along the beam and a parabolic shape of the thrust line have been assumed. Thus the following equation (Brady and Brown, 1993) is developed :

$$z_f/s = \sqrt{\{(z_o/s)^2 - f_c \cdot (1/3+n/4)[3/16+(z_o/s)^2]/E'\}} \qquad (3)$$

$$f_c/E' < 2/\ (2/3 + n/2) \qquad (3a)$$

$$f_c/E' < (z_o/s)^2 / \{(1/3 + n/4).[3/16 + (z_o/s)^2]\} \qquad (3b)$$

E': is the Young's modulus of the rock mass in compression.

Inequality 3b establishes an upper bound for the strain f_c/E'. There is a simple infinity of possible solutions that satisfy equations 1 through 3. Each solution corresponds to a specific thrust line which may be characterized say by the ratio n. For each thrust line, the values of z_f, z_o and f_c are determinate and may be evaluated by equations 1 through 3. The solution thrust line corresponds to a stationary minimum value of f_c.

In order to satisfy against the first mode of failure, the value of f_c has to be correlated to the uniaxial compressive strength q_u of the rock mass. The chosen values of the uniaxial compressive strength and the modulus of deformation have to consider any anisotropic behaviour of the rock mass, due to the existence of any weakness planes. The factor of safety against compression failure FS_c is given by :

$$FS_c = q_u/f_c \qquad (4)$$

In order to satisfy against the second mode of failure the derivative of the moment of resistance of the arch due to an increase in deflection has to be positive. For z_f less or equal to zero this is violated. Inequality 3b ensures that z_f is positive, but it is only necessary and not sufficient for preventing the second mode of failure.

The possibility against shear failure at the abutments may be evaluated directly. The frictional resistance force V_R that may be mobilized is :

$$V_R = (1/2)f_c.n.t.\tan\phi \qquad (5)$$

ϕ : the angle of internal friction of the rock mass.

The shear force V_a acting at the abutment is given by:

$$V_a = (1/2).\gamma.s.t \qquad (6)$$

Then the factor of safety FS_s against shear failure at the abutments may be defined by :

$$FS_s = V_R/V_a = [f_c.n/(\gamma.s)].\tan\phi \qquad (7)$$

3 IN SITU OBSERVATIONS

Bauxite deposits are abundant in the Greek mainland and intense mining operations expand within the geosyncline of Parnassos - Giona - Helicon -Oiti mountains in Central Greece, the main axis of which has a northwest - southeast orientation. The prevailing rocks in these regions generally consist of white, usually non crystalline limestones which stratigraphically belong to the Maestrichtian period. Apart from the upper strata which are rather blocky, the underlying limestone formations are usually clearly thin bedded. The majority of the bauxite deposits, which are intensively mined nowadays, are underlying dark coloured roudists bearing bituminous limestones of the Touronian-Senonian period.

Mechanized room and pillar mining is the usual mining method. Mining excavations are supported by abandoned ore pillars. Room layout and dimensions are designed under the general requirement that working stresses at any point within the surrounding rock should not cause under any circumstances roof or pillar failure, at least until the end of the depillaring stage, when gradual reduction of pillar initial dimensions takes place. The stability of the limestone roof is of dominant importance for the safe exploitation of the deposits, due to the quite frequent entry of personnel and equipment into the mined void. The main task of roof control depends on the unmined mineral which forms the pillars and on the self supporting ability of the rock mass of the roof, which is intensively reinforced by systematic and mechanized rock bolting.

The initial development (advancing) of the deposit involves the extraction of the ore by excavating 5 meter wide rectangular rooms and leaving 8 meter wide square pillars of ore. During the retreating stage, pillar sidewalls are reduced to 5 meters in each direction.

The Touronian limestones forming the immediate roof of the mining rooms, are generally considered to be relatively competent rocks. Hence, the mine layout and the reinforcement system selection is highly dependent on the persistence and the frequency of the main joint sets. However, despite the ordinary crossing of the rock mass of the roof by clear bedding planes and cross fractures, the roof appears to behave satisfactory.

Laboratory tests on intact rock specimens evaluated a uniaxial compressive strength of the limestone rock mass in the range of 70 to 150MPa (with a mean value of 100MPa), a Young's Modulus between 6-14GPa, a Poisson's ratio between 0.25-0.35, a unit weight of $26.5KN/m^3$ and a Point Load Index of about 4. The angle of internal friction is found to be between 33°-40°. The limestone has an RQD value within 50 and 75 and an average strata layer thickness of 0.2-1.0m. The usual condition of the main discontinuity sets may be characterized by the relatively rough surfaces, the separation of no more than 1mm, filled sometimes with soft clayey or calcite material. Most of the joints dissecting the rock mass are characterized as fair to favourable for the majority of the main access and development drifts, provided that they have been aligned properly. In general, groundwater appears as interstitial, with the exception of some winter months.

Due to the impermeability and deferrification of the bauxite orebodies, groundwater percolating through the limestone of the hangingwall, lays down clayey marl sediments, particularly in the areas of high joint density. These sediments have formed at the contact of the bauxite deposits and the limestone roof a relatively thin yellow clayey marl layer, 20-50cm thick, completely distinguished from the deposit and the hangingwall which is almost always scaled by mechanical means.

Besides the bedding planes of the roof, one or two (depending on the region) natural subvertical (75°-85°) main joint sets and sometimes a subordinate flat dipping subset may be distinguished. Limestone roofs are also suffering by nearby production blasting vibrations, while they are usually dissected by a karstic network containing numerous irregularly branching and interconnected channels. These features necessitate the modelling of the cracked roof beams as voussoir ones.

The overburden of the mining excavations varies with respect to the surface relief, but the majority of the mine workings takes place at a depth of about 300m under a gravitational stress field of more than 8MPa.

The Rock Mass Rating, according to CSIR, is evaluated to be in the range of 55 to 70, which characterizes the limestone as fair to good rock. This suggests a mean standup time of about 6 months for a 4m span. Rock support interaction analyses, as suggested by Hock & Brown (1980), have been implemented successfully to the main access drifts of the bauxite mines. The rock mass strength parameters m and s were estimated for these openings in the range of 2.0-0.5 and 0.05-0.0005 respectively (Economopoulos et al,1992).

Gradual sagging of the immediate roof layers (not always clearly visible, as these beds usually bend very little before they rupture), opening or propagation of critical fractures (usually from the center of the beam to the abutments) and shearing at the abutments renders the roof appropriate to voussoir beam analysis. Whenever the beam is relatively thin, it breaks through. In areas where the lateral stress is inadequate and the vertical load on the roof beam significant, the lower stratum is subjected to high tensile stress and usually fractures. This happens also when the roof beam consists of incompetent rock and the abutments may not provide sufficient lateral confinement. The arch action of the beam and any gradually growing non self supporting portion (weight) of the lower separated strata may overstress the rockbolts at the center of the room, which behave as drawbolts, thus encouraging further separation (Economopoulos et al, 1993).

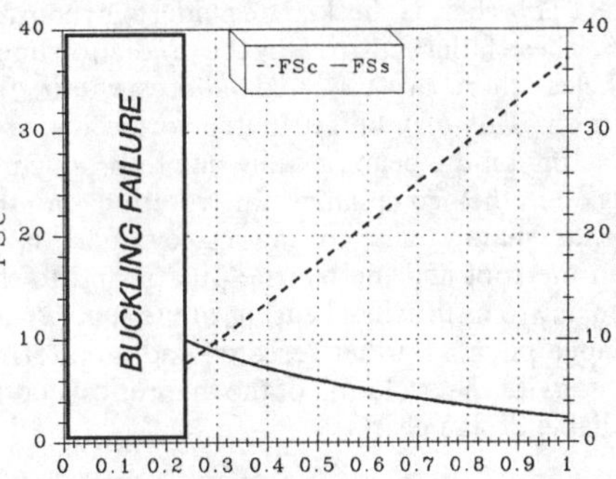

Figure 4. Factors of safety for compression and shear against strata layer thickness (E'=4GPa, q_u=40MPa, ϕ=35°).

Usually, there are very few cases where the visible transverse cracks of the roof are open initially, i.e. immediately after the complete exposure of the first roof layer. In most cases, although the tensile strength of the limestone rock mass has been exceeded and cracks are densely spaced, horizontal thrust holds the blocks tightly together.

The opening's span to bed thickness ratio varies within the range of 5 to 30, with a most characteristic mean value of 15. The maximum critical width, during the retreating stage, for almost all mining roofs with bed thickness of about 0.5m, is about 12m. Some critical transverse cracks may be observed in the external surface of the roof. Although small debris of rock are sometimes falling, the majority of such roofs remains stable for a long time, allowing the safe completion of the pertinent mining activities.

4 APPLICATIONS

A computer code is compiled in order to analyze the stability of a voussoir beam by using the equations described above. In this analysis a stationary minimum value of the maximum normal stress f_c is directly pursued for values of n greater than 0.5. For each thrust line, which corresponds to a value of n, stability against buckling is explicitly checked. Input parameters are the thickness of the strata layer t, the clear distance s between the abutments, the modulus of deformation E' and the unconfined compressive strength q_u of the rock mass, the angle of internal friction ϕ of the fractures and the unit weight γ of the rock mass. It is noted that the values of E' and q_u should be obtained either directly by in situ testings parallel to bedding or by laboratory tests after considering the appropriate scale effect and the directions of the main joint sets. Initially, the maximum normal stress f_c and the corresponding deflection Δ is calculated. At the same time, the evidence of buckling failure is investigated. Finally, the factors of safety against compression and shear failure are evaluated. The solution procedure involves sequential calculation of n, z_o, f_c, z_f and Δ. The unit weight of the limestone rock mass is taken as 26kN/m^3.

In Fig.4, the thickness t of the strata layer is correlated to the factors of safety against compression and shear and to the evidence of buckling. Input parameters are E'=4000MPa, q_u=40MPa, ϕ=35° and s=8m. Buckling occurs for values of t less than 0.2m. Increasing values of t cause a rapid increase in FS_c and a decrease in FS_s.

In Fig.5, the span s of the opening is correlated to the factors of safety against compression and shear and to the evidence of buckling. Input parameters are E'=4000MPa, q_u=40MPa, ϕ=35° and t=0.5m. Buckling occurs for values of s greater than 15m.

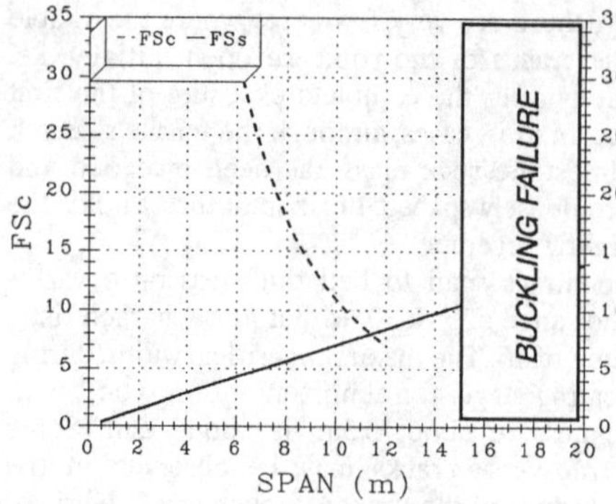

Figure 5. Factors of safety for compression and shear against the span of the opening (E'=4GPa, q_u=40MPa, t=0.5m, ϕ=35°).

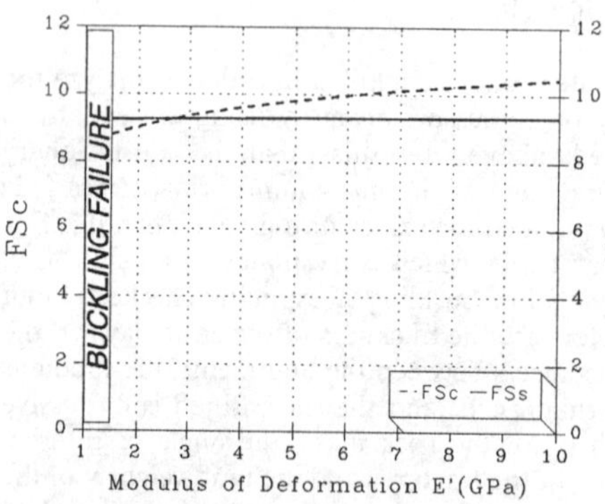

Figure 6. Factors of safety for compression and shear against the Modulus of Deformation (t=0.3m, s=8m, q_u=40MPa, ϕ=35°).

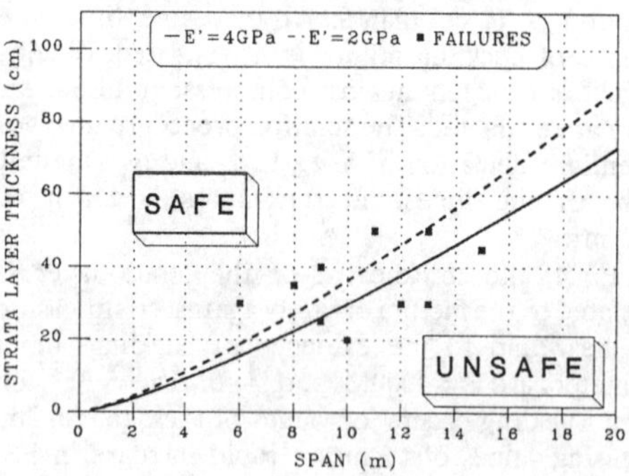

Figure 7. Stable and unstable conditions of Voussoir beam roofs against buckling.

Increasing values of s cause an increase in FS_s and a decrease in FS_c.

In Fig.6, the modulus of deformation E' is correlated to the factors of safety against compression and shear and to the evidence of buckling. Input parameters are q_u=40MPa, ϕ=35°, t=0.3m and s=8m. Buckling occurs for values of E' less than 1500MPa. Increasing values of E' cause a slight increase in FS_c, while FS_s remains constant.

In Fig.7, stable and unstable regions in buckling for values of s less than 20m and t less than 1m, are determined. This diagram assumes two values of E', i.e. 2000 and 4000MPa which are estimated to correspond to the long term value of E' for the particular limestone roofs and hangingwalls. The calculated value of f_c nowhere in the chart exceeds 17MPa. The thickness of the roof layers may be easily estimated at the site either during drilling operations or by obseving fallen layers of rock. Numerous observed failures in the mines, have been projected in the same figure. A comparison between the model results and the real failure conditions shows a satisfactory correlation.

5 CONCLUSIONS

The prevailing ground conditions in an opening, and especially in an actual bauxite mine network, are complicated due to the difficult to determine mechanical parameters and the involved unpredicted factors. The stability of the fractured roof beam layers is evaluated through an ad hoc compiled computer code which simulates them as voussoir beams. This modelling established correlations between the main design parameters, as encountered in Greek underground bauxite mines, and the three modes of failure involved.

A typical chart for the prediction of stable and unstable situations of the limestone beam roof in buckling, adapted to the real mining conditions, is provided. This chart is checked on numerous reported failures. These failures lie nearby the prediction lines; nevertheless there are reported failures within the safe region. This may be attributed to the dynamic forces acting on the beam, mainly due to production blasting, to the compliance in rotation at the abutments, mainly due to the clayey marl layer between the roof and the bauxite pillars, and to the rotation due to insufficient bearing, at the square (not rib) shaped pillars. Further research and monitoring is necessary for the evolution of the theoretical curves into reliable design charts.

ACKNOWLEDGEMENTS

This work is a part of a research project carried out by the Laboratory of Mining Technology of the

National Technical University of Athens, concerning ground control problems in bauxite mines and supported by the "BAUXITE PARNASSE Mining Co.". The authors express their sincere thanks to Mr. U. Kyriakopoulos, Managing Director, for his assistance and kind permission to present this paper. Partial support of the General Secretariat for Research and Technology is also acknowledged.

REFERENCES

Adler,L. & M.C. Sun 1976. Ground control in bedded formations. *Bull. Va. Polytech. Inst. Res. Div.*, No28.

Economopoulos, J.N., A.I. Sofianos & N.J. Koronakis 1993. Roof failure mechanisms in Greek underground bauxite mines. *Proc. I.S.R.M. Int. Symposium on Safety and Environmental Issues in Rock Engineering*, Lisbon, Portugal.

Economopoulos, J.N., N.J. Koronakis & J.G. Mastoris 1992. Study and application of rock mass-rockbolt interaction analysis in underground bauxite mining excavations. *Mining and Metallurgical Annals*, Vol.2, No1.

Evans, W.H. 1940-41. The strength of undermined strata. *Trans. Instn. Min. Metall.*, No50.

Beer, G.& J.L. Meek 1982. Design curves for roofs and hangingwalls in bedded rock based on "voussoir" beam and plate solutions. *Trans. Instn. Min. Metall.*, No91.

Brady, B.H.G. & E.T. Brown 1993. *Rock mechanics for underground mining*. Chapman & Hall, London.

Hoek, E. & E.T. Brown 1980. *Underground excavations in rock*, IMM, London.

Tunnelling and Ground Conditions, Abdel Salam (ed.) © 1994 Balkema, Rotterdam, ISBN 90 5410 363 9

Observational tunneling in squeezing ground

Toshiyuki Ichijo, Takahiko Okai & Mikio Moriyama
Nishimatsu Construction Co., Ltd, Japan

ABSTRACT: The Aioi Number 1 Tunnel was excavated initially using the top heading method, but large displacements occurred in the area inside the tunnel where there were changes in the ground quality between the tuffaceous rock and muddy schists. This displacement was controlled using appropriate auxiliary methods and by enhancing the support stiffness, and analysis was carried out on the variers kinds of measurement. Therefore, it was judged that the excavation by the top heading method was impossible, and was charged to the method of excavation by the side drift method. The resulting convergence was recorded as 1/5–1/10 of the convergence before.

1 INTRODUCTION

The Aioi Number 1 Tunnel, part of the national highway network in the south west of Japan, was designed as a road tunnel with a length of 733m, overburden of 50–100m, and excavated cross-section area of 95m^2. The ground consists mainly of Neocene tuffaceous rocks and Mesozoic muddy schists. At the boundary 285m from the start, the tuffaceous rocks change to muddy schists. These muddy schists are metamorphic by compression, slightly recrystallized, and accompanying the schistose surfaces considerable spalling is characteristic.

For the tunnel, excavation started using the top heading method but, there was excessive displacement in the tunnel at the geological boundary between the tuffaceous rocks and the muddy schists. Deformation of the rock bolts, spalling of the shotcrete, and buckling of the steel support occurred, so at the beginning of this stage (from the time that displacement occurred) appropriate auxiliary methods and the support stiffness were enhanced. Analysis was carried out on all the measured data generated, and important information on design and construction was collected.

This report details the appropriate auxiliary methods until the original tunneling method was changed, and gives conclusions relating to design and construction based on the results of the measurements.

2 TOPOGRAPHY AND GEOLOGY

This tunnel is situated 100–200m above sea level on comparatively flat hilly ground and, the overburden for the tuffaceous rocks is about 80–100m and for the muddy schists 50–80m. A geological profile is shown on Figure 1.

The ground conditions were set as between Grade B and Grade D by the Ministry of Construction for the standard design pattern. An explanation of the

Table 1 Constants for the muddy schists

	Value	Notes
Elastic Wave Rate	V_P=2.4km/sec	At the time of investigation V_P=4km/sec
Coefficient of Change	E_P=12–62MPa	Borehole Loading Test
Ultimate Bearing Stress	R_u=2.5MPa	Plate Loading Test
Adhesive Strength	c=0.25MPa	Estimated Values from the Occurrence of Foot Settlement
Angle of Internal Friction	ϕ=30°	Estimated Values from the Occurrence of Foot Settlement
Uniaxial Compressive Stress	q_u=1MPa	Estimated Value
Coefficient of Rock Strength	$q_u/\gamma h$<1.0	Estimated Value

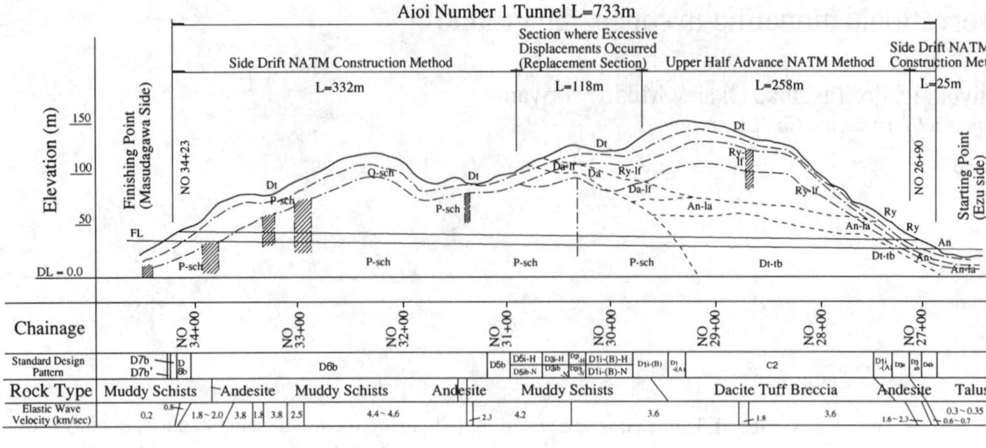

Figure 1. Cross-section of the geological conditions

basis for the standard design pattern is given below. The ground from the starting point was talus, Neocene lava, tuffaceous rocks and then Mesozoic Sangun metamorphic muddy schists. Within this the greater part of the tunnel excavation took place in the tuffaceous rocks and the muddy schists, the boundary between them being 285m from the entrance of the tunnel.

The tuffaceous rocks, where they were in contact with the lava in the vicinity of the entrance of tunnel appeared as deteriorated Grade D ground, while the comparatively hard rock more than 50m from the entrance was Grade C. The boundary with the muddy schists consists of areas of inclusions and these areas show the effects of deterioration, and after the action of weathering are Grade D. The muddy schists show a graphite-like state with 2–3mm phyllitic layers due to the action of low temperatures and high pressures, and accompanying the schistose surfaces considerable spalling is characteristic.

Also, the original rock being mudstone, the body of rock itself is soft, and on the face of the rock structure there are slip surfaces from the fragmented state. For this reason, the degree of consolidation is small and cores could not be taken even in-situ during the investigation, and it was impossible to carry out strength tests according to the rock core samples. A lot of effort was spent in understanding the rock constants, and plate loading tests, borehole tests, and simple elastic wave tests were carried out in-situ. The estimated rock constants and the measurements are shown on Table 1.

3 OUTLINE OF CONSTRUCTION DETAILS

On this tunnel construction, excavation was begun using the top heading method from the east entrance of the tunnel (Ezu side). The support materials employed were shotcrete, rock bolts, and steel support according to NATM.

For the tuffaceous rock part the advance was satisfactory mainly using the CI pattern; fissures developed but the rock quality was more or less comparatively hard and stable. From the area about 230m from the entrance, the effects of the boundary between the tuffaceous rocks and the muddy schists started to appear, and at about 280m the whole face of the top heading changed into muddy schist. The muddy schist which appeared on the face hardly resisted a hammer blow and contained locally thin layers of clay and veins of quartz. There was no seepage water on the face, the ground, with low strength in areas, supported itself and excavation could continue relatively smoothly.

However concerning continued excavation, it was confirmed that the convergence (the relative displacement inside the tunnel) after excavation, and the crown and foot settlement were too large. One week after excavation, both the convergence (horizontal measurement) and the crown settlement were 50mm, while the foot displacement exceeded 100mm; no tendency for the displacement to cease was seen, and it was increasing almost linearly. On examining the tunnel, changes to the shape of the rock bolt bearing plates in the sections built previously, and changes in the cracking in the shotcrete were detected. There was some possibility that the displacements violated the limits so, increased rockbolting was carried out in the section already built with the purpose of controlling the displacement. In addition, with the purpose of controlling, in particular, the settlement in the subsequent sections auxiliary methods were employed. Despite this there was no confirmation of any tendency for the displacements to cease, so excavation was suspended 42m into the muddy schist (325m from the entrance) and an upper half temporary concrete invert was placed.

The displacement rate decreased with the temporary closing up and the convergence ceased. Excessive displacement had occurred, and it was judged that the cause was insufficient support stiffness with insufficient bearing capacity, so after making changes to the support pattern (increase in support stiffness and use of a wing rib support construction) excavation was restarted. Figure 2 show the appearance of the support pattern.

After this, using the measured data, analysis was carried out, and excavation continued with three levels of changed support pattern based on the analysis results. However, depending on the enhancing of support stiffness the ground pressure action behaved like the hydrostatic pressure; with no temporary closing up the displacements did not stop, which is to say there was no change in conditions. In particular, even after the temporary closing up there was a gradual increase in the crown and foot settlements, that is no tendency towards an improvement was seen. Having therefore concluded it was not possible to excavate using the top heading method, excavation was stopped (at 401.65m from the entrance) and it became necessary to radically change the method of excavation.

From the measurements and the results of the investigations, the basic origin of the changes in the ground conditions was the decreasing coefficient of rock strength ($q_u/(\gamma \cdot h)$ where q_u is the uniaxial compressive strength, γ the unit volume weight, and h the overburden height), and thus the bearing strength was insufficient and it was clear that in particular the vertical displacement became bigger. For this reason the excavation method was changed to the side drift method. That the side drifts resulted in maintaining the bearing strength was confirmed, and excavation was completed satisfactorily without further occurrence of excessive displacement.

For the section where excessive displacements had occurred, the required cross-section within the tunnel had not been obtained so replacing of support became necessary. Even in the replacement section, after ensuring the bearing strength of the side drift excavated sections built previously, a replacing of support was employed.

4 RESULTS OF MEASUREMENTS AFTER CHANGES TO THE SUPPORT PATTERN

4.1 Basic policy for changes to the support pattern

Initially on entering the muddy schists there was the section where large displacements occurred and the following features were considered.

(1) Extrusion from the sides due to insufficient support stiffness.

(2) Structural stratification in the ground as the origin of uneven pressure.

(3) Low coefficient of rock strength as the origin of insufficient bearing strength.

(4) Sudden deterioration of the rock bed accompanying stress release after excavation.

The following countermeasures were chosen as standard, a margin of deformation (10cm) was set and construction continued.

(1) Support stiffness was increased to control displacement.

(2) Wing rib supports were used to increase the bearing strength.

(3) Measuring was strengthened and the state of the displacement was understood at an early stage.

(4) In cases where the displacement did not stop, a temporary closing up and laying of a temporary invert were carried out.

However, when the support stiffness was strengthened by one level while continuing construction, having not temporarily closed up, displacement stopped and although a thorough investigation of the pattern was carried out we could not find out why this was the best pattern.

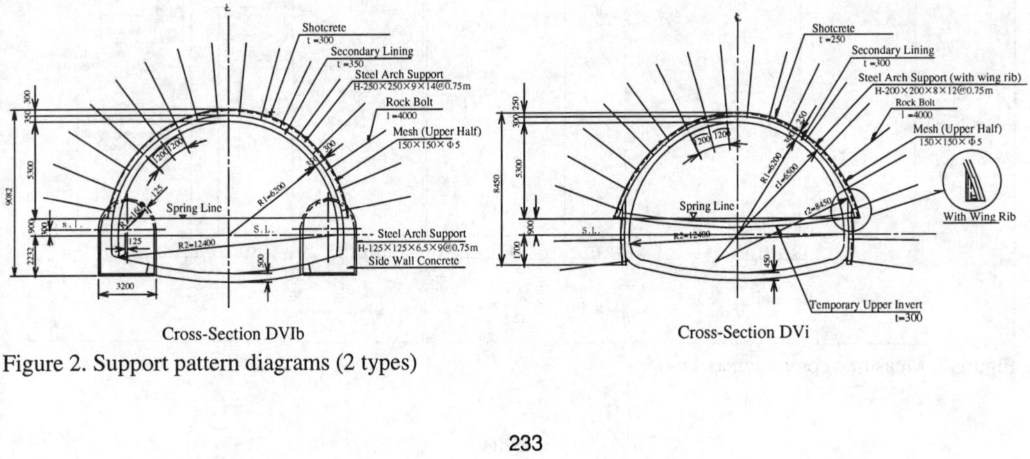

Cross-Section DVIb

Cross-Section DVi

Figure 2. Support pattern diagrams (2 types)

4.2 Measurement Work A results

This details the measurements carried out to monitor daily the convergence and the crown displacement (called Measurement Work A below). The Measurement Work A results for the section where the convergence occurred are shown in Figure 3.

(1) Behavior of displacement

The convergence increased linearly, and until the construction of a temporary invert no tendency for the displacement to slow was seen at all. For the DI section the diagonal displacement was small, and the horizontal displacement was large. Extrusions of the support in a horizontal direction became obvious, due it is thought, to the insufficient support stiffness. After that, due to increased steel in the support materials, the diagonal and horizontal behaviour was in good balance. The displacement stopped for a while due to the construction of a temporary invert, but on reopening excavation on the following section, the displacement near the face showed a tendency to increase. The crown and foot settlement, continued to show a tendency to gradually increase after the temporary closing up, and the results of the temporary closing up in relation to the settlement were not as clearly understood as the convergence.

The measured values for the point when the tunnel was temporarily closed up are shown on Figure 4 as the measured values (Japan Tunneling Association 1985) classified by support pattern for the convergence. On this figure, the value before displacement stopped at the point of the temporary closing up may be seen to be much bigger than the measured average value for DI. That is to say, the supports stiffness in this section were too small for the corresponding ground conditions.

(2) Rate of displacement

The horizontal relative displacement for each support pattern is shown in Figure 5. The oblique lines in the graphs show, from the actual figures obtained, the tendency (Japan Tunneling Association 1985) towards the rate of displacement, accompanying the change of displacement over time. The general control value rates of displacement (Nishimatsu Construction Co., Ltd. 1987) are shown in Table 4 for pattern D but, on this project the maximum measured values were more than 25mm/day, more than three times the control values. The rates of displacement, usually showed a tendency to decrease over time as shown in the graphs but, the values obtained tended to increase again over time and generally looked to be unstable. Additionally, relatively large

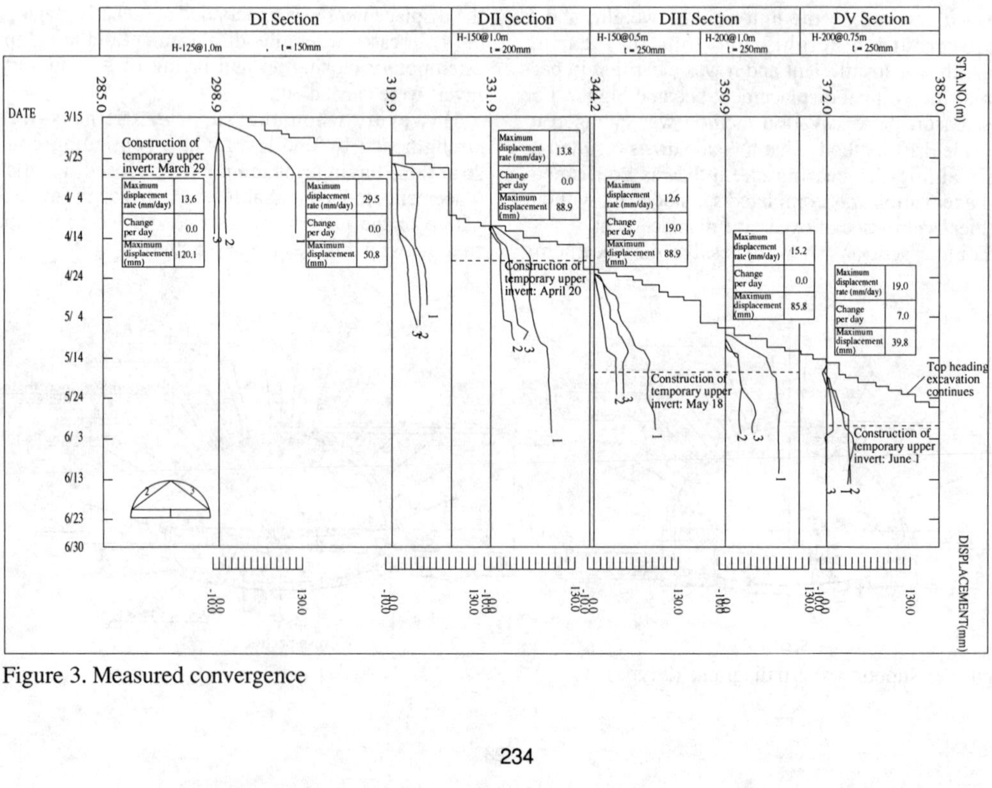

Figure 3. Measured convergence

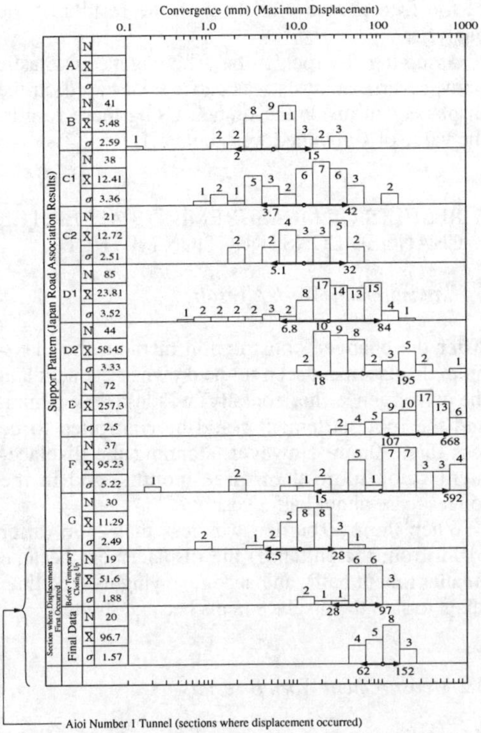

NOTE:
Before Temporary Closing Up — Only until before the placement of the temporary invert
Final Data — Includes all the data until before the replaceing and before the countermeasures of increased shotcreting etc.

Figure 4. Frequency distribution of the convergence (comparison with measured values) (Japan Tunneling Association 1985)

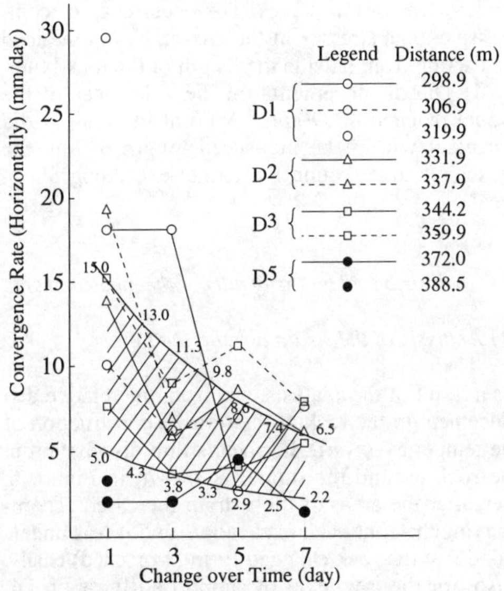

Figure 5. Displacement rate over time

Table 4. Control values for measurements

	Pattern C	Pattern D
Convergence	30mm	60mm
Initial Displacement Rate	4mm/day	8mm/day
Crown Settlement	15mm	60mm

(Nishimatsu Construction Co., Ltd. 1987)

rates of displacement continued and the displacement behaviour of the rock concerned is understood to be unique.

The rate of displacement after the changes to the support pattern became relatively small due to the upgrading in the support stiffness from DI to DV. However, even if the rates of displacement were controlled for a while, they tended to increase again when excavation continued, and there was no improvement due to the three levels of change in the pattern.

4.3 Measurement Work B results

According to what was stated before, the changes to the support pattern on the top heading method did not stop the displacements, and even with the increase in the level of support displacement was near the ultimate limits. Adding to the daily measurements of the condition in the section where displacement had occurred (Measurement Work A), a new set of measurements was initiated (Measurement Work B) to understand the rock behaviour in the vicinity of the tunnel and the stability of the support work, and great effort was spent in obtaining information which would be useful in design and construction. Measurement Work B included:

— Measurement of the relative displacement in the rock.

— Measurement of the stress in the shotcrete.

— Measurement of the ground pressure on the back of the face.

— Measurement of the axial stress in the rock bolts.

From the results of Measurement Work B the following points can be made.

(1) The shotcrete stress had extremely large values of 12–20MPa.

(2) The ground pressure on the back of the face averaged a large 0.5–0.6MPa. These values were in keeping with 30–35% of the corresponding total overburden pressure (overburden 80m).

(3) The rock bolt axial stress was more than

150kN (tension) in places. The occurrence of compressive axial stresses in the crown, shows the load was acting from outside the length of the rock bolt.

(4) The displacements on the side faces of the tunnel reached 40–90mm. A point of stability did not exist within the measured length of 8m, the loosening zone around the tunnel exceeding 8m in areas.

4.4 Analysis based on the results of the measurements

(1) Analysis of the strain around the tunnel

As a result of the analysis, and from the relative displacement in the rock at the time of construction of the temporary invert, the shear/strain distribution in the rock around the tunnel is shown in Figure 6. Hereafter the areas of high strain increased accompanying the continued excavation, and it was understood that the rock changed form (worsened) easily. Also, for the low level of support stiffness on DI, the strain distribution centered on one side, but depending on the increased support stiffness from DI–DⅢ the strain distribution became closer to concentric circles, and depending on changes to the support pattern a good balanced distribution was seen.

(2) Estimation of ground pressure action

Analysis was needed to determine the action of the ground pressure and the lateral pressure coefficient, due to the phenomenon of yield on the steel support. As shown on Figure 7, a temporary frame model was assuming for the steel support, values for the lateral pressure coefficient were set free, and a parameter study was carried out on the steel support condition. As shown on Figure 7, in the area of the steel support work yield the lateral pressure coefficient k may be seen to be less than 1.1, and the action of the load to average $p=0.5$–0.6MPa. In addition, the ground pressure measured on the back of the face agreed closely with the results of the analysis.

Assuming the rock to be a homogeneous elastic body, reverse analysis was carried out based on the displacement inside the tunnel. Using these results, the value of k obtained was about 1.1.

5 RESULTS OF MEASUREMENTS AFTER THE CHANGED CONSTRUCTION METHOD

5.1 Measurement Work A results

After the changed construction method, the maximum displacement seen in the crown settlement and the convergence (horizontally) was less than 60mm, and the foot settlement could be controlled to be less than 30mm. However, stopping the displacements took more than three months, and in the longest case about half a year.

When the overburden was less than 50m (after 610m from the entrance) the displacement became smaller (about half), and accompanying this the time taken to stop the displacements became shorter.

5.2 Measurement Work B results

After the changed construction method, the side wall displacement could be controlled to less than 16mm, the same as for the section where displacement had occurred, and it was confirmed that there was a large loosening zone.

Depending on the side walls, the settlement of the steel support was restrained and there was a large initial axial stress in the steel support; the change over ten days in the compressive stress was 200MPa and it reached the yield value at three weeks. The stress in the shotcrete was reduced on average 6–10MPa compared to that before the changed construction method, but it continued to be large. From these results it was estimated that the load action, p, in the section where displacement occurred was 0.6–0.7MPa.

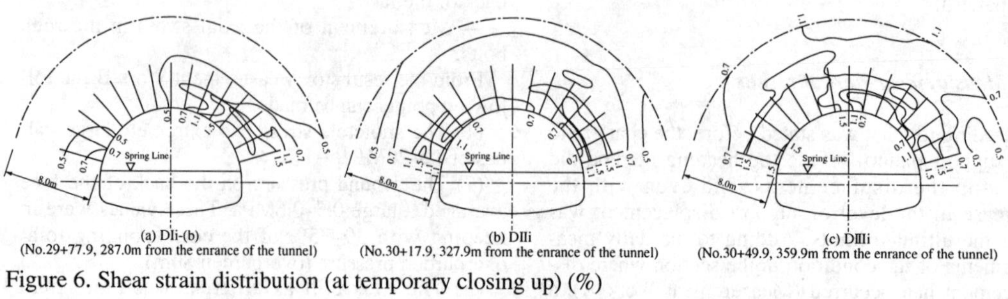

Figure 6. Shear strain distribution (at temporary closing up) (%)

In the steel support the stress that was seen was near to the ultimate state but, the displacements were not excessive and no cracking developed in the shotcrete so construction advanced while continuing measurements. Finally the displacement and the stress stopped, but this took a long time.

5.3 Results of measurements of stress in the secondary lining

In the sections which needed replacing, the rock around the tunnel was disturbed due to the original change of state and the work done in replacing, and it was considered necessary to monitor the stability of the tunnel for a long period. Also, extra reinforcement (D19 @200) cross-section was placed in the secondary lining in the replacement section, and the behaviour was monitored depending on the measured stress in the lining.

The reinforcement stress, was measured as 10–20MPa compression in the whole body of the lining, and these values hardly changed after the concrete hardened. From this, it is considered that there was almost no load action on the secondary lining after the establishment of a stable displacement. In short, the load action of the tunnel on the primary lining was supported and the stability of the tunnel was confirmed. Therefore the extra reinforcement was left out of the secondary lining after the change in construction method.

6 SUMMARY OF THE MEASURED RESULTS

6.1 Differences due to changed construction method and geological conditions

On the excavation of Aioi Tunnel Number 1 there was a big difference between the tuffaceous rocks and muddy schists. The measurements throughout the tunnel excavation are shown in Figure 8. From these it can be understood that the maximum displacement differed depending on the changes in the geological conditions.

The change in the displacement rate after the changed construction method in the muddy schists is shown in Figure 9. After the change in the construction method there were cases when the rate of displacement at the beginning was large then decreasing over time, and it is understood that the changed construction method was especially effective for the rock stability.

6.2 Relationship between displacement rate and level of displacement

The relationship between the initial displacement

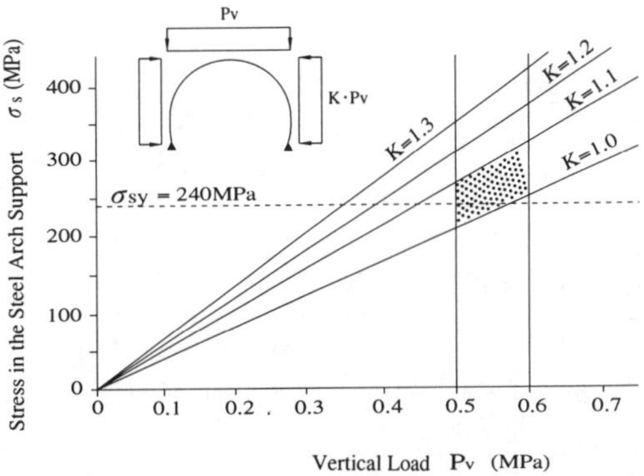

Figure 7. Correlation between the vertical load and the stress in the steel support (DIIi)

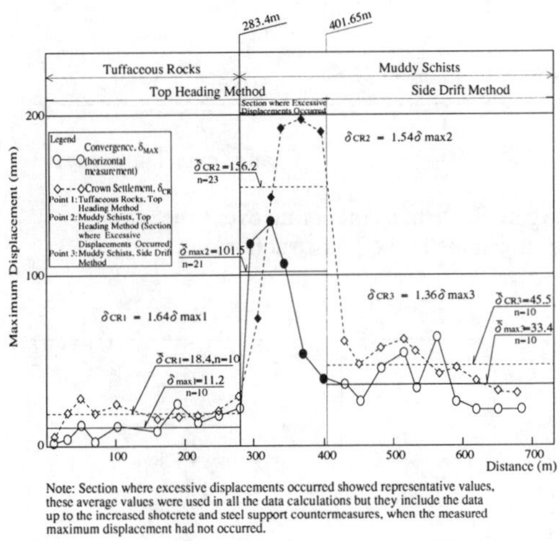

Figure 8. Maximum values for the convergence and the crown settlement

rate and the maximum convergence (horizontally) is shown on Figure 10. From these results the differences between sections where displacement was stopped and the other sections where displacement occurred can be clearly understood. The maximum values in the sections where displacements occurred are considered to include the values up until the increased shotcreting, and the ratio of the initial displacement rate compared to it finally is considered to be large. At the time of excavation of the sections where displacement occurred it was necessary to judge whether counter measures were required at the start of the displacement.

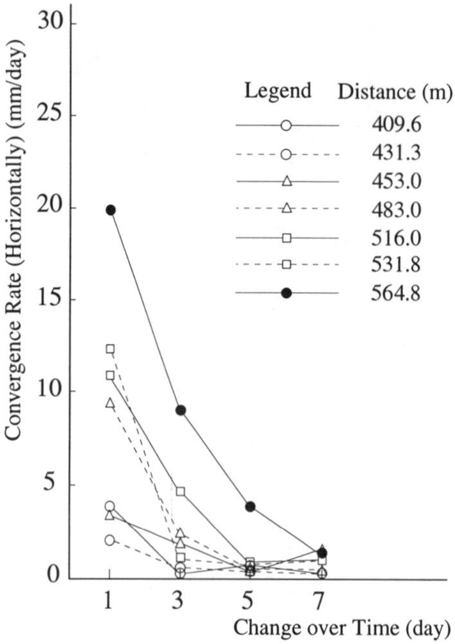

Figure 9. Displacement rate over time (after changed method of construction)

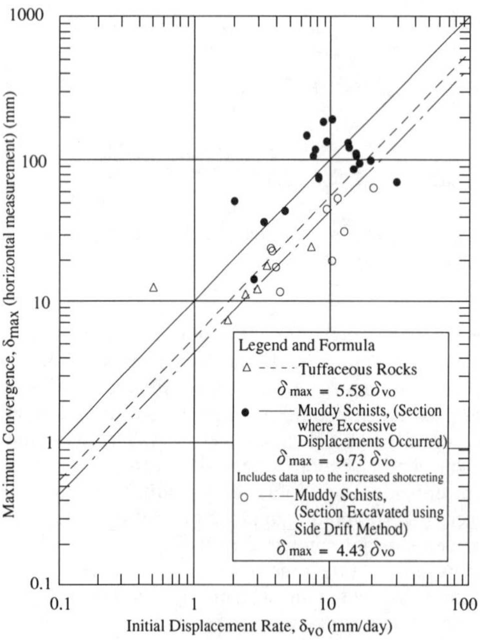

Figure 10. Relationship between the initial displacement rate and the convergence

7 CONCLUSION

For the conditions experienced here, it was decided initially that the cause was insufficient stiffness, and it was considered possible to manage a temporary closing up of the upper section and augmentation of the support stiffness. There was no change to the condition of the rockbed in the face,

the displacement of the rock concerned was bad and the condition of the rock was judged to be worsening with the advance of the face, so it was necessary to analyse the results over the period of measurement. Consequently after progress in the construction of additional measurements and changes to the support patterns, the side drift construction method was used.

On this project, we could complete the project according to the established measurement feedback of results characteristic of NATM.

It was necessary to note the initial displacement rate, especially for control of the measurements at the time of excavation, and according to the forecast rock conditions from the various measurements and analysis, it was possible to verify the design and construction.

For this report we were fortunate to be able to refer to other sources dealing with excavation in rock of similar condition.

Finally, we are very thankful for their help, with the construction counter measures for this tunnel, to each one of the officers of the Ministry of Construction Chugoku Regional Construction Bureau Hamada Office.

REFERENCES

Japan Tunneling Association 1985. *Study on the Evaluation of Ground Types*, p. 24–26

Nishimatsu Construction Co., Ltd. 1987. *A Manual for NATM Works*, p. 397

Sakurai, S. 1982. An Evaluation of the Results of the Measured Displacements in Tunnel Works. *Report for the Japan Society of Civil Engineers*

Sakurai S. 1983. Inverse Analytical Methods for the Results of Measured Displacements on Same Time Tunnel Excavations. *Report for the Japan Society of Civil Engineers*

Tunnelling and Ground Conditions, Abdel Salam (ed.) © 1994 Balkema, Rotterdam, ISBN 90 5410 363 9

Studies on the present state and the mechanism of trouble occurrence in tunnel construction in Japan

Akira Inokuma, *Public Works Research Institute, Ministry of Construction, Tsukuba, Japan*

Shouji Suzuki & Isao Maruyama, *Technical Development Department, Ohmoto-gumi Co., Ltd, Okayama, Japan*

Jun Mitsuo, *Civil Engineering Design Department, Tokyu Construction Co., Ltd, Tokyo, Japan*

Teruo Sakai, *Civil Engineering Design Department, Maeda Corporation, Tokyo, Japan*

Yutaka Tokunaga, *Technical Research Institute, Penta-Ocean Construction Co., Ltd, Japan*

ABSTRACT: The information on troubles obtained from individual tunnel construction cases in Japan is objectively verified from the point of view of data analysis based on the results of case investigations into troubles.The actual conditions of the troubles,geological survey tequniques and countermeasures are considered based on the cases. The multi-dimensional quantification theory is applied to the quantitative analysis, and difference between the trouble occurrence mechanisms is discussed.

1 Introduction

NATM is a construction method generally employed for mountain tunnels in Japan. The topography and geology of Japan vary so widely that rocks of different types and natures intricately coexist in even one and the same tunnel. For this reason, numerous attempts have been made for the development of geological survey techniques and auxiliary methods. In spite of the efforts, preventing troubles is still not a easy task. Possible causes of such troubles are geology, topography and construction conditions which interact each other. The studies on the "automatized technologies in tunnel construction" have been carried out as a part of the synthetic development of new construction technologies by the joint efforts of government and industry. In the course of the geological surveys into the ahead of the cutting face, the "case surveys into troubles in tunnel construction" were carried out by the working group of the geological survey into the ahead of the cutting face. In this paper, troubles are classified according to the case survey results to evaluate the occurrence process and the primary factors. The actual conditions of the troubles, geological survey prior to construction, trouble prediction and countermeasures were considered based on the cases. Meanwhile the multi-dimensional quantification theory was applied to the quantitative analysis of the troubles.

A collapse can be classified according to phenomena into a separation, run-off and stratum slippage. Meanwhile, accidents that occur in the supporting and supporting members such as the deformation in rock bolts are known generally as a deformed state. Abnormal phenomena such as a collapse, deformation and rock burst that occur in tunnel construction are known generally as a trouble.

2 Basic analysis

The total of 71 cases in 65 tunnels were taken for analyses. The excavation methods used for tunnel construction include 42 cases of blasting, 27 cases of mechanical excavation and 2 cases of a combination of blasting and mechanical excavation. The analytical results of survey were classified according to Table 1 "Variables". The "support pattern" shown in Table 1 indicates the standard support pattern[1] presented in Table 2. Both the "auxiliary method A" and "auxiliary method B" indicate the auxiliary methods that had been used before trouble occurrence. The "auxiliary method A" indicates a method that can be implemented with the use of facilities and machinery employed in ordinary construction. The "auxiliary method B", on the other hand, is a method that can not be implemented with such facilities and machinery.The "auxiliary method C" is a method not included in either of the above two methods. The "countermeasure A", "countermeasure B" and "countermeasure C" are the auxiliary methods taken either as a countermeasure after trouble occurrence

Table 1 Variables and categories used in this study

VARIABLES	CATEGORIES								
excavation method	blasting	mechanical							
tunneling method	full face	mini bench cut	short bench cut	long bench cut	drift method				
support pattern	A	B	CI	CII	DI	DII	DIII		
aux. method A	fore piling	steel timbering	diagonal rockbolt		face shotcrete		face rockbolt		
aux. method B	drainage	chemical grouting		pipe roof					
aux. method C	temporary closing								
overburden(m)	0-10	10-20	20-50	50-100	over 100				
geological type	mesozoic,paleozoic		tertiary deposit	igneous rock		diluvial,alluvium		metamorphic roc	
geological condition	rock boundaries	fracture zone	alternating strata	highly weathered	progressive fissures	clayish,sandy	dip slope	sound	swelling
ground water influence	main factor	influence factor		unrelated					
ground water condition	highly water content		perched water	highly confined					
spot of occurence	cutting face	crown	core	foot	bottom	side wall			
trouble type	collapse	support deform	earth deform	rock fall	RB deform	SC crack	rock burst		
location of trouble	cutting face	back of face							
occurence time	after excavation		after supporting						
collapsed vol.(m^3)	0-50	50-100	100-500	over 500					
prediction	possible with observation		possible with measurements		possible with geological survey				
geological survey	horiz. boring	observational	PS logging	probing	electric prospect		AE	seismic prospect	
countermeasure A	fore piling	steel timbering	diagonal rockbolt		face shotcrete		face rockbolt		
countermeasure B	drainage	chemical grouting		pipe roof					
countermeasure C	temporary closing		connection of supports		additional support		filler injection		

Table 2 Standard support patterns

support pattern	rock bolts			steel support		shotcrete
	length(m)	circum. interval	longi. interval	top heading	lower half	thickness(cm)
B	3.0	1.5(top.h)	2.0			5.0
CI	3.0	1.5	1.5			10.0
CII	3.0	1.5	1.2	H125		10.0
DI	4.0	1.2	1.0	H125	H125	15.0
DII	4.0	1.2	<=1.0	H150	H125	20.0

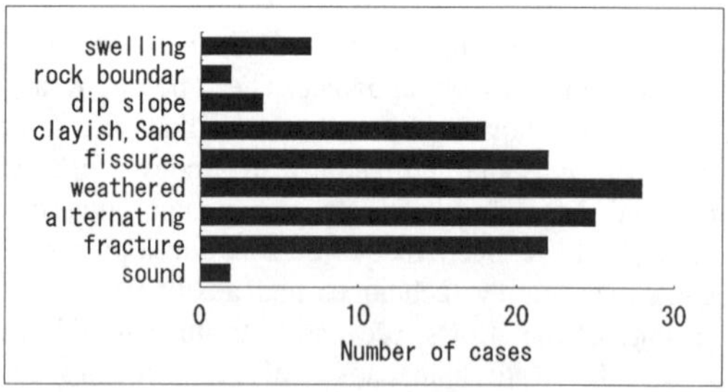

Fig.2 Geological conditions

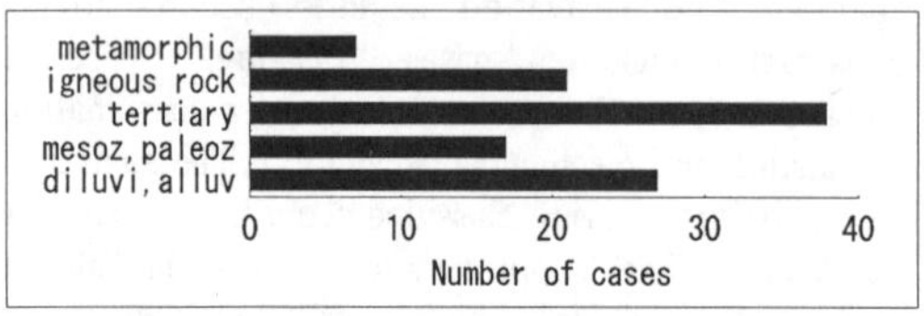

Fig.1 The classification of geology

or as a preventive measure for the succeeding excavation.

2.1 Geology

The classification of geology is shown in Fig 1. As multiple geologies coexist in the alternating strata and rock boundaries, the total number of geology and the number of cases do not coincide. As can been seen in the figure, trouble occurrences in the Tertiary deposit, as well as in the diluvial and alluvium deposits are conspicuous. Trouble occurrence can be also observed in the Mesozoic and Paleozoic sedimentary rocks, the igneous rock and the metamorphic rock.

Fig. 2 illustrates the geological conditions at the spots where trouble occurred. More troubles are observed in fracture and highly weathered zones as apparent from the figure. It must be noted, however, that a considerable number of troubles also occur in the alternating strata. It has been pointed out that the alternating strata, as well as the fracture zones, apt to cause troubles due to its geological property.

2.2 Trouble

Troubles can be roughly divided into 50 cases of collapses, 14 cases of deformed states, 5 cases of the simultaneous occurrence of collapses and deformed

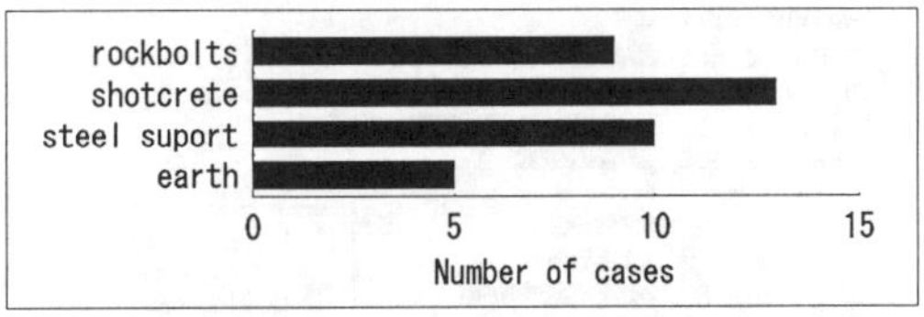

Fig.3 The types of the deformed states

states (hereinafter referred to as the collapse & deformed state) and 2 cases of rock burst. The types of the deformed states are shown in Fig. 3. The total number of the deformed states does not coincide with that of cases as multiple deformed states occur simultaneously in individual cases. In the "steel support" in the figure, deformed states in the shoulder and side wall are observed in three of all cases. In the remaining cases, settlements and thrustings are observed in the foot. In the "rock bolts", the failure in the head and a falling-off are observed in two different cases. Other cases show the deformed states in the metal washer. Large-scale deformed states such as the large deformation in the supporting and the failure in the rock bolts are relatively unusual as long as the ground discussed in this study is concerned.

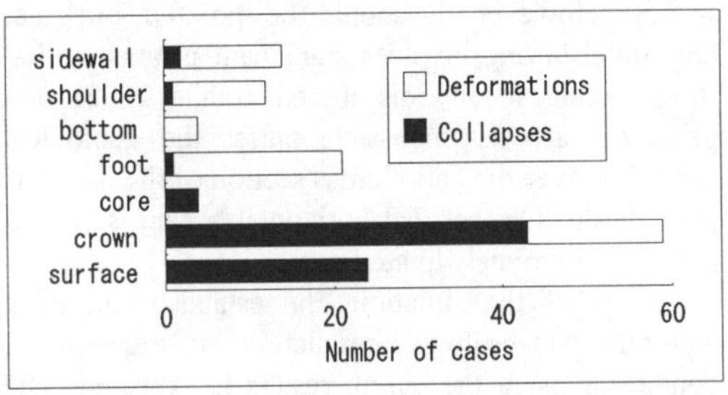

Fig.4 Spots of trouble occurence ("core", "crown" and "surface" are in the cutting face. "side wall", "shoulder", "bottom" and "foot" are in the back of the face)

From Fig. 4 in which the spots of trouble occurrence are shown, most collapses seem to have a tendency to occur in the surface, crown and the core of the cutting face. Meanwhile deformed states tend to occur in the back of the cutting face such as the foot, shoulder and the side wall. Such troubles most frequently occur in and the surrounding area of the crown, which is quite natural from the point of view of the structure and loading conditions of the tunnel.

Collapsed earth volumes, as can be seen in Fig. 5,

Table 3 The basic statistic of Tunnels and Collapses

	correlation coeff. (num. of cases=20)			
	sect. area	height	depth	collapsed vol
sect. area(m^2)	1.0	-0.35	-0.22	-0.11
max. height(m)		1.0	-0.04	0.57
max. depth(m)			1.0	0.09
collapse(m^3)				1.0
mean value	64.1(72.3)	5.2(5.5)	5.0(5.3)	152.1(102.5)
S.D.	25.0(21.0)	3.6(3.8)	3.5(3.8)	264.1(72.9)
cases	49	23	25	50

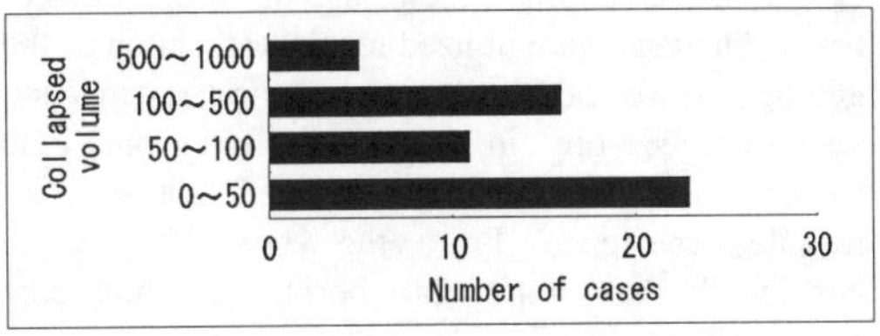

Fig.5 Collapsed volume

are less than 500 m^3 in most cases. No marked tendency is observed from the relationship between the geography and the collapsed earth volume. Table 3 presents the basic statistic of the excavated cross section area, the height, depth and the earth volume of the collapse. The correlation coefficients were obtained based on the 20 cases where all the above data was available. The mean values and standard deviation values given in the table were obtained based on the number of cases where the above data was available. The values in parentheses are limited to the cases from which the correlation coefficients were derived. From the mean values and deviation values shown in Table 3, the height and depth of the collapse were within the range of 5-10 m. The range of collapse in most cases is assumed to be within approximately 10 m irrespective of the dimensions of the excavated sectional area due to the fact that there is almost no correlation between a sectional area and other elements. Meanwhile there is a favorable positive correlation between maximum height and collapsed earth volume. It follows from the above that collapsed earth volume depends on the height of a collapse but not on the depth of a collapse. The above suggests that geological surveys should be carried out in the area 10 m ahead the cutting face. If trouble occurrence within this area becomes predictable, there will be enough time available for devising appropriate countermeasures, considering that supporting is usually arranged at an interval of 2-3 m even when the self-standing of the ground is favorable.

2.3 Trouble prediction

Geological surveys into the ahead of the cutting face are carried out in 31 cases (44 % of all cases) prior to construction. The geological survey methods employed before trouble occurrence are shown in Fig. 6. The percentage of individual surveys to the 31 cases is also shown in the figure. In the case of "probing", several experimental drillings were carried out using a drilling machine. The geology of the objective area was estimated from the results. In the case of "observational", the results of observations of the cutting face were utilized for the estimation of the geology. Horizontal boring and electric prospecting were carried out in 11 cases and one case respectively to survey the geological conditions after trouble occurrence. From the above, it can be concluded that horizontal boring is the most commonly employed geological survey method.

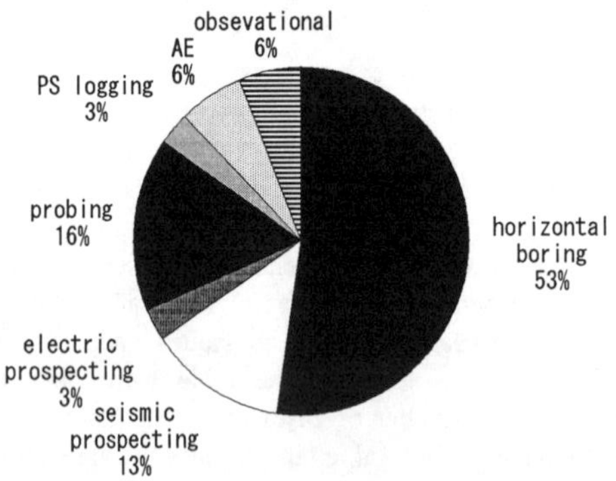

Fig.6 The percentage of employed geological surveys

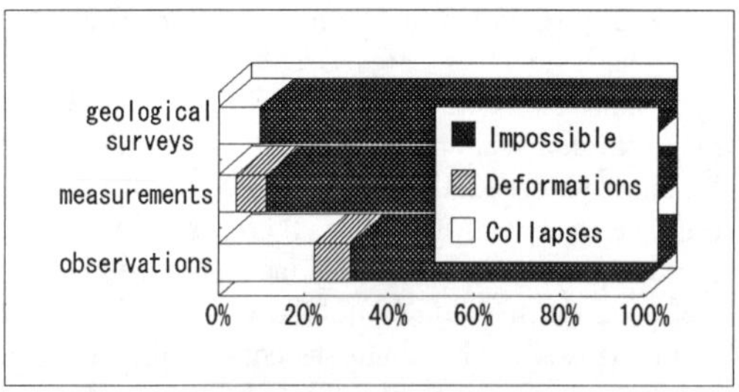

Fig.7 Possibility of trouble prediction

In Fig. 7, the number of cases, where the prediction of trouble occurrence was possible through observations and measurements prior to construction as well as through geological surveys carried out to estimate the geology of the objective area, are shown in percentage. The percentage of observations and measurements to the total of 71 cases, as well as that of geological surveys to 31 cases are also shown. From the figure, it can be assumed that the possibility of trouble prediction through observations is relatively high, while trouble prediction through measurements is difficult. Moreover, there is only 4 cases where trouble occurrence can be predicted through geological surveys, of which, in two cases AE (acoustic emission) measurements and visual observations were carried out to predict rock bursts. It follows from the above that at present horizontal boring is most commonly employed for geological survey. However, it should be pointed out that horizontal boring involves significant problems, that is, detecting locally distributed fragile zones and perched water is impossible unless the geological condition over the entire cross section of the tunnel is poor due to the fact that horizontal boring is carried out in the extremely limited areas.

In view of the situation, the establishment of a method that allows prediction of geological conditions using the boring results is expected. The significance of devising an appropriate approach that enables the objective description and evaluation of the variations in geological condition due to excavation to facilitate the routine observations of the cutting face, as well as the development of a method that permits the quantitative evaluation of geology and troubles is also recognized.

2.4 Ex post facto countermeasures

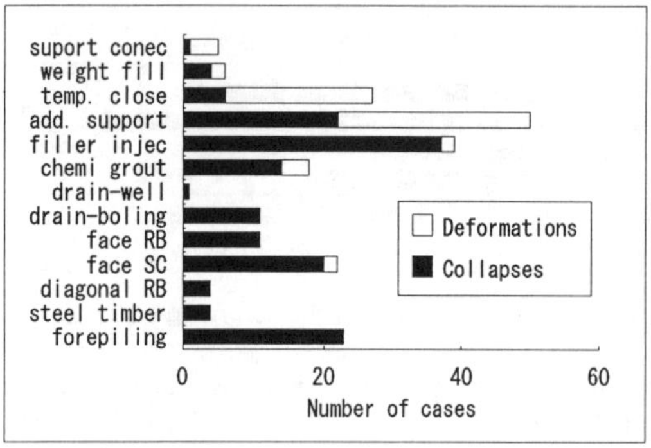

Fig.8 Countermeasures taken after trouble occurrence

Fig. 8 shows a histogram of the countermeasures taken after trouble occurrence. The diagonal bolts mentioned in the figure were employed in expectation of forepiling effects. The "Additional supporting" indicates additional support members and reinforcements such as additional rock bolts and increasing of shotcrete thickness. As auxiliary countermeasures against the deformed state, reinforcement of supporting as well as approaches that enable early closing of the tunnel section such as the additional supporting, temporary closing, and connection of supports are implemented. As the deformed state is considered to be induced by the excessive deformations of the tunnel, it is assumed to be quite natural that the above measures are employed to settle such deformations. The figure also suggests that the above countermeasures are effective enough against most deformed states. As can be seen in the case of the deformed state, most deformations could not be settled by the "additional supporting" but by temporary closing.

Various types of countermeasures were taken against collapses. It is not an unusual practice to carry out filler injection to fill cavities caused by such collapses. Urethane and air mortar are mainly used for a filler. Forepiling was employed in most cases. There are cases where collapses recurred between the piles. In other cases collapses recurred due to the deformation in the tip of the piles. For this reason, appropriate intervals of piles as well as the extension of forepiling arrangement must be considered. Grouting of a urethane filler is most commonly employed for forepiling.

3. Quantitative analysis using the multi-dimensional quantification theory

Obtaining quantitative data such as measured values is difficult at the time of, before and after trouble occurrence. Thus, most data available are limited to qualitative data. In this paper, obtained qualitative data is categorized according to Table 1 , to which the multi-dimensional quantification theory was applied.

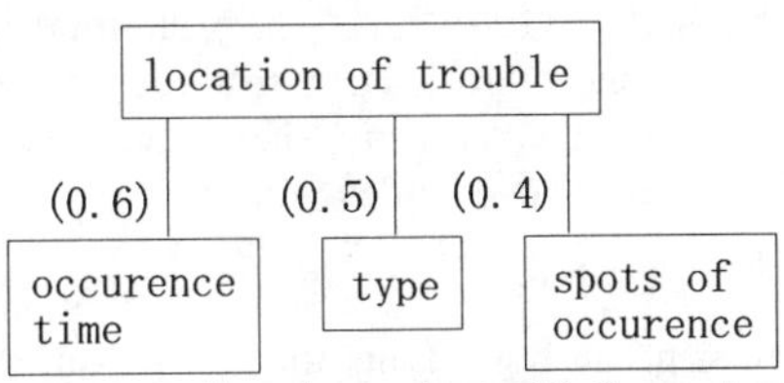

Fig.9 Indipendent values with Cramer's measure

In order to categorize troubles and to characterize each categorized groups, the quantification method of the third type was applied to all the variables excluding those included in the items "prediction" and "countermeasures" shown in Table 1. The variables whose independence are rejected by a level of significance of 5 % and whose Cramer's measure of association was found to be over 0.3 are shown in Figure 9. From Fig. 9, it is clear that "the location of trouble" is almost limited by "the type of trouble", "occurence time of trouble" and "the spots of trouble". Thus, "the location of trouble occurrence" was eliminated from the data because of its inappropriateness as a variable. As a result, the total number of cases and categories were 71 and 70 respectively.

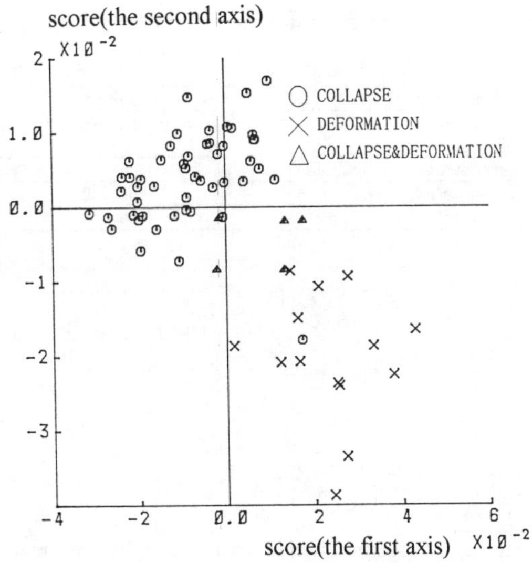

Fig.10 Distribution of scored values

Figure 10 shows a distribution of scored values with a transverse axis as the first axis and a longitudinal axis, the second axis. The cases of rock burst were excluded from the figure as their values are widely different from those of other cases. All the cases can be categorized into a collapse, deformed state and a rock burst. These three groups are clearly distinguishable, however, the cases where collapses and deformed states occur concurrently are very close to "collapse" and "deformed state" groups, making it possible to categorize them into an independent group. A cluster analysis was performed according to the Ward method using the resultant scored values up to the sixth axis obtained from the quantification method of the third type. All the cases were eventually classified into five clusters. The categories to which 60 % of all cases included the individual clusters responded are found and shown in

Table.4 The categories included the individual clusters

group	1.1	1.2		2	3
cluster	1	2	3	4	5
number of cases	23	16	11	19	2
excavation method	mechanical^{20}	blasting^{13}	blasting^{10}	blasting^{16}	blasting^{2}
auxiliary method A	forepiling^{15},face shotcrete^{13}	forepiling^{7},face shotcrete^{4}			
auxiliary method B	drainage^{14}				
geological type	diluvium&alluvium^{14}	mesozoic&palaezoic^{5},tertiary^{5},volcanic^{6}	tertiary^{6},volcanic^{6}	tertiary^{9},mesozoic&palaezoic^{7}	volcanic
geological condition	alternating strata^{14},highly weathered^{11},clayish&sandy^{9}	fracture^{8} zone,highly weathered^{6},progressive fissures^{7}	fracture zone^{4},highly weathered^{4}, clayish&sandy^{5}	fracture zone^{8},highly weathered^{7},alternating strata^{8},progressive fissures^{5},swelling^{5}	sound
influence by ground water		indirect factor^{7}	main factor^{8}		
condition of ground water			highly confined^{5}, concentrated^{4}		
spots of trouble	cutting face^{16},crown^{18}	crown^{16}	crown^{9}	crown^{11},core^{11},shoulder^{7},side wall^{7}	crown
type of troubles	collapse^{23}	collapse^{11},rock fall^{6}	collapse^{11}	collapse^{5},deform of steel support^{8},shotcrete^{12},rockbolts^{9}	rock burst
Spots of troubles	cutting face^{23}	cutting face^{16}	cutting face^{10}	back of the face^{18}	cutting face
Time of troubles	after excavation^{22}	after excavation^{16}	after excavation^{7} ,after support^{4}	after face advanced^{10}	after excavation
support pattern	DI^{9},DIII^{7}	DI^{7}			
Overburden(m)	50>h^{17}	50<=h^{14}	50<=h^{9}	50<=h^{17}	100<=h
Collapse vol.(m^3)	50>d^{12}	100>d^{13}			

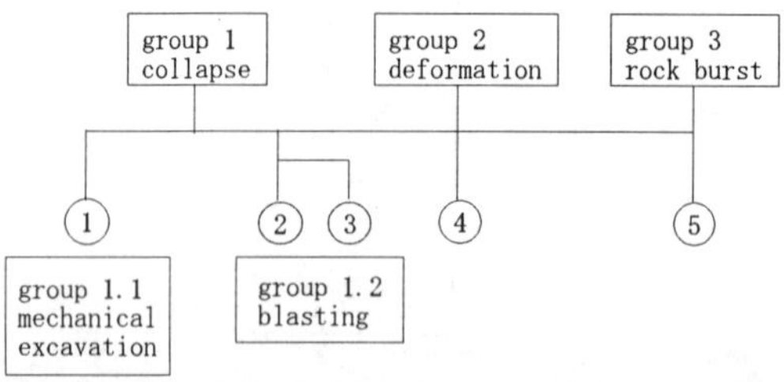

Fig.11 The relation within clusters and ther meanings

Table 4. The index appended to each category indicates the number of cases that responded to a specific category. Based on Table 4, these clusters can be categorized into the first group (a collapse), the second group (a deformed state) and the third group (a rock burst). A collapse is subcategorized into 1.1 group (machine excavation) and 1.2 group (blasting) according to an excavation method. The numbers within circle in Fig. 11 indicate clusters. From Table 4 and Fig. 11, the characteristics of individual groups can be summarized as follows.

1.1 group: In the diluvial and alluvium deposits with relatively shallow overburden where machine excavation is usually employed, collapses due to ground water and alternating strata occur all over the cutting face. In such a case, construction must be carried out with utmost care even countermeasures have already been taken against possible collapses.

1.2 group: In the Tertiary deposit, Mesozoic and Paleozoic deposits and igneous rock with relatively deep overburden where blasting is employed, collapses due to the fragility of the ground induced by fractured zones, highly weathered zones, and progressive cracks occur mainly in the crown. In such a case, either the geological condition over the entire ground is poor or fragile parts are locally encountered. In either case, collapsed earth volume is relatively large.

2nd group: In the Tertiary deposit, Mesozoic and Paleozoic deposits where overburden is relatively deep, deformed states due to the fragility, alternating strata and swelling property of the ground can be seen in the back of the cutting face. No such deformed state can be observed in the diluvial and alluvium deposits.

3rd group: Two unusual cases are observed where "the condition of the ground" is found to be "sound" and "overburden" is "100 $\leqq$ h", which illustrate a typical characteristic of rock burst that rock burst usually occur in the ground where geological condition is sound and overburden is deep.

3.1 Discriminant analysis using the quantification method of the second type

Based on the results obtained so far, the discriminant analysis of the first and second groups were performed using the quantification method of the second type. The variables used in the quantification method of the third type were employed in the analysis. Of which, however, "ground water influence","ground water condition" and "troubles" that directly explain troubles were eliminated. Two categories whose responses were limited to "sound" and "A pattern" were also eliminated. As a result of the discriminant analysis of the total of 43 categories and 64 cases, a maximum correlation ratio of 0.67 and an erroneous discrimination rate of 19 % were obtained.

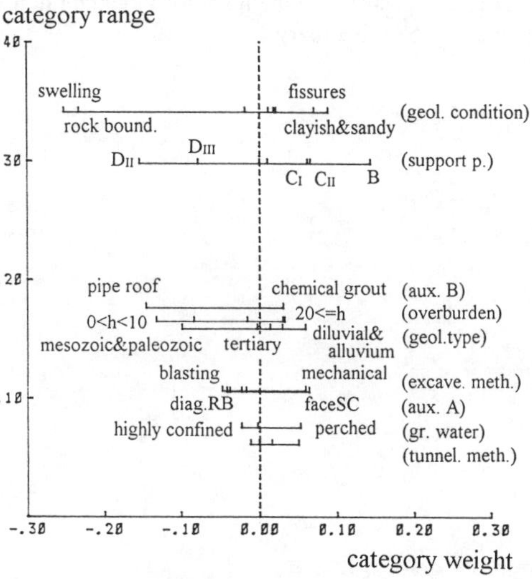

Fig.11 Distribution of variables and categories(Positive scored value of category weights indicates "collapse", and negative scored value indicates "deformed states")

Figure 11 shows the distribution of variables and categories with the category range on the longitudinal axis and category weight, on the transverse axis. In the above discriminant analysis, positive scored values of each category weights indicate "collapses" and negative scored values, "deformed states". As can be seen in the figure, categories "geological conditions", "support pattern", "auxiliary method B", "overburden" and "geological type" are shown in sequence of significance, suggesting that these five categories are primary factors in the discriminant analysis.Meanwhile as for the category weight, the two categories "clayish and sandy" and "progressive fissures" outweigh on the positive value side, illustrating the tendency of collapses to occur in the areas where the variations in the ground into clay and sand as well as fissures are induced by progressive weathering. From the fact that the patterns B and C as well as "perched water" take positive values, it can be assumed that in case overburden is relatively deep, supporting is often devised counting on the self-standing of the ground (thus no steel supporting is employed) until sudden changes in the ground such as fractured zones, local fragile zones and perched water are encountered. From Fig. 11, it is assumed that the type and method of excavation do not have a great influence on discrimination. As apparent from the figure, the categories "the Tertiary deposit", "the diluvial and alluvium deposits", "machine excavation" and the "heading method" exist on the positive value side. We may say from the above that collapses can easily occur when the self-standing of the ground is not assured. The ground water is rather on the positive side though the dependency of which is low. It may follow from the above that ground water can be a predisposing factor, if not a primary factor, in the causation of collapses. As can be seen in Fig. 11, no auxiliary countermeasure against ground water is included in the countermeasures taken against deformed state. This is due to the fact that ground water, which is an influential factor in the causation of collapses, has an extremely limited influence over the deformed states that occur in the back of the cutting face. It also shows that, in connection with the above, a cross section to which supporting is arranged rarely experiences the influence of ground water.[2]

The categories "swelling" and "rock boundaries" outweigh on the negative value side. It follows from the above that deformed states tend to be caused by stresses in the ground such as the swelling pressure of the ground due to its expansivity and the uneven pressures in rock boundaries. The "overburden" takes a positive value when the depth of which is under 20 m. It is often the case that when overburden is shallow, the geological condition of the ground is considerably poor. In such cases, rigid supporting systems such as D_{II} and D_{III} are employed. Carefully planned out countermeasure such as pipe roofing (as an auxiliary countermeasure) are taken against collapses and settlements for the prevention of troubles. These countermeasures are only effective for the prevention of collapses. Most deformed states can not be prevented with such countermeasures.

As already been understood, the discriminating factors exhibit a favorable correspondence with the characteristics of individual groups. It is assumed that the categorization of troubles and typical trouble occurrence mechanisms can be explained by what has been discussed.

4 Conclusions

The information on troubles obtained from individual cases is objectively verified from the point of view of data analysis based on the results of case investigations into troubles. The classification of such troubles are attempted as well as the difference between the trouble occurrence mechanisms is discussed. Some controversial points in trouble prediction and countermeasures are also presented.

1) Based on ground conditions and constructional conditions, troubles can be roughly divided into three groups: a collapse(1st group), deformed state(2nd group) and rock burst(3rd group). The collapse can be subdivided into two cases: machine excavation(1.1 group) and blasting(1.2 group).

2) Ground water relates mainly to collapses. In case the geological condition over the entire ground is poor, ground water can be a predisposing factor in the causation of collapses, while in locally distributed fragile zones, ground water becomes a primary factor. Ground water is assumed to have almost no relation to deformed states.

3) Most collapses occur in the cutting face. The range of the collapses has no significant correlation with geological conditions. Both the height and depth of the collapses are within 10 m.

4) At present, the geological condition of the ahead of the cutting face is predicted mainly through horizontal boring. However, the accuracy of this approach is extremely low. In view of the situation, the development of an appropriate method that enables the evaluation of the geological conditions based on the boring results is required.

5) Forepiling is the most commonly employed auxiliary countermeasure taken against collapses. The method is highly effective, nevertheless, the interval and extension of forepiling arrangement must be carefully designed.

6) A majority of deformed states are settled by the reinforcement of support members. In most of such cases, however, the ultimate stability is achieved through closing of the cross section with invert. Thus, early closing of the cross section is assumed to be most effective.

5 Acknowledgment

The authors would like to thank 18 private corporations for cooperation in providing us their valuable data through our questionnaire.

References:

1) Japan Road Association: Technical standard for highway tunnels (Structure) • Commentary, p.118, June 1989

2) Japan Tunneling Association: Research study report on construction methods for mountain tunnels (part 3), p.584, February 1990

Tunnelling and Ground Conditions, Abdel Salam (ed.) © 1994 Balkema, Rotterdam, ISBN 90 5410 363 9

A study on ground water behavior due to the construction of large-scale caverns at Kikuma underground oil storage plant

Hiroyasu Ohtsu & Kazuhiko Yamamoto
Taisei Corporation, Tokyo, Japan

Akio Okamoto
Japan Underground Oil Storage Co., Ltd, Tokyo, Japan

ABSTRACT : This paper is concerned with the investigation of ground water behavior due to the construction of large-scale caverns at Kikuma underground oil storage plant employing the concept of aquifer sealing method as one of the first attempts in Japan. The investigation describes the ground water-flow being heavily affected by the discontinuities present in rock masses based on field measurement results, and the applicability of the theory of crack tensor as a method to quantify the distribution tendency of discontinuities present in rock masses. The results show that higher frequency area of discontinuities calculated by the theory of crack tensor makes good agreement with the location of observed seepage area. This verification makes it clear that the theory of crack tensor has a great potential to predict quantitatively the higher permeable area.

1 INTRODUCTION

In early 70's, Japanese had such a bitter experience through oil-crisis due to Middle East War that a national oil stockpiling plan became an urgent task to stabilize national life. In response to this national demand, Japan government established the Petroleum Stockpiling Law and officially decided to promote construction of oil storage plants, and the Agency of Natural Resource and Energy established a study group on the new petroleum storage technology. Then, in addition to conventional oil storage methods, underground oil storage plant adopting the principle of aquifer sealing method, which is originally developed in Scandinavian countries, was selected as one of the feasible oil storage systems from a viewpoint of effective land utilization plan to keep surrounding environment intact.

Based on this national policy, a pilot project to construct small-scale underground oil storage plant and to investigate the applicability of the aquifer sealing method for geological condition in Japan was initiated in Kikuma town, Ehime prefecture.

After investigation considering several factors associated with design, construction, and operation at the pilot plant from 1979 to 1983, a study group concluded that the underground oil storage plant adopting the principal of an aquifer sealed method is possible in Japan safely with environmental conservation.

Japan government decided to initiate Kikuma project, which is one of the first attempts in Japan to construct large-scale underground crude oil storage plant, as a part of national petroleum storage plans in 1987 and the construction work was completed in 1993.

As developed by many researchers[1),2)], an aquifer sealing method is based on the simple principle that ground water-flow toward storage caverns prevents leakage of fuel and/or vapor gas. When fulfilling this requirement associated with an aquifer sealing method, it is necessary to employ the idea of ground water control. Hence, the observation of ground water-flow becomes a very important factor besides the stability of rock masses surrounding the caverns to construct a large-scale underground oil storage plant. As for the construction of an underground oil storage plant, the great attention is laid on this requirement associated with ground water-control, whose situation is completely different from that of ordinary cavern excavations such as underground electric power plants.

It is well-known that ground water-flow through rock masses is heavily affected by the discontinuities present in rock mass. However, the quantitative discussion associating the presence of discontinuities involved in rock masses with ground water-flow behavior due to construction have been insufficient yet. The reason for this is that proper examples are limited, which observe the feature of ground water-flow through rock masses due to excavation.

From these viewpoints, this paper presents ground water-flow due to the construction of large-scale caverns at Kikuma underground oil storage plant as an example to discuss the quantitative co-relationship between discontinuities present in rock masses and ground water-flow behavior. In order to quantify the distribution of discontinuities present in rock masses observed at construction site, the theory of crack tensor[3)] is employed.

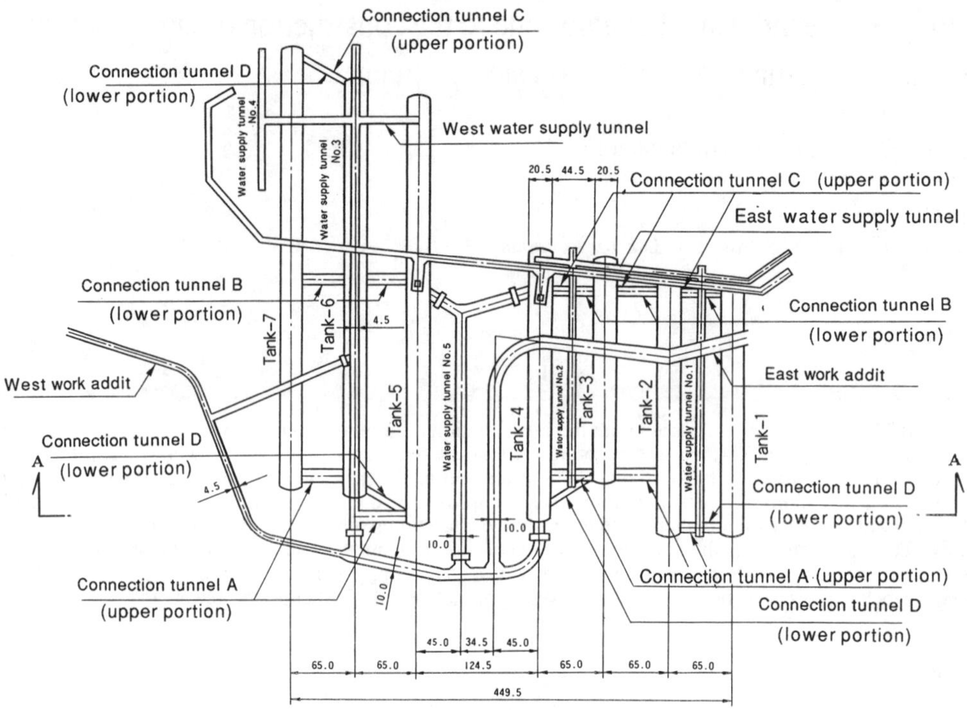

(a) Plane View

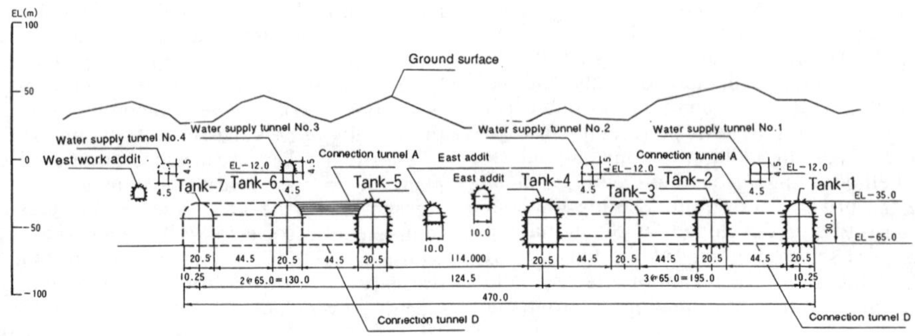

(b) Section A-A

Fig.1 Layout of rock Caverns

The contents of this investigation are summarized as follows:

1. the summary of ground water behavior such as the change of water table and discharge in rock caverns affected by the discontinuities present in rock mass.

2. the calculation of crack tensor to quantify the distribution of discontinuities observed at the exposed surfaces of caverns.

3. the co-relationship between the location of seepage area observed at the site and the higher frequency area specified by crack tensor.

Finally, results show the relationship between higher frequency area of discontinuities and the location of observed seepage area, and the applicability of the theory of crack tensor to predict the higher permeable area quantitatively.

2 THE OUTLINE OF KIKUMA UNDER-GROUND OIL STORAGE PLANT

It is the purpose of Kikuma project to construct underground oil storage plant, whose storage capacity is 1.5 million m^3. This chapter is concerned with the outline of Kikuma underground oil storage plant.

This project site is located in Ehime prefecture, southwest of Japan, and is close to the pilot plant constructed in 1980-1983 before this project.

The geological formation at the construction site mainly consists of upper Cretaceous granite, and the surface is covered with weathered deposit.

Typical rock properties are summarized as follows:

1. Averaged rock quality designation(RQD), the index to express frequency of existing fracture, is 80 %.

2. Compressive strength of rock core is about 80 MPa.

3. Averaged deformation modulus of rock masses is 6,500 MPa.

4. Seismic velocity of rock masses is 4.8 km/sec to 5.0 km/sec.

5. Averaged permeability coefficient is 10^{-6} cm/sec to 10^{-7} cm/sec.

Fig.1 shows the layout of Kikuma underground oil storage plant. As shown in Fig.1, the overburden of rock caverns is relatively shallow, which is from 80 m to 100 m, and the oil storage plant consists of seven adjacent rock caverns, connection tunnels, vertical shafts and water supply tunnels.

As for the dimensions of rock caverns, the feature of Kikuma plant is that the length of each cavern varies from 230 m to 450 m because of restrictions associated with above-ground surrounding facilities and housings, and that the size of each rock cavern is 20.5 m width and 30.0 m height, which is one of the biggest section of caverns constructed in Japan. Also, the distance between two adjacent caverns is 65m, whose dimensions were determined by means of stress analysis for the purpose of avoiding the stress-interference due to excavation of adjacent caverns.

Another feature of Kikuma plant is the employment of artificial aquifer sealing method with water supply holes drilled horizontally, whose dimensions were determined by means of unsaturated-saturated seepage analysis.

As water supply through the drilled holes was initiated during cavern excavation in order to prevent water table from lowering, cavern excavation was executed under water injection from upper portion. Consequently, this construction had become very rare case to apply the idea of the control of ground water behavior during construction artificially for large-scale cavern excavation.

3 GROUND WATER BEHAVIOR DUE TO CAVERN EXCAVATION

As known well, ground water-flow through discontinuous rock masses is heavily affected by the discontinuities present in rock masses, while the volume of discontinuities present is much smaller than intact portions.

Here, the feature of geological condition at construction site is that RQD is about 80 %, which means the volume of discontinuities present in rock masses is very small, and that the permeability of intact portions is very low, less than 10^{-7} cm/sec. The geological condition is fairy identical to general findings mentioned above.

Hence, in order to discuss the ground water behavior due to cavern excavation, it is inevitable to investigate the relationship between distribution of discontinuities and the location of seepage area.

3.1 The distribution of discontinuities

Many faults and cracks were observed at the exposed surface of rock caverns during excavation. Fig.2 shows a Schmidt's net to express the strike and dip of discontinuous planes observed at the exposed side wall surface of each rock cavern. As shown in Fig.2, there is no particular difference among rock caverns as to the distributions of strikes and dip of discontinuous plane.

The contours to express higher frequency of the strike and dip of discontinuous plane concentrate on two portions. One expresses the discontinuous planes, whose strike is from N40°E to N50°E. Another expresses the discontinuous planes, whose strike is from N30°W to N50°W. Both express the discontinuous planes, whose dip is from 60° to 80° toward the vertical direction. This feature of discontinuities described above is quite similar to those observed at Kikuma pilot plant, which is close to this construction site.

Comparing these strikes with the layout of rock caverns in horizontal plane, the discontinuities, whose strike is from N40°E to N50°E, intersect rock caverns with 90° toward the longitudinal axis of each cavern and the discontinuities, whose strike is from

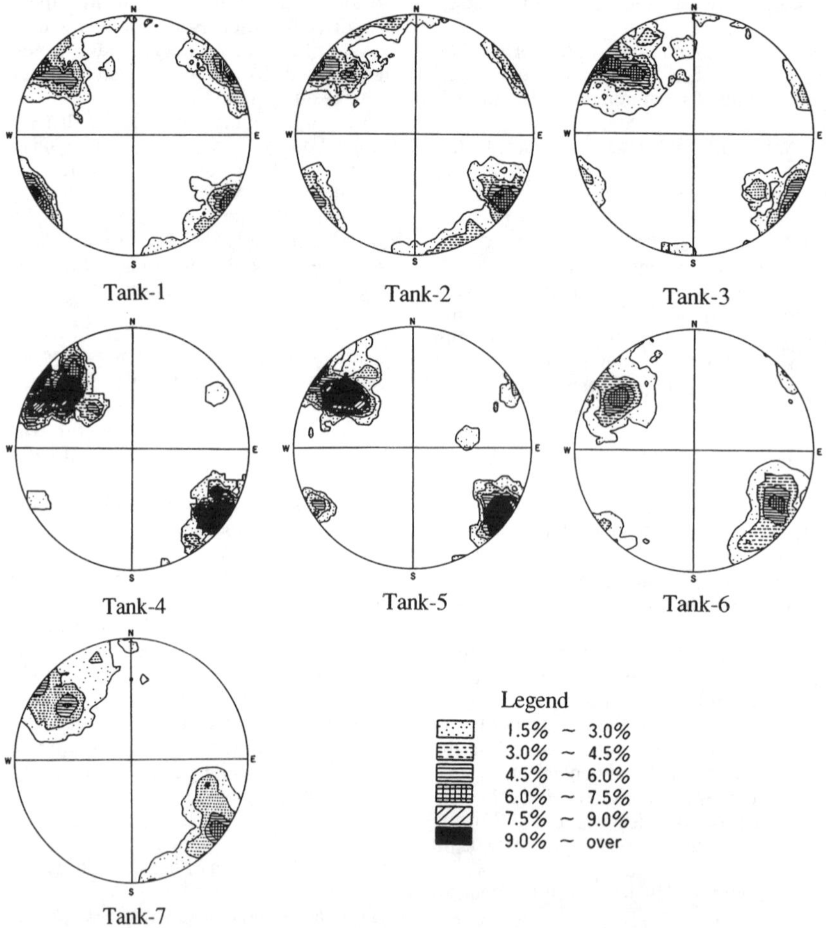

Fig.2 Schmidt's net projection of discontinuous planes at a surface of each rock cavern

N30°W to N50°W, intersect rock caverns with 30° toward the longitudinal axis of each cavern.

Hence, the distribution of discontinuities observed at the site are decisive one, and are classified into specific two groups by their strike.

3.2 The location of seepage area

When the dip of most discontinuous plane is from 60° to 80° toward the vertical direction, it is predictable that the vertical component of ground water-flow becomes dominant. Fig.3 shows the distribution of discharge points observed at the exposed surface of arch portions besides discontinuities. Among these discontinuities, four faults, which are indicated by F-1 to F-4 in Fig.3, are relatively continuous and large-scale. F-1 and F-2 belong to the group of discontinuities, whose strike is from N40°E to N50°E. On the other hand, F-3 and F-4 belong to the group of discontinuities, whose strike is from N30°W to N50°W.

As shown in Fig.3, there is a tendency that typical seepage areas concentrate on six specific zones, which are indicated by zone-1 to zone-6. Most discharge points included into each seepage zone are along the fault. Especially, discharge at zone-2, where F-1 crosses with F-2, is dominant.

Hence, there is a strong co-relationship between the location of discharge due to cavern excavation and discontinuities present in rock masses.

Fig.4 shows the change of water table due to cavern excavation as the example describing ground water behavior affected by discontinuities present in rock masses. The region that shows the typical change of water table concentrates on two specific areas as shown in Fig.4. One is the area connecting zone-1, zone-2 and zone-3, and the change of water table at the center of tank-2 is dominant among this specified area. Another is the area connecting zone-5 and

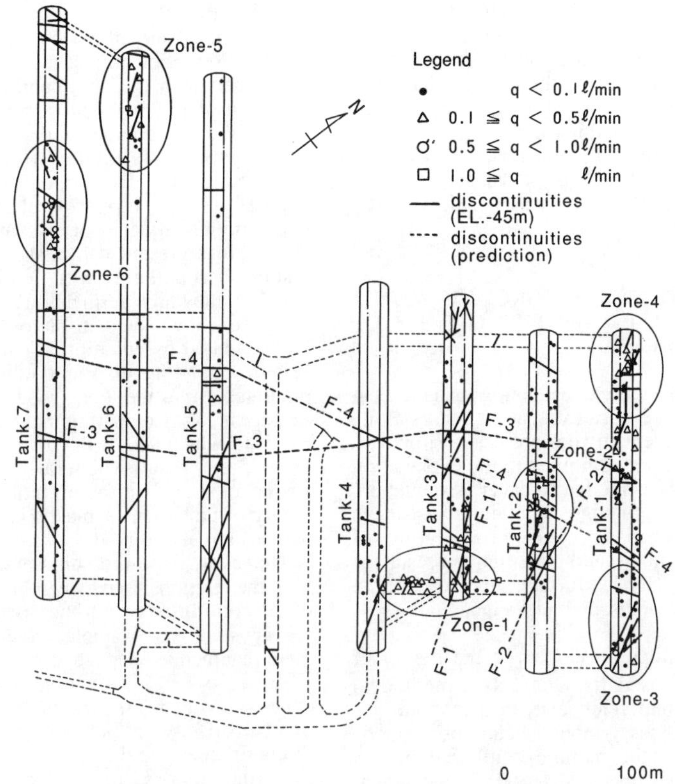

Fig.3 Seepage areas observed at the construction site

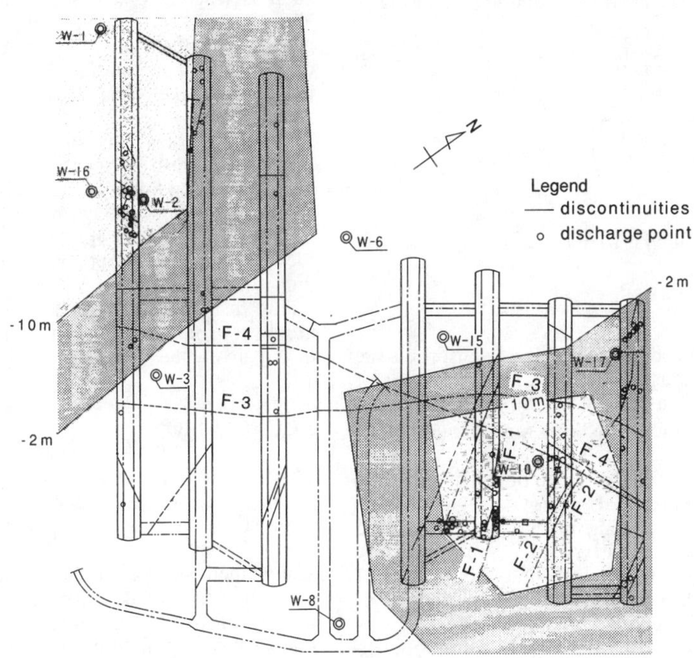

Fig.4 Change of water table due to cavern excavations

zone-6, and the change of water table at the edge of tank-7 is dominant among this specified area.

Consequently, there is a strong co-relationship between the region to show the typical change of water table and the location of seepage area based on discharge points expressed in Fig.4.

As a result, the investigation made it clear that ground water behavior such as the discharge and the change of water table are heavily affected by discontinuities.

4 THE INVESTIGATION OF GROUND WATER BEHAVIOR BASED ON THE THEORY OF CRACK TENSOR

As mentioned above, the investigation made it clear that seepage due to cavern excavation occurs along discontinuities present in rock masses. However, the previous discussion is limited to the explanation of phenomenon occurred due to cavern excavation. The reason for this is that the discussion to associate the location of seepage areas with feature of discontinuities based on field measurement results is insufficient yet. Hence, further investigation to specify distribution of discontinuities quantitatively is necessary.

Here, this investigation employed the theory of crack tensor proposed by Oda[3] as a method to quantify the distribution tendency of discontinuities present in rock masses. Among calculation methods presented by Oda, the method employed in this investigation is the one to calculate crack tensor based on the distribution of discontinuities observed at the exposed surface. The applied equation is shown as follows[3]:

$$F_{ij} = \frac{3\pi}{8} \frac{\langle t^2 \rangle}{\langle t \rangle} \frac{N^{(q)}/h}{\langle |\mathbf{n} \cdot \mathbf{q}| \rangle} N_{ij} \qquad (1)$$

$$N_{ij} = \int_{\Omega} n_i \, n_j \, E(\mathbf{n}) \, d\Omega$$

where F_{ij} are the components of crack tensor, t is the trace length of cracks, n_i (n_j) is normal vector of crack, and $E(\mathbf{n})$ is a density function expressing the distribution of cracks. $N^{(q)}/h$ means number of cracks per unit length, which cross with the measurement line with the length of h. The notation < > expresses the averaged values to each term.

Oda has also proposed the method to calculate the coefficient of permeability of discontinuous rock masses under the assumption that the feature of discontinuities is constant. However, feature of discontinuities observed at the exposed surface is generally various. Consequently, it is improper to calculate directly the coefficient of permeability by applying the theory of crack tensor for in-situ data.

From these viewpoints, this investigation is only concerned with the applicability of the theory of crack tensor as a method to associate the quantitative values of discontinuities with seepage areas. In this investigation, the index to determine higher frequency area of discontinuities is first invariant of crack tensor F_0. The reason selecting F_0 as the index to determine frequency of discontinuities is based on the findings presented by Oda, where he proposed that the discontinuities with first invariant F_0 greater than 7.0 show high permeability.

Fig.5(b) shows the distribution of the areas, where F_0 calculated by equation (1) are greater than 7.0. When depending on the previous findings, it is predictable that the areas specified in Fig.5(b) are seepage areas. Comparing those specified areas with Fig.5(a), which shows the distribution of discharge points, the results generally show that higher frequency area of discontinuities calculated by the theory of crack tensor makes good agreement with the location of observed seepage area.

However, there are some exceptions associated with the relationship between higher frequency area of discontinuities and the location of observed seepage area. For example, though zone-4 and zone-5 are regarded as seepage areas from a viewpoint of observation, first invariant F_0 corresponding to these zones is less than 7.0. The reason for the occurrence of the difference between the distribution of discontinuities and the location of discharge point is that the theory of crack tensor lacks the considerations for physical feature of discontinuities such as whether cracks are including clay or not. Though the values such as first invariant F_0 calculated by the theory of crack tensor are dependent on geometrical conditions of discontinuities, they are independent on physical feature of discontinuities.

Consequently, to apply the theory of crack tensor for in-situ problem, further investigation to consider physical feature of discontinuities besides geometrical conditions of cracks is necessary.

However, from engineering viewpoints, the result shows the strong co-relationship between the location of seepage area observed at the site and the higher frequency area specified by crack tensor.

This verification makes it clear that the theory of crack tensor has a great potential to predict quantitatively the higher permeable area.

5 CONCLUSIONS

This paper is concerned with the investigation of ground water behavior due to the construction of large-scale caverns at Kikuma underground oil storage plant employing the concept of aquifer sealing method as one of the first attempts in Japan.

The investigation described the ground water-flow heavily affected by the discontinuities present in rock masses based on field measurement results, and the applicability of the theory of crack tensor as a method

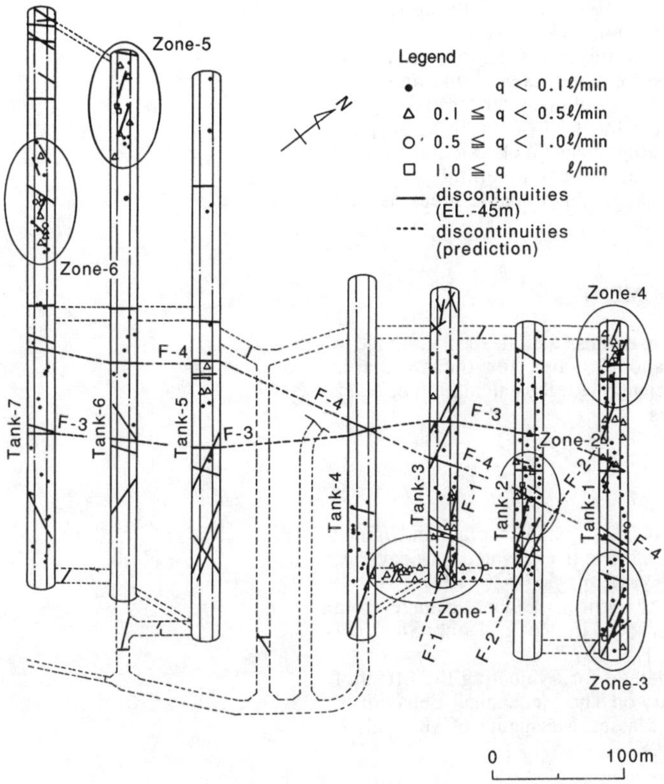

(a) Seepage areas observed at the construction site

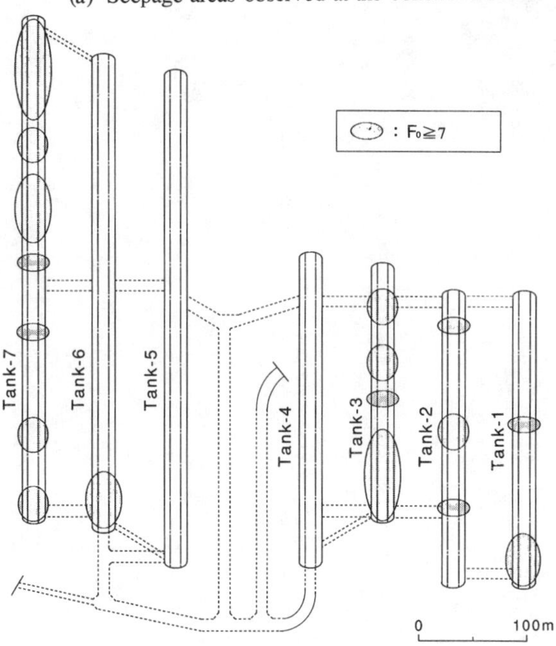

(b) Calculation results by crack tensor

Fig.5 Comparison of seepage areas and the areas where $F_0 \geqq 7$

to quantify the distribution tendency of discontinuities present in rock masses.

As a conclusion of this investigation, the results showed that higher frequency area of discontinuities calculated by the theory of crack tensor makes good agreement with the location of observed seepage area. This verification makes it clear that the theory of crack tensor has a great potential to predict quantitatively the higher permeable areas as first approximate.

6 ACKNOWLEDGMENT

The authors wish to express their deepest gratitude to Y. Hatsuyama and T. Ohya for their valuable cooperation associated with the calculation of crack tensor in this paper.

REFERENCES

1) Åberg, B. : Model Tests on Oil Storage in Unlined Rock Caverns, Storage in excavated rock caverns, Rock Store 77 Vol.2, pp.517-530, 1977.
2) Åberg, B. : Prevention of Gas Leakage from Unlined Reservoirs in Rock, Rock Store 77 Vol.2, pp.399-414, 1977.
3) Oda, M : A Method for Evaluating the Effect of Crack Geometry on The Mechanical Behavior of Cracked Rock Masses, Mechanics of Materials 2, pp.163-171, 1983.

Tunnelling and Ground Conditions, Abdel Salam (ed.) © 1994 Balkema, Rotterdam, ISBN 90 5410 363 9

Italy – Europe links across the Alps

M. Palliccia
Linea A.V. Torino-Milano-Venezia e Milano-Genova, ITALFERR-SIS.T.A.V. Roma, Italy

G. Viola
Linea A.V. Firenze-Bologna, ITALFERR-SIS.T.A.V. Roma, Italy

S. Pelizza
Tunnelling and Underground Space Center, Polytechnical University Torino, Italy

ABSTRACT: The Alps provide a natural barrier and in later ages, as well as at present, they hinder transportation, i.e. the connection between Italy and the rest of Europe. For this reason, the need for Alpine crossing through road and rail infrastuctures is a concern of much importance in Italy and in the Alpine countries The major problem is to provide adequate conditions for ever more rapid crossings. Consequently, very long road and railway tunnels have been or will be built. The paper will give a general overview of them through the Alps which have been constructed so far and which are planned in the future in Italy.

1. MOVING TO EUROPE

The physical boundaries between Italy and Europe consist of a natural barrier: the great Alpine mountain range which covers over 1200 km from East to West. Italy has always been a preferential traffic route along the European North to South axis and a direct link between South-West and Eastern Europe and it has thus always had to tackle the problem of crossing the Alps.

Goods travelling across the Alps recorded a net increase, moving from 18 MT in 1965 to 99.3 MT in 1990; 45% of these goods are shipped by railway and 55% by road (Table 1).

The average annual growth of trans-Alpine traffic through the whole mountain range, from Ventimiglia, in the West, to Villa Opicina, in the East, is estimated at around 3.25%; consequently freight traffic is expected to reach 114 MT in the year 2000.

A considerable shifting of traffic towards railway transport. Positive evidence of this is given by the number of foresees trans-Alpine railway tunnel projects against the number of road tunnel project.

The reason for this decision, mainly relates to the global low cost of rail trasport and to the reduction of environmental pollution. On the other hand this approach has already been adopted by two countries bordering Italy, Austria and Switzerland, who clearly prefer railway transport.

Table 1: Transport capacity (Mt/y) by rail along the main links without new long tunnels.

Pass name	1992	2000
Frejus	8.5	11.5
Simplon	5.5	7.3
St. Gothard	9.8	13
Brennero	7.2	9.5
Tarvisio	5	6.9
Villa Opicina	3.0	3.0

2 TUNNELS CROSSING THE ALPS

2.1 Technical problems of long and deep tunnels

The great road links across the Alps are progressively built at lower altitudes in order to reduce their length and gradient, and thus achieving higher speeds, saving energy and increasing transport capacity. At the same time, the access ways crossing the mountain valleys are shortened, reducing their environmental impact.

The reduction of mountain-pass altitudes is however a negative factor for tunnels, since these become longer (more than 10km as 60km), and are subjected to greater overburden (more than 500-600m as to 2.000m).

Greater overburden, on one hand, give rise to greater

stress on the tunnel, implying the need for a better knowledge of the rock mass to precisely calculate the supporting systems for the tunnel; on the other hand, however, it makes it difficult to know the geological, geohydrological and geomechanical situations at tunnel level.

The procedure adopted in the most recent projects for long and deep tunnels in Italy is the following (Pelizza, 1987):

- assessment of the geological - structural situation at tunnel level by means of on-the-surface surveys and, subsequently, deep geophysics, and long geological drilling aimed at ascertaining major structural discontinuities and separations between different lithotypes;
- sampling of different sub-emerging lithotypes with many and relatively short exploratory bores, and geomechanical classification of the samples;
- analytical evaluation of the geo-mechanical behaviour of the different rock masses at tunnel level, and structural project of the excavation and supports: the project is then verified during the work.

This procedure unavoidably leaves various uncertainties that can be clarified only when the tunnel is bored, specially when dealing with weak and complex rock masses, which today are still difficult to classify geo-mechanically: this must be accounted for in construction contracts, foreseeing a sharing of risks between the Client and the Contractor.

A precise means of investigation, which has been widely adopted in Italy in the last 10 years, is the pilot bore, drilled with a TBM (since 1982, 32 pilot bores for a total length of around 110 km have been drilled). In cases of deep tunnels, however, we have had to admit (with a few failures) that pilot bore drilling with a TBM in difficult geotechnical conditions is practically impossible.

In cases of long as well as deep tunnels, investigation times with a pilot bore along the tunnel axis become particularly long, and the cost is so high that this type of prospecting proves unfeasible, except for special cases: for example, the studies for the new San Gottardo railway tunnel in Switzerland (57 km long) foresee a 6 km TBM pilot bore, transversal to the main tunnel, in order to assess the geomechanical situation of a limited layer (200 m) of difficult material, and eventually to perform preventive consolidation on the layer itself (Anon., 1981).

Another very important aspect for deep tunnels is the geohydrological situation, for two reasons: on one hand flooding inside the tunnel due to violent water inflows during work causes danger for the workers, as in the case of the Mount Blanc tunnel, where the inflow was so great as to reduce the rock temperature, or requires costly and long operations to overcome the water area ,as in the case of the Gran Sasso tunnel where a large water table with a pressure of 60 bars suddenly had to be tackled (ANAS, 1989). On the other hand, the interception of large quantities of water causes alterations to the hydrological situation on the surface, with considerable damage to the environment. Geohydrological studies, like geomechanical studies, are very difficult and leave major uncertainties on detail situations, even though they allow to make some general forecasts: thus projects should always foresee long sounding bores performed from inside the tunnel under construction (Pelizza, 1991).

As for the length of great mountain pass tunnels, it has two main application consequences: long construction times and operational safety.

There are two ways to reduce construction times: to open intermediate accesses; to use TBMs for the construction. Both are negatively influenced by great covers. In relation to intermediate access, obviously there is a need for long and tilted tunnels or shafts, which represent bottlenecks, especially in case of eventual mechanised tunnelling.

As for boring machines, the problem is given by the required capacity to withstand great loads applied by the mountain, both on the structure of the machine (when difficult geotechnical conditions are tackled), as on the supporting structure of the tunnel (prefabricated or extruded lining), that has a rigid behaviour which allows to count only limited convergence values.

If, on top of a poor geomechanical situation, we were to add a considerable presence of water problems would become even more serious.

With such problems in mind, recent large tunnel projects in Italy adopt a conventional construction method based on the confinement-convergence criterion, with rock mass re-classification, even performed in advance through a short and repeated pilot drive: this system is flexible and may be adopted in any condition; though it is a low productivity system (less than 1 m/day) which requires a considerable amount of intermediate accesses.

The idea of using boring machines has not been completely given up, trusting that the TBM technology may rapidly improve and provide the sector with a "universal machine". In this view, the machine-tunnel pair is inseparable: the first important step to be made is, in fact, to reduce as much as possible the bore cross section in the project phase, and thus simplify the mechanical work.

The final major aspect affecting the design of long and deep tunnels is represented by operational safety. In road tunnels, a long drive in a closed environment, oppresses and bores drivers, who tend to pay less attention to their driving: this problem has been tackled on the motorway connecting Turin to the Frejus tunnel by improving the conditions inside its tunnels with artificial environments created with lights and coloured walls.

Another serious aspect of long road or railway tunnels is represented by accidents, that may be even more serious in case of fire: problems are linked to the need for rapid access of rescue vehicles to the accident spot, and for a swift evacuation of travellers.

Both of these problems are quite serious for railway tunnels, specially when they are over 20 km long. The gratest danger is represented by accidents followed by fire. The train will not be able to leave the tunnel on its own.

Furthermore, the consequences of accidents can become more serious due to the reduced space, the distances to travel in order to get outside or to rescue the victims and, finally, the panic of passengers in an underground environment.

For purpose of safety, we must bear in mind a few guiding principles, listed below, to direct decisions regarding both the cross sections and the overall architecture of long tunnels:

- reduce the possibility of accidents;
- limit the cause of fire (specially for passenger trains);
- reduce the length of tunnel sections between the outside and eventual intermediate underground stations, easily accessed from outside;
- avoid the concurrent crossing of trains;
- facilitate the entrance of emergency vehicles and tunnel evacuation;
- facilitate the access of rescue.

There are two other important issues:

- train stops caused by fire are to be avoided as much as possible;
- organization or construction measures should be adopted in order to ensure efficent rescue operations in shortest time in case of other accidents that might obstruct the movement of trains.

The solutions that could be assumed for a long tunnel are (Fig. 1):

1) single-tube with service tunnel;
2) double-tube with service tunnel;
3) double tube without service tunnel;
4) double-tube without service tunnel but with an intermediate emergency station with easy access to the outside where possible;

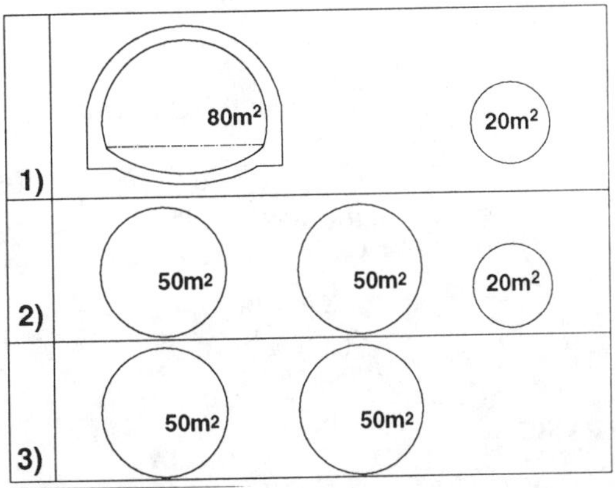

Fig. 1: Possible solutions for long tunnels (Areas as given as medium estimated values for net cross section).

Solution N° 1 offers least safety because of:

-the risk of head-on collisions;

-the evacuation of passengers from the track facing the service tunnel;

-the risk of the concurrent presence of other trains with the damaged train.

Solution N° 2 avoids the risks that exist for solution N° 1, but is considerably more expensive.

Solution N° 3 can be compared to solution N° 2 as for costs and it has the advantage that rescue teams have to use the second tube.

Solution N° 4 with an intermediate station offering easy access to rescue vehicles and appropriately equipped so that it may be considered equal to an open air station , offers the advantage of splitting the tunnel in two, reducing the time necessary to evacuate the tunnel.

Generally speaking, however, the double-tube solution proves to be better than a single-tube solution for the following reasons:

-it avoids the risks of head-on or lateral collisions;

-the accident affects one tube only;

-the fire cannot spread to trains travelling in the opposite direction;

-it avoids the risk of passengers being run over by trains travelling in the opposite direction;

-ventilation and smoke removing devices are simpler;

-the smaller section can make easier the excavation with a full-face boring machine.

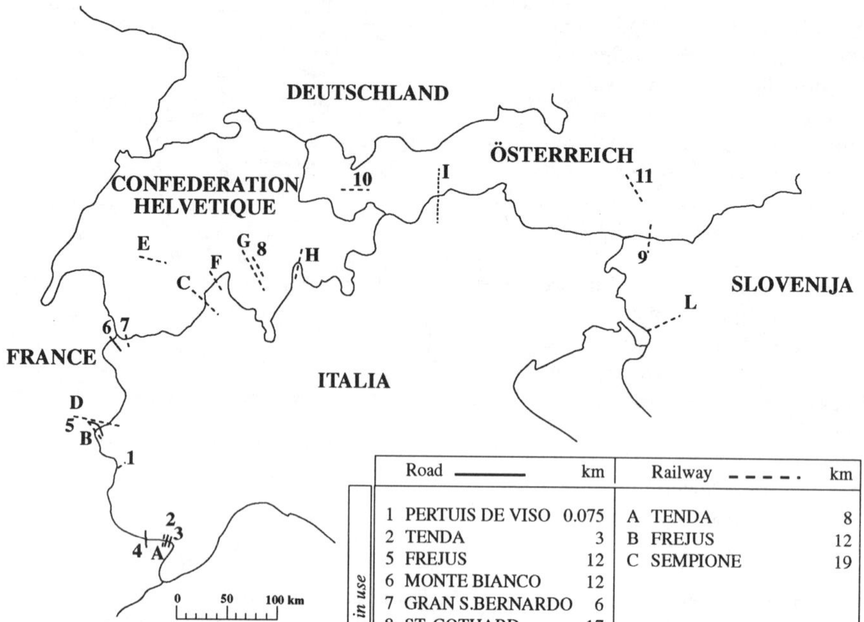

	Road ———	km	Railway - - - - -	km
in use	1 PERTUIS DE VISO	0.075	A TENDA	8
	2 TENDA	3	B FREJUS	12
	5 FREJUS	12	C SEMPIONE	19
	6 MONTE BIANCO	12		
	7 GRAN S.BERNARDO	6		
	8 ST. GOTHARD	17		
	9 KARAVANKEN	8		
	10 ARLBERG	14		
	11 TAUERN	6.5		
under planning	3 TENDA	15	D MONCENISIO	54
	4 MERCANTOUR	20	E LOETSCHBERG	32÷45
			F SEMPIONE	35
			G ST. GOTHARD	57
			H SPLUGA	36
			I BRENNERO	53
			L VILLA OPICINA	35

Fig. 2: Principal alpine tunnels.

2.2 Current tunnels through the Alps

Currently there are seven Italian tunnels across the Alps: four road tunnels, only two of which - Mount Blanc and Frejus - are of strategic importance for commercial traffic, three railway tunnels, two of which (Frejus and Simplon) are subjected to intense traffic (Fig. 2). All major tunnels are located in the western Alps and connect Piedmont and the Aosta Valley to France and Switzerland; these were true historical records (Table 2).

The reasons for this layout are mostly historical and morphological: in the western Alps it is unthinkable to accomplish crossings at relatively low altitudes, as is the case for the eastern side (Chiasso, Brennero, Tarvisio, Villa Opicina); the first tunnels were built, since the last century, under the reign of the Savoy in Piedmont, the only free Italian state bordering with Europe.

2.3 Tunnels of historical importance

The upsurge of industrial activity, which took place starting from the 19th century, determined an intensification of international trade, and highlighted the need for new means of transport and new communication channels.

The first tunnel to cross the Alps was the Frejus railway tunnel, also known as the Moncenisio tunnel.

In actual fact, the first Alpine tunnel was the Pertugio del Viso (Mount Viso hole), a 75 m long tunnel built in the Cozie Alps, close to Mount Viso, at 2859m above sea level, between the years 1475 and 1480, in order to facilitate trade between the marquisate of Saluzzo and France.

However, it is the 12,232m long Frejus tunnel, built between France and Italy in the years 1857 - 1871, that holds the record as the first tunnel across the Alps. The first section, 1646m long, saw the cut holes bored by hand. It was only in 1861 that compressed air drills were adopted for the first time in history in underground works; the prototypes, designed by G. Sommellier can be considered as the forefathers of percussion drills, still used today to bore blast holes.

In the following years the construction of railway tunnels spread along the Alps as well as in other regions, and in other countries.

Table 2: Italian Records in tunnel performance

Tunnel Name	Tunnel Purpose	Length (m)	Constr. period	Notes
Pertuis du Viso	Road	75	1475-1480	First tunnel of Western Alps
Frejus	Rail road	12.333	1857-1871	First tunnel of the Alps
Gran S. Bernardo	Road	5.854	1958-1964	First road tunnel of the Alps
Monte Bianco	Road	11.600	1959-1965	It was the longest roadtunnel world wide

2.4 Today's tunnels

We had to wait 100 years before the construction of a road tunnel, a requirement linked to the rapid evolution of road transport.

Italy can once again claim another record: the 11,660m long Mount Blanc motorway tunnel between Italy and France, the longest road tunnel in the world when it was built. The tunnel was constructed between 1959 and 1962 and was opened to traffic in 1965 (Bernieri,1965) (Cuaz,1984).

The next tunnel was the Frejus motorway tunnel, 12,878m long, located next to the existing railway tunnel (at a distance of 200 - 300 m); it also links Italy to France (Leonardi et al, 1981). The works, which started in 1975, finished in 1979 and the tunnel was inaugurated in 1980.

The two above mentioned road tunnels have become extremely important from a commercial point of view, and traffic through them has grown considerably (Table 3).

Table 3: Level of traffic during 1992 in highway tunnels Monte Bianco, Frejus and Gran S. Bernardo.

Tunnel	Vehicles		Total traffic increase compared to 1991 (%)
	Long vehicles	Total	
Monte Bianco	382,387	950,677	+1,3
Frejus	574,473	1,085,592	+5,0
Gran S. Bernardo	71,202	501,123	+3,0

3. NEW PROJECTS

Bearing in mind the prospects of intense evolution of commercial traffic between Italy and Europe, there are countless ideas, proposals and initiatives for the construction of Alpine crossings.

All initiatives foresee the construction of very long tunnels (20 to 60 km) with the aim of favouring access routes, reducing tunnel gradients and improving routes in order to allow to speed up traffic and increase transport capacities.

Attention is currently concentrated on railway lines, but there are also many road tunnel projects.
Amongst various ideas, there are 7 that attract most of the attention: the Mercantour - Ciriegia motorway tunnel and 6 railway tunnels, with their connected railway lines, on which the Italian Railway Company is performing feasibility studies: Moncenisio, Simplon, Gottardo, Spluga, Brennero and Villa Opicina.

3.1 The Mercantour motorway tunnel

This tunnel will connect western Italy to the South of France and Spain, and will represent the most important East - West Europe link, contributing to partially relieve traffic along the currently saturated coastal road (Crespo, 1992).

The route is currently being studied, yet no definite decisions have been made since this tunnel would represent an alternative to the Tenda road tunnel (almost 15 km long) which seems to fulfil mainly a local function.

The Mercantour tunnel, 20 km long, with a maximum cover of over 1500 m, is featured by almost monotonous geological conditions, since most of its

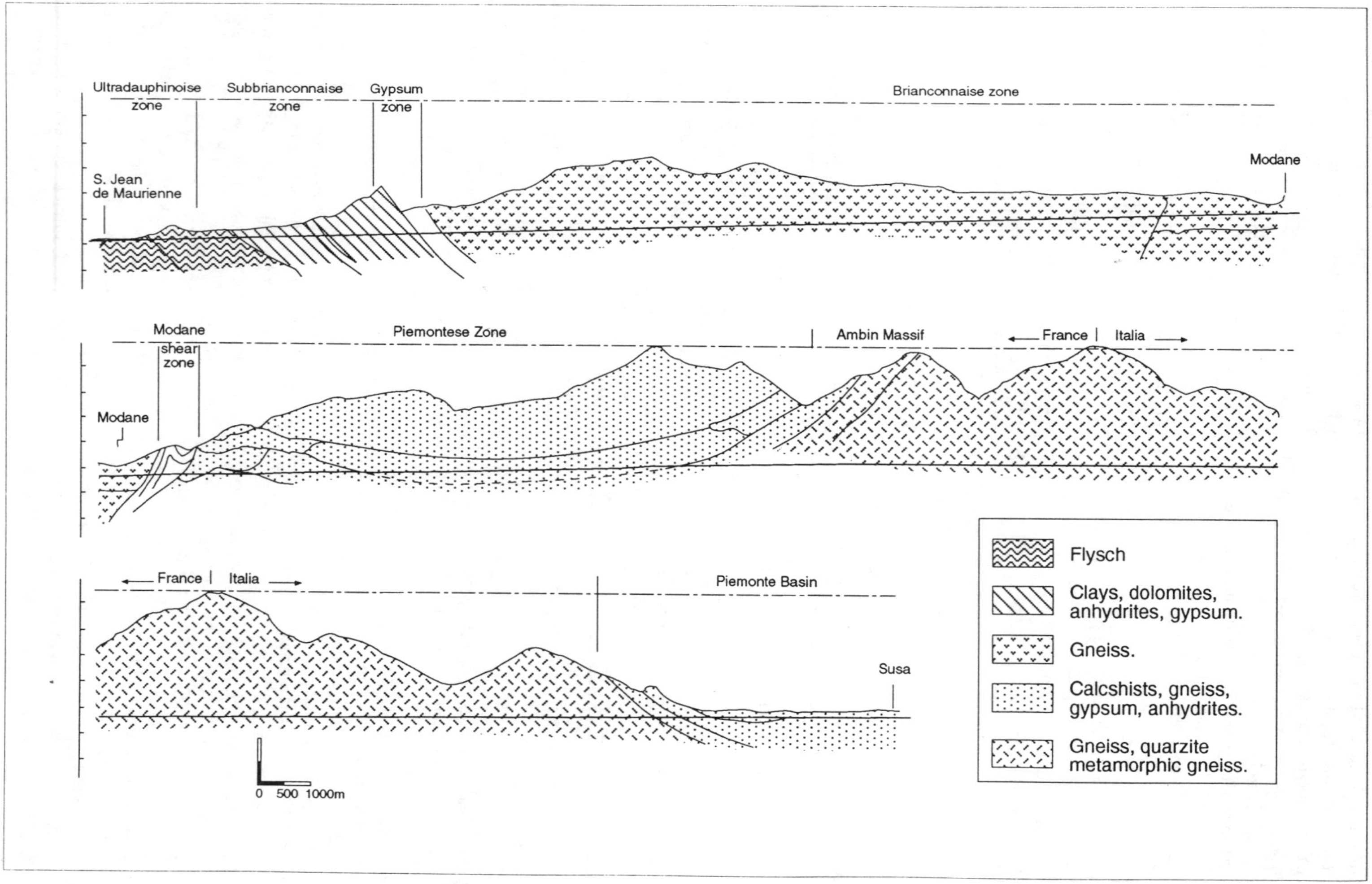

Fig. 3: Moncenisio tunnel - Geological section between S.J. de Maurienne and Susa.

route crosses the Crystalline Massif of Mount Argentera, consisting of granite intrusions located inside gneiss rocks.

From the tectonic point of view, the situation is relatively simple since the area is characterised by a number of vertical layers of mylonites correlated by slip or tectonic contact areas that are very difficult to cross with a tunnel.

During feasibility studies, surveys were performed thanks to a 3 km long pilot bore which supplied important data.

3.2 Moncenisio railway tunnel

Plans for the future high-speed train link between France and Italy, along the axis between Paris-Lyons-Turin-Milan- Venice and the Far East (Palliccia et al., 1992), include the crossing of the Alps through a base tunnel running from Saint-Jean-de-Maurienne (France) to Susa (Italy) (Fig.3).

The tunnel will be approximately 54 km long, of which 41 km lie in French territory and 13 in Italian territory. In relation to the length of the proposed work (54 km) and its overall perpendicular orientation to the major structural lines of the Alps, the project will cross the entire Alpine chain from west to east, from the "Ultradelfinese zone" to the "Piedmontese zone".

In general terms, the overburden is in excess of: 1000 m over 22 km, 1500 m over 10 km and 2000 m over 5 km of the tunnel.

The maximum overburden in the "Massiccio d'Ambin" zone is approximately 2300 m. Studies carried out (existing data research, 1/25,000 geological map, 5 km of boring, 40 km of seismic prospecting) have led to the elaboration of a 1/25,000 map and geological profile.

In addition to the geological aspects, the question of the internal temperature at the proposed altitudes is of particular interest due to the considerable thickness of the overburden.

Ground temperature increases with depth according to a mean increment (thermic gradient) of 1*C per 30 m, but this may be locally greater in particular in the presence of granites and geogneiss (Table 4).

On the basis of the bibliographical data available for existing tunnels in comparable geological formations, the temperature in the tunnel under Monte d'Ambin, namely in the zone with the greatest overburden, will be in the region of 50*C.

The main technical characteristics of the project are as follows:

a) Twin-tube construction (Fig.4): The twin-tube solution was chosen as opposed to the single tube solution because it facilitates:

- the prevention of possible faults as well as limiting their severity (active safety);
- the attenuation of the consequences of a possible fault (passive safety);
- construction conditions:
- smaller working faces and the possibility of using milling machines with an anticipated reduction in construction time and, as a result, costs (in relation to the times required by financing);
- the twin-tube solution limits the effects of ground convergence under large overburdens and reduces construction difficulties (one of the tubes will act as a pilot tunnel for the other).

b) No service tunnel:

The construction of a service tunnel, of a size which would give access to emergency and evacuation vehicles, implies a major added cost unless there are

Table 4 Some maximum temperatures in Alpin tunnels

Tunnel	Overburden (m)	Max. temp.	Rock type Medium (°)	Medium radient degree/km of over ...
Mont-Blanc	2500	≈40°	Granites	-
Simplon	2200	55.4°	Gneisses	27°
Arc-Isère	2100	42°	Gneisses+ Scistes	20°
Isère-Arc	1850	37°	Granites Gneisses	20°
St-Gothard	1750	30.7°	Granites	17.5°
Lötschberg	1630	34°	Granites+ Gneisses	20.8°
Fréjus	1610	29.5°	Schistes	18.4°
Tauern	1570	23.9°	Gneisses	15.2°
Le Bourget-du-Lac	1500	22°	Limestones	14.7°

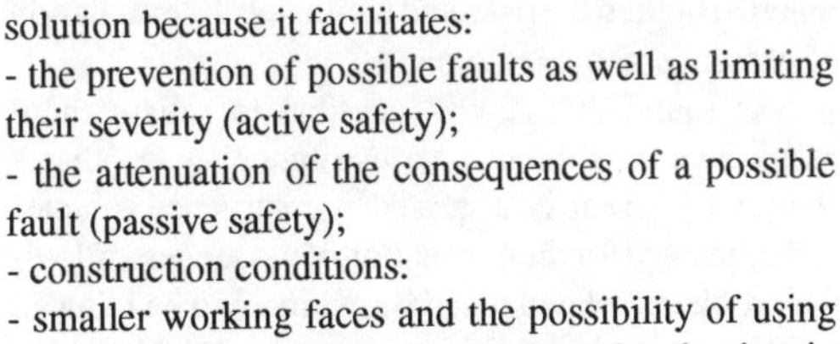

Fig. 4: Lyon-Turin High-Speed Railway line. Basic tunnel Cross section of a train communication branch.

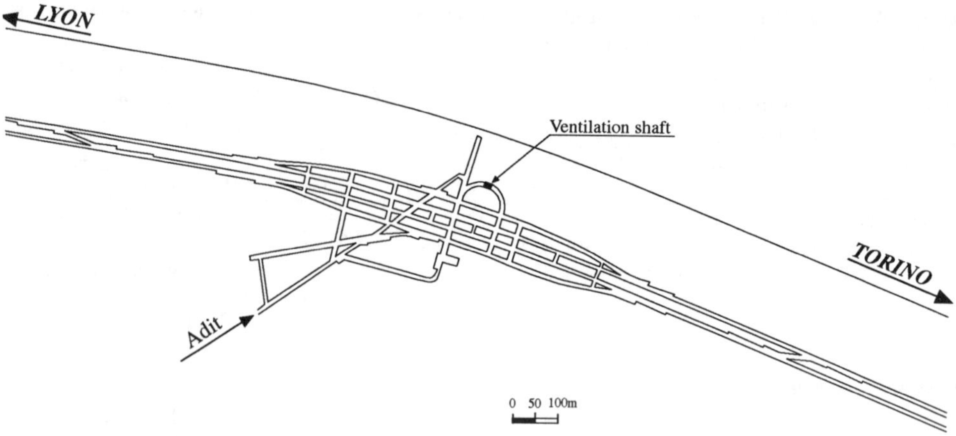

Fig. 5: Lyon-Turin High-Speed Railway line. Modane Underground Access Station.

other reasons for its construction (pilot tunnel, ventilation, site access...). For this reason it was preferred to use a "safety system" without a service tunnel given that bibliographical research showed that this also limited the risks and facilitated the evacuation of passengers to a "safe place".

It was found that it was possible to satisfy these requirements without a service tunnel given that a station, which can be accessed by emergency vehicles and organised for the evacuation of passengers, allows the problems related to a 54-km tunnel to be reduced to 30 km. A solution of this nature was also adopted in the tunnels forming part of the Shinkansen network in Japan. Moreover, the time allocated for maintenance work (4 hours each night) and the existence of the intermediate station, which can also be used by works equipment, avoid the need for a service tunnel for maintenance purposes.

c) Communicating passages between the two tubes In the proposed system based on the twin-tube solution, each tube acts as an evacuation tunnel for the other owing to the existence of communicating passages between the two tubes which are normally closed automatically. They can only opened using special operating procedures. Each communications passage lies 350 m from the next, has a usable length of 20 metres (between the doors) and is 3 metres wide. In addition, a 1.30 m-wide sidewalk will be constructed in each tube containing ducts for cables.

d) No displacement passages

On the basis of research performed to date, there are no technical or economic grounds for connecting passages between the two tubes to eliminate air displaced by trains.

e) A maximum slope of 12% in the tunnel

The maximum longitudinal slope in the tunnel is 12%. This level restricts the overburden to approximately 300 metres in correspondence with the Modane zone (internal station).

f) An underground service station near Modane (Fig.5)

Bibliographic studies and observations made have led to the proposal of a system to ensure the evacuation of passengers to a safe place within an acceptable time. The topography and longitudinal profile of the tunnel dictate that the highest point of the tunnel lies very close to the surface; this has resulted in the construction of an intermediate access and underground system below Modane (h=300 m) capable of receiving a train which has signalled a fault and in particular a fire aboard and evacuating the passengers in order to limit the risks of a stationary train in the tunnel and reduce the time taken to reach a safe place.

The station below Modane is the key point of the proposed solution. It basically provides rescue functions and enables:

- a train with an incipient fire aboard to halt with better protection than in the case of a train stopped in mid-tunnel;
- the evacuation of the train and the shelter of passengers in "safe places" before leaving the tunnel;
- the extinction of a fire aboard.

Lastly, the Modane service station, which is equipped with a precedence parking line in each direction and reciprocal points between the two lines, makes for a considerable increase in the tunnel's capacity (in the region of 20%) and improved operating elasticity, as well as guaranteeing regularity.

g) A net cross-section of 43 square metres

The proposed cross-section is of the circular type for two basic reasons:
- in the case of this particularly large project, a system of mechanised excavation works can be adopted over the greater part of the tunnel's length;
- the circular form is the most appropriate in view of the expected stresses (oriented in a non-symmetric manner) and the forces concentrated under large overburdens.
h) The installationof the following systems: ventilation, smoke disposal and cooling.
The expected schedules for the completion of construction work, in view of current knowledge regarding the geology and hydrogeology of the site, are 4.5 and 6 years for 5 and 2 intermediate connections (shaft or inclined) respectively. These schedules are based on the hypothesis that pilot tunnels will have been constructed beforehand and that both the preliminary administrative procedures and site plant will have been completed.

3.3 The Simplon railway tunnel

This new mountain pass tunnel should replace the old 19.8 km long twin bore tunnel, built, overcoming considerable difficulties, between 1898 and 1905, which was subsequently integrated in a new route at a lower altitude in order to increase the railway line potential.

This work must be considered as an alternative to other two tunnels - Gottardo and Spluga - within a project for developing traffic capacity in the central Alps.

The Simplon is located along the Zurich - Bern - Milan railway line which has links towards Genoa. Currently, highly articulated feasibility studies are still underway since around 110 km of tunnels are foreseen for a route which totals 441 km .

As for the geological aspects, we may reasonably state that the degree of uncertainty is quite low thanks to the existence of old tunnels; the Italian side is featured by crystalline rock (gneiss) with good geomechanical characteristics, and limited areas of tectonic disturbance (Ragni,1984).

Problems on the Swiss side are more complex due to the presence of micaschists, calcischists, limestones and gypsum; there are large possibilities of encountering great inflows of water that can reach high temperatures (as was the case for the old tunnels where an inflow of water at 45°C was encountered).

3.4 The St. Gothard railway tunnel

The tunnel is located on the Zurich - Bellinzona - Chiasso - Milan railway line; 186 km of the 645 Km long route is inside tunnels (Zuber, 1992 Gehriger, 1993). The main tunnel, located on the Swiss side, will be approximately 57 km long.

The route will be almost completely in a straight line, and the overburden will reach 2000 m in certain points in order to avoid critical geological areas, any interferences with surface infrastructures - dams - , and to allow the creation of 2 intermediate adits.

From the geological point of view, the tunnel will cross 3 different geological formations with areas featured by poor geo-mechanical conditions; a 5 Km long pilot bore will be drilled in order to visually check the behaviour of these materials and the foreseen presence of water.

Two intermediate boring fronts, that will divide the tunnel in three sections of approximately the same length, will be used in order to permit a reasonable construction time for the tunnel (around 10 years).

Two single-section tunnels, more suited for excavation with a boring machine, and easier to access for maintenance, are foreseen.

3.5 The Spluga railway tunnel

This tunnel fits in a project for a new North - South railway line called "Spluga integrato" (integrated Spluga) which will connect Briga - Chur (Switzerland) with Lecco - Bellinzona - Como - Milan, with the possibility of further extentions of the link to the East and towards the Ports of Liguria in the South; the railway line totals 543 km, 233 of which are inside tunnels (Anon., 1988).

The main tunnel is the Spluga tunnel, around 35 km long, between Soazza and Thusi, totally in Swiss territory.

At the moment this project seems to be outdated as the Swiss Government tend to prefer St.Gothard and Simplon pass.

The geology of the area will have to be investigated from scratch, since it has not yet been explored: thorough and specific investigations will thus be necessary.

3.6 The Brennero railway tunnel

The Brennero tunnel fits inside the works for a two-way development of the Verona -Innsbruck- Munich

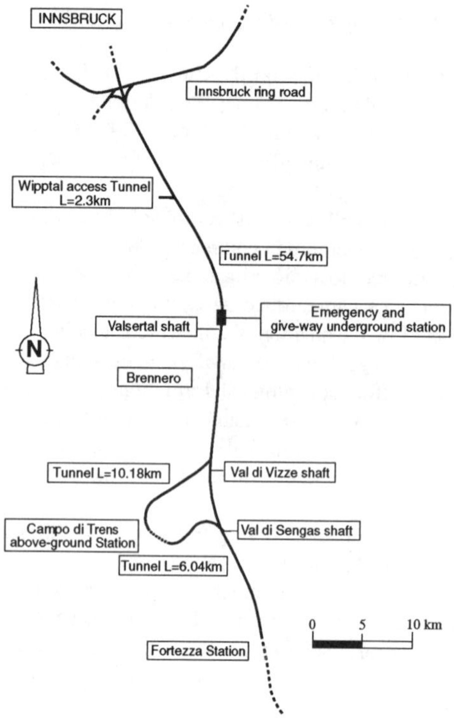

Fig. 6: Brennero tunnel - Plan of the tunnel between Innsbruck and Fortezza.

rrailway axis; The main operating features of the new railway line are: a total capacity of 400 trains per day, 20% of which are passenger trains; high-speed: 250 km/h for passenger trains and 100 Km/h for freight trains.

As for the tunnel, the results of the feasibility study foresee the construction of a tunnel between Innsbruck (Austria) and Fortezza (Italy), respectively located at 592 and 769 m above sea level. The tunnel will be approximately 53 km long and will have maximum cover of 1500 m; a link with Campo di Trens (927 m above sea level) is foreseen in order to warrant a suitable level of safety to the underground system (Fig. 6) (Potucek, 1992; Wessiak, 1993).

The Brennero tunnel crosses a complex series of quite different, and often poor, geological formations (phyllite, dolomite, limestone, micaschists, gneiss, porphyry and granite): a pilot bore will thus be needed (Fig. 7).

Four intermediate adit shafts will be located along the tunnel route to speed up its construction.

3.7 The Villa Opicina railway tunnel

The mountain pass of Villa Opicina connects Italy to the Balkan and Middle Eastern countries; currently this traffic channel is quite important for the apportionment of international traffic: transit totalling 2.1 MT in 1981 grew to 2.9 MT in 1990 with an increase of around 40%.

From this viewpoint, the construction of a new railway link between Lubljana and Mestre is currently being studied.

In conclusion, two possible alternative solutions are being evaluated: the first foresees the construction of a new link with High-Speed characteristic; the second evaluates the possibility of making significant improvements on the existing line in order to increase travelling speeds.

Varoius alternative solutions have been developed for the first hypothesis; the two most significant one foresee a common route in Italy but two different routes, A and B, in Slovenia.

Solution A has the shortest route (225 km) and the least gradient; these advantages are countered by a greater percetage of tunnels (37.6%) and the presence of three tunnels over 20 km long.

Solution B foresees 234 km of route, practically the same percentage of tunnels (36.7%) and a gradient of over 13% with two tunnels exceeding 20 km in length. Problems regarding the tunnels lies in the crossing of marly-arenaceous materials belonging to the Eocene Flysch and to rocks of the Alto Carso Calcareous Formation; and in particular, in the contact between these two lithologies, in the overthrust zones and in the fractured areas that may eventually be encountered. Karst phenomena and hydrogeological problems might be encountered inside the calcareous formation.

REFERENCES

ANAS, Cogefar 1979. Gran Sasso. Il Traforo Autostradale. Sefra, Milano, pp.267.

Anon. 1981. Le Tunnel Routier du Saint-Gothard. Ist. Grafico Casagrande, Bellinzona, pp.112.

Anon. 1988. Documentazione sul Progetto Spluga Integrato. Internal Report, Regione Lombardia.

Anon. 1992. Raccomandazione Comune delle Autorita' di Vigilanza sulle Ferrovie della Germania Federale, dell'Austria e della Svizzera Concernente la Sicurezza dei Viaggiatori nelle Gallerie Ferroviarie di Grandi Dimensioni. Internal Report.

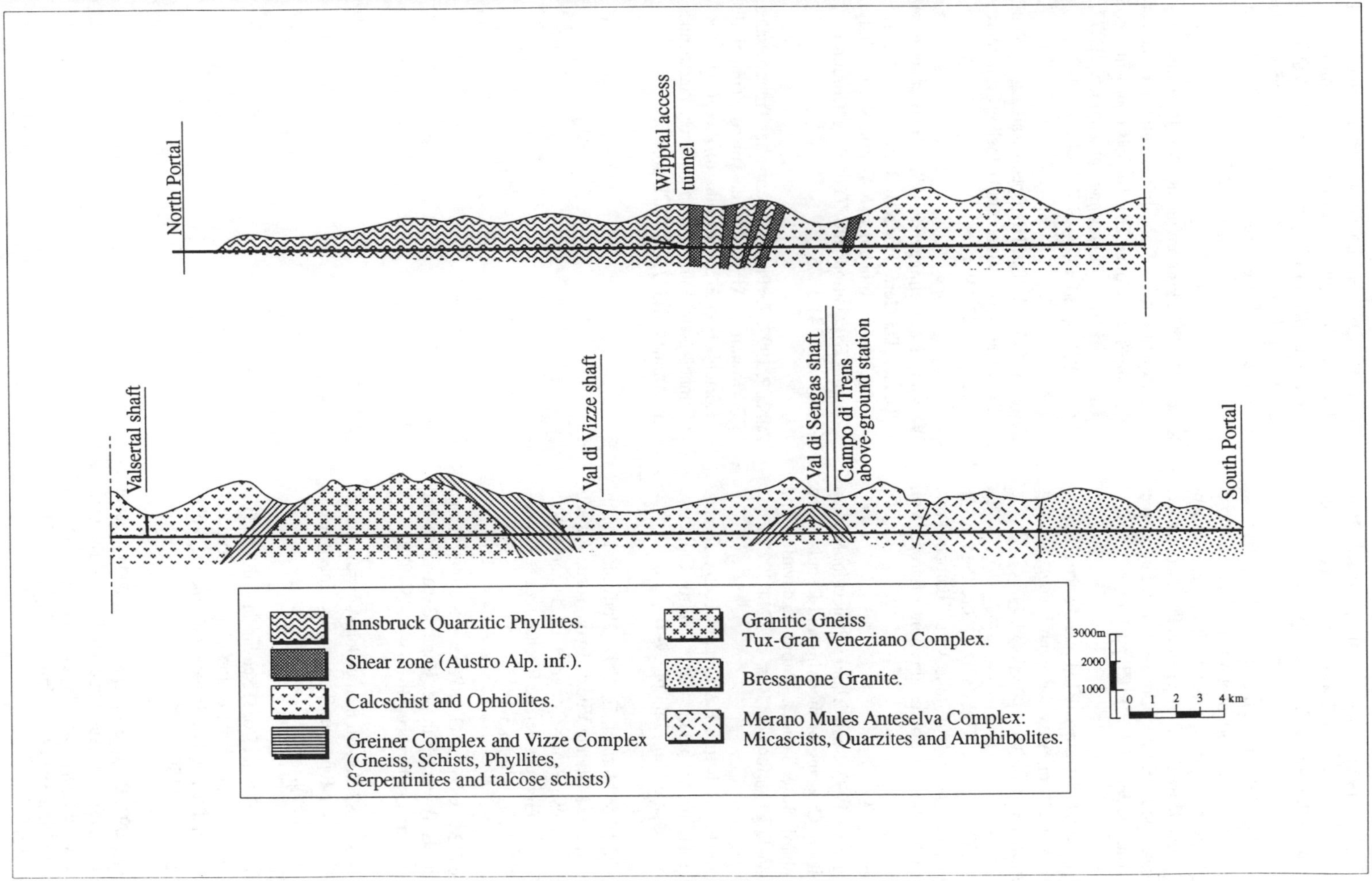

Fig. 7: Brennero tunnel - Geological section of the tunnel between Innsbruck and Fortezza

Bernieri, U.V. 1965. Una Colossale Opera che Onora il Lavoro Italiano: il Traforo del Monte Bianco, Trasporti Pubblici, N°. 10 oct, pp.1-20.

Crespo, A., 1992. Le Tunnel du Mercantour: un Ouvrage d'Art Important et Attendu pour Ameliorer les Liaisons Alpines entre l'Italie et la France dans la Nouvelle Realite Europeenne. Proc. Congr. Des Tunnels pour l'Europe, Grandes Percee Pireneennes et Alpines, JE377, CEIFICI, pp.1-12.

Cuaz, F., 1984. Il Traforo del Monte Bianco si avvicina ai Venti Anni. Atti e Rassegna Tecnica Soc. Ing. e Arch., N°.1-2, Jan.-Feb., pp.39-44.

Focacci, C.& Grasso, F.& Pelizza, S.& Pigorini, B. 1990. Les Tunnels d'Italie vers l'Europe. Proc. Franchissements Souterrains pour l'Europe, Vol. , pp.7-12.

Gehriger, W. 1993. Switzerland's Project for Modern Railway Crossing of the Alps and Details of the Gothard Base Tunnel. Rapid Excavation and Tunneling Conference, Boston: 1211-1223.

Leonardi, L.& Bellettini, F.& Lunardi, P. 1981. Frejus: Due Trafori tra un Secolo e l'altro. Gallerie e Grandi Opere Sott., Jul., pp.29-41.

Palliccia, M.& Maraini, E. 1992. Nouvelle Liaison Ferroviarie a Grande Vitesse Lyon-Turin: Aspects Techniques. Proc. Congr. Des Tunnels pour l'Europe, Grandes Percée Pirenéennes et Alpines, JE377, CEIFICI, pp.3-18.

Palliccia, M. & Viola, G & Pelizza, S. 1993. Italian Tunnels Toward Europe. Rapid Excavation and Tunneling Conference, Boston: 1225-1239.

Pelizza, S. 1987. Aspetti tecnici della costruzione dei trafori. Atti e Rassegna Tecnica Soc. Ing. e Arch., N°.4, Apr., pp.79-82.

Pelizza, S. 1991. Pilot Bore Excavation with TBM for the Design and Construction of Larger Tunnels. Tunn. and Undergr. Space Technology, Vol.6, N°.2, pp.185-189.

Potucek. 1992. Projet d'Amenagement de la Liaison Ferroviaire entre l' Autriche et l'Italie par le Brenner. Proc. Congr. Des Tunnels pour l'Europe, Grandes Percee Pireneennes et Alpines, JE377, CEIFICI, pp.1-12.

Ragni, U. 1984. Studio Geologico-Tecnico per il Progetto di Fattibilita' di un Nuovo Traforo del Sempione per il Collegamento Autostradale tra la Provincia di Novara (Italia) e il Cantone Vallese (Svizzera). Atti e Rassegna Tecnica Soc. Ing. e Arch., N°.1, Jan.-Feb., pp.62-71

Rocco, V. 1992. Programme de Developpement des Liaisons Routiers Transalpines entre la France et L'Italie. Proc. Congr. Des Tunnels pour l'Europe, Grandes Percee Pireneennes et Alpines, JE377, CEIFICI, pp.30-49.

Taille, J.Y. 1993. The Highspeed Rail Link Lyon-Turin. Rapid Excavation and Tunneling Conference, Boston: 1185-1193.

Wessiak, H. & Steininger, P. 1993. The Brenner Base Tunnel- The Core of the New Railway Axis Munich / Germany - Innsbruck / Austria-Verona Italy. Rapid Excavation and Tunneling Conference, Boston: 1241-1275.

Zuber P. 1992,. Le Projet de Nouvelles Transversales Ferroviaires Alpine de la Suisse (Tunnel du Loetchberg et du Saint-Gothard). Proc. Congr. Des Tunnels pour l'Europe, Grandes Percee Pireneennes et Alpines, JE377, CEIFICI, pp.4-5.

Tunnelling and Ground Conditions, Abdel Salam (ed.) © 1994 Balkema, Rotterdam, ISBN 90 5410 363 9

Improvement of stability conditions from half to full face excavation in difficult geotechnical conditions

S. Pelizza, *Tunnelling and Underground Space Centre, Turin, Polytechnic, University of Turin, Italy*
G. Corona & G. Graffi, *SIC, Civil Engineering Consultants, Turin, Italy*
F. Grasso, *University of Turin, Italy*
R. Raineri, *ANAS, Florence, Italy*

ABSTRACT: A 550 m long tunnel has just been excavated along the State Road SS 1 "Aurelia" in Tuscany (Italy). The tunnel overburden is 50 m and the rock mass is a very fractured and weathered flysch.
The rock mass is geotechnically very complex and it belongs to the Vb class of the Rabcewicz-Parker classification. The first part of tunnel was carried out at half section, supporting the tunnel with steel arches and shot-crete and, where necessary, with an umbrella steel pipe pre-reinforcement and a face reinforcement.

Severe problems of tunnel stability were encountered during the excavation. In the last 170 m of the tunnel a full face excavation was adopted in order to reduce the problems of stability.
In this excavation a systematic face reinforcement was applied, with fibre-glass pipes, steel arches and an inverted arch carried out close to the tunnel face.

The results of this method were very satisfactory, as tunnel stability was obtained and the rate of advancement was good.

1 ORGANISATION

The work concerns the 550 m road tunnel "Poggio Fornello". It is part of a new improved section of the National Road S.S.1 "AURELIA" under construction in Tuscany (Central Italy).
The tunnel has two tubes, but in order to save money only one has been built. The costs are high because of poor geomechanical conditions. The total area of the excavation is about 104 m^2; the maximum overburden is 50 m.
The tunnel has been excavated with only one advancing face proceeding from north to south.
The already poor geomechanical conditions have worsened in the last 170 m, with a maximum overburden of 25 m. In order to overcome the geotechnical difficulties, in this section we changed the advancing method from upper-half section to full section, and the results have been very good. This paper reports this type of work.
The tunnel passes through a rock mass formed by Cretaceous flysch; the strata are less than 10 cm thick. The predominant stratum is argillaceous slate (60-80% of the total), which is alternated with calcareous and arenareous layers.
The rock formation is intensely tectonic: the rigid materials (calcareous and arenareous) are subdivided into smithereens less than 10 cm in size; the poorest pieces much are the most capable of being deformed (argillaceous slate) are laminated into little chips that are even less than 1 mm in size. These chips have been repeatedly bent, and often the stratification is no longer visible.
The rock mass can therefore be defined as weak and complex.
Project studies have attributed the rock mass to the Vb class of the Rabcewicz-Parker classification.

2 TECHNIQUES AND BUILDING PROBLEMS

Excavation was carried out with the upper-half section method for about 60 m. The supports and the advancement methods were later adapted to the real behaviour of the rock mass, which was identified as excavation progressed.
We adopted the following technique for the excavation of the upper-halt section (Figs. 1 and 2):

- 75 cm excavation;
- coupled steel arches with double T of 200 mm, every 75 cm;
- widened steel arch foot;
- 30 + 20 cm thick shot-crete, with strong steel netting;

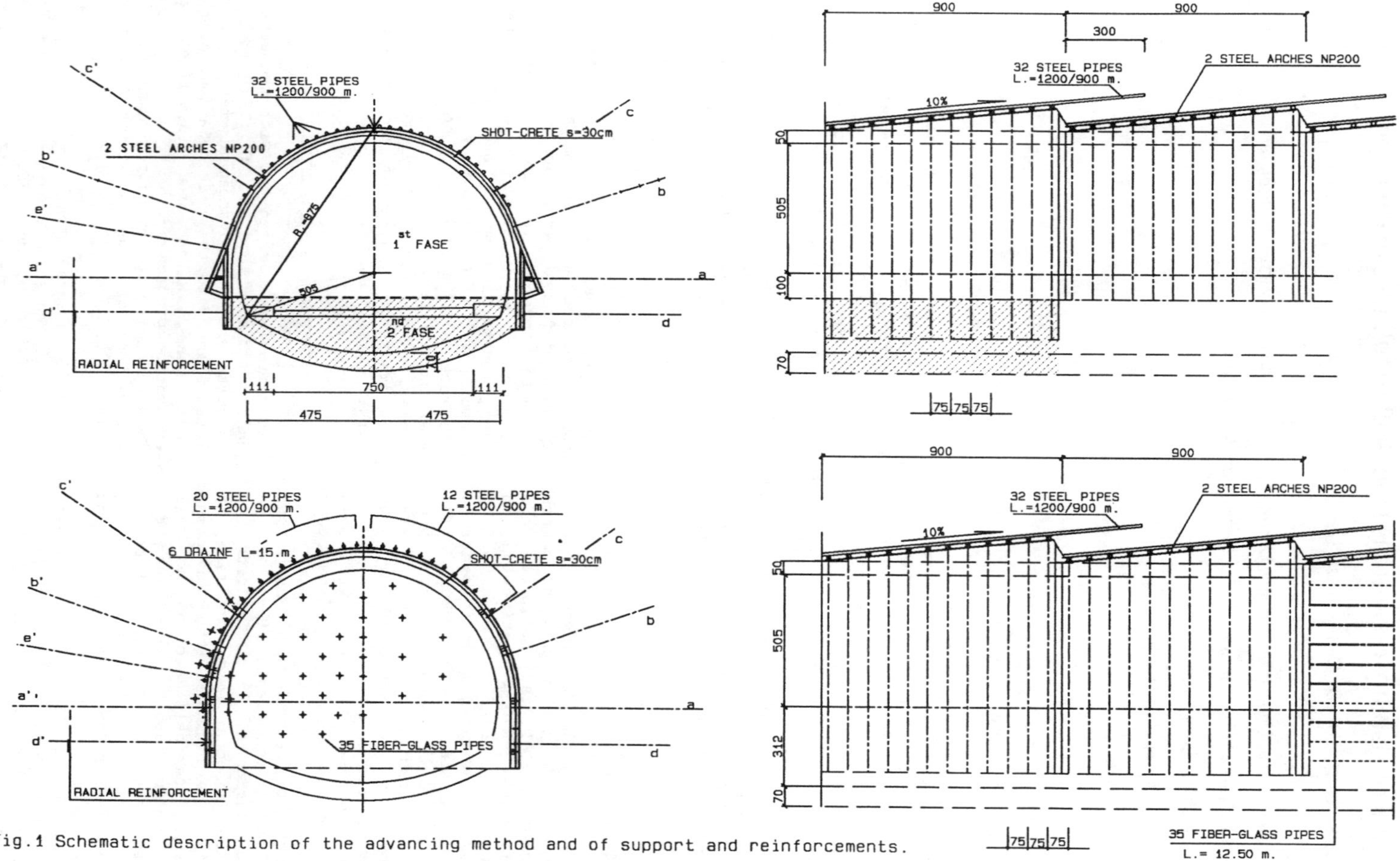

Fig.1 Schematic description of the advancing method and of support and reinforcements.

Above: upper-half section excavation. Below: full section excavation with reinforcements at the face.

- 6-8 m long radial reinforcements, placed every 8 m of advancement, with cemented and sometimes prestretched steel bars;
- umbrella with 32 steel pipes 12 metres long; with a 9 m usable excavation under the umbrella.

Tunnel behaviour was controlled by means of convergence and levelling techniques. We noted that:

- strains were very variable in various of the tunnel;
- the maximum speed of strain was reached at 2-3 diameters distance (2 to 3 weeks after excavation);
- convergence values in the upper section, without benching, 1 m and the steel arches deepened by 0,7-1 m (Figs 4 and 5).
- as we approached the last 170-130 m to the South, strains appeared with high values (20-30 m) which had already been registered at the face. It was impossible to carry on excavation with the methods described above.

These great and rapid strains (despite the low overburden of only 25 m) were due both to the worsening of the geomechanical characteristics of the rock mass and to the presence of water. This was not flowing water; it impregnated the rock. Above all on the left hand side of the excavation section the flysch was transformed by tectonic action into a limestone detritus mass that was quite lacking in cohesion, saturated with water and highly capable of being deformed.

The geotechnical characteristics of the rock concerned in this project were as follows:

- 50 KPa cohesion; 20° internal friction angle; modules of changeability: 2000 MPa.

When the tunnel buckled some analysis was carried out (with the FEM scheme) in order to assess the geotechnical parameters of the ground that had led to this collapse.

Fig.2 Realisation of umbrella pipes by means of a one arm drilling machine able to drill a 12 m hole with a monobloc drilling pipe and to introduce pipes into the hole.

The results of the analysis were: cohesion 10 KPa; internal friction angle 20%; modulus of changeability: 350 MPa.

Fig.3 Great unsymmetrical deformations of the upper section.

Fig.4 Great deformations of the upper section have required local propping to prevent the tunnel from buckling.

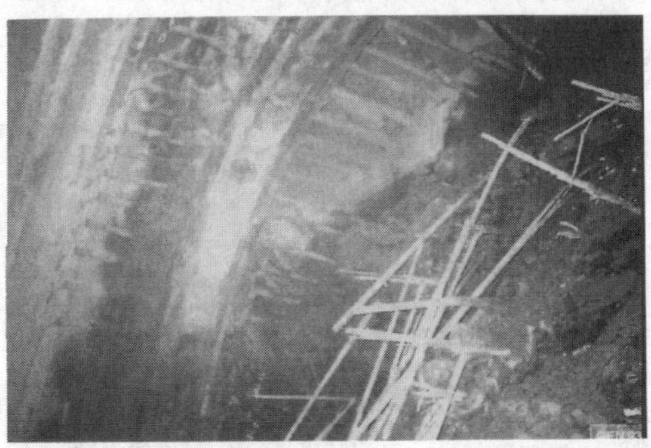

Fig.5 Full section excavation with reinforcing face.
In profile we can see the umbrella steel pipes and on the face (right) the fibre-glass pipes while the excavation is destroying them.

3 FULL SECTION EXCAVATION

Because of the conditions described above we decided to modify the excavation and support scheme. We followed the criteria below.
A rigid structure was laid. This had to be able to resist the geostatic loads of the 25 m overburden which was wholly resting on it.
The support ring had to be closed immediately at the foot. In order to do so it was necessary to make the definitive inverted steel arch immediately.
Excavation was carried out full section in order to install the steel arches that rested definitively on the earth foundation. This allowed the foot of the steel arch to be extended along inside the tunnel, which increased the support area of the lower steel arch. What is more it meant that the bottom part of the steel arch was incorporated into the inverted arch.
It is possible to work in this way as long as the face of the excavation can be kept steady and safe, in an absolutely vertical position. This geometry is possible in poor rock if a special face support is used.
Face stabilisation can be obtained with grouted bars that are inserted into the ground and which run parallel to the tunnel axis. These bars stiffen the ground and increase its resistance to shearing stresses because of the lateral confinement of the stresses.
Fibre-glass pipes are usually used because they are easily destroyed by the excavator blade when the earth is removed (Fig. 6).
The working schedule required to carry out 9 m of excavation is as follows:

* a face support must be made, using 35-40 fibre-glass pipes of 12,5 m in length. This is done by using a special ore arm drilling machine able to drill a hole up 14 m long with a monobloc drilling pipe. The pipes must then be cemented. This operation takes from 32 to 36 hours.

Fig.6 A tunnel stretch which is very deformed during the excavation phase.

* an umbrella of 32 steel tubes 12 m long has to be made (with the above mentioned machine) and then cemented. This takes about 32 hours.

* full face excavation (0,75 m advancement) with excavator laying of the steel arch (200 mm double) and laying of the shot-crete: 4 hours. The) m advancement under a 12 m umbrella necessitates the laying of 12 steel arches, so that the total time for excavation and support is 48 hours. A 9 m cycle requires about 5 days.

* the closing inverted arch at the base of the steel arch-shot-crete structure can be made according to the needs and the behaviour of the excavation. It may be used for sections 2-3 m from the face or for longer sections (up to 9 m) that may be 1 or 2 diameters distant from the face (this does not influence the cycle because this work can be carried out on the 6th day of the week.

4 CONCLUSIONS

The final section, the south end the tunnel, is about 170 m long. In this section the half-section excavation method has allowed for 20 to 30 m of tunnel to be built per month. The risk of big deformations remained, with the consequent need to carry out additional interventions in order to stabilise or the excavate structures under excessive strain (Fig. 7), and to make another benching excavation.
Otherwise, through full face excavation with face supports and reinforcements, we managed to produce on average 30 m of tunnel a month - safe from risk and problems - complete with inverted arch (and ready to receive waterproofing and final concrete covering).

REFERENCES

Carrieri, G & Grasso, P. & Mahteb, A. & Pelizza, S. 1991. Ten years of experience in the use of umbrella- arch for tunnelling. SIG Conference om Soil and Rock Improvement in Underground Woks, Milan, March 18-20: 99-111.

Grasso, P. & Mahtab, M. A. & Pelizza, S & Rabajoli, G. 1993. Consideration for design of shallow tunnels. AFTES International Conference on Underground Transportation Infrastructures, Toulou, Oct. 19-21.

Pelizza, S. & Barisona, G: & Campo, F. & Corona, G. 1989. Rapid umbrella-arch excavation of a tunnel in cohesionless material under an archaeologica site. International Congress on Progress and

Innovation in Tunnelling, Toronto, Spet.)- 14: 885-891.

Pelizza, S: & Paila, D. 1993. Soil and Rock Reinforcements in Tunnelling. Tunnelling and Underground Space Technology, Vol. 8, No. 3, 357-372.

Tunnelling and Ground Conditions, Abdel Salam (ed.) © 1994 Balkema, Rotterdam, ISBN 90 5410 363 9

Tieback anchors support record setting walls for Super Collider experimental halls in Texas

Birger Schmidt
Parsons Brinckerhoff, San Francisco, Calif., USA

Dale T. Moeller
Parsons Brinckerhoff, Dallas, Tex., USA

ABSTRACT: The Superconducting Super Collider particle accelerator is currently under construction near Dallas, Texas, USA. As part of this project, large sophisticated pieces of equipment will detect and measure subatomic particles resulting from high energy, proton-proton collisions. These detectors will be housed in large underground halls up to 69 m deep, founded in competent Austin Chalk. The walls, mostly in the weaker Taylor Marl, are permanently supported by tiebacks, and the roof is permanently supported by columns. This paper presents the design concepts and analyses for the halls.

1 INTRODUCTION

Construction is well underway on the Superconducting Super Collider (SSC) particle accelerator. This scientific machine will be the most powerful subatomic particle accelerator in the world and is comprised of an injector area, a collider tunnel oval of 87 km in circumference and two large underground experimental halls. See Richards and Robbins' paper at this conference for the general layout of the facility. The injector facilities begin the acceleration of protons, increasing their energy levels until injecting them into the collider tunnel. In the collider tunnels, two beams of protons are propelled in opposite directions at velocities approaching the speed of light before colliding them within huge detectors built in large underground experimental halls. These collisions, recreating energies that prevailed at the moment of creation, will provide data aiding scientists in their study of fundamental particles and forces of the universe.

The SSC detectors are very large, sophisticated pieces of equipment still under design. The largest of the detectors will weigh more than 35,000 metric tonnes, be about 30 m long and 26 m across. The experimental halls must provide room to assemble and maintain the components of the detectors. To meet these needs, the halls will measure about 34 meters wide, 110 meters long with the floor founded 66 to 69 meters below existing ground surface. Interior overhead cranes to be used in assembly of the detectors require a hook clearance of about 31 meters. Radiation shielding requires cover over the hall equivalent to 6 meters of concrete and an equal distance to adjacent ancillary shafts.

The large column-free space needed for the experimental halls posed a unique design challenge. The experimental halls are wider than typical underground caverns made for hydropower stations and about twice as wide as subway stations. They are taller than the typical hydro cavern, and twice as deep as a typical subway station. Only Scandinavian caverns for sports arenas have wider spans.

The first and most important decision was to choose between underground cavern mining and open cut construction. This choice had to be made based on constructibility and economics. The choice of open cut with backfilling above the roof was influenced by user requirements for equipment installation, operation and maintenance. It also afforded greater economy and shorter construction time. User requirements included two large permanent assembly shafts that straddle the halls at about the quarter points, surface and base slab load paths to move heavy detector components, and thick radiation shielding from adjacent and surface installations. The foundation slabs of the detector halls are subject to stringent performance criteria. The excavated vertical walls used to create the experimental halls are believed to be the World's tallest in soil or rock, permanently supported by tieback anchors.

2 GEOLOGY

One reason for selecting the North Texas site for the SSC was the simplicity and benign nature of the geologic formations. Only three significant geologic strata of the Upper Cretaceous are encountered at the site: Taylor Marl (TM), a shale with variable calcite content, outcropping to the east side of the SSC site above the Austin Chalk (AC). The Austin Chalk, a limestone of modest strength with few joints and faults, is found at the ground surface over the remainder of the site. Eagle Ford Shale lies below the chalk and is a weak claystone subject to deterioration and slaking upon exposure.

The Interaction Region (IR) 8 and Interaction Region 5 sites selected for the two large detector halls are on the east side of the main collider ring and have similar stratigraphy. Several borings were drilled at each IR site to determine the geologic conditions. Residual soil and weathered Taylor Marl comprise about 10 m depth. The fresh TM reaches a depth of about 43 m below ground surface at IR-8 and about 56 m at IR-5. This places the interface between the TM and AC below the middle of the wall of the halls. The Austin Chalk limestone forms the foundation material of both halls.

Interpretation of the boring data indicate that there are two steep faults with a vertical displacement of some 3 m at the IR 5 site. Additional faults of small displacements could be encountered during construction of either hall. These faults cause a break in the elevation of the interface between the marl and chalk. Based on experience elsewhere, however, the faults are not expected to result in significant weakness of the rock formations.

At the time when a western location of the experimental halls was contemplated, a 5 m diameter exploratory shaft was drilled and instrumented to determine the properties and performance of the Eagle Ford Shale (Brouillette et al., 1992). These results give insight into the behavior of the TM, which is similar though somewhat stronger.

Another shaft located on the southern part of the collider ring in marl was instrumented to assess the behavior of the TM. This shaft was 5.5 m diameter and 60 m deep. Inclinometer data measured the inward deflection of the shaft walls during drilling and generally confirmed the deformation modulus determined by laboratory tests. After initial excavation and support with about 75 mm of shotcrete, the displacements stopped after reaching a radial displacement of about 4 mm maximum in the upper, weathered TM. Creep in this material is very limited.

A laboratory testing program was conducted on samples of rock recovered from the boring program. Design properties of the two principal geologic materials are presented in Table 1.

The AC is generally characterized as alternating massive beds of blue-gray marly limestone 0.3 to 1.5 m thick with thin calcareous shale beds up to 0.3 m thick. Although the Austin Chalk is layered, the properties of the chalk are generally isotropic, except that the layering makes the horizontal permeability much greater that the vertical permeability.

The TM is a green-gray to blue-gray, fine-grained, laminated, calcareous shale, also known as marl, with horizontal to sub-horizontal interbedded thin calcite layers. It has an average rock composition of 70 percent clay (65% montmorillonite and 5% illite). Even though it has laminar bedding, it generally behaves as a massive material when fresh and raveling is not generally expected. The strength and deformation modulus of the TM increases with depth.

Table 1. Rock properties assumed for design

Parameter	Chalk	Marl	Remarks
Compressive Strength (MPa)	15.2	1.2+0.06z	z = depth (m) below top of fresh marl
Modulus (GPa)	3.45	0.69	
Cohesion (MPa)	-	0.34	(Drained)
Friction Angle (Degrees)	-	30	(Drained)
Permeability (cm/sec)			
Vertical	10^{-7} to 10^{-8}	10^{-8} to 10^{-9}	
Horizontal	10^{-6} to 10^{-7}	5×10^{-7} to 5×10^{-8}	

Both formations have significant in situ horizontal stresses. The TM has a K_O (horizontal to vertical total stress) of 1.3 to 1.9 while K_O ranges between 1.6 to 2.2 for the AC. These horizontal stresses are inferred from hydrofracture tests (Kim and Schmidt, 1992)

3 DEVELOPMENT OF STRUCTURAL CONCEPTS

For open-cut construction in sound rock, where backfill is not required, structural support elements to supplement the innate stability of the ground often are not needed. Safety, performance, economy and serviceability can be achieved with the prudent use of rock reinforcement, rock surface deterioration protection, grouting and dewatering. However, if a roof is required to create a covered space, additional structural support elements are needed. When the roof system must also support a great thickness of backfill, the structure becomes a formidable part of the underground structure and a new set of problems are imposed on the techniques used to stabilize the rock formation. The structural elements must transfer the heavy roof loads into the rock formations without creating rock instability that cannot be countered by additional rock reinforcement.

During the construction period and service life of a cut-and-cover structure, the structural elements are stressed concurrently with the rock reinforcing. The structural elements provide relatively stiff support when compared to the flexible nature of rock reinforcement elements. The redundant nature of the problem and the uncertainties in rock mass properties over time result in difficulty in evaluating the stress-strain compatibility.

Because of these uncertainties and difficulties, it is common for civil engineers to design a completely contiguous structural envelope. This usually consists of reinforced concrete roof, wall and floor slab, that make the initial ground reinforcement redundant once the installation is complete. In sound rock, this is not an economical solution.

The design optimization and value engineering process resulted in the structural configuration indicated in Figure 1. The roof system is a structural steel tied-arch frame constructed as a deep truss. Many true arch configuration were studied, but the selected solution minimizes the backfill and roof dead weight and avoids the need for high false work to support overhead forms for concrete placement.

Several alternatives were studied that employed an arch roof system supported directly on the TM rather than on columns. These solutions increased excavation, backfill and marl reinforcement to an extent that the construction became more costly. The final solution shown in Figure 1 was to transfer vertical loads through columns down to the reinforced concrete footing level in Austin Chalk.

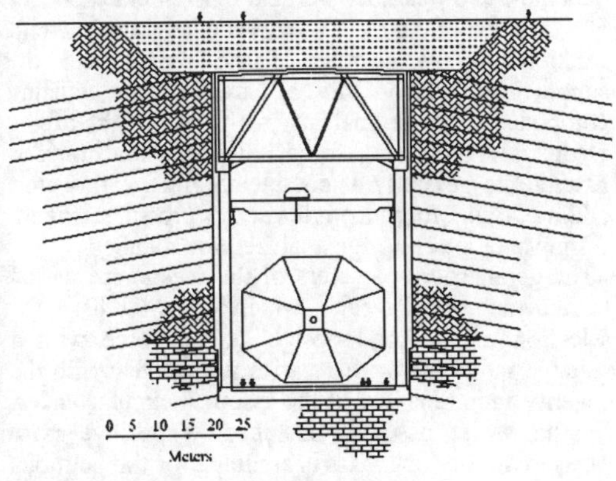

Fig.1 Experimental hall typical section

The selected design concept combines a structural roof system with the vertically excavated TM and AC side walls supported by tiebacks. These are constructed largely independent of each other. The structural stability of the roof is not dependent on the performance of the permanently tieback supported marl.

The important aspect of the structural design, aside from economy of construction, is the ultimate assurance of safety through redundant reserve between the geotechnical and structural support systems. This provides freedom for construction scheduling and sequencing, latitude for the unknown loads of future operations and maintenance, and allowance for future rock mass strain adjustments occurring with time.

The vertical reinforced concrete wall connecting the roof slab with the column cap beams ensure that non-uniform vertical loads are redistributed, both within the truss planes and between the main hall columns. Whether lateral rock pressures of the vertical walls are carried by the tieback anchors or by the rigid roof slab and horizontal bottom chord members of the structural steel roof frame, depends on the method, timing and sequence of construction as well as the relative stiffness of the geotechnical and structural elements sharing these loads. Where these uncertainties arise, the envelope of all potential loads that result in the highest forces and distortions are assumed.

Shafts exterior to the main hall shell are for utilities, detector electronic cables, and access for personnel, maintenance, equipment and detector components. The location, configuration, and method of construction for the shafts are mostly a function of the space required for these functions and the SSC operational needs.

The excavation techniques for shaft alternatives, that also accommodated the functional requirements, resulted in solutions that were economically close. Because of this, one detector hall will have circular shafts with the ground between the shafts and the hall providing the shielding. The other hall will have rectangular shafts that will be excavated simultaneously with the main hall, and then subsequently separated by a thick wall of lean concrete shielding.

4 WALL SUPPORT CONCEPT

The walls of the experimental halls are permanently supported by tieback anchors. The ground support has the following major components, as shown in Figure 2.

- Soldier piles in pairs, embedded in concrete caissons.
- Tieback anchors threaded through the soldier piles.
- Shotcrete placed in arches between soldier piles.

The tieback anchors are made up of seven-strand prestressing wire tendons, placed in boreholes inclined at 10 degrees from the horizontal and up to about 40 m long. At the hole bottom, the tendons are grouted in place, with grout filling the borehole;

this is the anchor zone. Between the anchor zone and the stressing head at the soldier piles, the tendon is isolated and free to stretch.

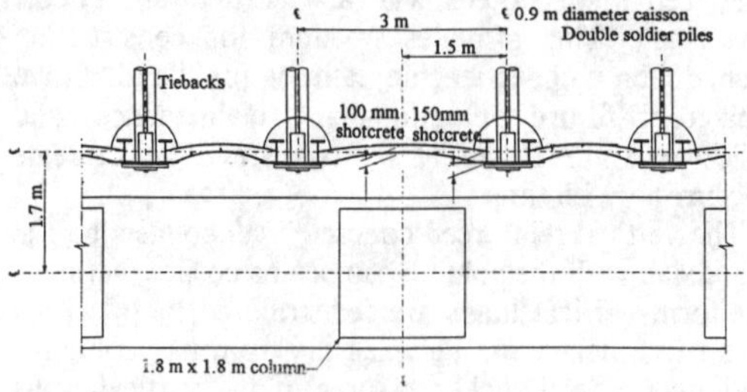

Fig. 2 Soldier pile and tieback arrangement

Corrosion protection is provided as a matter of course for these permanent tiebacks. The tendon length is surrounded by corrugated HDPE tube along the anchor length and smooth HDPE tube along the free length, and individual strands are covered by a plastic sheath in the free zone.

Horizontal drain holes will be installed in a pattern in both the TM and AC formations. The drains will be inclined up 10 degrees from horizontal and extend about 37 m from the face of the excavation. These drains will relieve potential water that may gather in tension cracks. Water collected in these drains will be collected along with the floor slab underdrains in a sump to be pumped to the surface for disposal.

Design of the anchoring system requires the determination of the following engineered parameters:

- Geometric parameters: vertical and horizontal spacing, and anchor inclination (chosen to be 10 degrees from horizontal)
- Design load for individual anchors
- Total design load for all anchors along a pair of soldier piles
- Anchor length and free length of tiebacks
- Stressing load, less than the design load for individual anchors
- Instrumentation, testing and monitoring requirements.

A tieback testing contract was undertaken to field-verify the selected design parameters. Additionally, the hall contractor will install test anchors to assure that the design anchorage forces can be achieved before proceeding with production tiebacks.

Construction will be performed in stages, which will affect the design parameters of the anchoring system. First a large excavation is made to 9 m depth, the level of the top of the roof structure. From this platform the caissons with twin soldier piles are installed by drilling 300 mm holes, inserting the piles and filling the holes with lean concrete. The caissons are approximately 3 m apart and must be drilled with great precision. Then the excavation of the main hall, with vertical walls, will proceed in stages. Only 3 to 5 m will be excavated at a time, with anchors installed in horizontal rows, tested and stressed. The rock exposed between soldier piles is covered with fiber-reinforced shotcrete, forming a horizontal structural arch. This shotcrete layer also protects the rock from deterioration due to atmospheric exposure.

As excavation proceeds, the walls will tend to move towards the open face, thereby adding load to the anchors by straining the free length. This is the reason why the initial stressing load of the anchors is less than the design load. This flexibility is required to permit the ground to adjust to the removal of existing horizontal stresses at the wall face.

A number of analyses are required to determine the anchor system parameters. Total anchor load and individual anchor lengths are determined by overall wall stability analyses of several kinds, and the initial stressing or lock-off loads are determined from numerical analyses simulating the excavation and anchor installation sequence.

A quick analysis is sufficient to determine that, in theory, the wall stresses will not exceed the initial, undrained rock mass strength. In other words, the walls can stand without support for a while. The long-term, drained strength, defined by a friction angle and a cohesion intercept, however, is insufficient for long-term stability with an adequate safety factor.

For the long term, the traditional stability analysis consists of determining the equilibrium of a wedge cutting into the wall, limited by a vertical tension crack. Observations of both natural and man-made cuts indicated that such vertical tension cracks occur at a distance of 0.4 to 0.5 times the height of the cut. Friction on the bottom of the wedge and the supporting anchor force are the stabilizing components of the analysis, with the weight of the wedge as the driving force. The tension crack is assumed to be partly or completely filled with water, adding driving force from the water pressure. Figure 3 shows a typical wedge analyzed for stability.

The geometric parameters of the wedge are varied to find the critical wedge, which forms the basis for selection of the total design anchor force with a suitable safety factor that varies subjectively with the severity and credibility of the assumed circumstances. In the worst assumed instance, a positive pore pressure in the rock mass is assumed for the fictitious failure plane along the wedge bottom.

Once the total required anchor force has been determined, individual anchor design loads are determined. The free length of the anchors are selected such that the anchor zones are outside the extent of the potential failure wedge. As check-analyses, standard slope stability analyses, assuming circular potential failure surfaces were also performed.

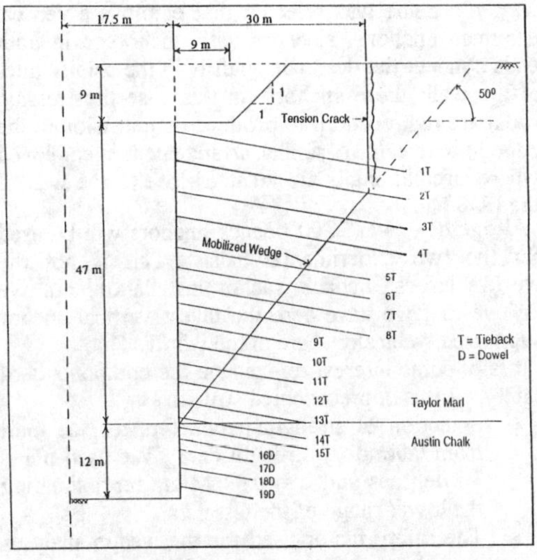

Fig. 3 Typical wedge analysis to check anchor lengths and loads

A finite difference, numerical analysis, using the FLAC code, was used to simulate the excavation and support sequence, using stages of construction some of which are shown in Figure 4. The program simulated the installation and stressing of the tieback anchors and produced final anchor loads at the completion of the excavation, verifying that when anchors were stressed to 80% of the design load, the final loads were less than 95% of the design loads. (Figure 5.)

The elastic free-field lateral wall displacement (without anchors) would be about 80 mm; the anchors reduce this to about 60 mm. This suggests that most of the lateral displacement originates from the effects of elastic lateral unloading at some distance away from the excavation.

The analyses also indicate that the TM at the AC interface tends to be overstressed some distance (21 m) from the face, a fact that should not result in instability. Elsewhere, the analyses showed that the effect of the anchors is largely to maintain the rock mass in the elastic range, without overstress or creation of shear zones.

These analyses are basically two-dimensional, tacitly assuming infinitely long walls. The actual excavation is of finite length and complicated by cutouts and adjacent shafts. The three-dimensional effects were analyzed approximately using a variety of three-dimensional wedges to determine anchor length and load requirements at outside and inside corners. Where tunnels enter the excavation, anchors, naturally, could not be placed. By placing continuous rods embedded in the shotcrete lining of these tunnels, the tunnels themselves were made to function as lateral ties.

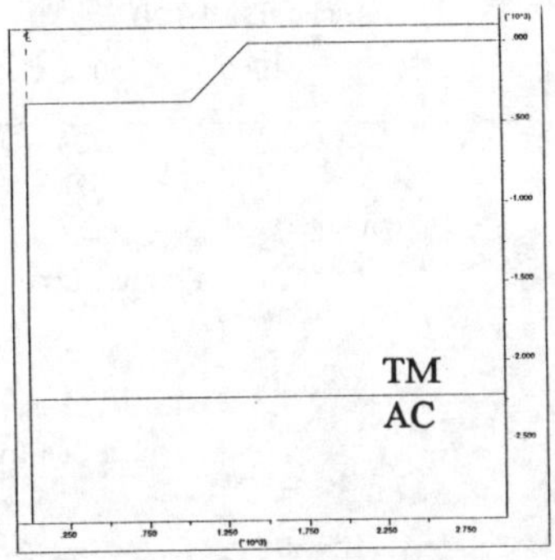

Stage 1 Open-Cut Portion Excavated

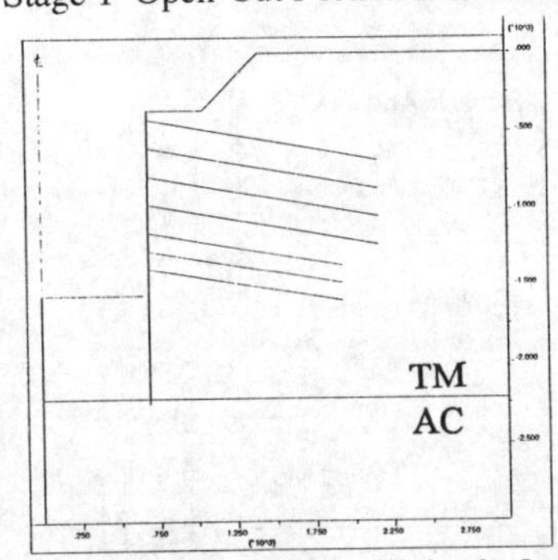

Stage 9 Excavation to 40m Tiebacks Installed

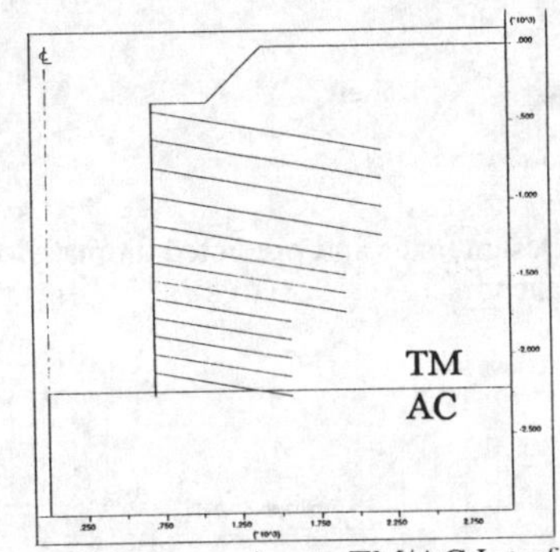

Stage 15 Excavation to TM/AC Interface

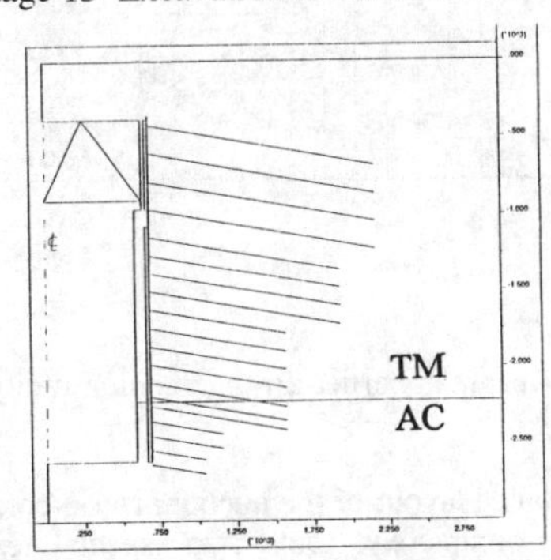

Stage 23 Hall Complete with Detector Load

Fig. 4 Stages of construction simulated in analysis

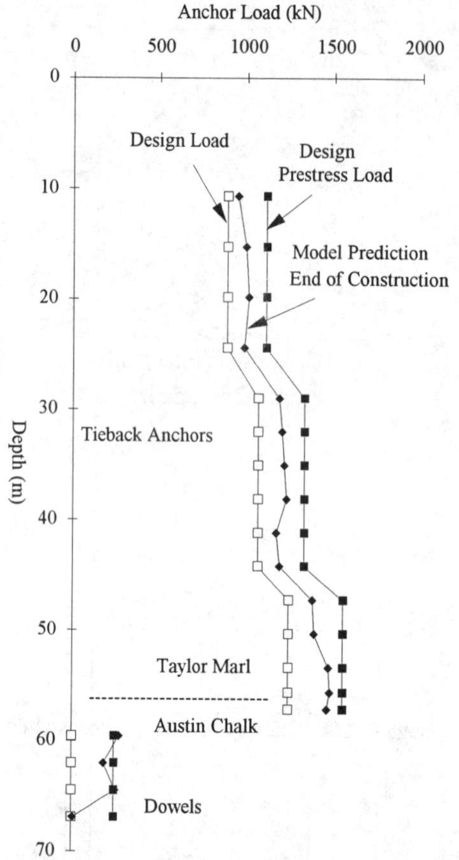

Fig. 5 Design loads and predicted ultimate loads in anchors

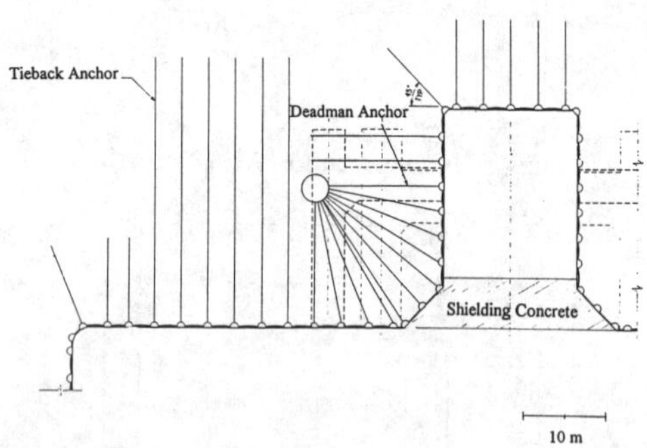

Fig. 6 Deadman corner arrangement at utility shaft

The detailed layout of the anchors is too complex to describe completely, but two special examples deserve some attention. Figure 6 shows the anchor arrangement at reentrant corners at the IR-5 hall. Because the anchors here would get in each others way, a scheme was selected that employs a vertical deadman anchor. Anchors with anchorage in both ends connect the deadman shaft with the soldier piles at the wall; these anchors (in this case prestressing rods) are tightened up to produce compression in the zone in between. A similar arrangement is employed where circular shafts are located close to the wall of the IR-8 hall.

Altogether, over 2300 tieback anchors will be used for the two experimental halls. This is not the world's largest tieback anchor installation, but we believe that we have here the tallest vertical anchor supported walls anywhere in the world.

It is of some interest to examine the components of safety for this unprecedented structure:

- Selection of strength parameters for the marl from laboratory strength data. We used phi = 30 degrees and c = 0.34 MPa, approximating the lower range of the test data.
- The safety factor used for the wedge analyses was 1.5 or greater for the most credible assumptions, and in excess of 1.2 for the worst assumed case (full water pressure in tension crack, gradually decreasing water pressure in the pores along the potential failure surface).
- The calculated anchor loads are less than the design loads, which include a safety factor on the tendons and the anchors of at least 1.67.
- The design includes horizontal drain pipes, intended to remove water from any tension cracks and reduce pore water pressures elsewhere.
- In the final condition, the roof truss will act as a compression member, a strut, which will take up load in the event that anchors relax over time.
- All anchors will be proof tested to 1.33 times the design load (a portion of them also subjected to more extensive performance tests) before being locked off.

An extensive instrumentation and monitoring program will be implemented to verify assumptions regarding loads and displacements, including instrumentation of some anchors, permitting corrective action should it become necessary.

5 CONCLUSION

This paper describes the design approach used to create large experimental halls for the SSC detectors. The use of open cut construction with permanently supported wall excavations and an independent roof and backfill support structural system was the result of a flexible approach to the design and analysis. This unique and challenging design is noteworthy given the dimensional requirements and geometry as well as the limited design time available. The efforts provided an economical and efficient design for the experimental halls. This was validated when the potential contractors three lowest bids were under the engineer's estimate.

6 REFERENCES

Hashash, Y. M. A., R. F. Cook, M. Hasek and B. Schmidt, 1994, "Design of Ground Supports for the Experimental Hall Excavations of the Superconducting Super Collider" submitted for publication in ASCE's Geotechnical Journal.

Kim, K. and B. Schmidt, 1992, "Characteristics of The State of In Situ Stresses for Design of the Superconducting Super Collider Interaction Halls", proceedings of the ISRM Symposium on Rock Mass Characterization, Chester, G. B.

Richards, D. P., and J. S. Robbins, 1994, "Geological Optimization of Underground Facilities for the Superconducting Super Collider, Texas, USA", International Congress on Tunnelling and Ground Conditions, Cairo.

Robinson, R. A., M. S. Kucher, and R. P. Brouillette, 1993, "Construction Behavior of the first Underground Opening of the Superconducting Super Collider Project", RETC, Boston, pp. 607-629.

Tunnelling and Ground Conditions, Abdel Salam (ed.) © *1994 Balkema, Rotterdam, ISBN 90 5410 363 9*

Analysis and interpretation of the geotechnical measurements for tunnels in south Turkey

Emam Soliman
Structural Engineering Department, Ain Shams University, Faculty of Engineering, Abbassia & Dar Al-Handasah Consultants, Cairo, Egypt

A.Amr Darrag
Civil Engineering Department, Cairo University, Faculty of Engineering, Giza, Egypt (Presently: Dar Al-Handasah Consultants, Cairo, Egypt)

ABSTRACT: A large tunneling project is being carried out in South Turkey. These tunnels have been driven using the New Austrian Tunneling Method (NATM). NATM is a construction method, where the type and the amount of primary tunnel support is designed on the basis of the anticipated subsurface conditions using empirical classification systems as well as the designer's experience. In-situ measurements provide an integral part of the method through providing information about the interaction of ground and tunnel support and foreseeing problems of lining and face stability. Numerical analyses provide useful means to support decisions taken during NATM design. This paper presents the results of the finite element analyses for the subject tunnels compared with actual geotechnical measurements. The paper also presents a three-dimensional approach for the analysis of this problem. Empirical design using the well-known NGI method (Q-system) is assessed based on field observations.

1 INTRODUCTION

Empirical design methods, which are based on comprehensive rock classification systems, have been extensively used for the design of tunnel primary support. The two most widely used methods in this respect are the NGI method (Barton et al., 1974), which has been revised by Grimstad and Barton (1988), and the geomechanical method (Bieniawski, 1984). Although such empirical methods provide useful guidelines for tunnel support selection, they are based on limited data which may not be compatible with the particular case of interest. Moreover, the effect of individual support elements is not quantified. The shortcomings of empirical design methods could be partly overcome by using the New Austrian Tunneling Method (NATM), which utilizes the results of monitoring deformations and stresses within the rock and liner as an integral part of the method. Successful utilization of NATM depends greatly on the experience of the designer. For this reason numerical analyses (either simplified as those recommended by Hoek and Brown(1980); Darrag(1993), etc.) or rigorous as those presented in this paper, are essential in order to support decisions taken during NATM design.

This paper presents methodologies for systematic two and three dimensional finite element (FE) analyses for NATM tunnel design. The analysis has been applied to a tunneling project that has just been completed in South Turkey. Both empirical and numerical analyses are assessed in light of the geotechnical measurements that have been taken during construction of one of the tunnels.

2 TUNNEL PROJECT

The Tarsus-Adana-Gaziantep six-lane dual Motorway is currently under construction in the South Turkey area. As part of the motorway, four twin highway tunnels are to be constructed. The length of the tunnels varies between 356 and 1470m. Typical tunnel cross sectional area is approximately 125 m2. Analyses given in this paper are related to one of the tunnels, which is located approxi-

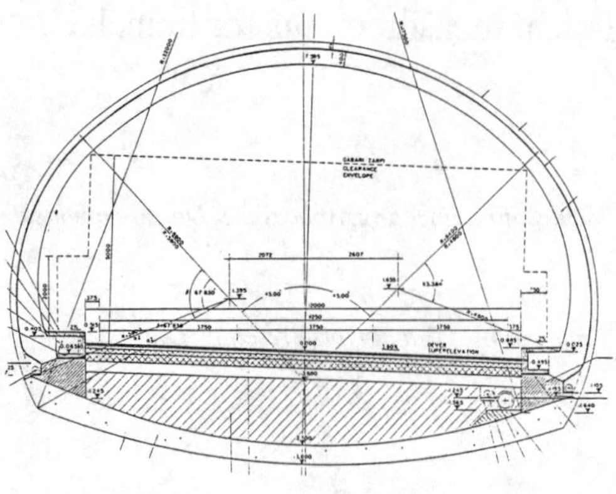

Figure 1: Typical cross section of the tunnel(with invert)

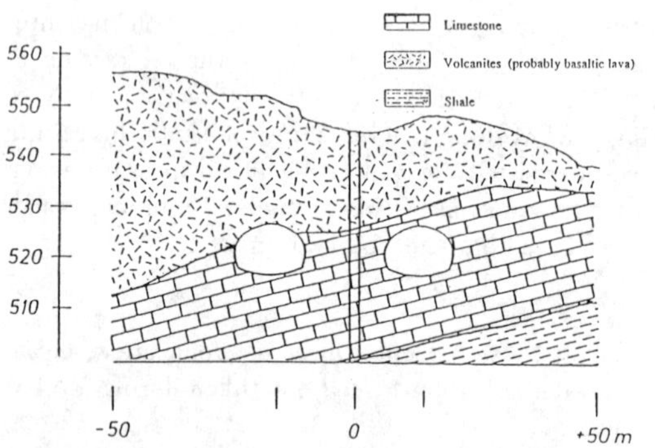

Figure 2: Geological cross section at station 192+368

mately 2.0 km west of Taslouck village. The subject tunnel is 361 m long, with an overburden that ranges between 10 and 45m. Typical cross section of the tunnel is shown in Figure (1).

3 GEOLOGY AND GEOTECHNICS

In the vicinity of the tunnel two major rock types, both of the Miocene age, are encountered. These are the Kalecick formation, which consists of basaltic volcanic rocks, and the Horu Limestone formation. Figure (2) shows a geological cross section at a location where one of the tunnel tubes is crossing the interface between the two layers.

The volcanic rock group is highly altered and irregularly fractured. The rocks are moderately weak to weak and contain some swelling minerals. Although at some short sections the water table may lie above tunnel level, in general water influx was expected to be low.

The Horu limestone is massive to thick bedded. Prominent karst features, steeply dipping master joints and slikensides are abundant in this formation. Consequently water influx will be concentrated along open joints or solution features. The rock is moderately strong to strong.

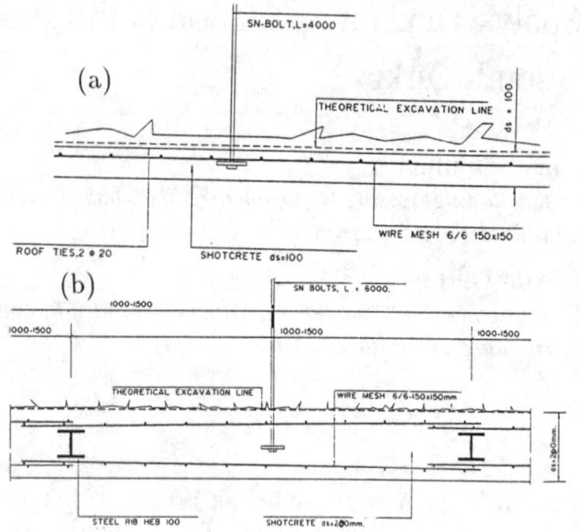

Figure 3: Sections of the primary rock support for class 3 (a) and class 5 (b)

4 EMPIRICAL SELECTION OF TUNNEL SUPPORT

The modified NGI empirical procedure (Grimstad and Barton, 1988) was used for selecting the primary tunnel support for the different rock classes (combinations of shotcrete, steel ribs, rockbolts and/or forepoling). Boundaries of the rock classes were assigned based on the Q values obtained for the existing rock strata following the NGI recommendations (Barton et al., 1974). For the limestone formation, the Q values range between 1.00 and 0.4, while for the volcanic rock group Q ranges between 0.10 and 0.01. Consequently, NATM rock mass class 3 (slightly friable) and class 5 (highly friable to pressure exerting) were assigned for the limestone and the volcanic rocks, respectively. Class 5 was also assigned to the limestone near the tunnel portals were the cover is relatively small. Based on the NATM classification, the primary rock support systems for the two classes were selected. These are schematically illustrated on Figures (3a and b) for rock mass classes 3 and 5, respectively.

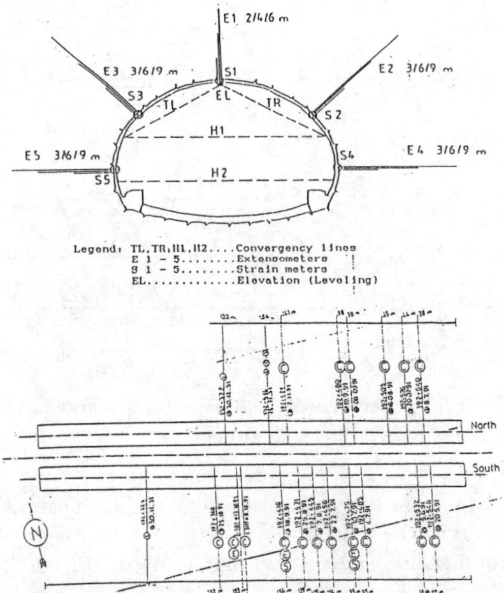

Figure 4: Layout of the monitoring-sections

5 GEOTECHNICAL MEASUREMENTS

The geotechnical monitoring program which was used for the subject site was intended to provide information on tunnel deformations, behaviour of the surrounding rock material and the effect of rock loading on the primary shotcrete lining by measuring of rock mass strains, normal forces and bending moments. The installed monitoring sections contained different combinations of the instruments (tape and borehole extensometers and shotcrete strain meters). The layout of the monitoring sections is given in Figure (4).

The relative displacement between two pins were measured with a tape extensometer. The pins were placed in prearranged patterns around the tunnel circumference. Both horizontal and diagonal convergencies were read. Borehole extensometers were used to record the values and distortion of movements and strains inside the surrounding rock mass. Shotcrete strain meters were used to record tensional as well as compressive strains in the shotcrete lining.

5.1 Monitoring Station 192 + 416

The following installed instruments were used at this station :

1.Three convergency bolts (for tape extensometer

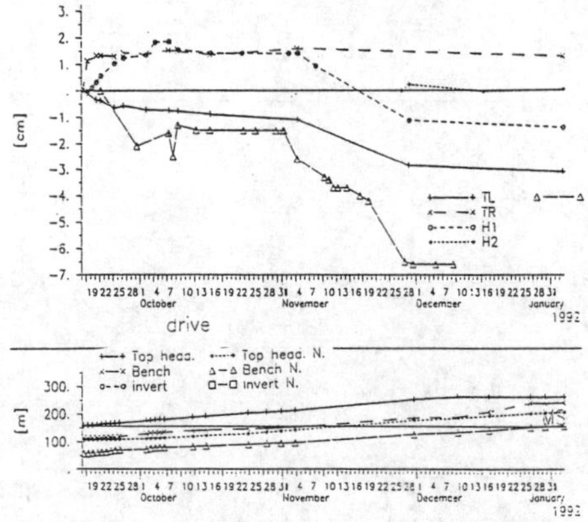

Figure 5: Convergency and leveling readings at station 192+416 (south tunnel)

measurements)

2.Two extensometers of 2 and 4 meters depth

3.Three strainmeters.

At this chainage limestone breccia was excavated and has shown few bigger cracks on the left side in vertical direction. Primary Tunnel support of 20cm shotcrete with rockbolts 4m long and steel ribs HEB 100 were installed. Convergency, leveling and strainmeter deformation readings for the south (right) tunnel are given in Figures 5, 6 and for the north (left) tunnel in Figures (7, 8 and 9).

5.2 Monitoring Station 192 + 368

At the right side of this station, clay including limestone bolders of a diameter between 0.2-0.5m was encountered. The majority of the ground mass at this station, however, consists of volcanic agglomerate. Three convergency bolts were installed at this station. The convergency and leveling readings are given in Figure (10).

6 TWO DIMENSIONAL ANALYSIS OF THE TUNNEL

The principles of the two dimensional finite element analyses for the subject tunnel followed guidelines given by Ahrens and Lindner(1982), Duddeck(1972), Duddeck and Erdmann(1985), Duddeck(1988) Müller(1978), Schulze and Duddeck (1964), Soliman(1991) and Soliman, Duddeck, and Ahrens (1993). The analysis has been carried out

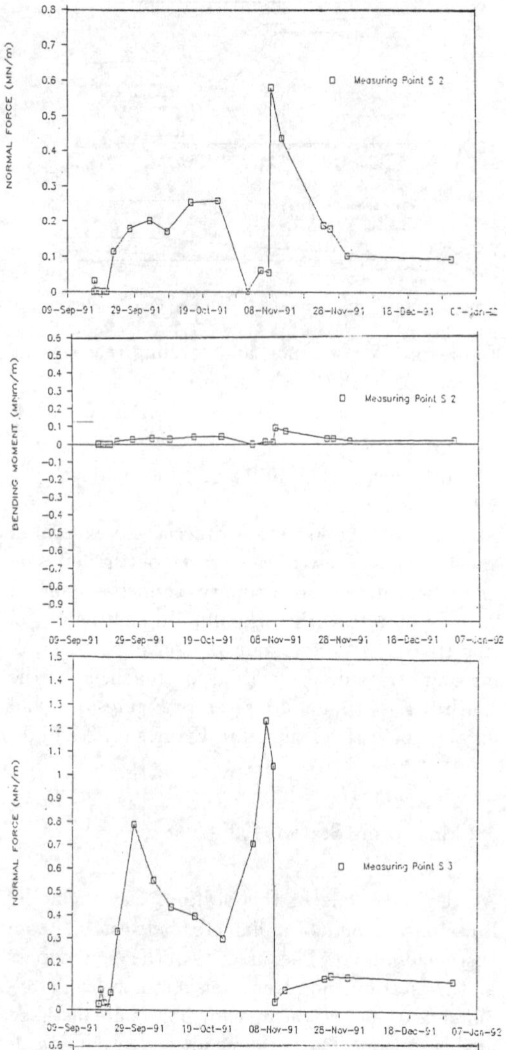

Figure 6: Strainmeter readings at station 192+416 (south tunnel)

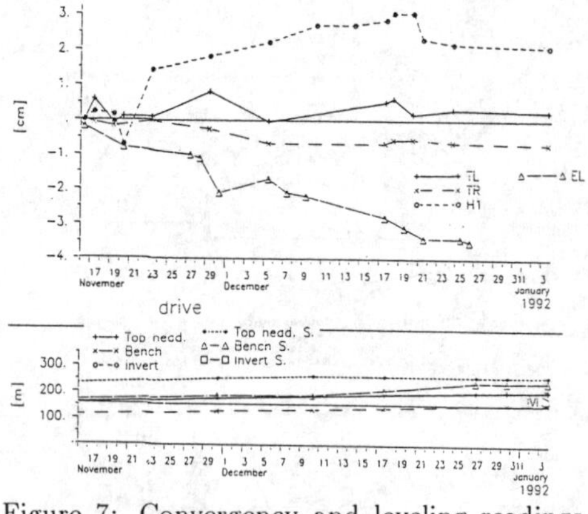

Figure 7: Convergency and leveling readings at station 192+416 (north tunnel)

at the a/m two stations. The numerical analysis followed the construction phases of excavating the top heading and the benching using the FE mesh shown in Figure (11).

6.1 FE-Results of Section at Station 192+368

The following rock parameters were used in the analysis :

Height of overburden above the crown = $25.0m$(class 3)
Young's modulus of the rock mass: $E_G = 5.0MPa$ (as interpreted from pressuremeter tests).
Poisson's ration: $\nu = 0.25$
Specific weight of the rock: $\gamma = 25kN/m^3$
Lateral horizontal pressure coefficient: $K_0 = 1.0$
Figure (12) shows the deformed shape due to excavation of top heading of the first tube. Normal forces, bending moment, horizontal and vertical displacements are given in Figure (13). Figures (14 and 15) show the results during the excavation of full section of the first and second tube.

6.2 FE-Results of Section at Station 192+416

For the rock between ground surface and tunnel crown ($25.0m$), the same rock properties as given in section 6.1 were used. Young's modulus E_G of $0.3MPa$ and Poisson's ratio ν of 0.30 were chosen for rock under the tunnel crown (class 5).
The results of the analysis during the installation

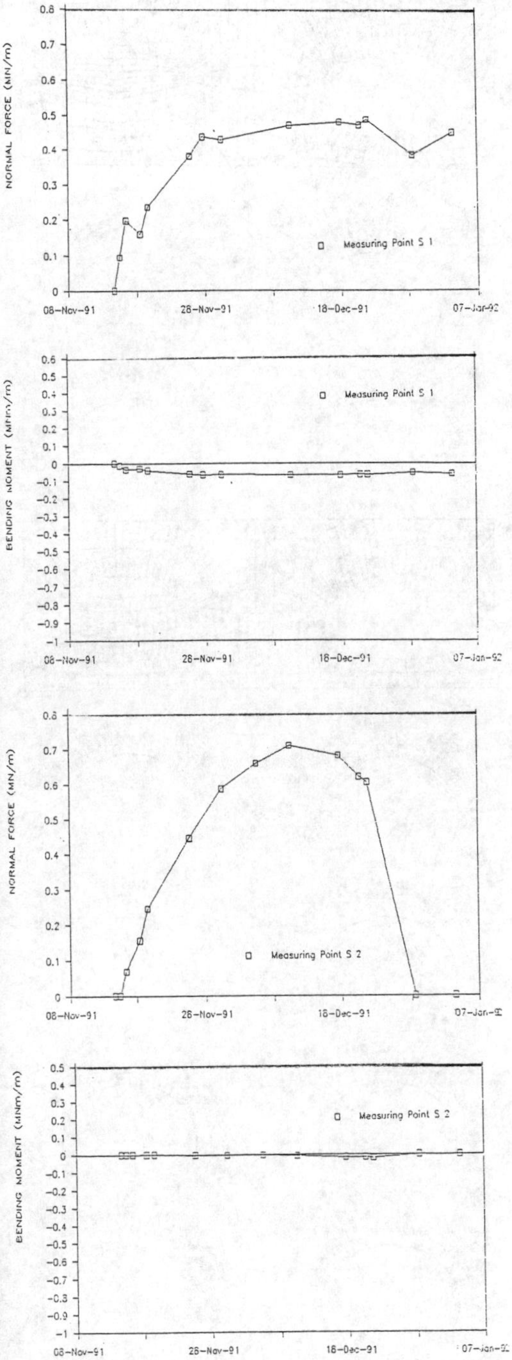

Figure 8: Strainmeter readings at station 192+416 (north tunnel)

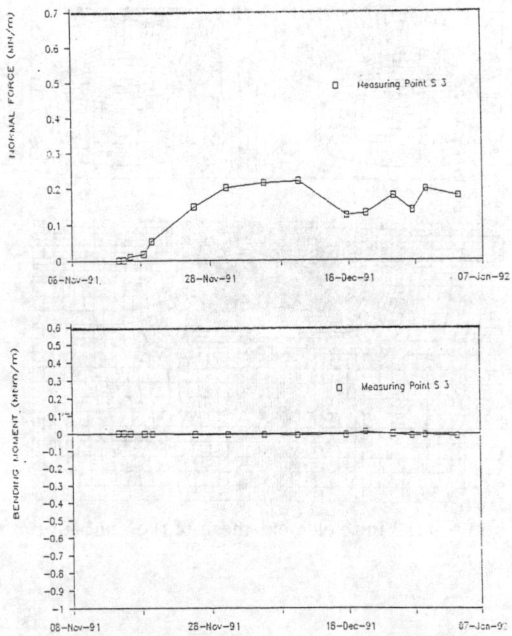

Figure 9: Strainmeter readings at station 192+416 (north tunnel)

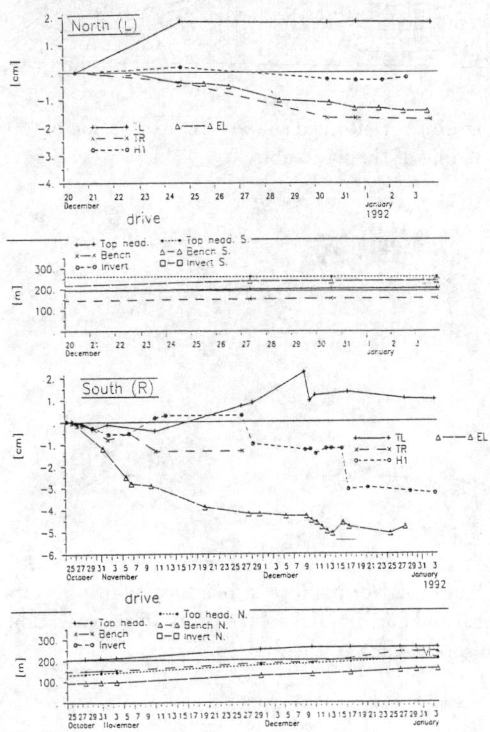

Figure 10: Convergency and leveling readings at station 192+368

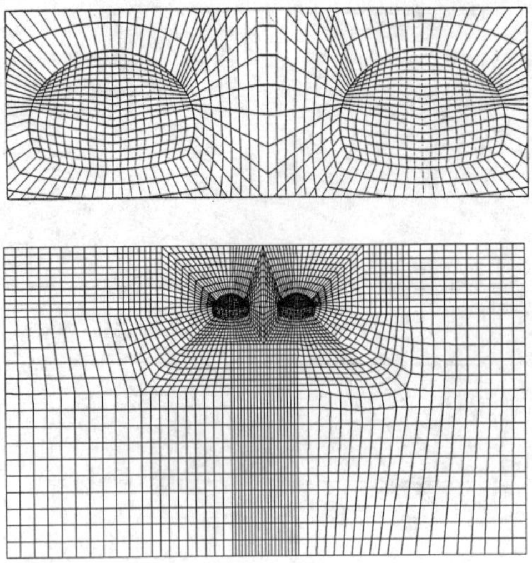

Figure 11: Finite element mesh of the double tube tunnel

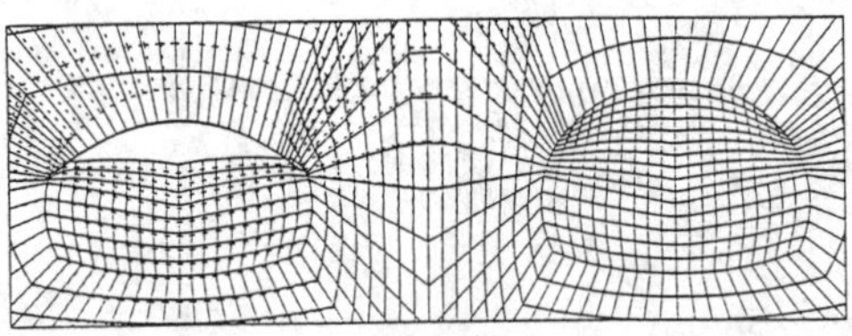

Figure 12: Deformed shape after excavation of top heading of the first tube

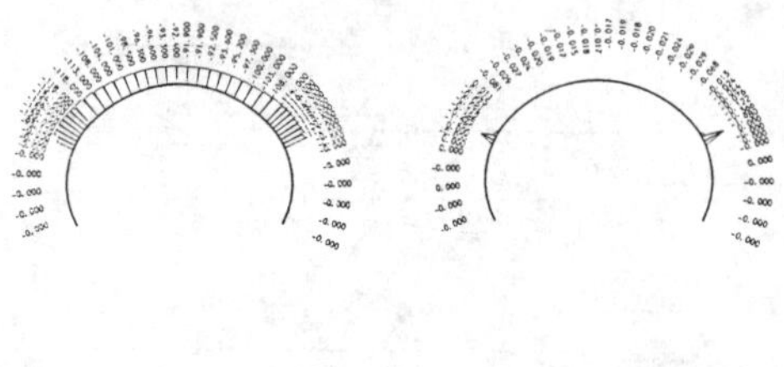

Figure 13: Normal forces, bending moments, vertical and horizontal displacements after excavation of top heading of the first tube

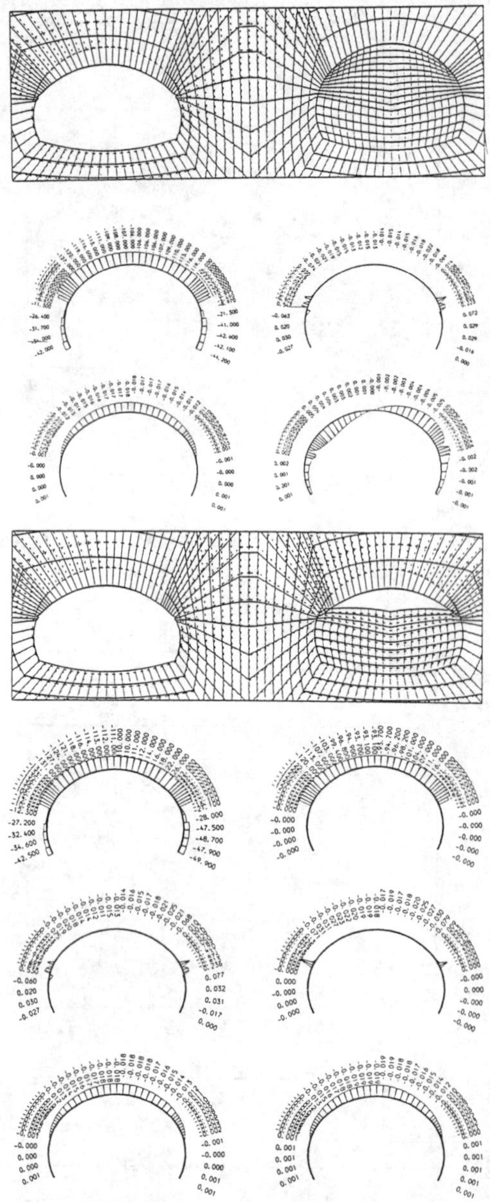

Figure 14: Normal forces, bending moments, vertical and horizontal displacements after excavation of benching of the first tube and top heading of the second tube

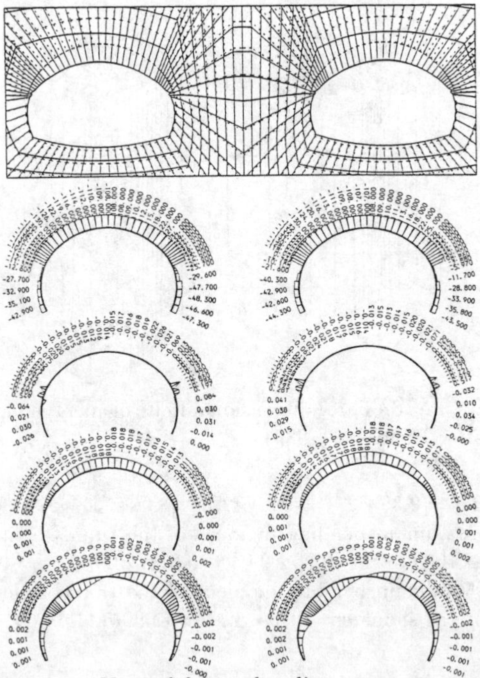

Figure 15: Normal forces, bending moments, vertical and horizontal displacements after excavation of benching of the second tube

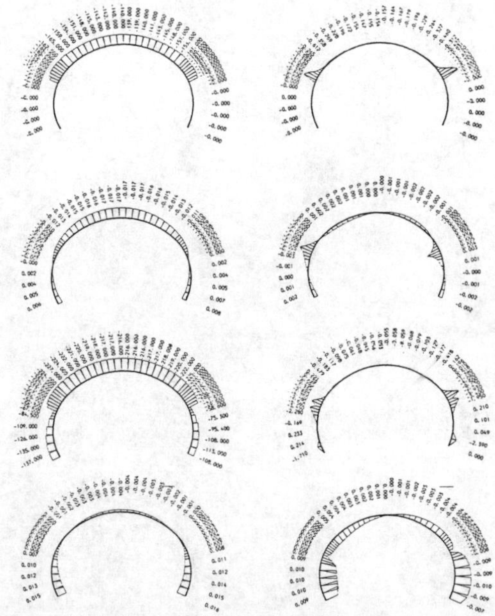

Figure 16: Normal forces, bending moments, vertical and horizontal displacements after excavation of top heading and benching of the first tube

of primary tunnel support are shown in Figures (16 and 17).

7 THREE-DIMENSIONAL ANALYSIS OF THE TUNNEL

The principal guidelines to use three dimensional models in the tunneling design have been described by Berwanger(1986), Dallmann(1988), Erdmann (1983), Kielbassa(1989), Kielbassa and Duddeck (1991) Soliman(1991) and Soliman, Duddeck and Ahrens(1993). Kielbassa (1989) has performed three dimensional analyses for non circular sections by using the FE Model as shown in Figure (19). He has developed diagrams to get the three dimensional results by using two equivalent 2-Dimensional models. One of the two models is used to calculate the predeformation of the surrounding ground during excavation, while the second model is to get the actual forces in the lining. Soliman (1991) has examined the case of advancing a second tube next to one that is also under construction and has shown the consequent effects on the analysis. Methodologies described by Kielbassa (1989) and by Soliman (1991) are utilized in analyses presented in this study.

For the analysis, the following three parameters are needed at station 192+416

$\alpha = \frac{E_G R}{EI} = \frac{(0.3*10^5*9.0)}{(1*10^6)*(7.43*10^{-4})} = 363.4$

$\beta = \frac{E_G R}{EF} = \frac{(0.3*10^5*9.0)}{(1*10^6)*(0.244)} = 1.107$

$\frac{L_u}{D} = \frac{1.5}{9.0} = \frac{1}{6}$

L_u is the unsupported length at the tunnel face. Therefore, the 3-D results at station 192+416 for the first tube are given as follows :

-Maximum radial displacement at the crown of the tunnel lining= $0.35 * 0.005 = 1.75mm$

-Maximum radial displacement at the shoulder of the tunnel lining= $0.35 * 0.009 = 3.15mm$

-Maximum radial displacement of the ground at the crown of the tunnel (including the predeformations)= $1.75 + 0.6 * 20 = 13.75mm$

-Maximum radial displacement of the ground at the shoulder (including the predeformations)= $3.15 + 0.6 * 14 = 11.55mm$

-Maximum normal forces of the tunnel lining at the crown= $0.35 * 242 = 84.7t/m$

-Maximum forces of the tunnel lining at the Shoul-

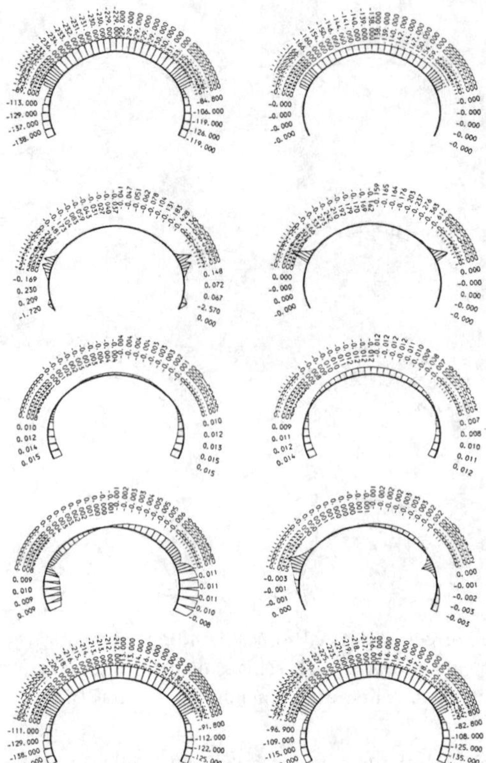

Figure 17: Normal forces, bending moments, vertical and horizontal displacements after excavation of top heading and benching of the second tube

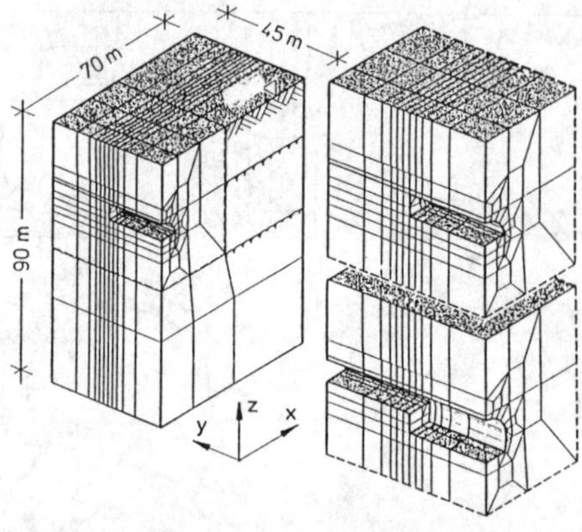

Figure 18: Three dimensional finite element model after Kielbassa (1989)

der $= 0.35 * 271 = 94.85t/m$
-Maximum bending moment of the tunnel lining at the crown$= 0.35 * 0.04 = 0.014m.t/m$
-Maximum bending moment of the tunnel lining at the shoulder$= 0.35 * 3.14 = 1.099m.t/m$

After the advancing of the second tube, we have :
-Maximum radial displacement at the crown of the first tube$= 1.75 * 1.06 = 1.855mm$
-Maximum radial displacement at the crown of the second tube$= 1.75 * 1.01 = 1.768mm$
-Maximum radial displacement at the shoulder of the first tube$= 3.15 * 1.2 = 3.78mm$
-Maximum radial displacement at the shoulder of the second tube$= 3.15 * 0.9 = 2.835mm$
-Maximum radial displacement of the ground at the crown of the first tube$= 13.75*1.015 = 13.96mm$

-Maximum radial displacement of the ground at the crown of the second tube$= 13.75 * 1.03 = 14.16mm$
-Maximum radial displacement of the ground at the shoulder of the first tube$= 11.55mm$
-Maximum radial displacement of the ground at the Shoulder of the second tube$= 11.55mm$
-Maximum normal forces of the tunnel lining at the crown of the first tube$= 84.7*1.05 = 88.94t/m$
-Maximum normal forces of the tunnel lining at the crown of the second tube$= 84.7*1.02 = 86.394t/$

-Maximum normal forces of the tunnel lining at the shoulder of the first tube$= 94.85*1.1 = 104.35t/$

-Maximum normal forces of the tunnel lining at the shoulder of the second tube= $94.85 * 1 = 94.85t/m$
-Maximum bending moment of the tunnel lining at the crown of the first tube= $0.014m.t/m$
-Maximum bending moment of the tunnel lining at the crown of the second tube= $0.014m.t/m$
-Maximum bending moment of the tunnel lining at the shoulder of the first tube= $1.099 * 1.1 = 1.209m.t/m$
-Maximum bending moment of the tunnel lining at the shoulder of the second tube= $1.099m.t./m$

Similarly, at Station 192+368

$$\alpha = \frac{E_G R}{EI} = \frac{(4.0*10^5*9.0)}{(1*10^6)*(8.33*10^{-5})} = 43217.28$$
$$\beta = \frac{E_G R}{EF} = \frac{(4.0*10^5*9.0)}{(1*10^6)*(0.10)} = 36.0$$
$$\frac{L_u}{D} = \frac{3.0}{9.0} = \frac{1}{3}$$

Therefore, the 3-D results at station 192+368 for the first tube are given as follows :

-Maximum radial displacement at the crown of the tunnel lining= $0.25 * 0.016 = 4mm$
-Maximum radial displacement at the shoulder of the tunnel lining= $0.25 * 0.006 = 1.5mm$
-Maximum radial displacement of the ground at the crown (including the predeformations)= $4 + 0.7 * 24 = 20.8mm$
-Maximum radial displacement of the at the ground Shoulder (including the predeformations)= $1.5 + 0.7 * 8 = 7.1mm$
-Maximum normal forces of the tunnel lining at the crown= $0.25 * 117 = 29.25t/m$
-Maximum normal forces of the tunnel lining at the Shoulder= $0.25 * 169 = 42.25t/m$
-Maximum bending moment of the tunnel lining at the crown= $0.25 * 0.014 = 0.0035m.t/m$
-Maximum bending moment of the tunnel lining at the shoulder= $0.25 * 0.614 = 0.1535m.t/m$

After the advancing of the second tube, we have :

-Maximum radial displacement at the crown of the first tube= $4 * 1.09 = 4.36mm$
-Maximum radial displacement at the crown of the second tube= $4 * 1.04 = 4.16mm$
-Maximum radial displacement at the shoulder of the first tube= $1.5 * 1.6 = 2.4mm$
-Maximum radial displacement at the shoulder of the second tube= $1.5 * 1.0 = 1.5mm$
-Maximum radial displacement of the ground at the crown of the first tube= $20.8*1.055 = 21.944mm$

-Maximum radial displacement of the ground at the crown of the second tube= $20.8*1.03 = 21.42mm$

-Maximum radial displacement of the ground at the Shoulder of the first tube= $7.1mm$
-Maximum radial displacement of the ground at the shoulder of the second tube= $7.1mm$
-Maximum normal forces of the tunnel lining at the crown of the first tube= $29.25*1.03 = 30.128t/m$

-Maximum normal forces of the tunnel lining at the crown of the second tube= $29.25 * 1.02 = 29.84t/m$
-Maximum normal forces of the tunnel lining at the shoulder of the first tube= $42.25 * 1.14 = 48.165t/m$
-Maximum normal forces of the tunnel lining at the shoulder of the second tube= $42.25 * 1.04 = 43.94t/m$
-Maximum bending moment of the tunnel lining at the crown of the first tube= $0.0035m.t/m$
-Maximum bending moment of the tunnel lining at the crown of the second tube= $0.0035m.t/m$
-Maximum bending moment of the tunnel lining at the shoulder of the first tube= $0.1535 * 1.12 = 0.172m.t/m$
-Maximum bending moment of the tunnel lining at the shoulder of the second tube= $0.1535 * 1.04 = 0.1596m.t/m$

8 COMPARISON OF RESULTS AND CONCLUSIONS

Results of analyses presented in this paper have been compared with available geotechnical measurements described earlier. The 2-D and 3-D results have also been compared. These comparisons rendered the following conclusions:

1- The preliminary selected tunnel primary support according to the modified NGI system were proven to be adequate since the maximum compressive stresses in the shotcrete were computed to be close to the allowable values. Consequently, the original NGI system is considered to be conservative.

2- The 3-D analysis was proven to be very effective in introducing the effect of predeformation of the ground during tunnelling (e.g. relaxation). Therefore, the analysis provides a more accurate comparison with the measurements than the 2-D analysis. This is clear from comparing the numerical 3-D results and the measured bending moments and normal forces shown in Figures 6,8 and 9. The displacements measured via convergency pins, which are more accurate than levelling readings, also compare well with the calculations, as shown in Figures 5,7 and 10.
3-Although the 3-D analyses provide more accurate results than the 2-D ones, the latter is still very useful during preliminary design stages.

REFERENCES

Ahrens, H., Lindner, E. , Lux, K.-H. 1982. Zur Dimensionierung von Tunnelausbauten nach den "Empfehlungen zur Berechnung von Tunneln im Lockergestein 1980", Die Bautechnik 8/1982, S. 260-273 und 9/1982, S. 303-311

Barton, N.,Lien, R. and Lunde, J. 1974. Engineering Classification of Rock Masses for the Design of Tunnel Support. Rock Mechanics, Vol. 6, 189-236.

Berwanger, W. 1986. Dreidimensionale Berechnung von tiefliegenden Felstunnel unter Berücksichtigung des rheologischen Verhaltens von Spritzbeton und des Bauverfahrens, Forschungsergebnisse aus dem Tunnel- und Kavernenbau, H. 10, Universität Hannover

Bieniawski, Z.T. 1984. Rock Mechanics Design in Mining and Tunneling, Balkema.

Dallmann, R. 1988. Kopplung von Rand- und Finite-Element-Methode beim Tunnelvortrieb im Voll- und Halbraum, Bericht 88-56, Institut für Statik der TU Braunschweig

Darrag, A. A. 1993. Integrated Analysis of Rock Support for NATM Tunnels. Engineering Research Bulletin, University of Helwan, January, 61-76.

Duddeck, H. 1972. Zu den Berechnungsmethoden und zur Sicherheit von Tunnelbauten, Der Bauingenieur 47 , H. 2, S. 43-52

Duddeck, H. and Erdmann, J. 1985. On structural design models for tunnels in soft soil. Underground Space, Vol. 9, 246-259.

Duddeck, H. 1988. Guidelines for the design of tunnels. ITA working group. J. Tunneling and underground Space Technology V.3 Vo.3 , 237-249.

Empfehlungen zur Berechnung von Tunneln im Lockergestein 1988, Die Bautechnik 57 (1980), H. 10, S. 349-356

Erdmann, J. 1983. Vergleich ebener und Entwicklung räumlicher Berechnungsverfahren für Tunnel, Bericht Nr. 83-40, Institut für Statik der TU Braunschweig

Grimstad, E. and Barton, N. 1988. Design and Methods for Rock support. Norwegian Tunneling Today, Norwegian soil and Rock Engineering Association, Pub. 5, 59-64.

Hoek, E. and Brown, E.T. 1980. Underground Excavations in Rock, Institutionof Minining and Mettallurgy, U. K.

Kielbassa, S. 1989. Spannungsumlagerungen an der Ortsbrust und Beanspruchung des Tunnelausbaus, Bericht Nr. 89-58, Institut für Statik der TU Braunschweig

Kielbassa, S. and Duddeck, H. 1991. Three-dimensional analysis of tunnels. Rock-Mechanics,115-132.

Müller, L. 1978. Der Felsbau, Bd. 3, Tunnelbau, Enke-Verlage, Stuttgart

Schulze, H., Duddeck, H. 1964. Spannungen in schildvorgetriebenen Tunneln, Beton- und Stahlbetonbau 59, H. 8, S. 169-175

Soliman, E. 1991. Ebene und räumliche Spannungs- und Verformungszustände von zwei nacheinander aufgefahrenen Tunnelröhren, Bericht 91-66, Institut für Statik der TU Braunschweig

Soliman, E., Duddeck, H. and Ahrens, H. 1993. Two- and Three-dimensional analysis of closely spaced double-tube tunnels. Tunneling and Underground Space Technology, Vol. 8, No. 1, pp. 13-18

4 Innovation, new researches and development of underground space

Tunnelling and Ground Conditions, Abdel Salam (ed.) © 1994 Balkema, Rotterdam, ISBN 90 5410 363 9

Seismic hazards for Greater Cairo underground metro lines

M.E.Abdel Salam
Egyptian National Authority for Tunnels NAT, Egypt
M.Abdel Fattah I., B.
Study Department NAT, Egypt

Abstract: The Greater Cairo underground metro lines were planned to cover and connect the over populated zones of Greater Cairo. Both cut & cover and bored tunnels are employed in the underground lines. Generally, the uncertainties which involve the computation of seismic resistance of underground structures are much more than the well defined parameters. These uncertainties can be classified into three main groups;

1. The parameters of the expected earthquake.
2. The characteristics of waves propagated at underground structure site.
3. The methodology of representing the dynamic response of the surrounding soil and soil-structure interaction.

This work presents a limited survey for the different approaches of the above mentioned factors in order to clarify the seismic hazard domain for the greater Cairo underground metro.

1 INTRODUCTION

There is a famous traditional saying that the underground structures are the safest structures with respect to the hazards caused from earthquakes. Allover the world, the underground metro lines are used as the ideal shelter to rescue and evacuate the waifs during destructive earthquakes. [Ref. 4, 13, and 14]. Therefore, that saying is practically, correct. In spit of the rarity of the literatures discussing the seismic resistivity of underground structures specially, that embedded in saturated sand, the objective of the present work is to establish a logic proof for this saying. Greater Cairo, with a present population of more than 10 millions, is one of the world's most dense cities. Therefore, metro system was the only solution to stand with the growing transportation problems. Studies proposed a network of three lines, [Ref. 1, 2, and 6], Fig.(1):

1. Regional line (L1), 42.5 km. long from Helwan to El Marg with underground cut & cover portion of 4.5 km. under the central business district of Cairo with 5 underground stations. Fig.(2), gives the common cross section of the under ground portion. (L1) is in service since 1987.
2. Urban line no.1 (L2), 18.3 km. long from Shoubra ElKhiema to El Giza railway station. The underground portion of (L2) is 16.9 km. with 13 underground stations. Both cut & cover and bored tunnels were employed, Fig.(2) shows the different cross sections of(L2) and typical section in underground station. (L2) has two interchanges with (L1) and one with (L3). The work in (L2) has commenced in February 1993.
3. Urban line no. 2 (L3) [proposed] 10 km. long from Imbaba to El Darrasa with 8 underground stations.

On October 12th 1992 an earthquake of magnitude 5.9 hit the middle part of Egypt. The surface epicenter was located about 40 km. south west of Cairo, Fig.(1), indicates the probable positions of the surface epicenter. Many buildings collapsed and numerous cracked. Although the distance from (L1) final station at Helwan to the surface epicenter was less than 20 km.,[Ref. 5], Fig.(1), all the surface and underground components of (L1) have not influenced and are completely intact. That earthquake was a practical full scale check test for (L1) seismic resistivity. The distinguished response of (L1) components, hasn't to under estimate the investigation of the seismic resistivity of underground metro lines of greater Cairo. The seismic study implies a conclusive presentation of greater Cairo soil formation and the local criteria for expected earthquake.

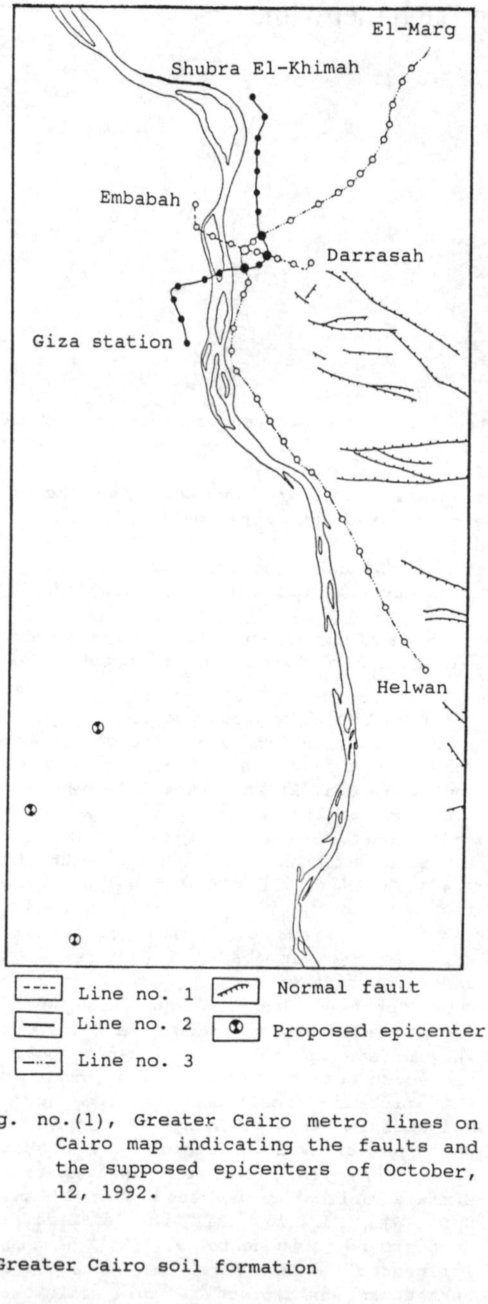

Fig. no.(1), Greater Cairo metro lines on Cairo map indicating the faults and the supposed epicenters of October, 12, 1992.

2 Greater Cairo soil formation

Extensive investigations were done to clarify the formation and properties of soil in which the underground lines exist. The investigation results can be concluded in the following main formation, [Ref. 6],:

1. A surface fill layer of thickness 3 to 10 meters. It was noted that the more over population areas the more thickness of fill layer.
2. The fill layer is followed by a clay or silty clay layer of depth between 2 to 12 meters according to the relative locations with respect to the formations of the deposits of Nile valley.
3. The above layers are resting over dense sand layer which extend downward to the limestone bedrock.
4. The water table is found at depths 1.0 to 2.0 meters from the surface with annual fluctuation of 0.5 meter.
5. No rock layers were observed in the results in spite of the existence of Mokattam mountain at Cairo south east and the Pyramids plateau at west. Fig.(3) illustrates (L2) sectional path and the surrounding soil formation.

3 Magnitude and location of the design earthquake

Accurate prediction of earthquakes parameters is so far, beyond the scientific attain. Extensive investigation were carried out in order to clarify the main three unknowns of expected earthquake. These unknowns are duration time, magnitude, and epicenter location. Moreover, the complete design of any structure has to include calibration of seismic resistivity. Therefore, codes and investigators set up the seismic design criteria for different locations according to two main principles ; the seismic history of the site, and economical considerations. Kebeasy & Maamoun, 1981, [Ref. 10] studied the historical seismic events of Egypt, as shown in table [1], and in accordance with the Egyptian geological formations in order to determine the magnitude and location of the design earthquake of the greater Cairo zone. They concluded that the design earthquake of the greater Cairo zone is of magnitude 6.7 and it will locate in Suze gulf about 130 km. far from Cairo. In another approach, they consider the acceleration of 32 cm/sec^2 (0.032 g) as an adequate value for the design earthquake in site effect. Fig. (4), presents the different approaches for getting the design earthquake. The figure proofs that the design earthquake of magnitude 6.7, located at Suze gulf, is so far, the max seismic hazard that might attack greater Cairo zone.

4 The in site hazards from local soil conditions

The released energy from the design earthquake has to travel about 130 km. from its proposed epicenter at Suez gulf to greater Cairo. The energy suffers during

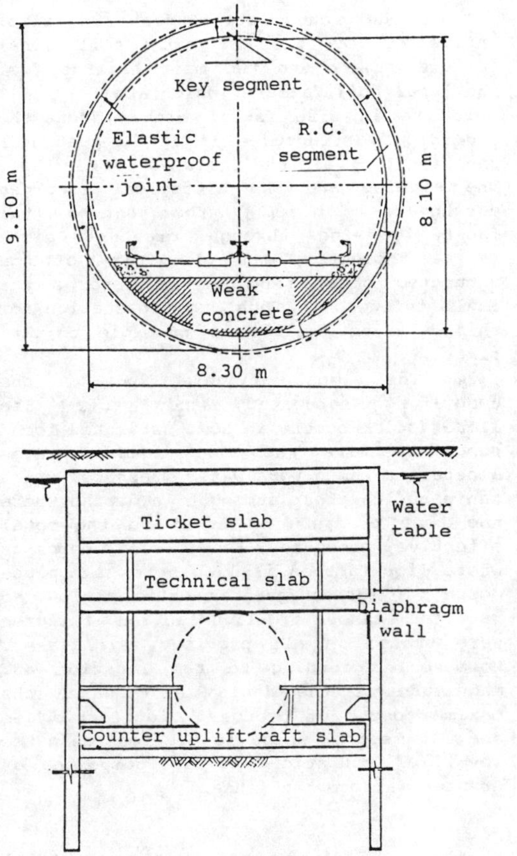

Fig. no.(2), Typical cross sections in board tunnel, and cut & cover 3 level station.

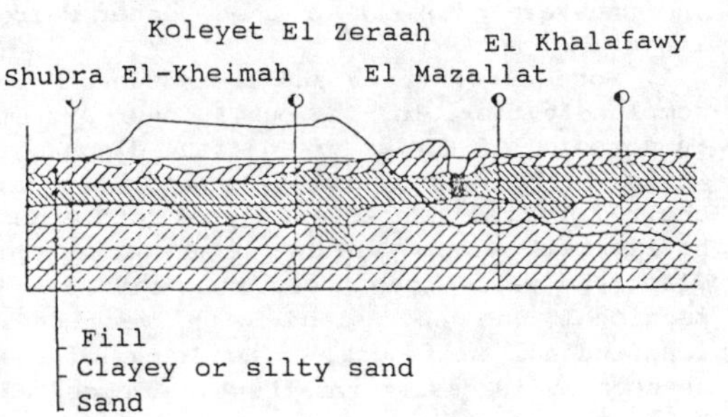

Fig no.(3), Geological profile of line no.2

its travel from geometrical dispersion, dissipation, damping, and absorption. All these factors will be expressed as the attenuation factors. Naturally, attenuation factors, are seriously, affected by the path conditions. The path from Suez gulf to

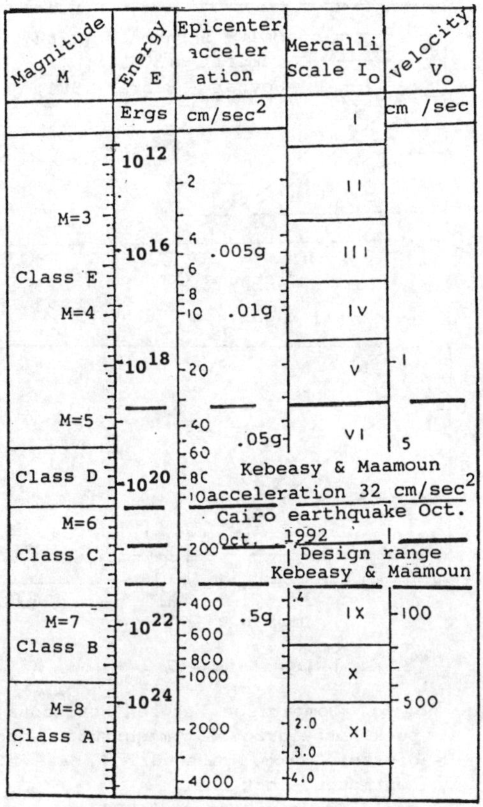

Fig. no.(4), Summary of rough relationship between magnitude energy, epicenteral acceleration, and ground velocity. after : Lomnitz & Rosenbieuth, 1976.

greater Cairo is composed from mixtures of rocky, sandy, and sandy clay soils with different degrees of saturations. That wide difference in the path soil formation, enlarge the attenuation of the propagated waves than that of single soil type. Continuous rock soil is the only soil type of elastic response to elastic waves propagation, [Ref. 4, 13, 14, and 16]. The response of other soil types varies between elastic plastic and completely plastic according to the soil type, plasticity, and degree of saturation. Elastic response allows waves to propagate with minimum attenuations. Consequently, elastic plastic or plastic response impede and weaken the released energy waves as it travel out from the epicenter. Donovan 1982, [Ref. 16] collect the various approaches for determining the in site accelerations due to earthquakes of different magnitudes propagated in continuous rocky soil, Fig. (5). Most of

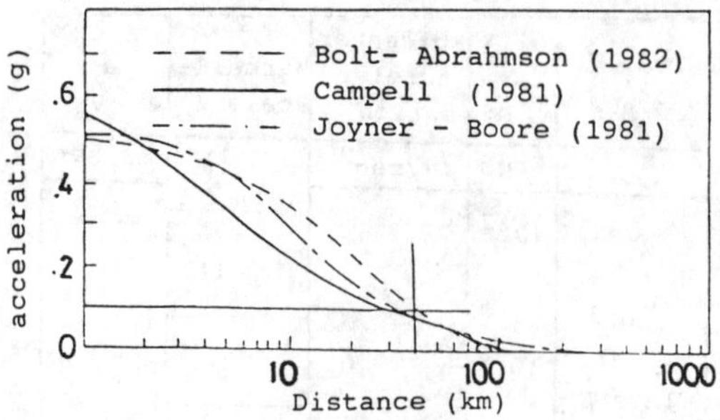

[a] Magnitude = 6.5

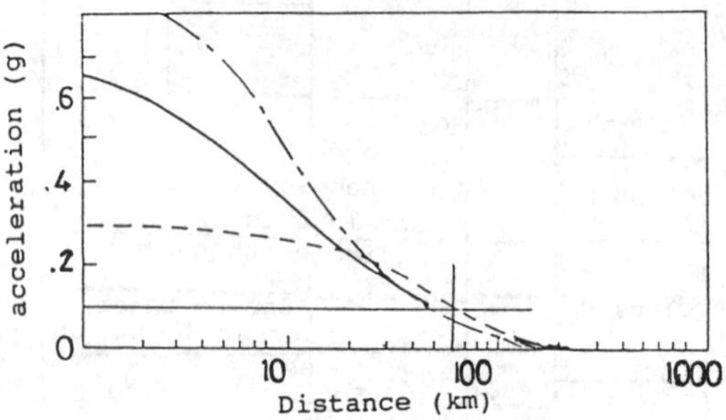

[b] Magnitude = 7.3

Fig. no.(5), Comparison between Horizontal peak acceleration, magnitude, and distance from epicenter. [case of continuous rock soil] modified from Donovan, 1982.

reliable references assign the in site acceleration to represent the hazard level of the site. Fig. (5), indicates that the in site acceleration at 130 km. (Greater Cairo distance from epicenter) for earthquake of magnitude 7.3 (which is much more than the design earthquake), is less than 0.04 g in all the indicated approaches. This result is based on assumption that the 130 km. path form epicenter to Greater Cairo is composed of continuous rock mass, i. e., the result overestimates the in site acceleration value for previously described mixture of soil formation of high attenuation potential. Therefore, it can be said that the maximum design value of the in site horizontal and vertical acceleration gh, gv in Greater Cairo zone is 0.04 g (about 40 cm/sec^2). Consequently, the maximum design value for ground velocity is about, 4.25 cm/sec., and the ground displacement is about 5.0 cm, [Ref. 4, 10, 13, and 16].

The above explained parameters of ground movement may cause local dynamic failure in some special soil formations. The dynamic local failure, increases the waves attenuation if it occurs in the wave path.

The underground structures are also, influenced by the in site soil local failure. There are two main hazards from the local soil failure, which are:

1. The in site fault which affects the underground structures if it is just over the fault, i.e. the fault is still vital. The studies and the historical recorded earthquakes, Table [1], proof that no vital faults passing through greater Cairo, (Fig.1) Moreover, the plasticity of the saturated dense sand layers absorbs the fault influence. Thus, no expectation for this type of failure is probable to take place.

2. The local liquefaction of the underlying saturated sandy soil. The liquefaction occurs in some saturated loose sands deposits when the induced waves generate dynamic pore water pressure, which vanishes its shear strength in depths where the pore pressure is equal to the total effective pressure. Loose sand deposits where liquefaction is liable to take place doesn't exist in greater Cairo region. No recorded cases of liquefaction failures were stated in all previous earthquakes. Moreover, according to the location and magnitude of the design earthquake, the maximum depth of proposed liquefaction is about 4 meters, which is too less than the level of foundations of any component of greater Cairo metro lines.

5. The main assumptions for the structural behavior of the underground parts of Greater Cairo metro lines.

The following assumptions are considered in evaluation of the seismic resistivity of the underground parts of the Greater Cairo lines:

1. Both cut & cover and bored tunnel are completely flexible. The cut & cover system is composed of series of adjacent diaphragm walls with no structural connections in their joints. Moreover, the different horizontal slabs which form with the diaphragm walls the underground space, are sectional one way slabs with very weak longitudinal stiffness. The logic of the assumption is easier for the ring segmented bored tunnel where no structural connection in the joints between the segmented rings.

2. The bored tunnel ring section with its segments is also, completely flexible in the transverse direction.

3. All the additional structures such as accesses, and technical structures are structurally, independent.

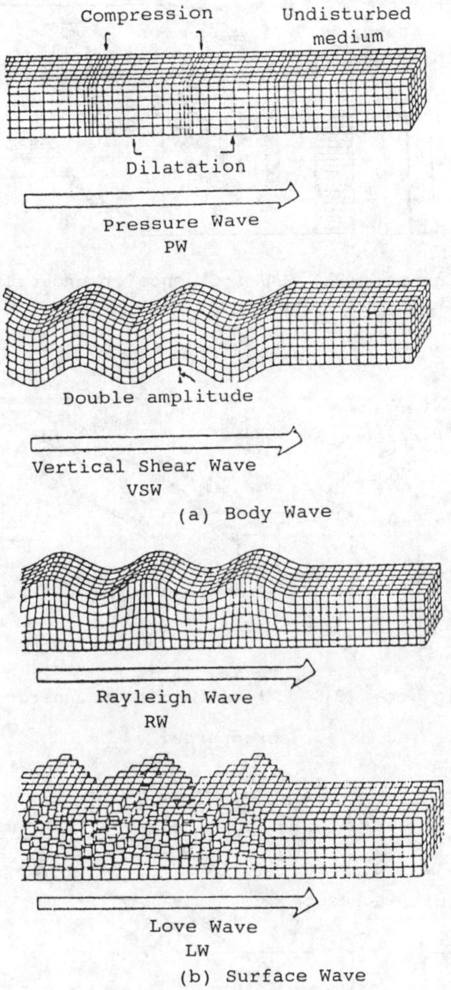

Fig. no.(6), Schematic diagram Showing the different types of waves induced from an earthquake, after St. Jhon & Zahrah 1987.

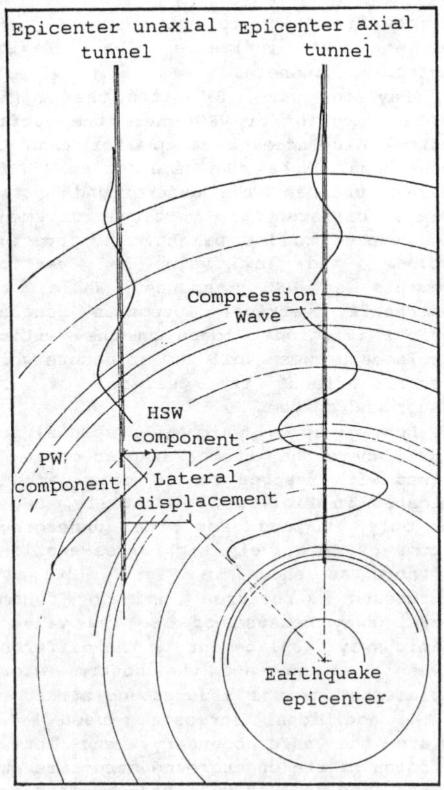

Fig. no. (7), PW propagation and the generated HSW in unaxial longitudinal buried structures.

6. The induced In- Site dynamic loadings

The released energy from the earthquake causes in site wave. The analysis of this wave in its general form is too difficult. In order to facilitate the analysis, the in site wave is resolved into the following known types of propagated waves, Fig. (6):

1. Pressure wave PW, which is usually, accompanied by the horizontal shear wave, HSW, specially for the long extend structures. HSW is the perpendicular component , and PW is the axial component of PW. PW generates axial reciprocating compression and tension in the underground structures, while, HSW oscillates the soil structure system laterally. Fig. (7) explains the difference in effect of the pressure or horizontal shear wave for long extent underground structures. HSW has a serious effect for high rise super structures, but it has no appreciable influence for underground structures.

Tunnels and flexible longitudinal underground structures, overcome HSW by the flexibility of ring connections as explained in Fig.(8). PW is the fastest wave propagated from earthquake. Therefore, it's the first to affect and hit the site of soil structure system.

2. Vertical shear wave, VSW. It is the main wave type. It represents about 2/3 of the total released energy. VSW causes dynamic vertical displacements of soil structure system, which is also, very dangerous to super structures, but it has no global effect for tunnels and underground structures which absorb its effect through its flexible joints. VSW is slower than HSW thus, the time lag between VSW and HSW, depends mainly, on the

distance from site to epicenter, Fig. (9) explains VSW influence for flexible longitudinal tunnels.

3. Rayleigh wave, RW. It's the initial and last portion of VSW where the surface vertical displacement is greater than the deeper ones. RW has the same act as VSW for super structures. The underground systems undergo differential vertical movements with respect to its height. The structure responses the low value of vertical movements as VSH response, While, the differential vertical movements generate vertical frictional force on the vertical interfaces between soil and structure which doesn't affect the safety of the underground system.

4. Love wave, LW. which is a special form of HSW, where the lateral displacements are reduced with respect to the soil depth as indicated in Fig.(10). Generally, LW, is the only, hazard for the underground systems of relatively high cross-sectional stiffness as shown in Fig. (10). The underground structures undergo dynamic lateral sway because of LW. The value of dynamic sway displacement is the difference between the top, and the bottom lateral displacement of the underground structure. If the additional stresses caused by LW violate the safe boundary, the lateral stiffness of the underground structure must be increased to suit that case of loading.

Fortunately, each of the above mentioned waves has its own velocity. Thus, it doesn't hit systems simultaneously as long as that systems are reasonably far from the earthquake epicenter. Obviously, the power of each wave depends mainly, on the earthquake magnitude, the distance between the epicenter and the system site, and the soil properties in this distance. RW, and LW, are damped in granular soil at relatively, short distances from epicenter. Therefore, some of these waves are liable to hit the soil- underground structure system.

The analysis of LW effect on Greater Cairo metro lines, leads to the following results :

4.1 Considering the maximum lateral displacement of 5.0 cm.,

4.2 Assuming a conservative shallow depth of LW @ 2000. m.

4.3 Hence, for underground structure height of @ 20. m., the total lateral sway value is 0.05 cm. [1/2 mm.] considering zero soil structure interaction.

Therefore, the design value of the LW lateral sway has no significant effect on underground structure global safety.

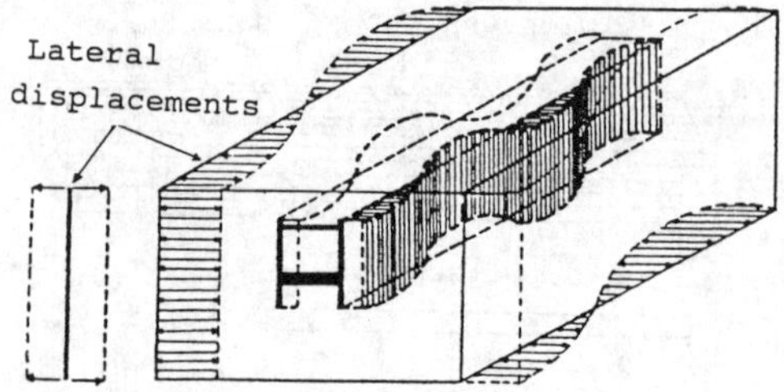

Fig. no. (8), HSW influence on underground structures.

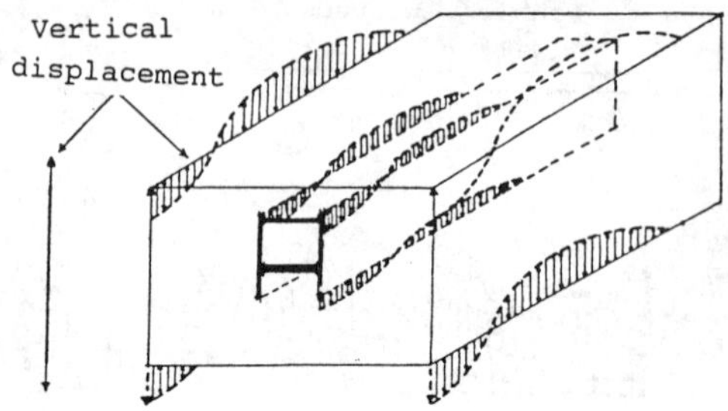

Fig. no. (9), VSW influence on underground structures.

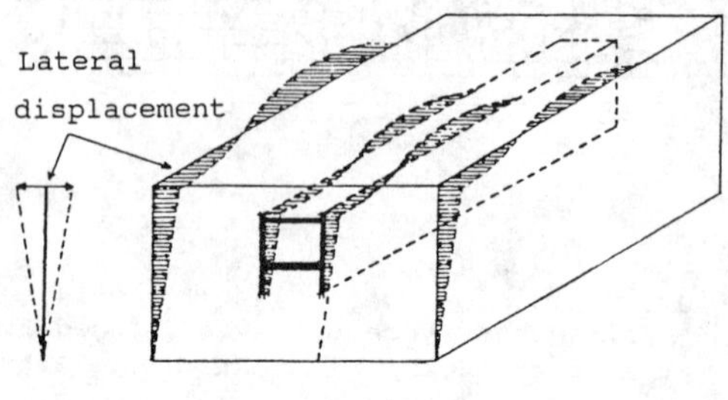

Fig. no. (10), LW influence on underground structures.

7. Approaches to estimate the seismic resistivity of the soil-structure system

The estimation of seismic dynamic response of the soil-structure system implies a previous determination of full informations about the in-site dynamic loading, and the local soil-structure interaction. Both dynamic and pseudo static approaches are used in that determination.

Dynamic approach depends upon the principle of dividing the in site time dependent force caused by the earthquake into damped part, elastic or stiffness part, and the inertia part. Thus, the analysis can expressed in the basic fundamental following equation : [Ref. 4, 12, 13, 14, and 16]

$F(t) = F_i + F_d + F_s$ where:

$F(t)$: The total applied dynamic seismic force,
F_i : The inertia force, $F_i = m u''$,
F_d : The damping force, $F_d = c u'$,
F_s : the elastic force, $F_s = k u$,
m : mass,
c : critical damping coefficient,
k : system stiffness,
u : dynamic movement displacement,
u' : dynamic velocity, $u' = d u / dt$
u'' : dynamic acceleration,
$u'' = d u' / dt$.

The dynamic response analysis suffers from a lot of difficulties. It can be summarized in the following items:

1. It necessitates the accurate determination of the above mentioned parameters for both the soil and the underground structure. Most of the dynamic parameters are calculated from empirical formulas, which significantly, vary according to the approach of each formula derivation.

2. The determination of the responded soil mass in the dynamic equation is too difficult specially for different soil formations, and degrees of saturation. The published literatures in this topic are mainly, hypotheses.

3. The dynamic approach implies, practically, the assignment of parameters of certain soil type allover the analysis, which isn't simulate the in site soil formation.

4. The dynamic analysis cannot be performed for the random applied dynamic force F(t). It has to be modified to a defined mathematical form such as the simplified general form: [same refs.]

$F(t) = A \sin w t + B \cos w t$ where:

w : the earthquake induced frequency.
A & B : constants depending on the earthquake input parameters.

This form overestimates the earthquake magnitude because it envelop the random vibration of the earthquake. Also, F(t) is highly affected by the path from the epicenter to the analyzed structure. The propagated waves are strongly attenuated due to refraction because of faults, zones of liquefaction in the path of the waves. The %age of attenuation isn't the same for HW, SW, RW, and LW. Moreover, the consideration of the elasticity or plasticity of the soil in the path influence the in-site F(t) characteristics.

Therefore, it can be easily said that the ambiguities involve dynamic analysis lose the practical dependability of its output results.

The pseudo static approach is so far, the only available alternative for determining the seismic resistivity of the underground structures. It is easier and faster than the dynamic analysis. Although it also, based on empirical formulas and coefficients, the practical design dependability of its output results are more conservative than that of dynamic analysis. Most of international codes recommend pseudo static procedure in evaluation of seismic resistivity. The dynamic load in pseudo static is taken as the product of the vibrated mass by the specified fraction of gravitational acceleration "g_h or g_v ". That fraction is a site property based on the location and magnitude of the design earthquake, Fig.(5). Applying any of the pseudo static techniques (like Mononobe & Okabe equation for retaining structures or Westergaard hydrodynamic lateral water pressure ....etc), the equivalent static load is obtained. Generally, pseudo static approach under estimates the seismic resistivity of the analyzed systems.

8. Conclusions and Recommendations

The presented surveying study of seismic hazards of the Greater Cairo underground metro lines, can be summarized in the following main items ;

1. The underground metro lines, are generally, located in saturated dense sandy soil.

2. The maximum design earthquake magnitude is 6.7, and located at Suze gulf [130 km. from Greater Cairo].

3. The expected design value of horizontal & vertical site acceleration due to the design earthquake is less than 0.04 g.

4. Two main seismic hazards are expected to hit the Greater Cairo underground metro lines:

4.1 The wave propagated from the design earthquake epicenter.

The study has proved that the known five wave types, PW, HSW, VSW, RW, and LW have no effect on the under ground structural safety. The waves influence are absorbed through the flexible joints of the underground structure.

4.2 The probable local soil dynamic failure because of faulting, and liquefaction, have no possibility to occur because of the nature of soil formation and the foundation depth of the underground structure.

5. The pseudo static approach for seismic resistivity is so far, the more dependable technique for design purposes.

6. The Greater Cairo underground metro lines are the safest structures with respect to the proposed design earthquake of 6.7 magnitude and located at Suze gulf. Its seismic resistivity need no additional structural stiffners or reinforcements than that of primary cases of loadings.

7. The flexibility of connections between sectional unit elements must be checked for the transverse dynamic response specially, the water tightness of these connections.

Table no. 1, Catalogue of historical large earthquakes.
(Abbreviated from Kebeasy & Maamoun, 1981.)

Date	Region	I_o	M	Remarks
2200 BC	Tell Basta	VII	5.8	Deep Fissures
24-20BC	Alex. off sh.	IV	4.0	strong waves
320 AC	Alex. off sh.	VII	5.9	Many houses destructed
553	Alex. off sh.	V	4.8	
4. 796	SE Med. sea	VI	5.2	Felt at dif. locations of Egypt, damage light houses in Alex.
859	Belbeis	VI	5.5	Felt in Delta & Alex.
5.1111	East Cairo	VIII	5.8	Destruction in Rehachope temp
8.1303	Fayoum	VIII	6.5	Sever earthqu. damage from Cairo to Alex.
1326	Alex. off sh.	V	4.8	Light houses shocked
3.1687	Alex. off sh.	VI	5.2	Alex vibr. @ 10-12 days
10.1698	Rosetta	VI	5.5	Nile valley
9.1754	Tanata	VIII	6.0	2/3 of Cairo bldgs.damaged& thousands were killed
8.1847	Fayoum 29.5o N 30.5o E	VIII	6.2	3042 bldgs. destr.& 85 killed & 62 injured.
6.1870	East Medit. 32oN 30oE	VIII	6.5	Sever, felt in vast areas.
7.1879	Alex. off sh.	V	4.8	Strongly felt.
12.1908	Alex. off sh.	V	4.8	Felt earthq.
12 Oct. 1992	SW Cairo	VII	5.9	damage in old bldgs Greater Cairo zones.

I_o = maximum intensity at the epicenter
M = body wave magnitude

ACKNOWLEDGMENT

The authors express their deep gratitude for the head and engineers of the NAT's, studies departments for the voluntary dedicated efforts in presenting this work.

REFERENCES

1. Abdel Salam, M.E., 1991.
TUNNELLING IN CONGESTED CITIES, CAIRO METRO NETWORK.
International Symposium, Egyptian Society of Engineers & National Authority for Tunnels, NAT, Cairo, Egypt.
2. Abdel Salam, M.E., 1992.
TUNNELLING THE CAIRO METRO LINE NO.2
Tunnelling and Underground Space Tech. Vol.7, No. 4, pp. 407 - 413.
3. Douglas, W.s. & Warshaw,R., 1971.
DESIGN OF SEISMIC JOINT FOR SAN FRANCISCO BAY TUNNEL.
J. Str. Eng. ASCE, Vol.97, No.ST4, pp. 1129-1141
4. Dowrick, D.J., 1987.
EARTHQUAKE RESISTANT DESIGN.
Jhon Wiley & Sons, U.K.
5. Egyptian Geological Survey and Mining, EGSMA.
EGSMA bulletin No. 3, October 1992.
6. Egyptian National Authority for Tunnels, NAT.,(L1), (L2), documents., 1982-1993 .
7. Egyptian Society for Earthquake Engineering, ESEE 1988.
REGULATIONS FOR EARTHQUAKE- RESISTANCE DESIGN OF BUILDINGS IN EGYPT.
ESEE publications.
8. Indian Standard, 1986.
CRITERIA FOR EARTHQUAKE RESISTANT DESIGN OF STRUCTURES.
Indian Standard Institution, New Delhi 110002.
9. Japanese Aseismic Code. 1991.
STANDARDS FOR ASEISMIC CIVIL ENGINEERING CONSTRUCTIONS. PP. 23-1 , 23-87.
10. Kebeasy, R.M. & Maamoun, M. 1981.
SEISMICITY AND EARTHQUAKE RISK OF THE PROPOSED SITE OF SHOUBRA AL-KHAIMAH ELECTRIC POWER STATION.
Bulletin of IISEE, Vol. 19 Special Issue, pp21-33.
11. McLean, C.D. & Gribble, C.D., 1990.
GEOLOGY FOR CIVIL ENGINEERS.
ELBS, Unwin Hyman London.
12. Park,Y.Ji. & Ang, A.H.S., 1985.
MECHANISTIC SEISMIC DAMAGE MODEL FOR REINFORCED CONCRETE.
J. Struct. Eng. ASCE Vol.111, No. 4, STRUCTURES.

Tunnelling and Underground Space Tech. pp. 722-739.
13. Richart, F.E.Jr., 1975. FOUNDATION VIBRATIONS. Proc. Foundation Engineering Handbook, Edited by Winterkorn, H.F. & Fang, H.Y., Von Nostrand Reinhold Co., N.Y., pp. 673-699
14. Seed, H.B., 1975. EARTHQUAKE EFFECTS ON SOIL FOUNDATION SYSTEMS. Proc. Foundation Engineering Handbook, Edited by Winterkorn, H.F. & Fang, H.Y., Von Nostrand Reinhold Co., N.Y., pp. 700-732
14. Steedman, R.S., & Zeng, X., 1992. THE SEISMIC RESPONSE OF WATERFRONT RETAINING WALLS. Proceeding of Conference " Design and Performance of Retaining Structures" Ithaca Sec. ASCE, No. 25., pp. 873-887
15. St Jhon & Zahrah, T.F., 1987. ASEISMIC DESIGN OF UNDERGROUND Vol.2 pp.165-197.
16. Uang,C.M. & Bertero,V.V., 1991. UBC, SEISMIC SERVICEABILITY REGULATIONS: CRITICAL REVIEW.

Tunnelling and Ground Conditions, Abdel Salam (ed.) © 1994 Balkema, Rotterdam, ISBN 90 5410 363 9

Analysis of ground water seepage towards tunnels

M. Atwa & E. Leca
Laboratoire Central des Ponts et Chaussées, France

Abstract : Tunneling in water bearing soils affects water conditions in the surrounding ground and tends to produce seepage towards the opening. This may result in water inflows into the tunnel, face instability due to high hydraulic gradients and long term settlements induced by the dissipation of pore water pressures. The purpose of the work described in this paper is to analyze seepage towards tunnels under steady conditions by means of two-dimensionnal and three-dimensionnal finite element models. The influence of ground conditions, such as soil anisotropy and layered ground profiles on the water flows are analyzed for permeable tunnels and lined tunnels of finite permeabilities. The paper also considers the three-dimensional flow pattern towards the face of a tunnel under construction, with emphasis on the influence of soil anisotropy as well as the presence of a grouted soil mass at the face, on seepage towards the tunnel.

1 INTRODUCTION

The construction of a tunnel in water-bearing soils modifies the hydraulic equilibrium of the surrounding ground, and tends to generate seepage towards the opening. Moreover, because of defects in the water-tightness of the tunnel liner, and/or the presence of pervious materials around the liner, tunnels are often found to act as drains after they have been constructed, therefore producing long term seepage in the surrounding ground mass.

Seepage towards a tunnel may result in several detrimental consequences, including :

- water inflows into the tunnel, during and after construction,
- face instability, due to high hydraulic gradients,
- long term surface settlements, when tunnels are excavated in or underneath compressible soil layers.

The problem of seepage towards tunnels have been addressed by several authors. An overview of previous works is provided in Leca et al. (1993). Analytical solutions have been derived, based on Dupuit's asumptions by Bear (1979), Averjanov (1949), Romanova (1956), Romanov (1959), to compute the water flow, and, in some cases the water draw-down produced by seepage towards the walls of a tunnel cross-section, for unconfined aquifers under steady conditions. Solutions for estimating water flows towards horizontal drains have also been proposed, based on experimental work (Keller and Robinson, 1959) or analytical derivations (Polubarinova-Kochina, 1962).

Additional numerical work has been accomplished by Chishaki and Yang (1977), Chishaki et al. (1978), Chishaki (1984), Sato (1983), Sato and Iizawa (1983), and analytical solutions were derived by Rat (1968 and 1973), as part of the hydrological studies for the Channel tunnel project. Most of the more recent works also included consideration for transient flows towards the tunnel. In some cases, comparisons were also made with experimental results obtained from model tests (Hele-Shaw models). The case of confined aquifers have also been studied by Chishaki (1984), Polubarinova-Kochina (1962) and Rat (1968 and 1973).

Leca et al. (1993) analyzed these solutions and compared their results with those obtained from finite element calculations. These methods can be considered to provide adequate background to analyze seepage towards tunnels, but are mostly limited to idealized conditions, and usually cannot address the three-dimensional flow of water towards a tunnel under construction.

In the present paper, the capabilities of the finite element method to analyze complicated problems of interest for tunnel construction are considered, including the presence of heterogeneous soil conditions, soil anisotropy, as well as the influence of liner permeability and drainage conditions at the face. Only steady flow conditions are considered. Additional background on the transient analysis of water flow towards tunnels is provided in Atwa et al. (1993).

2 COMPUTATIONAL METHOD

The Finite Element program CESAR-LCPC was used in this study to analyze the influence of various aspects of the geological and construction conditions on seepage towards shallow tunnels constructed in water-bearing soils. This program has the capability of analyzing the main aspects of geotechnical and structural design. In particular the subroutine NSAT was designed to compute seepage through saturated and unsaturated porous media. The input of data and analysis of results are processed through a pre- and post-processor. Both two- and three-dimensional problems can be analyzed.

The geometry considered in this study is described in Figure 1. The tunnel, of circular cross-section, is located at a depth to axis of H = 10m. Its diameter is equal to D = 5m. The impervious bedrock is located 10m below the tunnel axis. The natural water level (out of the tunnel area) is located at the ground surface. The boundaries of the model are set at a distance to center-line of 50m. Most of the computations were carried out for the tunnel cross-section described in Figure 1, which is adequate to analyze seepage towards the walls of a constructed tunnel under working conditions. The following aspects were considered :

- influence of soil anisotropy,
- influence of heterogeneous soil conditions,
- influence of the tunnel liner's permeability.

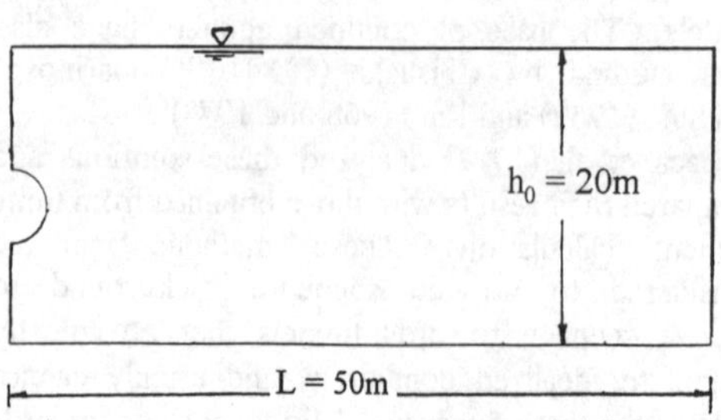

Fig.1 Two-dimensional model

Additional computations were also carried out in three-dimensional conditions (Figure 2), to study seepage towards the face of a tunnel under construction. In this latter case, the boundaries of the model were set 50 m away from the tunnel face along the longitudinal direction, and at 80 m from the tunnel axis in the lateral direction.

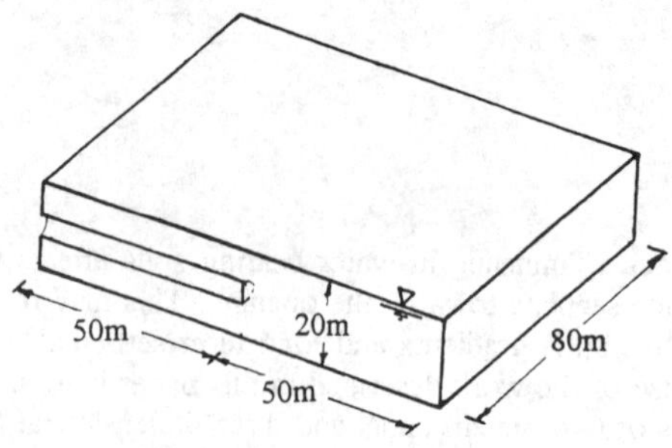

Fig.2 Three-dimensional model

3 TWO-DIMENSIONAL ANALYSIS

Both unconfined and confined hydrological conditions were analyzed, with more emphasis put on the former. The soil's permeability was typically chosen equal to $k = 10^{-4}$ m/s in the homogeneous case, which corresponds to a relatively pervious material.

Typical results obtained from the finite element analysis are displayed in Figure 3 for a pervious tunnel in a homogeneous unconfined aquifer : distribution of hydraulic heads (Figure 3a), distribution of water pressures (Figure 3b) and hydraulic gradients (Figure 3c). The head lines drawn in Figure 3a indicate a strong concentration of seepage at the tunnel invert, where hydraulic gradients in excess of 1 develop (Figure 3c). Maximum draw-down of the water table is equal to 13 m and the water flow towards the tunnel is equal to $Q = 0.36\ 10^{-3}\ m^3/s.m$.

The network of head lines obtained for a confined aquifer (Figure 4), is significantly different from that for an unconfined aquifer, with the head lines tending to circle around the tunnel, whereas they are almost vertical at a distance of one to two diameters from the tunnel springline in the former case. The water flow towards the tunnel is equal to $Q = 1.19\ 10^{-3}$ $m^3/s.m$ and the maximum hydraulic gradient to $i_{max} = 3.86$.

For the following part of the paper, only unconfined aquifers will be considered, i.e. seepage towards the tunnel may result in a draw-down of the water table surface. A similar analysis could be carried out for the confined aquifer case (Figure 4), which better matches the conditions for an underwater tunnel.

Part of our study was aimed at analyzing how the seepage conditions shown in Figure 3 would be altered by the presence of anisotropic and heterogeneous soils around a permeable tunnel, as well as the installation of a liner of finite permeability. The results of this study are summarized in Table 1, where the computed water flows towards the tunnel, maximum water draw-downs and maximum hydraulic gradients are listed for each computed case.

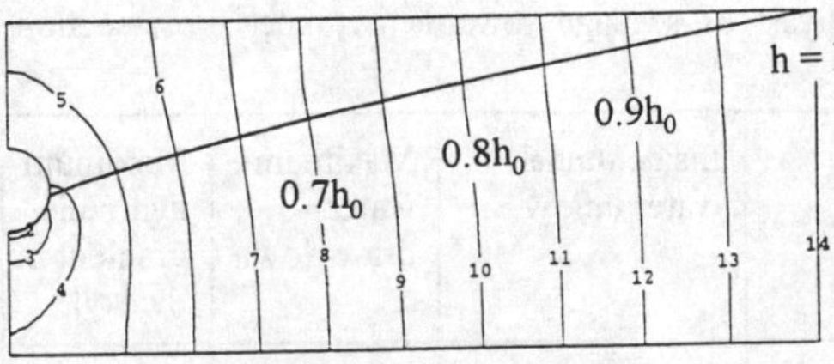

(a) Hydraulic head distribution

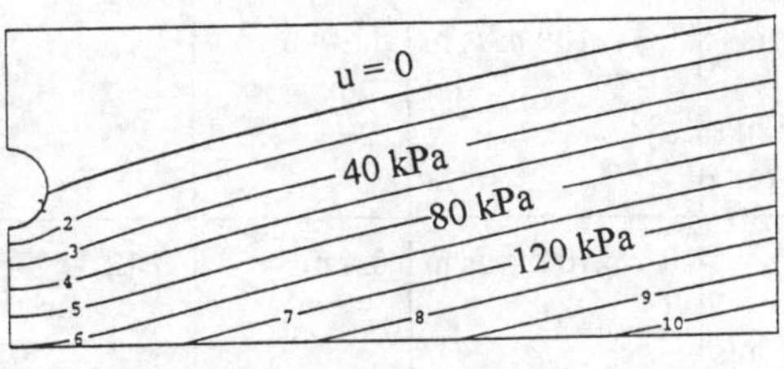

(b) Water pressure distribution

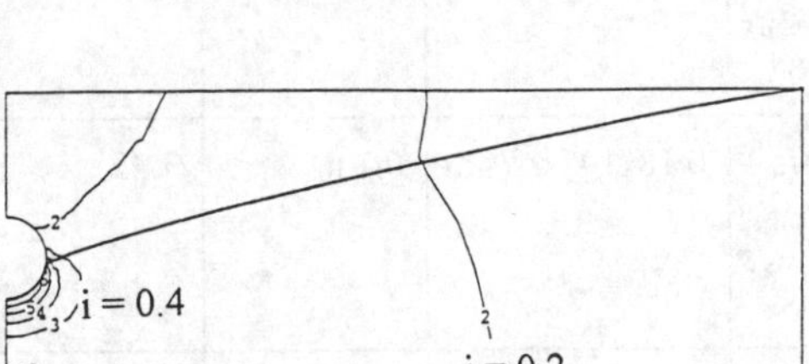

(c) Hydraulic gradients distribution

Fig.3 Main features of unconfined flow towards a permeable tunnel

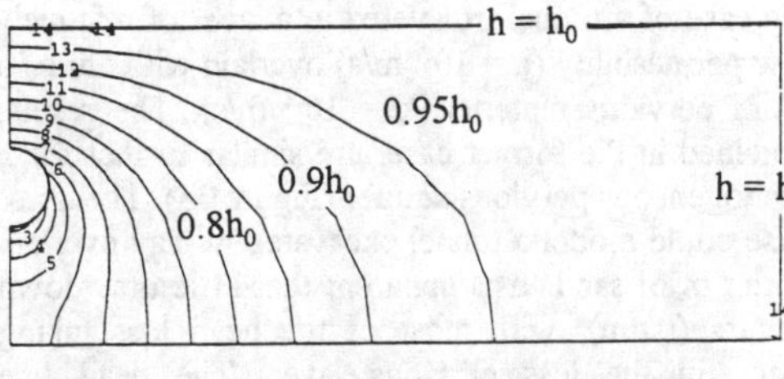

Fig.4 Hydraulic head distribution for confined flow towards a permeable tunnel

3.1 Influence of soil anisotropy

The influence of the soil's anisotropy was analyzed for a ratio of 10 to 1 between the horizontal permeability, k_x and the vertical permeability, k_y, with the average equivalent permeability being kept equal to that used for the study with homogeneous soil conditions : $k_{eq} = \sqrt{k_x . k_y} = 10^{-4}\,m/s$.

The network of head lines obtained for this case is shown in Figure 5 : the head lines are more elongated along the horizontal direction in the tunnel vicinity than in the isotropic case and less water draw-down is computed (8.2 m, to be compared to 13.0 m obtained with isotropic conditions). The water flow ($Q = 0.99\ 10^{-3}\ m^3/s.m$) and the maximum hydraulic gradient ($i_{max} = 3.89$) are much higher than for the isotropic case (Table 1).

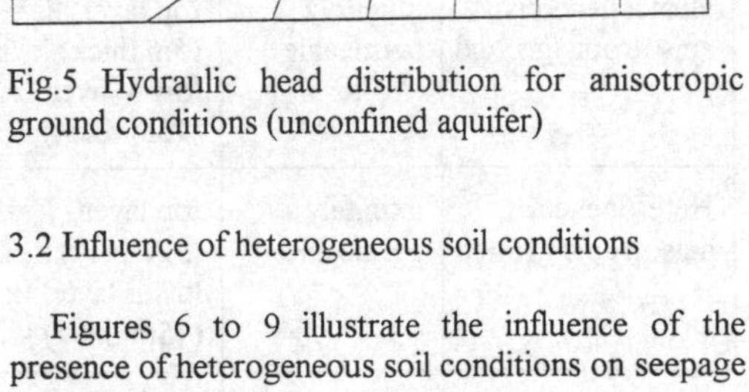

Fig.5 Hydraulic head distribution for anisotropic ground conditions (unconfined aquifer)

3.2 Influence of heterogeneous soil conditions

Figures 6 to 9 illustrate the influence of the presence of heterogeneous soil conditions on seepage towards the tunnel.

Figure 6 corresponds to the case of a tunnel excavated in a layer of relatively high permeability (k = 10^{-4} m/s) overlain with a 5 m thick more pervious layer (of permeability k = 10^{-7} m/s), and Figure 7 to the case of a tunnel excavated in a layer of relatively low permeability (k = 10^{-7} m/s) overlain with 5m of a more pervious material (k = 10^{-4} m/s). The results obtained in the former case, are similar to that for a homogeneous pervious aquifer (Figure 3a). The latter case could model a tunnel excavated in clay overlain with 5m of sand. It is apparent that little drawdown occurs (0.4m), with most of the head loss taking place in the less pervious layer. The head line network displayed in Figure 7 resembles that obtained for a confined aquifer (Figure 4).

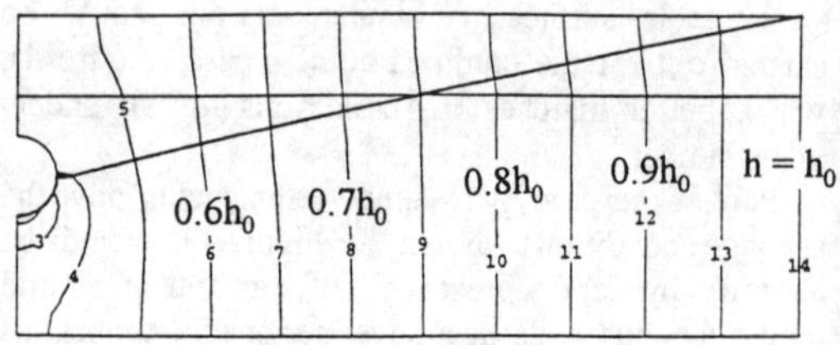

Fig.6 Hydraulic head distribution for layered isotropic ground conditions - tunnel in the higher permeability layer

Table 1 : Results obtained for the two-dimensional analysis of seepage towards a tunnel cross-section, (unconfined aquifer)

Soil configuration	Tunnel liner permeability	Permeability, k	Inside tunnel water inflow	Maximum water draw-down	Maximum hydraulic gradient in the soil
Homogeneous, isotropic ground	infinitely permeable	k = 10^{-4} m/s	0.36×10^{-3} m³/s.m	13.0 m	1.15
Homogeneous, anisotropic ground	infinitely permeable	$k_{eq} = 10^{-4}$ m/s $k_x/k_y = 10$	0.99×10^{-3} m³/s.m	8.2 m	3.89
Heterogeneous, isotropic ground	infinitely permeable	top layer : k = 10^{-7} m/s (5m thick) : lower layer : k = 10^{-4} m/s (15m thick)	0.31×10^{-3} m³/s.m	11.0 m	1.05
Heterogeneous, isotropic ground	infinitely permeable	top layer k = 10^{-4} m/s (5m thick) : lower layer : k = 10^{-7} m/s (15m thick)	0.16×10^{-5} m³/s.m	0.4 m	3.12
Heterogeneous, anisotropic ground	infinitely permeable	top layer : k = 10^{-7} m/s (5m thick) $k_x/k_y = 10$ lower layer : k = 10^{-4} m/s (15m thick) $k_x/k_y = 10$	0.88×10^{-3} m³/s.m	8.8 m	2.85
Heterogeneous, anisotropic ground	infinitely permeable	top layer : k = 10^{-4} m/s (5m thick) $k_x/k_y = 10$ lower layer : k = 10^{-7} m/s (15m thick) $k_x/k_y = 10$	0.18×10^{-5} m³/s.m	0.1 m	3.42
Homogeneous, isotropic ground	$k_l = 2.\ 10^{-9}$ m/s	k = 10^{-7} m/s	0.18×10^{-6} m³/s.m	4.5 m	0.25
Homogeneous, anisotropic ground	$k_l = 2.\ 10^{-9}$ m/s	$k_{eq} = 10^{-7}$ m/s $k_x/k_y = 10$	0.26×10^{-6} m³/s.m	1.4 m	0.60

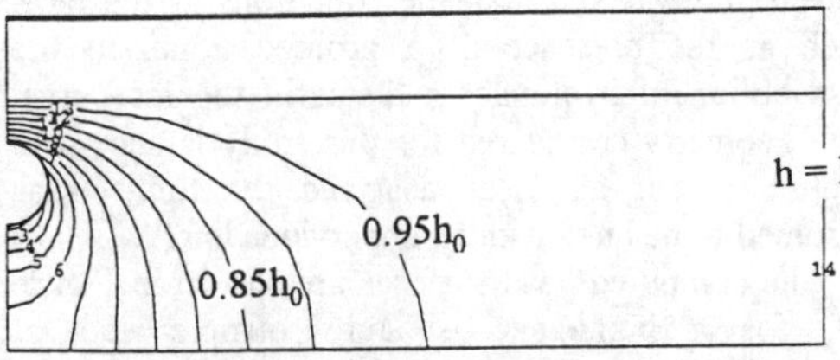

Fig.7 Hydraulic head distribution for layered isotropic ground conditions - tunnel in the lower permeability layer

Figures 8 and 9 show the results obtained with combined effects of soil anisotropy and heterogeneity, which is usually the case of natural ground conditions. Both computations were carried out for two layers of equivalent permeabilities equal to 10^{-4} m/s and 10^{-7} m/s respectively, with the ratio of the horizontal to the vertical permeabilities being equal to 10 for both layers. In comparison with the isotropic case, it is found that higher water flows and hydraulic gradients are computed when the tunnel is excavated in the layer of lower permeability, whereas the results are almost unchanged when the tunnel is excavated in the more pervious layer.

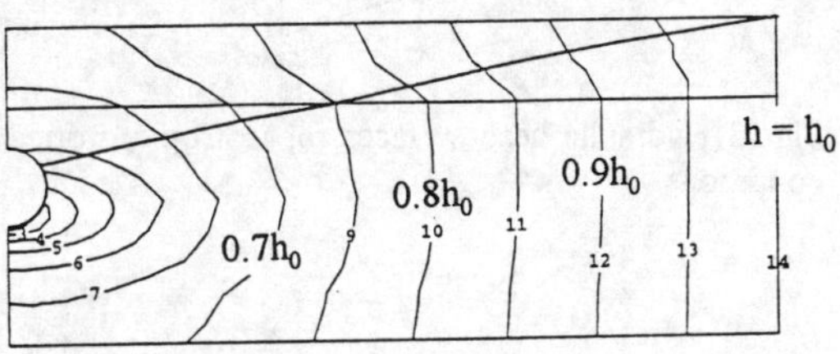

Fig.8 Hydraulic head distribution for layered anisotropic ground conditions - tunnel in the higher permeability layer, (k_x/k_y = 10)

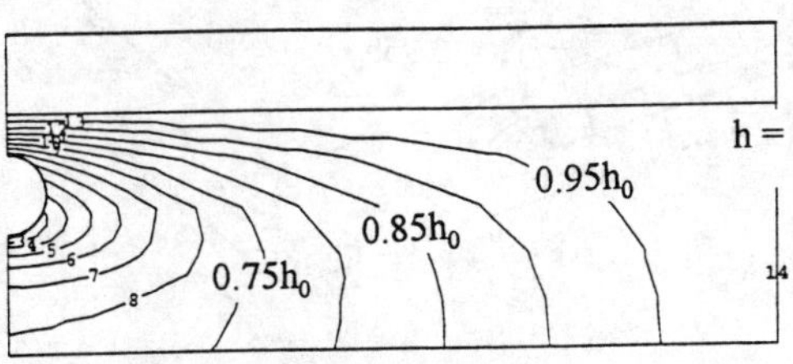

fig.9 Hydraulic head distribution for layered anisotropic ground conditions - tunnel in the lower permeability layer, (k_x/k_y = 10)

3.3 Influence of the liner's permeability

A preliminary estimate of the seepage pattern towards a tunnel can be obtained by assuming that the tunnel is completely impervious (tunnel liner permeability is significantly small in comparison to that of the soil) or that the tunnel is completely pervious (unlined tunnel or pervious lining system). Site conditions usually fall between these two extreme cases. In the present study, we tried to analyze how seepage conditions are modified when a finite value of the liner's permeability is accounted for.

The results obtained with a lined tunnel are shown in Figure 10, for isotropic soil conditions and Figure 11, for anisotropic soil conditions. In both cases the permeability (or equivalent permeability) of the soil was chosen equal to 10^{-4} m/s, with the ratio of the liner's permeability to the soil's permeability being equal to 1/50. The lined tunnel case results in less seepage and water draw-down and lower hydraulic gradients in the soil mass, in comparison to the unlined tunnel case, with most of the head loss taking place through the liner.

It is also of interest to compare the water pressures obtained with a permeable liner to that corresponding to an impervious tunnel. In this latter case, the water pressures around the tunnel are equal to the hydrostatic values, since no seepage towards the tunnel occurs.

The distribution of water pressures acting on the tunnel wall, obtained for both cases are plotted in Figure 12. Lower water pressures are computed for the permeable liner than for the impervious tunnel ; however, this latter case results in a more uniform distribution of pressures acting on the liner.

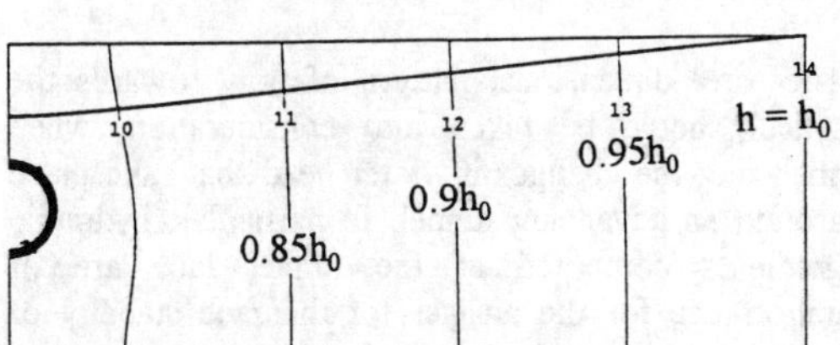

Fig.10 Hydraulic head distribution for a lined tunnel of finite permeability ($k_l = 2.10^{-9}$ m/s)- homogeneous, isotropic ground conditions

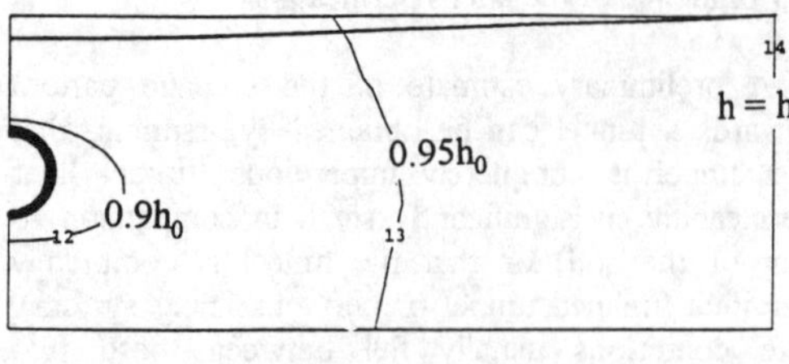

Fig.11 Hydraulic head distribution for a lined tunnel of finite permeability (k_l = 2.10-9 m/s)- homogeneous, anisotropic ground conditions

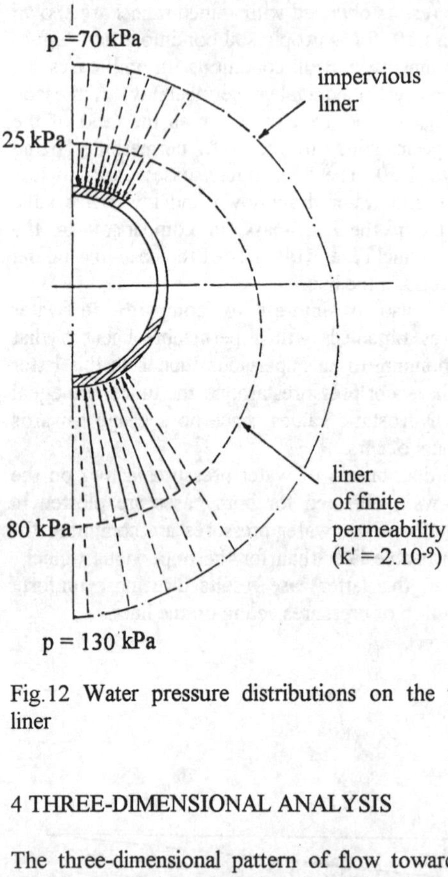

Fig.12 Water pressure distributions on the tunnel liner

4 THREE-DIMENSIONAL ANALYSIS

The three-dimensional pattern of flow towards the tunnel should be taken into consideration, when analyzing the changes in water head that take place around an advancing tunnel. In particular, hydraulic gradients computed at the tunnel face are of importance for the analysis of the face stability of tunnels constructed in water-bearing soils. Some of the benefits of three-dimensional finite element computations of seepage towards a tunnel face have been described in Leca et al. (1993). The present study is aimed at analyzing the influence of anisotopic soils and drainage conditions at the face, such as the presence of a grouted zone, in the distribution of hydraulic gradients in the face area. The geometry considered for this study is shown in Figure 2. For all cases analyzed the tunnel was assumed to be lined with an impervious liner.

The computed head surfaces are shown in Figure 13, for a homogeneous and isotropic soil of permeability k = 10^{-4} m/s, and in Figure 14, for the anisotropic soil conditions considered in section 3 (k_x = 3.16 10^{-4} m/s, k_y = 0.32 10^{-4} m/s). In the latter case, the head surfaces are more elongated along the tunnel axis and of more limited extent in the lateral direction than for homogeneous soil conditions.

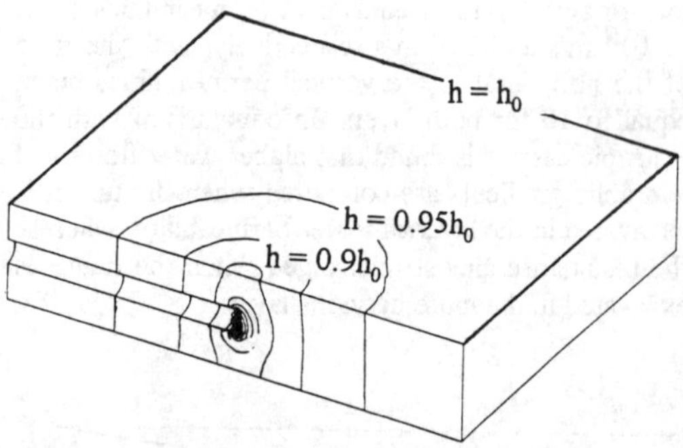

Fig.13 Hydraulic head surfaces for isotropic ground conditions

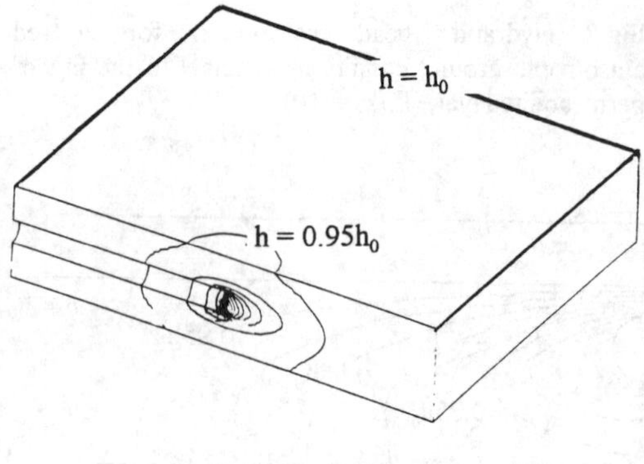

Fig.14 Hydraulic head surfaces for anisotropic ground conditions

The hydraulic gradients in a tunnel cross-section cut through the tunnel face are shown in Figure 15, for the isotropic case, and in Figure 16, for the anisotropic case. It is apparent that higher gradient values are concentrated in the lower edges of the tunnel face in the latter case.

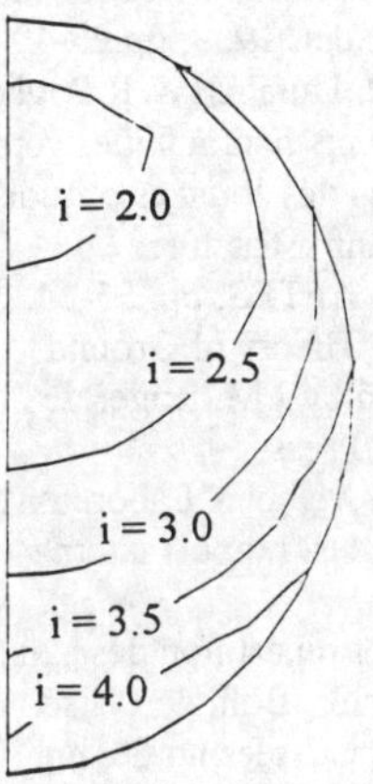

Fig.15 Distribution of hydraulic gradients on the tunnel face, for isotropic ground conditions

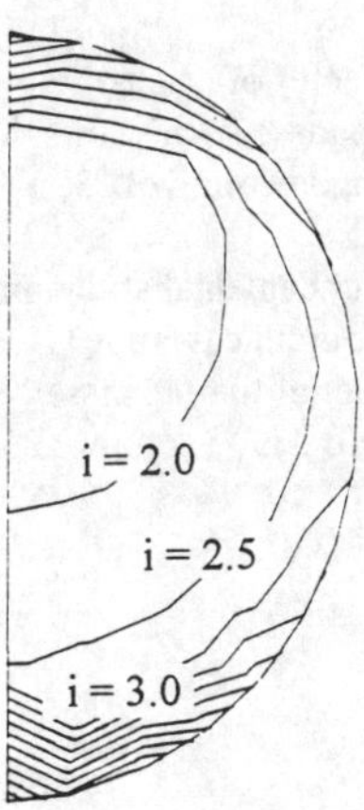

Fig.16 Distribution of hydraulic gradients on the tunnel face, for anisotropic ground conditions

The construction of tunnels in water-bearing pervious soils may result in high values of water flows into the tunnel. To overcome this problem, grouting can be used to create a zone of less pervious ground in front of the tunnel face. This was studied, as part of the present work, by considering a 5m long cylinder of grouted material in front of the tunnel face. The permeability of the grouted material was chosen equal to $k' = 5.10^{-6}$ m/s. The computed head surfaces and hydraulic gradients are shown in Figures 17 and 18. In comparison to the case of an ungrouted front, it is found that the head losses are more concentrated in the face area, and that higher hydraulic gradients develop in the lower part of the tunnel face ; in this latter case, however, the high hydraulic gradients occur in an area which should be strenghtened by the grouting works.

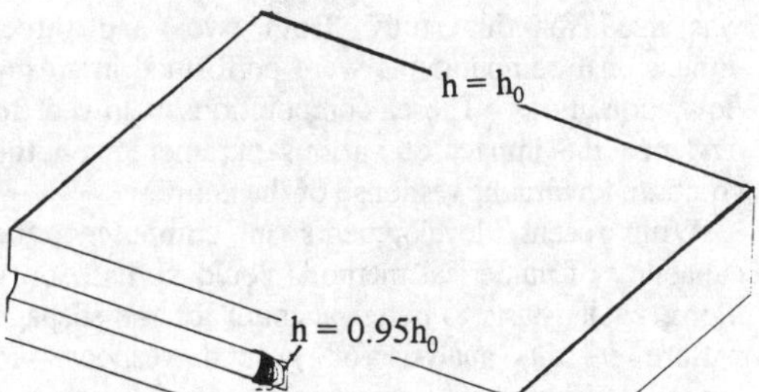

Fig.17 Hydraulic head distribution for grouted front

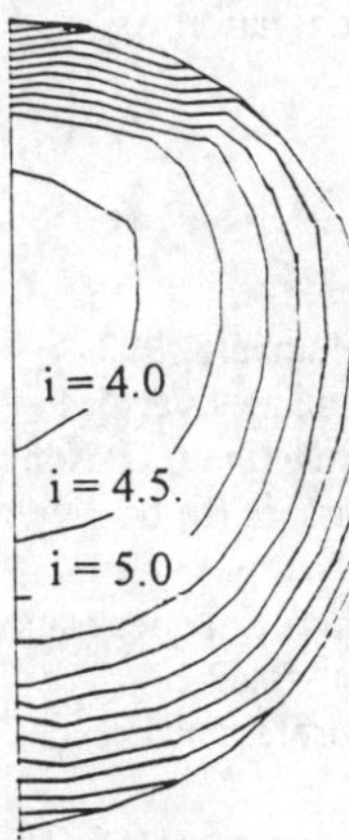

Fig.18 Distribution of hydraulic gradients on the tunnel face for a grouted front

5 CONCLUSIONS

The construction of a tunnel in water-bearing grounds produces seepage towards the opening, and may result in detrimental effects to both construction operation and surrounding structures.

Based on analytical, numerical or experimental approaches, solutions have been proposed to analyze seepage towards underground openings, but these methods are usually limited to seepage towards tunnels under working conditions in homogeneous isotropic grounds.

In the present study, the potentials of the finite element method to better account for the complexity of various ground and construction conditions have been investigated. The following aspects were considered : soil anisotropy and heterogeneity, liner permeability, use of grouting to limit seepage towards the working face of a tunnel under construction. The computer program CESAR-LCPC was used for this study. Both two- and three-dimensional computations were performed in steady flow conditions. These computations allowed to evidence the impact of various parameters on the computed hydraulic response of the aquifer.

With recent developments in computers, the capability of numerical methods could be used on a more regular basis to better account for the seepage pattern in the analysis of ground response to tunneling. This could be particularly useful for cases where limited support is used at the face during excavation, or where large consolidation settlements can be expected to take place as a result of tunnel construction.

However, the accuracy of numerical predictions are strongly dependent on the knowledge of the actual in situ hydrological conditions and rely on data obtained from adequate site investigation.

REFERENCES

Atwa, M., E. Leca, M. Rat, P. Humbert 1993. Analyse des écoulements hydrauliques vers les tunnels en régime transitoire. Bulletin de Liaison des Laboratoires des Ponts et Chaussées (to be published)

Averjanov, S.F 1949. Réglage du régime des eaux souterraines au voisinage des ouvrages hydrotechniques (en russe). Gidrotekhnicheskoe Stroitelstvo, n° 7, 6p.

Bear, J. 1979. Hydraulics of Groundwater. McGraw Hill Inc.

Chishaki, T. 1984. A study on confined flow of ground water through a tunnel. Kyushu University, Ground Water, Vol.22, No 2, pp. 162-167.

Chishaki, T.& S.D. Yang, 1977. Two dimensional analysis of unconfined groundwater flow from a tunnel opening. Memoirs of the Faculty of Engineering, Kyushu University, Vol.39, No 1, pp.37-64.

Chishaki, T., S.D. Yang, & K. Numata 1978. Non steady flow of ground water through a tunnel. Memoirs of the faculty of engineering, Kyushu University, vol.38, n°1, pp.61-81.

Keller, J.& A.R. Robinson 1959. Laboratory research on interceptor drains. Journal of the Irrigation and Drainage Division, ASCE, Sept. 1959, pp.25-40.

Leca, E., M. Atwa, M. Rat,P. Humbert & R Puglisi 1993. Analyse des écoulements hydrauliques vers les tunnels. Comptes Rendus des Jounées d'études Internationales de Toulon, Infrastructures Souterraines de Transports, AFTES, pp. 55-64.

Polubarinova-Kochina 1962. Theory of Ground Water Movement, translated by J.M. Roger de Wiest, Princeton University Press.

Rat, M. 1968. Tunnel sous la Manche. Laboratoire Central des Ponts et Chaussées, rapport interne Juillet, 68.

Rat, M. 1973. Ecoulements et répartition des eaux souterraines autour des tunnels. Bull. de Liaison des Laboratoires des Ponts et Chaussées,no 68, pp. 109-124.

Romanov, A.V. 1959. Méthodes de calcul de l'écoulement vers les ouvrages de drainage (in russian). Voprosy filtratsionnykh raschetov, Recueil III.

Romanova, E. J. 1956. Influence de la fissuration du tapis d'étanchéité amont sur les écoulements sous un barrage (in russian). Voprosy filtratsionnykh raschetov, Recueil II.

Sato, K. 1983. Hydraulic character of discharge hydrograph for tunneling. Japanese society of soil mechanics and foundation Engineering,vol.23, n°4, pp. 27-33.

Sato, K. & M. Iizawa 1983. Fundamental study on unsteady flow around underground cavern in unconfined groundwater. Proc. of the Japanese society of civil engineering, No.337, Sept., pp. 213-220.

Tunnelling and Ground Conditions, Abdel Salam (ed.) © 1994 Balkema, Rotterdam, ISBN 90 5410 363 9

Towards optimum design of bored tunnel section

G.S.Ebaid, M.E.Hammad & S.A.M.Ibrahim
Ain Shams University, Egypt

A.A.Maharram
Cairo University, Egypt

ABSTRAC: Bored tunnels excuted using shield driven machines are widly used in Egypt during the last few years. Precast concrete reinforced concrete segments is the most commonly way for lining such tunnels. Many factors should be considered in the design of these segments. The ultimate strength envelop of tunnel segment is the proper way to include such large number of variable in the design of tunnel liner. The given study deals with the procedure of using the strength envelop to reach an optimum design.

1. INTRODUCTION

Bored tunnels are used widely in Egypt during the last ten years. Most of these tunnles are excucted using shield driven machine. Examples of these projects is the Ahmed Hamdy tunnel, Greater Cairo waste water project and the Cairo Metro project phase II.
In all these projects, or simillar ones, reinforced concrete precast segments are widely used as linear to these tunnels.
Ribbed or solid R.C segment can be applied with different numbers of segments per ring. Also tapered or universal shape segments can be applied according to the case under consideration. The less the number segment per ring, the less number of segments shape, the lessweight of segments the much easy construction will be. Many factors affect the choice of type of segments to be applied for a certain project.

2. FACTORS AFFECTING BORED TUNNEL LINER DESIGN

Many factors take place in the design of bored tunner liner, which can be summerized in the following:

2.1. The tunnel function, e.g.roadway,railway, waste water, hydro power ...... etc. The requirements will define the net dimensions of cross section. Also the method of excution has an influence on this choice. Large diameter tunnels can be replaced by more than one of smaller diameter.Twin or more tunnel can be proposed. Also the location of tunnel center line from the ground surface has an influence in the choice of tunnel dimensions. Bored tunnel driven using the shield machine should be at depths not less than certain values particularly when it passes through residential areas .

2.2. Location of tunnel center line from the ground surface and consequently the vertical load above the tunnel, including loads existing at the ground surface.

2.3.The geological profile of soil through which the tunnel is passing, types of soil layers, existance of ground water and the soil characteristics at various levels and values of the soil parameters for these layers,particularly surrounding the tunnel section, and their maximum and minimum values. From these parameters the coefficient of subgrade reaction and the soil modulus of elasticity E_s.

2.4.The rigidity of the tunnel liner $E_l I_l$, which is the product of the elasticity modulus of liner material E_l , and the moment of interia/unit length of liner I_l. This will include the existance of ribs or solid segments, and the number of segments taken/ring.Murwood/3/,and /4/gives methods for including this parameter in the modified moment of inhertia of section. The larger number of segments/ring the more flexible liner will behave thus the smaller modified moment of inertia of the segment will be.Also the relative rigidity of liner with respect to surrounding soil,

$$\alpha = \frac{E_s R^3}{E_l I_l},$$

which also includes the radius of the tunnel R. /2/.

One or more types of segments can be proposed for the tunnel liner, and each type can have different types of reinforcement as shown in Fig. (1,2). Each proposed section will have its relative rigidity parameter.

2.5.The resulting straining actions in the tunnel section including the maximum bending moment and normal force. Such values will vary along the length of tunnel since it depends on the over burden pressure on the tunnel and the design consept used. In /2/ a universal graph represents the maximum bending moment in the section by four different design methods. By applying such curve one can get the maximum bending moment at various critical sections along the tunnel route and the results of various design methods.By using such graph it is very easy to consider the maximum and minimum values of for each specific location; since the soil parameters may change within a certain range in the same location, as a result of the possible change in soil condition at the same site. Also it can be noted that if it is proposed sections of different number of segments per ring, a representative value of α will be different from one proposal to the other. Most of the design methods give nearly the same thrust in the tunnel liner, bendig moments vary too much from one design method to the other. But by using such graph shown in Fig.(3,4), one get the value of the maximum bending moment for each location under a given type of segments and number of segments per ring and soil characteristics.

2.6 Some other factors affect the design of the tunnel liner such as the degree of tigntness required, type of bolting system for segments, maximum weight of element according to lifting equipment available, type of shield used and experience of workmanship, possibility of using more than one type of segments per ring .. etc. All such factors and others can be included in the design of the segments and the overall cost of blocks and tightness.

3. OPTIMUM TUNNEL LINER DESIGN

It has been clearly stated how so many factors affect the design of tunnel segments in the transverse direction.To check the stresses in the proposed section under wide range of values of straining actions will be inconisent.The best way is to draw the ultimate strength envelop for each type of section proposed, and even with various percentages of reinforcements used as shown in Fig. (5), (6). Such type of curves can be reached using the normal ultimate strength analysis for reinforced concrete sections. By plotting the values of the resulting straining actions (M & N) on the ultimate strength envelop one can get the best section and the resulting safty factor,and what is the best required percentage of reinforcements needed in the section.

By getting the results of these analysis one may take the actual parameters of the number and to be included in a more advanced design method, such as finite element program which simulate the behavior of the soil and liner to check this section.

REFERENCES

K. SZECHY: The art of tunneling, Akademia Kiado Budapest 1967.

G.S.Ebaid, M.E.Hammad: Aspects of circular tunnel design, Tunnels&Tunneling Journal July/Aug., 1978.

Murwood A.M.: The circular tunnel in elastic ground, Geotechnique 25, No.1, 1975.

G.S.Ebaid: Effect of Joints on circular tunnel design, Bull. of Faculaty of Eng. Ain Shams Univ. No. 9, 1978.

M.E.Hammad: Design of circular syptens bored in elastic ground Bull.of Faculaty of Eng. Ain Shams Univ. No. 9, 1978.

Peck, R.B. et al: State of art of soft ground tunneling RETC proceedings Vol.1, Illinois 1972.

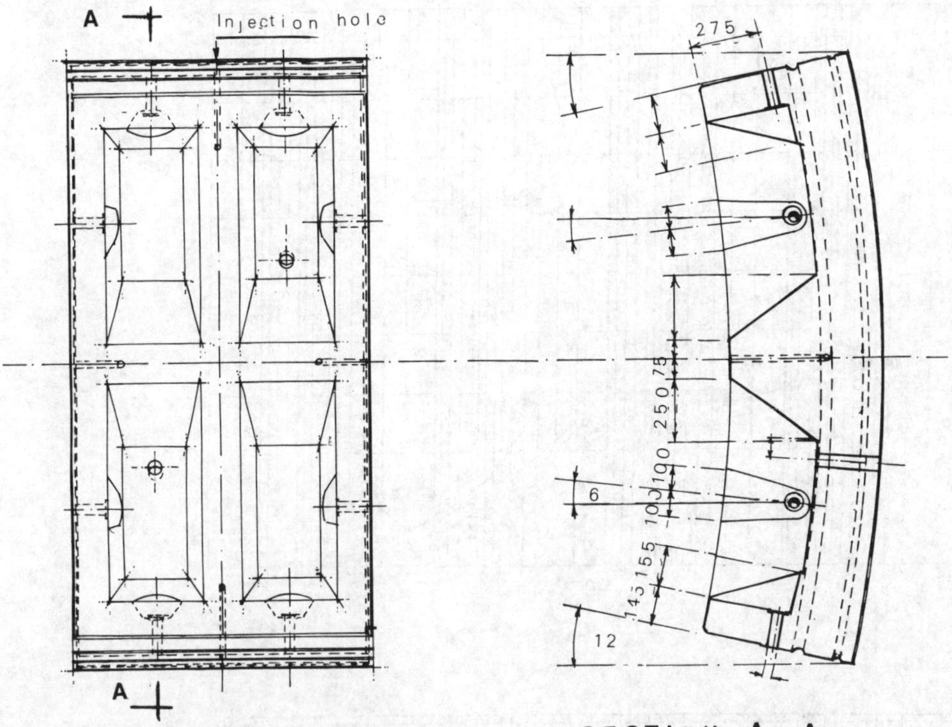

ELEVATION OF SEGMENT

SECTION A – A

FIG (1)

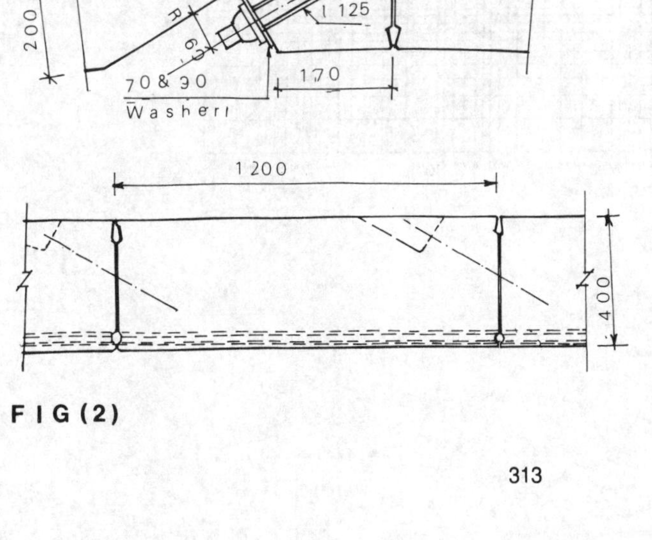

FIG (2)

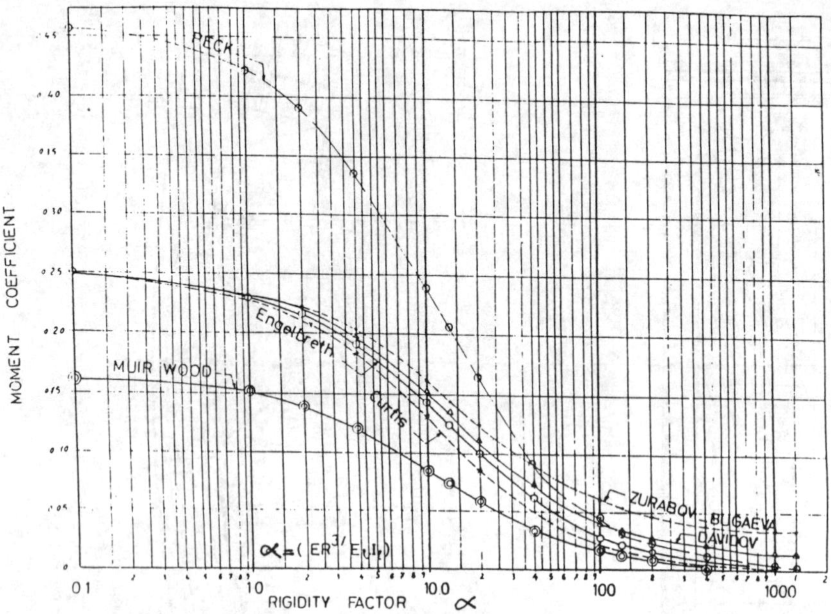
PECK
Engelbreth
MUIR WOOD
Curtis
ZURABOV - BUGAEVA
DAVIDOV
MOMENT COEFFICIENT
RIGIDITY FACTOR α
0.1
1.0
10.0
100
1000

FIG (3)

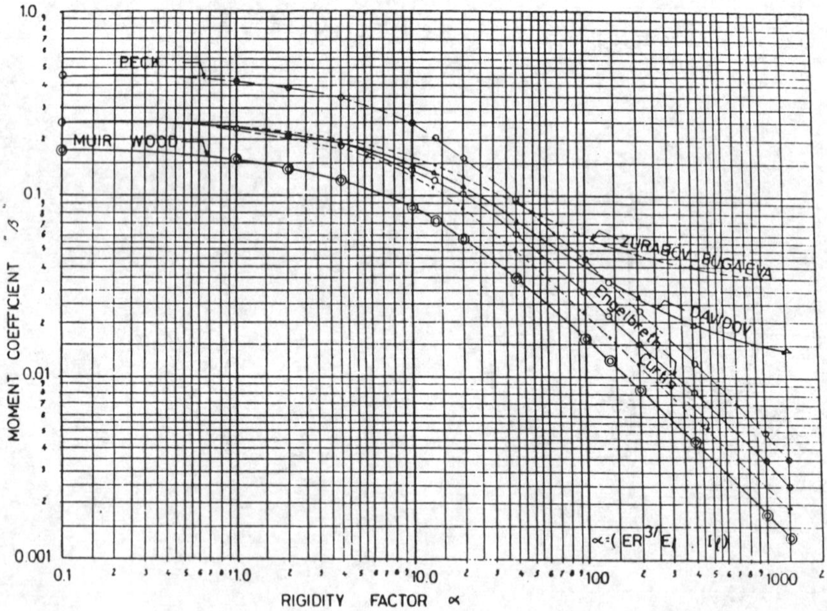
1.0
0.1
0.01
0.001
PECK
MUIR WOOD
ZURABOV - BUGAEVA
DAVIDOV
Engelbreth
Curtis
MOMENT COEFFICIENT β
RIGIDITY FACTOR α
0.1
1.0
10.0
100
1000

FIG (4)

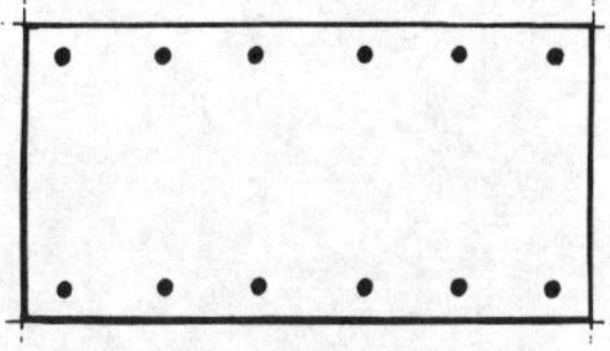

SOLID SEGMENT WITH DIFFERENT RFTS

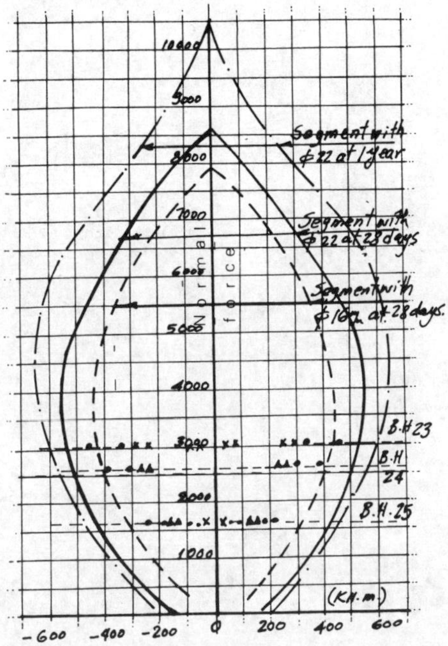

FIG (5)

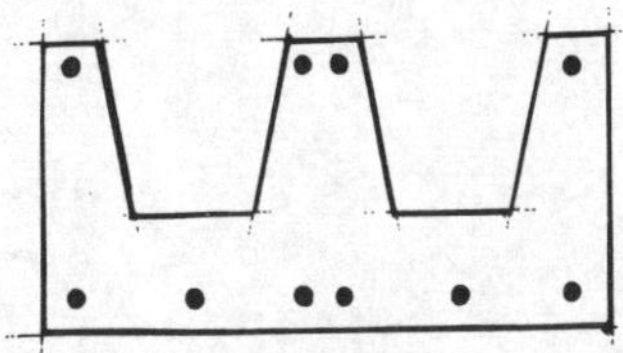

RIBBED SEGMENT WITH DIFFERENT RFTS

FIG (6)

Tunnelling and Ground Conditions, Abdel Salam (ed.) © 1994 Balkema, Rotterdam, ISBN 90 5410 363 9

Fire protection in traffic tunnels – Initial findings from large-scale tests

Alfred Haack
STUVA, Cologne, Germany

ABSTRACT: As the number and length of road and rail tunnels increases and more people use them, fire safety issues take an ever greater importance (Haack 1992). In order to improve fire protection in subsurface traffic facilities, eight Western European nations have joined Germany in undertaking a comprehensive study of fire in such facilities. Laboratory tests and tests conducted in existing tunnels are being carried out in Germany, Finland and Norway, with the overall objective of investigating the possibilities for optimal protection of people and preservation of property in the event of fires in subsurface transportation facilities.

1 PROBLEMS TO BE FACED

Fire in traffic tunnels (road or rail), are an international problem. They are characterised by the danger they present to the persons affected and, in many cases by the considerable amount of material damage that they cause (see Figure 1). Serious accidents resulting in injuries have been individually reported from Great Britain, Japan, Canada and the U.S.A. Various major fire accidents have also occured in tunnel facilities in Germany. These cases have served to draw attention to the possibilities of escape and rescue, which are made more difficult in a tunnel scenario.

Fig. 1 Destroyed rail cars and interior furnishings at an underground stop following an urban commuter train fire on April 8, 1980, in Hamburg, Germany.

Serious fire cause considerable material damage, not only to rolling-stock or vehicles, but often to the tunnel facilities as well. This damage is brought about by the massive development of heat and aggressive combustion gases, among other causes. Although the effects of fire seldom threaten the stability of a tunnel structure, they often reduce the availability of the tunnel for traffic, at least for a certain length of time.

The effects of this latter aspect should on no account be underestimated. In some cases, redevelopment can take weeks or months. If as a result, service in important tunnel sections of a metro or urban commuter network has to be discontinued, the inevitable result will be serious disturbances in their everyday operation in built-up areas (such as Berlin, Hongkong, London, New York or Paris). As a consequence, in Germany special "Guidelines for Furnishing and Operating Road Tunnels" (1985) expressly point to the effects of a case of fire and consequently call for suitable measures already within the scope of preventive fire protection.

Generally speaking, it has been determined that the risk of fire in traffic tunnels is on the increase. The reasons for this are:

1. The growing density of traffic, especially on the roads.
2. The increasing travelling speeds in rail traffic (see Figure 2).

Fig. 2 Fast train leaving a tunnel.

3. The growing number of tunnels, with ever greater individual lengths.
4. Increasing of vandalism, including increasing a trend towards terrorism (witnessed, e.g., in cases of arson).

Such developments must be taken into consideration in developing a safety concept for tunnel traffic.

2 TRAFFIC TUNNEL STATISTICS

Numerous rail and road tunnels are in operation (Table 1) or are being built in various European countries. As examples, the following major projects are either under construction or in the planning stage:

1. The Channel Tunnel (rail) between Britain and France, which is approximately 50 km long.
2. The Great Belt Tunnel (rail) in Denmark, between the islands of Funen and Seeland, which is about 7 km long.

Table 1. Rail and road tunnel operations in various West European countries.

	Traffic tunnels (route km)			
	Metro	Rail	Road	Total
Austria	15	105	210	330
Switzerland	-	360	140	500
Germany	550	380	70	1000
France	270	650	180	1100
Great Britain	200	220	30	450
Italy	60	1150	600	1810
Norway	20	260	370	650
Spain	200	750	100	1050
Total	1315	3875	1700	6890

3. The Alpine transit routes for rail traffic in Austria/Italy (the Brenner Base Tunnel) and in Switzerland (where the Gotthard Base Tunnel and the Lötschberg Tunnel are being contemplated), each approximately 40 to 50 km long.
4. The Gibraltar Tunnel between Spain and Morocco, which will probably be about 50 km long.
5. The Mont Cenis Tunnel between France and Italy, again with a length of about 40 to 50 km.
6. A large number of tunnels within the framework of the planned pan-European high-speed rail links, scheduled to be completed by the year 2015 (UIC 1989).
7. Various tunnels planned in Norway beneath straits in order to connect islands with one another or with the mainland; the overall length will be greater than 100 km. A number of these tunnels will be more than 10 km long and will be constructed at depths of 600 m and more below sea level (Ovstedal and Melby 1992).

Taking all these new construction measures into account, the pan-European traffic tunnel network is likely to far exceed 10,000 km by the year 2000.

3 CASES OF FIRE IN TRAFFIC TUNNELS

3.1 Subsurface metro and urban commuter systems

At the beginning of the 1980s, the German Federal Ministry of

Transport in Bonn commissioned a comprehensive survey in order to improve fire protection in tunnels for tram, urban railway and metro systems (Heffels et al. 1984).

Within the framework of this STUVA-research project, a survey relating to cases of fire in tunnels that had occurred during the last years was carried out in 86 cities worldwide. Some detailed results were subsequently updated and presented by Haack 1992.

Furthermore, a large number of literary sources were evaluated with respect to the frequency, cause, spread, parallel phenomena, fire-fighting methods, and the resultant damage.

The most important findings obtained in this survey of fire accidents in rail tunnels can be summarized as follows:

1. Given the rail-bound vehicles customarily used today for commuter transportation, the fire flash-over point must be reckoned with after some 7 to 10 minutes.

2. The fire duration fluctuates considerably depending on external circumstances, and can last between 30 minutes and several hours.

3. Considerable quantities of fumes can evolve even in the case of relatively small fires, leading to a critical situation for the passengers and the train crew (see Figure 3).

4. In a number of cases, the entire tunnel cross-section was filled with smoke after a short time to such an extent that visibility was reduced to less than 1 m, even when searchlights were switched on.

5. In many cases, communication problems between the train driver and the central control room or between the control room and the fire brigade, inaccurate data pertaining to the position of the train in the section of tunnel, as well as poor radio telephone connections between the rescue and fire-fighting crews in the tunnel and the command centre on the surface, resulted in critical time losses.

6. Combatting the fire is made considerably more difficult if the train remains stuck in a section of tunnel rather than at a stop underground.

7. In a number of fire accidents, there were either no wall hydrants available in the tunnel or they were set up too far apart. This meant that valuable time (roughly 10 minutes required per 100 metres; Günther 1985) was lost in setting up a supply of water to extinguish the fire.

8. The considerable material damage (examples are shown in Figures 1 and 4) that resulted in some cases as a result of major fires came about through the substantial fire load of the rolling-stock. In the case of the rail-bound vehicles currently in

Fig. 3 Major amounts of combustion gas were released and attempts to extinguish this urban commuter train fire (on April 8, 1980, in Hamburg) were futile.

Fig. 4 Burned-out metro car resulting from a fire accident in Cologne on October 24, 1978.

use, this normally amounts to between 50 and 80 kg of combustible material (wood equivalent value) per square metre of passenger accomodation area*).

9. Damage caused to tunnel structures made of concrete largely arises from flaking as well as the condensation of fumes on the tunnel wall, tunnel ceiling and the interior furnishings of the stops (see Figure 1).

By and large, the evaluation of various fire accidents with regard to rolling-stock revealed that the type of vehicle had a considerable influence on damages suffered. The examples shown in Figure 5 support this statement. In case (a), the roof of an urban railway car completely constructed of steel was preserved. In contrast, the fibre-glass reinforced plastic roofs in case b) - installed in order to save weight - burned completely. If an aluminium construction is used, then the total loss of the roof components can be expected, as in case (c).

The extent of roof damage caused to the vehicle body influences the extent of fire load to the tunnel lining in the roof zone. In this respect, case (c) undoubtedly produces the most unfavourable conditions.

3.2 Tunnel facilities for long-distance traffic

The situation pertaining to tunnel facilities for long-distance

Fig. 5a All-steel body; fire accident on July 24, 1979, on the Hamburg urban commuter system.

Fig. 5b Fiber-glass-reinforced plastic in the roof section; fire accident in the Hamburg metro on April 11, 1979.

Fig. 5c Aluminium body; fire accident on April 8, 1980, on the Hamburg urban commuter system.

Note

*) *For comparison (also wood equivalent values):*
- *For housing on average 30 to 60 kg combustible material/m² of story area.*
- *For stores, on average approx. 100 kg combustible material/m² of sales area (varies pronouncedly between the various goods departments).*
- *For administrative offices, on average from 15 kg combustible material/m² of floor area to 300 kg/m² of filing space area.*

traffic is similar. A number of major fire accidents (for details see Haack 1992) - resulting, in some cases in considerable damage to persons and property - are known from this field. In the case of the not more than 243 km long Moorfleet motorway tunnel near Hamburg, for example, the pre-stressed concrete roof of one of the two tunnels had to be entirely renovated (see Figure 6). A particularly tragic incident occured when a touring coach burned out in Paris (see Figure 7). As a result, 44 schoolchildren on an excursion lost their lives because they were unable to escape fast enough from the burning bus.

Fig. 6 Completely destroyed pre-stressed concrete roof following a lorry fire in the Moorfleet/Hamburg motorway tunnel in 1968.

Fig. 7 A touring bus that caught fire on July 31, 1982, in Paris; 44 schoolchildren died as a result of the accident.

The statistical report on the first years of operation for the federal motorway link Elbe Tunnel in Hamburg (see Table 2) is informative in estimating risks for road traffic in tunnel facilities.

Table 2. Fires occurring in the Federal Motorway Elbe Tunnel, Hamburg, during the first 15 years of operation.

Year	Traffic 10^6 Veh.	Total Overall	Fires Car	Lorry
1975	19.9	7	5	2
1976	23.3	17	15	2
1977	25.0	8	5	3
1978	25.2	5	3	2
1979	25.4	9	8	1
1980	26.9	7	4	3
1981	26.6	7	6	1
1982	27.9	5	3	2
1983	28.7	5	4	1
1984	29.5	6	3	3
1985	29.4	12	7	5
1986	30.2	8	7	1
1987	32.1	19	10	9
1988	34.5	13	9	4
1989	36.5	9	7	2
Total	421.1	137	96	41

The data show that, on average, one vehicle fire per month occurs within the tunnel. Although lorries account for only about 15 % of total traffic on weekdays, they cause approximately 30 % of the fires. In other words, lorry fires are considerably more likely to happen than car fires. So far, these fire accidents in the Elbe Tunnel have not had major consequences, mainly because one fire brigade vehicle is on call at the southern portal and another at the northern portal and the control room is manned around the clock with personnel who are trained in firefighting technicalities.

At present, an average of just over five major fires involving passenger-carrying rolling stock occur on the German Federal Railway each year. Fortunately, such incidents have not occured in tunnel facilities.

3.3 Probability of fire accidents

Taking the results of the above mentioned research work and according to statistics of PROGNOS 1987 the probability of fire accidents can be roughly assumed to be as follows when the main fields of passenger traffic are compared:

railway	1
bus	2.5
private cars	24

With concern to the driven km the following situation is given for Germany as an example:

rail: 1 tunnel fire case per 0.5×10^9 km

road: 1 tunnel fire case per 10×10^9 km

Again there is the factor of about 20 to 25 between rail and road. In other words, travelling by railway can statistically be accepted as about 20 to 25 times safer than travelling on road.

4 LARGE-SCALE TESTS FOR IMPROVING FIRE PROTECTION

4.1 Research objectives

Against the background described above, and in consideration of the fact that subsurface traffic facilities are increasing in number, length and number of users, ministries in Germany and other European countries have commissioned a comprehensive survey in order to improve fire protection in subsurface traffic facilities - something urgently required throughout Europe and the world. In Germany, STUVA (the Research Association for Underground Transportation Facilities, Inc., Cologne) together with the Institute for Construction Materials, Massive Construction and Fire Protection at the Technical University of Braunschweig (Germany) is responsible for the scientific side of the project. The Technical Research Centre of Finland, Fire Technology Laboratory at Espoo and the Norwegian Road Research Laboratory in Oslo as well as SINTEF in Trondheim, Norway are also involved in the project. Other European countries such as Austria, Great Britain, France, Italy, Switzerland and Sweden have joined the project which got the EUREKA status in June 1991.

The overall objective of the project is to investigate the possibilities for optimal protection of people and transportation facilities; and to examine the alternatives open to the fire brigade with respect to saving people and combatting fires in such tunnels.

The research programme is intended above all to provide information on the following aspects of tunnel fires in subsurface traffic facilities (Haack 1986 and Kordina 1981):

1. Fire phenomena: Chronological and spatial distribution of temperature and combustion gases; combustion rates; liberation of energy; fire distribution speed; air and hot gas discharge volumes; and the influence of ventilation plants.

2. Escape, rescue and fire-extinguishing possibilities (in causal relationship with the results from 1, above): Functioning of fire-warning systems; efficacy of emergency lighting and escape guidance systems; overall fire-fighting services and equipment; and, in particular, available and required clearance times for vehicles and tunnels. Also, the communication possibilities between the fire-fighting and rescue crews and the command centre, and the nature and requirements of heavy respiratory apparatus.

3. The effect of the surrounding structural parts on the fire: Transfer of heat from the tunnel structure to the rock; temperature curves and supporting behaviour of different types of tunnel support systems (e.g. tunnel shells made of concrete, cast iron, steel, natural stone or tunnel in rock without shell-type supports).

4. Reusing the structure: Damage; residual supporting behaviour; possibilities and time required for redevelopment (operational periods of immobilization).

5. Accumulation of theory: Extending and improving the basic principles of energy and mass balances; investigating the validity ranges of existing domestic and foreign heat balance models and, if necessary, redevelopment or modifying the models; expanding existing heat

balance computing programmes; extending supporting load calculations for structures in the event of fire accidents.

4.2 Test programme

The test programme is divided into tunnel tests and laboratory tests:

a) Tunnel tests: Several types of tunnel tests were conducted during 1986 to 1992 and are planned respectively:

1. Wood fire tests, as well as fire-fighting and rescue attempts, conducted in a 3.2-km-long, non-operational railway tunnel in Germany with a mouth profile of approx. 45 m^2 clear section.

2. Wood fire tests conducted in the branched tunnel system of an abandoned subterranean quarry in Finland with a clear rectangular cross-section of approximately 25 m^2. These tests simulate conditions in an underground train system.

3. Vehicle, wood and heptane fire tests in an approx. 2.3-km-long transport tunnel in Norway with a flattened mouth profile of about 35 m^2 clear section (Figure 8).

Fig. 8 Reflection tables in the Norwegian test tunnel for determining the spread of combustion gas with the aid of video cameras.

All the tests in Finland and Norway have been finished (see Table 3) and are now being evaluated.

b) Laboratory Tests: The laboratory tests, with relevant model calculations, were conducted at Braunschweig (Germany) Technical University. These tests and their evaluation have been completed.

5 FIRST IMPORTANT FINDINGS

The evaluation of the international test programme is still being continued. But some first important findings have been established already:

1. The influence of damage both to the vehicles and the tunnel lining, especially in the crown area, depends on the type of car. This fact, already derived from numerous fire accidents (see chapter 3.1) was entirely confirmed by the tests in Norway. The roof of those vehicles constructed of steel resisted the heat, whereas the roofs of the public bus and of the metro car F4 both made of aluminium were completely destroyed at a rather early stage of the test fire. The same happened, as preservation property in the event of fires in subsurface expected, to the roof of the private car with a plastic body (see tests No. 7, 14 and 20 in Table 3).

2. The temperatures during most of the rail car and bus fires reached maximum values of about 800 to 900 °C, in one case about 1000 °C. Against this the temperature during the test with the heavy good vehicle loaded with 3 t of modern furniture climbed up to more than 1300 °C. Along the tunnel, temperatures decrease over a relatively short section. For escape, the situation is worse downwind than upwind (see Figure 9).

3. The size of the fires has been recalculated on the basis of the Swedish and Norwegian measurement of heat release. The rail car fires mostly amounted to between 15 and 20 MW, the burning of the heavy good vehicle was measured at about 100 MW.

Table 3. List of fire tests conducted in Repparfjord Tunnel, Norway

No.	Date of test	Fire load	Conducting countries[1)]	Other countries taking part[1)]	Air velocity during test [m/sec]
1	07.12.90	wooden cribs No. 1	D	-	0.3
2	24.07.91	wooden cribs No. 2[5)]	D	-	0.3
3	08.08.91	private car[5)]	D	-	0.3
4	19.08.91	metro car F3 (steel)[5)]	D	-	0.3
5	29.08.91	half railcar F5	D	-	0.3
6	04.09.91	half railcar F6	D	-	0.3
7	23.08.92	public bus	D	A,S	0.3
8	28.08.92	wooden cribs No. 3	D	A,S	0.3
9	30.08.92	wooden cribs No. 4	D	A,S	3-4
10	31.08.92	wooden cribs No. 5	D	A,F	6-8
11	13.09.92	1.5 rail cars F2Al+F7	D, I[2)]	A,F	6-8/3-4
12	25.09.92	rail car F2St (steel)[5)]	D, A[3)]	F	0.5
13	07.10.92	rail car F1	D, A[4)]	F,S	0.5
14	14.10.92	metro car F4 (Alu)	D	A,F,S	0.5
15	23.10.92	mixed load simulating truck load	F,A	D,S	0.5
16	27.10.92	heptane pool No. 1	F,A	D	0.6-1.0
17	28.10.92	heptane pool No. 2	F,A	D	1.5-2.0
18	29.10.92	heptane pool No. 3	F,A	D	1.5-2.0
19	29.10.92	heptane pool No. 4	F,A	D	2.0-2.5
20	04.11.92	private car with plastic body[5)]	A	D,F	0.5
21	12.11.92	heavy good vehicle	GB	A,D,F,N	6-8/3-4

Notes:
1) Actively in Norway participating countries:
A = Austria, D = Germany, F = France, GB = Great Britain, I = Italy, N = Norway, S = Sweden
2) Performance test of an Italian refuge box
3) Test of Austrian rail car, positioned in a clear distance of 100 m behind the burning car
4) Concerning Austrian rail car as 2), but clear distance 50 m
5) Additional measurement of dioxin and furane emission

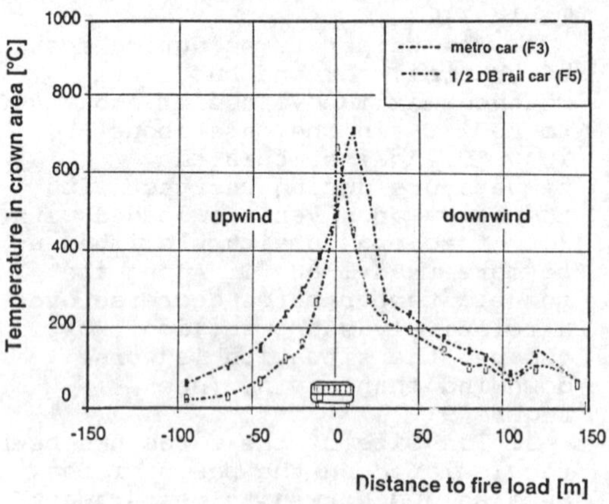

Fig. 9 Temperature curve in longitudinal axes during fire test with a metro car and half of a German Railway rail car (Kordina and Richter 1992).

4. All the rail and road car fires registered a fast development during the first 10 to 15 minutes. So, seen from the aspect of temporal development and heat emission, the hydro carbon curve of RABT covers the reality much better than the unit temperature curve of DIN 4102 (see Figure 10). But, in this connection, there is one question not yet answered. The European countries have to discuss and think it over, if the Dutch design curve - especially taken for the static calculation of underwater tunnels - with its maximum temperature value of 1350 °C, is closer to reality than the RABT curve in Germany with its highest temperature value of 1200 °C. This question arose as a consequence of the test with the heavy good vehicle.

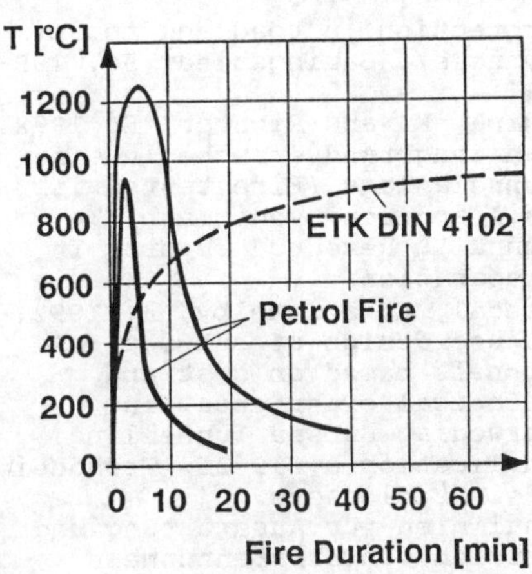

Fig. 10a Comparison between real petrol fires and the unit temperature curve of DIN 4102

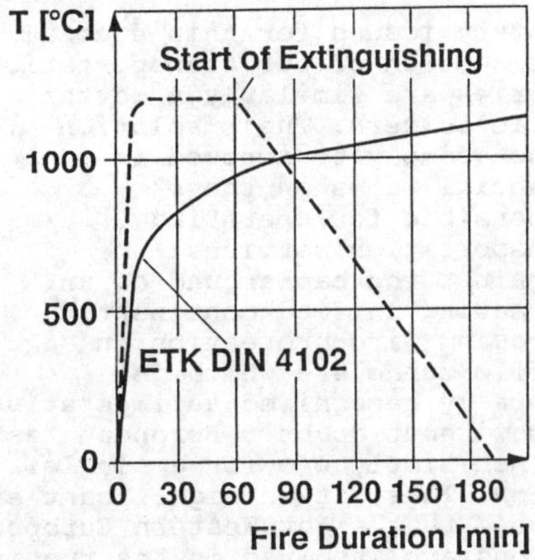

Fig. 10b Comparison between the RABT-curve and the unit temperature curve of DIN 4102

5. Modern outfitting of rail cars makes them much more resistant against ignition than that of earlier vehicle generations. This could be proved by the fire tests with two halves of German Federal Railway passenger coaches, which both had all-steel bodies but with different lining materials for their walls and roof. The two latter-mentioned test vehicles were stripped off seats and other internal furnishings in order to obtain a true comparison of the influence of the roof and wall lining materials.

The coach lined with materials based on fibre-glass reinforced, unsaturated polyester, in accordance with the latest state of technology (Table 3, line 5), developed a considerable fire (see Figure 11). However, the other coach, lined with materials based on a phenol resin basis, which releases energy at a lower rate (Table 3, line 6), proved to be flame-resistant. In both cases, an ignition energy of approximately 6 l isopropanol (= 7,293 cm^3, corresponding to 200 MJ = fire load of a seat) was applied. The phenol resin lining did not catch fire, even after the ignition energy was doubled.

6 CONCLUSION

Improving fire protection in traffic tunnel facilities is an important task, particularly given the increasing importance of tunnels in the infrastructure of modern society. In recent decades, tunnelling has been championed by the public, and in turn, by the politicians - both for ecological reasons and as means of revitalizing our cities. Consequently, tunnels have been financed almost exclusively with public funds.

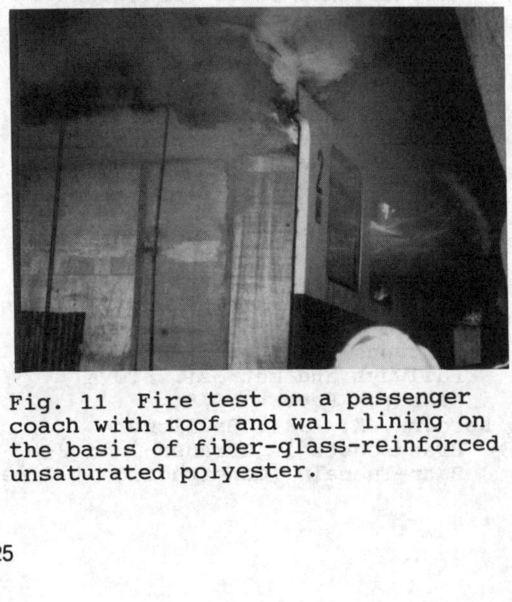

Fig. 11 Fire test on a passenger coach with roof and wall lining on the basis of fiber-glass-reinforced unsaturated polyester.

Safety issues for this special part-section of our transportation arteries are similarly a matter of public concern. Their solution cannot simply be imposed in financial terms on those responsible for operating transportation services.

Against the background of an increasing unified continent, improving fire protection in traffic tunnels - which is generally recognised as imperative - is without doubt a European task and not simply one for a single nation. Thus, it is significant and welcome that eight Western European nations are involved in the running research project. Given the current state of progress, the test series should be completed by mid-1994 including the tremendous volume of evaluation work.

REFERENCES

DIN 4102: Brandverhalten von Baustoffen und Bauteilen

Günther, K.-P. 1985. 10 Jahre Brandschutz im Hamburger Elbtunnel (10 years fire protection in the Hamburg Elbe-Tunnel). Brandschutz/Deutsche Feuerwehr-Zeitung 39 (7), 263-264, as well as information from the City of Hamburg.

Haack, A. 1986. Geplante Brandversuche in einem stillgelegten Eisenbahntunnel (Planned fire tests in an abandoned railway tunnel). In Transport gefährlicher Güter (Proceedings of an International Conference), 25.-27.7.1986 in Innsbruck, Österreich, 90-101.

Haack, A. 1992. Fire Protection in Traffic Tunnels - Initial Findings from Large-scale Tests; Tunnelling and Underground Space Technology, vol. 7 (1992), No. 4, 363-375

Heffels, P. Marquardt, H.-J.; and Staub, L. 1984. Verbesserung des Brandschutzes in Tunnelanlagen für Straßen-, Stadt- und U-Bahnen (Improvement of fire protection in tunnels for trams, commuter railways and metros). STUVA Research Report 18/84.

Kordina, K. 1981. Baulicher Brandschutz in Straßen- und U-Bahn-Tunneln (Constructional fire protection in road and metro tunnels). Bauingenieur 56, 189-191.

Kordina, K. and Richter, E. 1992. Fahrzeugbrandversuche in einer Tunnelanlage (Fire tests with rail and road vehicles in a tunnel). Research report. In preparation.

Ovstedal, E. and Melby, K. 1992. Future design of sub-sea road tunnels based on cost and technical experience. In Norwegian Subsea Tunnelling, Publication 8, 19-26. Trondheim: Tapir Publishers.

Richtlinien für Ausstattung und Betrieb von Straßentunneln (RABT). (Guidelines for Furnishing and Operating Road Tunnels). Editor, Forschungsgesellschaft für Straßen- und Verkehrswesen, Köln.

Vorschlag für ein europäisches Hochgeschwindigkeitsnetz (Proposal for a European high-speed railway network). 1989. UIC-Gemeinschaft der Europäischen Bahnen (Association of the European State Railways), January 1989.

Tunnelling and Ground Conditions, Abdel Salam (ed.) © 1994 Balkema, Rotterdam, ISBN 90 5410 363 9

Assessment of hydraulic efficiency of grouted plug between diaphragm walls

Ahmed A. Hassan & Fathalla M. El-Nahhas
Ain Shams University, Faculty of Engineering, Cairo, Egypt

ABSTRACT: Construction of cut-and-cover tunnels and subway stations within water-bearing soil layers in congested cities requires good control of groundwater levels adjacent to the deep excavation in order to minimize settlement of near-by existing buildings. Grouted plugs were used between the toes of the diaphragm walls during the construction of the Regional Line of Cairo Metro and will be used for the subway stations of Line (2) of the same project. This paper contains the results of a parametric study which was carried out to investigate the effect of geometrical configuration of the grouted plug on the hydraulic efficiency of the system for controlling groundwater levels. The results were compared with field measurements compiled during construction of the Regional Line of Cairo Metro. Optimization of grouted plugs is discussed.

1 INTRODUCTION

Construction of underground structures in congested cities located on water bearing soil deposits is considered one of the difficult tasks facing civil engineers. The presence of buildings and different types of structures in the vicinity of such a project puts some constrains on the designers and requires special construction techniques to be utilized. Controlling the groundwater level under the surrounding structures during construction is essential for minimizing their settlement and hence avoiding their damage.

Five stations of the Regional Line of Cairo Metro were constructed through the Nile alluvial soils using diaphragm walls provided with grouted plugs. Groundwater table was present at shallow depth of less than 3 meters while the depth of excavation was about 10 meters. The same system was used for constructing the cut-and-cover tunnel connecting these stations (El-Nahhas et. al, 1988). Seven deeper subway stations (about 22 m deep) are presently being constructed using the same procedure for Phase 1 of the second line of Cairo Metro.

This paper focuses on the role of grouted plugs in controlling groundwater table in the area surrounding deep excavations supported by diaphragm walls. Hydraulic efficiency of such grouted elements is assessed based on results of numerical modelling of the groundwater flow around diaphragm walls.

2 GROUTED PLUGS

The permeability of sand deposits can be reduced to 1/100 and up to 1/1000 of its in-situ value if grouted efficiently, AFTES, (1991). Grouting is carried out simply by filling the voids between sand particles. The grouting materials can be injected in stages. For instance, in the Cairo Metro project, injection of bentonite-cement slurry is followed by a second stage of grouting using soft-silica gel in order to reach the specified permeability of 10^{-6} m/sec. Conventionally, the elevation of the plug and its thickness are selected to fulfill an average factor of safety against heave of 1.1 with maximum value of average hydraulic gradient within the plug of 3, Salam (1984).

The combined hydraulic action of water-tight diaphragm walls and low-permeability grouted plug develop semi-impervious large container. Therefore, excavation of soil and pumping-out of groundwater within these boundaries should not result in any disturbance to the groundwater table in the surrounding area during construction. Once the reinforced concrete base-slab of the station (or cut-and-cover tunnel) was placed, permanent control of water table is achieved.

3 THEORY AND NUMERICAL SOLUTION

The 2-D simulation in vertical planes is successfully applied to many groundwater flow problems such as flow through earth dams, flow under continuous floors of hydraulic structures, cofferdams,...etc. All these problems can be considered as steady state problems for which the unsteady state conditions exist only for a very short time.

In order to simulate mathematically the two dimensional groundwater flow problems in an anisotropic and nonhomogeneous porous media under the steady state conditions, Laplace equation is used. This equation should be associated with the boundary conditions which represent the physical boundaries of the studied domain. The general form of Laplace equation may be written in the form, (Bear, 1972),

$$\frac{\partial}{\partial x}\left(K_x\frac{\partial h}{\partial x}\right)+\frac{\partial}{\partial y}\left(K_y\frac{\partial h}{\partial y}\right)=0 \quad (1)$$

where,
h is the potential head [L];
K is the permeability coefficient [L/T];

Alternatively, a two dimensional finite element model can be used to solve this problem numerically, Hassan, et.al (1989, 1992). This model gives the possibility of simulating a wide variety of groundwater flow problems. It facilitates the simulation of the anisotropic and non-homogeneous aquifers because it enables the user to input the hydraulic characteristics in elementwise way. The model output is in the form of equipotential lines, velocity vectors, hydraulic gradients and seepage forces.

4 CASE STUDY

Figure 1 illustrates the configuration of Cairo Metro cut-and-cover tunnel and because of symmetry only one half of the studied domain will be considered. This domain lies between the center line of the tunnel and a known hydraulic head boundary. Figure 2 shows the boundary conditions of the problem. A known hydraulic head value is applied at the boundary 1-2. The boundary 2-3 is a free surface where the elevation of any point equals the hydraulic head. The boundary 3-4-5-6 is an impervious boundary at which the normal hydraulic gradient vanishes. The boundary 8-1 is assumed the lowest flow line at the domain bottom. The boundary 6-7 is a known hydraulic head boundary at which the construction should be under dry condition. The problem was solved for different cases using a finite element mesh consisting of 746 elements and 415 nodes.

5 PARAMETRIC STUDY

The parametric study was designed to investigate the effect of configuration of the grouted plug on its hydraulic efficiency. Five cases were studied assuming the same permeability of the grouted plug used in Cairo Metro cut-and-cover tunnel that was of 10^{-6} m/sec. Constant thickness of the grouted plug is assumed for the first three cases. The central portion (4.86 m wide) of the grouted plug of the remaining two cases was reduced from 3 m to either 2.25 or 1.5 m as shown in Table 1.

Table 1. Cases of study

Case	Thickness of plug (m)
1	3.00
2	2.25
3	1.50
4	3.00 to 2.25
5	3.00 to 1.50

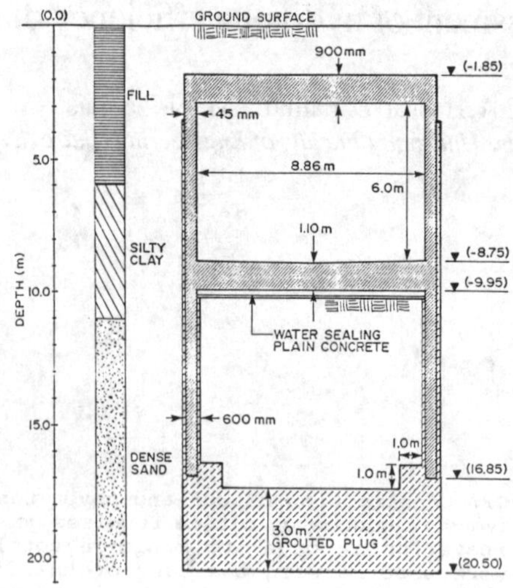

Fig.1 Cairo Metro Tunnel: configuration and soil profile

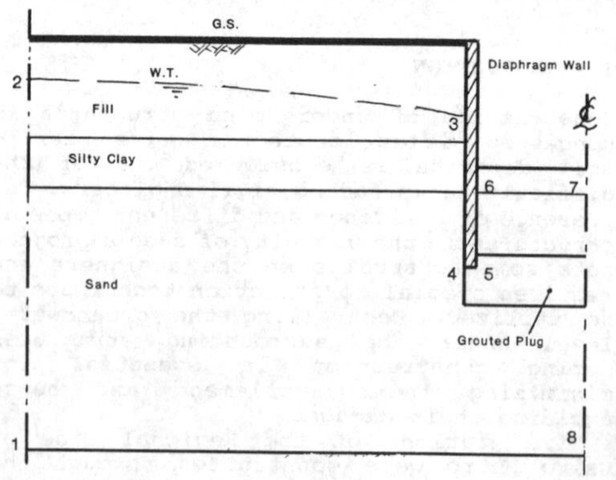

Fig.2 Problem definition and boundary conditions

6 RESULTS AND DISCUSSION

For each case, the distribution of water pressure as well as the velocity field within the plug are determined. In addition, the hydraulic gradient under the tunnel and at the inlet and exit of the grouted plug was investigated. The uplift force that acts on the plug bottom was calculated and compared against the weight of the soil column above this level. The ratio between these two forces was defined as the factor of safety against heave. Furthermore, the pore water pressure distribution adjacent to the diaphragm wall as well as below the tunnel was checked against the hydrostatic pressure distribution.

Figure 3 shows the velocity field overall the studied domain for case 1. To give a clear idea about the flow conditions within the grouted plug, a zoom view is illustrated in Figure 4 showing the flow concentration and the steep change in the hydraulic gradient near the toe of the diaphragm wall. The distribution of potential head within the plug is given in Figure 5.

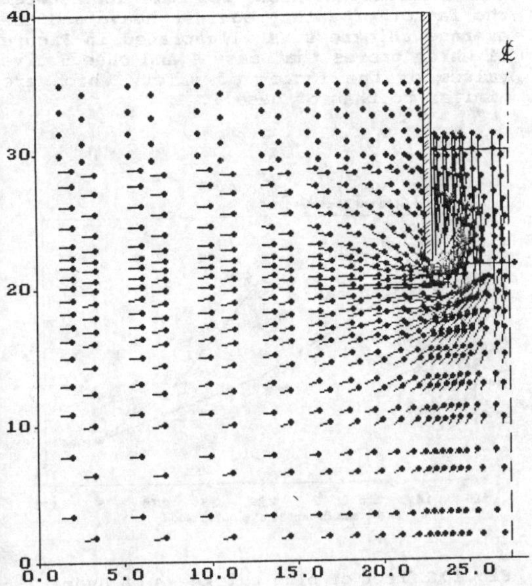

Fig.3 Velocity field (case 1)

Figures 6 and 7 show the potential head distribution and the velocity field, respectively, within the plug for case 5 at which the plug thickness is reduced in its central portion from 3 m to 1.5 m. Comparison between these two figures and figures 4 and 5 shows the effect of the thickness reduction on the flow conditions and the hydraulic gradient within the plug.

To investigate the effect of the plug thickness on the pressure distribution below the tunnel, a comparison between case 1, case 2 and case 3 referring to the hydrostatic pressure distribution is shown in Figure 8. In this figure, it is noticed that case 1 gives the lowest pressure distribution on the diaphragm wall. Also, this figure shows a comparison between the numerical result and field measurements compiled during the construction of the Regional Line of Cairo Metro. It is found that the difference between the numerical result and the field measurements is very small. In addition, the obtained result gives a detailed pressure distribution within the different layers below the tunnel including the grouted plug.

The trend of the hydraulic gradient change within the plug was investigated for the five cases. Figure 9 illustrates the trends of the hydraulic gradient change within the 3 m plug of case 1. Figure 10 presents the relation between the average hydraulic gradient along the center line of the plug and its average thickness. This relation

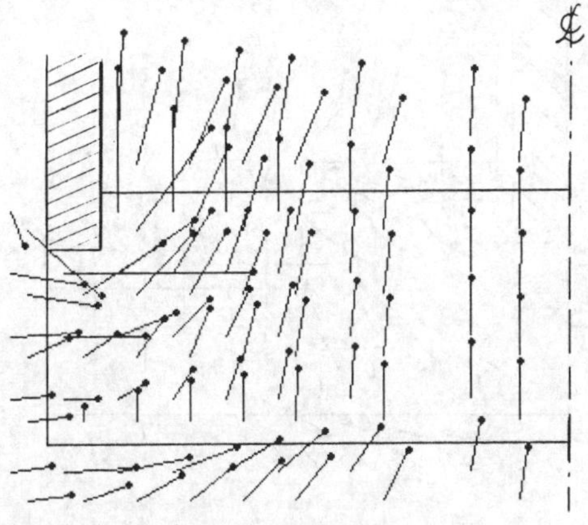

Fig.4 Velocity field within the plug (case 1)

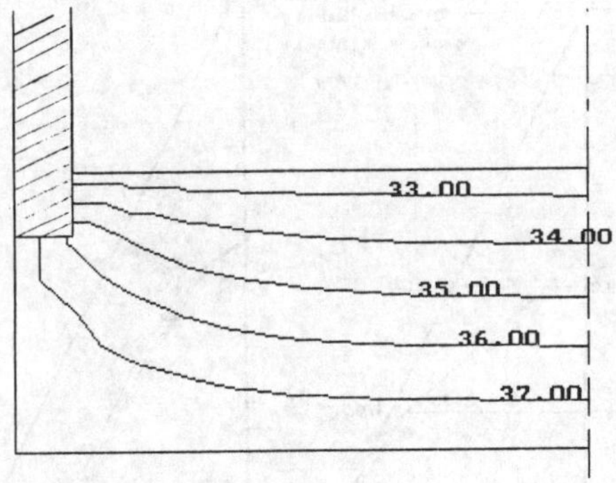

Fig.5 Equipotential lines within the plug (case 1)

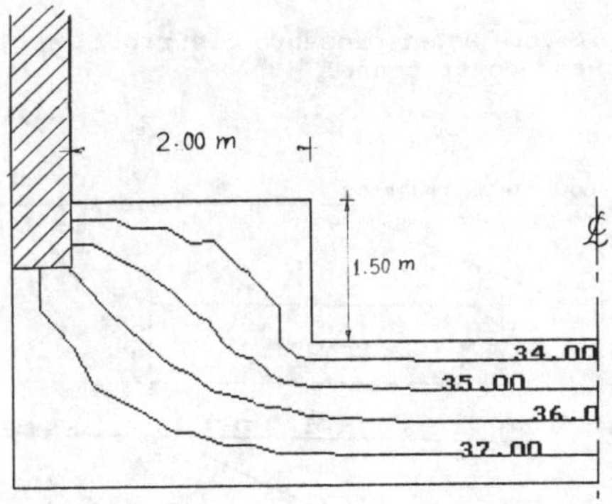

Fig.6 Equipotential lines within the plug (case 5)

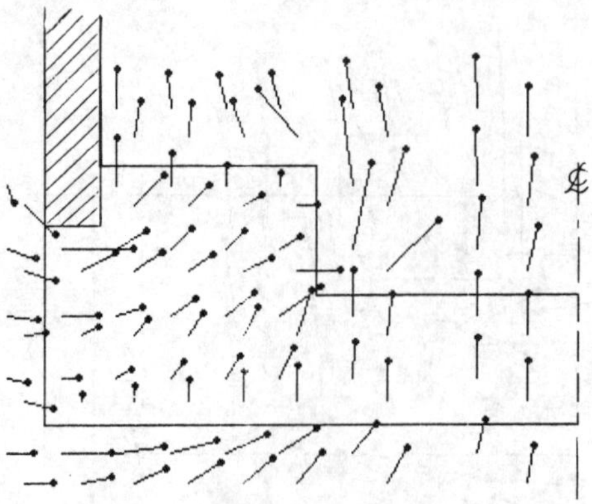

Fig.7 Velocity field within the plug (case 5)

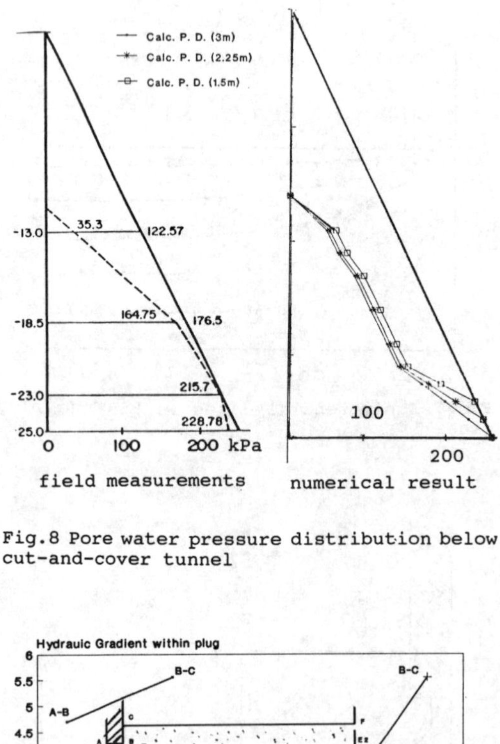

Fig.8 Pore water pressure distribution below cut-and-cover tunnel

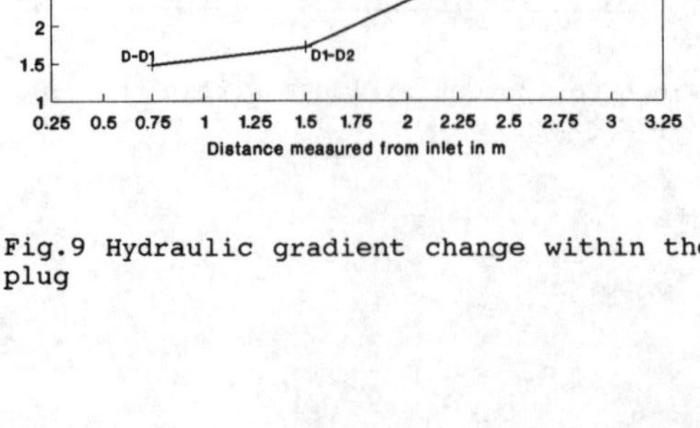

Fig.9 Hydraulic gradient change within the plug

shows that the 3 m plug gives the best gradient recommended in the conventional design of the plug that is less than 3. It shows also the possibility of obtaining a value of the hydraulic gradient near the recommended one if the plug thickness is optimized taking results of case 2 and case 4 as a guideline. Also, the relation between the factor of safety against heave and the average thickness is illustrated in Figure 11 which proves that case 4 and case 5 give values of the factor of safety which are similar to that of case 1.

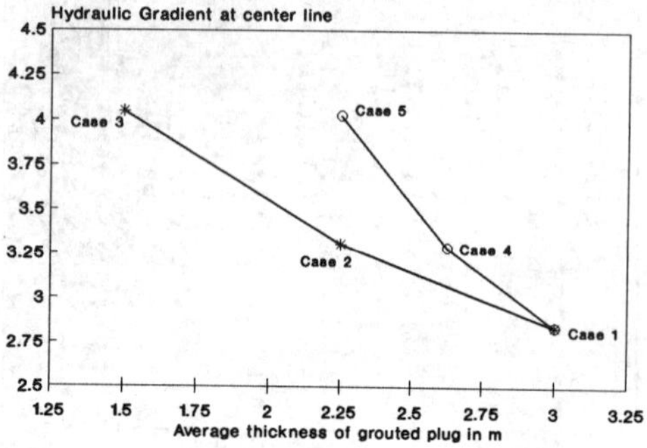

Fig.10 Effect of plug thickness on hydraulic gradient

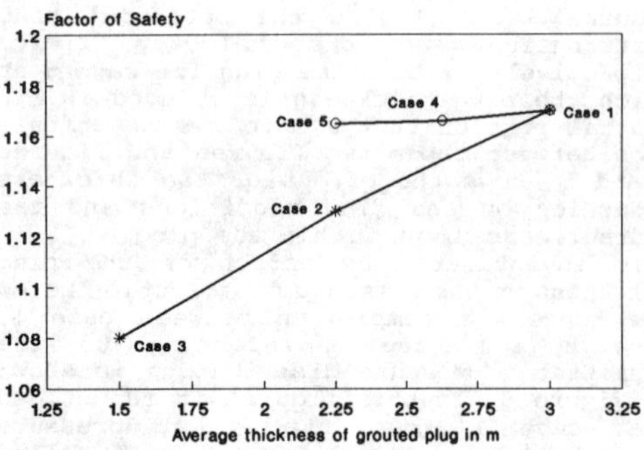

Fig.11 Effect of plug thickness on factor of safety against heave

7 CONCLUSIONS

1. This paper confirms the successful numerical modeling of groundwater flow around cut-and-cover tunnels and subway stations provided with grouted plugs. This was evident by good matching between measured and calculated behaviours.
2. Conventional design of grouted plugs is oversimplified utilizing assumption of one dimensional equilibrium and flow conditions. On the other hand, two dimensional analysis

represents the actual behaviour more realistically to calculate the variation of both hydraulic gradient and potential head in the two dimensional domain.
3. Numerical results given in this paper indicate the need for optimization of configuration of grouted plugs. Such optimization will require new design guidelines taking into account the two-dimensional nature of the problem.

REFERENCES

AFTES, 1991. Recommendations on grouting for underground works. Tunnelling and Underground Space Technology, Vol. 6, No. 4: 383-461.

Bear, J., 1972. Dynamics of fluids in porous media, New York, American Elsevier.

El-Nahhas, F., Eisenstein, Z, Shalaby, A., Salam, M. and El-Bedaiwy, S., 1988. Geotechnical performance of diaphragm walls during construction of Cairo Metro tunnel. Proc. Tunnels and Water: 259-266. Rotterdam, Balkema.

Hassan, A. A. 1989. Hydrology of wadis in arid regions with application to quality models. Ph.D. thesis, Ain Shams University, Cairo, Egypt.

Hassan, A.A., Shalaby, G.A. and Ahmed A.A., 1992. Coffer-dam-flow hydraulics interaction in porous medium. Soil mechanics & foundation, Vol. 7: 45-49.

Salam, M., 1984. Grouting of alluvial soil in Cairo Underground Metro. Proc. Tunneling in Soft and Water-Bearing Grounds: 135-142, AFTES.

Tunnelling and Ground Conditions, Abdel Salam (ed.) © 1994 Balkema, Rotterdam, ISBN 90 5410 363 9

Transferring the European experience using Mixshields for the employment in Cairo

Martin Herrenknecht
Herrenknecht GmbH, Germany

B. Maidl
Universität Bochum, Germany

ABSTRACT: By the early 1980's, a significant amount of work had been carried out using slurry tunnelling machines for advancing tunnels in various water bearing sands and gravels. However, in the mid 1980's it was realised that such technology had its limitations. This led to the decision that a new system, capable of tackling projects where water bearing sands, gravels and silts, clays and hard rock conditions co-existed, was needed. These conditions would best be fulfilled by developing a tunnelling machine in which conversion from one running mode into another be achieved relatively simply from within the tunnel when ground conditions changed during advance. Thus, the idea of the Mixshield was born.

1 INTRODUCTION

Since 20 years the Hydroshield is successfully used for tunnelling in noncohesive, waterbearing soils. More than 25 km of tunnels of all kinds are constructed by this system; now with diameters ranging from 2 to 11 m , and covers between 0.5 and 3 times the diameter. By segmented concrete lining, rates of progress from 24 m in 24 hours were achieved with an average of 15 m per day. The reason for the widespread use of the Hydroshield is the reliable, clearly arranged and simple system. The air cushion behind the diaphragm-wall guarantees

- the supporting pressure all over the face within very narrow limits
- a simple, partial or total evacuation of the working chamber by pumping, giving easy access to the front
- and, consequence, the fast removal of obstacles or the fast exchange of the tools

The experience led to a perfection of the Hydroshield system; but they also revealed the limits of application determined by geological conditions.

Which are those limits?

Hydraulical mucking used as a rule with the Hydroshield requires a separation plant which basically comprises mechanical screens and hydrocyclones. By this, it is possible to separate grain sizes as small as 62 microns approx. from the liquid economically. With high investment expenses the separation limit may be lowered to 20 microns by adding a second stage of hydrocyclones with smaller diameters or even by a centrifuge. The investment costs of hydrocylones with smaller diameters or even by a centrifuge. The investments costs of the additional hydrocyclon stage are about DM 200.000,--; for a centrifuge producing 30t/h of solids at a flowrate of $120m^3/h$, DM 900.000,-- must be paid; the operating costs are to be added in both cases. It must be remembered that the chances of a breakdown grow with the complications of the system which may affect the progress of the tunnelling machine.

With the additional equimpent silty ground may be included in the scope of the Hydroshield; however this makes sense only with a part of silt not to large; it should not exeed 20 %. In the case of a more important part of silt the separation costs become uneconomic: the use of a dry mucking system under compressed air may be the better solution in the case of a tunnel laying part of its length in clayey, silty soils and the remainder part in water-bearing sands and gravels. That this is not a theoretical example, shows the construction of well known project "Galleria Aurelia" in Rome; a two-rack railway tunnel with an outer diameter of 10.64m. There, the first 1,200 m were in stable clay; the remainder in waterbearing sands and silts. The clay was cut by the open wheel, loaded with lobster arms on the belt-conveyor and carried out by train. It took months to change the machine into a Hydroshield underground; parallel to that a separation plant was installed on the trailers. With this adequate system tunnelling in the sands took place with difficulties.

2 The MIXSHIELD

Obviously it was reasonable to develop a tunnelling machine for those or similar geological conditions; this type should be adaptable on site to the variations of the ground with a minimum of work and expenses.

Based on this preconditions, the company Wayss & Freytag AG developed together with the manufacturer Herrenknecht GmbH a machine which we called "Mixshield" because of their different possibilities of application.

2.1 Requirements

The Mixshield is designed for the following requirements:

1. It has to work as a shield machine with
   - liquid supported face
   - face-support by compressed air
   - face-support by earth
   - hard rock TBM with conveyor mucking systems.

The basic machine should be the same for all four versions, that means steel construction, the cutting wheel with gear, the erector and the trailers.

2. The conversion between the different types has to be possible with little amount of time and costs.
3. The conversion has to be possible underground.

Together with the additional demand for strictly separating of the functions "cutting of the ground" and "supporting of the face" the design allows the cutting of the grour in all 4 versions in the same manner; the variations concern the medium to support the face and the mucking.

2.2 The basic machine

To meet the before mentioned conditions is only possible with a "drum type" machine, that means a peripheral drive which allows to install the conveyors for dry mucking in the open centre. Here we see one of the main differences between Mix- and Hydroshield. The first has a peripheral drive; the second is a centre shaft machine. In addition, there is an other advantage of the drum type machine: it is possible to install a higher torque. However we did not intend to do without the well-proven air cushion. It is located in the annular space limited by pressure wall and diaphragm wall in one direction and the steel cylinder covering the gear and the shield's skirt in the other.

Part of the basic machine is also an integrated double chamber air lock which provides acces to the working chamber. It is possible to mount different cutting wheels on the driving drum; usual is an open cutting wheel with 3 to 7 spokes.

2.3 Mixshield version "liquid supported face"

If no stones bigger than 100 mm are expected, the centre passage may be closed. Suction pipe and feeding nozzles can be placed without problems. The supporting pressure is regulated by the air cushion. In this version, the construction of the machine is similar to that of the Hydroshield.

If there are stones bigger than 100 mm to be expected, a stone crusher will be assembled or a closed wheel with disks will be used to breake the boulders in such pieces, that with the slurry system the fractured stones can be handled.

In this version, the mucking place by a screw conveyor, located in the centre of the machine. A rotary lock delivers up the muck inthe atmospheric part of the tunnel were it is carried out by train. The capacity of the screw conveyor

and the rotary lock is to be co-ordinated with the progress of the machine and the expected grain size of the ground. The screw conveyor is fed by buckets on the cutting wheel. To avoid connection between compressed air at the face and the tail sealing it is useful to inject the space between shield and earth by some injection nozzles in the shield's skirt.
Only the working chamber is under air pressure; the link to the tunnel being under atmospheric pressure is the rotary lock. Access to the face is possible by an air lock. A separation plant is not necessary any more.

2.4 Version "earth pressure balanced shield"

For effective stabilizing of the face by earth, in this version the construction of a drum in connection with the cutting wheel is necessary, the only outlet for the ground is the hole which leads to the spiral conveyor. The spokes of the cutting wheel are connected by a wheel rim. The triangular spaces between the spokes were closed by steel plates in such a manner that the material flows naturally to the spiral conveyor. In principle, the ground should be suited to close the working chamber pressure-tight by complete filling of the drum and the screw conveyor. Especially in Europe's geological conditions, this will be rather seldom. In this case, the pressure-thightness has to be achieved by injection of the screw; injection of the material in the drum is also possible. The pressure is controlled by pressure gauges in the working chamber and the screw conveyor. The pressure is controlled by pressure gauges in the working chamber and the screw conveyor. The pressure can be regulated by the speed of the screw. At the outlet of the screw conveyor a regulation valve as well as an automatic weighing plant is located. In the tunnel, the material is transported by train; with sufficient water-content mucking by pumps is possible. If the natural composition of the ground is not altered, a separation plant may be avoided.

3 SITE EXPERIENCES

3.1 HERA circuit in Hamburg

In underground construction work, one should not do too many steps of development at the same time. So we decided to use the Mixshield with liquid supported face at first.

The première took place in Hamburg, where an underground circuit was to be constructed for the "Deutsche Elektronen-Synchroton", abbreviated "DESY", in which for microcosmic investigation electrons and protons where brought to collison at very high energy. The tunnel, which is not a true circle, is 6.3 km in circumference which corresponds to a diameter about 2 km. Four underground hols subdivide the tunnel in 4 equal parts; the experiments will take place there. The tunnel has a cover from at least 10 m for radiation protection reasons; the maximum cover is 30 m.

The borings revealed mostly medium and fine sands with layers of marl up to 1.6 m thickness. Partially the tunnel touches the underlaying very compact and cohesive marl. Maximum groundwater pressure will be 1.4 bar The obviously small layers of marl did not indicate the conversion of the machine as an economic solution; so we chose the Mixshield in the special design with liquid supported face.

3.2 Mixshield, special design "HERA"

The machine has an outer diameter of 5.95 m. The pressure in the front chamber is regulated by the air cushion. The cutting wheel may be turned with a maximum torque of a 130 mt; the rotational speed is regulated dependent on the load. To facilitate steering the machine is articulated which was not too urgent having a 780 m radius. Dependent on the number of segements of the lining 12 thrust rams are employed with 160 to each.

Shifting of the cutting wheel in axial direction for 300 mm eases tool-exchange. A complicated sealing system prevents the pressurized liquid from coming into the tunnel. Openings in the diaphragm wall, which can be locked, allow access to the front chamber to remove obstacles or to change the tools. The trailers run on rubber wheels, the so-called "elephant" feet, so the usual installation of rails is avoided. The system was mounted on the innert of the experimental hall South in April 85; the boring begun in the middle of Mai. The encountered sand had a large part of silt; so we were forced to install a decantation basin as well as a centrifuge. After 300 m we found imbedded in a marl layer cobbles and stones in an unexpected number and density which diminished the rate of progress to a low 6 m per day in a 3-shift operation. 2/3 of the time were required to remove the obstacles; blocks with a volume of about 2.5 m were split hydraulically. After passing those obstacles, rates of 18 m per day and 80 m per week were achieved.

4 GRAUHOLZ TUNNEL

4.1 Tunnelling System; E alternative Proposal

The alternative proposal for heading the tunnel from one side with a Mixshield with an excavated diameter of 11.6 m was chosen. The tunnel was then to be lined with a monocoque shell with an internal diameter of 10.6 m and a wall strength of 40 cm. A ring of segments with a width of 1.8m comprising 6 elements and a keystone. The elements have to be sealed by means of all-round neoprene membranes.

In lose deposits, in rock with low overcap and in mixed-face conditions the Mixshield was operated as a slurry shield with fluidsupported face. In the rock sections, the Mixshield functioned as a normal TBM.

4.2 The tunnelling machine

The machine can operate in the slurry and the dry modes. In the slurry mode the machine has to be able to withstand waterpressures up to 40 m. The machine has also to cope with varying ground and in particular boulders and cobbles. Large boulders shall be cut still in the soil matrix by the disc cutters placed on the six arms of the open cutting wheel. In the center of the machine a stone crusher with an opening width of close to 1 by 1 meter is located. The oversize stones will be lifted by the arms of the cutting wheel and drop into the crusher.

A suction pipe removes the slurry from the base of the shield. The slurry is then transported to a first separation system on the trailing equipment of the shield where separation by sieves and hydrocyclone takes place. A secondary circuit goes to the surface where the finer particles shall be separated by a centrifuge and a settling pond. The mucking sheme is similar to the one described by Becker (84) but needs to be adapted to detailed local ground conditions.

In the dry mucking mode a conveyor belt will be placed in the central opening and the muck will be transported on a long conveyor belt through the complete tunnel to the outside.

5 MIXSHIELD METRO CAIRO

In order to decide on the design for the Mixshield Cairo, we took into consideration the following machines which have already been used in other projects:

a) Metro Lille
outer diameter 7.7 m
S1 tailskin seal
hydrostatic drive
double erector
b) U-Bahn Nürnberg/Metro Moscow
cutterwheel with double equipment - discs and flat chisel-type teeth
c) Duisburg U-Stadtbahn
stone-breaker
injection of the grouting mortar through the tailskin
d) Sevres-Chêres
Paris sewers in the most difficult geology
e) Stadtbahn Mülheim
double equipment
hard rock machine with slurry transport
f) Tramway Strasbourg
g) Metro Lyon
h) Metro Rouen

We were able to use the experience we had with these machines to decide on the design of the Cairo machine. However, it became clear that this information was not sufficient for the Cairo project and therefore we drew up a list of specifiations together with the Joint Venture.

5.1 List of Specifications

After the order had been placed with Herrenknecht GmbH in April 93 the Joint Venture and Herrenknecht formed a technical commission in order to draw up these technical specifications.

Basics for the list of specifications:

technical data of the TBMs up to 1993
values derived from previous experience/use
parameter
geology
tunnel profile
hydrological conditions
logistic

5.2 Rough draft of the Cairo Mixshield

After approximately 3 months' consultation the general rough draft of the TBM was submitted to the Joint Venture and approved by them.

Mixshield which could be used with slurry transport
assembly of the shield in parts, can only be screwed not welded
cutterwheel which can be used for cutting in both directions, left and right
hydrostatic drive
vacuum erector (weight of the segments approx. 6 tonnes)
final calculation of the shield and the cutter wheel

5.3 Innovations in the Cairo Mixshield

- vacuum erector able to pick up segments weighing up to 6 tons
- airconditioned control cabin
- agitator
- injection through shield tail
- automatic program for slurry circuit

5.4 Final remarks

With a design and development of machines by experienced engineers as in the case of the Cairo Mixshield it is possible to obtain the best results in the design, construction and assembly as well is in the actual construction of the project in itself.

Machines which involve a high degree of innovation can only be realised by cooperation and experimentation in the design phase.

The example of Cairo shows that this cooperation between the building industry, the contractor and the machine manufacturer is absolutely necessary if the thorny problems that can arise are to be solved.

At this particular point I would like to express my special thanks to the Joint Venture Cairo for their particularly fruitful cooperation during the 4 years in which we have worked together on research and development, manufacture and construction

Tunnelling and Ground Conditions, Abdel Salam (ed.) © 1994 Balkema, Rotterdam, ISBN 90 5410 363 9

Prefabricated lining, conceptional analysis and comparative studies for optimal solution

Teodor Iftimie
Civil Engineering Institute, Bucharest, Romania

ABSTRACT: The paper tries to achieve a synthesis of the main factors which influence the achievement of an economic lining, in the conditions of preserving the entire safety concept. At the begining it is achieved a systematised analysis of all methods and structural models for circular linings, precast or cast in place. An hierarchy of the main factors which influence the behaviour of the lining is established and the most important factors are analised in detail. The type of joint between the precast segments, the configuration of the segments in cross-section and the way of achieving the injection at the extrados of the lining are the essential factors which are analized in interconnection to determine an optimal solution.

1 INTRODUCTION

In the tunnelling history the use of prefabricated lining is closely connected with the circular shield method.

The obvious advantages of this method imposed it to the most diverse soils and in the most different dommains, from railway tunnels to metropolitan line and from urbane sewers to hydrotechnic galleries.

For more than a century of existence, this method and therefore the prefabricated linings have conquered the whole world, representing at tis moment the main proportion in the total volume of tunnels building.

The advent of schield machines (TBM) with bentonite sludges or air pressurised brought to a new and spectacular developpment of this domain, giving the impression of a quick and safe solution in tunnelling in any soil conditions.

In a simplistic way, the problem resumes at the finding of the most adequate excavating equipment and the most economical lining which should suit the geological situation.

If the problem of equipment is one of choise, that of lining remains open, considering the complexity of factors that are influencing it. The choice of the lining represent the most important factor in a tunnel project. The lining ensures not only the stability of the excavated opening but also the use in safety service conditions, influencing also the advancing speed during execution.

The lining is also the most important singular element in the total cost of tunnelling representing, according to several statistics, between 15 and 40%.

The importance of this element in the general proces of tunnel designing, made it one of the most studied elements.

Because of their degree of use, prefabricated lining have always been the first subject of this researches, many important improvements and innovations both in the domain of calculation and in that of constructive solutions, being recorded in time.

The present paper tries to systematise the experience that was accumulated and to highlight some of the problems that are specifically to this kind of linings.

2 BRIEF ANALYSIS OF STRUCTURAL METHODS

The establishing of the structural model for tunnels is a more difficult problem than that of overground structures. Tunnel linning is a foreign body in to system (soil) of wich initial stress state was disturbed through the excavation of the hole.

The surrounding ground is the source of applied forces to the lining, but is at the same time bearing element which cooperates with the lining for the stabilisation of stress state.

The real model of tunnel is extremely complex, because of lots of parameters and of the complexity

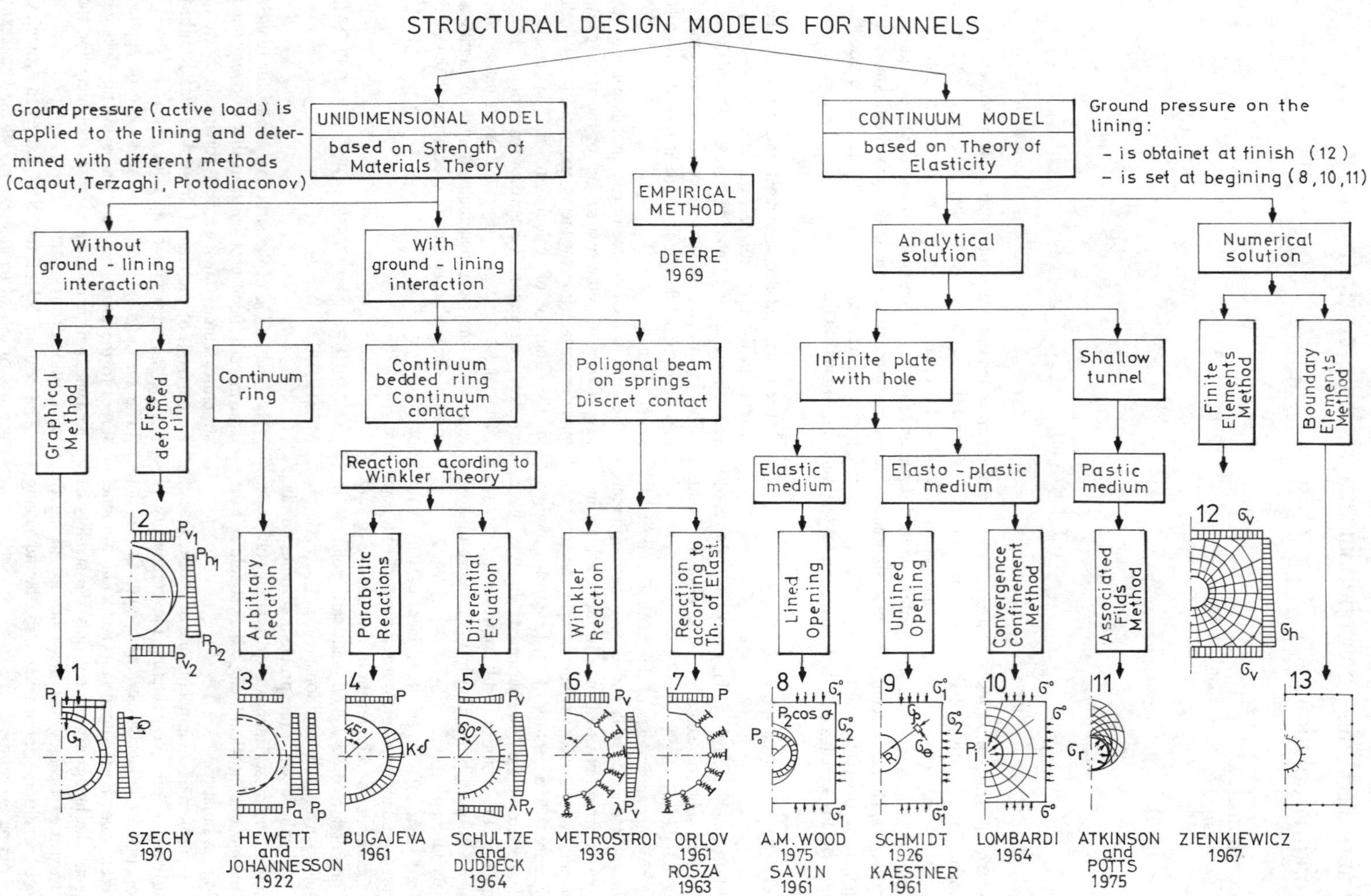

Fig.1 STRUCTURAL DESIGN MODELS FOR TUNNELS

of system behaviour.The tunnels designing engineer most frequent, has to create a structural model which behaves most fitted to the real system.

This model must include the statical calculation scheme, the behaviour material laws and the concept of safety.

The I.T.A. Guidelines for the Design of Tunnels, present the general design procedures for tunnels, from site investigation to in-situ monitoring, and from design criteria and structural design models to structural detailing of the lining.

The present paper tries to detail some problems specifical for circular prefabricated lining.

In Fig.1 is an attempt of systematisation of calculation model for circular lining, as much as is posible in an evolutional order. The general division corresponds at the two domains of Deformable Mediums Mechanics.

The unidimensional model considers the lining a curved beam and the loads are established separately with diverse theories from which the more known and more comunly used are those of Terzaghi and Protodiaconov.

The two or three dimensional continuum model determines the stress and strain state in the ground and in the lining using methods of the Theory of Elasticity and numerical methods.

The tunnels designer has to choose a method from the existing lots and he will be bewildered a little.

What is the most adequate method?.

The first temptation is to make a comparison

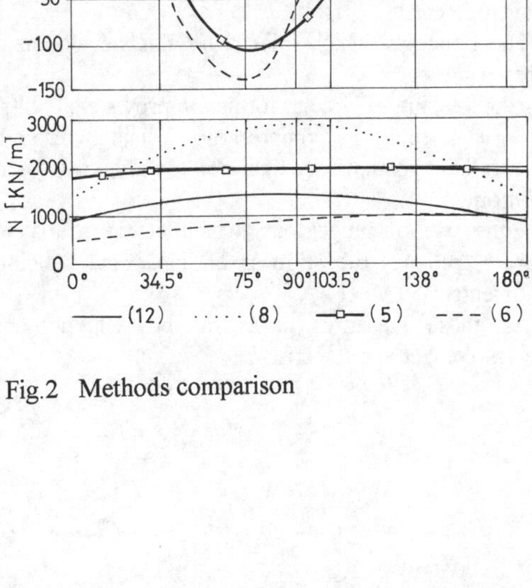

Fig.2 Methods comparison

between the best known methods.

In Fig.2 is attempted such a comparison, for the calculation of lining made of prefabricated elements with articulated joints.

Four methods were used, three clasic between the best known and F.E.M. with a fine discretisation in the zone of the hinges. The first two are from the first group (unidimensional model), considering the cooperation between lining and ground according to Winkler theory and with a detaching on the zone of calotte. The first method of Schulze and Duddeck is with continuum contact between lining and ground and with its own establishing of loads.

The second method is of poligonal beam on springs, with discrete contact at the junctions and establishing of loads according to Protodiaconov.

The third method, of Muir-Wood and Curtis, as well as F.E.M., belongs to the second group (continuum model), but on the subgroupe with analytical solutions, the infinite plate with lined opening.

In order to homogenize the assumptions for loads of methods Schulze/Duddeck and M.Wood with F.E.M., the assumption of relaxation of initial stresses with 60%, in the moment of lining assembly, was admitted.

The results obtained are far to be near and far to dispel doubts and to give a definitive option for a certain method.

Nevertheles, a similitude between the methods of the same group for moments, as order of size, and between F.E.M. and poligonal beam method as forme of moments diagram is perceived.

In this comparison, is not looked for the order of size of sectional stresses, but the form and their sign counterpart of joints position between segments.

The existance of joints hinge type between segments is a certainty which can not be neglected and which is taken into consideration only in poligonal beam method and F.E.M. The method of poligonal beam on springs is perfectly adapted to electronic calculation and capable to catch the peculiarities of behaviour of this lining type, given by the type and rigidity of joints and the number and their pasition in section.

The use of F.E.M. with its increased posibilities of analysis, elasto-plastic and time-dependent, can be an arbitrage for the clasic methods, pointing out the deficiencies and qualities of diverse methods.

But, the F.E.M. is a more laborious method and therefore less used.

The method of poligonal beam of springs, improved with conclusions of F.E.M. analysis, especially for loads, can still be a simple, efficacious and rapid

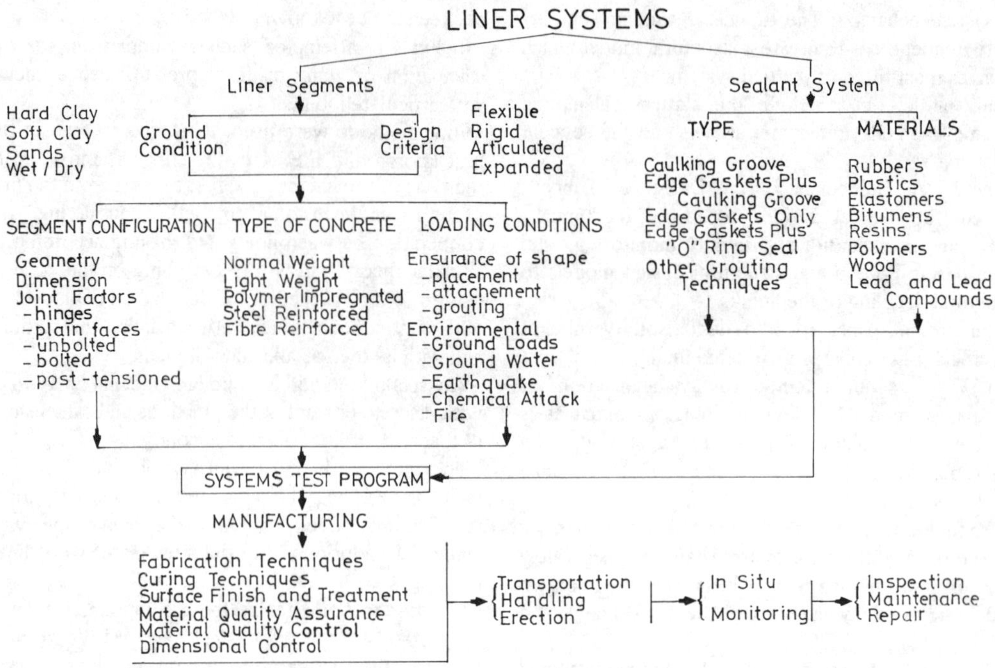

Fig.3 Tunnel Linner System Design Flow Chart (Tartaglione, 1977)

analysis instrument of prefabricated lining.

The efficiency can be ensured only through a thorough knowledge of the main factors which influenced the behaviour of this lining type.

3 ELEMENTS OF PREFABRICATED LINER DESIGN

In the shield method some constructive systems are used:

- the system with two lining, a prymary prefabricated lining and a secondary lining of cast in place concret;
- the system with a cast in place pressed lining;
- the system with a prefabricated lining.

The prevalent tendency is to use only one prefabricated lining which serves as both immediate and permanent lining.

The main requirements for tunnel lining segments are:

-immediate bearing capacity against external earth and water pressures without detrimental deformation or leakage;

- resistance to inpact stresses due to rough handling, transport or erection operations;
- resistance to high axial stresses produced during the advancement of the shield by the propulsion jacks;
- resistance to moisture and ground water effects on the segment itself, on its joints;
- economy in construction and maintenance.

In a simplified scheme (Fig.3) it is tried to resume the most important factors which intervene in designing of satisfactory lining system.

Between the elements presented in this scheme, the type of lining, the configuration of segments and the keeping of forme, appear to be among the most important and they will have been briefly analised further.

A schematic diagram is given in Fig.4, taked and actualised, illustrating the sequence in lining development.

The concepts which infer from this development are:

- the keeping of circular form with provisional bolts;
- the keeping in compresion of lining, through controllabe grouting or expansion towards excavated contour;
- the use of hinge joint which centre the axial forces, reduces the stiffness of lining and bending moments.

All these measures guide to the reduction of reinforcement and the cost.

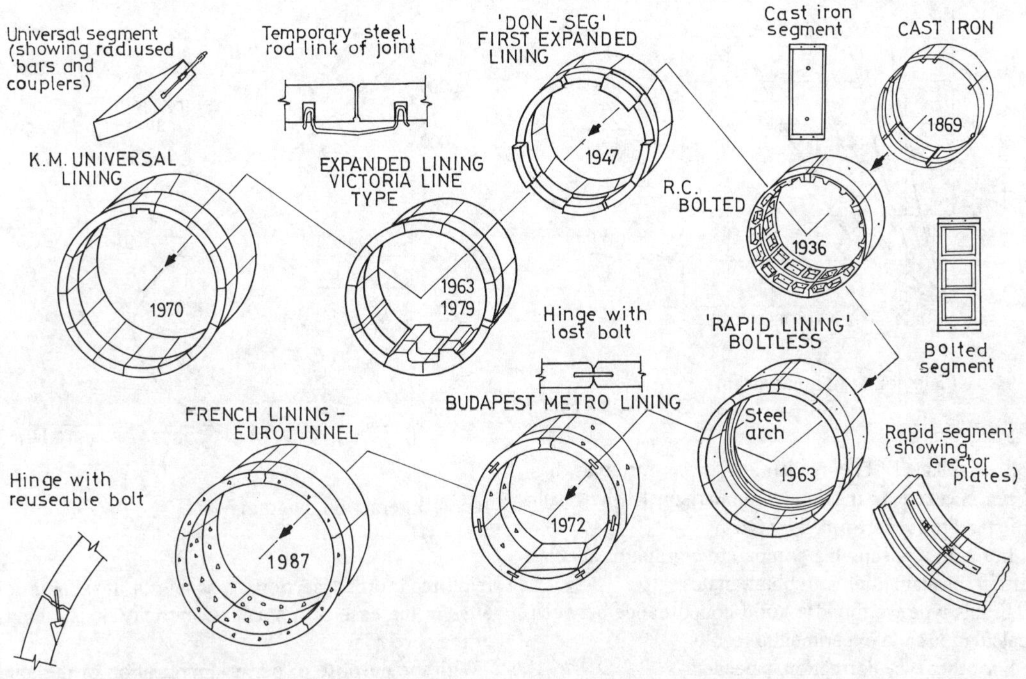

Fig.4 Development of Prefabricated Tunnel Linning (Golder Associate, 1976)

4 THE STUDY OF JOINTS BETWEEN SEGMENTS

The existence of joints between segments is the main characteristic of prefabricated lining.

Mechanic behaviour analysis of hinges with convex-concave radius and of plane joint stiff at bending, must highlight the infuence of eccentricity and stiffness rotation on bending moments.

Also, must analyse the stress-strain state, all around of contact zones, for the establishmentof the influences on neighbouring zone and on all structure.

In joint zone, the ring compresion force is transmitted from one segment to an ather through a surface, less than that of usual cross section.

For plane joint we start from the assumption, confirmed by practice, about the impossibility to materialize an perfect contact between the two faces, because of the following causes:

- the impossibility to make some geometric perfect segments. The tolerances are admited in the most cases;
- the difficulty to keep the circular form both in the montage phase and in service;
- the elastic deformation of ground;
- deformability of prefabricated elements.

The manufacturing inaccuracy may cause bearing on the edges (open joint), with stress concentration and rotation between elements, which appear at erection or under the loads, can cause similar effects or more to those existant.

In a static scheme, the contact zone between segments may be assimilated to an elastic clamping with a rotation stiffness Cd.

This stiffness may be calculated using a simple scheme (Fig.5), keeping in mind the fact that the joint can only take over compression stresses.

Considering a parabolic distribution of stresses, the rotation angle α and the stiffness at rotation Cd can

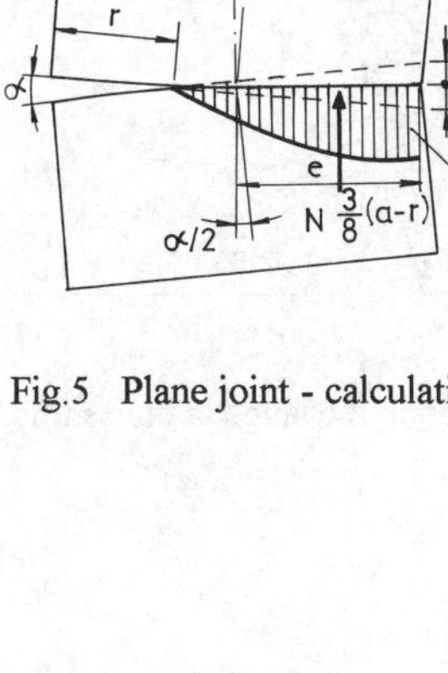

$$\alpha = \frac{27\,N}{32\,a b (1-2m)^2 E_o}$$

$$M = C_D \cdot \alpha$$

$$C_D = \frac{N \cdot a(1-2m)}{4\,\alpha}$$

Fig.5 Plane joint - calculation elements

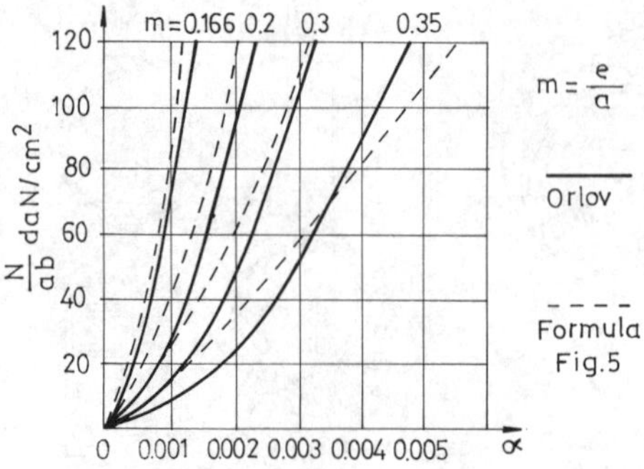

Fig.6 Plane joint. Angular strain

be established. The realism of these formulas has been checked by making a comparison to the result of stand tests, performed by Orlov.

In Fig.6, it is shown a comparative graph of angular strain for plain joint with blunt angles.

In this type we found a good concordance between calculations and experimental results.

For other type correction is needed.

The influence of angular strain in joints over the static action of linings depends greatly of the dimensions of its elements and the distribution of joints on the contour of lining. Using this formulae, the bending moment and axial forces inside the joints in case of various eccentricities, can be calculated, with the purpose of drawing the M-N interaction diagram.

Fig.7 present the diagrams of interaction in case of eccentric compresion, in usually cross section and in plane joint with blunt angles, for a prefabricated lining 30cm thick.

It was noticed that the bending bearing capacity of the joint is always smaller than that of the usual

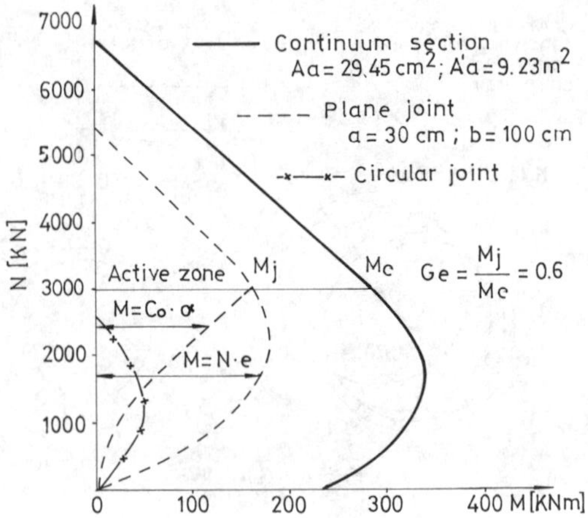

Fig 7 Interaction diagram M - N

reinforced cross section, reaching a maximum of 60% in the case of a small eccentricity, in the limits of the core zone.

With the purpose of better appreciation of the joint effect, in the case of using electronic calculation of lining, the degree of elastic clamping Ge of a such joint was conventionable defined, as the ratio between the moment of the joint and that of continuum cross-section, considered perfect clamping.

Ge can take values between 1.0, for perfect clamping and 0 for hinge joint, but practically in the case of the studied joint between o.6 and 0.

Therefore, the stiffness matrix of a beam with elastic clamping can be written, multiplying the elements of the stiffness matrix for a double clamping beam by the same coefficients, function of Ge.

In this way the values of moments in lining can be

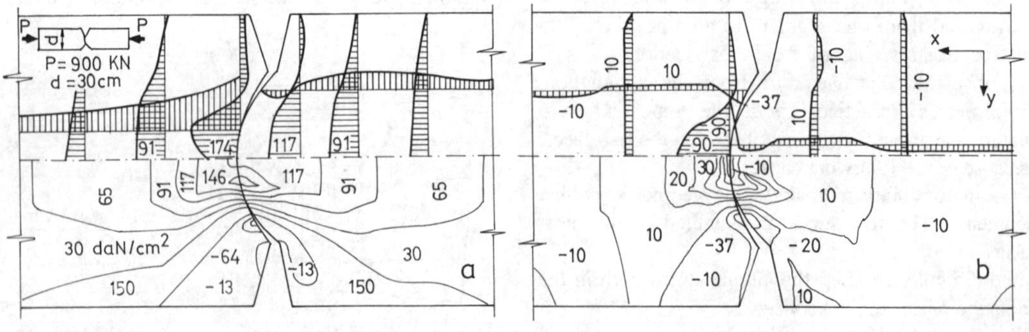

Fig.8 Stresses distribution in the hinge at concentric load . a) σx stresses b)σy stresses (- tension)

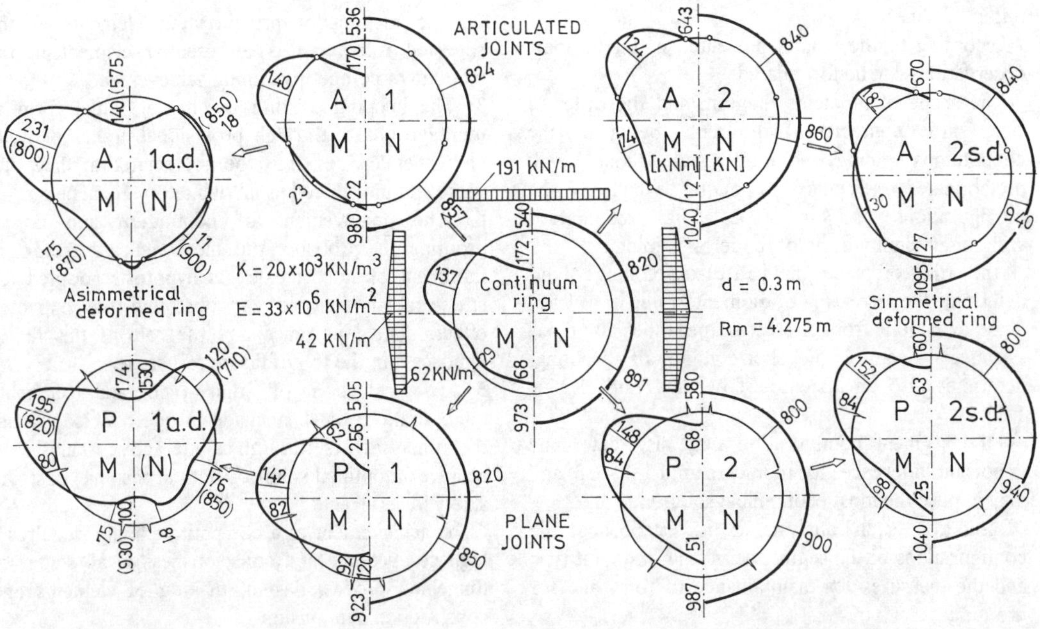

Fig.9 Influence of joint type (A, P), segments configurations (1, 2) and initial deformation (s.d., a.d.) on sectional stresses M-N.

calculated, considering the effect of the eccentricity of the axial force in joint and influence of degree of elastic clamping Ge.

This judgement may be extended also upon the fissured section, approximating the degree of elastic clamping corresponding to the new rigidity.

Therefore, it can be calculated, in a simple way, but through iteration, the effect of rigidity reduction by fissuration upon the value of moments in the lining.

The cas of joints with bolts, may also be considered in the same way. For hinge joint with convex-concave faces, mechanic behaviour analysis is more simple than for plane joint.

Jansen, in his dissertation, analyses the main aspects of this joint type behaviour, from rotation phenomenon to that of sliding and flattening.

A test to check the relations establishing the height and wideness of flattening, as well as the stress distribution in the contact zone, was made using F.E.M.

With this purpose were discretized both components of hinge, and contact surface was established through iteration. A exemplification is given in Fig.8, and present a good concordance with experimental results of Rosza.

5 ANALYSIS OF THE MAIN TYPICAL FACTORS INFLUENCING PREFABRICATED LINING BEHAVIOUR

In the specialised literature there are numerous attempts to establish, analyse or classify the factors influencing lining behaviour, in generaly or for those prefabricated in particular. This paper is trying to bring in the first plane some typical factors, which are considered esentials for a good structural analysis of prefabricated lining.

We propose the following factors:

- the type of joint;
- the keeping of circular form in initial phaze;
- the configuration of segments on ring.

The influence of these factors over the size of sectional stresses, is marked further, with a structural analysis on tunnel lining considered successively continuous, with hinge joint and with plane open joint (Fig.9).

Two configuration of segments on ring, with both type of joints and also the influence of the initial simetric and assymetric deformations are analysed.

Evidently, other configurations (with 4, 6 or 8 segments on ring), as well as other joint opening or deformation variants can be analysed. Only the qualitative influence of these three factors are pointed out in this lines.

According to the results presented in Fig.9, some observations can be formulated:

- the shape of moments diagram and the order of size, are appreciably different between the conntinuous lining solution and those with prefabricated elements, no matter which is the configurations or joint type. That recommend without ambiguity using of model with joints.
- the using of plane joints in place of hinge joint, determine the increasing of moments value until 1.5;
- the deviations from circular shape, especially those assymetric, caused by shortcoming of grouting, determine also an increase of bending until 1,35 - 1.65;
- the segment configuration may also determine important differences for moments until 1.4 - 1.8, and may impose the most economical solution.

Based on such analysis it is established the configurations of the segments on ring, the joint type and the measures for maintainance of form and for grouting.

Afterwards we can get to a parametric study for selected solution, varying diverse parameters as the earth pressure coefficient ,the reaction modulus and other.

Finally we can draw a wrapper of maximum moments wich gives a better safety but increase the cost.

A comprehensive analyse , in this sense, make Duddek.

The analysis of all factors and parameters influencing the behaviour of prefabricated lining serves finally at the establishing of safety concept.

This domain remains still open for underground constructions, because of the complexity, scattering and lots of parameters which should be considered in an interacting unity.

An inportant part of future research effort must concentrated on this domain.

6 CONCLUSIONS

1. It is obvious that as for as tunnels are concerned no matter what method is used, the structural analysis has still an orientative character, being at the same time unreplacable. Here the engineer art, intuition and experience intervene decisively compensating the failings of the method used.

Nevertheless a conclusion is abave doubt: the structural model adopted must take into consideration the existent joint, their type and distribution on the ring.

2. The making of prefabricated elements at the sugested tolerances is an esential element in the strategy of competitive lining achievement.

3. The keeping of ring in circular initial form at montage phase, through provisional arches or bolts and other devices, must be appreciated at they true value and introduced as an obligatory element.

4. The achievement of continuum and corect grouting, is probable the most delicate factor, in fulfilment or not of calculus hypotesis adopted for structural model and for obteining the expected effects. The inportance and the role of this factor imposes it in the top of future researches.

5. The selection of joint type and segments configuration, must be made in interconnexion with the other factors through a interactive analysis, to achieve an optimal solution, both at the level cost and safety in service.

The achievement of a competitive lining, requires a high competence and consciousness in all stages of tunnelling proces, design, making of elements and construction supervision.

In all these stages the human factor is decisive.

REFERENCES

A.F.T.E.S.- Working Group nr.7, 1993. Temporary supports and permanent lining.Tunnels et Ouvrages Souterains. Supplement au nr.117,Mai-Juin.

Duddek,H. 1972. Zu den Berechnungsmethoden und zur Sicherheit von Tunnelbauten. Der Bauing.,47.

Golder Associates, James F. Mac Laren Limited. 1976. Tunnelling technology - an appraisal of the state of the art for aplication to transit systes. Ministry of Transportation and Communication, Ontario, Canada.

I.T.A. - Working Group.1988. Guidelines for the Design of Tunnels. T.U.S.T. Nr.3.

Jansen, P.1984. Tragverhalten von Tunnelausbauten mit Gelenk Tubbings. Brauschweig Techn. Univ. Dissertation.

Muir Wood,A.M. 1975. The circular tunnel in elastic ground. Geotechnique 25,No.1.

Orlov,S.A.1961.Metodi staticeskogo rasciota sbornih jelezobetonih obdelok tonnelei.Moscow. Gostroiisdat.

Rozsa,L. 1979. Precast concrete segment lining of the Budapest Metro. Tunnels and Tunnelling, Dec.

Schulze,H., Duddek,H. 1964. Spannungen in schild-vorgetriebenen Tunneln.Beton und Stahlbetonbau,8.

Tartaglione,L. 1977. Segmental Concrete Liner for Soft Ground Tunnels. Journal of the Construction Division, 6.

Tunnelling and Ground Conditions, Abdel Salam (ed.) © 1994 Balkema, Rotterdam, ISBN 90 5410 363 9

Design improvement of reaction force in Fronte-Jacking Method

Katsuyasu Ikuta, Shoichi Furuyama, Isao Abe & Hikosaburo Yoshida
East Japan Railway Company, Tohoku Construction Office, Japan

ABSTRACT: At railway sites, there are many kinds of methods to construct underpasses under busy railway lines. In the case of Fronte-Jacking Method, which is a nonopen cutting method, protective work for tracks called Pipe Roof Method is generally done to keep train services safe.

In this work, steel pipes (about 60 cm in diameter) were put side by side horizontally to make them into a roof or hat-like shape. It bears overhead load and acts as lagging. After this protective work, a prestressing steel strand and Fronte-jacks were used to haul an under-pass structure constructed by surface work and put it in the prescribed position.

This time skin friction force was expected between soil and pipes, and planning, design and construction of Pipe Roofs were improved so as not only to protect tracks but also to use them as reaction forces. This report has made clear that Pipe Roofs for reaction forces of haulage can be fully expected as tractive reaction force from the result of measurement, and can reduce the construction cost.

1 INTRODUCTION

Many methods are used for constructing railway-crossing structures such as the passage under railway lines. Among these constructions, there is a nonopen cutting method in which a structure is constructed by excavating from both sides of the track or subgrade surface, instead of starting excavation from the surfaces of the track or subgrade, and since former method requires no closing of railway lines, it is frequently used in recent years. In such a nonopen cutting method, reaction facilities for propulsion or traction are provided at a location which will not hinder railway lines. These reaction facilities may differ depending upon the sizes of structures, crossing lengths, etc., but sometimes these facilities become large-scale ones and account for a large part of construction expenses.

In Fronte-Jacking Method (hereinafter abbreviated to "Fronte Method"), which is one of the nonopen cutting method and which has comparatively many results of adoption, a track protecting work called the "Pipe Roof Method" is generally carried out in order to ensure the travelling safety of trains.

This time, the authors have newly planned, designed and executed this Pipe Roof work, not only for using it for the purpose of track protection but also for making the Pipe Roof into the tractive reaction of Fronte Method, expecting skin frictional resistance between the Pipe Roof and earth. The present paper 1) introduces the outline of the Pipe Roof Reaction Method, 2) indicates that the Pipe Roof can be expected to work fully as the tractive reaction force in view of the measurement results of Pipe Roof axial force and 3) reports other findings.

2 PIPE ROOF REACTION METHOD

2.1 Securement of Tractive Reaction (refer to Figure 1).

In the Fronte Method, the following basic conditions must be satisfied in order to haul the culvert. Namely, "tractive reaction > maximum tractive force." Maximum tractive force is ordinarily a sum of penetration resistance of the cutting edge and skin frictional resistance, etc. at the 1st calvert. On the other hand, explanation is given below regarding what will become of the tractive reaction, by dividing it into the former period and the latter period.

At the former period of traction, the 1st culvert is in the position outside of the passive slip line of earth in front of the reactive culvert, and for the tractive reaction in such a case, "(passive pressure of earth in front of reactive culvert) + (frictional resistance at the bottom surface of reactive culvert)" can be expected, and traction is fully possible.

In the latter period of traction, on the other hand, the 1st culvert comes inside the passive slip line of earth in front of the reactive culvert, and therefore, it becomes impossible, contrary to the case of the former period of traction, to expect the passive pressure of earth in front of the reactive culvert as a tractive reaction. Then in order to compensate this shortage of tractive reaction, generally the following means were used in the past to cope with this situation.

(1) To provide a protrusion on the bottom floor slab of the reactive culvert and to expect passive pressure of earth in front of the protrusion.

(2) To construct a reactive wall on the outside of the reactive culvert and to expect the passive pressure of earth in front of this reactive wall.

(3) To load temporarily soil on the reactive culvert and to expect an increase in frictional resistance at the bottom surface of the reactive culvert.

(4) To make the reactive culvert larger and longer, thereby expecting an increase in frictional resistance at the bottom surface of the reactive culvert.

(5) To increase the number of division of traction, to shorten the length of the tractive culvert and to make smaller the tractive force itself.

However, (1) and (3) have naturally their limits and, in (2), the shaft has to be made larger. In (4), it is necessary to make the reactive culvert larger or longer than is necessary. In (5), the number of division increases, and the number of preparation for traction also increases. Consequently, it can be imagined that these methods may not sometimes necessarily be said to be very rational.

2.2 Pipe Roof Reaction Method

Therefore, the authors have conceived an idea in which the reactive culvert is connected to the Pipe Roof, that is executed as a track protective work at the time of traction in Fronte Method, so that tractive reaction, which is applied to the reactive culvert, will be propagated to the ground by way of the Pipe Roof. This means that expectation is entertained of the skin resistance between Pipe Roof and the ground, and this forms the image of placing the friction pile in the pile foundation in a lateral direction. Figure 2 shows a conceptual drawing of the Pipe Roof Reaction Method.

Since this method uses the Pipe Roof which has already been put, it has the following advantages: It uses no special facilities and does not require to make the reactive culvert longer or larger than is necessary. Consequently, there is no need of making the shaft larger than is

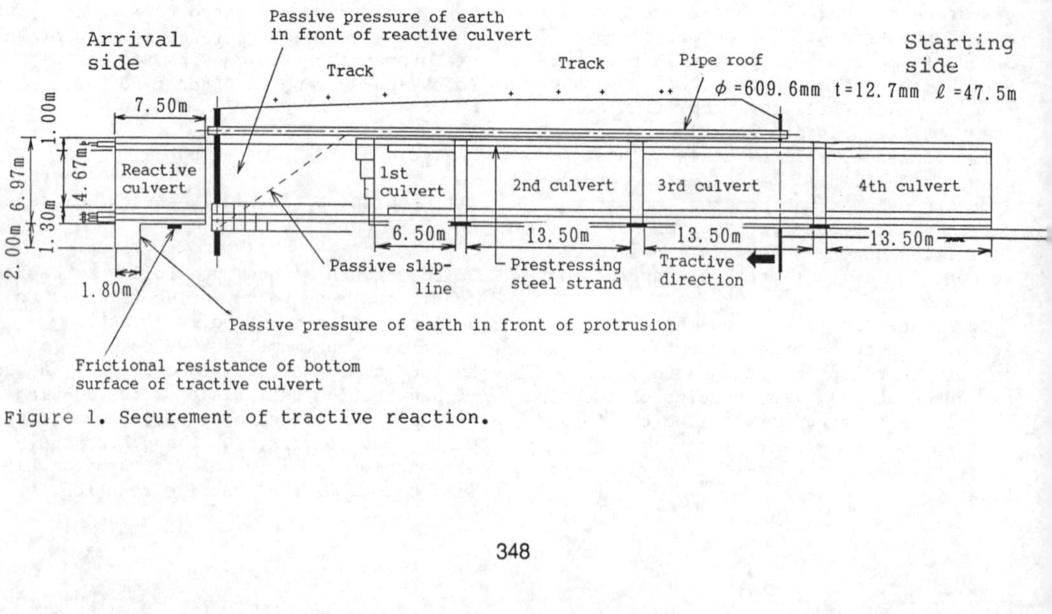

Figure 1. Securement of tractive reaction.

necessary, and it can secure tractive reaction.

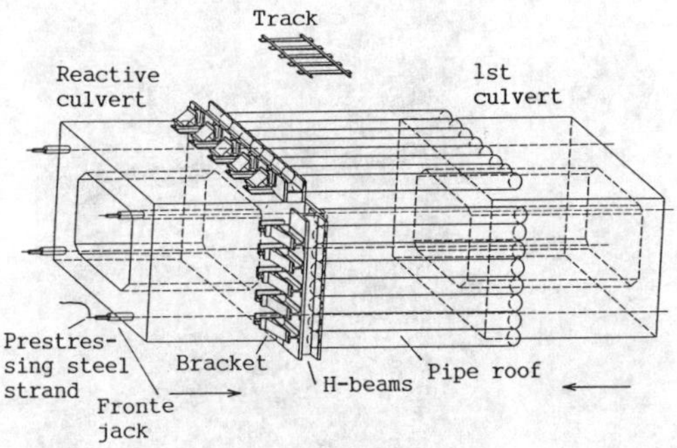

Figure 2. Conceptual drawing of Pipe Roof Reaction Method.

3 DESIGN OF PIPE ROOF REACTION METHOD

3.1 Outline of Construction Site

Figure 1 shows the General View of the viaduct which has adopted the Pipe Roof Reaction Method for the first time. The size of the culvert is 19.00-m wide, 6.97-m high (3 lanes and 2 footpaths) and 56.00-m long in total. The culvert has a cross section which is larger for the culvert to be constructed by the Fronte Method.

Although size is small, construction of a reaction culvert is possible; hence a four-division single-sided sliding traction system has been adopted. Culvert length is 6.5 m for the 1st culvert which is affected by penetration resistance of the cutting edge and skin resistance, etc., of the culvert, and is decided to be 13.5 m for the 2nd to 4th culverts which are affected by only the skin resistance of the culvert.

Since a straight-line-type Pipe Roof for track protection is difficult to construct due to the locational conditions, etc., it has been planned to construct the gate type as shown in Figure 3.

As for the geological conditions, the area concerned is in a banking section, and its geology is gravel-mixed clay having an n-value of about 5. Unconfined compressive strength is 0.98 kgf/cm^2. Earth covering is 1.4-m thick on the arrival side and 2.3-m thick on the starting side.

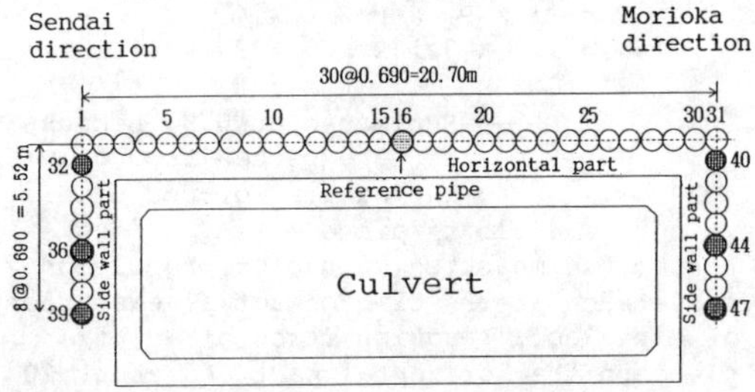

Figure 3. Layout drawing of Pipe Roof.

3.2 Securement of Tractive Reaction

For tractive reaction of the latter period of traction, the present work site can give the expectation of the following values:

(1) Passive pressure of earth in front of reaction walls (retaining walls on the right and left reactive culvert): R_1=210 tf

(2) Passive pressure of earth in front of protrusion provided on the bottom floor slab of reactive culvert: R_2=980 tf

(3) Frictional resistance at the bottom surface of reactive culvert: R_3=630 tf

Even if all the above are totalled, it is estimated that 1,180 tf will be in short against the necessary tractive reaction of 3,000 tf. Therefore, it has been decided to make good this shortage by Pipe Roof Reaction Method. In calculating how much tractive reaction can be expected from Pipe Roof Reaction Method, it has been decided to estimate it from the following eq.(1), which has been obtained by transforming the calculation formula (skin resistance fraction) of tractive resistance at the time of Pipe Roof execution, and to confirm the estimated value by the measured value of tractive resistance at the time of actual execution of the Pipe Roof.

Estimation Formula of Pipe Roof Reaction:

$$R = \mu_0 \cdot P_m \cdot F \cdot L \cdot \alpha \qquad (1)$$

where R : Pipe Roof reaction (tf)
μ_0: Frictional coefficient between earth and pipe
P_m: Average earth pressure around pipe (tf/m^2)
F : Surrounding area per unit length of Pipe Roof (m^2)
L : Length of Pipe Roof (m)
α : Coefficient decrease $\mu 0 \cdot P_m$ is 1.0 tf/m^2 (empirical value).

$R = \mu_0 \times P_m \times F \times L \times \alpha$

$= 1.0 \times 3.142 \times 0.6096 \times 46.1 \times \underline{0.5}$

└── Estimated at 0.5, because it is expected from only the natural-ground side.

$= 44.15$ tf/piece

From the measured values of propulsion resistance at the time of actual execution of Pipe Roof, reaction force of 30 tf/piece on the horizontal part and about 20 tf/piece (calculated as 1/2 of skin resistance fraction out of propulsion resistance) are supposedly expected as reaction force, and when these two values are totalled, the total becomes 1,250 tf, which can, in calculation, satisfy the shortage of 1,180 tf.

3.3 Connection of Pipe Roof and Reactive Culvert (refer to Photos 1, 2, and 3)

In order to propagate tractive reaction to the ground by way of reactive culvert, a method has been adopted, as shown in Figure 4, to connect the edge part of Pipe Roof, a bracket fitted to the reactive culvert and H-beams (waling material in the longitudinal direction).
The edge part of Pipe Roof has shown slight irregularity due to propulsion accuracy, but this has been coped with by filling-up of packings made of steel plates. This connection method has been also used for the Pipe Roof on the horizontal part.

Photo 1. General view of reactive culvert.

Photo 2. Connection of Pipe Roof and reactive culvert.

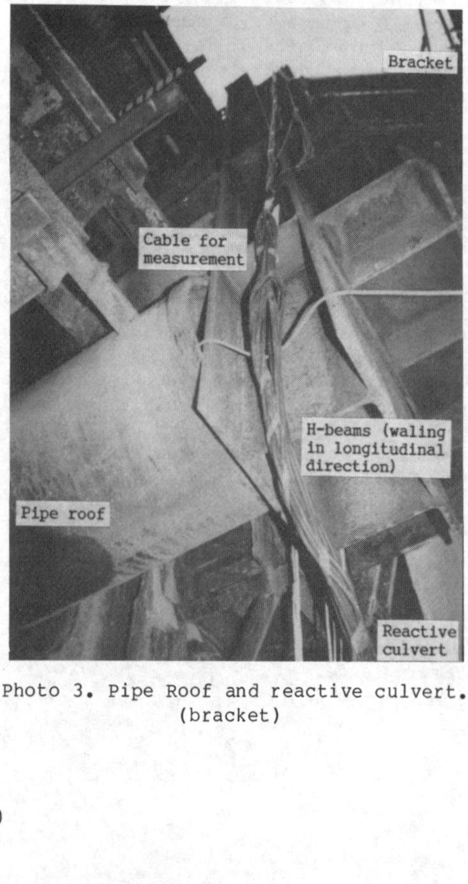

Photo 3. Pipe Roof and reactive culvert. (bracket)

4 MEASUREMENT AND ANALYSIS OF AXIAL FORCE

4.1 Measurement of Axial Force

Measurements have been made twice. One is when the 1st culvert came to 5-m away from the reactive culvert (when earth is still remaining in front of the reactive culvert), and the other is when the 1st culvert came to 0 m from the reactive culvert (when no earth is remaining in front of the reactive culvert).

In order to measure the axial force, two strain gages were stuck horizontally per cross section and strain has been measured at the same time of traction.

Parts of measurement results are shown in Table 1 and Figure 5. Further, Table 2 shows a comparison between skin resistance fraction out of propulsion resistance during Pipe Roof execution and measurement result at the present time.

From the above, the following can be considered:

(1) At the side wall part, Morioka Direction shows larger axial force than Sendai Direction. In this way, axial force shows a pretty wide dispersion depending upon the position of Pipe Roof.

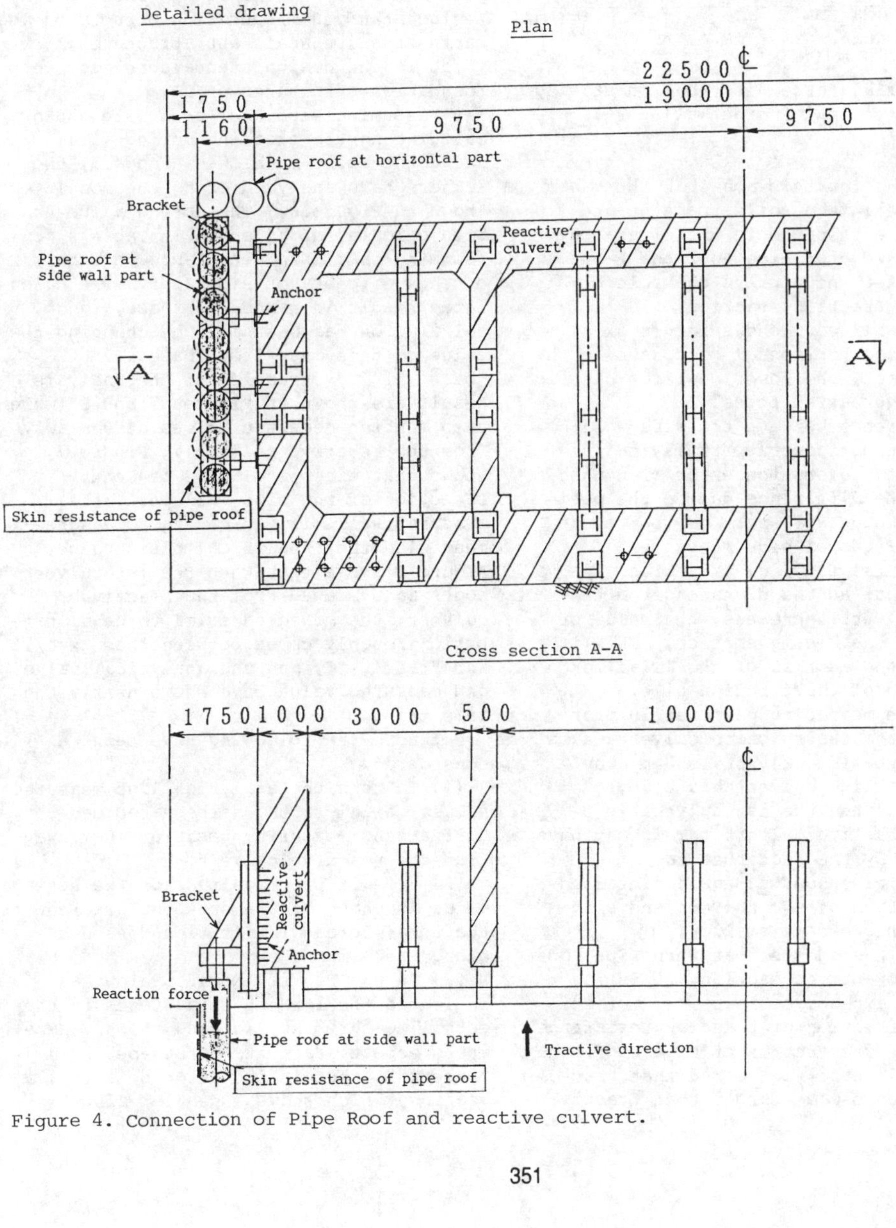

Figure 4. Connection of Pipe Roof and reactive culvert.

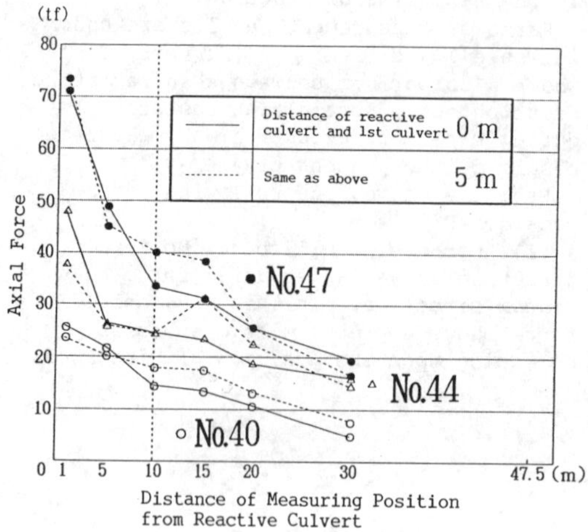

Figure 5. Axial force distribution drawing (Morioka direction of side wall part).

The reasons for this are that there was a difference in the soil strength, and there was a difference in the clearance between the edge of Pipe Roof and H-beams of the bracket which plays the role of propagating tractive reaction.

(2) Side-wall part shows larger axial force than the horizontal part, and at the side wall part, the lower position part is showing larger axial force.

The difference between the side wall part and the horizontal part may be related to the execution error of connection, and the difference due to the position of the side wall part may be related to the magnitude of side pressure.

(3) The axial force distribution indicates that as the distance from the reactive culvert increases, the value of axial force has become smaller. This is the same trend as that of the axial force distribution of the friction pile.

(4) At the measuring position 10 m or less away from the reactive culvert, axial force when the 1st culvert is 0-m away from the reactive culvert has become smaller than when the 1st culvert is 5-m away from tractive culvert. This may have been caused by the fact that earth surrounding Pipe Roof is again disturbed by the traction of the culvert and skin resistance has become smaller.

(5) Table 2 suggests that when Pipe Roof is used as tractive reaction, it is possible to expect, at least, 1/2 of skin resistance fraction of tractive resistance at the time of execution of Pipe Roof. At the side wall part, it seemed that tractive reaction became larger than tractive resistance at the time of work execution. This may be due to the fact that the surrounding ground which was disturbed at the time of Pipe Roof traction has been so compacted as to permit the ground to be used as reaction force. This trend is conspicuous at Morioka Direction.

4.2 Analysis

As mentioned in 4.1 (3), the axial force distribution of Pipe Roof in the Pipe Roof Reaction Method resembles the case of the friction pile; hence, the authors made an analysis using, as a model, the friction pile mentioned in "Building Construction Design Standard (Foundation Structures and Earth-Pressure Resistant Structures," which is the design standard related to foundation structures of JR.

By assuming Pipe Roof as a pile and by calculating the shearing spring value using the above-mentioned standard, the authors have analyzed using the model shown in Figure 6. The spring value has been calculated, as shown in Table 3, by assuming that skin resistance acts on 1/2 of the periphery of Pipe Roof. The calculated result is named the "base," and calculation has been made by changing the value of this base from "base x 2" to "base x 1/2." Examples of the analysis result are shown in Figures 7 and 8 (which also mention measured values at 0 m away from the reactive culvert). Figure 7 shows that excepting No.32 of Sendai Direction of the side wall part, the measured value is indicating good agreement with the "analytical basic value." Figure 8 shows that when the 1st culvert comes to 0 m away from the reactive culvert, the measured axial force distribution roughly comes between "base x 1/2" and "base x 2", and the analytical value and measured value have shown nearly the same trend.

Further, the following have been considered.

(1) In both the analytical and measured values, axial force has been reduced to half at the measuring position 10-m away from the reactive culvert.

(2) Even if the magnitude of the spring value is changed, the steepest gradient of the axial force distribution does not change very much.

(3) In the Figure, No.47, which has indicated the largest axial force in the actual measurement, is also shown. Here, the steepest gradient was steeper than "base x 2," but in its later behavior, a section, which shows a gentler slope is

Table 1. Measurement results of Pipe Roof axial force (in units of tf).

Gage position / Pipe Roof No.	Distance of measuring position from reaction culvert								
	1 m	2.5m	5m	10m	15m	20m	25m	30m	40m
Center of horizontal part No.16	20.3 [13.8]	15.3 [14.5]	24.5 [16.5]	21.8 [20.5]			12.8 [6.5]		
Morioka direction of side wall part 1st pc No.40	23.5 [25.5]		20.0 [21.5]	17.8 [14.3]	17.3 [13.3]	13.0 [10.5]		7.5 [4.8]	
Morioka direction of side wall part 5th pc No.44	37.8 [48.0]		25.8 [26.3]	24.5 [24.5]	31.3 [23.5]	22.5 [18.8]		14.5 [16.0]	
Morioka direction of side wall part 8th pc No.47	73.3 [71.0]		45.0 [48.8]	40.0 [33.5]	38.3 [31.0]	25.5 [25.5]		16.5 [19.3]	
Sendai direction of side wall part 1st pc No.32	25.0 [15.8]		30.8 [23.3]	25.5 [19.0]	20.5 [10.0]	12.8 [7.8]		8.3 [6.0]	
Sendai direction of side wall part 5th pc No.36	25.3 [23.0]		- [16.3]	24.0 [17.3]	23.3 [10.8]	20.3 [11.8]		15.0 [7.8]	
Sendai direction of side wall part 8th pc No.39	35.5 [29.0]		29.8 [23.5]	32.5 [21.3]	36.0 [12.8]	24.5 [13.5]		13.0 [9.3]	7.5 [3.0]

Note) Upper Stage: Condition in which earth is remaining in front of reactive culvert: Tractive force = 2,960 tf

Lower Stage: Condition in which there is no earth in front of reactive culvert: Tractive force = 3,122 tf

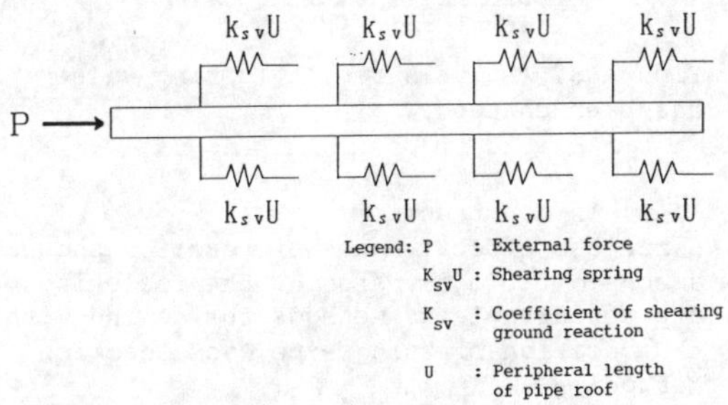

Figure 6. Analytical model.

seen. Later the gradient has become steeper again, thereby showing a value different from the analytical value.

The reason for the existence of a section with a gentler gradient may be that shearing spring reaction around Pipe Roof has exceeded the elastic region and has yielded. Further, the reason for the difference in the magnitude of the gradients may be that the spring value is not uniform throughout the Pipe Roof section.

5 CONCLUDING REMARKS

The present attempt, which is aiming, for the first time, at causing the ground to positively bear tractive reaction by way of the reactive culvert and Pipe Roof, has clearly shown that it is possible to expect Pipe Roof to act fully as tractive reaction force, from the result of measurements of Pipe Roof axial force, which have been carried out at the time of construction.

Matters of caution in the planning, design and construction when adopting the Pipe Roof Reaction Method are shown below.

(1) In Planning

At the work site where a certain degree of earth covering exists and the effect on the track is considered minimal, Pipe Roof Reaction Method is an effective construction method of securing tractive reaction. When this method is used, it is possible to carry out traction work without making the reactive culvert larger and without making the reaction wall larger.

(2) In Designing

Out of tractive resistance during the Pipe Roof work, about 1/2 of skin resistance is likely to be secured, even though it is the lowest as reaction force. Axial force

Table 2. Comparison with time of Pipe Roof propulsion.

Pipe Roof Position \ At the Time of Measurement	At the time of Pipe Roof Propulsion A(tf)	At the time of using Reaction B(tf)	Ratio (B/A)
Center of horizontal part No.16	30	20	0.67
Morioka direction of side wall part No.40	30	26	0.87
Morioka direction of side wall part No.44	50	50	1.00
Morioka direction of side wall part No.47	60	70	1.67
Sendai direction of side wall part No.32	20	25	1.25
Sendai direction of side wall part No.36	50	25	0.50
Sendai direction of side wall part No.39	60	36	0.60

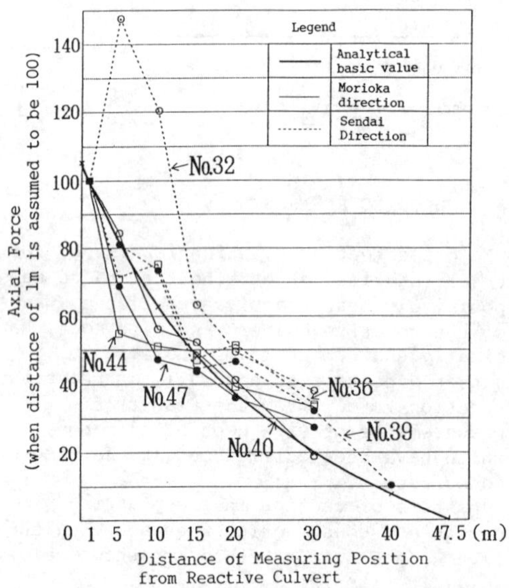

Figure 7. Analysis result (0-m away from reactive culvert).

distribution in the present construction method can be calculated using the same model as the friction pile. For the shearing spring value to be used at this time, the value calculated by Building Construction Design Standard (Foundation Structures and Earth-Pressure Resistant Structures), which is the design standard related to Foundation Structures can be used as it is.

Table 3. Spring value.

Case \ Spring value	Shearing spring value surrounding Pipe Roof
Base	1.25×10^3 tf/m^2
Base x 2	2.50×10^3 tf/m^2
Base x 1/2	6.25×10^2 tf/m^2

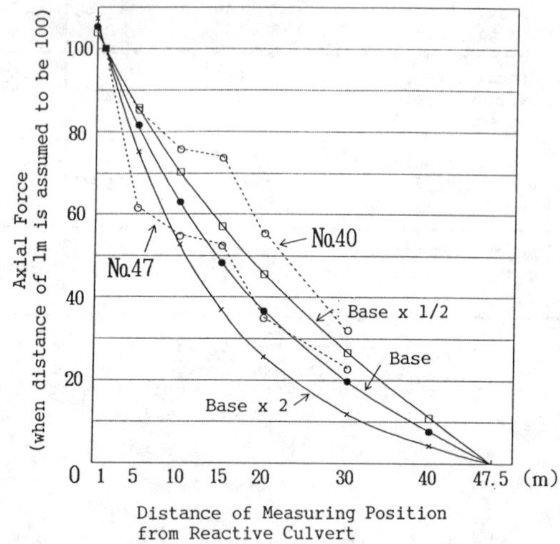

Figure 8. Analysis result (spring value has been changed).

(3) In Construction

Since the effectiveness of reaction showed a considerable dispersion, the following are conceivable as methods for coping with the occasion of using Pipe Roof Reaction Method:

i) Clearance of the bracket, which plays the role of propagating force between the reactive culvert and Pipe Roof, should be adjusted by diligently filling it up using thin steel plates.

ii) Holes have been driven on Pipe Roof beforehand, and after propulsion of Pipe Roof, injection should be given from the periphery direction of the Pipe towards the natural ground, thereby increasing the loosened skin resistance between the natural ground and Pipe Roof.

iii) Injection should be given to the joint between Pipe Roofs, thereby making the spring value of each Pipe Roof uniform and making the behaviors at the time of tractive reaction the same.

Tunnelling and Ground Conditions, Abdel Salam (ed.) © 1994 Balkema, Rotterdam, ISBN 90 5410 363 9

Development of vertical-horizontal shield machine

Hiroyuki Itoh, *Ishikawajima-Harima Heavy Industries Co., Ltd, Japan*
Norio Kainuma, *Tokyo Electric Power Company, Japan*
Yoshio Fukawa, *Kawasaki Sewerage Bureau, Japan*
Ken-ichi Kaneko, *Taisei Corporation, Japan*

ABSTRACT

A vertical-horizontal shield machine has been developed to carry out efficient excavation of shaft and adit tunnels deeper underground.

The vertical-horizontal shield machine consists of a main shield for vertical excavation with a rotatable sphere incorporated in sub-shield for horizontal excavation. After vertical excavation, the sphere rotates 90° to start the sub-shield. This newly developed machine is capable of excavating shafts and adits continuously using only a single machine.

This report describes the vertical-horizontal shield machine, an implementation example of the right-angled rotating shield machine, the first sphere shield machine, and the vertical- horizontal shield machine planned to be the second machine.

1. INTRODUCTION

City underground space is being effectively used to accommodate a large number of city facilities such as power and communication tunnels, gas lines, waterworks and sewage pipes, subways, etc. As the underground space at the shallow and medium-shallow levels becomes more limited, newly planned facilities must be located deeper underground.

Construction work deeper underground involves several problems that cannot be solved by conventional technology because of the high pressure of underground water. The problems of shield excavation construction methods for a 50 m deep tunnel can probably be solved using conventional technology. However, several technical problems remain concerning the shield excavation shaft.

In the case of the pneumatic caisson construction method, although unmanned operation has largely increased due to improvement in remote control technology, there is work that still must be done by man power in a compressed air environment, so that a working depth of 50 m is the limit set by the Labor Sanitation Law. Depending on the geological features, an additional precautions to prevent air leakage are needed even at shallow levels. Furthermore, since excavated sediment is discharged using buckets, excavation efficiency deteriorates as the depth increases, thus lengthening the construction period.

When carrying out shaft excavation using the cast-in-situ diaphragm wall method, a wall length of nearly 100 m is generally needed for floor construction of 50 m because of the deep impermeable ground. This increases both construction period and cost. Furthermore, there are always problems of boiling and heaving, requiring additional precautions to prevent them.

Whichever method is adopted, it is necessary to improve the foundation and cut the diaphragm for protection at the time of advancement. So, the greater the depth at which the work is performed, the more dangerous it becomes. Under such circumstances, the development of advanced shield technology for safe, quick and secure operation is required.

The "Vertical-Horizontal Shield Machine" described here is a machine that excavates the shaft before performing continuous excavation of the adits, and has been developed with the purpose of rationalizing shield excavation shaft construction at greater underground depths.

This report describes the vertical-horizontal shield machine, the implementation example of the right-angled rotating shield machine, the first sphere shield machine, and the vertical- horizontal shield machine planned to be the second machine.

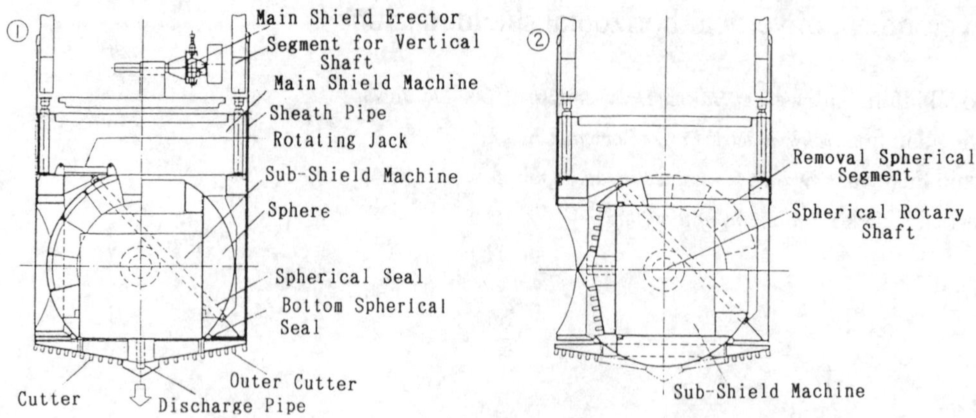

Figure 1. Outline drawing of vertical-horizontal continuous shield machine

2. VERTICAL-HORIZONTAL SHIELD MACHINE

2.1 Basic structure

As shown in Fig. 1, the vertical-horizontal shield machine consists of a main shield for vertical excavation with an incorporated sphere that houses a sub-shield for horizontal excavation.

When the main shield reaches the specified depth, the sphere is rotated 90° to start the sub-shield for horizontal excavation. During vertical excavation or at the time of sphere rotation, stopped water inside the machine is held by a seal.

The slurry shield system is used for the main shield, and the reverse circulation drill method, with an excellent record in large aperture shaft construction, is used for vertical excavation. Excavated sediment is collected at the center of the cutter before being discharged by the pump. Furthermore, the cutter and the cutter driving device used in the main shield are designed to be used jointly for the sub-shield.

The main shield consists of a sphere with the built-in sub-shield, a vertical shield frame to house the sphere, sheath pipes to protect the frame, and an outer cutter for vertical excavation as shown in Fig. 2.

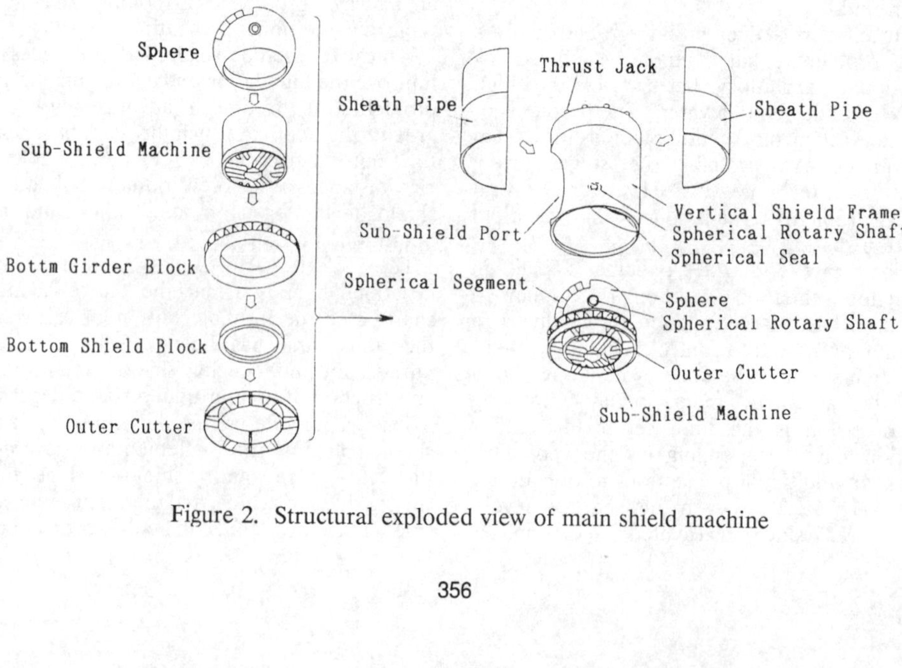

Figure 2. Structural exploded view of main shield machine

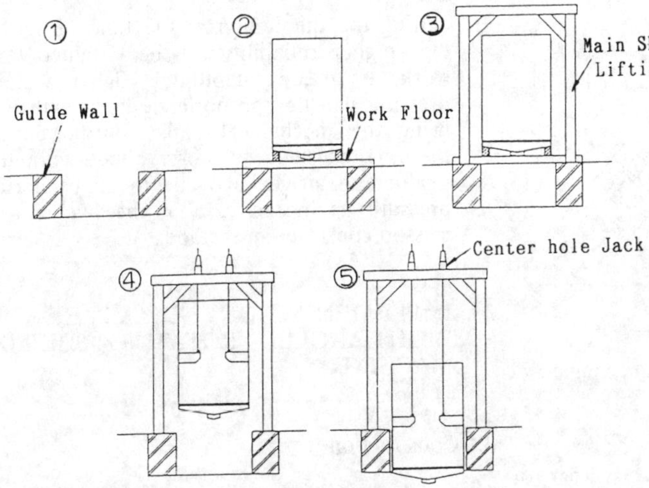

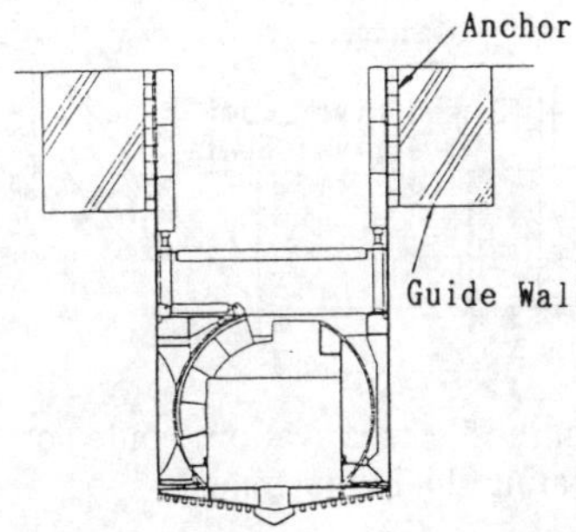

Figure 3. Excavation procedure of main shield machine

Figure 4. Shaft excavation condition

2.2 Excavation procedures

Vertical-horizontal shield excavation procedures are described below.

(1) Install the main shield lifting frame on the ground as shown in Fig. 3 and assemble with the main shield at the inverted position. At the time of initial excavation, suspend the shield machine with the main shield lifting frame and let the shield machine lower under its own weight while preventing rolling and pitching of the machine by jack operation.

(2) As shown in Fig. 4, after the tail of the main shield passes through the guide wall, start excavation using the thrust jack while applying inverse force to the segment in the same manner as for an ordinary shield machine, when the buoyancy overcomes the weight of the shield machine. To make use of the weight of the main shield lifting frame as the inverse force, keep the segment and the main shield lifting frame firmly tied with an anchor, etc.

(3) Once the main shield reaches the adit crown position, fix the sheath pipe to the segment, lower the main shield with the thrust jack, and expose the cutting part of the vertical shield frame as the opening for the sub-shield.

(4) Upon completion of vertical excavation, pull out the connecting pin that connects the outer cutter to the cutter head of the sub-shield using the hydraulic jack to remove the outer cutter from the sub-shield machine. Leave the outer cutter, which is no longer needed, at the bottom of the vertical shaft, use the thrust jack of the sub-shield to slide the entire sub-shield upward before storing it into the sphere rotating locus.

(5) Turn the sphere 90° with the rotating jack so that the cutter head comes from the downward position to the horizontal position. Then successively remove the no longer needed sphere segments by regaining the rotating jack.

(6) After the sphere is rotated, press the thrust jack of the sub-shield against the steel wall for inverse force inside the sphere to push down the sub-shield from the sphere, and then weld the tail plate, mount the erector, etc., to complete the sub-shield. Install the temporarily assembled steel segment inside the sub-shield, and use it as the inverse force to start the sub-shield.

2.3 Shield method features

(1) The size of the shaft method used with this method makes it possible to reduce the thickness of the lining by using the segment in the shaft, with the outer diameter being reduced 50-70% as compared to conventional

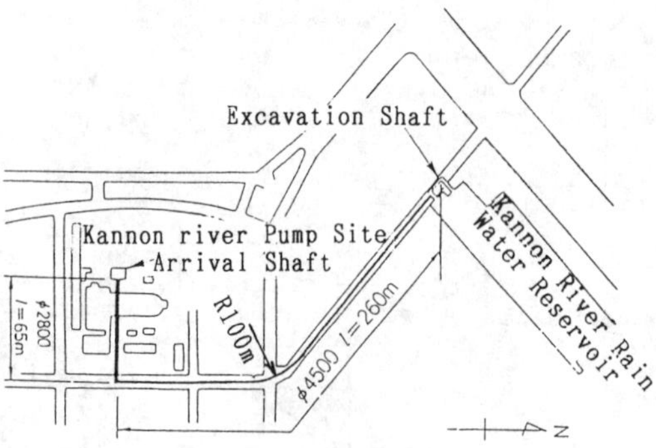

Figure 5. Excavation route of right angled rotating shield machine

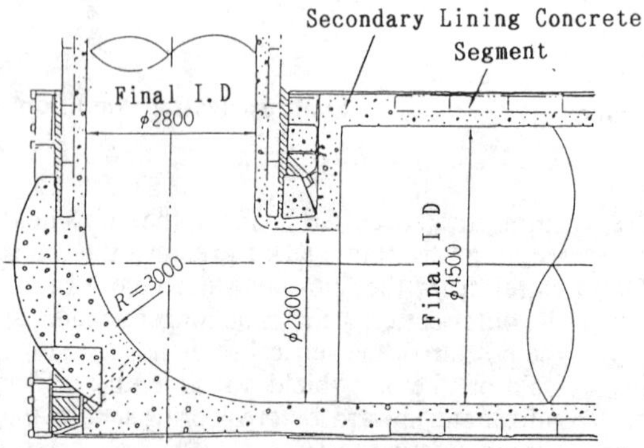

Figure 6. Final tunnel shape of right angled rotating shield machine

methods. The reduction in shaft size is more economical not only because it leads to the reduction of land cost but also because it reduces the quantity of dug up soil.

(2) The construction of the shaft by the shield method shortens the construction period by 50% as compared with conventional methods, reducing field expenses and alleviating public inconvenience, such as road blockage, etc.

(3) The construction cost includes the fixed cost of the shield machine and variable costs depending on the depth of the shaft. The shaft excavation cost using this method is more economical as the depth increases, with the depth of 30-40 m being the diverging point.

(4) Because there are no dangerous jobs such as cutting the diaphragm at the time of start, etc., higher reliability can be obtained for work at greater depths. Higher water pressure can be appropriately dealt with by using the mechanical seal. Furthermore, there is no need to worry about workers health and air leakage due to atmospheric pressure as in the case of the pneumatic caisson construction method.

3. IMPLEMENTATION EXAMPLE OF RIGHT-ANGLED ROTATING SHIELD MACHINE

3.1 Construction work

The right-angled rotating shield machine has basically the same structure as the vertical-horizontal shield machine. However, instead of excavating a shaft, it excavates a tunnel at a right angle from the original tunnel and parallel to the surface.

One example of a tunnel made by the right-angled rotating shield machine is the conveyance ditch in Kawasaki City, connecting the Kannongawa rain water reservoir and the pump site, which also works as a rain water retaining pipe. The outline of the construction work is shown below.

Finished tunnel inner diameter x total length

* Main shield: ø4500 mm x 260 m
* Sub-shield: ø2800 mm x 65 m

Soil characteristics: Silt (N value: 2 - 4)
Soil depth: 11.5 m (average)
The right-angled rotating shield machine was used for the following reasons.

(1) Low construction cost

(2) Shortest route and shortest construction period

(3) No need of work on the road

The excavation route is shown in Fig. 5. Because of the foundation pile driven in the site where the tunnel bends vertically, it was difficult to penetrate the sharp curve. Furthermore, because the site was the entrance of a factory and the pipes buried under the ground had to be shifted, it was impossible to make the rotating shaft from the ground.

Note: The tunnel diameter of the right-angled rotating shield machine is reduced from the main

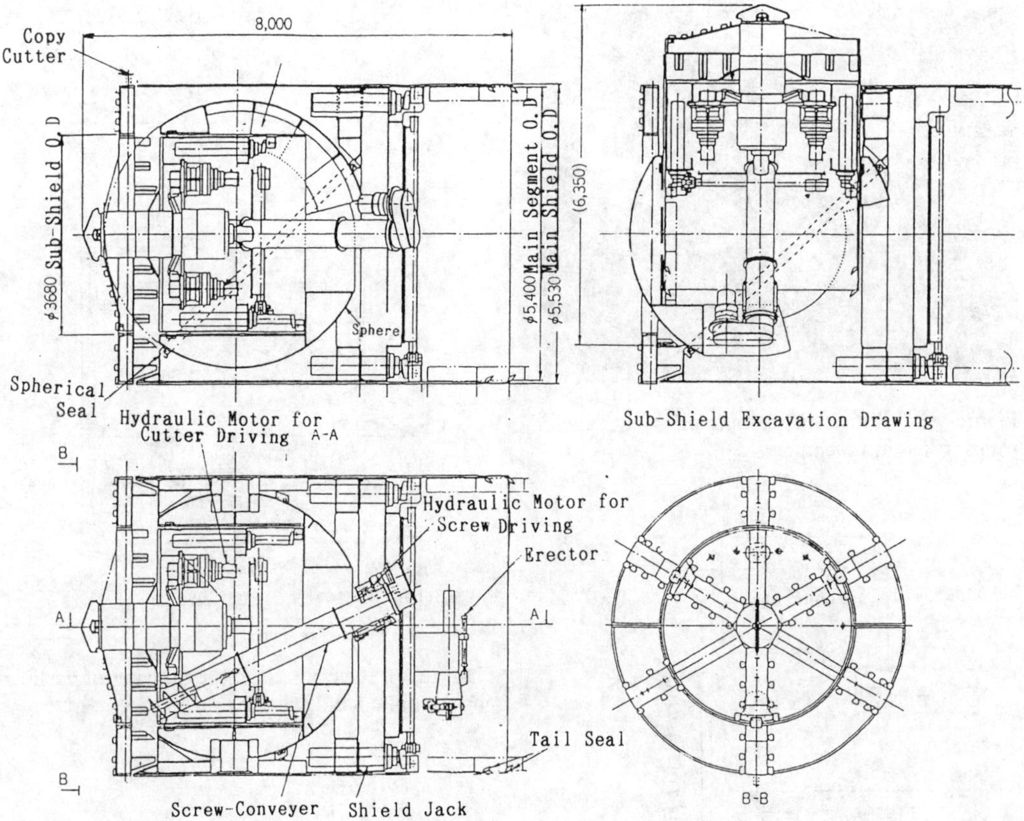

Figure 7. Overall assembly drawing of right angled rotating shield machine

shield to the sub-shield, but since this is a storage pipe, there are no functional problems. The final shape is shown in Fig. 6.

3.2 Structural features

As shown in Fig. 7, the outer diameters of the right-angled rotating shield machine are main shield: ⌀5530 mm, sub-shield: ⌀3680 mm. The right-angled rotating shield machine has basically the same structure as the vertical-horizontal shield machine, but the mud-pressure type machine is used because of the adit excavation carried out using the main shield, considering soil condition, installment space for slurry treatment plant, etc. Hence, the sub-shield is mounted with the screw conveyor to be used also as the muck discharging device of the main shield.

Moreover, the cutter driving device is also used for the sub-shield and main shield as in the case of the vertical-horizontal shield machine. However, since the torque of the mounted cutter should enable excavation with the main shield, the cutter must have a higher excavation capacity than a vertical-horizontal shield machine with the same diameter.

Therefore, the sub-shield cutter torque mounted may be excessively large device for the sub-shield for adit excavation, but the torque may prove insufficient for the main shield excavation capacity. In this construction work, however, the soil was soft and there was no need to mount the excessively high torque for the main shield, so that appropriate measures could be taken. So, compared with the vertical-horizontal shield machine, the right-angled rotating shield machine has a narrower selection range in terms of cutter torque, such as the outer diameter ratio between the main shield and sub-shield, the excavated soil characteristics, type of shield machine, etc.

Fig. 5 shows a curve of R 100 m in the

Figure 8. Photo of completed right angled rotating shield machine

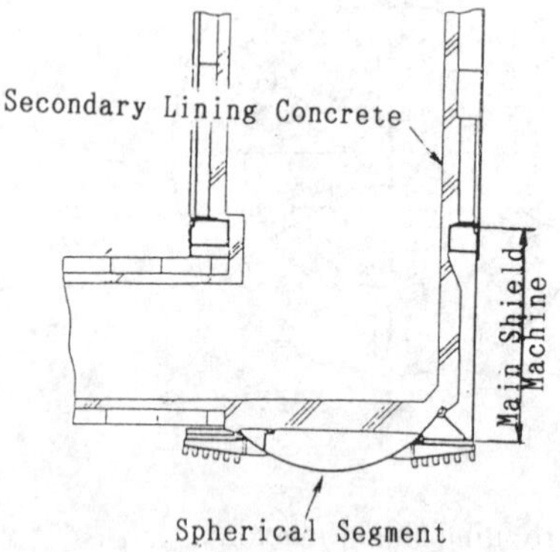

Figure 9. Final shaft shape of vertical-horizontal continuous shield

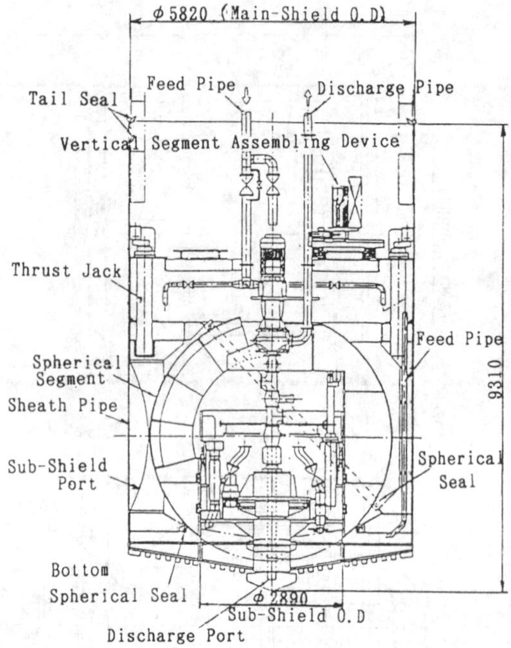

Figure 10. Overall assembly drawing of vertical-horizontal shield machine

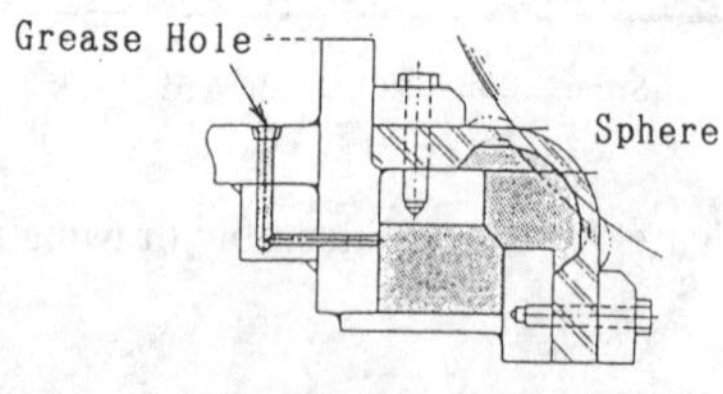

Figure 11. Bottom rotary seal

excavation route, requiring measures for curve excavation using the main shield, which is quite unlike the vertical-horizontal shield machine.

The over cut device is designed to have the slide rod, located in the outer cutter, connected to the copy cutter jack installed on the sub-shield cutter, lowing the bit to go in and out at the end of the slide rod to reach the outside of the outer cutter.

Furthermore, since the sphere is installed inside the shield, the machine length becomes longer than the normal shield, calling for measures to keep the machine length short. The shield jack of the main shield machine is located here at the front most position. However, placed at this position, the jack interferes during sub-shield start. The jack is, therefore, stored in the outer drum, and is designed to slide backward together with the outer drum by securing the air seal of the machine after the completion of main shield excavation. The structure is shown by the four cylindrical bodies in Fig. Fig. 8

4. INPLEMENTATION EXAMPLE OF VERTICAL-HORIZONTAL SHIELD MACHINE

4.1 Construction work

The first vertical-horizontal shield machine planned to be adopted in the sewage construction work in Tokyo is described here. The outline of the construction work is shown below.

shaft inner diameter x shaft depth (Main shield excavation): ⌀5000 mm x 39 m

Finished tunnel inner diameter x Total length (Sub-shield excavation): ø2450 mm x 433 m
Soil characteristics: Silt, sand layer, gravel layer
Tunnel minimum curve: R 10 m

Furthermore, the final structure of the vertical shaft is designed to have a steel structure by combining the main shield frame with the sphere bottom as shown in Fig. 9, while dismantling the internal structural materials of the sphere, etc. inside the machine but partially leaving the sphere at the bottom of the vertical shaft, in order to withstand the rising pressure of the underground water. This allows, unlike the final shape of the right-angled rotating shield machine, an effective passage space to the adit from the shaft.

4.2 Structural features

As shown in Fig. 10, the outer diameters of the vertical-horizontal shield machine are main shield: ø5820 mm, sub-shield: ø2890 mm.

The slurry shield machine is also used for the sub-shield so that a muck discharging system can be used with fluid transportation during main shield shaft excavation, the same as in the reverse circulation drilling method.

As for the seals of the sphere rotating section, the bottom sphere seal is installed to dismantle the internal structural materials such as the sphere, etc., in the machine after the start of the sub-shield. Also, a sphere seal is installed diagonally at the upper part to stop water during main shield vertical excavation and sphere rotation.

The inflating-type rubber seal, shown in Fig. 11, is adopted as the bottom sphere seal and is designed to stop water by operating the internal pressure of the seal, into which grease is injected. The reason why the seal has this shape is because it is necessary to prevent the cutting part of the sphere of the cutter from damaging the seal when passing through the seal. The durability and water stopping performance have been fully confirmed through an element performance confirmation test, with the water stopping performance of the cutting part of the sphere found to be 0.98 MPa.

Furthermore, it is necessary to stop water at the opening for the sub-shield starting port during dismantling inside the machine. Water is stopped by the entrance segment. The entrance segment, as shown in Fig. 12 and Fig. 13, is a special segment installed with the partition plate required for injecting back-filling material to stop the water at the tail clearance between the adit segment and the main shield shaft opening after the sub-shield passes out of the main shield.

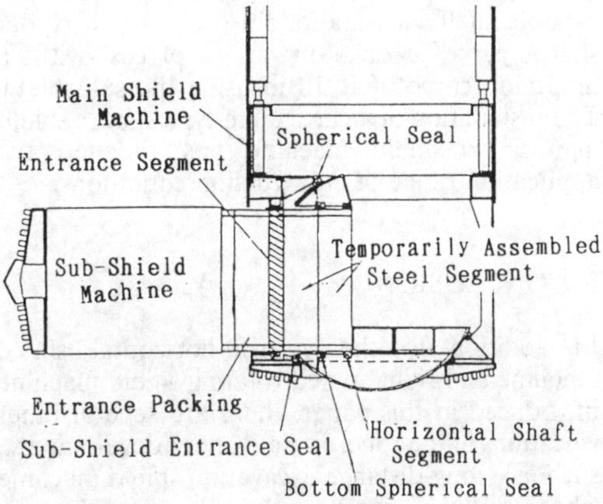

Figure 12. Installation drawing of entrance segment

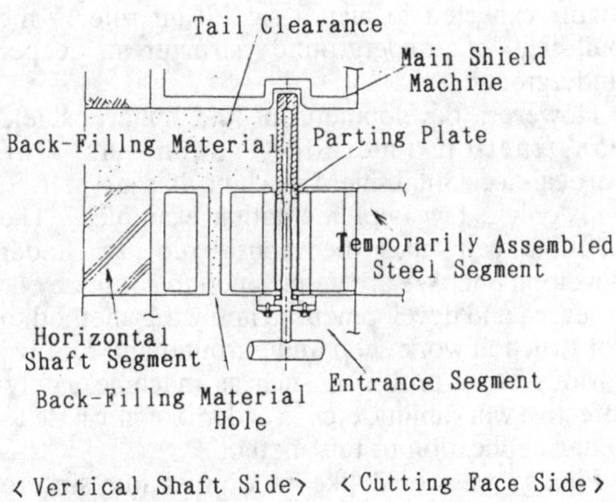

Figure 13. Detailed drawing of entrance segment water sealing section

Before excavation, the temporarily assembled steel segment is fully assembled from inside the shaft, and the entrance segment is erected before the sub-shield passes through the main shield, while the partition plate is pushed down immediately after the passage of the sub-shield to inject into the tail clearance the back-filling material, which is then solidified to stop the water.

The cutter driving device for the sub-shield is also used for the main shield, but the torque of the cutter has a smaller excavation capacity than that of the right-angled rotating shield machine used for adit excavation. It has been confirmed through the vertical shaft excavation experiment that the cutter torque needed for shaft excavation is approximately one-tenth of the cutter torque needed for adit excavation.

The sub-shield is designed as an articulate type

capable of 9° articulation in order to carry out sharp curve excavation at 4 places with a minimum curve of R 10 m using the sub-shield. The installation of an articulate-type sphere shield in the sub-shield machine has widened the application range of construction conditions.

5. CONCLUSION

In addition to the vertical-horizontal shield machine and right-angled rotating shield machine introduced in this paper, there are several other variations of the sphere shield under development, such as a long-distance excavation shield machine whose cutter bit can be replaced several times in the tunnel by using the sphere. Various other ideas concerning construction methods with a sphere shield have also been thought up, so that the sphere shield construction method is now highly expected to play a significant role in the building of underground structures deeper underground.

However, development of the sphere shield construction methods, such as the vertical-horizontal shield machine, has just started with only a few implementation examples. The method can thus be considered as under development. We are determined to continue our research and development to apply this method to construction work deep underground and to solve various other problems such as machine safety, effective workability, etc., in order to contribute to wider application of this method.

Finally, we would like to express our sincere thanks to all concerned people for their cooperation throughout the process of this development.

Tunnelling and Ground Conditions, Abdel Salam (ed.) © 1994 Balkema, Rotterdam, ISBN 90 5410 363 9

Study on the design of large-cross-section highway tunnels

Katsuhiko Iwai, Tatsuro Iwasaki & Kazuya Takekuni
Japan Highway Public Corporation, Japan

1 INTRODUCTION

Japan Highway Public Corporation (JH) plans to construct the New Tomei and New Meishin Expressways which are to serve, along with the existing First Tomei and Meishin Expressways, as an artery of Japan in the twenty-first century and beyond. Designed as six-lane motorways with full shoulders, the New Tomei and Meishin Expressways are to be spacious enough for safe and comfortable driving and to share the task of carrying future motor traffic with the First Tomei and Meishin Expressways. Tunnel sections of the planned expressways will also have the same width as the earthwork sections and bridge sections will.

Tunnels to be constructed on the New Tomei and Meishin Expressways will have a large oval cross-section (roadway width of 16.5 m, excavation width about of 20 m). To mention underground structures in Japan having such large cross-sections, there are only caverns for hydroelectric power stations and for oil storage tanks constructed in relatively sound rocks, but no tunnels as linear structures with such a large cross-section as on the New Expressways ever constructed. This paper deals with studies on some problems concerning the design and construction of the planned large-cross-section tunnels, such as the inner cross-sectional shape, final concrete lining, primary ground supports, occasional preset supports, and foot reinforcement.

2 CROSS-SECTIONAL SHAPE

2.1 *Width of roadway*

The total width of 16.5 m, consisting of three lanes (3.75 m × 3), and left and right shoulders (3.25 m and 2.0 m respectively), has been adopted for the New Tomei and Meishin Expressways. The cross-section will also include a passageway for inspection personnel on both sides. To cope with the expected enlarging of marine containers (4.1 m high), the construction gauge has been set as high as 4.8 m.

2.2 *Conception of inner cross-sectional shape*

The inner shape is to accommodate space for such facilities as ventilation systems, interior panels, inspection passageways, in addition to the above-mentioned construction gauge.

The ratio of the height (including the invert) to the width of the inner cross section of previously designed and constructed tunnels is 0.65 or above. Considering the past records of design, the height–width ratio of around 0.65 was judged appropriate for the large-cross-section highway tunnels. As the shape of cross-section of the tunnels, a three-arc conjoint arch was adopted taking account of stability and economy.

2.3 *Analysis of mechanical characteristics of proposed shapes*

Six inner shapes, each of which had a different three-arc conjoint arch cross-section, were proposed. The height–width ratio varied around 0.65 with the ratio R1/R2 (refer to Fig. 1) and the height of side wall as shown in Fig. 1. First, the deflection of lining was calculated for each of six by frame analysis with the lateral pressure coefficient of 0.5 and 1.0 as a parameter. Then, selected two shapes superior to others were put to finite element analysis. After that, the best one (Case 2) was chosen to be adopted.

The frame analysis revealed that for the lateral pressure coefficient of 0.5, the bending moment at the crown, at the side wall, and at the invert was minimized in Case 2, while for the lateral pressure coefficient of 1.0, the bending moment in Case 1,

	Case 1	Case 2	Case 3	Case 4	Case 5	Case 6
Section data	h (height below spring line) = 1.25 m R1 (radius of top heading) = 9.85 m R2=R3 = 7.55 m R2/R1 = 0.77 Cross-sectional area of excavation = 226.2 m² Height–width ratio of inner shape = 0.55 Height–width ratio of tunnel = 0.67 Circularity of top heading = 0.9	h (height below spring line) = 1.50 m R1 (radius of top heading) = 9.90 m R2=R3 = 7.20 m R2/R1 = 0.73 Cross-sectional area of excavation = 225.5 m² Height–width ratio of inner shape = 0.55 Height–width ratio of tunnel = 0.67 Circularity of top heading = 0.88	h (height below spring line) = 1.75 m R1 (radius of top heading) = 10.00 m R2=R3 = 6.75 m R2/R1 = 0.58 Cross-sectional area of excavation = 224.1 m² Height–width ratio of inner shape = 0.55 Height–width ratio of tunnel = 0.67 Circularity of top heading = 0.85	h (height below spring line) = 2.00 m R1 (radius of top heading) = 10.15 m R2=R3 = 6.20 m R2/R1 = 0.61 Cross-sectional area of excavation = 221.9 m² Height–width ratio of inner shape = 0.55 Height–width ratio of tunnel = 0.67 Circularity of top heading = 0.82	h (height below spring line) = 2.25 m R1 (radius of top heading) = 10.35 m R2=R3 = 5.35 m R2/R1 = 0.55 Cross-sectional area of excavation = 220.4 m² Height–width ratio of inner shape = 0.55 Height–width ratio of tunnel = 0.64 Circularity of top heading = 0.78	h (height below spring line) = 2.50 m R1 (radius of top heading) = 10.55 m R2=R3 = 5.25 m R2/R1 = 0.50 Cross-sectional area of excavation = 221.0 m² Height–width ratio of inner shape = 0.55 Height–width ratio of tunnel = 0.64 Circularity of top heading = 0.76
Cross-sectional shape	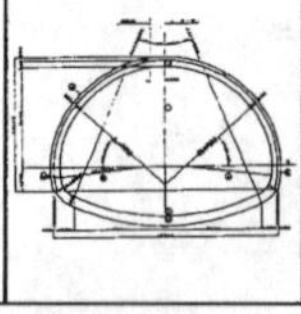	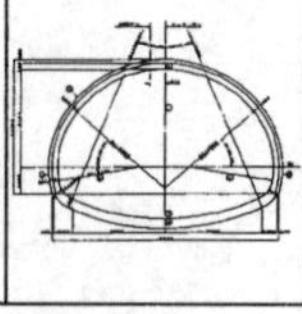	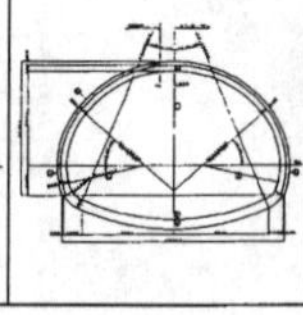	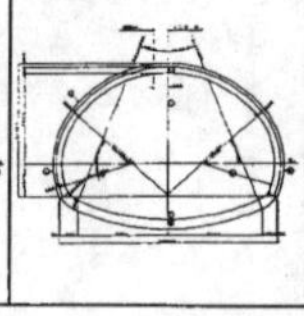	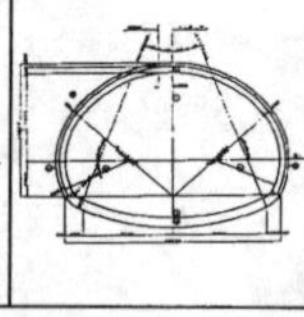	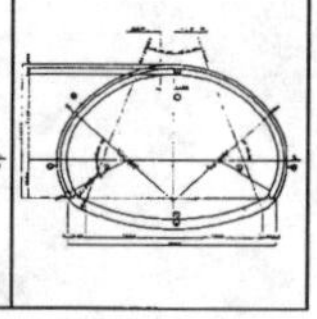

Fig. 1 Proposed cross-section for analysis

Case 2, and Case 3 showed similarly small values, indicating that these cases are more advantageous than others. For the lateral pressure coefficients of 0.5 and 1.0, both axial force and stress in the lining were minimized in Case 2 and Case 1, respectively.

The frame analysis for the radii of curvature at the transition part R5 from the side wall to the invert of 2.0 m and 2.5 m revealed that the bending moment at the arch and the side wall in the two cases were almost the same, but at the radius of curvature at R5 of 2.5 m the bending moment at the invert became significantly small.

Finite element analysis was conducted for Case 2 and Case 6, these two cases being judged superior to others through the frame analysis. As a result, it was confirmed that Case 2 and Case 4 showed similar levels of displacement, stress, safety factor for loosening in the surrounding ground, and stresses in support members.

2.4 *Adopted cross-sectional shape*

The finite element analysis did not reveal any significant difference between Case 2 and Case 4 in stress and deformation in the ground and support members during excavation. It was confirmed, however, that as a result of the frame analysis, the cross-sectional shape of Case 2 is the most favorable.

Consequently, the cross-sectional shape of Case 2, as shown in Fig. 2, was finally adopted for the planned large cross-section.

3 STUDY ON FINAL CONCRETE LINING

3.1 *Purpose*

As mentioned earlier, the tunnels to be constructed on the New Tomei and Meishin Expressways will have such an excavated cross-section as more than 200 m²). The stability of the final lining, therefore, becomes important not only during excavation but also over a long period of time after completion. Studies were then carried out on the following items.

(1) Evaluation of the structural stability of final lining using the allowable stress method

(2) Evaluation of the structural stability of final lining in terms of the ultimate strength given according to the classification of ground by grade

(3) Study on the timing of removal of forms for the final lining

Based on the results thus obtained, the stability of the final lining for the large-cross-section tunnels was to be evaluated.

As for the classification of ground, this paper follows that of Japan Highway Public Corporation. The grades in this classification are arranged, in the order of the soundness of ground, as A, B, CI, CII, DI, DII and E. The grade A represents a very hard ground, and E a very soft ground. The grades dealt with in this paper are from B to DII, so that the grades B, CI and CII represent a hard to medium-hardness rock, and DI and DII a soft rock.

3.2 *Method of analysis*

Analytical methods applicable to the final lining are the frame analysis, the finite element analysis, the crack extension model, and the discontinuous deformation analysis (DDA), etc. For this study, the widely used frame analysis was employed.

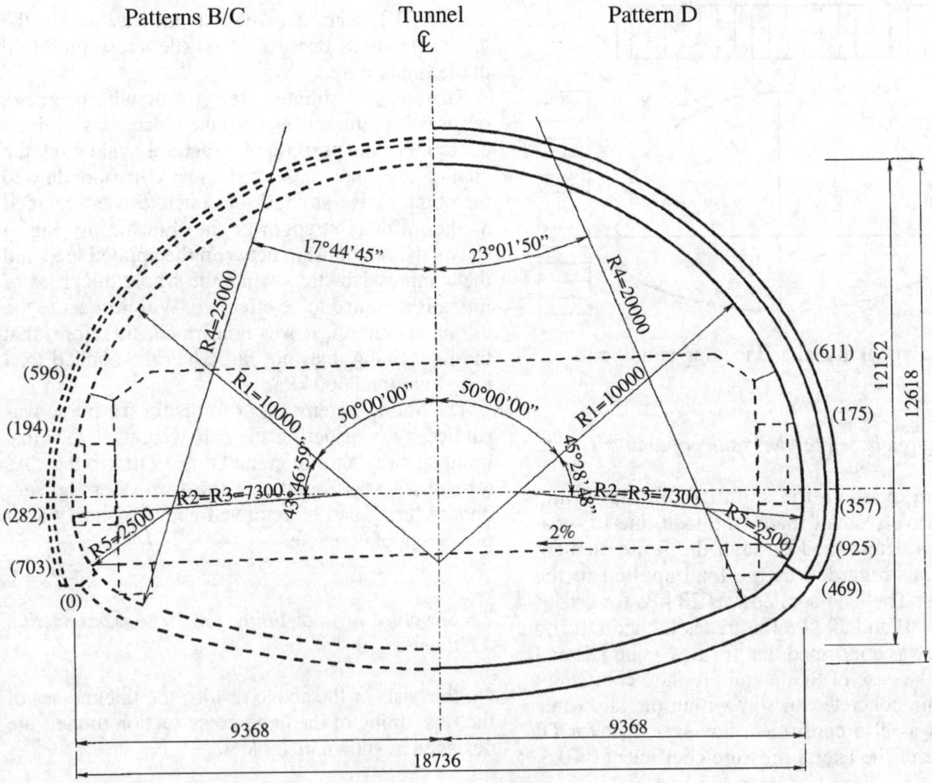

Cross-sectional area of inner shape	181.3 m²	R1 = 10 m R2 = R3 = 7.3 m	Cross-sectional area of inner shape	187.3 m²	R1 = 10 m R2 = R3 = 7.3 m
Cross-sectional area of excavation	212.7 m²	R4 = 25 m R5 = 2.5 m	Cross-sectional area of excavation	233.1 m²	R4 = 20 m R5 = 2.5 m
Height–width ratio	64.9%		Height–width ratio	67.3%	
Circularity	88.2%		Circularity	88.2%	

Fig. 2 Cross-sectional shape of large-cross-section tunnel

3.3 *Cross section and conditions for analysis*

Referring to experiences in the past with two-lane tunnels, a rule was thumb was applied to give the thickness of final lining as 50 cm for the grade B, CI and CII grounds, and 60 cm for the grade DI and DII grounds. Likewise, the modulus of elasticity of ground has been given by Japan Highway Public Corporation according to the grade of ground. The ground spring constant was provided based on the "Calculation of Ground Spring Constants for Direct Foundations" in the Specification for Substructures of Road Bridges—Japan Road Association." These values in terms of the ground grade are shown in Table 1. In addition, the ground spring is assumed to act perpendicularly on support members, and to have a nonlinearity which is valid for valid for compression only.

3.4 *Analytical model and load applied*

Fig. 3 shows the analytical model and loading conditions. In the analysis, overburden load and lateral pressure were increased step by step from 2 to 800 kPa. Since a ground with thick earth cover was assumed, the lateral pressure coefficient was set at 1.0 and uniformly distributed load was applied.

Table 1. Physical properties of ground

Grade of ground	Modulus of elasticity (MPa)	Spring constant (MPa)
B	5000	2240
CI	2000	898
CII	1000	448
DI	500	225
DII	150	67

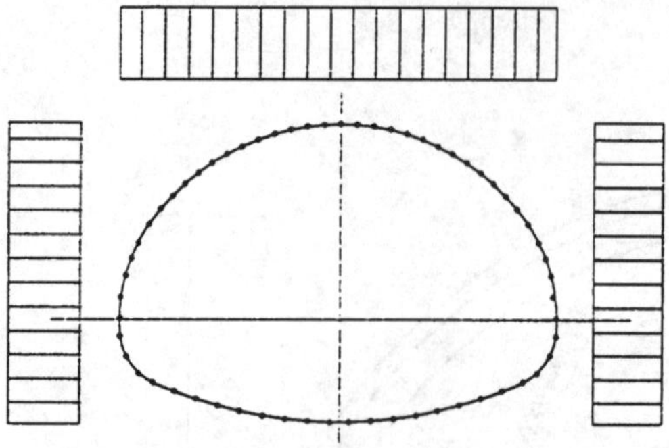

Fig. 3 Analytical model and loading conditions

3.5 *Evaluation based on the results of analysis*

In order to evaluate the stability of final lining under relatively small loads, the load due to loosened zones determined through the finite element analysis was regarded as the load applied to the final lining. The load applied was 20 kPa for grades B, CI and CII and 50 kPa for grades DI and DII. As a result, it was confirmed that for B, CI and CII and for DI in the case of the lateral pressure coefficient of 1.0, plain concrete can stay within the allowable stress. It was also confirmed, however, that for DI in the case of the lateral pressure coefficient of 0.5, and for DII, plain concrete cannot stay within the allowable stress because of tensile stress produced in the final lining.

Structural ultimate strength, at which cracks occur but a failure does not take place, was evaluated, too. For the purpose of structural evaluation, the applied load that causes the stress corresponding to the compressive strength of concrete was regarded as the ultimate strength of the final lining. Fig. 4 shows the relationship between the applied load and the compressive stress when the whole thickness of lining is assumed to be effective. With respect to the ultimate strength, it was confirmed, therefore, that the final lining does not fail when the applied load is up to around 600 kPa.

The timing of removal of forms for the lining was studied by considering the self-weight of the final lining at early ages in grade DI and DII grounds. As a result, it was found that the forms for the secondary lining can be removed about 24 hours after placement of concrete.

3.6 *Structure of final lining for large-cross-section tunnels*

On the basis of the above results, the thicknesses of the final lining of the large-cross-section tunnels are designed as shown in Table 2.

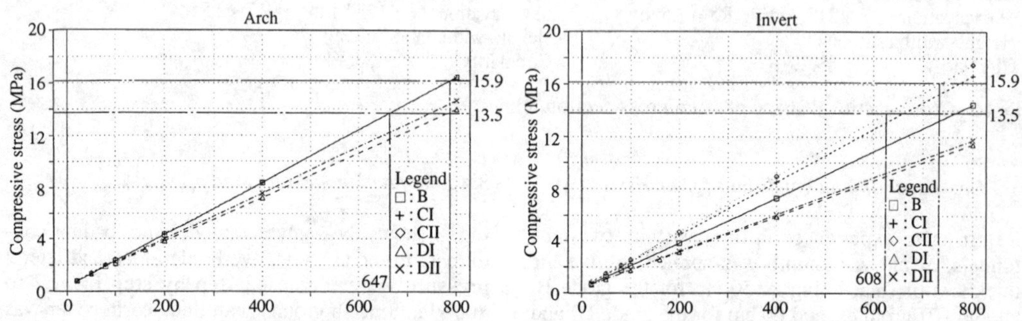

Fig. 4 Relationship between load applied and compressive strength (whole thickness of lining assumed effective)

Table 2 Proposed design thicknesses of final lining

Grade of Ground	Thickness of Arch	Thickness of Invert	Structure	Remarks
B – CII	50 cm	50 cm	Plain concrete	Application of steel fiber reinforcement to be considered
DI	60 cm	80 cm	Single reinforcement	
DII	60 cm	80 cm	Double reinforcement	

4 STUDY ON PRIMARY SUPPORT STRUCTURE

4.1 *Analysis on stability of tunnel supports*

Two-dimensional, finite element, elastic sequential analysis was carried out, assuming typical ground conditions and physical properties for each of the grade B, CI, CII, DI, and DII grounds. The items studied here are as follows:

(1) Adequacy of assumed standard support patterns

(2) Adequacy of cross-sectional shape

(3) Comparison of excavation methods (top heading short bench method vs. center drift method) (refer to Fig. 5)

(4) Verification of the effectiveness of shotcrete on the invert

Table 3 shows the cases analyzed for the verification for the items (1) to (4), and Table 4 shows the physical properties of ground used in the analysis.

The support patterns used in the analysis were determined by upgrading patterns used for previously constructed tunnels with relatively large cross-sections.

4.2 *Results*

1) *Adequacy of assumed standard support patterns*

Referring to past measurements which had revealed the depth of the plastic zone to be less than 2 m and the convergence of top heading less than 0.3% in two-lane tunnels, the proposed support pattern, for a ground of grade B to DII, is verified to assure the

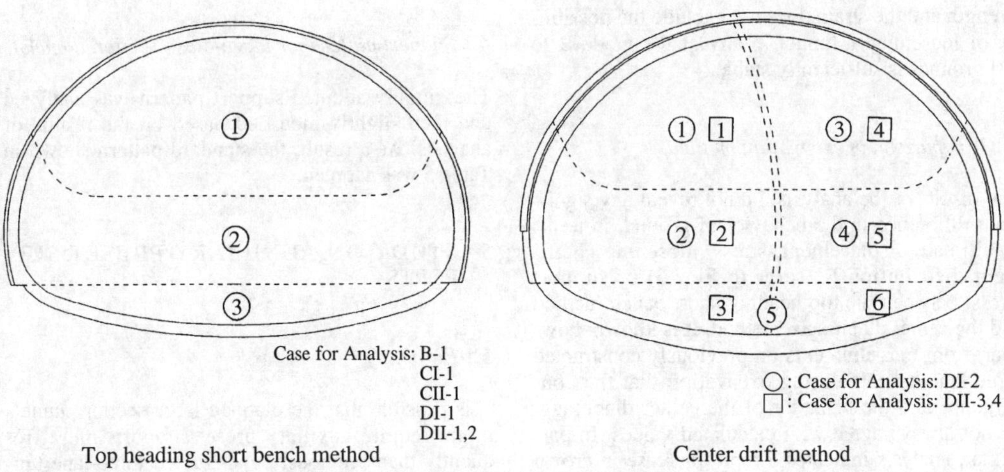

Fig. 5 Proposed excavation method

Table 3 Cases for analysis

Case	Grade of ground	Cross-sectional shape	Earth covering (m)	Initial lateral pressure coefficient	Excavation method	Preset supporting	Foot reinforcement	Invert shotcrete	Analytical model
1	B		300						
2	CI				Top heading short bench	Not used			Axisymmetric model
3	CII			1.0					
4			100					Not used	
5	DI	See Fig. 3.			Center diaphragm		Not used		Entirety model
6					Top heading short bench	Forepoling (ignored in analysis)			Axisymmetric model
7								Used	
8	DII		40	0.7	Center diaphragm			Not used	Entirety model
9								Used	

Table 4 Physical properties of ground (assumed values)

Grade of ground	Unit weight γ(g/cm^3)	Modulus of elasticity E (MPa)	Poisson's ratio μ	Cohesion C (MPa)	Internal friction angle ϕ (degree)	Tensile strength σ_t (MPa)	Elastic limit parameter k
B	2.5	5000	0.25	3	50	0.6	6
CI	2.4	2000	0.30	2	45	0.4	6
CII	2.3	1000	0.30	1	40	0.2	6
DI	2.2	500	0.35	0.5	35	0.1	4
DII	2.1	150	0.35	0.2	30	0.04	4

stability equivalent to that of two-lane tunnels. It was also confirmed that stresses in shotcrete and steel-arch supports and strains in rock bolts were all within the limits of standard design values. Evaluation of the results of the analysis based on the critical strain revealed that, although the relationship between the unconfined compressive strength and the strain does not exclude the possibility of loosening, a tunnel in any of the grade B to DII grounds is sufficiently stable.

2) *Comparison of excavation methods*

The results of the analysis did not reveal any significant difference in the behavior of ground, in terms of ultimate displacement, stress, loosening coefficient distribution R (refer to Fig. 6) or support stress, between the top heading short bench method and the center diaphragm method. It is known, however, from measurements on previously constructed three-lane tunnels during excavation that the convergence and the behavior of the center diaphragm do not always agree with calculated values. In particular, analysis indicates a rapid increase in crown settlement during the delayed excavation of the concaved D-section, while measured values tend to be smaller. Thus, the prediction by analysis is not always perfect.

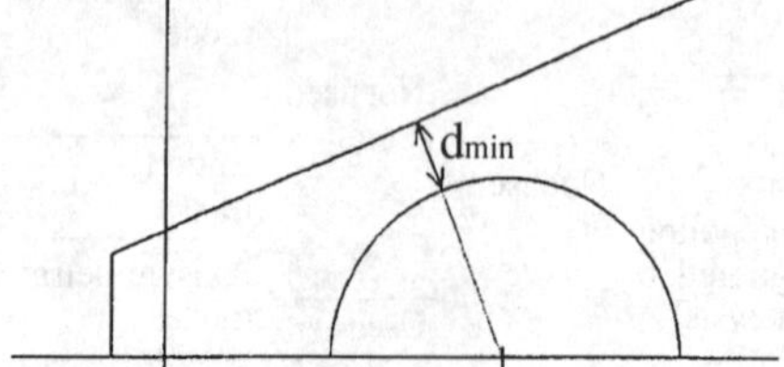

Fig. 6 Definition of loosening coefficient

Considering the advantages of the center diaphragm method, such as working face stability and smaller settlement, it was decided that the center diaphragm method should also be adopted as a standard construction method in conjunction with the top heading bench method.

4.3 *Support pattern for large-cross-section tunnels*

The initially adopted support pattern was analyzed and then slightly modified based on the results of analysis. As a result, the standard pattern shown in Table 5 was adopted.

5 STUDY ON AUXILIARY PRESET SUPPORTS

5.1 *Purpose*

It is possible that the planned large-section tunnels might require auxiliary preset supports more frequently than previously constructed three-lane tunnels did in poor ground conditions. The effect and necessity of preset supporting, and the relations between preset supports and tunnel supports were studied through two-dimensional elastic sequential analysis. The analysis dealt with the following points:

(1) Effects of preset supports

(2) Comparison of the effects of preset supports with different peripheral bounds

(3) Effects of preset supports at different ground stiffnesses

(4) Effects of preset supports with different excavation methods (top heading short bench method and center diaphragm method)

Table 6 shows the eight cases analyzed, and Table 7 shows the physical properties of ground. The lateral pressure coefficient of 0.5 for the earth covering of 20 m was assumed. The support pattern in this case is DII as shown in Table 6.

Table 5 Proposed standard support patterns for large-cross-section tunnels

Grade of ground	Excavation method	Rock bolt			Steel arch support	Shotcrete (cm)
		Length (m)	Spacing (m)			
			Circumferential	Longitudinal		
B	Top heading method	1.5	4.0	1.5	H150 (top heading only)	15
CI	Top heading method (Center diaphragm method)	1.5	6.0	1.5	H200	20
CII	Top heading method (Center diaphragm method)	1.2	6.0	1.2	H200	25
DI	Top heading method Center diaphragm method	1.0	6.0	1.0	H200	30
DII	Top heading method Center diaphragm method	1.0	6.0	1.0	H200	30

Allowance for deformation will be studied separately.
Initial supporting for DII will be studied separately.

Table 6 Cases for analysis

Number	Modulus of deformation of ground (MPa)	Excavation method	Support pattern	Preset supporting	Foot reinforcement
1	50	Top heading short bench	DII	Forepoling (injection forepoling)	Not used
2	50	Top heading short bench	DII	Horizontal jet grout (180° of top heading)	Foot piling with jet grout
3	50	Top heading short bench	DII	Horizontal jet grout (120° of top heading)	Not used
4	20	Top heading short bench	DII	Horizontal jet grout (180° of top heading)	Foot piling with jet grout
5	50	Center diaphragm method	DII	Forepoling (injection forepoling)	Not used
6	50	Center diaphragm method	DII	Horizontal jet grout (180° of top heading)	Foot piling with jet grout
7	50	Center diaphragm method	DII	Horizontal jet grout (180° of top heading + center diaphragm)	Foot piling with jet grout
8	20	Center diaphragm method	DII	Horizontal jet grout (180° of top heading)	Foot piling with jet grout

Table 7 Physical properties of ground

Unit weight γ(t/m^3)	Modulus of deformation E (MPa)	Poisson's ratio μ	Cohesion C (MPa)	Internal friction angle ϕ (degree)	Tensile strength σ_t (MPa)	Elastic limit parameter k
2.0	50 20	0.42	0.05	30	0.01	2

5.2 *Evaluation of results*

The analysis indicated that in a ground with a modulus of deformation of 20 to 50 MPa, the ground excavated by the top heading short bench or CD method without preset support causes the extension of the plastic zone and the augmentation of convergence and settlement at the foot of the side wall, thereby spoiling the support stability. It was also estimated that preset supports such as a rigid shell and foot reinforcement effectively restrain ground deformation and plasticization.

According to the change of distribution of loosening coefficient in the analysis, forepoling is effective for ground stabilization but not significantly effective in restraining plasticization of ground below the invert, which extends to the depth of 7 to 9 m. In a ground where heaving of the bottom occurs, some kind of restraining works such as a temporary invert is necessary.

In a ground of a small modulus of deformation, preset supporting should cover the entire periphery of the top heading arch. For such conditions, the addition of a center diaphragm with preset support hardly contributes to tunnel supporting. It can be concluded, therefore, that there is no significant difference between with and without a center diaphragm. In this connection, there is also a need for further study on the division of heading.

The analysis indicated that stiff preset supporting and foot reinforcement resulted in increases in sectional force and stress in the steel arch. This tendency is more pronounced when ground stiffness is low or unsymmetrical earth pressure takes place. In such cases, it is necessary to use stiffer support structures to adopt preset supporting and foot reinforcement so as to restrain ground loosening.

6 CONCLUSION

As mentioned earlier, this study is confined to desk works such as a survey of past records of construction and computer analysis. Since the reliability of the present numerical analysis still remains insufficient to predict actual behavior of tunnel supports and ground, it is necessary to carry out a full-scale test construction so that the analytical methods will be improved.

For the upcoming large-scale construction of the New Tomei and Meishin Expressways, a large number of stable and durable large-cross-section tunnels should be constructed safely and economically in the shortest possible period of time. In order to solve various technical problems regarding large-cross-section tunnels and obtain information necessary for thorough investigation and study, experimental construction is indispensable.

It is also necessary to conduct various performance tests for tunnel facilities to ensure the safety of various during high-speed running.

Tunnelling and Ground Conditions, Abdel Salam (ed.) © *1994 Balkema, Rotterdam, ISBN 90 5410 363 9*

Total work control system for tunnels using satellite GPS and automatic laser control

Tadashi Kanzaki, Koichi Nakano & Shuichi Nishizawa
Construction Technology Development Department, Technology Division, Taisei Corporation, Japan

ABSTRACT:
Our proposed system applies electronics technology to conventional geographical surveying for tunneling inside mountains. Called the Total Tunneling Work Control System, the proposed system reduces labor and enhances the precision and efficiency of surveying work. The system is composed of two sub-systems. One is the GPS satellite precision fixed-point surveying system which is used to check reference points at both ends of a tunnel prior to commencement of construction work. This method, replacing conventional triangulation or traverse surveying, uses satellites to determine three-dimensional positions on the ground. Relative positioning is achieved by using more than two receivers to simultaneously survey reference points and measurement points. Quicker and more precise fixed-point positioning is now possible. The other system is the three-dimensional marking system. In this system, a scanning device automatically controls a laser beam to continuously illuminate points on the excavation face and conducts face-marking more quickly and precisely. This feature has greatly improved face marking work by reducing labor and "outbreak", thereby enabling construction of high quality tunnels. The system was developed using the latest technology to meet the demanding needs of the construction industry. Its efficiency has been proven through actual application to construction work and we expect it will be put to much wider use.

1 INTRODUCTION

During the construction of the longest tunnels, control point surveys often have to operate over a distance of some 10 km. Prior to work commencement, control points are accurately surveyed at both ends of the planned tunnel using several triangulation points. This requires highly precise and efficient surveying.

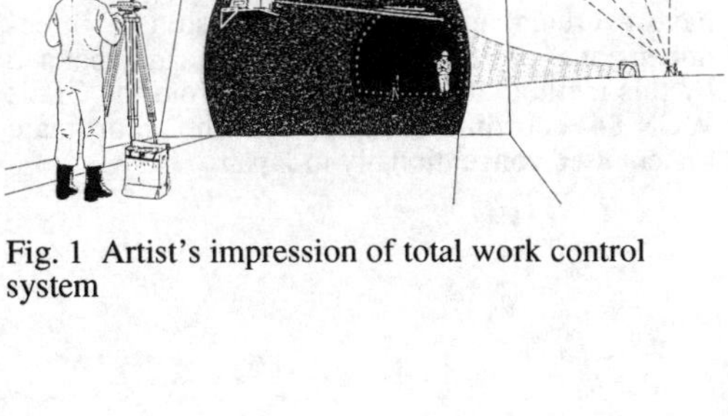

Fig. 1 Artist's impression of total work control system

During routine construction operations, the alignment and excavation face must be maintained continually under accurate and up-to-the-minute control.

The system proposed here is a systematized method of tunnel construction control in which the GPS precision positioning technique is integrated with a laser-based in-tunnel positioning method, as shown in Figure 1. The system has been put to practical use in various construction projects, including the Ichinose Tunnel work for the Hokuriku Shinkansen route.

2. GPS PRECISION FIXED-POINT POSITIONING SYSTEM

2-1 *Problems with conventional positioning systems*

When large-scale construction work has to be aligned accurately, such as in the case of a dam or tunnel, or when work takes place in a remote location, surveys often have to make use of control points that extend for some 10 km. This, it is well known, requires a tremendous amount of time and effort.

Conventional survey methods, which include the electronic tachometer, suffer from a number of

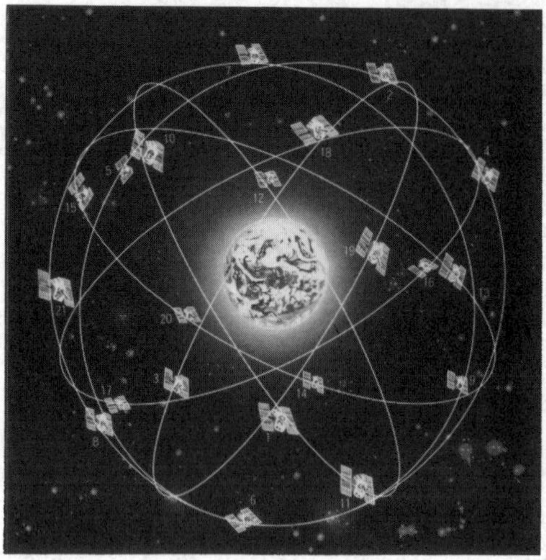

Fig. 2 Artist's impression of GPS satellites

problems as outlined below.

(1) Visual contact between pairs of points is necessary.

(2) The quality and precision of the results depends greatly on the expertise and experience of the surveyor.

(3) The work is susceptible to weather conditions.

Another major problem is the uncertainty in meteorological correction techniques, leading to unsatisfactory precision.

With the aim of developing a more accurate survey method, we carried out a large number of experiments and studies in which survey precision was quantified, the result being a new fixed-point positioning method of great precision.
Survey activities in a remote mountainous area are schematically illustrated in Figure-1.

2-2 *Outline of GPS high-precision positioning*

The GPS (Global Positioning System) was developed by the U.S. Department of Defense to allow the accurate determination of the coordinates of a point on the earth using satellites. The system is outlined in Fig. 2. Currently, 24 satellites make up this system, 4 circling the earth in each of six 20,000 km orbits. There are plans to increase the total inventory of satellites.

These GPS satellites continually transmit quasi-micro waves at 1.5 GHz and 1.2 GHz, and these are called the L1 and L2 waves, respectively. These transmissions contain codified data from atomic clocks, ionospheric correction data, and satellite orbital information. The codes fall into two categories, known as P and C/A. The former are for military use while the latter are available free of charge to anyone.

The distance between a satellite and a measuring point is determined by the arrival time difference of radio waves. The four distance data from four satellites are used for three-dimensional positioning of the measuring point. There are two positioning methods. One that uses one receiver is called the single positioning method, which is already used for vehicle/marine navigation. However, this cannot delete error factors at the time of observation, resulting in an error of about 30 m. Hence its application to construction work was considered impractical. The other one, called the relative positioning method, uses two or more satellites to obtain vectors between receivers. This enables high precision positioning in a three-dimensional manner over a wide area.
To improve survey accuracy, there was a need for a relative positioning method based on measuring the phase of the satellite carrier waves using many receivers. Development of such a method can be considered a major breakthrough in the elimination of errors.

The major sources of error in GPS surveying include the following:

(1) Satellite clock errors

(2) Receiver clock errors

(3) Carrier wave delay during passage through the ionosphere and atmosphere

In Fig. 3, the various symbols have the following meanings: R: Distance traveled by the signal between satellite and receiver, n: Number of wavelengths, λ: Wavelength, $\Delta\phi$: Phase.

The relative positioning system is shown in Fig. 4. Suppose the carrier wave from one satellite is received by two receivers and the phase difference obtained. If this is done for two satellites, the difference between the two phase measurements can be calculated. This double calculation of phase difference enables errors to be eliminated. As illustrated in Fig. 5, if the true path differences (the difference in distance from each end of the baseline to the satellite) between Receivers 1 and 2 are $\Delta\rho_1$ and $\Delta\rho_2$ respectively and the positioning error is δ, the difference between the phase measurements from the two satellites, Σ, may be expressed as follows.

$$\Sigma = (n_{1\lambda} + \Delta\rho_1 - \delta) - (n_2\lambda + \Delta\rho_2 - \delta)$$
$$= (n_1 - n_2)\lambda + \Delta\rho_1 - \Delta\rho_2$$

This indicates that double error calculation will have a crucial effect on the precision of three-dimensional positioning. Positioning data obtained by this method are then converted from the GPS's WGS-84 coordinate system to the coordinate system used conventionally in Japan.

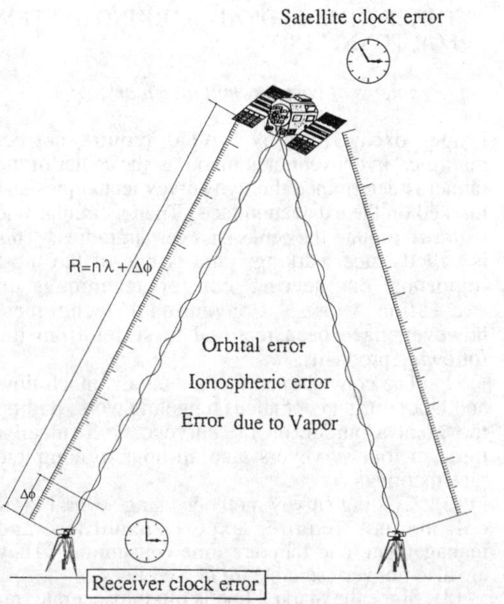

Fig. 3 Carrier distance and sources of error

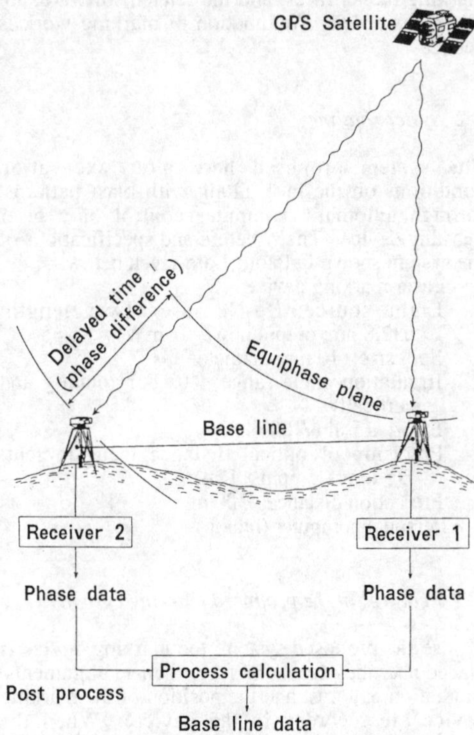

Fig. 4 Schematic drawing of relative positioning system

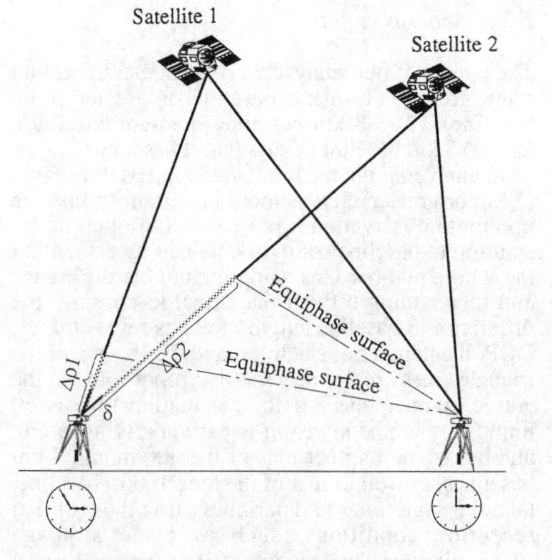

Fig. 5 Schematic drawing of double phase difference

Table 1. Determinants of baseline solutions

Baseline length classification	Number of baselines	Observation time classification (min)	Number of satellites	P-DOP
Long baseline (8 km or greater)	3	10, 20, 30, 45, 60, 75, 80, 120, 180	4-7	2.5-6.4
Medium baseline (3-8 km)	6	Same as above	4-6	2.5-8.0
Long baseline (8 km or greater)	7	Same as above	4-5	2.3-7.5

2-3 *Technical problems with GPS and their solution*

2-3-1 *Experimental*

The major purpose of our experiments was to quantify the three issues that were thought to have a great effect on survey precision: survey method and distance between survey points; the number of satellites and their pattern; and the survey time.

These issues were analyzed according to the following procedure:

(1) Carrier phase data received at each survey point was transferred to a computer

(2) The rough position of each survey point was fixed, and the residual between each point and survey result was calculated

(3) Major cycle slippage was corrected based on carrier phase measurements

(4) Double difference calculations were carried out and cycle slippage corrected

2-3-2 *Analysis results*

The results of our analysis have been classified into three groups: (1) 8 km baseline or greater (long baseline), (2) 3-8 km baseline (medium baseline), and (3) 3 km baseline or less (short baseline).

In analyzing the medium and long baseline cases (3 km or greater), it was more important to find the minimum observation time required to obtain a fix solution (a baseline solution obtained by estimating the whole number bias using the double difference and then adding a fixed bias) than to consider the difference in baseline length. Results reveal that P-DOP, the number of satellites, and the lengths of the triangle sides are not necessarily proportional to the whole number bias. If the calculation time is 60 minutes, this phenomenon is particularly apparent, and the whole number bias of the 500 m and 4 km baselines, as well as that of the long 10 km baseline, take the same time to determine. It is thought that reception conditions, such as cycle slippage (discontinuous data reception) and noise, have a great impact on positioning precision during the double measurements of carrier phase difference.

In analyzing the results for baselines of 3 km or less (short baselines), a fix solution was obtained in a significantly shorter observation time than in the case of the medium and long baselines with the same arrangement of satellites. It is thought that P-DOP and the number of satellites considerably influence the observation time required.

These experiments give a quantitative evaluation of the three factors which have an effect on survey precision, and this has led to a successful standardization of precision fixed-point positioning. Upon this foundation, a new system is now being used for control point survey activities in construction of the Kuki Tunnel for the High-speed Linear Train project and Chubu Electric Co., Ltd.'s Nikengoya waterway.

2-4 *Results of GPS positioning*

The studies for a precision fixed-point positioning system provided, through a series of experiments, quantitative data on the fundamental problems of positioning techniques, including observation method and distance between survey points, the arrangement of satellites, and observation time. The resulting data were systematized and documented in the form of a manual and used to standardize a positioning method. The system is now being used for control point surveys at three tunnel construction sites, including the Kuki Tunnel for the High-speed Maglev Train, with the aim of further refining it into a commercially effective product. The precision now achievable with this method is in the millimeter range.

3. THREE-DIMENSIONAL MARKING SYSTEM FOR TUNNELING

3-1 *Problems of conventional face marking*

Proper excavations by NATM require correct marking. In conventional marking, the center of the tunnel is determined through survey techniques and marked on the excavation face. Then a circular line is drawn around the center at a certain radius. This is called "face marking," and is one of the most important engineering control techniques in excavation work. Conventional techniques, however, have been reported to suffer from the following problems:

(1) The conventional method entails much time and labor during continuous tunneling work, putting too great a burden on the surveyors. Typically, three to four surveyors take an hour to complete face marking.

(2) Existing survey methods using several laser collimators require expert handling and management, and they are time-consuming. They are also not very suitable for curved lines.

(3) Since the marked line is often inaccurate, re-touching work, such as chiseling, is required and as a result, outbreaks often occur.

To solve these problems, a device for efficiently marking tunnel faces, and the related software, are urgently needed. Automation of marking work is needed.

3-2 *System outline*

The system proposed here plots excavation conditions on the face, along with blast patterns, through automatic computer control of a laser marking device. The structure and specifications of the system shown in Photo 1 are given below.

- Laser marking device
  - Light source: He-Ne laser (wavelength: 632.8 nm; output power: 1 mW)
  - Spot size: 10 mm diameter
  - Irradiation angle range: 40° horizontally and vertically
  - Error: ±5 mm/100 m
  - Precision of optical distance measurement: ±(5 mm + 5 ppm x Dm)
  - Projection distance: 300 m
- Personal computer (laptop)

3-3 *Features of the proposed marking system*

To use the proposed system, the marking device is placed near the excavation face. Tunnel alignments, irradiation patterns, and the position of the marking device are recorded in the device. When the coordinates of an area to be irradiated are input (that is, the station), three-dimensional coordinates of each irradiation point are calculated. The laser beam is then directed to the calculated irradiation

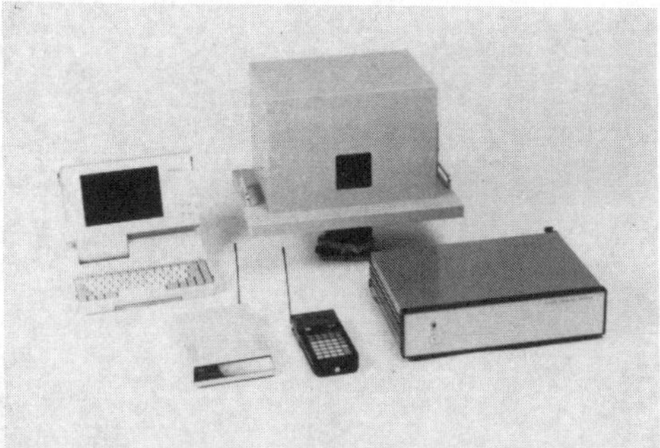

Photo 1 Laser marking device

Photo 2 Placement of the device

Photo 3 Simultaneous two-cross-section irradiation

points. This three-dimensional positioning system, which has already been committed for actual applications, has a number of novel characteristics. These are outlined below.

3-3-1 *Characteristics of overall structure*

(1) After registration of alignment, vertical alignment, and irradiation pattern in advance, inputting the face (or station) position automatically initiates laser marking of the registered pattern.

(2) Up to 200 irradiation patterns can be stored in memory, and any one may be selected.

(3) The marking device, when correctly placed, allows high-precision laser irradiation (Photo 2).

3-3-2 *Equipment*

(1) The galvanometer scanner uses two computer-controlled mirrors to scan the x and y axes for accurate laser irradiation. High-speed control of the mirrors enables after-image marking to be adopted. This type of mirror is capable of 24-hour operation, whereas the motor-driven type is not.

(2) Downsizing, adoption of a light-weight structure, and full integration has made the system compact in size.

3-3-3 *Special functions*

(1) Continuous high-speed scanning enables "afterimage" marking.

(2) The built-in wavelength-based distance meter enables the face to be measured at the same time as it is marked.

(3) Simultaneous double cross-sectional irradiation of the upper and lower halves is possible, making the system compatible with the micro bench method (Photo 3).

In addition to these major functions, the system also allows the size of the pattern section and the speed of irradiation to be adjusted.

3-4 *Improvement over conventional face marking*

Since completing this new system, it has been used in construction of the Shin Akechi Tunnel (Tochigi Prefecture), the Hokuriku Shinkansen Ichinose Tunnel (Gunma Prefecture), and the Shinkannon Tunnel (Kyoto Prefecture). The following improvements have been demonstrated:

(1) The system is effective in terms of both energy and labor saving during marking work.

(2) High-speed, continuous refreshing of the irradiated pattern means no physical marking required, reducing work time.

(3) Improved drilling precision has eliminated "outbreaks" and "chiseling," consequently improving quality and cost-efficiency.

(4) The system is applicable to any cross section, but is especially effective in the case of large cross sections and curved tunnels.

(5) Constant refreshing of the markings enables instant analysis of rock projections on the blast surface, allowing chiseling to be carried out during excavation.

(6) The system can also be used where conventional timbering is employed.

(7) The system reduces danger at the face when fragile rock conditions prevail, and thus ensures safe tunneling.

The great merit of the system has been demonstrated, especially as regards energy saving, economy, and construction management.

3-5 *Conclusion*

The new system has been proven to have excellent functionality, and successful results were achieved in construction of the Kuki Tunnel for the High-speed Linear Train, a tunnel for Chubu Electric Co. Ltd.'s Nikengoya power station, and the Ichinose Tunnel for the Hokuriku Shinkansen.

As an example of the success of these systems, consider the Nikengoya power station tunnel. The site of this watercourse tunnel is a mountainous region 120 km away from Shizuoka city. Because of the terrain, surveying would have taken a month using the conventional method. With the new system, it was completed in only a few days.

In the case of the Ichinose Tunnel, systematized implementation of surveying work involving the establishment of control points using GPS, tunnel alignment control, and laser marking led to labor saving and improved efficiency. There was also a reduction in outbreaks during face excavation.

Tunnelling and Ground Conditions, Abdel Salam (ed.) © 1994 Balkema, Rotterdam, ISBN 90 5410 363 9

Small-diameter automatic shield tunnel cast-in-place lining method realizes long-distance tunnel construction – Development of ACE MOLE 1200-M2 construction method

Yukio Kimura & Tsunenori Saito
NTT Telecommunication Field Systems R&D Center, Tukuba, Japan

ABSTRACT: Traffic is paralyzed on the main roads of large cities, resulting from the fact that the construction of underground channels, pipes and conduits utilized in water supply, power and telephone systems on which the inhabitants' life depends is on the increase, which situation leads to the tightening of regulation on the frequent recurrence of excavation on the main roads by all the proprietors; therefore, at present demand for a long-distance trenchless method of laying these installations underground is increasing day by day.

On the other hand, we are confronted with difficult problems such as shortages and aging of skilled workers, and sharp rises in labor costs.

At NTT (Nippon Telegraph and Telephone Corporation), smaller diameter tunnels have been required since the adoption of more optical fiber telecommunication cables. Along this line, a strong desire for development of a shield tunneling construction method has risen which enables construction of small inner-diameter tunnels of about 1.0 meter extending about one kilometer.

An automatic shield tunnel cast-in-place lining method has been developed, instead of the conventional segment assembling type, that uses high-early-strength resin mortar as its lining material and allows lining work in the presence of water. NTT has established the automatic shield tunneling system "ACE MOLE 1200-M2 construction method" that can select free alignment according to the alignment of a road and the condition of underground facilities and constructs a 1.2-meter small inner-diameter tunnel with a targeted driving distance of one kilometer.

The resin mortar lining material satisfies high-strength and high-durability requirements, and therefore attracts the attention of those engaged in sewerage systems who have corrosion problems with cement concrete pipes caused by wastewater.

The M2 method features not only completely unmanned processes in the tunnel, from excavation to tunnel lining work, but also contributes to efficiency, improved working environment and safety Extruded Concrete Lining (ECL) automation construction technology not employing reinforcement which has a bright future.

This paper outlines the development of the M2 method.

1 INTRODUCTION

In Japan, development of automation tunnel construction technology is being promoted to realize efficiency at construction sites and save manpower in order to respond to the scarcity and aging of job site workers, rising labor cost, and more businesses resulting from expanding public investments.

Construction work accidents at job sites must be prevented; construction workers must be released from hard physical labor so that attractive construction businesses may be created. From this point of view, automatic construction technology must be developed to tackle such issues as improving work efficiency, saving manpower and assuring safety at job sites.

In large cities, in addition to its worsening road traffic congestion, the control over frequent road excavation work conducted by public utility companies is being strengthened. These conditions have brought about rising needs for long-distance non-excavation type construction work.

To meet sophisticated, diversified information telecommunication needs, NTT is constructing telephone cable tunnels that accommodate multiple cables in Tokyo and other large cities, employing the shield tunneling construction method.

The introduction of optical fiber cables which might be called a dramatic technological innovation is on a large scale underway. This involves digging small diameter tunnels which have inner diameters of one meter or so and long distances: therefore it is strongly desirable that a new shield tunnelling method for those tunnels be put to practical use.

In this case, it seems that establishing a fully automatic system for digging them at high speed on a technical basis provides a right answer to the problems we are confronted with, because of their limited working space.

2 TACKLING OUR TASKS

Little technology for shield tunnel construction method has been established for construction of 1.0 meter inner-diameter tunnels, because it is difficult for construction workers to work in tunnels and to secure space within the shield machine for implementing automatic operation. Thus, hume-pipe jacking method has been used frequently, a pipe-end-pushing construction method in which its driving distance is limited. The longest distance achieved was about 400 meters.

Thus NTT worked on the development of an automated method for lining small-diameter shield tunnels with cast-in-place resin mortar, whose strength is developed soon after placement, instead of prefabricated segments and which can be used in underwater construction. And then, it experimentally built a 170 meter long small-diameter tunnels with a radius of curvature of 200 meters so far, and successfully developed an automated tunneling system using the "ACE MOLE 1200-M2 Method".

The development of such a method by which excavation of tunnels with inner diameter of 1.0 meters or so and with long distances of one kilometer or so according to the course of existing main roads or street is made possible, and by which proper control of the alignment is also available under any circumstances, for example, to avoid underground installations which might be buried is strongly required to satisfy the stricter environmental conditions of urban construction.

The following technical requirements had to be met to implement a construction method for the one-kilometer driving length and high-speed operation:

(1) Improved shield machine system reliability

(2) Higher speed transport vehicle to transport resin mortar lining material and excavated earth; improved stable machine angle capabilities

(3) To implement construction workability at 80-meter radius curvature (including improved shield machine direction control at curved sections)

(4) Improved long-distance traction capabilities for control cables

Various improvements and developments were made on the basis of these requirements.

The ACE MOLE 1200-M2 makes practical use of the lining material itself, automatic lining unit, lining material wagon and overall system control. Testing of the sewerage system construction work in Yokohama City (1.2-meter inner-diameter, 674-meter driving length, and 100-meter radius curvature at three locations) reveals that technical capability for installation of a 1.0-meter small-diameter shield tunnel extending over a long-distance of one kilometer has been assured.

3 SYSTEM OUTLINE

3.1 System outline

Photo-1 is a machine system image using the M2 construction method. Figure-1 is the system diagram for the machine system, and Table-1 provides the shield machine specifications. The system has a shield machine consisting of four sections: excavation apparatus, local electrical control apparatus, power supply apparatus and lining apparatus; unmanned transport vehicle that transports excavated earth and lining material; one-man controlled operation house that monitors and controls the entire system, and mixing plant that kneads the lining material.

Photo-1 Machine system image diagram

Table-1 Shield machine specification

Outside diameter	φ1, 480mm
Machine length	12, 145mm
Shield jacks	30tf×750mm×6pcs.
Thrust jacks	30tf×750mm×6pcs.
Type	Muck pressure type
Other	Four - divided machine

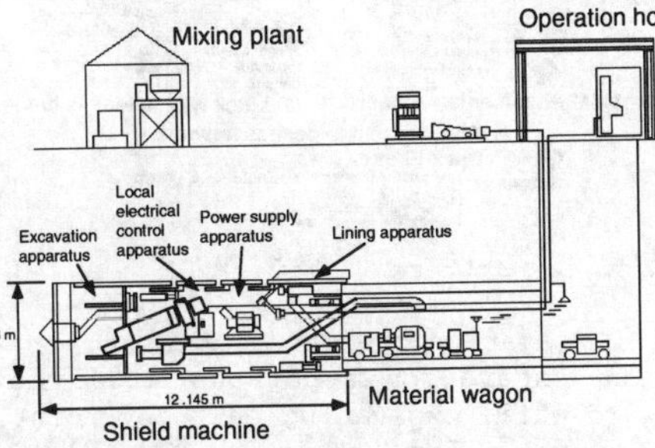

Figure-1 System diagram

3.2 Tunnel lining material

The ordinary shield tunnel construction method uses steel segments or reinforced concrete segments for lining work. Recently, the ECL construction method that assembles reinforced bars in the tunnel and places normal concrete has been put under trial. Development of automatic lining work in both construction methods remains to be a difficult issue for small diameter shield tunneling.

The target of the M2 construction method is to enable unmanned lining work in narrow tunnels, and features automatic building of a non-reinforced lining wall by simply feeding lining material. For this purpose, resin mortar was developed and implemented as lining material that assures high-early-strength.

(1) Resin mortar features

The general features of this lining material are as follows:

(i) Advantages: high-strength, quick setting, excellent chemical resistibility and abrasion resistibility

(ii) Disadvantages: large shrinkage when hardening, and the cost tends to be expensive

To develop new resin mortar with features of extremely low shrinkage and quick-setting which permits adhesion-in-water, improvements on the mix combination of lining materials were made as follows:

Improvements made in resin mortar:

(i) The strength before the improvement dropped about 30% when soaked in water, but after increasing the special surface processing density applied to sand which is the aggregate (see Table-2), the strength drop ratio was significantly improved.

(ii) More additive was applied to the anti-segregating agent which helped achieving a mix combination that does not segregate the materials for three days after kneading. Table-3 shows the mix combination of resin mortar, and Figure-2 indicates the strength properties of the lining material.

Table-2 Mix combination of special surface processing

Material	Weight (phs ※)
Silane coupling agent	0.15
Meth	2.0
Water	0.15

※ parts per hundred sand

Table-3 Mix combination of lining material

Item	Material
Resin	Unsaturated polymer
Aggregate	Sand
	Calcium Carbonate powder
Anti-segregating agent	Fine powder of silica
Hardener	Methyl Ethyl Keton Peroxide

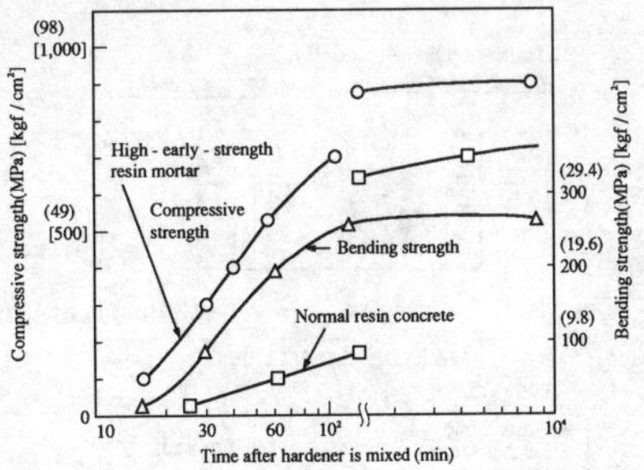

Figure-2 Lining material strength properties

(2) Basic characteristics test of mortar

In view of the basic characteristics of resin mortar, various tests were conducted to compare the test results with the lining material requirements.

The results were satisfactory as shown in Table-4.

Figure-3 illustrates the production of adhesion-in-water specimen. The lining material temperature when kneading is an important factor that determines the material segregation, and this is particularly important when the material is used for cast-in-place. The material mix combination is adjusted to retain the material temperature within a certain range. Because the construction work covers considerable distance, a temperature adjustment unit is provided in the plant mixer (Figure-4), with which resin mortar kneading is completed at a constant temperature throughout the four seasons. The sprinkler unit automatically adjusts the water temperature so that it retains a preset temperature. This development does not affect the working environment but constantly permits stable lining work.

Table-4 Requirements and test results of lining material

Requirments	Method of Experiment	Method	Results
Quick setting Compressive strength above 6.6 MPa within 30 min, 31.6 MPa within 50 min, 32.3 MPa within 7 days after casting Flexural stength above 7.6 MPa within 50 min, 21.5 MPa within 7 days after casting	Compressive test JIS A 1181 JIS A 1182 Flexural test JIS A 1181 JIS A 1184	Promoter is added	**Compressive strength** 43.3 MPa after 30 min, 56.2 MPa after 50 min, 73.1 MPa after 7 days **Bending strength** 24.1 MPa after 50 min, 25.6 MPa after 7 days
Low shrinkage Shrinkage rate within ±0.1% after 24 hours after casting	Measurering by micrometer with comparator method	Poly-vinyl-acetate is added	**Shrinkage rate** within -0.028% after gel within +0.019% after 24 hours
adhesion strength in the presence of water Bending strength above 49 MPa within 3 hours after casting	Adhesion strength test in the presence of water	Special surface processing is applied to sand	**Adhesion Bending strength** 58.9 Mpa after 3 hours
Material segregation After 3 days after mixing no segregation of polymer mortar	JIS A 1123	Anti-segregating agent is added	**Material seregation** no segregation after 3 days

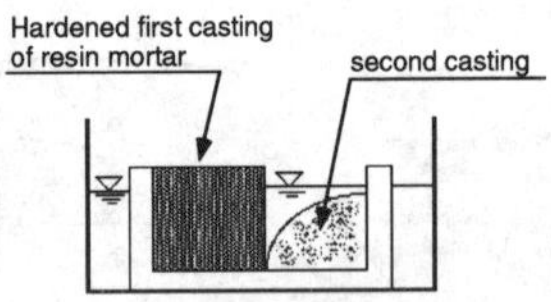

Figure-3 Production of adhesion-in-water specimen

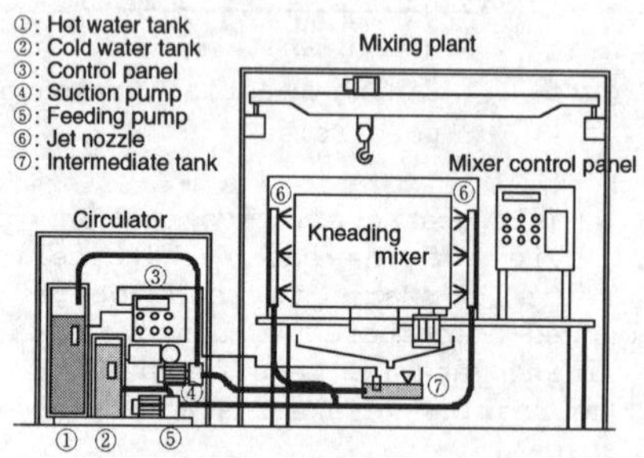

Figure-4 Temperature adjustment unit

3.3 Work cycle

It takes about two hours to complete one work cycle, and then repeated. As shown in Figure-5, one work cycle consists of earth excavation, lining form removing, pushing forward, back-filling, lining form setting and lining work.

In the four-section shield machine, a shield jack used for excavation is provided at the excavation apparatus section, while a driving jack is provided at the lining apparatus section. A multiple driving method that operates these jacks by turns has been adopted.

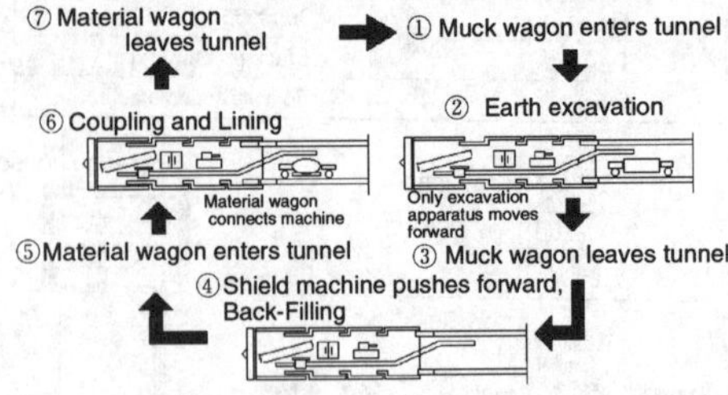

Figure-5 Work cycle

(1) Excavation process

The excavation apparatus alone moves forward while excavating earth by rotating the cutter wing and extending the six shield jacks forward. The screw- conveyer hauls proper amount of the excavated earth into the machine. The pressure pump then sends the earth into the soil packet on the muck wagon waiting at the rear end of the machine. The radio-controlled muck wagon starts running and transports the earth to the shaft. The shield machine is equipped with a gyro compass, rolling meter, pitching meter and stroke-angle meter to control the machine position with tolerable values according to the planned alignment. It is extremely difficult to precisely control the shield machine directivity by simply relying on the thrust of six jacks. In addition, considerable time is spent to maintain the cross-section and plane alignments. Because of these disadvantages, a compound oil-pressure control mechanism that provides tension and fixed capabilities was added to each shield jack (Figure-6).

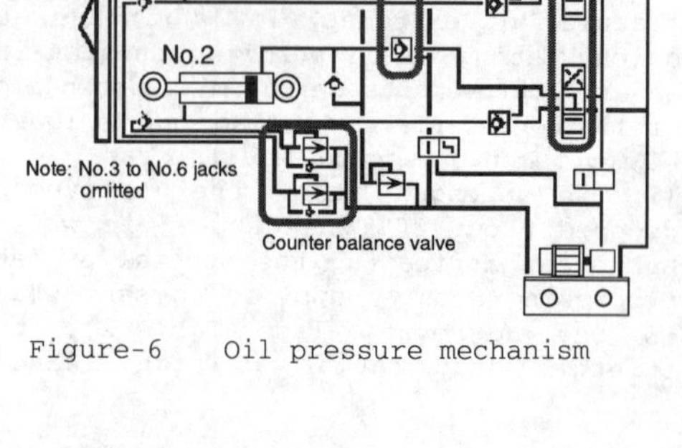

Figure-6 Oil pressure mechanism

(2) Shield machine moves forward

The local electrical control, power supply and lining apparatuses move forward by contracting the shield jacks and extending the thrust jacks, which push the resin mortar lining cast previously in the form. This forward move connects the excavation apparatus to the local electrical control apparatus. The construction work is seriously affected if shield machine rolling occurs, especially when it covers considerable distance or involves curved sections. To prevent the rolling, a stabilizer equipped with corrective rolling capabilities is provided at the local electrical control apparatus.

The electric and control cables that control the shield machine are pulled along, using pulleys and curtain rails with low friction, as the shield machine moves forward.

(3) Lining work

The material wagon loaded with resin mortar automatically moves into the tunnel and is, through radio-control, connected to the rear of the shield machine. The resin mortar is then sent with pressure to the lining form at the lining apparatus. One lining form completes a 70-centimeter wide lining wall.

3.4 Automatic lining system

The lining system is shown in Figure-7.

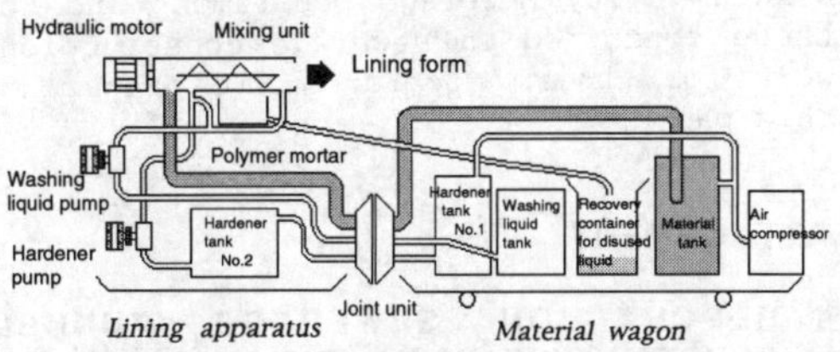

Figure-7 Lining system outline

(1) Mixing unit

At the mixing unit, resin mortar and hardener are equally mixed while the screw operates at 700 rpm, and continuously performs lining work. Because it is important that proper amount of hardener is mixed with resin mortar, the material wagon informs (by wireless unit) the machine-side of the weight of resin mortar supplied within a specific time frame. The machine-side adjusts the amount of hardener for mixing, using the sequencer in the machine. For quality control purposes of the lining wall, the optimum mixing rate of resin mortar and hardener is within the range of 2.5 - 3.5 phr (phr: parts per hundred resin). The hardener control flow, as shown in Figure-8, provides the optimum mixing rate. The mixing rate below 1.5 phr results in lack of hardening, while the mixing rate more than 5.0 phr means excessive contraction rate which leads to hair cracks. To avoid this situation, this system adopts a control method that injects the hardener according to the amount of resin mortar supplied every 20 seconds to achieve the targeted 3.0 phr mixing rate. When lining is completed, the resin mortar remaining in the mixing unit is washed and recovered.

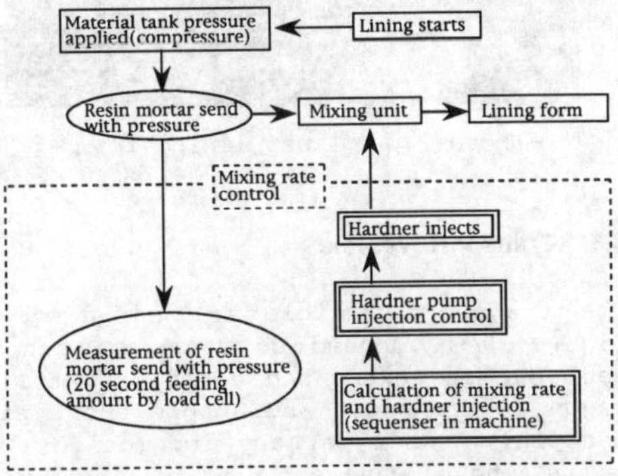

Figure-8 Mixing rate control

(2) Lining form unit

The lining form unit is composed of an outer-lining form, inner-lining form, end lining form, removable jack and thrust jack. The inner-lining form is divided into eight pieces. When lining, the lining form is completely closed to prevent resin leakage and is extended and contracted by the two jacks equipped on each piece, as shown in Figure-9 and Photo-2. A very durable fluorine-contained resin is coated over the lining form surface to avoid the problem of adhesive feature between the lining form and resin mortar.

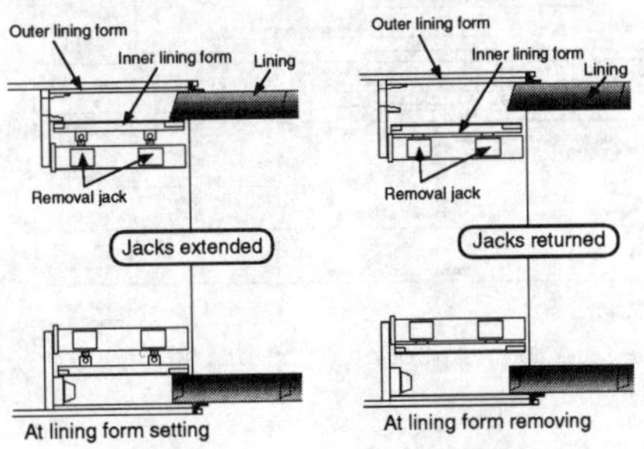

Figure-9 Lining form unit

(3) Control of lining thickness

The load cell equipped on the material wagon measures the material tank weight to check the amount of lining material used. The lining thickness is confirmed, using a supersonic sensor.

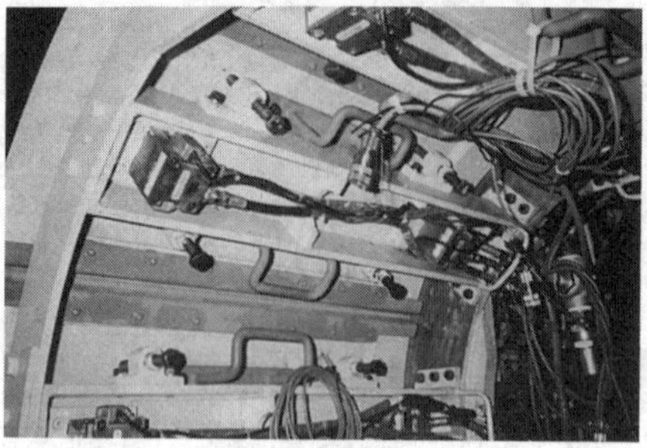

Photo-2 Inner lining form

3.5 Transport vehicle

The transport vehicle is, as shown in Figure-10, composed of a muck wagon to haul out the excavated earth, a material wagon to transport and supply the lining material, and a battery locomotive that pulls these wagons. A driving unit, control unit and power unit are housed in the battery locomotive.

A sequencer and remote radio control are combined for the vehicle to run automatically in the tunnel. Because the M2 system is an unmanned tunneling system, rails for the wagons, as used for normal shield construction methods, are not provided. To control the wagon's running position, a tilt angle sensor provided on the wagon constantly monitors the rolling angle and maintains it to be within ±5°. To allow wagons smooth running, a steering method combining two oil jacks has been devised and adopted to provide the wagons with a stable neutral position.

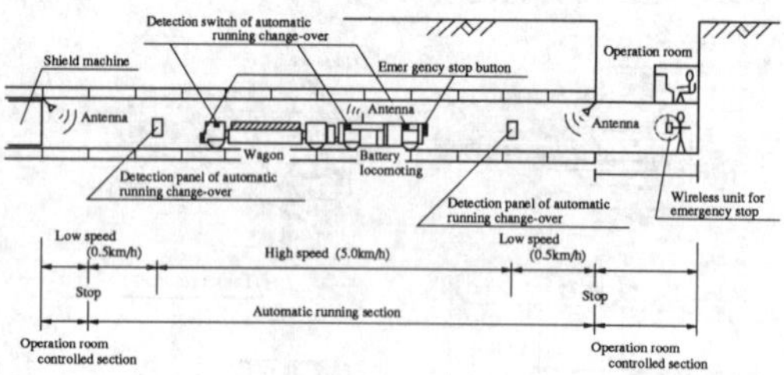

Figure-10 Automatic running system

CONCLUSION:

Hitherto, the M2 construction method was improved to realize a 1.2-meter small-diameter, long distance (one kilometer), curved (R=80 meters) tunnel construction method. Practicability of the method was then studied for application to a 1.2-meter inner-diameter, 674-meter long sewerage construction work (for details, see attached "Results of implementing M2 construction method to sewerage construction"). As a result, the M2 method is now considered a new type ECL construction method proving that it possesses the basic capabilities of an automatic tunneling system using resin lining material (excavation and lining work), and without reinforcement.

In order to assure quicker construction work and to further improve system reliability, we must promote technical development in the following areas:

(1) Improvement of adhesive property of resin lining material

(2) to establish an automatic simultaneous back-filling method

(3) to atomize in-tunnel measurements

(4) to devise a lining form unit with no resin leakage, and to assure quicker mortar filling into lining forms

We must also concentrate on enhanced functions while positively applying the construction method to long-distance, sharp-curved construction work.

We are very indebted to the persons in the Yokohama City Sewerage Department and to those concerned in sewerage construction work for their cooperation in preparing this paper.

ATTACHMENT

1 REPORT ON SEWERAGE TUNNEL CONSTRUCTION USING THE M2 METHOD

1.1 Outline of the construction

The M2 method was utilized for a sewerage construction in Yokohama city. The construction is a sewage main trunk rehabilitation work ordered by Yokohama municipal sewerage department. Table-1 shows the construction conditions. The construction route, passes a housing area from north to south and crosses a national highway bypass which carries a high traffic of 100,000 vehicles per day. The route is downsloped because the departure shaft site was difficult to obtain at the lower end and the shaft was installed at the upper end of the route.

1.2 Construction results

The construction was executed in sequence of shield machine carrying-in, initial driving, lining moldability check and main driving.

The shield machine completed the lining of 962 rings with the driving length of 674 m in October, 1993. Photo-1 shows a part of the completed tunnel. Analysis of data obtained from the construction indicates the follows:

Photo-1 Completion

(1) Finishing of the lining walls

No cracks were found on the completed lining walls and the walls were uniformly structured. It is confirmed that mixing ratios between resin mortar and hardener are controlled within a range of 3.0 phr ±0.5 phr (see Figure-1), which satisfies a required quality level. Also, the top-end lining thickness, which is an important factor of lining, satisfies the management criteria of 12 cm (see Figure-2).

On the other hand, further study is required for improving the adhesive property-in-water of lining materials and the material leakage from forms.

(2) Directional control

The construction accuracy, which is important particularly for sewerage tunnelling, is within ±35 mm apart from the baseline as shown in Figure-3, sufficiently meeting the management criteria of ±50 mm. This accuracy can be rated high, considering that secondary lining for meandering adjustment is not performed in this M2 method. Also, we confirmed that a required accuracy for curvature lining is fully controllable with six jacks.

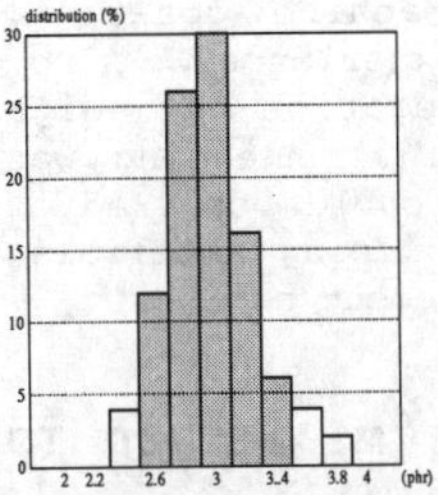

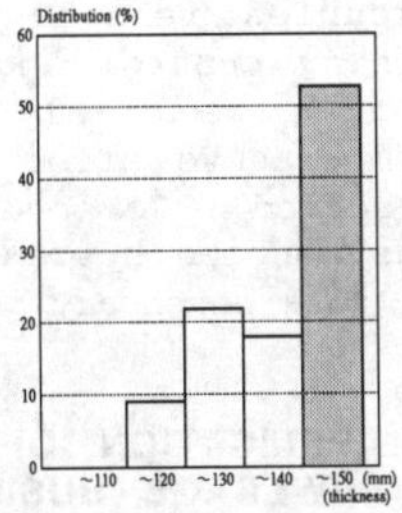

Figure-1 Mixing rate distribution

Figure-2 Top-end lining thickness

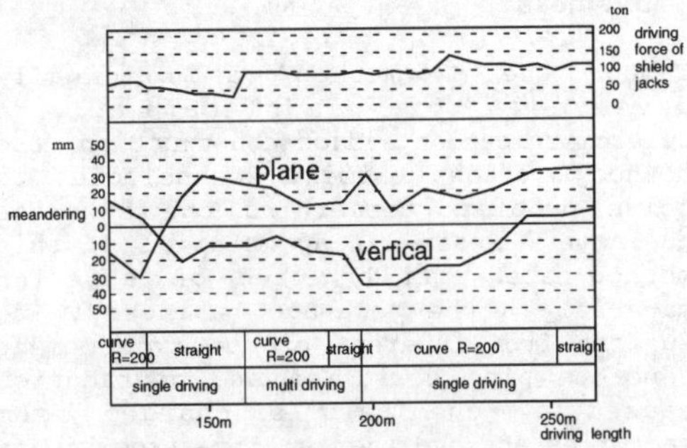

Figure-3 Construction accuracy (plane, vertical section)

(3) Work efficiency

The average driving length per day was 3 m, which was below the target of the average driving length (4.2 m per day). After improving the machine, the maximum driving length was extended to 4.6 m per day (6.5 rings) in several operations.

In the present system, 4.9 m to 5.6 m per day (7 to 8 rings) is the limit due to the cycle time. We must challenge higher speed operations by shortening the curing time for lining.

1.3 Troubles and actions

This was the first time for the M2 method to execute a long distance (674 m) construction with 100 m curvature radius, though the method was used for a shorter distance construction in the past. Before

Table-1 Work outline

Client	Yokohama City Sewerage Department	Driving length	674m
Construction site	Yokohama - City Kanagawa - Prefecture	Planimetric alignment	R=100m : three, R=200m : four
Purpose	sewage tunnel	Longitudinal alignment	-0.29% slope
Driving period	1991.10～1994.3	Earth covering	GL-6～-15m,Water pressure:4.9Pa
Inner tunnel diameter wall thickness	1.2 m 10 cm	Soil conditions	Silty Clay (Value N=2～30)

the construction started, various tests were conducted to confirm functions of the machine. At the initial stage of the construction, however, we observed troubles related to machine position, lining control and inexperience. Also, lining material segregation at high temperature after material kneading was observed. For every trouble occurred of the machine system and lining materials, we took necessary improvement actions.

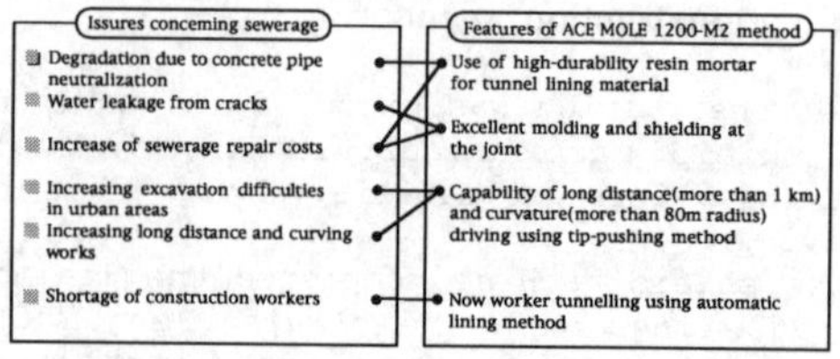

Figure-4 Sewerage problems and M2 method features

2 APPLICATION OF THE M2 METHOD TO SEWERAGE BUSINESS

2.1 Problems concerning the sewerage business

Though the M2 method was originally developed for application of telecommunication's duct construction, the method provides element technologies to solve problems concerning the sewerage business. Resin lining material of this method also can be used as superior material in the sewerage business which suffers from inherent problems of cement concrete pipe corrosion and degradation caused by chemicals discharged from factories and hydrogen sulfide from sewage water.

2.2 Chemical resistance test results of resin mortar

To evaluate chemical resistance of resin mortar when used for sewerage facilities, tests were conducted to check strength changes before and after specimens of resin mortar were soaked in the test liquid of acid solution (sulfuric acid, hydrochloric acid and acetic acid) and alkali solution (hydroxide solution). For comparison, the same test procedure was also applied to concrete.

As given in Figure-5, the tests clearly indicate that resin mortar has higher chemical resistance against acid than concrete. And the two materials showed the similar level of chemical resistance against alkali. Photo-2 shows degradation of resin mortar and concrete after soaking them in sulfuric acid for two weeks.

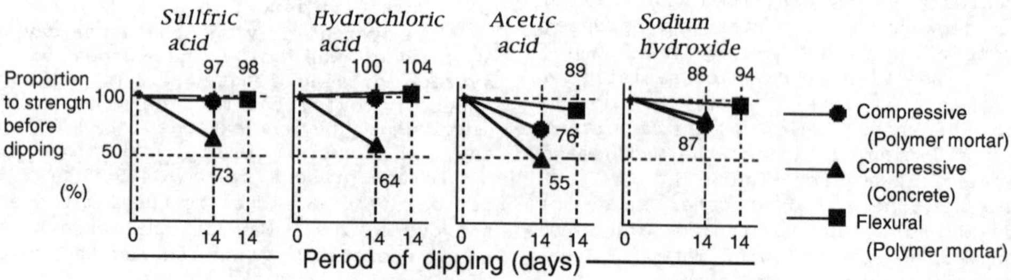

Figure-5 Test results of chemical resistance

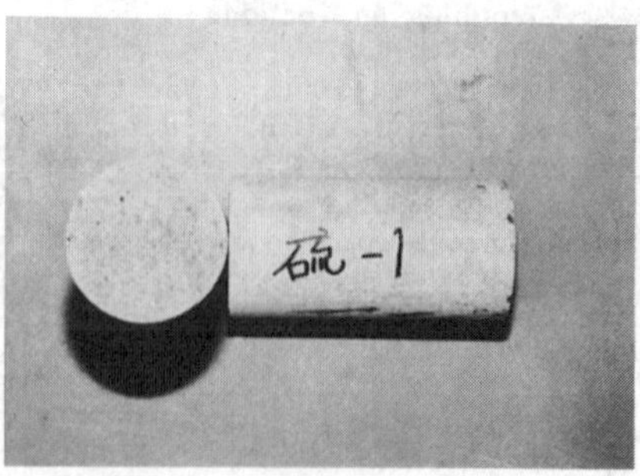

Resin mortar

Cement concrete

Photo-2 Specimen degradation (two-week sulfuric acid soaking)

Tunnelling and Ground Conditions, Abdel Salam (ed.) © 1994 Balkema, Rotterdam, ISBN 90 5410 363 9

A 3D CAD geological analysis system

S. Koyama & R. Morita
Construction Technology Development, Taisei Corporation, Japan

E. Hisakawa
Information Systems, Taisei Corporation, Japan

ABSTRACT: This system is intended for the visualization of information on subsurface geological conditions to facilitate underground design and construction work. For this purpose, the accurate estimation of geological structure and ground properties based on the data obtained in the investigation and construction phases, as well as the visualization of the modeling results and underground structures using a 3D CAD system are required. This report presents the results of applications of this system to field data.

1 INTRODUCTION

With the recent increase in the development of the underground space, underground structures have become larger in scale and more complex. As a result, technologies that enable the automatization, systematization and enhancement of design and construction works to ensure the safety and improved quality of construction, as well as to reduce labor are required. In this report, the rationalization of design and construction works through the visualization of information on various aspects of underground conditions is discussed. As visual observation of geological conditions three-dimensionally distributed underground is impossible, engineers are obliged to judge the unseen geology through imaginary three-dimensional data which is created based on limited information such as bore hole data and two-dimensional data obtained from the various analytical results. Such three-dimensional data, however, often depends on an engineer's subjective point of view, resulting in oversights and inconsistent judgements.

Under this situation, it is important to prevent oversights that were unavoidable with the use of conventional methods, as well as to expedite design and construction works and to simulate various possible imaginary situations using a three-dimensional model of a geological structure objectively formed by computer, based on the obtained underground data. In order to enhance efficiency of design and construction works, it is also important to comprehend the geological structure and the locational aspect of the structure in the objective site, as well as to quickly and accurately visualize the data on the progress of construction works.

In view of the above, this system was developed for the accurate estimation of geological distribution and the easier visualization of obtained data. This report presents a description of the system along with the results of applications of the system to field data.

2 3D CAD GEOLOGICAL ANALYSIS SYSTEM

The technical background to this system for its application to civil engineering, the software configuration, the processing functions of the system, as well as the data estimation method for modeling will be discussed below.

2.1 Background

The geological structure dealt with in civil engineering consists of three-dimensionally distributed strata and rock masses. Non-conformities and discontinuous planes, such as faults and fracture zones often encountered in these strata and rock masses along with folding and erosion, make the geological structure even more complex.

Data that is available in engineering works is usually limited to linear or two-dimensional data such as bore hole data and cutting face investigation data. This data, however, is available only scatteredly, so methods that enable the accurate estimation of unknown values between known values are required. The triangulation, inverse distance and the geostatistics methods can be mentioned as a few of such methods, however, they still require some background that can attach physical significance to the estimation.

This system was developed, based on the results of reviews on an estimation method that enables the enhancement of the estimation accuracy of geological distribution and ground properties, to visualize data on geological structure and underground structures using a 3D CAD system, as well as to ultimately facilitate efficient design and construction works.

2.2 System configuration

The core of this system consists of VULCAN, a three-dimensional underground supervision system developed by KRJA and geostatistics. The software called VULCAN consists of the following three functions: supervision of the geological data base, modeling of geological structure using the inverse distance and triangulation methods and three-dimensional display. Geostatistics allows the estimation of data on stratum boundary, and ground properties based on the geological data base established by VULCAN. The obtained results are visualized and displayed by VULCAN's three-dimensional graphic function. A display of the analytical results transferred from a three-dimensional analytical program such as the FEM program is also possible with the use of VULCAN. The software configuration of the system is shown in Fig. 1.

2.3 Processing functions

In this system, a geological data base is firstly established utilizing bore hole data obtained during investigation at a limited number of bore hole sites and geological data obtained through geological reconnaissance. Secondly, based on the established data base, geological structure is modeled utilizing geostatistics, the triangulation and inverse distance methods to evaluate the estimated data. Thirdly, the geological structure is displayed three-dimensionally using the estimated data. A structure can be formed with the CAD system, and a real-time display of a birds-eye view scaled up or down to any desired level is possible by shifting the viewpoint.

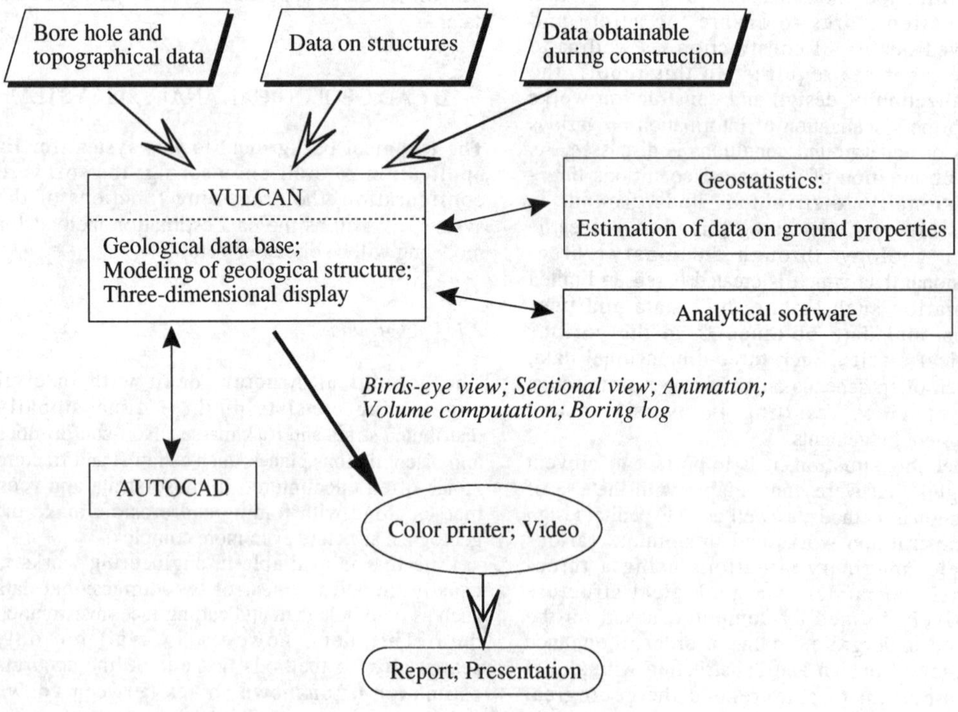

Fig. 1 Software configuration

The two-dimensional section of any cross section can also be drawn up. Adding of data as it becomes available during construction (such as pre-boring and cutting-face investigation data in tunnel construction) allows the remodeling of the geological structure, enabling enhancement of the accuracy of the predicted data as well as the prediction of the geology in areas where succeeding construction is to take place. The process flow diagram of the system is shown in Fig. 2.

This system can be utilized as an effective tool which provides useful information during the comparison of design values to a geological distribution, as well as consideration of the location of anchor arrangement in the design phase. The system is also effective in the construction phase for safety supervision and devising appropriate countermeasures against deformations.

2.4 Data estimation methods

Due to the fact that only a limited amount of bore hole investigation is carried out in engineering works, the estimation of geological structure has to be done by adding geological data obtained from the results of geological reconnaissance to those obtained in construction. Cutting face investigation data and pre-boring data that is available in tunneling works are examples of such data. In this report, the triangulation and inverse distance methods as well as geostatistics are discussed to predict stratum boundary distribution.

Triangulation method: This method uses a triangulation algorithm called the Delauney method which determines the most equilateral set of triangles starting essentially from the center of the map and working to reduce elonate triangles along the edges.

After this is done, the results of the triangulation are used to form a smooth grid using the Bicubic spline function.

Inverse distance method: This method is frequently used to estimate a value between the values already known. A set of values obtained from a set of sampling data located close to each other tends to present greater similarity than that obtained from a set of sampling data which was far apart. In view of the above tendency, the degree of contribution to a mean interpolated value is weighted as a function of the distance between an unknown value and a value already known. This method, however, has the disadvantage that the obtained data is subject to the influence of the overall tendency of a specific surface.

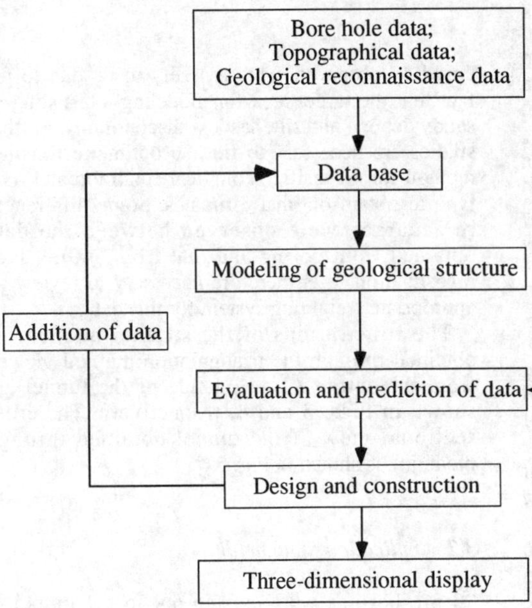

Fig. 2 Process flow diagram

Geostatistics: A stochastic method that was developed particularly for the study of mineral deposits. This method is distinguished from other methods by the fact that unknown values can be estimated based on the stochastic theory when irregular space fluctuation makes modeling impossible with the use of smooth mathematical functions. An advantageous characteristic of this method is that the estimation of errors is possible. However, computation requires a lot of time, which is a disadvantage.

The applicability of the above mentioned methods to a specific stratum and ground property was examined by comparison using actual field data.

3 EXAMPLE OF APPLICATION

3.1 Application to tunnel A

The construction site of this tunnel is located in an urban area where overburden is as shallow as 5-24 m, and the area immediately above the tunnel is densely populated. For this reason, consideration had to be given to environmental impacts such as subsidence and drought that directly affect residents in the adjacent areas. In order to devise effective countermeasures against possible trouble as well as to ultimately ensure safe construction, this system was employed, firstly to estimate stratum boundary from the data obtained through boring and cutting face investigations carried out near the surface, and secondly to estimate the locational relationship between the tunnel and anticipated stratum boundary, together with the geological structure at the excavation site.

The system was applied to an analysis of the area with the shallowest overburden of 5 m. Stratum boundaries were modeled based on the data obtained from the 16 borings carried out within an area of 200 m x 100 m in the XY plane as well as that obtained from cutting face investigation. The stratum boundary was classified into the following five geological types: embankment, alluvium loam, diluvium sand, diluvium silt and mudstone. The grid for the stratum boundaries was divided at a pitch of 2 m for modeling. A three-dimensional display of the ground surface, the stratum boundaries, the groundwater level and the tunnel gave a clear illustration of their location. A cross-sectional view and a profile of the tunnel were also drawn up.

The stratum boundaries were estimated through the triangulation and inverse distance methods using the data obtained from boring and cutting face investigation. No marked difference between the results obtained through the above two methods could be observed. This is likely to be due to the fact that the target area for modeling, consisting of sandy deposit and silt, has less discontinuity. Further studies are necessary to find an optimal estimation method for a specific geological condition and rock type to obtain optimal estimates. Some differences in accuracy were observed between the data obtained from boring and that from cutting face investigation. It is therefore necessary to review an appropriate weighting system for this data.

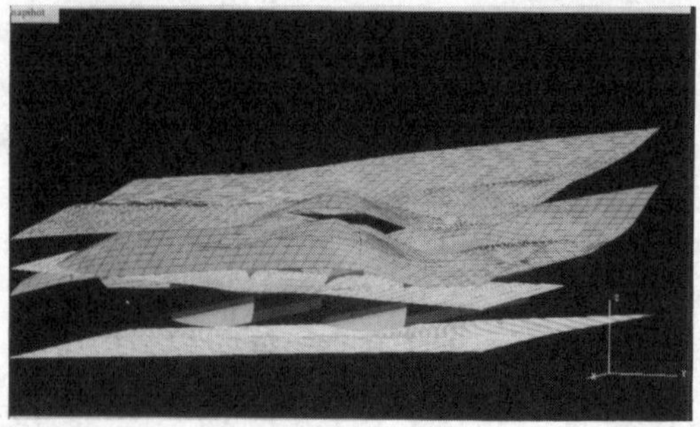

Fig. 3 View from outside of the tunnel

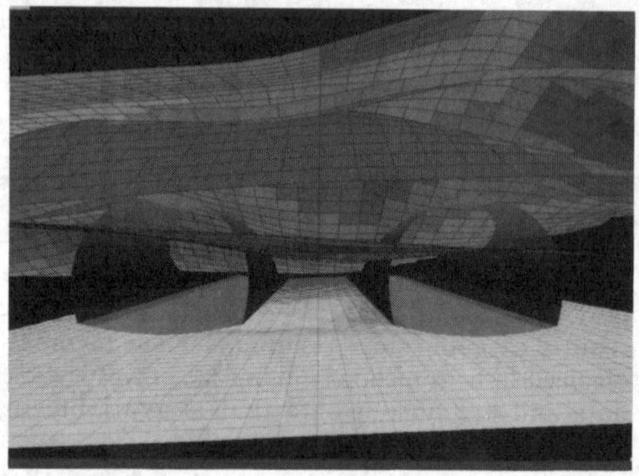

Fig. 4 View from the entrance of the tunnel

The illustrations of the stratum boundaries obtained through the triangulation method viewed from the entrance and outside of the tunnel are shown in Figs. 3 and 4, respectively. The cross sectional view of the tunnel obtained through modeling is shown in Fig. 5.

3.2 Application to tunnel B

23 pre-borings were carried out in the tunnel to comprehend the geological structure of the area where two tunnels cross each other. The stratum

boundary distribution of granite and sedimentary rocks ahead of the cutting face was estimated three-dimensionally using the obtained bore hole data. The triangulation and inverse distance methods as well as geostatistics were employed for comparison.

The analysis was performed within an area of 34 m x 60 m in the XY plane. The center line of the tunnel runs along the y-axis at x = 0. The positive direction of the y-axis shows the direction of progressive excavation. Borings were carried out in all directions centering around the origin.

The contour maps of stratum distribution are shown in Fig. 6. In the figure, map (a) was obtained through geostatistics, (b), through the triangulation method and finally (c), the inverse distance method.

In the inverse distance method, modeling was carried out with 1.0, 1.5 and 2.0. degrees of distance. No marked difference was observed among them. Modeling using the degree of 2.0 presents the greatest similarity to modeling through the triangulation method. The triangulation and inverse distance methods develop a similar tendency. In the illustrations (a) and (b) in Fig. 6, projections can be observed in the extrapolated sections between 0-10 m and 45-60 m on the y-axis. This is due to excessive weighting of known values, which is a disadvantage.

In geostatistics, weighting was carried out using the Ordinary Kriging method. In this method, the formation of such projections was not observed, showing its distinctive tendency from the other two methods.

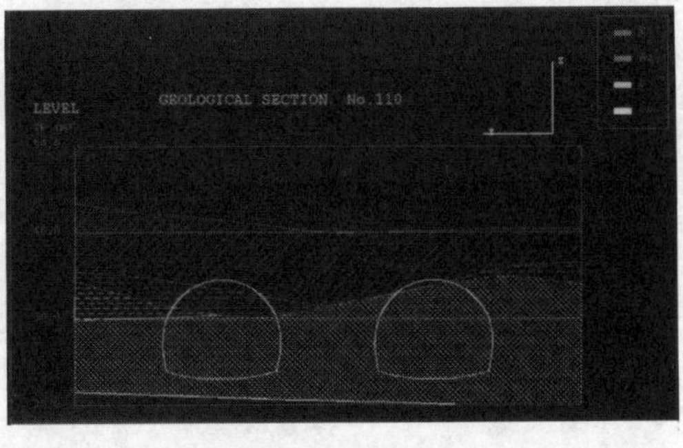

Fig. 5 Example of sectional view

3.3 *Application to tunnel C*

The ground at and around the entrance of this tunnel was soft and thus a considerable amount of water inflow was anticipated. To prevent loosening of the ground as well as settlement of the ground surface, the pipe-roofing method was employed as an auxiliary method. The arrangement and number of roof pipes as well as the stratum boundaries of alternating layers of sand stone and sand and gravel were three-dimensionally displayed. As only a limited amount of bore hole data was available (from 6 bore holes), the stratum boundaries were estimated using the triangulation method. The geological structure at the entrance of the tunnel along with the roof pipe arrangement is shown in Fig. 7.

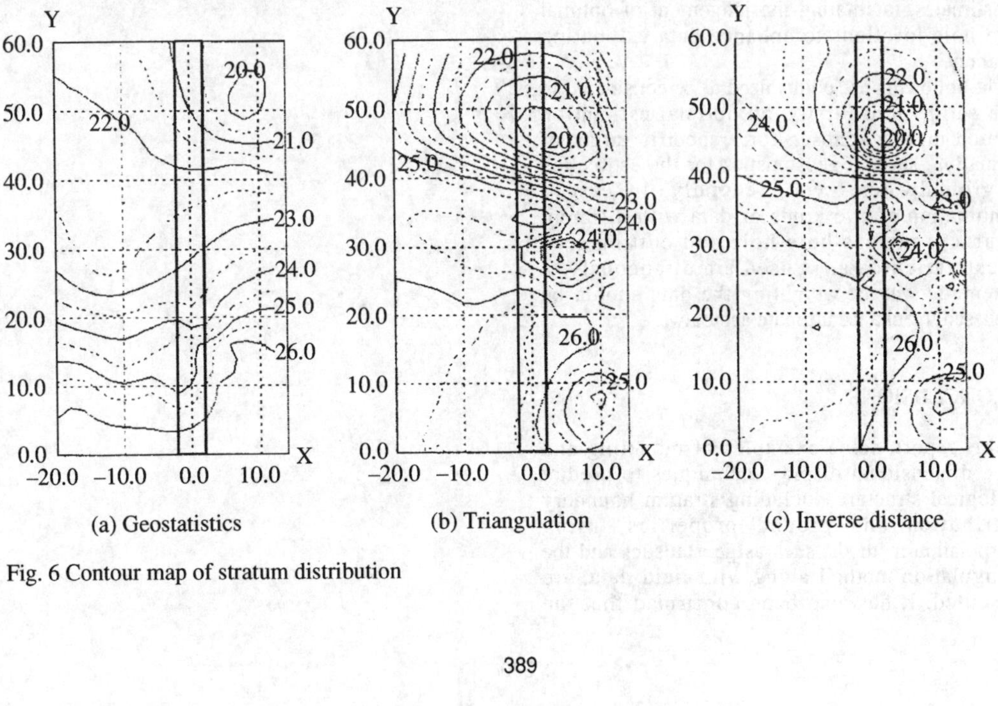

(a) Geostatistics (b) Triangulation (c) Inverse distance

Fig. 6 Contour map of stratum distribution

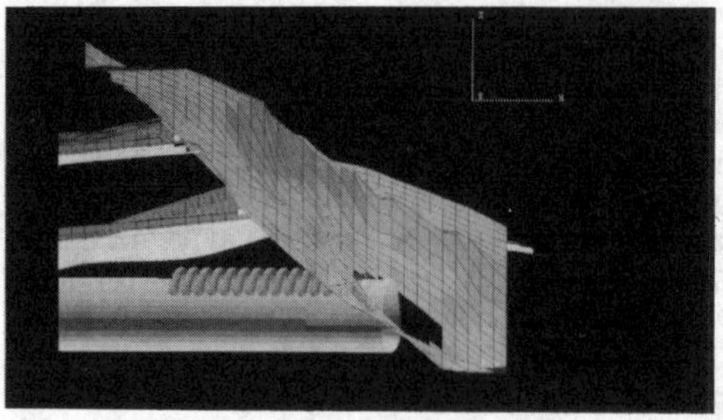

Fig. 7 Birds-eye view of tunnel entrance

4 DISCUSSION

Various methods for estimating data on stratum boundaries through bore hole data and cutting face investigation data obtained from tunnels, have been discussed. As only a limited amount of boring is carried out in mountain tunnels, geological structure can be estimated only through geological reconnaissance and cutting face investigation data. In such a case, data estimation using the triangulation method is the most appropriate, as other methods such as geostatistics which is based on stochastic theory, is less reliable in terms of accuracy due to the limited amount of sampling data. On the other hand for tunnels in urban areas, geostatistics can be a reliable estimation method as enough data to support a stochastic ground is available through bore holes, geological reconnaissance and cutting face investigation. Moreover, geostatistics allows computation of errors to estimates, facilitating the judgement of optimal bore hole locations to enhance data estimation accuracy.

The following are points need to be considered in data estimation. Firstly, an optimal estimation method must be chosen for a specific target of estimation, such as geostatistics for the estimation of groundwater level. Secondly, in case a combination of two kinds of data with different accuracy, such as bore hole and cutting face investigation data, is used simultaneously, a systematic way of weighting the data should be discussed to enhance ultimate accuracy.

5 CONCLUSIONS

In this report, some examples of modeling and three-dimensional display techniques to predict geological structure, including stratum boundary distribution and ground properties, using interpolation methods, such as geostatistics and the triangulation method along with field data, are presented. It has also been confirmed that the visualization of data on underground geological conditions and underground structures at the construction site using the CAD system is essential from the view point of the reconfirmation of design data as well as to increase efficiency and safety.

To enhance the accuracy of modeling through adopting geostatistics to the estimation of geological structure and ground properties, continuous verification of the estimation accuracy with field data is important.

Further, in order to expand the function of the system in the future, the development of additional systems that enable the display of predicted data with improved accuracy by integrating the results of geophysical exploration as well as the prediction of the location of faults is expected.

REFERENCES:

Journel, A.G. and Huijbregts, Ch.J.: Mining geostatistics, Academic Press, 1991

Tunnelling and Ground Conditions, Abdel Salam (ed.) © 1994 Balkema, Rotterdam, ISBN 90 5410 363 9

A system for presentation of engineering geological data for tunnel design

F. Løset, N. Barton, R. K. Bhasin, C. G. Rawlings & G. Vik
Norwegian Geotechnical Institute, Oslo, Norway

ABSTRACT

In two recent, major projects, NGI has collected a large amount of engineering geological data. Both were geotechnical design projects, one comprising underground caverns for the disposal of solid, low and intermediate level radioactive wastes in Great Britain, the Nirex project; and the other, the Olympic (1994) Ice Hockey Cavern at Gjøvik, Norway which is the largest rock cavern for public use in the world (span width = 61m). In both of these projects, an orderly, well arranged presentation of engineering geological data was very important. A new system has been worked out to facilitate this. Engineering geological data can be collected during geological mapping at or below the surface, by logging drillcores or by carrying out laboratory testing. These data are important in planning and designing tunnels and rock caverns, and in the evaluation of reinforcement for tunnels. Such data are also necessary for numerical analyses. To make these data easy to interpret, a special geotechnical chart has been developed showing the data as histograms. The chart has also been constructed as a spreadsheet in Lotus, thus making it easy to store the data and to combine different data groups. In the Nirex-project this system has been used for the logging of 6 km of 100mm diameter core as well as during earlier surface mapping. In this way, it is practical to compare data from different locations or from various depths in different boreholes. This is very important in the design of tunnels and rock caverns.

INTRODUCTION

During planning and construction of tunnels and rock caverns, a large quantity of engineering geological data is often collected. The source of these data will, during the planning stage, be surface mapping, core logging or mapping in nearby underground excavations. These data will be important for the design of the new project.

During the construction of a rock cavern a lot of data will usually be collected by mapping in the current cavern. These data will be used for designing the final rock support, but are also stored as a data base for later use.

For major underground excavations the quantity of engineering geological data may be very large and it is therefore important that the data are systematically recorded and stored in such a way as to enable easy access and easy use.

In two recent, major projects, NGI has collected a large amount of engineering geological data. Both were geotechnical projects, one involving underground caverns for radioactive waste disposal in Great Britain, the Nirex project; and the other, the Olympic Ice Hockey Cavern at Gjøvik, which is the largest rock cavern for public use in the world (span width = 61m). In both of these projects, an orderly, well arranged presentation of the data was necessary.

A new method for recording, storing and manipulating the data has been developed at NGI. The geotechnical data is recorded on a logging chart displaying histograms which are subsequently entered in a PC based spreadsheet (Lotus) which allows the data to be combined or manipulated according to the user's wishes. A description of the logging chart was given by Barton *et al.* (1992).

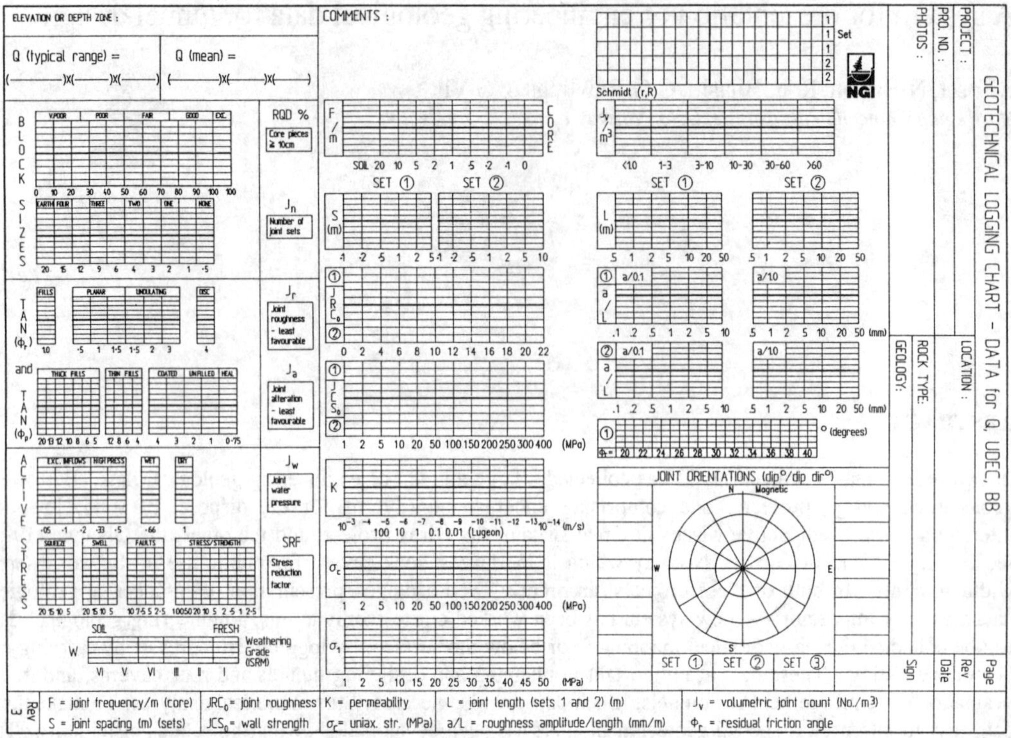

Figure 1   NGI's Geotechnical Logging Chart

The basis of the engineering geological data are the six different parameters in the Q-system: RQD, J_n (joint set), J_r (joint roughness), J_a (joint alteration), J_w (joint water) and SRF (stress reduction factor). Histograms for these parameters are found on the left hand side of the chart in addition to weathering. In the middle section of the chart, there are histograms for joint spacing, joint roughness and strength parameters JRC and JCS, permeability, rock strength and rock stress. On the right hand side of the chart there are histograms showing joint length, joint roughness amplitudes and joint orientation.

By mapping the data from one special location, large or small, the information can be recorded on the chart, and using Lotus, the data from several locations can be easily combined. This facilitates a review of the different data in the form of depth logs.

THE GEOTECHNICAL LOGGING CHART

The geotechnical logging chart highlighted in this paper can be used during geological mapping underground or at the surface, and during core logging. Figure 1 shows the chart used during surface or underground mapping. Rectangular graduated areas are given for making numerous individual observations of the various parameters. The frequency of observations will depend on the type of mapping. During field mapping, for example, one rectangle may correspond to one exposure. During core logging one rectangle will correspond to a fixed length of core. The histograms will in this way show the percentages of the different qualities for each parameter.

After mapping, the data is entered in a PC based spreadsheet (Lotus) and the logging chart can be printed out. Figure 2 shows a finished logging chart typical of that obtained from field mapping. The Lotus programme will automatically calculate mean values for the different parameters and can be used to combine several charts when information on specific depth intervals is required.

The basis of the engineering geological data is the six different parameters in the Q-method which are found on the left hand side of the chart. In the Q-

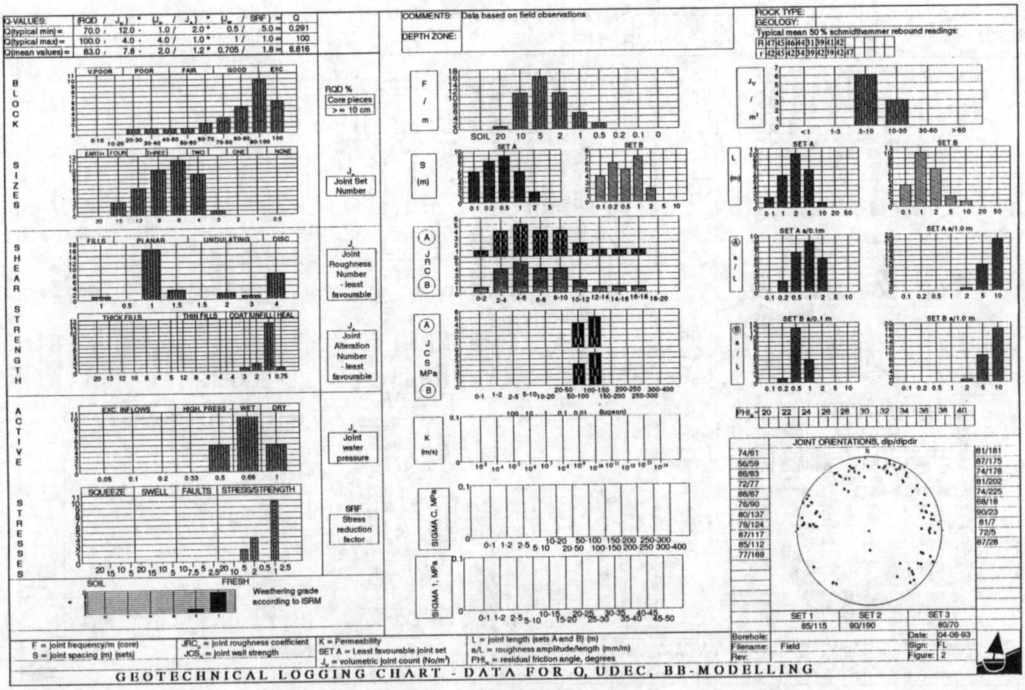

Figure 2 *Example of logging chart based on field mapping.*

method the stability of a rock mass is described by a Q-value based on six parameters (Barton, *et al.*, 1974):

$$Q = \frac{RQD}{J_n} \times \frac{J_r}{J_a} \times \frac{J_w}{SRF}$$

RQD = Rock Quality Designation (Deere, 1963)

RQD is the total length of core pieces over 100mm (between natural joints) expressed as a percentage of the core run length. During field mapping the RQD-values can easily be calculated by using the number of joints per cubic metre.

J_n = Joint Set Number

These numbers vary from 0.5 for unjointed rocks to 20 for completely crushed or earth-like material. RQD and J_n together give a description of the joint pattern and block size in the rock.

J_r = Joint Roughness Number

These numbers vary from 4 for discontinuous joints to 0.5 for planar slickensided joints. For joints with soft infilling a nominal value of J_r = 1 is used.

J_a = Joint Alteration Number

These numbers vary from 0.75 for healed joints to 20 for discontinuities with thick fillings of swelling clay. The J_a numbers depend generally on the thickness and mineralogy (friction properties) of the fillings. J_r together with J_a give an approximate indication of the friction angle along the joints or filled discontinuities. Other parameters which are supplementary to this are JRC, JCS and ϕ_r.

J_w = Joint Water Reduction Factor

This parameter varies from 1.0 for dry rock masses to 0.05 for rock masses with very high inflow, as recorded or expected at the proposed excavation depth.

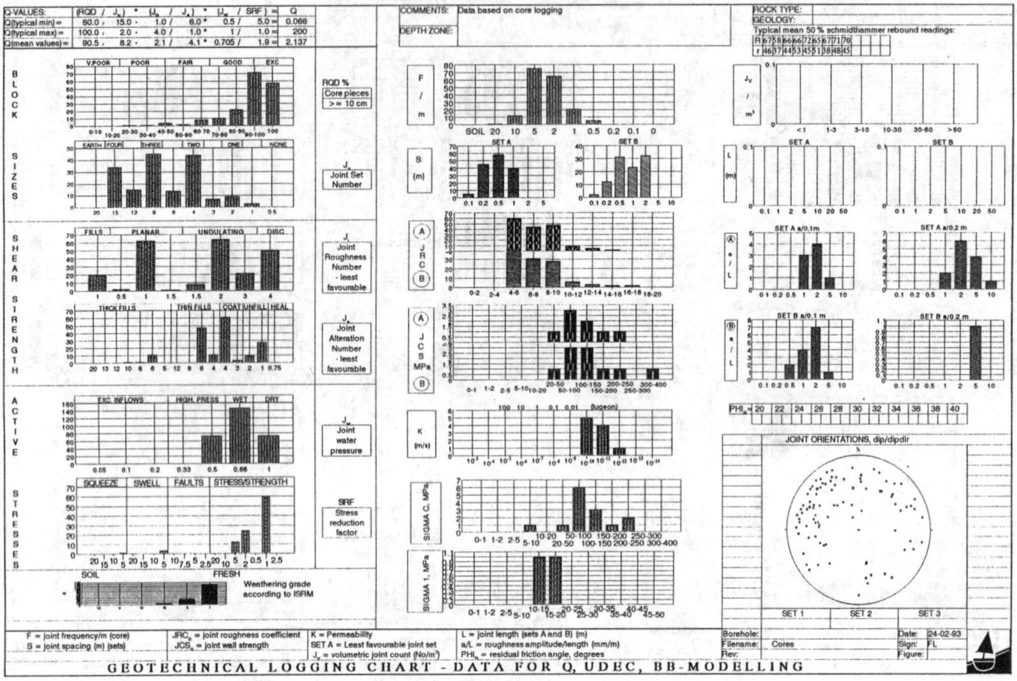

Figure 3 Example of logging chart based on core logging.

SRF = Stress Reduction Factor

For the case of competent rock, the SRF-value is based on the ratio between rock strength (σ_c) and principal stress (σ_1) at the tunnelling depth. Special cases are swelling and squeezing rocks and fault zones. The factor may vary from 1 for hard rocks with moderate stress to 20 and more for rocks with extremely high stress, or for cases of extreme squeezing or swelling. J_w together with SRF gives a description of the active stress situation in the rock mass at the tunnelling depth.

The typical minimum, maximum and mean Q-values are automatically calculated and are found in the upper left corner of the chart.

In the lower left corner of the chart, histograms concerning weathering grade are found. Six different classes, I to VI, according to ISRM are used.

In the middle section of the chart the following parameters are found:

F/m = Frequency of joints per metre along a line of sample (*i.e.*, along core).

S(m) = Joint spacing. Spacing for the most prominent joint set (A) and the second most prominent joint set (B) are given.

JRC = Joint Roughness Coefficient. This number may be estimated visually or estimated by measuring joint roughness amplitude over length.

JCS = Joint Wall Compression Strength. This value is calculated from Schmidt hammer readings along joint surfaces.

K(m/s) = Permeability of rock mass. The numbers may be based on Lugeon test measurements in boreholes (where 1 Lugeon $\approx 10^{-7}$m/sec).

σ_c = Rock compression strength. These numbers may be based on laboratory testing or calculated from Schmidt hammer readings.

σ_1 = Major principal stress. These numbers are based on *in situ* stress measurements.

On the right hand side of the chart the following parameters are found:

R = Schmidt hammer rebound number on fresh rock.

r = Schmidt hammer rebound number on joint surface.

J_v/m^3 = Number of joints per cubic metre.

L(m) = Length of joints in metre (joint set A and B).

a/L = Maximum amplitude of roughness in millimetres estimated on joint profiles of two different lengths (for field mapping usual lengths are 0.1 and 1.0m).

PHI_r = Residual friction angle (ϕ_r). These numbers may be estimated in the field using J_r/J_a, or measured more exactly by laboratory testing.

In the lower right corner of the chart there is space for joint orientation data. These data can be given both as dip/dip direction numbers or as a stereo plot (the lower hemisphere of the Schmidt net is usually used).

EXAMPLES OF THE USE OF THE CHART

The geotechnical logging chart was developed for the Nirex project with respect to the design of underground caverns for radioactive waste disposal in Great Britain. Examples of data presentation are shown in Figure 2 (field data) and Figure 3 (core logging data). Using these charts the ranges and mean values of the different parameters can easily be compared. The accuracy of the value for each parameter will depend on the way in which the parameters were obtained. Parameters such as J_w and SRF are directly linked to a specific tunnel and the same values are estimated for these two parameters in Figures 2 and 3. During core logging it is usually impossible to measure joint length, but J_a (joint alteration number) can often be directly observed. J_a may often be difficult to estimate during field mapping because of surface weathering. It may also have a larger rating at the surface for the same reason.

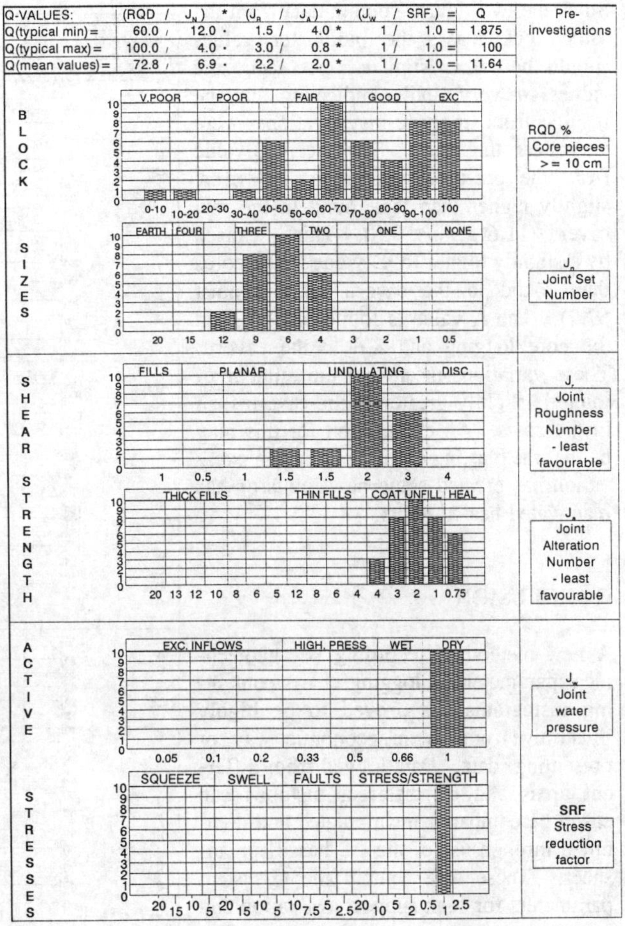

Figure 4 *Data from pre-investigations at Gjøvik Ice Hockey Hall*

This new method of data presentation has been used extensively at the Olympic (1994) Ice Hockey Hall at Gjøvik, Norway. During excavation, mapping was often carried out for each round of blasting. Geotechnical logging charts for about 40 small areas inside the cavern therefore exist. By combining different logging charts it is possible to get exact information about the different engineering geological parameters for the various sections of the cavern.

Figure 4 shows the data concerning the Q-system from the pre-investigations. This data can be compared to the data from the whole cavern given in Figure 5.

Since the two figures contain data from the same rock mass, the parameter values should be rather similar. As seen, the values are very similar indicating that the results from the pre-investigations were very near the truth. The mean Q-value from the pre-investigations is, however, slightly higher than that found in the real cavern (11.6 against 9.5). This is caused by a slightly higher RQD value in the cores than found in the cavern (72.8 against 67.2). The J_n value is found to be 6.9 in the core logging and 8.6 in the cavern. These variations mean that the number of joints is slightly greater in the cavern than in the cores. An explanation for this may be that the blasting carried out during excavation has created some new joints or has reactivated healed joints.

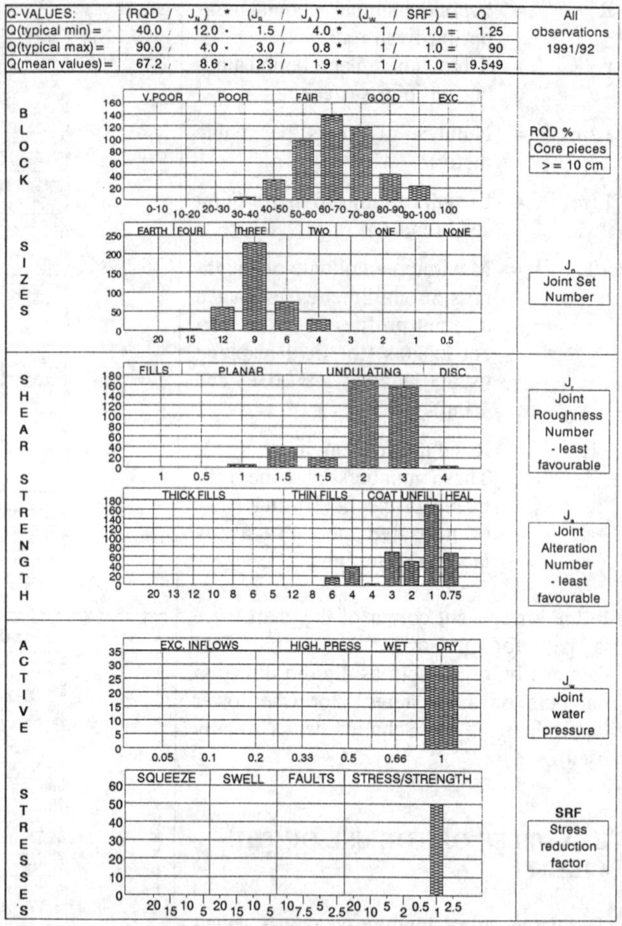

Figure 5 *Data from engineering geological mapping during the excavation of the Gjøvik Ice Hockey Hall.*

CONCLUSION

A new method of recording key geotechnical parameters in logging charts containing histograms has proved to be highly illustrative, and is an economic way of presenting data. Data logged from different areas and by different methods can easily be compared, manipulated and combined through use of the PC based spreadsheet. These data contain the Q-system parameters for rock support evaluation and information for setting up input data files for numerical modelling of critical sections of underground excavations.

REFERENCES

Barton, N., R. Lien and J. Lunde, 1974, "Engineering classification of rock masses for the design of tunnel support". Rock Mechanics, Vol. 6, No. 4, pp. 189-236.

Barton, N., F. Løset, A. Smallwood, G. Vik, C. Rawlings, P. Chryssanthakis, H. Hansteen and T. Ireland, 1992, "Geotechnical Core Characterisation for the UK Radioactive Waste Repository Design". 1992 Proc. of ISRM Symp. EUROCK, Chester, UK.

Deere, D.U., 1963, "Technical description of rock cores for engineering purposes". Felsmechanik und Ingenieurgeologi, Vol. 1, No. 1, pp. 16-22.

Tunnelling and Ground Conditions, Abdel Salam (ed.) © 1994 Balkema, Rotterdam, ISBN 90 5410 363 9

A singular point in the cut and cover works for the Bilbao underground
The bottleneck in the old town

J.R.Madinaveitia
Bilbao Underground 'IMEBISA', Spain

R.Anero
Agroman, Construction Company Ltd, Madrid, Spain

ABSTRACT: The construction of the Bilbao underground is carried out to a large extent by a tunnel bored in rock. Nevertheless, the areas next to the river lay across quaternary and recent fillings that oblige to build a cut & cover. When the underground gets to the old part of the town the layout of the works goes through a bottle-neck where all the difficulties of the project get together in hardly 20 m: geometric, geologic, presence of monumental buildings and concentration of commercial and social life that have compelled to use different techniques and controls that are described in this report.

1 SITUATION

Bilbao, main industrial centre of the Basque Country, is placed in the North of Spain at 14 km from the Cantabric sea and it streches out on both banks of the Nervion estuary. The valley is narrow and surrounded by hills, with a population of over a million of inhabitants that together with the industries and harbour installations spread over all the low lands.

Since the end of the 19th century the railway transport has two lines that join together the coast and the centre of Bilbao, each of them following a bank of the estuary but with two important problems, none of them cross over the estuary and both of them leave the more important areas in travel generation: the business and finantial areas, out of their service.

Moreover these railway lines do not follow a design according to the necessities of the population, for example, the railway in the left bank goes along a height of +6 above the sea level, while the population in this area lives on the hills in a height of +80 above the sea level, what hinders the access to and from the station.

Traditionally these orographic difficulties have had priority in the road transport but the traffic in the approaches to the centre of Bilbao is saturated and the traffic-jams are usual in the rush hours. The underground presently under construction is the answer to these problems.

2 DESCRIPTION OF THE WORKS

The Bilbao underground -in its line 1- profits from the existing railway substructure on the right bank of the estuary, modifying the surface layout by burying the stations in the most important population centres, what improves urbanistically the environment and changes its layout when it gets to the centre of the city crossing the estuary at several times. The new layout of 8,7 km is totally subterranean. This new line of the Bilbao underground is built in bored tunnel in marly rock, what locally is known as ''cayuela'', but when it gets to the proximities of the river quaternary fillings appear and oblige to modify the construction

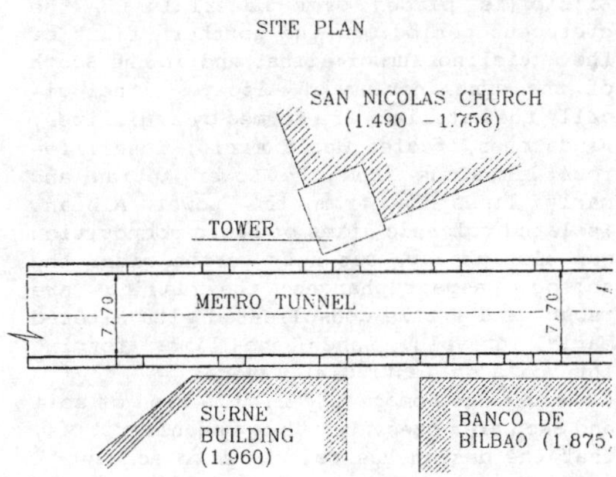

Figure 1

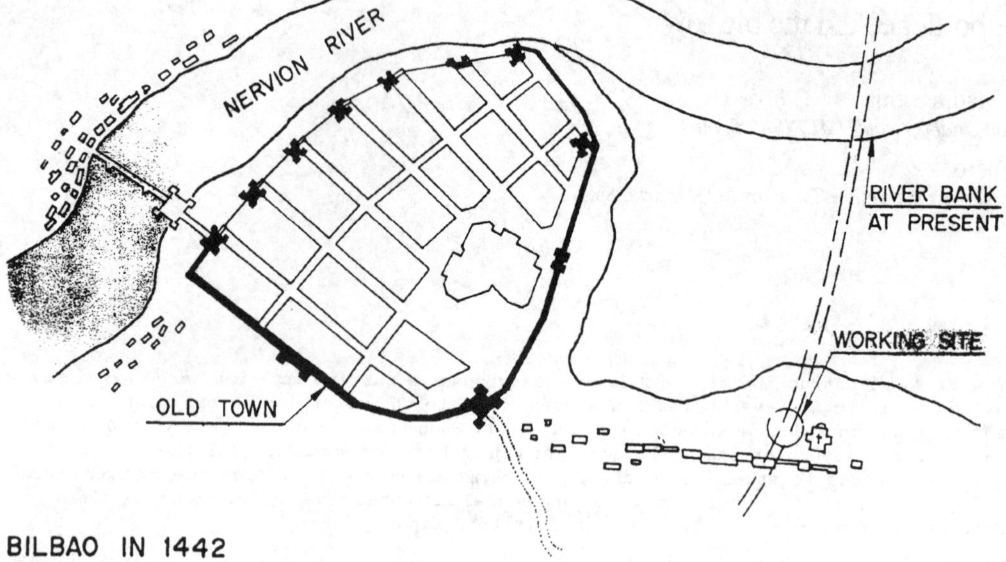

BILBAO IN 1442

Figure 2

typology that becomes a cut & cover using as timbering reinforced concrete diaphragm walls, sometimes in a provisional way and others with a definitive structure responsability.

A singular point in this layout is the passing of the underground through the medieval town, where in hardly twenty metres the greatest difficulties of the project are gathered. This bottle-neck during the carry out of the works is the aim of this report. (Figure 1).

3 GEOLOGY AND GEOTECHNIQUE

Bilbao is placed over materials of the cretaceus period, in the northern flank of the anticlinorium of Bilbao and at the South of the synclinorium of Biscay. Lithologically the materials are formed by argillites, sandstones, facies Weald marls, fossiliferous limestone from the Lower Aptian and marly limestone from the Lower Albian. Isolated volcanic dikes of basic composition can also appear. But next to the river the geologic aspect changes; the fillings are recent and little consolidated with altered marls, gravels, sands and limes forming intercalated lenticular layers.

There is no homogenity in this kind of soil and each area presents its own peculiarities, that the design has to take into account.

As the ancient plan dated in 1442 shows (Figure 2), the city was surrounded by walls and the San Nicolás Church -that is presently disturbed by the underground works- was still and outskirt hermitage. So, now it is easy to understand the heterogenity of the fillings that the works have to pass through (Figure 3) and why the cut & cover typology is the most adequate.

It is important to add that the river proximity secures a high water table and also a high speed gradient in perpendicular direction to the works, above all when crossing the gravel layers, what is very frequent.

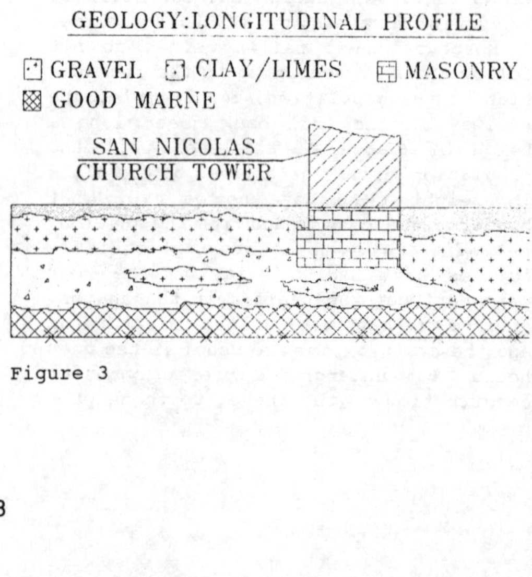

Figure 3

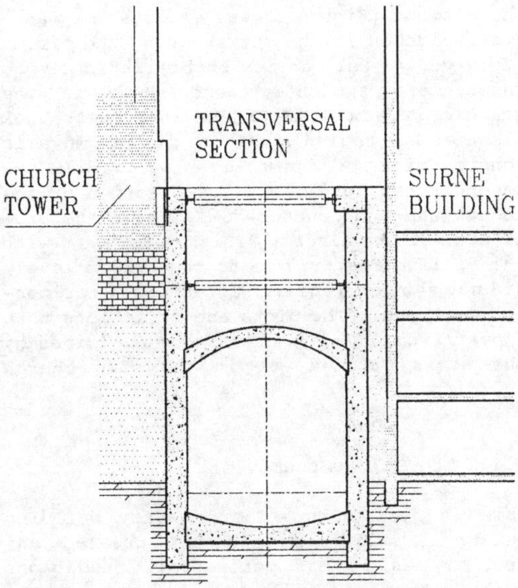

Figure 4

4 BUILDING PROCEDURE

The building procedure consists on the construction of a timbering formed by two one-metre-thick reinforced concrete diaphragm walls, embedded in the good marne rock, for what a trench cutter has been used. This machinery breaks the rock by the rotation of roller bits and extracts the detritus by pumping up the bentonite mixture that fills the trench. Thus it avoids all kind of vibration that would have doubtless affected the close buildings, catalogued like monumentals.

Due to the little room spared between the present buildings and the underground layout, the closing of the section with a dome and a counter-dome has been solved by directly embedding these structural elements, also built on reinforced concrete, in the diaphragm walls. (Figure 4).

The process of excavation has been discontinuous and in different phases in order to place in the critic zone two levels of struts (one at street level in the rest of the layout) and get to the height of the dome. This was concreted after securing the embedding of its extremes and the filling was made some days later in order to get a third level of struts. The cut & cover excavation and the covering with concrete of the counter-dome has been made advancing in a longitudinal direction under the dome shelter, that is to say, a cut and cover has been finally constructed as a tunnel.

5 DIFFICULTIES OF THE NEIGHBOURHOOD

The works would not have any singularity if it were not for the neighbourhood conditionings which are multiple and varied.

5.1 Geometrical conditionings

Figure 4 shows the geometrical conditionings to fix the section of the underground works into the section of the street. This difficulty would have been easily solved by lowering the level of the railway tracks about 20 m, then the tunnel would have been bored in rock and the affections to the foundations would have been minimal, but this implied a penalization to all the future users of the next ''Casco Viejo station'' due to the fact that its platforms would be 18 m deeper than the present ones.

5.2 Geological conditionings

With the selected solution the tunnel has to be bored in rock and partially in loose soil that as it has been described in point 3 is specially changeable and heterogeneous. Moreover, this decision of keeping the layout as more superficial as possible, in order to make easier the access of the users to the platforms, has obliged to modify the boring system in this area, that has been made with a roadheader to a large extent.

5.3 Hidrogeological conditionings

The proximity of the river and the presence of a water table at less than five meters from the surface with a transversal hydraulic

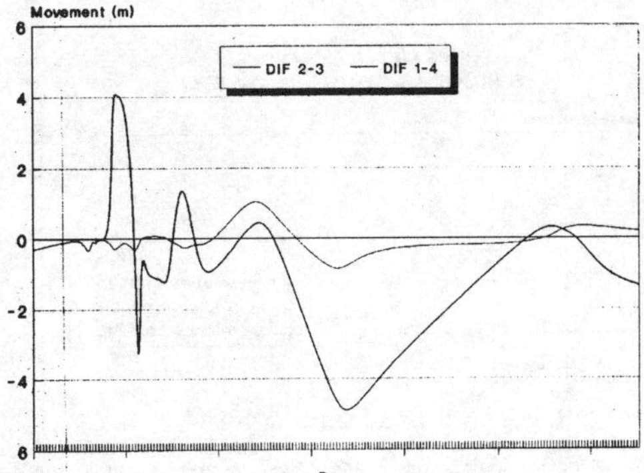

Figure 5

gradient oblige to consider the need of ''breaking'' the diaphragm walls, which are acting as a dam, in some areas between the street surface and the bottom rock.

The works have been carried out in different phases leaving important sub-rivered areas free and also controllling the variations of the water table (Figura 5) that could modify the foundations of some close buildings.

5.4 The side conditions

These conditions were the most worrying since the underground construction affected three buildings -two of them monumental- to a greater or lesser extent.

5.4.1 San Nicolás Church

A barroc building whose present configuration dates back from 1756 but which keeps elements from a gothic building erected in 1490 over the remains of a hermitage from the 13rd century.

Its foundations in the southwest tower, the one closest to the works, have been formed by breakwater sections contained in perimetral masonry walls placed over the gravel possibly reached in the low tides. These foundations have suffered the effects of no less than fifteen catastrophic floods throughout their history and the west façade of the church showed a group of crannies, before the beginning of the works, that denoted the existence of differential settlings between the tower and the rest of the building.

In order to improve the foundations of this tower and before constructing the underground cut and cover, it was underpinned with micropiles and jet-2 treatment, programming as well the control of possible movements.

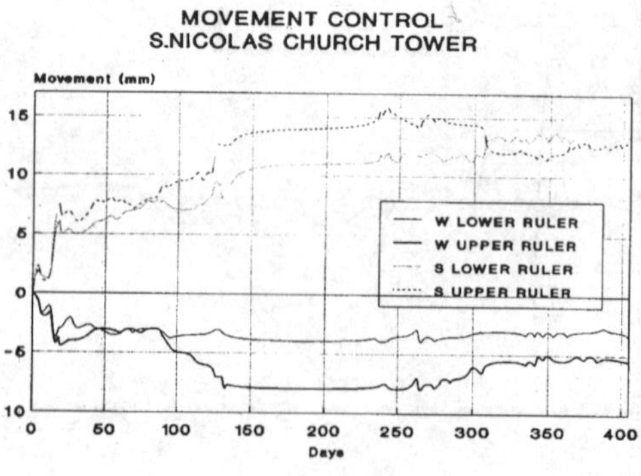

Figure 6

The outcome (Figure 6) was satisfactory and its incidence in the works during the first 120 days, as well as its stabilization, once these were finished, were shown in the graphics.The most important increments took place at the beginning of the diaphragm walls construction and when excavating to place the second strut level. The construction of the tunnel and the counter-dome, under the protection of the already finished dome and with the filling restored up to the street level, did not show appreciable affections (horizontal sections). The picks and variations that appear from 230 days on must be explained by the works carried out in the same church building.

5.4.2 The Bilbao Bank

This is just a two-floor masonry building built in 1875 that has not created any problem, as it has perimetral foundations over quite clean and reasonably homogeneous gravel layers. In view of its monumental nature the façade has been protected from possible strokes of the machinery.

5.4.3 The Surne building

This is a relatively modern building, dating from 1960, that belongs to an insurance company. From the underground works point of view this building presents two problems both related to the condition of the diaphragm walls that form its foundations, which are at the same time the closing walls of four subterranean floors of a car park.

These diaphragm walls were some of the first built in Spain with this technique and their quality was not perfect, showing some deficiencies in the concrete compacity. These deficiencies have favoured the aggression of the subsoil waters and the seasonal variation of the water table, slightly saline by the influence of the sea tides.

Moreover, these diaphragm walls were provisionaly anchored to the soil during the construction process of the car park basements. Although these anchors were distightened when constructing the horizontal iron wrought, they were not removed, so the underground works had to cross a heap of partially concreted wires. The solution chosen, that has been proved to be correct, consisted in identifying from the inside of the car park basements all the anchors hidden behind the brick partitions and make sure that they were inactive to clean them in the greatest possible length in order to avoid they could occasionally work together with the old foundation diaphragm walls. The later

Figure 7

Figure 8

carrying-out of the underground timbering using a trench-cutter did not have any problem even if in some occasions the residual wires left in the soil got caught in the roller bits but without any consequences in the foundations of the Surne building.

5.5. Transport conditionings

The site, where the underground is now being constructed, has been the entrance of the bus public transport for the last ten years. During the works the bus routes have been changed and some streets for pedestrians only have been opened to the wheeled traffic. Moreover, at hardly 50 meters from the works the Bilbao-Customs railway station terminal is located and thousand of passengers have to walk cross the working area. (Figure 7).

The presence of so many pedestrians, above all in the rush hours, has been the determining factor that has compelled to take special security measures. Actually, in some occasions the workers formed a line to guide the pedestrians in view of the impossibility, for lack of space, of installing a permanent closure (figure 8).

5.6. Citizens conditionings

The old part of Bilbao is an important commercial centre so, the difficulties of access occasionally caused by the works had, as a consequence, the isolation of certain areas, the ''cul de sacs'' in some streets, and physical impossibility of access of public vehicles in others. This problem has been solved by negotiating with merchants and citizen associations that have introduced changes in the working programme to make coincide the most important disturbances with the lesser economic activity of the year.

In this way, the neighbourhood has conditioned not only the typology employed but also the constructive proceedings and its adaptation to social necessity and economic activities, what has been achieved by keeping an extensive sounding of the works and a high level of public information.

6 CONCLUSION

A singular point in the access of the Bilbao underground to the old town, the actual bottleneck of the project has been solved thanks to the application of coordinated techniques in space and time, not specially complex techniques applied one by one but techniques that as a whole implied and important hindrance to the development of the project. This hindrance was greater when the public opinion was opposed to the cutting off the main vehicles access to the old town and at the same time the works endangered the monumental buildings located at a few centimetres.

In the summer of 1993 these difficulties have been overcome and nobody in town remembers what initially was considered an unsolvable problem.

Tunnelling and Ground Conditions, Abdel Salam (ed.) © 1994 Balkema, Rotterdam, ISBN 90 5410 363 9

Construction of a road tunnel next to the harbour in Barcelona

Juan Ramón de Clascà Marín
EUROPROJECT, S.A., Spain
José Manuel Serrano Herreno & Carlos López Carreras
EUROGEOTECNICA, S.A., Spain

ABSTRACT: The road tunnel of Columbus goes by the below ground to the harbour area. The characteristic of this road corresponds as a concrete frame under the street level with two way direction. The presence of the high level ground water in which saturates the sandy materials of the proximity of the sea, has motivates the adoption of constructive procedures that will assure the water lightness. The communication describes the project as apply methods for the execution work.

1 INTRODUCTION

The new Arterial System of Roads in Barcelona has been configurated through the ages by successive contribution to it that have tried to satisfy the communication needs that every period creates.

Some of the elements belonging to Barcelona's new road system, completed in 1992, were previously conceived by Ildefonso Cerdá in 1860. The arterial scheme was revised by León Jausely in 1906 and complemented by Romeu i Porcel in 1917. In the course of years, this scheme underwent adjustment and reconfiguration until year 1963, in which the resulting Arterial Road System Scheme conceived the layout of the city by-passes essentially as it is now.

This scheme could finally be carried out in 1992, by the time Olympic Games were held, and the works to build the arterial system of roads that surrounds Barcelona were fully completed.

The Coastal By-Pass, which integrates into Barcelona's arterial system of roads, as it passes through Columbus Avenue implies the execution of a 450 m long underground section with a 9.80 m clear height and on overall width of 22 m. This section, because of its closeness to the Harbour, creates a number of constructive difficulties arisen from the fact that highly–permeable sandy-saturated materials can be found there.

The building project conceived in this section implies the execution of the tunnel by means of prefabricated retaining walls. This technique, put forward by the building company, was chosen because of the benefits it provided: Due to the fact of being prefabricated modules, their sizes could be optimized and reduced to a minimum, bearing in mind not to bar for a long time the underground water flow to the sea.

The situation of the underground section is shown in Fig. 1, in which can also be seen how close the sea is.

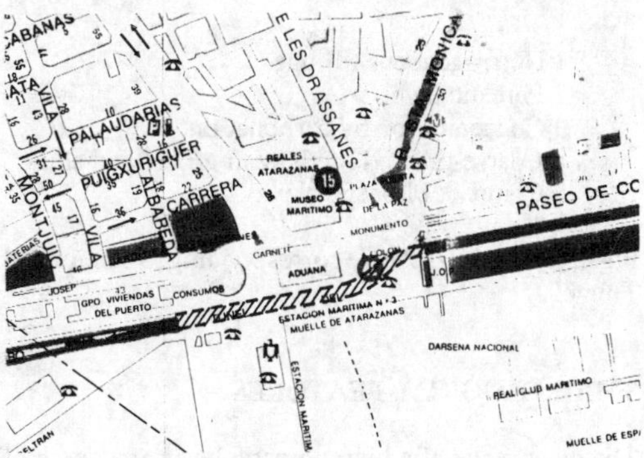

Fig. 1 The situation of tunnel

2 GEOLOGY

The studies that have been carried out show that there is a variety of soils in the tunnel area. The upper-level soil consists of a filling composed of clay and sand, and blocks abound. It is a heavily heterogeneous set of materials between 3 and 5 m wide. The intermediate level consists of quaternary sands of a medium-to-coarse grain size, and irregular gravel layers are inserted in it. This formation is about 15 m thick. The permeability of these materials has been assessed by means of pumping tests and on the whole is estimated to be between 10^{-2} and 10^{-1} cm/sec.

An impermeable layer lies underneath theses sands, some 20 m deep, and consists of Pleocene clays tath are highly-compacted and continuous along the whole underground section.

The sandy layer is locally cemented due to carbonatation processes, not always completed, which determines scarcely continuous compacts layers.

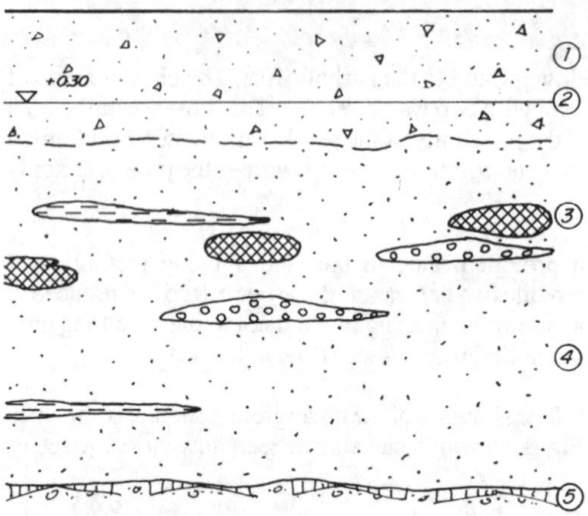

1 Heterogeneous fillings
2 Phreatic layer
3 Concentration by carbonation
4 Coarse-grained sands with gravel and slime
5 Compact Pliocene clays

Fig. 2 Geological scheme corresponding to Columbus' Tunnel

3 HIDROLOGICAL FEATURES

The occurrence of a high phreatic layer at a depth of 3 m and the occurrence of permeable sandy materials in the whole area pose very specific hydrological problems.

The underground work goes parallel to the Harbour, so that, to a certain extent, it means barring the natural flow of underground water which goes from the city down to the sea. The sandy aquifer formation in the tunnel area lies in the direction of the inner part of the city and gets thinner and thinner and eventually disappears, some 500 m away from the tunnel. Its hydraulic gradient rate is 0.5%.

The storage rate of this aquifer is 10^{-2}, which means a certain degree of confinement provided by the filling. A hydrological study by means of a fint differences simulation has been carried out aimed at estimating the growth of the phreatic level generated upstream due to the building of the tunnel. The study carried out have shown that the growth of the phreatic level is likely to be 30 cm, over a tolerance limit not higher than 50 cm, so this margin for safety was considered to be wide enough.

4 THE BUILDING OF THE TUNNEL

4.1 *Impermeabilization*

The technique is based on prefabricated retaining walls and entails the previous excavation of the ground and its backfill with bentonite-cement, which creates a plastic shield into which the concrete prefabricated walls are inserted. The functions of bentonite-cement can be summed up in two respects :

1. Impermeabilization of the tunnel, thus eliminating the possibility of lickage passing through the wall components.

2. The ability to build watertight areas or cofferdams aimed at reducing the phreatic level and making it possible to carry out the excavation.

As far as the percentage of bentonite sludge is concerned, it represents 10% of the overall weight of the cement. This percentage makes it sure that there will be enough resistance against the pressure generated by the prefabricated components, as soon as sludge becomes solid.

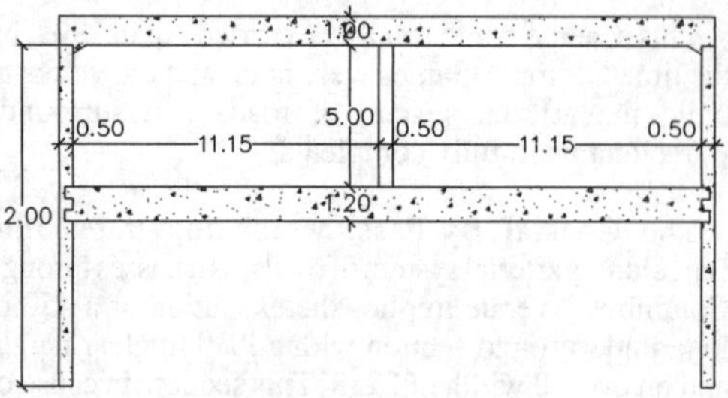

Fig. 3 Section type

4.2 *Execution process*

The execution of the tunnel section has to be carried out in several stages which go as follows :

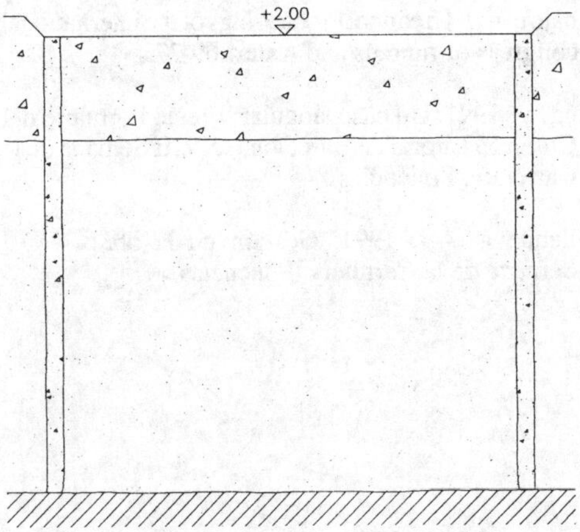

Fig 4 Construction of the plastic walls by means of bentonite-cement.

Fig 5 Insertion of the prefabricated walls and construction of the upper flagging.

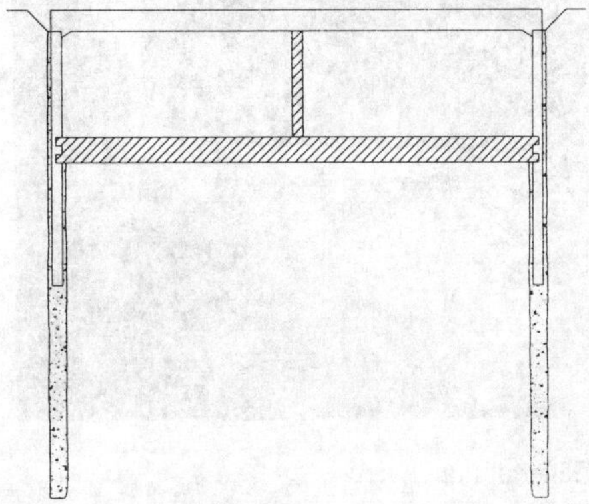

Fig 6 Lowering of the phreatic level and construction of lower flagging and central support.

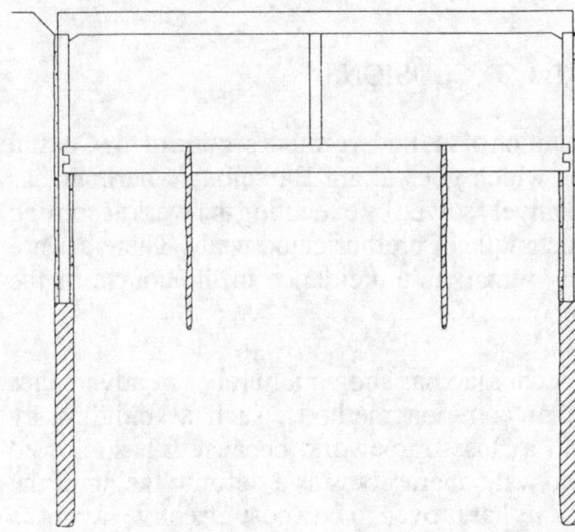

Fig 7 Anchorage of lower flagging through micropiles and removal of the plastic walls.

1. Building of the plastic walls

The first stage consists of the excavation prior to the insertion of the walls. This excavation must be 22 m deep so that it goes through the impermeable clay layer. After that, the resulting excavation is backfilled with bentonite-cement with the purpose of achieving fully impermeabilization of the tunnel as well as enough resistance for the walls to rest on it.

2 . Insertion of the walls and the upper flagging

The next step is inserting the prefabricated walls taking care that their components are accurately aligned. Theses components mus be kept suspended until the bentonite-cement has become had. After that, the upper flagging is concreted. This upper flagging rests directly on the walls.

3. Reduction of the phreatic level and construction of the lower flagging

Once the upper flagging has been constructed, the water stored in the plastic wall cofferdams is pomped. Next, the ground is excavated and the assembled lower flagging is built as well as the central piles.

4. Anchorage of lower flagging and removal of the plastic walls

In some places subject to higher hydrostatic pressures anchorage of the lower flagging has been carried out by means of micropiles which have provided a higher resistance against these pressures. One more task within this last stage is the removal of the plastic walls which lay below the prefabricated components. This operation will restore the underground water flow below the

Fig 8 General tunnel view.

prefabricated components and, thus, will limit the risk tath the phreatic level comes up in those areas with a higher occurrence of water.

5 FINAL CONCLUSIONS

The execution of the underground section of the Coastal By-Pass which goes along Barcelona's harbour has been positively solved by executing the work according to the technique of prefabricated walls. There has not been any remarkable incidence in the course of the building process.

This technique has shown a number of advantages over other common methods, such as building by means of a close framework, because it has allowed using the walls themselves as a definite facing of the tunnel. This has proven to be a cost-effective saving as far as concrete is concerned.

The success in using these prefabricated walls is based on the fact of having achieved watertightness for a short time during the building process and adopted dessigns that guarantee a full impermeabilization when the tunnel goes into operation.

REFERENCES

Alonso, E. & Gens, A. 1985. Proyecto de Soterramiento de la Ronda Litoral frente al monumento a Colón. Reconocimiento del terreno. Síntesis e Interpretación.

Clascà, J.M. 1987. Soterramiento de la Ronda Litoral frente al Monumento a Colón. Proyecto constructivo.

Clascà, J.M., Sharma, D., Martí, J. & Ruiz, J. 1988. Análisis del impacto de la construcción de un túnel bajo el nivel freático. Proceedings of the international congress of tunnels and water: 69-77.

Roig, V. 1991. Un caso singular: Frente Portuario del Cinturón Litoral de Barcelona. XVIII Semana de la Carretera. Ponencias.

Villanueva, A.J. 1991. Gestión de la obra. XVIII Semana de la Carretera. Ponencias.

Tunnelling and Ground Conditions, Abdel Salam (ed.) © 1994 Balkema, Rotterdam, ISBN 90 5410 363 9

Utilization of waterway tunnel survey system using ground penetrating radar

T. Nakamura & H. Kubota
Tokyo Electric Power Co., Inc., Japan

M. Inagaki
Geo Search Co., Inc.

ABSTRACT: Ground penetrating radar(GPR)is a non-destructive survey device using electro-magnetic waves. GPR can produce structural information on waterway tunnels continuously and quickly. Information needed for maintenance of such tunnels has mainly been obtained by boring from inside the tunnels. But boring requires a long time, and does not produce sufficient information. We have improved GPR to be suitable for tunnel maintenance work through laboratory and field tests. Consequently it is proved that the improved GPR can be well applied to actual maintenance work of waterway tunnels instead of boring.

1 INTRODUCTION

Because Japan is a mountaneous country with a high rate of precipitation, a lot of water power plants have been actively constructed. For these plants, water conduit types are commonly used rather than dam types because a greater water head can be obtained.

Tokyo Electric Power Company(TEPCO)has approximately 660km waterway tunnels, 70% of which are more than 50 years. Several cave-ins have ever occured in several times due to poor quality of lining concrete, the existence of voids behind the lining and shortage in lining thickness. We have made an effort to investigate the direct reasons of the cave-ins quantitatively with schematic boring surveys. However, boring locations must be limited to one every 100 to 200 metres because of restrictions on downtime in power generation and investigation budget. It was the only way to develop an image of the tunnel condition by global interporation with long-distance borings.

Recently we focused on ground penetrating radar(GPR)as a continuous, quick and non-destructive method in order to give a more accurate diagnosis and to make a more suitable maintenance plan. GPR was basically recognized to be a highly effective method to investigate the lining and void thicknesses of waterway tunnels with several field tests from 1985 to 1986.

Continuous efforts have been made to improve the system by increasing the survey depth and accuracy, by making clear the electro-magnetic properties of lining concrete and by improving the skill of interpreting GPR records. Consequently, not only the thicknesses of lining and void but also the thicknesses of collapsed soil and water reservoir behind the lining have come to be able to be obtained from GPR records.

The GPR tunnel survey system has now reached the practical level. This paper describes the outline of GPR with examples.

2 THEORY

GPR is a survey device using electro-magnetic(EM)waves. Once EM wave pulse is radiated from the antenna, it hits the lining surface at first and is partially reflected. The transmitted pulse propagates until it hits the next boundary as shown in Fig.1.

A series of reflected pulses are continuously plotted by a chart recorder to form a visual record as shown in Fig.2.

The strength and polarity of the reflected pulses depend upon the contrast of dielectric constants between both interfacing layers. Variations in the strength and polarity observed on the charts help to interprete the records.

EM wave velocity in a medium varies according to its dielectric constant. This means that it is necessary to convert the layer thickness on the chart to the real thickness and to get the profile of the layer structure.

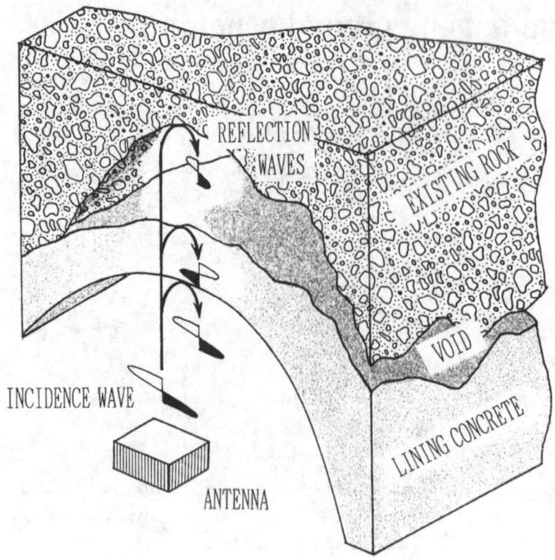

Fig.1 Illustrated view of wave reflection

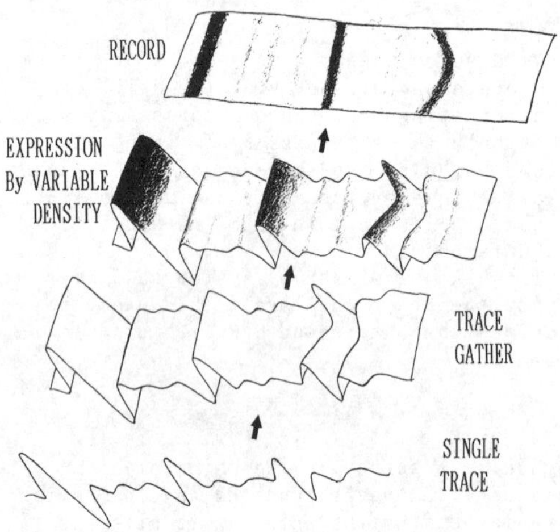

Fig.2 Building-up of visual record

3 SURVEY SYSTEM

The survey system consists of GPR equipment and a vehicle on which the equipment is carried. The system block diagram is shown in Fig.3.

Three types of vehicles are available depending upon tunnel heights as shown in Table 1.

The survey team basically consists of four operaters : a driver, a machine operater, an antenna operater and a technical assistant. An example of a survey in process in a large scale tunnel is shown in Fig.4.

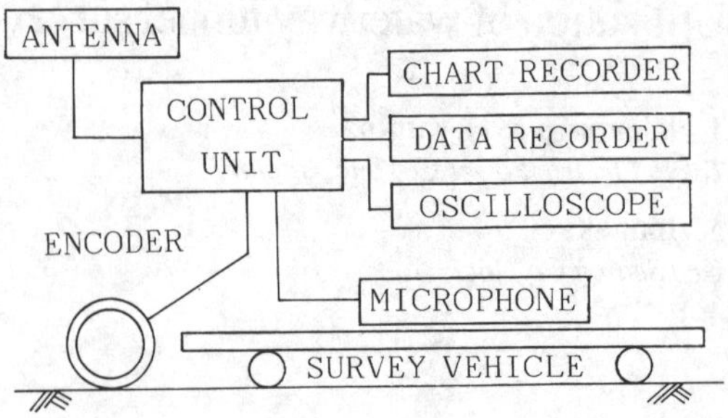

Fig.3 System block diagram

Table 1. Vehicle specification

Tunnel size	Small	Medium	Large
Tunnel height	1.5～2.5m	2.5～5.0m	5.0～8.0m
Vehicle	Electro-driven car	Diesel driven car	Truck
Speed	1～2km/hr	2～3km/hr	3～5km/hr
Continuous survey distance	5km	More than 10km	More than 10km
Antenna support	Absorbing box	Boom	Boom

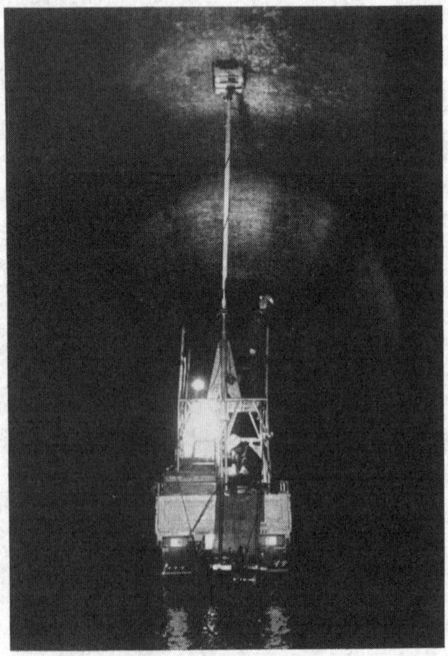

Fig.4 Vehicle in survey

4 EFFORTS AIMED AT DEVELOPING PRACTICAL USE LEVELS

GPR was to be found effective means of waterway tunnel investigation by a basic demonstration test. There were, however, still some problems that had to be overcome in order to attain a practical level of use. The following tasks were picked up to solve the problems.

1. To increase the maximum survey depth by developing a higher power type or a pulse widened type.
2. To investigate and analyse the dielectric constant of lining concrete for interpretation of the records.
3. To improve the interpretation skill
4. To develop data processing software and profile drawing software.
5. To make the equipment lighter and more compact.

4.1 Power increases and pulse widening

The GPR system should increase the survey depth to be applicable to tunnels with thicker lining. It can be attained by increasing the transmitted power or extending the pulse width. Each effectiveness has been evaluated through laboratory and field tests. Fig.5 shows the simulated amplitude change due to thickness variation of concrete slab. Field test agreed with this simulation.

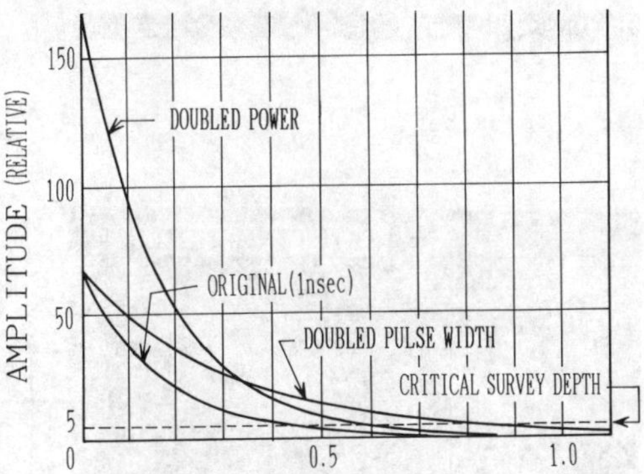

Fig.5 Simulated amplitude due to concrete thickness

It is recognized that extending the pulse width is more effective than increasing the power for increasing the survey depth. It has been revealed that unwelcome signal clutter also increases as the power increases and it creates confusion in identifing the reflected pulse record from the real boundary.

While extending the pulse width would in principle lower the resolution, of records the test result has showed that pulse length less than 2 nano sec(60cm)does not practically impair the interpretation.

4.2 Electro-magnetic properties (dielectric constant)of lining concrete

Dielectric constants of media have an important role in GPR survey. Dielectric constants are supposed to be much dependent on water content of the media. We measured the dielectric constant of lining concrete on 300 samples taken out of actual tunnels. Fig.6 shows the relationship between the dielectric constant and the water content.

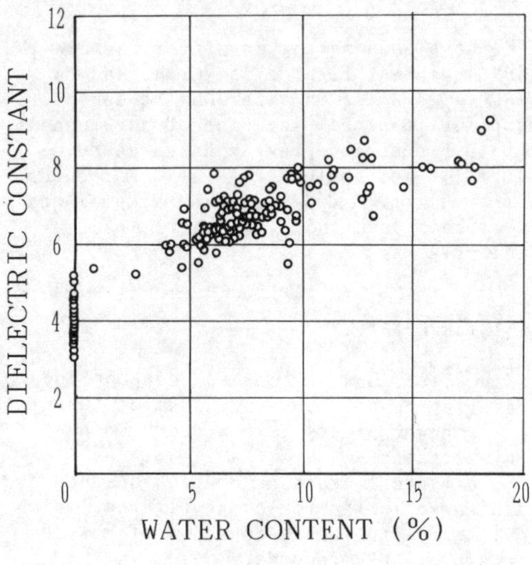

Fig.6 Dielectric constants of lining concrete samples in waterway tunnels

Although the dielectric constant is much more dependent upon the water content than other properties such as density and gravel content, the dielectric constant for each medium is determined as a constant value because the dependence on the water content seems weak(see Fig.6)and difficult to integrate into the system.

4.3 Improving skill level of data interpretation

Radiated EM waves travel through various

paths where the waves are subject to divergence, absorption, diffraction and reflection. Thus, such a travel gives the recorded waves a complicated appearance, which makes it very difficult to interprete the recorded waves quantitatively in terms of the strength of reflected waves. Furthermore, there are not only simple structures such as lining/existing rock or lining/void/existing rock but also more complicated structures including collapsed soil or water reservoirs. Therefore, the interpretation should be carefully carried out with a theoretical approach. Also, we tried to extract and accumulate various interpreting knowledge from the boring data taken all through the surveys. As a result,more complicated structures have come to be able to be successfully interpreted.

4.4 Improvement of equipment and data processing

We have been making an effort to make the GPR equipment lighter in weight and to make it smaller in size. Now it can be applied to almost any size of our tunnels.

Data processing program has been developed to get better image of GPR records. In addition, a data handling program has also been developed for complicated analysis.

5 INTERPRETATION

Since 1987 GPR system has been applied to waterway tunnel surveys. 258km of tunnels in 49 power plants have already been investigated.

Some typical examples of all the records are shown in the following figures, which points out the basic concepts of how to interprete GPR records.

Fig.7 is a typical case where the lining concrete makes contact with the existing soil or rock. Reflection waves are much weaker than those in the following example with voids, although observable.

Fig.8 shows an example with the continuous voids behind the lining. The unique and easy to recognize signature of the void, a strong and inverse polarity reflection, appears in record.

Fig.9 shows an example with reinforcing bars inside the lining. The strong black band near the surface and periodical thin comb-like strings stretched from the band indicate the existence of reinforcing bars. Reflection at the boundary between the lining and the rock is quite difficult to recognize because it is weakened by

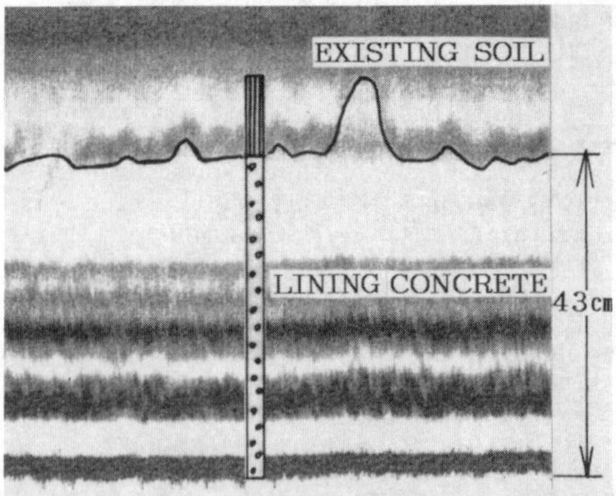

Fig.7 Lining concrete in contact with existing rock

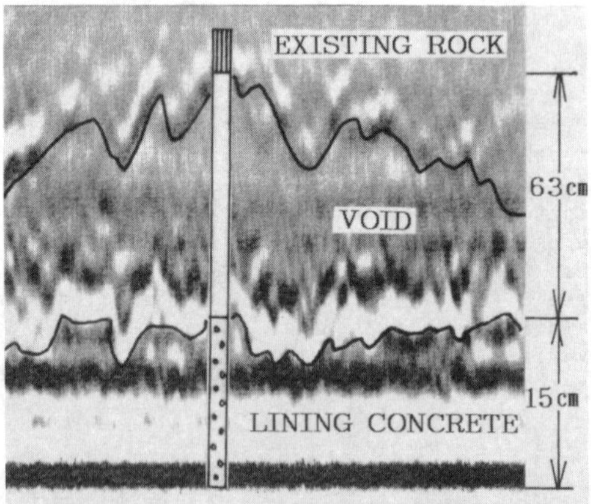

Fig.8 Voids behind the lining

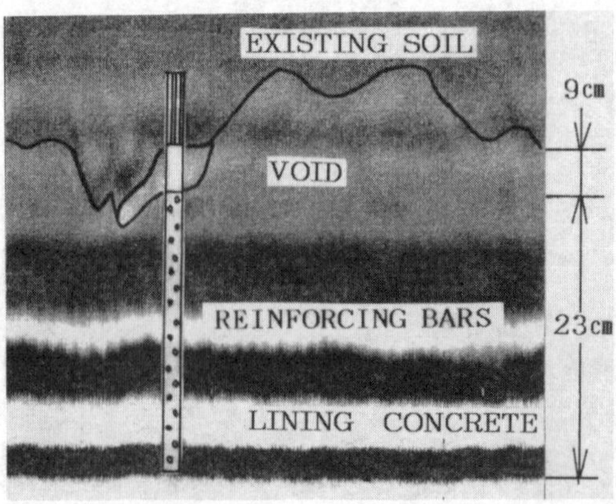

Fig.9 Reinforcing bars in the lining

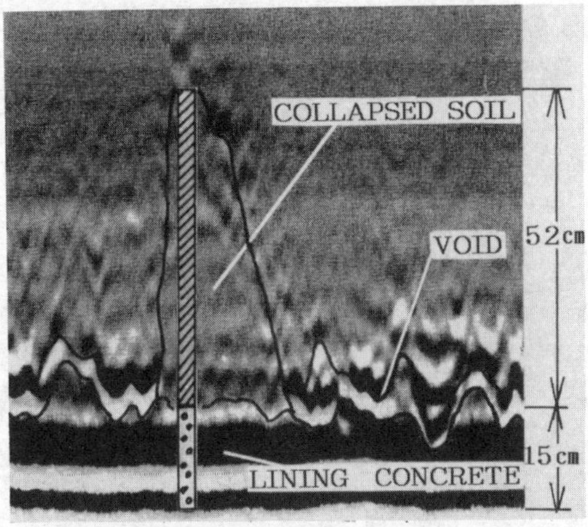

Fig.10 Collapsed soil behind the lining

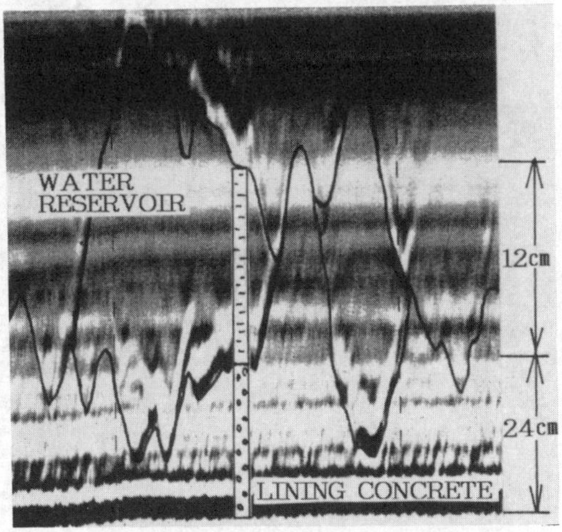

Fig.11 Water reservoir behind the lining

reinforcing bar reflections.

Fig.10 shows an example with collapsed soil behind the lining. Continuous reflections form the steep triangle zone. This is the typical signature of a pile of collapsed soil or rock debris.

Fig.11 shows an example with water reservoir inside the void. The dielectric constant of water is very large, approximately 80. That means the EM wave velocity is very low. Thus the thickness of a water reservoir in record is exaggerated almost three times thicker than those of concrete and soil. Consequently the water reservoir is likely to form a lense shape reflection continuity as shown in the record.

6 APPLICATION LIMIT

Based on the results of application to the actual surveys and of the research for practical use, the application limits of GPR is shown as follows in terms of the survey depth and the accuracy.

6.1 Survey depth

The strength of radiated waves is gradually weakened by the effects of divergence, absorption and reflection. Thus, survey depth depends upon the dielectric constant of media as well as the layer structure of media. Although survey depth cannot be determined theoretically, survey depths of typical layer structures are shown in Table 2 which has been obtained by a statistical analysis of all the data.

Table 2. Survey depths in typical layer structures

Layer structure	Survey depth
Lining/Existing rock	30～ 40cm
Lining/Void/Existing rock	50～170cm
Lining/Collapsed soil/Void /Existing rock	30～100cm
Lining/Water reservoir/Void /Existing rock	50～100cm

The survey depth for the lining varies approximately 30 to 50cm. Therefore, GPR can be succesfully applied to almost all of our waterway tunnels.

6.2 Accuracy

The relation between the GPR-observed lining thickness and the measured thickness is shown in Fig.12. About 90% of the dispersion falls within the range of ±5cm. The case of the void thickness is also shown in Fig.13. About 80% of the dispersion falls within the range of ±10cm. Two reasons are considered for the dispersion.

1. Local difference in the dielectric constant of lining concrete
2. Location error of boring. Location error is considered to be dominant, especially in such a case that the reverse surface of the lining is rough. However the accuracy of this system is acceptable for the application to waterway tunnel surveys.

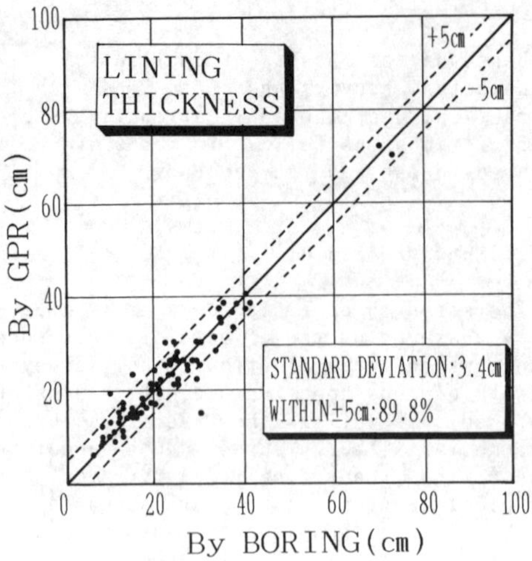

Fig.12 Lining thickness comparizon between by GPR and by boring

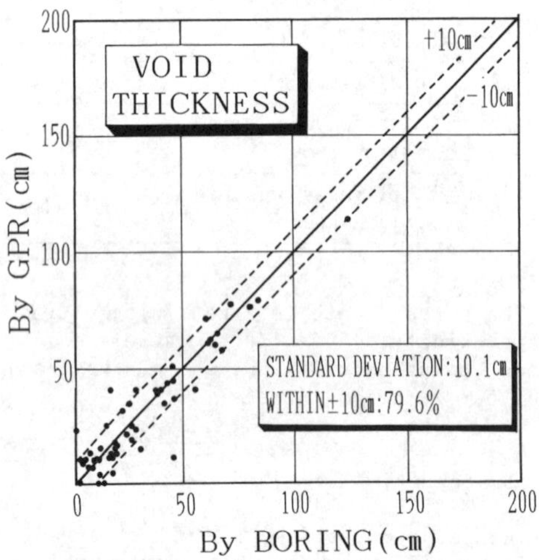

Fig.13 Void thickness comparizon between by GPR and by boring

7 CONCLUSION

The GPR survey system has proved to be very effective for waterway tunnel surveys and its application scope has been obtained through various kinds of field and laboratory studies.

In terms of how quickly, continuously and non-destructively to obtain the essential information all through the tunnel, The GPR system is superior to convensional boring survey. This system is very helpful in collecting and accumulating the information necessary to manage waterway tunnels systematically.

REFERENCES

Egawa,K. et. al. 1986. Study on the application of ground penetrating radar to the waterway tunnel survey. Electric power civil engineeving No.203:13-22(in Japanese)

Kondoh,S. and Inagaki,M. 1986. Application of ground penetrating radar to water conduit lining and void survey. Annual meeting SEGJ:110-111(ditto)

Okada,H., Inagaki,M. and Haida,T. 1987. Electrical property of concrete. Annual meeting SEGJ:204-205(ditto)

Tsuda,M., Inagaki,M. Shin,K. 1991. Interpretation of collapsed soil in ground penetrating radar record. Annual meeting SEGJ:309-311(ditto)

Kubota,H., Inagaki,M. and Shin,K. 1991. Interpretation of water reservoir in ground penetrating radar record.Annual meeting SEGJ:401-404(ditto)

Kubota,H. and Inagaki,M. 1991. Study for practical use of water conduit survey system by means of ground penetrating radar. JSCE:211-216(ditto)

Kubota,H., Inagaki,M. and Fujiyama,Y. 1992. Effect of radiated wave power for penetrable depth. Annual meeting SEGJ:313-316(ditto)

Tunnelling and Ground Conditions, Abdel Salam (ed.) © 1994 Balkema, Rotterdam, ISBN 90 5410 363 9

Underground road crossing of an archaeological area

S. Pelizza
Tunnelling and Underground Space Centre, Polytechnic University of Turin, Italy

G. Corona & G. Graffi
SIC, Civil Engineering Consultants, Turin, Italy

P. Zola
Itinera S.p.A.

ABSTRACT: During the construction of the A32 Turin-Bardonecchia motorway, which projected in order to speed up the road link between Italy and France via the Frejus tunnel, an area of great archaeological interest was discovered near the town of Rivoli. In order to respect the characteristics of that area the project for the construction of an artificial tunnel was modified and a natural tunnel was built instead.

1 BACKGROUD

The Susa Valley, a crossroards of Europe, has been used since antiquity by armies, pilgrims and merchants travelling between the Po Valley and the Rhône basin.

The Turin-Bardonecchia motorway represents the final stage in the major communications project, beginning with the Frejus Tunnel, between north-western Italy and southern France, encouraging traffic flows from the Channel to the Mediterranean and south-east Europe.

In terms of its technology and methods of construction the Turin-Bardonecchia A32 is a "new generation" motorway. It runs along the entire Susa Valley through extremely complex orographical, morphological and anthropological conditions.

The Susa - Rivoli section, which is approximately 35 kilometres long, runs through a densely urbanised valley characterised by numerous infrastructures, so much so that the motorway route was conditioned by a number of limiting factors: namely, the bed of the Dora river, the Turin-Modane-Lyons and Turin-Bussoleno railway lines, two major roads (S.S.24 and S.S.25).

The roadway was constructed with two carriageways, separated by a 2-metre unpassable median strip; each carriageway comprises two 3,75-metre lanes flanked by a 3-metre emergency lane and 0,75-metre shoulder, giving a roadway total width of 24,50 metres.

During construction work on the motorway historically and culturally important finds were made relating to Roman and medieval settlements.

During excavation works for the construction of an artificial tunnel (photo no.1) within the ambit of lot 31,

photo no.1

between progressive Km 32+520 and km 32+628, an archaeological zone of considerable interest was identified, including the first known tract of the Imperial Roman road to Gaul along which stood at least thirty-six "tombs", part in earth and part with monumental structures, dating back to a period between the 4th century A.D. and the Longobard epoch. In agreement and collaboration with the Archaeological Superintendence of Piedmont a project was drawn up which, although enabling the motorway to be constructed along the planned route, would also respect the archaeological zone.

2 STRUCTURAL CHOICE

The solution considered most appropriate to resolve the construction of a subway under a zone abounding in archaeological finds, while at the same time offering a sufficient margin of safety, consisted in the construction of two natural tunnels, one for each carriageway, with an interaxis of 23,50 metres. The choice of structure and the geotechnical characteristics of the soil in which excavations would be made conditioned the methods and machinery used to realise the project.
The rectangular cross-section of the natural tunnels has an internal width of 12.10 metres and a height of 5.45 metres; the tunnel lining is in a reinforced concrete with a constant thickness of 1,00 metre (photo no. 2). The choice of the

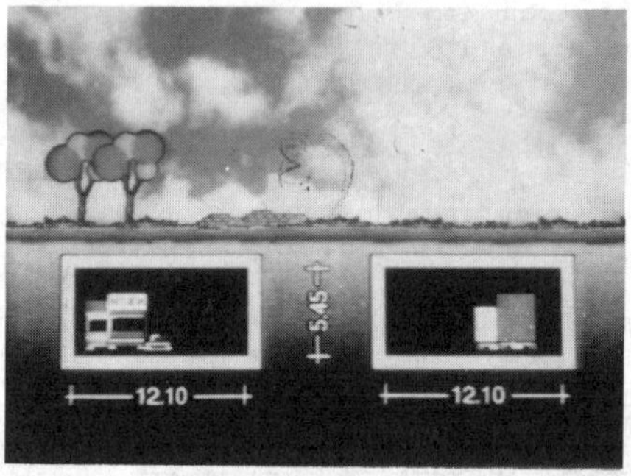

photo no.2

rectangular cross-section was imposed by the limited existing overburden and the necessary to preserve the gradient of the roadway; the overburden measured from the extrados of the finished work has a thickness of between 0.80 to 1.30 metres.
In order to define the methods of executing the project it was necessary to make a detailed geophysical and geotechnical study to identify the geomechanical characteristics of the soil in order to evaluate subsidence and the behaviour of the soil during the excavation stage.

3 GEOLOGY AND GEOTECHNICS

Extensive boring was carried out to collect undisturbed samples, and SPT penetrometric tests were performed at varying depths up to a depth of 16.00 m from the surface plane.
The average stratigraphy of the soil, which was also controlled using exploratory boreholes sunk to a depth of approximately 6 metres, can be defined as follows:
from 0.00 m to 1.00-1.50 m:
sheets of grey-blue coloured silty clay
from 1.00-1.50 m to 8.00 m:
more or less silty sand alternating with fine gravel levels
from 8.00 m to 16.00 m:
level consisting of well thickened gravel.
With regard to SPT (N) values, it was found that there was a variability of between 3:7/8 blows in the silty-clayey level reaching on average between 18-20 blows in proximity of the gravel stratum where refusal was reached.
In the exploratory boreholes "pocket" tests were also carried out using vane tests to calculate the non-drained cohesion (Cu) of the clayey level and the behaviour of the walls of the hole were checked at various depths.
The results of the tests and field observations were used to calculate the values of maximum subsidence and acceptable pressures at the contact between the tunnel floor slab and in situ soil. With reference to settlement it was found that the values were incompatible with the protection of the integrity of the "tombs" lying approximately 0.30-0.10 m from the surface plane and therefore at a distance of approximately 0.40-0.90 m from the intrados of the tunnel.
An analysis of the geomechanical characteristics of the soil in immediate contact with the archaeological zone revealed the need to carry out preconsolidation operations in the entire area involved in the construction of the two tunnels so as to make both ground settling in contact with the upper edge of the tunnel during the excavation phase and the overall long-term settlement of the tunnels compatible with preserving the integrity of archaeological finds.

4 METHODS OF EXECUTION

The choice of the methods of execution and machinery was aimed to minimise all ground settling, above all differential settlement, during excavation works.
The tunnels were therefore constructed using soil preconsolidation, taking samples as works progressed, and operating according to the following stages:

4.1 Realisation of preliminary support works and soil consolidation around the archaeological zone

The need to preserve and not decompress the soil in the archaeological zone meant that it was necessary to construct support bulkheads, consisting of micropiles and ties-back (Berlin-type), around the edge of the four openings; this operations is indispensable for the layout and consolidation of the two underground excavation faces (photos no.7 and 8).

photo no.3

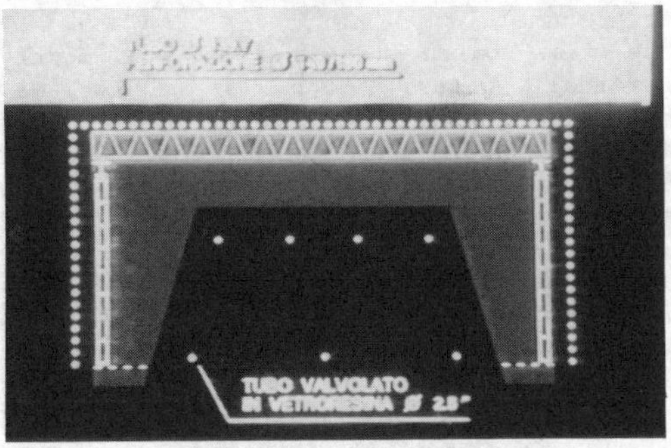

photo no.4

4.2 Preconsolidation works for soil involved in tunnel excavation work

The preconsolidation of the soil which was indispensable for the completion of underground excavations was planned taking account of both the geomechanical characteristics of the soil and the shape of the rectangular cross-section and considerable dimensions of the excavation. Construction work was divided into the following operational stages (photos no.3 and 5).

a) execution of horizontal micropiles fitted with valves (sub-horizontal grout treatment technology) around the edge of the hole, working from the two openings. Soil consolidation was completed using PVC "manchette"-type pipes.

b) consolidation of the internal core at the excavation profile using a "manchette" pipe in plastic reinforced by fibre glass to guarantee the stability of the face during the advance of work.

4.3 Realisation of excavations and completion of underground consolidation works

The poor geomechanical characteristics of the soil together with the need to avoid causing any alterations to the overlying "tombs" dictated the controlled progression of underground excavations and their limited advance, also including the shutting-down of the section. The executive sequences took the form of the following phases (photos no.3-10):

a) Preliminary excavation, carried out exclusively around the edge of the hole profile, to a depth of not more than 1.50 m and shaping of the central core into a shoe compatible with the geomechanical characteristics of the soil acquired after consolidation operations to guarantee face stability; followed by the casting of the invert support, using class 300 concrete, realised with metal sections and placed at an interaxis of 1.00 m (photo no.9), capable of supporting the horizontal trusses;

b) laying of horizontal reticular structures (photo no.10) realised using

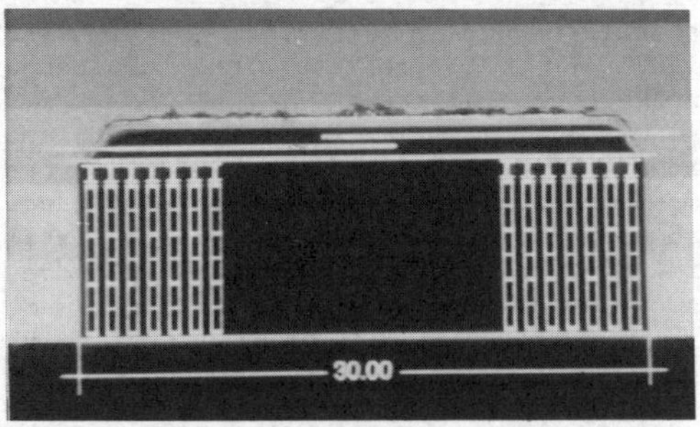

photo no.5

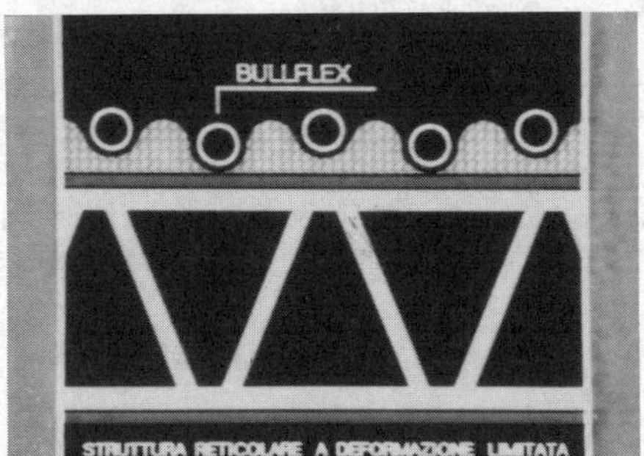

photo no.6

photo no.7

photo no.8

controlled deformation steel structural work, to support the advance of excavation work.
Reticular beams are fitted with DYWIDAG tighteners to control and annul subsidence in the overlying soil.
In order to avoid dangerous localised ground settling, caused by the impossibility of achieving a continuous and uniform contact between pre-horizontal treatment devices, placed around the edge, and the reticular beams, a "mattress made from reticular polypropylene" filled with highly resistant additive-enriched grout was used (photo no.6).

c) execution of fibre-reinforced shoot-crete sprayed around the edge of excavation and onto the face.

d) casting of a reinforced concrete bed to act as a strut.

e) consolidation of inverts and foundation bed using micropiles fitted with valves (photo no.3).

4.4 Casting of reinforced concrete casing

Having completed the excavation the final cross-section was realised using concrete compensated for contraction in order to obtain a waterproof structure.

5 OPEN-AIR CONNECTION WORKS

The tunnels were then completed by the construction of two 0.75 m sidewalks protected by metal barriers and four external openings using a flute-mouthpiece structure.

6 CONCLUSIONS

Given that the subway below the archaeological zone is the last section required to make lot no.31 usable, it was important to carry out works using three 8-hour shifts a day, operating contemporaneously from the two openings of each tunnel.
The time taken to complete preliminary external works, such as bulkheads, micropiles, preconsolidation of the core was 40 consecutive solar days, whereas excavation operations, carried out using an excavator fitted with a bucket with two ripper teeth, took fifteen working days, equivalent to a tunnelling speed of 1.00 m from each opening.
The final casing ad finishing works

photo no.9

photo no.10

required to complete the tunnels were completed in a further twenty days, making a total of seventy-five days.
The precautions adopted and operating methods used enabled the works to be duly carried out without even causing marginal damage to the archaeological complex.

Tunnelling and Ground Conditions, Abdel Salam (ed.) © 1994 Balkema, Rotterdam, ISBN 90 5410 363 9

Micro-multi-box tunneling method (MMB Method)

T. Sugito, *Taisei Corporation, Japan*

M. Ukegawa, *Toda Corporation, Japan*

H. Hamamura, *Tone Corporation, Japan*

M. Mohri, *Mitsui Engineering & Shipbuilding Co., Ltd, Japan*

ABSTRACT: Micro-multi-box Tunneling Method is a shield tunneling method for constructing a large underground space accommodating such structures as roads, railroads or parking lots. The method, whose development began in 1988, has been completed through a number of basic experiments and design reviews followed by demonstrative execution. The method is designed for a tunnel cross section measuring 15 m in height and 25 m in width and is intended for a road having three or more lanes. At present, use of circular shield tunneling machines or multi-face circular shield tunneling machines to construct tunnel cross section of this scale is impractical because of difficulties involving the manufacture and handling of shield tunneling machines and segments, and the developed tunneling method is more practical in this regard.

1 INTRODUCTION

Urban development started with the Egyptian and Mesopotamian civilizations in large river basins. The fertile alluvium was productive and human civilization developed by leaps and bounds. There are, however, examples of European style cities built on hard base rock, but in Japan with many rivers flowing from steep mountains, one hundred and ten million people live on top of alluvium. In this small space, houses, roads, railways and all other facilities are jumbled together, and it is reasonable to say that great care is paid to the effective use of space underground in order to preserve living space.

However excavation of underground spaces in the weak sand and clay alluvium is not easy, and this initiated the development of the shield tunneling method. Circular shield tunneling machines of up to 14m in diameter have been used in projects up to now but, coping with using a single body shield tunneling machine is difficult, and these days effective underground use involves large volume excavations. With the MMB Method the external sections are progressively excavated with a 2-3m small/medium sized box shield, and having built the whole structure with connecting box shields, the internal large space is open excavated.

Development of the box shield method commenced in the autumn of 1988, and since then there has been much study into its practical application. Using a 1.0m square test machine (pipe jacking type) various tests were carried out: excavation performance, investigation into an agitator for preventing soil adhesion in the chamber, testing the crushing of stones, test confirmation of grouting coverage around the rectangular cross-section, and development of the concrete segments, etc. In addition, a practical scale 2.85m square machine was used on this project with a rectangular section pressurized slurry shield method, to confirm the theory.

Another reason was that for the box shield method we obviously could not confirm curve excavation and the operation of the erector, full scale face stability and position control, excavation speed and other functions of excavation by testing only. In addition, starting with the development of the excavation method for the large scale underground tunnel TUBE structure (Figure 1), element tests relating to the structural materials were partly finished. So on this trial project, another objective was to test the possibility of using the box shield tunneling as a prelining in the MMB method.

In this report, after discussing factors in the MMB method, there is a report on the demonstration carried out at the Toda Corporation Tsukuba Research Center. The four companies, Taisei Corporation, Toda Corporation, Tone Corporation, and Mitsui Engineering & Shipbuilding Co., Ltd., are all presently further researching the M-M-B method together.

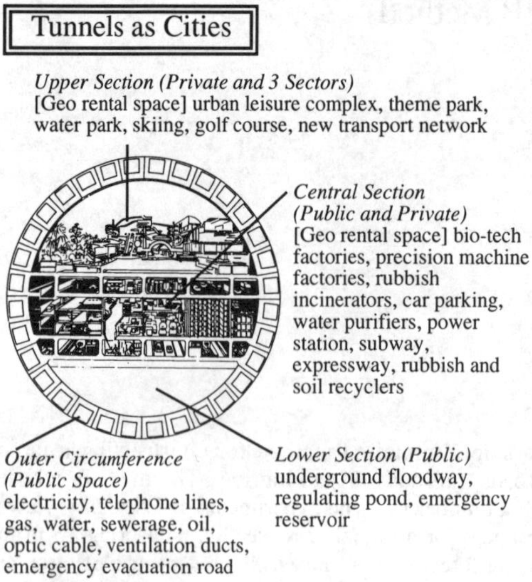

Figure 1 TUBE Concept

2 MMB METHOD

2.1 Examples of Application of MMB Method

Here are some examples of ideas for the use of underground excavated spaces in present-day Japan.

Figure 2 shows an example where use of cut and cover is not possible. It shows a cross-section through an underground railway, but it could also be used where there were important historic sites or areas of important animal and plant life on the surface.

Figure 3 shows an example of underground space where above ground space is scarce, e.g. for parking, subway train depot, oil and gas storage, rubbish management facilities etc.

Figure 4 shows an example of an underground urban large scale roadway but this could also be a subway station or a pedestrian area.

These are not agreed projects for the near future but examples of actual requests received where the strengths of the MMB method may be seen.

2.2 Features of MMB Method

Figure 5 shows an example of the sequence of construction using the MMB method. It is a design for a two story two lane road tunnel with a height of 22m and a width of 15m.

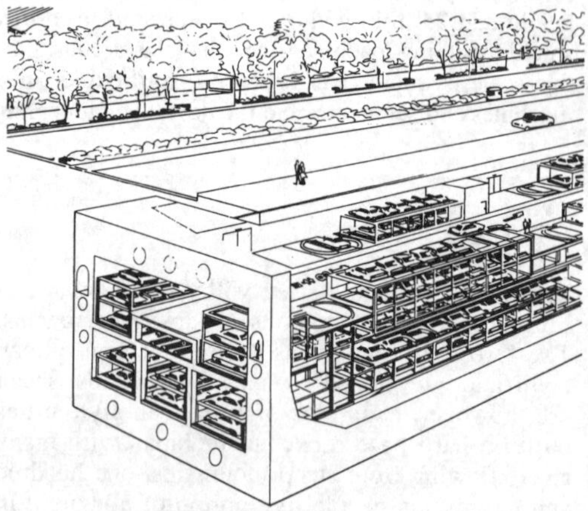

Figure 3 Example of Tunnel (2)

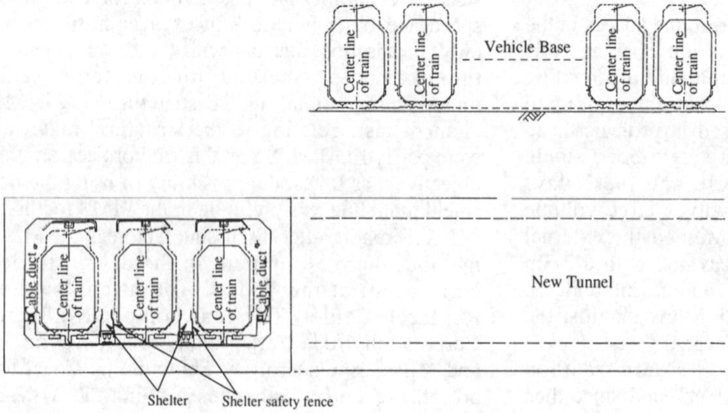

Figure 2 Example of Tunnel (1)

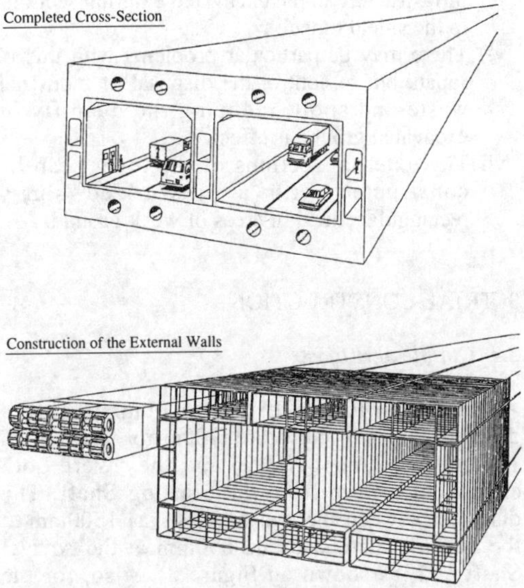

Figure 4 Example of Tunnel (3)

1) The whole structure is divided into rectangular box sections excavated in sequence by box shield machines (Nos. 1 and 2 in the Figure).

2) The connecting sections in the tunnel are excavated having done ground improvement works (No. 3). Figure 6 shows both excavation and support methods, but these are selected depending on the ground conditions, water pressure, length of connecting sections, and excavating length etc.

3) Construction of the whole body (No. 4). Some segments are partially removed and part of the whole body is constructed, and based on the steel concrete structure, concrete is placed in the segments.

4) Open excavation of the inner sections using the bench cut method (No. 5).

5) Finishing works for the inner sections, and construction of the roadbed (No. 6).

Depending on the sequence of construction this method has the following features.

i) When using small or medium sized box shield machines a deep overburden is not necessary, compared to excavation with a large section shield tunneling machine.
ii) The total construction time depends on the number of shield machines decided upon and how they are used.
iii) For forks and slip roads in the road tunnels, and energy related storage tanks, adjustments can be made in the connecting section dimensions for the shield tunnel part.
iv) Even in cases where there is no leeway in the width of the site for the underground space, in order to excavate using the box shield directly, it can be designed with the necessary minimum cross-section.
v) It is possible to locate electric, telephone, potable and sewerage water and other

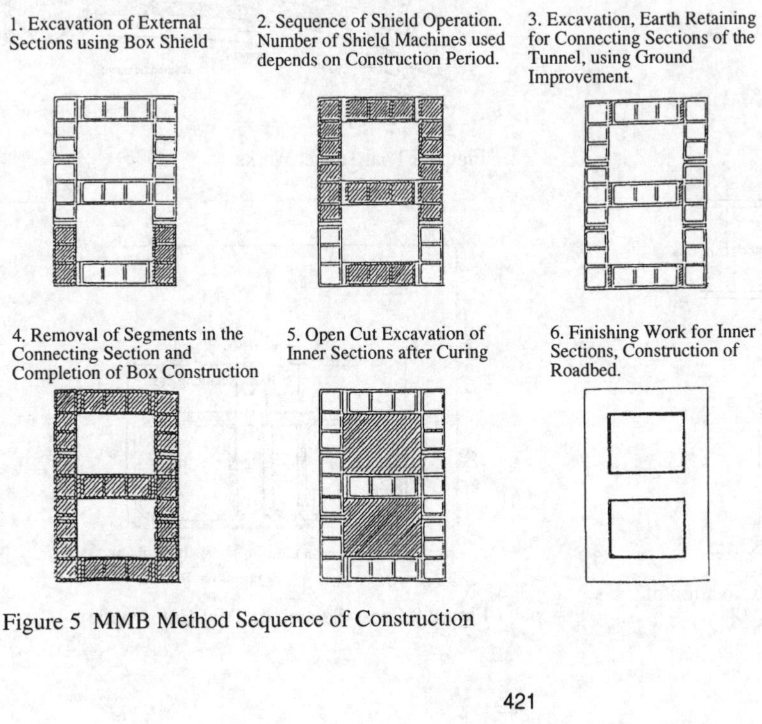

Figure 5 MMB Method Sequence of Construction

infrastructure in the excavated external sections in the shield tunnel.

vi) There may be particular problems with this in Japan, but to control the disposal of industrial waste and spoil, reducing the quantity of excavated ground is effective.

vii) The external sections of the tunnel can be constructed with a pressurized slurry rectangular shield in areas of weak ground.

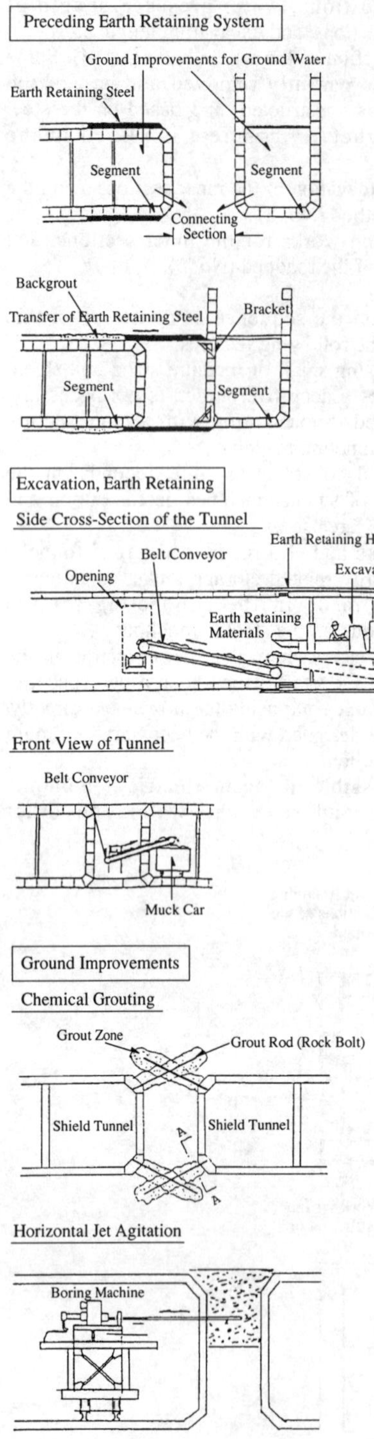

Fig. 6 Support Method for Excavation of Connecting Section

3 TRIAL CONSTRUCTION

3.1 Outline of Project

As shown on Figure 7, the two tunnels have a different radius of curvature (100m for Tunnel No. 1, 80m for Tunnel No. 2) and they were both excavated in turn from the Starting Shaft. The distance between the two tunnels was 1500mm at the Starting Shaft end and 850mm at the Arrival Shaft end (as shown on Figure 8). Also, for the confirmation of the MMB method, out of a 40m length, the segments were excavated for a 20m section from the Starting Shaft (called the connecting section), and after removal of the segment wall the two tunnels were joined into one body.

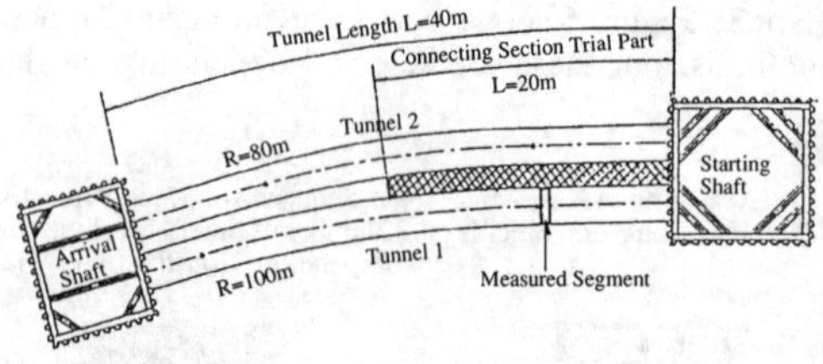

Figure 7 Trial Tunnel Works

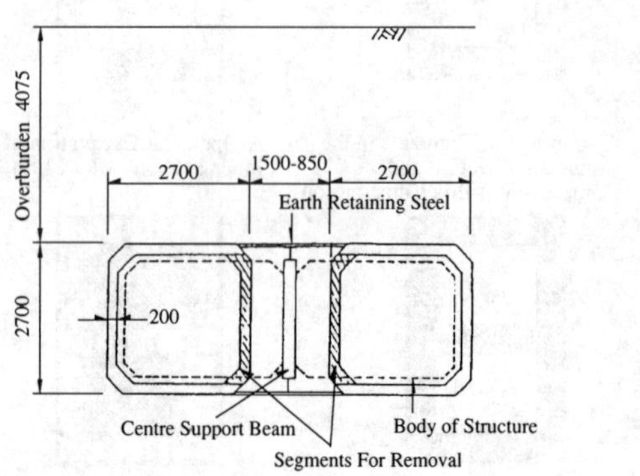

Figure 8 Cross-Section of Trial Construction

In reality for construction of a large cross-section tunnel, the empty space between the segments would be buried as construction materials, but in this test in order to keep this space empty a centre support beam was constructed, as shown on Figure 8. Moreover, the alignments of both tunnels incline upwards at an inclination, i = 0.5%. As for the trial construction, after having basically confirmed the excavation ability on a full scale test machine, including construction on a curve using a box shield (previously stated as fundamental points of the MMB method), the MMB method was employed on the excavation of the connecting section, removal of the segments, and structural construction.

3.2 Box Shield Method

3.2.1 Box Shield Machine

Figure 9 shows the complete pressurized slurry box shield machine used on this trial project. Its special features are explained below.

(a) Drum Cutter
Having considered the effects of a cutter on the natural ground, cutter torque, costs etc. it was decided to use two (an upper and a lower) 1,425mm diameter drum cutters. The cutters were spoke type with the cutter and gear case action separate to prevent earth adhesion, and equipment was installed to clean the cutters with charge slurry blown onto them.

(b) Chamber
This was different from that of a circular type, the bottom being flat as retention and sedimentation of the ground was expected. The haunches on the corner sections, and the dead space being small, to protect against sedimented ground the cleaning equipment was placed inside the chamber in the lower part of the corner sections.

(c) Curve Construction
Depending on the over cutter (stroke 100mm) being controlled within the cutter spokes and the articulating equipment in the shield (articulation angle 3°), it was decided that the minimum radius of curvature for construction would be 40m (on this project R=80m).

(d) Control of Position
For rolling correction, eight of the twelve shield jacks can incline up to ±3° during construction. For additional correction of the shield machine position, clay paste can be injected into the ground at the bottom of the shield.

(e) Backgrouting
Using the space between the segment haunch and the tail of the machine to establish grout pipes, it was possible to backgrout from within the excavated section at the same time.

(f) Manhole
For taking apart the ring cutter, if there were unexpected conditions, it was decided that there should be access to the working face from inside the chamber.

(g) Erector
Usual circular shield machines are equipped with a standard action rotation, thrust erector; depending on the screw jack moving horizontally (vertically), it was possible to erect the rectangular segments (8 per ring).

3.2.2 Conclusions

The conclusions from the results of this trial construction are listed below.

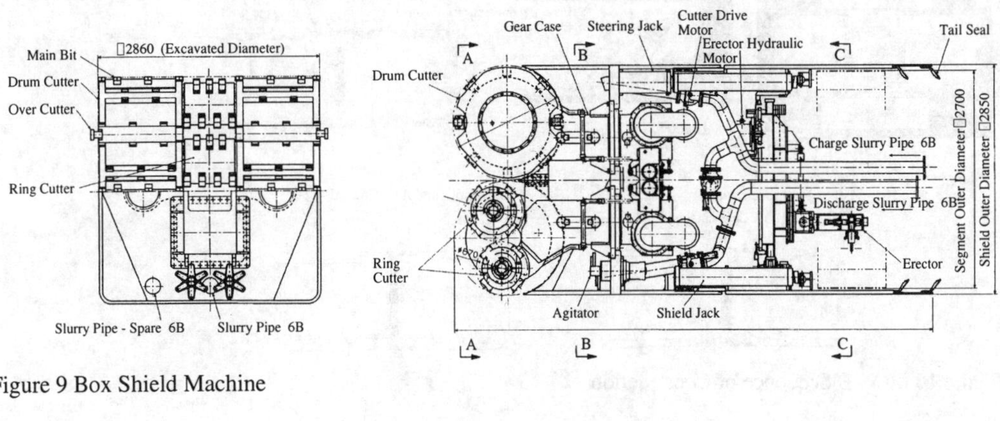

Figure 9 Box Shield Machine

1) Concerning the stability of the working face, and the looseness of the ground, measured values for the change in conditions and the volume of dry spoil are not different from the calculated values; excavation is possible while maintaining stability at the working face, and the box shield method used compares in scale to the test machine.

2) Excavation was basically carried out at a speed of 30 mm/min but it was possible to excavate at 50 mm/min. The thrust used was 25% of the installed capacity and was no different from that of regular circular slurry shield machines. There was almost a linear relationship between the excavation speed and cutter torque, and between the excavation speed and thrust.

3) It was possible to control the position easily with the shields jacks and the inclining jacks. In addition, the direction of rotation of the cutters affected the machine, and by understanding their influence they could be used to inversely control the pitching.

4) Stable pressure grouting was possible from the corners of the machine at the same time, in the same way as for circular shields, with the advantage that there was no grout coming out of the carcass of the shield machine. Also with grouting rates of 120-130% good coverage was possible.

5) A radius of curvature of 80m (minimum design radius 40m) was employed for the articulating equipment and the over cutter controlled inside the cutter spokes. On the trial, construction was carried out with care and the curve cutting response was good. However, when constructing the curve, a tendency was seen for rolling as a result of the partial choice of the shield jacks. For use of the inclining jacks, some care was necessary, but there was sufficient confirmation that curve construction with a rectangular cross-section was possible.

3.3 MMB Method

3.3.1 Sequence of Construction (Figure 10 shows each stage)

(1) Injection Grouting
In advance from the grout holes established in the segments, and from within the tunnel in the upper and lower sections of the connecting section grout injection was carried out to improve the ground in the connecting section.

(2) Excavation / Earth Retaining
In relation to the construction of the connecting section various methods were considered; however it was constructed from the shaft in the direction of the axis of the tunnel. For the excavation a mini excavator with a spoil bucket was used. For earth retaining, 300mm steel channels connected to each other by bolts were employed. The "Excavation → Spoil Removal → Connection of Earth Retaining → Welding of Segment and Earth Retaining" construction cycle was carried out repeatedly with 300mm steps for the 20m excavation. On this demonstration project, a leased mini excavator was used and, as hand labour was used on the earth retaining, a pace of 900mm/day of excavation was adopted. In reality, using an excavator equipped with an erector for the construction of the earth retainer, it would be possible to proceed more quickly.

(3) Construction of Central Bulkhead
Before removing the segments, beams (300mm RSJ) were put on the earth retaining, and were packed tight with the central columns (300mm RSJ) [for 9m out of the 20m excavation in the connecting section]. To pack the central columns tight, an attachment on the mini excavator was used.

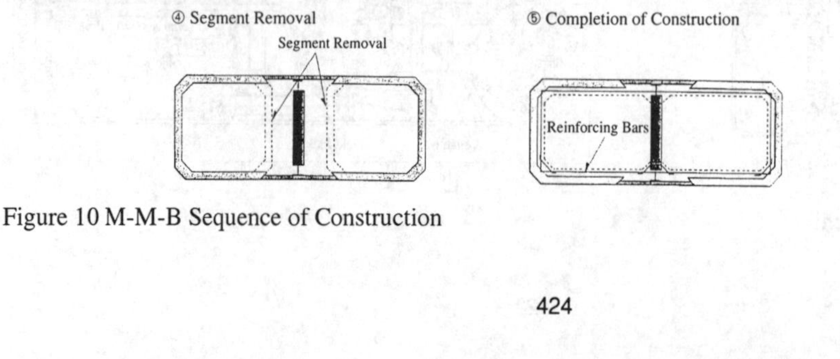

Figure 10 M-M-B Sequence of Construction

(4) Segment Removal
Having taken out the bolts in the part to be removed beforehand, the segments could be taken apart, but in order not to have difficulties removing the taper, part of the segment was removed by burning. Using a chain block the segments were removed .

(5) Construction of the Main Structure
Stress reinforcement was placed on the inside of the cross-section of the structure and, the segments on the outside were included in the design as part of the composite structure (Cross-Section in Figure 11). For placement of the concrete, a two pour (base, and wall/crown) was decided upon. For the wall/crown pour, in order to reduce the pressure on the side formwork and to give good coverage, high performance concrete was used.

3.3.2 Segments

Figure 12 is a comparison of calculated and actual measured values where the lateral loading on the right side was removed for the connecting section excavation.

Figure 13 shows the change of stress over time on the side axis for the connecting section working face. Accompanying the excavation in the connecting section, depending on the stress relief in the ground, the changes in ground pressure and the transmission of bending moments between the segment rings for the measured segment concerned (segment 10), did not appear at the same time as the excavated section of the working face. There was some change in the stress level but the ground on the right side of segment 10 was excavated, and according to the change in the acting load there were no big changes in the final force redistribution in the cross-section. In cases where there is no lateral load on the right hand side, there will be large bending moments especially on the right hand side center, and the bending moment loss will tend to be big (compression inside, expansion outside).

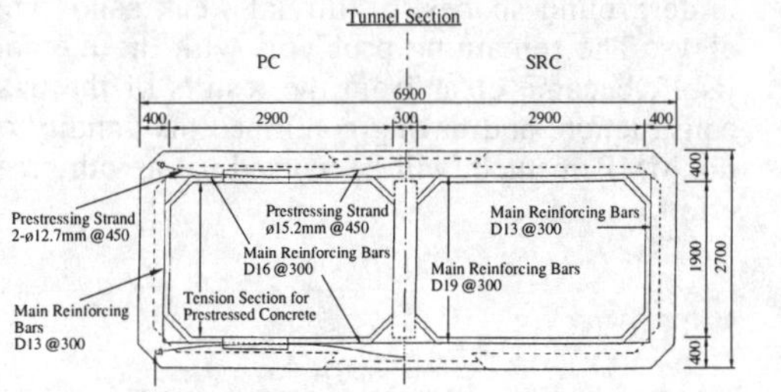

Figure 11 Construction Section

3.3.3 Conclusions

For the construction on this trial project, even though there was not enough time for preparation, we choose to excavate the connecting section and backgrout it, and it was decided that use of the MMB method was a prime objective. Steel reinforced concrete and prestressed concrete were used in the body construction, and having received the first results, considering the construction stages individually there are many problems. In particular, the excavation of the connecting section, the construction efficiency of the earth retaining, the safety viewpoint, and conditions with high water pressures. However, having investigated a construction method for the connecting section running in parallel with the actual construction and having analyzed the use of construction materials, for example the idea of using a sliding steel plate for the earth retaining in the connecting section, investigations are planned to continue.

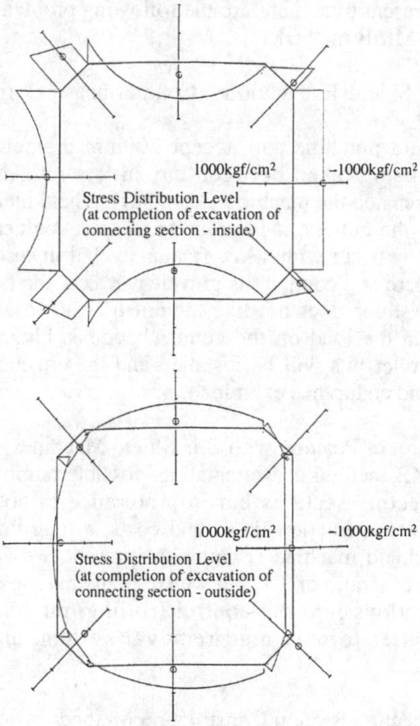

Figure 12 Stress Distribution Level

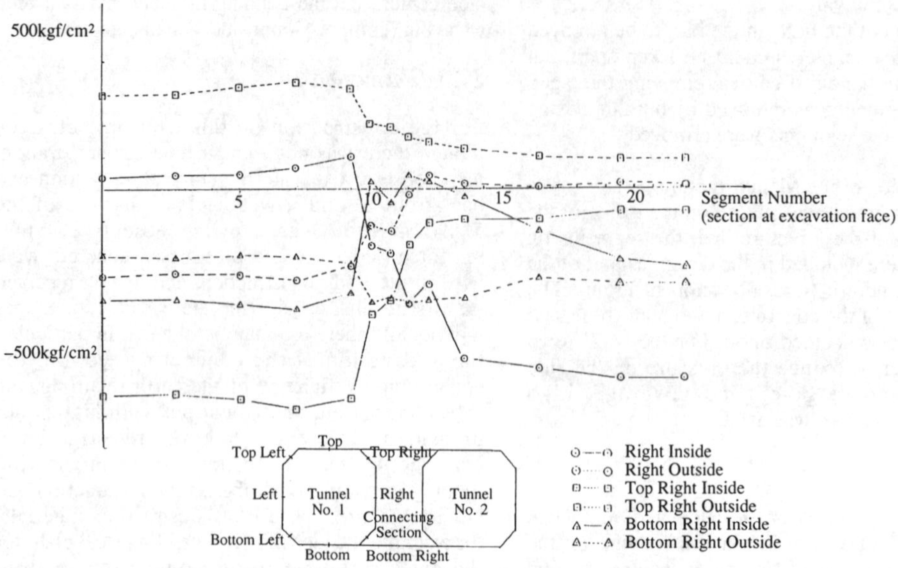

Figure 13 Influence of Excavation in the Connecting Section on Segments in Tunnel No. 1

4 PROBLEMS

At the present time there are the following problems with the MMB method.

(1) Box Shield Excavation - Countermeasures for Stones
The shield machine can accept 200mm diameter objects being taken between the cutters into the chamber inside the machine. Also, having held them between the cutters, it is possible to split boulders where they occur with a low frequency, but in cases where there is a continuous gravel river bed, the top of the machine does not like the crush on the tooth bits when the load on the cutters becomes bigger, and so roller bits will be installed and the crushing ability and endurance examined.

(2) Control of Position with Box Shield Machines
The MMB method uses a small as possible number of connecting sections but in preference to both favorable construction times and costs, a long thin shaped shield machine is better. At present, we are thinking of a ratio of 3:1, but we are in the middle of investigations into the control (rolling, pitching etc.), cutter torque, mud removal system, and erector.

(3) Connecting Section Construction Method
Having done tests at Tsukuba, using the methods explained before, two tunnels can be linked. However with a deep overburden, ground and water pressures will be high. Weak ground is often seen in Japan, and if ground improvements are carried out construction will be possible but the cost will be extremely high. For this reason ground improvements are rarely done. In order to confirm construction and safety, we are investigating, for example, installing sliding steel plates on the upper and lower parts of the connecting section, ground retaining methods and system technology construction methods.

5 CONCLUSION

The MMB method, and not cut and cover, is an effective method for constructing large scale underground spaces in alluvial weak sands and clays. The remaining problems with the method itself, became clear from the results of the trial construction, and having confirmed the validity of the MMB method, will be worked upon with great effort.

REFERENCES

Tachikawa, Watanabe et al. 1993.9. Report on Box Shield Demonstration Construction, 48th Annual Conference Summary, *Japan Society of Civil Engineers* VI pp.96, 97

Taniguchi, Tada et al. 1993.9. Actual Construction using the MMB Method, 48th Annual Conference Summary, *Japan Society of Civil Engineers* III p 54

Tunnelling and Ground Conditions, Abdel Salam (ed.) © 1994 Balkema, Rotterdam, ISBN 90 5410 363 9

Numerical modelling of shotcrete and concrete tunnel linings

Gunter Swoboda
Innsbruck University, Austria

Ahmed Moussa
Helwan University, Cairo, Egypt

ABSTRACT: Prediction of the behavior of the tunnel lining during construction is a complicated problem due to the complexity of tunnel modelling and material nonlinearity. The few existing numerical models for shotcrete in tunnelling are not able to predict the actual behavior of the lining. In most cases material nonlinearity and the distribution of the stresses and strains along the thickness of the shotcrete lining were ignored. In this paper the authors attempt to give a more realistic simulation of the partial excavations used in NATM which produce incremental loading on the shotcrete lining with the progress of excavation with time, while the shotcrete strength is increasing. The model takes into consideration nonlinearity, cracking, plastic deformation history, and the time dependent strength of shotcrete. A practical application of the model in the analysis of a subway tunnel with a double lining system is shown. The finite element coupled with boundary elements are used. A special six nodded beam element is used to model both shotcrete and final linings. An interface element is introduced to model the possible relative deformation between the two linings. Material nonlinearity and cracking of concrete are considered.

1 INTRODUCTION

The driving of tunnels according to the New Austrian Tunnelling Method (NATM) is performed by partial excavation and shotcreting. In some cases rock bolts are associated with shotcrete to provide a rapid support for the cavity. To date there is no accurate method for designing the thickness of shotcrete lining or the amount of any additional reinforcement which should be applied to suit the particular ground condition of the tunnel. Semi-empirical approaches using Degrees of Rock Quality Designation, RQD, have been found to be a useful guide to rock quality and the support required for loosening ground conditions [4]. To design the tunnel supporting system adequately needs both great experience and great theoretical knowledge, and these qualities are even more important when applying this system correctly during construction [10]. Field measurements are efficiently applied in order to confirm whether the flexible support system is satisfactorily functioning and to indicate the proper timing for operations such as final tunnel lining installation [1].

The prediction of shotcrete behavior during tunnel construction is a complicated problem because of material nonlinearity and the progress of both load with excavation and shotcrete properties with time. Few numerical models for shotcrete in tunnelling have been developed. To illustrate of the principles of NATM, a closed form solution for a circular tunnel in an elasto-plastic media under hydrostatic state of stress is often used [12]. To determine the pressure on the lining, the characteristic curve method has been developed and used. In this analytical method, the shotcrete is generally modelled as a thin-walled or thick-walled tube. The support action of shotcrete is simulated by a radial pressure. The accuracy of the results is greatly reduced as the surrounding soil or rock becomes irregular and as the tunnel cross section becomes different from the circular shape. In addition, only the axial deformation of the shotcrete tube is considered; this means that both bending action and tension cracks are ignored. For these reasons, we cannot apply this method to study the actual behavior of shotcrete in tunnelling under different constructions and loading conditions.

It was found that the finite element method is the most suitable method for analyzing tunnel excavations, since the varieties in soil and loading conditions can be considered. Due to the complexity and high cost of three dimensional analysis, a simulation of the three-dimensional effect near the tunnel face by a two-dimensional model is often used. Much attention was given to two-dimensional FE modelling of soil and rock mass by many researchers, but very few models for shotcrete in tunnelling have been developed.

The major difficulties in the nonlinear analysis of shotcrete in tunnelling arise in selecting appropriate finite element models and constitutive relations for the materials under different loading and cracking conditions. Since the shotcrete lining is not an independent structure, the choice of the FE model for shotcrete was mainly constrained by two factors. The first is that the shotcrete element should be able to combine with two-dimensional plane strain elements used for soil, and the second is that the FE mesh used should not be very fine, otherwise the compution time will be too large.

Till now there are no established constitutive relations for shotcrete because of the lack of experimental data describing its mechanical behavior. The tests on shotcrete cubes done by Sezaki [11] show a much softer behavior of shotcrete than that of concrete under compression. Complete stress-strain curves for shotcrete under both tension and compression are not available. Also, available field measurements of stresses and strains in tunnel lining cannot be properly used because the depth of the measuring points is usually not recorded, and the strain distribution along the shotcrete thickness cannot be neglected.

Because of these constraints, different types of approximations were used in the existing models. The shotcrete lining was modelled by a normal plane strain element like that used for the surrounding soil but with a shotcrete modulus of elasticity or by a two-nodded beam element. These models for shotcrete lining are not accurate, since they take into consideration only the axial and shear forces. The bending action cannot be modelled, unless a great number of 2D elements are used, and if we consider the nonlinearity and the cracking of shotcrete, the shotcrete thickness must be divided into a larger number of elements. This means that a very fine mesh is needed for shotcrete and this also leads to a fine mesh for the surrounding soil. The required CPU time will increase too much, while the final results may not be satisfactory. A coupled beam-boundary-element model had been developed by Pöttler [9]. The model gives better simulation of the shotcrete lining but it cannot consider plasticity and the variations in the surrounding soils. The BEAM 6 element developed by Swoboda [15] provides an acceptable solution for the FE modelling problem as it considers all the possible deformations of the lining.

In this paper the authors attempt to give a more realistic simulation of the partial excavations used in NATM which produce incremental loading on the shotcrete lining with the progress of excavation with time, while the shotcrete strength is increasing. The model takes into consideration: (a) the nonlinearity of shotcrete under compression and cracking in tension, (b) plastic deformation in shotcrete due to early loading, (c) the development of shotcrete strength with time, (d) presentation of the axial, shear and bending deformation of the lining shell.

As an application, the finite element method coupled with boundary elements is used to analyze a critical cross-section in a Munich subway tunnel. In this tunnel, the final lining is designed to be water-proof. To guarantee the impermeability of the final lining a minimum compression zone is required. The objective of the study is to predict and hence control the distribution of cracks in the concrete lining under extreme loading conditions. External loads on this lining system are the full water pressure, a possible failure of the shotcrete lining and the expected surface load due to future constructions. The tunnel excavations are simulated by a two-dimensional finite element model. A special six nodded beam element is used to model both shotcrete and final linings. An interface element is introduced to model the possible relative deformation between the two linings. Material nonlinearity and cracking of concrete are considered.

2 SIMULATION OF TUNNEL EXCAVATION

For two-dimensional FE modelling of tunnels, a special beam element (BEAM 6) has been used to model shotcrete in order to simulate the combined action of the tunnel lining and the surrounding soil which is modelled by plane strain elements (LST). The advantage of this beam element is that it can describe the real behavior of the lining as an arched frame using a small number of elements in addition to achieving combined action with LST elements used for the surrounding soil. The beam element used has six nodes, each with two degrees of freedom (u_i, v_i). The element nodes are arranged to form three couples, each couple with two nodes connected by a rigid truss element as shown in Fig. 1. The derivation of the stiffness matrices of both LST and BEAM 6 elements is given in reference [15].

To simulate the three dimensional effect near the tunnel face, a stiffness reduction model had been used [14] as shown in Fig 2. This model uses a support core with a modified modulus of elasticity, E_s, where

$$E_s = \alpha_a E \tag{1}$$

and $0 < \alpha_a < \infty$. The full stress field in front of the tunnel face is applied to this core. Two steps have been assumed to simulate the advance of the tunnel face due to excavtion; the first

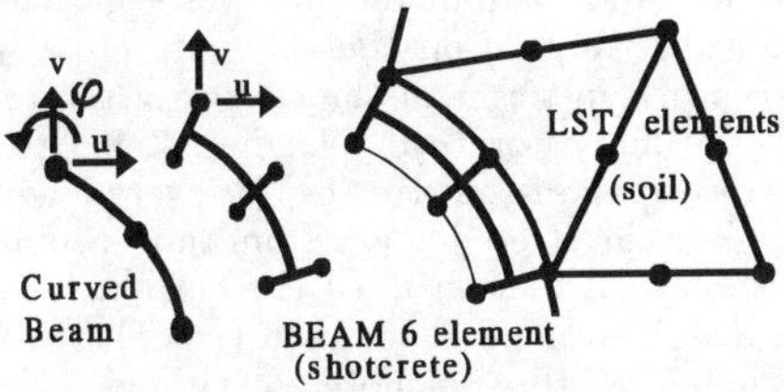

Figure 1: BEAM 6 element

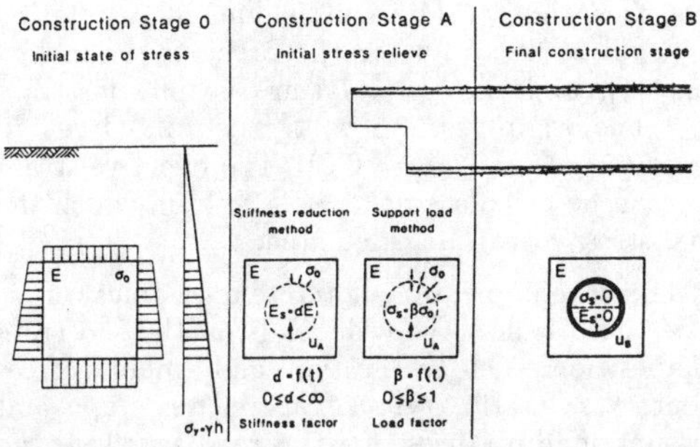

Figure 2: 2D simulation of the 3D effect [14]

consists of eliminating the elements of the excavated area by reducing their elasticity modulus to zero or by deleting these elements, and the second step consists of eliminating the primary stresses in these elements at the same time by integration over the element to get the equilibrium nodal forces $\{f\}$

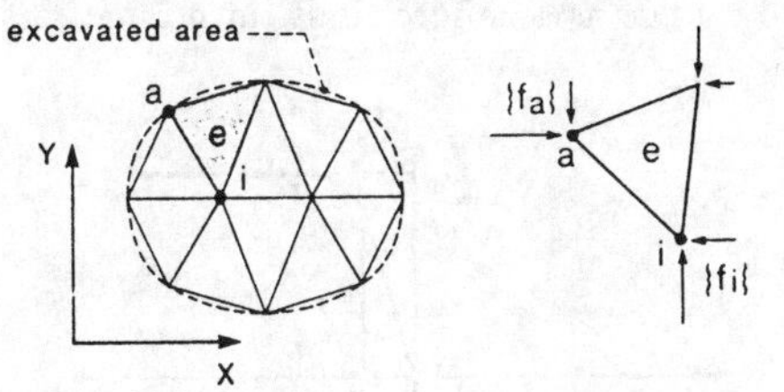

Figure 3: Simulation of excavation [14]

$$\{f_e\} = \int_v [B]^T \{\sigma_o\} dv \tag{2}$$

The sum of the forces in the (n) excavated elements gives the total load vector $\{F\}$

$$\{F\} = \sum_{e=1}^{n} \{f_e\} \tag{3}$$

This load vector is divided into two parts:

$$\{F\} = \left\{ \begin{array}{c} F_a \\ F_i \end{array} \right\} \tag{4}$$

where $\{F_a\}$ represents the loads on the excavated boundary, and $\{F_i\}$ are the forces in the interior part which are not to be considered as is clear from Fig. 3, so that the load vector $\{F\}$ is transformed into a load vector $\{F_a\}$

3 CONSTITUTIVE MODELS

Particularly in the case of large initial displacements in tunnelling, the shotcrete can experience loads that do not otherwise occur in concrete structures. The shotcrete is already subjected to very heavy loads during its initial curing phase. Based on the available experimental data and the various constitutive relations proposed by different codes, it was found that a parabolic relation is the most simple and representative one for the stress strain relationship for both of shotcrete and concrete under compression up to the maximum stress point ϵ_1, f_c. This relationship can take the form

$$\sigma_c = \frac{f_c \epsilon_c}{\epsilon_1}(2.0 - \frac{\epsilon_c}{\epsilon_1}) \tag{5}$$

The value of the strain at maximum stress ϵ_1 depends on the concrete type and quality. For normal concrete ϵ_1 can be taken as 2‰. Tests on shotcrete showed a much softer behavior. Based on these tests an average value for ϵ_1 can be taken as 5‰. After reaching the maximum stress point ϵ_1, f_c a softening behavior was recorded in the tests. This is also suggested by CEB-FEP model code 1990. This softening behavior depends mainly on the state of stress of the shotcrete and the degree of confinement. Because of the large variation of the experimental test results in softening, which is largely affected by the testing condition and strain rate, we can say that there is no accurate stress-strain relationship for this softening. A linear relation would be enough if the values of the strain and stress at failure ϵ_u, σ_f are properly assumed. Based on this discussion we propose the stress-strain relationship shown in Fig. 4 for shotcrete under compression.

Because of the lack of experimental data describing the behavior of shotcrete in tension, the present model is based on the experimental data and proposed models for concrete. It is found that the CEB-FEP model can be taken as a representative one. This model describes the behavior of concrete in tension in two parts; the first one is a stress-strain relationship up to the maximum tensile stress reached, and the second is a bilinear stress-crack opening relationship. For the purpose of numerical analysis, we assume that the crack is smeared on the spacing between each two cracks. The stress-crack opening relationship must be transformed to an equivalent smeared stress-strain relationship. An alternative simplified model can be assumed by using a backward extension of the parabolic relation for compression up to the tensile strength of shotcrete followed by a linear softening up to complete crack opening. The cracking strain ϵ_{cr} can be calculated as the strain corresponding to the maximum tensile stress of shotcrete

$$\epsilon_{cr} = \epsilon_1(1 - \sqrt{1 - f_t/f_c}) \qquad (6)$$

and the maximum tensile stress f_t can vary from $0.2f_c^{\frac{2}{3}}$ to $0.3f_c^{\frac{2}{3}}$ The ultimate tensile strain ϵ_{tu} is assumed to be 0.3‰ at which complete crack-opening is assumed.

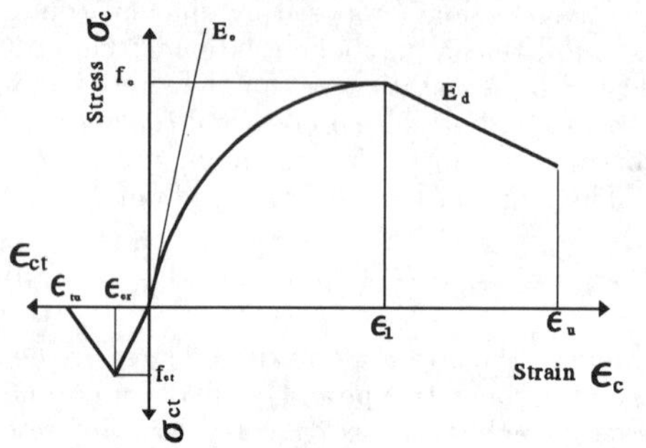

Figure 4: Stress-strain curve for shotcrete without preloading

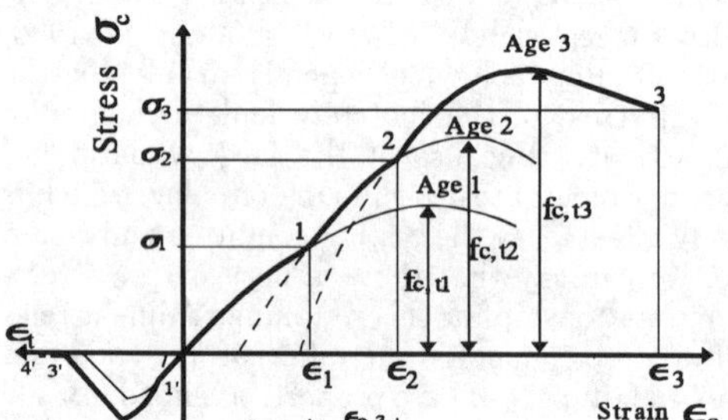

Figure 5: Constitutive model for preloaded shotcrete

For the case of incremental loading with time, which normally takes place in a shotcrete lining during tunnel construction, a strain history based model is proposed. For calculating the stress at a certain location in a shotcrete section which has a previous strain history we suggest the model shown in Fig. 5. This constitutive model is based on combining the short-term $\sigma - \epsilon$ curves for shotcrete at the different loading times, in such a way that the plastic deformation due to early loading and shotcrete hardening with age are taken into consideration. In this combination the creep of the shotcrete in the mean time between each two successive load increments was neglected. The strain distribution on the shotcrete cross section is recorded in terms of the strains of the top and bottom fibres from every loading step. To avoid the recording of the stresses, for every loading we calculate the stress history from the strain history in a loop calculation until the current loading is reached. For the first loading, ($i = 1$), the stress-strain relationship for shotcrete without preloading is used. For any other loading i, the new stress-strain curve is defined in terms of the initial stress, σ_{i-1}, the initial strain ϵ_{i-1}, and the new compressive strength f_{c,t_i}. This can be done by calculating the equivalent strain for the previous strain history $\epsilon_{o,i}$, and then calculating the plastic strain $\epsilon_{pl,i}$. The origin of the new stress-strain curve will be at ($\epsilon_c = \epsilon_{pl,i}$, $\sigma_c = 0.0$). The effective strain $\epsilon_{ce,i}$ can be calculated as $\epsilon_{ci} - \epsilon_{pl,i}$ from which the final stress σ_i can be calculated.

The development of shotcrete mechanical properties with time largely depends on the shotcrete composition, such as the type and amount of cement, W/C ratio, type of aggregates, type and amount of admixtures. For this reason we kept the shotcrete compressive strength as an input parameter. The initial modulus of elasticity is calculated as the tangent for the parabolic $\sigma - \epsilon$ relation at its origin. The progress of shotcrete strength with time is generally as given in reference [5].

Since a shotcrete lining is a temporary supporting structure, a trilinear stress strain curve is assumed for cold twisted steel used as reinforcement as in Fig. 6. The ultimate tensile strength of the steel is used instead of the yield strength which is usually used, as a failure limit, in normal structures.

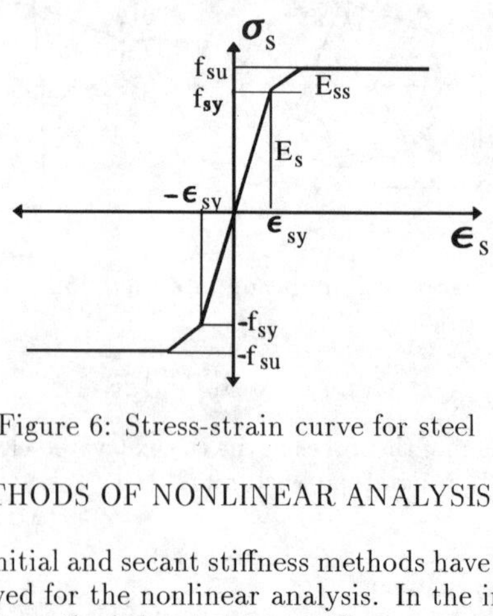

Figure 6: Stress-strain curve for steel

4 METHODS OF NONLINEAR ANALYSIS

Both initial and secant stiffness methods have been employed for the nonlinear analysis. In the initial stiffness method the structure stiffness matrix $[K_o]$ is calculated only at the beginning from the initial stiffness parameters, which are E_o, A_o and I_{zo} for the beam 6 element, and it is kept constant for all iterations. For each iteration the internal forces are calculated and the differences between external and internal forces, i.e. the unbalanced forces, are applied as external load until equilibrium is achieved.

In the secant stiffness method the structure stiffness matrix $[K_i]$ is updated for every iteration, using the actual secant stiffness parameters for beam 6 element ($E_{sec,i}, A_i, I_{z,i}$). These stiffness parameters are calculated depending on the load level from the iteration before, based on the true strain distribution.

4.1 Internal forces

The internal forces of a R. shotcrete cross-section under known strain history are calculated using the following steps:

1. Divide the total section into n micro-layers.
2. The internal force in any micro-layer, i is calculated as follows:

   • Define the strain history of this micro-layer, $\{\epsilon_{i,j}\} = \{\epsilon_{i,1}, \epsilon_{i,2}, ..., \epsilon_{i,k-1}\}$, from the saved strains from the previous loadings as:

$$\epsilon_{i,j} = \epsilon_{T,j} + (\epsilon_{B,j} - \epsilon_{T,j}).\frac{y_i}{h_j} \qquad (7)$$

   where k is the number of the current loading, and j is any loading number from 1 to $k-1$.

   • Calculate the stress history from the strain history using the proposed constitutive model.

   • Calculate the stress increment due to the current loading as:

$$\Delta\sigma_{i,k} = \sigma_{i,k} - \sigma_{i,k-1} \qquad (8)$$

   • Calculate the force increment for micro-layer i due to current loading as:

$$\Delta F_{i,k} = \Delta\sigma_{i,k}.dA \qquad (9)$$

3. Calculate the increment in the forces in the steel layers $\Delta F_{si,k}$ in a similar way.
4. The total cross-sectional forces after loading k are calculated as:

$$N_{t,k} = \sum_{i=1}^{n} F_{ci,k} + \sum_{i=1}^{2} F_{si,k} \qquad (10)$$

$$M_{t,k} = \sum_{i=1}^{n} F_{ci,k}.(y_i - h/2) + \sum_{i=1}^{2} F_{si,k}.(c_i - h/2) \qquad (11)$$

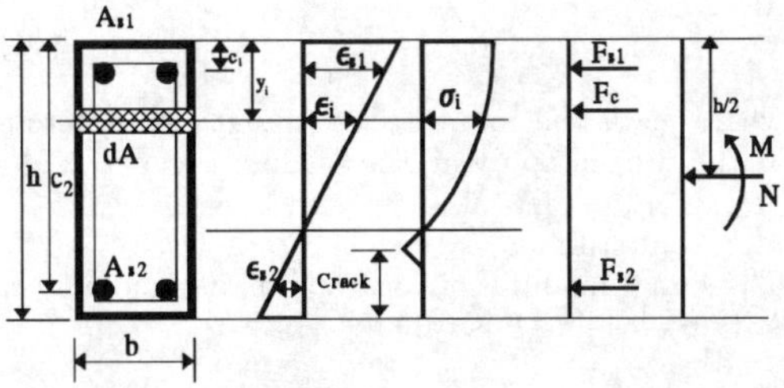

Figure 7: Internal forces for shotcrete cross-section

5. The increments in the cross-sectional forces due to loading k are calculated as:

$$N_k = \sum_{i=1}^{n} \Delta F_{ci,k} + \Delta \sum_{i=1}^{2} F_{si,k} \qquad (12)$$

$$M_k = \sum_{i=1}^{n} \Delta F_{ci,k}.(y_i - h/2) + \sum_{i=1}^{2} \Delta F_{si,k}.(c_i - h/2) \qquad (13)$$

4.2 Stiffness parameters

The stiffness parameters for the beam element are calculated by numerical integration over the cross section using a strain history based modulus of elasticity for each micro-layer i

4.2.1 Stiffness parameters for micro-layers

The stiffness parameters for any shotcrete micro-layer i are calculated according to the method of analysis employed and considering the strain history as follows:

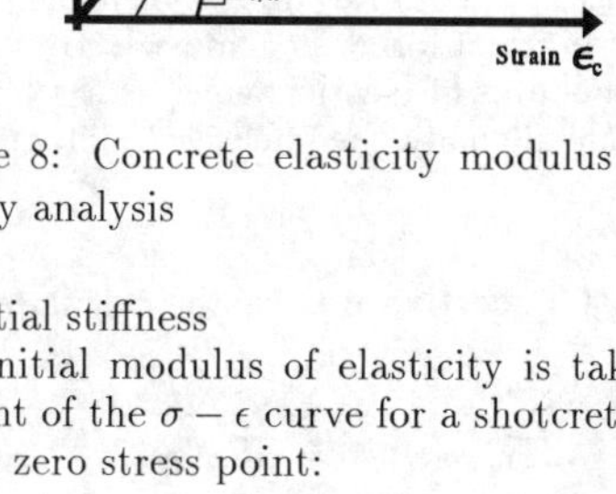

Figure 8: Concrete elasticity modulus for strain history analysis

a. Initial stiffness

The initial modulus of elasticity is taken as the tangent of the $\sigma - \epsilon$ curve for a shotcrete of age t_k at the zero stress point:

$$E_{o,k} = \frac{2\, f_{c,k}}{\epsilon_1} \tag{14}$$

where $f_{c,k}$ is the compressive strength of shotcrete at the time of applying the loading number k.

b. Secant stiffness
The secant modulus of elasticity of any micro-layer i at any loading k is calculated as:

$$E_{sec,i,k} = \frac{\Delta\sigma_{i,k}}{\epsilon_{i,k}} \tag{15}$$

where $\Delta\sigma_{i,k}$ is the stress increment in layer i due to loading k and $\epsilon_{i,k}$ is the strain increment in layer i due to loading k. It should be noted that the same equation is used to calculate the secant modulus of elasticity for steel layers $E_{s,sec,i,k}$. This modulus is constant for steel before yielding E_s, and has a different value for yielded steel, depending on the strain level.

c. Tangent stiffness
The tangent modulus of elasticity of any micro-layer i at any loading k is calculated from the tangent to the parabolic $\sigma - \epsilon$ curve for the considered layer as:

$$E_{tan,i,k} = \frac{f_{c,k}}{\epsilon_1}.(2 - \frac{\epsilon_{ef,i,k}}{\epsilon_1}) \tag{16}$$

where $\epsilon_{ef,i,k}$ is the total effective strain for micro-layer i at the beginning of load increment k and ϵ_1 is the strain corresponding to the maximum stress. It should be noted that the tangent elasticity modulus is never negative, so it is taken as zero for both cracked concrete and concrete that has compression strain greater than ϵ_1. The modulus of elasticity of steel is calculated in the same way.

4.2.2 Actual stiffness of the cross-section
The stiffness parameters of the whole cross section are determined by integration of the shotcrete micro-layers and steel layers. For both secant and tangent stiffness methods the stiffness parameters of the beam cross section are based on the true strain distribution due to current loading in addition to the strain history from the previous loadings. This true strain distribution is calculated using iterations until equilibrium between the external and the internal forces is achieved. By taking the initial modulus of elasticity, E_o, as a reference value, we can calculate the effective axial stiffness as:

$$E_o\, A_e = E\, A \tag{17}$$

where the total effective area can be calculated as:

$$A_e = \sum_{i=1}^{n} \frac{E_i}{E_o}\, b\, \Delta h + \sum_{j=1}^{2} \frac{E_{sj}}{E_o}\, A_{sj} \tag{18}$$

The effective total bending stiffness, EI_z, is also expressed in terms of the initial modulus of elasticity, E_o, as a reference value and the effective moment of inertia as

$$E_o\, I_{z,e} = E\, I_z \tag{19}$$

where the effective moment of inertia of the cross section around z axis through the effective centre of the cross section, $I_{z,e}$, is calculated as:

$$I_{z,e} = \sum_{i=1}^{n} \frac{E_i}{E_o} [\frac{b\,(\Delta h)^3}{12} + b\, \Delta h(\frac{h}{2} - y_i)^2] + \sum_{j=1}^{2} \frac{E_{sj}}{E_o}\, A_{sj}\, (\frac{h}{2} - c_j)^2 + A_e\, Y^2 \tag{20}$$

where A_{sj} is the area of the steel layer j at c_j from the top surface of the section, E_i is the modulus of elasticity of the shotcrete micro-layer number i, E_{sj} is the secant modulus of elasticity of the steel layer number j, Δh is the micro-layer thickness, and Y is the distance between the effective centre of gravity and the middle of the cross section

$$Y = \frac{1}{A_e} [\sum_{i=1}^{n} \frac{E_i}{E_o}\, b\, \Delta h(\frac{h}{2} - y_i) + \sum_{j=1}^{2} \frac{E_{sj}}{E_o}\, A_{sj}(\frac{h}{2} - c_j)] \tag{21}$$

5 APPLICATION TO TUNNELLING

In most cases, after completion of the excavation operations, tunnels are lined by a final concrete lining. The function of this lining is to ensure the safety and the serviceability of the tunnel during its working life and to give the tunnel the required shape. Two different methods exist to prevent water seepage. The first method is the installation of a water-proof membrane between shotcrete lining and final lining. In this case, the full water pressure is acting on the final lining. Additional loads on the final lining may occur through failure of the shotcrete lining. In the second method, the final lining is designed to be a water-proof lining. In this case, no water-proof membrane is necessary. To guarantee the impermeability of the final lining, a minimum compression zone is required. External loads on this lining system are again the full water pressure and a possible failure of the shotcrete lining.

5.1 Finite element model

For finite element modelling of the final lining the BEAM 6 element can be used. An interface element should be introduced to model the possible relative deformation between the final concrete lining and the shotcrete lining. The proposed two-dimensional finite element model is shown in Fig. 10.

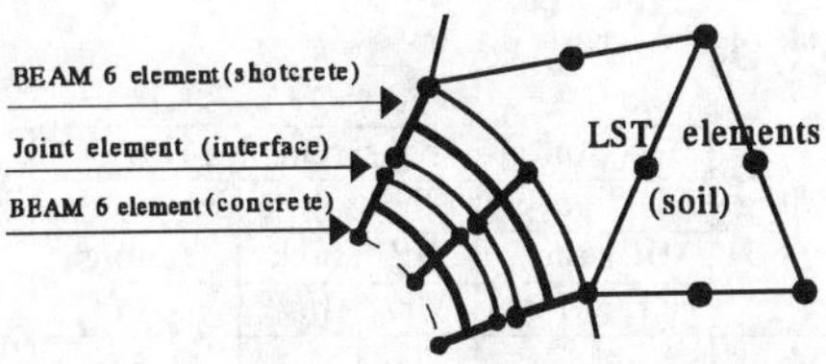

Figure 9: Two-dimensional finite element model for double tunnel lining.

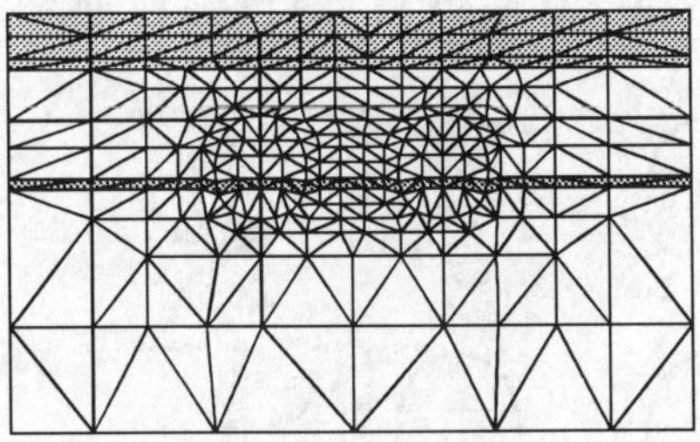

Figure 10: Two-dimensional finite element mesh for Munich subway tunnel.

The proposed model for the final lining (using the second method) is introduced to the Munich subway tunnel to predict and hence control the distribution of cracks in the concrete lining under extreme loading conditions. The project consists of two parallel tunnels of 5.6 m final diameter and 6.0 clear height. The centre line to centre line distance of the tunnels is 13.5 m. The overburden above the two tunnels is 7.3 m. The calculated tunnel cross-section is located under the IBM building, Candidplatz 9 and 11. It was assumed that the driving of the two tunnels is not parallel. This means that unsymmetrical load conditions are considered.

The simulated two-dimensional finite element mesh is as given in Fig. 10. Both shotcrete and final concrete linings were modelled by BEAM 6 elements using an interface joint element in between. The surrounding soil is modelled by LST elements. Boundary elements are introduced to model the infinite domain. The soil classification shows different soil layers starting with 4 m gravel layer followed by 7.8 m silicon clay and 0.8 m sand layer followed by continuos silicon clay. The properties of the different soil layers are given in Table 1.

5.2 Loading steps

The loading steps assumed to simulate the partial excavations of the tunnel are explained in Fig. 12, and can be summarised as follows:

- Loading -0- Initial stress
  This loading represents the initial state of stress in the soil. The sources of the initial stresses are the weight of submerged soil, according to γ' given in Table 1, load of the surface building which was assumed to vary from 80 to 100 Kn/m^2, and the ground water pressure with a ground water table of -2.0 m. A subsoil drainage with a maximum pressure of 1 bar in the sand layer is performed.

- Loading -1- Initial stress relief for crown of the inbound track. The initial stress relief factor is given in Table 1.

- Loading -2- Excavation and shotcreting of the crown and initial stress relief for the bench and invert of the inbound track. At the same moment an internal air pressure of 1.0 bar is applied on the excavation contours. The ground water pressure in the sand layer is reduced to 1 bar. Inside the tunnel, the internal air pressure is acting on the shotcrete lining and on the outer contours the reduced ground water pressure is applied. The shotcrete lining is assumed to be water-and gasproof.

- Loading -3- Excavation and shotcreting of the inbound track and initial stress relief for the crown of the outbound track.

- Loading -4- Excavation and shotcreting of the crown and initial stress relief for bench and invert of the outbound track. Internal air pressure and water pressure act as described in loading 2.

- Loading -5- Excavation and shotcreting of the outbound track.

- Loading -6- Removal of the internal air pressure; full water pressure act on the shotcrete lining.

- Loading -7- Final lining cast in place in both tracks under dead load conditions. The water was assumed to penetrate the shotcrete lining and the full ground water pressure was applied directly to the inner lining.

- Loading -8- Demolition of the existing buildings and simulation of an unsymmetrical excavation stage, starting from the vertical axis of the outbound track on a level of -7.0 m below ground surface.

- Loading -9- Symmetric excavation conditions on a level of -7.0 m below ground surface.

Table 1: Soil parameters for Munich subway tunnel

Type of soil		Gravel	Silicon clay	Sand
E (MPa)	E_1	120.0	90.0	120.0
	E_2	180	120	180
Unit weight KN/m^3	γ	22.5	21.0	22.5
	γ'	13.0	11.0	11.0
Poisson's ratio ν		0.35	0.15	0.35
Cohesion C (KPa)		4.0	30.0	4.0
Friction angle ϕ (°)		37.5	20.0	37.5
Stress relief factor α		0.2	0.4	0.4

Table 2: Construction steps and compressive strength of shotcrete lining

loading	Compressive strength MPa				
	Shotcrete lining				Concrete Lining
	IB Track		OB Track		
	Cr.	Inv.	Cr.	Inv.	
2	7.5	-	-	-	-
3	15.0	7.5	-	-	-
4	20.0	20.0	7.5	-	-
5	20.0	20.0	15.0	7.5	-
6	25.0	25.0	25.0	25.0	-
7 to 11	25.0	25.0	25.0	25.0	25.0

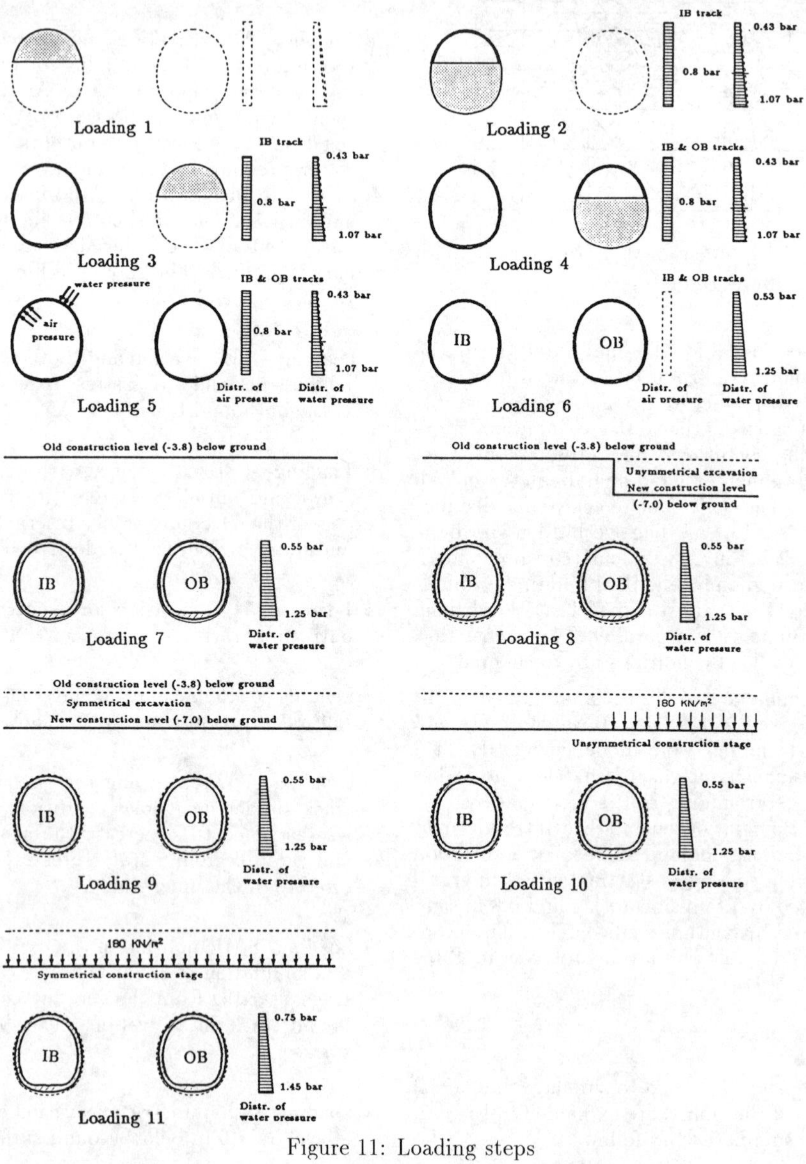

Figure 11: Loading steps

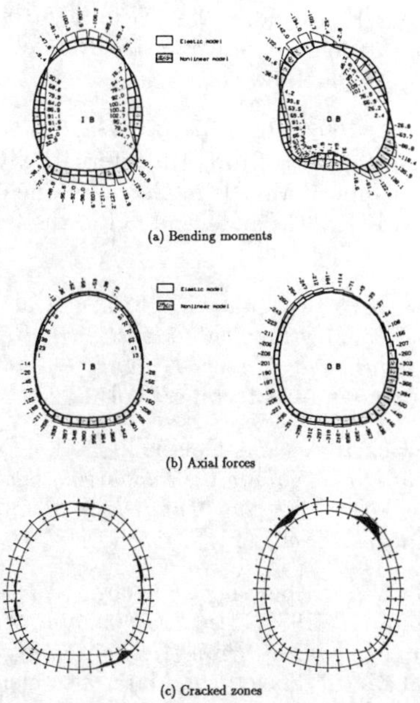

Figure 12: Internal forces and cracked zones in the concrete lining after loading 10.

Table 3: Parameters used for $\sigma - \epsilon$ curve for shotcrete and concrete in compression

Type	ϵ_1	ϵ_u	E_d
Shotcrete	0.005	0.01	0.0
concrete	0.002	0.0035	0.0

- Loading -10- Unsymmetrical construction stage of a new building starting from the vertical axis of the outbound track with assumed building loads of 180 KN/m^2.
- Loading -11- Symmetric construction stage of a new building. The ground water table is increased to flood level.

The construction steps of the shotcrete safety lining and of the concrete final lining and the development of shotcrete strength according to the progress of excavations are explained in Table 2, and the parameters of the stress-strain curve for shotcrete are given in Table 3.

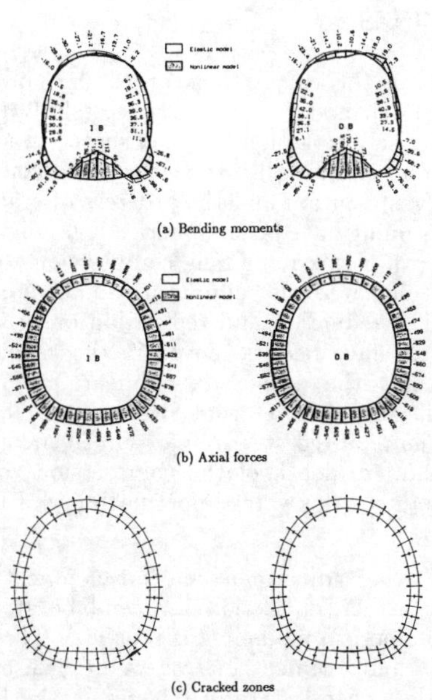

Figure 13: Internal forces and cracked zones in the concrete lining after loading 11.

5.3 Numerical results

Cracks in the concrete final lining of the outbound track started to develop in loading 8 due to unsymmetrical surface excavation. After the symmetric excavation stage, a reverse action could be observed and the cracks in the inbound track started to close again, while new cracks developed in the outbound track. The worst loading stage for the final lining is loading 10, the unsymmetrical construction stage. Under these conditions the final lining is not waterproof. Because of the temporary nature of this load case, the cracks in the final lining can be ignored. In the following load case these cracks are closed again; the minimum compression zone is always higher than the required 15 cm, which proves the water impermeability of the final lining. The internal forces resulting from the nonlinear analysis and the cracked zones are shown in Figs. 12 and 13.

CONCLUSIONS

The results of the calculations prove the ability of the proposed model to predict successfully the behavior of the interaction between shotcrete and concrete linings under all investigated load conditions. The distribution and the progress of cracks in the inner lining have been numerically predicted and the depth of the remaining compression zone has been calculated all round the tunnel cross-sections. The reduction and redistribution of the linings forces have been achieved in the calculations by using the proposed nonlinear material model. The resulting internal forces are used to calculate the required steel reinforcement area to keep a minimum depth of the compression zone which guarantees the water impermeability of the tunnels.

For further adjustment and development of the model, further interaction between laboratory, field, and numerical modelling is required. Three-dimensional finite element analysis would be useful to define the general state of stresses in the lining near the tunnel face. Extensive experimental work is required to define the complete behavior of shotcrete under different loading conditions, such as compression, tension, and creep. The results of these tests must be formulated for numerical use.

REFERENCES

1. Adachi T. 1992. *Simulation of Fenner-Pacher curve in NATM.* Proc. 4th International Symposium on Numerical Models in Geomechanics, held in Swansea,U.K., G.Pande and S.Pietruszczak, eds., Vol 2, 663-673.

2. Aydan Ö., Sezaki M., Kawamoto T. 1992. *Mechanical and Numerical modelling of shotcrete.* Proc. 4th Internatl. Sympos. on Numer. Models in Geomech., held in Swansea,U.K., G.Pande and S.Pietruszczak, eds., Vol 2, 757-764.

3. CEB-FIP Model Code 1990, Final Draft. 1991. Bulletin D'Information No. 203 Chapter 2.

4. Gullan G.T. 1975. *Shotcrete for Tunnel Lining.* Tunnels & Tunnelling, Vol. 22 37-47.

5. Kusterle W. 1985. *Fruhfestigkeiten des Spritzbeton.* Berichtsband der Internationalen Fachtagung Spritzbetontechnologie. Editor: W.Lukas, Innsbruck.

6. Moussa A. 1993. *Finite Element modelling for shotcrete in Tunnelling.* Ph.D. thesis, Innsbruck University, Austria.

7. Nilson A. H. et. al. 1982. *Finite Element Analysis of Reinforced Concrete.* State of Art Report, ASCE, New York.

8. Pöttler R. 1992. *Influence of Rock Bolts on shotcrete Stresses.* Proc. 4th Internatl. Sympos. on Numer. Models in Geomech., held in Swansea,U.K., G.Pande and S.Pietruszczak, eds., Vol 2, 739-746.

9. Pöttler R., Swoboda G. 1986. *Coupled Beam-Boundary-Element Model (BE-BEM) for Analysis of Under Ground Openings.* Computers and Geotechnics, Vol. 2, 239-256.

10. Rabcewicz L. V. and Golser J. 1973. *Principles of Dimensioning the Supporting System for the "New Austrian Tunnelling Method".* Water Power 88-93.

11. Sezaki M., Kibe T., Ichikawa Y. and Kawamoto T. 1989. *An Experimental Study on the Mechanical Properties of Shotcrete.* Journal of the Society of Materials science, Japan, Vol. 38, no. 434, 1336-1340.

12. Sezaki M., Aydan Ö., Kawata T., Moussa A. and Swoboda G. 1992. *Numerical modelling for the representation of shotcrete hardening and face advance of tunnels excavated by bench excavation method.* Proc. 4th Internatl. Sympos. on Numer. Models in Geomech., held in Swansea,U.K., G.Pande and S.Pietruszczak, eds., Vol 2, 707-716.

13. Swoboda G. and Moussa A. 1992. *Numerical modelling of shotcrete in tunnelling.* Proc. 4th Internatl. Sympos. on Numer. Models in Geomech., held in Swansea,U.K., G.Pande and S.Pietruszczak, eds., Vol 2, 717-727.

14. Swoboda G. 1990. *Numerical Modelling of Tunnels.* in C. S. Desai and G. Gioda: Numerical Methods and Constitutive Modelling in Geomechanics, Courses and Lectures, No.311, 277-318 Springer-Verlag Wien.

15. Swoboda G. 1979. *Finite Element Analysis of The New Austrian Tunneling Method.* Third International Conference on Numerical Methods in Geomechanics, Aachen, 581-586.

16. Swoboda G. 1993 *Programmsystem FINAL - Finite Element Analysis Program for Linear and Nonlinear Structures.* Ver 6.6, Print University of Innsbruck.

Tunnelling and Ground Conditions, Abdel Salam (ed.) © 1994 Balkema, Rotterdam, ISBN 90 5410 363 9

Development of steel diaphragm wall method

Kazuyuki Tazaki & Noriyuki Hirosawa
Nippon Steel Corporation, Japan

Hiroshi Furube
Hazama Corporation, Japan

Akira Nakazawa
Shimizu Corporation, Japan

ABSTRACT: In overpopulated cities in Japan, the construction of large underground structures at great depth is inevitable for better use of limited space available. The steel diaphragm wall method is a method developed to enable the construction of such underground structures by the excavation method. This method, featuring space saving and manpower saving, has many similar points with the slurry wall system, except the employment of steel products of special shapes instead of reinforcing bars. The present paper outlines this method, describes the results of performance evaluation tests, and reports the results of its application to a practical project.

1. INTRODUCTION

In Japan, the cast-in-place concrete underground walls have been widely used for underground construction at great depth. In this method, however, the wall thickness must be increased with increasing depth, resulting in increased workload for processing of reinforcing bars. In Japan's large cities, the area of construction site is restricted. To utilize the available underground space more effectively, needs for reduced wall thickness and new construction methods, such as prefab method, which can eliminate the transportation and erection of large frames of reinforcement have become increasingly stronger. Moreover, the lack of labor required for proceeding of reinforcement is a serious problems in Japan.

For effective utilization of the underground at great depth, the development of a construction method in which the thickness of retaining walls can be minimized to ensure the maximum utilization of available space is required.

To solve these problems, a construction method, called "steel diaphragm wall method", which can reduce the wall thickness and achieve manpower and space saving by using high-strength prefabricated members has been devleoped. Success has been achieved in the practical application of this method through various performance evaluation tests.

2. OUTLINE OF STEEL DIAPHRAGM WALL METHOD

The steel diaphragm wall method is a kind of underground diaphragm wall method. It differs from the slurry wall system in that special steel members (NS-BOX produced by Nippon Steel) are used as structural members instead of reinforced concrete. In this method, NS-BOX members are erected and the space around them is filled with concrete or a slurry-cement mixture.

NS-BOX members can be classified into several types depending on the design and construction conditions. The present paper describes the GH-R type developed for the construction of underground diaphragm wall at great depth.

The structures of GH-R type are shown in Fig. 1 and their uses are given in Table 1.

GH-R type is produced by rolling the flange with large joint and freely combining it with web element.

As described above, NS-BOX is characterized in that joints are provided at the ends of flanges of H-section and the continuous wall structure is formed by connecting the members with each other.

3. APPLICATION AND FEATURES OF STEEL DIAPHRAGM WALL METHOD

The steel diaphragm wall method is applied to the construction of temporary plus permanent underground walls for underground roads and stations, start and arrival shafts in the shield tunneling method, etc. This method is particularly advantageous for underground construction at great depth in urban areas.

1) Because of high quality reliability, the steel diaphragm walls can be used as permanent underground walls.

2) As continuous thin walls with high strength can be formed (wall thickness: 1/2~2/3 of concrete walls), the steel diaphragm wall method is advantageous for construction in limited space (Fig. 2).

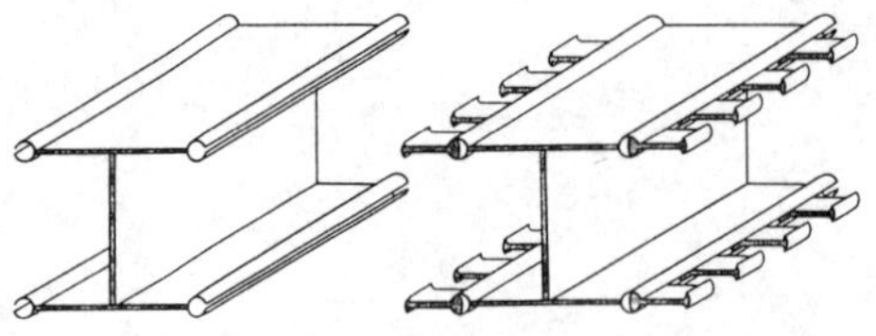

Fig. 1. Structure of GH-R

Table 1. Users of GH-R

Item		GH-R
	Design method	Steel structural design method
	Applicable depth	Up to 80m
Dimensions (mm)	Member height	350 - 1,300
	Effective width	800 - 900 (Standard : 850)
	Length	3,000 - 20,000
Fitting allowance (mm)	In the wall direction	40 - 50
	In the directoin at right angles to the whe wall	15 - 20
	Material (JIS)	Equivalent to SS400,SM400,SM490

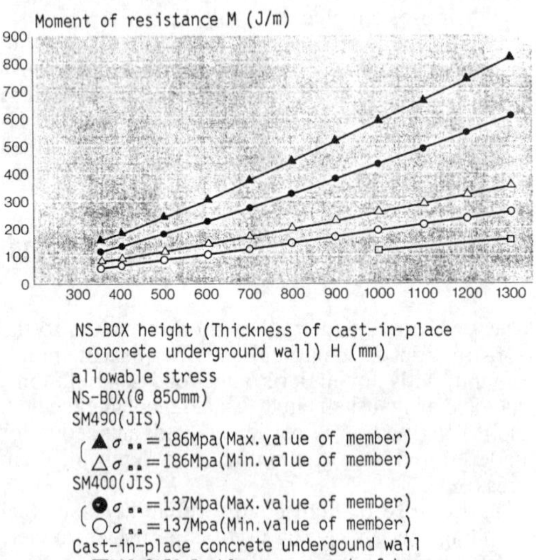

Fig. 2. Moment of resistance of GH-R

3) Due to heavy section performance, the workload for timbering can be reduced, resulting in shorter construction term.

4) The prefabrication can be reduced the number of skilled workers, such as welders, and make it possible to achieve manpower saving at site.

5) As NS-BOX members are shop-assembled products, the reinforcement processing yard is not required at site. As the thin diaphragm walls are built, the number of slurry disposal pits can be reduced, resulting in saving of space at construction site. As the amount of slurry to be disposed is reduced, the amount of treated industrial waste can be reduced.

6) As the members to be erected are small in size, their handling is easy. NS-BOX is suitable for the construction of diaphragm walls under the road where the height of underground space is restricted.

4. DESIGN METHOD

The section force to be produced with respect to the earth and water pressures is calculated by the conventional method or elasto-plastic method in the same manner as for the retaining wall.

For GH-R type, the stress is calculated in accordance with the conventional steel structural design method. As a rule, it is assumed that the stress is borne by the steel structure only and concrete does not bear the stress at all.

An example of connection between the NS-BOX wall and the floor concrete, the inside concrete wall is shown in Fig. 3. The connection between the inside concrete wall and the base concrete is accomplished by using reinforcing bars or studs.

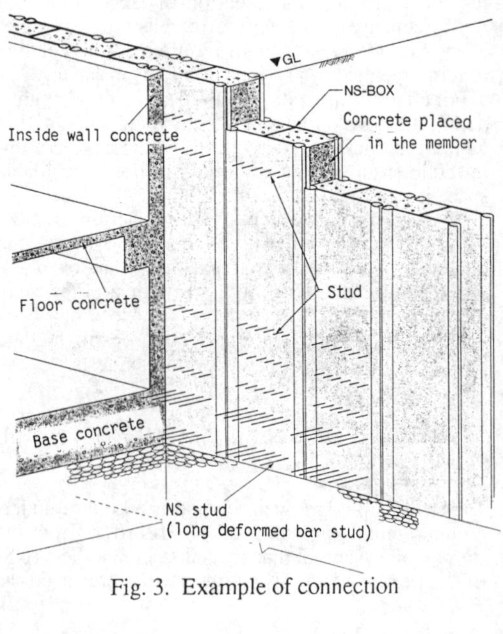

Fig. 3. Example of connection

5. CONSTRUCTION METHOD

NS-BOX is erected and concrete or a slurry-cement mixture is placed, after excavation in the similar manner as that in the slurry wall system. A general construction procedure for steel diaphragm wall is shown in Fig. 7.

6. RESULTS OF PERFORMANCE EVALUATION TEST (CONCRETE PLACING TEST)

6.1 Purpose

For permanent utilization of the steel diaphragm wall, it is necessary to reliably cut off water by placing homogenous concrete around the steel diaphragm wall members. The purposes of concrete placing test are to establish the concrete property which ensures an excellent filling performance, to determine the workability, filling performance of the concrete and the strength of concrete, and to check the appropriateness of the sizes of the steel diaphragm wall, including the slit width, joint shape and the shape of opening in the web.

6.2 Test method

This test is schematically shown in Fig. 4. The test specimen (steel diaphragm wall member) was installed in a form which was used as a substitute for the trench. Concrete was placed from a tremie pipe into the form filled with water at a rate of 5 m/hr and the concrete placing condition was checked through the acrylic boards installed at the both sides of the form. After curing, the form was removed, the cores were drilled and the specimen was cut. Moreover, the condition of concrete at the joint after hardening was checked.

The arrangement of the test specimen and the top view of the form used for the test are shown in Fig. 5. Openings were provided by intermittently arranging the joints to connect the GH-R members, and concrete was placed in such a manner that it flowed out from the inside of the member to the outside (cover). The concrete property is shown in Table 2.

6.3 Results of test

The top surface of the placing concrete was raised nearly in a horizontal state. Concrete could be placed with a maximum difference of about 1.0 cm (flow gradient: 1/9~1/50) in the height of top surface. The joints on the compression and tension sides were satisfactorily filled with concrete. The separation of aggregate and mortar or poor concrete placing did not occur (Photo 1). Compressive strength test revealed that the compressive strength of the core was more than 40 MPa. In a word, concrete of homogeneous high quality could be placed.

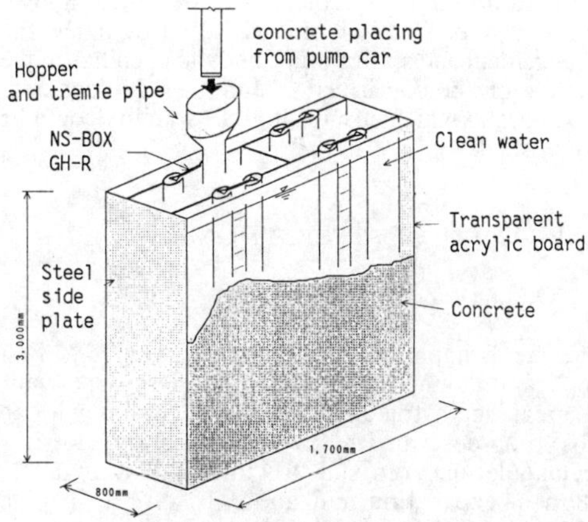

Fig. 4. Schematic drawing of test

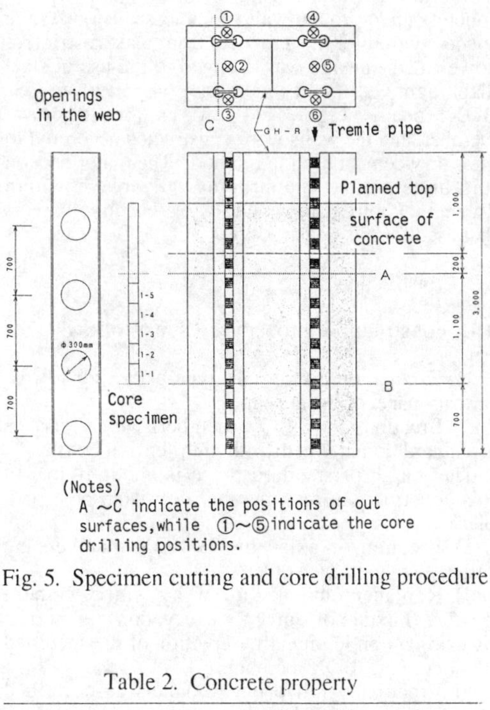

(Notes)
A ~C indicate the positions of out surfaces,while ①~⑤indicate the core drilling positions.

Fig. 5. Specimen cutting and core drilling procedure

Table 2. Concrete property

Kind of cement	Max. aggregate size (mm)	Range of slunp flow (cm)	Range of air amoumt (%)	Water cement ratio W/C(%)	Fine aggregate ratio s/a(%)
BB	20	55±5	±1	39.6	51.4

Quantity of material per unit volume of concrete (kg/m³)

Water W	Cement C	Fine dggregate S	Coarse aggregate G	High-performance AE water-reducing agent
182	460	831	830	8.28

Based on the results of the test described above, the size specification of GH-R, particularly the arrangement and size of joints and the openings in the web, could be confirmed. Moreover, the concrete mixing ratio which ensures high flowability could be established.

7. EXAMPLE OF APPLICATION

7.1 Outline

The steel diaphragm wall method found its first application to the construction of underground walls at great depth for an underground parking lot in Tokyo as described below. This parking lot is a rectangular underground structure of 49 m in depth, 38 m in excavation and about 10 x 12 m in plan which was built under an office building as shown in Fig. 6.

At first, this structure was designed with temporary earth-retaining wall by cast-in-place concrete underground wall (thickness: 1 m). As the space available for construction was restricted, however, the design was changed to the use of steel diaphragm wall (wall thickness: 64 cm) using NS-BOX (beam height: 40 cm) as the structural member, and the walls were constructed according to the slurry cement mixing method. The major data on this underground diaphragm wall are shown in Table 3. Photo 2 shows GH-R of the same type as that used for this construction.

7.2 Construction

The construction procedure is as follows (see Fig. 7).

1) Excavation by a rotary type excavator for the primary panel of linear shape

2) Erection of NS-BOX members which serve as a standard, joining them with high-tension bolts

The length of standard piece is about 10 m. In this construction, one member consisted of 4 or 5 pieces.

3) Erection of adjacent NS-BOX members by fitting them with each other

4) Replacement of slurry by slurry-cement mixture (mixture of slurry for excavation and mortar) by using a tremie pipe after erection of the specified number of NS-BOX members

5) Erection of all primary panels as described above

6) Excavation for the secondly panel in the corner

7) Erection of NS-BOX member in the corner and replacement of slurry by slurry-cement mixture by means of the tremie pipe

8) Erection of all panels by repeating the steps described above

Compared with the conventional cast-in-place concrete underground wall, the steel diaphragm wall is characterized in that steel members with joint (NS-BOX) are erected. When erecting NS-BOX, special attention was directed to the verticality of NS-BOX

Photo 1. Concrete placed in GH-R joint

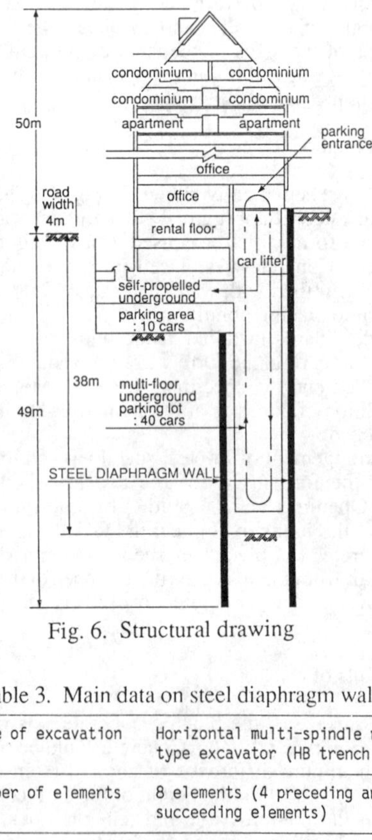

Fig. 6. Structural drawing

Table 3. Main data on steel diaphragm wall

Type of excavation	Horizontal multi-spindle rotary type excavator (HB trench cutter)
Number of elements	8 elements (4 preceding and 4 succeeding elements)
Wall thickness	640mm (NS-BOX : 400mm)
Total wall length	43.6(11.8×2÷10.0×2)m
Design excavation Length	49m
Excavated arca	2,049m²
Quantity of NS-BOX	83set (3,665m)

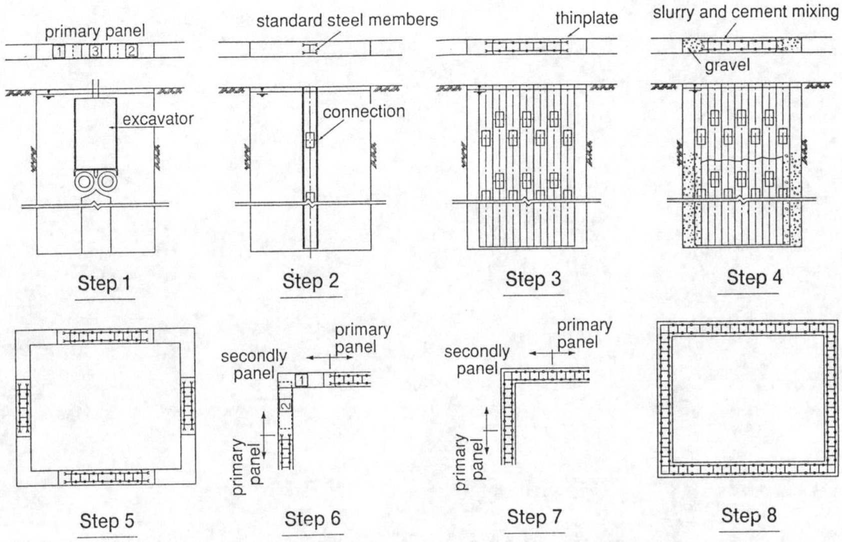

Fig. 7. Construction procedure

Photo 2. GH-R

(field joint in particular), but discrepancy of the joints was not caused.

In this construction work, the average number of working days was 7~8 days when the standard type preceding elements were used. The efficiency of this method stands comparison with that of the conventional concrete wall construction method.

The behavior of the retaining wall at excavation was observed. The results obtained from this measurement were nearly the same as the calculated values (steel structural design). Accordingly, the appropriateness of this design method could be confirmed.

The construction of steel diaphragm wall could be completed smoothly as described above. It was confirmed that the quality of the wall as the retaining wall, particularly the water cut-off property, was very high.

8. CONCLUSION

The outline of the steel diaphragm wall method, the results of performance evaluation tests and the application of this method to a practical project have been described. As a result of performance evaluation tests and application of NS-BOX (GH-R), the following were confirmed.

1) The concrete mixing ratio which ensures a high filling property was established and the filling property at the joint was confirmed.

2) This method was applied to a practical project. As a result, the applicability of this method and the high water cut-off property of the constructed underground wall could be confirmed.

In parallel to the activities for development as described above, the Steel Diaphragm Wall Association was organized in November 1992 in Japan to promote the application of this method and to establish the technology. Nippon Steel and 31 major construction companies are participating in the Association.

The steel diaphragm wall method is a new type of diaphragm wall construction method for application in urban areas, aiming at space and manpower saving. It is expected that this method will find wide application for the construction of underground walls for underground roads, stations, tunnel shafts, etc.

Tunnelling and Ground Conditions, Abdel Salam (ed.) © 1994 Balkema, Rotterdam, ISBN 90 5410 363 9

Development of a faster and longer drive shield tunnelling method

Hiroji Tokoh & Tadashi Suzuki
Tokyo Branch, Okumura Corporation, Japan

Kenji Arakawa
Head Office, Okumura Corporation, Osaka, Japan

ABSTRACT: There are so many roads, railways and other structures built underground as well as on ground surface in large cities of Japan. In order to continue expansion of the social infrastructure such as traffic and utility networks in the congested cities, the space of underground has to be more efficiently utilised than ever before. A construction technology to build very long tunnels is required to this end. Under such a circumstance, we have been tackling research and development of a new shield tunnelling method in which operation of segment erection can be performed while a shield machine advances in soft ground, and eventually a longer tunnel can be bored at least twice faster than conventional methods. In this paper, we report three of new techniques developed to realize the faster and longer drive shield tunnelling method. The primary technique is the hexagon "Honeycomb" segment so named after a unique graphic pattern of the precast concrete lining.

1 DEVELOPMENT OF HONEYCOMB SEGMENT

1.1 General

In order to fast build long tunnels by shield machines, the separation of two operations, namely segment erection and shield advance, should be first reviewed. It is a key point how these two operations are performed simultaneously. In case ordinary segments rectangular in shape are used in such a simultaneous operation, the erected segments are structurally unstable until they have formed a closed circle and are not able to sufficiently resist shove force for the sake of shield advance. Techniques of the rock tunnelling by TBM, in which resistance against shove force relies on the surrounding ground, may be a good reference for the simultaneous operation. From the ground conditions which is generally soft and incompetent in Japan, however, application of such techniques of the rock tunnelling is much limited.

Hexagon is a shape that we first thought of for segments. As shown in **Figure 1**, hexagon segments have a nature to be firmly placed with their 3 sides of a hexagon restricted by other segments beside. They are able to resist shove force for shield advance as soon as being placed. This nature enables a shield machine to advance while segments are being erected. Also, shove force can be resisted by the lining, like the conventional methods. Thus, hexagon segments was considered in this R&D project as the starting point for the said simultaneous operation.

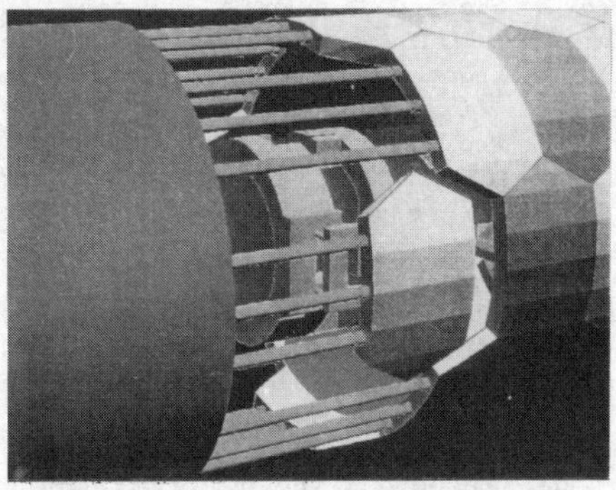

Fig.1 Concept of Honeycomb segments

Stability in structure, which hexagon segments generally have, seems to have been appreciated in European countries from old days, particularly in Germany where some Metro projects were completed using hexagon segments in 1960s. Apart from these examples in the past time, this R&D project started in a firm anticipation that hexagon segments would be utmost suitable for the said simultaneous operation.

Between 1991 and 1993, a series of full scale load tests were conducted to check if the Honeycomb segments could be applicable to the incompetent and changeable ground in Japan. As the R&D project proceeded, several features and benefits that the Honeycomb segments had were clarified. They had sufficient rigidity at their joints. Also they showed the following merits technical and economical in comparison with ordinary rectangular segments: (1) to be particularly suitable for automation in erection owing to their complete identity in size and dimension, (2) to provide tunnel with a so-called smooth bore without any bolts or bolt boxes imposed inside tunnel, which eliminating necessity for a secondary lining whose main purpose is rust or corrosion control of the lining.

1.2 Structure of Honeycomb segment

1.2.1 Shape and joint

Among a variety of shapes generally called as hexagon, such a hexagon was selected for the Honeycomb segment as the regular hexagon is made a little taller with each of six corners kept at 120 degrees. The selection was made in consideration of two technical aspects contradictive each another: the top corner of the hexagon should be narrower from a view point of structural strength; it should be wider for easier handling.

A series of bending tests were carried out, using the Honeycomb segments of the dia 4050mm model as shown in **Figure 2.** Comparison among various type of joints was made. Based on the results of the bending tests, the joints of Honeycomb segment were decided to be as shown in **Figure 3.** Particular considerations were made during the design of joints for the sake of smoothness of the internal tunnel face. Two of each of the through-bolts and the oblique bolts were provided. Also, two pairs of sockets and plugs were provided on each of oblique sides to ensure accuracy and less disalignment in erection.

From a consideration only with regard to stresses developed at joints of any type of segments, there is no doubt that four pieces of segments per ring is ideal. However, the division (number of pieces) of the Honeycomb segments will be six, eight or larger even numbers, taking

Fig.2 Dia 4050mm model joints bending test

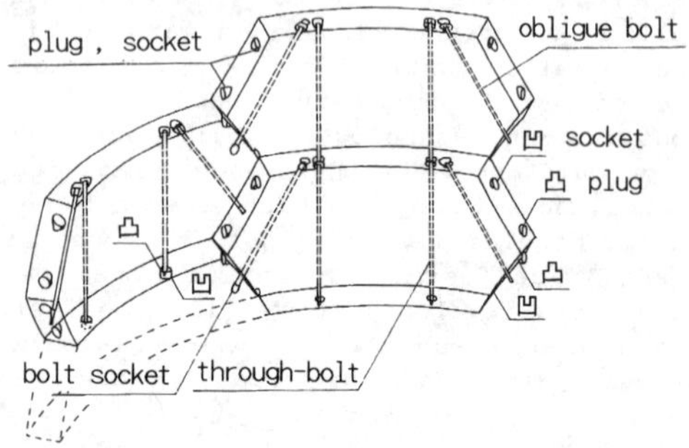

Fig.3 Structure of Honeycomb segments

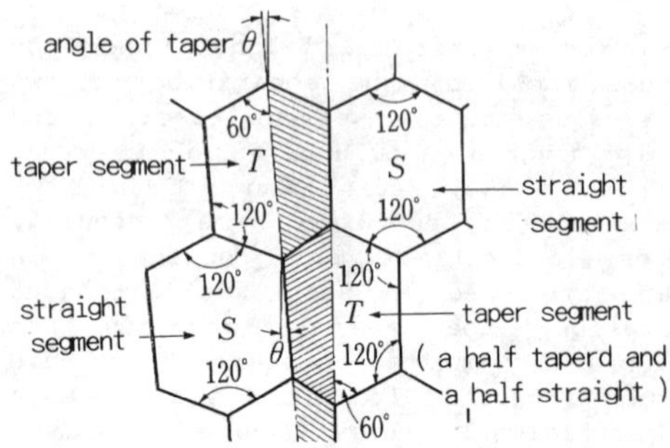

Fig.4 Arrangement of tapered segments

the facility in handling and transportation and applicability to curve alignment into considerations.

1.2.2 Taper segment

As each of Honeycomb segments is ahead of other segments beside by one half of its width when it is erected. The front end of segment lining is always an uneven line. Therefore, design of taper segment is not easy but it is very important since most of shield tunnels included curves on their alignment in Japan.

It was found during the design that taper could be well provided by making only a half section of the Honeycomb segment tapered as shown in **Figure 4.** The miniature model of the lining confirmed that the design concept of taper segments was correct in details.

1.3 Stress and deformation

Being different from ordinary rectangular segments, the bending moment (Mj), which is acting at the joints of the Honeycomb segments, is divided into two moments (M1) and (M2) as shown in **Figure 5.** The moment (M1) causes the gap between segments at the joints while the moment (M2) forces the segments to rotate and the top corner of the segment to bow inside. **Figure 6** shows such movements. The bolt resists the moment (M1) as a tensile member and restrains the gap between two segments. Under a normal loading situation, the moment (M2), specific to the Honeycomb segment, is overcome by the friction developed between two segments and thus such rotational movement of the Honeycomb segments stays negligible. **Figure 7** is the result of the three dimensional FEM analysis to show how the Honeycomb segment is deformed particularly around the oblique sides. The stress caused by the moments (M1) and (M2) are so transmitted as to twist the oblique sides and to slightly warp the part between the oblique sides outward.

With such friction and division of stress at the joints, which are related to the unique shape of Honeycomb segment, the joints of Honeycomb segment showed a sufficient rigidity in the dia 4050mm bending test and the following dia 5300mm full scale load tests.

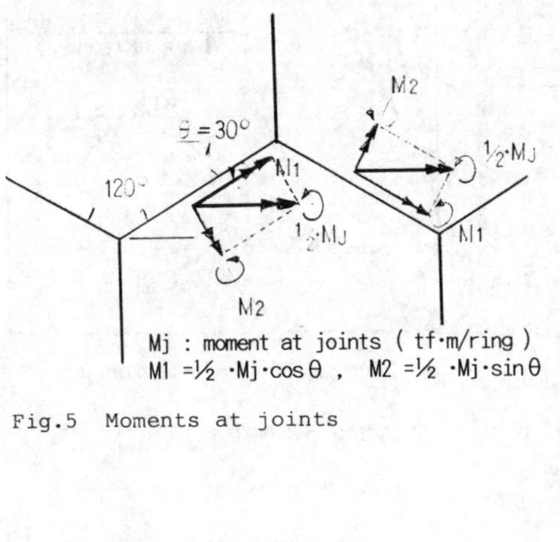

Fig.5 Moments at joints

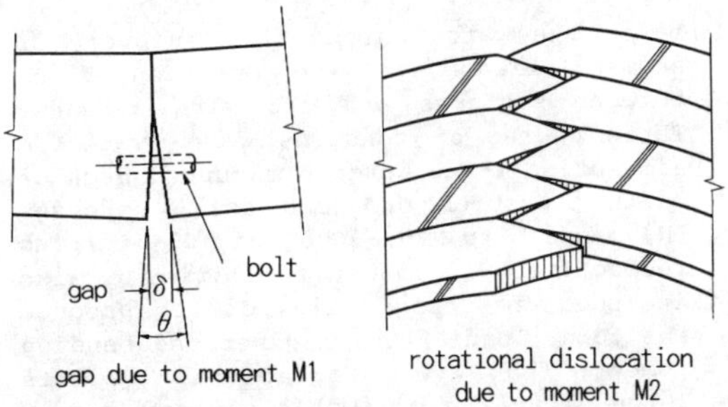

Fig.6 Dislocation at joints

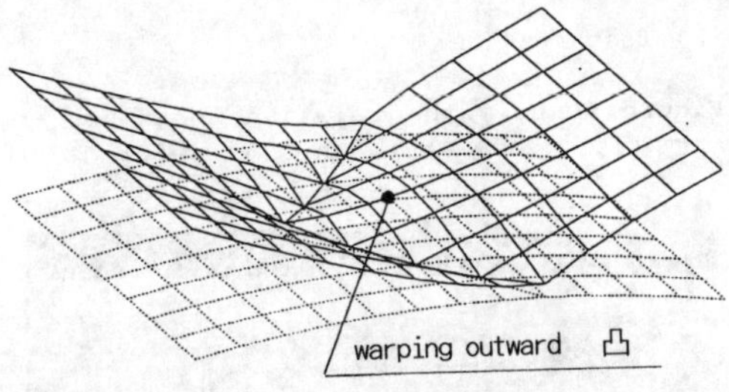

Fig.7 Deformation around oblique sides

1.4 Dia 5300mm full scale load tests

Between January and March 1993, the full scale load tests were carried out. The Honeycomb segments o.d. 5300mm, thickness 250mm, width 1000mm, which formed one ring by 6 pieces, were used. The diameter of 5300mm was selected in anticipation of a future application to the small diameter Metro projects in Japan. In these tests, the behaviours of the Honeycomb segments were verified under permanent and temporary loading conditions.

1.4.1 Segment rings load test

(1) Loading facilities and load control

Purpose of this test was to verify the stress and deformation at the joints on oblique sides of the hexagon segments and the behaviour of the whole lining. **Figure 8** shows the segments sample which included two of complete rings, and was erected upright in a ring and the loading facilities which included 4 number of prestressed wire strands to tie up around the segment rings. Tension of the wire strands was to accurately simulate the earth pressure which would act onto the tunnel lining. The tension was transmitted to the segment rings through thirty six (36) number of dia 30mm round bars. They stood along the periphery of segment rings to omit friction force between segments and the wire strands. Three degree of tensions were available simulating three earth pressures. Each of earth pressures produced axial stresses (N) equal to 40, 60 and 80 tf/ring respectively. The facilities also included prestressed tie rods to produce the point load PV which generated bending moment. The PV was varied until moments larger than the allowable maximum moment (Ma') were developed at the joints on the oblique sides.

(2) Test results

Figure 9 shows the relation, obtained in the load test, between the deformation

Fig.8 Dia 5300mm segment rings load test

(mm) of the segment ring in diametre and the bending moment (tf-m/ring) along with the PV at the points No. 3 and 7. The axial stress (N) was taken as a parameter for the relation chart. From the chart, the deformation was as small as 7 to 9 mm i.e. about 1/1500 of the diameter of lining even when the allowable minus bending moment (Ma) was acting at the points 3 and 7.

Figure 10 shows the relation between the gap (mm) of segments and the bending moment (Mo) at the joints which caused the gap. The figure indicated that the gap became smaller as the axial stress (N) was getting larger. The gap measured 0.15 to 0.20mm almost negligible even when the allowable bending moment was acting at the joints. The deformation of the top corner of the segment toward the tunnel center was also negligible.

Deformation under a free load state, where the tension of the wire strands was released, was also measured as small as 3.2mm. This meant the Honeycomb segments could be erected inside the tail of shield machine without any difficulty.

Measurement of strains of the concrete and reinforcing bar indicated that there was no stress concentration undue at any part of the segment and that the measured strains reasonably met the bending moment which acted at the joints.

It was also confirmed visually that there was no crack or damage on the

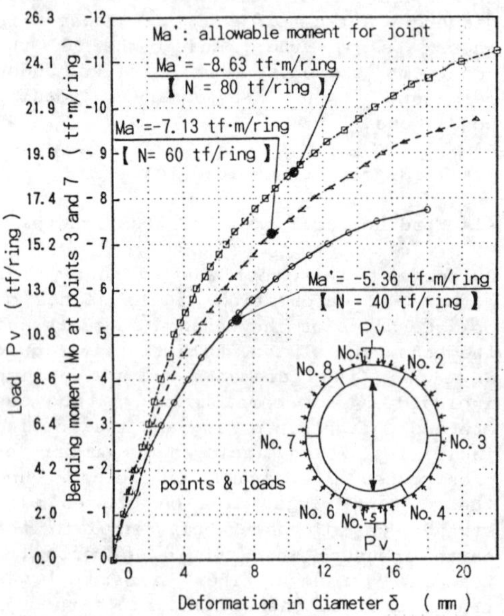

Fig.9 Deformation - moment relation

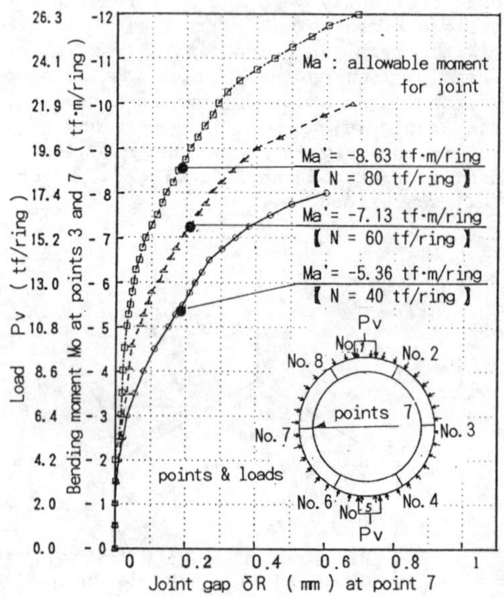

Fig.10 Joint gap - moment relation

segments even under the load exceeding the allowable one. In comparison with ordinary rectangular segments, the Honeycomb segments had a sufficient strength owing to their shape and configuration with no sectional voids by bolt boxes.

1.4.2 Shove force loading test

(1) Loading facilities and load control

The Honeycomb segments have to resist the shove force for shield advance in such a state that one of them is always ahead of others by its half of width as shown in Figure 1. Also it should be expected that the shove force eccentricly acts on the segment in actual tunnelling. Even in such circumstances, the Honeycomb segment is required to ensure stability of the tunnel lining and not to cause any hindrance for shield drive. Therefore, the shove force loading test was carried out using the three shove jacks 170 tonne per each as shown in **Figure 11.** The jacking force was loaded 25mm off the center line on the side face to simulate such a state of eccentric shoving.

(2) Test results

Figure 12 indicates how the segment was deformed and dislocated. Measuring only about 1.0 mm of dislocation, it was proved that the erected Honeycomb segments well resisted shove force. Also from the strain of the concrete, the Honeycomb segment always stayed in an acceptable state of stress.

Fig.11 Dia 5300mm shove force load test

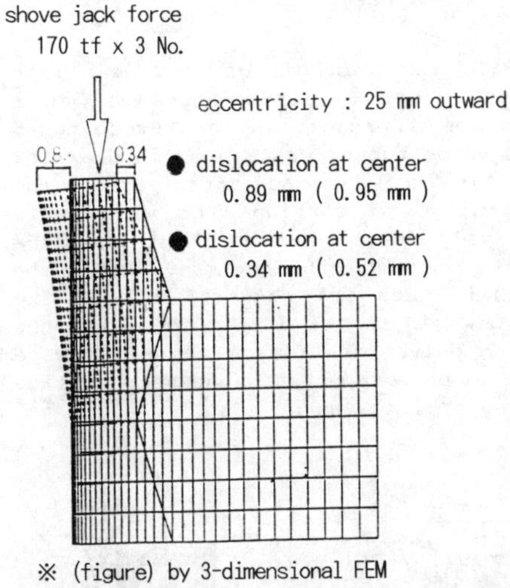

Fig.12 Results of shove force load test

2. DEVELOPMENT OF LONG WEAR LIFE BIT

Wear of cutting bits is a large concern in shield tunnelling in a distance and exchange of worn bits at the tunnel face is a big problem. The exchange has often been performed by men's hands after providing an additional shaft or ground treatment at the tunnel face, which is costly and time consuming. In order to achieve the faster and longer shield tunnelling, required is a technique to make exchange of bits easy or a technique to dispense with such exchange.

2.1 Outline of a new long wear life bit

In this R&D project, a new type of cutting bit was developed. A system to detect wears is incorporated in the bit. As the wear becomes serious with excavation slowing down, the bit is raised outward by a hydraulic jack to correspond the wear (mm). Mechanism of the bit and raising system is outlined in **Figure 13.** The bit has carbide tips embedded at multiple stages from the top end to the below. When the bit is worn down to the first wear line, the hydraulic jack installed at the cutterhead arm raises the bit to recover the height of wear and the next new tips start to cut the ground. This system is generally simple in mechanism and is much useful for safety by dispensing with bit exchange at the tunnel face. Thus the bit raising system ensures long wear life of the bit and saves large cost and time which are inevitable for conventional bit exchange.

2.2 Trial application

In 1991, the said bit of long wear life was used for its performance test in a sewer tunnel project in Nara Prefecture, Japan whose tunnel was 2130mm in diameter and 1116.3m long. As anticipated, the first stage of carbide tips which were relatively short were worn out when the tunnel reached 1000 meters. Then the hydraulic jack was actuated to put the bit forward. In the inspection after the 1116.3m drive, the first stage tips were found to be worn out and the second stage tips were little worn.

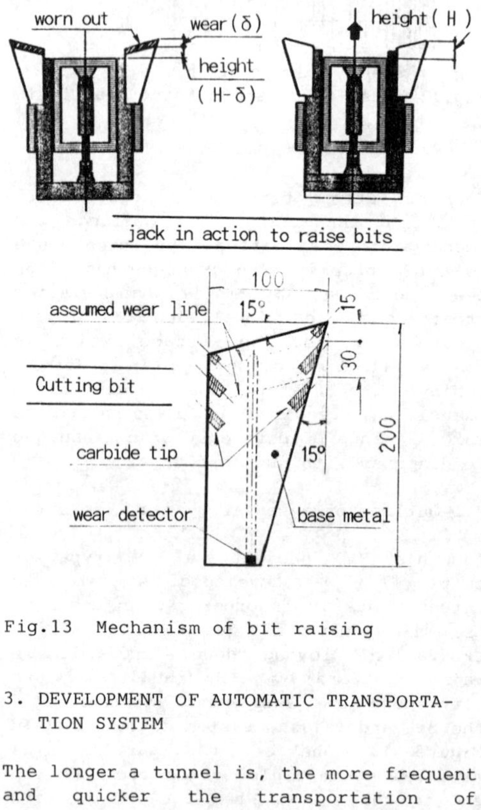

Fig.13 Mechanism of bit raising

3. DEVELOPMENT OF AUTOMATIC TRANSPORTATION SYSTEM

The longer a tunnel is, the more frequent and quicker the transportation of excavated soils and segments should be. In this R&D project, a new automatic system has been developed to safely and efficiently perform the transportation of excavated soils and segments. As shown in **Figure 14**, the transportation was highly automated using a non-operator battery train. The automatic transportation system was tested for efficiency in a water supply project in Nara Prefecture, Japan from 1992 to 1993, whose tunnel was o.d. 2910mm and 1080m long.

3.1 Major components of the system

The followings feature the system : (1) constant indication of position of the train, (2) speed down and up at the sharp curve sections and (3) auto stop at the tunnel face and the shaft entrance. Several control sensors are installed to check up linkage among train and cars and actuation of brakes. An immergency stop system is also incorporated for a time when any obstacle is found on the track or the derailment of train occurs.

In addition, the back and forth movement of mucking cars, which usually depends on operator's perception, was tried to be automated. Eventually, the auto-controlled mucking cars smoothly received the soils from belt conveyor. This operation was based on the collected data such as volume of loaded soil, position of mucking cars and penetration speed of shield machine. As another benefit of this automatic system, the volume of excavated soil was monitored in a real time.

3.2 Trial application

The automatic transportation system worked well to complete the 1080m long EPB shield tunnelling work in which 4 number of sharp curves each with a radius ranging 15 to 40m were included on the route. The operation proceeded smoothly except in a short section where extremely hard clay was excavated, which came down to the mucking cars in the lump 20 to 30cm.

4. SUMMARY

4.1 The hexagon Honeycomb segments were developed to realise the simultaneous operation of segment erection and shield advance. Their shape and structure were proved to be reasonably designed and to well negotiate the curve alignment. Through the various loading tests, they were found to be rigid, strong and good for cost saving. It is assumed that the Honeycomb segments will be more widely adopted in shield tunnelling in future.

4.2 The long wear life bit was developed. Multiple stages of carbide tips were embedded. A system to detect the wear of the bit was incorporated. The bit was raised outward by a hydraulic jack to correspond the progress in wears. Bit exchange at the tunnel face can be eliminated. A great deal of cost saving and acceleration in construction time

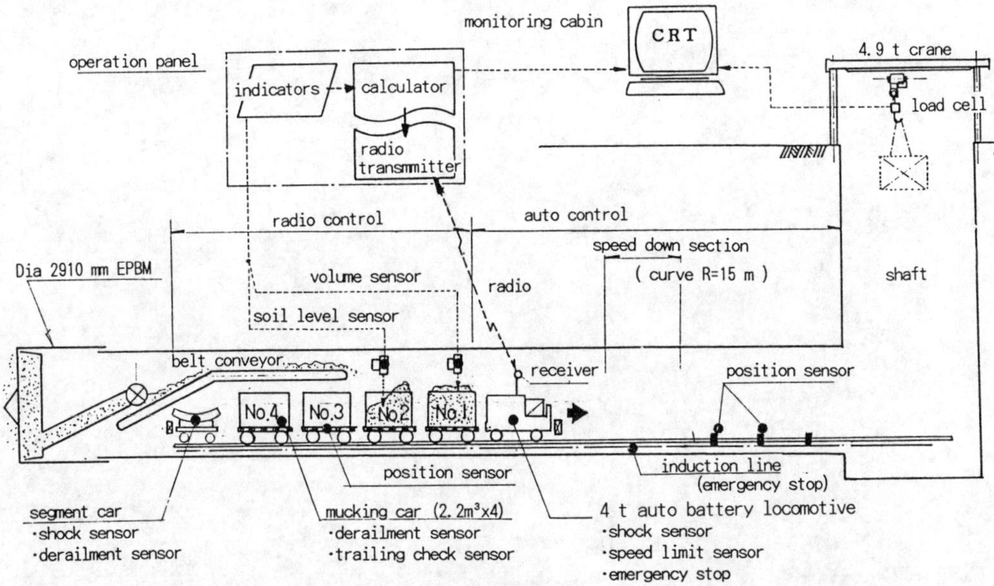

Fig.14 Outline of automatic transportation system

will be ensured in construction of long tunnels. Structure and mechanism of the bit will be further improved to cope with various ground conditions through more trial applications in future.

4.3 The automatic transportation system was developed. Smooth and safe transportation of excavated soils and segments was achieved in the long tunnel work. Shield tunnelling will be improved in respect of man power saving and safety by this system.

5. ACKNOWLEDGEMENT

We sincerely thank Okumura Machinery Corporation and Ishikawajima Kenzai Kogyo Co., Ltd. for their cooperation in this reserch and development. We also wish to express our gratitude to Dr. Minoru Yamamoto, the honorary professor of Tokyo Metropolitan University and others concerned for their valuable advices.

Tunnelling and Ground Conditions, Abdel Salam (ed.) © 1994 Balkema, Rotterdam, ISBN 90 5410 363 9

New tunneling technique, PASS Method

Y.Tsuchiya & S.Kenmochi
Japan Railway Construction Public Corporation, Tokyo, Japan

T.Hara & K.Nakayama
Fujita Corporation, Tokyo, Japan

ABSTRACT: A new technique, the "Pre Arch Shell Support Method", which involves construction of an arched pre-lining ahead of the tunnel face prior to excavation, in order to stabilize the face and restrict surface settlement to an absolute minimum in unconsolidated shallow tunnels in the city regions, has been developed. After field experimentation and testing, this method was employed on an actual city tunnel project.
The results are summarized below:
1) It was possible to drive the tunnel safely since the cutting face was in a stable condition under the umbrella of the pre-lining.
2) Houses along the tunnel route were not affected by the tunnel construction.

1. INTRODUCTION

In recent years, NATM tunneling has been on the increase, even in unconsolidated shallow ground in cities in Japan.

Many difficulties have been encountered however, in the construction of those tunnels, because of problems in ensuring work safely and of minimizing the impact on buildings, utilities and services.

Various techniques such as division of the cutting face, chemical injection, fore-poling etc. have been adopted, although these measures were not really considered to be satisfactory from the point of view of reliability, practicability and economy.

In this paper, we report on the development of a new technique to overcome these difficulties on city tunnel projects. This is the Pre-Arch-Shell-Support (PASS) Method.

2. OUTLINE OF THE METHOD

In the PASS Method a thin mortar lining in the form of an arched shell is cast ahead of the face prior to excavation (pre-lining), enabling the tunnel to be excavated safely, economical and speedily.

The outstanding characteristic of this method, which distinguishes from conventional fore-poling methods, is the support provided by the arched shell structure.

For practical application of the method, the size and range of the pre-lining were determined on the basis of model experiments, numerical analysis and research of practicability as follows:

Transverse coverage: 120° arc of upper section
Depth : 4m from the face
Thickness of liner : 17cm

3. DEVELOPMENT OF TECHNOLOGY

3.1 Sequence of working

The process of forming the pre-lining is as follows:

1. A boom-mounted 5-shaft auger drill is set up in the calculated position using a computer controlled surveying and positioning system.
2. The auger then drills a slot 17cm thick, 81cm wide and 4m deep into the natural ground ahead of the tunnel face.
3. As the auger is gradually withdrawn, mortar is injected through a nozzle in the center of each auger to fill the drilled slot to a depth of 2.5m.
4. To form a continuous arched lining, steps 1 to 3 are repeated laterally overlapping the edge of each segment along the circumference of the tunnel.

The tunnel is driven by repeating this process after every 2m of excavation.

3.2 Pre-lining machine

The following were prerequisites for the

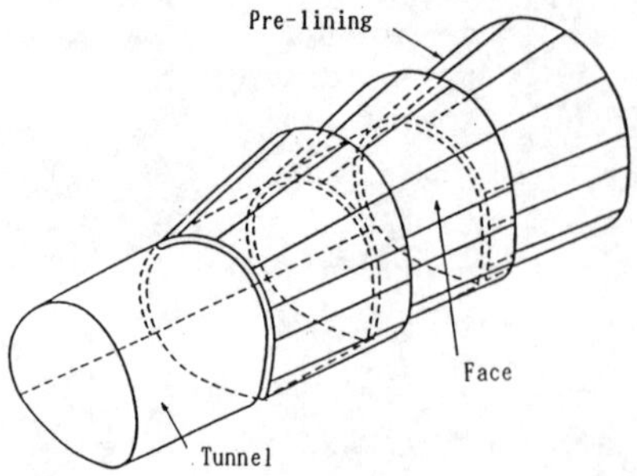

Fig 1. Conceptual illustration of the pre-lining

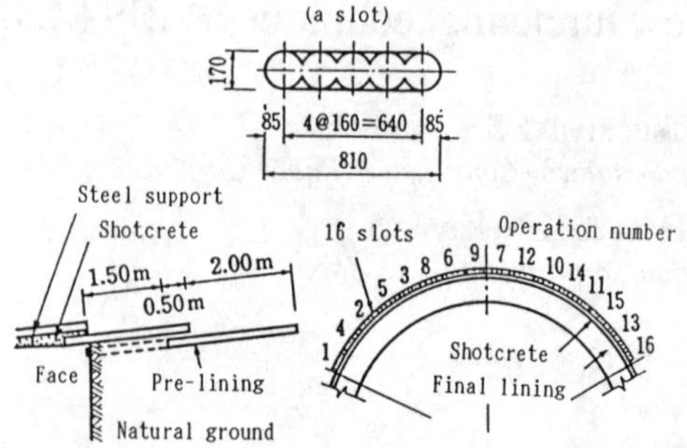

Fig 2. Sequence of the prelining operation

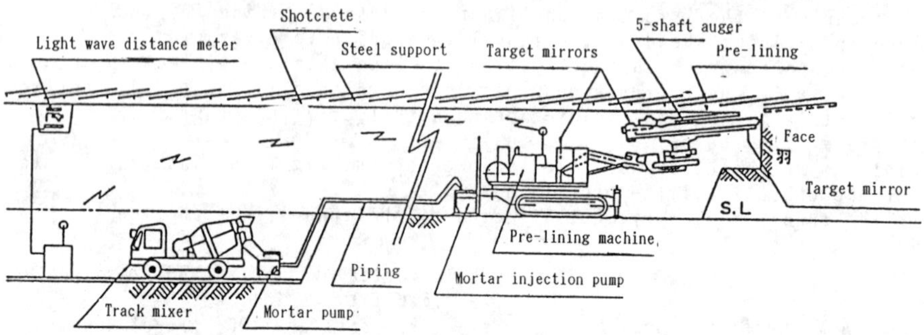

Fig 3. View of the pre-lining operation

pre-lining machine:

1. Ability to form a liner with high accuracy and speed.
2. Applicable to various shapes of tunnel section.
3. Applicable to a double track tunnel without requiring the base machine to be moved.

After detailed research and design, a crawler type machine, with a 5-shaft auger, multi-joint boom, injection pump and automatic control equipment, was manufactured.

3.3 Automated control system

In order to form the pre-lining with high accuracy and speed, the following automated control system was developed:

1. Automated measuring system: Using a special purpose light-wave distance meter with automatic target searching, the true coordinates of the auger are measured.
2. Automated auger setting system: Using a specially developed control program for the multi-joint robot, each joint's adjustment value is calculated, to set the auger in the required position.
3. Automated drilling system: Drilling is executed to the predetermined depth while rotational torque and driving speed are automatically controlled.
4. Automatic injection system: The rate of withdrawal of the auger is computer controled to balance it with the rate of the injection of mortar, to ensure correct filling.

3.4 Material for lining

The following were prerequisites the lining material:

1. Economy, superior pumpability and easy of handling.
2. Equal to shotcrete in early compressive strength.
3. It was necessary that the wet lining material should not flow back out of the slot.

After a number of material tests, a plain mortar was selected.

Table 1. The specification of the mortar

Specified strength	W/C	S/C
80kgf/cm² (24hrs.)	40%	2

Photo 1. The pre-lining machine

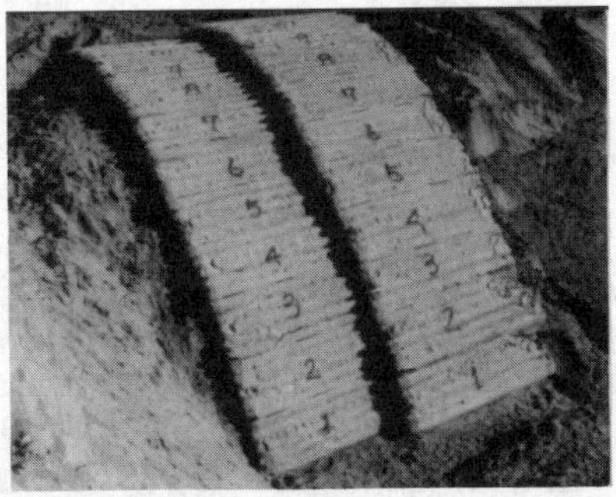

Photo 2. The view of the pre-lining

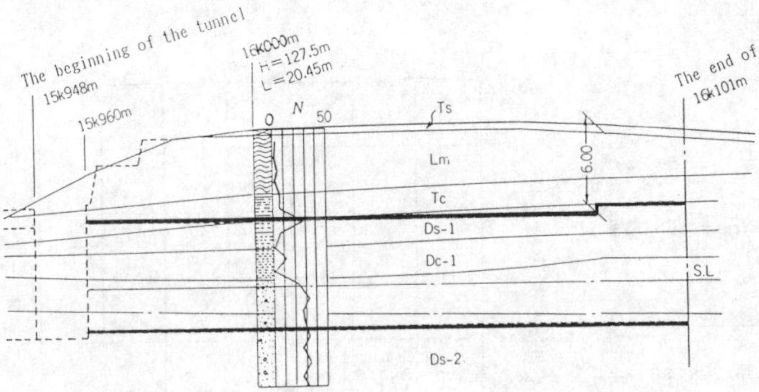

Geological name	Geological symbol
Top soil	Ts
Loam	Lm
Tuffaceous clay	Tc
Silt	Dc
Clayey silt	
Sandy silt	
Sand	Ds

Fig 4. Geological profile of the Katsutadai tunnel

4. FIELD EXPERIMENTAL WORK

A full scale experimental was carried out in a sandy ground slope for the purpose of confirming the pre-lining accuracy, construction speed and reliability of the machinery.

As a result, the PASS method was shown to be suitable for application to an actual tunnel project, as the continuity of the pre-lining was achieved and the construction was completed with no particular problems.

5. FEATURES OF THE METHOD

Distinguishing features of the PASS method are:

1. The pre-lining umbrella creates a safe working environment at the tunnel face.
2. The pre-lining minimizes ground surface settlement and ensure face stability.
3. Large sectional excavation is made possible by the pre-lining, with resultant higher productivity and safety.
4. Entire pre-lining operation is automated.

6. APPLICATION TO A TUNNEL PROJECT

KATSUTADAI tunnel is a 10m diameter railway tunnel, located in a residential suburb of Tokyo.

Many houses are located very close to the tunnel, although the ground cover to the tunnel is only 6～7m.

Moreover, the tunnel overburden is comprised of loam and tuffy clay formation(N-value less than 10). The tunnel passes through a diluvialsand and silty clay formation (N-value10～20). The sandy formation at the upper section is in particular aquifer and not fine.

Given these restrictive conditions, there was concern that the face would be unstable and that the ground surface settlement would result from driving the tunnel.

So, the PASS method was adopted in order to drive the tunnel safely without affecting the houses along the route. The result was as follows:

1. The houses along the tunnel suffered no adverse effects, since the surface settlement was limited to 30mm due to top heading work and 40mm in total.

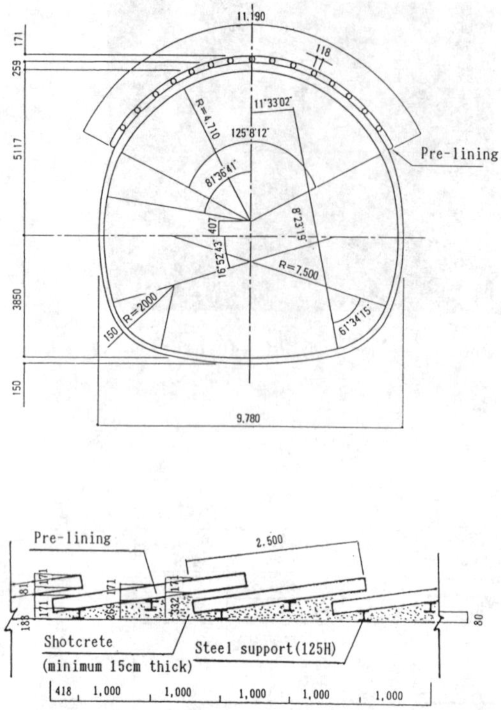

Fig 5. The structural plan of the Katsutadai tunnel

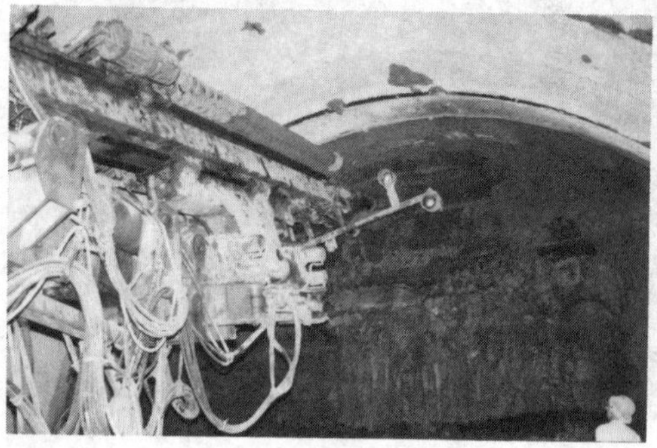

Photo 3. Pre-lining operation in the Katsutadai tunnel

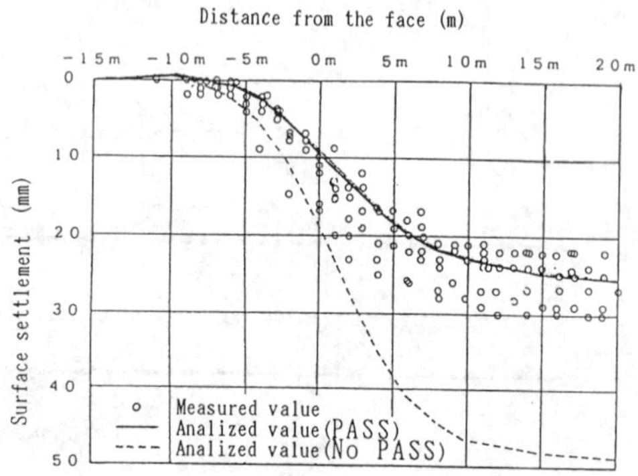

Fig 6. The surface settlement curve

2. The tunnel was driven safely, since the face was kept in a stable condition under the pre-lining umbrella.

3. It is considered that the ground surrounding the tunnel was in an elastic state during tunneling, since the measured surface settlement curve coinsided well with the results of the 3-dimensional elastic FEM analysis.

4. A cycle of pre-lining work, 4m long, was completed in 5～6 hours. This contributed greatly to reduction in the project cost.

The PASS method section, 110m long, successfully broke through in 3.5 months despite the difficult site conditions had many restrictions.

7. FUTURE DEVELOPMENT

The PASS method was awarded the 1992 Technology Development Prize by JSCE(the Japan Society of Civil Engineers). The following tasks are now being tackled in order to make this method more popular.

1. Development of a drilling system for application in sandy gravel and weathered rock etc.

2. Development of a pre-lining machine with the capacity to drill to greater depths, required for larger section tunnels.

3. Reduction of the cycle time for the pre-lining operation.

This method will be applied on a road tunnel project in Karuizawa, a popular summer resort area of Japan, toward the end of 1993, where it will be necessary to stabilize the tunnel face in unconsolidated volcanic ash ground.

The ground conditions are as follows:

1. The whole length of the tunnel route is Asama volcanic ash, composed of a mixture of loam, pumice, sandy soil and cohesive soils etc.

2. The ground cover is 3 to 20m

3. The ground water table is below the proposed tunnel route, however there is the possibility of permeating into the tunnel because of the thin ground cover.

Monitoring and analysis will be performed to confirm the effecttiveness of support of the PASS method.

8. CONCLUSION

By using the PASS method, it is firmly believed that NATM tunneling can achieve considerably wider application in soft ground where the shield tunneling method has primarily been used.

In closing, we would like to express our profound gratitude to Prof. Adachi for his appropriate advice and cooperation.

5 Numerical and experimental modelling of tunnels and underground structures

Tunnelling and Ground Conditions, Abdel Salam (ed.) © 1994 Balkema, Rotterdam, ISBN 90 5410 363 9

Dynamic analysis of tunnels

Sayed Abdel-Salam & Osman Elsayed Shallan
Structural Design Department, Faculty of Engineering, Zagazig University, Egypt
Adel Y. Akl
Department of Structural Engineering, Faculty of Engineering, Cairo University, Egypt

ABSTRACT— Dynamic analysis of tunnels is very complicated, where the equation of motion may be solved by many procedures , also the infinite domain of soil must be represented for the wave propagation ,and the nonlinear properties of soil must be taken into consideration for a good representation. The aim of this research is to develop a new type of infinite element and use it as a numerical tool to study the seismic behavior of tunnels where a comprehensive program with many capabilities is introduced.

1 INTRODUCTION

Tunnels are increased all over the world and also Egypt in recent years. Tunnel analyses are faced by the problem of infinite domain specially for dynamic loads. Roller or fixed supports may be used at enough distance for static analysis but not for dynamic analysis due to wave reflection. In the past few years, research effort has been expanded in developing new types of elements. These elements are the *infinite elements* which are arranged at the boundary of the finite element. There are many types of infinite elements such as [1 to 12] which are different from each other in the number of nodes, mapping functions, or type of the decay term. A new type of infinite element has many advantages is introduced to represent the infinite domain of soil.

Two main procedures are known for solution of dynamic equations of motion, the first is the modal analysis procedure (*frequency domain*) ,and the second is the step by step integration procedure (*time domain*). The modal analysis can solve linear systems only but the step by step integration can solve any arbitrary nonlinear system. The step by step integration can be obtained by different algorithms such as *average acceleration method, Wilson-θ method,* or *Newmark-β method* [13 & 14]. A new accurate algorithm known by *Beta-m method* is used in this research [15].

A parametric study is made on a circular reinforced concrete tunnel which has the same diameter and cross section of the cairo tunnel (phase two). The seismic behavior of tunnel is studied for different depths of soil overlying the tunnel. Two soil models are used in this research, linear model and hyperbolic Duncan model.

2 METHOD OF ANALYSIS

A computer program is introduced for solving static and dynamic problems using finite or coupled finite-infinite elements. In the dynamic analysis, the equation of motion is solved in the time domain using *Beta-m* method. *Duncan* model is introduced into the program to solve the nonlinear soil problems. Many types of infinite element are contained into the program in addition to the proposed type.

2.1 *Equation of Motion due to Earthquake Excitation*

The formulation of the equation of motion of a *MDOF* system under earthquake excitation can be carried out in a matrix form using the equation :-

$$[M]\{\ddot{U}\}+[C]\{\dot{U}\}+[K]\{U\}=-[M]\{\ddot{U}_g\} \quad (1)$$

in which $\{U\}$ is the column vector of nodes displacements relative to the moving base. $\{\dot{U}\}$ and $\{\ddot{U}\}$ are the first and the second derivatives of the displacement vector $\{U\}$ with respect to time. [M], [C] and [K] are the mass matrix, damping matrix and stiffness matrix of the system respectively. In the above equation the value $\ddot{U}_g$ is the component of the ground acceleration in the direction of the corresponding degrees of freedom, so that if the motion of the base in the horizontal direction only , the all degrees of freedom in the same direction have the base acceleration but the acceleration U_g corresponding to the vertical degrees of freedom equal zero [16].

The system of the above equation is solved by using *Beta-3* algorithm. *Beta-m* method is considered as extension of the *Newmark-β method*, where m is the highest derivative included in the relation between last and current time steps. Two algorithms with m = 2 and m = 3 are introduced. It is found that *Beta-3* method is unconditionally stable and yields most accurate results [15].

2.2 *Proposed Infinite Element*

In spite of infinite elements are investigated in many researches , they are not completely defined. The decay term and the wave number may take different forms. Also the decay constants which give a good result for one problem may failed for another one.

A new type of infinite element is introduced, this element has two nodes only and extend to infinity in the X direction as shown in figure (1).

The conventional Lagrange interpolation functions are used for Y coordinates and special shape functions are used for X coordinates. The coordinate of any point within the element can be related to the nodal coordinates as the following :-

$$X = (1+L\xi)\,[N_1 X_1 + N_2 X_2] \quad (2)$$

$$Y = N_1 X_1 + N_2 X_2 \quad (3)$$

Where $N_1 = 1/2\,(1-\eta)$

$N_2 = 1/2\,(1+\eta)$

L = Scaling Factor

It is shown from equation (2) and figure(1) that, X coordinate of any point within the element depends on L value as the following :

$$X = X_o\,(1 + L\,\xi)$$
$$= X_o + L\,X_o\,\xi \quad (4)$$

The above relation (4) is shown in figure (2),

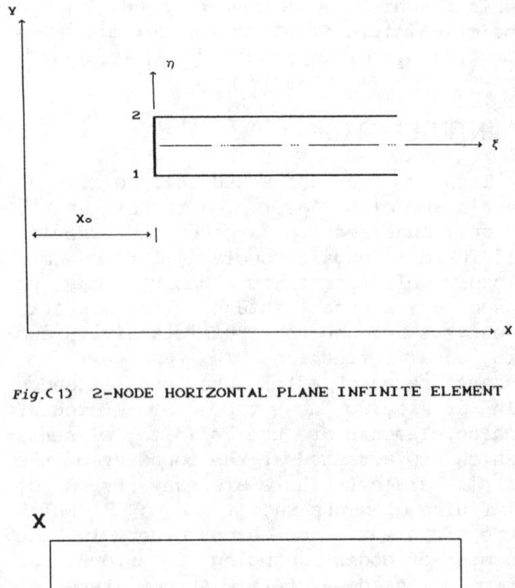

Fig.(1) 2-NODE HORIZONTAL PLANE INFINITE ELEMENT

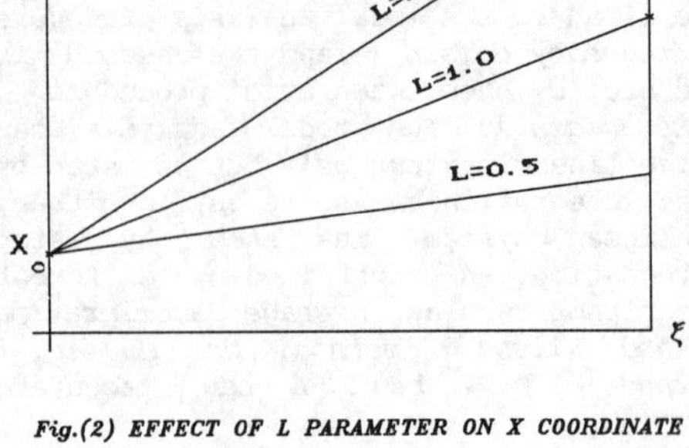

Fig.(2) EFFECT OF L PARAMETER ON X COORDINATE

From the above figure (2), it is shown that :

I As ξ increase X increase up to ∞ .
II The rate of increasing of X coordinate depend on the location of finite boundary X_o and on the scaling lengthL.

The displacement functions of the element is introduced by using the conventional Lagrange polynomial multiplied by the decay terms as the following :

$$U = F_d F_s (N_1 U_1 + N_2 U_2)$$
$$= M_1 U_1 + M_2 U_2 \qquad (5)$$

$$V = F_d F_s (N_1 V_1 + N_2 V_2)$$
$$= M_1 V_1 + M_2 V_2 \qquad (6)$$

Where :-

$M_1 = F_d F_s N_1$ and $M_2 = F_d F_s N_2$

The first term F_d representsthe wave shape and equals to $\cos(b\xi)$, and the second term F_s is the decay of the wave with increasing ξ and equal to $e^{-A\xi}$, A is the decay factor and b is the wave number.

Solving equation (4) for ξ yields :

$$\xi = (X-X_o)/L \qquad (7)$$

Then :

$$F_s = e^{-A\xi} = e^{-A(X-X_o)/L} \qquad (8)$$

It is clear from equation (8) that, as ξ (or X) increases, F_s deceases up to zero at infinity. So, as X tends to infinity U and V approach to zero.

The element stiffness matrix and mass matrix are introduced by the same procedure of the finite element [17] but the integration domain in ξ direction is from 0 to ∞ . After satisfying the strain-displacement and stress-strain relationships, the element stiffness matrix [k] can be calculated using the *virtual-work* principles or the *potential-energy* theorem to obtain the relation :

$$[k] = \int_v [B]^T [D] [B] \, dv \qquad (9)$$
$$= \int_{-1}^{+1} \int_{-1}^{\infty} [B]^T [D] [B] \, |J| t \, d\xi \, d\eta$$

Where :-

[B] = strain-displacement matrix .

[D] = stress-strain matrix .

$|J|$ = determinate of jacobian.

t = element thickness.

Also, the consistent mass matrix [M] may be calculated using the relation :

$$[M] = \int_v [N]^T [N] \rho \, dv \qquad (10)$$

$$= \int_{-1}^{+1} \int_{-1}^{\infty} [N]^T [N] \, |J| \rho \, t \, d\xi \, d\eta$$

Where ρ is the mass density, and

$$[N] = \begin{bmatrix} M_1 & M_2 & - & - \\ - & - & M_1 & M_2 \end{bmatrix} \qquad (11)$$

The integration is carried out numerically by using Gauss-Legender quadrature in the finite direction η and by using Gauss-Laguerre quadrature in the infinite direction ξ .

The introduced infinite element has these advantages :-

I- The element has two nodes coincide with the nodes of the finite element at the boundary , so the number of degrees of freedom of the finite element mesh not increase.

II- Using a proper value for scaling length L decrease the breadth of the finite element mesh than any type of other infinite element.

III- This element gives good results in the time domain analysis where the previous researches demonstrate that, the infinite element is used with the frequency domain analysis.

2.3 *Soil Models*

Two soil models are used in this research, Linear and *Duncan* models. Linear model is simple but not accurate because the soil is considered as a linear elastic material. The nonlinear behavior of soil is studied by *Duncan* [18], where he represent the stress-strain behavior of soil by a semi-hyperbolic .

2.3.1 *Linear model*

This model represents the soil as a linear elastic material. Two independent constants are needed to describe the soil, they are Young's modulus and bulk modulus or Young's modulus and Poisson's ratio.

2.3.2 *Duncan model*

The *stress-strain* relationship is introduced by Duncan and his colleagues using a triaxial tests with constant confining pressure. This relation is assumed to be approximated by a hyperbola. They introduced that the tangential *yang's modulus* E_t is as the following [18] :-

$$E_t = \left[1 - \frac{R_f\,(1 - \sin\phi)\,(\sigma_1 - \sigma_3)}{2\,C\cos\phi + 2\,\sigma_3\sin\phi}\right]^2 \times K\,P_a\left[\frac{\sigma_3}{P_a}\right]^n \qquad (12)$$

Where :-

σ_3 = minimum principle stress which is compression.

σ_1 = maximum principle stress which is compression.

P_a = atmospheric pressure.

K = modulus number, nondimensional.

n = modulus exponent, typical range -1.0 to 1.0 .

R_f = failure ratio, typical range 0.5 to 0.9 $= \dfrac{(\sigma_1 - \sigma_3)_f}{(\sigma_1 - \sigma_3)_f}$.

C = cohesion intercept, units as P_a.

$\phi = \phi_o - \Delta\phi \log_{10}(\frac{\sigma_3}{P_a})$

ϕ_o = friction angle, radians.

$\Delta\phi$ = reduction in ϕ for 10-folded increase in σ_3.

Duncan assumed that the bulk modulus B can be calculated from the following equation :

$$B = K_b\,P_a\left[\frac{\sigma_3}{P_a}\right]^m \qquad (13)$$

Where :-

K_b = bulk modulus number, dimensionless.

m = bulk modulus exponent, typical range 0.0 to 1.0 .

Eight parameters are required to evaluate the tangential *Young's modulus* E_t and the *bulk modulus* B . The required parameters are K, n, R_f, C, ϕ_o, $\Delta\phi$, K_b, and m . The first six parameters are required to obtain E_t and the last two parameters are required to obtain B.

3 PARAMETRIC STUDY

The introduced program is used to study the seismic behavior of a circular reinforced concrete tunnel which has the same diameter and cross section of the cairo tunnel (phase two). In both linear and nonlinear soil model analysis, the effect of depth of soil overlying the tunnel is studied. The typical dimensions of the structure is shown in figure(3). The middle diameter of the tunnel ϕ is 8.35 m and the tunnel thickness is .40 m . The height of the soil medium overlying the rock base H is 35.0 m. After studying the performance of the proposed infinite element , the optimum width of soil block B is introduced to be 3ϕ. To study the effect of depth of soil overlying the

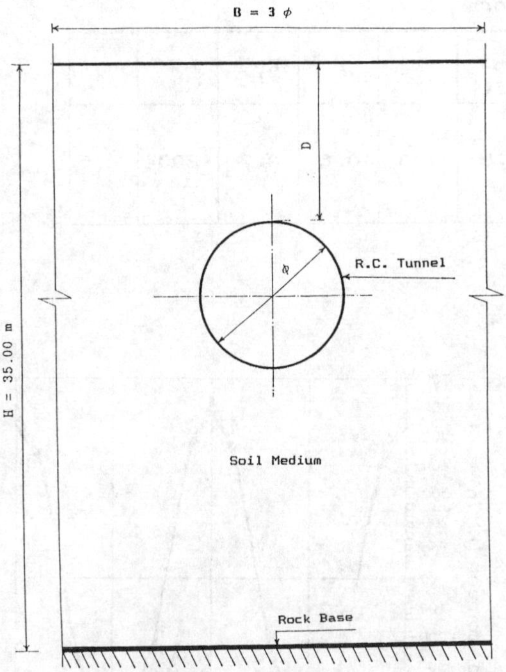

Fig. (3) SCHEMATIC REPRESENTATION OF SOIL-TUNNEL MODEL

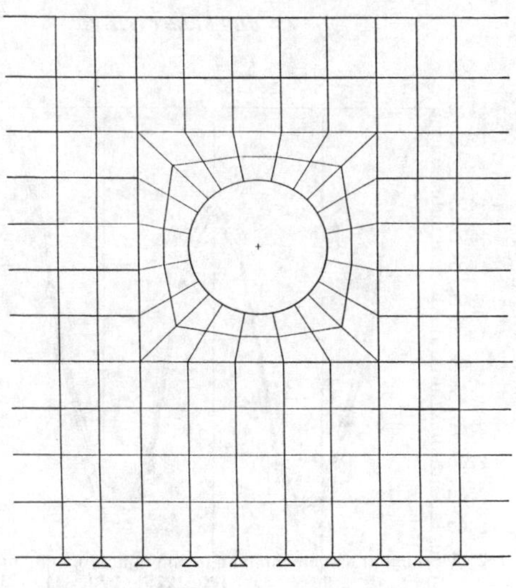

Fig. (4) HYBRID MESH OF FINITE AND INFINITE ELEMENT

tunnel, four embedment depth ratios D/H = 0.15, 0.30, 0.45, and 0.60 are chosen. The finite and infinite elements mesh which representing the problems are as shown in figure(4). The near field is represented by four-node isoparametric elements and the effect of the far field is taken by the proposed two-node infinite elements. The tunnel is modeled by twenty frame elements. The Young's modulus of the tunnel is constant for all cases and equals to 2.10E7 KN/m^2. Since the tunnels are long, the problems are idealized as plane strain analysis.

October 17, 1989 Santa Cruz Mins (Loma Prieta) earthquake is selected as a design earthquake. The effect of earthquake excitation is represented by applying its two component acceleration at the bedrock base of the different systems. The horizontal and vertical components of the earthquake accelerations are shown in figure(5) for the first twenty seconds of the earthquake duration.

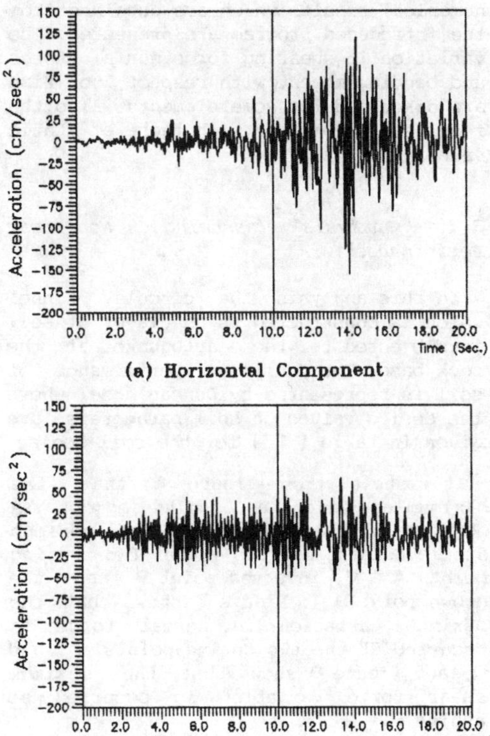

Fig.(5) GROUND ACCELERATION RECORDS OF EARTHQUAKE

TABLE (1) Standard Hyperbolic Parameters

Unified Classification	RC Stand-AASHTO	γ Kip/ft^3	ϕ_o deg.	$\Delta\phi$ deg.	C Kip/ft^2	K	n	R_f	K_b	m
Silty Clayey Sand (SM-SC)	100	.135	33	0	0.5	400	0.6	0.7	200	0.5

RC = Relative compaction

γ = Weight density

The integration of the dynamic equation of motion must be stable and accurate. The stability of solution depends on the used algorithm of integration and the accuracy depends on the type of algorithm and on the time step size Δt. It is found that, using the time step increment Δt of .02 second with β-3 [15] algorithm gives sufficient accurate results.

In the following sections some of the numerical results which are obtained from the introduced program are presented. The variation in shearing force, normal force and bending moment with respect to time are discussed for some elements. Also, the results are compared with those of static values.

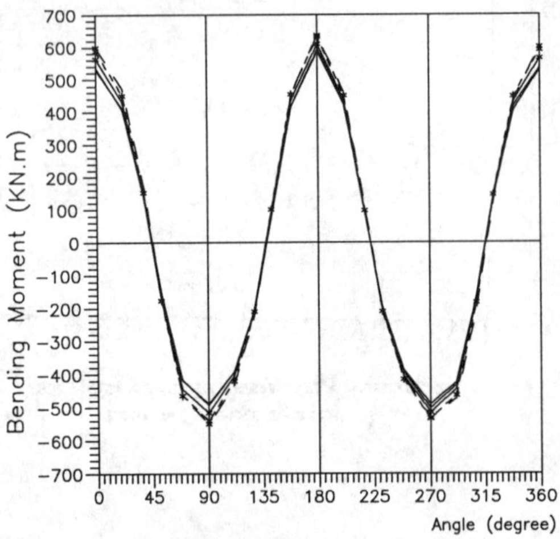

Fig.(6) BENDING MOMENT DIAGRAM AT DIFFERENT TIMES

3.1 *Analysis assuming Nonlinear Performance*

In this analysis the circular tunnel which is constructed in one layer of soil is subjected to the earthquake at the rock base. The nonlinear performance of soil is represented by Duncan model where the design values of soil parameters are given in table (1) for the soil media.

It's shown from figure 6 that the maximum variation in bending moment is occurred at the same points of maximum static bending moments (the two spring points A & C, inverted point B and the crown point D). Figure 7 shows that the maximum variation in normal force is occurred at the two spring points A and C, and figure 8 shows that the maximum shear force variation is occurred at angles $\theta = \mp 45^\circ$.

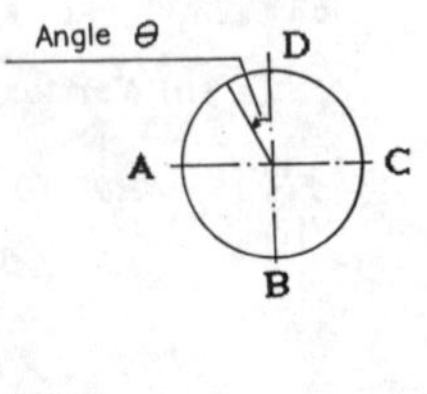

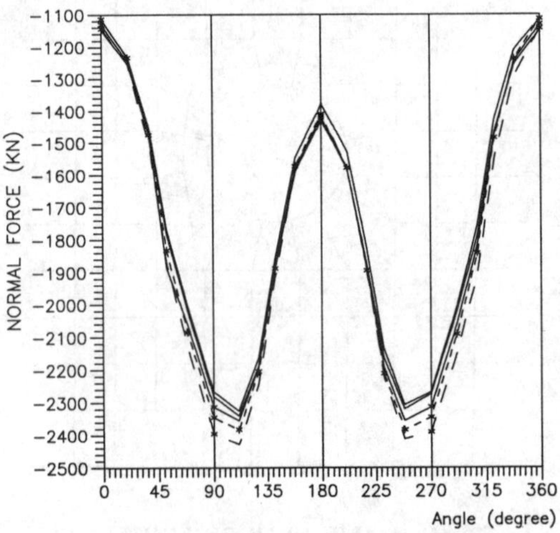

Fig.(7) NORMAL FORCE DIAGRAM AT DIFFERENT TIMES

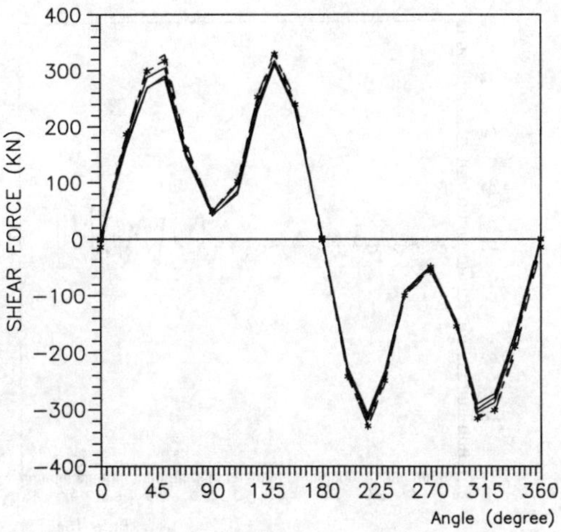

Fig.(8) SHEARING FORCE DIAGRAM AT DIFFERENT TIMES

Figure 9 shows that as the depth of soil overlying the tunnel decreases the variation in shearing force increases. Also, figure 10 shows that as the depth of soil decreases the variation in normal force increases. Figure 11 shows that the variation in moment due to earthquake increases as the depth of soil overlying the tunnel decreases. Since the variation in shear force, normal force and bending moment increase as the depth of soil decreases, the effect of earthquake excitation is more significant for shallow tunnels than deep tunnels.

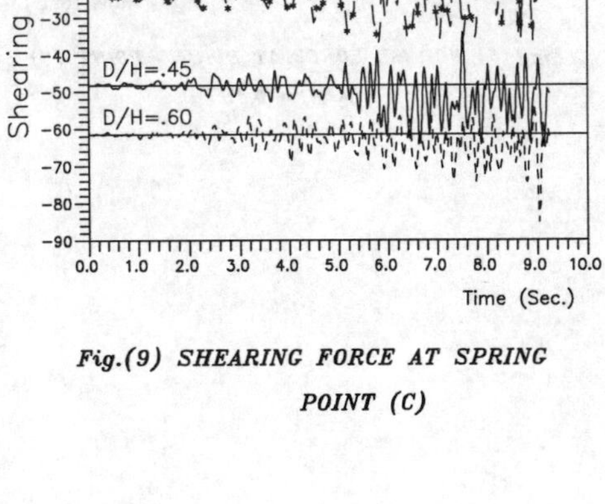

Fig.(9) SHEARING FORCE AT SPRING POINT (C)

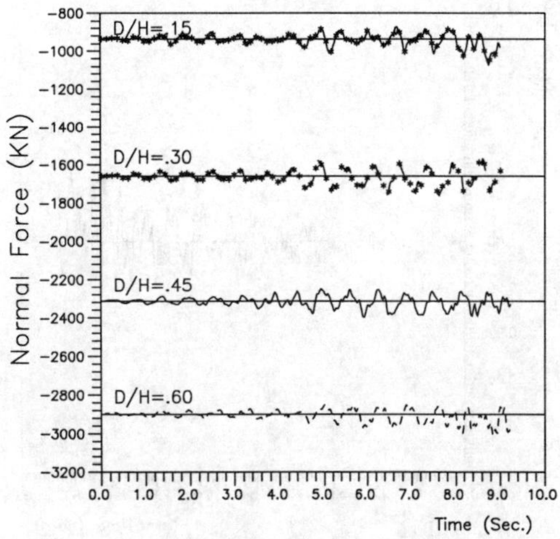

Fig.(10) NORMAL FORCE AT SPRING POINT (C)

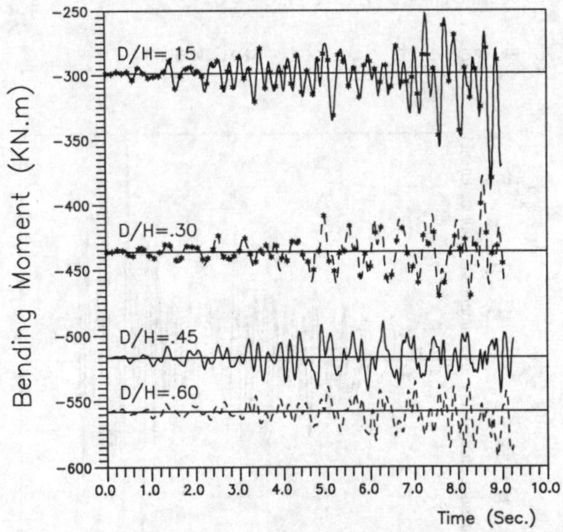

Fig.(11) BENDING MOMENT AT SPRING POINT (C)

By comparing figures 12 and 13 , they introduced that the maximum value of the moment ratio M_t/M_o at the spring point C is increased from 1.05 at the depth ratio D/H=0.6 to 1.19 at the depth ratio 0.15 . Also, it is found from figures 14 and 15 that the maximum value of the normal force ratio N_t/N_o is increased from 1.03 to 1.10 at the same point.

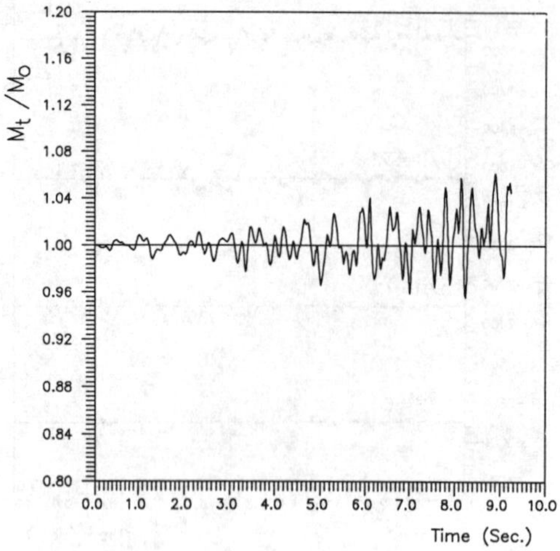

Fig.(12) BENDING MOMENT AT SPRING POINT (C)
D/H = 0.60

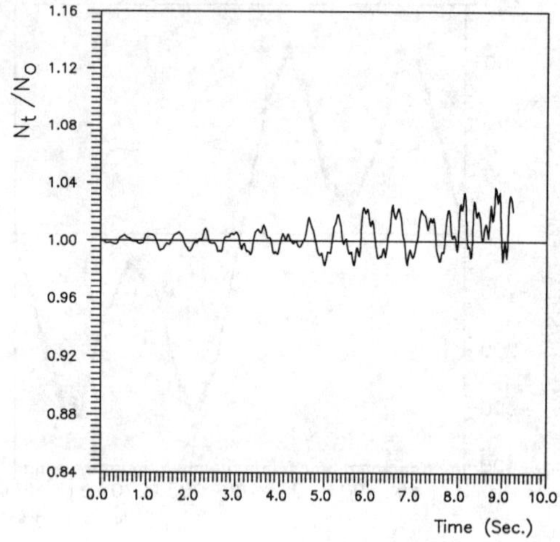

Fig.(14) NORMAL FORCE AT SPRING POINT (C)
D/H = 0.60

M_t = moment at time (t)

M_0 = moment due to own weight

N_t = normal force at time (t)

N_0 = normal force due to own weight

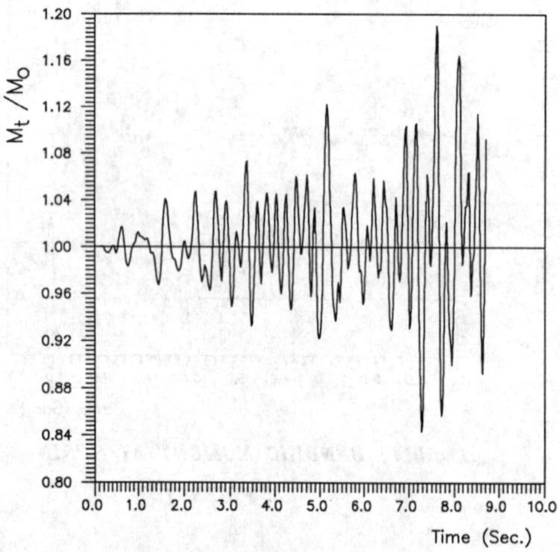

Fig.(13) BENDING MOMENT AT SPRING POINT (C)
D/H = 0.15

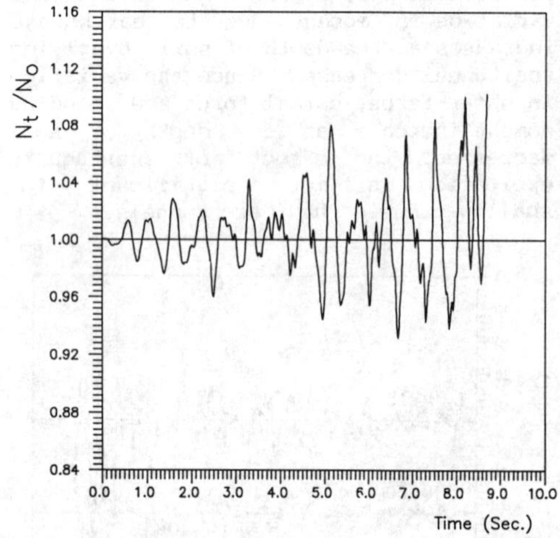

Fig.(15) NORMAL FORCE AT SPRING POINT (C)
D/H = 0.15

At time t equals 0 ,the results represent the own weight analysis. Figure 16 shows that as the depth of soil overlying the tunnel increase the values of bending moment increase. It is also shows,as the depth ratio D/H increases the rate of increasing the bending moment decreases.

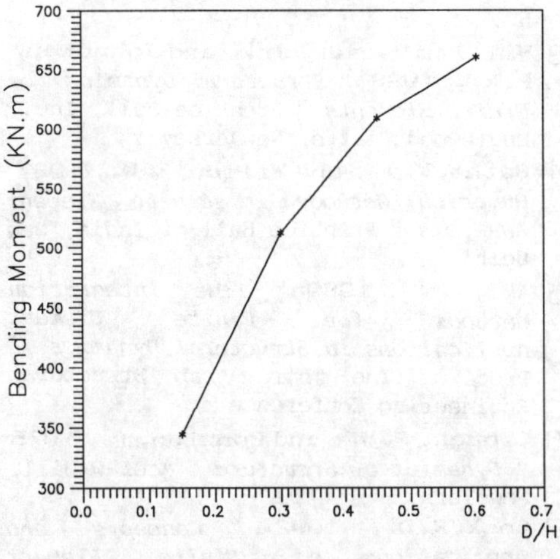

Fig.(16) BENDING MOMENT AT INVERT POINT (B)

(t=0.0)

3.2 *Analysis assuming Linear Performance*

In the nonlinear performance,Duncan model is failed before the maximum earthquake acceleration is reached,where tension failure is showed at the top elements after 9 seconds for most problems. So linear soil analysis is done to know the effect of earthquake at the maximum accelerations. The applied earthquake duration is increased to reach twenty seconds. The soil constants used are 13000 KN/m^2 for the Young's modulus and 0.33 for the Poisson's ratio.

The effect of the depth ratio D/H on the performance of tunnels under seismic excitations is the same as nonlinear soil analysis, where the earthquake is more significant as the depth ratio D/H decreases.

Figure 17 and 18 show that the seismic excitations is very significant where the maximum value of of the moment ratio M_t/M_o at the spring point A is increased from 1.20 at the depth ratio D/H=0.6 to 1.72 at the depth ratio 0.15 .

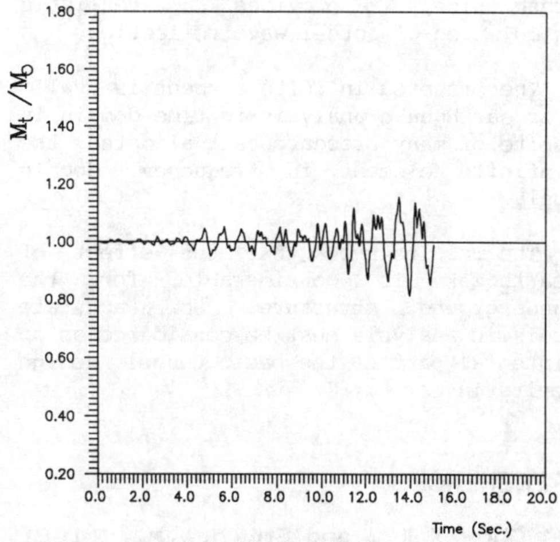

Fig.(17) BENDING MOMENT AT SPRING POINT (A)

D/H = 0.60

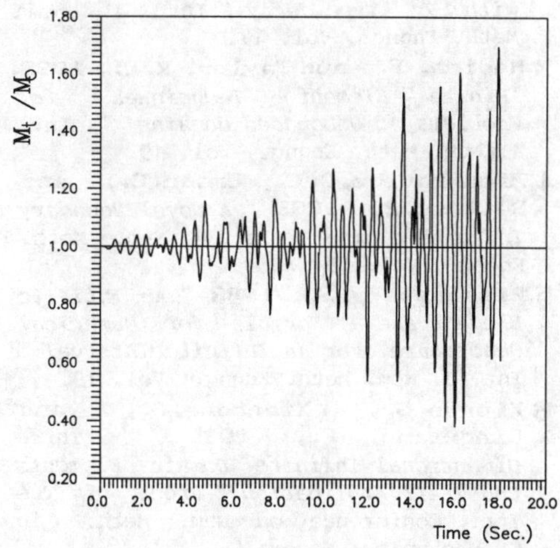

Fig.(18) BENDING MOMENT AT SPRING POINT (A)

D/H = 0.15

4 CONCLUSIONS

From the previous analysis,the most important conclusions can be summarized as follows :

Using the proposed infinite element as a numerical tool is very efficient in soil structure interaction problems. In static analysis the infinite element decreases mesh size which leads to minimum labor, time and computer memory. In dynamic analysis using the infinite element technique is more significant because in addition to the reduction of mesh size, it provides a realistic simulation of actual wave reflection.

The proposed infinite element is valid for earthquake analysis in time domain in spite of many researches validated the infinite element in frequency domain only.

The results show that the effect of earthquake is considerable for the underground structure. So, accurate seismic analysis must be considered as an integral part of the basic tunnel design criteria.

REFERENCES

1 Chow, Y.k., and Smith, I.M., 1981 *"Static and Periodic infinite Elements "* Int. J. Num. Meth. Engng. Vol. 17.

2 Lynn, P.P., 1981 *" Infinite Elements with 1/r^n Type decay "* Int. J. Num. Meth. Engng., Vol. 17.

3 Medina, F., and Taylor, R.L., 1983 *"Finite Element Techniques for Problems of Unbounded domains "* Int. J. Num. Meth. Engng., Vol. 19.

4 Zienkiewicz, O.C., Emson, C., and Bettess, P., 1983 *" A Novel Boundary Infinite element "* Int. J. Num. Meth. Engng., Vol. 19 .

5 Pissanetzky, S., 1983 *" An Infinite Element and a Formula for Numerical Quadrature over An Infinite Interval "* Int. J. Num. Meth. Engng., Vol. 19.

6 Xiurun, G., Xianrong, G., and Dingxiang, F., 1985 *" Three Dimensional Infinite Domain Elements and Their Application "* Proc. of The Int. Conference on Num. Meth. in Geomechanics, Nagoya .

7 Sadek, F.H.F., 1987 *" Employment of Infinite Elements in Static and Dynamic Soil Structure Interaction Problems "* M.Sc. Thesis, Cairo University .

8 Karpurapu, G.R., 1988 *" Composite Infinite Element Analysis of Unbounded Two-PhaseMedia"* Adv. Eng. Software, Vol. 10, No. 4 .

9 Hodhod, H.A., 1989 *" Employment of Infinite Elements in Soil Structure Interaction analysis "* M.Sc. Thesis, Cairo University .

10 Selvadurai, A.P.S., and Karpurapu, R., 1989 *" Composite Infinite Element for Unbounded Saturated-Soil Media "* Journal of Geotechnical Engineering, Vol. 115,No. 11,.

11 Godbole, P.N., Viladkar, M.N., and Noorzaei, J., 1990 *" Nonlinear Soil-Structure Interaction Analysis Using Coupled-Infinite Elements "* Computers & Structures , Vol. 36, No. 6 .

12 Angelov, T.A., 1991 *" Infinite Elements-Theory and Applications "* Computers & Structures , Vol. 41, No. 5 .

13 William Weaver, Jr., and Johnston, P.R., 1988 *" Structural Dynamics by Finite Elements "* Prentice-Hall, Inc., Englewood Cliffs, NewJersey .

14 Bathe, K.J., and Wilson, E.W., 1987 *" Numerical Methods in Finite Element Analysis "* Printice Hall of India, New Delhi .

15 Akl, A.y., 1989 *" Time Integration Methods for Finite Element Applications in Structural Dynamics "* Proc. of The Third Arab Structural Engineering Conference .

16 Clough, R.W., and Penzien, J., 1975 *" Dynamics of Structure "* McGraw-Hill, New York.

17 Cook, R.D., 1992 *" Concepts and Applications of Finite Element Analysis "* John Wiley and Sons Inc. , New york, 1974.

18 Duncan, J.M., Byrne, P., Wong, K.S., and Mabry, P., 1980 *" Strength, Stress-Strain and Bulk Modulus Parameters for Finite Element Analysis of Stresses and Movement in Soil Masses "* Report No. UCB/GT/80-01, Department of Civil Engineering, University of California, Berkeley, CA .

19 William Weaver, Jr., and Johnston, P.R., 1984 *" Finite elements for Structural Analysis "* Prentice-Hall, Inc., Englewood Cliffs, New jersey .

20 Cheung, Y.K., and Yeo, M.F., 1980 *" A Practical Introduction to Finite Element Analysis "* Pitman Publishing Limited, London .

21 Wolf, J.P., 1985 *" Dynamic Soil-Structure Interaction "* Prentice-Hall, Inc., Englewood Cliffs, New jersey .

22 Chen, Y., and Krauthammer, T., *" Seismic Effects on Large Reinforced Concrete Lifelines-I. Theory "* Computers & Structures , Vol. 42, No. 2 .

Tunnelling and Ground Conditions, Abdel Salam (ed.) © 1994 Balkema, Rotterdam, ISBN 90 5410 363 9

Analysis of deck road tunnels

A.A.Ahmed
Ain Shams University, Cairo, Egypt

ABSTRACT: In soft ground tunnels, the choice of the lining system is always considered as a major task in the tunnel project. Stiff lining ensures the stability of tunnel opening, the integrity of overlying surface structures and it also affects the rate of advance of the tunnelling operation. Accordingly, great improvements have been made in recent years in the design of lining systems and new lining techniques have been developed.

In road tunnels, adjusting the deck slab stiffness in the global stiffness of the ground-lining system is considered a more realistic idealization or the ground-lining interaction, particularly in cases of soft ground tunnelling conditions.

This paper evaluates some factors affecting the analysis of road tunnels driven in soft ground such as; the relative stiffness between tunnel lining and the surrounding ground, the intensity of lateral earth pressure, and the adopted continuum constitutive relationship. Also, one of the main objectives of this work is to determine the contribution of the deck slab in the ground-lining interaction feature and its effect on the final stability of ground-lining system.

1 INTRODUCTION

During tunnel excavation, the state of stresses in the ground continuum are significantly altered, specially for soft ground conditions. In soft ground tunnels, or soils characterized with relatively small "stand-up time", immediate support must be provided to the surrounding ground or the tunnel will collapse. In such ground conditions, the use of stiff lining systems is always required. In such a case, the ground usually participates actively by providing resistance to outward deformation of the lining (Duddeck, 1988).

It is considered, particularly for circular tunnel lining, that the idealization of deck slab stiffness in the global stiffness of the tunnel lining system, may produce a more rigid tunnel lining system, which could lead to more stable tunnel. Accordingly, more rigid deck systems may be needed. Ground-lining interaction analysis are usually complicated because it must fullfill the different factors affecting the interaction process. These factors can be divided into two main groups. The first group concerning the materials characteristics of both ground and lining and the details of excavation process. While, the second group concerning the compatibility and equilibrium conditions at the interface medium between ground and lining.

2 CURRENT LINING DESIGN APPROACHES

A significant number of tunnel lining design methods are readily available. These methods are divided into analytical and empirical methods. The analytical methods such as the finite element requires accurate and detailed information about the different elements affecting the nature of the ground-lining interaction. The finite element method is a very powerfull technique provides all the ground and structural elements needed for the interaction simulation. It can handle some interaction features of great complexity such as materials nonlinearity and time dependency.

The simplicity of the circular geometry and the much simpler interaction between a stiff structure and soft ground are the reasons why, internationally, the design models for soft ground tunnels (specially for shield-driven tunnels) are fairly well-developed (Duddeck and Erdmann, 1985). A brief discussion about the currently used lining design approaches are given by (El-Nahhas, 1980; Duddeck, 1988 and Ahmed, 1991). These approaches can be classified as:

- Methods based on subgrade reaction (bedded beam model).
- Relative stiffness approach (continuum model).
- Convergence-confinement approach.
- Finite element approach.
- Observational approach.

Among these approaches, it should be noted that the finite element method as a tool for tunnel design indicated promising results. It enable modelling the different elements affecting the soil-lining interac- tion. It can model the materials nonlinearity, creep, time dependency and effect of lining hinges. Some investigators recommended the use of combination of the finite element method and the convergence - confinement concept which may lead to more realistic simulation of tunnel excavation (Eisenstein and Negro, 1985 and El-Nahhas et al., 1992).

3 IDEALIZATION OF THE PROBLEM

The ground-interaface-lining interaction idealization are simulated and analysed employing the finite element method. The ground continuum is simulated using constant strain triangular elements, and the ground constitutive relationship is treated using both linear elastic and hyperbolic nonlinear volume change technique proposed by Duncan & Chang, 1970 and Wong & Duncan, 1974. The nonlinear analysis is carried out following an iterative technique by which the continuum yielding zones in the vicinity of tunnel are determined and adjusted in the global ground- inteface-liner stiffness to relatively low tangent modulus values.

The liner is simulated using the conventional beam-column finite elements. Each element consists of two nodes and each node has of three degrees of freedom. The beam-column element is more convenient to simulate any lining system, whatever its shape, specially in cases of hinged linings.

In order to link elements having three degrees of freedom (liner elements) to elements having two degrees of freedom (ground elements) an interface finite element is strongly needed. This type of shear elements considers the bending moments develops at the neutral axis by the friction at the interface surface (King & Pandey, 1978 and El-Nahhas et al., 1992). The interface finite element is characterized by two stiffness components, known as normal and tangential stiffness (K_n & K_s). The nonlinear behaviour of the interface can be programed using a special technique of iteration after performing a series of direct shear tests to evaluate the components of the element stiffness.

The ground-interface-liner system is evaluated and analysed considering external loading condition equivalent to the forces due to excavation following the stress reversal approach (Kulhawy, 1975). In this approach, the excavation process is simulated by computing the initial stresses, which are in equilibrium and at rest condition, acting on the excavation surface. These stresses are integrated in order to determine the nodal point forces which are applied in the opposite direction to the nodal points of ground-interface -liner system.

Although the interaction between tunnel lining and ground is tunnelling technology dependent, but in the present study, no reduction in the gravity forces (in-situ stresses) have been considered due to excavation process.

4 GROUND CONSTITUTIVE BEHAVIOUR

Ground constitutive relationships, of geologic media, determined in the laboratory usually demonstrate that nonlinearity can not be ignored for proper prediction of deformation. In the present work, the ground continuum constitutive behaviour is updated using both the classical linear elastic relationship and the hyperbolic nonlinear model which is formulized by Duncan & Chang (1970) and Wong & Duncan (1974) as;

$$E_t = \left[1 - \frac{R_f(1-\sin\phi)(\sigma_1 - \sigma_3)}{2C\cos\phi + 2\sigma_3\sin\phi} \right]^2 K P_a(\sigma_3/P_a)^n$$

$$\nu_t = \frac{\dfrac{G - F\log(\sigma_3/P_a)}{d(\sigma_1 - \sigma_3)}}{\left[1 - \dfrac{}{K P_a(\frac{\sigma_3}{P_a})^n \left\{ 1 - \dfrac{R_f(1-\sin\phi)(\sigma_1 - \sigma_3)}{2C\cos\phi + 2\sigma_3\sin\phi} \right\}} \right]^2}$$

E_t, ν_t tangent modulus and the tangent value of Poisson's ratio,
C, ϕ cohesion and angle of shearing resistance of soil,
R_f failure ratio and found to be between 0.75 and 1.,
K, n modulus number and modulus exponent,
G, F initial tangent Poisson's ratio and the reduction of initial tangent Poisson's ratio for a tenfold increase in (σ_3),
d rate of change of Poisson's ratio with the radial strain,
P_a the atmospheric pressure, and
σ_1, σ_3 the major and minor principle stresses.

5 CASE STUDY

The case study utilized in the analysis is shown on Figure 1. It is a relatively large diameter tunnel (6.20m) driven at a depth to centerline of about 11.70 meters below ground surface. The properties of the lining system, as well as the ground properties and nonlinear parameters are summarized on Figure 1. The interface element stiffness components K_n and K_s are idealized as 2×10^8 & 2×10^4 kN/m^3 respectively.

To study the effect of relative stiffness between the tunnel lining and the surrounding ground on the tunnel stability, linear and non-linear ground idealization, a wide values of modulus of elasticity ranging between (10^4-10^5 kN/m^3) and modulus number of (100-1000) are assigned in the analysis. Also, a wide value of the coefficient of lateral earth pressure (K_o) are adopted to determine the effect of lateral pressure coefficient on the straining actions developed in the tunnel lining.

As mentioned before that, the main objective of this study is to determine the contribution of the tunnel deck slab, as a member of the lining system, on the final stability of tunnel lining. To enlighten such point, the lining is assigned in the global system twice, with and without deck slab. The deck slab properties and configuration are taken as those of lining.

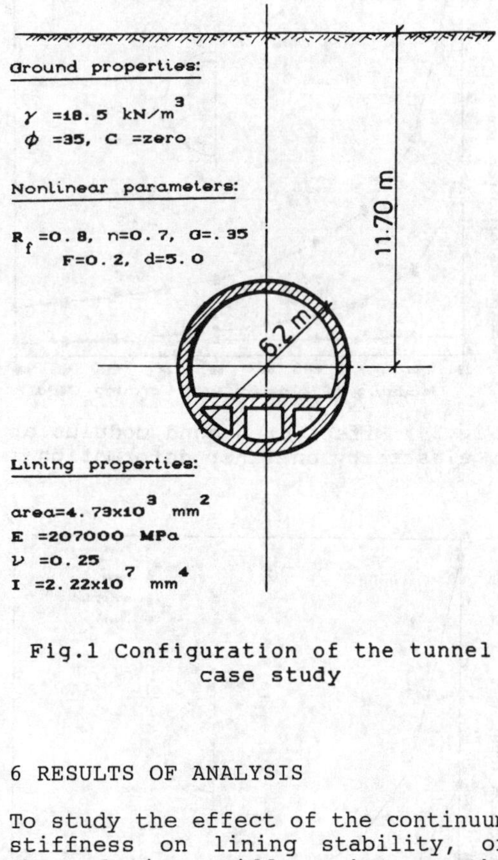

Fig.1 Configuration of the tunnel case study

6 RESULTS OF ANALYSIS

To study the effect of the continuum stiffness on lining stability, or the relative stiffness between the continuum and the lining, the lining is assigned in the analysis without deck slab. Figure 2 illustrates the effect of ground continuum stiffness (modulus of elasticity in case of linear elastic and the modulus number in case of nonlinear analysis) on the inward vertical movement at both tunnel crown and invert, and the outward horizontal displacement at the springline. The results of the lining deformation, indicate an elliptical mode of lining deformation. The results also, prove that increasing the ground continuum stiffness (in both cases of linear elastic and non-linear analysis) resulted in a considerable reduction of the radial deformation at both tunnel invert and springline, while the vertical displacement at the tunnel crown remained unchanged in case of linear elastic and slightly decreased in case of nonlinear analysis.

Figure 3 shows the relationship between the continuum stiffness parameters for (linear elastic and nonlinear idealization) and the lining thrust at crown, springline, and invert. This figure indicates

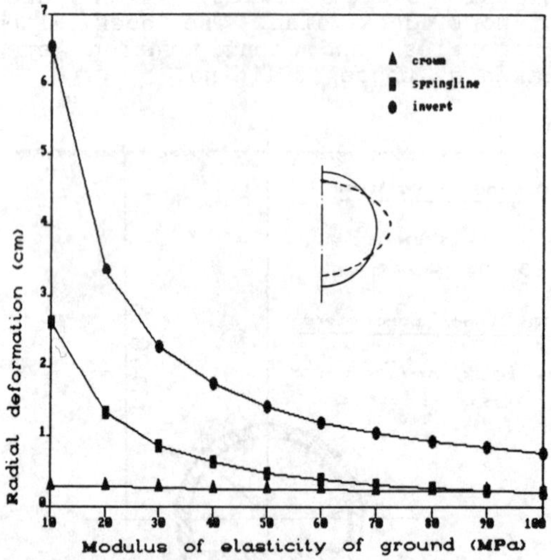

Fig.2-a Effect of ground modulus of elasticity on liner deformation

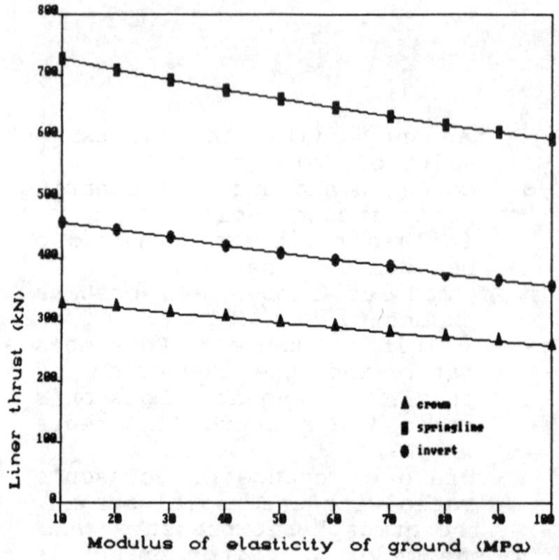

Fig.3-a Effect of ground modulus of elasticity on liner thrust

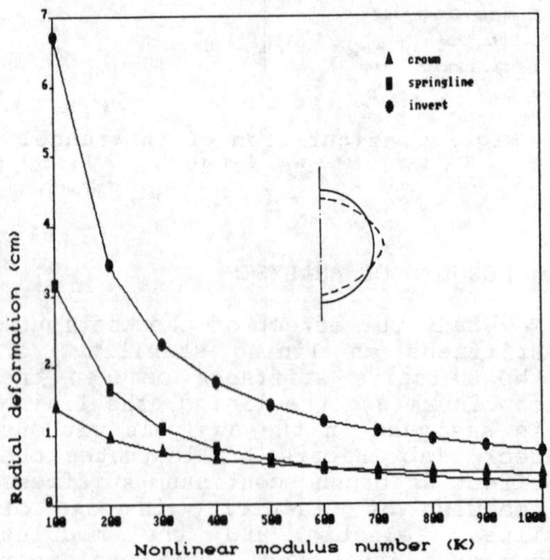

Fig.2-b Effect of ground nonlinear modulus number on liner deformation

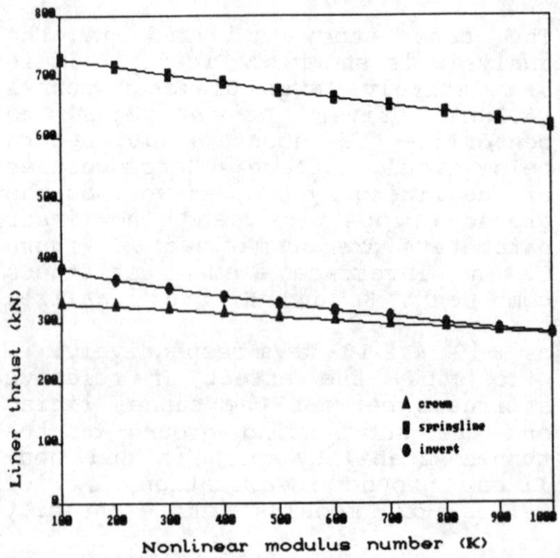

Fig.3-b Effect of ground nonlinear modulus number on liner thrust

that the distribution of the lining thrust is non-uniform with a maximum value at tunnel springline and minimum value at tunnel crown. The figure also, proves that decreasing the relative stiffness between ground and lining, or increasing the ground stiffness (for both cases of linear and nonlinear) resulted in some reduction of the lining thrust.

To clarify the main objective of the study, which is the contribution of deck stiffness as a member of the lining system, the tunnel lining is assigned in the analysis with and without adjusting the deck slab stiffness. The ground continuum constitutive relationship is also idealized as linear elastic and nonlinear. Figure 4 illustrates the relationship between the ground lateral pressure coefficient and the lining radial deformation (with and without adjusting the deck slab). The results of the lining deformations indicate that considering the deck slab stiffness in the global stiffness of the lining system could lead to substantial reduction of the lining deformations at tunnel crown, springline, and invert. The results also prove that, as the ground lateral pressure coefficient increased, the lining radial deformation decreased. The barchart shown on Figure 5 illustrates quantitively the reduction of lining radial deformation associated with assigning the deck slab stiffness in the global stiffness of the lining system.

Figure 6 shows the relationship between the lining thrust (with and without deck slab) and lateral pressure coefficient. The results illustrate that the lining thrust distribution is non-uniform with a maximum value at the tunnel springline. It is also noted that, as the ground lateral pressure coefficient increased the lining thrust at both crown and invert increased, while the lining thrust at the springline remained constant. Figure 6 also indicates that, due to assignment of the deck slab, the lining thrust at both tunnel crown and springline are relatively unchanged. The figure also shows that the lining thrust at the tunnel invert increased with deck slab due to the tension forces develop in the deck slab.

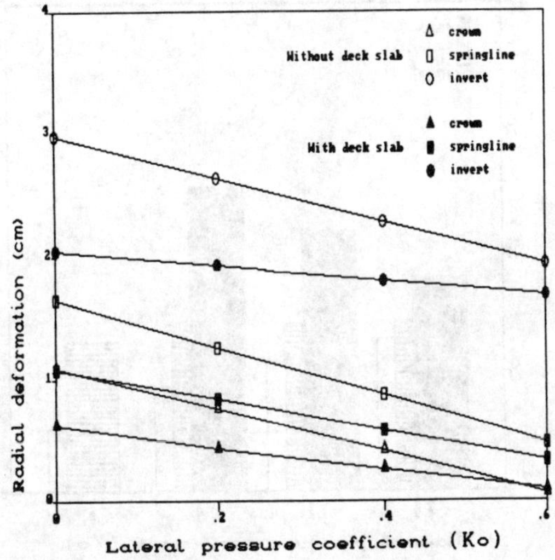

Fig.4-a Effect of lateral pressure coeff. on liner radial deformation (case of linear elastic)

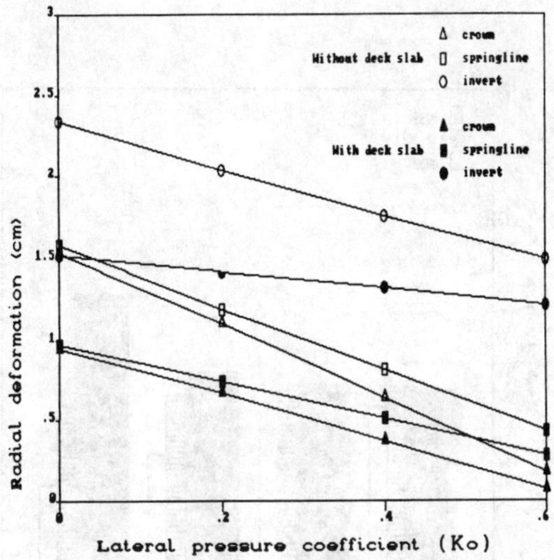

Fig.4-b Effect of lateral pressure coeff. on liner radial deformation (case of nonlinear)

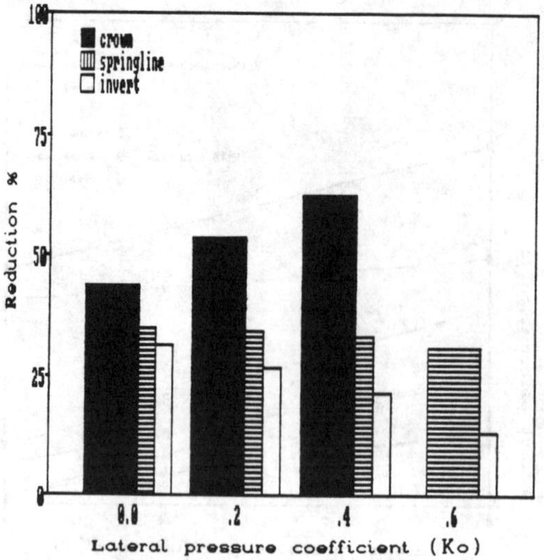

Fig.5-a Reduction of liner radial deformation associated with deck slab (case of linear elastic)

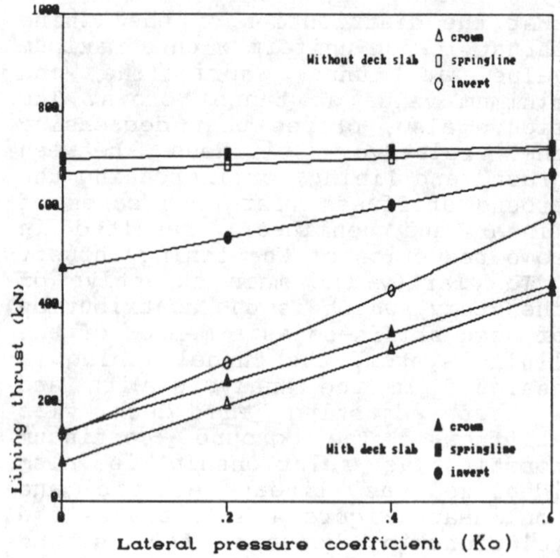

Fig.6-a Effect of lateral pressure coefficient on liner thrust (case of linear elastic)

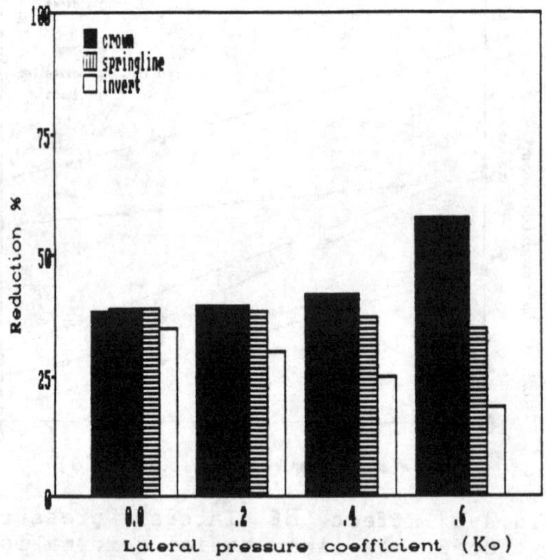

Fig.5-b Reduction of liner radial deformation associated with deck slab (case of nonlinear)

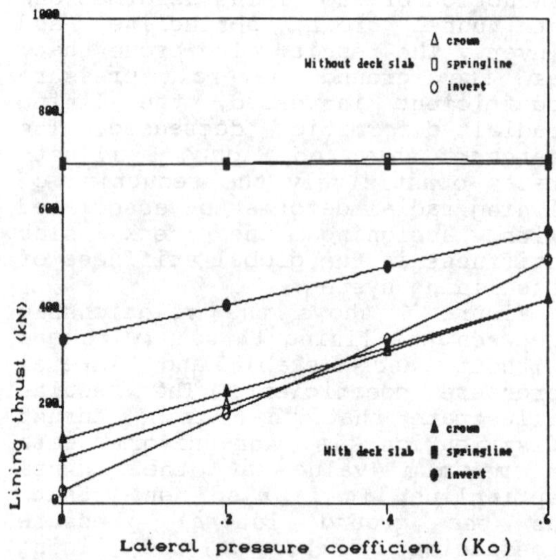

Fig.6-b Effect of lateral pressure coefficient on liner thrust (case of nonlinear)

Figure 7 indicates the bending moment and normal force distribution in the lining system due to in-situ forces only, with and without the deck slab. The figure indicates the tension forces developed in the deck slab and its effect on increasing lining thrust in the bottom portion of tunnel. Figure 8 illustrates the distribution of bending moments in the lining system (with and without deck slab) due to traffic load. This figure indicates that, the developed moments in case of adjusting deck slab are reduced to about 50%.

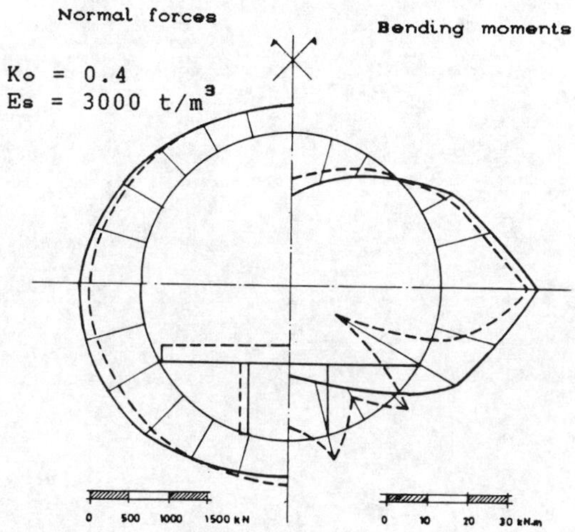

Fig.7-a Lining straining actions (case of linear elastic)

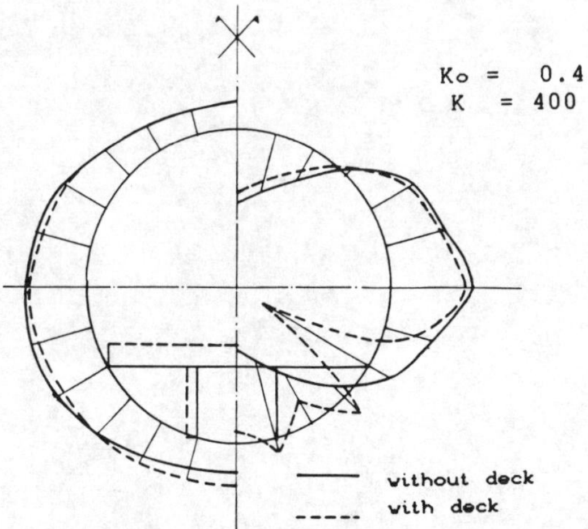

Fig.7-b Lining straining actions (case of nonlinear)

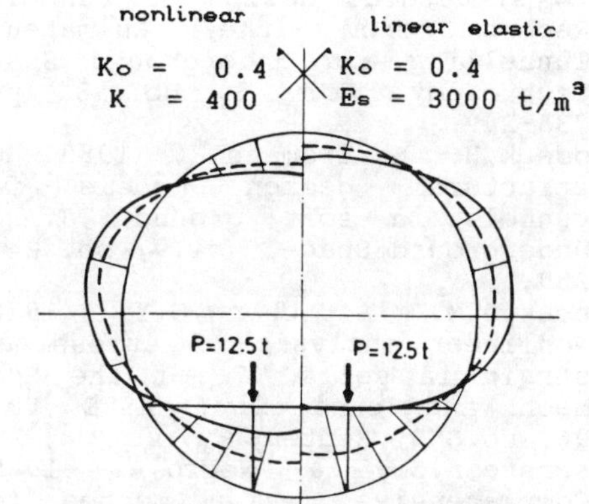

Fig.8 Lining straining actions due to traffic load

7 CONCLUSIONS

It is obvious from the results of the present study, that idealizing the deck slab stiffness in the global stiffness of tunnel lining system can be considered as a good technique for improving the reliability of interaction between the tunnel lining and the surrounding ground. The idealization of deck slab is always characterized by significat reductions in the tunnel lining deformations.

Also the results indicate that considerable reduction in the tunnel lining straining actions is expected due to assignment of deck slab stiffness particullarly with traffic loading condition.

The results also demonstrate that, decreasing the relative stiffness between the tunnel lining and ground resulted in substaintial reductions in lining deformations as well as lining internal forces.

REFERENCES

Ahmed,A.A. 1991. Interaction of tunnel lining and ground. Ph. D. Thesis, Dept. of Civil Eng. Ain Shams Univ., Cairo, Egypt, 216p.

Duddeck,H. 1988. The ITA's approach to structural design of tunnels. Report from the animateur, Tunnelling and Underground Space Technology, Vol. 3, No. 3, pp. 235-249.

Duddeck,H. & Erdmann,J. 1985. On structural design models for tunnels in soft ground. J. of Underground Space, Vol.9, pp. 246-259.

Duncan,J.M. & Chang,C.Y. 1970. Nonlinear analysis of stress and strain in soils. J. of the Soil Mech. and Found. Div., ASCE, Vol. 96, No.SM5, Septemper.

Eisenstein,Z., & Negro,A. 1985. Comprehensive design method for shallow tunnels. Proc. Underground structures in Urban Areas, ITA, Praug, Vol.1, pp. 375-391.

El-Nahhas,F. 1980. The behaviour of tunnels in stiff soil. Ph.D. Thesis, Dept. of Civil Eng. Univ. of Alberta, Canada, 305p.

El-Nahhas,F., El-Kadi,F. & Ahmed,A. 1992. Interaction of tunnel linings and soft ground" J. of Tunnlling and Underground Space Technology, Vol.7, No.1, pp.33-43.

King,G.J.W. & Pandey,P.C. 1978. The analysis of infilled frames using finite element. Proc. Instn Civ. Engrs, Part 2, Des., pp. 749-760.

Kulhawy,F.H. 1975. Stresses and displacements around openings in homogeneous rock. Int. J. for rock mechanics, Mining Sciences and Geomech. Abstracts, Vol. 12, pp. 43-57.

Wong,K.S. & Duncan,J.M. 1974. Hyperbolic stress-strain parameters for nonlinear finite element analysis of stresses and movements in earth masses. Report No. TE 74-3, Univ. of California, Berkeley.

Tunnelling and Ground Conditions, Abdel Salam (ed.) © 1994 Balkema, Rotterdam, ISBN 90 5410 363 9

Dynamic analyses of large underground caverns using discrete-element codes Verification and reliability

M.R.Akky, D.Rosidi, M.N.Madianos & J.Y.Kaneshiro
Geomatrix Consultants, Inc., San Francisco, Calif., USA

ABSTRACT: This paper discusses the seismic loading conditions, seismic response and stability of rock caverns based on discrete-element modeling, limit-equilibrium analyses, and historical observations of underground cavern performance during earthquakes.

1 INTRODUCTION

Although they depend on the veracity of the constitutive model and on good engineering judgement, verification and justification of static numerical models of underground struc- tures can utilize closed-form and analytical solutions. Static numerical models ultimately are verified based on field measurements made during construction.

On the other hand, dynamic numerical models of underground structures, although also dependent on the veracity of the constitutive model and on good engineering judgement, are less easily verified. It may be years before the models are tested by an earthquake event. In this paper, a combination of numerical methods, pseudo-static method, and observed data from underground structures during historical earthquakes were used to examine the reliability of the seismic design of large underground caverns. Two-dimensional, discrete-element computer codes were used to perform dynamic analyses of the caverns: the Universal Distinct Element Code (UDEC) and the Discontinuous Deformation Analysis (DDA).

Probable and credible event earthquake loading scenarios were developed from a seismic hazard analysis for the design of rock caverns in a heavily jointed granite. The discrete element analyses are compared with each other and are contrasted with limit equilibrium analyses, and empirical observations. Limitations of the various analyses and numerical methods, as well as the amount of conservatism involved, are discussed. This somewhat random method of verifying dynamic analysis of underground caverns, combined with good engineering judgment, appears to be the best method for evaluating the reliability of such structures given the data base on critical underground structures and their performance during historical earthquakes.

2 SEISMIC LOADING CONDITIONS

The source characterization effort comprised of a suite of interrelated geologic, seismotectonic, seismologic, and earthquake engineering studies to characterize and model potentially significant seismic sources in terms of their geometry, capability, and earthquake recurrence. Potential seismic sources in the study area were identified through published literature, and the magnitudes and locations were cross-checked and compared with surface geologic structures.

Seismic hazard analysis was based on assessing the recurrence of earthquakes on potential seismic sources and on the ground motion attenuation relationships appropriate for the site. Results of the analysis were expressed as hazard relationships between amplitudes of response spectral values and the annual frequencies or return periods of exceedance for those amplitudes. Spectrally matched time histories for annual frequencies of 10^{-4} and 10^{-6} (probable and credible) were generated for the site under study. Such typical velocity time histories are shown on Figure 1. These spectrally matched time histories were used in the rock stability analyses.

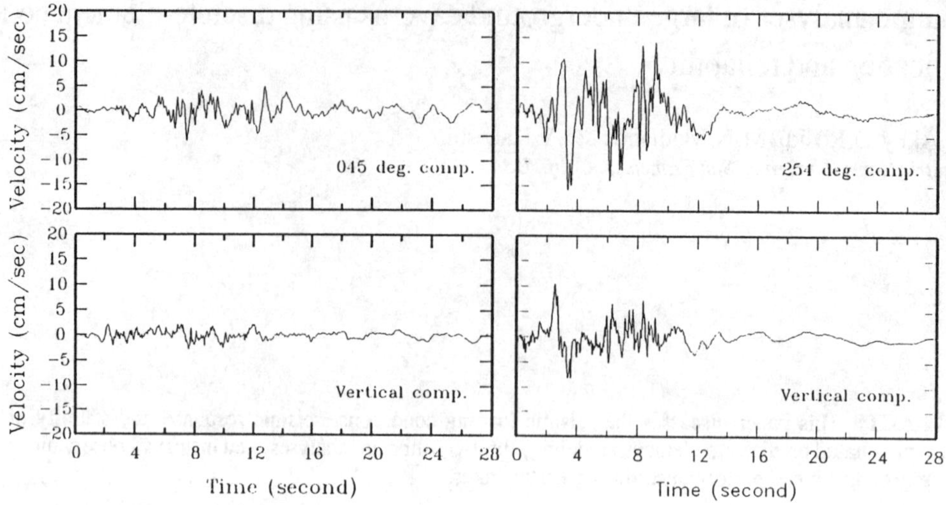

a) Probable event.

b) Credible event.

Fig. 1. Spectrally matched velocity time histories.

3. SEISMIC RESPONSE AND STABILITY EVALUATIONS OF ROCK CAVERNS

Seismic response analyses and stability assessments under earthquake loading conditions were conducted for a typical cavern with prescribed rock joint patterns and rock quality. The analytical techniques included the two-dimensional "discrete-element" methods developed by Cundall (1971) and Shi (1989), as well as "pseudo-static" wedge analysis. In many continuum mechanics formulations, discontinuous or interface elements are provided to allow for some small displacement. Such formulations generally are incapable of simulating large displacements and/or rotations. Total separation and creation of new contacts as blocks come together are possible in discrete element methods. Many problems in rock mechanics involve analyzing rock masses that interact with each other, in the process creating new contacts and gaps. Discrete-element methods, unlike continuum mechanics, are formulated specifically for this type of problem.

Three, two-dimensional, analytical models were used to assess the dynamic response and stability of the rock cavern. These models include (1) the Universal Distinct Element Code (UDEC) model, (2) the Discontinuous Deformation Analysis (DDA) model, and (3) a pseudo-static, limit-equilibrium wedge model.

3.1 Analyses using the Universal Distinct Element Code (UDEC)

The Universal Distinct Element Code (UDEC, Version 1.82; Itasca, 1992) is a numerical program where the rock mass is presented as an assemblage of distinct blocks having joints between blocks. Blocks may be modeled as either rigid or deformable materials. Deformable blocks are subdivided into a mesh of finite elements, with each element behaving according to a prescribed linear or nonlinear stress-strain constitutive relationship. The relative motions of the blocks along discontinuities is also governed by prescribed linear or nonlinear force-displacement relations (springs) for movement in both the normal and shear directions. UDEC uses time-marching, finite-difference schemes to solve the force equations of motion in the system. At each time step, new contact forces are calculated from the known displacements; Newton's second law is then used to calculate the new block positions. The model also incorporates nonreflecting boundaries for performing dynamic analyses. Dynamic analyses were performed using bi-linear (elastic, perfectly plastic) deformable blocks and joints. The joint set distribution for the caverns were simplified from a complex randomly generated and mapped joint distribution using the two key blocks (wedges) as shown in Figure 2 for computational ease and faster analysis under earthquake loading conditions. The key block assumption was verified by static analysis of the complex case and field confirmation. The ratio of the horizontal to vertical stress was initialized at 1.0. The strength of the materials and discontinuity is governed by the Mohr-Coulomb model. The deformation moduli assumed were based on poor rock conditions to simulate the complex joint distribution actually present.

3.1.1 UDEC Static Analyses

The caverns were excavated in four stages. Rock bolts were installed after each bench was excavated, and sufficient iterations were allowed to reach an equilibrium state. A conservation shear joint friction angle of 25° and no cohesion was assumed in the analyses.

Mohr-Coulomb strength/stress ratio contour plots after completion of excavation is shown on Figure 2 as well. When the strength/stress ratio is 1 or less (i.e., the induced stress is equal to greater than the strength), the excavation is not necessarily unstable, but rather the strength of the zones, where the strength/stress ratio is less than unity, will drop to a residual strength value and the "over-stress" will be redistributed to the adjacent rock mass (Fairhurst and Pei, 1990). This interpretation is substantiated since less than 1% of the intact elements yielded.

The major and minor principal stresses of the intact rock range from 7 to 12 MPa and from 2 to -0.5 MPa, respectively. The major principal stresses around the cavern are shown on Figure 3. The maximum value of the principal stresses are in agreement with other theoretical models; that is, the overburden stress at the center of the excavation is approximately 1.7 MPa, and the maximum principal stresses around the cavities are about 7 to 12 MPa. This represents stress concentration factors between about 4 and 7 around the cavity and at corners, consistent with boundary-element solutions for underground openings in an elastic medium (e.g., Hoek and Brown, 1980). The major principal stresses indicate that the maximum compressive stresses in the intact rock (7 to 12 MPa maximum) are localized and are much less than the estimated strength of the intact rock (assuming a uniaxial compressive strength of 50 to 100 MPa). When the strength of the intact rock is greater than five time the major principal stress, new fractures generally will not develop around the opening, and stress concentrations around the excavation will be relieved rapidly with small inward displacements (Hendron and Fernandez, 1983). The minor principal stresses show that the maximum tensile stresses in the intact rock (< 1.4 MPa) are less than the estimated tensile strength of the rock (5 to 10 MPa), so the excavation is also stable from the stand-point of tension.

In this discontinuum analysis where the strength of the rock greatly exceeds the induced stress in the intact rock, the most important parameter is the movement among the blocks, in particular the critical wedge movements. The major principal stresses show a lower stress regime within the wedges than in the intact rock surrounding the opening. This is because the critical wedges are not coupled with the surrounding free-field in situ stresses due to wedge movement. At the completion of excavation, the maximum displacements are due to wedge movements plus intact rock displacements and are on the order of 5 to 7.5 centimeters. Maximum differential displacements (i.e., between the wedge and the

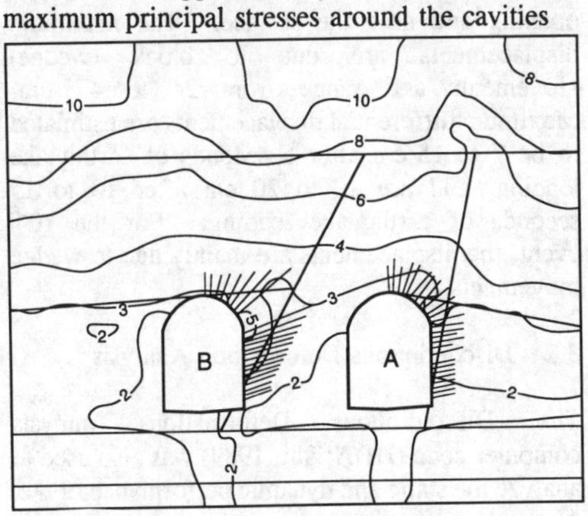

Fig. 2. Static, UDEC strength/stress contours, rock bolt pattern.

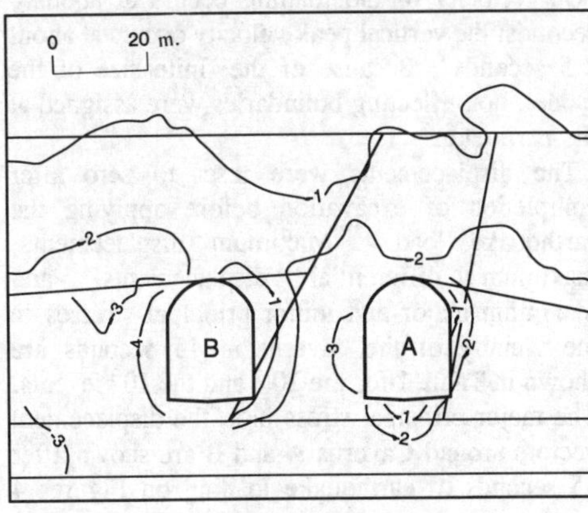

Fig. 3. Static, UDEC major principal stress.

adjacent and surrounding rock mass) are estimated to be on the order of 3 to 5 centimeters.

The model shows what is happening from a phenomenological standpoint and identifies critical wedges that may require additional support. The static results showing the principal stresses and displacements in the vicinity of the caverns are summarized in Table 1.

Table 1. Summary of discrete element analyses.

Analysis	UDEC						DDA	
	Static		10^{-4}		10^{-6}		Static	10^{-4}
Cavern	A	B	A	B	A	B	--	--
Max. Total Disp. (cm)	5.4	7.5	5.9	8.7	46	45	Not Ob-served	Col-lapse
Est. Differential Wedge Disp. (cm)	3	5	3-6 cm		10-20 cm			
Max. Major Principal Stress (MPa)	4.8	6.8	5.0	7.0	7.1	8.4	3.5	6
Max. Minor Principal Stress (MPa)	-0.5	-0.9	0.5	-0.8	-1.5	-1.4	1	1

3.1.2 UDEC Dynamic Analyses

The simplified model was considered appropriate for modeling phenomena for a preliminary risk assessment. Dynamic analyses were performed at two levels of shaking for the annual exceedance frequencies of 10^{-4} and 10^{-6} and incorporating both horizontal and vertical components of the earthquake velocity records. The analyses were run for 15 seconds, corresponding to the primary duration of strong ground motion. The horizontal peak velocity of the loading occurs at about 4 seconds; the vertical peak velocity occurs at about 2.5 seconds. Because of the finite size of the model, nonreflecting boundaries were assigned at the perimeter.

The displacements were reset to zero after completion of excavation before applying the earthquake load. Maximum displacements, maximum differential displacements, and maximum major and minor principal stresses in the vicinity of the caverns at 15 seconds are shown in Table 1 for the 10^{-4} and the 10^{-6} events. The major principal stress plot, the displacement vectors around Caverns A and B are shown after 15 seconds of earthquake loading on Figures 4 and 5a and b for the 10^{-4} event, and on Figures 6 and 7a and b for the 10^{-6} event. These figures show typical results.

For the 10^{-4} event, the maximum major principal stresses vary from 5 to 7 MPa, and the maximum minor principal stresses vary from -0.4 to -1 MPa. These are consistent with the principal stresses discussed for the static case. However, the strength/stress ratio, the major and minor principal stresses, and the shear stresses do not differ significantly from the static case, indicating a relatively small additional stresses induced by the earthquake loads. This is explained by the fact that in a discontinuum model, movement along joints is allowed, and most of the shear strain energy is dissipated by movement of the blocks along joints. The maximum displacements, which are due primarily to block (wedge) rigid body movements, range from 2.5 to 8.7 cm. Maximum differential displacements are estimated to range from 1 to 3 cm after 5 and 10 seconds of earthquake loading, and from 3 to 6 cm after 15 seconds of earthquake loading.

For the 10^{-6} event, the maximum major principal stress varies from 7 to 9 MPa, and the maximum minor principal stress varies from -1.2 to -1.5 MPa. The Mohr-Coulomb strength/stress ratio and the shear stresses do not change significantly from the 10^{-4} event. However, the major and minor principal stresses are higher than for the 10^{-4} event. A comparison of the maximum major and minor principal stresses in Table 1 shows a 20% to 75% increase in stress from the 10^{-4} to the 10^{-6} event. The strength of the intact rock still greatly exceeds the stresses generated by the earthquake. However, with the higher loading, about 27% of the elements yielded around the opening and near the surface. The maximum displacements are due to block (wedge) movements and range from 24 to 47 cm. Maximum differential displacements are estimated to be 7 to 15 cm after 5 seconds of earthquake loading, and are 10 to 20 cm after 10 to 15 seconds of earthquake loading. For the 10^{-6} event, the displacements are mainly due to wedge movements.

3.2 Discontinuous Deformation Analysis

The Discontinuous Deformation Analysis computer code (DDA; Shi, 1989) was also used to analyze the static and dynamic performance of the

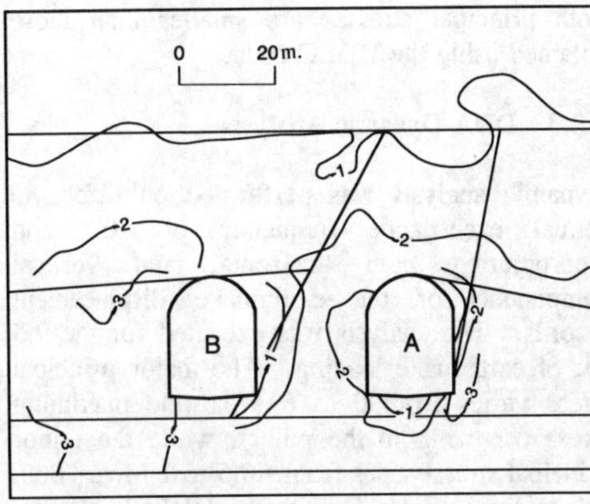

Fig. 4. 10^{-4} event, UDEC maj. principal stress.

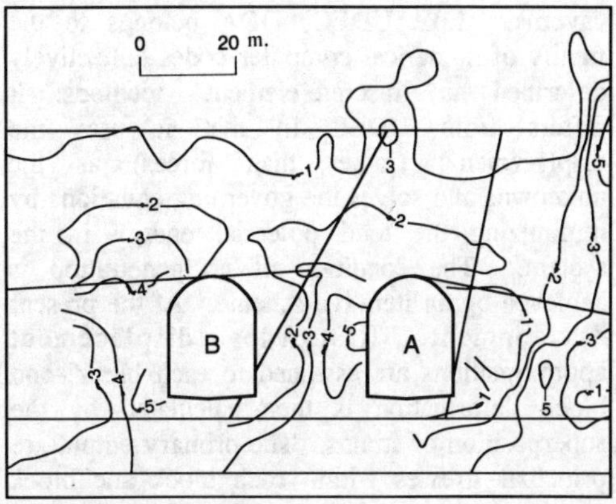

Fig. 6. 10^{-6} event, UDEC major principal stress

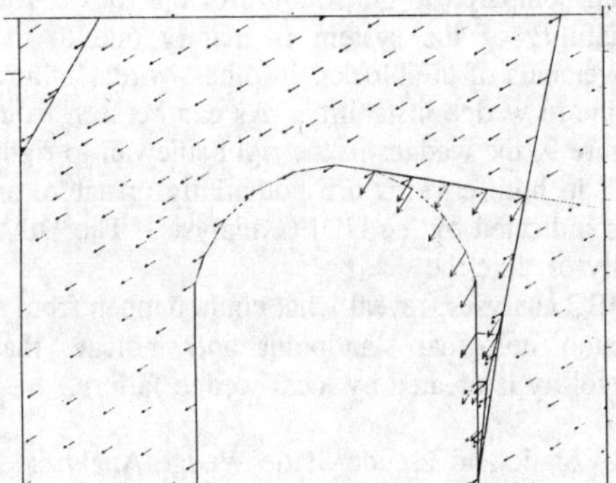

Fig. 5a. 10^{-4} event, UDEC disp. for cavern A.

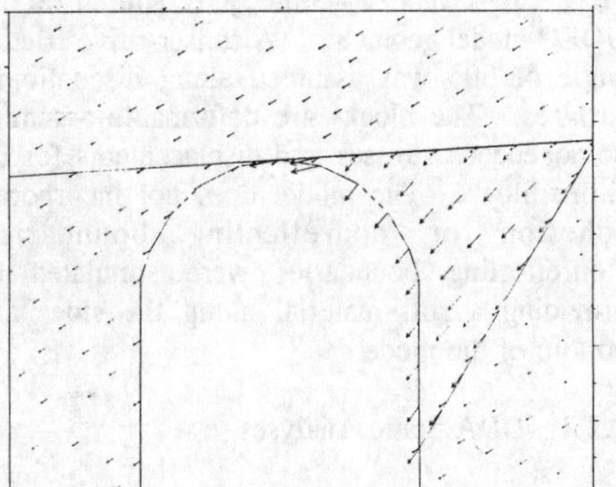

Fig. 7a. 10^{-6} event, UDEC disp. for cavern A.

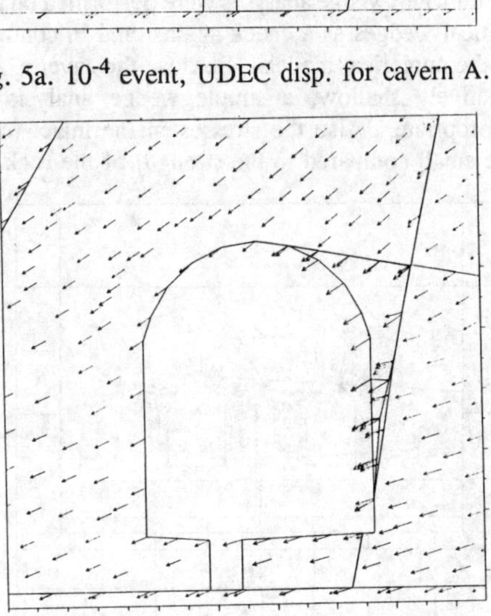

Fig. 5b. 10^{-4} event, UDEC disp. for cavern B.

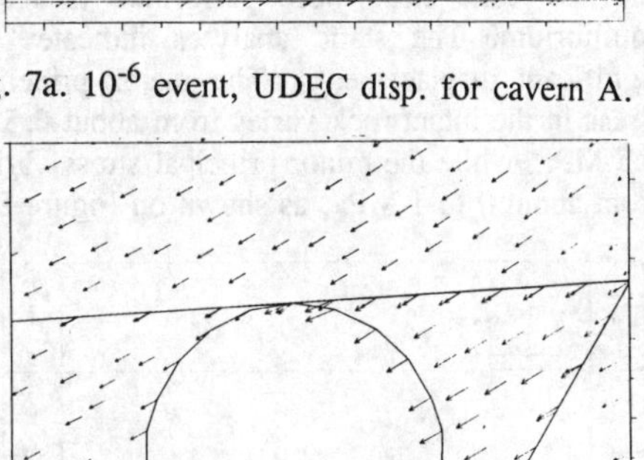

Fig. 7b. 10^{-6} event, UDEC disp. for cavern B.

caverns. Like UDEC, DDA belongs to the family of numerical computer codes collectively described as "discrete-element" methods. It differs from UDEC in that it uses the displacements (rather than forces) as the unknowns and solves the governing equations by minimizing the total potential energy in the system. The condition of no penetration is achieved by an iterative scheme. At the present development, first-order displacement approximations are assigned to each block, and block deformation is then calculated by the superposition of strains. The primary output are principal stresses within each block and block positions at selected time steps. Consequently, our DDA interpretation and discussion is limited to these parameters.

The DDA model geometry is similar to the UDEC model geometry. A conservative friction angle of 30° was assumed along discontinuity surfaces. The blocks are deformable assuming homogeneous stresses and displacements for the entire block. The model does not incorporate cohesion or nonreflecting boundaries. Nonreflecting boundaries were simulated by providing a soft material along the sides and bottom of the model.

3.2.1 DDA Static Analyses

Unlike in UDEC analyses, the excavation was performed by removing the material all at once and 250 time steps were performed to reach equilibrium. The static analyses indicates no significant displacement. The major principal stress in the intact rock varies from about 0.5 to 3.5 MPa, while the minor principal stress varies from about 0 to 1 MPa, as shown on Figure 8.

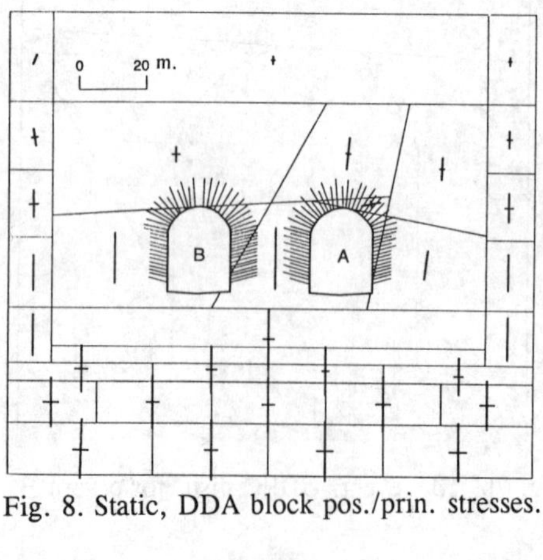

Fig. 8. Static, DDA block pos./prin. stresses.

Both principal stresses are smaller than those obtained using the UDEC code.

3.2.3 DDA Dynamic Analysis

Dynamic analysis was performed only for the annual exceedance frequency of 10^{-4} and incorporating both horizontal and vertical components of the earthquake displacement records. The analysis was executed for the 9.5 sec of earthquake loading. The major principal stress varies from 0 to 6 Mpa (the maximum stress occurring in the pillar), while the minor principal stress varies from - 0.5 to 1 Mpa. Both values are comparable with the UDEC's results. As with the results from UDEC analyses, the earthquake induced stresses in the intact rocks are small compared to the strength of the rocks. The instability of the system is mainly due to the movements of the blocks. In other words, failure is due to wedge instability. As can be seen from Figure 9, the wedges in the right sidewall to right arch in both caverns are potentially unstable, as also indicated by the UDEC analysis. The DDA analyses, like the UDEC analyses, reveal what could happen from a phenomenological standpoint and indicate that instability is created by local wedge failure.

3.3 Static and Pseudo-Static Wedge Analyses

Static and pseudo-static, two-dimensional, limit-equilibrium wedge analyses were performed on the critical wedges as a check against and in contrast to the numerical models. Because the caverns are relatively shallow, a simple wedge analysis is appropriate. Also the stresses in the intact rock are small compared to the strength of the rock.

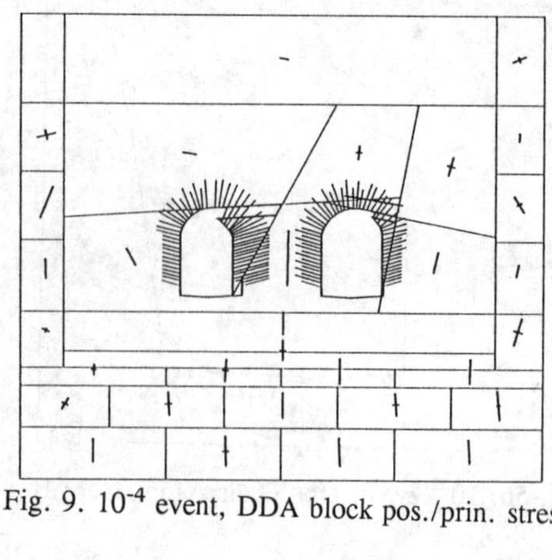

Fig. 9. 10^{-4} event, DDA block pos./prin. stresses.

Failure of the excavation is governed by local failure; hence limit-equilibrium methods may be applied. The pseudo-static method is also a simplified approach readily assessing the capacity of the support system to support the wedge during earthquake loading. However, unlike discrete-element modeling, displacements cannot be estimated.

The shear strength of the joints was conservatively calculated by using a joint friction angle of 25°and no cohesion. The rock bolt pattern assumed is shown in Figure 2. It was assumed that 80% of the ultimate strength of the bolt/cable in tension and 40% of the ultimate strength in shear across the minimum cross-sectional bar area were available for mobilization and that these values were within the working stresses of the bar.

The critical wedges were analyzed under the 10^{-4} and 10^{-6} earthquake loading events. In a pseudo-static analysis, typically only the horizontal accelerations (k_h) are used. This is because the vertical accelerations (k_v) are smaller, and rarely do the peak horizontal and vertical accelerations occur at the same time. However, for comparison with UDEC and DDA, which incorporated time histories for horizontal and vertical motion, the wedges were also analyzed to include the vertical acceleration. The pseudo-static acceleration was based on two-thirds of the mean peak ground acceleration to account for the transient nature of ground motions and to represent the average repeatable accelerations. The method for calculating the factor of safety of the wedges is shown on Figure 10. The results of the wedge analysis are summarized in the Table 2 below.

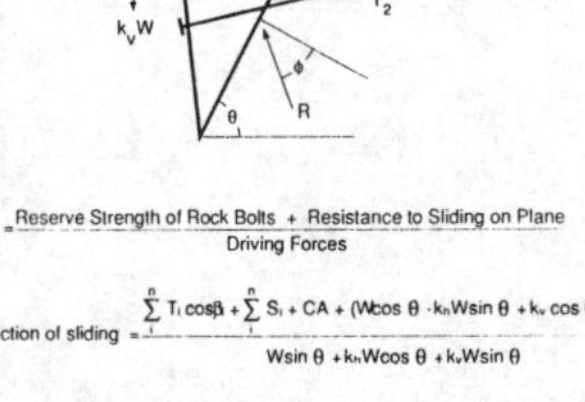

$$F.S. = \frac{\text{Reserve Strength of Rock Bolts} + \text{Resistance to Sliding on Plane}}{\text{Driving Forces}}$$

$$F.S \text{ in direction of sliding} = \frac{\sum_{i}^{n} T_i \cos\beta_i + \sum_{i}^{n} S_i + CA + (W\cos\theta - k_h W\sin\theta + k_v \cos\theta)\tan\phi}{W\sin\theta + k_h W\cos\theta + k_v W\sin\theta}$$

Fig. 10. Pseudo-static model.

Table 2. Factor of safety, limit equilibrium method.

Cavern	Static	k_h 10^{-4}	$k_h + k_v$ 10^{-4}	k_h 10^{-6}	$k_h + k_v$ 10^{-6}
A	2.0	1.9	1.7	1.7	1.1
B	1.5	1.2	1.1	0.8	0.6

Factors of safety for Cavern B for the 10^{-6} event were less than 1; 0.8 and 0.6 for k_h and $k_h + k_v$, respectively, and are in general agreement with DDA and UDEC results.

3.4 Behavior of underground structures during historical earthquakes

Empirical studies show minor damage to tunnels for mean peak ground acceleration (mpga) from 0.2 g to 0.5 g and damage to tunnels for mpga from 0.5 g and greater (Dowding and Rozen, 1978; Sharma and Judd, 1991). It should be noted that the input motion for UDEC was based on earthquake time histories recorded at the free field of a rock site. Few data exist on subsurface earthquake records. However, as shown by Sharma and Judd (1991), consider-ably less damage occurs to underground struc-tures that are at depths greater than 50 meters, and no heavy damage occurs below 300 m. Therefore, these surface earthquake records must be deconvoluted to the appropriate depth before being input to UDEC type analyses. This deconvolution may alter some frequency contents of the ground motions.

4 CONCLUSIONS AND LIMITATIONS

The static analyses show that for the intact rock stability, the caverns are safe. The analyses show that the intact rock does not result in failure by compression or tension. Induced stresses in relatively shallow caverns in competent rock are of less concern than is wedge movement. The UDEC and DDA dynamic analyses shows that instability is created by local wedge or block failures. These failures can be prevented by systematic and timely installation of support (i.e., rock bolts and shotcrete) before plastic deformation occurs. The pseudo-static, limit-equilibrium wedge analyses, and empirical studies contrast with the discrete-element modeling of UDEC and DDA. They cannot be compared easily, but can be used to assess the reasonableness of the UDEC and DDA results and the phenomena observed in the numerical models. The results are in general agreement with UDEC and DDA. Lower factors of safety and values less than 1 were calculated for the 10^{-6} event, indicating potential instability and need for additional rock bolt support.

One limitation imposed on all the analyses is that they involve a two-dimensional representation of a three-dimensional problem. In three-dimensional analysis, failure can occur on one or more failure planes. Furthermore, in lieu of actual rock joint data, simplified geometries were assumed in all the UDEC, DDA, and limit-equilibrium analyses. Finally, the discrete-element models do not model the formation of new cracks or shear planes due to high internal stresses. In addition, the joint model assumed an elastic, perfectly plastic Mohr-Coulomb behavior. Under dynamic loading, it will be more critical to look at others non-linear joint models, such as the Barton-Bandis joint model (Bandis et al., 1985). This model allows for degradation of joint shear strength during cyclic loading, a phenomenon important to any geological materials. However, the analyses performed used conservative joint and shear plane friction values to simulate residual and degraded joint properties.

5 REFERENCES

Bandis, S.C., Barton, N.R., and Christian son, M., 1985, Application of a new numerical model of joint behavior to rock mechanics problems: Proc. Inter. Sym. on Fundamentals of Rock Joints, Bjorkliden, Sweden, September 1983, pp. 345-356.

Cundall, P.A., 1971, A computer model for simulating progressive large scale movements in blocky rock systems: in Proc. of the Sym. of the Inter. Soc. for Rock Mech., Nancy, France, v. 1.

Dowding, C.H. and Rozen, A., 1978, Damage to rock tunnels from earthquake shaking, J. Geotech. Eng. Div., Amer. Soc. of Civil Eng., 104 (GT2), 175-191.

Fairhurst, C. and Pei, J., 1990, A comparison between distinct element method and the finite element method for analysis of the stability of an excavation in jointed rock, tunnelling and Underground Space Technology, Pergamon Press, vol. 5, n. ½.

Hendron, A.J., and Fernandez, G., 1983, Dynamic and static design considerations for underground chambers in Howard, T.R., editor, Seismic Design of Embankments and Caverns, Proc. Amer. Soc. of Civil Eng. Geotech. Div., Philadelphia, Penn., May.

Hoek, E. and Brown, E.T., 1980, Underground excavations in rock, Inst. of Mining and Metallurgy, London

Sharma, S., and Judd, W.R., 1991, Underground opening damage from earthquakes, Eng. Geology, 30: 263-276.

Shi, G.-H., 1989, Discontinuous Deformation Analysis - A new numerical model for the static and dynamics of block systems, Ph.D. Thesis, University of California, Berkeley, August.

Tunnelling and Ground Conditions, Abdel Salam (ed.) © 1994 Balkema, Rotterdam, ISBN 90 5410 363 9

Semi-quantitative estimation of water shielding requirements and optimization of rock cover for sub-sea road tunnels

Per Hagelia
Norwegian Road Research Laboratory, Oslo, Norway

ABSTRACT: An important task related to planning of sub-sea road tunnels is to estimate the extent and cost of water sealing. This paper presents a new conceptual approach and a preliminary empirical equation. The model was developed from simple theory, practical experience and quantitative tunnel data. It is shown that the requirement for water panels is a function of hydraulic gradient, rock cover and a new joint frequency parameter. For the first time we may obtain a semi-quantitative estimate of the water shielding requirement using geological field data. The equation provides an option for optimization of rock covers for a range of rock conditions and sea-depths.

1. INTRODUCTION

An important research topic in engineering geology is to develop pre-investigation methods which can be used to convert field data to parameters of direct relevance for the cost analysis. With respect to tunnelling, the rock mass should preferably be classified in quantitative terms, because cost factors such as rock reinforcement and water protection measures are functions of the rock mass variability on-site.

Several methods are already available for the estimation of rock reinforcement. The most famous ones are perhaps the Rock Mass Rating method, (Bieniawski 1973) and the Q-system (Barton et al. 1974). These systems account for important theoretical constraints, and provide empirical relationships between quantified joint characteristics and extent and type of ground support. Moreover, they represent standardized engineering routines, an important prerequisite for efficient communication of experience between rock excavations.

However, no equivalent system has as yet been introduced for water sealing.

1.1 *Current engineering practice*

Pre-investigations for sub-sea road tunnel projects has up to now focused mainly on stability. This important part of the planning involves geological field mapping, localization of fault zones, as well as identification of the rock surface underneath the sea. Characterization of the rock mass in terms of seismic velocity provides important on-site data from which costs of rock support may be estimated.

In contrast, the potential for water leakage is usually not considered in much detail at the planning stage. Costs are generally inferred from completed tunnel projects, without making use of quantified geological and hydrolological data. Thus, the method involves subjective judgement factors.

Experience from previous projects, indeed, demonstrates this problem. Data from eight Norwegian sub-sea road tunnels show that the extent of water shielding and frost protection was variably underestimated at the planning stage (Øvstedal and Melby 1990). In average the protected area had to be increased by ca. 40 % compared to the original plans, in some cases to more than 100 %. Significant differences between planned and executed grouting were also reported. Only in a few instances, the actual costs of water sealing corresponded well with the forecast costs.

The discreapancies between forecast costs and actual costs may of cause have several reasons. Yet, a quantitative approach is necessary in order to put firm constraints on the total costs. Surprisingly, a study of 19 completed tunnel projects demonstrated that the reported tunnel data (geological e.t.c) were not sufficiently quantified to give a proper explanation for the actual extent of water sealing (Hagelia 1991a, 1991b). In particular, very little attention has been drawn to "little- to moderately jointed rocks", although water panels are frequently installed in tunnel sections sharing such characteristics. Seven of the projects were, according to present standards, well documented sub-sea tunnels. Undoubtedly, therefore, the current engineering practice merely reflects the present state of knowlegde.

1.2 *Objectives*

The main motivation for a water seepage analysis related to sub-sea tunnel projects is to estimate:

1. the potential for localized large leakages and associated requirement for grouting.
2. the pump capacity.
3. the total area of minor seepages which requires water shielding/frost protection.

If this can be done, on-site data may be used to find a tunnel depth "at which the tunnel length is weighed against costs of water inflow prevention measures" (Grønhaug 1978).

This paper provides important new constraints on such modelling, and deals primarily with problem 3. A preliminary empirical equation, which incorporates the most significant variables, has been developed. This allows us to estimate semi-quantitatively the requirement for water panels on the basis of joint frequency statistics. The influence of rock cover can also be modelled.

2 A CONCEPTUAL MODEL

2.1 *Important variables and problems*

Fig. 1 shows a vertical section of a typical sub-sea tunnel. In accordance with established flow laws, water seepage into the tunnel depends significantly on the hydraulic gradient (h = tunnel depth beneath sea level), the thickness of the rock cover (h_r) as well as soil cover (h_s), and the permeability of rock and soil, respectively.

Typically the main challenge is to obtain representative estimates of the variation in permeability. This may be achieved by direct measurements of hydraulic conductivity, or indirectly by using joint frequency data.

The hydraulic conductivity of a given rock mass is a function of its joint network. Direct estimates of hydraulic conductivity may be obtained from exploratory drilling (pumping tests e.t.c.). It should, however, be noted that such tests may be representative of only small volumes, sometimes giving results of dubious practical value. This is due to the fact that test lengths intersect a relatively small number of fractures, thereby being sensitive to scale effects (Neuman 1987; Carlsson et al. 1990).

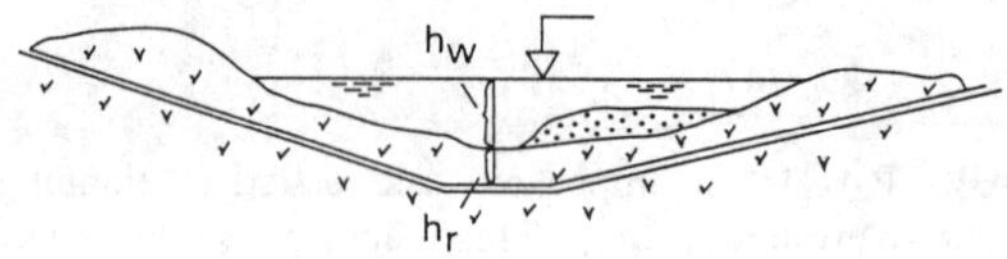

Fig. 1 Vertical section parall to sub-sea tunnel. h_r and h_w correspond to thickness of rock cover and sea depth, respectively. $h = (h_r+h_w)$ is the distance from sea level to tunnel, an expression of hydraulic gradient. Dotted area represents soil cover (h_s). The key issue is to identify a joint parameter which simulates variations in hydraulic conductivity. Note that this depends on an evaluaton of scale effects (see text for details).

It is well established that fluid flow through a fracture is strongly dependent on the size of the joint aperture, but also on joint roughness and tortuousity. Experiments show that the volumetric flow rate through natural joints is related to, and mostly greater than, the cube of the aperture size (cf. Pyrak-Nolte et al. 1990). A single large-aperture joint (some mm) may result in a very large water leakage, and demand for grouting.

However, an important feature of worldwide geological observations is that, in each region, joints with relatively large apertures are very few (i.e. covering a very small area of a tunnel) compared with the number of joints with small apertures (cf. Barton and Zobach 1990). Thus, statistically the overall requirement of water shielding, or areal distribution of leakages in sub-sea tunnels, is:

1. little sensitive to the number of large apertures (locally high hydraulic conductivity).
2. very sensitive to the number of joints with small size apertures (generally low hydraulic conductivity).

Indeed, experience from tunnelling shows that the small water seepages which typically require water panels are located in rocks with ordinary small apertures (<< 0.5mm).

Therefore, we aim at characterizing the *gross scale hydraulic conductivity* in relatively large volumes of rock. It is noteworthy that classical water seepage modelling only focuses on volumetric flow rate (leakage in litres/minute). In contrast, water shielding in a particular tunnel section depends both on the volumetric flow rate (and therefore on the gross scale hydraulic conductivity of the rock mass), as well as the areal distribution of leakage points in that section. Herein we investigate the possibility of using some kind of joint frequency data in order to meet these requirements. The present modelling involves:

1. identification of a joint parameter which simulates the variation in hydraulic conductivity, and at the same time expresses the potential for spread leakage.
2. application of Darcy's law using an approach which minimizes the influence of scale effects and associated non-linear terms.

2.2 *Darcy's law, scale effects and Representative Elementary Volumes*

According to Darcy's law:

$$Q = k\ A\ \Delta h\ /\ L \qquad (1)$$

where Q = volumetric flow rate; k = hydraulic conductivity; A = cross sectional area normal to flow direction; Δh = hydraulic gradient; and L = length of traversed medium (cf. Chapman 1981). Darcy's law applies only to saturated ground, and assumes laminar flow in homogeneous and isotropic media. Numerical studies by Long et al. (1982) has shown that only when a rock mass contains a "sufficient number of cracks" it may be treated as a Darcy medium. This scale effect is simply a result of the inhomogeneous nature of fractured crystalline rocks. The fever joints, the greater is the influence from each individual joint.

Hence, the first step in our modelling is to estimate the critical number of joints, or rather; the size of a rock volume which is likely to contain at least this number. This problem is related to the determination of a Representative Elementary Volume (REV), a minimum volume beyond which there is no influence from scale effects.

Oda (1988) found that the size of a REV is at least three times larger than the typical joint trace length. Drilling tests at Aspø in Sweden indicate that hydraulic conductivities in rocks with very low joint frequencies (ca. 1 fracture per metre) were not significantly influenced by scale effects for test scales in excess of ca. 50-100 m (cf. Carlsson et al. 1990). Taken together this suggests that if the typical joint length is 10 m, a 50 m tunnel section should be close to a REV, particularly if the joint frequency is > 1/m.

Recognizing that joint frequencies in most cases are well above this figure, we assume that 50 m tunnel sections define a REV. Average data from 50 m sections (each corresponding to ca. 1000 m^2 of exposed rock surface), therefore, make up the empirical basis in our modelling (see below). Data collection at this scale has the advantage that we now can be quite sure that the flow rate is linearly correlated with hydraulic gradient. Moreover, in view of the small apertures of most joints, water seepages from individual joints should in general be small and characterized by low velocity and laminar flow. High velocity flow promotes a turbulent regime (cf. Soyeaux 1990).

Thus, having minimized the probability of deviations from Darcy's law, we attempt to identify a suitable "leakage sensitive" joint parameter.

2.3 *Significance of steep to vertical joints, and definition of a new joint parameter*

In a study of several deep-seated tunnels, Selmer-Olsen (1981) demonstrated statistically that 95 % of large and permanent leakages (1-100 l/min; large apertures) were associated with joints dipping 60-90° (i.e. steep to vertical). Moreover, the number of such leakages was systematically increasing with increasing dip angle (Fig. 2). The reason for this is probably that the flow direction in most tunnels is sub-vertical. Thus, the steep to vertical joints represent the shortest pathway along the hydraulic gradient.

Similar relations have also been observed from water seepage along small-aperture joints. Indeed, the smaller water seepages are typically related to sub-vertical joints (Storås 1988; Blindheim and Øvstedal 1990; Hagelia 1992). Such a consistency is expected, because there is no conceptual difference between joints with small and large apertures.

This experience is extremely significant for the identification of a leakage sensitive joint parameter. Obviously, incorporation of all joints would seem to introduce a large uncertainty. We, therefore, infer that the number of steep to vertical joints per m^2 of rock is the most relevant joint frequency parameter in the present context. Hence, we define:

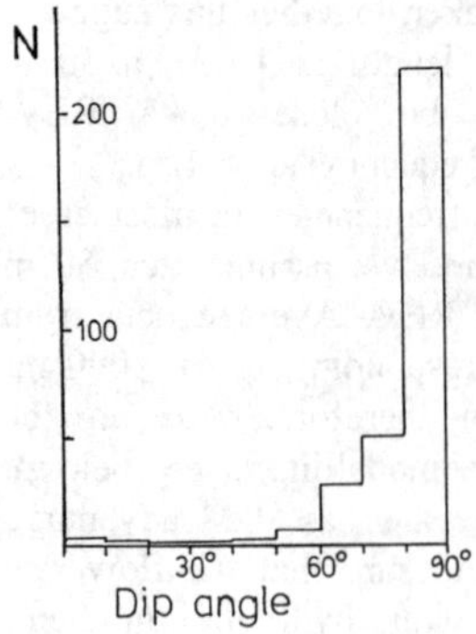

Fig. 2. *Influence of dip angle on the occurence of large leakages. N is number of leaking joints. Note that 95 % of leakages are associated with steep to vertical joints (dipping 60-90°). Data of Selmer-Olsen (1981).*

J = average number of joints/m^2 dipping 60-90°. The size of the apertures are < 0.5 mm.

Average J-numbers representative of 50 m tunnel sections should, therefore, to a great extent be a reflection of the effective hydraulic conductivity. It is here implicitly assumed that the large scale invoked, tends to cancel out effects of small range aperture variations, as well as the most significant effects of variable joint roughness and tortuousity. Joints may be considered to be well distributed rather than clustered at this scale. Hence, in a broad sense the potential for spread leakage should be positively correllated with J.

It is, however, possible that J slightly overestimates the permeability in situations where water leakage is restricted to one of several sub-vertical joint sets. Yet, the presence of additional steep and "dry" joint sets contributes to seepage at intersection points. Indeed, when the number of joint connections increases this should lead to enhanced permeability (cf. Zhang et al. 1992).

2.4 *Simple mathematical approach*

Darcy's law (equation 1) may now be developed in terms of sub-sea tunnel parameters. Δh corresponds to the vertical distance (h) between sea-level and tunnel, and L corresponds to the rock cover (h_r) when the soil cover can be neglected. Thus, considering tunnel sections of constant areas (ca. 1000 m^2 = A = unit tunnel area), equation (1) may be rewritten:

$$Q = k\, h / h_r \tag{2}$$

Since J approximately simulates the difference in hydraulic conductivity between unit tunnel areas, we assume:

$$k \approx \alpha J \tag{3}$$

where α is a constant. Hence,

$$Q \approx \alpha J h / h_r \tag{4}$$

According to Darcy's law (eq. 1) the term k/h_r represents the rock conductance (c), which states that at constant hydraulic conductivity and gradient, water inflow (l/min) will decrease as the thickness of rock cover increases. Thus, when k/h_r is constant the water inflow (Q) increases linearly with hydraulic gradient (cf. Fig. 3a). Based on (2) and (3) we define conductance (C) in terms of the geological parameter J:

$$C = J/h_r \tag{5}$$

Similarly, resistance to flow (R) is expressed:

$$R = h_r / J \tag{6}$$

We further define X as the requirement of water shielding in percent of a unit tunnel area. X is linearly correlated with the actual leakage area, but somewhat larger (NOTE: we do not include water panels which have been installed for reasons unrelated to actual water seepage, for example in order to maintain good symmetry)

It is important to note that X may not readily be substituted into (4), for two reasons:

1. the volumetrical smallest leakages do not require shielding due to evaporation and some tolerance.

2. above a certain level of leakage the shielding requirement reaches 100 % of a unit tunnel area, irrespective of volumetric inflow.

This is illustrated in Fig. 3b, showing an h vs. X diagram. We may assume that when X = 0, then Q > 0, and that Q may reach q_1 l/min (small!) before shielding is required (at X=0 %). The X = 0-100 % range is expected to be accompanied with a steady increase in water inflow. Q may also increase linearly beyond the value which requires 100 % shielding.

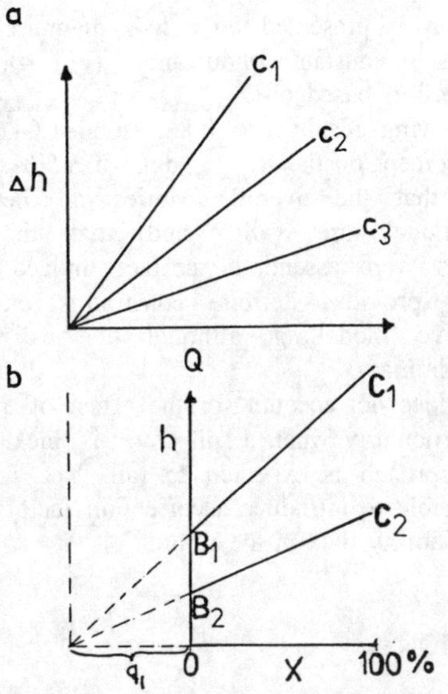

Fig. 3a. Darcy's law; the volumetric flow rate (Q) correlates with hydraulic gradient (Δh) when conductance (c) is constant; $c_1 < c_2 < c_3$. Fig. 3b. Modification of Darcy's law applied to water shielding requirement (X % of unit tunnel areas) in sub-sea tunnels at variable depths (h) beneath sea level. Rock conductance $C_1 < C_2$ (where $C = J/h_r$; J is a joint parameter defined in text and h_r= rock cover). Very small leakages up to q_1 are accepted before water shielding is required (i.e. X = 0 when $Q = q_1$).

A consequence of this model (Fig. 3b) is that lines of constant conductance (C) have intercepts (B) on the h-axis. Thus, for a given conductance (C_1), a critical hydraulic gradient ($h = B_1$) must be reached before water shielding is required (X = 0 % up to B_1). Tunnel sections with no requirement for water shielding, may be treated as domains where this critical hydraulic gradient has not been reached. Both intercepts and slopes (herein called A) decrease as a function of increasing conductance.

From this we infer that the shielding requirement, X, for tunnel sections with variable C, is related to tunnel depth, h, by the following general equation:

$$h = AX + B \tag{7}$$

If the model given in Fig. 3b is strictly valid, linear correlations of rock resistance (R) with A and B are expected (from detailed inspection of Darcy's law). Accordingly, equation (7) may be rewritten:

$$h = \beta RX + \gamma R \tag{8}$$

where β and γ are constants which relate the slopes and intercepts of lines on the h-X diagram to the numerical values of R.

However, in view of the approximations we have made, empirical data are necessary in order to test this further.

3. CALIBRATION

Using empirical data from two sub-sea tunnels (Maursund and Flekkerøy tunnels) it was found that the general model (eq. 7) closely resembles reality. Hence, J is a joint parameter which can be used with success in practical modelling. It should be noted that conventional joint parameters such as RQD value and Q-value were tried, but no good systematics appeared. Herein we include the most significant results pertinent to calibration. The full documentation is given in Hagelia (1992), and will be published together with additional data elsewhere.

3.1 *Empirical data*

Average tunnel depths (h) and rock covers (h_r) for 50 m tunnel sections were obtained from vertical tunnel profiles. J-numbers and water shielding requirement (X) were obtained from detailed tunnel mapping. J for a complete 50 m section was obtained by counting the average J-number on both sides in 10 m sub-sections. Thus, each J-number included in the calibration is the arithmetic mean of 10 individual estimates. This proceedure was thought to improve the accuracy. A scale bar was used as a tool during counting.

Investigations were restricted to the sub-sea sections, because here the water level was well constrained. The data set, covering 1650 m of tunnel, is totally dominated by sections without soil cover (in a few instances it was assumed that rather thin soil covers have permeabilities similar to rock).

The rocks in both tunnels are gneisses, usually showing little or no evidence of deep whethering along the joints. Joint roughness (J_r in the Q-system) ranges from 1-2. There is no general correlation between J-number and RQD value.

3.2 *An empirical equation*

From field data conductance (C) and resistance (R) were computed according to equations (5) and (6). Discriminating between conductance groups, it was found that tunnel sections with similar C, show good linear correlations when plotted on h vs. X diagrams (correlation coefficients, r = 0.92-0.96). From linear regression of each group it was possible to estimate A and B in equation (7). Plotting R versus A and B, respectively, provided a basis for further calibration.

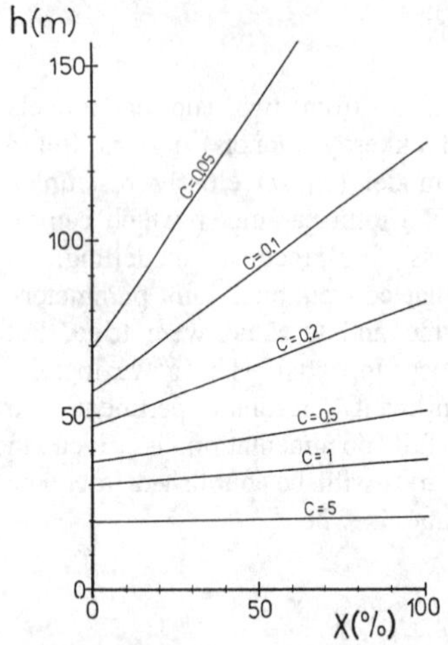

Fig. 4. h-X plot showing the calibration (equation 11) for variable rock conductance (C).

Slopes (A) were found to vary linearly with R. Linear regression gave a good fit (r = 0.97) with slope = 0.0713 (i.e. β in equation 8), and intercept = 0.0475. Thus, approximately;

$$A = 0.0713\,R \qquad (9)$$

Intercepts (B) are related to R by a power law, thus deviating from equation (8) at this point (r = 0.81):

$$B = 30.17\,R^{0.2741} \qquad (10)$$

Upon rearrangement of equation (7) and substitution for A and B, we obtain the following empirical equation:

$$X = \frac{h \div 30.17\,(h_r/J)^{0.2741}}{0.0713\,(h_r/J)} \qquad (11)$$

This calibration is presented on a h-X diagram, including lines of constant conductance (Fig. 4). A linear regression based on $X_{calculated}$ vs. $X_{measured}$ gave the following result; r = 0,86, slope=1.04, with a low intercept on the $X_{calculated}$-axis of 3.5 %. This proves that the overall features of the conceptual model are valid, and that the approximations were essentially correct. Indeed, equation (11) provides definite constraints on semi-quantitative modelling, although the calibration in preliminary.

The model does not account for the effect of a soil cover. Particularly when a soil cover is thick, the present approach is expected to fail. Yet, it should be possible to introduce a correction factor based on permability data in the future.

4. APPLICATIONS

The calibration provides a completely new tool for the estimation of water shielding requirements in sub-sea tunnels. For the first time we are permitted to undertake a semi-quantitative modelling of this important cost factor already at the planning stage.

4.1 Pre-investigation and cost analysis

It is suggested that J-number statistics should be made available at the earliest posible stage of pre-investigation. It is of particular interest to observe how the J-number varies regionally. At Flekkerøy it was found that J was dependent on rock type (Hagelia 1992). Otherwize it is expected that regional trends may occur in a single rock unit, for example when approaching fault zones.

When a tunnel corridor has been proposed, detailed counting within 50-100 m sections may be done. Combined with constraints on water level it should be possible to estimate the water shielding requirement under land.

Interpretation of the sub-sea tunnel section is more complicated, because we to a great extent must rely on indirect observations (acoustic sounding, refraction seismics, tomography e.t.c.). However, it is frequently possible to locate the main lithological boundaries approximately, together with the main weakness zones. Joint statistics from field work may then be used in combination with constraints on rock quality from seismics or other methods. Barton et al. (1991) found that Q-values may be determined indirectly from seismic velocity (V_p). Thus, based on field data of joint characteristics in general, indirect Q-estimates may also give information relevant for J-number approximation.

Having calculated X for sub-sections covering the entire tunnel corridor, the price of water shielding may be estimated:

$$\text{Price} = \frac{\Sigma (X_1, \ldots X_n)}{n\ 100} P\ A \qquad (12)$$

where n=number of tunnel sections, P = price/m^2 for a given water panel, and A = total tunnel area.

4.2 Optimization of rock cover

The empirical equation (11) can be used to model how the shielding requirement varies with rock cover for any fixed value of J. It is possible to construct such X vs. h_r diagrams with curves for different sea depths ($h_w = h \div h_r$). This provides an option for modelling optimal rock covers (i.e. the rock covers which result in the lowest possible shielding requirement), for a range of rock conditions and sea depths.

Fig. 5 illustrates the dependence of shielding requirement on rock covers for sea depths ranging from 0 to 100 m. In this example all curves are calculated with a constant joint frequency (J = 5/m^2). The optimal rock cover for a tunnel section in such rocks underneath 35 m of sea is approximately 25 m, corresponding to a water shielding requirement of ca. 35-40 % (stippled line). In the case of 50 m sea depth the optimal rock cover is approximately 60 m and the theoretical shielding requirement increases to ca. 60 % (dotted line).

Similar diagrams calculated for increasingly higher J-numbers (not illustrated herein) show that curves for constant sea depths become systematically displaced towards higher shielding requirement. At the same time minima occur at thinner and thinner and rock covers. In view of potential stability problems associated with increasing joint frequencies, many of the thin optimal rock covers from this modelling should be treated with causion.

The sub-sea environment is corrosive and may damage various types of installations. Chloride penetration may reduce the strength of shotcrete, thereby reducing the tunnel stability. Thus, the present account has a direct bearing on maintenance costs as well.

A complete optimization proceedure involves calculation of the appropriate diagram for each 50-100 m sub-section, identification of the optimal value for each individual section, followed by weighting of these optima. Together with a stability analysis, this should provide essential information for the identification of a safe and cost efficient rock cover.

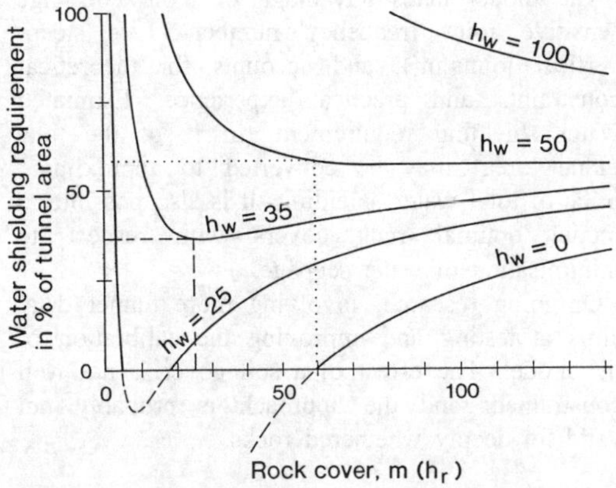

Fig. 5. Water shielding requirement in % of tunnel area (X) as a function of rock cover (h_r) when J= 5. Note that the optimal rock cover depends on sea-depth (h_w).

5. PROBLEMS AND FUTURE DEVELOPMENTS

The calibration given by equation (11) is preliminary and the uncertainty is, therefore, as yet not well constrained. On-going research aim at improving the calibration by including more empirical data.

The model is probably not valid for rocks characterized by extensive and deep whethering. It should be recalled that calibration was based on data from gneisses where whethering was only weakly developed.

If joint material is likely to be eroded or dissolved as a consequence of excavation, leakage will inevitably increase. Future research should focus on this problem, because it might shed light on maintenance costs, and put further constraints on the quantitative modelling.

With respect to investigations of the sub-sea part of a tunnel, the present author has found that zones of high J-numbers can be detected on available resistivity profiles (Hagelia unpubl. data). This should be investigated further.

6. CONCLUSIONS

Current pre-investigation methods for evaluation of water seepage into tunnels involve subjective judgement factors. In an attempt to improve the methods, a preliminary empirical equation for water shielding requirements in sub-sea tunnels has been developed.

The model takes advantage of a new leakage sensitive joint frequency number (J = steep-vertical joints/m^2), and accounts for theoretical constraints and practical experience. Estimated water shielding requirement (in % of the total tunnel area) may be converted to approximate price of total water shielding. It is also possible to model optimal rock covers with respect to minimisation of water seepage.

On-going research, involving more tunnel data, aims at testing and improving the calibration of the model. The effect of a soil cover is not well constrained, and the approach is probably not valid for deeply whethered rocks.

REFERENCES

Barton, C.A. & M.D. Zobach 1990. Self-similar distribution of macroscopic fractures at depth in crystalline rock in the Cajon Pass scientific drillhole. In N. Barton & O. Stephansson (eds.), *Rock Joints*, p.163-170. Rotterdam, Balkema.

Barton, N., R. Lien & J. Lunde 1974. Engineering classification of rock masses for the design of tunnel support. *Rock Mechanics* 6: 189-236.

Barton, N. 1991. Geotechnical design. *World tunnelling*, November 1991.

Bieniawski, Z.T. 1973. Engineering classification of jointed rock masses. *Transactions, South African Institution of Civil Engineering* 15 (12): 335-344.

Blindheim, O.T. & E. Øvstedal 1990. Water control in subsea road tunnels in rock. In J. Krokeborg (ed.), *Strait Crossings*, p. 223-230. Rotterdam, Balkema.

Carlsson, A., G. Gustafson, U. Lindblom & T. Olsson 1990. Scale effects in the determination of hydraulic properties in rock masses. In A.P. Cunha (ed.), *Scale Effects in Rock Masses*, p. 103-117. Rotterdam, Balkema.

Chapman, R.E. 1981. *Geology and water. An introduction to fluid mechanics for geologists*. The Hague, Martin Nijhoff/Dr. W. Junk Publ.

Grønhaug, A. 1978. Requirements of Geological Studies for Undersea Tunnels. *Rock Mechanics*, Supplement 7: 27-40.

Hagelia, P. 1991a. Vasslekkasje i landtunnelar og undersjøiske tunnelar (1). *Internal report* 1498. Oslo, Norwegian Road Research Laboratory.

Hagelia, P. 1991b. Vasslekkasje i landtunnelar og undersjøiske tunnelar (II). *Internal report* 1499 Oslo, Norwegian Road Research Laboratory.

Hagelia, P. 1992. Semi-quantitative estimation of water shielding requirements for sub-sea road tunnels using geological and hydrological data. *Internal report* 1569. Oslo, Norwegian Road Reasearch Laboratory.

Long, J.C., J.S. Remer & P.A. Witherspoon 1982. Porous media equivalents for networks of discontinous fractures. *Water Resources Research* 18/3: 645-658.

Neuman, S.P. 1987. Stochastic continuum representation of fractured rock permeability as an alternative to the REV and fracture network concepts. In I.W. Farmer, J.J.K. Daemen, C.S. Desai, C.E.Glass & S.P. Neuman (eds.), *Rock Mechanics, Proceedings of the 28th US Symposium*, p. 533-561. Rotterdam, Balkema.

Oda, M. 1988. A method for evaluating the representative elementary volume based on joint survey of rock masses. *Canadian Geotechnical Journal* 25: 440-447.

Pyrak-Nolte, L.J., D.D. Nolte, L.R. Myer & N.G.W. Cook 1990. Fluid flow through single fractures. In N. Barton & O. Stephansson (eds.), *Rock Joints*, p. 405-412. Rotterdam, Balkema.

Selmer-Olsen, R 1981. Betraktninger over store vannlekkasjer i dyptliggende tunneler. *Fjell sprengningsteknikk, Bergmekanikk, Geotek*nikk 1981. Trondheim, Tapir.

Soyeaux, R. 1990. Seepage flow under dams with jointed rock foundations - A discrete hydraulic model with respect to laminar and turbulent flow. In N. Barton & O. Stephansson (eds.), *Rock Joints*, p. 745-752. Rotterdam, Balkema.

Storås, I. 1988. Ingeniørgeologisk analyse av vannlekkasjer og injeksjon i Hvaler-tunnelen. Thesis, Trondheim, The Norwegian Institute of Technology.

Zhang, X., R.M. Harkness & N.C. Last 1992. Evaluation of connectivity characteristics of naturally jointed rock masses. *Engineering Geology* 33: 11-30.

Øvstedal, E & K. Melby 1990. Future design of sub-sea tunnels based on cost and technical experience. In J. Krokeborg (ed.), *Strait Crossings*, p. 79-86. Rotterdam, Balkema.

Tunnelling and Ground Conditions, Abdel Salam (ed.) © 1994 Balkema, Rotterdam, ISBN 90 5410 363 9

A non-destructive assessment of tunnel lining stability

M. Hisatake
Department of Civil Engineering, Kinki University, Osaka, Japan

J. Murakami & T. Toki
Asanuma Corporation, Osaka, Japan

ABSTRACT: A non-destructive assessment of a model tunnel and an actual road tunnel under construction has been conducted using a back analysis method. The applicability of the method has been confirmed through comparison of analytical and measured results.

1 INTRODUCTION

In the construction of a tunnel, it is very important to maintain the safety of the tunnel by monitoring its movement. For this purpose, a back analysis method, coupled with field measurement results, has been recognized as a very powerful tool to aid in the construction of a tunnel.

In conducting an ordinary back analysis, it is necessary to input many tunnel displacement measurements. In practice, however, only a few of the displacements are measured at the site, and this restricts the applicability of the method. If the number of the displacements is fewer than that of unknowns, such as external loads, unlimited sets of solutions exist which can satisfy the displacements measured. In this case, therefore, it is impossible to uniquely obtain one set of solutions.

Recently, a new back analysis method has been proposed (Hisatake et al., 1992), in which just a few displacement measurements, as shown in Figure 1, are enough for the calculation of the external loads, strains and displacements of a tunnel lining.

In this paper, for the practical usage of the new back analysis method, the applicability of the method is firstly examined through model experiments on a tunnel lining. In order to find the beginning of the instability of the lining, a method to interpret the displacement measurement results is also proposed here. It requires no sophisticated theories.

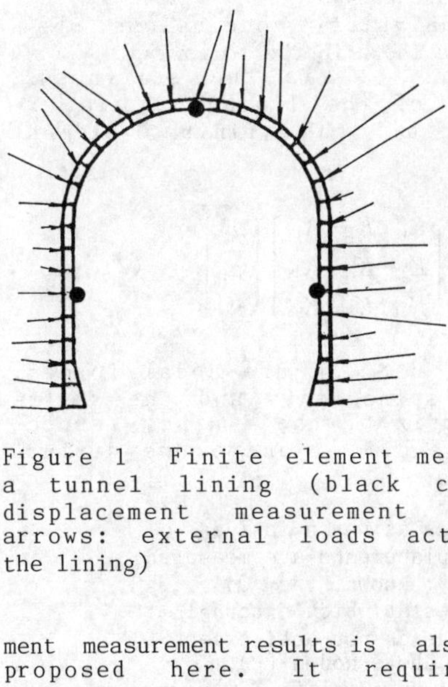

Figure 1 Finite element meshes of a tunnel lining (black circles: displacement measurement points, arrows: external loads acting on the lining)

Secondly, the new back analysis method is applied to an actual tunnel site for a non-destructive assessment of the tunnel stability.

2 BACK ANALYSIS

External loads acting on a tunnel lining can be back analyzed by an ordinary method if the following condition is satisfied (Hisatake et al., 1983):

$$Nu \geq Nf \tag{1}$$

where Nu is the number of lining displacements measured and Nf is the number of unknowns, such as external loads. Many values for the external loads are supposed in practical problems, as shown in Figure 1. Thus, according to equation (1), displacement measurements of a greater number must be conducted at the site when applying an ordinary back analysis. It is difficult to satisfy this condition, however, and it is unrealistic from an engineering point of view.

In order to overcome this theoretical shortcoming, a new back analysis method has been proposed by theoretically restricting the problems. The method gives one set of solutions and guarantees reasonable accuracy, even though there are far fewer displacement measurements than unknowns.

The analytical conclusions obtained are shown in the following.

Equation (2) is the stiffness equation of the lining. It is expressed by the finite element method,

$$\begin{Bmatrix} f^* \\ f_i \\ f_0 \end{Bmatrix} = \begin{bmatrix} K_{11} & K_{12} & K_{13} \\ K_{21} & K_{22} & K_{23} \\ K_{31} & K_{32} & K_{33} \end{bmatrix} \begin{Bmatrix} u^* \\ u_i \\ u_0 \end{Bmatrix} \quad (2)$$

where f, u and K are nodal forces, nodal displacements and the stiffness matrix of the lining, respectively, and *, i and o are defined as follows:

- * : nodes at which inside displacements are measured (u*: known, f* = 0)
- i : nodes at which external loads do not act, except the above nodes (*), (ui: unknown, fi = 0)
- o : nodes at which external loads act, (uo: unknown, fo: unknown)

By minimizing the sum of the square of the external loads (unknowns, fo), the displacements (unknowns, uo, ui) and fo are analyzed by the following equations:

$$u_0 = -A_{33}^{-1}b_3 + A_{33}^{-1}H_{13}^T[H_{13}A_{33}^{-1}H_{13}^T]^{-1} \cdot \{H_{13}A_{33}^{-1}b_3 - F_1^*\} \quad (3)$$

$$u_i = -K_{22}^{-1}[K_{21}u^* + K_{23}u_0] \quad (4)$$

$$f_0 = H_{33}u_0 + F_3^* \quad (5)$$

where

$$A_{33} = H_{33}^T H_{33}, \quad b_3 = H_{33}^T F_3^*$$

$$H_{13} = [K_{13} - K_{12}K_{22}^{-1}K_{23}],\ F_1^* = [K_{11} - K_{12}K_{22}^{-1}K_{21}]u^*$$

$$H_{33} = [K_{33} - K_{32}K_{22}^{-1}K_{23}],\ F_3^* = [K_{31} - K_{32}K_{22}^{-1}K_{21}]u^*$$

All displacements and external loads have been determined. Strains and stresses of the lining can be easily calculated by using stress-strain and strain-displacement relationships. The stability and the accuracy of the results have been confirmed through numerical analyses(Hisatake et al.,1992).

3 MODEL TESTS

3.1 Experimental device and lining

Figure 2 and Photographs 1 and 2 show a two-dimensional experimental device and a 1/20 scale model tunnel lining which is made of cement mortar. External loads are applied to the outer boundaries of the lining through 9 rods as the reaction forces from the steel frames, which are introduced by fastening bolts. Displacements and strains are measured at 5 points and 15 points, respectively, on the inner boundaries of the lining. Vertical and horizontal displacements are measured for each measuring point, and strains are measured in a circumferential direction.

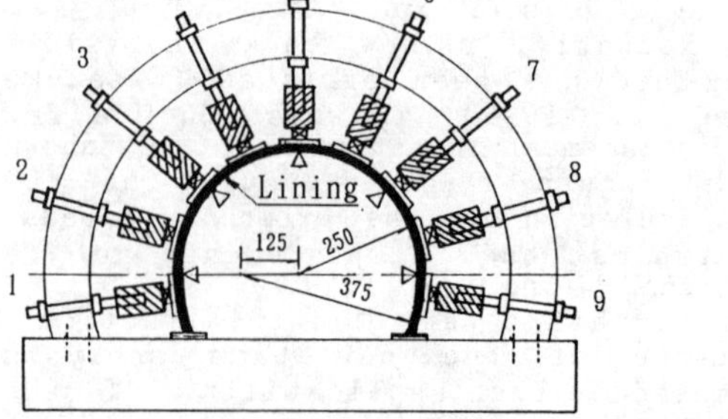

Figure 2 Experimental device and model lining

Photograph 1 The state of the experiment

Photograph 2 Lining models

3.2 Experimental results

Figure 3 shows the finite element meshes, in two dimensions, used in the analysis. The numbers written in the figure denote the nodal (inside) and the element (outside) numbers, respectively. Only the

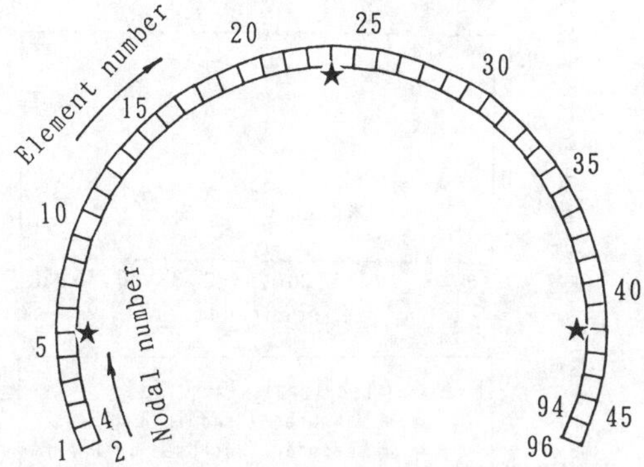

★:Nodes that input displacements are given

Fig.3 Finite elements used

displacements measured at the 3 points marked with stars in Figure 3 are fed into the back analysis as input data, and displacements (horizontal and vertical directions) and strains in a circumferential direction are analyzed. The analytical results are compared with the measured ones, as shown in the following.

Figure 4(a) shows the external loads applied to the lining. The maximum external load (Fmax) is nearly 30% of the load (Fcrack) at which cracks occur in the lining.
The analytical displacements are compared with the measured ones in Figure 4 (b)(horizontal) and Figure 4(C)(vertical). They agree very well with each other. In Figure 4(d), the analytical strains are compared with the measured ones.

When the maximum force (Fmax) increases to 90% of the force (Fcrack), almost the same results are shown in Figure 5. The analytical strains of an element are only available when the element is sandwiched between two points of a displacement measurement. In this experiment, therefore, the analytical strains between element numbers 6 and 41 are available and almost coincide with the measured ones.

Figure 6 shows the same results when the maximum load is applied in the inclined direction from the vertical line. In this case, the same results are obtained except analytical strains near the lining edges, which show some error. The maximum strain, however, is reasonably accurate.

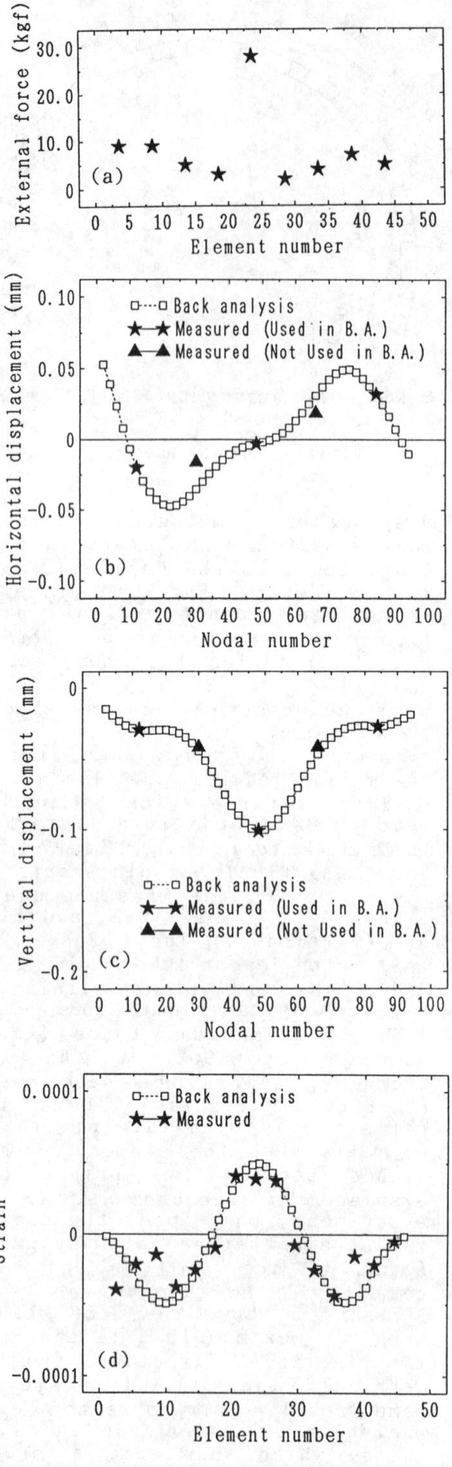

Figure 4 Comparison of analytical and measured results(Fmax=0.3Fcrck)

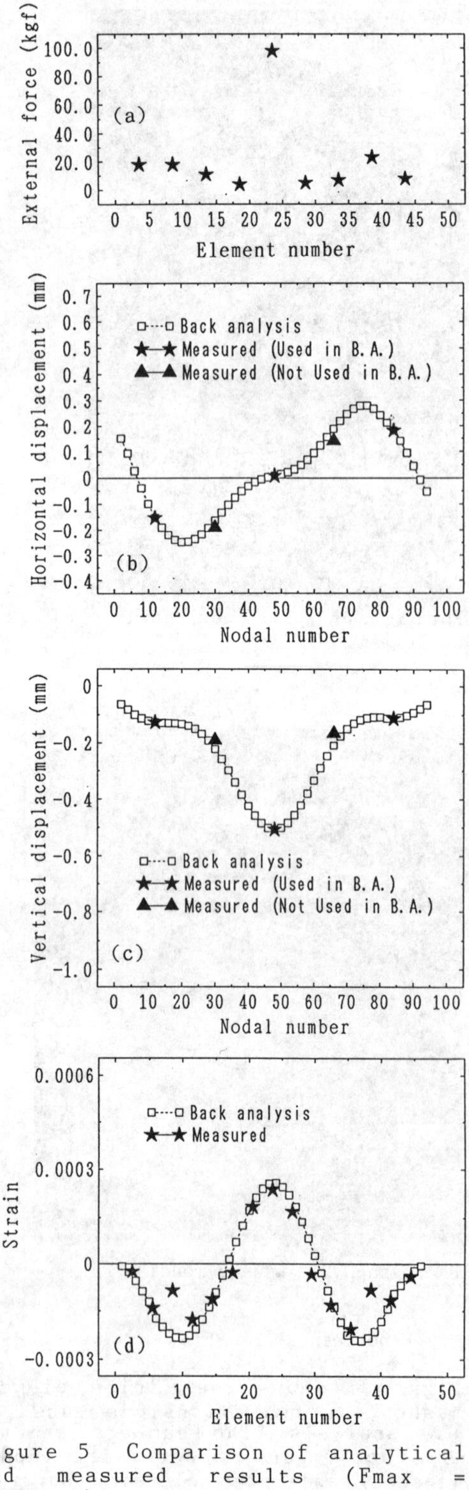

Figure 5 Comparison of analytical and measured results (Fmax = 0.9Fcrack)

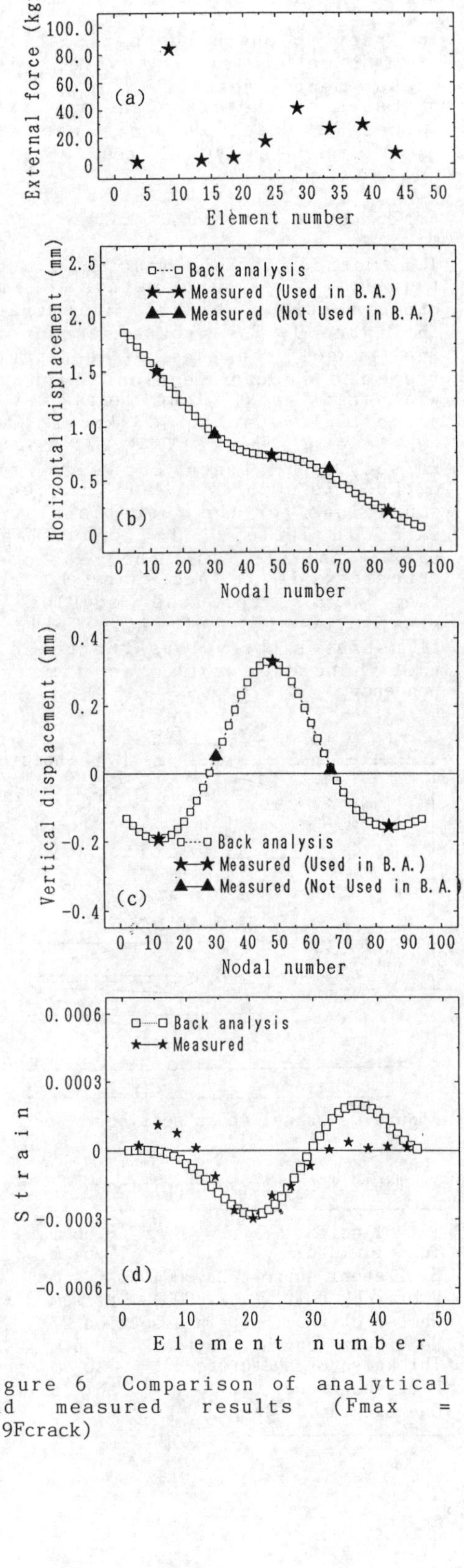

Figure 6 Comparison of analytical and measured results (Fmax = 0.9Fcrack)

From a comparison between the analytical and the experimental results, it is recognized that this back analysis method is applicable to practical problems.

3.3 Finding the beginning of the instability

The most important thing is to know the beginning of the instability of the lining. However, the back analysis method mentioned previously can only be applicable to a lining which is kept at a continuous state without cracks. In short, the method should not be used after occurrence of cracks in the lining.

By conducting the model experiments, however, we have noticed some very interesting facts, as shown in the following.

Figures 7 and 8 show changes in the strains for the respective convergence measurements when the

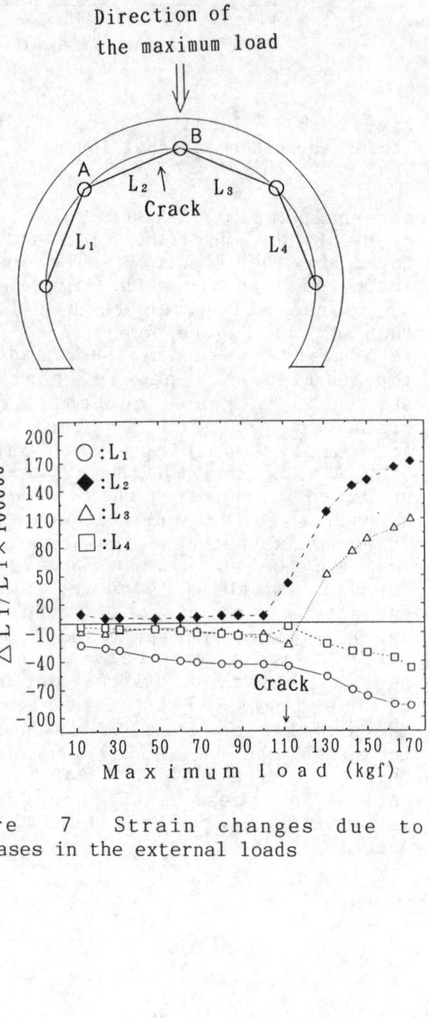

Figure 7 Strain changes due to increases in the external loads

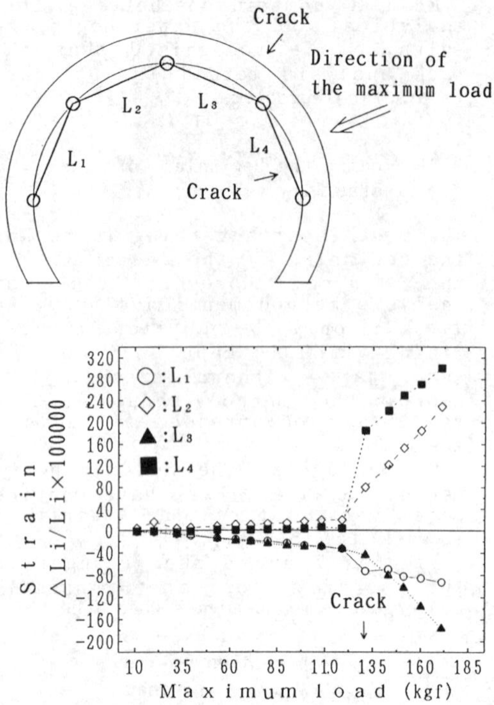

Figure 8 Strain changes due to increases in the external loads

external loads increase. The direction of the maximum load is indicated in the figures. The external loads at 9 points in Figure 2 are of course all given to the lining, such as in Figure 4(a). Very interesting results can be read from these figures, namely that some strains increase suddenly when a crack is produced in the lining. Accurately speaking, one part of the lining in which a crack is produced shows this tendency strongly. For example, when a crack is produced between points A and B in Figure 7, the convergence results in this span are strongly subject to the effect of the occurrence of the crack. Results of the other convergence results are, of course, affected by the occurrence of the crack. This fact leads to the conclusion that it is possible to catch the beginning of unstable phenomena without any sophisticated back analyses, if convergence measurements are continuously conducted.

4 APPLICABILITY TO AN ACTUAL TUNNEL

In order to check the applicability of the method at the site, field measurement results for lining stresses of shotcrete and for axial forces of steel supports are compared with the analytical ones.

4.1 Executive conditions

The method is applied to a road tunnel, 131 square meters of its cross section, which is illustrated in Figure 9. The ground surrounding the tunnel consists of a stratum from the Neocene period which has weathered rock (longitudinal wave velocity V = 0.6 to 1.5 km/sec) and conglomerate rock (V = 2.3 to 2.8 km/sec). The bench cut excavation method is employed and executive conditions for the tunnel are indicated in Table 1. In conducting a back analysis of the shotcrete stresses, it is necessary to know the values for the modulus of elasticity (E) of it. Table 2 illustrates the values for the E of the shotcrete which has time dependency.

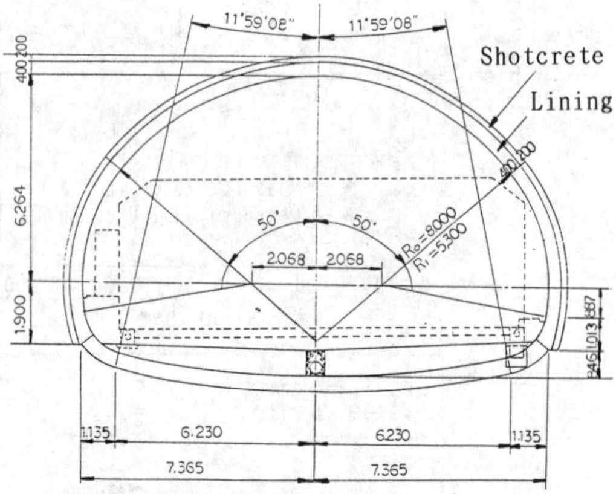

Figure 9 Tunnel cross section

Table 1 Executive conditions

Tunnel height	8.56 m
Tunnel width	15.54 m
E of steel support	21,000,000tf/sqm
E of rock bolt	21,000,000tf/sqm
Number of rock bolt in a section	27
Rock bolt length	6 m
Thickness of shotcrete	20 cm
E of shotcrete	Table 2
One excavation length	1 m

Table 2 Modulus of elasticity for shotcrete with time dependency

Day	tf/square meters
1	1,601,800
3	1,619,200
5	1,861,300
8	2,029,800
16	2,243,400

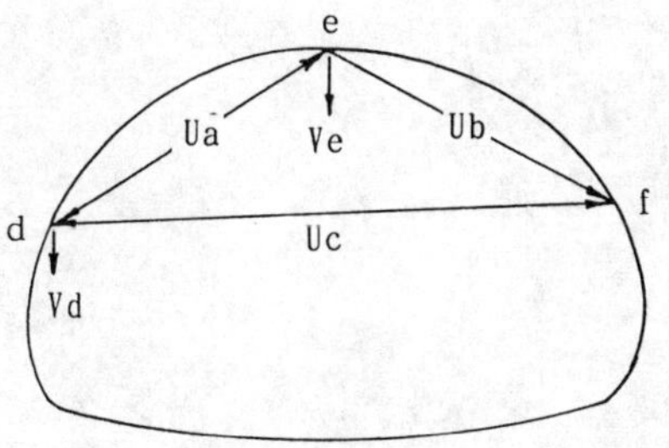

Figure 10 Positions of the displacement measurements

4.2 Displacement measurements

Tunnel convergence (Ua, Ub and Uc) and settlement (Ve and Vd) at the crown and the side wall are measured as shown in Figure 10. Relative displacements at points "d", "e" and "f" are calculated from the above data. They are then fed into the back analysis to calculate stresses and axial forces of the upper part of the lining. Figure 11 shows the finite element meshes used in the analysis for the upper part of the lining, and the respective element numbers are written in Figure 11.

4.3 Comparison of the results

Figure 12 illustrates analytical and field measurement results of shotcrete stresses in a circumferential direction. The shotcrete stresses are directly measured with pressure cells which have been buried in the shotcrete. As the modulus of elasticity of the shotcrete increases with time, as shown in Table 2, analytical stresses are obtained by summing up the respective incremental stresses.

The analytical results of the axial forces of the steel support are compared with the field measurement results in Figure 13.

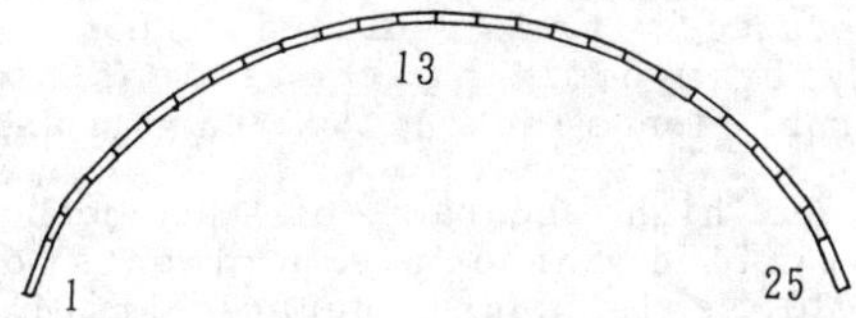

Figure 11 Finite elements used and element numbers

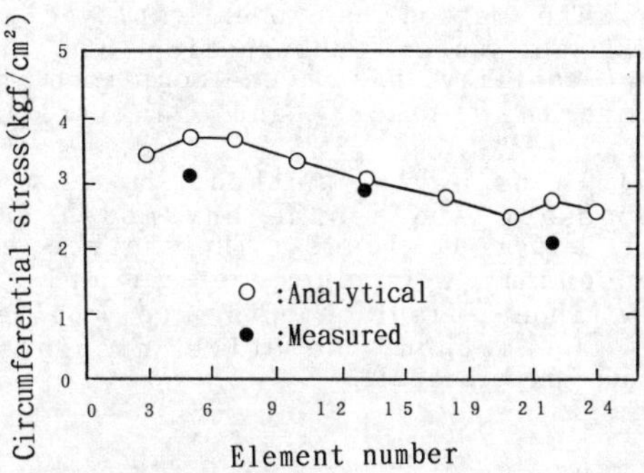

Figure 12 Analytical and field measurement results of shotcrete stresses

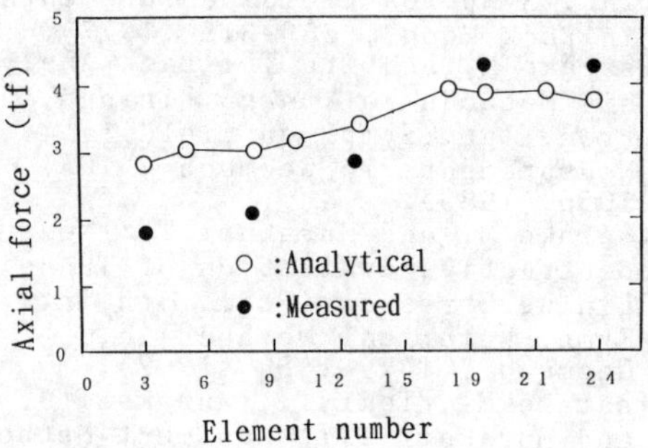

Figure 13 Analytical and field measurement results of axial forces of a steel support

From Figures 12 and 13, the stresses and axial forces obtained by the back analysis compare favorably with those measured at the site.

5 CONCLUSIONS

The conclusions obtained in this paper are as follows.

(1) The back analysis method pro-

posed by Hisatake et al. is briefly introduced. It is derived theoretically by minimizing the sum of the external loads acting on the tunnel lining.
(2) The high accuracy of the method is verified through experiments on a 1/20 scale model tunnel lining, made of cement mortal.
(3) It is possible to catch the beginning of the unstable phenomena of a lining with convergence data, without conducting any sophisticated back analyses.
(4) The method has been employed on a tunnel under construction and its applicability has been confirmed by comparing field and analytical data.
(5) In using the method, it is not necessary to assume any modes for the external loads. The method is reasonably accurate even if only a few input displacements are applied to the method. Relative displacements are available.

REFERENCES

kovari K., Amstad ch. and Fritz P.: Integrated Technique for Rock Pressure Determination, Proc. Int. Symp. on Field Measurements in Rock Mech., 289-316 (1977).

Hisatake M. and Ito T.: Back Analy sis Methods to Tunnel Linings, Proc. Int. Symp. on Field Measurements in Geomech., 1047-1056 (1983).

Hisatake M. and Murakami T.: Non-destructive Evaluation of Tunnel Lining Stresses, Proc. 4th Int. Symp. Numerical Methods in Geomech., 1479-1484 (1992).

Hisatake M., Toki T., Murakami J. and Murakami T.:Assessment Method of Tunnel Lining Stability and Model Experiments, Proc. the 25th Symp. Rock Mech. in Japan, 201-205 (1993).

Tunnelling and Ground Conditions, Abdel Salam (ed.) © 1994 Balkema, Rotterdam, ISBN 90 5410 363 9

Design for umbrella method based on numerical analyses and field measurements

N. Kotake, Y. Yamamoto & K. Oka
Toyo Construction Co., Ltd, Tokyo, Japan

ABSTRACT: Design for the umbrella method, the preliminary stabilization method by the long, rigid, and shell-shaped structure, is discussed based on the field measurement data obtained from an earth tunnel where the sub-horizontal jet grouting was adopted, and the results of three-dimensional boundary element analysis conducted for the purpose of understanding the fundamental behavior. It suggests that the range of stabilization and the reinforcement of the foot are examined depending on the purpose to apply the method. Simplified approach for the structural design is also examined.

1 INTRODUCTION

Umbrella method, a preliminary ground improvement for tunnelling by the conical structure created in the core ground ahead of the face, has been originally developed in Italy. The method has been recently confirmed to be effective to achieve stability for large tunnels with poor geological conditions in Japan as well. Instead of dividing the tunnel cross section and excavating in stages in the past, the large face excavation becomes possible, consequently leading to rapid excavation and improvement of productivity by mechanization with large-scale machinery. The method is one of the efficient alternatives to be examined when operations from the surface are impossible or inappropriate, and especially when the disturbance of the environments such as surface settlement is estimated to be not permissible. Thus, the method is considered as a principal solution of engineering problems expected in construction projects of very large tunnels as well as the urban tunnels.

To date, a unified design method with mechanically definite bases has not been established yet, due to insufficient practical experience and difficulty to evaluate three-dimensional effects. Since the umbrella method is generally considered to be expensive as a large-scale supplementary supporting method for NATM, the preliminary investigation becomes essential to find a reasonable balance of economic and technical factors, preferably through computational safety evaluations. It could be done by clarifying the reinforcing mechanism, so as to define the scope of application, to determine the design specification, and to grasp the requirements for the engineering characteristics of the stabilized zone for design and further development.

The stabilization method by the long, rigid, and shell-shaped structure is referred to as the umbrella method in this paper. The corresponding representative methods are: the sub-horizontal jet grouting, the umbrella-arch method with injected steel pipes, and the combination of steel pipe and jet grouting. There are substantial difference in the execution procedure, machinery and equipments, and reinforcing materials. However, the basic concept and design philosophy seem to be common to these methods.

This paper deals with design aspects of umbrella method based on the field measurement data obtained from an earth tunnel where sub-horizontal jet grouting was adopted, and the results of three-dimensional boundary element analysis conducted for the purpose of understanding the fundamental behavior of the umbrella structure and the surrounding ground. Simplified approach for the structural design is also examined.

2 CASE HISTORY OF AN EARTH TUNNEL

Shoryou No.1 and No.2 Tunnels were driven through shallow and loose volcanic sandy deposits near Mt. Fuji for a three lane motorway. Little ground arch effect could be expected because of its thin covering of up to 18 m as compared to the excavation width of 16 m and cross-sectional area of 140 m². Injected forepiles were firstly adopted at Shoryou No.2 Tunnel, driven by side drifts heading. As a more effective way to reduce loosening of the ground, the sub-horizontal jet grouting was applied at No.1 Tunnel and made possible to drive by top heading (see Figures 1 and 2). Partially stabilized patterns were also adopted in the range where bearing capacity was considered to be sufficient based on the preliminary analysis.

Figure 3 shows the surface and crown settlements monitored in the sections, whose geological conditions are similar such that the deformation modulus of the ground ranges 15~40 MPa, and earth covering 0.3~1.1 D. It should be noted that the surface settlement is monitored along with the face advance, while the crown settlement is measured after the primary support is installed.

As for the effects of the umbrella method, the following can be stated;

1) The pattern with stabilized range of 180° by jet grouting shows a clear effect to reduce the surface settlements, while the settlements observed in partially stabilized patterns by jet grouting are as large as those in injected forepiling.

2) It is seen that the difference of the crown settlements among various patterns is relatively small compared to that of the surface settlements. In fact, more settlements occurred in patterns of jet grouting than injected forepiling during bench and inverted arch excavation. Therefore, the effect of jet grouting after the passage of heading face and during bench and inverted arch excavation can not be clearly observed.

3) The surface settlement occurred during bench and invert excavation was nearly equal to the crown settlement. It was probably because the whole tunnel structure settled in the loose ground.

4) More effects could be expected for reduction of deformation by introducing additional measures to prevent the settlements occurring after the top heading.

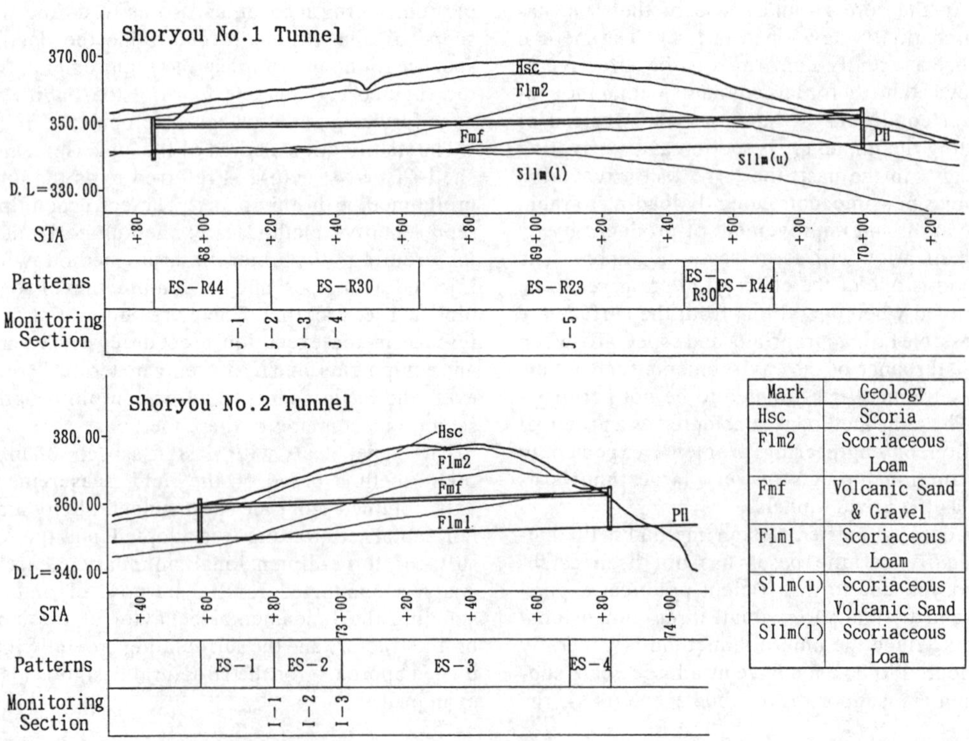

Mark	Geology
Hsc	Scoria
Flm2	Scoriaceous Loam
Fmf	Volcanic Sand & Gravel
Flml	Scoriaceous Loam
SIlm(u)	Scoriaceous Volcanic Sand
SIlm(l)	Scoriaceous Loam

Fig. 1 Geological profile of Shoryou No.1 and No.2 Tunnels

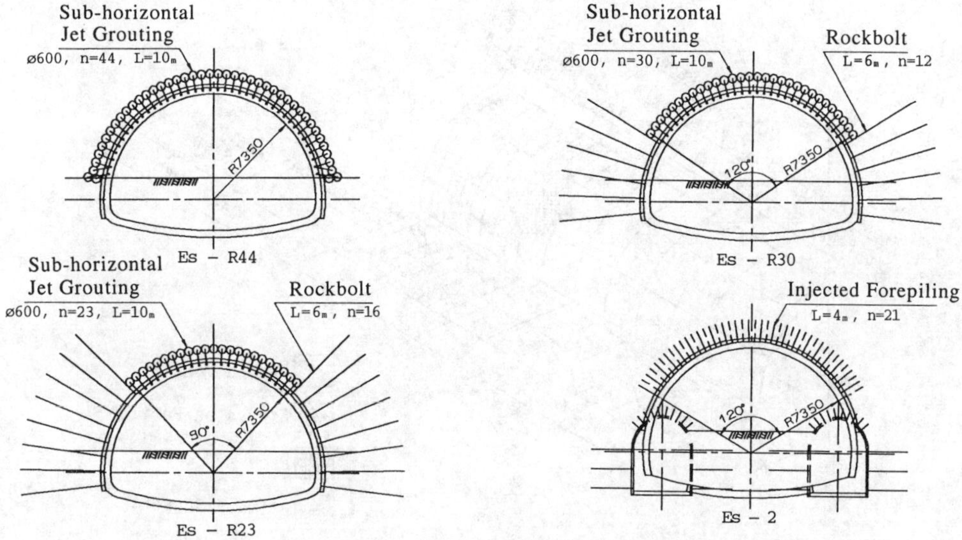

Fig. 2 Stabilized patterns

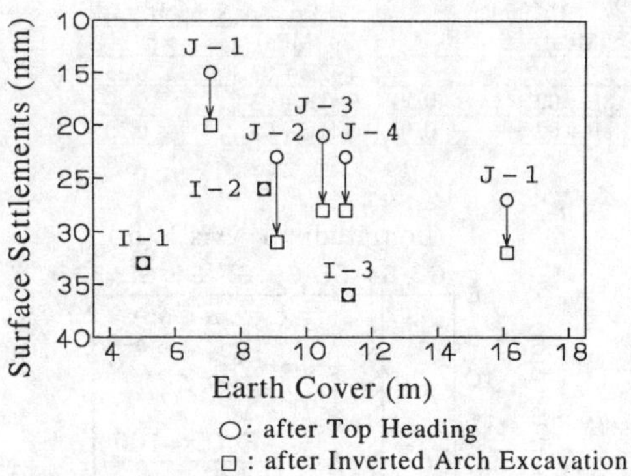

Fig. 3 (a) Surface settlements

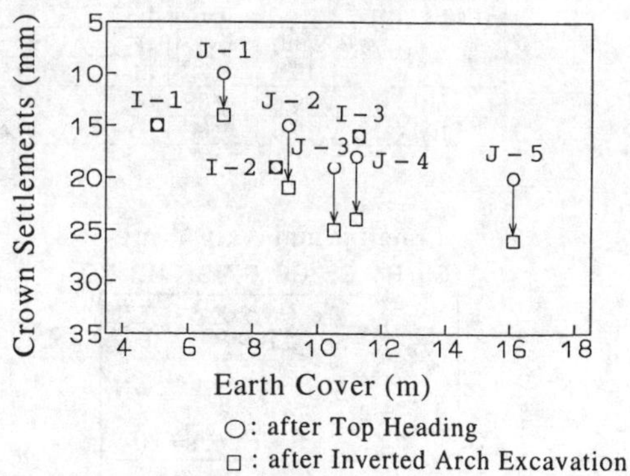

Fig. 3 (b) Crown settlements

3 THREE-DIMENSIONAL BOUNDARY ELEMENT ANALYSIS

Three-dimensional boundary element method (3-D BEM) is applied to the numerical analysis of tunnel excavation with the umbrella method to clarify the fundamental behavior of the umbrella structure and the surrounding ground along with the face advance by the incremental step approach. The hypothetical geological conditions are assumed by considering a typical shallow tunnel in a loose ground.

3.1 *Boundary element model*

The situation to be analyzed is to drive a tunnel by top heading together with sub- horizontal jet grouting under the earth covering of 1D. The BEM model of tunnel excavation part and analysis steps are shown in Figure 4. The ground is excavated incrementally with the installation of the primary support at one step behind. The excavation diameter is 10 m, and the stabilized range is 180° along the periphery of the heading section with the effective thickness of 0.5 m.

The whole model, a closed domain symmetrical model, consists of 3 regions modelled as continua: the ground, the umbrella structure, and the primary support with the umbrella structure. The properties for each region are displayed in Table 1. The initial stress is specified due to the weight of the overburden and the coefficient of lateral earth pressure K_h of 0.5.

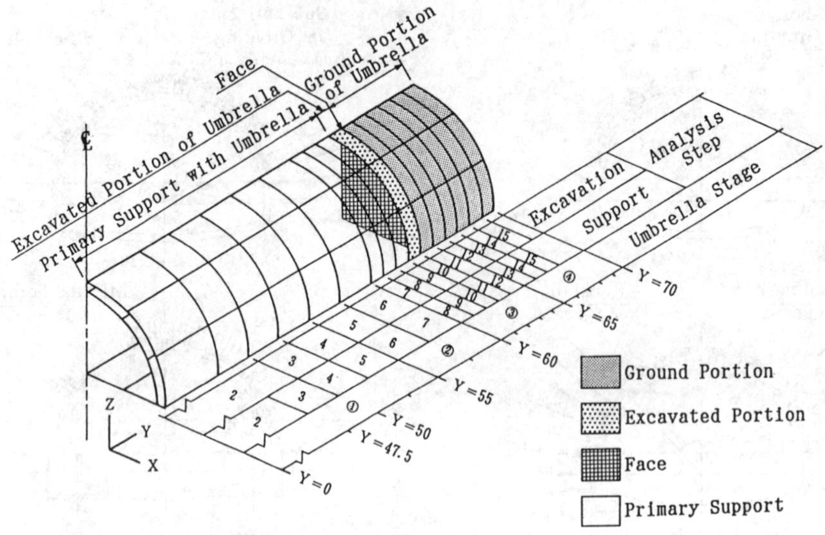

Fig. 4 Three-dimensional boundary element model and analysis steps

Table 1 Properties for Boundary Element Analysis

Ec/Es	Ground			Umbrella		Primary Support	
	γs (kN/m³)	Es (MPa)	νs	Ec (MPa)	νs	Ep (MPa)	νp
1	19.6	9.8E + 01	0.35	–		4.32E + 03	0.35
10	19.6	9.8E + 01	0.35	9.8E + 02	0.30	6.41E + 03	0.30
100	19.6	9.8E + 00	0.40	9.8E + 02	0.30	6.41E + 03	0.30

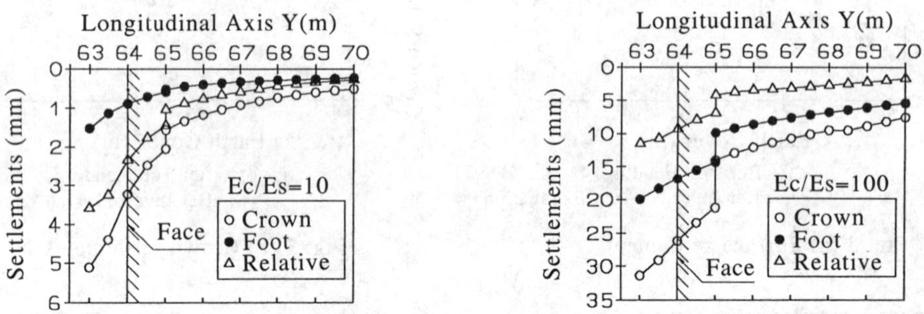

Fig. 5 Deformation profile

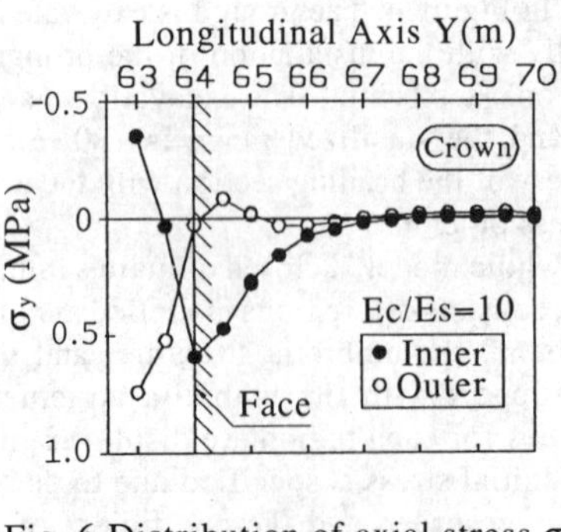

Fig. 6 Distribution of axial stress σ_y to the longitudinal direction

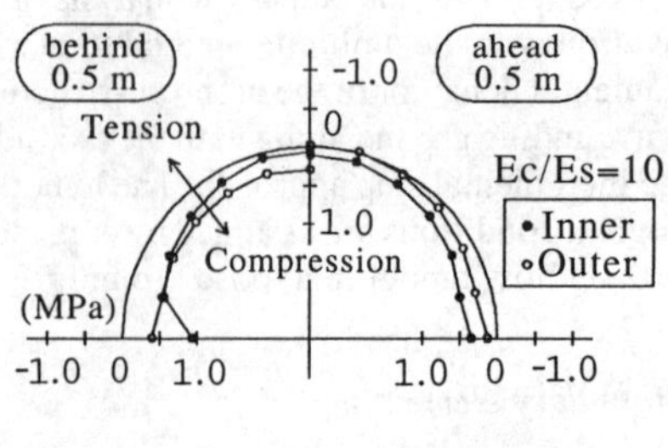

Fig. 7 Distribution of circumferential stress σ_θ in transverse sections

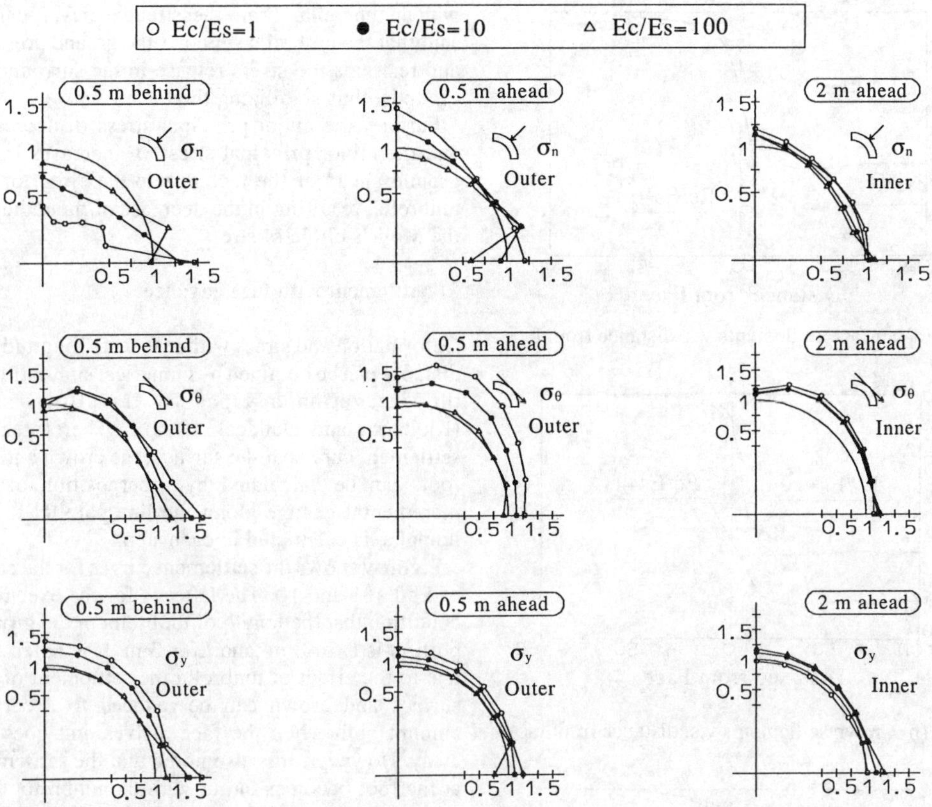

Fig. 8 Stress acting on the umbrella structure and the internal ground

The stiffness ratio of the umbrella structure to the ground, Ec/Es, is considered to range approximately from 10 to 100, where Ec and Es are the deformation moduli of the umbrella structure and the ground, respectively. Regarding Ec/Es as a significant parameter, the analysis was conducted for Ec/Es of 1 (without umbrella), 10, and 100.

3.2 *Results and evaluation*

1) Deformation and stress patterns

The results are arranged in relation to the deformation and stress which occur in step 10 after the umbrella structure is created. The deformation is represented as the longitudinal profile of settlement along the crown and the foot, and the vertical relative displacement between the crown and the foot as shown in Figure 5. It can be seen that the settlement along the crown increases when Ec/Es increases, in which the proportion of the foot settlement increases.

Figure 6 shows the longitudinal distribution of axial stress σ_y along the crown of the inner and outer surface of the structure. Figure 7 shows the circumferential stress σ_θ along the inner and outer surface of the transverse sections; at 0.5 m ahead of the face in the ground portion and at 0.5 m behind the face in the excavated portion, respectively.

From these results, it can be said that the behavior of the umbrella structure is expressed as a beam along the axes of the crown and the foot, and as an arch in the transverse section of the excavated portion which primarily undergoes compression, and further the whole umbrella structure showing behavior as a beam on an elastic foundation.

2) Face stability effect

The transverse distribution of the ground stress is shown in Figure 8 for the normal stress σ_n, the circumferential stress σ_θ, and the axial stress σ_y, which occur along the boundary between the ground and the structure. They are expressed as fractions of the initial stress.

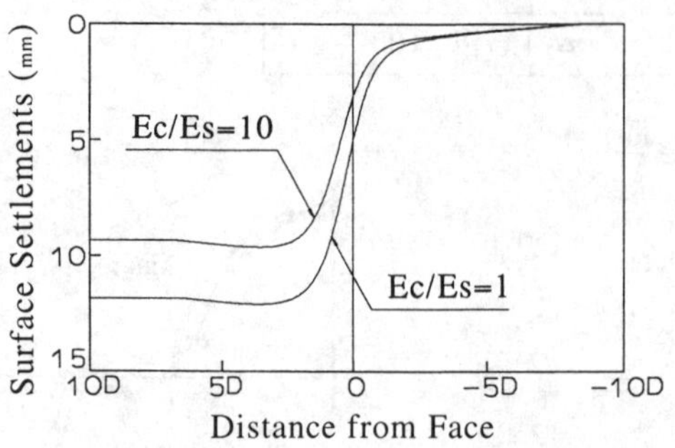

Fig. 9 (a) Surface settlements v.s. distance from face

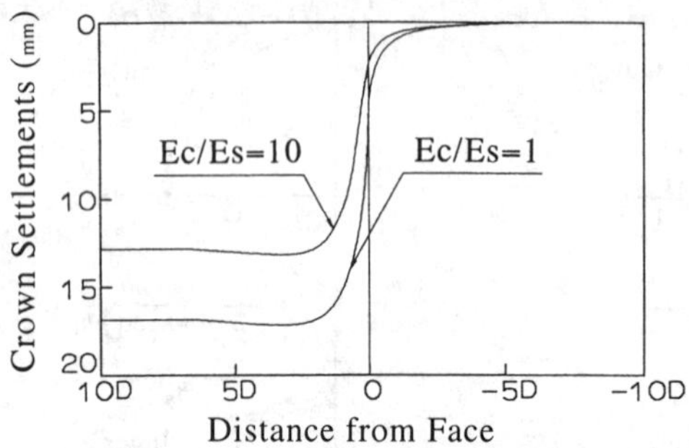

Fig. 9 (b) Crown settlements v.s. distance from face

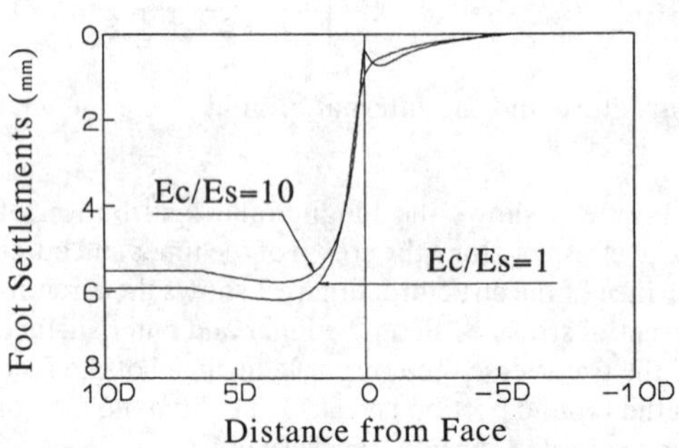

Fig. 9 (c) Foot settlements v.s. distance from face

The ground stress along the outer side of the umbrella is thought to be equivalent to the ground pressure which acts on it, and that along the inner side of the structure to be equivalent to the external force which acts on the internal ground ahead of the face through the umbrella.

The presence and absence of the umbrella being compared, the change from the initial stress appears to be smaller at the outer side of 0.5 m ahead and behind the face. Also, the greater the stiffness ratio, the closer it is to the initial stress. At the inner side of 2 m ahead, σ_y increases in compression for the cases with an umbrella. From these tendencies, it can be said that the umbrella supports the ground pressure and restrains the stress release in the surrounding grounds, thus stabilizing the face.

Further, the major principal stress σ_1 decreases and the minor principal stress σ_3 increases in the ground ahead of the face due to the effect of the umbrella, resulting in the decrease of the radius of the Mohr's circle of stress.

3) Settlement with face advance

Deformation and stress with respect to longitudinal distance can be obtained by numerical simulation of the excavation advance by travelling mesh (Kielbassa and Duddeck 1991). Therefore, the settlement curves at the surface, the crown and the foot, can be calculated by superposition of the incremental change along the line parallel to the tunnel axis calculated in each step.

Figure 9 shows the settlement curves for the cases of Ec/Es=1 and 10. The latter is for the execution condition that the length of umbrella in the ground portion is L_{max}=8 m ,and L_{min}=3 m. It is noted that due to the effect of umbrella the settlement of the surface and crown can be reduced to a certain amount both when the face arrives and goes far away. However, it is also noted that the settlement at the foot becomes larger with the umbrella than without it. It is explained that the increase in rigidity of the primary support by adding the umbrella structure behind the face leads to the stress concentration at the foot ground of the primary support.

4 SIMPLIFIED APPROACH

The three-dimensional analysis, despite the most suitable for the tunnelling problem, can not be used on a regular basis because of the high cost and processing time associated with it. Thus, the possibility of applying simplified approach is examined for practical purpose as well as for a more extensive parametric study of various conditions.

In this study, the stability of the umbrella structure is focused on, from the standpoint that it is necessary to be structurally stable for the purpose to function satisfactorily.

4.1 *Simplified models*

The shell, beam and arch models are examined to find an equivalent structural model to the 3-D BEM under the corresponding loading and structural conditions. The stability of the umbrella structure has been evaluated with regard to the developed stress due to excavation force.

(1) Shell model

To simulate the behavior of the shell-shaped structure in a more simplified fashion, the umbrella structure is modelled by shell elements to be analyzed by three-dimensional finite element method. The reaction force from the ground is modelled by the corresponding outer and inner springs of discrete form, which act non-linearly only on compression to the vertical and horizontal directions. The foot and the end of the structure are supported by springs, either. (See Figure 10)

(2) Beam and arch models

Beam models are examined with relation to the longitudinal direction along the crown. A beam simply supported on one end and built-in on the other end, is applied according to the conventional design of pipe roof method. Considering the reaction force from the internal ground, bedded- beams are examined to have the ground pressure charged on the excavated portion and beyond.

Arch models are examined with relation to the transverse section of the excavated portion; one is a bedded-arch-frame, and the other is a 2-D plane strain finite element method (FEM) with the umbrella modelled by beam elements.

4.2 *Comparison with 3-D BEM*

Table 2 shows the maximum and minimum stresses calculated by these models, for comparison with the incremental stress occurred in step 10 of 3-D BEM. It is, first of all, noted that the approximate critical stress can be assessed by the shell model for both axial and circumferential stresses.

The developed stress by bedded-beams are relatively close to the axial stress by 3- D BEM, while those by other beam models are several times larger. When the beam effect is dominant and the axial stress along the crown is critical, the bedded-beam model might be possibly used as a simplified design method. The developed stress by simplified arch models are several times larger than the circumferential stress by 3-D BEM.

The umbrella structure created in shell-shape is thought to exhibit effects as both a beam and an arch. However, the results by these models will tend to be too safe if straightly applied in design, since both characteristics can not be considered at the same time.

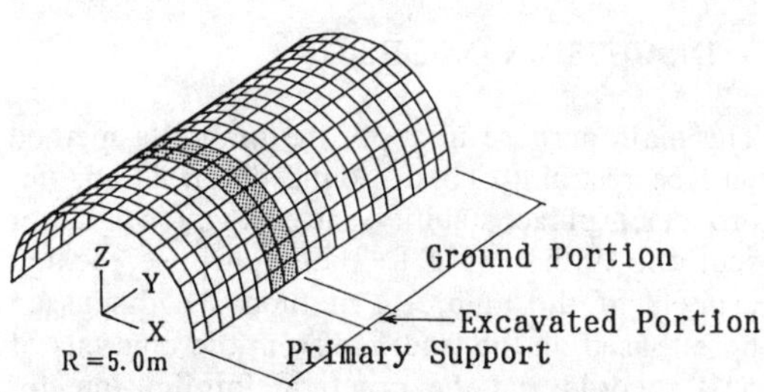

(a) Finite element model

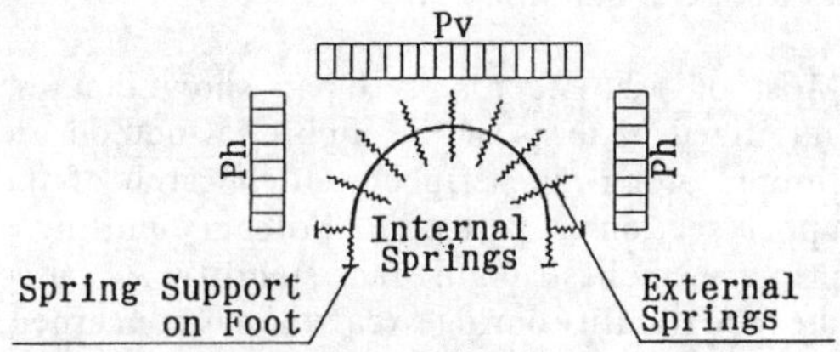

(b) Loading condition

Fig. 10 Shell model

Table 2. Comparison with simplified approach

	Model		Developed Stress (MPa)			
			Es/Ec=10		Es/Ec=100	
			max	min	max	min
Longitudinal (Axial)	Simple Beam	P	6.5	-6.5	6.5	-6.5
	Built-in on One End	P	3.7	-3.7	3.7	-3.7
	Bedded-beam (1)	P	1.0	-1.0	1.4	-1.4
	Bedded-beam (2)	P	1.5	-1.5	3.0	-3.0
	Shell Model	Fig. 10	1.5	-0.9	2.7	-0.8
	3-D BEM	Fig. 4	0.7	-0.9	1.1	-0.7
Transverse (Circumferential)	Bedded-arch -frame	P, P/2, P/2	2.9	0.6	3.7	-0.4
	2-D Plane Strain FEM		2.6	1.2	2.1	0.2
	Shell Model	Fig. 10	0.6	-0.3	0.9	-0.5
	3-D BEM	Fig. 4	0.6	-0.2	0.7	-0.4

5 DISCUSSION ON DESIGN

The main purpose to apply the umbrella method will be restraint of the ground deformation, improvement of face stability, and excavation of larger section. With regard to these purpose, design aspects of the umbrella method are discussed here, based on the findings from the analysis of past records and the practical implication derived from the numerical analyses.

1) Range of stabilization

Most of past records in Japan show that the stabilized patterns of the umbrella method are limited along the periphery of the arch of the upper section or part of it, probably putting a primary emphasis on the face stability. As far as the face stability during excavation is concerned, the stabilization of such range might be sufficient for the tunnels of ordinary scale, such as two lane motorways, even driven in poor geological conditions.

It is recognized in this study that the stabilized range of the whole periphery of the arch will be effective to constrain the settlements during top heading. However, it is stressed that due to the foot settlements occurring after the passage of heading as well as during the succeeding excavation of bench and inverted arch, the final deformation of the tunnel and the surrounding ground can not be properly restrained. Therefore, to restrain the unfavorable ground deformation or the damage to neighboring buildings during the whole period of tunnelling, any additional measures to prevent settlements at the foot of the primary support will be essential.

2) Reinforcement of the foot ground

The reinforcement at the foot is required to be operated as close to the excavation face as possible to reduce the settlement at the foot by resisting the concentrated stress. It can be done by sub-vertical jet grouting, micropiles, combination of jet grouting and steel pipes, etc. (SIG 1992)

3) Structural stability

Structural stability of the umbrella structure is considered to be one of main factors to fulfill its functions and will be required for the reasonable safety assessment, especially for unprecedented geological conditions and scale of excavation. It is desired to accumulate the results of numerical analyses for more realistic situations in company with analysis of field measurement data. This will help us determine the suitable loading and structural conditions for design.

4) Creation of arch

Though the beam effect is predominant, contribution of the arch effect is important to mobilize the load supporting mechanism as a shell structure, particularly for sub-horizontal jet grouting. For the case of pipe umbrella, it would be judged from both economical and technical points of view whether to increase the stiffness of steel pipes for beam effect or to stabilize more by grout injection for arch effect.

5) Application to large tunnels

As for application to large tunnels, with wider and higher excavation section and flatter shape, severe conditions such as difficulty in face stability and the large developing stress in the structure are expected (Kotake et al. 1992). To cope with this, the load supporting capacity must be improved, e.g. by the further reinforcement of materials to form the structure or by increase of thickness of the stabilized zone.

REFERENCES

Kielbassa,S., Duddeck,H. 1991. Stress-strain fields at the tunnelling face -Three-dimensional analysis for two-dimensional technical approach: *Rock Mechanics and Rock Engineering* 24:115-132

Koizumi,M., Imamura,O., Oka,K., Gouda,K. 1990. Construction of 3-lane tunnel with thin earth covering -Shoryou No.1 Tunnel on Tokyo-Nagoya Expressway- (Part2): *Tunnels and Underground*, Vol.21 No.6, June:463-471

Kotake,N., Yamamoto,Y., Oka,K. 1992. A study on applicability of horizontal jet-grouting in very large tunnel. Proc. Annual Conf. JSCE:692-693

SIG 1992. Soil and rock improvement in underground works. Proc. Int. Cong. Towards New Worlds in Tunnelling Vol.2:865-872. Acapulco

Tunnelling and Ground Conditions, Abdel Salam (ed.) © *1994 Balkema, Rotterdam, ISBN 90 5410 363 9*

Face instability in tunnelling through thick fault zones

T. Ogawa
Taisei Corporation, Yokohama, Japan

K.Y. Lo & R.K. Rowe
The University of Western Ontario, London, Ont., Canada

K. Sekiguchi
NKK Corporation, Kawasaki, Japan

ABSTRACT: This paper deals with the problem of face stability of tunnelling in weak, thick fault zones. The results of three-dimensional elasto-plastic analysis of stresses and displacements around the advancing face are presented. Using the results, together with a plane strain elasto-plastic closed-form solution, the concept of load sharing ratio between rockmass and support system is developed. Two mutually consistent criteria for face stability are introduced and design guidelines are suggested.

The method of analysis is then applied to case histories of tunnelling in which large displacements and instability were repeatedly encountered. The results of analysis are generally consistent with field observations.

1. INTRODUCTION

A most difficult problem in rock tunnelling is crossing thick, highly sheared and altered fault zones. Cases have been reported where heavy support systems failed leading to TBM's being frozen in place, and construction delays for years. A currently popular design concept, the characteristic line approach, by which rock strength is mobilized through tunnel deformation, works well when the face is stable and deformation is moderate. It is however ineffective when large deformations occur at and around the advancing face.

This paper provides a solution to the problem of face instability of tunnelling in weak, wide fault zones. Based on three-dimensional elasto-plastic finite element method, stresses and displacements around the advancing tunnel face resulting from initially isotropic stress conditions are presented. These results enable the development of the concept of Load Shearing Ratio (LSR) between the unexcavated rockmass at the tunnel face and support system. Two mutually consistent stability criteria are established by consideration of limiting displacement and of formation of a continuous plastic zone. Design guidelines for tunnels constructed through thick, weak fault zones are also suggested.

This method of analysis is then applied to case histories of tunnelling in thick fault zones where large displacements and instability were repeatedly encountered. The results of analysis of these case records indicate that the predicted performance is generally consistent with the results of field observations, particularly for the condition of face stability.

2. CONCEPT OF THE LOAD SHARING RATIO

Consider a circular tunnel driven in an elasto-plastic medium under isotropic initial stresses. As the tunnel face advances, the initial stresses are relieved and displacements occur at and around the tunnel face.

The displacements of interest are shown in Figure 1. These are the axial (maximum) displacement at the centre of the face u_{fm}, the radial displacement at the crown of the face u_{fc}, and the plane strain displacement at the crown u_{ps} at the same section under consideration after the face has advanced a sufficient distance to develop plane strain condition at this section.

The excavation process may be conceived as a reduction in the radial stress in the rock at the face from the initial value of P_0 to zero. A temporary support pressure provided by the tunnel face gradually disappear when the face is advanced beyond the section under consideration. Part of this pressure is then carried by the support system provided close to the face.

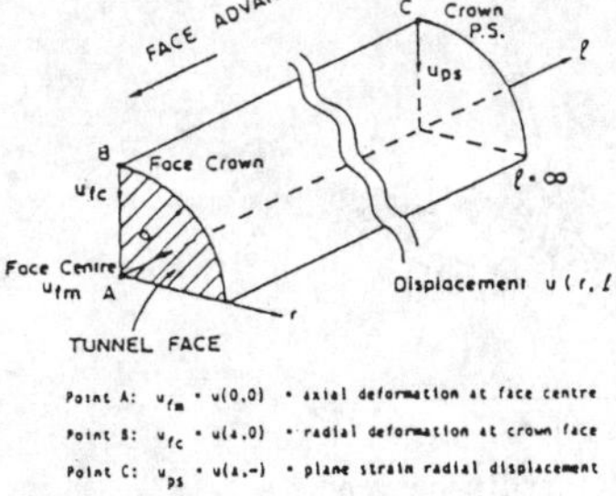

Figure 1: Definitions of radial and axial displacements

The radial stress thus simulates this face effect, and this fictitious temporary support pressure may be expressed as

$$\sigma_r = LSR \cdot P_0 \quad (0 \leq LSR \leq 1) \tag{1}$$

where LSR is the load sharing ratio, representing the amount of initial stress relieved, or percentage initial stress carried by the rockmass prior to the installation of the support system. If the rockmass is elastic after excavation, no support is necessary and LSR = 1. In general, however, a plastic zone will be present and some support is required ($0 \leq LSR \leq 1$).

Considering stability alone, in principle, it is possible to design an appropriate support system to limit the radial deformation behind the face. However, it is difficult to control the face displacement. The stability condition at the face therefore requires careful consideration.

3. STABILITY CONDITION AT THE ADVANCING FACE

The first consideration is to investigate how much load the rockmass can carry without excessive deformation leading to instability at the face. It will be shown that at any point between the face and the plane strain condition, a general elasto-plastic deformation increases with load sharing ratio LSR as illustrated in Figure 2. The maximum displacement at the centre of the face may be represented by the dimensionless displacement $\Omega_{fm} = u_{fm}E/(aP_0)$ in which E, a and P_0 are, respectively, the elastic modulus, radius of the tunnel and the initial stress. The relationship between Ω_{fm} and LSR is illustrated in Figure 2. (The relationship is quantified in subsequent sections.) If the rockmass remains elastic after excavation, the deformation is linear with LSR, and the tunnel is stable everywhere. In general, however, the deformation increases rapidly with increasing LSR concurrent with plastic zone development. One stability criterion is therefore to limit the deformation within a

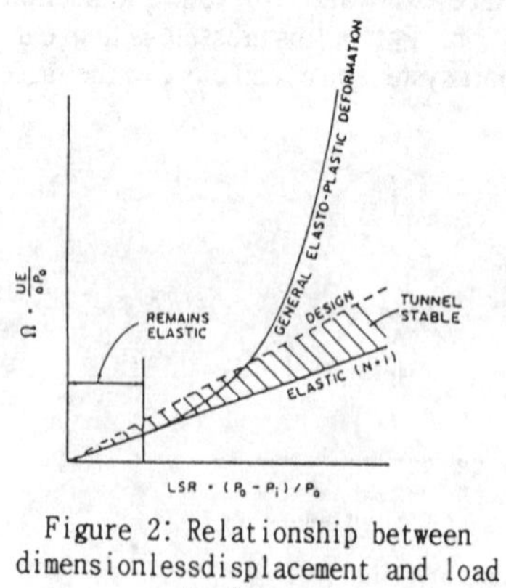

Figure 2: Relationship between dimensionlessdisplacement and load sharing ratio

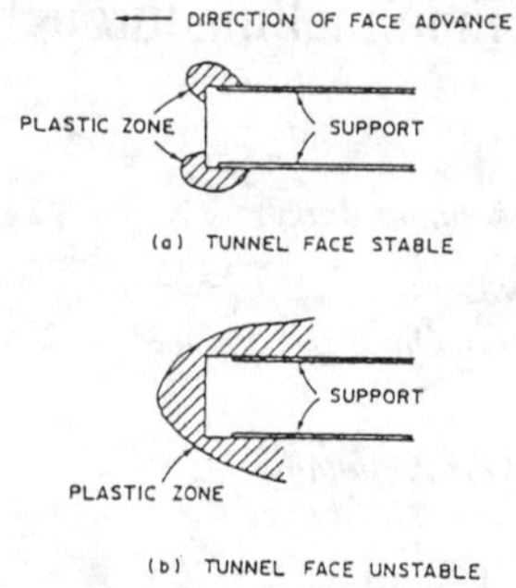

Figure 3: Illustrations of plastic zone formed around an advancing face: (a) face stable; (b) face unstable

maximum allowable (design) value, such that an excessive deformation will not lead to instability at or near the face.

Another stability criterion may be represented by the development of plastic zones around an advancing face as illustrated in Figure 3. Plastic zones will start to develop from the corner of the tunnel face as the support pressure decreases, i.e., the value of LSR increases. The tunnel is considered to be stable at this stage. As the support pressure decreases further, the plastic zone develops progressively and finally arrives at a stage where a continuous plastic zone is formed at and in front of the tunnel face (Figure 3b). The tunnel is considered unstable if the plastic zone reaches or surpasses this stage.

In the subsequent sections, the displacements at and behind the face , and the development of plastic zones will be analyzed so as to obtain a solution for the stability condition at the face.

4. LOAD SHARING RATIO IN PLANE STRAIN CONDITION

For a complete solution of the elasto-plastic problem, the equation of equilibrium, compatibility condition along with the stress-strain relationship, yield criteria, plastic potential, and flow rule are required. General concept necessary for the solution of elasto-plastic problem may be found elsewhere(Ogawa and Lo 1987). The solution within the plastic region will depend upon the assumptions of (a) yield criterion, (b) use of associated or non-associated flow rule, and (c) dilatancy angle ψ.

4.1 *$c \neq 0$, $\phi = 0$ (frictionless material)*

From the plane strain closed-form elasto-plastic solution, the plastic radius b, around a circular tunnel of radius a, under isotropic initial stress P_0 is given by the following equation (Lo et al. 1984, Ogawa and Lo 1987):

$$b = a \cdot \exp(\frac{P_0 - P_i - c_u}{2c_u}) \quad (2)$$

where P_i is the support pressure and c_u is the undrained shear strength.

The condition for a plastic zone to occur is given by

$$P_i < P_{i_{crit.}} = P_0 - c_u \quad (3)$$

Using the conventional stability number,

$$N = P_0 / c_u = \gamma z / c_u \quad (4)$$

where, γ = unit weight of rock, and
z = depth of the tunnel

Equation (2) may be written as

$$b / a = \sqrt{\exp(N \cdot LSR - 1)}, \quad N \cdot LSR > 1 \ and \ 0 \leq LSR \leq 1 \quad (5)$$

where LSR = $(P_0 - P_i)/P_0$, the load sharing ratio.

Equation (5) indicates that the development of plastic zone in the plane strain condition is governed by the material strength and the amount of initial stress relieved. This equation may be used as a preliminary means of evaluating what percent of the initial stress can be relieved before an extensive plastic zone develops. (A critical value of b/a in practice may be about 3 to 4.)

The radial displacement in the plastic zone is given by the closed-form plane strain solution (Ogawa and Lo 1987) viz.:

$$u = \frac{1+\nu}{E} \cdot \frac{1}{r}[2b^2 c_u (1-\nu) + r^2(1-2\nu)\{2c_u \ln(r/a) + P_i - P_0\}] \quad (6)$$

where r = radial distance from the tunnel centre ($a < r < b$)
E = elastic modulus of rock
ν = Poisson's ratio of rock

Putting N = P_0/c_u, LSR = $(P_0-P_i)/P_0$, and $r = a$, the surface radial displacement in a dimensional form is

$$\Omega_a = \frac{1+\nu}{N}[2(1-\nu)\exp(N \cdot LSR - 1) - (1-2\nu) N \cdot LSR] \quad (7)$$

For the elastic case, Ω_a = LSR(1+ν).

Equation (7) gives a convenient expression of tunnel surface displacement in terms of stability number and load sharing ratio. For a constant Poisson's ratio and a known stability number, the tunnel surface displacement is governed solely by the load sharing ratio, i.e., the amount of initial stress relief due to excavation.

4.2 $c \neq 0, \phi \neq 0$

The plastic radius from the closed-form plane strain solution (Ogawa and Lo 1987) is given by

$$b = a[\frac{2}{N_\phi + 1}\{\frac{N(N_\phi - 1) + 2\sqrt{N_\phi}}{N(N_\phi - 1)(1 - LSR) + 2\sqrt{N_\phi}}\}]^{\frac{1}{N_\phi - 1}} \quad (8)$$

where $N_\phi = (1 + \sin\phi)/(1 - \sin\phi)$.

The condition for the plastic zone to occur is

$$P_i < P_{i_{crit.}} = (1 - \sin\phi) P_0 - c \cdot \cos\phi \quad (9)$$

The dimensionless tunnel surface displacement at r = a in this case is

$$\Omega_a = \frac{u_a E}{a P_0} = (1+\nu)[2(1-\nu)(b/a)^{N_\psi + 1} \times \frac{N(N_\phi - 1) + 2\sqrt{N_\phi}}{N(N_\phi + N_\psi)} + \frac{(1-N_\phi) + (N_\psi + 1)\{(1-\nu)N_\phi - \nu\}}{N_\phi + N_\psi} \times \{\frac{N(N_\phi - 1)(1 - LSR) + 2\sqrt{N_\phi}}{N(N_\phi - 1)} - (1-2\nu)\frac{N(N_\phi - 1) + 2\sqrt{N_\phi}}{N(N_\phi - 1)}] \quad (10)$$

Where $N_\psi = (1 + \sin\psi)/(1 - \sin\psi)$, and
ψ = dilatancy angle (= 0 when there is no plastic volumetric strain ; = ϕ when an associated flow rule is satisfied.)

It may be seen in equations (8) and (10) that the plastic radius and the tunnel surface displacement are expressed in terms of stability number and load sharing ratio. For a constant Poisson's ratio, friction angle, and dilatancy angle with a known stability number, the tunnel surface displacement is controlled by the load sharing ratio, i.e., the amount of initial stress relief due to excavation.

It has been shown in the above discussions that both plastic radius given by equations (5) and (8) and elasto-plastic displacement given by equations (7) and (10) can be described in terms of the stability number N , and the load sharing ratio LSR in the plane strain condition. It may also be shown that the plastic radius and the elasto-plastic deformation at any point between the face (3-D condition) and the plane strain condition can be expressed in terms of N and LSR. Since there is no known closed-form 3-D closed-form elasto-plastic solution, this will be studied from the results of 3-D (axisymmetric) finite element analysis.

5. 3-D FINITE ELEMENT ANALYSIS OF THE ADVANCING FACE OF A TUNNEL

To obtain the stresses and displacements at and around the face, a finite element code NAPG-2D was used (NKK 1988). The code employs a pseudo-viscoplastic algorithm with the Euler's time-marching scheme to acquire elasto-plastic solutions (Zienkiewicz and Cormeau 1974).

In the 3-D (axisymmetric) analysis, a circular tunnel in an elasto-perfectly plastic material is modelled for the boundary conditions and geometry shown in Figure 4. The rock strength is defined by cohesion c, and angle of internal friction ϕ with a dilatancy angle ψ. Using two different limiting 3-D analyses (Ogawa 1986, Lee 1989), it was verified that the outer boundaries were far enough away from the tunnel so that the boundary conditions (rough or smooth) did not affect the calculated stresses and displacements.

As discussed in the previous section, the normalized plastic radius (b/a) and the elasto-plastic displacement ($\Omega = uE/aP_0$) are expressed in terms of stability number and load sharing ratio. Since the dimensionless displacement Ω already involves E, a, P_0, it is only necessary to perform the parametric study for N (effect of c_u on b/a and Ω) and for LSR (effect of P_i on b/a and Ω). The results of analyses are discussed in the next section.

6. RESULTS OF 3-D ANALYSIS

6.1 $c\neq0,\phi=0$

In presenting the results of 3-D analyses, the effect of load sharing ratio on dimensionless displacements for different stability number will be examined first for the case of $c\neq0,\phi=0$. The development of the plastic zone and the effects of load sharing ratio on plastic radius are then studied. It may be noted here that since $\phi=0$, there is no plastic volumetric strain (i.e., $\psi=0$). From these investigations, the stability criterion at the tunnel face may be quantified.

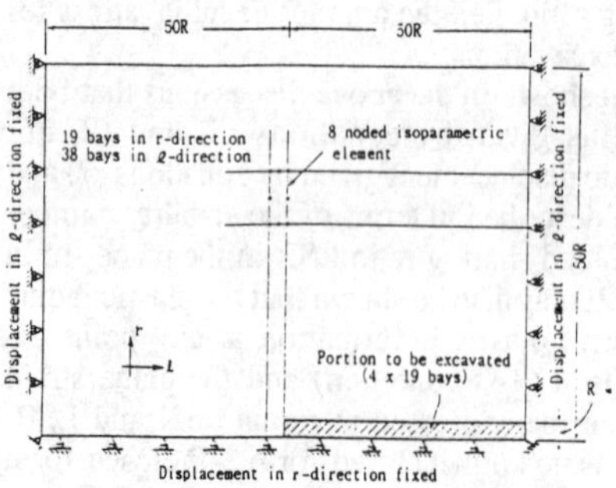

Figure 4: Finite element mesh used in the analysis

(a) Relationships Between Displacements and Load Sharing Ratio

The relationship between LSR and the normalized displacements at the face centre (u_{fm}), and under plane strain condition (u_{ps}) for different stability number (N) are shown in Figure 5(a). Similar relationships between the normalized displacement at the crown of the face (u_{fc}) are shown in Figure 5(b). The elastic response (N=1) is also shown in these figures. It may be observed that:

(i) for a given value of stability number, the normalized displacements increase with increasing LSR,

(ii) for a given LSR, the rate of deformation increases with increasing stability number,

(iii) the normalized displacements depart from the elastic state at smaller LSR with increasing N.

The physical meaning of these results is clear. For a rockmass with a given strength and initial stresses, larger deformations occur if the rockmass carries more load. Conversely, if the design is such that the LSR is, say, 0.4, deformation may become excessive for stability number greater than 6. It is also significant to note that although u_{ps} will be the largest of the three displacements, u_{fm} is the critical quantity to consider for face stability. This is because it is possible, in principle, to design an adequate support system behind the face to limit the deformation, while it is difficult to limit the face deformation directly.

An approach to the design for face stability is therefore to design for an LSR such that the face displacement is within tolerable limits. A suggested design line, based on 1.5 times elastic displacement at the face centre is shown in Figure 5(a). It remains to be examined whether this suggested design line satisfies the second stability criterion.

(b) Effect of LSR on Plastic Zone

The development of plastic zones with increasing LSR for a stability number of 15 is shown in Figure

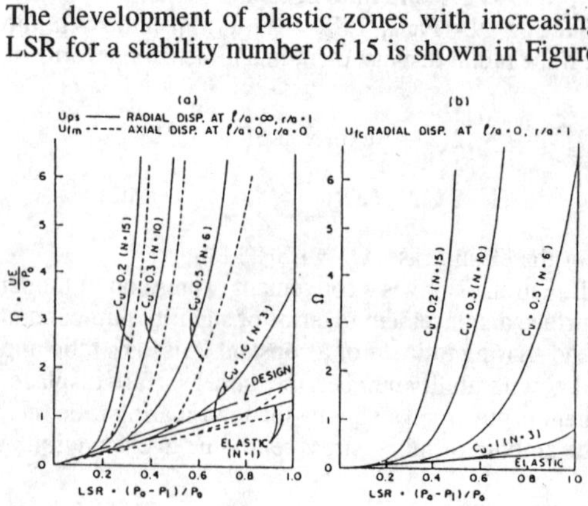

Figure 5: Effect of LSR on dimensionless displacements: (a) u_{ps} and u_{fm}; (b) u_{fc}

6. The theoretical values of the plastic radius calculated from Eq. 5 under plane strain conditions are also shown by the arrows. From the figure the following observations may be made:

(i) Plastic zone increases with the amount of initial stress relieved. For large values of l/a, the radii of the plastic zones calculated by the finite element analysis are consistent with those calculated from the closed-form plane strain solution.

(ii) The growth of the plastic zone starts from the corners of the tunnel face and fans out into the interior of the rockmass. The plastic zone occurs not only behind but also in front of the tunnel face. These results are generally consistent with those reported (Harada 1989a,b,c, Panet and Guenot 1982, Lee and Rowe 1990).

(iii) It may be seen in Figure 6 that a continuous plastic zone develops around the face at a value of LSR between 10 and 20%, establishing a situation where instability of the face becomes possible.

To characterize the development of a continuous plastic zone around the face which may lead to instability, the normalized plastic radius at the tunnel face $(b/a)_{fc}$ and under plane strain conditions $(b/a)_{ps}$ may be used. These quantities are plotted against LSR in Figure 7. In the figure, the line for the elastic state (N=1) is also shown. It may be observed that

i) for a given value of stability number, the normalized plastic radius increases with increasing LSR;

ii) for a given LSR, the rate of development of plastic zone increases with increasing stability number;

iii) the normalized plastic radius departs from the elastic state at smaller LSR with increasing N. The observations mentioned above are consistent with those made for displacements in Figure 5.

It follows, for face stability, that it is possible to design for an LSR such that the plastic radius is within tolerable limits. A suggested design plastic radius, $b/a=1.6$, based on the development of continuous plastic zone around the face, is shown in Figure 7. It may be noted that normalized plastic radius increases rapidly for a small increase in LSR above this line.

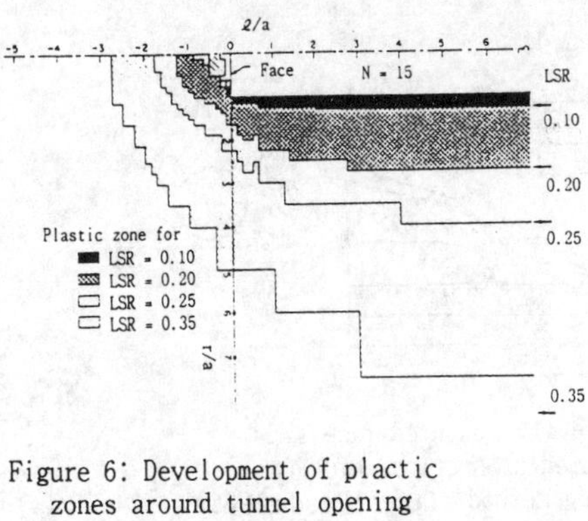

Figure 6: Development of plactic zones around tunnel opening

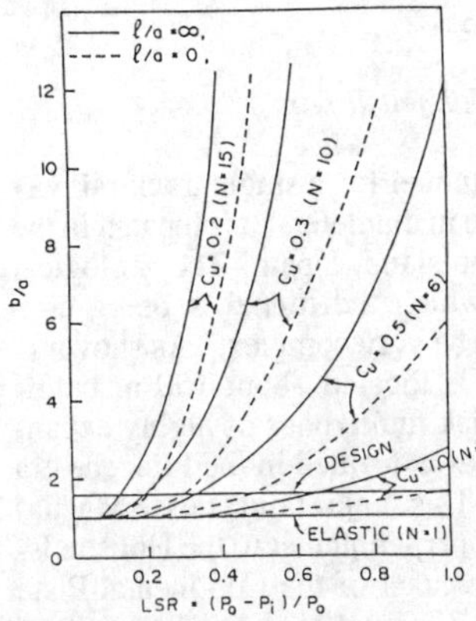

Figure 7: Effect of LSR on plastic radius

6.2 $c\neq0,\phi\neq0$

The effects of load sharing ratio and stability number on dimensionless displacements and development of plastic zone have been studied. Since similar results to those discussed in the previous section were obtained, effect of friction angle and dilatancy angle on the displacement is shown here.

Normalized radial displacement at the crown of the face, Ω_{fc}, is plotted in Figure 8 against the reciprocal of N for LSR = 100 %. For $\psi=0$ (no plastic volumetric strain), the relationship between Ω_{fc} and 1/N are shown by the solid lines, while the dotted lines are for $\phi=\psi$ (associated flow rule). It may be observed that:

i) the normalized displacements increase with decreasing values of the friction angle;

ii) the normalized displacements become very large, when there is a plastic deformation in the plastic region; and

iii) for smaller values of 1/N (large values of N), the effect of friction angle on the deformation is relatively small compared with that of N.

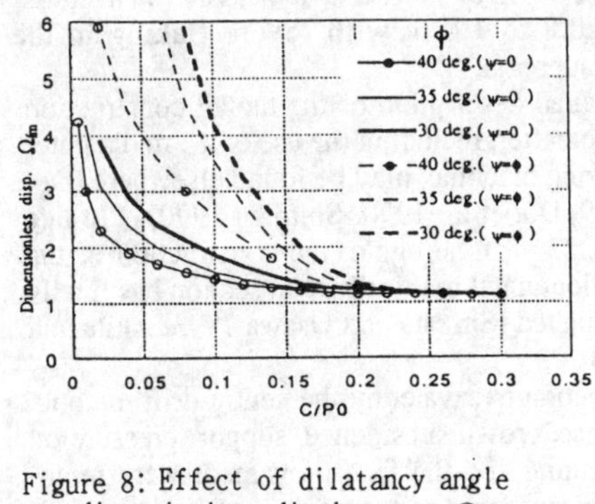

Figure 8: Effect of dilatancy angle on dimensionless displacement Ω_{fm}

7. CASE HISTORIESs

7.1 *Nabetachiyama Tunnel, Japan*

The Nabetachiyama tunnel for a single track railway is 9.1 km long and 6 m in diameter. It is located in the central part of Honshu Island, Japan. The geological conditions through which the tunnel is being constructed are known to be very complex. As shown in Figure 9, the tunnel is located about 150 m below ground surface, in thick fault zones of highly expansive mudstone which was formed in the Pliocene era of the Tertiary Age. This major fault (Fossa Magna) strikes approximately perpendicular to the Honshu Island, formed by the contact of the Continental Plate and the Pacific Plate. The mudstone consists of alternating layers of thin tuff and tuffaceous sandstone. The composition of an individual layer of the mudstone varies widely from rock fragments scattered in weathered clay to flaky brittle rock. The undrained shear strength c_u reported is as low as 125 kPa(Doi et al. 1989).

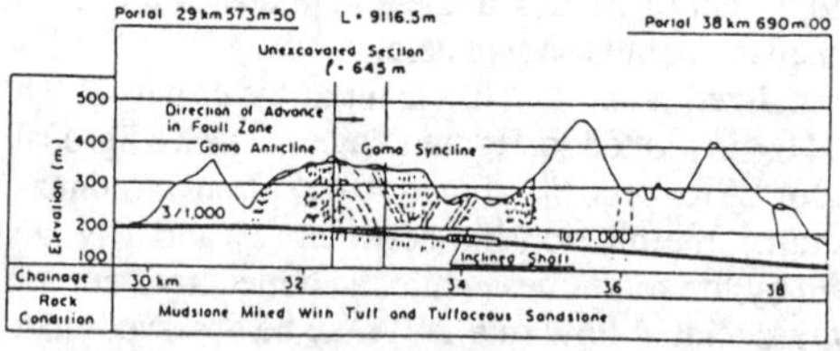

Figure 9: Longitudinal section of Nabetachiyama Tunnel (after Harada 1989a)

By mid 1985, tunnel excavation was 93 % completed by the short benchcut method with shotcreting and rock bolting. No major problems were encountered until the tunnel face reached a section of 645 m in the Shiiya formation (Figure 9). The geology in this section was extremely unfavourable for tunnelling. The floor of the excavated tunnel heaved as much as 0.8 to 1.0 m with severe flaking in the shotcrete supports.

The detailed description of the tunnel construction history thereafter, including the use of a 3 m diameter tunnel boring machine, may be found elsewhere (Doi et al. 1989, Doi et al. 1990, Shimizu 1990). In late 1992, after 7 years struggle, the construction of the 3m dia. pilot tunnel in this difficult section has finally been completed (Shibata and Osawa 1993, Akita and Sato 1993).

In the section excavatedby the centre drift method, convergence, crown subsidence, support pressure on the steel lining and the face movement were monitored. The measured convergence and crown subsid-

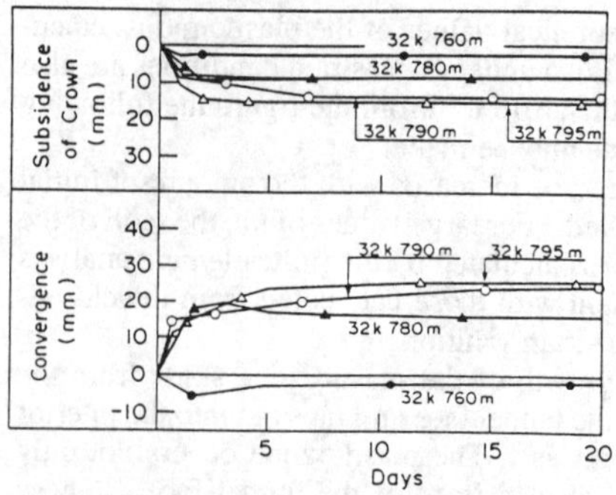

Figure 10: Measured displacements at sections of centre drift heading method (after Doi et al. 1989)

ence between Stations 32k760m and 32k775m (stable section) are shown in Figure 10. It may be seen that the deformation due to excavation developed very quickly in a short period of time and generally stayed constant for larger values of time, suggesting that the rock-lining interaction is an elasto-plastic type, and that most deformation was caused by plastic flow in rockmass. Maximum crown subsidence of the support is about 15 mm, whereas convergence reached 25 mm and measured support pressure was 2.4 to 3.0 MPa, close to the insitu initial stress P_0 deduced from the unit weight of rock and the depth of the tunnel and indicates $K_0 = 1$. It may be noted that in this section the tunnel stability was meintained successfully by the support system used.

The measured face displacements at Station 32k798m are shown in Figure 11. Four measurement pins were installed on the stationary face and the inward deformation of the tunnel face was monitored with time. It may be seen that the face deformation was large, amounting to 1 to 3 m within 2 days. Pin 5, installed at the upper right section of the face, showed the smallest face deformation of about 1 m among the 4 pins. The face displacements indicated by points A and B for pin 5 will therefore be used for the analysis, since the diaplacements measured at the other pins show a more unstable state.

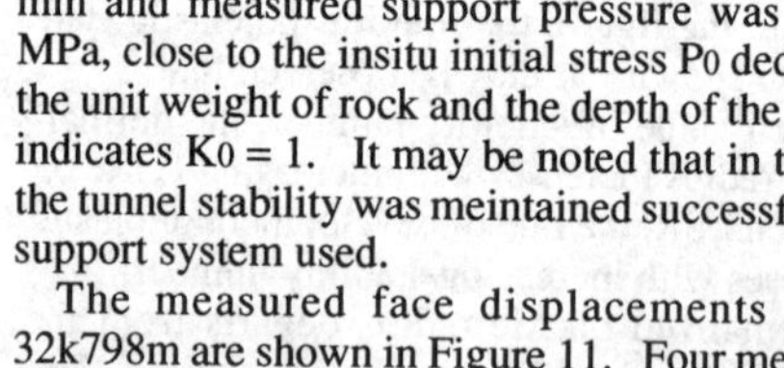

Figure 11: Measured face displacemen at sections ofdrift heading method (after Harada 1989a)

The computed displacement-LSR relationships are shown in Figure 12 using rock parameters and tunnel geometry of the centre drift. Measured face displacement at pin 5 are markd by A and B on Figure 12. It may be seen that both points lie in the region of instability. From Figure 12 it is inferred that dimensionless plastic radius b/a is already as large as 4.6 for point A (see Figure 7 for N=15 and LSR=0.34), indicating instability.

The theoretical results are therefore consistent with the field measurements and the observations that the face could not be advanced (Doi et al. 1989). For stability, it appears that a design value of LSR not exceeding 0.18 is required.

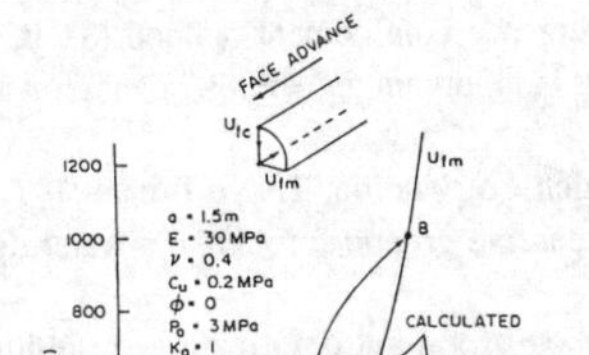

Figure 12: Analysis of tunnel performance of Nabetachiyama Tunnel (centre drift, station 32K798m)

7.2 *Akakura Tunnel, Japan*

The Akakura Tunnel is 10.3 km long and 8 m in diameter. It is located about 35 km east of the Nabetachiyama Tunnel. The geological condition through which the Akakura Tunnel was constructed is very similar to that at the Nabetachiyama Tunnel. The depth of the tunnel varies from 100m to 400m (Figure 13).

The construction of the tunnel started in 1969, and was completed in 1976. During construction, the following two problems were encountered. The first problem occurred when the tunnel was excavated through a sand/sandy conglomerate layer. The construction was very slow and difficult due to large inflow of water at STN 4K234m. The second problem appeared as instability of the tunnel walls due to large earth pressure at STNs 5K170m and 8K009m. The earth pressure was as large as 1.5 MPa at STN 5K170m.

During construction of the tunnel, mudstone samples were recovered for laboratory tests from the boreholes in the tunnel and from the ground surface, as the tunnel face advanced. Comprehensive mineralogical analysis was performed together with uniaxial compression tests, uniaxial creep tests, and

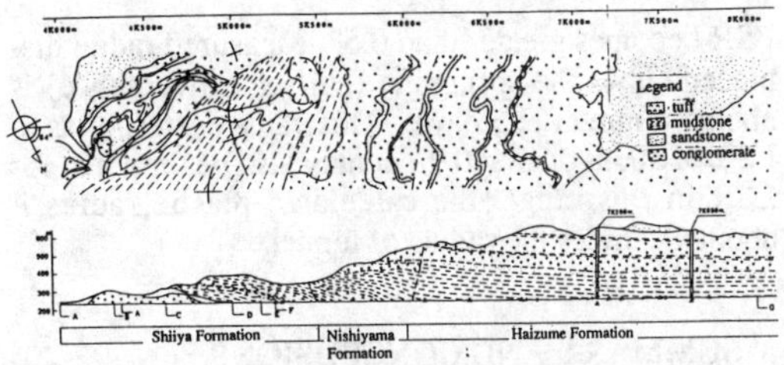

Figure 13: Geological condition at Akakura Tunnel (after JRCC 1977)

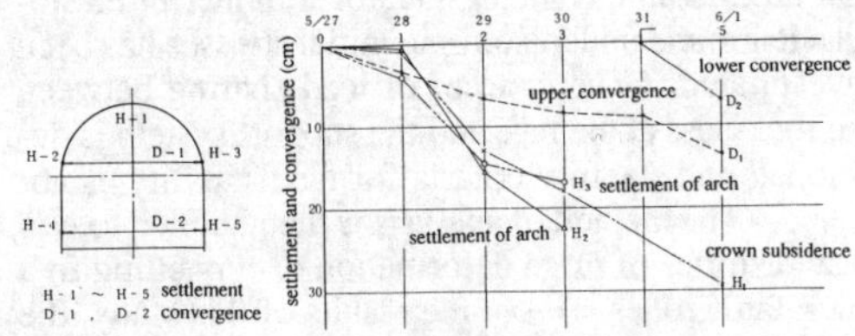

Figure 14: Measured displacement at 5K200m (after JRCC 1977)

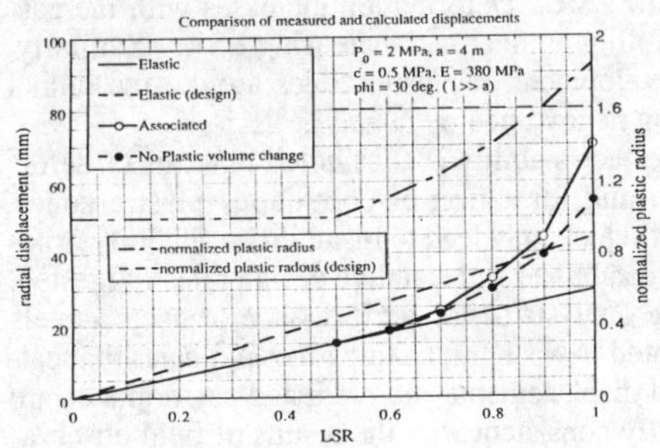

Figure 15: Comparison of Measured and calculated displacements

triaxial compression tests. It was found that main constituent of the mudstone was Na-montmorillonite which was very susceptible to swelling when wetted.

The measurements of convergence and subsidence of the arches at STN5K200m are shown in Figure 14. It may be seen that whole tunnel subsided about 200 mm in the first 3 days with a wall closure of about 80 mm. To demonstrate the method of analysis developed, the convergence measured at the upper section is compared in Figure 15 with the calculated results for a circular tunnel with equivalent cross section and the strength and deformation parameters determined on the rock cores recovered from the site.In the analysis, two cases were considered:associated flow rule (φ=ψ), and no plastic volume change (ψ=0).

It may be seen in the figure that for the parameters used, development of a plastic zone commences when

LSR becomes greater than 0.5. Measured radial displacement at the springline (half of convergence) is about 40 mm after 3 days. This value corresponds to LSR of about 0.85 to 0.9. It may also be seen that for LSR in this range, the calculated plastic radius is about 1.5 times the radius of tunnel.

8. SUMMARY AND CONCLUSIONS

Using the results of elasto-plastic 2-D closed form solution and 3-D (axisymmetric) finite element analysis, the deformations and formation of plastic zones at and around an advancing face of a tunnel in elasto-plastic ground under isotropic initial stresses has been investigated. The concept of load sharing between the rockmass at the face and the support system is developed and stability criteria for face stability established. The method of analysis is then applied to two case histories of large deformation of tunnelling in a thick fault zones. From the results of this study, the following conclusions may be drawn:
(a) The inward displacements at the face and the radial displacements at any distance from the face are governed by the stability number N and the load sharing ratio LSR. Deformation increases with increasing stability number and increasing LSR. Similarly, the development of plastic zones around the tunnel opening is governed by N and LSR.
(b) The face stability is considered in terms of deformation and formation of continuous plastic zones, both of which may lead to instability. Stability criteria are establihed and tentative design lines suggested.
(c) The analysis of the performance of deep tunnels excavated in weak thick fault zones indicates that calculated displacements and predicted performance are generally consistent with the results of field observations, particularly conditions of face stability.

It is therefore suggested that the design concept and the methods of analysis developed may be useful in the problem of deep tunnels crossing weak thick fault zones.

9. ACKNOWLEDGEMENT

The study was performed using a finite element code NAPG-2D (Nonlinear Analysis Program for Geotechnical Engineering-2 dimensional version) developed by the NKK Corporation, Japan. The authors are grateful to Messieurs N. Doi, S. Hattori, J. Hayashi, M. Minemoto, Y. Takahashi, K. Akita and R. Shimizu of the Japan Railway Construction Public Corporation for allowing us to publish the results.

10. REFERENCES

Doi, N., Hattori, S., and Hayashi, J. 1990. Challenge in highly-expansive ground (Nabetachiyama Tunnel). *Tunnels and Underground.*

Doi, N., Minemoto, M.., Takahashi, Y., and Akita, K. 1989. Tunnel through highly-expansive mudstone zone (Nabetachiyama Tunnel). *Proceedings, International Congress on Progress and Innovation in Tunnelling.*

Harada, N. 1989a. Convergence confinement method (1), in a series "New techniques in mountain tunnelling," *Tunnels and underground.*

Harada, N. 1989b. Convergence confinement method (2), in a series "New techniques in mountain tunnelling," *Tunnels and underground.*

Harada, N. 1989c. Convergence confinement method (3), in a series "New techniques in mountain tunnelling," *Tunnels and underground.*

Japan Railway Construction Corporation, Tokyo Branch 1977. Overall analysis of expansive ground for Akakura water diversion tunnel.

Lee, K. M. 1989. Prediction of ground deformations resulting from shield tunnelling in soft clays. *Ph.D. thesis*, The University of Western Ontario, London, Ontario.

Lee, K. M. and Rowe, R. K. 1990. Finite element modelling of the three-dimensional ground deformations due to tunnelling in soft cohesive soils: Part I-Method of analysis; Part II-Results. *Computers and Geotechnics.*

Lo, K. Y., Ng, M. C., and Rowe, R. K. 1984. Predicting settlement due to tunnelling in clays. *Proceedings of GEOTECH '84, ASCE*, Atlanta, GA.

NKK Corporation, Steel Research Center, Civil and Building Technology Research Department: NAPG-2D, Version 3.0, *Theoretical Manual*, 1988.

Ogawa, T. 1986. Elasto-plastic, thermo-mechanical and three-dimensional problems in tunnelling. *Ph.D. thesis*, The University of Western Ontario, London, Ontario.

Ogawa, T. and Lo, K. Y. 1987. Effects of dilatancy and yield criteria on displacements around tunnels. *Canadian Geothechnical Journal.*

Panet, M. and Guenot, A. 1982. Analysis of convergence behind the face of a tunnel. *Tunnelling '82, The Institution of Mining and Metallurgy.*

Sekiguchi, K. Rowe, R. K., Lo, K. Y., and Ogawa, T. 1990. Time step selection for the 6-node non-linear joint element in elasto-viscoplasticity analysis. *Computers and Geotechnics.*

Sekiguchi, K. Rowe, R. K., Lo, K. Y., and Ogawa, T. 1992. Time step selection for the 6-node non-linear joint element in elasto-viscoplasticity analysis. *Soil and Foundations.*

Shimizu, R. 1990. Personal communication on Nabetachiyama Tunnel.

Zienkiewicz, O. C. and Cormeau, I. C. 1974. Visco-plasticity-plasticity and creep in elastic solid-A unified numerical solution approach. *International Journal for Numerical Methods in Engineering.*

Tunnelling and Ground Conditions, Abdel Salam (ed.) © 1994 Balkema, Rotterdam, ISBN 90 5410 363 9

A new approach for ground reinforcing design in tunnelling

S. Pelizza, D. Peila & P.P. Oreste
Dipartimento di Georisorse e Territorio, Tunnelling and Underground Space Center, Politecnico di Torino, Italy

ABSTRACT: the convergence confinement method has been widely applied to tunnel design due to the simplicity of the used numerical approach. A solution, able to take the existence of an area with different mechanical properties around the tunnel into account has been proposed in this paper. The ground behaviour considered for this analytical formulation includes a strain softening law and a Hoek and Brown yield criterium both for the reinforced area and for the natural ground.

A new approach for the design of bolt reinforcement has been proposed using this convergence-confinement formulation. The bolted ground has been modelled by changing, the gradient of the softening branch of the natural ground stress-strain curve and the residual strength criteria parameters in a zone around the tunnel as thick as the bolt length.

This approach has been tested on the basis of the physical model results proposed by Indraratna and Kaiser (1989,1990) and has been applied to the case of a railway tunnel in weak rock reinforced with bolts and steel cables.

1. INTRODUCTION

The use of rock mass reinforcing techniques such as fully-grouted bolts or cables and grouting injections has been frequently applied to many tunnels in recent years, with good results in terms of safety and speed of operations, even in weak ground conditions (Pelizza and Peila, 1993).

The design of these reinforcing schemes with analytical axisymmetric models (convergence-confinement approach) has encountered many theoretical problems as it is difficult to take the properties of the new material, obtained with the application of the reinforcing technique, into account with the usual proposed models.

For this reason a new convergence-confinement approach which is able to model a reinforced rock mass zone with different properties, has here been developed.

This new approach allows one to study different reinforcing techniques applied in tunnelling such as: bolting and grouting injections. It is also possible to simulate the blasting damage to the rock mass since this effect can be modelled as a zone, around the tunnel with worse geotechnical properties than those found in the natural rock mass.

Since the reinforcing effect of rock bolting appears to be directly linked to the post peak behaviour of the rock, this reinforcing effect has been simulated in this paper by varying the gradient of the softening curve of the original rock mass stress-strain behaviour and increasing the properties related to the residual behaviour of the original rock mass.

2. PROPOSED MODEL

The studied geometry is that of a circular tunnel, driven in a rock mass where a zone with different geomechanical properties exists around the tunnel, initially subjected to undisturbed hydrostatic stress field, P_0 (figure 1). The problem is studied in plane-strain conditions, therefore the tridimensional effect of the tunnel face is disregarded.

The two materials are both considered homogeneous and isotropic with the peak yield criteria proposed by Hoek and Brown (1980):

$$\sigma_1 = \sigma_3 + \sqrt{mC_0\sigma_3 + sC_0^2}$$

where: σ_1, σ_3: the maximum and minimum principal stress at failure; C_0: the uniaxial compressive strength of the intact material; m, s: constants which depend on the rock mass nature.

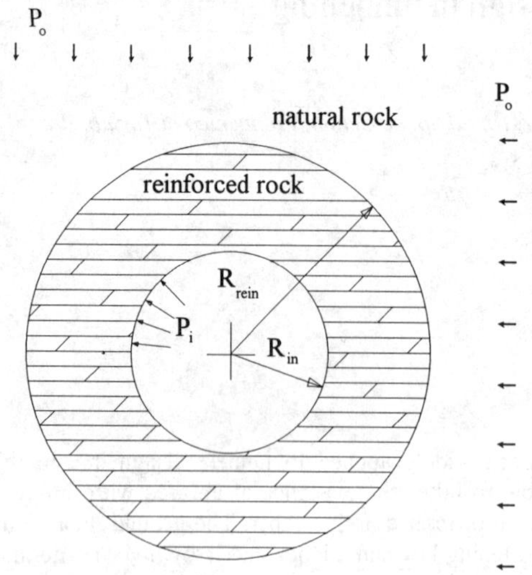

Figure 1. Scheme of the adopted model

It is also assumed in the model that the residual yield criteria is expressed by the same analytical formulation: $\sigma_1 = \sigma_3 + \sqrt{m_r C_0 \sigma_3 + s_r C_0^2}$ where m_R and s_R are the residual strength criteria parameters.

The adopted stress-strain law is with strain softening and three parameters: f (F), h (H), and α describe the plastic behaviour of the mass after failure, as can be seen in figure 2.

In order to describe the stress-strain behaviour of the rock mass in the strain-softening zone, the formulation proposed by Brown et al. (1983) has been used and it is assumed that both m and s decrease from peak values to the residual values m_r, s_r, linearly with the tangential strain:

$$m_{(i)} = m + (m_r - m) \cdot \frac{(\varepsilon_{\vartheta_{(i)}} - \varepsilon_{\vartheta_{(R_{plas})}})}{(\alpha - 1) \cdot \varepsilon_{\vartheta_{(R_{plas})}}}$$

$$s_{(i)} = s + (s_r - s) \cdot \frac{(\varepsilon_{\vartheta_{(i)}} - \varepsilon_{\vartheta_{(R_{plas})}})}{(\alpha - 1) \cdot \varepsilon_{\vartheta_{(R_{plas})}}}$$

where: $m_{(i)}$ and $s_{(i)}$ are the m and s values at a point where the tangential deformation is $\varepsilon_{\vartheta(i)}$.

Because of the algebraically complexity of the equations, it is not possible to obtain a closed form solution for the definition of the stresses and deformations around the tunnel. A numerical solution based on a finite difference numerical scheme is therefore adopted. The computational procedure involves the discretization of the plastic zone in annular rings starting from the elastic-plastic interface, towards the tunnel.

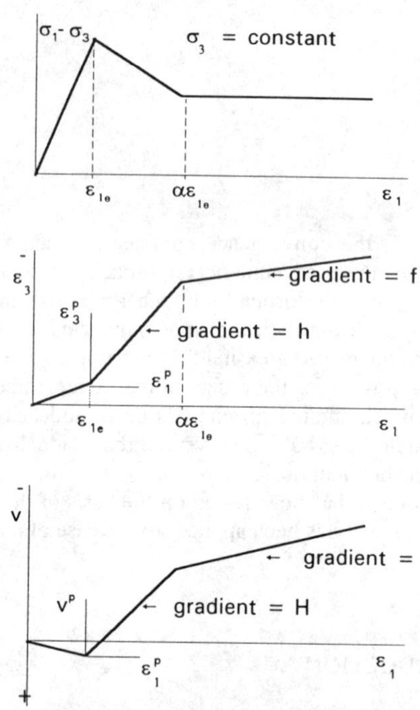

Figure 2. Adopted material behaviour (ν is the volumetric strain, ε_1^p, ε_3^p, ν^p are the total post-peak strain increments (plastic+elastic) in the plastic zone. The compressive stress and related strain are considered positive. (redrawn from Brown et al., 1983).

For the simulation of the reinforced rock mass zone, the computation must be carried out as follows: for $R = R_{rein}$ the displacement and the radial pressure have been calculated by disregarding the existence of the reinforced zone and considering a tunnel of radius $R_{in} = R_{rein}$. Starting from these couples of values (radial pressure and radial displacement) at R_{rein}, stresses and displacements are calculated in the reinforced zone by a step by step procedure toward the internal radius of the tunnel for both elastic and plastic material.

A more detailed description of this numerical approach has already been presented by Peila and Oreste, (1993).

3. ROCK BOLTING SIMULATION IN CONVERGENCE CONFINEMENT MODELS

Many different models for the simulation of the bolted rock mass in axisymmetric convergence-confinement models have been proposed in technical literature: Bischof and Smart (1975) proposed designing fully bonded bolts as a pressure on the tunnel boundary while Grasso et al. (1989) introduced the concept of "reinforced rock mass" and modelled the bolt action as a fictitious increase of the rock cohesion. This last approach has been applied, with good results, to the design of a tunnel excavated in a rock mass with poor geotechnical properties and reinforced with steel-cables and fully grouted rebars.

Stille et al. (1989) developed an analytical elasto-plastic formulation which is able to simulate the bolted rock mass response with four different hypothesis of bolt performance and compared the results with the measured data of the Kielder experimental tunnel.

Indraratna and Kaiser (1988,1990a,1990b) introduced the global behaviour of bolted rock mass (passive fully grouted bolts) in the basic equation of the axisymmetric problem. In this last approach the bolt influence is simulated by a reinforced rock mass zone with better geomechanical properties (cohesion (c_{rein}) and friction angle (ϕ_{rein})): the new properties of the rock mass are linked to the bolt density, $\beta = \pi d \lambda a / S_L \cdot S_T$ (d : bolt diameter; λ : friction coefficient between bolt and rock; a : tunnel radius S_L and S_T: longitudinal and transversal distance between the bolts) and are defined as:

$$\sin\phi_{rein} = \frac{\beta \cdot (1 + \sin\phi) + 2\sin\phi}{\beta \cdot (1 + \sin\phi) + 2}$$

$$c_{rein} = \frac{2 \cdot c \cdot \cos\phi \cdot (1 - \sin\phi_{rein}) \cdot (1 + \beta)}{2 \cdot \cos\phi_{rein} \cdot (1 - \sin\phi)}$$

The friction factor λ is related to the mobilised shear stress acting on the grouted bolt and it is analogous to a coefficient of friction. The magnitude of λ falls in the range of tan (ϕ/2)< λ < tan (2ϕ/3) for smooth steel bolts and approaches tang(ϕ) for rough surfaces while the dimentionless ratio (β/λ) has been found to vary between 0.1 and 0.4 (Indraratna and Kaiser, 1988,1990a, 1990b).

The Indraratna and Kaiser method has been applied to the back analysis of some physical models and has shown to agree well with the measured data but this solution required some simplifications: the linear yielding criteria (Mohr-Columb criterium) and the elasto-brittle-plastic stress-strain law.

Some of the previously described models have proved that the bolt effect can be simulated with good results with an overall variation of the rock mass properties in the bolted zone around the tunnel.

This approach is also considered in this work and the bolt action on the rock mass has been simulated with new, improved properties.

To develop this special application of the proposed general convergence-confinement scheme, as described in paragraph 2, a new model which is able to properly asses the influence of the bolts on the rock mass properties, has been adopted.

Since the fully grouted bolts start their action when the rock mass deforms they mainly act during the plastic displacements of the tunnel, that is to say when there is a plastic radius around the tunnel . Therefore, in this work, the bolt influence on ground properties has been related with the softening and residual behaviour of the natural ground while it has been assumed that bolts do not modify the natural ground peak properties.

The parameters which are found to be the best in the simulation of the new, reinforced rock mass are the gradient of the softening branch of the stress-strain rock mass curve and the residual strength parameters, as can be seen in figure 3.

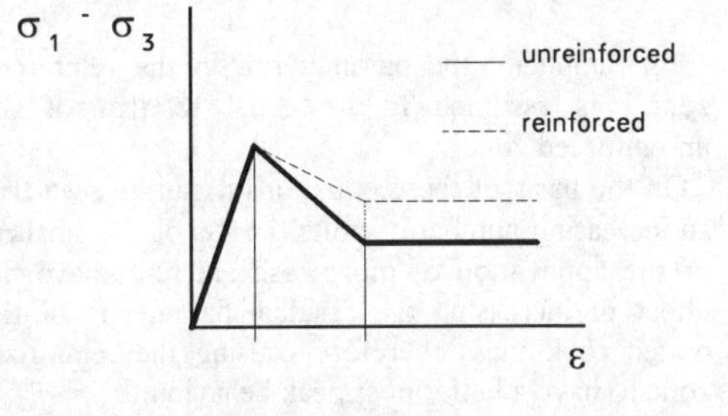

Figure 3. Adopted elasto-plastic model of the reinforced rock mass.

Since around the tunnel, in the plastic zone, the $\sigma_3 = \sigma_r$ variation is between zero (at tunnel surface) and P_{crit} (at the elastic-plastic interface) the improvement is considered only for this range of stress values.

The parameter $\Delta\sigma_3$, which defines the translation of the residual strength criteria (Mahtab et Al., 1993), is therefore considered to be a function of the bolt density, the bolt properties and stress condition around the tunnel. The following equation describes the proposed parameter: $\Delta\sigma_3 = k \cdot \beta \cdot L / a$ where β is the bolt density (Indraratna and Kaiser 1988, 1990a, 1990b), L is the bolt length, a is the tunnel radius and k is a costant with the same dimensions of Po . This value of $\Delta\sigma_3$ is assumed to be constant in the considered stress interval.

The new residual strength criteria envelope is therefore defined and the residual ground properties are modified based on the above described variation of the stress-strain law. The following equations reports the new residual value m_r and s_r and the new plastic parameter f if an associate flow rule is chosen.

$$m_r^{rein} = m_r + \frac{1}{P_{crit}}\left[\left(s_r - s_r^{rein}\right)\cdot C_0 + \Delta\sigma_3 \cdot m_r + \frac{\Delta\sigma_3^2}{C_0} + \frac{2\Delta\sigma_3}{C_0}\sqrt{m_r \cdot C_0 \cdot \left(P_{crit} + \Delta\sigma_3\right) + s_r \cdot C_0^2}\right]$$

$$s_r^{rein} = \frac{\Delta\sigma_3}{C_o}\cdot\left[\frac{\Delta\sigma_3}{C_o} + m_r + 2\cdot\sqrt{\frac{\Delta\sigma_3 \cdot m_r}{C_o} + s_r}\right] + s_r$$

$$\frac{f^{rein}}{f} = \left[\frac{P_{crit} + \sqrt{m_r^{rein}\cdot P_{crit}\cdot C_o + s_r^{rein}\cdot C_o^2} - C_o\sqrt{s_r^{rein}}}{P_{crit} + \sqrt{m_r \cdot P_{crit}\cdot C_o + s_r \cdot C_o^2} - C_o\sqrt{s_r}}\right]$$

For simplicity the parameter h in the reinforced zone is assumed to be equal to that of the unreinforced zone.

On the basis of these equations it can be seen that an increasing number of bolts (i.e. S_L or S_T smaller) or the application of more resistant bolts have the effect of increasing the residual parameters of the bolted rock mass therefore causing the reinforced zone to have a better post peak behaviour.

4. CALCULATION EXAMPLE

Some calculations have been carried out to study the effectiveness of the proposed model for the simulation of rock bolting reinforcement around a tunnel.

The laboratory physical model results, as presented by Indraratna and Kaiser (1988,1990a, 1990b), have been compared with the values predicted with the proposed approach.

Figure 4 reports the geometry of the physical model. The experiments were performed with an artificial rock, 'Gypstone' and four sets of weakness planes were inserted in the model to simulate joints.

The values of the unreinforced material mechanical parameters have been obtained by means of a back analysis of the tunnel wall displacement measured for various stress levels as reported in figure 5 in order to carry out a consistent analysis.

The measured tunnel displacements for various stress levels with a bolt density of β=0.22 and a bolt length of 100mm have then compared with the predicted values. Table 1 reports the material parameters used in the analysis and figure 6 presents the results of the comparison. The calculated data closely overlap the measured data showing a good agreement.

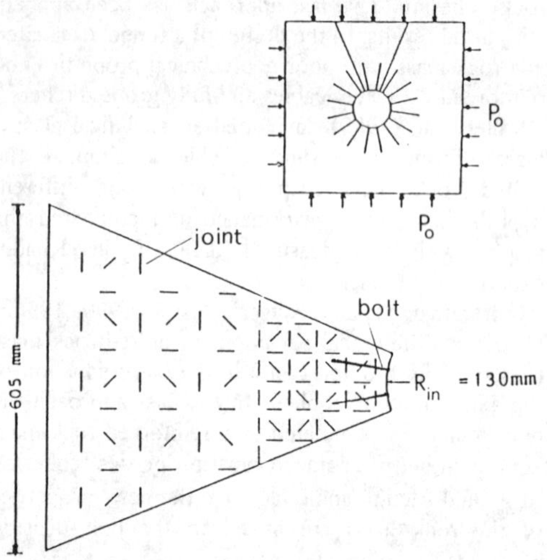

Figure 4. Geometry of the physical model (redrawn from Indraratna and Kaiser, 1988)

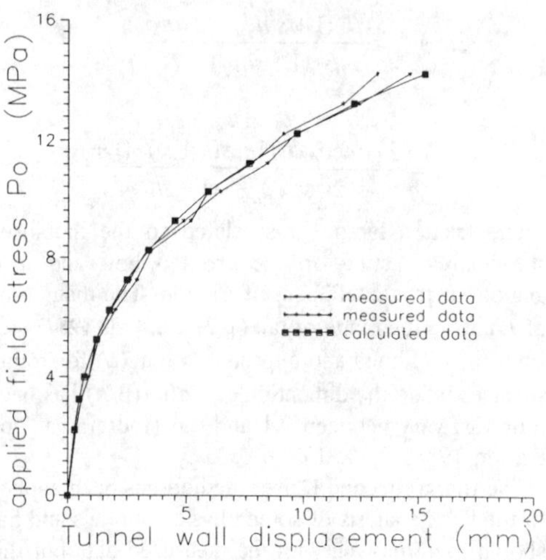

Figure 5. Comparison of measured and predicted values of the convergence response of the jointed sample

Table 1. Geomechanical parameters of the unreinforced and reinforced rock mass for the simulation of the physical model presented by Indraratna and Kaiser, 1988,1990

PARAMETER	unreinforced ground	reinforced ground
deformation modulus (MPa)	1350	1350
Poisson modulus, ν	0.25	0.25
m (peak value)	0.82	0.82
s (peak value)	0.017	0.017
m (residual value)	0.75	0.809
s (residual value)	0.013	0.017
f	1.75	1.75
h	1.75	1.75
α	2	2
Co (MPa)	21.5	21.5

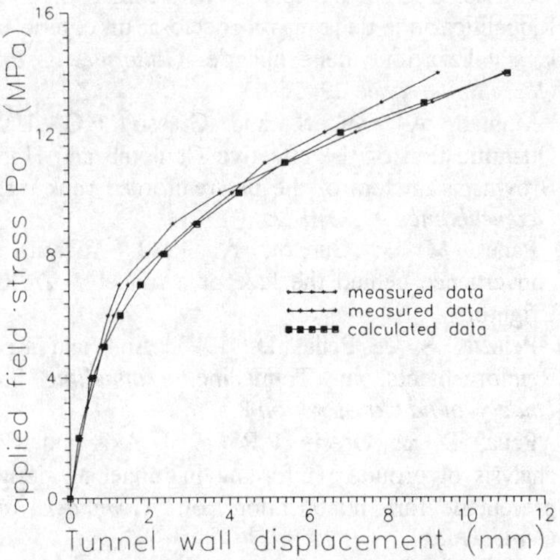

Figure 6. Comparison of measured and predicted values of the convergence responce of the bolted sample β=0.22

The numerical model has also been applied to study the behaviour of a real tunnel which has been excavated in a poor rock mass (V class of Bieniawski classification) using systematic reinforcement with steel grouted rebars and cables (Grasso et al., 1989). The tunnel has an equivalent diameter of 13 m and the reinforced zone has a thickness of 8 m . Various reinforcing geometries were adopted in the real tunnel but in the simulation it was assumed to have a bolt density of β=0.1865 (λ=0.35, $S_L=S_T=1$m). The tunnel depth assumed in the simulation is of 150 m .

Table 2 reports the used unreinforced and reinforced rock mass geotechnical properties (Grasso et al. 1989). The adopted values of the parameters f and h are those of a non associate flow rule with ΔV/V=0.

Table 2. Geomechanical parameters of the unreinforced and reinforced rock mass

PARAMETER	unreinforced ground	reinforced ground
deformation modulus (MPa)	800	800
Poisson modulus, ν	0.3	0.3
m (peak value)	0.4	0.4
s (peak value)	0.001	0.001
m (residual value)	0.35	0.502
s (residual value)	0.00025	0.0305
f	1	1
h	1	1
α	3	3
Co (MPa)	3.48	3.48

Figure 7 reports the obtained convergence-confinement curves with and without bolts. It can be observed that the model predicts a reduction of the final convergence from 2.94m to 0.9m which is in good agreement with the convergences measured in the real tunnel (Grasso et al. 1989)

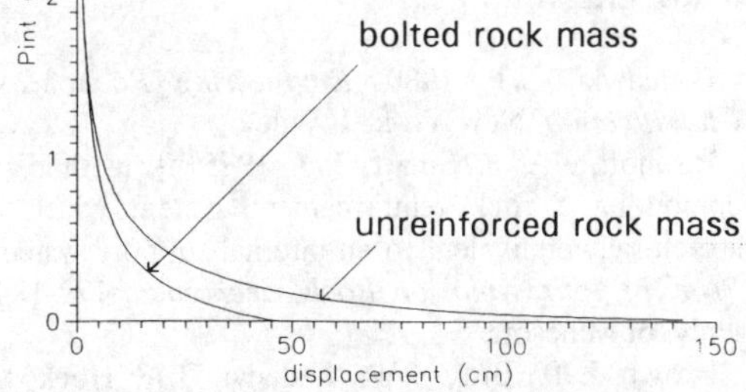

Figure 7 - Calculated convergence-confinament curves with and without bolting

5. CONCLUSION

A new convergence-confinement model which is able to simulate the presence of a zone around a tunnel with mechanical properties different from the natural rock mass, has been presented.

The stress-strain law which has been considered in the analytical formulation of the model is a strain-softening law with an empirical non-linear peak and residual yield criteria.

This new model has been applied to the simulation of tunnel reinforcing with bolts by varying the properties of the bolted area.

This simulation of the zone reinforced by bolts has been carried out under the hypothesis that the bolt action can be described by varying the gradient of the softening branch and the residual parameters of the unreinforced ground mechanical behaviour.

The proposed approach has been applied to simulate some of the results of the physical model presented by Indraratna and Kaiser (1988,1990a, 1990b). This comparison has proved that the measured and predicted results are in good agreement.

The case history of a tunnel excavated in weak rock and systematically reinforced with bolts and steel cables has also been analized with the proposed model. A significative reduction in the tunnel displacement has been calculated, starting from the known geomechanical parameters and from the bolting geometry.

The calculated result has been found to be in good agreement with the measured convergence in the observed tunnel.

On the basis of the carried out tests the proposed model has proved to be a good design tool for rock bolting around tunnels. However it is necessary to carry out further tests in order to confirm these results

6. REFERENCES

Bieniawki, Z.T. 1989. *Engineering Rock Mass Classification*. New York: J. Wiley,

Bischoff, J.A. & Smart, J.D. 1975. A method of computing a rock reinforcement system which is structurally equivalent to an internal support system. *Proc. of 16th Symp. on Rock Mechanics*: 179-184. Univ. of Minesota

Brown, E.T., Bray, J.W., Ladanyi, B. & Hoek, E. 1983. Ground Response Curves for Rock Tunnels: *Journal of Geotechnical Engineering* 109: 15-39

Hoek, E. & Brown, E.T. 1980. *Underground Excavations in Rock*. London: IMM

Indraratna, B. & Kaiser, P.K. 1988. Stabilization of Underground Opening in Jointed Rock by Fully Grouted Boltes. *Proc. of 15th Canadian Rock Mechanics Symposium*. Toronto

Indraratna, B. & Kaiser, P.K. 1990a. Design for Grouted Rock Bolts Based on the Convergence Control Method. *Int. J. of Rock Mech. Mining Sci. & Geomech.* 27 :269-281

Indraratna, B. & Kaiser, P.K. 1990b. Analytical model for the design of grouted rock bolts. *Int. J. for numerical and analytical methods in geomechanics* 14: 227-251

Kaiser, P.K. 1980. Effect of stress history on the deformation behaviour of underground openings. *Proc. of the 13th. Canadian Rock Mechanics Symposium*:133-140. Montreal: The Canadian Institute of Mining and Metallurgy (ed.)

Lembo-Fazio A. & Ribacchi R. 1986. Stato di sforzo e deformazione intorno ad una galleria. *MIR - Recenti sviluppi e nuovi orientamenti nel campo delle gallerie*:11.1-11.36. Torino: G. Barla (ed.)

Grasso, P.G., Mahtab, A. & Pelizza, S. 1989. Riqualificazione della massa rocciosa: un criterio per la stabilizzazione delle gallerie. *Gallerie e Grandi Opere Sotterranee* 29: 35-41.

Mahtab, A., Xu, S. and Grasso, P.G. 1993. Quantification of the effective Coulomb and Hoek-Brown parameters of the pre-reinforced rock mass. *Geomechanics 93*, Ostrava

Panet, M. & Guenot, A. 1982. Analysis of convergence behind the face of a tunnel.: 197-204: Brighton

Pelizza, S. & Peila, D. 1993. Soil and Rock Reinforcements in Tunnelling. *Tunnelling and Underground Construction* 8

Peila, D. & Oreste P.P. 1993 Axisymmetrical analysis of ground reinforcing in tunnelling design. presented for pubblication on *Computer and Geotechnics*

Ribacchi, R. & Riccioni, R. 1977. Stato di sforzo e di deformazione intorno ad una galleria circolare. *Gallerie e Grandi Opere in Sotterranee* 5: 7-18

Stille, H., Holmberg, M. &Nord G. 1989. Support of Weak Rock with Grouted Bolts and Shotcrete. *Int. J. of Rock Mech., Mining Sci. & Geomech.* 26: 99-113

ACKNOWLEDGEMENTS: the authors would like to thank Prof. N. Innaurato for the help given during the preparation of this work. The financial support for this work was provided by the 60% MURST (Italian Ministry for University and Science).

Tunnelling and Ground Conditions, Abdel Salam (ed.) © 1994 Balkema, Rotterdam, ISBN 90 5410 363 9

Analysis of the effects of the two new twin tunnels excavation very close to a big diameter tunnel of Caracas subway

Giafranco Perri
Engineering School of the Central University of Venezuela

ABSTRACT: A numerical analysis by finited element method is carried out to simulate the excavations of two new twin tunnels very close to a big diameter tunnel of Line II in Caracas sub way. Later the structure of the big tunnel already built is verifyed and is indicated consecuently the measures to be implementes to reduce to aceptable values all influence of the new excavations on the operation of the two way Line II tunnel.

1. INTRODUCTION

Due to the abrupt increase of the number of passengers in the Line I of the Caracas Subway, the construction of a new line paralel to the South of the first in the central section of the latter, is been projectes. this new protect anticipates the construction in its beginnig, of two twin tunnels of aproximately 6 m in diameter parallel to the two way tunnel of 11 m in diameter that belong to another line operation: the Line II.

These new twin tunnels, in the most critical sector, will be located very close to the tunnel that now exist passing by at a distance, between the floor of this tunnel and the roof of the new ones, of aproximatly 5 m and with relative horizontal asymetric location betwen the twin tunnels an the tunnel in operation.

It is to be expected, that due to the new excavations, line deformations of the tunnel already built, will occur wich can demage its structure and consecuently affect the operation of the Line II trains, that must not be interrumped.

2. GEOMETRY

Between the progressions 0+000 and 0+100, the twin tunnels, parting from the "Capuchinos" station with a parallel alignment from the double track tunnel, will be excavated sideways to the latter with an aproximate separation of 5 m therefore. During the excavation of the 6 m in diameter twin tunnels, a decompression will take place in the soil that holds the two way tunnel walls, causing deformations and stress in the line structure of that tunnel (Fig. 1).

Between the progressions 0+350 and 0+460, both twin tunnels tend to get closer to the double track tunnel as they go deeper and the "PVI" twin tunnel continues to get closer to the double track tunnel until it lines up under the latter, with a minimun separation of 5.5 m between the roof and the bottom of both tunnels respectively, givin way to a new critical configuratigon in relation with the settlements of the soil, caused by the excavation of the twin tunnels.

Line the taking into account that the most critical conditions will take place when the twin tunnels line up under the double track

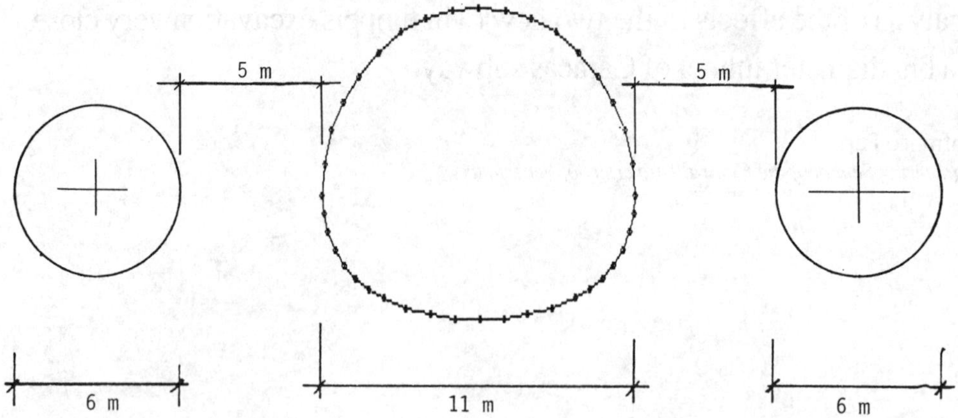

FIGURE 1

tunnel and in an assymetrical way respect to the latter, it is considered as the critical section of the analysis the one located in the progressive 0+400, since a strong assymetry takes place in the localization of the "PVI" twin tunnel in relation to the double track and because the separation between both tunnels is aproximately of 15.0 m, being the separation between the roof of the "PVI" and the bottom of the double track of around 5.90 m (Fig. 2).

3. GEOTECNIQUE

Basically, the subsoil of the sector is characterized by a soil of colluvium-alluvium origin, of a predominantly granular nature and of a compactness from medium to dense, that rest directly on the rocky layers in place, characterized as quartz-micaceus, meteorized, fractured, from soft to hard, schists.

The contact between the soil with a coluvium-aluvium origen and rocky stratums in place, can be seen all along the sector, in a very erratic way, and wich the tunnels pass through those two geological formations in an alternate way and, consequently, the geotecnical conditions vary as the tunnels meet each other totally or partially in one or the other formation.

Considering that the most unfavorable geotechnical condictions will be when the tunnels, or a great percentage of them, is located in the weak coluvium-alluvium layers, it was decides that the relevant analysis were carried out under the following geotechnical conditions:

- Progressives 0+000 to 0+300

a) The double track tunnel is located in coluvium-alluvium soils.

b) The bottom of the double track tunnel is supported on rock and the walls and the bottom in colluvium alluvium soils.

- Progressives 0+300 to 0+460

a) The entirely of the tunnels is found in colluvion-alluvium soils.

The geotecnical parameters of the subsoil for the analysis that were carried out, were the following:

Module of elasticity (E): 200 MPa
Module of Poisson (v): 0.40
Module of soil reaction (Ks): 3000 MPa/m
Module of rock reaction (Ks): 150MPa/m
Soil friction (ϕ): 25°
Rock friction (ϕ): 35°
Soil cohesion (c): 0.01 MPa
Rock cohesion (c): 1 MPa

4. NUMERICAL ANALYSIS

4.1 Progressives 0+000 to 0+300

By means of the unidimensional finite elements the definitive line of the double track tunnel was modeled (Fig. 1), and the conditions of the existing load corresponded to the geostatic and hydrostatic pressures were applied: a coverage of 8 m and phreatic level at -10 m.

For each analysis carried out, two load conditions were taken into account, the first only with the geostatic pressures and in the second with the geostatic and the hydrostatic pressures.

The soil decompression, resulting from the excavation of the twin tunnels, was simulated by means of a reduction by an 83.33% of the magnitud of the module of soil reaction.

This value simulate sufficiently enough the excavation of the twin tunnels by means of a mechanic hield and the assembly of prefabricated reinforced concrete line on the tail of the shield, according to the practice constantly utilized in the metro of Caracas (Perri and Centeno, 1987).

4.2 Progresive 0+300 to 0+460

To simulate the excavation of the twin tunnels and, thus, be able to study the interaction with the double track tunnel, a model of planes finite elements made up of 378 nodes and 323 isoparametric quadrangular elements (Fig. 2) was elaborated.

Geometrically speaking, the model include the two twin tunnels and the double track tunnel, and it extends from the soil surface to a depth of 2 diameters under the bottom of the

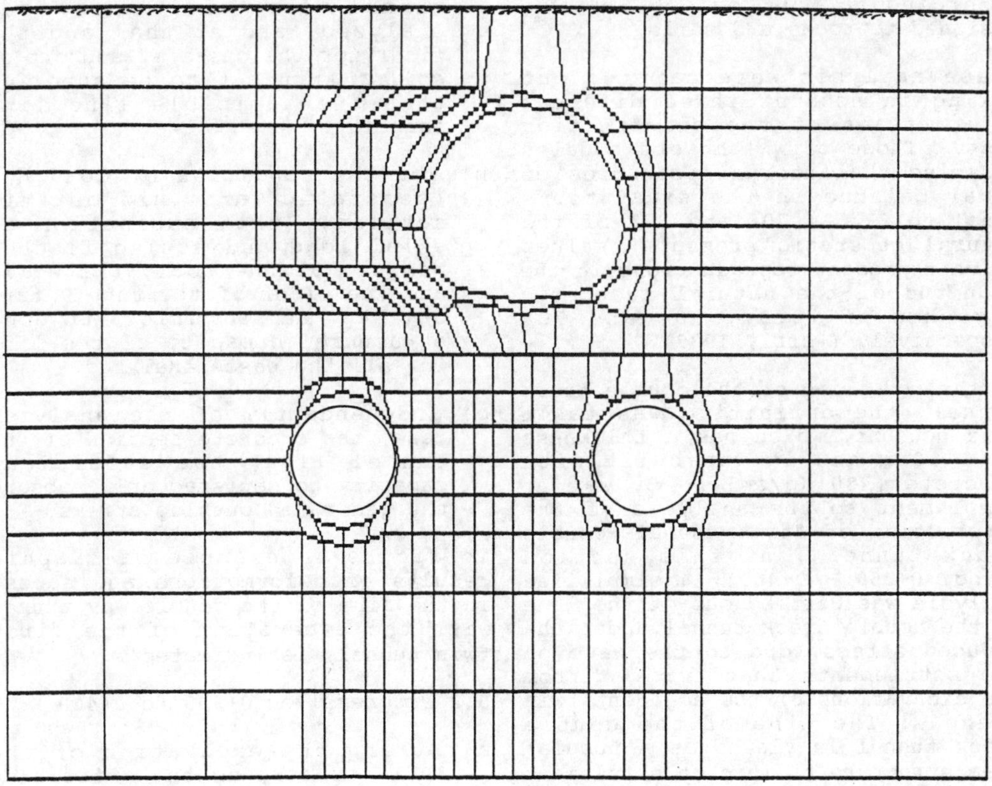

FIGURE 2

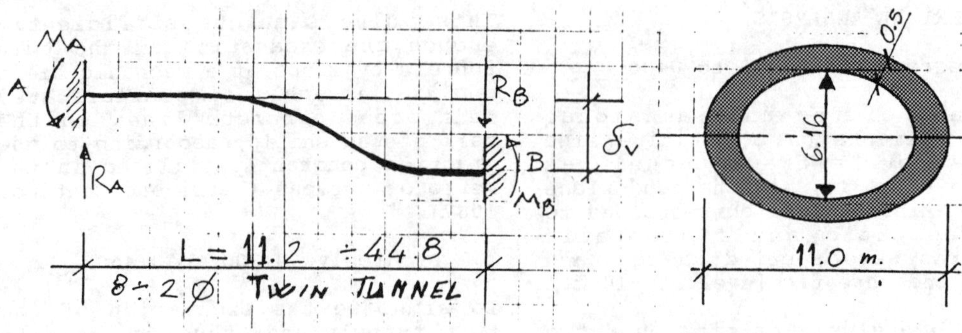

FIGURE 3

twin tunnels and up to a minimun distance of 2.5 diameters from the sideways of the wall of each twin tunnel.

Subsequently, the excavation of the twin tunnels is simulates. In the first place, the excavation of the "PVI" is simulated and inmediatelly after, the excavation of the "PVD". A model of elastic plasticity according to Mohr Coulomb law was assigned to rocks and soils.

These analysis were carried out taking in account three diferent efficiencies of the construction cess. Those of which were modeled assuming the internal pressures of final balance in the excavations equal to 87.5%, 70% and 30% of the natural geostatic pressure; values which represent reproductions in the magnitud of the natural geostatic pressure of 12.5%, 30% and 70% respectively (Perri, 1989).

In the modelling of the double track tunnel, the definitive was taken into account, which has a thickness of 0,50 m and it was built with concrete (280 Kg/cm²). By way of complement to the analysis of the interaction of "twin tunnels-double track tunnel", at this section (Prog. 0+300 - 0+460), a simplified analysis was carried out of the line of the double track tunnel under the induced stress, due to the wave on the settlements that derived from the excavations of the adjacent twin tunnels: The line of the double track tunnel is viewed as a tubular beam structure to which a settlement of contrary supports was imposed (Fig. 3).

5. ANALYSIS OF THE RESULTS

5.1. Progresive 0+000 to 0+300

a) For the condiction of load N°1 (only geostatic actions), the combination of axial load and flector moment at the roof of the tunnel for all the cases analyzed, and at the and the floor for the cases of support on soils, turn out to be superior than the admisible line line capacity.

b) For the condition of load N°2 (geostatic and hidrostatic actions), the combination of axial load and acting flector moment, at the roof, the walls and the floor of the tunel, turn out to be inferior to the structural capacity of the line for all the cases analyzed.

c) Independently of the analyzed case, the concrete section of the tunnel line, has sufficient capacity to satisfactorily absorb the generated cutting stress.

In figure 4, example of graphic results of deformations and stress in the line of the double way tunnel with the excavation of the first twin tunnel, is indicated.

5.2 Progressives 0+300 to 0+460

a) During the excavation of the "PVI" tunnel, as the efficiency

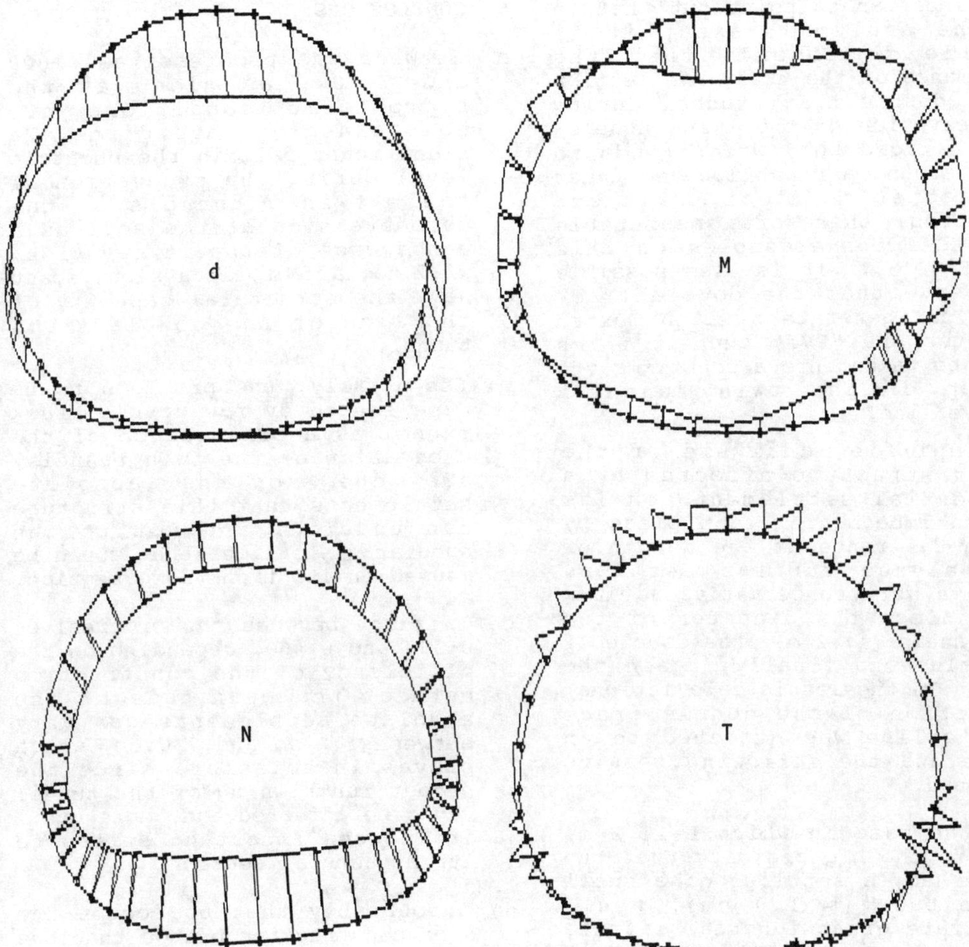

FIGURE 4

of the (magnitude of the balance pressure) construction process reduces and the magnitude of settlements experienced by the double track tunnel, tends to incresae and the distribution of those settlements being slighty asymetrical and of greater magnitude in the zone located above the "PVI" tunnel. The estimated angular distortions for the tranversal section of the double track tunnel, calculated between consecutive nodes, tends to increase up to slightly superior values to the maximum tolerable limite of 1/500, in the case of less efficiency of the construction process.

b) As the excavation of the "PVD" tunnel advances, and the efficiency of the construction process reduces, the magniture of the settlements tends to increase, but their distribution tends to be aproximately symetrical with respect to the axis of the double track tunnel.

Consequently the magnitudes of the angular distortions of the transverse section of the line are negligible.

It is importan to point out that in all the analyzed cases, it is possible to abserve that the performance of the tranverse section of the double track tunnel during the excavation of the twin tunnels, is of a rigid body which tends to rotate around a point located under the latter's floor, without generating, therefore, measurable magnitude coonvergencies on this line (Fig. 5). It is also possible to observe that the double track tunnel can generate a slight extra load over the "PVI" tunnel, which had to be taken into account for the checkup of the twin tunnel's definive line.

The simplified analysis of the tubular structure affected by a differential settlements on its finished embedments, took place by varying the range of the length of the analyzed tubular section, between a very conservative minimum of two times the diameter of the twin tunnels (11.2 m), the double of that value and finally, again the double. In theelastic calculations, the section of the double track tunnel's line was attached to an ellipse and the following results were found:

a) In the case in which L=11.2 m, due to the big rigidity that the tube had, a settling of a small magnitude (≈ 1.0 cm), would generate great internal effects in the section of the tunnel that would exceed the admissible capacity of the line. It most be taken into consideration that the line does not have a significan logitudinal metallic reinforcement capable of absorbing the traction efforts that are generate.

b) In the case in which L=22.4 m, the magnitude of the stress in the line section continue to be umbearable for the structure (the maximun traction effort nearly of 100 Kg/cm^2).

c) In the case of L=44.8 m, the stress is reduce to compatible magnitudes with the resistance of concrete (traction effort and maximum compression of about 25 Kg/cm^2).

6. CONCLUSIONS

a) Between the progressives 0+000 and 0+300, as along as the hydrostatic conditions are maintained, avoiding a significant fall in the pheatric level during the excavation of the twin tunnels, the decompression of the soil as a conseuence of that excavation, will not affect, in a significant way, the structural capacity of the line of the double track tunnel.

b) Essentially, the problems to be experienced by the double track tunnel as a consequence of the excavation of the twin tunnels, are due to the absolute settlements that this structure can undergo, and not to the angular distortions that could be caused in its transverse section.

c) In fact, between the progresives 0+300 and 0+460, depending on the efficiency of the construction method, the magnitude of the absolute settlements can vary between 1.6 cm and 7.0 cm, which proves inadmissible since the proper functioning of the tunnel would be affected and, more over, it induced important stress to its structural section.

d) Undoubtably that to accept the same construction method that has been so successfully used in the building of the twin tunnels of the I, II and II Lines of the Caracas Metro, a previous consolidation of the soil that surrounds the bottom of the double track tunnel and little more under the roof of the twin tunnels, must be carried out with the purpose of minimizing the magnitude of the settlements that can take place.

e) However, to be able to carry out an adequate and effective consolidation of that strip of soil, it must be done from within the double track tunnel which would make it into a very slow method since it would have to be done during the night (when the double track tunnel was closed down) and during a long period of time due to the limited

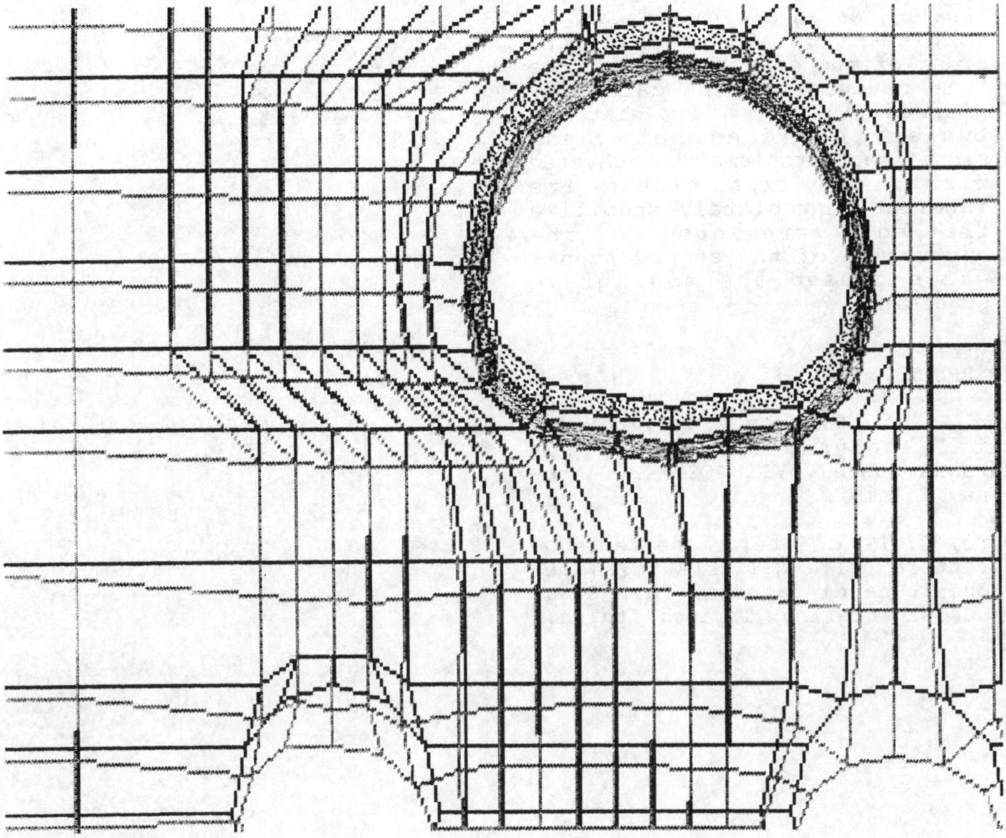

FIGURE 5

productivity of the work.

f) Another possibility would be to act, during the excavation of the tunnels, from the same front by means of extremely difficult in sections of consolidation, and with uncertain results and big delays in time for the progress of the excavations.

g) These circuntances suggest the need to use the most advanced and technical construction method which must be based o the use of close front excavation shield capable of maintaining the natural geostatic pressures in the soil without causing significant decompression on underground water flow: a "hydroshield", that through the hydraulic systems generates a preassure all around istelf at a magnitude that equals that of the soil and the flow of the underground water.

h) By means of the use of this type of shield, the efficiency of the construction method could be achieves so that the reduction of the magnitude of the vertical geostatical preassure can be found under the 10%, by which the stress in the double track tunnel can be minimized and so that the decrease of the phreatic level can be avoide, there fore guaranteeing the confinment conditions required for that tunnel.

i) Finally, as an alternative and as complement of the previus, the order of the construction of the twin tunnels must be carefully planned, in such a way that the tunnel that does not cause significant problems is excavated first, and from within its interior, apropiately stabilize the soil previous to the excavation of the second tunnel which caused the most grave problems.

REFERENCES

Perri, G. & Centeno, R. 1897. Numerical Analysis of interaction Caracas Subway. VII PCSMFE. Cartagena, Colombia.

Perri, G. 1989. Interazione terreno-struttura nelle gallerie della metropolitana di Caracas. Int. Cong. on Geoengineering. Torino, Italia.

Tunnelling and Ground Conditions, Abdel Salam (ed.) © 1994 Balkema, Rotterdam, ISBN 90 5410 363 9

Effect of bench length on stability of tunnel face

J. Seki, *Engineering Section of Maiko Tunnel Naka-kita J.V. (Maeda Construction Co., Ltd), Japan*
K. Noda & E. Washizawa, *Geotechnical Laboratory, Maeda Construction Co., Ltd, Japan*
T. Suzuki, *Civil Engineering Division, Maeda Construction Co., Ltd, Japan*
K. Nishino, *Tunnelling Group, C.E. Division, Maeda Construction Co., Ltd, Japan*

ABSTRACT: The tunnelling method with a shorter bench length ranging from 2 to 6 m has seen more extensive use in recent years. This technique provides an advantage of simultaneous excavation of the upper and lower sections. Since the bench is short, it is vital to evaluate the effect of the bench length upon the stability of the face. We investigated this problem by means of case history studies and three–dimensional finite element analysis. The investigation demonstrated that, with a bench longer than 5 m long, the displacement at the face is limited to a range similar to that in the case of the top heading method.

1. INTRODUCTION

In mountain tunnelling, there is an increasing number of projects using the method by which, with a shorter bench from 2 to 6 m, the upper and lower sections are excavated at the same time.

Maeda Construction Co., Ltd., calls the method "Maeda Fuzzy Bench"(MFB), making an effort to improve it and aiming at its wider application. The greatest feature of the method is to perform excavation at a time on a single face, while, in response to the ground condition encountered, appropriately selecting a length and a shape of the bench sufficient to assure the stability of the face. In addition, auxiliary methods are used as required. When the ground is good enough (hard rock mass), having enough self–supporting properties, the bench length is reduced to 0 m (full–face attack). In the case of a weak ground, a bench about 6 m long is made, leaving the core.

The MFB method requires less labor and allows more rapid excavation. What makes it possible is more concentrated excavation, compared with the conventional top heading practice (with a bench longer than about 20 m). Problems to be tackled when applying this method are (1) effect of a shorter bench on the stability of the face, and (2) influence exerted by sudden distressing of the ground upon the stability of tunnel. Up to now these problems have not been fully studied or discussed and no definitive strategies optimum for different ground conditions have been established. This paper reports the results of our investigation upon the stability of the face through study of case history data and numerical analysis.

2. STUDY OF CASE HISTORY DATA

With respect to eleven tunnels bored by the MFB method, the authors et al. investigated in 1991 the length and shape of the bench, and collapses of the tunnel face. The total length investigated was 8,050 meters, of which 7,030 meters (87 %) were excavated by the MFB method. At tunnel entrances where the overburden is low, or in weak soils which have experienced a collapse, the top heading method was employed. Looking to records of tunnelling operations after the construction of the four tunnels, some cases are found where even the tunnel entrance was bored by the MFB. As shown by these cases, a trend has emerged to apply the MFB even in more adverse conditions.

2.1 *Records of bench length and shape*

Shown in the Fig.1 are data of length and shape of the bench classified by the ground quality. The MFB method was applied up to Class DI in the case of two–lane–road tunnels, whereas for Shinkansen railway tunnels it was used in lower ground class, up to I_N. The method may be applied, according to some field engineers, to poorer grounds, e.g., DII, I_s, I_L, if a suitable auxiliary method is used for stabilizing the face.

Longer benches are made in poorer grounds. It is 5 to 6 m long (5 to 6 support–intervals) in the case of CI, DI, IV_N, I_N. In some locations of CII or DI or I_N Class, the core is left.

(See Table 5)

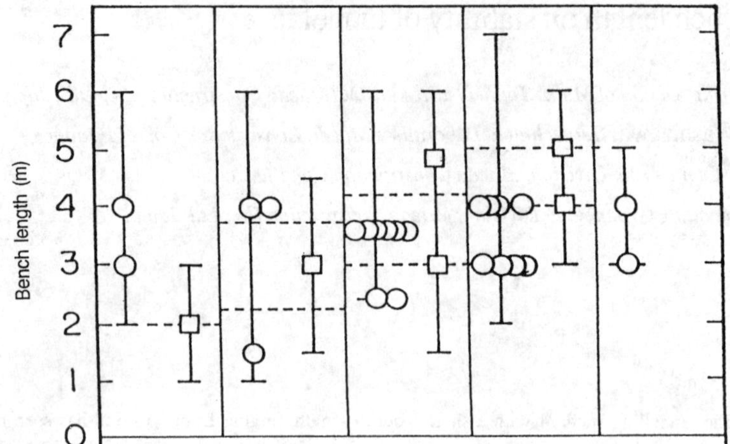

Ground class	Road	B	C_I	C_{II}	D_I	D_{II}
	Railway	IV_N	III_N	II_N	I_N	I_S, I_L
Ground quality		Good ⟵⟶				Poor
One cycle length (m)		2.0	1.5	1.2	1.0	--
Remarks		Legend: ⊢○⊣ : Road; ⊢□⊣ : Railway; ------ : Standard length		Partially leaving the core; an auxiliary method is used.	Leaving the core; auxiliary methods are fully used.	With fore-piling

Fig. 1 Bench lengths adopted in tunnelling operations by the MFB method, for different ground classes

(Remarks) ⊥ represents the range of bench length.

○ and □ represent the bench lengths usually selected.

2.2 *Records of instability of the face*

Fig. 2 illustrates three patterns of face collapse which may be encountered when the bench is short. We found cases where collapse occurred at the crown only (a) and cases where collapse reached the upper section face (b), but found no cases where collapse reached the lower section face due to the short length of the bench. When the collapse of the upper section face as shown in (a) and (b) is not serious, the MFB was still used. When there is a risk of significant collapse because of groundwater inflow, the conventional top heading method may be applied instead of the MFB method. This is because, when significant water inflow exists, the MFB may cause greater damage with a larger amount of collapse and make it harder to perform countermeasures.

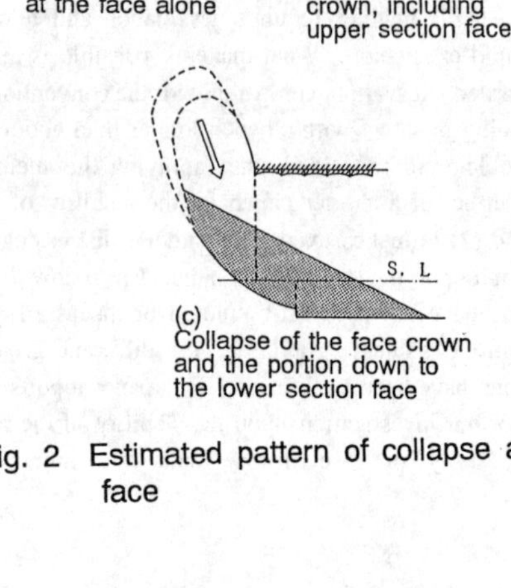

Fig. 2 Estimated pattern of collapse at the face

The following are typical collapses.

(1) Collapse from the upper section of the face up to the crown

During excavation of a two–lane–road tunnel with the support pattern CII and about 4 m long bench, collapse occurred at the top of the upper arch section (Fig. 3). The clay in the fracture zone within the granite layer, and the part over the clay, flowed out with a huge volume of water.

(2) Collapse of the upper section face

In the course of construction of a Shinkansen railway tunnel with the support pattern II_N and a bench about 6 m long, a collapse occurred along the slickenside over about one meter in the depth direction of the central portion of the upper section face (Fig. 4). The ground at the face was composed of tough breccia and mudstone which could be easily crushed by hammering.

It may be possible to judge that the investigated cases of collapse would occur even if the conventional top heading method had been used. There is a possibility, however, that the collapses may have been induced by the shortness of the bench. Therefore, numerical study was implemented to grasp the relationship between bench length and stability of tunnel face.

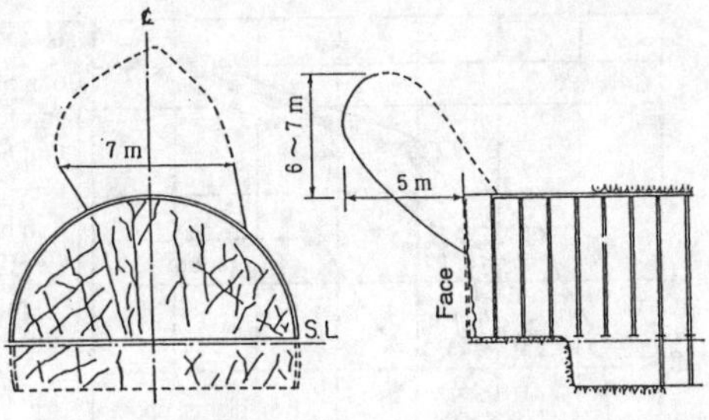

Fig. 3 Cases of collapse from the upper part of the face to the crown

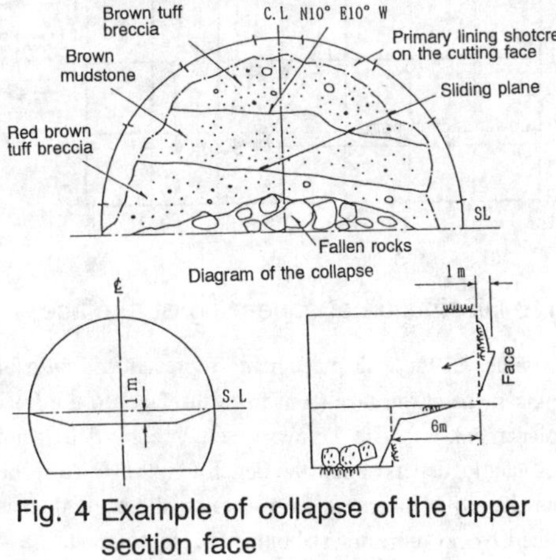

Fig. 4 Example of collapse of the upper section face

3. THREE–DIMENSIONAL FINITE–ELEMENT ANALYSIS

3.1 *Analytical scheme*

By three–dimensional finite–element elastic analysis, the bench length and shape were studied in relation to squeezing and displacement of the tunnel face and stress distribution in the vicinity of the tunnel face.

The tunnel model used was based on a two–lane–road tunnel with horseshoe shape, 5.5 m in radius, excavated without support. Fig. 5 shows the model. The analytical conditions are given in Table 1 supposing the ground to be homogeneous and the initial stress distribution to be uniform. The initial stress was given by external force only, ignoring the dead weight effect.

The cases for analysis with different bench lengths and shapes are shown in Table 2.

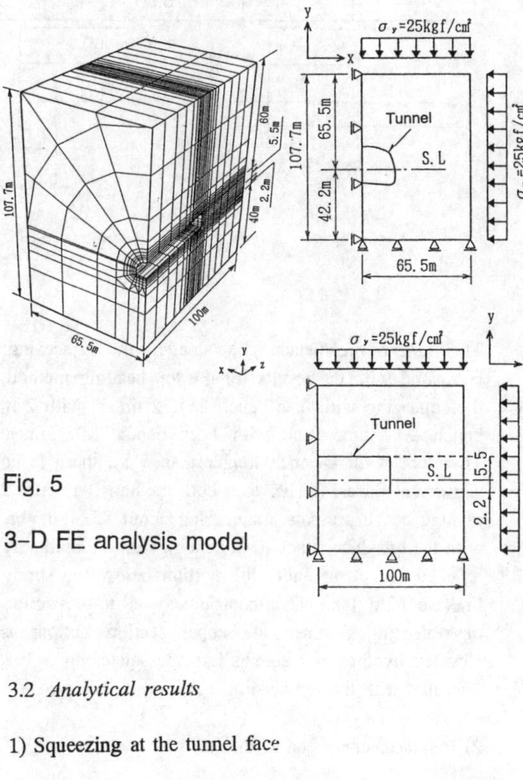

Fig. 5

3-D FE analysis model

3.2 *Analytical results*

1) Squeezing at the tunnel face

When a large displacement is created due to squeezing at the tunnel face, toppling or collapse by slipping of the tunnel face is very likely to take place. Fig. 6 shows, for each bench length and shape, the distribution of squeezing (in the Z direction, i.e., tunnel driving direction) at the tunnel face in the section cut along the tunnel centerline. Fig. 7 is the relationship between bench length and maximum squeezing displacement.

Table 1 Conditions of 3-D FE analysis model

Ground properties	Elastic wave coefficient E (kgf/cm^2)	10,000
	Poisson's ratio ν	0.3
Initial stress σ_0 (kgf/cm^2)		25.0
Remarks: $\sigma_x = \sigma_y = \sigma_z = \sigma_0$ · Supposing an overburden of 100 to 125 m in height.		

Table 2 Cases studied by means of 3-D FE analysis

Cases	(1)	(2)	(3)	(4)	(5)	(6)	(7)
Bench length (m)	0.0	2.0	5.0	10.0	∞	5.0	5.0
Bench height (m)	0.0	2.2	2.2	2.2	2.2	2.2	3.2
Remarks	Full-face attack	Upper section bottm S.L.	Ditto	Ditto	Top heading	Leaving core	1 m-higher upper section bottom

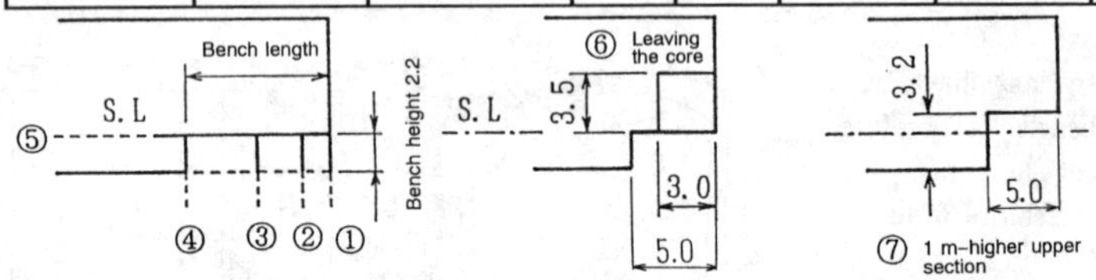

The longer the bench, the smaller the squeezing: compared with the results for the top heading method, the squeezing with 0 m bench is 1.32 times, with 2 m bench 1.12 times and with 5 m bench 1.05 times. Therefore, with a bench longer than 5 m, there is no significant difference between both methods. Even a 2 m long bench presents a squeezing about 85 % of that with the full-face attack (0 m bench), with a remarkably reduced squeezing under the portion below the spring line. So a 2 m long bench contributes well to prevention of squeezing. When the upper section bottom is elevated by 1 m or a core is left, the squeezing is less than that with the top heading method.

2) Displacement ahead of the face

When a large displacement of the crown at the face is produced, the tunnel face or its vicinity are prone to failure. Fig. 8 illustrates the crown settlement and horizontal displacement at the upper section foot in relation to the distance from the face, for different lengths and shapes of benches. In Fig. 9 the displacement ahead of the face is plotted against the bench length.

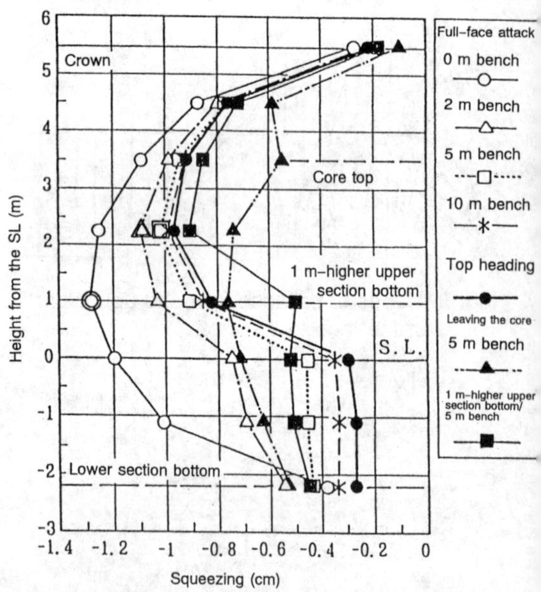

Fig. 6 Distribution of squeezing at the face

The values of these displacements represent the ratio of displacement at the face (0 m from the face) to the final displacement. The crown settlement does not significantly depend upon the bench length. Even if the bench length is increased from 0 to ∞, the settlement is reduced by no more than about 3 %. In contrast, the

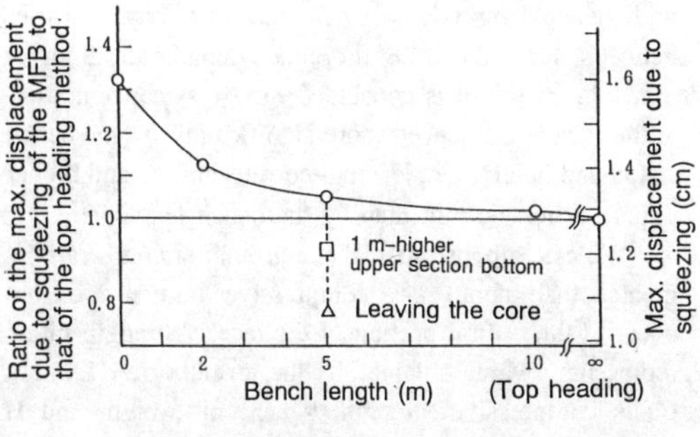

Fig. 7 Relationship between bench length and maximum displacement due to squeezing at the face
(variation ratio = u max/u5,max
u max=each maximum displacement
u5, max=max displacement in the case of the top heading method)

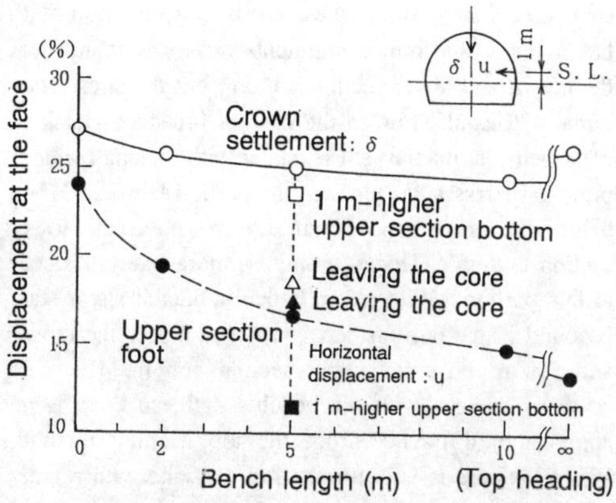

Fig. 9 Relationship between bench length and displacement ahead of the front (crown settlement and horizontal displacement at the upper section foot)

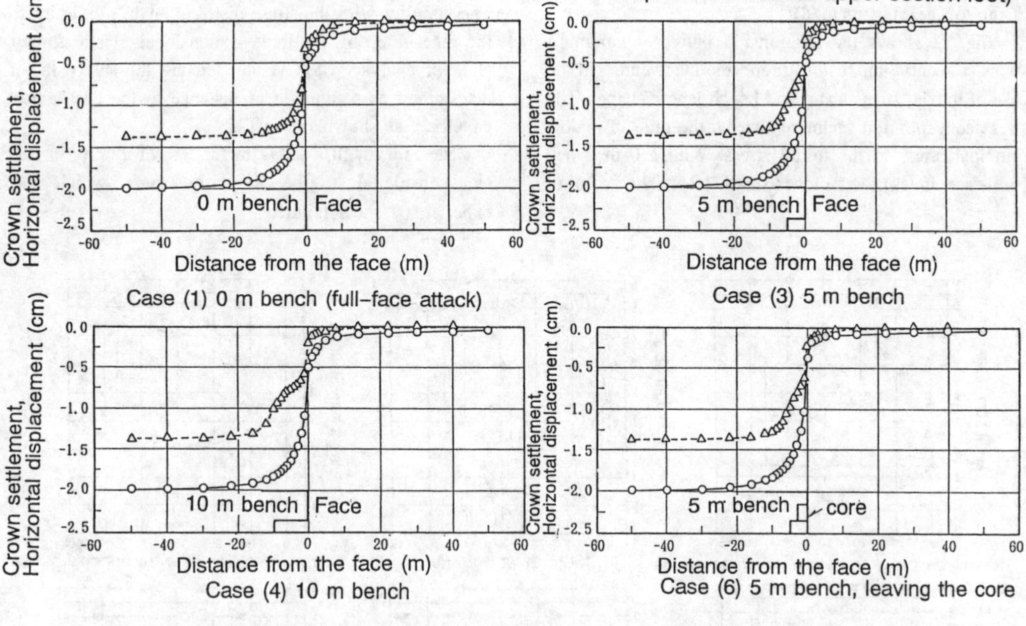

Fig. 8 Crown settlement and horizontal displacement vs. distance from the face
(Legend)
△–△ : Horizontal displacement
○—○ : Crown settlement

horizontal displacement at the upper section foot varies substantially with the bench length. With a longer bench, the displacement is smaller and its increase after excavation along the distance from the face becomes more gentle. Leaving the core is effective exclusively for settlement ahead of the crown. Elevating the upper section bottom by 1 m is more effective for preventing the horizontal displacement at the upper section foot than for preventing the crown settlement.

3) Stress distribution at the face

A state near biaxial stress condition occurs in the ground of the tunnel face where the stress in the tunnel axis direction (σz) is small. As the major principal stress becomes larger, the face is more prone to compressive failure. Fig. 10 illustrates the distribution of major and minor principal stresses at the face around the tunnel centerline for different bench lengths and shapes. In each case for short benches, stress concentration with larger major and minor principal stresses is found near the crown and lower section bottom, but the stress state remains triaxial. But, in the case of full–face attack (0 m bench), a biaxial stress state with a small minor principal stress takes place in the portion between 1.5 m below the crown and about 1.5 m above the lower section bottom. Hence a failure is more likely to occur at this portion. With a 2 m bench, a biaxial stress state is found in the portion between 1.5 m below the crown and 1.5 m above the upper section bottom. The state scarcely changes with the bench length, so there is no appreciable difference from the top heading method. When the core is left, a triaxial stress state with a large minor principal stress remains, offering preventive effect against failure of the face. The 1 m–higher upper section bottom has an effect to limit the zone of biaxial stress state to an area spanning 1 m (2.5 m in the case of the top heading method).

Fig. 11 shows the relationship between maximum stress concentration factor around the face center (in the case of biaxial stress state) and bench length/shape. The stress concentration factor represents the ratio of major principal stress to the initial stress. With a 0 m bench (full–face attack) the factor is small, 1.16. The variation with bench length is so small, that is to say, it varies between 1.25 and 1.28 for bench lengths of 2 m or more. From this it is concluded the stress concentration at the face is influenced more significantly by the bench shape and height, i.e., leaving–core technique and higher upper section bottom than by the bench length.

We can suppose that, if the ground strength ratio is greater than about 1.5, compressive failure does not occur at the central portion of the face. "Strength ratio" represents a value obtained by the formula $\sigma c/\gamma H$ where σc is compressive strength, γ per unit weight and H overburden height. We will discuss such destruction in the following section.

4. ANALYSIS OF THE STABILITY OF THE FACE

4.1 *Plastic zone in the vicinity of the face*

The stability with the stress distribution in proximity to the face given by the three–dimensional elastic finite–element analysis was evaluated using the safety factor determined as a function of distance to the destruction envelope, as shown in Fig.12.

Elements with a safety factor (Fs) of a unity or less are considered to be a plastic zone which has experienced shear failure.

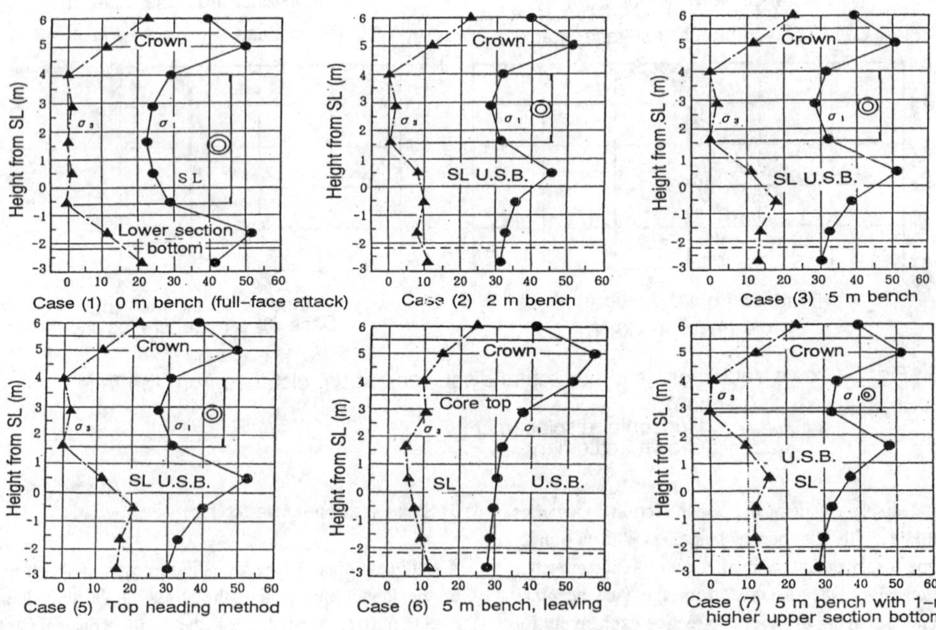

Fig.10 Distribution of major and minor principal stress

(Remarks)
Abscissa: Principal stress $\sigma 1$, $\sigma 3$ (in kg/sq cm)
U.S.B.: upper section bottom
The zones marked with ◎ are in biaxial state.

The strength characteristics are summarized in Table 3. We referred to the values of physical properties estimated for each class of the rock mass classification of dam foundations[2], took the values corresponding to the different strength ratios of the ground, and tabulated against the scale of ground classification of road tunnels. Shown in Fig. 13 are the distributions of safety factor in the longitudinal direction of the tunnel with a ground strength ratio being a unity, for different lengths and shapes of bench. In the case of full–face attack (0 m bench), the plastic zones presenting a safety factor Fs less than a unity appear 1 to 2 m below the crown and in an area of 1 m in front of the face 1 to 2 m above the lower section bottom. By contrast, with a bench, the plastic zones are located 1 to 2 m below the crown and 1 to 2 m above the upper section bottom. In both cases, lower safety factors are found at the central portion of the face. However, the differences in safety factor in front of the face by the bench length are small. The safety factor distribution is similar to that of the top heading method. In the case of the leaving–core technique, no plastic zone appears in the face, and the safety factors in the area ahead of the face are greater than those of the other cases. With the 1 m–higher upper section bottom, plastic zones are generated at the central portion of the upper half section face, but larger safety factors are obtained in the area ahead of such portion compared with the case with the lower upper section bottom. For different bench lengths with different ground strength ratios, a similar trend is found in the distributions of safety factors. For a ratio of 0.5, plastic zones appear up to 2 to 3 m ahead of the face. For a ratio of 2 and 4, no plastic zone is generated. The safety factor distribution with a 5 m bench is quite similar to that of the collapse case (Compare the area with $F_S \leq 1.2$ in Fig. 13 with the collapse zone shown in Fig. 4).

Fig. 14 shows the relationship between bench length and minimum safety factor of the face. From this diagram, it is known that the minimum safety factor does not varies with the bench length, but increases when the core is left.

4.2 *Sliding failure at the tunnel face*

We are going to discuss, considering each strength ratio of the ground given in Table 3, the relationship between bench length and stability against sliding failure of the whole face including the upper and lower sections. It can be supposed that the stability of the upper section alone could be achieved by using an auxiliary method suitable for the ground condition encountered, and that, against sliding below the spring line, the bench in the lower section would behave like a retaining bank. The analysis model and sliding line are illustrated in Fig. 15.

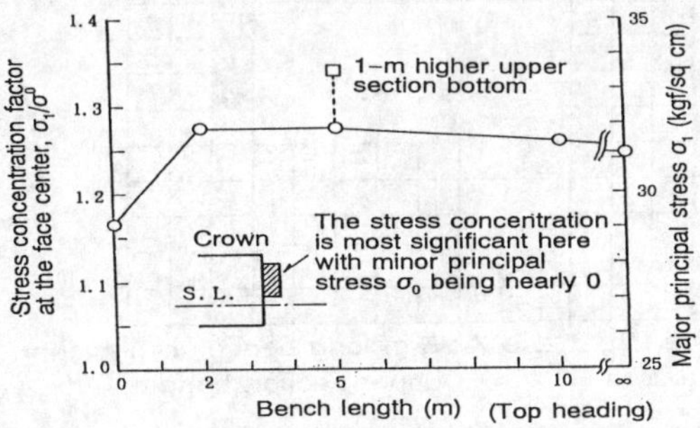

Fig. 11 Relationship between major principal stress at the face center and bench length

Table 3 Strength characteristics for different ground strength ratios with the ground classification scale

Ground class for road tunnel	CI ~ CII		DI	DII	E		
Strength ratio σc/γH (kgf/cm²)	--	4.0	2.0	1.0	0.5		
Unconfined compressive strength σc (°)	--	100.0	50.0	25.0	12.5		γ·H = 25 (kgf/cm²)
Internal friction angle φ (kgf/cm²)	--	45.0	38.0	30.0	30.0	15.0	Set up on the basis of classification by Central Research Institute of Electric Power Industry.
Cohesion C (kgf/cm²)	--	20.7	12.2	7.2	3.6	4.8	$C = \frac{\sigma c \cdot (1-\sin\phi)}{2 \cdot \cos\phi}$

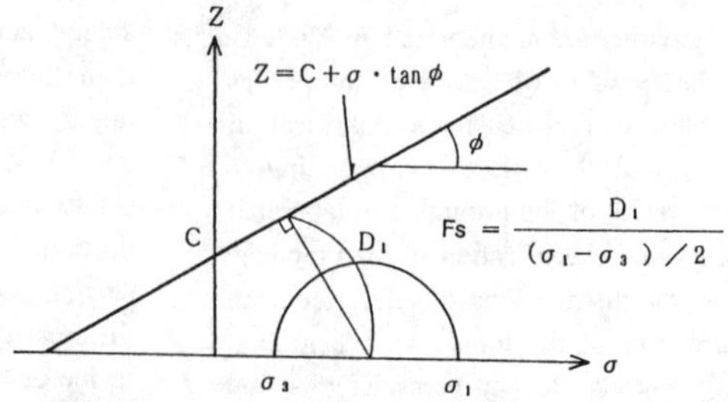

Fig. 12 Diagram of failure

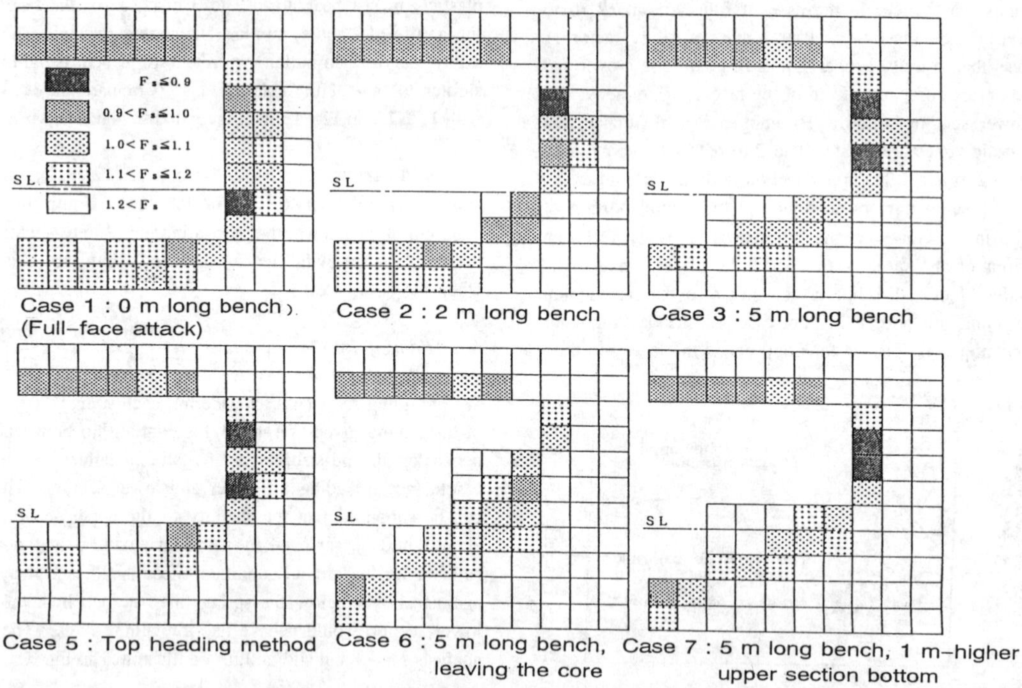

Fig. 13 Distribution of safety factors at the tunnel center for each bench length and shape, in the case of the ground strength ratio being a unity

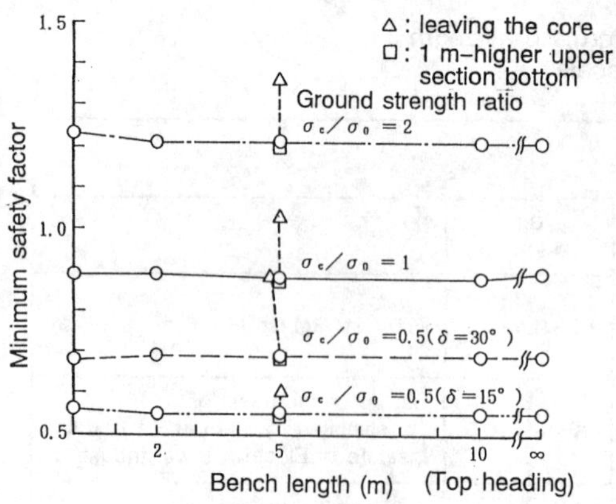

Fig. 14 Minimum safety factor of the face against shear failure vs. bench length

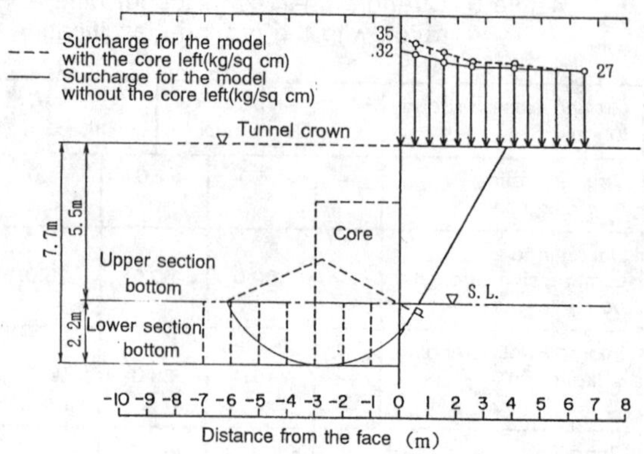

Fig. 15 Analysis model of sliding and sliding line

The surcharge considered is the vertical stress at the crown located at the tunnel center given by the finite element analysis. First, for a 0 m bench (full-face attack), the sliding line passing about 1 m below the spring line was determined (this sliding line is straight). Next, a compound sliding line was drawn, which is the straight line connecting to a circular arc contacting the lower section bottom (2.2 m below the spring line).

The safety factor is plotted against the bench length in Fig. 16. As the bench becomes longer, the safety factor is greater. But, with a bench of 6 to 7 m or longer, the safety factor remains constant, because the sliding line is located at the level of the upper section bottom (spring line). When the ground strength ratio is a unity, the safety factor becomes 1.2 or more for a bench length of 4 m or more. In the case of the ground strength ratio being 2 or more, the safety factor is 1.4 or more even with a 0 m bench (full-face attack). The leaving-core technique increases the safety factory by about 3 to 6 %, while the increase due to the 1 m-higher upper section bottom is no more than about 1 %.

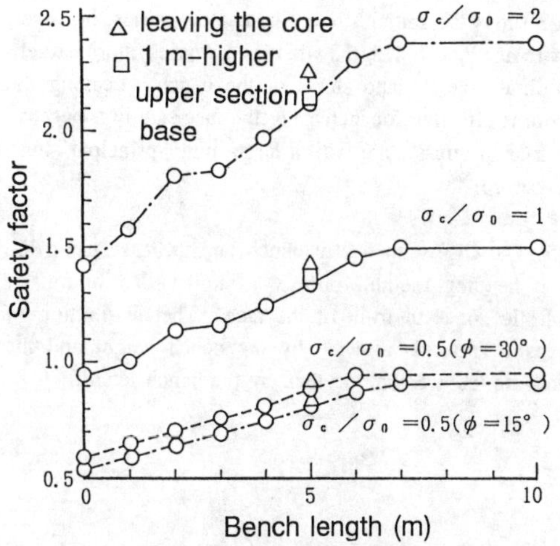

Fig.16 Factor of safety against sliding vs. bench length

5. STABILITY WITH THE BENCH LENGTHS FOUND IN CASE HISTORIES

From the case studies discussed here and the records in published documents, the actual bench lengths and shapes were determined as shown in Table 4 to estimate the stability of the face for different ground strength ratios. Two kinds of safety index were obtained: the minimum safety factor of the face against shear failure from Fig. 14 and the safety factor of the whole face against sliding from Fig. 16. The values of these indices were determined for each of the ground strength ratios given in Table 4. But, for the ratio of 4, no safety factor is shown.

As known from Table 4, as the ground strength ratio gets smaller, the safety factor against shear failure and that against sliding failure become lower. When the ground safety ratio is a unity, a safety factor against shear failure more than 1.0 can be obtained by leaving the core with a 5 m bench. For the ground strength ratios of 4 or more, the safety factors are large, that is, 1.6 for the shear failure safety factor and 2.7 or more for the sliding safety factor. It should be noted that, however, the stability of the face largely depends upon the discontinuity of the rock mass which is not considered in this paper. There are almost no tunnelling cases in a ground with a strength ratio of 1 or less. It is conceivable that the tunneling practice in such a ground would accompany swelling which would be quite different from the elastic behavior treated in the analysis discussed in this paper.

The stability against squeezing at the face is difficult to evaluate quantitatively. But it is probable that, for cases with a smaller ground strength ratio, some preventive effect against squeezing would be expected by making a longer bench and leaving the core.

6. CONCLUSIONS

The tunnelling method with a shorter bench, by which the upper and lower sections are excavated at the same time, has found wider use in recent years. We studied the bench length and shape in relation to the stability of the face, through case history studies and numerical analysis. This has revealed the following.

(1) The longer the bench, the smaller the displacement due to squeezing at the face. But with a bench length of 5 m or more, the displacement is almost the same as that found in the top heading method. Even a bench about 2 m long presents sufficient effect for preventing squeezing.

(2) The bench length scarcely exerts influence upon settlement of the crown. To limit settlement ahead of the crown, leaving the core is an effective strategy.

(3) There is a close relationship between bench length and horizontal displacement ahead of the face at the foot of the upper section. Longer benches will cause smaller displacement.

(4) The major principal stress at the face center is about 1.3 times the initial stress. It scarcely varies, however, with the bench length, whereas it varies more largely with the height and shape of the bench. Leaving the core is effective for increasing the face stability, because a triaxial stress state with a large minor principal stress is set up.

(5) The case with a 2 m bench and the case using the top heading method present a similar distribution of plastic zones in front of the face. The distribution of plastic zones is affected by the bench height and the leaving–core technique than by the bench length.

(6) A preventive effect against sliding failure of the whole of the face, like a retaining bank, can be expected of the bench. Longer benches tend to offer a greater safety factor. When the bench length exceeds 6 to 7 m, the safety factor remains almost constant, which is similar to that of the conventional top heading method.

(7) From the study of the stability of the tunnel face in relation to the actual bench length and shape adopted in different ground classes, it is known that the smaller is the strength ratio, the lower are the safety factors against shear failure of the face and sliding failure of the whole of the face. So, in the case of a ground strength ratio ranging from 2 to 1, they try to increase the safety factor by leaving the core.

Table 4 Estimated safety factors for the actual bench lengths and shapes (initial stress of 25 kgf/sq cm, overburden about 100 m high)

Ground class	Road	B	CI	CII	DI	DII	E	Remarks
	Railway	IV_N	III_N	II_N	I_N	I_S, I_L, Special S, Special L		
Estimated groung strength ratio		4 or more			4~2	2~1	1~0.5	
Cycle length (m)		2.0	1.5	1.2	1.0	1.0	1.0 or less	
Bench length (m)		2.0	2.25~3.75	3.0~4.2	4.0~5.0	4.0~5.0	20 or more	
No. of support–intervals for a bench		1.0	1.5~2.5	2.5~3.5	4.0~5.0	4.0~5.0	--	
Bench shape					(Partially leaving the core)	Leaving the core	Leaving the core	
Auxiliary method				Partially used	Fully used	Fully used		Fore–piling Face bots, etc.
Shear failure safety factor		1.6 or more (Attention should be paid to the discontinuity of the rock mass.)			1.6~1.2	1.3~1.0	0.9~0.6	Estimated from Fig. 14
Sliding failure safety factor		2.7 or more (Attention should be paid to the discontinuity of the rock mass.)			2.7~2.0	2.0~1.3	1.5~0.9	Estimated from Fig.16
Remarks		The safety factors given here are estimated, supposing an initial stress of 25 kgf/sq cm and the physical properties being equal to those given in Table 3 for each graound strength ratio.						

Table 5 Standard Support Pattern

For road tunnels (specified by the Japan Highway Public Corporation)

Ground classification	Tunnelling method	Cycle length (upper section) (m)	Rock bolt			Steel arch support			Shotcreting thickness (cm)	Lining thickness (cm)		Margin for deformation (cm)		
			Length (m)	Interval		Upper section	Lower section	Interval (m)		Arch ·side wall	Invert	Upper section	Lower section	Invert
				Circumferential (m)	Longitudinal									
B	Top heading	2.0	3.0	1.5 (only for upper section)	2.0	Not provided	Not provided	--	5	30	0	0	0	0
CI	Top heading	1.5	3.0	1.5	1.5	Not provided	Not provided	--	10	30	0	0	0	0
CII	Top heading	1.2	3.0	1.5	1.2	H-125	Not provided	1.2	10	30	0	0	0	0
DI	Top heading	1.0	4.0	1.2	1.0	H-125	H-125	1.0	15	30	45	0	0	0
DII	Top heading	1.0 or less	4.0	1.2	1.0 or less	H-150	H-125	1.0 or less	20	30	50	10	0	0

For railway tunnels (specified by the Railway Technical Research Institute)

Support / Ground classification	Rock bolt			Shotcreting thickness (cm)		Steel support
	Location	Length (m) × No.	Longitudinal interval (m)	Arch · Side wall	Invert	Kind
V_N	----	----	----	5 (Average)	----	----
IV_N	Arch	2 × 0 ~ 8	(As required)	5 (Average)	----	----
III_N	Arch	2 × 12	1.5	10 (Average)	----	----
II_N	Arch · Side wall	3 × 16	1.2	Hard rock 10 (Average) Soft rock 7 (Minimum)	----	----
I_N	Arch · Side wall	3 × 20	1.0	15 (Minimum)	----	(125 H)
I_S	Arch · Side wall	3 × 14 4.5 × 8	0.8 ~ 1.0	15 (Minimum)	15 (Minimum)	150 H
I_L	Arch foot · Side wall	3 × 12	0.8 ~ 1.0	20 (Minimum)	----	125 H

These results have revealed the fact that with a bench length of 5 m or more, the stability of the face similar to that of the top heading method can be achieved, and that even about 2 m long bench would contribute considerably to the stability of the face. It should be noted, however, that the discussion in this paper is a fundamental study on the basis of the analysis supposing an elastic continuous mass, without considering slackening, sliding and toppling due to discontinuity of the rock mass or swelling in a poor rock mass. In actual tunnelling work, these factors should be taken into full consideration with regard to the stability of the face. We have intention of studying farther referring to records of future tunnelling work. We are also studying the tunnel stability which may be influenced by sudden distressing of ground in the case of a short bench. We hope to have an opportunity to report on this subject in another paper.

REFERENCES

1) Case histories of mountain tunnelling by the MFB method, September 1991, In-house document of Maeda Construction Co., Ltd.

2) K. KIKUCHI, M.FUJIE, N.OKA, T.KOBAYASHI, 1984, Geotechnical evaluation of the load bearing capacity of rock mass for dam foundation, Application Geotechnology; Special Issue

Tunnelling and Ground Conditions, Abdel Salam (ed.) © 1994 Balkema, Rotterdam, ISBN 90 5410 363 9

A simple numerical model for analysis of cut-and-cover tunnels

A.M.Shalaby
Department of Civil Engineering, Shoubra Faculty of Engineering, Zagazig University, Banha Branch, Cairo, Egypt

ABSTRACT: The use of numerical models, such as the finite element method, that incorporate sophisticated numerical techniques, is becoming a part of modern geotechnical engineering design. Such models are expected to simulate the field behavior considerably better than the conventional methods. However, they need large capacity computers. On the other hand, the proposed analytical model is simpler and needs small capacity computers to provide the deeply needed aid at least for preliminary designs. Site problems may be evaluated on the spot without running a highly sophisticated, time consuming and costly finite element program, i.e. the structure can be back-analyzed using the measured deformed shape. This study also includes comparison of results of this simple model with actual field measurements and results of sophisticated finite element analyses.

1. INTRODUCTION

Observational procedure have been accepted as a realistic tool in geotechnical engineering particularly when ground conditions have a significant effect on the behaviour of the project under consideration.

During construction of cut-and-cover tunneling and diaphragm walls with the use of the observational approach, engineers may face some problems which usually require reasonable and quick decisions to be made.

This paper proposes a simple analytical model which can be used to deal with such situation. This model was tried to analyze four different case histories and comparison between calculated and measured lateral deflection or distribution of earth pressure is also presented and discussed in this paper.

2. MODEL DESCRIPTION

The proposed simple model simply considers each soil layer as a thin tie having length and properties representing soil characteristics of this layer. The proposed numerical model can be used to calculate the internal forces and displacements in the underground structure for any given lateral pressure distribution or vice versa. The required program to run this model can be any plane frame truss program which is a two-dimensional analysis using the direct stiffness method and a linear elastic model.

The simplified simulation of the geotechnical aspects and the geometry of the problem are the most important input data of this analysis. In this work, the diaphragm wall was simulated by a beam, fixed at depth and supported by rollers at roof and raft elevation or any other kind of support such as struts. The soil was represented by a series of horizontal linked members. These varied in length, with the length decreasing with increasing depth. The displacement is varied according to field measurements taken from the measuring device installed in the wall (such as the slope indictor). The force required to cause the strain increases as the length of the member decreases. The simplified geometry is shown in figure (1) as an example.

Properties of the wall were assigned to the beam and soil properties were assigned to the linked members. The linked members are assumed to be hinged at the wall and at a point located away from the zone of influence. This zone is represented by a 45° line drawn from the fixed end of the wall as shown in figure 1. This ensured that the free end of the linked member was not influenced by the deformation of the soil.

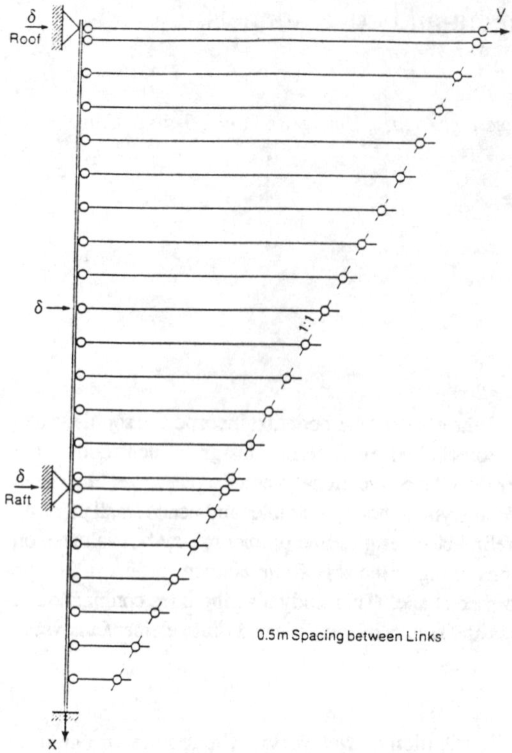

Figure 1: Simplified Geometry for the Model

The measured horizontal displacements were prescribed to the two rollers supports and a point midway between them. The program calculates the forces in each link member which represent distribution of earth pressure for comparison with the measured and those calculated by complicated finite element programs. The advantages of such model are its simplicity, economy and speed. Alternatively, the distribution of earth pressure can be used as input data to calculated the corresponding deformed shape of the wall. The program may be run on a microcomputer which lends itself to on site use. The lateral earth pressure or the displacements can be calculated within minutes of the reading being taken.

This will provide an evaluation for the daily site problems aiding in proposing the suitable solution without the need to run a highly sophisticated, time consuming finite element program. Soil and structural material parameters must be ready for use and should present the materials used on each site. It is advisable that these parameters are carefully verified before their use.

3. CASE HISTORIES

The proposed model was applied successfully in many cases. In this paper, four case studies will be presented with comparison of results of the proposed model, with those of complicated finite element programs and field measurements when it is available.

3.1 CASE HISTORY 1

This study was conducted during the construction of the first line of the Greater Cairo Metro project. Full description of the work is shown by A. Shalaby (1989). The running tunnel and station section consists of R.C. boxed section, the diaphragm wall works as the sides and the raft and roof are the bottom and the top of the section. The geometry of the problem is shown in figure (1) and the parameters for the R.C. and soils are presented in table (1).

Table 1: Soil Parameters for Greater Cairo area

Parameter/Material	γ_{sat} kN/m^3	γ_{dry} kN/m^3	σ_t kpa	E Mpa	ν
Fill	18.0	13.04		15,000.0	0.4
Silty Clay	18.3	13.04		20,000.0	0.4
Medium Density Sand	18.0			20,000.0	0.4
Dense Sand	20.0			22,000.0	0.4
Grouted Dense Sand	22.0		1,200.0	200,000.0	0.4
Concrete		25.0	3,350.0	31.25E7	0.2
Steel*		1.91	36,000.0	5.034E7	0.4
Wood*		0.6		10,000.0	0.4

* = equivalent or ficticious parameters

γ_{dry} = dry unit weight σ_t = tensile strength

E = elastic modulus ν = Poisson's ratio

γ_{sat} = saturated unit weight

These parameters for Greater Cairo area were verified by A. Shalaby (1989), A. G. Shalaby (1990), El-Nahhas, et al (1989) and other researchers. Figure (2) presents the results of the proposed model showing the soil stratification. The program used to conduct the work for the proposed model was PFT by Simmonds (1988) and the complicated finite element program chosen for this study was SAFE,"Soil Analysis by Finite Element" developed by Chan (1986). The hyperbolic model, introduced by Duncan and Chang (1970) was used in this program. Full details of the model and parameters are given by A. Shalaby (1989). Figure (3) provides a

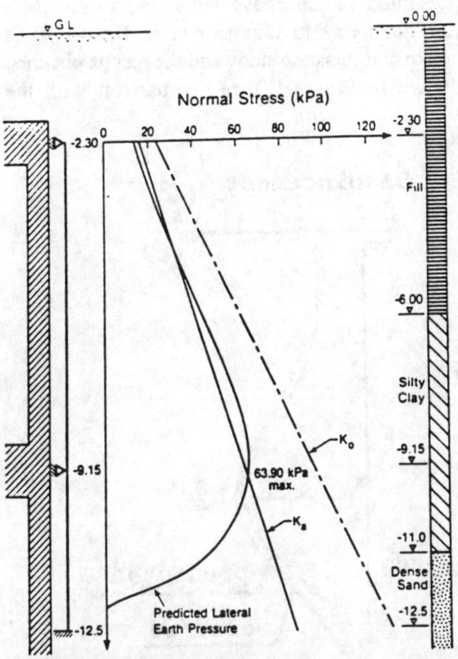

Figure 2: Net Lateral Earth Pressure as Predicted by the Model for Case History 1

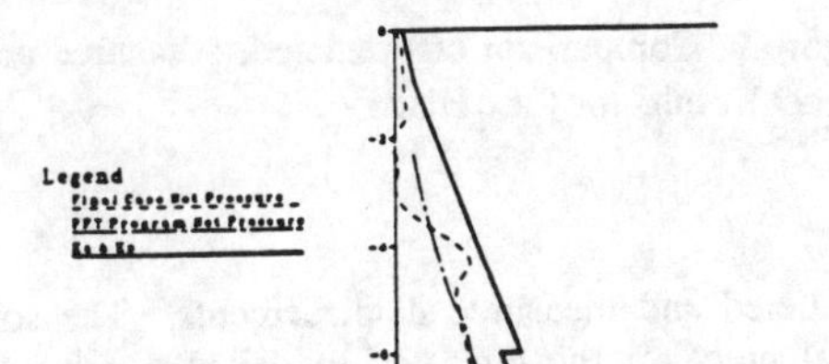

Figure 3: Comparison of SAFE and Simple Model

comparison between the results obtained by the proposed model, F.E.M. and the classical ones. From figure (2) and (3), the agreement between the results can be concluded and verifing the proposed model and supporting it. Results of the station section indicated the same finding.

3.2 CASE HISTORY 2

This case history is presented by Lee et al., (1989). It is a deep excavation in Singapore which is commonly carried out in marine clay. In order to identify a constitutive model which can realistically represent the behavior of the used marine clay in deep excavation, the finite element program CRISP which incorporates the modified cam clay model was used for post-construction analysis. The obtained results were compared with those measured in the field.

This study was based on an excavation to construct deep basements for the City Telecommunications Centre building. This excavation went through the

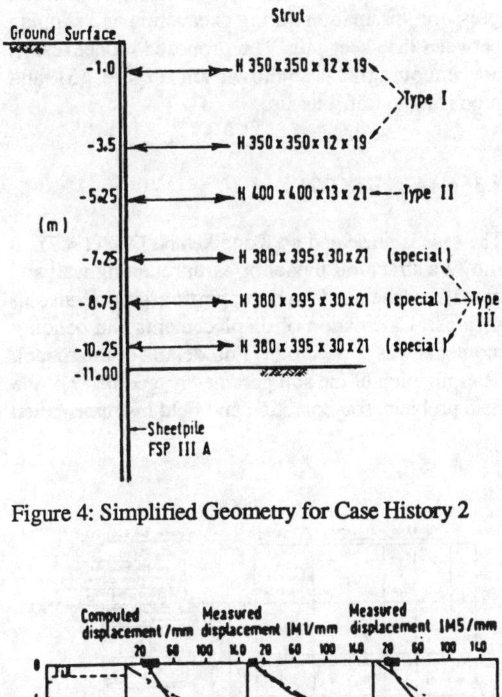

Figure 4: Simplified Geometry for Case History 2

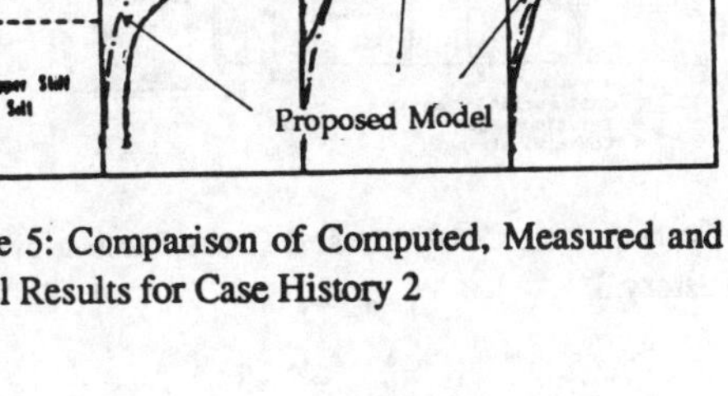

Figure 5: Comparison of Computed, Measured and Model Results for Case History 2

first layer of fill with about 2m depth underlaid by a deep layer of a normally consolidated marine clay up to 20m depth, while the sheet piles were driven to a depth of about 30m below ground surface into stiff silt layer. Figure 4 shows the geometry and strut levels, the proposed model geometry is similar to figure (1) of the case history 1. The parameters used for the proposed analysis is presented in the table included in the paper by Lee et al., (1989). This study presents the computed wall displacement with the results from inclinometers. The field results were compared with the finite element analysis using CRISP and the proposed model. It was concluded that the F.E.M. analysis under estimate the maximum lateral sheet pile displacement and overestimates base and toe movements, while the proposed model results showed a reasonable agreement with field measurments. The authors attribute this discrepancy to the uncertainty of soil parameters, pore water pressure dissipation during excavation and slipping between the sheet pile. The proposed model results are encouraging as shown on Figure (5) and promising to fulfill its aim.

3.3 CASE HISTORY 3

The case is presented by Konoike and Ono (1987). It shows a structural model of earth retaining wall and modelling the strut and supporting ground also as springs. Calculation of displacements and bending moments was carried out. Full details of the model, determination of the soil parameters, geometry of the field problem, soil condition and field instrumentation

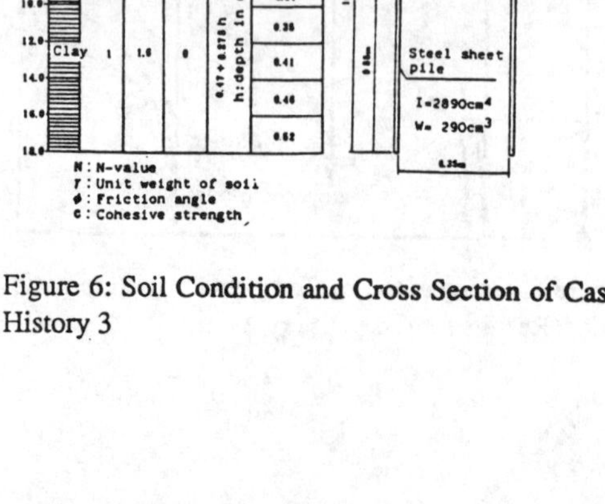

Figure 6: Soil Condition and Cross Section of Case History 3

are presented in the above work and figure (6). Similar geometry to that shown in figure (1) is adapted to suit this case study and the results obtained are shown in Figure (7), in comparison with the

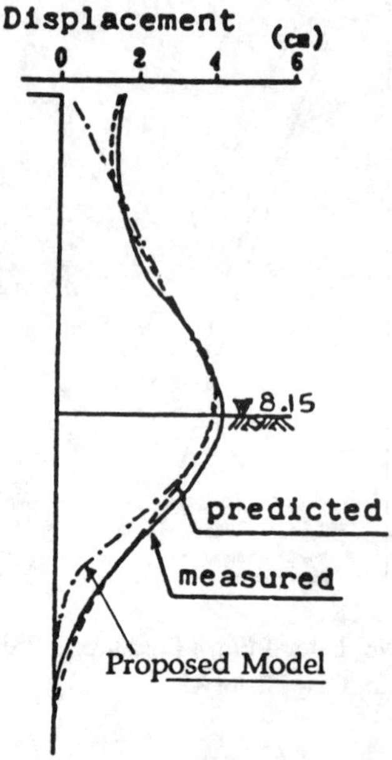

Figure 7: Comparison of Predicted, Measured and Model Results for Case History 3

predicted and measured displacements. The soil parameters for the proposed model were adapted according to the data and procedure presented by Duncan et al., (1980). The study presented another case with reinforced concrete wall. Figure (7) shows the strong agreement between the results of the study and the proposed model. This seems to give a good prediction of the displacement and prove that this simple model could be employed as a tool for site control of cut-and-cover excavations and also would give a useful information for the design of earth retaining structures.

3.4 CASE HISTORY 4

Broms et al., (1986) carried out finite element analysis for braced excavations in Singapore with subsequent field measurements. The analytical work was done using a modified version of the computer rogram EXCAV, which was originally developed at the University of California at Berkeley by Chang and Duncan utilizing a non-linear hyperbolic soil model. They stated the major difficulty in choosing soil parameters for the use in the analysis, with presenting the soil parameters were used in this study for the proposed model. Their work was carried out utilizing three case records. The choice of the third case for our study was based on the fact that they used the knowledge gained from the other two cases, to complement analysis of the third case. The depth of excavation was 15m with width varies from 6m to 12m in mainly marine clay. The inclinometer was installed to measure the lateral displacement of the wall. Figure (8) presents the geometry of the problem and the soil formation for the third case record.

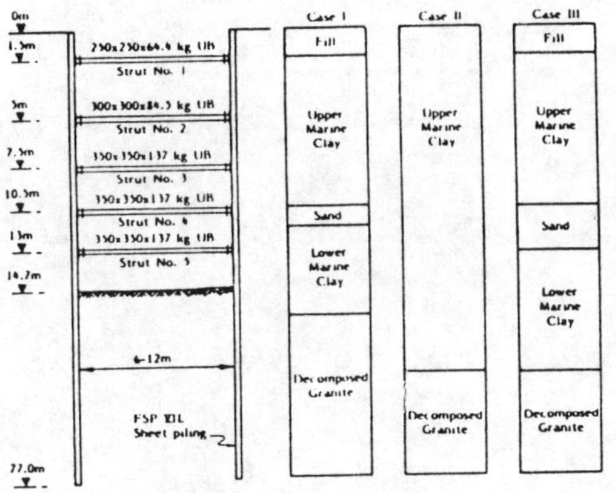

Figure 8: Soil Profile and Transverse Section for Case History 4

In order to conduct the analysis required for the proposed model, the previous procedure used in Case History 1,2 and 3 was carried out by adapting the model geometry, soil parameters and properties. A comparison of measured and computed lateral displacement of the wall provided by Broms at al., (1986) with the results obtained from the analysis using the proposed model is shown in figure (9) for the final stage of excavation. The calculated 75mm of max. displacement using the Finite Element Method provided only about half of the observed max lateral displacement of about 150mm The proposed model results showed a max. of about 123mm. Back-analysis of this case using F.E. analysis considering the depth of the marine clay layer is extended up to 20m gave results in good agreement with the measured values. This modification of soil profile was based on the results of cone penetration

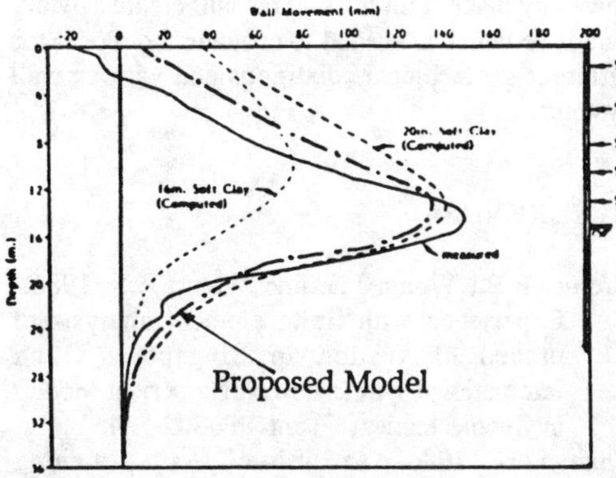

Figure 9: Comparison of Computed, Measared and Model Results for Case History 4

test. It was concluded that the computed values smaller than the measured ones by about 20%.

The results showed that the proposed model results in good agreement with the measured considering the time, effort, speed of obtaining the results to provide a valuable tool for preliminary designs and solving site problems.

4. CONCLUSIONS

Comparing the results of the proposed model with different case histories, it is shown that a reasonable agreement between the measured values and those obtained by the model is possible. The results are extremely encouraging and promising to provide a simple tool to aid the preliminary design with considerable saving in time and cost. Also to serve in providing important on-site information to help engineers to face and solving almost their daily problems during construction which usually require reasonable quick decisions to be made.

The proposed numerical model can be used to calculate the internal forces and lateral displacements in the underground structure for any given lateral pressure distribution. Alternatively, if the observational approach is used, the structure can be back-analyzed using the measured deformed shape.

In order to confirm these finding, more verification of the model results are required to provide the extremely needed support. The researchers are encouraged to use the model in their analysis to compare the results with the computed and measured results. This can be done in very simple way using different computer programs which are cabable to solve any plane fram truss. This will create a wider data base which is needed to evaluate the model for different geotechnical conditions and various wall systems.

REFERENCES

Broms, B.B.; Wong, I.H. and Wong, K.S. 1986. Experience with finite element analysis of braced excavation in Singapore. 2nd International symposium on Numerical Models in Geomechanics / Ghent PP. 309-324.

Chan, D.H., 1986. Analysis of strain softening material. Ph.D. Thesis, Departement of Civil Engineering, University of Alberta, Canada.

Duncan, J.M., and Chang, C.Y. 1970. Non linear analysis of stress and strain in soils. Journal of Soil Mechancis and Foundation Division, ASCE, Vol. 96, No. SM5, Sept.

Duncan, J.M., Byrne, P., Wong, K.S., and Mabry, P., 1980. Strength, Stress-Strain and bulk modulus parameters for finite element analysis of stresses and movements in soil masss, Report No. UCB/GT/80-01, College of Engineering, University of California, Berkeley, California, U.S.A.

El-Nahhas, F.; Eisenstein, Z. and Shalaby, A.; 1989. Behaviour of diaphragn wall during construction of Cairo Metro. Proc. of the 12th. Int. Conf. on Soil Mech. and Foundation Eng., Rio de Janeiro, Brazil, pp 1483-1486.

Konoike, K. and Ono, K. 1987. Prediction of the behaviour of earth retaining walls. Proc. 9th Southest Asian Geotechnical Confrence, Sept. PP 2-13 to 2-22, Bangkok, Thailand.

Lee, F.H.;Yong, K.Y.; Lee, S.L. and Toh,C.T. 1989. Finite element modelling of a strutted excavation. Proc. of the third International Symposium on Numerical models of Geomechanics (NUM OGIII),May, Niagra Falls, Canada, PP 577-584.

Shalaby, A.G. 1990. Behaviour of tunnels in some Egyptian Soils, Ph.D. thesis, Ain Shams University Cairo, Egypt.

Shalaby, A.M.1989. Geotechnical behaviour of diaphragm walls, Ph.D. thesis, Alberta University, Canada.

Simmonds, S.H. 1988. Personal communication and the user manual for computer program (PFT). Internal Note, Deptement of Civil Engineering, University of Alberta, Canada.

Tunnelling and Ground Conditions, Abdel Salam (ed.) © 1994 Balkema, Rotterdam, ISBN 90 5410 363 9

Effects of development of shotcrete stiffness on stresses and displacements of single and double tunnels

Emam Soliman
Ain Shams University & Dar Al-Handasah Consultants, Cairo, Egypt

H. Duddeck & H. Ahrens
Institut für Statik, Technische Universität Braunschweig, Germany

ABSTRACT: Tunnel construction applying the shotcrete method becomes more and more popular in tunneling. It consists of two shells. The first has to act as a primary or temporary support during the building time. The second one is considered as the final support. The primary tunnel support is designed to carry 100% of the primary stresses of the ground. It is formed by shotcrete, steel ribs or lattice arches and welded wire meshes. In contradiction with Reinforcement Concrete in buildings, shotcrete in tunneling carries the load soon after its application. So, the effect of the nonlinear change of the shotcrete stiffness with the time on the tunnel analysis must be studied. The paper shows the results of the three dimensional analysis of single and double tube tunnels taking into consideration the nonlinear change of the shotcrete stiffness with time. The results are presented in diagrams that show the relative changes in stresses and deformations, in comparison with the case of constant shotcrete stiffness.

1 PROPERTIES OF SHOTCRETE

Sprayed concrete is a mixture of cement, aggregate and water projected at high velocity from a nozzle into place to produce a dense homogeneous mass. Gunite is a term used for sprayed concrete where the maximum aggregate size is less than 10 mm. Shotcrete is a term used for sprayed concrete where the maximum aggregate size is 10 mm or greater, see the concrete society (1979). The shotcrete process has grown into an important and widely used construction technique. Shotcreting is usually classified according to the process used, wet-mix or dry-mix, and the type of aggregate used, coarse or fine. Although information on shotcrete construction and application is available in the literature, there is presently a scarcity of useful engineering data and the information available shows a wide range of values. This is attributable, in part, to a lack of standard testing procedures, variations in constituent material quality and gradation, nonuniformity of application techniques, the absence of testing standards, and difficulties in correlating factors between specimens and in-place shotcrete cores. The physical properties of sound shotcrete are comparable to those of conventional mortar or concrete having the same composition. It should be stressed that the properties and performance of shotcrete are largely dependent on the conditions under which it is placed and ultimately on the competence and experience of the application crew. Most published values for 28-day strength are in the range of 200 to 480 kg/cm^2, although dry-mix shotcrete in some tests has developed strengths in excess of $690 kg/cm^2$. The modulus of elasticity generally is between 170000 to 410000 kg/cm^2, about the same as conventional concrete, see ACI (1992). Composites of shotcrete and poly- styrene or polyurethane foam, special fiberglass and other plastics forming a sandwich construction is another possibility. Steel fibers has been used in shotcrete to reduce crack propagation, to increase flexural strength and to improve ductility, toughness and impact resistance, see Maidl and Dietrich (1990 and 1992).

2 TIME DEPENDENT BEHAVIOUR OF TUNNEL LINING

For the analysis of stresses and displacements in tunnel linings and in surrounding rock or soil, the material properties must be considered as realistically as possible, to prove the safety and stability of construction. The behaviour of the lining and the rock or soil is dominated in almost all cases by non-linear material properties. The analytical methods can not deal with non-linearities. Numerical methods are needed as that of the finite element methods. Iteration processes included herein can take into account the non-linear properties. Many kind of rock and soil show creep behaviour. For many tests the characteristic deformation behaviour of such materials is qualitatively well known. On applying a constant compressive stress to a creep specimen, an initial strain results instantaneously. This strain is mainly elastic. Time-dependent deformation then occurs at a rate which decreases with increasing time. This is often called primary creep phase or viscoelasticity. If the compressive stress exceeds the yield limit, the period of decreasing creep rate if followed by a period in which the creep rate is constant or nearly constant (steady state creep phase). The described time-dependent behaviour can be represented by the well known rheological model, see Müller (1972), Duddeck and Vollstedt (1979) and Huang (1991). Creep behaviour of concrete can be described quite well by a single Kelvin body. Figure 1 shows normal forces in the lining against time (t) by using a two-dimensional tunnel model, after Müller (1972), where α is the percentage of the elastic relaxation of the rock due to the excavation before the shotcrete lining acts. Figure 2 shows the vertical stresses in a cross-section of the rock near the tunnel.

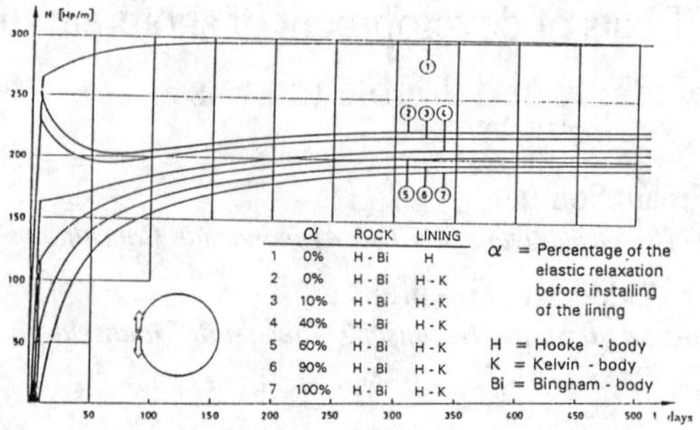

Figure 1: Normal forces in the tunnel lining considering creep properties after Müller (1972)

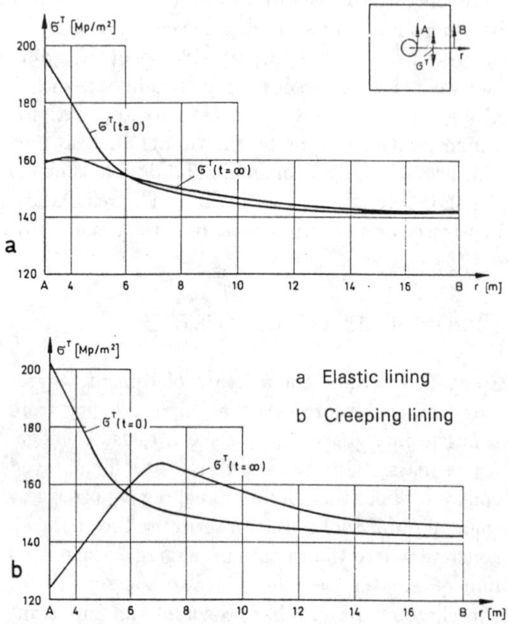

Figure 2: Vertical stresses in a cross-section of the rock after Müller (1972)

3 THREE-DIMENSIONAL ANALYSIS OF SINGLE AND DOUBLE TUNNEL

For the determination of the stresses and deformations for a tunnel, two different approaches may be chosen. If the ground is soft and the support of the lining acts very close to the excavation face (as in shield driving), a two-dimensional approach (plane strain model) may be justified, see Duddeck and Schulze (1964), Duddeck (1972), Ahrens and Lindner (1982), Duddeck (1988), Duddeck and Erdmann (1985) and Soliman (1991). When shotcrete is used for tunnel support, where the lining is acting only a certain distance behind the excavation face, a three-dimensional finite element model may be valid. The ground responses to each round of excavation by a release of stresses at the excavation zone and corresponding inward displacements. By using a three dimensional model, the continuous geometrical changes at the excavation face can be taken into account. For further information about the three dimensional analysis see Kielbassa (1989), Soliman (1991), Erdmann (1980), Berwanger (1986), Dallmann (1988), Kielbassa and Duddeck (1991) and Pierau (1980).

4 DEVELOPMENT OF THE ELASTIC MODULUS OF SHOTCRETE WITH THE TIME

Many authors have studied the effect of the nonlinear behaviour of the shotcrete with the time on the analysis of the tunnel, see Soliman (1991), Kielbassa (1989), Berwanger (1986) and Erdmann (1980). Weber (1979) has given an empirical formula for the nonlinear change of the elastic modulus of the concrete with the time:

$$E(t) = E(28).a.e^{\left(\frac{c}{t^{0.6}}\right)} \quad (1)$$

Where:
$E(t)$ is the elastic modulus of concrete at time (t)
$E(28)$ is the elastic modulus of concrete after 28 days
t is the time in days
a and c are constants depending on the type of cement and water-cement ratio.
Figure 3 shows the development of the elastic modulus of the concrete $E(t)$ with the time t for $E(28) = 24000 MN/m^2$, $a = 1.062$ and $c = -0.446$.

For the analysis, the stepped line in Figure 3 is used. Excavation advance steps equal to two meters every five hours is assumed.

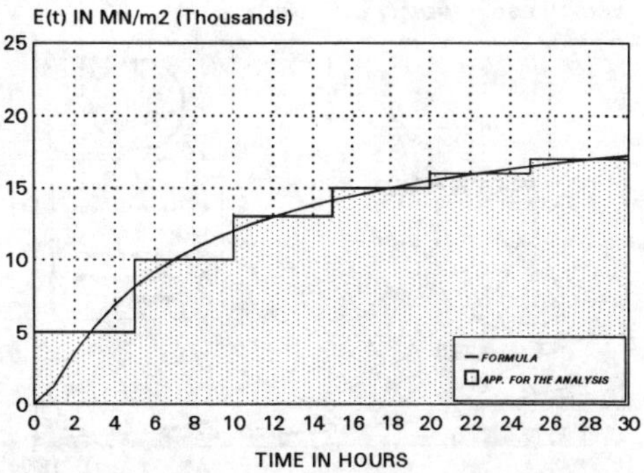

Figure 3: Development of the elastic modulus of the concrete with the time

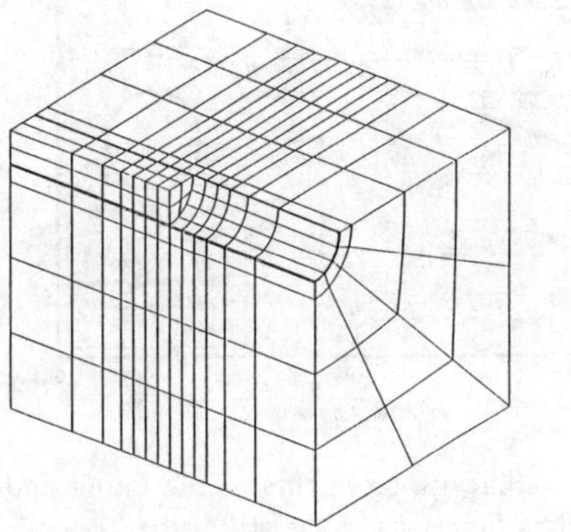

Figure 4: Three-dimensional finite element model for the analysis of single tunnel

5 RESULTS OF SINGLE TUBE TUNNEL

To take into account the stress changes at the tunneling face prior to the lining support, it is necessary to apply a three-dimensional finite element model. Figure 4 shows a FEM-mesh for the analysis of single tube tunnel. The results are obtained after calculation of 15 excavation advances. The following properties of the ground and the tunnel lining are chosen:
Young's modulus of ground: $E_G = 100 MN/m^2$
Young's modulus of shotcrete after 28 days: $E_L = 15000 MN/m^2$
Thickness of tunnel lining: $t = 0.30m$
Radius of the tunnel: $R = 5.00m$
Height of overburden above the crown: $h = 35.00m$
Poisson's ratio: $\mu = 0.30$
Specific weight of the ground: $\gamma = 0.02 MN/m^3$
Ratio of lateral pressure: $k_0 = 0.5$
Vertical primary stresses of the ground: $\sigma_{v_{prim}} = -.70 MN/m^2$
Horizontal primary stresses of the ground: $\sigma_{h_{prim}} = -.40 MN/m^2$
Unsupported length: $L_u = 2.0m$
The results are given as functions of the stiffness relation $\alpha = \frac{E_G R^3}{E_L I}$.
For comparison, the results of a two-dimensional model are given also.
Figure 5 shows the radial displacements of the lining and the ground for both cases of constant and variable stiffness of shotcrete. In case of stiffer

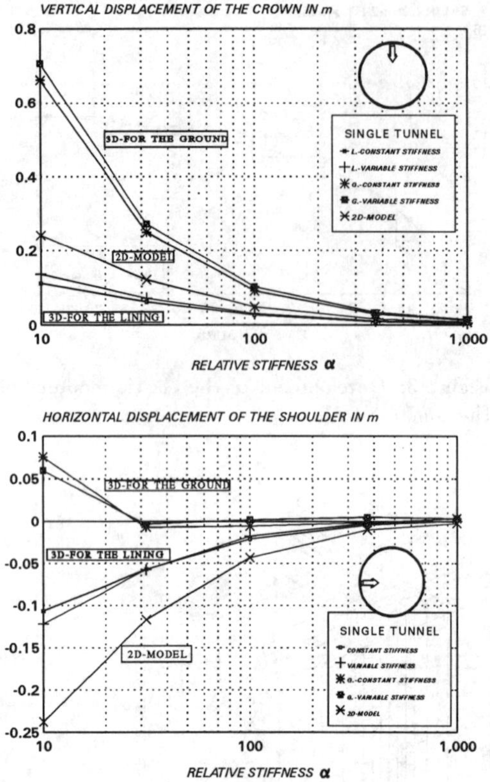

Figure 5: Radial displacements of the lining and the ground as functions of the stiffness ratio α

ground the radial displacements of the lining at the tunnel shoulders are only -0.17 of those valid for the constant shotcrete stiffness. At the tunnel crown, they are as much as 23% larger than those values calculated by assuming constant shotcrete stiffness. In Figure 6, the hoop forces and the bending moments of the lining are given. They are smaller than those for constant shotcrete stiffness.
The final ground stresses around the tunnel lining can be determined from Figure 7.

6 RESULTS OF DOUBLE TUBE TUNNEL

For the analysis of double tube tunnel, a three-dimensional finite element net is used, as shown in Figure 8. The first tube has already been excavated and the second tube is driven consecutively. The tunnel spacing (pillar) is taken as 0.5 times the tunnel radius. The results are given in Figures

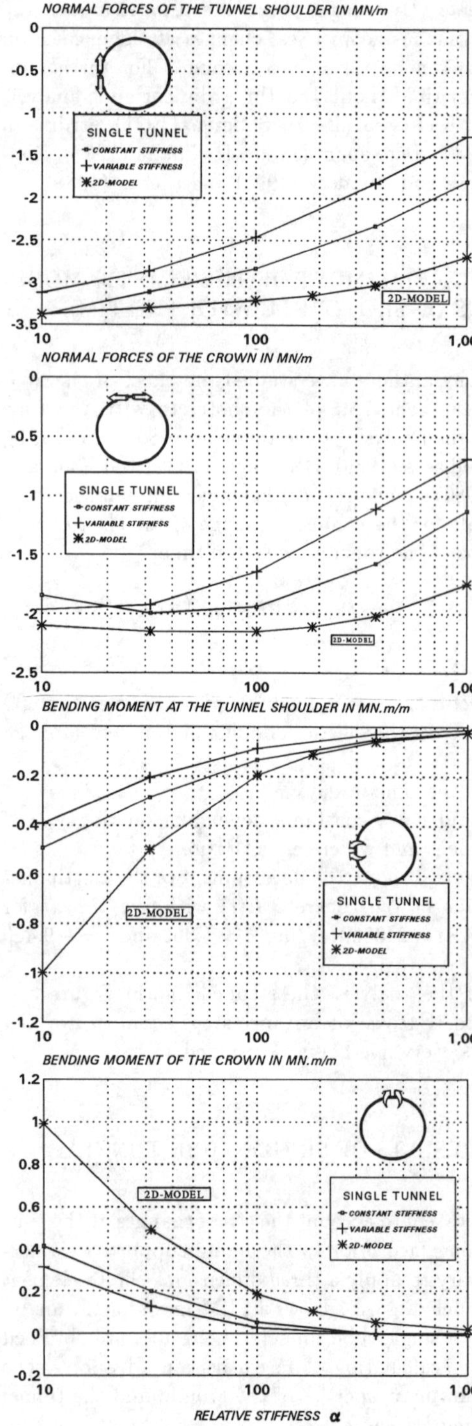

Figure 6: Normal forces and bending moments of the tunnel lining as functions of the stiffness ratio $\alpha = \frac{E_G R^3}{E_L I}$

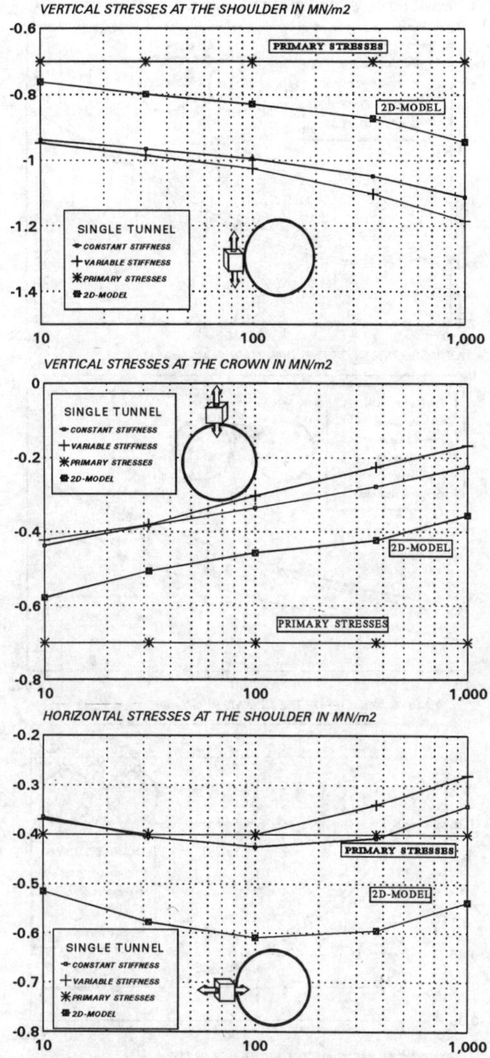

Figure 7: Final stresses of the ground around the tunnel lining as functions of the stiffness ratio α

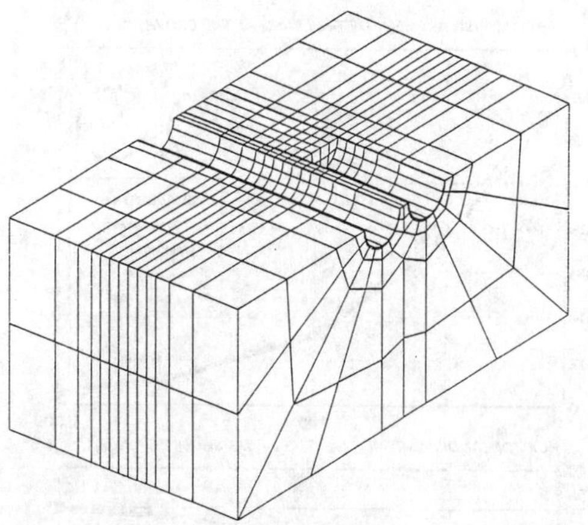

Figure 8: Three-dimensional finite element model for the analysis of double tunnel

9, 10, and 11. The horizontal displacements of the lining at the shoulder of the first driven tube are affected during the excavation of the second tube. Ahead of the excavation face of the second tube, they are increased to 228% of those of single tube tunnel (max. value in the diagrams) and then decreased again. The vertical displacements at the crown are less affected. The normal forces and bending moments in the lining shoulder of the first driven tube, especially those closer to the second tube, are larger than in case of a single tube tunnel.

REFERENCES

ACI Manual of Concrete Practice, Part 5, 1992. Guide to shotcrete

Ahrens, H., Lindner, E. , Lux, K.-H. 1982. Zur Dimensionierung von Tunnelausbauten nach den "Empfehlungen zur Berechnung von Tunneln im Lockergestein 1980", Die Bautechnik 8/1982, S. 260-273 und 9/1982, S. 303-311

Berwanger, W. 1986. Dreidimensionale Berechnung von tiefliegenden Felstunnel unter Berücksichtigung des rheologischen Verhaltens von Spritzbeton und des Bauverfahrens, Forschungsergebnisse aus dem Tunnel- und Kavernenbau, H. 10, Universität Hannover

Construction and Formwork Committee of The Concrete Society, 1979. Specification for sprayed concrete,ISBN 0721011322

Dallmann, R. 1988. Kopplung von Rand- und

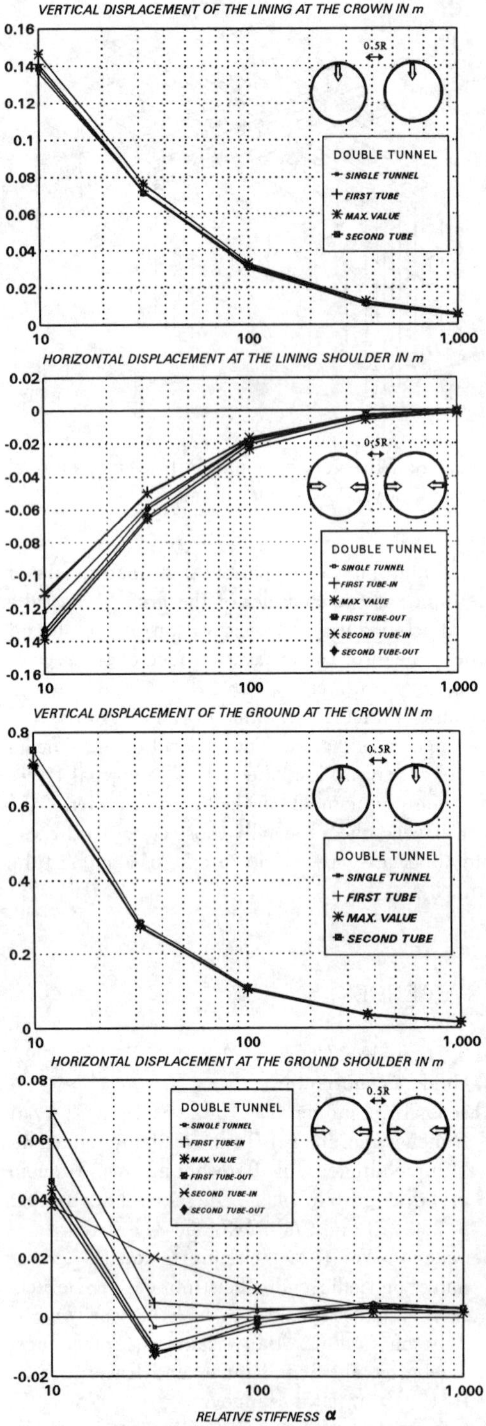

Figure 9: Radial displacements of the lining and the ground of double tunnel as functions of the stiffness ratio $\alpha = \frac{E_G R^3}{E_L I}$

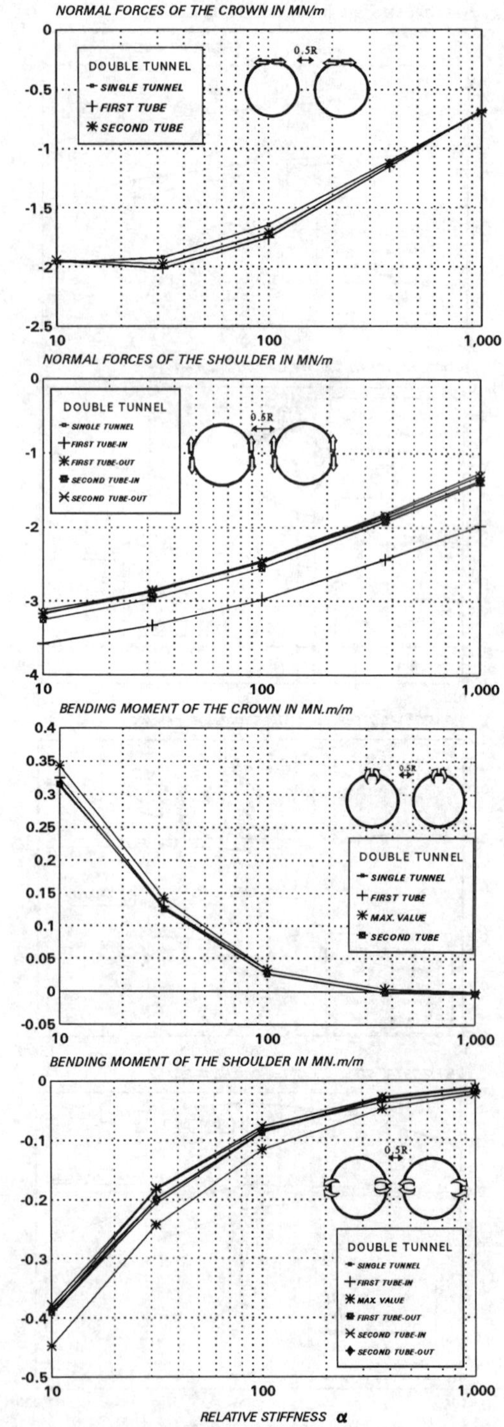

Figure 10: Normal forces and bending moments of the tunnel lining of double tunnel as functions of the stiffness ratio $\alpha = \frac{E_G R^3}{E_L I}$

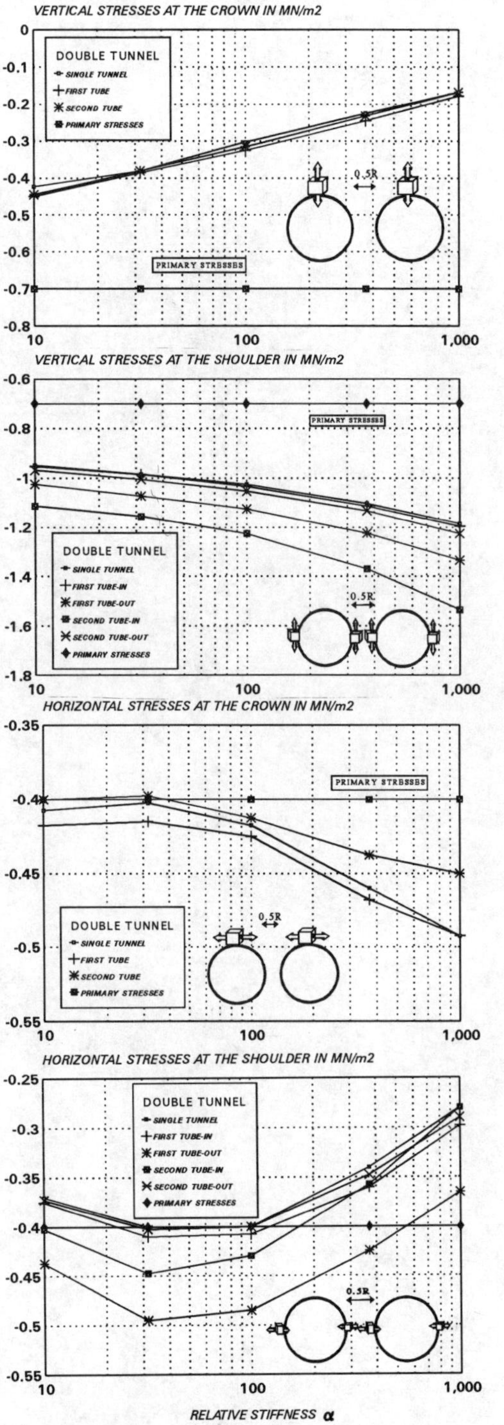

Figure 11: Final stresses of the ground around the tunnel lining of double tunnel as functions of the stiffness ratio $\alpha = \frac{E_G R^3}{E_L I}$

Finite-Element-Methode beim Tunnelvortrieb im Voll- und Halbraum, Bericht 88-56, Institut für Statik der TU Braunschweig

Duddeck, H. 1972. Zu den Berechnungsmethoden und zur Sicherheit von Tunnelbauten, Der Bauingenieur 47 , H. 2, S. 43-52

Duddeck, H. and Erdmann, J. 1985. On structural design models for tunnels in soft soil. Underground Space, Vol. 9, 246-259.

Duddeck, H. 1988. Guidelines for the design of tunnels. ITA working group. J. Tunneling and underground Space Technology V.3 Vo.3 , 237-249.

Empfehlungen zur Berechnung von Tunneln im Lockergestein 1988, Die Bautechnik 57 (1980), H. 10, S. 349-356

Erdmann, J. 1983. Vergleich ebener und Entwicklung räumlicher Berechnungsverfahren für Tunnel, Bericht Nr. 83-40, Institut für Statik der TU Braunschweig

Kielbassa, S. 1989. Spannungsumlagerungen an der Ortsbrust und Beanspruchung des Tunnelausbaus, Bericht Nr. 89-58, Institut für Statik der TU Braunschweig

Kielbassa, S. and Duddeck, H. 1991. Three-dimensional analysis of tunnels. Rock-Mechanics,115-132.

Maidl, B. and Dietrich, J. 1990. State of technology for using steel fiber concrete or steel fiber shotcrete in tunnel constructions, 3rd International colloquium on the fixed link Europe-Africa, Morocco

Maidl, B. and Dietrich, J. 1992. Verification of serviceability for steel fiber reinforced concrete in tunnelling. Structural Engineering International, Vol. 2, pp. 114-117

Müller, K. 1972. Zeitabhängige Spannungsumlagerungen beim Felshohlraumbau, Bericht 72-4, Institut für Statik der TU Braunschweig

Müller, L. 1978. Der Felsbau, Bd. 3, Tunnelbau, Enke-Verlage, Stuttgart

Schulze, H., Duddeck, H. 1964. Spannungen in schildvorgetriebenen Tunneln, Beton- und Stahlbetonbau 59, H. 8, S. 169-175

Soliman, E. 1991. Ebene und räumliche Spannungs- und Verformungszustände von zwei nacheinander aufgefahrenen Tunnelröhren, Bericht 91-66, Institut für Statik der TU Braunschweig

Soliman, E., Duddeck, H. and Ahrens, H. 1993. Two- and Three-dimensional analysis of closely spaced double-tube tunnels. Tunnelling and Underground Space Technology, Vol. 8, No. 1, pp. 13-18

Weber, J.W. 1979. Empirische Formeln zur Beschreibung der Festigkeitsentwicklung und der Entwicklung des E-Moduls von Beton, Beton- und Fertigteiltechnik, Heft 12

Tunnelling and Ground Conditions, Abdel Salam (ed.) © 1994 Balkema, Rotterdam, ISBN 90 5410 363 9

The role of face pressure for shields with positive ground control

Z. Eisenstein & O. Ezzeldine
The University of Alberta, Alb., Canada

ABSTRACT: Among the most effective methods of face stability and settlement control for tunnels constructed in soft ground are the Bentonite Slurry Shield method, the Earth Pressure Balanced Shield Method, and the Compressed Air Method. To optimize the design for these methods, it is required to balance soil movement at the face and at the ground surface with the reduction of the necessary pressure providing the support which is ultimately transmitted to the lining system. A numerical analysis using the finite element method evaluates the minimum requirements of face stability. Comparison of the results with actual case histories shows the effects of the method of soil support and those of the ground condition on the displacement field around the excavation and on liner loading system.

1 INTRODUCTION

The development needs of modern cities requires the construction of drainage and sewer works, mass rail transit and road tunnels beneath densely developed areas. Although the problems encountered in the construction of these projects are diverse, some common trends may be identified and the methods used to solve these problems can be added to the designer's experience. For obvious practical reasons which include accessibility, serviceability, and economy, tunnels in urban areas are constructed at shallow depths. Problems of ground water control are commonly present and concerns are raised about the soil's capability to sustain the imposed stress changes related to the excavation. A successful excavation scheme would preserve the integrity of the project against shear failure and seepage forces and would preclude excessive ground movements that may endanger neighbouring structures.

The principle of ground control using shield methods is to provide an uninterrupted support to the excavation boundaries. A rigid shield followed by an adequate lining system restricts soil movement at the circumferential boundary of the tunnel. However, face support remains a major concern in the excavation design. A variety of methods are used to improve the face stability either by dividing the face into partially self supported subdivisions (New Austrian Tunnelling Method, also known as Sequential Excavation Method) or by ameliorating soil conditions using external agents such as freezing, grouting, dewatering, etc. . Positive ground control methods such as the Compressed Air, the Bentonite Slurry Shield (BSS), or the Earth Pressure Balanced Shield (EPBS) methods rely on minimizing the disturbance at the face so that (1) induced stress changes conform to the resisting capacity of the existing soil conditions and (2) the resulting ground movements are within acceptable limits imposed by the site conditions.

2 METHODS OF POSITIVE GROUND CONTROL AT THE FACE

Three basic methods of face support are used to achieve the objectives of positive ground control; compressed air, balanced earth pressure shield, and bentonite slurry shield methods. A description of the technologies adopted in theses methods is provided by Eisenstein and Ezzeldine (1992: 2). The use of compressed air for sub aquatic constructions has been practiced since the sixteen century. Application of the method to tunnelling projects started in 1871 during the construction of the New York tunnel under the Hudson River. Successes and failures of the methods have been noted during the construction of London's underground railway system between 1886 and 1900 (Kirkland 1984). Until now, compressed air has been considered an acceptable face support alternative considering its major advantage in controlling ground water pressure. On the other hand, health and safety concerns impose restrictions on the application of the method.(Jardine et al. 1991 and Anderson et al 1991).

The Earth Pressure Balanced Shield method consists on providing a rigid support to the face by regulating the amount of excavation. The rate of advance of the shield is related to the volume of transported muck through a controlled screw conveyor from a closed spoil retaining area. This retaining area acts as a buffer zone and controls the movement of the face toward the excavation. A more thorough description of the earth pressure balanced shield method is provided by Finno (1983). In order to transmit the required pressure to the face from the shield jacks located at the tail, the shield has to overcome the adhesion developed along its skin by keeping its relative movement toward the advancing direction. This yields, in general a thrust pressure at the face higher than the initial horizontal pressure and results in an upward movement in front of the face. Control measures employed in this method are described by Xi (1990). As a development of the earth pressure balanced shield, the muddy soil pressure balanced shield was developed, or the D. K. shield (Hagimoto and Kashima 1985). Instead of controlling earth pressure by regulating muck displacement, mud making agent is added to the soil in the retaining area in order to provide the excavated soil with fluidity and impermeability, and to improve the uniformity of the support pressure.

Slurry methods for face support have been used for a wide variety of soil conditions during the last fifteen years with a substantial degree of success and their advantages render them attractive alternatives to either of the above two methods. Bentonite slurry is injected under pressure at the face of an excavation. In order to fulfill its function as a counterbalance to earth and water pressure, slurry pressure has to be built up over a layer of finite thickness. This layer, called "the cake", which is formed from a mix of bentonite slurry and soil particles has an improved shear strength properties and low permeability. As the progress of the excavation requires a continuous process of forming and destroying of the cake layer, the design of the composition of the slurry must take into account the variability of cake thickness, Mohkam and Bouyat (1985 and 1989). The slurry is used also as a transporting agent of the excavated muck through a specially designed cycle. German and Japanese technologies in building Bentonite Slurry Shields represent two distinctive ways in applying the method. German boring machines, also known as the Hydroshield, relay on compressed air reservoir to maintain constant the support pressure inside a slurry chamber. A description of the technicalities of the method is provided by Babendererde (1991) and by Anheuser (1985). Japanese boring machines are used with comparable degree of success. While air pressure is not used to regulate the bentonite pressure, the continuity of face support is guaranteed by the effectiveness of the face openings and the adequacy of cutter bits in use. Kurosawa (1985) provides a description of the recent Japanese technology.

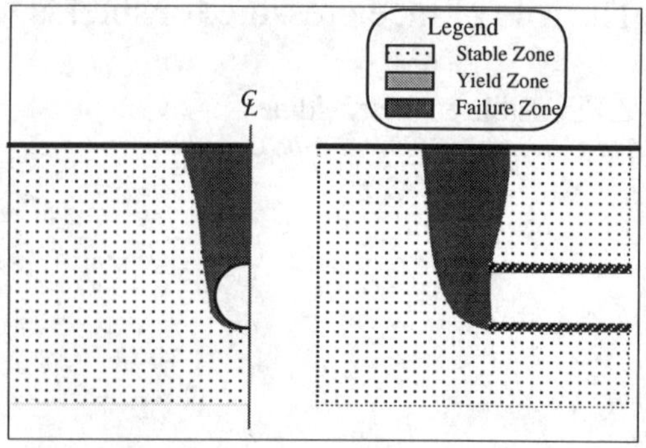

Figure 1 a: failure mechanism in cohesive material (after Kimura and Mair 1981)

3 CHARACTERISTICS OF FACE FAILURE BASED ON EXPERIMENTAL MODELS

Face stability of cohesive material under undrained loading is quantified using *stability ratio* , *N:*

$$N = \frac{p_o - p_f}{c_u} \tag{1}$$

where

p_o = total overburden pressure at the tunnel axis level; p_f= applied pressure at the face of excavation; and c_u= undrained cohesive resistance. Broms and Bennermark (1967) suggest, through a semi empirical derivation that a value of $N = 6$ to 8 could be considered as a safe limit for the design of face stability.

Experimental models are shown to be useful as a realistic tool of investigating the characteristic of face failure. The use of centrifuge models has the advantage of permitting the study of gravitational stress field. At the same time, reservations could be made concerning the particle size effect and the authenticity of the representation of the actual details of various excavation methods.

Centrifuge model test of face stability problem on soft clay soils is performed by Kimura and Mair (1981). For a fully supported shaft circumference, failure zone is found to take the shape shown in Figure 1(a). The resultant stability ratios proved to be dependent, up to a certain limit, on the *depth ratio* (the ratio between the cover depth , *H*, and the tunnel diameter , *D*) as shown in Figure 2. Physical models are used by Casarin and Mair (1981) to study the same problem. The rate of face movement is found to increase gradually as the face pressure decreases and failure is considered to have taken place when the ratio of face displacement to the tunnel diameter reaches a considerable high value (4%). Stability Ratios obtained from the analysis are presented in Figure 2.

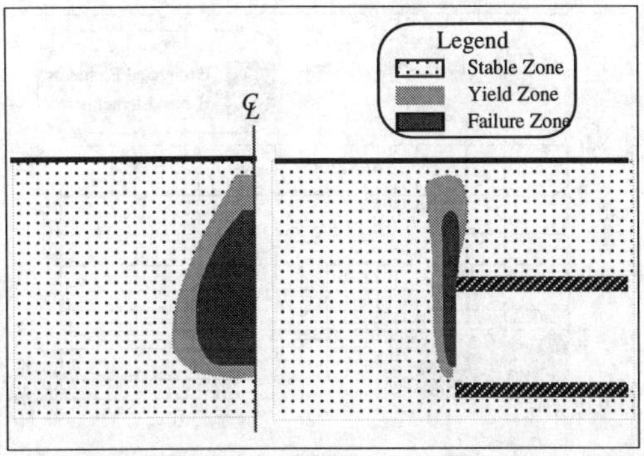

Figure 1 b: failure mechanism in frictional material (after Chambon et. al. 1991)

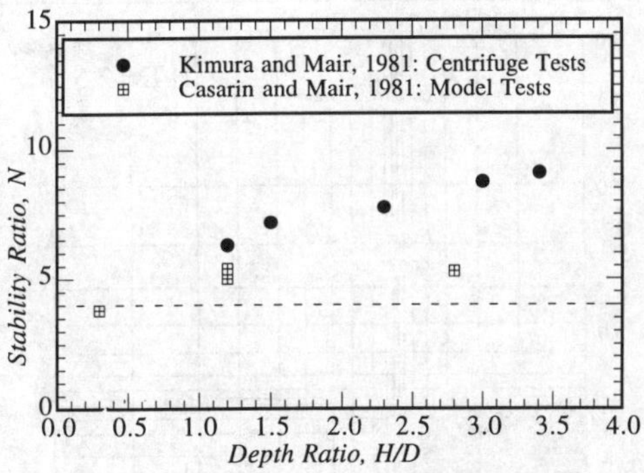

Figure 2: stability ratio versus depth ratio from physical models on cohesive soils

Chambon et al. (1991) present results of centrifuge tests for face stability in granular soils performed in Germany (Bochum) and France (LCPC[1]). The failure zone described in the analysis is reproduced in Figure 1(b). Failure mechanisms are found to exhibit a consistent degree of similitude, Figure 3(a), and could be described as follows: as face pressure decreases, face movements are almost negligible until a certain pressure is reached, the yield pressure p_y, where the rate of face deformation increases constantly until collapse pressure p_c is reached. At this point uncontrolled face displacement takes place characterized by sudden rise of air pressure inside the model tunnel. Figures 3(b) and 3(c) show the obtained collapse and yield pressures with respect to the various tunnel diameters and to the cover ratios. Based on these figures, it may be concluded that the effect of depth ratio and the tunnel diameter is more obvious on the collapse pressure than it is on the yield pressure. As shown in Figure 1(b), yield starts in a region of limited thickness in the longitudinal direction and take a bulb shape in the transversal direction. Unless the depth ratio is very low, collapse occurs before failure zones reach ground surface and then failure propagates rapidly upward. Beside that, results showed that the depth ratio has little influence on the limit collapse pressure.

1Laboratoire Central des Ponts et Chaussées

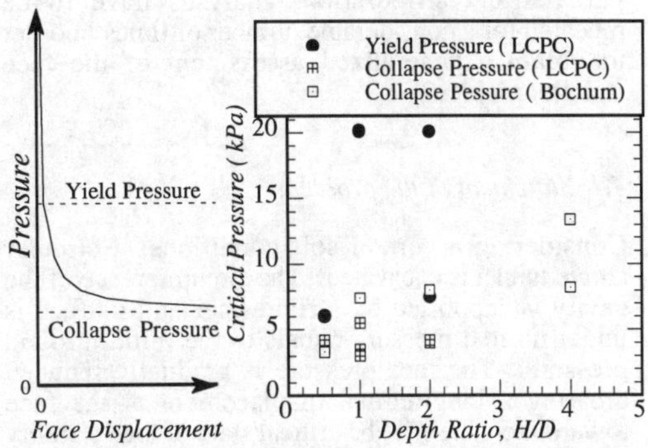

Figure 3 a-b: critical face pressure versus depth ratio from centrifuge tests on non cohesive soil (Chambon et. al. 1991)

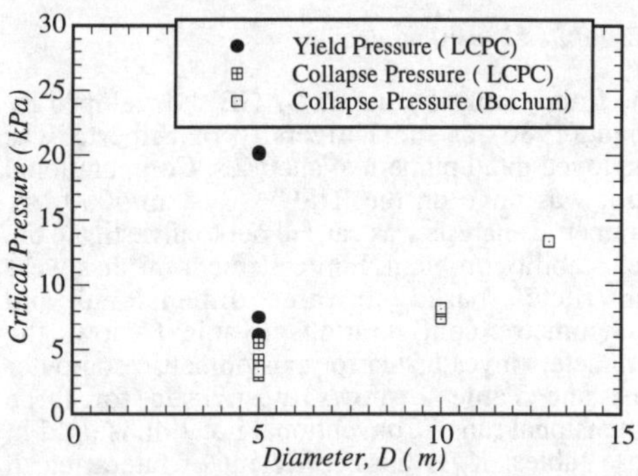

Figure 3 c: critical face pressure versus tunnel diameter from centrifuge tests on non cohesive soil (Chambon et. al. 1991)

4 APPLIED NUMERICAL MODELS

The purpose of the study carried out is to investigate ground behaviour with respect to stress reduction at the face of an excavation. Numerical analysis using the finite element method is found to be an appropriate tool of investigation as it provides a satisfactory simulation accuracy of the geometry and the constitutive model of the problem. Three dimensional analysis, although exhaustive in its computational effort, has become more affordable due to recent advances in computer systems both regarding their speed and their storage capacity. The ability of the method to account for nonlinear material behaviour through iterative processes allows the simulation of a realistic material behaviour. Meanwhile the ability to divide the excavation process into several loading steps allows the inspection of the various stages of excavation and the observation of stress paths followed by different elements in the soil mass. As the obtained results are related to the specific geometrical and parametric configuration, analyses have to be repeated for a considerable number of times in order to obtain a generalized assessment of the face stability problem.

4.1 Statement of the problem

Considering a certain soil conditions, a circular single tunnel is excavated. The circumference of the cavity is supported by a rigid liner and its face is under normal pressure equals to the initial ground pressure. The face pressure is gradually reduced creating a longitudinal displacement at the face toward the gallery. The critical state of face stability is reached when the magnitude of face pressure is reduced such that the rate of increase of face displacement associated with face pressure reduction follows an accelerating trend. At this point, yield pressure at the face is recorded.

4.2 Finite element models

The finite element program SAGE™ developed by Chan (1986) at the University of Alberta was employed in all numerical analyses. Computational work was done on the RISC™ System/6000. A parametric analysis was carried out to investigate the face stability problem. Finite element meshes were constructed based on three dimensional and axisymmetric configurations. Table 1 shows the parameters investigated for axisymmetric computer runs and Table 2 shows those used for three dimensional runs. Conventional notation is used in these tables: K_o is the coefficient of lateral earth pressure at rest, γ is soil density, ν is Poisson's ratio, ϕ is the angle of internal friction, and c is the cohesive resistance.

Axisymmetric models

An elasto-plastic model based on Mohr- Coulomb failure criterion with non associated flow rule is employed in the axisymmetric analyses. Figure 4 shows three meshes of the five adopted meshes for the analyses and their boundary conditions. Eight node elements are selected for all meshes. A constant tunnel diameter of 5.0 m is chosen for all analyses. Young's modulus for the elastic state is chosen to be 10.0 MPa.

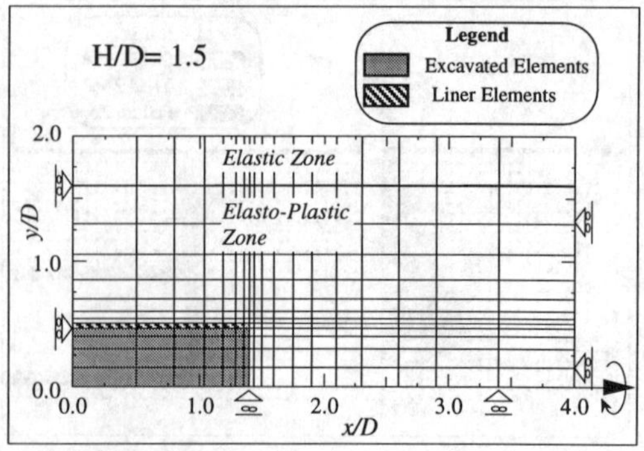

Figure 4 a: axisymmetric finite element mesh for tunnel face stability analysis (H/D= 1.5)

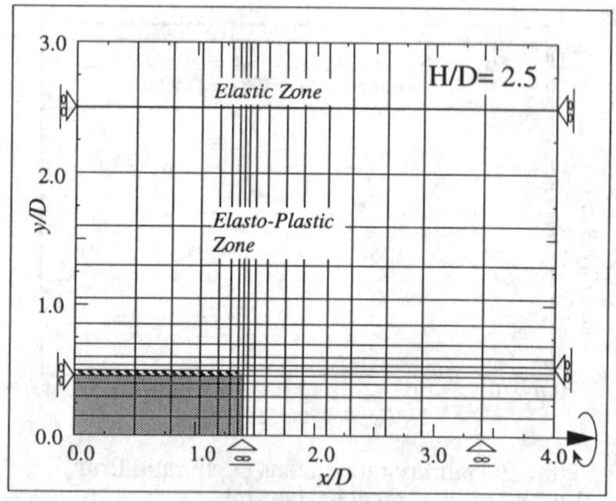

Figure 4 b: axisymmetric finite element mesh for tunnel face stability analysis (H/D= 2.5)

Each run includes a number of steps ranging between 20 and 30. The first step consists of applying initial state of stresses. Linear elastic model is assumed with a Poisson's ratio, ν, calculated based on the chosen coefficient of lateral earth pressure at rest K_o. The second step consisted of applying the actual material properties. In case of pure frictional material, an elastic zone is assumed at the top elements of the mesh in order to save computer effort in the iteration process as this region is under low confining pressure and, thus, is expected to fail. However, this assumption has been found not to interfere with the calculated values of yield pressure at the face. The excavation of tunnel region is simulated by eliminating the excavated elements from the global stiffness matrix and then by applying nodal forces at the boundaries of the excavation that are equivalent to the stresses at one side of the boundary. The supporting system at the circumferential boundary of the tunnel shaft (the shield and the liner) is assumed to be rigid and high

Table 1: Parameters used for Axisymmetric Finite Element Analysis of Tunnel Face Stability

#	Ko	H/D	γ	ν	φ	c
1	0.60	1.50	20	0.20	30	0
2	0.60	1.50	20	0.25	30	0
3	0.60	1.50	20	0.30	30	0
4	0.60	1.50	20	0.35	30	0
5	0.60	1.50	20	0.40	30	0
6	0.60	1.50	20	0.45	30	0
7	0.80	1.50	20	0.20	30	0
8	0.80	1.50	20	0.25	30	0
9	0.80	1.50	20	0.30	30	0
10	0.80	1.50	20	0.35	30	0
11	0.80	1.50	20	0.40	30	0
12	0.80	1.50	20	0.45	30	0
13	0.70	1.50	20	0.20	30	0
14	0.70	1.50	20	0.25	30	0
15	0.70	1.50	20	0.30	30	0
16	0.70	1.50	20	0.35	30	0
17	0.70	1.50	20	0.40	30	0
18	0.70	1.50	20	0.45	30	0
19	0.50	1.50	10	0.30	20	0
20	0.60	1.50	10	0.30	20	0
21	0.60	1.50	10	0.30	25	0
22	0.60	1.50	10	0.30	30	0
23	0.60	1.50	10	0.30	35	0
24	0.60	1.50	10	0.30	40	0
25	0.60	1.50	10	0.30	45	0
26	0.70	1.50	10	0.30	30	0
27	0.80	1.50	10	0.30	20	0
28	0.80	1.5	10	0.30	25	0
29	0.80	1.5	10	0.30	30	0
30	0.80	1.5	10	0.30	35	0
31	0.80	1.5	10	0.30	40	0
32	0.80	1.5	10	0.30	45	0
33	0.50	1.5	20	0.30	30	0
34	0.60	1.5	20	0.30	20	0
35	0.60	1.5	20	0.30	25	0
36	0.60	1.5	20	0.30	30	0
37	0.60	1.5	20	0.30	35	0
38	0.60	1.5	20	0.30	40	0
39	0.60	1.5	20	0.30	45	0
40	0.70	1.5	20	0.30	30	0
41	0.80	1.5	20	0.30	20	0
42	0.80	1.5	20	0.30	25	0
43	0.80	1.5	20	0.30	30	0
44	0.80	1.5	20	0.30	35	0
45	0.80	1.5	20	0.30	40	0
46	0.80	1.5	20	0.30	45	0
47	0.80	2.0	20	0.30	20	0
48	0.80	2.0	20	0.30	25	0
49	0.80	2.0	20	0.30	30	0
50	0.80	2.0	20	0.30	35	0
51	0.80	2.0	20	0.30	40	0
52	0.80	2.0	20	0.30	45	0
53	0.60	2.0	20	0.30	30	0
54	0.65	2.0	20	0.30	30	0
55	0.70	2.0	20	0.30	30	0
56	0.75	2.0	20	0.30	30	0
57	0.80	2.5	20	0.30	20	0
58	0.80	2.5	20	0.30	25	0
59	0.80	2.5	20	0.30	30	0
60	0.80	2.5	20	0.30	35	0
61	0.80	2.5	20	0.30	40	0
62	0.80	2.5	20	0.30	45	0
63	0.80	2.5	20	0.20	30	0
64	0.80	2.5	20	0.40	30	0
65	0.40	2.5	20	0.30	30	0
66	0.50	2.5	20	0.30	30	0
67	0.60	2.5	20	0.30	30	0
68	0.70	2.5	20	0.30	30	0
69	0.80	2.5	20	0.30	30	0
70	0.90	2.5	20	0.30	30	0
71	0.80	1.0	20	0.20	20	0
72	0.80	1.0	20	0.30	20	0
73	0.80	1.0	20	0.40	20	0
74	0.80	1.0	20	0.20	30	0
75	0.80	1.0	20	0.30	30	0
76	0.80	1.0	20	0.40	30	0
77	0.80	1.0	20	0.20	40	0
78	0.80	1.0	20	0.30	40	0
79	0.80	1.0	20	0.40	40	0
80	0.80	3.5	20	0.20	20	0
81	0.80	3.5	20	0.30	20	0
82	0.80	3.5	20	0.40	20	0
83	0.80	3.5	20	0.20	30	0
84	0.80	3.5	20	0.30	30	0
85	0.80	3.5	20	0.40	30	0
86	0.80	3.5	20	0.20	40	0
87	0.80	3.5	20	0.30	40	0
88	0.80	3.5	20	0.40	40	0
89	0.80	1.5	20	0.30	0	10
90	0.80	1.5	20	0.30	0	20
91	0.80	1.5	20	0.30	0	30
92	0.80	1.5	20	0.30	0	40
93	0.80	1.5	20	0.30	0	50
94	0.80	1.5	20	0.30	10	10
95	0.80	1.5	20	0.30	10	20
96	0.80	1.5	20	0.30	10	40
97	0.80	1.5	20	0.30	10	60
98	0.80	1.5	20	0.30	20	0
99	0.80	1.5	20	0.30	20	5
100	0.80	1.5	20	0.30	20	10
101	0.80	1.5	20	0.30	20	20
102	0.80	1.5	20	0.30	20	30
103	0.80	1.5	20	0.30	20	50
104	0.80	1.5	20	0.30	30	5
105	0.80	1.5	20	0.30	30	10
106	0.80	2.0	20	0.30	0	10
107	0.80	2.0	20	0.30	0	30
108	0.80	2.0	20	0.30	0	20
109	0.80	2.0	20	0.30	0	60
110	0.80	2.0	20	0.30	10	10
111	0.80	2.0	20	0.30	10	20
112	0.80	2.0	20	0.30	10	40
113	0.80	2.0	20	0.30	10	60
114	0.80	2.0	20	0.30	20	10
115	0.80	2.0	20	0.30	20	20
116	0.80	2.0	20	0.30	20	40
117	0.80	2.0	20	0.30	20	30
118	0.80	2.0	20	0.30	20	60
119	0.80	2.0	20	0.30	20	50
120	0.80	2.0	20	0.30	30	10
121	0.80	2.5	20	0.30	0	20
122	0.80	2.5	20	0.30	0	30
123	0.80	2.5	20	0.30	0	40
124	0.80	2.5	20	0.30	10	10
125	0.80	2.5	20	0.30	10	20
126	0.80	2.5	20	0.30	10	40
127	0.80	2.5	20	0.30	20	10
128	0.80	2.5	20	0.30	20	20
129	0.80	2.5	20	0.30	20	30
130	0.80	2.5	20	0.30	30	5
131	0.80	2.5	20	0.30	30	10

Dimensions:

c: *kPa*

γ: *kN/m3*

Table 2: Parameters used for Three Dimensional Finite Element Analysis of Tunnel Face Stability

#	Ko	H/D	γ	ν	φ	c
1	0.80	2.0	20	0.30	20	0
2	0.80	2.0	20	0.30	30	0
3	0.80	2.0	20	0.30	40	0
4	0.80	2.0	20	0.30	0	40
5	0.80	2.0	20	0.30	10	10
6	0.80	2.0	20	0.30	10	20

stiffness parameters are selected for elements at this region. The following steps consist of gradually reducing the pressure applied at the face following about 25 equal steps and allowing an iteration strain tolerance of 10^{-4}. In cases where convergence is not met the loading step is subdivided into smaller subload steps. Soil movement at the center of the face is monitored and a non dimensional parameter of face displacement, Ω, is calculated as:

$$\Omega = \frac{\delta_x E}{D((K_o \gamma H_o) - p_f)} \tag{2}$$

where δ_x= horizontal soil movement at the centerpoint of the face; and H_o= depth of the tunnel axis. It should be noted that the parameter Ω is not the face displacement but it is defined as a dimensionless ratio of soil movement at the face to the amount of stress reduction. As such the parameter represents the slope of a pressure- face displacement curve. The rather complex form of definition of parameter Ω is used to obtain a dimensionless term and thus to make the results more applicable.

Figures 5, 6, 7, and 8 show typical results of the relationship between parameter of face displacement and the *face pressure ratio PF* for different cases of non cohesive soil where

$$PF = \frac{p_f}{K_o \gamma H_o} \tag{3}$$

At the face pressure ratio of 100% the parameter of face displacement is an undefined number, therefore a certain stress relief is required until the parameter reaches a constant value. At a certain degree of stress relief, the yield point, Ω starts to increase at an accelerating rate. From the figures the yield point is well defined and the accelerating rate of increase of Ω is clear in cases of frictional material while the increase is gradual for cohesive materials and geometrical construction is needed to select the yielding points. This behaviour is expected as has been shown from experimental results described above. The increase of shear strength parameters , c and/or ϕ, will result in decreasing the face pressure ratio required at yield point. Smaller influence is detected for the effect of K_o and ν. The increase of either of theses two parameters results in decreasing the required face pressure ratio due to the increase of confining pressure.

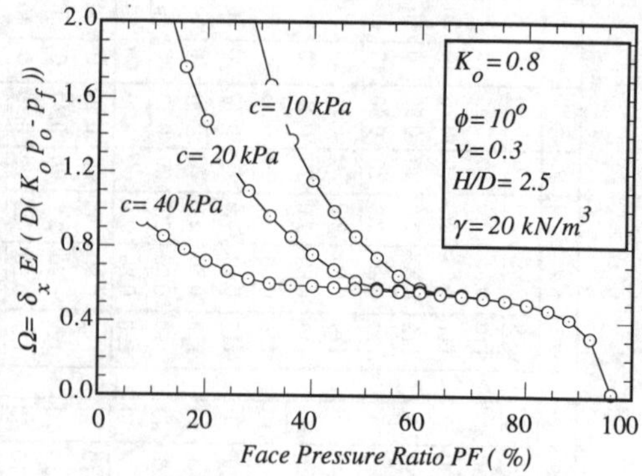

Figure 5: pressure ratio versus parameter of face displacement for various values of cohesion

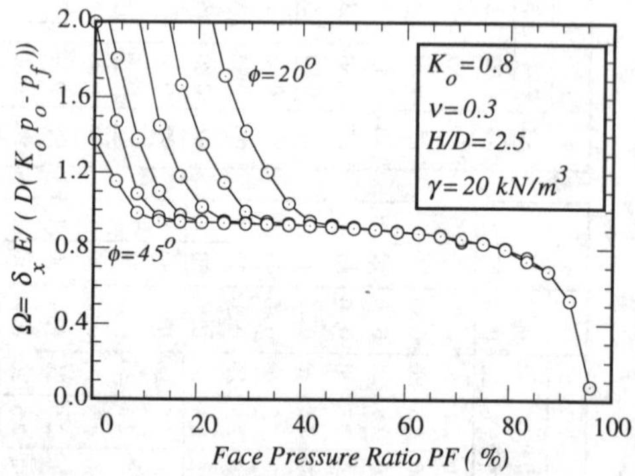

Figure 6: pressure ratio versus parameter of face displacement for various values of angle of internal friction ϕ

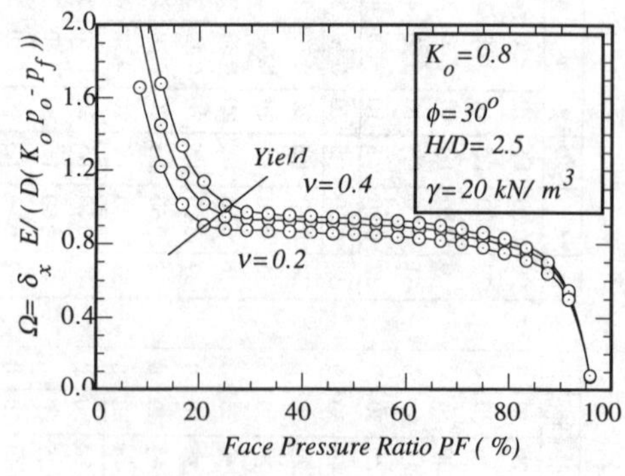

Figure 7: pressure ratio versus parameter of face displacement for various values of Poisson's ratio ν

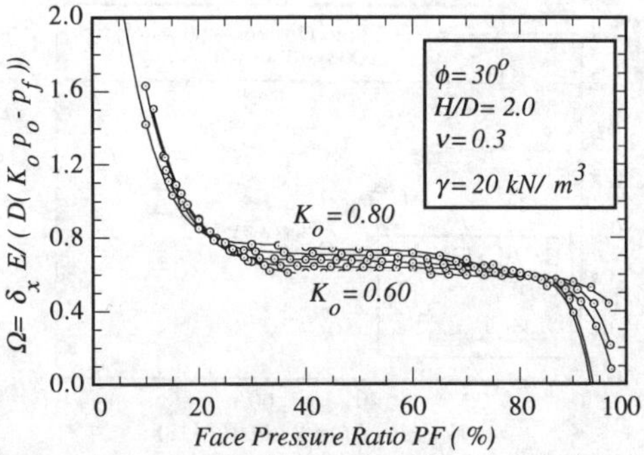

Figure 8: face pressure ratio versus parameter of face displacement for various values of coefficient of at rest lateral pressure K_o

Three dimensional analyses

The same constitutive model and loading process are used as in the axisymmetric analyses. The considerable computational effort needed for this type of analysis requires the reduction of the number of loading steps, total number of steps ranges between 8 to 16 is used for the same strain tolerance as in the axisymmetric cases. Figure 9 shows the constructed finite element mesh: it consists of 711 elements each of 20 nodes which resulted in a total nodal points of 3,622 and a global stiffness matrix size of 14,429,094. As in axisymmetric analyses, elements at near the ground surface were assigned linear elastic material properties in case of frictional material.

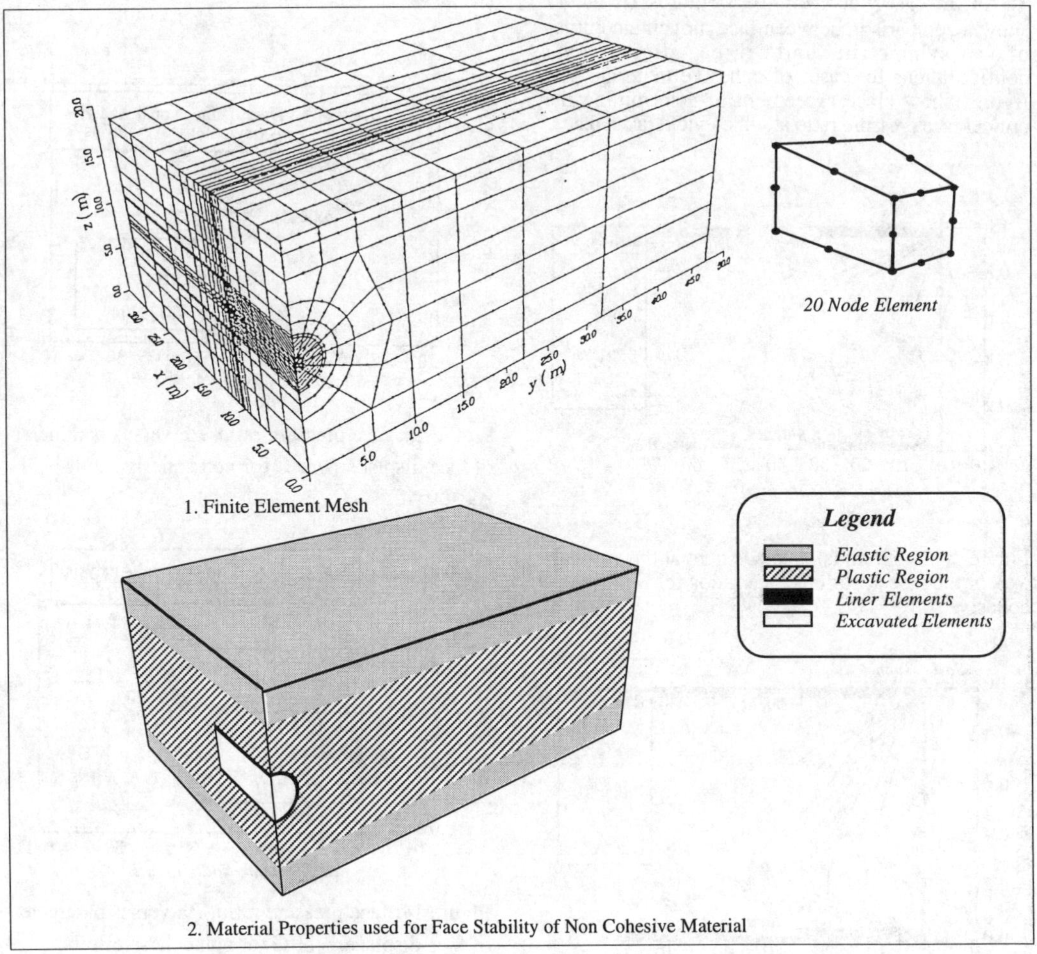

Figure 9: Three Dimensional Finite Element Mesh used for Tunnel Face Stability Analyses

Figures 10 and 11 show a typical deformation pattern at the centerline of the supported face for frictional and cohesive soils, respectively. Soil movement increases as *PF* decreases with the maximum displacement taking place at the lower half of the face and its location gets higher as the face pressure is reduced. As shown by the figures, the elevation of the point of maximum displacement is close to the lower third of the face in case of cohesive soil while it is closer to the centerpoint in case of non cohesive soil. This is due to a higher vertical component of displacement in the cases of cohesive soil. Figures 12 through 14 show a comparison between face movement curves of axisymmetric and three dimensional configurations for cases of frictional material. The simple, non-dimensionless relationship between pressure reduction and face displacement is shown on Figure 15 for the case where $\phi = 40^o$. Figures 16 and 17 show a comparison between face movement curves of axisymmetric and three dimensional configurations for cases of cohesive material. The figures show close agreement in determining the critical face pressure ratio at which yield takes place.

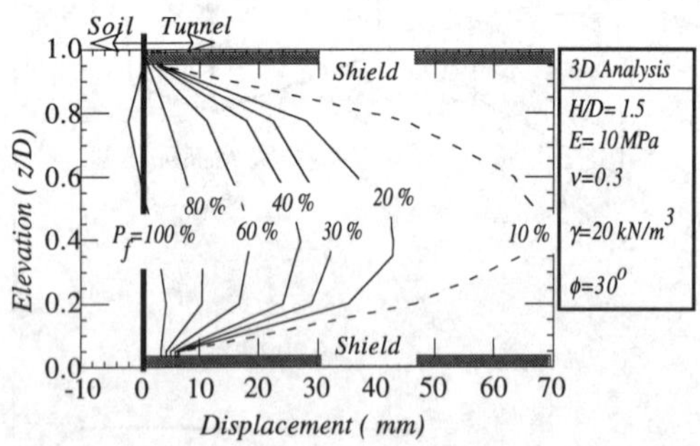

Figure 10: longitudinal displacement at the vertical axis of the face at various face ratios for non cohesive soils

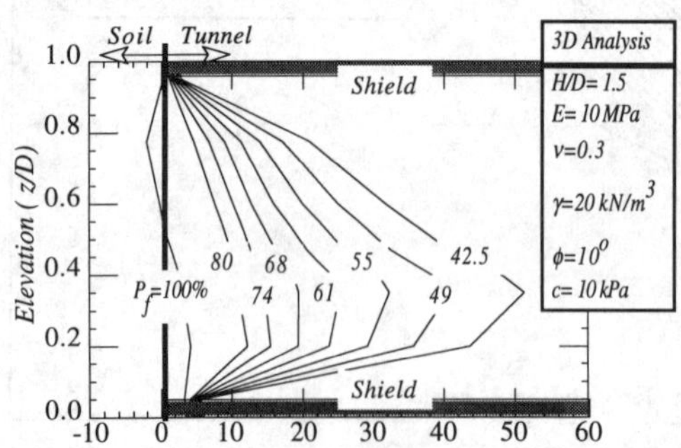

Figure 11: longitudinal displacement at the vertical axis of the face at various face ratios for cohesive soils

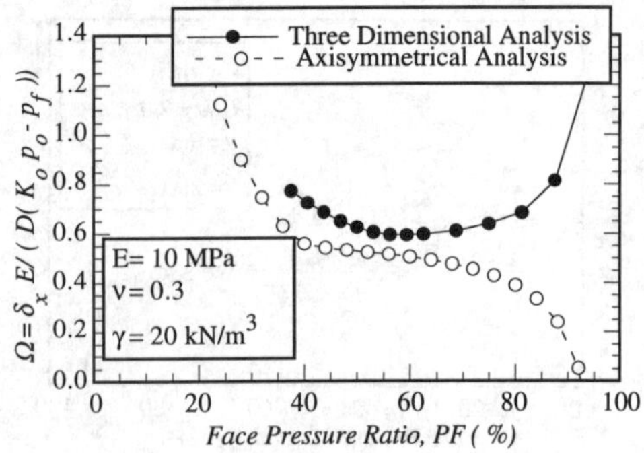

Figure 12: face pressure ratio, *PF* versus parameter of face displacement Ω for non cohesive soils, $\phi = 20^o$

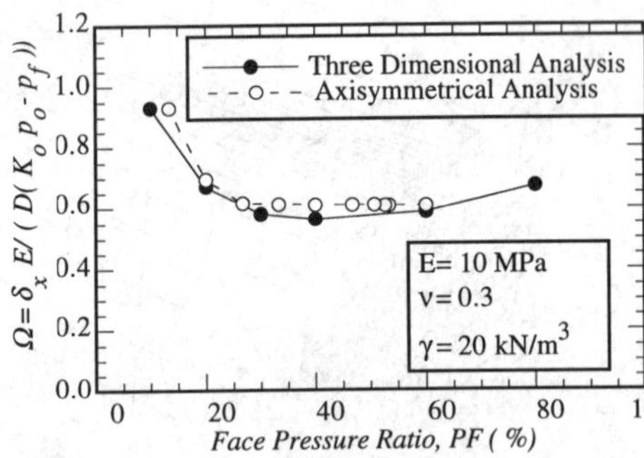

Figure 13: face pressure ratio, *PF* versus parameter of face displacement Ω for non cohesive soils, $\phi = 30^o$

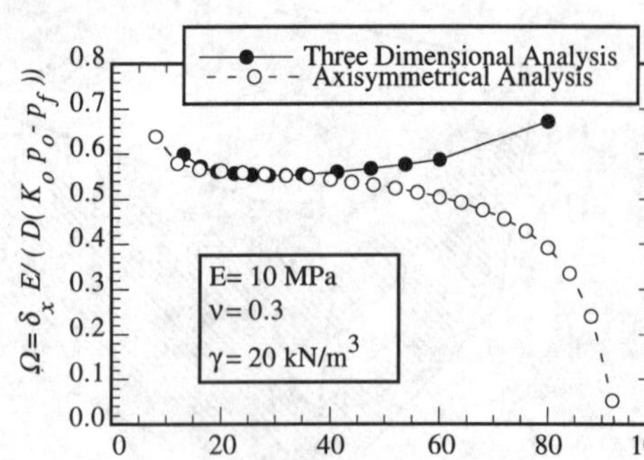

Figure 14: face pressure ratio, *PF* versus parameter of face displacement Ω for non cohesive soils, $\phi = 40^o$

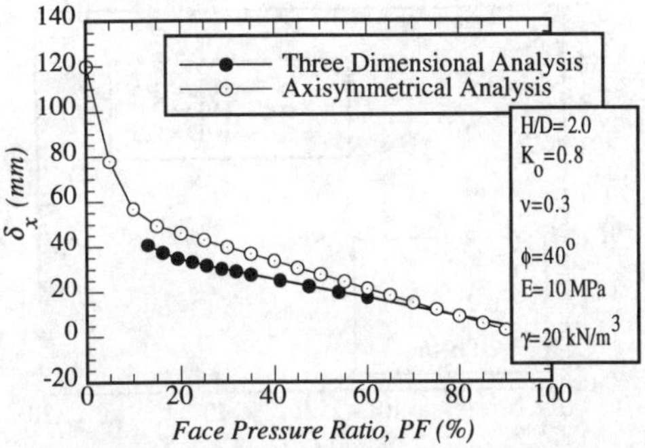

Figure 15: horizontal deformation at the centerpoint of the tunnel versus the face pressure ratio *PF*

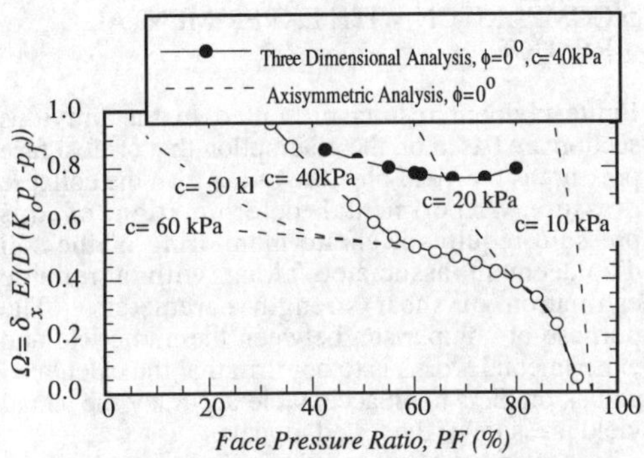

Figure 16: face pressure ratio, *PF*, versus parameter of face displacement, *Ω*, for frictionless soil

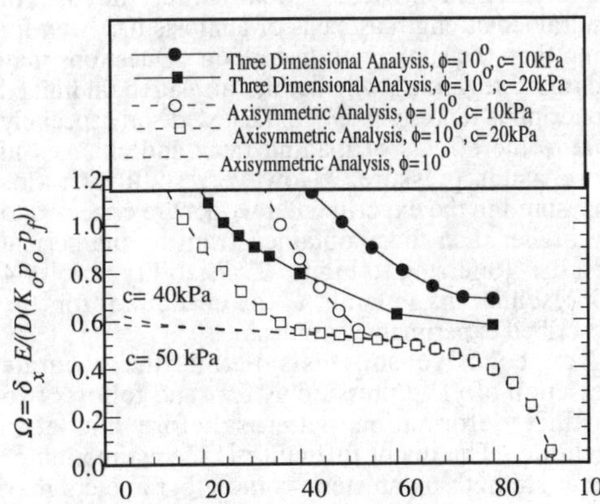

Figure 17: face pressure ratio, *PF*, versus parameter of face displacement, *Ω*, *φ*= 10°

Design criterion

Processing of the results obtained from the axisymmetric analyses shows that the conventional Rankine expression for active earth pressure fits to a high degree the magnitudes of the required face pressure at yielding points with the introduction of two influence factors as follows:

$$p_f = I_\phi(K_a\ \gamma\ H_o) - I_c(2c\sqrt{K_a}) \tag{4}$$

where I_ϕ= influence factor of frictional resistance; I_c= influence factor of cohesive resistance; and K_a= coefficient of active lateral earth pressure.

$$K_a = \frac{1-\sin\phi}{1+\sin\phi} \tag{5}$$

Figures 18 through 20 show changes of I_ϕ for non cohesive material with respect to the angle of internal friction, the coefficient of lateral earth pressure at rest, and Poisson's ratio. The figures indicate that the angle of internal friction has affects to a high degree the magnitude of the influence factor while K_o and ν lower impact on it. The scatter of the obtained values of I_ϕ may be referred to the tolerance used in the iterative process and to the geometrical construction employed to define the value of critical face pressure. Figures 21 and 22 show the effect of the angle of internal friction on the influence factors for cohesive material. By inspection of Equations and 4, for a purely cohesive soil (K_a= *1.0* and I_ϕ= *1.0*), I_c is equivalent to one half the stability ratio (Equation 1). Figure 22 indicates that the critical stability ratio for this type of soil is equal to 4.0.

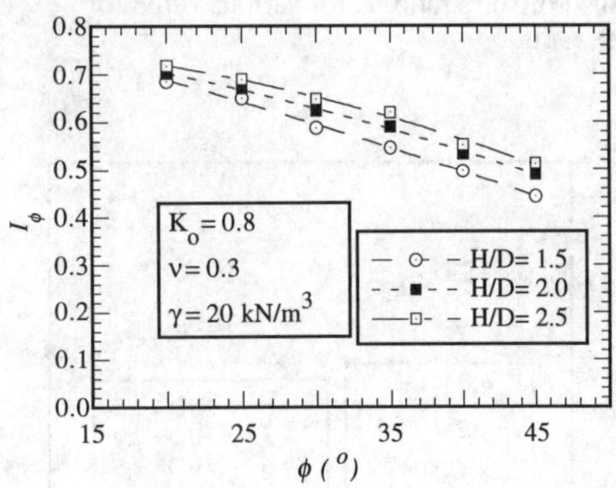

Figure 18: influence factor of frictional resistance versus angle of internal friction, *φ*, for various values of depth ratio

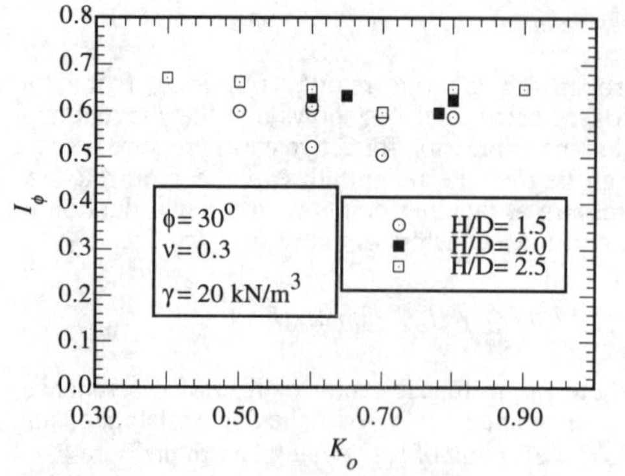

Figure 19: influence factor of frictional resistance versus coefficient of at rest lateral pressure, K_o, for various values of depth ratio

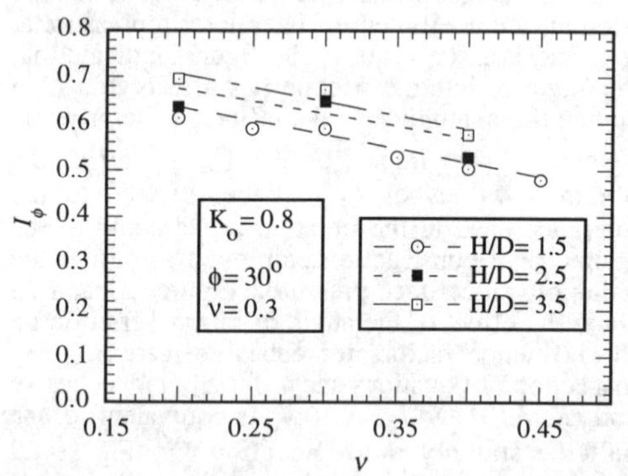

Figure 20: influence factor of frictional resistance versus Poisson's ratio, ν, for various values of depth ratio

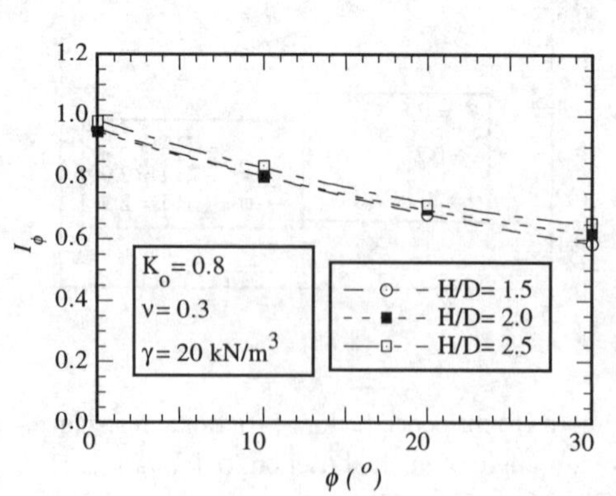

Figure 21: influence factor of frictional resistance versus angle of internal friction, ϕ, for various values of depth ratio

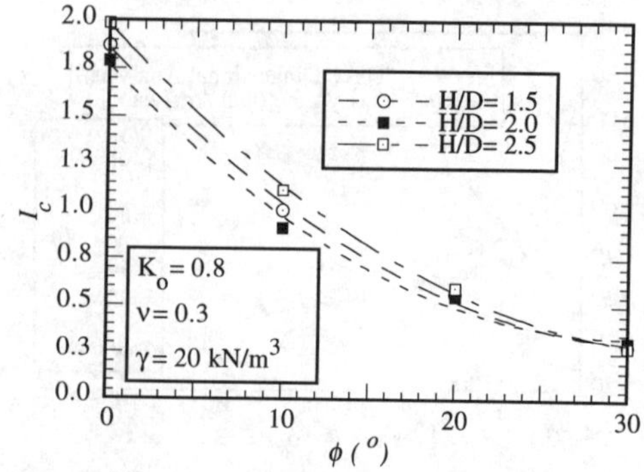

Figure 22: influence factor of cohesive resistance versus angle of internal friction, ϕ, for various values of depth ratio

5 COMPARISON WITH EXPERIMENTAL RESULTS

Finite element results presented in the previous section are based on the assumption that critical face pressure is the yield pressure rather than the collapse pressure. Experimental determination of this pressure requires accurate monitoring of the soil displacement associated along with a reliable estimation of shear strength parameters. The purpose of comparison between the numerical and experimental works is to confirm that the calculated values predicts to an acceptable accuracy the actual yield pressure of the tested models.

In case of cohesive soil difficulty arises from the gradual increase in the rate of displacement of the face centerpoint. Therefore, the point at which the face displacement starts to follow an accelerating rate with respect to the reduction of face pressure cannot be clearly identified. Meanwhile, the use of undrained strength as basis of analysis has been for long time the subject of discussion. Questions may be raised regarding whether the measured undrained cohesion is representative to the tested soil especially due to the effects of loading rates and changes in pore water pressure. However, stability ratios measured in the experimental works are expected to be higher than those obtained from the numerical results. Referring to Figure 2, a stability ratio of 4 represents, in general, a lower bound for the described experimental results.

Non cohesive soil tests permit the accurate detection of yield pressure as the sand follows two distinct deformation patterns before and after yielding. The major difficulty is the estimation of shear strength parameters as these parameters may be influenced by a number of factors such as the changes of dry density, capillary, homogeneity, etc.. As stated by Chambon et al. (1991), the tested soil may be considered purely frictional sand of $\phi = 40^o$. Figure 23 shows a comparison between the experimental results presented by the authors and the performed numerical results. From the figures numerical analysis results are close to the yield pressure while collapse pressure are of lower magnitude as expected.

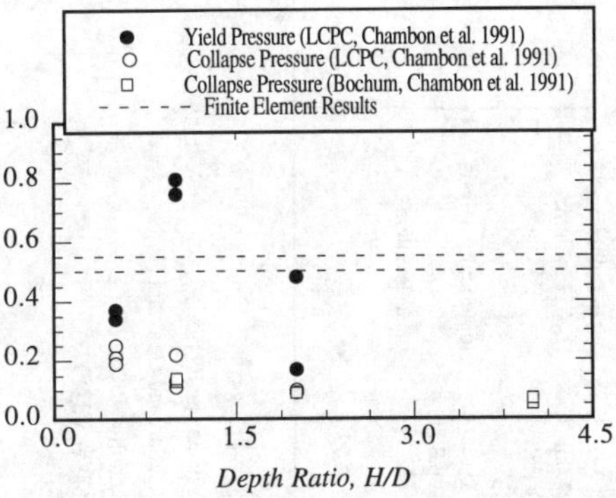

Figure 23: comparison between centrifuge results and finite element results for frictional soils, $\phi = 40^{o}$

6 COMPARISON WITH CASE HISTORIES

The objective of comparison between numerical results and actual case histories is to indicate that practical experience rely on the yield pressure as the minimum required pressure to support the face. As previously mentioned, methods of face support are of various types and each of these types generates a stress and a displacement field which is categorically different than those generated by the other types and some face support measure are used for purposes beyond the integrity of the face such as controlling water flow by compacting soil particles at the face, reducing surface settlement by pushing the ground forward as excavation proceeds, or improving the cutting performance of the machine.

In case of cohesive soil, Kirkland (1984) and Peck (1969) confirm that a stability ratio of 4 represents a general requirement in the use of compressive air to support a cohesive soil in an undrained case. As no reference was mentioned by the authors about the basis of this selection indicate that it is of empirical nature. Tables 3 and 4 show case histories compiled from different projects using various types of face support for cohesive and non cohesive soils respectively.

3.1 and 3.2) Belfast: Glossop and Farmer (1979) present an investigation of the effect of compressed air in supporting the face of a tunnel excavated for sewerage purpose in Sydenham, Belfast, Ireland. Monitoring soil movement around the excavation gives clear indication of a consolidation process resulting from applying and, then, releasing the supporting air pressure. Ground conditions around the excavation consists mainly of sleech: dark gray organic clay. Undrained shear strength is measured using field vane and triaxial tests and a typical ground stratigraphy is presented by the authors. During construction, Initial applied air pressure is found to be unsatisfactory (Case 1) and has been increased (Case 2). As the table shows, the safe support of the excavation is only reached when the average I_c is lower than 2. Defining the *factor of safety* (*FS*) as the ratio between the actual face pressure and face pressure required to support the face as calculated from Equation 4 and to neutralize ground water pressure, an average factor of safety of 1.2 is proven to be satisfactory.

3.3 and 4.1) Cairo: Greater Cairo wastewater project, as described by Safwat (1983), extends 33 km using Bentonite Slurry Shields, Earth Pressure Balanced Shields, and Compressed Air Shields. Compressed air method is used for a construction length of 2.5 km. At this part, the tunnel is excavated in a mixed face ground condition consisting of a water bearing sandy layer underlying cohesive clayey silt soil. El-Nahhas et al. (1991) describes results of monitoring program executed at this part. Measurements include Piezometer and shallow and deep displacement readings. Pore water pressure has risen in the cohesive material due to the use of compressed air which produces upward movement in front of the tunnel followed by downward displacement after the release of compressed air. The final settlement has been found to be within the tolerable settlement limits. If the tunnel section was considered totally imbedded into the cohesive layer, the actual I_c based on estimated undrained strength would be equal to 0.88. Using a value of $I_c = 2$, no face pressure is required to maintain the face stability and the only purpose of the compressed air is, then, to control seepage forces which results in factor of safety of 1.233. If the section was considered to be totally imbedded in sandy soil, then a factor of safety of less than unity would be obtained. It is believed that the stability of the lower half of the face is enhanced by the increase of shear strength of the sand due to capillary forces.

3.4, 3.5 and 4.3) Grimsby: a sewer tunnel has been constructed in Grimsby, as described by O'Reilly et al. (1991), from 1979 to 1981. Initially, the tunnel has been excavated under free air conditions. Ground conditions consist of soft to very soft clay (Marine warp) and in some instances a mixed face of Marine warp and Glacial till (Stony clay). If undrained cohesion of the Marine warp is used to calculate I_c relative to unsupported face, then a value exceeding 2 is obtained. Actually, the project suffered excessive settlements at ground surface and major difficulties in maintaining the face stability. Air pressure was then introduced and no further difficulties were encountered. Field monitoring described by the authors shows results of two sites A and B. Factors of safety calculated for the two sites are about *1.22*. As pore water pressure measurements and effective shear strength parameters are provided, Influence factor for frictional material, I_ϕ, is verified at Site B (4.3). Ground water table ranges between 1.6 to 3.8 which corresponds to a factor safety ranging between 1.0 and 2.66.

3.6 Mexico City: a sewer tunnel of 5.3 km was constructed in Mexico City using a Bentonite Slurry

Table 3: Assessment of Factors of Safety of some Cases History in Cohesive Soils

#	Place	Time	D (m)	Method	*Length* *(km)*	H/D	c (kPa)	ϕ	Ic	$I\phi$	*F.S.*	Remarks	Reference
3.1	Belfast		2.74	C.A		1.32	8.8-10.6	0	1.86-2.24	-	0.91-1.08	Some Problems	1
3.2	Belfast 2		2.74	C.A		1.32	8.8-10.6	0	1.43-1.72	-	1.11-1.32	No Problems	1
3.3	Cairo(up)		5.15	C.A		2.23	60	0	0.88	-	1.233	Mixed Section	2, 3
3.4	Grimsby(B)	79-81	3	C.A		2.21	18	0	1.23	-	1.23	No Problems, Total Stresses	4
3.5	Grimsby(A)	79-81	3	C.A		2.21	26	0	1.87	-	1.21	Problems when Ic= 2.66	4
3.6	Mexico City		4	BSS	5.34	2.43	23-40	0	0.81-1.4	-	-	No Problems	5
3.7	Nagoya	76-79	3.85	BSS	2.1	5.71	40-45	0	0.26	-	-	under sea	6
3.8	San Francisco	81	3.7	EPBS	0.915	2.46	27.4	0	(-)	-	3-4.23		7, 8, 9
3.9	Shanghai- up		4.33	EPBS	2	0.92	10	10.5	0.95	0.78	3.44	Upward movement	10
3.10	Shanghai-low		4.33	EPBS	2	0.92	3	12.5	0.82	0.76	3.29	Upward movement	10
3.11	Tokyo		3.73	BSS	1.6	10.27	78.48	0	1.325	-	1.36	No Problems	11
3.12	Unknown		4.13	D.K.	0.888	2.55	22.56	0	0.69-1.12	-	1.37		12

C.A.: Compressed Air
D.K.: D.K. Shield

Table 4: Assessment of Factors of Safety of some Cases History in Frictional Soils

#	Place	Time	D (m)	Method	*Length* *(km)*	H/D	c (kPa)	ϕ	Ic	$I\phi$	*FS*	Remarks	Reference
4.1	Cairo(low)		5.15	CA		2.23	-	30	-	0.55	0.88	Mixed Section	2, 3
4.2	Edmonton, SLRT	89-90	5	BSS	0.28	2	-	33.5	-	1.38	2.3	No Problems	13
4.3	Grimsby- Site B	79-81	3	CA		2.21	-	32	-	0.53- 1.41	1- 2.66	No Problems	4
4.4	Hamamatsu		2	D.K.	0.526	3.85	-	40	-	0.33- 1	0.6- 1.82		12
4.5	Kaoshung-Taiwan		3.914	D.K.	1.224	1.98	-	30	-	1.06	3		12
4.6	Lyon_ D	1984	6.5	BSS		1.87	-	33	-	0.59	1	Numerical Analysis	14, 15, 16, 17, 18, 19
4.7	Osaka	84-85	11.22	BSS		1.96	-	45	-	0.72- 1.68	1.44- 3.36		20
4.8	Rome	89	10.6	BSS	1.5	2.83	-	36	-	1.18	1.852	No Problems	21, 22, 23, 24, 25
4.9	Takaido		2.4	BSS	0.785	10.83	-	45	-	0.48	1		26
4.10	Unknown		5.22	D.K.	0.499	3.54	-	36	-		0.63- 2.72		12
4.11	Yotsugi		2.4	BSS	0.829	2.63	-	20	-	1.39	2		26

References

1:Glossop and Farmer (1979)
2:El-Nahhas et. al. (1991)
3:Safwat (1983)
4:O'Reilly et. al. (1991)
5:Schmitter et. al. (1988)
6:Ohta et. al. (1978)
7:Clough et. al. (1983)
8:Clough et. al. (1985)
9: Finno (1983)
10:Maidl et al. (1990)
11:Murakami et. al. (1990)
12:Hagimoto and Kashima (1985)
13:Eisenstein and Ezzeldine (1992:2)
14:Ferrand and Bouyat (1985)
15:Bouyat et. al. (1985)
16:Ferrand and Morcrette (1988)
17:Mohkam and Bouyat (1989)
18:Morcrette and Bouyat
19:Chaffois et. al. (1988)
20:Hashimoto et. al. (1988)
21:Anheuser (1985)
22:Becker (1985)
23:Casisnis et. al. (1985)
24:Capellari and Ottaviani (1982)
25:Pandolfo (1986)
26:Miki et. al. (1977)

Machine. As described by Schmitter (1988), soil condition consisted of highly compressible clay. Face pressure is almost equal to the initial pore water pressure at tunnel axis. Considering total stresses, I_c has a value between *0.81* and *1.4* due to variation of shear strength. Effective stress calculations showed that for $I_c = 2.0$, no face pressure is required to support the face and the only purpose of the slurry pressure is to neutralize water pressure.

3.7 Nagoya: the construction of a water supply tunnel under Nagoya Harbor, Japan, is designed to pass under the sea. Soil conditions, as described by Ohta (1978) at the face include in many places of the project highly permeable layers. As the tunnel level is considerably deep under sea level, high bentonite slurry pressure is used to neutralize hydrostatic pressure. In spite of the high variation of shear strength along the tunnel trajectory the calculated pressure required for soil support has always been very small compared to the total face pressure.

3.8 San Francisco: the first application of EPBS was used in the construction of San Francisco Clean Water Project (N2). Description of the project and numerical analysis of the obtained data is found in Clough et al. (1983 and 1986) and Finno (1983). The entire face has been immersed in a normally consolidated clayey soil known locally as the Bay Mud. A number of wooden piles passed through the tunnel section and had to be cut by the machine. As mentioned before, face pressure of the EPBS is usually higher than the initial horizontal pressure. Negative values of I_c is calculated for this project as the face pressure is higher than the initial horizontal pressure. The value of factor of safety was higher than 3.0 and reaches 4.23 at the section where face pressure had to be increased to cut the wooden piles.

3.9 and 3.10 Shanghai: a sewage tunnel has been driven in Shanghai, China through soft and saturated soil. Soil conditions as described by Maidl et al. (1990) indicate that the tunnel section consisted to two layers of silty clay divided by a thin layer of silty sand passing almost at the tunnel axis. Average applied face force is calculated by subtracting the friction and adhesion forces along the shield skin from the total applied jacks force. Face pressure is calculated by dividing the net face force by the shield cross section. As in the previous case, high value of factor of safety is obtained. The values of I_c and I_ϕ presented in Table 3 are calculated assuming that fully shearing resistance is mobilized.

3.11 Tokyo: Bentonite slurry shield machine has been used in the construction of a cable tunnel in Tokyo passing under two rivers. As described by Murakami (1990) the tunnel has a high depth ratio and is passing mainly through a diluvium strata of stiff clay. Total stress analysis resulted in a value of I_c of 1.325 and a factor of safety of 1.36.

3.12 D.K. Shield: D.K. Shields may be considered a transition technique between the EPBS and the Bentonite Slurry Shield. Face stability is not directly achieved through shield jacking at the liner as in case of EPBS. The mix of mud with the spoil provide a better distribution of the support pressure inside the bulkhead and, thus, at the face. Hagimoto and Kashima (1985) show a case where D.K. shield was constructed in a cohesive soil with a factor of safety considerably less than that of the EPBS.

4.2 Edmonton, LRT: the extension of Edmonton Light Rail Transit system has included the construction of a tunnel using the Hydraushield; a Bentonite Slurry Machine based on German technology. Soil conditions at the tunnel section, as described by Eisenstein and Ezzeldine (1992) consists for a part of the project of a post glacial outwash of sand and gravel deposits with a very high degree of heterogeneity while on the other part of the project the lower half of the cross section passes through Edmonton Formation, a Cretaceous Bedrock. Ground water table is below the tunnel level. The function of the slurry pressure, therefore included, beside face support, the formation of a permanent and reliable cake thickness and mucking of the excavation spoil. A factor of safety of 2.3 is calculated.

4.4, 4.5, and 4.10 Hamamatsu: D.K. Shield described above by Hagimoto and Kashima (1985) has been used in non cohesive soils in Hamamatsu, Japan, and in Kaoshung, Taiwan. The presented information from the authors shows a wide range of values of factor of safety which may be attributed to the specific working conditions at these projects.

4.6 Lyon- D: the construction of the Lyon Subway is described by Ferrand and Bouyat (1985) and Ferrand and Morcrette(1988) and Morcrette and Bouyat (1990). The tunnel trajectory was constructed in 1984 under the Rhône and the Saône in a sandy gravely soil. Due to the importance of the project, a number of analyses were initiated in different domains. Bouyat et al. (1985) investigates experimentally the interaction between the slurry and the sandy soil and Chaffois et al. (1988) have used a three dimensional analysis to determine the yield pressure. The resultant pressure compares very well with the current analysis.

4.7 Osaka: a large diameter tunnel has been constructed in Osaka City, Japan, for stormwater management using the Bentonite Slurry Shield. The project is described by Hashimoto et al. (1988) passes through a predominantly very hard diluvium sand and gravel of Standard Penetration Number exceeding 50. In some parts of the project, ground freezing has been used to insure the face stability and therefore low slurry pressure is needed. In the part were no additional face protection was used, assuming an angle of internal friction $\phi = 40^o$, a value of I_ϕ very close to that suggested in Figure 18 is obtained.

4.8 Rome: a large diameter railway tunnel is constructed in Rome, Italy. Excavation works as described by Casinis et al. (1985) and Pandolfo (1986) have been carried out using the Hydroshield in a competent hard clay in a part of the project and in silty sand layers in other parts and the entire project has been constructed under ground water table. Face pressure has been found necessary for the part constructed in the sandy layer and a factor of safety of 1.852 is calculated.

4.9 Takaido and Yotsui: Miki et al. (1977) presents a description of the Bentonite Slurry Shield technology in Japan and shows that the method of calculation of the supporting face pressure in cohesionless soils is close to Equation 4, except that they use K_o instead of K_a and an empirical factor instead of I_ϕ. The authors justified their method of calculation by three case histories of projects in Japan, Yotsugi, Ushiku, and Takaido. The three projects were carried out in sandy soils with, respectively, low, variable, and high Standard Penetration Test Numbers at the excavated section. While it is difficult to estimate a reliable value for the angle of internal friction of Ushiku project, $\phi = 20^o$ and $\phi = 45^o$ may be considered a realistic estimation for Yotsugi and Takaido projects respectively which yields in the factors of safety shown in Table 4.

7 CONCLUSIONS

The study presented shows the influence of the face pressure, as applied with the Bentonite Slurry Shield, the Earth Pressure Balanced, and the compressed air shields, on face stability and ground movements. The magnitude of the face pressure is required to be optimized to limit the displacements at the face and at the same time to control the loads imposed by the shield on the lining. Thus the problem studies the careful balance between these two requirements.

REFERENCES

Anderson, J.M. and Lamont, D.R., 1991. A Comparison of International Legislation Concerned with Tunnelling in Compressed Air. *'Tunnelling 91', 6th Int.Symposium the Institution of Mining and Metallurgy*, London, England, pp. 55- 64.

Anheuser, L., 1985. Technologie et Applications Récentes des Boucliers à Boue Allemandes. *Tunnelling in Soft and Water- Bearing Grounds, Proceedings of an International Symposium organized by Association française des Travaux en Souterrain*, Lyon, France, pp. 15- 20.

Babendererde, S., 1991. Tunnelling Machines in Soft Ground: a Comparison of Slurry and EPB Shield Systems. *Tunnelling and Underground Space Technology* , Vol. 6, No. 2, pp. 169- 175.

Becker, C., L., 1985. Galleria Aurelia Roma - Europe's Largest Hydroshield. *Tunnelling in Soft and Water- Bearing Grounds, Proceedings of an International Symposium organized by Association française des Travaux en Souterrain,* Lyon, France (M. Legrand, editor), pp. 69- 72.

Bouyat, C., Mohkam, M., and Morcrette, J. P., 1985. Le Soutènement Liquide, une Recherche Appliquée pour le Bouclier à Pression de Boue du Métro de Lyon. *Revue Française de Geotechnique*, N. 30, pp.31- 45.

Broms, B.B and Bennermark, H., 1967. Stability of Clay at Vertical Openings. *Journal of Soil Mechanics and Foundation Engineering Division, American Society of Civil Engineers*, Vol. 93, pp. 71- 94.

Casarin, C. and Mair, R.J., 1981. The Assessment of Tunnel Stability in Clay by Model Tests. *Soft Ground Tunnelling, Failures and Displacements* (edited by D. Resendiz and M.P. Romo), Balkema, Rotterdam, pp. 33- 44.

Casinis, C., Nisio, P., and Capata,V., 1985. Settlement and Face Stability Boring a Large Tunnel. *Eleventh International Conference on Soil Mechanics and Foundation Engineering*, San Francisco, pp. 2057- 2062.

Chaffois, S., Laréal, P., Monnet, J., and Chapeau, C., 1988. Study of Tunnel Face in a Gravel Site. *Numerical Methods in Geomechanics, Innsbruck,* (Swoboda, editor), pp. 1493-.1498.

Chambon, P., Corté, J-F., Garnier, J., and König, D., 1991. Face Stability of Shallow Tunnels in Granular Soils . *Centrifuge 91* (edited by Hon-Yim Ko and Francis G. Mclean), Balkema, Rotterdam, pp. 99- 105.

Chan, D., 1986. Analysis of Strain Softening Material. *Ph.D. Thesis*, The University of Alberta, Edmonton, Alberta.

Clough, G. W., Shirasuna, T., and Finno, R., J., 1985. Finite Element Analysis of Advanced Shield Tunneling. *Fifth International Conference on Numerical Methods in Geomechanics*, Nagoya, pp. 1167- 1174.

Clough, G. W., Sweeney, B.P. and Finno, R.J., 1983. Measured Soil Response to EPB Shield Tunneling. *Journal of Soil Mechanics and Foundation Engineering Division, American Society of Civil Engineers*, Vol. 109, pp. 131-149.

Eisenstein, Z. and Ezzeldine, O., 1992. Ground Control by Shields with Positive Soil Support. *Cairo second International Symposium on Tunnelling; Current Experiences in Tunnelling*, January 27- 28, pp. 75-88.

Eisenstein, Z. and Ezzeldine, O., 1992. The Effect of Tunnelling Technology on Ground Control. *Tunnelling and Underground Space Technology*, Vol. 7, No. 3, pp. 273- 279.

El-Nahhas F., El-Kadi F. and Shalaby, A., 1991. Field Measurements During Construction of a Compressed Air Tunnel in Cairo. *Tunnelling and Underground Space Technology*, Vol. 6, No. 1, pp. 123- 127.

Ferrand, J. and Bouyat, C., 1985. Le Métro de Lyon et son Bouclier à Pression de Boue. *Tunnelling in Soft and Water- Bearing Grounds, Proceedings of an International Symposium organized by Association française des Travaux en Souterrain,* Lyon, France, pp. 21- 28.

Ferrand, J. and Morcrette, J.P., 1988. Métro Ligne D: Premiers Résultats et Bilan des Chantiers de Génie Civil. *Tunnels and Water*, Balkema, Rotterdam, pp. 869- 875.

Finno, R.J., 1983. Response of Coheesive Soil to Advanced Shield Tunneling. *Ph. D. Thesis*, Stanford University.

Glossop, N. H., and Farmer, I. W., 1979. Settlement Associated with Removal of Compressed Air Pressure during Tunnelling in Alluvial Clay. *Geotechnique*, Vol. 29, No.1, pp. 67- 72.

Hagimoto, H. and Kashima, Y., 1985. D.K. Shield Method. *Tunnelling in Soft and Water- Bearing Grounds, Proceedings of an International Symposium organized by Association Française des Travaux en Souterrain,* Lyon, France (M. Legrand, editor), pp. 53- 60.

Hashimoto, K., Kato, T., Watanabe, Y., Matsuda, M, Tamiya, Y, Shima, M, Fujihara, Y,, and Yoshikawa, T, 1988. *Underground Stormwater Reservoir constructed by f 11.22 m Slurry Shield. Tunnels and Water* (J. Manuel Serrano, editor), Balkema, Rotterdam, pp. 69- 72.

Jardine, F. M., Evans, A., and McCaig, R.H., 1991. Studies of the Incidence of Decompression Sickness on United Kingdom Compressed-Air Projects. *'Tunnelling 91', Sixth International Symposium organized by the Institution of Mining and Metallurgy,* London, England, pp. 29- 40.

Kimura, T. and Mair, R. J., 1981. Centrifugal Testing of Model Tunnels in Soft Clay. *Tenth International Conference on Soil Mechanics and Foundation Engineering*, Stockholm, pp. 319- 322.

Kirkland, C.J., 1984. Design and Construction of Urban Tunnels- Compressed Air. *Advances in Tunnelling Technology and Subsurface Use,* Volume 4, No. 4, pp. 191- 194.

Kurosawa, S., 1985. Up-to-date of the Art of Japanese Shield. The Selection Criteria of Slurry Shield and Earth Pressure Balanced Shield. *Tunnelling in Soft and Water- Bearing Grounds, Proceedings of an International Symposium organized by Association française des Travaux en Souterrain,* Lyon, France (M. Legrand, editor), pp. 9- 13.

Maidl, B., Hou, X., and Yi, X., 1990. Field Measurement of Ground Movements and Variations of Pore Pressure caused by EBP Shield Construction. *Tunnel and Underground Works Today and Future. Proceedings of the International Congress, the International Tunnelling Association Annual Meeting 1990*, pp. 335- 349.

Miki, G., Saito, T. and Yamazaki, H., 1977. Theory and Praxice of Soft Ground Tunneling by the Slurry Mole Method. *Tunnelling in Soft Ground, Ninth International Conference on Soil Mechanics and Foundation Engineering*, Information from the Specialty Session No.1 in Tokyo, pp. 6- 15.

Mohkam, M, and Bouyat, C., 1989. Simulation en Laboratoire du Creusement de Tunnel avec des Techniques Diverses et des Conditions Hydrogéologiques Différentes. *Tunnels and Water*, Balkema, Rotterdam, pp. 1239- 1246.

Mohkam, M. and Bouyat, C., 1985. Research Studies for Slurry Shield Tunnelling. Tunnelling 85, *Fourth International Symposium organized by the Institution of Mining and Metallurgy*, London, England, pp. 235- 242.

Morcrette, J. P. and Bouyat, C., 1990. Métro de Lyon. Les Difficuletés du Creusement sous le Rhône avec le Bouclier à Pression de Boue. *International Congress on Large Underground Openings*, Firenze, Italy, pp. 822- 828.

Murakami, K., Nakano, M., and Suzuki, S., 1990. Measurement and Analysis of Ground Deformation for Shield Tunneling undercrossing a Class A River. *Tunnel and Underground Works Today and Future. Proceedings of the International Congress, the International Tunnelling Association Annual Meeting 1990*, pp. 357- 364.

O'Reilly, M.P, Mair, R.J., and Alderman, G.H., 1991. Long-Term Settlements over Tunnels: an Eleven-year Study at Grimsby. *'Tunnelling 91', Sixth International Symposium organized by the Institution of Mining and Metallurgy,* London, England, pp. 55- 64.

Ohta, Y., Matsui, T., Matsubara,H., Akiyama, M., and Baba, T., 1978. Slurry Shield Method in Complex Formation under Sea- Construction of Water Supply Tunnel under Nagoya Harbor. *Tunnelling under Difficult Conditions, Proceedings of the International Tunnel Symposium,* Tokyo, (I. Kitamura, editor), pp. 203- 208.

Pandolfo, A., 1986. IlSottoattraversamento Ferroviario di Centri Urbani con Scarsa Copertura ed in Ambienti Idergeologici Particolarmente Difficili. *International Congress on Large Underground Openings*, Firenze, Italy, pp. 252- 263.

Safwat, A., 1983. The Greater Cairo Wastewater Project. *World Water '83, the World Problem*, pp. 165- 172.

Schmitter, J.M., Lopez Portillo, M.V., and Orozco, J.C., 1988. Slurry Shield Tunneling for Mexico City. *'Tunnelling 88', Fifth International Symposium organized by the Institution of Mining and Metallurgy*, London, England, pp. 289- 294.

Terzaghi, K., 1943. *Theoretical Soil Mechanics.* John Wiley, New York.

Xi, Z., 1990. Control Measures Employed in Earth Pressure Balanced Shield. Tunnel and Underground Works Today and Future. *Proceedings of the International Congress, the International Tunnelling Association Annual Meeting 1990*, pp. 445- 451.

6 Repair and management of underground projects

Tunnelling and Ground Conditions, Abdel Salam (ed.) © *1994 Balkema, Rotterdam, ISBN 90 5410 363 9*

Examination of passages of Aswan high dam tunnels

E.A.Abdel-Hafiz
Faculty of Engineering, Ain Shams University, Cairo, Egypt

A.M.El-Molla
Faculty of Engineering, Al-Azhar University, Cairo, Egypt

ABSTRACT: The purpose of this investigation is to examine the submerged parts of the inlets and outlets of the Aswan High Dam tunnels and to locate any damage or defect in addition to propose the method of remedy for that damage.

1 INTRODUCTION

The Aswan High Dam was designed to serve three basic functions; to store water from the annual flood, to generate hydroelectric power, and to control floods in the Nile. Construction began in 1960 and the dam was completed in 1968.

The massive amount of water needed annually, passes to the downstream of the dam via the diversion canal and the power house which is located at the eastern bank of the River, Fig. (1).

The construction of the tunnels started in 1964 and they have been in operation since 1967. Maintenance programs have been carried out regularly for these elements which are of vital importance to the durability and performance of the structure.

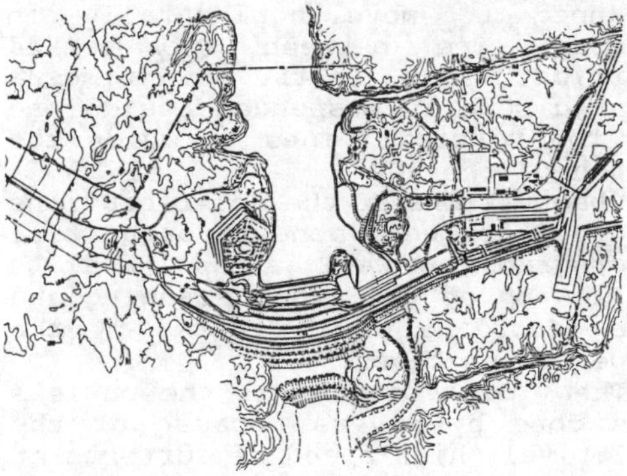

Fig.1 General layout

The inlets and outlets of the tunnels are greatly affected by the dynamic loading resulting from operation. The maintenance programs usually overlooked these parts of the structure, because it is not possible to uncover them. Thus, the inspection of these parts is an important task in order to check its safety against any defects or damages which may cause problems for the power house.

In 1991, the authors were involved in a task to explore the status of the tunnels inlets and outlets. This paper presents the results of this task.

2 AREA UNDER INVESTIGATION

The power house consists of six reinforced concrete tunnels of circular cross section of a diameter 15.00m each. These tunnels are lined by a reinforced concrete layer of one meter thickness.

The six tunnels in the upstream are divided into two inlets for each tunnel, giving a total of 12 inlets in the upstream. Before reaching the power house in the downstream, each tunnel is divided into two branches, giving 12 branch. After leaving the power house, each branch is divided into two outlets, giving a total of 24 outlet.

The inlets of the tunnels are provided with trash racks, maintenance gates, and main gates, while the outlets are provided with maintenance gates. Figure (2) shows a section in one of the tunnel, while figures (3) and (4) show the inlets and outlets of the tunnels, respectively.

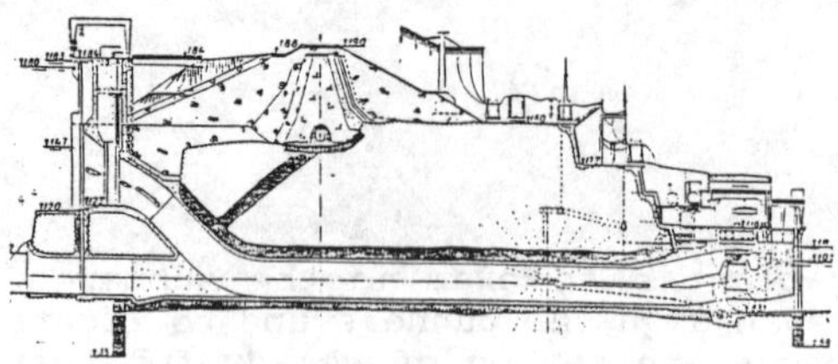

Fig.2 Tunnel section

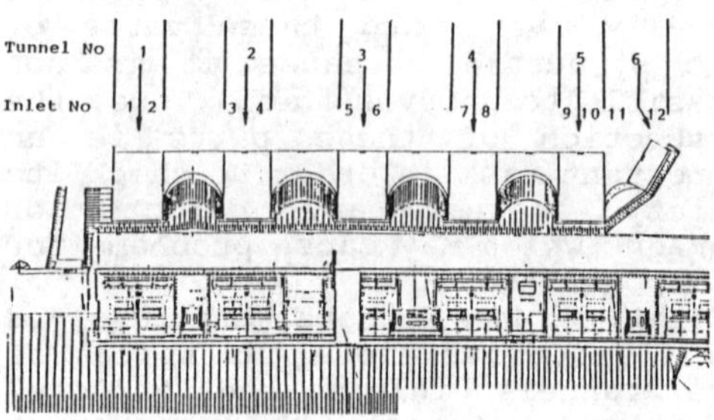

Fig.3 Tunnels'inlets

3 WORK PROCEDURE

The survey of the tunnels' inlets and outlets had to be done under water since it was impossible to get a "dry case". The techniques used to monitor the under water parts were chosen according to the inspected part. Thus, for the upstream section of the structure, where the water pressure cannot be tolerated by divers, a Remotely Operated Vehicle (ROV) was used. On the other hand, in the downstream segment, where the water velocity is very high and the ROV could not stand the currents, professional divers were in charge of the monitoring program.

Due to the very high velocities emerging from the tunnels in the downstream side, the monitoring program was scheduled to be performed during the time of minimum demand for electricity from the power house in order to close the tunnel, thus the work took place usually at night. For some tunnels, the eddies in the downstream were very severe that, not only the inspected tunnel had to be closed, but the few adjacent ones as well.

The ROV, works as a robot for video scanning under water to overcome the difficulties of high pressure. It is connected to a T.V. monitor to enable the working team from following, correcting, and modifying the process.

Special precautions have been considered in planning of the ROV work, where datums in all directions were taken in order to correlate all the locations. In the meantime, to ensure that the scanned area is always uniform, the distance between the ROV camera and the scanned surface was kept fixed. For more elaborate study for a certain area, the ROV camera was put close to the scanned surfaces in order to view the shape of the defect and its nature. Meanwhile, overlap of scanned areas was assured.

In order to avoid missing of scanning any part, the ROV was planned to move horizontally in zigzag line between two marked vertical lines. In the mean time, a marked rope is suspended beside one of the vertical lines to guide the pitch of the ROV.

Observations of the deteriorations and defects were done through still photographs as well as the immediate follow up of the video scanning and the recording of the notes by the supervision team.

The video scanning at the outlets was done by divers because of the relatively high speed of currents at the outlets which makes the control of the ROV real difficult. The observation of the downstream scanning was immediate through watching a T.V. monitor connected to the camera through cables. Communications, comments, requests for close-up views and other instructions were given directly to the divers through a special telephone connection.

Stand-still photographs were also taken during the recording in order to help more in identifying the damages and their causes.

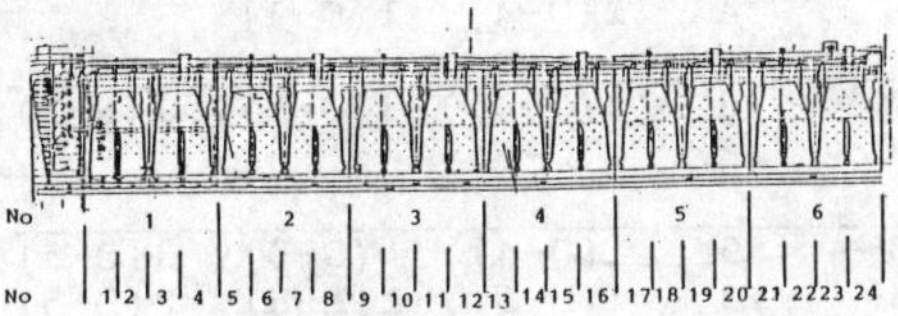

Fig.4 Tunnels' outlet

Due to the vast number of surfaces required to be scanned, the wall

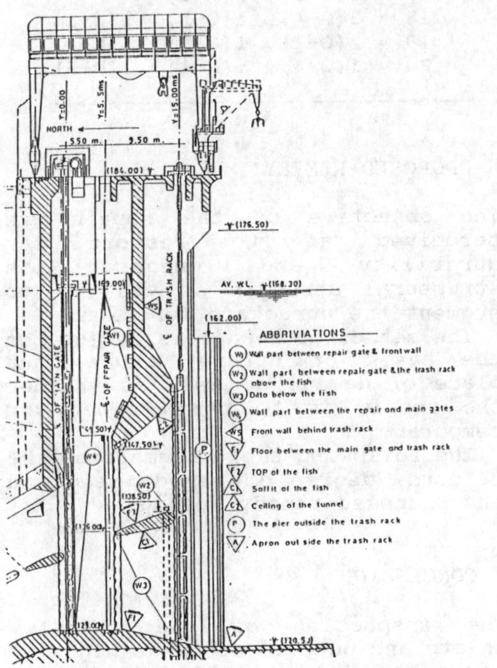

Fig.5 Sectional elevation in tunnels' inlet

surfaces in the upstream were divided into portions as shown in Fig. (5). In the meantime, the surfaces of the walls of the outlets were divided into portions as shown in Fig. (6).

4 CLASSIFICATION OF DEFECTS

The observed deteriorations cover a wide variety of nature, locations, and priorities. In order to arrange these defects according to their importance and nature, the following categories were set:

4.1 Degree of remedy need

The defects were classified according to importance, severity, and location as follows:

1. Optional (O): Defect with no danger what so ever on the structure's safety, operation and durability. It is up to the authorities to repair it.
2. Normal (N): Defect which is not necessary to be repaired as it poses no danger on the structure's integrity, but it is preferable to be repaired.
3. Essential (E): Defect that should be repaired, but it has no immediate danger on the structure's integrity.
4. Dangerous (D): Defect that must be repaired soon to maintain the structure's integrity.

4.2 Description of defects

Due to the vast nature of defects, they were classified as follows:

1. Loose debris. 2. Fixed debris.
3. Falling parts of racks.
4. Surface erosion, honeycombs with no exposed reinforcement.
5. Missing concrete parts, erosion, with exposed reinforcement.
6. Cracks in concrete.

5 OBSERVATIONS

Tables (1) and (2) summarize the remarks on the tunnels' inlets and outlets, respectively. Inlets or outlets with no defects were not shown.

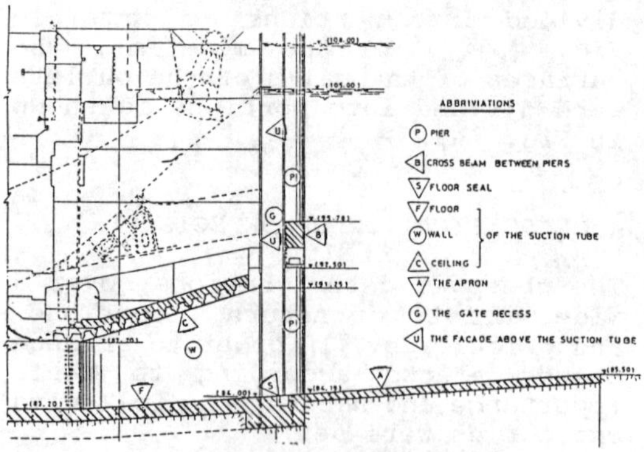

Fig.6 Sectional elevation in tunnels' outlet

Table 1. Defects in tunnels' inlets

Tunnel	Inlet	Category
1	1	2(O-4), 1(O-1), 1(O-2)
	2	1(O-4)
2	4	2(O-4), 1(O-1)
3	5	4(O-4), 2(O-2), 2(N-4)
	6	2(O-4), 1(O-5), 1(O-3)
4	7	2(O-3)
	8	1(O-3), 2(N-5)
5	9	2(O-3), 1(N-4), 1(N-5)
	10	4(O-5), 1(N-5), 1(N-6)
6	11	1(O-1), 1(N-5)
	12	2(O-3), 1(N-2), 3(N-5)

The tables identify the tunnel as well as inlet or outlet number, then the defects are categorized with the order of: number of defects found then the degree of remedy need as given in article 4.1 (O, N, E, and D), and finally the description of defect as given in article 4.2 (1 to 6).

The tables show that the deterioration found in the outlets of the tunnels is more than that in the inlets. In the meantime, while most of the defects in the inlets are minor, few in the outlets must be considered for immediate repair, especially in outlet number eight which had exposed reinforcement and eroded concrete around the pier and on the floor.

Table 2. Defects in tunnels' outlets

Tunnel	Outlet	Category
1	1	1(O-4), 1(O-1), 1(N-5)
	2	1(O-1), 1(E-5)
	3	1(O-1), 1(N-4), 1(D-5)
	4	1(O-4), 1(O-1)
2	6	2(O-4), 3(O-5), 1(N-5)
	8	4(D-5)
3	9	1(O-4), 1(O-5), 3(D-5)
	10	4(N-5), 1(E-5), 2(D-5)
	11	1(N-1), 2(E-5), 1(D-5)
	12	2(E-5), 1(N-1), 1(D-5)
4	13	1(N-1), 1(O-1), 1(D-5)
	14	1(O-1), 1(E-5), 2(D-5)
	15	1(O-1), 1(E-5)
	16	1(O-5), 1(N-1), 1(D-5)
5	17	2(O-1), 1(N-1)
	18	3(O-4), 1(O-5), 1(N-1)
	19	2(O-1), 1(O-5), 1(N-1)
	20	1(O-4), 1(N-1), 1(D-5)

6 PROPOSED REMEDY

The objective of the repair was perceived as to restore the durability and enhance the structural strength as well as to augment the structure function.

The method of repair is decided on the basis of nature, cause, and place of damage as well as whether the repair is to be permanent or temporary.

The following table summarizes the defects, degree of remedy need and the proposed remedy method.

7 CONCLUSION

The inspection of the tunnels' inlets and outlets of Aswan high dam power house demonstrated that they are generally in excellent condition. Few imperfections were found where they need to be treated.

The authors recommend that similar monitoring programs should take place for the submerged parts of all major hydraulic structures in Egypt.

Table 3. Defects and remedy method

defect type	degree of need O	N	E	D
1	I	I	-	-
2	I	I	-	-
3	I	-	-	-
4	II	II	-	-
5	II	II	II	III
6	I	I	-	-

The proposed methods indicated in the table are:

I : removal.
II : patch repair
III: casting by tremie.
IV : injection

REFERENCES

Allen, R.T.L. & Edwards, S.C. 1987. The repair of concrete structures. Blackie & Son Limited.

Anderson, J.M. 1983. Remote controlled hydrocrete. Concrete 17(11), 12-15

McLeish, A. 1983. Development of methods of rehabilitating damaged offshore concrete structures. Taylor Woodrow Report No. 014H/83/2504.

The High Aswan Dam construction authority 1970. Project bulletins (1-9).

Tunnelling and Ground Conditions, Abdel Salam (ed.) © 1994 Balkema, Rotterdam, ISBN 90 5410 363 9

Preconditions and concepts for professional qualification

M. Bandmann
Tiefbau-Berufsgenossenschaft, Munich, Germany

ABSTRACT: Tunnelling and specialized civil engineering are characterized by a high risk of accidents and unfavourable working conditions. As part of the project, two courses are being pursued in the endeavour to find ways of improving conditions at the work place, increasing work safety and reducing health hazards. After the accident situation has been assessed, areas requiring action defined and accidents analyzed, training programmes will be developed and tested for various target groups. These will result in more intensive and more broadly based qualification of employees. They will also provide the basis for implementing training programmes and carrying out research work into safety questions at the Tunnelling Training and Experimental Institute (TTEI). This combined approach will improve safety at work and reduce health hazards.

1 OVERALL OBJECTIVE OF THE PROJECT

Tunnelling and specialized civil engineering are characterized by a high risk of accidents and unfavourable working conditions. Work has to be carried out largely under inadequate visual conditions, with intense dust pollution, noise and vibration and under considerable physical stress. The work places and the working environment on construction sites change constantly and the potential stress and potential hazards are in some cases subject to very sudden changes. An on-going problem is that large numbers of semi-skilled workers, workers unused to learning, foreigners etc. have to be employed.

As part of the project, various approaches are being tried in order to find ways of improving conditions at the work places, increasing work safety and reducing the health hazards in tunnelling and specialized civil engineering.

Two courses have been selected for achieving these objectives:

- Development and application of programs to improve professional qualifications

The definition and quantifying of the main danger points and the health risks in tunnelling and in specialized civil engineering serve as a starting point for ascertaining deficiencies in the present qualification structure and in further education and training programmes at company and national level and for determining their causes. Ways of overcoming obstacles to becoming professionally qualified or of creating suitable incentives to learning are shown. This forms the basis for defining the relevant target groups in this sense and for investigating ways of reaching and addressing these groups. It is particularly important to include all employees as far as possible and not just to involve a small number of safety experts. Transparent training concepts and learning programmes to help secure jobs are being set up, tried and introduced specifically for the various groups, taking into account the time factor and local conditions. The aim is to achieve professional qualifications which reduce the accident and health risks, to promote the deployment of employees in a wider range of activities and to increase safety awareness and health consciousness.

- Providing the basis for developing and testing improvements in technical aspects of construction work and for implementing further training measures

The establishment of a Tunnelling Training and Experimental Institute (TTEI) is planned as the best possible way of implementing measures to improve professional qualifications and for the development, testing and demonstration of new work techniques. The concept of a facility of this type and programmes for R&D are being developed. This will form the basis for the actual planning and installation of the TTEI which will be followed by the start-up phase for training and research.

The existence of the Tunnelling Training and Experimental Institute TTEI will for the first time offer facilities for trying out new technical ideas and developments with special emphasis on improving working conditions in the construction industry. It will also provide ideal conditions for implementing further training measures in a practical environment aimed at instilling safety awareness on a broad front, demonstrating and teaching safety-relevant modes of behaviour and offering wider qualifications for different fields of work.

The project is of model character in two respects. Firstly, social aspects of working life (wider qualification, greater safety awareness and sense of responsibility) are combined with technical aspects ("humanizing" of technology). This will directly improve safety at work and reduce health risks.

The second outstanding feature is that the TTEI will combine the measures described for acquiring professional qualifications with further advances in technical aspects of the working environment. Apart from the positive benefits of being able to test technical developments on a large scale under the conditions encountered in practice, the involvement of tunnelling workers will for the first time also allow features for humanizing working life to be incorporated in technical innovations from the outset.

2 OBJECTIVES OF THE ASSISTANCE PROGRAMME

The project is being backed by funds provided by the German Federal Ministry for Research and Technology (BMFT) and the Tiefbau-Berufsgenossenschaft.

The furthering of research and development projects to improve working conditions in the construction industry has the following among its objectives:

- Identifying starting points for "humanizing" technology
- Ascertaining and preparing new forms of professional qualification
- Acquisition of knowledge about the effects of pollution and stress and methods of dealing with them and on preventive measures for the protection of employees.

Tunnelling and specialized civil engineering is a sector which has great economic significance. The structures (tunnels for roads, railways, supply and drainage, power stations etc.) are mainly commissioned by public authorities (Federal or state governments, local authorities, transport administrations and utilities, etc.) for the benefit of the general public.

On the other hand, working conditions in this sector of the construction industry are characterized by above-average accident risks and health hazards and considerable physical stress for personnel. Public authorities therefore have a very great responsibility to ensure that projects carried out on their behalf do not automatically involve exceptional risks and hazards for those working on them.

The aim of the project is to help toughen preventive measures for the protection of workers in line with the objectives of the assistance programme. It will do this by improving technical aspects of work processes and by increasing the qualifications of personnel and so achieve a considerable improvement in working conditions in tunnelling as a whole.

3 SCIENTIFIC AND TECHNICAL WORK OBJECTIVES OF THE PROJECT

The object of this work is firstly to develop and try out concepts and programmes for improving the professional qualifications of those working in tunnelling. Special importance is attached to gaining qualifications which reduce the accident risk and to wider qualification for various fields of work. It will include both horizontal and vertical further training for operatives and management level respectively. The first objective will be to transfer existing know-how from projects in the "Humanizing of working life" (HWL) programme to all employees, rather than just passing on the results to a limited number of safety experts. It is also intended that the Tunnelling Training and Experimental Institute (see below) should correlate basic and further training on the one hand with improvements in work techniques on the other. The ultimate aim is that safety awareness and health consciousness should be promoted on a broad front so that employees are able to estimate risks in advance and consequently reduce them, for themselves as well as for other persons.

The research project also provides for the planning and establishment of a Tunnelling Training and Experimental Institute (TTEI). At present no experimental and further training establishment exists specifically for this branch of the construction industry. There are to be two sides to the work of this institute, firstly basic and further training sessions under the real-life conditions of a "construction site" and secondly research and development aimed at technical improvements in tunnelling, directly including ways of improving working conditions.

The TTEI is to have its own tunnelling "construction sites" featuring the widest possible range of different characteristics. It is planned if possible to use an existing disused tunnel system. This will enable the various activities which take place on site to be demonstrated and practiced in the experimental and training establishment with the above objectives in view and also provide sufficient free space for testing technical improvements.

The concept of the TTEI is to be developed as part of this project and various possible locations examined for their suitability. Courses and basic and further training sessions are to be held as soon as the basic facilities have been established.

The entire project will be realized in two parts which will initially be independent:

- Part A: Development and application of programmes for improving the professional qualification of employees in tunnelling
- Part B: Creating the conditions for developing and testing improvements in technical aspects of construction work and for implementing further training measures

Parts A and B will be merged into Part C as further training courses, safety training sessions, research and development start up at the TTEI.

4. PROSPECTS OF SUCCESS, BENEFITS

The Tiefbau-Berufsgenossenschaft has been providing basic and further training for safety experts for many years. It is also involved in research work to improve working conditions in the tunnelling industry.

The STUVA is predominantly active in research and development in tunnelling, a major part of its work being within the HWL programme. The STUVA also plays a central role in the planning and running of courses for further professional training and for the improvement of working condi-tions.

Both institutions have the expertise and capacity in terms of technical facilities and personnel to carry out the project in line with its objective. The technical and scientific success of the project can therefore be taken for granted. The syllabuses and training programmes for 9 different tunnelling activities have already been largely completed after detailed consultations in technical committees consisting of representatives of the Tiefbau-Berufsgenossenschaft, the STUVA and of experts from the construction industry and universities.

The project involves no technical or financial risk. Preliminary investigations have revealed the existence of several tunnel systems which are available in principle for the purposes of this project. One of these systems would be assigned by the German government free of charge to the Tiefbau-Berufsgenossenschaft for its use subsequently for training purposes. Rights and obligations arising in the future from the takeover of this tunnel system would devolve upon the Tiefbau-Berufsgenossenschaft. This means that after the end of the subsidized phase, the Tiefbau-Berufsgenossenschaft would take over the tunnel system as a training and research facility and would thus also assume responsibility for all consequential encumbrances. Research will be carried out in permanent cooperation with the STUVA.

The existence of the TTEI will also provide opportunities for commercial exploitation at a later stage. One option which might be examined is whether parts of the tunnel complex, which will be a major component of the TTEI, can be used for industrial research. This would help enable the TTEI to pay its own way after the end of the subsidized phase. Fundamental interest in the existence of a facility of this kind has already been expressed by various branches of the industry.

Apart from fulfilling the social obligation to improve working conditions in a sector of working life which is afflicted by high accident risks and health hazards, the following positive affects are expected from this project:

- A reduction in the risk of accidents and health hazards in a branch of working life which is notorious in this respect will contribute in the short, medium and long term to a reduction in the costs of the health service.
- Increased safety at the work place and professional qualification of workers employed in tunnelling will have direct positive effects on the construction schedule and ultimately on construction costs as well. In the final analysis this will have a positive effect on the national economy since public authorities are the clients for most tunnelling and specialized civil engineering projects.
- The improved conditions for research and development in the technical sector and more especially the possibility of experimenting in a real-life environment will firstly enable the relevant work techniques to be made more efficient. Secondly it will be possible to carry out work and follow objectives which have not been thoroughly addressed or pursued intensively enough because of the present lack of adequate test facilities. A considerable innovation effect is expected as a result.

5 SCIENTIFIC AND TECHNICAL STATE OF THE ART

Even today the construction of underground cavity structures still involves greater danger and more physical stress for the persons employed on this work than other industrial activities.

The civil engineering accident statistics reveal that out of an average of 450,000 persons insured by the Tiefbau-Berufsgenossenschaft, one in eight suffers a notifiable accident each year. The accident rate in tunnelling is seven times as high as the ave-rage for the industrial sector. A major reason for the high risk of accident is the predominant use of semi-skilled labour and the consequent low level of training of the employees in this sector. Further professional training for the various branches of activity in the form of safety courses is needed to improve protection against accidents.

The annual statistics on current tunnelling activities in the Federal Republic of Germany which the STUVA has kept since 1977 with the assistance of the Tiefbau-Berufsgenossenschaft still reveals a high proportion of and requirement for tunnels driven underground. For example, about 80% of the traffic tunnel construction projects currently in progress with a total length of about 170 km are being constructed by the enclosed method. A planned construction volume for traffic tunnels with a total length of about 300 km can probably be expected by the year 2000. Of this total, about 150 km will be accounted for by city underground railway lines, 110 km by planned new routes of the Deutsche Bundebahn and about 40 km by road tunnels. It follows that the number of employees in tunnelling will rise in the future from the current figure of around 10,000.

The Tiefbau-Berufsgenossenschaft has a legal mandate to train work safety experts. The training of all employees in safety matters is not included in this mandate. Safety experts are usually given largely theoretical instruction at training establishments.

Development of work techniques is usually undertaken in research departments or by companies. Some companies have experimental and test facilities (e.g. shotcrete test sites) which allow "laboratory testing" of new processes and equipment. However, the extent to which real-life con ditions can be simulated is naturally restricted so that realistic trials are generally only possible directly on site on actual projects and are thus subject to time and cost restraints.

The establishment of a Tunnelling Training and Experimental Institute (TTEI) is desirable for further training purposes as well as for technical development. At present there is no facility of this kind in existence specifically for this branch of the construction industry.

This brief description of the actual state of affairs leads directly to the following starting points for improving the situation in tunnelling.

- All employees working in tunnelling should take part in basic or advanced safety training.
- The syllabus of basic and advanced training courses must be greatly enlarged beyond pure work safety aspects in order to meet the objective of "Improvement of working conditions".
- In the Tunnelling Training and Experimental Institute - which is still to be established - training courses can be held and technical developments tested under optimum conditions similar to those encountered in practice. These two partial objectives can be pursued jointly. This will also enable employees to be closely involved in the innovation process as "users" of new techniques.

There is an experimental tunnel installation in Switzerland (Hagebach experimental tunnel) in which developments in equipment, etc. for driving tunnels in rock are tested. This facility is, however, operated by a private firm and used practically exclusively by industry. There is no basic training or further training of employees. Experience acquired there will be used in as far as it is relevant and transferable to individual areas of this project.

6 TRAINING MANDATE OF THE TIEFBAU-BERUFSGENOSSENSCHAFT

Pursuant to the law relating to company doctors, safety engineers and other industrial safety experts of 12.12.1973, the Tiefbau-Berufsgenossenschaft (TBG) in common with all other statutory accident prevention and insurance institutes in industry has a mandate to provide the necessary training for persons entrusted with implementing measures for the protection of employees and accident prevention measures at companies.

For many years the TBG has been training industrial safety experts, safety officers, managers, apprentices and students in addition to members of works councils in seminars lasting several weeks held at their training facilities. It therefore has comprehensive experience of its own in the field of further professional training for this group of safety experts. Doctors from the TBG's own industrial medicine department are employed as teachers for this basic and advanced training.

Since 1972 the Tiefbau-Berufsgenossenschaft has been visiting civil engineering sites throughout Germany with a mobile training centre in which it holds 1 1/2 hour safety training sessions for the employees there. The latest teaching and lecturing techniques are used, e.g. video films of work procedures on construction sites, in order to highlight, for example, accident hazards.

The Tiefbau-Berufsgenossenschaft has collaborated with the STUVA to draw up a booklet "Tunnelling - Safety at Work" as a guide for tunnelling workers. The results of the "Protection against Accidents" research project carried out jointly by the STUVA and TBG have also been incorporated in this booklet.

The Tiefbau-Berufsgenossenschaft is a state-recognized test centre for all civil engineering equipment, including tunnelling and specialized civil engineering plant. The TTEI would be a suitable location for performing equipment safety tests in accordance with the EC Machinery Directive and CEN safety standards. The purpose of the equipment tests is to improve safety at work and thus also to humanize working conditions for employees on underground construction sites.

The TGB and STUVA participate intensively in the work of the "Health and Safety" study group of the International Tunnelling Association (ITA). Know-how gained from this international exchange of experience will also be applied to the project.

7. WORK SCHEDULE

The work of the entire project will be implemented in two main parts which will initially run parallel although they will be correlated:

- Part A: Development and application of programmes to improve professional qualifications in tunnelling
- Part B: Creating the conditions for developing and testing improvements in technical aspects of construction work and for implementing further training measures
- Part C: Merging of the teaching, training and research work at the future TTEI.

The two parts A and B are subdivided into the following work packets (WP):

- WP A1: Definition and quantification of safety and health hazards in tunnelling and specialized civil engineering, definition of target groups and determining how they can be reached
- WP A2: Establishing training concepts and learning programmes

Draft programmes for basic and further training of employees in tunnelling will be developed in this work stage - according to the results of WP A1. In general a multistage programme in the following form would be feasible:

- Instruction and information sessions (individual lectures, short courses)
- Seminars (training and discussion sessions over several days with practical exercises and demonstrations on machines)
- Training courses. These include courses (possibly lasting several weeks) for special professional groups and/or groups of employees. For example,

machine operators (excavators, loaders, tunnelling machines, shotcrete spraying machines) are to be trained on the most up-to-date equipment in realistic operating conditions.

All concepts and documents must be drawn up so that they can reach specific groups and address their different requirements. Three programmes are planned for the following groups:
- Trainees (apprentices, students)
- Tunnel construction workers
- Management personnel (e.g. supervisors)

Programmes of instruction are being developed for the following work areas and professional groups in tunnelling and are being tried out in half-week and one-week courses:
- General aspects of tunnelling
- Safety in tunnelling
- Excavation work in tunnels
- Transport in tunnels
- Maintenance work in tunnels
- Shotcreting
- Blasting work under ground
- Working in compressed air
- Specialized civil engineering

Part of the teaching programme has been completed and tried out and improved in 2 pilot courses. All teaching concepts will be ready by the end of 1994. The courses will be held by the Tiefbau-Berufsgenossenschaft at their own training facilities with the support of experts from the STUVA, from the fields of research and science and from the construction industry. These courses are to be combined subsequently with a practical training programme at the new Tunnelling Training and Experimental Institute TTEI.

Part B: Creating the conditions for developing and testing improvements in technical aspects of construction work and for implementing further training measures

Establishing the Tunnelling Training and Experimental Institute TTEI for basic and further training of tunnel construction workers and for performing research and development functions in tunnelling equipment and process technology.

8 CONCLUSION

Efforts are being made to increase work safety and at the same time reduce health hazards in tunnelling by combining basic and advanced safety training for tunnellers with improvement of conditions at work places in this industry. The creation of a Tunnelling Training and Experimental Institute TTEI is intended to provide the basis for implementing training programmes in line with practical needs to improve the qualifications of employees. The TTEI will also provide facilities for practical research and development work for equipment and work techniques applied in tunnelling.

Further professional training combined with improved working conditions will make work under ground safer and make the profession of tunneller more attractive again.

Tunnelling and Ground Conditions, Abdel Salam (ed.) © *1994 Balkema, Rotterdam, ISBN 90 5410 363 9*

Role of geotechnical monitoring in quality management of tunnelling projects

Abdel-Hady H. Hosny & Fathalla M. El-Nahhas
Ain Shams University, Faculty of Engineering, Abbassia, Cairo, Egypt

ABSTRACT : Tunnelling projects are considered as high-risk projects because of the uncertainties involved in the underground works, and hence need high level of controls. "Feedback Control System" is considered an ideal model for quality management of tunnelling projects which require continuous interaction between design and construction from the planning stage and up to commissioning. Geotechnical monitoring during construction can be implemented as a technical quality element of the feedback control system for major tunnelling projects. This paper describes a flow chart of the feedback control system and identifies the role of geotechnical monitoring within its framework. Some examples of construction monitoring programs utilizing geotechnical instrumentations for tunnelling projects in Egypt are also described.

1. INTRODUCTION

Managing for quality of civil enegineering projects requires adequate control tools to achieve the intended objectives of the project in the best possible way and with the best possible results. Monitoring of different quality elements (technical, financial, ... etc) consitutes an essential aspect within the internal cycle of th control system of any project.

Tunnelling projects are characterized by their high sensitivity to ground conditions. Hence, geotechnical monitoring has been considered as a major item in their technical control program. On the other hand, there have been growing trend and changes of attitudes of decision - makers of tunnelling projects to invest in more geotechnical reconnaissance to minimize the uncertainties (Etienne and Constantin, 1993). The more information available on subsurface ground conditions and quantitative peformance of full-scale underground structures in these conditions to a prospect contractor, the more informed its bid.

In-situ geotechnical monitoring programs can, therefore, play a remarkable role in the management control system; before, during and after construction. Their results can be utilized to minimize uncertainties, to control quality of production, and to guide new plans of action (or modify preliminary ones) as well as to confirm performance and to provide data-base for similar future projects. This paper focuses on this role of geotechnical monitoring within the framework of total quality management of tunnelling projects.

2. FEEDBACK CONTROL SYSTEM

Figure 1 illustrates a flow chart which models the operations, flow of information and decision making process which constitute a feedback control system appropriate for a medium to large sized engineering construction project. Although this flow chart applies equally well in the conventional design and construction process, where the two phases are largely separated, a control system of this type can

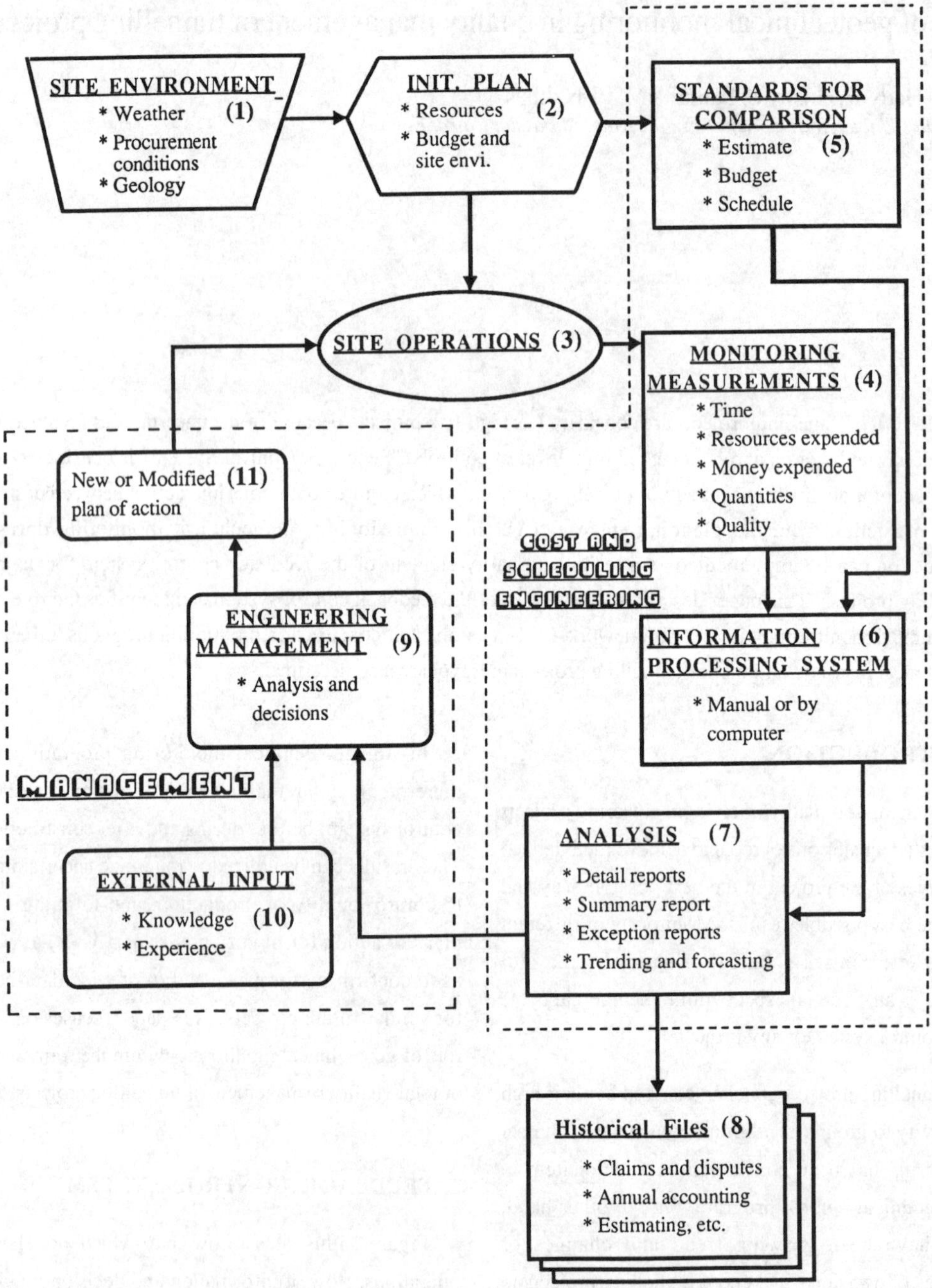

Figure 1 : Flow chart of "Feedback Control System"

have its greatest impact in the professional construction management approach where there is a strong interplay between all aspects of the system concept; design; procurement; construction and commissioning. This approach especially prevalent in large heavy-industrial projects, such as tunnels refineries, power stations ... etc. In projects of this type, engineers involved in design and those in construction interact continuously to optimize the facilities from both points of view.

Site environment (box 1 - Figure 1), is considered regarding major conditions that can affect the initial plan, such as geology, preliminary geotechnical data, weather and procurement delays or even unexpected good facilities on site. The initial plan (box 2) is made with consideration of the available resources, proposed budget and site environment. The initial plan is fed to (box 3) and site operations get underway. This initial plan becomes also reference standards for control purpose (box 5). The site operations generate indicators or progress which are monitored and measured (box 4, quantities in place, elapsed time, money expended, resources consumed and quality) and fed as data into a system (box 6) to produce information for decision makers.

The feedback system operates continuously throughout the life of a project. Associated with it is a feedback time. Ideally, the time through parts 4,5,6 and 7 should be as short as possible so that engineering managers and supervisors can receive quantitative, accurate and up-to-date information in time to make decision and formulate plans of action so as to have maximum impact in controlling those operations which are generating the information in the first place (Hosny, 1992).

3. GEOTECHNICAL MONITORING

Monitoring of tunnels and underground works during construction using in-situ geotechnical instruments plays multiple essential technical roles. During early construction stages, monitoring results are used as factual basis for quantitative assessment of the proposed design and selected construction technology. If needed such field measurements also can be utilized to guide modifications or refinement of the construction procedures, particularly at critical portions of the underground project. On the other hand, systematic field measurements taken at reasonably spaced control sections along the proposed alignment of a project can be utilized to document the overall performance. Such valuable records also can be used in a database for the design of similar future projects (El-Nahhas, 1992).

A successful geotechnical monitoring program can be designed for a tunnelling project according to the following sequence :

(1) detailed examination of surface and subsurface conditions along the proposed route, and suggested methods of construction.

(2) Identification of the main elements of behaviour pattern and expected general response of surrounding soil deposits, groundwater, and the lining system (Figure 2).

(3) Choosing the variables to be monitored and estimating the expected range to be measured.

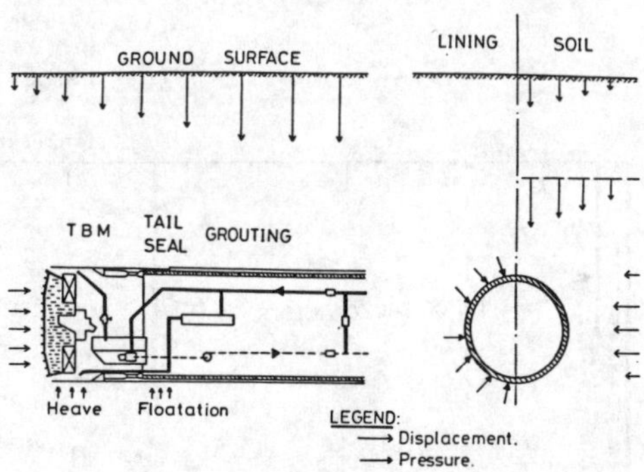

Figure 2 : Behaviour pattern of a soft ground tunnel

(4) Choosing the appropriate type and number of instruments for each variable.

(5) Time schedulling of the monitoring program and its correlation with the project overall time table.

(6) Formation of monitoring teams and setting up of the reporting mechanism of measurements and status of construction progress.

(7) Immediate evaluation of the measurements and fast interpretation of the data.

(8) Formluating, in advance, the possible course of actions to be taken when measurements are significantly different from predicated values.

(9) Detalied analysis of compiled measurements to document overall performance.

Several geotechnical monitoring programs have been recently implemented during construction of tunnelling projects in Egypt. Figure 3 and 4 show layouts of the instrumentation arrays for the regional line of Cairo Metro project and the spine tunnel of Cairo Wastewater project respectively. Similar programs have been proposed for the tunnels and subway stations of the second line of Cairo Metro and for the new tunnels under Suez Canal for El-Salam Syphon as well as for the rehabilitation works of Ahmed Hamdi tunnel..

4. GEOTECHNICAL MONITORING IN CONTROL SYSTEM

Quality management of tunnelling projects utilizing the feedback control system is outlined in Figure 1. It is interesting to compare the nine steps of geotechnical monitoring listed above with the flow chart given in Figure 1. It is observed that these steps can be incorporated within the different boxed of the feedback control system. In particular, the field data

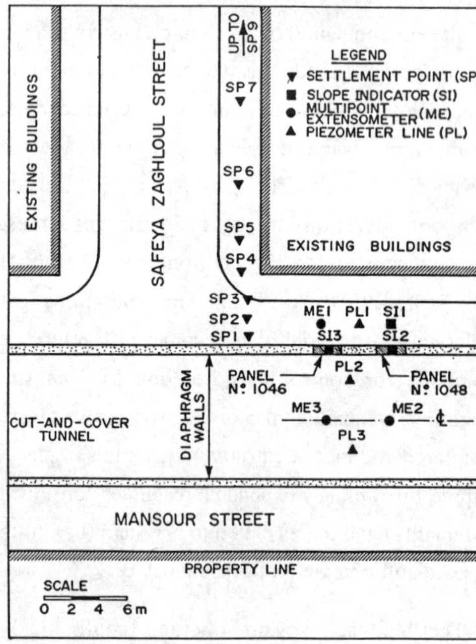

Figure 3 : Instrumented section of Regional Line of Cairo Metro

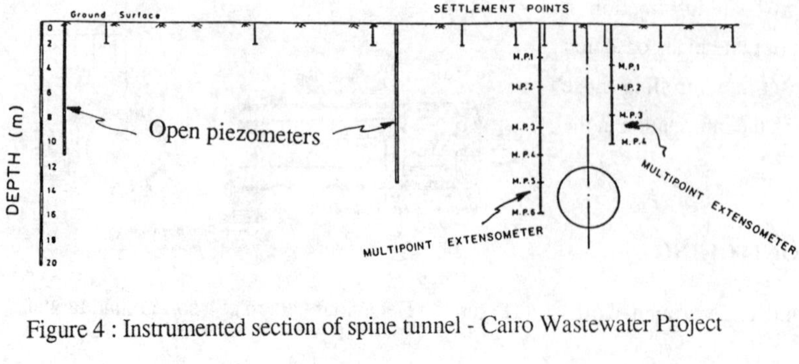

Figure 4 : Instrumented section of spine tunnel - Cairo Wastewater Project

compiled during different stages of construction is considered as a main item of box (4) which is processed and analyzed (box 6 and 7). The outcome of these factual measurements is one of the technical elements employed by management (box 9) to decide on new or modified construction procedures (box 11 and 3).

Therefore, geotechnical monitoring of tunnelling projects is regarded as one of the valuable technical elements within the "Feedback Control System", particularly within the internal loop including boxes 3,4,6,7,9,11 and 3 as shown on Figure 1.

As a matter of fact, there is also a new trend for relying on in-situ instruments to assess durability and long term behaviour of tunnel lining (Schiessl and Raupach, 1992).

5. CONCLUSIONS

In-situ geotechnical instrumentation programs are qualified to play a major role in the quality management of tunnelling projects. Quantitative results of these programs can be used to guide the continuous interaction between design and construction within the "Feedback Control System". The role of geotechnical monitoring of tunnelling projects in Egypt have been expanding from pure research work to fulfillment of an active element within their quality management.

REFERENCES

El-Nahhas, F. 1992. Construction monitoring of urban tunnels and subway stations. Tunnelling and Underground Space Technology, Vol. 7, No. 4, pp 425-439.

Etienne, M. and Constantin, B. 1993. Controlling and optimizing highway tunnel costs. In Proc. of Int. Congress on Options for Tunnelling Amsterdam, pp. 817-827.

Hosny, A.H. 1992. Control techniques between theory and practice for tunnelling projects. In Proc. of Int. Symposium on Current Experiences in Tunnelling, Cairo, pp 399-423.

Schiessl, P. and Raupach, M. 1992. Monitoring system for the corrosion risk of steel in concrete structures. Concrete International, July, pp. 52-55.

Tunnelling and Ground Conditions, Abdel Salam (ed.) © 1994 Balkema, Rotterdam, ISBN 90 5410 363 9

Rises in subway construction costs and specific examples of reducing the costs

Jiro Inose & Yukihiko Ishikawa
Teito Rapid Transit Authority

ABSTRACT: Subways are the main means of transportation in large cities. In Tokyo, construction work on four new subway lines is now in progress and it is hoped that the construction of some other planned lines will start as soon as possible. On the other hand, subway construction costs are rising rapidly and there arise the problems of how to secure construction funds and the general profitability of subway operation. These problems are the factors that have delayed the construction of new subway lines.

This paper reports the results of our studies and analyses of the trend towards rises in construction costs of the lines constructed by the Teito Rapid Transit Authority (TRTA). It also deals with the shield tunneling methods, which are considered to have economic advantages and as a result have been recently used more, and describes the up-to-date shield tunneling methods the TRTA has recently implemented or is planning to implement to save on the construction costs.

1 CHANGE IN SUBWAY CONSTRUCTION COSTS

Since the end of World War II, the TRTA has completed and opened to traffic seven new subway lines with a total length of 152.5 km. It is now continuing the construction of a 14.6-km section of a new line. At present, the TRTA's total operating length of subway lines is 162.2 km, including the Ginza Line which was transferred to the TRTA by a private owner before the end of the war.

Table 1 Construction period and length by line

Line	Construction period	Length
Marunouchi Line	1951–1962	27.4 km
Hibiya Line	1959–1964	21.1 km
Tozai Line	1962–1969	31.8 km
Chiyoda Line	1969–1979	25.1 km
Yurakucho Line	1970–1988	29.4 km
Hanzomon Line	1973–1990	10.9 km
Nanboku Line	1986–	21.4 km

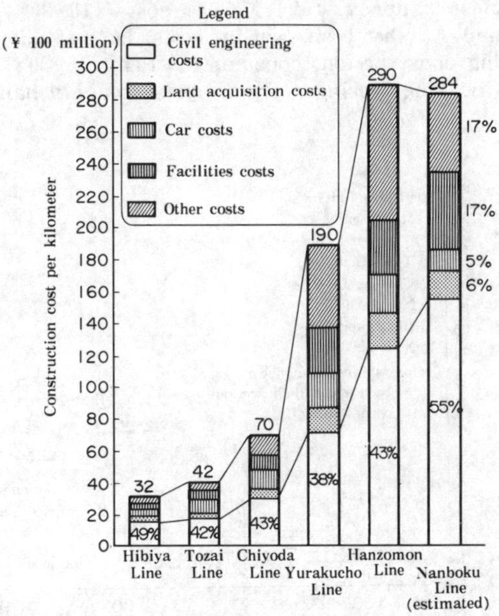

Fig. 1 Construction cost of subways by line

Of these lines, a 6.8–km section of the Nanboku Line was opened to traffic in 1991, and the remaining sections are now being constructed.

Fig. 1 shows the changes in the construction costs of these lines. Since these lines differ from one another in construction period, construction method, scale of facilities, geological condition, urbanization and land condition, we cannot make the direct comparisons. But the Fig.–1 will be sufficient for us to observe general cost trend changes. According to the figure, while the construction cost of the Hibiya Line which started operation about 30 years ago was ¥3.2 billion per kilometer, that of the Hanzomon Line which was completed in 1990 was ¥29.0 billion and that of the Nanboku Line scheduled for completion in 1997 is estimated at ¥28.4 billion. The cost of the two newer lines is 9.1 and 8.9 times, respectively, as much as that of the Hibiya Line, indicating substantial increases in costs.

According to the purpose of construction, subway construction costs can roughly be classified into ① land acquisition costs (approx. 10%), ② civil engineering costs (40–55%), ③ facilities' construction costs (railroads, finishing work, electric facilities, signaling and communications facilities, train shed facilities) (approx. 15%), ④ car costs (10–20%), and ⑤ other costs (sundry costs, interest during construction, consumption tax) (15–30%). Civil engineering costs comprise roughly half of the total construction costs, and their rising trend is similar to that of construction costs as a whole. Thus, it is clear that if it is possible to reduce civil engineering costs, that will greatly contribute to savings in subway construction costs.

2 ANALYSIS OF FACTORS CONSTIBUTING TO RISES IN SUBWAY CONSTRUCTION COSTS

The factors responsible for increasing subway construction costs can roughly be divided into ① price increases and ② increase in construction work. We studied the relationship between construction costs and prices by comparing the construction costs per kilometer of the TRTA's subway lines using the figures of the Hibiya Line as indexes, and the consumer price index in the wards of Tokyo (1955 = 1) (Fig. 2).

According to Fig. 2, while subway construction costs have generally been rising in step with price increases, those of the recent three lines are increasing at far higher rates than the consumer price index. Then we compared construction costs with changes in the wages of construction workers, which forms the greater portion of labor costs. The wages began to rise sharply around 1970 and this coincides with the sudden increase in construction costs. But the increase rate of the wages is much greater than

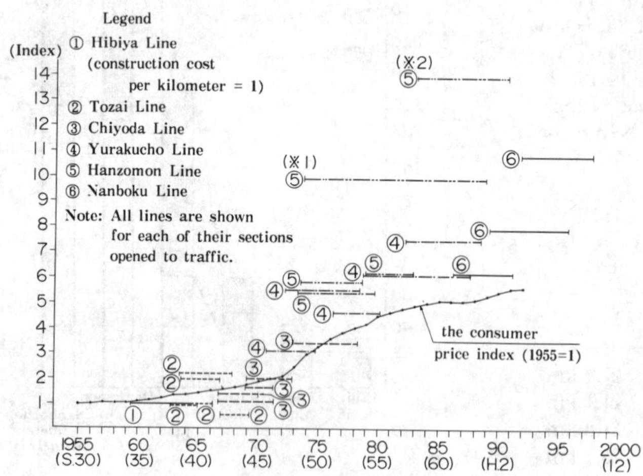

⑤•1) 4.4 km: Because this section passes below privately owned land, land acquisition problems arose and the construction work was suspended for a long period of time. This resulted in increase in construction costs and interest during construction. It took 15 years and three months from thestart of construction work to the opening to traffic.

⑤•2) 0.9 km: Because this section also passes below private owned land, special construction methods, such as underpinning of high-rise buildings, causing rises in construction costs; land acquisition costs increased, too.

Fig. 2 Comparison of subway construction cost by line and prices

that of construction costs, the former becoming more than three times as high as the latter in the 1990s (Fig. 3). As noted above, civil engineering costs account for 40 to 55 % of the subway construction costs, and labor costs make

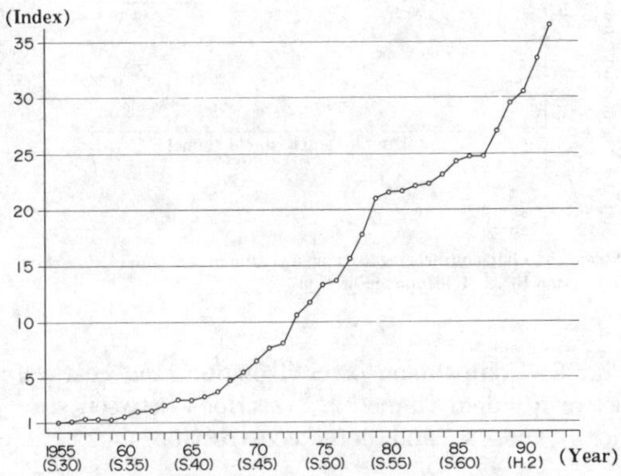

Fig. 3 Changes in the wages of construction workers (1955 = 1)

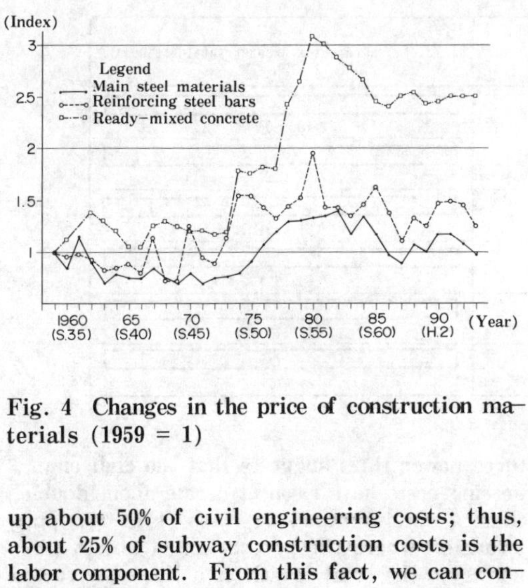

Fig. 4 Changes in the price of construction materials (1959 = 1)

up about 50% of civil engineering costs; thus, about 25% of subway construction costs is the labor component. From this fact, we can conclude that rises in labor costs affect increasing subway construction costs. Since no noticeable increase is observed in the price of construction materials, we can consider that this has not had a great effect on subway construction costs (Fig. 4).

Another factor contributing to increased subway construction costs is an increase in construction work. A typical cause is the deepening of tunnels. In many cases, recent subways have to be built under existing underground facilities: underground transportation facilities such as existing subways and expressways; increasingly larger and deeper underground facilities, such as water supply, sewerage systems, electric supply and telephone lines; and deep foundations of buildings, bank protections and bridges. Because of this, newer lines must be constructed deeper underground (Fig. 5). While the average excavation depth of the Hibiya Line is 12.6 meters, that of the Hanzomon and Nanboku Lines is roughly twice that.

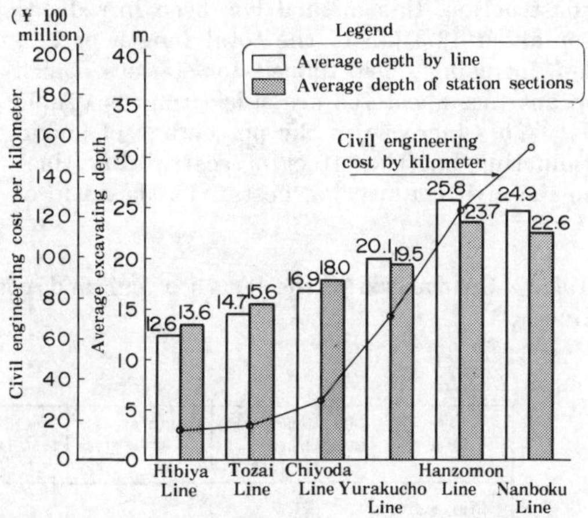

Fig. 5 Relationship between excavating depth and civil engineering costs by line

As Fig. 6 shows, civil engineering costs increase in direct proportion to the depth of excavation in the case of cut-and-cover construction. Moreover, there is the need to underpin other facilities and carry out protective works. As a result, the increase in excavation depth has had a serious influence on rising civil engineering costs.

3 CONSTRUCTION WORK COST OF SHIELD TUNNELS

In the present subway construction, the TRTA generally adopts the cut-and-cover method for stations and shield tunneling methods for sections between stations. The shield tunneling method is better than the cut-and-cover method in that it does not obstruct the road transport and causes fewer environmental problems during construction. The shield tunneling method is especially suited to subway construction in deeper places; while the construction work costs of the cut-and-cover method increase in proportion to the excavation depth, those of shield tunneling method are scarcely affected by depth. Thus, this shield tunneling method is a very good option for the recent subway projects which have to be carried out at increasingly greater depths (Fig. 6).

The TRTA began to employ the shield tunneling method for the Tozai Line, which was constructed in the 1960s. After that, it has increased the ratio of shield tunneling for new lines; at last in the case of the Nanboku Line now under construction, this method has been introduced for about 16.3 km of the total length of 21.4 km, including shield tunnels for stations, which means that about 76% are shield tunnels (Table 1). This increase in the percentage of shield tunneling has the effect of restraining rises in the civil engineering costs; as the ratio of

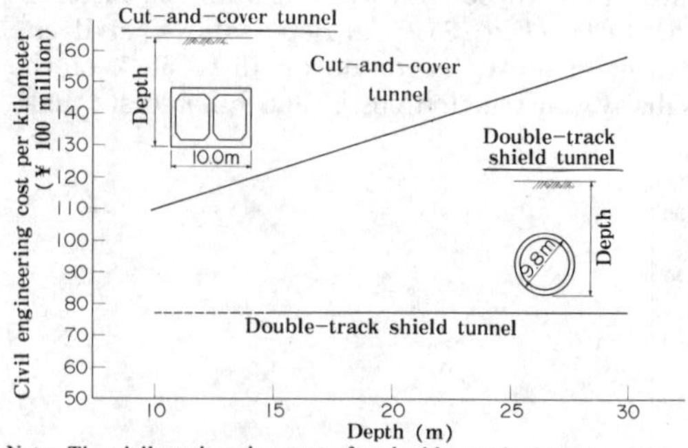

Note: The civil engineering cost of a double-track shield tunnel is one for a 1 kilometer section.

Fig. 6 Comparison of civil engineering costs of between shield tunneling (sections between stations) and cut-and-cover construction

Table 2 Comparison of the length of cut-and-cover construction and shield tunneling sections by line

Line	Total length (km) (underground length)	Length of cut-and-cover tunnels (km)	Length of shield tunnels (km)	Ratio of length:cut-and-cover shield tunneling
Hibiya Line	21.1 (17.3)	17.3	0	(100 : 0)
Tozai Line	31.8 (17.1)	15.4	1.7	(90 : 10)
Chiyoda Line	23.0 (19.1)	15.4	3.7	(81 : 19)
Yurakucho Line	29.4 (26.0)	17.0	9.0	(65 : 35)
Hanzomon Line	10.9 (10.9)	3.3	7.6	(30 : 70)
Nanboku Line	21.4 (21.4)	5.1	16.3	(24 : 76)

shield tunneling rises, the shield tunneling costs will naturally influence the civil engineering costs to a great extent.

Figs. 7 and 8 show the changes in shield tunneling costs as well as the trends of various costs consisting of shield tunneling costs and their ratios.

The comparison of civil engineering costs and shield tunneling costs, which are part of civil engineering costs, between the Chiyoda Line and three newer lines suggests that the civil engineering costs have risen at a rate about double that of shield tunneling costs (Fig. 7). In a situation where the percentage of shield tunneling to the total length of a line constructed is rising, this fact shows that the increase in civil engineering costs largely reflects rapid rises in works' costs other than shield tunneling costs, mainly rises in cut-and-cover construction work costs. Thus, we can conclude that the first point we should do to cut civil engineering costs is to reduce the station construction costs, which are mostly cut-and-cover construction work costs.

In the 1980s, the TRTA began to adopt the slurry shield method, switching from the traditional pneumatic shield method, and the slurry shield method has been introduced to all sections of the Nanboku Line. Even in this situation, the shield tunneling costs of the Nanboku Line are higher by only about 15% than those of the Hanzomon Line, compared to the total civil engineering costs, which have increased about 25%. This suggests that the greater use of the shield tunneling method has been very effective in checking increases in civil engineering costs. Because of this, efforts to cut shield tunneling costs are an important step towards making savings in the total civil engineering costs.

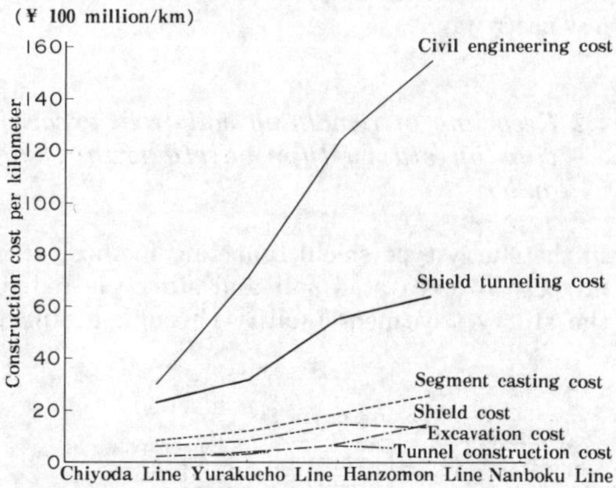

Fig. 7 Comparison of rises in civil engineering cost and in shield tunneling cost(per kilometer)

According to the purpose of construction works, the shield tunneling costs are divided into ① segment casting costs, ② shield machine costs, ③ excavation costs and ④ tunnel construction costs. In the Nanboku Line, partly because of the introduction of the slurry shield method, shield machine costs and segment casting costs have increased and make up 64% of the total shield tunneling costs. Thus, it will be an important point to control rises in these two costs to slash shield tunneling costs.

The TRTA is studying strategies for cutting civil engineering costs from the above-mentioned viewpoints, and is trying to improve traditional design and construction methods and to develop new methods. The next section will outline several measures we are now adopting or planning for shield tunneling work.

4 EXAMPLES OF MEASURES TO SAVE CONSTRUCTION WORK COSTS

4.1 *Development of the three-centered slurry type station shield tunneling method: long-distance application of a shield machine*

At present, shield machine is fabricated for each section of a stations and sections between

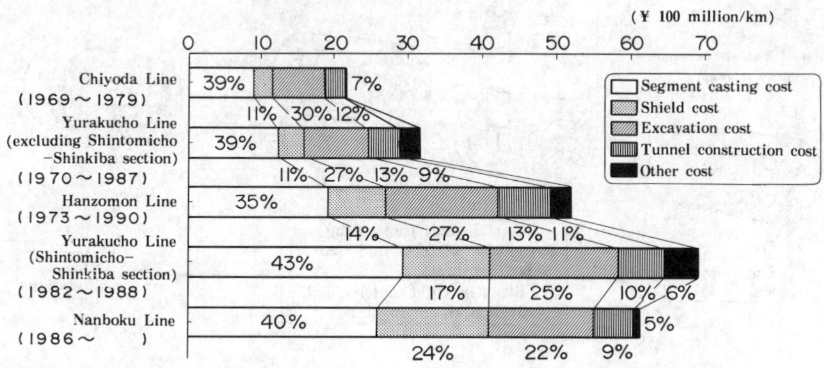

Fig. 8 Changes in shield tunneling cost by line

stations. An ordinary shield machine is capable of excavating a section of 2 km or more. But stations are built about every one kilometer on average, and so the ordinary usage of the shield machine is uneconomical.

To deal with this problem, the TRTA devised a new method; this method is named the "three-centered slurry type station shield tunneling method" and has the following characteristics: ① An ordinary slurry type double-track shield machine (main shield machine) is used to excavate a tunnel between stations; ② at the station, a smaller shield machine is mounted on both sides of the main shield machine in a shaft constructed in advance. These smaller shield machine will make platforms of the station; ③ the station section is excavated by a three-centered station shield machine;④ after the station tunnel is completed, the two smaller shield machines are removed in the shaft; and ⑤ the main shield machine is used again to excavate the next section between stations (Fig. 9).

This method has the following economic effects:

1. To construct a subway line, the traditional tunneling method needs two double-track shield machines and one station shield machine (or the construction of the station by the cut-and-cover method). The new method requires only one double-track shield machine and two smaller shield machines (costs of one smaller shield machine are roughly equal to the costs of one single-track shield machine). The double-track shield machine is capable of excavating a two-kilometer or longer section.

2. The new method requires only one supporting station at the shield starting base where a soil and sand treatment plant, a segment storing yard and other facilities are established.

3. This method using station shield machines is especially economical in the case where the station section is very deep and there is a need to protect important structures at ground level which needs high construction costs.

The TRTA plans to introduce this new method in the construction of the Nanboku Line which is now under way.

4.2 *Recycling of remaining soil from excavation by slurry type shield trunneling method*

In the slurry type shield tunneling method, the mixture of excavated soil and slurry is fed to the slurry treatment facility through a slurry

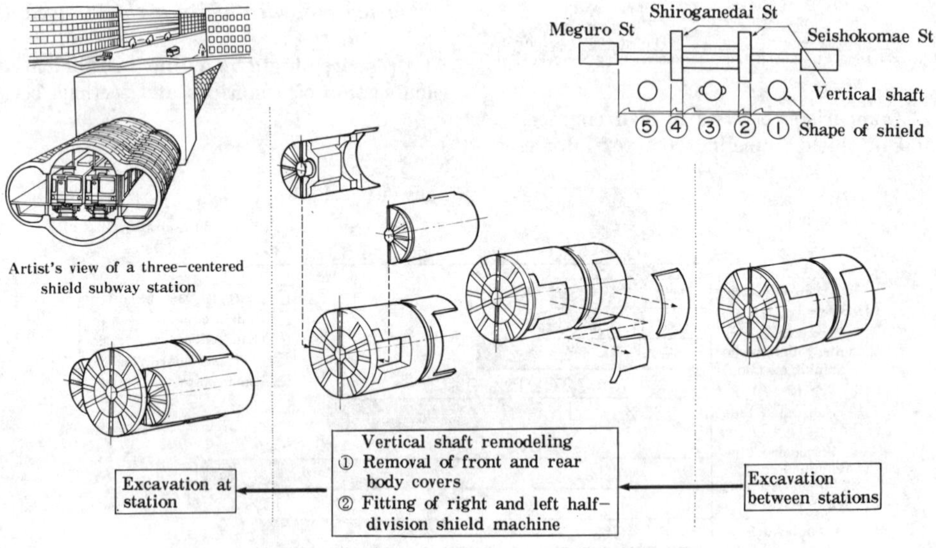

Fig. 9 Procedure of the Method of the Three-Centered Section Slurry Type Shielding of Tunnel Sections at Stations

drainage pipe, where it is separated into slurry and soil. During primary treatment, sandy soil, such as sand and gravel is discharged, and secondary treatment discharges cohesive soil in a cake state from the excess slurry.

The remaining soil has usually been dumped to reclaim the land from the sea or develop the building land, but these locations have decreased recently. The pollution of the environment by the disposal of excess soil is also becoming a serious issue. As a result, the dumping grounds are becoming farther and harder to get, causing the costs involved to rise rapidly. Most of the excess soil generated from slurry type shields is regarded as industrial waste, and the number of dumping sites are limited, which makes the cost even higher.

The TRTA has developed a new system for dealing with this exess soil. In this method, the soil from slurry type shield is improved and recycled as a construction material, thus reducing the cost of slurry treatment and construction materials. The flowchart of this method is shown in Fig. 10.

(1) Backfilling of improved sandy soil

After the primary treatment, sandy soil is improved by adding a limestone-based solidifying agent. This improved soil is used to backfill cut-and-cover tunnels, mainly in the sections where no facilities are buried.

(2) Backfilling of treated slurry

In the case where the excavated soil is mainly of a cohesive type, a large quantity of excess slurry is generated. A cement-based solidifying agent is added to this excess slurry, which is used in a fluid state for backfilling purposes. When it is used around a buried pipe, this improved slurry completely fills any spaces around it and helps prevent the damage to the pipe and collapse of the road surface resulting from the gaps around the pipe. This process saves the cost of the secondary treatment of excess slurry.

(3) Reuse of remaining soil as invert material for shield tunnels

The inverts of the shield tunnels form the foundation of the rail structure that supports the load of trains via the rail bed, and ready mixed concrete is now used to make inverts.

We devised a method of using the mixture of sandy soil, and excess slurry or water with a solidifying agent instead of ready mixed con-

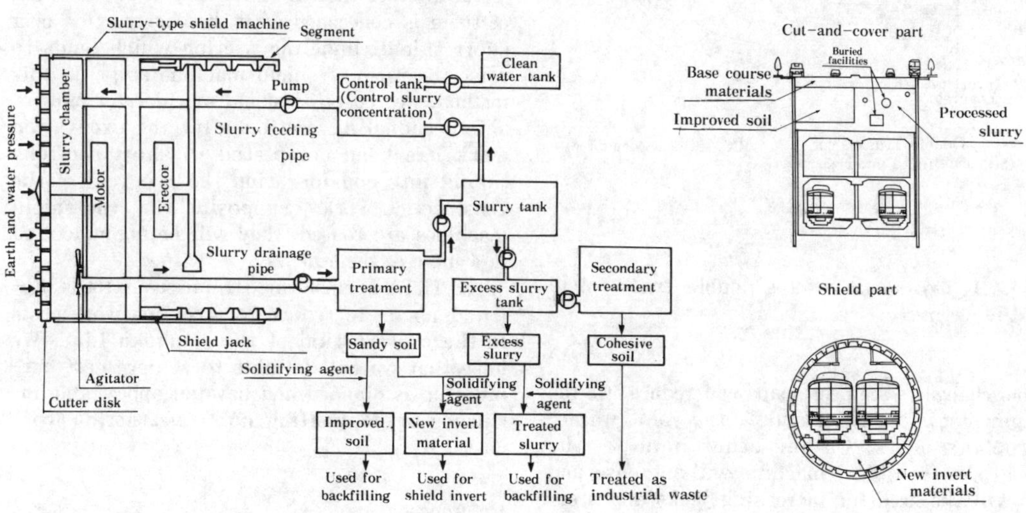

Fig. 10 Recycling methods of remaining shield soil

crete. This material is transported in a fluid condition and supplied to the invert section.

The above methods of recycling the remaining soil are now used in the construction of the Nanboku Line.

4.3 *Use of larger shield segments*

One method of saving the segment casting costs is to enlarge the segment size–increase in segment width and fewer number of segments–. In a traditional double-track shield, eight RC segments, with an outer diameter of 9800 mm and width of 1000 mm, are used. As a result of our studies on larger segments, we concluded that we would be able to achieve a larger size taking into consideration our facilities at the segment casting plant, lifting capacity of the shield machines, enduring capacity of the segments and transportation of the segments. Thus, we decided to increase segment width to 1200 mm without changing the number of ring segments (Fig. 11).

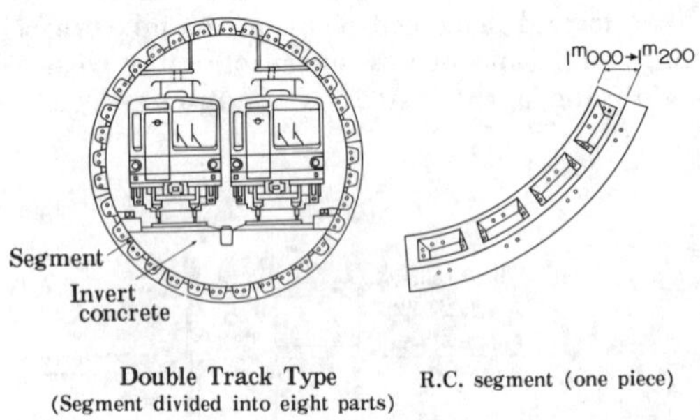

Fig. 11 Cross section of a double-track shield and RC segment

This change in segment shape will reduce ① the segment casting costs and ② the joint waterproofing costs. On the other hand,③ the length of the shield machine will increase and so will the need for increasing jacks and erectors.

But when we combine these three factors, we will be able to achieve a cost reduction of about 4% compared with the traditional method. We will also expedite excavation by over 7%, which will also lead to a shorter construction period and a further cost saving. We are using these wider segments in the construction of the Nanboku Line.

5 CONCLUSION

This paper discussed rising subway construction costs and ways to reduce these costs, placing a major emphasis on civil engineering costs. We found that the major part of cost increases has occurred in cut-and-cover construction, and cost reduction should be focused on this factor. But the construction by the cut-and-cover method has recently had to cope with the deepening of construction sites, environmental conservation, safety measures and the increase in special works, which leads to the increased amounts of construction work. Because of this, it is very hard to reduce the cut-and-cover construction work costs merely through technical research and development. Thus, in order to cut construction work costs, the strategies of having beneficiaries in the areas along the planned line bear part of the construction work costs – for example, having regional developers bear part of the costs of a station, entrances, exits and passages or having the owner of a building construct the entrance and exit – are one of possible approaches.

We are also studying several cost saving methods other than those described above. One of them is concerned with the construction of a short shield tunneling section which tends to raise the ratio of shield machine costs. In this method, we reuse the shield machine for another shield tunneling section after the excavation work has been completed in short sections taking into consideration the progress of the construction work. Supposing that the shield machines are reused, they will be manufactured in a more economical way.

The TRTA is employing the various methods outlined above to reduce construction work costs in the construction of the Nanboku Line. We hope that we will be able to achieve cost curtailment as planned and have an appreciable influence on controlling our construction work costs.

References:

Hirato Irie, "Substantiative Study of Subway Line Improvement Plans in Large Cities", 1990.

Tokyo Statistical Association, "Tokyo Statistical Yearbook", 1991.

Tunnelling and Ground Conditions, Abdel Salam (ed.) © 1994 Balkema, Rotterdam, ISBN 90 5410 363 9

Rehabilitation of Ahmed Hamdi Tunnel under the Suez Canal – Part 1: Study and design

Takayoshi Otsuka
Japan Shield Engineering Co., Ltd, Japan

Isis A. Kamel
Engineering Department, Suez Canal Authority, Egypt

ABSTRACT: This paper deals with the investigation of a tunnel damaged by salt water as well as the design for its rehabilitation. Based on the study and analysis of the causes of the deterioration of the existing tunnel, a design for a new tunnel lining structure with fleece and waterproofing membrane is proposed with the aim of avoiding future deterioration and obtaining a maintenance free tunnel.

1 INTRODUCTION

Ahmed Hamdi Tunnel, located about 17km to the north of Suez City, is the only tunnel connecting the Sinai Peninsula with Egypt and plays an important role in the development of the peninsula.

The construction of this tunnel completed in 1983 with the Open Shield Tunnelling Method, is no longer reliable after only 10 years due to deterioration of the Reinforced Concrete Segment (hereafter RC Segment) caused by saline water which is supposed to have been seeping since immediately after completion.

Results of investigation and analysis on the causes of damage and design study for the rehabilitation of the tunnel is reported hereafter.

2 OUTLINE OF EXISTING TUNNEL

The tunnel is 1640m long and 11.4m in its diameter as shown in Fig.1. The ground condition along its route is stable with unconfined compression strength of qu=10-20kgf/cm^2 and is considered a impermeable layer with coefficient of permeability of k=10^{-5}-10^{-7} cm/sec. Each section of the tunnel lining is of 60cm-thick RC Segments and consists of 16 segments including one key-segment. The segments are set in place and joined together by straight joint method for longitudinal of tunnel.

3 MECHANISM OF DETERIORATION

According to the site survey and analysis of the findings, deterioration mechanism of the tunnel can be explained as follows.

(1) Based on the seepage of sea water found at a side slope of the tunnel entrance, it is assumed most likely that sea water is flowing through a decomposed mudstone layer of high permeability. This was confirmed by fibre-scope observation through grout-holes in segments with show formation of water-course around segment rings. This is probably due to incomplete backfilling mainly of pea-gravel without sufficient grouting of cement milk.

(2) Seeping water reached inside the tunnel because of the opening of the joints between the segments, spreading over the segment surfaces, road deck and ceiling panels. It is believed that such opening of segment joints resulted from installation of bearing pads for making the flexible segmented ring.

As the result, compressibility of the foam-gasket was reduced, decreasing water-tightness substantially. Furthermore, division of a segement ring into as many as 16 pieces increased the number of joints and increased the water coming through.

The amount of seeping water is **45±5m^3/day** this volume has not changed for these ten years.

(3) As Photo-1 and Fig.2 shows, seeping saline water has penetrated into segments, road deck and ceiling panel, corroding the reinforcing steel bar and causing cracks and tearing off of covering concrete. Compressive strength of the the concrete fc is 415-650kg•f/cm^2 and its elastic modulus Ec is 3.2-3.8 x 10^5kg•f/cm^2.

Carbonation depth of the segment concrete is 5-11mm. The reason the reinforcing steel bars were corroded under these conditions is that:

- Actual concrete covering depth varied from 0 to 25mm through the design specified 25mm covering.
- Chloride was accumulated at the depth of 2 to 3cm as shown in Photo-2.

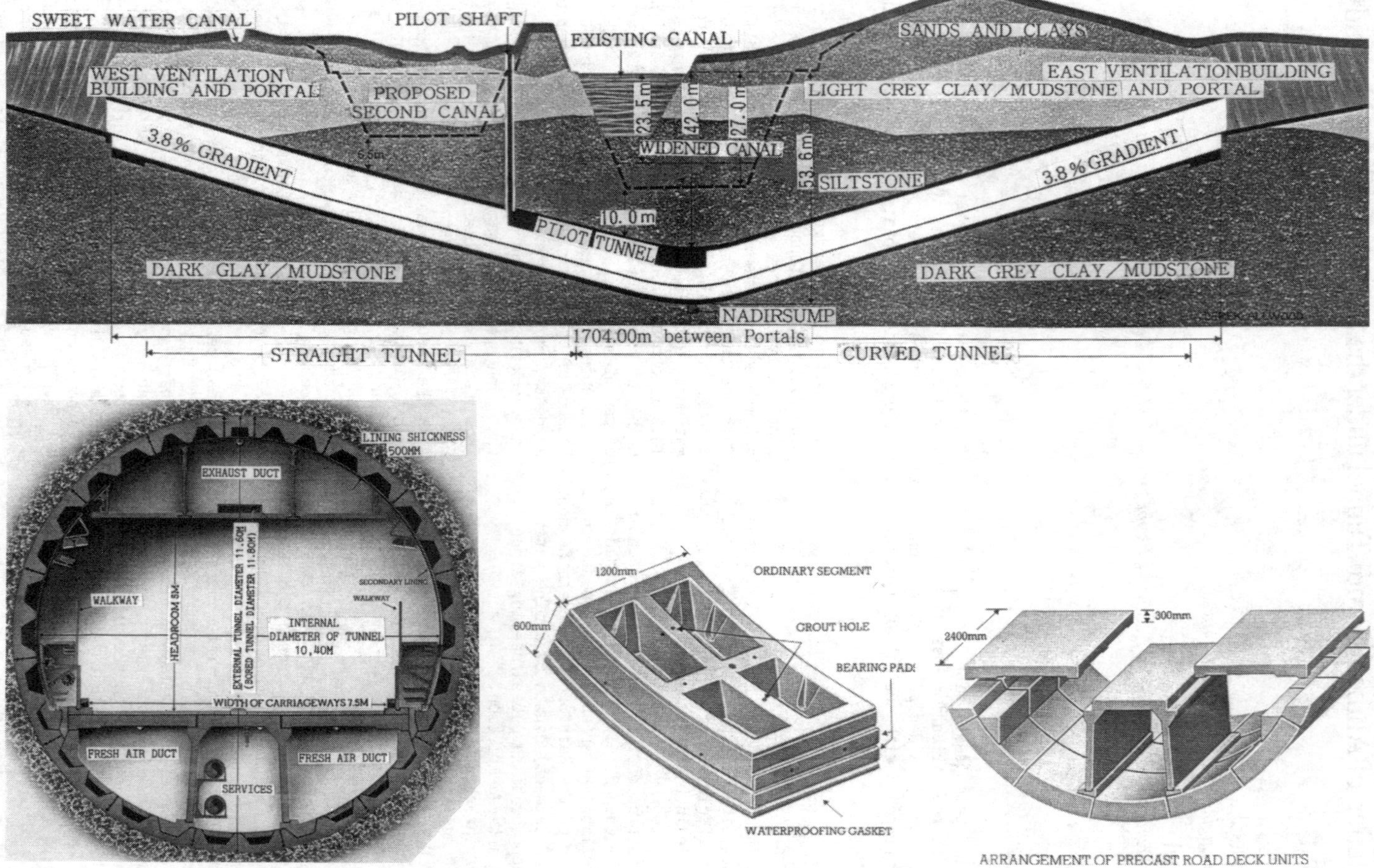

Fig.1 Outline of Existing Tunnel

Photo-1 Deterioration of Segment

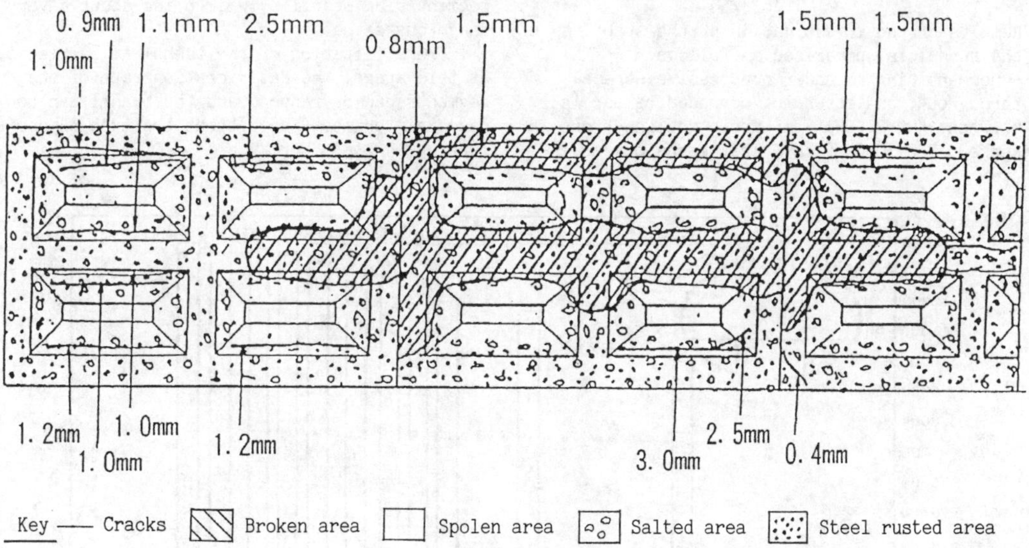

Fig. 2 Sketch of Segment Showing Extent of Deterloration and Crack Width

- Drying by ventilating air might have had an adverse effect (by making cracks in the concrete).

On the other hand, high inclusion of chloride in the original concrete may have been the cause of re-bar corrosion as well as the cause of low density solid cement by the separation of calcium (Ca) and the yielding of calcium hydroxide ($Ca(OH)_2$) in large quantities. In addition, dolomitic limestone used as coarse aggregates for the concrete could also be the cause by yielding $Ca(OH)_2$.

Concrete with low density solid cement facilitates permeation of water and air.

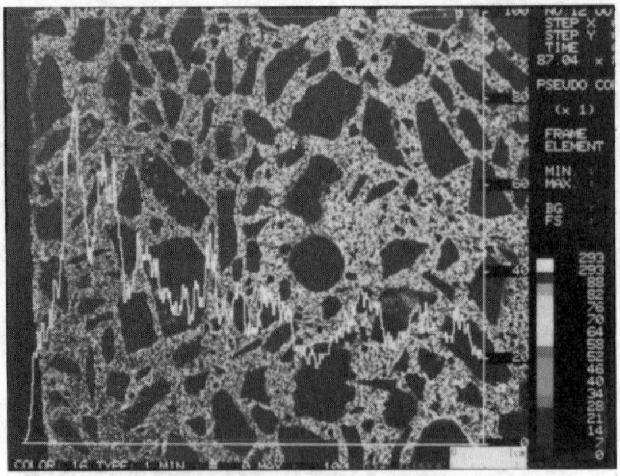

Photo-2 Chlorine (CL) Distribution of Segment obtained from Electron Probe Micro Analyzer

It was observed that saline water that had permeated into the segments was concentrated by evaporation of the water, and had crystallized into solid salt at its limit of solubility, thus causing splitting off of the concrete.

4 NECESSITY OF REHABILITATION

Results of the investigation of the stability of the tunnel is summarized as follows.

- Segment ring is under compressive stress throughout, and stress of corroded re-bar is between 1400-2100kgf/cm² where design stress is between 700-1000kgf/cm². That means stress is some of the re-bars is close to the yielding stress of 2400kgf/cm².
- The tunnel is in stable condition under integrity with in-situ mudstone. However, it may become unstable if the K-segments fall, which is quite possible because they are not wedge-shaped, or if the surrounding rocks are weakened.
- Deterioration of the segments shown in Fig.2 progressed year by year as shown in Fig.3. Splitting off of the segment-rib concrete and disappearance of re-bars are remarkable and in a dangerous condition. The segments supporting the road deck are also severely damaged with covering concrete falling off everywhere as shown in Fig.4.

If deterioration proceeds further to the inner re-bar corrosion, the road deck could collapse.

The above observations lead to the following technical assessment which strongly recommends implementation of thorough rehabilitation of the tunnel.

- Although the tunnel is presently stable due to integration of the segments and in-situ mudstone, further progress of segment deterioration may cause collapse of the tunnel if it is left unrepaired.
- If canal expansion work is carried out without rehabilitation of the tunnel, possible loss of segment ring stability will bring about a very dangerous situation.
- If rehabilitation of the tunnel is carried out at this stage, the deterioration rate of the segments can be reduced and the tunnel can be made maintenance free without too much expenditure.

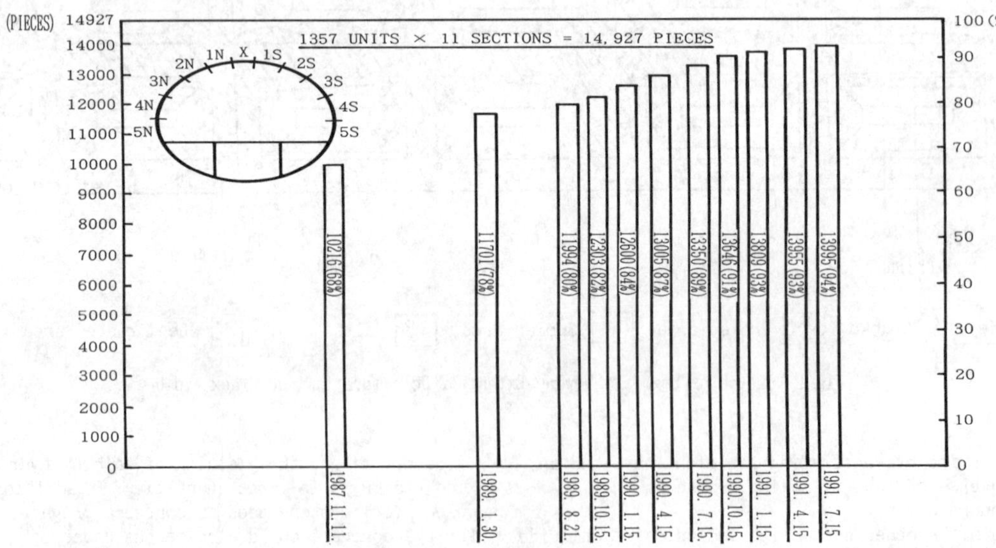

Fig.3 Yearly Progress of Deterioration of Segment Rings (SCA data)

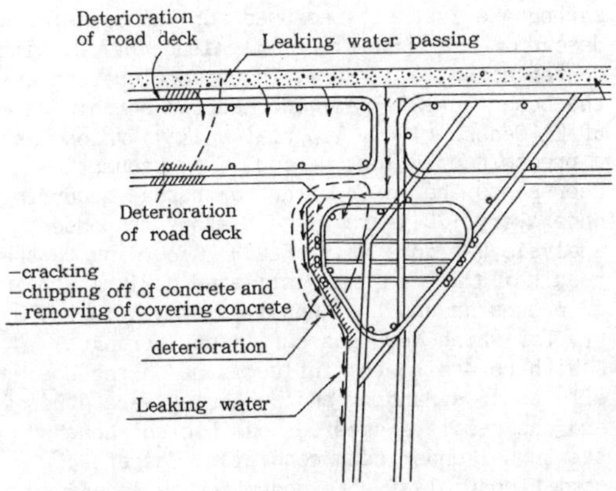

Fig.4 Deterioration of Road Deck Support

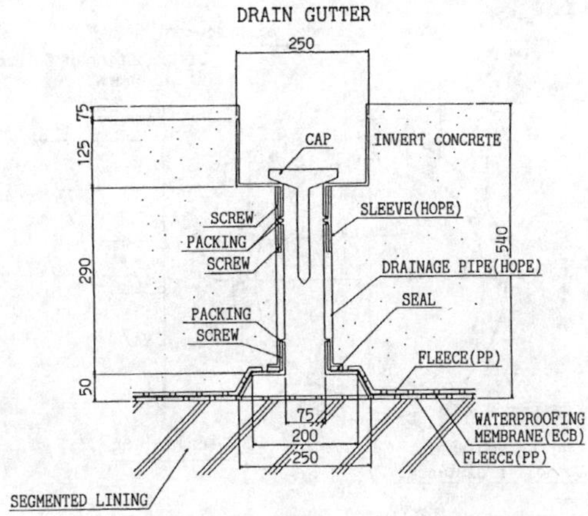

Fig.5 Drainage System of Leaking Water

5 OUTLINE OF THE REHABILITATION STUDY

5.1 Design policy and selection of construction method for the rehabilitation design policy for the rehabilitation work are summarized as follows:

- Present functional capacity of the tunnel (Construction gauge, height: 5.0m, width: 7.5m) shall be secured.
- New Reinforced Concrete Lining (hereafter NRCL) will be designed for the tunnel structure taking the following conditions into consideration with the target of avoiding similar problems of deterioration after completion of the rehabilitation work.
- Predicted external load will be borne by NRCL alone.
- Concrete covering depth of re-bars will be secured.
- NRCL will be free from seeping water and oxygen exposure in order to provide protection against deterioration.
- Water drainage system will drain water by the fleece as shown in Fig.5, and the waterproofing membrane (hereafter WPM) will guide the water to the lower part of the tunnel.
- Coarse aggregates of river gravel with low lime content and fine aggregates with low chloride content will be selected as the materials for NRCL.
- Structural design of NRCL will take the expansion plan of Suez Canal into consideration.
- The tunnel will be maintenance free after it is rehabilitated.

Although conventional repair methods suggest application of mortar or shotcrete after removal of the deteriorated concrete, we recommend installation of NRCL inside the segment ring and renewal of the road deck as the most suitable rehabilitation method satisfying the policy as mentioned above.

5.2 Study on NRCL

5.2.1 Design Load and Concept

Factors studied in the structural design of NRCL include: a) earth pressure, b) water pressure, c) self weight, d) buoyancy, e) influence of canal expansion work, f) seismic load, and g) influence by ships.

The most influential of these is e), and the design concept for it is described below.

- Temporary increase of water pressure due to an increase of canal depth is taken into consideration. Such a situation can be expected because of the existence of the impermeable layer in the ground above the tunnel.
- Study of buoyancy is required because minimum thickness of the earth cover will become as small as 10.5m, almost equal to the tunnel diameter, by the expansion of the canal. In that study, the ground above the tunnel was studied and checked for safety relative to every method of construction, with the emphasis on assessment of shear-resistance mechanism, cohesion and internal friction angle of the ground.
- Rebound that might arise from canal expansion was estimated by the quasi-linear method, the multi-dimension method (Skemyton Bjerrum) and by FEM. giving the design value of 200-300mm at the bottom of the canal. The effect of rebound on the NRCL must be studied in two directions, longitudinal and transverse.

The design model assumed transmittance of ground deformation according to the following sequence: Segment Lining - Fleece - W.P.M. - NRCL. The design model for these are shown in Figs.6.

Only 16 earthquake with magnitude M=5.0 or more have taken place in Egypt near the tunnel location during the past 90 years, according to the Meteorological Agency of Japan. However,

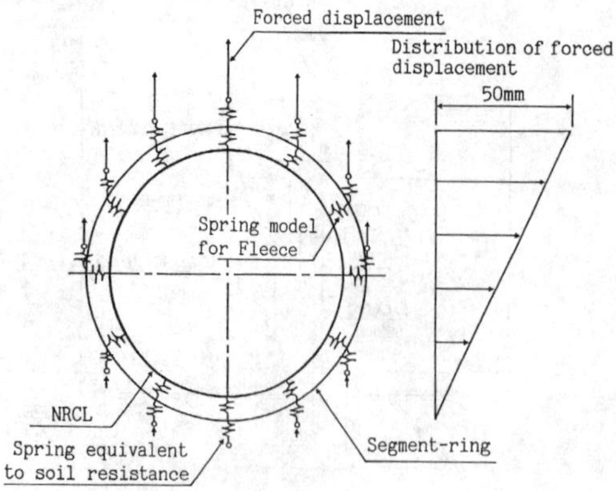

Fig.6 Structural Analytical Model of Rebound Study for Canal Expansion

earthquake intensity assumed for the seismic design of the tunnel is of magnitude M=7.0, with center distance $\triangle$=30km, input acceleration at the base α=107.5gal, and adopted seismic wave of El Centro (1940, Imperial Valley) which has a predominantly long period. This assumption represents the scale of an earthquake occurring once every 100 years. Time-history response analysis was done using FLUSH. According to the result of the analysis, estimated maximum strain of ground around the tunnel area is εmax= 4500 μ, for which NRCL has sufficient strength.

With regard to the influence of ships, the effects of stranding and anchoring were studied and the result suggested possible collapse of the tunnel under such condition. Therefore, strict control of ship operation is necessary. Safety of NRCL was assessed according to the following procedure with view to the reasonable design of the structure.

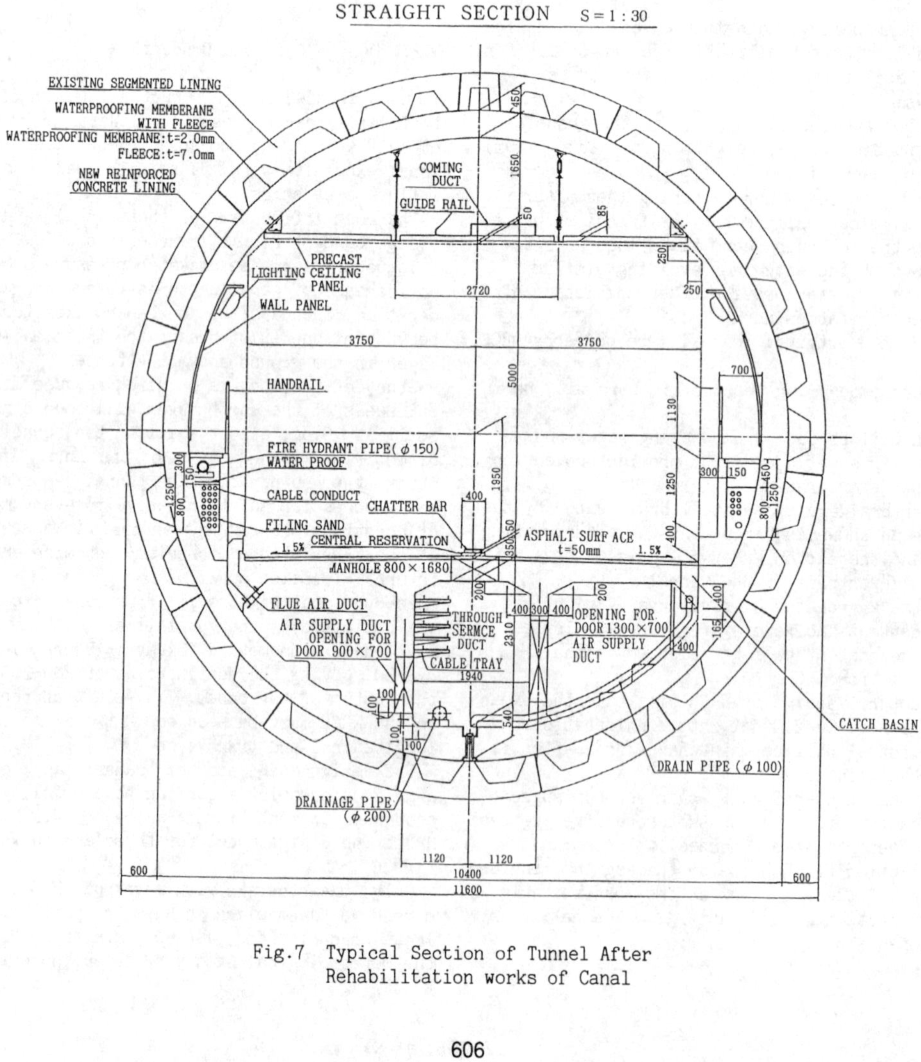

Fig.7 Typical Section of Tunnel After Rehabilitation works of Canal

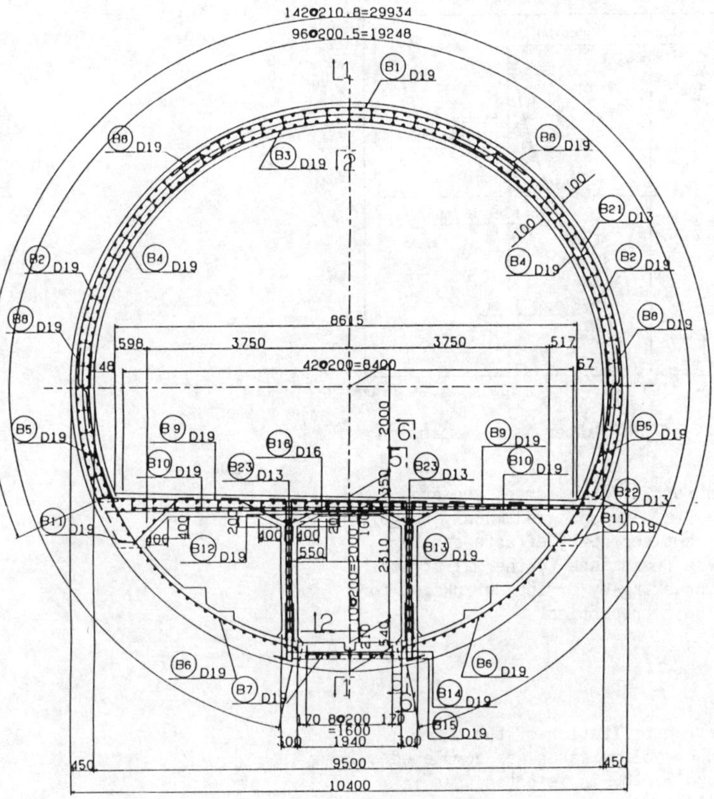

Fig.8 Reinforcement Arrangement of NRCL and Road Deck

Allowable stress design method

This is to be applied for design conditions occurring frequently.

Limit state design method (Ultimate Limit State)

This is to be applied for design conditions occurring rarely. For example, NRCL is usually loaded with earth pressure and its own dead load without inclusion of water pressure because seeping water guided to the lower part will be discharged normally. (Allowable Stress Design)

On the other hand, its safety was studied with the addition of water pressure in order to assure that the NRCL would not collapse in case of drainage trouble. (Limit State Design)

5.2.2 Results of Design

Outline of the results of the design is described with a drawing of the typical section of tunnel and the reinforcement arrangement of NRCL shown in Fig.7 and 8.

- It was possible to design NRCL so that loads a) to G) can be borne by itself.
- Design thickness of NRCL is 40cm against the thickness of 45cm as constructed because the segment ring has already been deformed. To ensure durability, the minimum covering of re-bars is 70mm (no part of re-bars within 70mm from the surface).

Although epoxy coated re-bars have been used in some cases, they were not adopted for use here because some of the previous uses were not successful.

- To the counter the influence of canal expansion, the following was studied backfilling of segment-ring, reinforcement of in-situ mudstone by grouting thickening of fleece, increase of re-bar quantity; use of steel fiber concrete and use of glass fiber concrete. As the result, it was decided to increase the fleece and the re-bars as shown in Fig.9 taking into consideration functional efficiency, ease of construction, construction period, and cost. Fleece is

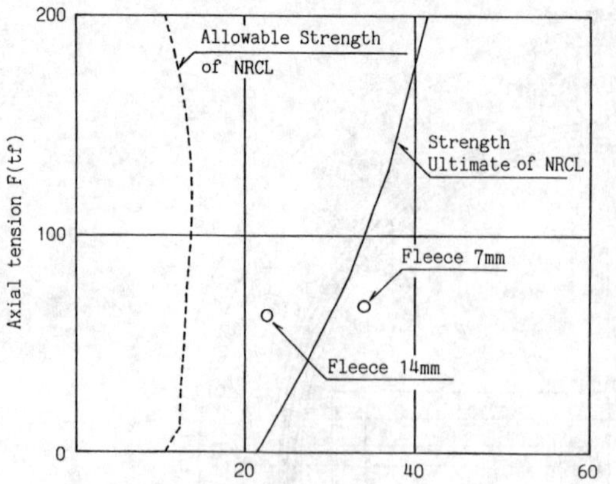

Fig.9 Safety of NRCL after Canal Expansion

effective to absorb deformation of the ground.
- Fleece and W.P.M. are designed mainly to guide seeping water, but secondary effects can be expected such as resistance to thermal stress, prevention of cracking by drying shrinkage, for NRCL and a seismic properties.

6 CONCLUSION

The design for rehabilitation of the shield-tunnel damaged by saline water was completed using fleece and W.P.M. as is often used in mountain tunneling.

Deterioration of tunnels is seldom announced and is not understood sufficiently.

The general cause of tunnel deterioration is seeping water, and lack of strength of segment rings is not responsible is most cases.

Damage by salt is a common problem of tunnels running in the seabed, and the result of this survey was referenced for the design of Trans-Tokyo Bay Highway Shield Tunnel.

Finally, we hope for successful completion of this tunnel in 1995, to facilitate safe passage between Egypt and the Sinai Peninsula, and to serve international needs through regional development of Sinai and expansion of the Suez Canal.

REFERENCES

1) ARAB REPUBLIC OF EGYPT;
Introduction of The Ahmed Hamdi Tunnel.
2) Japan International Cooperation Agency, Basic Design Study Report on The Project for Rehabilitation of Ahmed Hamdi Tunnel in Arab Republic of Egypt, Feb.1992.
3) AHMED HAMDI TUNNEL TECHNICAL COOPERATION TASK TEAM, JAPN;
The Recommendation on Rehabilitation Works for The Ahmed Hamdi Tunnel Final Report, May, 1990.

Tunnelling and Ground Conditions, Abdel Salam (ed.) © 1994 Balkema, Rotterdam, ISBN 90 5410 363 9

Rehabilitation of Ahmed Hamdi Tunnel under the Suez Canal – Part 2

Yukio Aratani
Kajima Corporation, Tokyo, Japan

ABSTRACT: This is a project to rehabilitate an existing tunnel by installing a waterproofing membrane to cover the entire surface of the existing segmented primary lining and place a reinforced concrete secondary lining while maintaining one lane open for general traffic during the daytime hours. Because of this special condition on the work, the tunnel is divided into upper and lower parts by means of a temporary road deck which is set above the existing road deck, and an attempt has been made to shorten the construction schedule by employing movable type scaffolds and concrete forms which were designed to secure the one lane of traffic. A mix of high slump, high strength concrete was designed for the reinforced concrete secondary lining and a severe control standard has been set for the total salinity content of the concrete. This paper presents a report on the construction work which is now in progress with completion scheduled for October 1995.

1 FOREWORD

This project is to rehabilitate an existing tunnel. As described in Part 1, this rehabilitation is being performed by newly constructing a reinforced concrete secondary lining within the existing segmented primary lining. Also, because the existing tunnel must remain available for use during the construction period as the one and only land transportation route between the two sides of the Suez Canal, selection of the method of construction to complete the work within the construction period was subject to severe restrictive conditions.

Further, strict control is required in the installation of the waterproofing membrane which is provided in the design to prevent damage due to saline water which caused the necessity for the present rehabilitation. Likewise, thorough quality control is also required regarding salinity in the concrete to be placed.

This paper presents a report on the rehabilitation works which are now in progress on this tunnel under the above described restrictive conditions with emphasis placed on schedule control and quality control.

2 OUTLINE OF WORKS

The work of rehabilitation of Ahmed Hamdi Tunnel was commenced in November 1992 and is scheduled for completion in October 1995.

Figure 1. Rehabilitated tunnel.

In outline, the work to be performed is as follows:

- Total length of tunnel: 1,642m
- Lining concrete: 21,100m^3
- Road deck concrete: 11,500m^3
- Reinforcing steel, D10 to D20: 3,950 ton
- Waterproofing membrane, 2mm thick: 53,580m^2
- Ceiling: 11,480m^2
- New walkway: 1,640m x 2
- Electrical systems: power cable, tunnel lighting, closed circuit television, carbon monoxide monitoring, visibility monitoring, fire alarm, communication and other systems
- Piping works: tunnel drain, fire hydrant, etc.
- Relocation of fresh water supply pipelines: 3,310m x 2 lines

3 CHARACTERISTICS OF THE WORK

The special characteristics of the work can be described as follows:

1) One lane of traffic shall be maintained throughout the work during the hours from 6:00 am to 8:00 pm with vehicle clearance of 4.5m in height and 3.5m in width.

2) The configuration of the existing tunnel shall be surveyed and new construction made to conform with the existing tunnel configuration.

3) The new reinforced concrete lining has an internal diameter of 9.5m, an average thickness of 45cm and a significant thickness of 40cm. In addition to being thin, this concrete lining has double reinforcement and is designed as a continuous structure for the full length of 1,640m. Consequently, in order to achieve satisfactory filling of concrete, concrete of excellent workability is required.

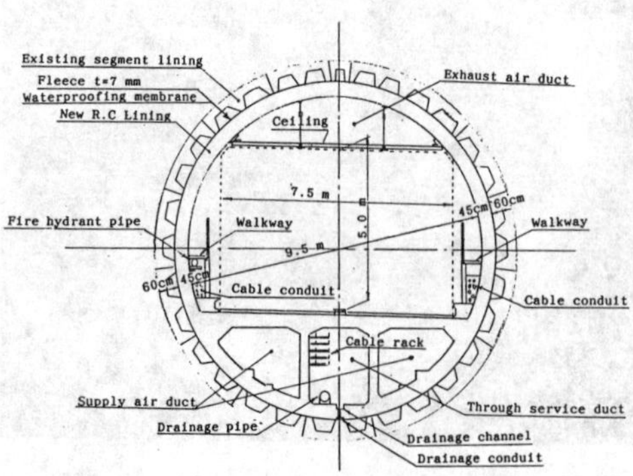

Figure 2. Typical cross section of rehabilitated tunnel.

4) Because the new reinforced concrete lining must be protected from high concentration saline seepage water, the design includes a layer of fleece applied to the inner surface of the segmented lining to lead seepage water to the drainage system in the tunnel invert, and a waterproofing membrane applied over the fleece to protect the new concrete lining. Consequently, the segmented lining and the new lining concrete must be isolated except for the anchor bolts which are installed at the bottom of the tunnel.

5) From the standpoint of securing traffic safety within the 1,642m long tunnel which will remain open as an arterial highway during the work, it is not desirable to have work sites dispersed. Consequently, it is necessary to secure the safety of general traffic within the tunnel, and at the same time facilitate management of the work, by concentrating the work as much as possible.

4 METHOD OF CONSTRUCTION

Considering the special characteristics of the work described above and the sequence of construction operations shown in Figure 3, the following method of construction was adopted.

1) Employ a temporary road deck system to secure the passage of general traffic during construction.

2) Divide concrete placing into four stages; base concrete, lower lining, and

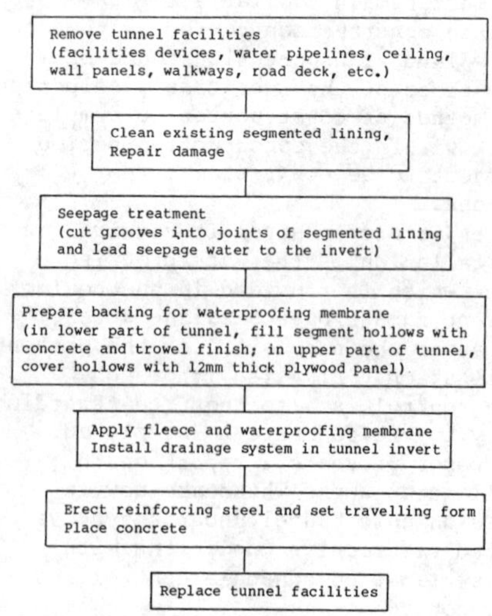

Figure 3. Sequence of operations in rehabilitation works.

road deck below the temporary road deck, and arch lining following removal of the temporary road deck. (Figure 4.)

The temporary road deck is constructed at a height of 1.6m above the existing road deck by placing cross beams so as to be supported in the hollows of the segmented lining on each side of the tunnel and placing steel deck slabs on these beams.

The advantages of this temporary road deck system are the fact that construction operations can be performed below this road deck without being affected by the passage of general road traffic, and the fact that construction operations are concentrated within the range of the temporary road deck, thus facilitating management of the work.

The existing road deck is removed as the temporary road deck is advanced. The road deck is removed three spans (2.4m per span x 3 = 7.2m) at a time during the limited nighttime hours by raising the deck slabs by jacking and then lifting and carrying them out by 18 ton forklift.

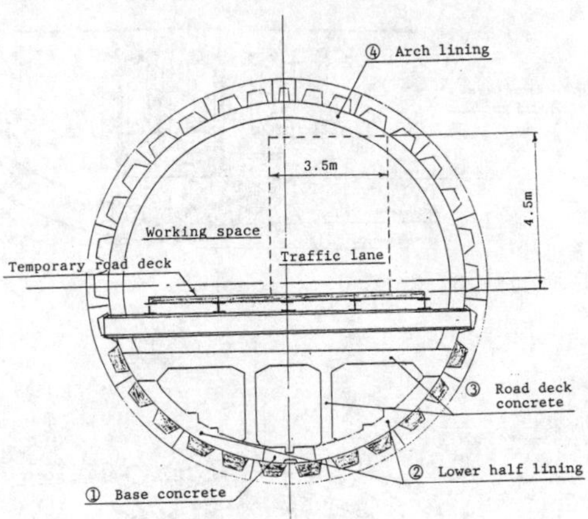

Figure 4. Stages of concrete placing.

5 CONSTRUCTION SCHEDULE

In order to complete the work within the prescribed construction period, it was necessary to establish an appropriate length of block for concrete placing, and study how to shorten the cycle time of operations.

The line of the tunnel includes profile grades of up to 3.823% with a 2,700m radius vertical curve at the center, and a 1,000m radius horizontal curve in the eastern half. Therefore, to secure the required precision in construction, and considering the quantify of concrete to be placed at one time (not more than 100m^3 at most), the length of block for concrete placing was set at 10.5m.

In determining cycle time, the age at which concrete forms can be stripped is an important factor. Having confirmed that the required strength of concrete can be obtained in 12 hours for lining concrete and 36 hours for road deck concrete, a cycle time of 2.5 days was determined. (Road deck concrete has a cycle time of 5 days, but two sets of forms are provided.)

In order to maintain the 2.5 day cycle, and to secure general traffic, all scaffolds and concrete forms were planned as movable type units as follows:

- Ceiling panel scaffold (rail mounted)
- Lower lining reinforcement erection scaffold (suspended)
- Lower lining waterproofing membrane scaffold (suspended) Figure 6.
- Segmented lining surface treatment scaffold (rail mounted)
- Lower lining concrete form (suspended) Figure 7.
- Road deck concrete form (rail mounted) Figure 8.
- Arch lining reinforcement erection scaffold (rail mounted)
- Arch lining waterproofing membrane scaffold (rail mounted)

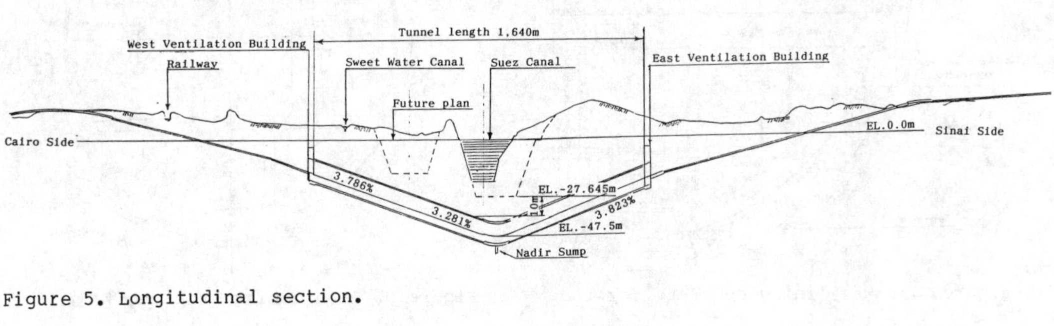

Figure 5. Longitudinal section.

· Arch lining concrete form (rail mounted) Figure 9.

The construction schedule is shown in Figure 10.

6 PRINCIPAL CONSTRUCTION MATERIALS

6.1 Waterproofing membrane

There are various types of waterproofing membrane; plastic group, rubberized asphalt group, synthetic rubber group and others. On this project, ethylene copolymer bituminous membrane (ECB) of 2mm thickness was selected.

This membrane possesses superior qualities of resistance to weathering and chemicals, but additionally it has a wide range of welding temperature for making joints and a signal layer which changes color when the membrane is damaged, thus making it a superior material from the standpoint of quality control during construction. (Figure 11.)

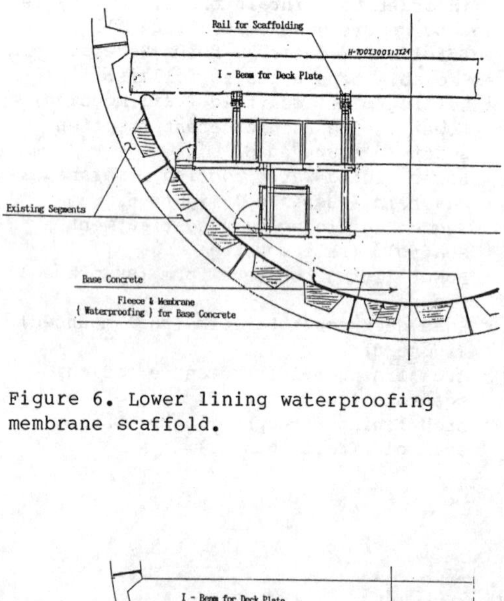

Figure 6. Lower lining waterproofing membrane scaffold.

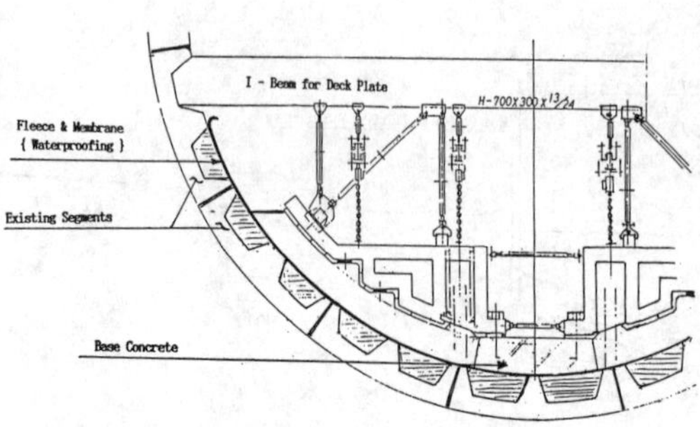

Figure 7. Lower lining concrete form.

6.2 Fleece

The fleece was chosen for its permeability, protection of the waterproofing membrane, and reduction of friction between the segmented lining and the waterproofing membrane after the lining concrete has hardened. It is made of high alkali-proof polypropylene, non-woven, 7mm thick.

6.3 Drainage channel and drainage conduit

For the drainage system to be installed in the tunnel invert, HDPE was selected from among PVC, HDPE and stainless steel, for the reason that it possesses superior qualities of resistance to weathering and chemicals.

In the drainage channel which forms the upper part of this drainage system, NAP

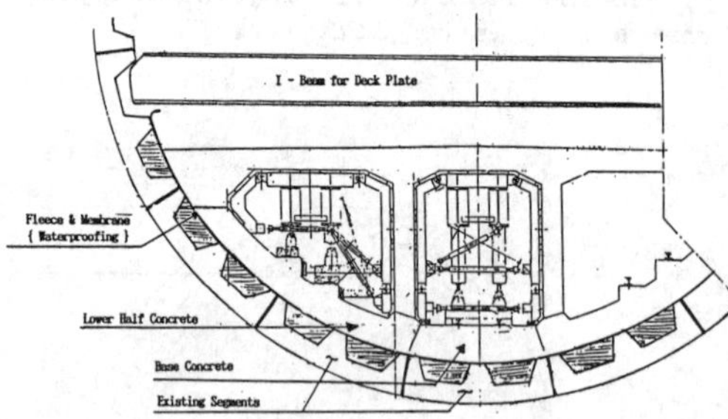

Figure 8. Road deck concrete form.

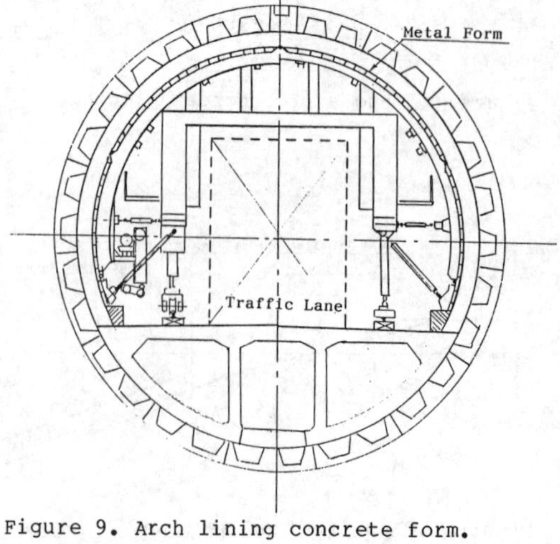

Figure 9. Arch lining concrete form.

membrane (3mm thick) manufactured by Huls Troisdorf AG of Germany, which has projections on its back side, was used to secure bond with the concrete.

For the drainage conduit which forms the lower part of this drainage system, 10mm thick HDPE sheet is being fabricated into trapezoidal form. This fabrication of the HDPE sheet is being performed by hot melt welding with polypropylene resin. At the joints with the ends of the waterproofing membrane, perfect watertightness is being

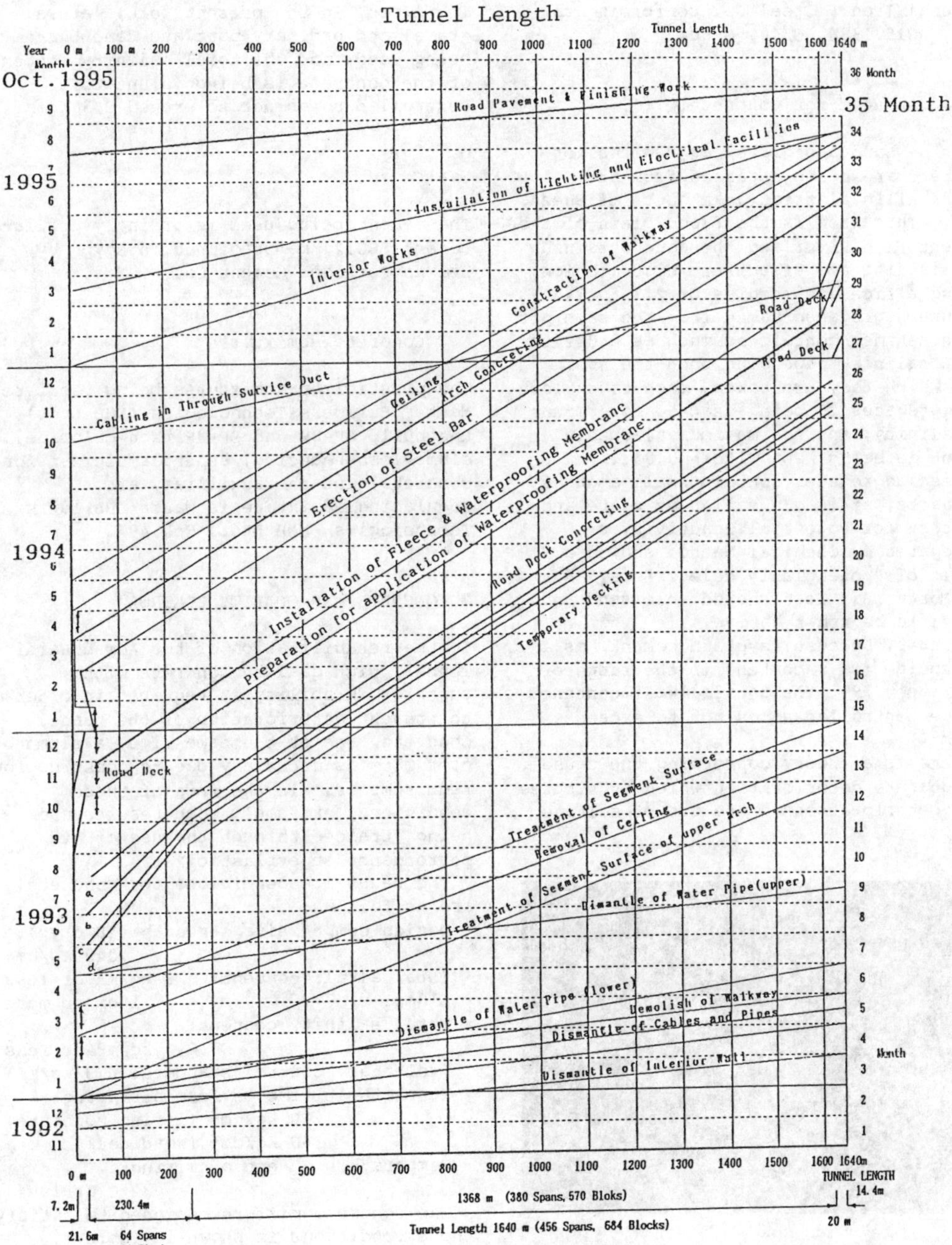

Figure 10. Construction schedule.

obtained by using water swelling rubber sandwiched under pressure from both sides.

6.4 Reinforcing steel

The reinforcing steel being used is manufactured in Egypt by Alexandria National Iron & Steel Co. conforming to ASTM, A615 1984, G/40, G/60.

6.5 Aggregates for concrete

Both the sand and gravel aggregates for concrete are being obtained from the Fanara hills district 30km north of Suez City. This area is the flood plain of the ancient Nile River and the sand obtained here is fine and of uniform distribution, having a fineness modulus of 2.3 to 2.5. Because aggregates taken from the same pit may sometimes contain as much as 0.04% to 0.06% salinity depending upon the spot from which they are taken, each truck load of aggregates is being checked for amount of salinity, and the maximum allowable amount is being kept within 0.03%.

These aggregates contain small amounts of quartz, siliceous schist or agate and the test for Potential Reactivity of Aggregates by Chemical Method showed a result of "potentially deleterious," but the Mortar Bar Test showed this sand and gravel to be non-reactive.

However, because Egyptian cement has a low sodium (Na_2) content in the range of 0.65% to 0.79%, the total alkali content can be controlled so as not to exceed 3kg/m^3.

There is a theory concerning the cause of concrete deterioration which attributes this deterioration to the use of sulfate resisting cement and calcareous dolomite aggregate, and the use of water or aggregates which contained salinity. (According to this theory, when limestone or dolomite are used, calcium within the cement separates out, promoting the formation of $Ca(OH)_2$ and increasing the permeability of the hardened concrete.) Therefore, in the present work, natural gravel and ordinary portland cement are being used, and the total salinity content of the concrete is being voluntarily controlled so as not to exceed 0.3 kg/m^3.

Figure 11. Installation of waterproofing membrane.

6.6 Cement

The cement being used is ordinary portland cement (BS12) manufactured in Egypt by Helwan Portland Cement Co.

6.7 Concrete admixtures

The plasticizer admixtures being used are Master Builders Technologies (NMB Ltd.) Rheobuild SP-8S and SP-8R (Non-chloride, slump-loss-reducing, superplasticizer for concrete) and the auxilliary air regulating admixture is Master Builders Technologies (NMB Ltd.) No. 404.

7 CONCRETE AND QUALITY CONTROL

In the rehabilitation of the Ahmed Hamdi Tunnel, high quality concrete with superior durability is required in order to prevent deterioration of the tunnel concrete, and at the same time, a slump of over 18cm is necessary due to construction requirements. In the face of these conditions, extremely good results are being attained through the use of high performance superplasticizer.

The requisite conditions for concrete are as follows:

- Design compressive strength (28 days): 240 kg/cm^2
- Total alkali content: 3 kg/m^3 or less
- Slump: 18cm or more
- Total salinity content: 0.6 kg/m^3 or less (voluntary control standard: 0.3 kg/m^3)
- Strength at time of form stripping: 10 kg/cm^2 (lining concrete) 140 kg/cm^2 (road deck)
- Temperature at end of mixing: 32°C or less

The concrete mixture designed to satisfy these conditions is shown in Table 1.

Testing of concrete materials and quality control of concrete is being

Table 1. Concrete mixture.

Class			Aggregate			Admixure	
	Cement Kg	Water Kg	Sand 0-5 Kg	G1 Agg 5-25 Kg	G2 Agg 3-9 Kg	SP-8S 2% Kg	No. 404 Kg
S1	378	160	692	957	106	7.56	3.78*

* 1% Solution

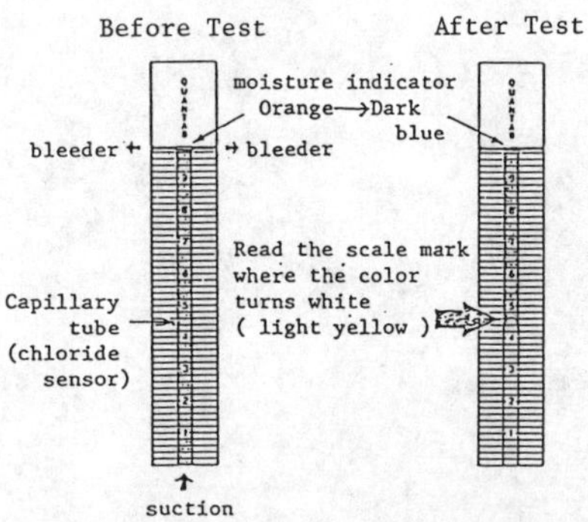

Figure 12. Quantab salinometer.

performed doubly at the site laboratory and at the Suez Canal Authority Research Center. Special emphasis is being placed on control of the total salinity content, voluntarily testing the amount of salinity in each truck load of sand, and measuring the salinity content at each placing of concrete. (Measurement of the salinity content of concrete aggregates, mixing water and fresh concrete is being performed by "Quantab" salinometer manufactured by Onoda Cement Co. (Figure 12.) by which measurements can be performed in 20 minutes or less.)

Because the temperature of the concrete at the end of mixing in the summer has a great affect on the amount of slump lost before placing, emphasis is being placed on temperature control. The temperature of mixing water is being lowered by about 5°C by cooling tower, and concrete is being placed during the nighttime or early morning when the humidity is high and the air temperature will exceed 30°C.

Because the water/cement ratio of the mixed concrete is a low 42.3%, it forms an extremely dense concrete which is not subject to bleeding. On the other hand it shows a tendency to form air voids on the surface of upper face forms, and various measures are being tried to prevent this, including improvement of concrete vibrators and improvement in the method of placing concrete.

8 CONCLUSION

By the time that this report is published the construction work will have advanced to the stage where placement of arch concrete has been commenced, but at the present time of writing, relocation of the water pipelines has been completed, and the first three blocks of the base concrete of the new tunnel lining have just been placed.

In addition to the fact that construction is being performed while securing the passage of general traffic, this project to rehabilitate the Ahmed Hamdi Tunnel is by nature a project of special character in which the work must constantly be studied in relation to the existing structure and tunnel facilities as the work is advanced. It should be added that, the work is being advanced through the united effort of the owner, the Suez Canal Authority, the engineer and the contractor regarding not only quality control in construction, but all phases of the work including traffic control, safety control, tunnel illumination and ventilation.

Tunnelling and Ground Conditions, Abdel Salam (ed.) © 1994 Balkema, Rotterdam, ISBN 90 5410 363 9

Workplace environmental conditions and innovative tunnel driving techniques: Measurement and control*

S. Pelizza
Tunnelling and Underground Space Center, Politecnico di Torino, Italy

M. Patrucco
CNR, Centro di Studio per i Problemi Minerari, Politecnico di Torino, Italy

G. Benedetto
I.E.N. Galileo Ferraris, Torino, Italy

ABSTRACT: The paper refers on the results of a data collection on workplace environmental conditions at the faces of some Italian tunnels, where the primary drifting is carried out by means of High Energy Hydraulic Hammers (H.E.H.H.).

Thanks to the substantial improvements in the machine size and efficiency, the use of H.E.H.H. has dramatically grown since some years in Italy, mainly due to the quite stringent Italian regulations on the use and transportation of explosives: as a consequence the blasting techniques can sometimes prove uneconomical.

But at the work places some problems arise, mainly due to the exposure conditions of workers to noise and airborne particulates, since, to win the same quantity of rock won by traditional drilling and blasting techniques, the H.E.H.H. machine must work much longer than a multiple arm boomer.

Moreover, it is a common practice to carry on the mucking operations together with the H.E.H.H. machine working. As a consequence, the resulting overall environmental pollution at the face can rise to concerning values.

The paper summarizes some of the possible control techniques suitable both to reduce the pollutants production and to lower the exposure of workers, with special reference to the results of the research work which is at present carried on by the machine producers

1 FOREWORD

The environmental pollution due to noise and airborne dust in underground mining and tunnelling operations is with no doubt a concerning problem, and substantial efforts are being spent in order to achieve an effective technical control of the situation, side by side with the progress of researches on occupational lung disease and hearing impairement.

Substantial improvements in the ventilation design, and the use of dedusting apparatuses (mainly based on water sprays and coanda effect) provide an effective reduction of the airborne particulate concentration even in the most critical situations (usually at the faces, both when drilling and blasting, and mechanical driving techniques are used, such as auger or drum type winning machines, or full face Tunnel Boring Machines).

Also the risk of hearing impairement is now commonly recognized, and substantial efforts are carried on in order to copy with the exposure limit values suggested by the E.E.C. 86/188 Standard (enforced in Italy since 1991 by the D.L.277/91). Even if the noise reduction of some sources (e.g. drilling machines, rockheaders) is a very difficult task, substantial exposure reduction is possible thanks to the use of specially designed operators cabs: useful information on the actual noise levels both inside the cabs and in the proximity of the operating machine is furthermore now available (e.g. for loading and hauling machines), in accordance to the 86/662 EEC Standard.

Nevertheless, it is common knowledge that the

* This research is being carried on by the Centro per la Sicurezza del Lavoro Minerario e di Scavo (a Research Center on Safety and Health in Underground Operations), Georisorse e Territorio Department of Politecnico di Torino, in cooperation with the Centro di Studio per i Problemi Minerari, C.N.R., care of the same Department

introduction of innovative technologies can involve some unespected risks: the present paper refers on a research work about the environmental impact at workplace due to the growing use of H.E.H.H. machines in tunnelling operations.

2 THE USE OF H.E.H.H. IN TUNNELLING OPERATIONS

H.E.H.H. are since more then 20 years commonly used in oversized stoneblock popping in quarrying.

The technical progress makes today available very highly powered machines (up to 150 kW for hammers weighting more than 7 tons) with impact energy values up to more than 12 kJ/blow, often with automatic regulation of both the percussion rate and the blow energy. The largest sized machines proved to be quite suitable to extend the use of the mechanical driving techniques to difficult winning conditions (e.g. where the hydraulic excavators, even if equipped with ripper tools, cannot work profitably).

Due to the reasons summarized in Tab.1, these machines are regarded with particular interest in Italy, where they are widely used in the primary winning at tunnel faces, in the enlarging of pilot tunnels, and in scaling, cleaning and auxiliary operations.

Tab.1. Main reasons supporting the selection of mechanical driving techniques in Italian tunnelling operations and in particular of the H.E.H.H. machines.

-a-the usual length of italian tunnels -except of course the main transalpine tunnels- ranges from some hundreds to a few thousands meters;

b- the tunnel driving work has often to be performed in quite altered, superficial formations, and special care is needed to prevent a decrease in the overall stability and the risk of landslides;

c-tunnels often and often have to be driven in urbanized areas and in the proximity of buildings of significant artistic or technological value;

d-owing to the fact that the rock geomechanical characteristics can remarkably vary along the tunnel axis, tunnel boring machines and continuous miners can show discontinuous or uneconomical performances;

e-the Italian regulations on the use and transportation of explosives are so stringent that, sometimes, the use of explosives may become uneconomical.

Since the middle '70 an extensive research work has been carried out by the Authors on the H.E.H.H. machines technology and use, both with reference to the maximum achievable production rates as a function of machine size and job conditions, and to the expectable enviromnental impact at the surface (the main results are summarized in /1,2,3,4/).

3 H.E.H.H. AT TUNNEL FACES AND ENVIRONMENTAL CONDITIONS: MEASUREMENT TECHNIQUE

Where the health conditions of workers in the underground are considered, and compared with the situation typical of other winning techniques, some remarkable technological and operative differences can be pinpointed.

In order to get experimental data a series of measurements have been carried out at the faces and/or at the enlargement site of some Italian road and railway tunnels, where H.E.H.H. are used for primary driving or to fulfill auxiliary operations.

3.1. *Noise measurements*

Tab.2 Layout of the noise measurement apparatus.

A-*in situ recordings:*

1. Sony TCD-D10 PRO II digital audio tape-recorder (Sony DT-90P tape)

2. two channel microphone feeder/conditioner Larson Davies LD 2200 C

3. microphone preamplifiers Larson Davies LD 900

4. 1/2" microphones Larson Davies 2559.

B-*laboratory analysis*

B1- *sound pressure vs time:*

1. Sony TCD-D10 PRO II digital audio tape-recorder
2. preamplifier Larson Davies LD 826 B
3. integrating sound level meter Larson Davies LD 800 B
4. personal computer and software Larson Davies LD 800/3.07

B2-*noise frequency distribution:*

1. Sony TCD-D10 PRO II digital audio tape-recorder
2. two channel real time analyzer Bruel & Kjaer B&K 2144.

Noise recordings have been systematically carried out, on a time base of more than 20 min during the machine operation, by means of a two channel digital audio recorder: the mycrophones were positioned in the proximity of the head of the machine operator, and at 1-1.5 m distance from the side of the carrier (on a trypod) respectively.

The recordings were subsequently analyzed in the laboratory in order to obtain the values of Leq(lin) and Leq(A), the time evolution of the sound pressure level and the 1/3 octave band distribution of noise in the frequency range 20 Hz-20kHz.

The layout of the measurement apparatus is summarized in Tab.2.

3.2. *Airborne dust measurements*

The airborne dust pollution was measured at the same positions, on a 1-2 h time base. The samples were collected by means of battery powered Dupont P4000 and AMETEC Alpha 1 samplers (flow rate 2 $dm^3/min.$, open holders for 25 mm dia., 0.8 μm pore diameter cellulose filters). The mass dust concentration of the samples was systematically determined.

Where a significant quartz content in the rock was feared, also respirable dust samples were collected, by means of the aforesaid sampling devices equipped with Dorr-Olver 10mm nylon and aluminum cyclone separators. At this stage of the research work, the microscope analysis technique (phase contrast illumination, 500x magnification) was considered acceptable to get preliminary results.

4 H.E.H.H. AT TUNNEL FACES AND ENVIRONMENTAL CONDITIONS: MEASUREMENT: RESULTS AND COMMENT

Even if the environmental conditions in tunnels where H.E.H.H. machines are used (either for primary winning or for auxiliary operations) are, as expectable, quite different from site to site, on the basis of a global analysis of the environmental situation at the workplaces, supported by the results of the direct measurements, the following considerations can be drawn:

1. The pure blowing time of H.E.H.H. is at least similar, but often quite longer than the pure drilling time, where similar driving situations are compared. For instance, the pure drilling time required by a double arm boomer to prepare a "blast event" by which, say 300 m^3 of rock are exploited, is of the order of 100 minutes. When the same quantity of rock must be exploited by means of a suitable H.E.H.H. machine, a blowing time of 180-250 minutes can be expected.

2. Even if, as formerly discussed in /1/, and confirmed in Fig.1, in common rock winning at tunnel face the ratio between the effective blowing time and the overall working time of H.E.H.H. usually ranges from 0.3 to 0.5, all the same the equivalent sound pressure level in the proximity of

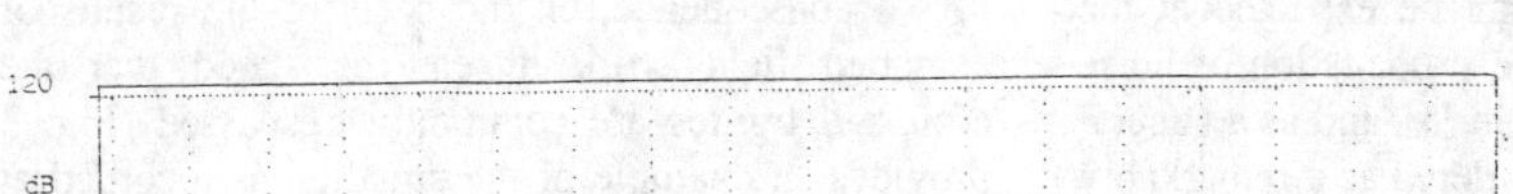

Fig.1 Time variation of the sound pressure level in dB(lin) as measured in a gallery with 80 m^2 section, generated from a H.E.H.H. of 5000 kg operating in primary drifting in fractured gneiss.

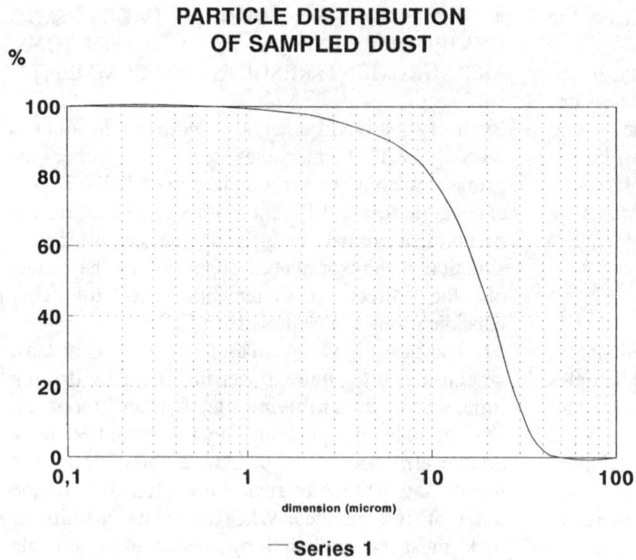

Fig. 2 Typical airborne dust size distribution drawn from the samples (total dust) collected at a tunnel face at the outside of the carrier cab. The upper particle dimension limit is a function of the sampling velocity.

the machine is high (a rough mean value can be assumed in approximately 105-110 dB(A)). Moreover, as in the case of drifters, impulsive components of sound pressure exceeding 140 dB(lin) can commonly be identified, in particular if uneffective strokes occur. As a consequence, also where H.E.H.H. machines are used instead of the drilling and blasting winning technique, a remarkable noise pollution can be expected at the face and along the tunnel (the propagation of noise in tunnels has been discussed in /5/, and as a general statement tunnels can be considered as a remarkably reverberant environment).

3. Since H.E.H.H. machines leak of dust control devices, during the blowing and ripping actions large amounts of airborne dust can be produced, due both to the breakage effect of the rock, and to the fall of the winned rock from the face. Airborne dust at workplaces, as directly collected in different tunnels, show an average value size distribution mainly in the field of particle diameters larger than 25 μm (see Fig. 2, which provides a typical example), but, where the quartz content in the rock is high, concerning concentrations should be feared also with reference to the respirable limit values for special dust.

4. It is a common practice to carry on the mucking operations while the H.E.H.H. machine is working. As a consequence, the resulting overall environmental pollution at the face can be furtherly increased with reference to both the airborne dust concentration and noise levels

5. The soundproofing of the carrier cab -the H.E.H.H. is commonly installed on an hydraulic excavator instead of the standard rear bucket- is not specially designed to copy with such a situation (the information on the noise levels inside and ouside the cabs officially required refers to measurements carried on during standard loading operations, which are remarkably less noisy).

As a consequence, on the basis of the results of extended field work, it can be stated that the aforesaid figures are commonly trespassed (Fig. 3 provides an example of the situation as recorded at the inside of a carrier cab at tunnel face during the primary winning operation: as it can be seen both the overall noise level and the frequency distribution show a crytical working condition).

Moreover, only in less than 15% of the analyzed cases the carrier cabs were fitted with forced circulation of conditionned and filtered air, this explaining some dust concentration values measured at the inside (exceptionally up to 25 mg/m^3 total dust on a short period).

6. Finally, the presence at the face of a second man is often required, in order both to provide the machine operator with some visual help on the optimal positioning of the tool against the rock -the visibility from the operator place can become very poor during the machine operation- and to intend a water flow derived from a hose to the face and the muck, in order to reduce the airborne dust pollution.

As a matter of fact the working conditions of a man standing at the side of the carrier proved

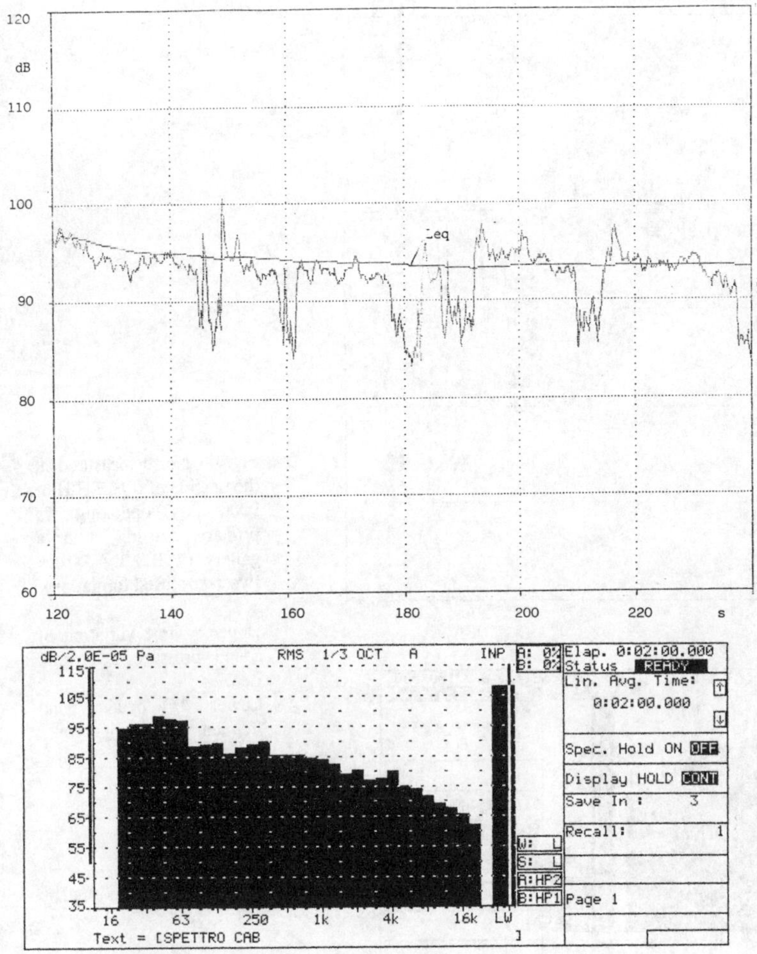

Fig. 3 Noise measured inside the operator's cab as generated from a H.E.H.H. of 9000 kg operating in primary drifting in a gallery of 100 m^2 (rock flysch stratified formation).

Upper: Time variation of sound pressure level and Leq.
Lower: 1/3 octave band frequency spectrum.

sometimes to be quite unconfortable -remarkably worser than that at the operator position- both with reference to the dust concentration: (dust pollution values widely exceeding the A.C.G.I.H. limits -e.g. more than 45 mg/m^3 total dust- have sometimes been recorded at the ouside of the carrier cab, and to the exposure to noise (see Fig.4).

5. H.E.H.H NOISE AND DUST EMISSION CONTROL: PRESENT SITUATION AND FUTURE TRENDS

The producers of H.E.H.H. machines are obviously conscious of the environmental problems afore discussed, and, as a result of their effort, the new generation machines are fitted with noise dampeners or with a continuous carter; to provide an example, Fig.5 reproduces the results got by a primary firm as to noise control of a machine of a size suitable for road construction application. But no solution can anyway be expected with regards to the noise generated by the action of the tool against the rock.

The dust problem is mainly due to the fact that at present generally the only attempt of reduction is based on the use of a water flow derived from a hose which a worker intends to the muck. Whilst the drilling machines for underground uses are in general provided with water injection systems (commonly a water flow -sometimes with wetting agents- is forced through the drilling rod and special holes drilled in the bits), and special nozzles are commonly used to control the dusts in the immediate proximity of cutters and tools of the mechanical rock winning machines (see e.g. /7/), the reduction of dust formation where H.E.H.H.

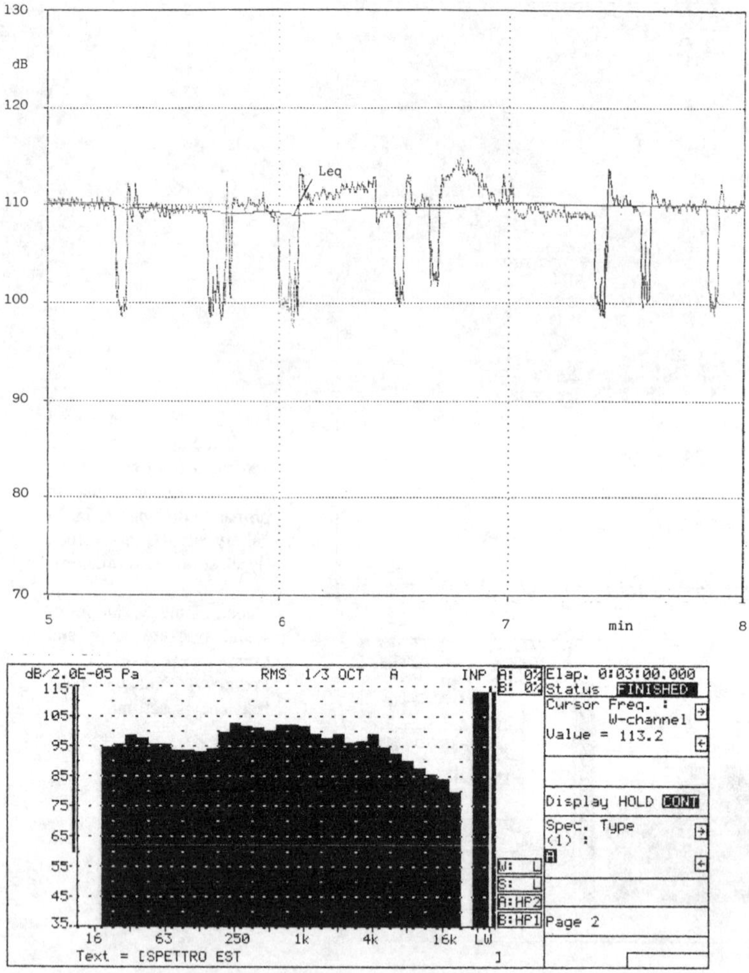

Fig. 4 Noise measured at the outside of a H.E.H.H.of 9000 kg operating in primary drifting in a gallery of 100 m^2 (rock flysch stratified formation).

Upper : Time variation of sound pressure level and Leq.
Lower: 1/3 octave band frequency spectrum

machines are used seems to be a target quite difficult to get: the water nozzles widely used for instance in the mechanical coal winning machines proved to be insuccessful: experimental tests have shown that the sudden cooling of the tool dramatically reduces the tool duration.

A more encouraging approach, but limited to the protection of the machine operator, can be based on the use of cabs specially designed for the H.E.H.H. mountings: the cabs should be extra sound proofed, and feeded with filtered and conditionned air. In this case (but in general for the cabs of every mining and tunnelling machine) windshiels wipers of improved efficiency are quite necessary.

CONCLUSION

The measurement instrumentation and the operative technique used confirmed their efficiency; asides from the Leq(A) workers exposure values, the knowledge of the frequency distribution of the emitted noise can provide a basis both for the noise control and particularly for the cab soundproofing.

Similarly, a further data collection on the dust concentration and particle distribution could be useful to provide input data for control attempts.

The measurement results confirmed that, at least in some operative situations, the exposure conditions of workers both at the inside and, much more, at the ouside of the carrier cab, during the operation of H.E.H.H. machines, are rather concerning, with

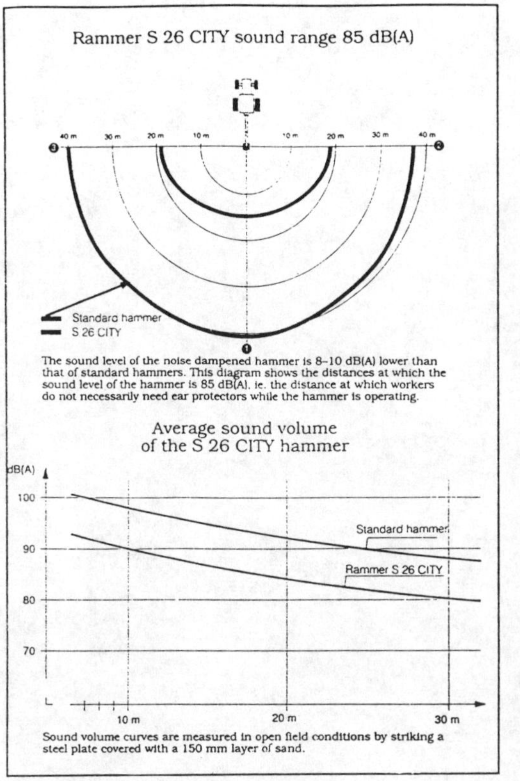

Fig.5 Example of the results of noise control of a H.E.H.H. machine (drawn from 6).

reference both to noise level and airborne dust concentration.

The main Producers of H.E.H.H. are, since some years, developing new ideas in order to reduce the emitted noise, and some incouraging results are now available, even if the noise levels are still high. As to the airborne dust pollution, experimental tests carried out on machines equipped with water nozzles clearly shown quite uneconomical results, due to the dramatically reduced time duration of the tools. In the autors opinion the possibility of dust control by means of exhausting systems, similar to those installed on wagon drills, should be carefully considered.

It can be expected that in future the presence in the underground of workers other than the machines operators vill no longer be necessary control of the emissions.Therefore the extensive use of cabs specially designed as a function of the machines equipment, together with the technical improvements in the emission control, will provide safe and confortable environment for the operators.

ACKNOWLEDGEMENTS

The authors are very obliged to the Contractors and Entreprises that made possible the measurements at the tunnel faces, and to the machine Manifacturers and Dealers for the information kindly provided. A special thank to Mr.F.Bronuzzi of the Acoustic Laboratory of the Energetica Dept. and to Prof.C.Clerici of the Georisorse Dept. of the Politecnico di Torino for their help, respectively in the analysis of the noise recorded data and of the collected dust samples.

REFERENCES

/1/ Armando E., M.Clerico, G.Moroni, M.Patrucco, S.Pelizza & L.Sambuelli 1989. Application of high energy hydraulic hammers to large section tunnels driving: Case histories in italian road and railway tunnels. *International Congress on Progress and Innovation in Tunnelling*. Toronto (Canada).

/2/ Godio A., M.Patrucco, G.Ranieri & L.Sambuelli 1992. Analysis of ground vibration due to the use of High Energy Hydraulic Hammers in tunnel drifting. *International Congress 'Towards New Worlds in Tunnelling*. Acapulco (Messico).

/3/ Armando E., C.Baratelli, M.Patrucco, S.Pelizza & L.Sambuelli 1992. On the possibility of using a seismic waves attenuation parameter for forecasting the performances of High Energy Hydraulic Hammers when used in tunnel drifting. *International Congress 'Towards New Worlds in Tunnelling*. Acapulco (Messico).

/4/ Fornaro M., M.Patrucco & L.Sambuelli 1993. Vibrations from exlosive, high energy hydraulic hammers and tunel boring machines: some experiences in Italian Tunnels. *International Congress on Options for Tunnelling*. Amsterdam (The Netherlands).

/5/ Clerico M. & M.Patrucco 1990. Noise propagation in underground workings, as a function of rock and support characteristics. *International Conference on Minesafe International 1990*. Perth (Australia).

/6/ No blast, no damnation 1993. *World Mining Equipment,* 17, 2.

/7/ Colinet J.F., J.J.McClelland & R.A.Jankowski 1991: Interactions and Limitations of Primary Dust Controls for Continuous Miners.*U.S.B.M. RI* 9373, Pittsburgh, USA.

Tunnelling and Ground Conditions, Abdel Salam (ed.) © 1994 Balkema, Rotterdam, ISBN 90 5410 363 9

Large scale mechanized repair work of a railway tunnel using precast concrete lining segments

Seiji Tamitsu, Mitsuharu Mizuno & Masanori Nasu
East Japan Railway Co., Japan

ABSTRACT : Approximately 20 years have now elapsed since the Tokyo Tunnel was constructed as a shield tunnel. But, with the rise in the ground water level, it is presently submerged.Rust caused by seepage of water from the segment joints and various other troubles have therefore arisen. The secondary tunnel lining method was opted for as a countermeasure. But, development of technology became necessary because of the many problems involved principally in relation to manpower with the conventional cast-in-place concrete method. The problems arose from the severe restrictions in time and space involved with operating lines. The PCW method developed at this time by prefabricating and mechanizing the secondary tunnel lining construction is therefore reported here.

RESUME : Vingt années environ se sont écoulées depuis la consruction du tunnel de Tokyo par tunneliers. Toutefois, en raison de la hausse du niveau des eaux, ce tunnel est aujourd'hui submergé. Un certain nombre de problèmes se sont par conséquent produits comme l'apparition de rouille provoquée par l'infiltration d'eau à partir des joints des segmnets. Une méthode de revêtement secondaire du tunnel a été adoptée comme solution au problème. Toutefois, la mise au point d'une certaine technologie fut nécessaire en raison des nombreux problèmes rencontrés, principalement au niveau de la main d'oeuvre, avec la méthode conventionnelle de béton coulé sur place. Le problème était dû aux limitations rigoureuses en matière de temps et d'espace sur les lignes d'opération. La méthode PCW, mise au point lors de la prébabrication et de la mécanisation de la construction du revêtement secondaire du tunnel, est par conséquent indiquée dans le présent rapport.

1. FOREWORD

The Tokyo Tunnel connecting the Yokosuka line and the Sobu line was constructed to alleviate congestion and assure transportation for the rapidly increasing number of commuters from the outskirts of metropolitan Tokyo. Construction of this urban shield tunnel, which was the first of its kind in Japan, with an overall length of about 9.5 km was commenced in 1965. The tunnel, which is actually two single-track tunnels side by side, was opened for service in 1976 (Fig. 1).

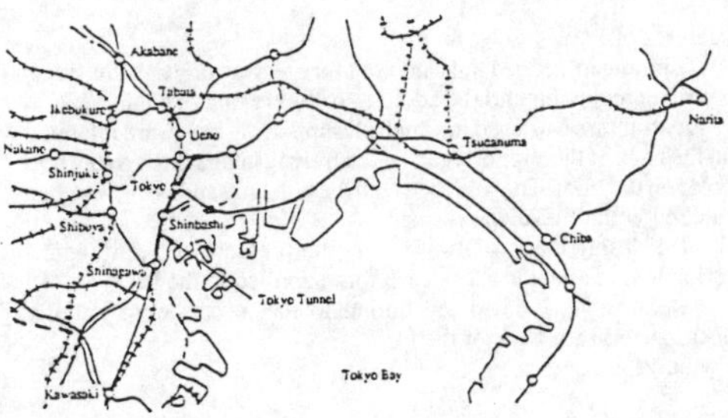

Fig. 1 Location plan view

The tunnel presently connects the Keiyo district and Keihin district with about 280 trains a day, and its importance continues to increase as a route for the Narita Express connecting Narita International Airport with the metropolitan area.

2. PRESENT AND PAST CONSTRUCTION METHOD OF SECONDARY TUNNEL LINING

Since the geology of the tunnel consisted principally of sandy alluvial layers, the ground water level was below the tunnel at the time of construction. But, with the strengthening of ground water pumping regulations in Tokyo in 1971, the ground water level sharply rose above the tunnel level. Although the secondary tunnel lining construction was partially completed, about 5.5 km still remained in the extension of the single track line in the primary tunnel lining section (Figs. 2 and 3).

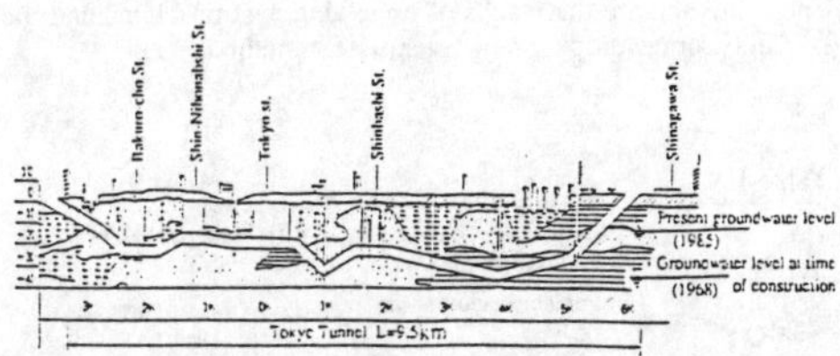

Fig. 2 Longitudinal drawing

For this reason, rust developed in the segments, bolts, and other steel parts of these sections as a result of the fresh air supplied by forced ventilation from the highly conductive leakage water containing dissolved salts. Leakage water itself caused early deterioration of overhead lines, rails, and signal cables in addition to creating sludge from oxidation of iron parts. There was also fear of the deterioration reducing the withstanding strength of the tunnel itself.

Although the injection method was used in the past as a means of stopping water, the secondary tunnel lining method was opted for as a radical measure at this time to prevent rust and leakage since the tunnel strength was sound. Secondary tunnel lining by the cast-in-place concrete method was implemented on a test basis from 1985 to 1986. Since the tunnel was in regular use, this was done under severely restricted conditions of time and space during the three hours from the last train at night to the first train in the morning. Since it was necessary to work while maintaining the safety of daily train operation, efficient work was not possible and the need to develop new methods was pressing (Fig. 4).

Construction methods developed from 1990 to 1992 and their results will now be explained.

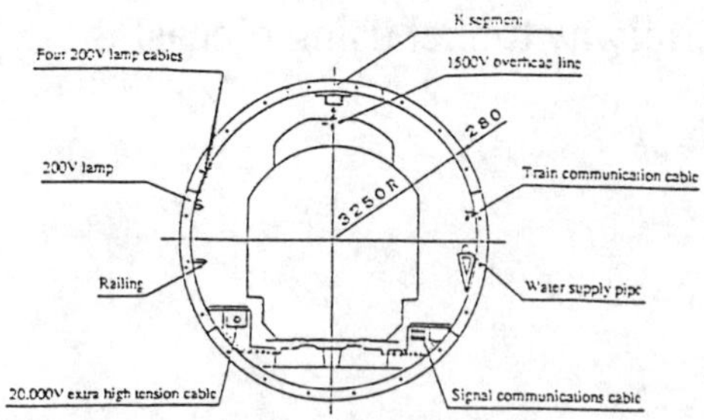

Fig. 3 Cross sectional drawing of a tunnel

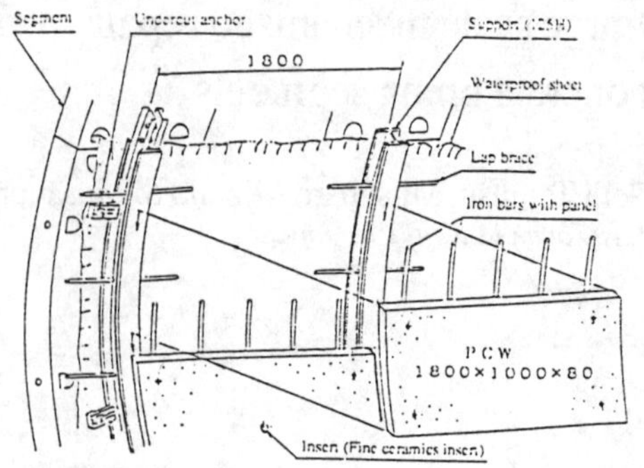

Fig. 5 Concept drawing of PCW construction method

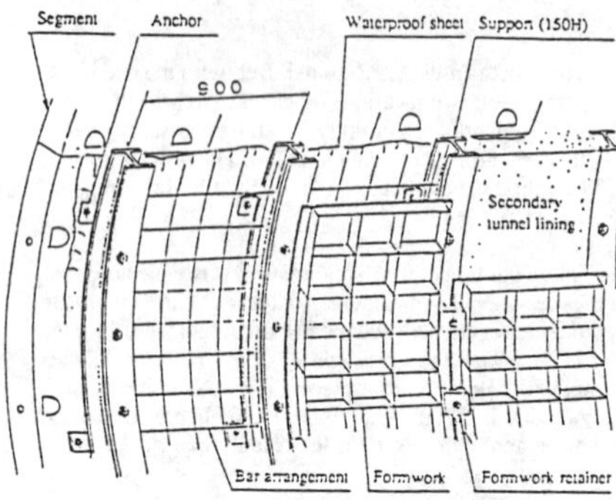

Fig. 4 Conceptual drawing of conventional method

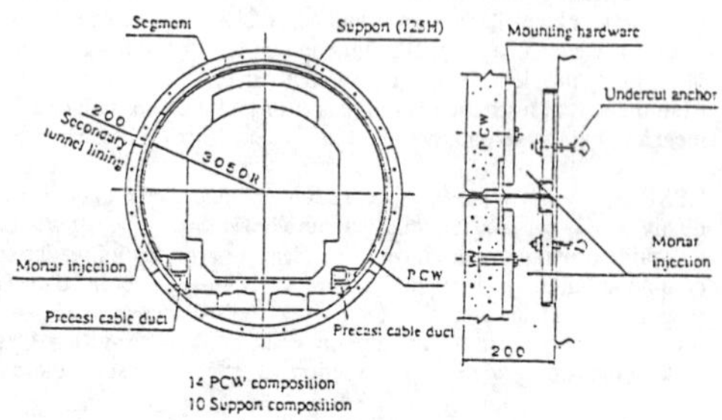

Fig. 6 Structural diagram of PCW construction method

3. PROBLEMS OF CONVENTIONAL CONSTRUCTION METHODS AND NEED TO DEVELOP NEW METHODS

The following are the results of reviewing past problems and the concepts for developing new construction methods.

Table 1

Item	Problems in conventional construction methods	Basic concepts for developing new methods
Investment efficiency	(1) Implementation speed is slow	(1) Increase in construction speed with prefabrication
Aging of the labor force	(2) Dependence on heavy midnight labor	(2) Labor saving by mechanizing construction
Lack of skilled people	(3) Securing large numbers of experienced workers is difficult	(3) Simplification of work by assembly of secondary products
Quality control	(4) Supervision of concrete work is difficult	(4) Improvement of quality by using secondary products
Need for maintenance	(5) Development of rust because of exposed supports	(5) Steel materials such as supports will not be exposed from the lining.

4. OUTLINE OF THE NEW CONSTRUCTION METHOD

As shown in Figs. 5 and 6, the tunnel lining structure was changed from the conventional steel framework to Precast Concrete Walls (PCW). Waterproof sheets were installed on the back of the walls to drain leakage water at the bottom, and the gap with the PCW was filled with mortar to assure waterproofing and prevent rust. A curtain wall structure in which supporting structures are covered with PCW was also used to prevent development of rust on the secondary tunnel lining.

Compared to the margin assured at the start of construction, a margin of over 200 mm was provided for the tunnel lining structure and the composition was as follows.

Precast concrete wall	70 mm
Supports	125 mm (125H)
Waterproof sheet	5 mm

Features of PCW

- Ceramic imbedded nuts are used for erection of PCW considering rust prevention and the addition of future tunnel equipment.
- A structure was used in which all supporting hardware mounted on the back of the supports can be securely fastened through the bolt holes on the front. The joint between the bolt hole and PCW was also caulked with pliable epoxy resin.
- The design load used was the mortar pressure of 5 t/m² (application height about h = 2 m/each time) applied to the back.
- Reinforcing bars for anchoring to the mortar were initially positioned on the back of the PCW.
- PCW data

Table 2

Item	Data	Description
Dimensions	1800 × 1000 × 70	14 layers per section equivalent to six steel frames
Weight	About 500 kg	
Strength of concrete	430 kg/cmt²	Reinforcing bars SD295 (D16, D13)

5. DEVELOPMENT OF CONSTRUCTION MACHINES

Since maintenance work on operating lines is conducted under severely restricted conditions and is a field in which mechanization is greatly delayed, mechanization was introduced as much as possible with the implementation of the new construction methods.

The object of development was to utilize general construction vehicles as much as possible to save labor and hold down development costs while improving safety and the working environment.

(1) Railroad vehicles

Although these are limited to 2 ton vehicles in Japan because narrow gage tracks (G=1067mm) are used, minimum investment was made possible by placing the front wheels of general construction vehicles on a bogie and changing the rear wheels to iron wheels. This enabled the vehicles to travel on tracks thus improving efficiency in transporting materials and construction machinery.

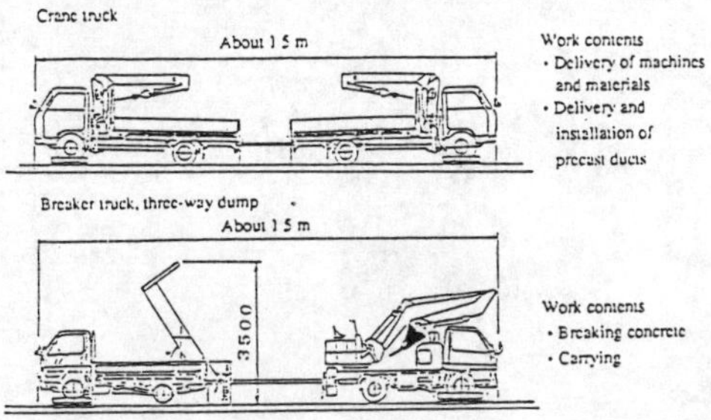

Fig. 7 Railroad vehicle

(2) Work staging bogie

A bogie capable of coping with the entire tunnel section with a slide and lift mechanism when working, while fitting into the car clearance limit of operated line tunnels when travelling empty, was developed to improve efficiency and safety by such measures as installing safety hand rails.

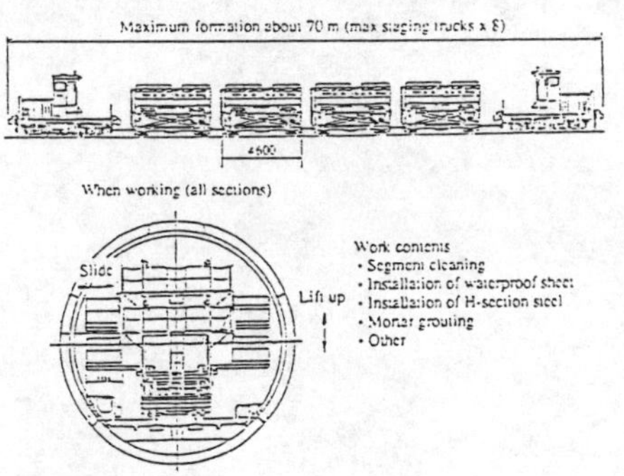

Fig. 8 Work staging truck

(3) Self-propelled bogie

Special towed type concrete agitator cars for tracks have been manufactured in the past. However, a self-propelled bogie on which a regular agitator car is placed, with the iron wheels of the bogie driven by the drive wheel of the car, was developed since work efficiency was poor in midnight work on operated lines and development costs were high.

A mortar grouting pump car was also coupled between the two cars thus enabling efficient grouting to be conducted while moving over a wide range.

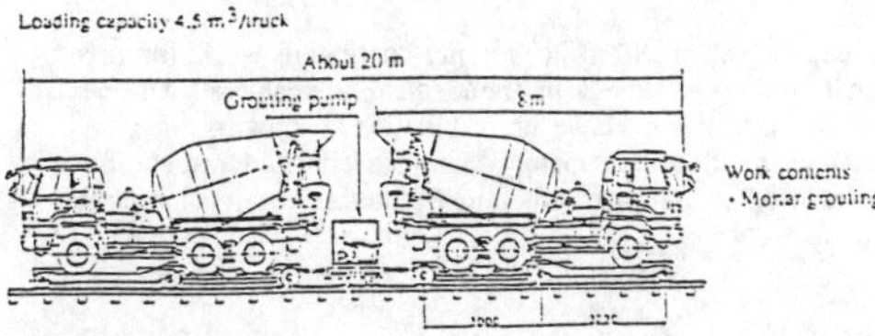

Fig. 9 Self-propelled truck

(4) PCW erection bogie

The mounting arm for safe and efficient erection of PCW over the full face of the tunnel has 360° rotating functions. Extension and retraction functions were also developed together with an erector having fine adjustment functions. These were mounted on the end together with a PCW hauling bogie.

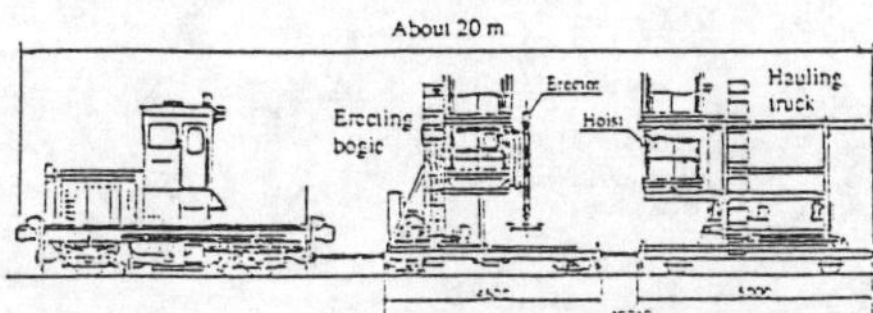

Fig. 10 PCW erector

6. RECORDS AND EVALUATIONS

The conventional method of pouring concrete in forms required much manpower and installation speed was about 100 m/yr. However, it was possible to simplify the work cycle with the newly developed PCW construction method which introduces a wide range of mechanization and prefabrication as shown in Fig. 11. As a result, it was possible to reduce the number of workers to about one half of those required in the conventional method and to satisfactorily

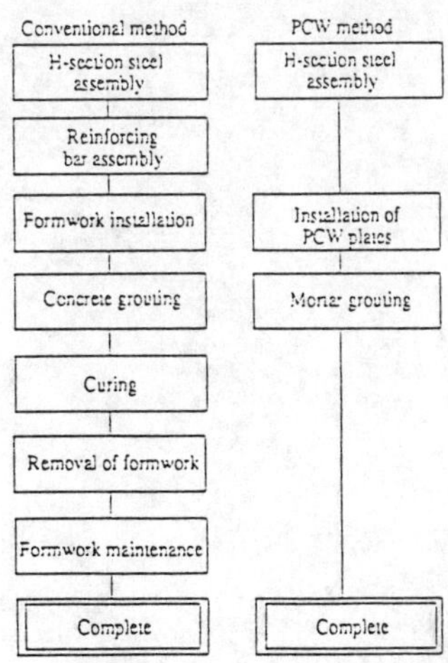

Fig. 11 Comparison of construction cycle

complete about 650 meters of a difficult section at about three times the normal speed. Problems that require study in the future include the application to other tunnels and underground structures and the application of the machines developed at this time to other construction work.

Since the construction sites require continuous work, improvement in working efficiency may be adequately considered in the next cycle of construction. However, installation of supports and waterproof sheets still depends mainly on human labor and further efforts are required to improve structures, materials, and construction methods.

7. SUMMARY

Remodeling the construction of railroad tunnels on operating lines involves work within a limited time and space while protecting train operating equipment such as electric power, signal and communication equipment. Practically all of the work has been dependent on manpower and has been very difficult to mechanize. Development of the PCW construction method is therefore believed to have a great impact on future mechanization including construction on operating railway lines.

Tunnelling and Ground Conditions, Abdel Salam (ed.) © *1994 Balkema, Rotterdam, ISBN 90 5410 363 9*

An effective repair system for lining concrete or power cable tunnels

Jyunichi Yahagi
Civil Engineering Department, Kandenko Co., Ltd, Japan

Yoshitada Tuburaya
Underground Power Civil Engineering Department, Tokyo Electric Power Services Co., Ltd, Japan

Akira Nagata
Construction Equipment Technology Department, Kumagai Gumi Co., Ltd, Japan

ABSTRACT: In Japan, power cables are encased in cylindrical or box-shaped reinforced concrete tunnels with cross-sectional openings of 3 to 20 m^2. At present, the space inside these tunnels, except for passageways, is congested with cables and supporting struts. When a concrete shell deteriorates as a result of chloride contamination, these cables must be removed so that repair work can be carried out safely and efficiently. However, transferring the cables is a difficult and arduous process, and a method that allows tunnels to be repaired without transferring cables has long been awaited.

Repairing a deteriorated concrete structure mainly consists of chipping off the deteriorated part and filling it with new material. Our research into the chipping process was presented at the 15th International Conference in Toronto in 1989.

This paper discusses our subsequent research into a method of automating the filling and finishing work.

1 INTRODUCTION

In repairing power cable tunnels with salt damage, the affected concrete wall is chipped off and new material is filled in the chipped part. This paper describes an automated repair system that improves the efficiency and quality of the filling and finishing work.

The major repair methods for reinforced concrete structures include surface treatment to prevent deterioration, crack repairing, and section repairing. The section repairing method is employed when the surface concrete is stripped off over relatively a large area, thereby necessitating a new wall surface. In this method, the lining concrete which is deeply contaminated with salt or embrittled concrete is chipped off, and new material is filled. This improves the durability of the overall structure by recovering the soundness of the lining concrete. Filling material can be by batching, prepacking, forming, or spraying. In selecting an automated method that can be used to repair any section of a power cable tunnel, we considered specifically the application region (corners in a rectangular tunnel, areas around cable supporting struts, and areas behind cables) and workability, including material filling and surface finishing behind reinforcing steels. We concluded that the spraying method offered the best application performance, although we know that certain problems must be solved before the method can be used for all repairing conditions. We then proceeded to develop a filling system using the spraying method.

The properties of the filling materials, functions of the shotcrete robot, and specifications of the backup rear components are outlined below.

2 DEVELOPMENT CONDITIONS

Power transmission tunnels in Japan are generally circular or rectangular, as shown in Figures 1 and 2. Tunnels are constructed with precast concrete or cast-in-place concrete.

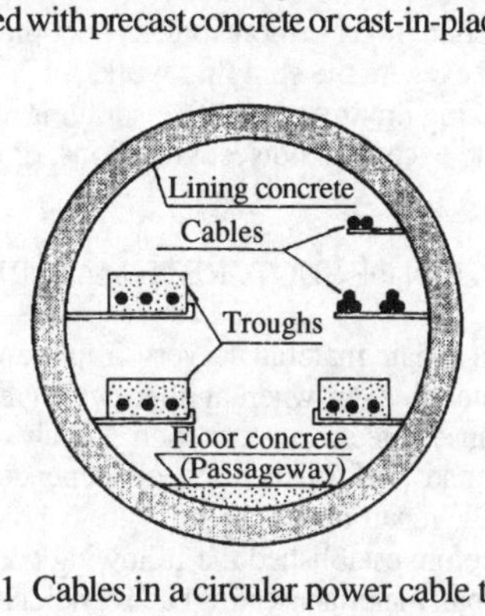

Fig. 1 Cables in a circular power cable tunnel

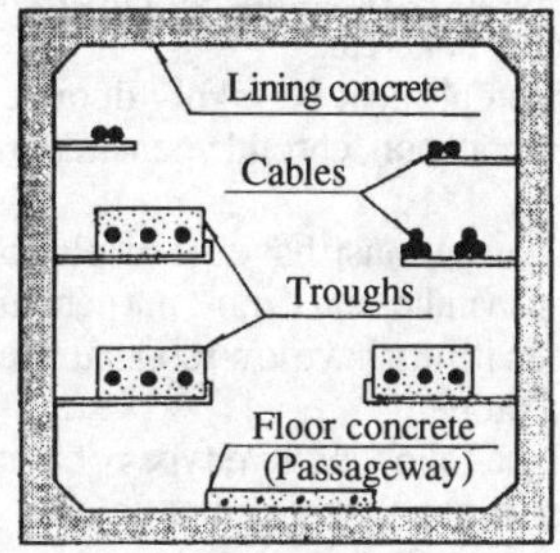

Fig. 2 Cables in a rectangular power cable tunnel

Since power transmission tunnels located near the coast are susceptible to deterioration from the salt-laden air, a robot capable of efficiently repairing damaged concrete has been needed.

The robot we developed is intended to repair rectangular tunnels. First, a chipping robot breaks up the deteriorated concrete wall using water jets. The new robot then sprays shotcrete material on it. The filling robot was developed with the following conditions in mind.

1) Transportability: The robot must be of block and assembly construction so that it can be passed through a 750 mm diameter manhole. Moreover, the assembled robot must be less than 600 mm wide so that it can move through tunnel passageways.

2) Operability: The robot must work automatically or semi-automatically (manually) from a centralized control source.

3) Flexibility of operating range: The robot must be able to automatically spray side walls and ceiling corners where no obstacles are present. For other areas where obstacles such as cable troughs exist, spraying can be performed through a nozzle semi-automatically. The minimum clearance between the sprayed area of the wall and the cables is 100 mm.

4) Work smoothness: The shotcrete plant on the rear dolly must be capable of easily manufacturing shotcrete material inside a tunnel.

5) Spraying capacity: The spraying capacity must be 5.0 m^2/h.

6) Labor-saving: No more than five people should be required to execute the spraying work.

7) Working environment: The spraying work must not generate excessive noises, vibrations, or dust.

3 SELECTION OF SHOTCRETE MATERIAL

Selecting a repair material is very important. A low-quality material, even when applied with great care, is likely to cause defects to occur soon. Repair work duration, cost, and performance are all dependent on the quality of the repair material.

We therefore established the following conditions to evaluate repair materials, testing each candidate material.

1) The material must be strong: it must have better bonding, flexural, tensile, and compressive properties than the existing concrete.

2) The material must be highly durable: it must be resistant to carbonation, chrolide penetration, and cracking.

3) The material must have good placing ability: it must be easy to handle, and it must not generate excessive dust. Moreover, it must have low rebound ratio, and fill the intended area properly.

We tested the following five types of materials.

1) Rubber-base polymer cement mortar
2) Resin-base polymer cement mortar (A, B)
3) Cement mortar with acrylic fiber
4) Plain mortar (c:s = 1:3)

The test results are shown in Table 1.

As the table indicates, each material tested was stronger than the existing concrete, all had similar strengths relative to each other. However, cement mortar with acrylic fiber is superior in durability and placing ability to the other materials tested.

We therefore selected it as the repair material.

4 STUDY OF REPAIR SYSTEM

4.1 *Properties of the finished surface*

(1) Development of a spraying nozzle
A conventional spraying nozzle requires at least a 300 mm clearance between the wall to be sprayed and the nozzle head. This makes it impossible to insert the nozzle head into the 100 mm clearance behind the cables and spray the surface there. Moreover, the structure of a conventional spraying nozzle does not allow walls to be sprayed at a right angle, which means that the shotcrete material does not bond well, and is highly likely to break away or rebound.

The nozzle we developed eliminates these disadvantages and enables walls to be smoothly and efficiently sprayed. The elbow-shaped head of the new nozzle is connected to the end of the nozzle pipe and the nozzle opening is perpendicular to the nozzle pipe axis. This structure allows the nozzle pipe to be kept parallel with the wall. Furthermore, the air nozzle for the new spraying nozzle is designed to be inserted from the back of the nozzle head so that air is directed towards the center of the spraying opening.

Figure 3 shows the nozzle head we developed.

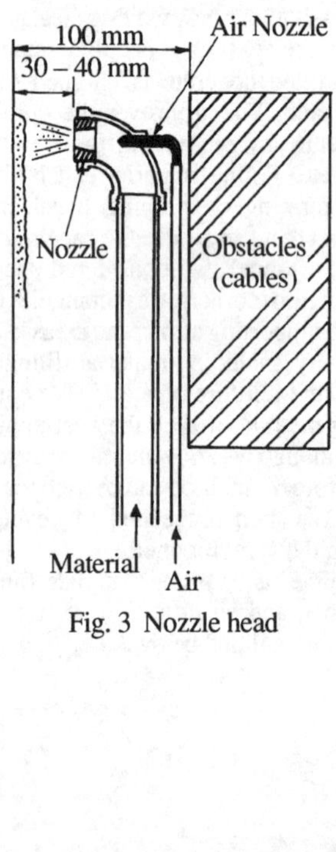

Fig. 3 Nozzle head

Table 1 Test results of repair materials by splaying methods

Item	Material	Rubber-base polymer cement mortar	Resin-base polymer cement motar A	Resin-base polymer cement motar B	Cement mortar with acrylic fiber	Plain motar (c:s = 1:3)	Note
Property	Bonding Strength (kgf/cm²)	22	31	23	23	19	Standard air curing at the age of 91 days: Temp. 20° C Moisture 60%
	Flexural Strength (kgf/cm²)	123	75	101	87	87	
	Compressive Strength (kgf/cm²)	571	398	603	750	677	
	Tensile Strength (kgf/cm²)	58	31	48	37	30	
Durability	Depth of Carbonation (mm)	4	16	9	0	2	Age: 180 days
	Depth of Chrolide penetration (mm)	11	18	11	8	15	
	Cracks	Yes	No	Yes	No	No	
Placing ability	Level of material filling around rebars	many voids	many voids	many voids	no void	many voids	
	Level of dust: Necessity of mask	Yes	Yes	Yes	No	Yes	
	Rebound ratio (%)	29 – 46	16–32	14–53	1	20 – 30	
	Completion time of hand trowel finish (min)	10	20	20	60	20	
Spraying method		Dry	Dry	Dry	Wet	Dry	

░░ : Unsuitable as repair material

(2) Optimum conditions for shotcrete on each area of the wall

Even when the new nozzle is used, the spraying method does not always produce a smooth finished surface. It has some irregularities. We therefore performed experiments intended to establish the optimum shotcrete conditions for producing the smoothest possible finished surface.

To study the optimum spraying conditions, we divided a tunnel wall into side walls, ceiling, and corners, and evaluated the effect of nozzle movement, nozzle travel speed, distance between nozzle and wall, and air spraying pressure. We found that nozzle movement has the greatest effect on the finished surface. The experiments proved that the best finished surface can be obtained by three-layer spraying (the nozzle moves in a transverse direction and sprays the wall three times) in which the final layer is sprayed with the pitch of the nozzle reduced by one-half. The finished surface we obtained in the experiments had an average roughness of about 5.0 mm.

Figure 4 and Photo 1 show the shotcrete condition and the finished surface, respectively. Table 2 shows the optimum conditions for shotcrete determined as a result of the experiments.

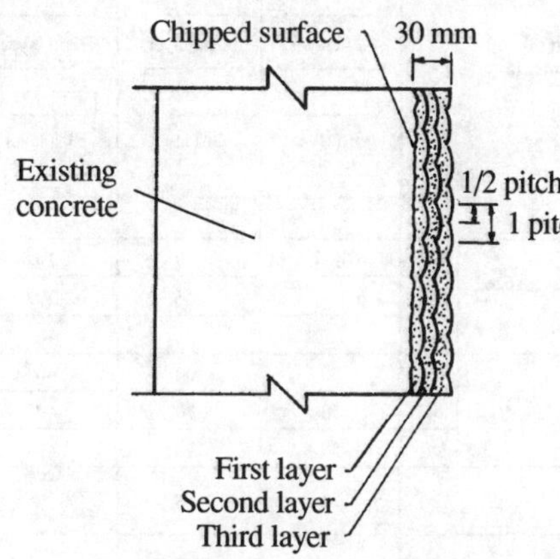

Fig. 4 Shotcrete condition

4.2 *Development of shotcrete robot*

(1) Development criteria

A shotcrete robot must satisfy the following field working conditions.

1) It performs spraying work with the tunnel's cables kept in place.

2) It ensures labor-saving and high-efficiency execution.

3) It can cope with tight working environments.

4) It ensures high-quality execution.

5) It can enter into a tunnel through a manhole (750 mm diameter).

6) It can spray every surface of a tunnel, excluding the floor.

7) It's spraying capacity is more than double the

Photo 1 Finished surface

Table 2 Optimum conditions for shotcrete

		Side walls	Ceiling	Corners
Movement of shotcrete nozzle		Moving route •1st layer •2nd layer •3rd layer	Moving route •1st layer •2nd layer •3rd layer	Moving route •1st layer •2nd layer •3rd layer
Nozzle travel speed (m/min)	1st	6.0 – 7.5	10	10
	2nd	6.0 – 7.5	6	10
	3rd	7.5	6	10
Distance (mm)	1st	50	100	40
	2nd	50	70	40
	3rd	50	70	40
Air pressure (kgf/cm^2)		4	5	5
Pump volume (l/min)			3.6	
Shotcrete thickness (mm)			30	
Shotcrete ability (m^2/h)			5.0	

chipping capacity, and it can spray coats as thick as 30 mm.

(2) Construction and functions of the robot

The robot we developed is a three-axis orthogonal automated robot that can automatically spray the side walls, ceiling, and corners of a rectangular tunnel. The robot

Table 3 Specifications of shotcrete robot

Item		Specification		
Dimension		L x W x H 2,600 x 930 x 2,460 mm		
Axis		X axis	Y axis	Z axis
	Motor	AC Servo 500W	AC Servo 500W	AC Servo 500W
	Stroke	1,500 mm	1,560 mm	500 mm
	Speed	12 m/min	12 m/min	12 m/min
	Driving	Rack & Pinion	Ball screw	Ball screw
	Weight	150 kgf	45 kgf	30 kgf
Nozzle	Speed	Max. 10 m/min		
	Pich	1 mm		
	Weight	5 kg (with hose)		
Control system		Teaching & playback		
Total weight		300 kg		
Protection		Waterproof		
Detector		Distance detector, Touch detector		
Power supply		200 V		
Carriage		4-CASTER, outrigger		

Table 4 Rear Dolly Components

Carrige	Components
Water tank — Compressor	Water tank (for mixing and washing) Shotcrete compressor
Shotcrete plant	Mortar pump Mortar mixer Belt conveyor Water measure (for mixing) Operation panel Control panel
Material transporter	Materials (package)

consists of an x-axis mover, a y-axis mover, a z-axis mover, a nozzle locking device, and an electric controller.

The robot has a nozzle head which is designed to work with cable-to-wall clearances as small as 100 mm, and can be operated remotely to allow spraying in tight work spaces. The robot moves maximum 1,500 mm longitudinally, 500 mm transversely, and 1,560 mm vertically within a tunnel.

Figure 5 is the schematic view of the shotcrete robot.

(3) Robot specifications

The shotcrete robot is designed to continuously spray all wall surfaces except the floor in each 1,500 mm space between the power cable supporting struts.

Table 3 shows the specifications of the robot.

4.3 *Backup rear components*

The spray method of concrete repair requires that shotcrete material be fed to the robot. The material is first mixed with water in a mixer, and the resulting mixture is sent to the hopper of a mortar pump. The mortar pump pressure feeds the mixture into the shotcrete robot.

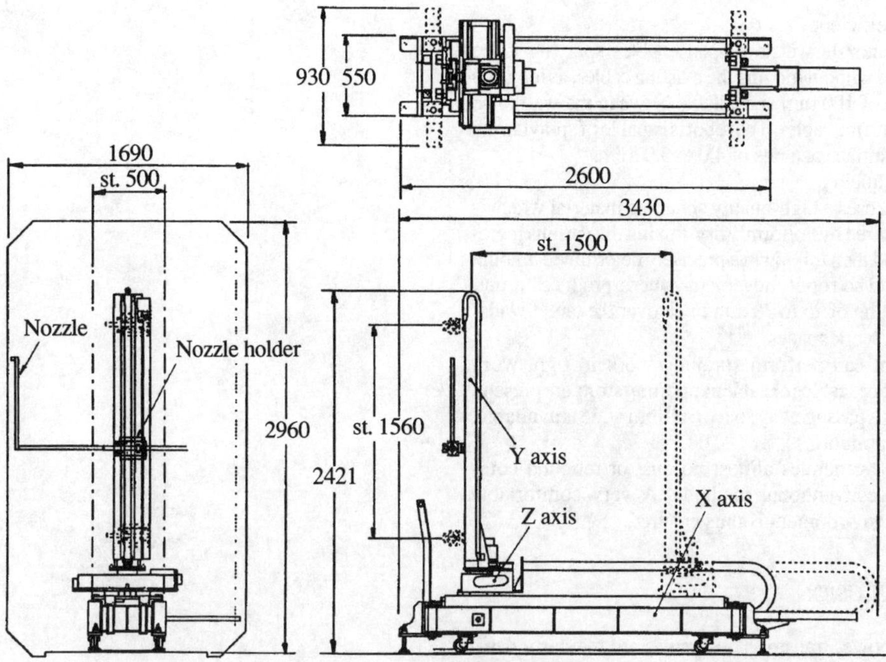

Fig. 5 Shotcrete robot

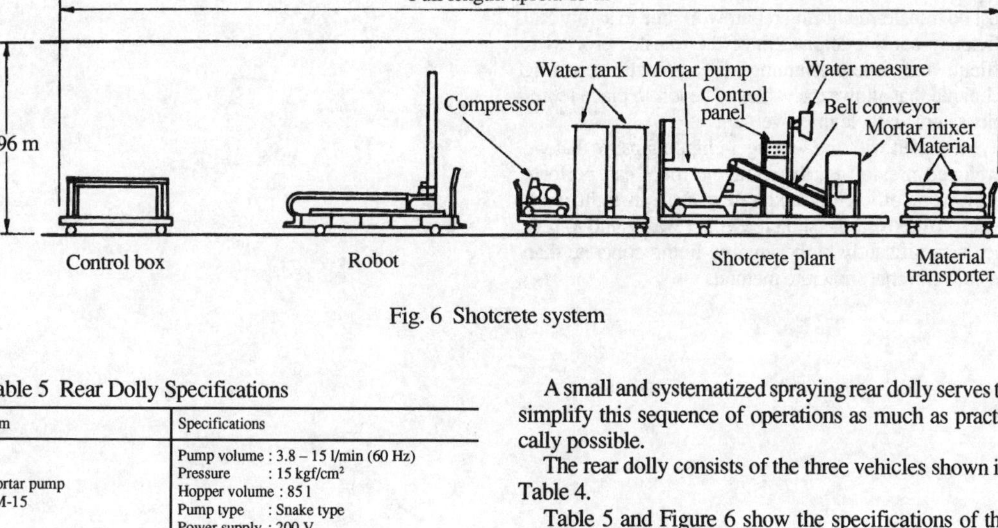

Fig. 6 Shotcrete system

Table 5 Rear Dolly Specifications

Item	Specifications
Mortar pump DM-15	Pump volume : 3.8 – 15 l/min (60 Hz) Pressure : 15 kgf/cm² Hopper volume : 85 l Pump type : Snake type Power supply : 200 V
Mortar mixer	Volume : 50 l Power supply : 200 V
Compressor SU-125PB	Volume : Max. 205 l/min Pressure : Max. 7 kgf/cm² Power supply : 200 V

A small and systematized spraying rear dolly serves to simplify this sequence of operations as much as practically possible.

The rear dolly consists of the three vehicles shown in Table 4.

Table 5 and Figure 6 show the specifications of the major components and a general view of the shotcrete system, respectively.

5 FEATURES OF NEWLY-DEVELOPED ROBOT

The shotcrete robot we developed has the following features.

(1) High efficiency
The new nozzle we developed enables spraying to be performed without moving the existing cables, as long as a clearance of 100 mm is available between the wall to be sprayed and the cables. The robot is capable of spraying the concrete surface at a rate of 4.0 to 5.0 m^2/h.
(2) High quality
The robot uses a high-quality shotcrete material which is manufactured by uniformly dry-mixing the various ingredients, and then mixing in a precisely determined amount of water. The robot movement pattern produces a uniform coating of 25 to 30 mm thick over the target walls.
(3) Tight work spaces
The robot can perform spraying work in tight work spaces where cables or cable supporting struts are present, as long as a passageway space 600 mm wide is available.
(4) Low pollution
The robot generates almost no dust or rebound compared to conventional methods. A very comfortable working environment is thus ensured.

6 CONCLUSION

Seaside power transmission tunnels are regularly damaged by the saline air, and a high-efficiency and high-performance repairing technique has long been sought.

The restricted working environment of these transmission tunnels has meant that the power cables they house must be transferred during repair work due to safety and efficiency considerations. Since the transfer process is difficult and time-consuming, an automatic repairing technique that allows the cables to be left in place represents a considerable improvement.

When used with the water-jet-chipping robot that we developed previously, the shotcrete robot can perform repair work efficiently and safely even in tight work spaces. Moreover, the shotcrete robot was found to produce a significantly higher-quality lining concrete than the conventional shotcrete method.

7 General tunnelling topics

Tunnelling and Ground Conditions, Abdel Salam (ed.) © 1994 Balkema, Rotterdam, ISBN 90 5410 363 9

Insurance control of tunnelling projects

A.F.Abouleid
Faculty of Engineering, Cairo University, Egypt

Y.A.El-Kady
General Organization for Housing, Building and Planning Research, Egypt

ABSTRACT : The control sequence of the Egyptain Insurance Pool has been recently applied to the main tunnel of the east bank section of the Greater Cairo wastewater project . The procedure required the submission of four types of reports, namely ; Stability report , Risk Definition report , Progress report , and Final report . Standrad formats for these reports were filed for contract 3 tunnel between Ameria and Souk El-Samak, contract 4 tunnel from Souk EL-Samak to Abdeen , and contract 12 between Abdeen and Ein El-Siera . This paper presents some observations and offers some recommendations concerning efficient control of insurance of large tunnelling projects . More information which could be included in the insurance reports is also recommended particulary for tunnels and underground structure .

1 INTRODUCTION

The license to start works for building or construction shall not be granted unless the applicant presents an insurance policy . The insurance policy covers the liabilty of engineers for the damages inflicted on the owner of the project or others due to any drawbacks that threaten the safety of the structure .

The application of the license shall be submitted to the Egyptain pool for civil liability insurance for building works . Enclosed with it shall be all the documents and data concerning administrative information about the designing and supervising engineers, architectural and structural executive drawings and the calculation notes of the structural design.

The insurance pool shall entrust one of the engineering consulting offices approved by it (control office) with reviewing the designs of the project subject matter of the insurance policy, and controlling the works during the construction period. The control operations of the consulting office require the submission of four types of reports. These reports are ; the Stability report, Risk definition report, Progress report, and Final report.

The sequence of control actions of the insurance pool has been applied to the main tunnel of the east bank section of the Greater Cairo wastewater project. This paper presents some observations and remarks on the work and the problems that arose during the progress of the works, which were noticed while carrying out periodical inspection of the execution of the tunnel. Recommendations are presented for the identification of the possible risk sources or unusual construction risks which may be faced in the construction of a high-risk projects such as tunnels or underground structures. These recomendations provide more information about the risk sources which could be included in the insurance pool reports to allow the insurance pool to appreciate fully the risks relevant to the construction of tunnels, leading to efficient technical control of the work.

2 EXPLANATION OF INSURANCE REPORTS

As mentioned above the consulting control office is required to submit to the insurance pool four types of reports during the control operations. Special format for each type of reports shall be forwarded to the insurance pool:

1. Stability report; this report is to be established before the very begining of the works. The report should be forwarded to the insurance pool within one week from

the moment the control office is in possession of all the necessary data. The control office, after having examined and analyzed in detail the information about the soil characteristics and all elements of the foundations and the structure, has to give his opinion about the risks of collapse during both the construction and decennial period.

2. Risk definition report; this report is to be prepared in order to complement and expand on the information contained previously in the stability report. This report will allow the insurance pool to appreciate fully the risks relevant to the policy subscribed. More particularly, the insurer will be able to estimate the value and nature of any item permanently fixed to the structure, and likely to be damaged by losses occurring to the works. In the conclusions, which constitutes the essential part of the report, the control office, after having analyzed in detail all the elements of the report and contributing to the risks, shall give his opinion about the risks of collapse and losses during both the construction and decennial period.

3. Progress report; concerning the technical control of work, the control office shall draw up several progress reports to inform the insurance pool about the progress of work. However, the control office is bound to inform the pool, in any case and without delay, whatever constitutes a major alteration of the risks,stated in either the stability or risk definition reports, during the construction period. The major alteration of the risk shall be fully described and the control office gives his detailed opinion and conclusions expressing any technical reserves.

4. Final report; this report may be considered as a report that brings to an end the task and duty of the control office. The control office shall indicate in the report that the risks relevant to the clauses of the insurance policy can be considered as normal, or suspend some guarantees of the policy .

3 DESCRIPTION OF WORKS CONCERNING THE INSURANCE POLICY

The works to be insured include the construction of the main wastewater tunnel of the east bank section of Greater Cairo. The construction of the tunnel was carried out under three contracts ; contract 3 between Ameria to Souk El Samak, contract 4 between Souk El Samak to Abdeen and contract 12 between Abdeen to Ein ElSiera.

The primary lining of the tunnel was constructed by using bolted pre-cast reinforced concrete segmental rings. Bentonite slurry machine was used to drive the tunnel, together with the use of compressed air in some parts of the tunnel.

The internal diameters of the tunnel are 5.0 and 4.0 meters. Shafts have been located along the tunnel route at the changes of direction, diameter and gradient. Shafts were numbered 1 to 5 in contract 3 , 6 to 13 in contract 4 , and 14 to 20 in contract 12. Both shafts and tunnel were lined with acid resistant bricks pointed with acid resisting epoxy mortar.

4 REMARKS AND PROBLEMS CONCERNING CONSTRUCTION WORKS

Some remarks and problems were observed during the construction of the tunnel while carrying out the insurance control operations.

4.1 Summary of the remarks

The remarks and observations can be summarized in the following:

1. The underground water table rised at various areas along the route of the tunnel. As a result, water flooded in the basements of some neighbouring buildings.

2. Subsidence of the subsoil occurred at some places, particularly where leakage of compressed air at the face of the tunnelling machine took place, creating collapse of soil at the ground surface.

3. Leakage of ground water into the tunnel through the primary lining. The quantity of water leakage was relatively large near shaft No. 16. The leakage of water was also noticed at some places after the tunnel has been lined with the acid resistant bricks.

The above remarks were reported in the insurance progress reports. The contractor and the project consultants were notified by the above remarks in order to provide the control office with the studies about the ground movement around the tunnel, underground water table changes, and the remedial measures to be carried out to prevent water leakage into the tunnel.

4.2 Summary of execution problems

The important problems that arose during the construction are:

1. A distortion (bulge) occurred in the pre-cast reinforced concrete lining of the tunnel with a length of 32 meters between shaft No. 3 to shaft No. 1 (contract 3). The difference between the levels of the bottom points of the lining rings reached about 65 mm , and the vertical diameter of the tunnel decreased by an amout ranging between 125 to 190 mm. The remedy was carried out by strengthening of the invert of the tunnel section. This strengthening involved heavy reinforcing the cast in situ concrete at the invert and making it integral with the primary lining.

2. Cracks appeared in the walls of a Mosque and also in a wall under the ramp of the fly-over bridge in the vicinty of El Azhar street near shaft No.11 (contract 4).This cracks occurred since the Mosque and the bridge wall suffered settlement damages during the drive of the tunnel under this region. Settlement ceased soon after the tunnelling machine had passed, and remedial measures have been carried out later on according to the project insurance cover.

3. A depression in the tunnel lining occurred near shaft No. 16 (contract 12). This problem occured due to sudden escape of compressed air (blow) creating collapse of the subsoil at the front of the machine shield and a depression in the tunnelling machine. The solution adopted was to drive the shield through a vertical curve to reach the correct level and inclination of the tunnel. Moreover, the cast in situ reinforced concret, which is used in the secondary lining of the lower part of the tunnel lining, was increased.

4. Cracks of about 5 mm were noticed in the upper part of the primary tunnel lining near the bolt holes between shaft No. 19 and shaft No. 20. This part of the tunnel was executed through limestone rocky formations. The cracks occurred due to that the upper part of the tunnel was subjected to concentrated loads, because of the movement of the rock blocks as a result of not completing the process of injection of all gaps outside the lining. Remedial measures have been carried out by making additional injection at the critical parts of the tunnels and injection of the cracks with epoxy resin.

In the progress reports, the contractor was requested to follow up the remedial works , after being executed, by carrying out field measurements of the levels of the upper and lower parts of the tunnel and any changes in the diameter of the tunnel section at the sites where the problem arose, in order to verify the structural stability of the tunnel.

5 APPRECIATION OF THE RISKS RELEVANT TO THE WORKS

Regarding the technical control of the works, several progress reports were presented to inform the insurance pool about the progress of the work together with detailed reports about the remarks which were noticed during periodical inspection of the tunnel.

The contractor and the project consultants were also notified by the remarks included in these reports in order to provide the control office with the detailed studies and field measurements previously requested in the progress reports.

While inspecting the tunnel, we obeserved that the water leakage into the tunnel had been remedied at most places. However, there are some places still need to be remedied to prevent water leakage. We recieved some measurements of the ground surface subsidence and the changes of the underground water level at some places, but we have not recieved the measurments for other remaining areas.

Therefore, it was concluded in the progress reports that the risks related to the works can be considered as normal, except for the observed remarks and problems, for which the technical information collected does not allow us to appreciate the relevant risks. Consequently, the insurance pool notified the concerned parties of the project that these incidents are considered as a violation of the conditions of the insurance policy and shall invalidate the insurance cover for the project.

Several meetings were held with the project consultants to review all the information and studies previously requested by the control office, relevant to the whole of works, in order that the control office can express the final report of the project.

Later on, the control office recieved records of the dimensional checks of the tunnel section, which demonstrates that no movement occurred to the tunnel structure. The information obtained about the ground water levels shows no significant changes in the vicinity of the tunnel. Regarding the water leakage into the tunnel, the tunnel has been finally inspected by representatives of the contractor, project consultants, and wastewater authorities. It was found that all leaks were sealed and that the tunnel is in an acceptable condition in accordance with the specifications.

Accordingly , the control office

forwarded the final report to the insurance pool, which brings to an end the duty of the control office.

6 CONCLUSIONS

As a result of efficient control actions of insurance applied to a large tunnelling project, the following conclusions can be drawn from this study:

1. The most important source of risk for tunnelling projects is the variation of the nature of the subsoil formation along the tunnel route. It is necessary to investigate fully the physical and engineering properties of the soil that may be encountered during construction. Geological study is also needed to determine the extent of the soil strata and its stability below tunnel.
2. The fluctuation of the underground water level in the vicinty of the tunnel should be recorded during construction.
3. Field measurements of the subsoil movement around the tunnel and the displacement of the tunnel should be carried out during the construction stages.
4. Periodical inspection during the execution of the tunnel is neccessary to observe any leakage of water into the tunnel.
5. The insurance pool allow the control office to establish the stability report of the project within one week. In view of such high risk project, adequate time should be allowed which will lead to better appreciation of the risks relevant to the works.

Tunnelling and Ground Conditions, Abdel Salam (ed.) © 1994 Balkema, Rotterdam, ISBN 90 5410 363 9

Chemicals for tunnel waterproofing

H.F. Fenner
Sika AG, Zurich, Switzerland

ABSTRACT:
Old railway-tunnels had a life expectancy of approximately 100 years prior to rehabilitation work. Todays traffic-bearing tunnels, including those for highspeed-railways with their numerous electro-mechanical installations, have increased the required quality level of watertightness. This paper describes the historical evolution and todays state of the art in tunnel waterproofing. Definitions of permissible leakage rates, related to different uses of tunnels and underground structures are being documented. Shotcrete and sheet membrane systems are particularly focused on.

INTRODUCTION

For the past 80 years, SIKA has been developing a range of technologies for the waterproofing and sealing of tunnels.

SIKA has supplied and continues to supply the principal waterproofing product systems, while in some countries the Company also offers a full contract service.

- Rigid waterproofing	Mortar admixtures Joint sealing compounds Gunite admixtures (preliminary sealing)
- Bituminous waterproofing	Bituminous felt systems Polymer-modified bituminous felts, paints and coatings
- Integral concrete water- proofing	Waterproofing admixtures Waterstops Injection resins Repair mortars
- High polymer water- proofing membranes	Polyvinylchloride (PVC) and Hypalon (CSM) membranes
- Liquid-based systems	Bitumen, polymer-modified bitumen, polyurethane, epoxy resins, polyester resins

With all these various options, we are able to offer a much wider range of products than most of our competitors.

2. A BRIEF HISTORY OF TUNNEL WATERPROOFING TECHNOLOGY

- Even in the days of ancient Babylon underground workings were often waterproofed bitumen.
- **In the 19th century:**
  Mining engineers used timber or cast-iron ring linings to seal their shafts and tunnels. The Museum of Mining in Bochum, Germany, shows that cast-iron ring linings ere in common use in both Germany and England, the only difference being that the Germans plugged the joints with timber while the English used lead.
- **Late 19th century:**
  In the new railway age, railway tunnels were lined with quarry stone or brick with mortared joints. Water was drained away via slots or channels.

- **Early 20th century:**
The first underground railway tunnels were constructed in Budapest, London and Moscow, lined with cast-iron ring segments.

- **Pre-1916:**
Early remedial works in railway tunnels (e.g. St. Gotthard line) were designed to afford protection against:
1) Soft waters from granite deposits, low in calcium carbonate, which attack and dissolvelime compounds.
2) Waters low in calcium carbonate with critical levels of aggressive free carbonic acid.
3) Calcium sulphate (gypsum) in concentrations larger than 0.03% (0.3 g/litre of water) and magnesium sulphate (Epsom salts) in concentrations larger than 0.2% (2 g/litre of water), which cause mortars to break up through crystalline expansion.

- **Remedial measures (pre-1916)**
1) Pointing joints with rapid-hardening cement.
2) Rendering with rapid-hardening cement.
3) Packing joints with lead wool.
4) Raking out joints. Filling with a layer of rapid-hardening cement and a special organic filler material; packing with lead wool, tar-impregnated rope and sealing mastic.
5) Attaching asbestos cement sheets or zinc plates to the tunnel arch.

- **1916-1922:**
Electrification of the St. Gotthard line. Large-scale use of quick-setting cement admixture Sika-4a and waterproofer Sika-1, particularly for sealing tunnel roofs.
All tunnels between Lucerne and Luino were treated. Following extensive trials, three principal techniques were adopted, all requiring the use of specially trained personnel:

1) Sound masonry linings: the joints were cleaned out with a pressurized water hose (4-6 bars), then repointed with a cement mortar containing Sika-3.
2) Linings of random rubble or small brick: a waterproof layer of cement rendering incorporating Sika-4 was applied over the whole surface, often reinforced with wire mesh. The substrate was thoroughly cleaned to remove all soot and foreign bodies, and non-rusting hooks were used to anchor the wire mesh.

3) In the Coldrerio and Massagno Tunnels, where the combination of frost action and hot gases from the locomotives had seriously damaged the original linings, guniting was used to effect repairs.

-**Post-1921:**
Waterproof renderings and preliminary sealing of tunnels in Germany, Austria, Italy, Spain and France.

-**1934-1935**
Construction of Hochtor and Mittelörl Tunnels on the Grossglockner route. Outer face of arch waterproofed with mortar containing Sika-1, with selective use of corrugated iron sheeting to drain areas subject to heavy inflows of water.

- **circa 1935:**
Some use of polyisobutylene membranes to seal sections of the Siegfried Line in Germany.

- **1942-1944:**
First use of the "Oberhasli technique", which is still used in tunnel construction today. Named after the power station project in Switzerland for which it was first developed, the technique involves the use of rubber hoses to channel away inflowing water. The hoses are grouted in place by hand with a mortar incorporating the quick-setting waterproofing compound Sika-4a, then withdrawn to leave a preformed drainage channel.

- **1949-1959:**
The Semmering Tunnel in Austria was waterproofed throughout using "Oppanol" sealing membranes. This system suffered a setback following numerous instances of water infiltration.

- **1945-1965:**
Multiple layers of torched-on bituminous felts, applied in strips 1 m wide, often with the aid of special travelling rigs. Following a fire in a Munich underground railway tunnel in 1965, the

system was abandoned on safety grounds and superseded by other systems developed for nimed tunnels.

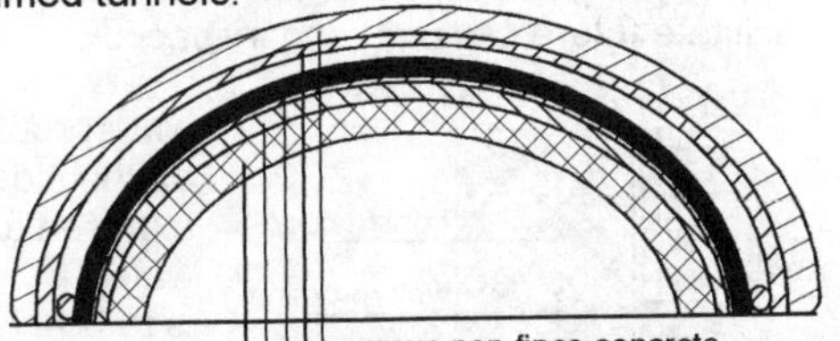

(From a 1964 Sika handbook on waterproofing techniques for tunnels and galleries:)

- **1956:**
Kaptenberg road tunnel in Austria. Pioneering use of a polymer-modified cement-based waterproofing slurry.

- **1958:**
Mechanized installation of channel drainage system as a more advanced variant of the "Oberhasli technique". Flat-section hoses or half-channels were gunited in place using a cement gun supplied by Aliva.

- **1959:**
Dürnstein road tunnel. Waterproofing system based on use of bituminous felts.

- **1959-1963:**
Jauntal railway in southern Austria: selective use of porous concrete drainage elements in areas subject to heavy water infiltration.

- **1960-1965:**
20-30 road tunnels constructed over this period in Germany, Austria and Switzerland were waterproofed with a spray-applied coating of Sikaplan Colmasyn glass fibre reinforced polyester resin.

- **circa 1960-1972:**
The inner face of the finished concrete tunnel lining was sealed with a polymer-modified cement-based waterproofing slurry such as Sika-101.

- **1965-1968:**
First use of the system that is now standard practice for road tunnels, i.e. polymer waterproofing membranes mechanically anchored at selected points. The work was carried out by trained Sika personnel. In its early form the system did not incorporate a protective padding layer of drainage quilt, and the membrane itself was too thin (1 mm); consequently it tended to puncture easily.

- **circa 1965-1972:**
Liquid waterproofing coatings (e.g. neoprene, latex, bitumen) were used on tunnel construction projects in West Germany and Switzerland.

- **1969:**
The Rofla Tunnel in the Via Mala gorge in Switzerland was waterproofed with a polymer membrane continuously bonded to the substrate with a special spray-applied hot seal adhesive.

- **1972:**
Application of polymer membranes using travelling "up and over" installation rigs which can cover the whole tunnel arch floor-to-floor in a single pass (mechanized installation).

St. Gotthard Tunnel. Use of Sikaplan tunnel waterproofing membrane incorporating a "tell-tale" layer to aid detection of mechanical damage.

- **1981-1982:**
Continuous adhesive bonding technique used for the first time with a Sikaplan tunnel waterproofing membrane incorporating an integral drainage quilt.

The history of waterproofing in Austrian road tunnels

Period	Tunnel length (m)	Treated length (m)	%	Comments
pre-1950	4524	0	0.0	No effective
1950-1960	2265	324	14.3	waterproofing Incl. Seehof Tunnel in Tyrol
1960-1970	17'363	3669	21.1	Gunite/spray-applied coatings.
1970-1975	22'217	11'911	53.3	Using NATM and polymer membranes
1975-1980	58'890	47'185	80.0	Improved technology
1980-9185	38'394	32'699	85.2	
Total	**143'653**	**95'788**	**66.5**	**(mean figure)**

Source: "Basic principles for the installation and testing of tunnel waterproofing systems", Publication No. 365, Highways Research Department of Austrian Ministry for Economic affairs, Vienna 1988.

3. CONCEPTS OF WATERPROOFING

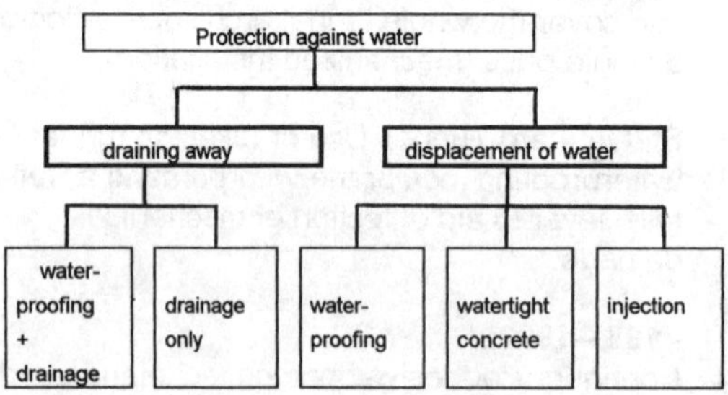

- Draining the water away is prefered to displacement of water.

- Mistakenly, sometimes drainage products are marketed as waterproofing.

- For driven tunnels, as much attention should be paid for the drainage as for the waterproofing. In case of Geotextiles, this includes chemical composition (possible alkality in contact with concrete) and longterm behaviour regarding hydraulic transmissivity under load (clogging).

Furthermore their fire-rating should be verified. A very important criterion for most Geotextiles is the weight / m². Diameters of less than 15cm for horizontal drainage pipe are often too small. The proper design of control shafts and pipes should facilitate future cleaning and inspection.

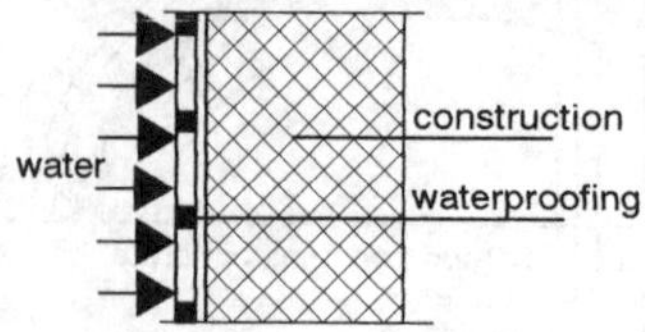

waterproofing on the side exposed to water- is to be prefered (outside)

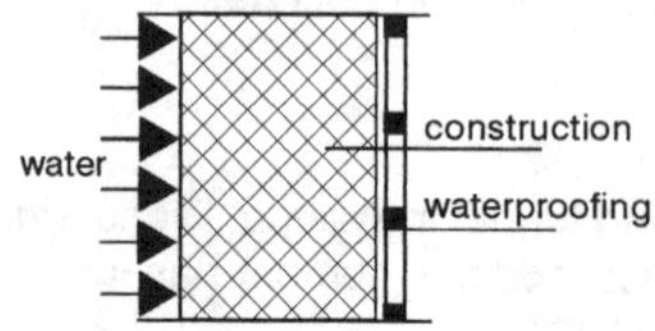

waterproofing on the water-averse side (inside)
-> aesthetical problems possible
-> problem with water pressure possible
-> construction is exposed to attack by chemically aggressive water

- blistering possible
- peeling off or mechanical damage possible
- in case of frost, problems possible

an important distinction for protection against water

- water pressure
- chemical agressivity of water

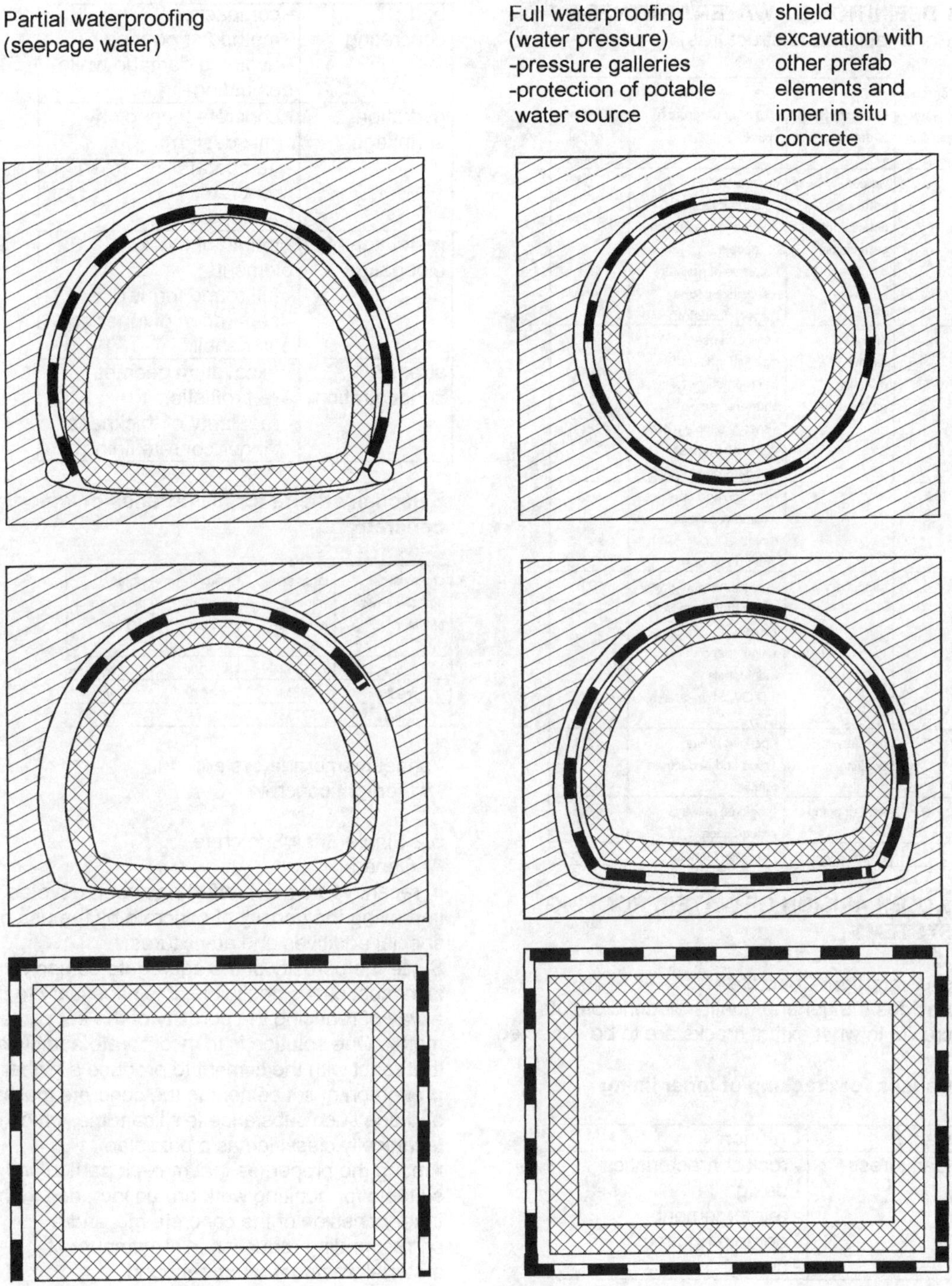
Partial waterproofing
(seepage water)
Full waterproofing
(water pressure)
-pressure galleries
-protection of potable
water source
shield
excavation with
other prefab
elements and
inner in situ
concrete

4. DEFINITION OF WATERTIGHTNESS

(For underground structures)

Degree of water-tight-ness	Characteristics of damp	Use of underground space	water inflitration l / m² per 24 hours
1	vapour diffusion from outside not permissible	dry rooms - clean air rooms - prolonged presence of people - storage of moisture sensitive goods (paper, foodstuff)	0
2	dry vapour diffusion permissible	- civil defense installations and its wet rooms like toilets, showers, etc. - energy supply rooms - general purpose underground space	0
3	dry	- storage and otherwise commercialized rooms - metro stations	< 0,001
4	almost dry	- road tunnels highway - road tunnels in frosty zones - high speed traintunnels (TGV, Shimkansen, ICE)	< 0,01
5	capillary soaking	- parking areas - road and pedestrian tunnels	< 0,1
6	light dripping	- regional railways - metro lines	< 0,5
7	dripping water	- sewage tunnels	< 1,0

5. COMPARISION OF WATERPROOFING SYSTEMS

5.1 Watertight Concrete

Besides the general quality of concrete, its crucial to what extent cracks are to be expected.

Causes for cracking of inner lining

Cause	Criterion
load stress	- rock characteristics - design - reinforcement - demolding time - curing
bad concreting	- compaction - method of pouring - avoiding damage while demolding
hydration shrinkage	- concrete technology (mix design) - temperature of fresh concrete - air-temperature
restriction of elongation	- length of block elements - full round formwork - separation of inner and outer shell
stress concentration	- excavation geometry - re-profilation - regularity of thickness of inner conrete lining

Selectian criteria for membranes or watertight concrete

chemical agressive-ness of water	inner shell without reinforce-ment	water pressure ≤ 3 bar	water pressure ≥ 3 bar
low	M	C/M	M
medium	M	C/M	M
high	M	M	M

- sheet membrane system M
- watertight concrete C

5.2 Single Shell Shotcrete

A) General

In recent years much effort has been devoted to improving the density of concrete by the use of special additives and admixtures.

Since the porosity of the aggregate components is in most cases negligible, it is necessary to find a way of reducing the porosity of the set cement matrix. One solution is to incorporate additives that react with the cement to produce a higher proportion of set cement in the concrete mass: and one such substance is silica fume, which is technically classified as a pozzolan.

Among the properties that make it particularly suitable for guniting work are an increase in the inner cohesion of the concrete mix and a corresponding reduction in the amount of material wasted through rebound.

B) Pozzolans

Pozzolans are substances which, when combined with calcium hydroxide, harden to form waterproof products. It may be useful to briefly look into the historical background. Both the ancient Greeks and the Romans were able to make a cement by adding volcanic ash - a natural pozzolan - to burned lime.

A suitable type of volcanic ash was found near the Italian town of Pozzuoli (Latin name: Puteoli) - hence the name "pozzolan".

In more recent times the role of synthetic pozzolans has become increasingly important. Fly ash is a waste product from the combustion of coal in thermal power stations;

slag is a by-product of steel production; and silica fume is a by-product of the metallurgy used to make silicon metal and silicon alloys. The main differences between silica fume and fly ash lie in the reactive component of silica fume and in its much smaller particle size, as expressed in the specific surface in m^2 / g.

While fly ash and slag have been used for many years, silica fume technology is still at a relatively early stage in many countries.

C) Silica Fume

Silica fume is obtained by reducing quartz (SiO_2) with carbon in an electric arc furnace. A portion of the partially reduced quarz evaporates in the form of silicon monoxide, which oxidizes again into SiO_2 as soon as it comes into contact with oxygen in the cooler zone of the furnace.

This SiO_2 condenses as amorphous silicon dioxide in the form of microscopically fine spherical particles. With an average diameter of around 0.1 μm, silica fume particles are approximately 100 times finer than cement particles.

The physical properties of silica fume, as compared with fly ash and cement, are summarized in Figs. 1-4.

The principal differences lie in the SiO_2 content and the relative fineness of the materials, characterized here by the specific surface.

D) Chemical reaction of cement with Pozzolans

The hydration of cement results in the release of calcium hydroxide, which, being readily soluble in water, constitutes the weakest component in the set cement. This manifests itself in the form of efflorescence - the white, chalky deposits that are a familiar sight on the surface of concrete and masonry.

However, if a pozzolanic material is added to the conrete mix, the SiO_2 present in such substances reacts with the calcium hydroxide to form a solid calcium silicate hydrate, which contributes significantly to the strength and density of the set cement.

Based o n this technology, a range of gunite admixtures has been developed for specific applications, giving the following improved performance characteristics:

- higher density
- lower porosity
- higher impermeability
- better resistance to chemical attack
- superior resistance to frost and deicing salts
- reduced rebound

All these properties are related in some way to the compressive strength of the conrete, which is normally adopted as the characteristic value.

The improved adhesion of the fresh gunite brings with it another advantage - and perhaps the most significant from a commercial point of view - namely a dramatic reduction in rebound wastage (half the normal quantity or less).

E) Applications

Various tests were carried out using the dry-spray method of guniting. In one series of tests a slurry of silica fume was fed via a prewetting nozzle into the stream of dry materials at the main nozzle, while in other tests silica fume was factory-mixed in powder form with the dry ingredients. Rebound wastage was reduced by up to 70%. Measurements of dust levels based on the use of silica fume in powder form or as a slurry dosed in the manner described revealed no measurable differences. The quality of the gunite is measurably imporved, particularly in terms of lower capillary porosity/water permeability and - depending on the product - higher strength.

However, really significant improvements in these parameters are only observed when specially formulated products are used, such as Sikacrete-PP1, a polymer-modified silica fume admixture. Designed for convenience of use on site, Sikacrete-PP1 is simply added to the other dry ingredients in the conrete mixer drum and mixed in the normal way. The following performance data, based on comparative measurements taken in the Hagerbach test gallery, highlight the gains in quality very clearly.

5.3High Polymer Membranes

see chapter 6

5.4Bituminous

- no longer in use in driven tunnels due to large ires in 1965
- In open cut tunnels the national roofing felts ith oxydized bitumen have been replaced by a newer generation of polymerised felts
- while with the old felts an addtitional layer of elt for every 3 meter of water pressure has been state of the art, the new polymerized bitumen need few layers
- the full bonding and hence the restriction of ater flowing behind the waterproofing is an advantage. Some points to watch are, imited load bearing capability, (bitumen is at approx. 100°C a "frozen liquid"), the need for dry substrate

5.5Combinations

Combinations of different groups of products are applied in special situations ,eg. as:

- watertight concrete construction and full circle sheet membrane
- horseshoe sheet membrane and watertight concrete at base slab
- watertight concrete and injection
- watertight concrete and in-built repair or drainage system (e.g. pipes)
- liquid applied or sheet membrane with U-shape steel

6 WATERPROOFING WITH SYNTHETIC SHEET MEMBRANES

6.1Which lining system - where

Lining system	Sealing against water	
Type of tunnel	without pressure	under pressure
Open-cut excavation	Partially lined (horeseshoe lining) A B C	Fully lined (A) B C D
driven tunnels smooth, even surface	partially lined / Fully lined A (B) C	Fully lined (A) (B) C D
driven tunnels irregular, uneven surface	partially lined / Fully lined A (B) C	Fully lined (A) (B) C D

6.2Lining systems

	Loose-laid	Fully-bonded
Single-layer	A	B
Single-layer, laid in sections, confined by waterstops with or without precautionary injection tubes	C	-
Double-layer, laid in sections, confined by waterstops with tubes for: - testing and eventually injection of a repair medium	D	-

6.3Single-ply installation with compartmentalization

Example: Floor slab construction joint, open tunnel construction method. The waterproofing surface is segmented to follow the concreting phases. For each project, an installation plan is normally drawn up for the water bars, taking into account the construction joints, expansion joints, installation sequence (concrete first then sheeting, or vice versa), waterproofing detail, mechanical protection measures against damage etc.

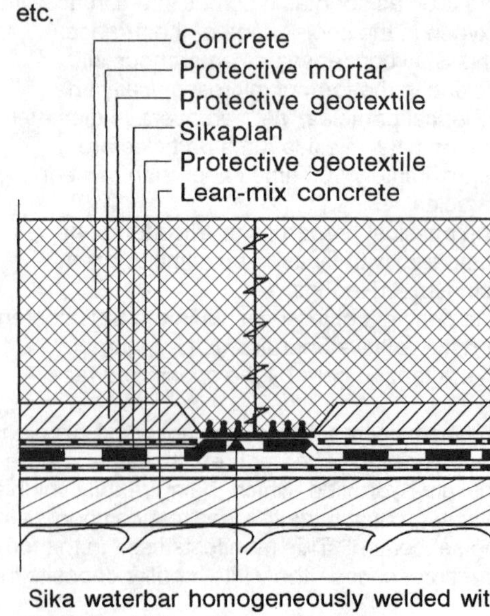

Sika waterbar homogeneously welded with Sikaplan high polymer sheeting

6.3a Open cut tunnel (Single ply high polymer sheet with compartmentalization)

1) Waterbar
2) hot-air welding of PVC membrane onto PVC waterbar
3) Installation sequence
4) Injection from inside

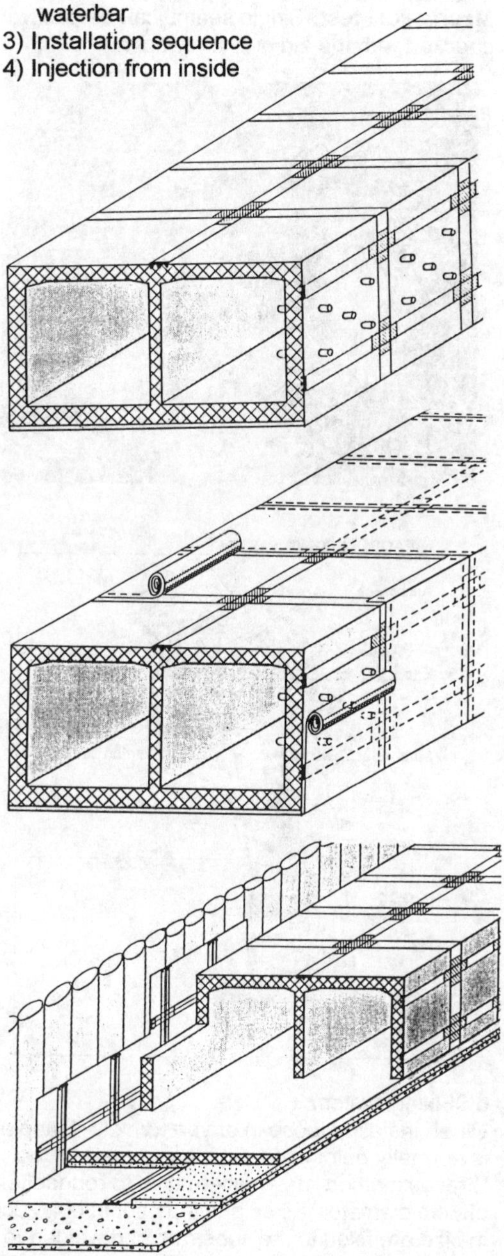

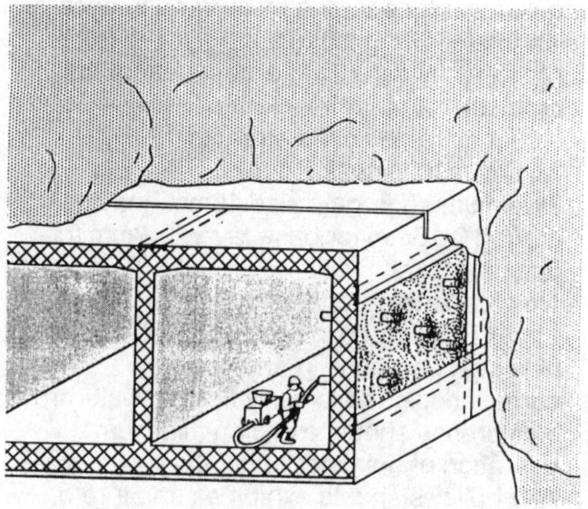

6.4Two-ply installation
The gaps between the 2 layers are for inspection and possible injection in the event of damage:
A. After installation of the high-polymer sheeting
B. After installation of the reinforcement
C. After conreting
D. At any later date

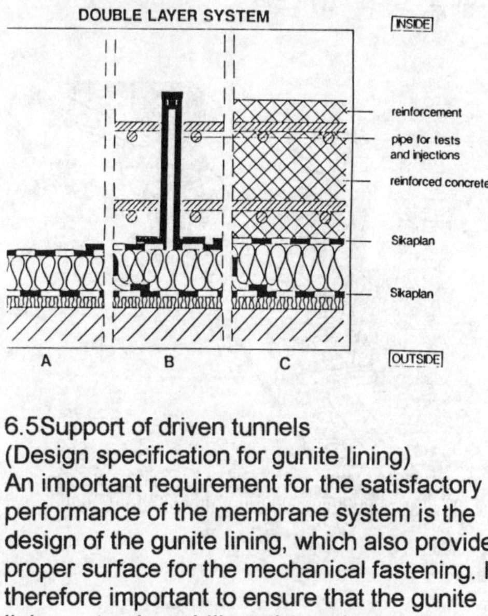

6.5Support of driven tunnels
(Design specification for gunite lining)
An important requirement for the satisfactory performance of the membrane system is the design of the gunite lining, which also provides a proper surface for the mechanical fastening. It is therefore important to ensure that the gunite lining not only stabilizes the rock mass effectively, but also serves to even out any hollows, sharp edges and other irregularities in the rock surface, as well as to cover any steel girders, rock bolts, etc., in order to provide a

backing surface for the membrane in accordance with the following criteria:

- No running water (wet areas present no problem)
- Adequate rigidity and strength
- Minimum thickness 5 cm
- Maximum aggregate size 16mm; may be graded 0-16mm for base layer, 0-8mm for finishing layer
- No crushed aggregates
- Localized surface irregularities left by mining or blasting operations in any backing surface designed to carry flexible waterproofing membranes should exhibit a minimum 5:1 ration of diameter to depth
- Any hollows should exhibit a curvature of minimum radius 20cm

6.6 Jointing system

Sikaplan tunnel sheeting or membranes are homogeneously heat welded at the overlap. Joints made on site are either double seams with an inspection conduit in between or single seams.

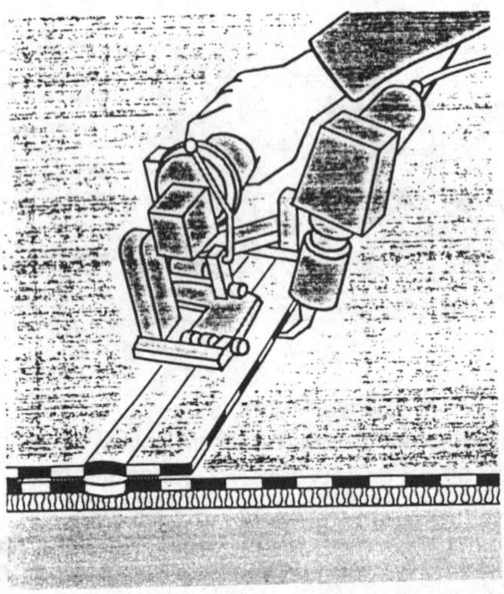

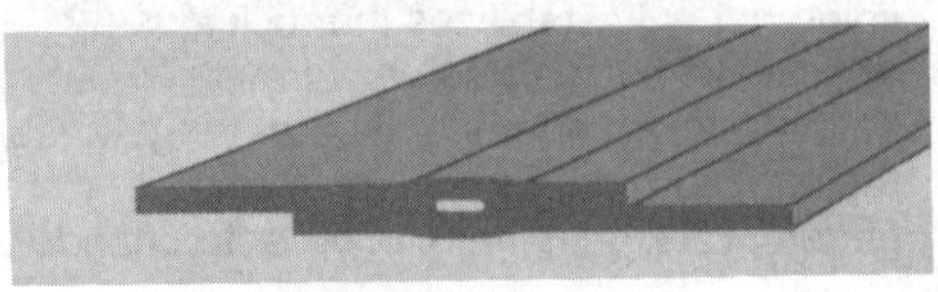

6.7 Inspection technology

Compressed air test: Double seams are checked at a pressure of around 2 bar. They pass inspection if the pressure falls by less than 20% in 5 minutes.

Mechanical test: Single seams can be easily checked with the aid of a screwdriver.

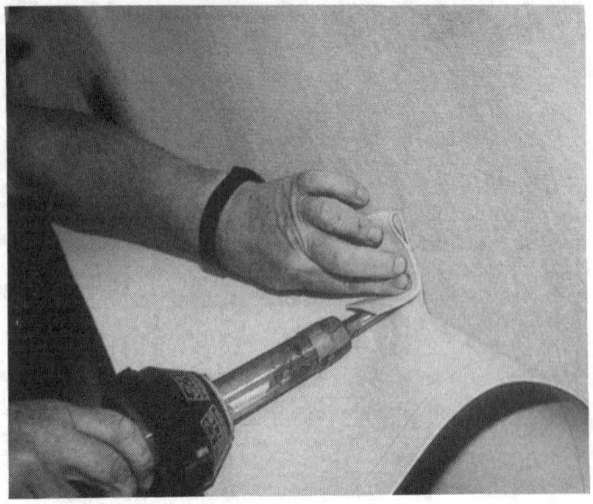

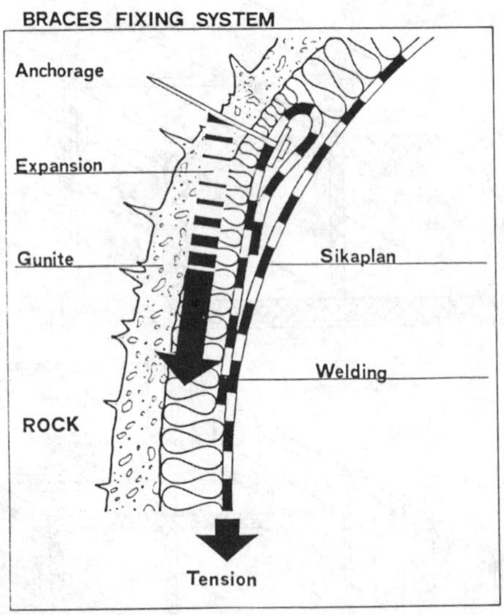

6.8 Fixing system

When installed loose in driven tunnels, Sikaplan is normally point-fixed. Sika has develped the "braces method" as an alternative to roundelles of 8cm diameter. Sikaplan strips approx. 4x20cm in size are fixed to the substrate and welded to the waterproofing membrane. This method absorbs the tensile stress which occurs during concreting without weakening the Sikaplan waterproof sheeting. The system is patented.

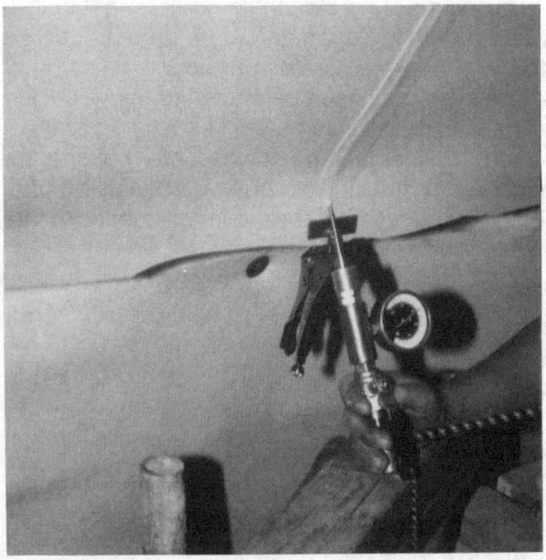

6.9 Installation technology
Special installation machines with mobile platforms have been developed with Sika's assistance over the past 40 years.
The diagram above shows the sheeting being installed prior to concreting of the inner ring.
Sika have helped to set standards for materials and systems in Switzerland, Germany, Austria, France, Japan and elsewhere.

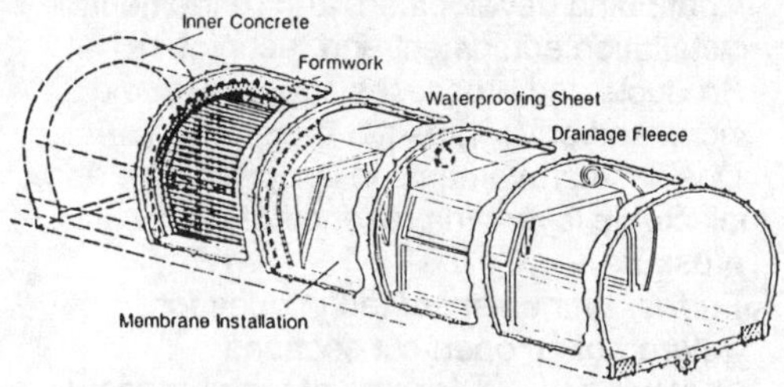

INSTALLATION SEQUENCE

6.10 The Materials

6.10.1 Classification of synthetic materials

Thermoplasts
-> Change of form by heat

PVC Polyvynilchlorid
PE Polyethylene
HDPE High-density-Polyethylene
LDPE Low-density-Polyethylene
LLDPE Linear-low-density Polyethylene
ECB Ethylene Copolymer Bitumen
EVA Ethylene-Vinyl-Acetate

Elastomers / Synthetic Rubbers ->rubber - elasticity

IIR Butyle Rubber
EPDM Ethylene-Propylen Diem Monomer

Thermosets / Duromers

Initially two component, when mixed one time hardening materials (no change of form)
EP Epoxy
PES Polyester

Worlwide well above 80% of membranes for driven tunnels are based on PVC

6.10.2 Geotextiles for drainage + protection

Raw material	Weight per m²	Characteristics
PP Polypropylene PES Polyester	300 500 800 1000	- length of fibres - thermal bonding of fleece - synthetic bonding of fleece - behaviour in fire - hydraulic transmissivity under load - chemical resistance - thickness
PA Polyamide	500 - 1000	

7. REHABILITATION

7.1 General
For this wide field from small repairs to major rehabilitation programs under traffic wich cost nearly as much as a new tunnel, 4 different examples are given.

7.2 Cracks and block joints
A solution for new and old joints: **Combiflex**
The system consists of a high polymer strip of Hypalon membrane bonded in place with a two component epoxy-adhesive as shown in figure (21/22). It works on wet substrate; with water underpressure special precautions have to be taken.
One of the reasons why Combiflex was selected was the difference in air-pressure pushing out an ordinary sealant.
- Other applications in tunnels are expansion joints, cracks and crevices, construction joints (including angled configurations).

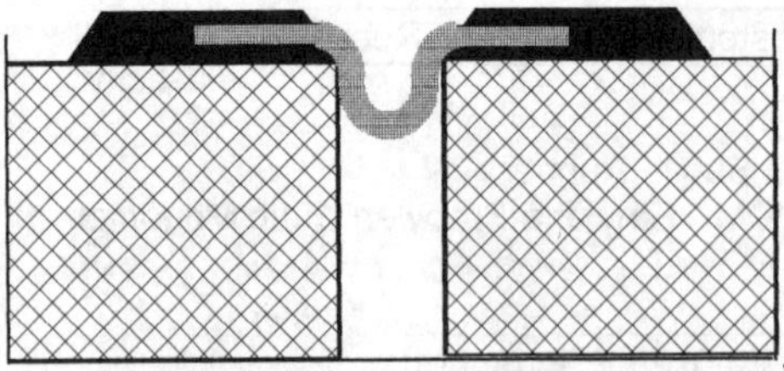

21

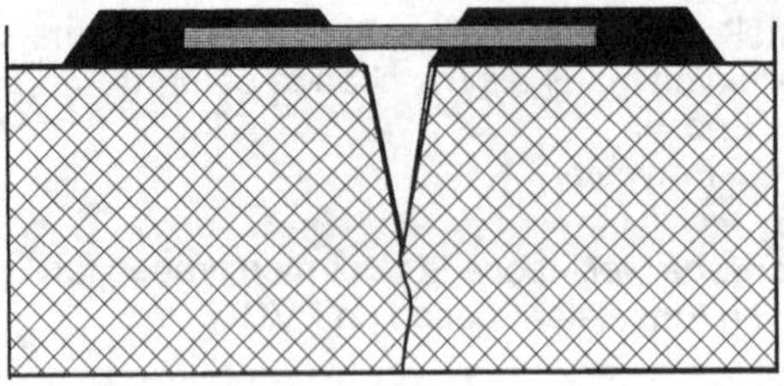

22

- Figure (23) shows the application in ventilation ducts with prefab elements.

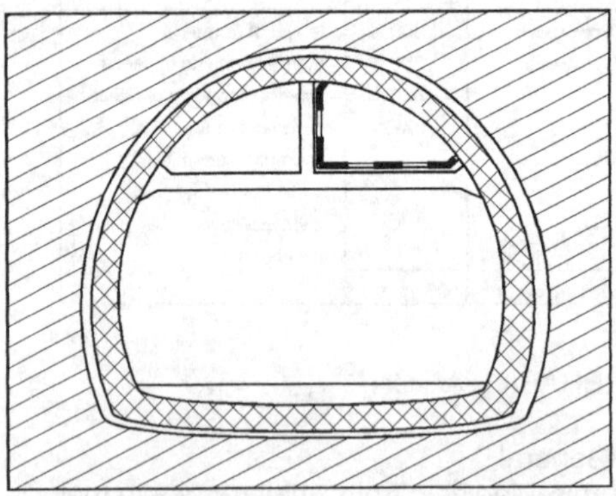

23

7.3Strengthening and waterproofing the old conrete and brick arch over a large area

- After thorough substrate preparation by wet sand blasting, a steelmesh reinforced gunite lining was applied in two coats.
  The 0-6mm gunite with Sigunite accelerator contained 400kg of sulphate-resistant cement per 1000 litres aggregate and had to achieve a strength of 35 N/mm² after 28 days.
- To reduce dust formation and rebound, antidust compound was added to the dry mix. For the rehabilitation of the Hauenstein Tunnel a total area of more than 40'000 m² was repaired using gunite. This railway tunnel (8'135m long) was built between 1912 and 1916. This major rehabilitation program under heavy traffic took place around 1990.

7.4Sheet membrane for railroad tunnel rehabilitation under traffic

- The repair of an old single track tunnel in the Akita region in Japan had to be done under traffic of approx. 2 trains per hour.
- The system chosen as shown in figure (24) consisted of installing steel H section girders where the inside flanges were later holding the wooden formwork. The build-up consisted of Geotextile-PVC-membrane inner concrete ring.

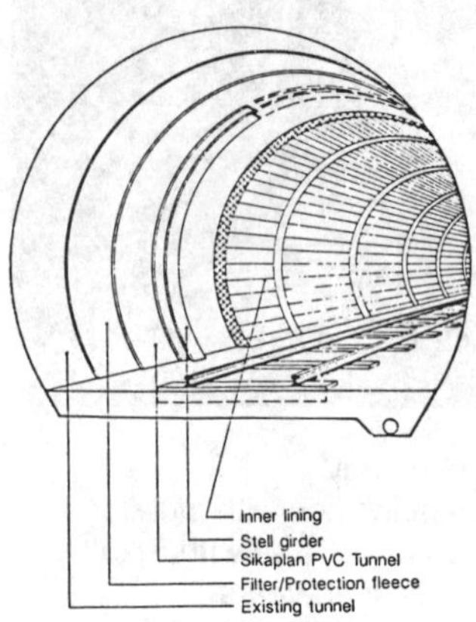

24

8. PROSPECTS FOR THE FUTURE

We see scope for further development in the following areas (examples):

- Continuing development and refinement of installation equipment and techniques
- Products and processes coping with the increased environmental requirements.
- Design improvements in the protective and injectable layers in presence of high water pressure
- Further refinement of techniques for waterproofing open-cut sections
- Improving the resistance of membranes to perforation, thereby doing away withthe protective mortar
- Improving the ease of handling and installation (general)
- Participation in trials and new tunnel projects
- Continuing efforts to reduce cost
- Continuing research and development on single shell tunnels

Tunnelling and Ground Conditions, Abdel Salam (ed.) © 1994 Balkema, Rotterdam, ISBN 90 5410 363 9

CAD and its application to railway infrastructure engineering for urban transport site

P.Girbal, P.Reiss & L.Gratiot
The Paris Independent Transport Authority, France

ABSTRACT : Within the scope of its railway engineering work for public transport, the Paris Independent Transport Authority (RATP) has added certain special tools to its computer aided design system, which performs the computings to define a tunnel section, taking into account the specifications of the rolling stock using the tunnel, and also to draw up a rail layout using interactive site plans and elevations. This document sets out the range of functions offered by this system and the working methods associated with its use.

1. INTRODUCTION

Endless problems are encountered in the engineering of underground infrastructures in urban public transport. These difficulties arise because of the space constraints of the underground area where the infrastructures are required. Hence, to develop this type of project, it is essential to use high performance tools which allow, among other design features, a three-dimensional approach.

With this aim, RATP/SOFRETU have added two specific applications to their CAD CATIA system (software produced by IBM and DASSAULT SYSTEMES). The first is designed to define the tunnel section and the second to create a rail layout using a site plan and elevation.

2. GAUGE CALCULATION SOFTWARE

2.1 Sources of information

An outline of the essential features of a tunnel, which are necessary to allow the passage of guided rolling stock is drawn up using informations obtained from four sources: the rolling stock itself, the guidance system, the specifications for the line layout and various administrative requirements.

The parameters which are set as a result of these informations form part of the data which is required by the software for calculation purposes.

1. Models of the rolling stock are made either from data obtained from the manufacturer or from standard specification details of theoretical rolling stock.

A first category of data relates to the geometrical dimensions of the rolling stock's body shell (the straight-forward length of the body, the distance between bogie pintles, the static envelope of the shell including its construction and distortion tolerance levels.)

A second category of data bears on the characteristics of the suspension and running gear (flexibility, operation tolerances, setting or working tolerances).

These details are required to evaluate the potential shifts in position of the rolling stock, transversally or vertically, given different stress levels, which may affect it.

This data is completed by details of the acceptable wear levels between two overhauls. This information helps to create a behavioural model of the rolling stock over a period of time and also helps to take account of the consequences of the rolling stock ageing process.

When studying construction for a major network project, defining the rolling stock precisely is not always feasible. Theoretical rolling stock can, however, be defined from principle data.

For contract purposes, the real rolling stock should be more favourable from the width gauge point of view than the theoretical rolling stock.

2. The guidance mechanism is set in accordance with the track laying tolerance levels, shifts in position and wear and tear between overhauls.

3. The track layout plan sets out the dimensions of the structure's gauge based on the following four parameter markers: the plan radius, the actual cant, the cant deficiency related to speed and the connecting elevation radius.

The plan and elevation radii cause geometrical throws.

The actual cant causes a leaning of the body of the vehicle on its suspension towards the inside of the curve when the rolling stock is at rest, whereas the cant deficiency related to the anticipated speed of the rolling stock as it rounds the curve, causes the body of the vehicle to lean towards the outside of the curve.

4. The arbitrary administrative variants which need to be considered are the result of regulations or guide-lines related to safety and modes of operation.

Three types of gauge are required to define the section.

The dynamic elements of the gauge, together with the track wear and tear factors and margins enable us to calculate the limit and nominal gauge measurements.

2.2 Defining the Gauge

The abilities of the software are explained in several stages, with the methodology and associated functions set out in parallel.

1. Presentation of a methodology table relating the constraints leading to the definition of the inner arc of the outlined tunnel section.
2. Giving cant to a track section with regards to the limitation value imposed by the production programme of the project.
3. Initialisation of the parameters related to the track section set with possibility of duplication, thus saving data capture for the study of parameter variants regarding the track section.
4. Presentation of the parameters related to the rolling stock:

The number of parameters varies according to the type of suspension with which the rolling stock is equipped or the level of complexity of the model studied.

It is therefore possible to use a model in principle to establish pre-dimensions or to carry out control procedures with regard to already built rolling stock.

As before a possibility of duplication, saving on data capture allows the study of variants for a predetermined model.

5. The calculation of a group of gauges on two adjacent tracks with the choice of automatic inter-axis calculation (inter-axis capture is also possible).
6. Production of linked listing and design supporting notes.
7. Analysis of the physical conditions involved in drawing up the gauge for a defined section of track: radius, cant, cant deficiency, speed limitation.
8. Rapid definition of the inner arc of a circular tunnel section.
9. Presentation of actual research examples: cross-section of the Channel tunnel (use of a complex checking model,) and study of the gauges for the Cairo Metro.

3. RAILWAY LAYOUT DESIGN SOFTWARE: PLAPRO

3.1 General introduction

The realisation of an urban site railway project confronts the designer with all sorts of necessary administrative, technical and financial considerations. The final project must be a compromise between the relevant factors.

Finding a solution to relieve these considerations and to achieve the best possible realisation of the project makes the research process crucially important. We are required to find a way to deal with the many variants and solutions, including repetitive tasks of calculation and printing of documents, both of which being possible to automate.

Inspired by the classical working method of defining the project by way of a site plan and elevation, the PLAPRO software, designed by the RATP, undertakes these jobs which assist the design process. It has the following special features which help to fulfil its role:

1. a structured approach to modelling track construction;
2. the ability to analyse and modify variants;
3. an interactive link between the plan and the elevation;
4. a wide range of features for dressing site plans and print listings.

3.2 Defining the Plan Layout

1. Positioning track connections (modifications and variants)

The positioning of track connections for the plan layout is done interactively from basic geometrical input: main directions, length of connecting curves, radius of connected circles, offset of axes and capture of constraint points.

These factors are kept with the model, so that it is possible to refer back to them to modify any layout or create a variant.

The interaction between the various functions of the PLAPRO software allows us to draw up repeat studies regarding the geometry of the plan layouts quickly and easily.

2. Ensuring that track connections conform to operational passenger comfort standards

Taking the trains' speed, when they pass the different connections, into account permits us to check the geometric compatibility of the pre-defined plan layout with the standards of comfort and safety laid down in the project specifications: uncompensated transverse acceleration, variation in transverse acceleration, and maximum warping.

Memorising the pass-by speed at each connection allows us to make repetitions between the modification of the geometric characteristics of the plan layout and the control checks. These repetitions lead to the best possible realisation of the project.

3. Producing the resulting layout

The production of the different project plan layouts (inter track axis, track axes and centre line of the structure) follows a precise logic and clear progression.

The plan layouts can be drawn up each one from the others according to the rules of calculation which are programmed into the software: off-set axis and calculations around a fixed centre point.

4. Printing listings and supporting design notes

The listings and supporting design notes which are produced from the different layouts which have been drawn up with the PLAPRO software, allow us to justify the geometrical positioning and comfort and safety specifications of the layouts which form the design base.

5. Automatic dressing of the plan layout

Different functions permit the plan layouts to be labelled or dressed up: co-ordinates, overall distances, type of elements and specifications used in the layouts.

The labelled details remain tied to the layouts for which they provide information so that automatic updating can take place in the event of any modifications to the geometry of the associated plan layout.

3.3 Defining a side elevation

1. Defining the site elevation

The definition of the height along a layout is conditionned by the vertical modelling of the site.

This job is carried out with the help of 3D modelling routines used in the CATIA software, which are back up tools to PLAPRO. It works by intercepting or interpoling objects in three dimensions along the vertical surface following the plan layout defined previously.

2. Wedging up the side elevation

The wedging up of the side elevation is achieved according to different considerations for which interactive control is possible due to the PLAPRO software, such as:

Minimum distances which need to be respected in relation to certain elements of the site.

Maximum up and down gradients permitted by safety standards and rolling stock specifications.

Coincidences to avoid between different characteristic elements of the plan and side elevation (connection bends, points, entrances or exits of stations.)

Numerous interactive functions, used at the project design or modification stage, are intended to achieve the best possible geometry for the side elevation.

3. Cutting up the project into work sheets and dressing of grids

Defining the project's side elevation needs the issue of one or more work sheets at a given level which overlap each other.

These work sheets are dressed up with grids containing numerical information, (level of the site or of the project, overall distances) or alphanumerical data (position of stations, location of intersected or folowed roads, type of track laying) allow the definition of the project and its environment.

The PLAPRO software cuts up and automatically labels the grids, the contents of which can be parametricaly defined, and updates the labels and plan dressing if the project is modified.

3.4 Defining the 3 dimensional framework

1. Defining of the plan and side elevation layouts enables a three dimensional framework to be drawn up on which, according to the different areas of the project, a section of the tunnel can be applied.

The 3D model which links the site and the project works is used to confirm, a posteriori, the validity of the positioning and can be used to produce realistic perspective pictures, for information or project promotion.

2. The 3D modelling of the site and works can also be used to cut up the project for more detailed analysis and verification of constraints in the most "sensitive" areas.

3.5 Examples of Projects undertaken by the RATP/SOFRETU using PLAPRO

These projects cover most urban railway transport systems:

1. Metro.

The Paris METEOR Line: 7.2km - 7 stations
Line 2 of the Cairo Metro: 18.5km - 19 stations
Line 2 of the Monterrey Metro : 9.5km - 10 stations

2. Regional Lines

The Paris RER Line D: 5km between two existing stations

3. Val

Line 2 Lille: 12km - 17 stations
Rennes: 3.8km - 7 stations

4. Tramways

Lille Mongy: 16km - 28 stops
Rouen: 11km - 24 stops

4. CONCLUSION

The specific applications which have been developed at the RATP/SOFRETU under CAD-CATIA, have brought qualities of homogeneity and precision to projects. Claiming that traditional technical drawing and design methods would do the same is impossible.

More fundamentally, the speed at which it is now possible to study a solution or variant, reduces the overall delay in study and analysis, provides the best possible solution to design the works, and to take into account most of the requests for design modifications.

Tunnelling and Ground Conditions, Abdel Salam (ed.) © 1994 Balkema, Rotterdam, ISBN 90 5410 363 9

Design of the shield tunnel crossing under the bottom of the Yodo River, Osaka by the pressured slurry shield method

Toshinobu Imanaka, Toshimasa Kamata & Masatomi Okuno
Nippon Telegraph and Telephone Corporation (NTT), Osaka, Japan

PREMISE: This theme is concerned with the Planning and Design which are now under consideration by NTT Osaka pertaing to a tunnel for the purpose of telephone works crossing under the bottom of the Yodo River, Osaka, Japan.
The tunnel designed is 3.7 meters in diameter and 2465 meters in length. The construction is planned at the location of the underground point of approximately 50 meters in depth. The construction method adopted is the pressured slurry shield method and the relative construction works are scheduled to begin before the end of 1993.

1 INTRODUCTION

NTT has been aiming to construct the most reliable telecommunication facilities and providing the best services as one of sophisticated leading companies in the telecommunication industries in Japan.
The tunnels for telecommunications are ones of facilities which NTT has construcred in various urban areas in Japan.
Now, this paper is the description of the plan and design of the tunnel for telecommunications by the Shield construction method crossing under the bottom of the Yodo River, one of the principal rivers situated in Osaka, Japan.
Osaka city divided into north and south by the Yodo River, shown in Fig.1, is the central city of the western part of Japan and has been flourished particularly in commerce. There are many civil facilities such as bridges acrossing this Yodo river for road, railway service, etc. NTT also has one bridge crossing the Yodo River used for telecommunication however the telecommunication traffic concentrated at this point which is not a favarable shape in view of countermeasures against natural disasters and terrorism.
NTT, under those circumstances, has decided to plan and renew the present facilities in Osaka into tunnel loop line, as shown in Fig.2. There are no tunnels crossing under the bottom of the Yodo river in Osaka, until this plan executed.

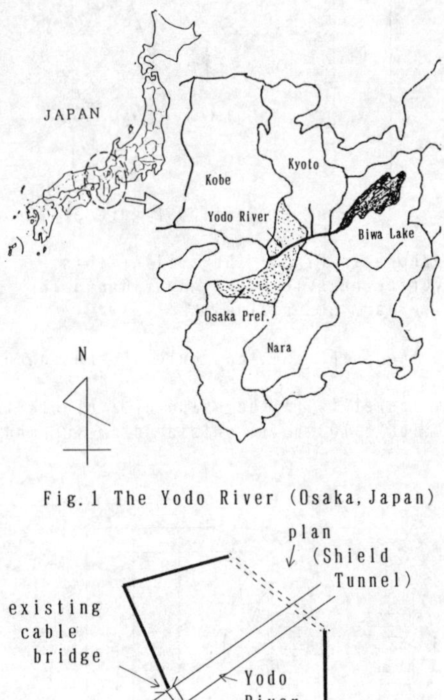

Fig.1 The Yodo River (Osaka, Japan)

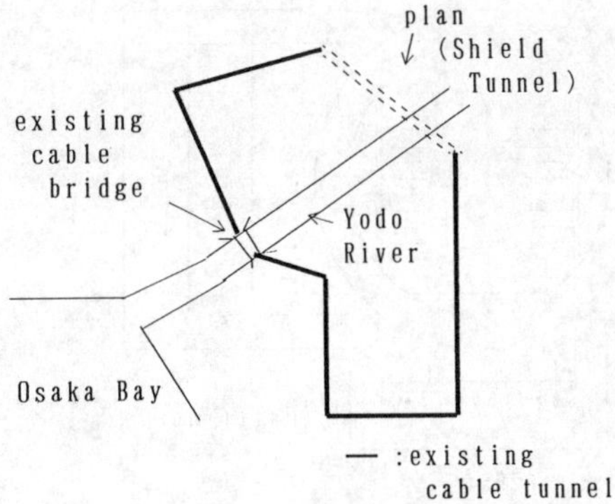

Fig.2 Telecommunication tunnel network in Osaka area

2 PLAN OF THE SHIELD TUNNEL

2.1 Soil

The soil formed under the Yodo River, from the surface to the bottom, they are, sand layer, clay layer in holocene deposit, grave layer in pleistocene deposit called "Temma Layer" respectively.
The depth of the tunnel passes is from Osaka Peil(O.P.) -30 meters to -40 meters as shown in Fig.3. The soil of this layer belongs to Osaka Layer Group in quaternary generation.

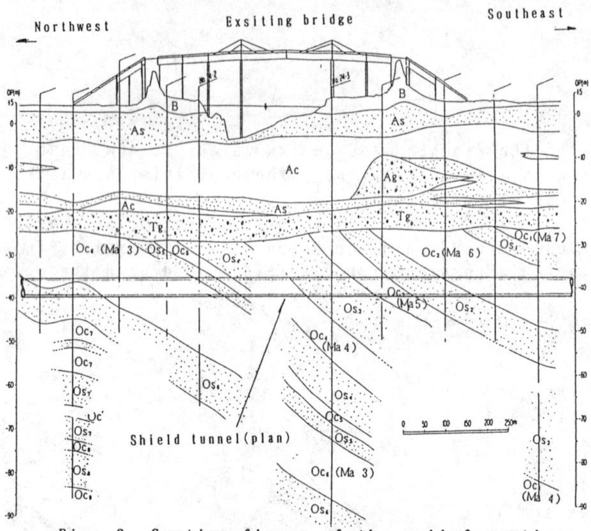

Explanatory notes

O.P. : Osaka Peil

* in holocene deposit
- B :Banking
- As:Sand layer
- Ac:Clay layer
- Ag:Gravel layer

* in pleistocene deposit
- Tg:Gravel layer (called "Temma Layer")
- Oc:Clay layer (Ma : Marine clay)
- Os:Sand layer
- (called "Osaka layer group")

Fig. 3 Section figure of the soil formation

2.2 The shape and the route of the tunnel

The tunnel is in the shape of a circle in section, 3,700 mm in outside diameter and 2,950 mm in inside diameter as shown in Fig.4. The total length to be forwarded by pressured slurry shield is 2,465 meters and the length accross the river out of it is 560 meters as shown in Fig.5.
The plane linear of tunnel across the river has been designed to be straight line as much as possible 8.5 meters in horizontal distance so as not to affect on the existng caisson foundation of the existng bridge. The tunnel shall be dig-forwarded in stable pleistocene deposit as shown in Fig.3. As a result, the depth of the earth covering from the bottom of the river become 33 meters and 47 meters from the top of the levee.

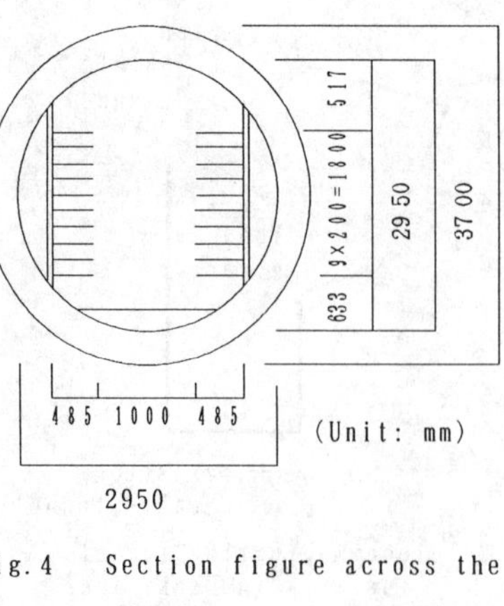

Fig.4 Section figure across the Yodo River

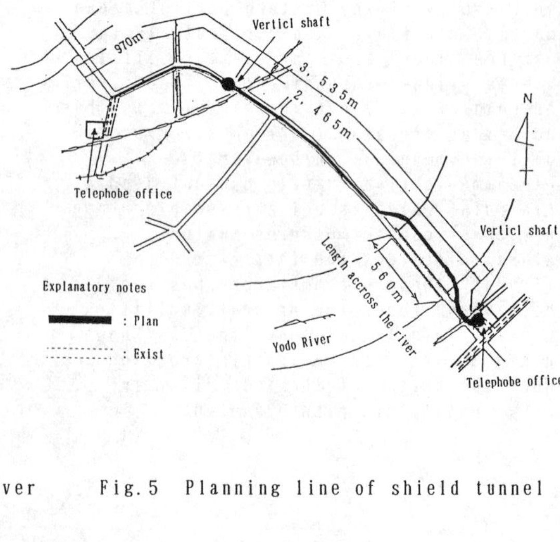

Fig.5 Planning line of shield tunnel

3 SHIELD TUNNELING METHOD

The pressured slurry shield method is chosen taking the following conditions into account.

1. Environmental conditions

The underground of the bottom of principal river, the Yodo River has to be crossed by this method. There is an exisiting caisson foundation close by. The construction of big scale levee is planned in the right side of the river.

2. Ground conditions

The tunnel will go through in the mutual layers of clay and sand in Osaka layer group, the maximum pore water pressure is 0.31 MPa which is quite high.

3. Design conditions

long digging distance of 2465 meters has to be considered. There are two acute curvatures that the radious of curvature is 30 meters. The maximum depth of the earth covering is 49.2 meters taking account of the big scale levee.

The pressured slurry shield method make it possible to apply for any soil conditions such as stable ground, a low N-value of clay layers, a high ground water level aquifer, aler-learning sand stratum aquifer water-bearing gravel stratum etc, stabilizing tunneling face by the effects of both the pressuring of slurry and the character of muddy water.

4 DESIGN OF THE SHIELD MACHINE

4.1 Evaluation of cutter bits

Taking account of environmental conditions that the change of the bits will be difficult, the height of the ahead bits in the size of 80mm(60mm) are made and the height of the cutter bit in the size of 65mm(50mm) are made respectively.
The height of bits both for ahead bits and cutter bits made of super alloy in the size of 50mm(30mm) are designed.
The figures in the brackets are in ordinaly circumstances. Due to the special production of the cutter bits, no inspection and change for the bits shall be performed though the tunneling distance of 2,465 meters is rather long.

4.2 Evaluation of the tail seal

3 steps of wire brush and putty grease automatic impregnation type are adopted as the tail seal to hold out high pore water pressure(0.31 Mpa) and long distance tunneling(2,465 m)

4.3 Evaluation of the works at the acute curvature points

one-stage mid turn type shall be adopted to catch up with 2 points of acute curvature points (R=30m) so that the volume of the outbreak can be decreased as shown in Fig.6.

Taking the above 3 factors into account, the Shield Machine shall be designed as shown in Fig.7.

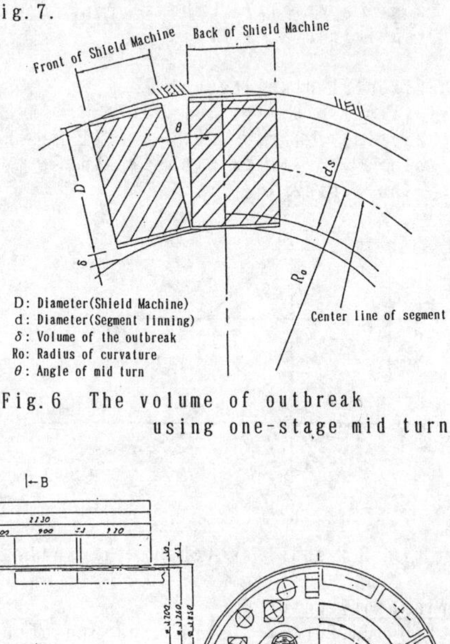

Fig.6 The volume of outbreak using one-stage mid turn

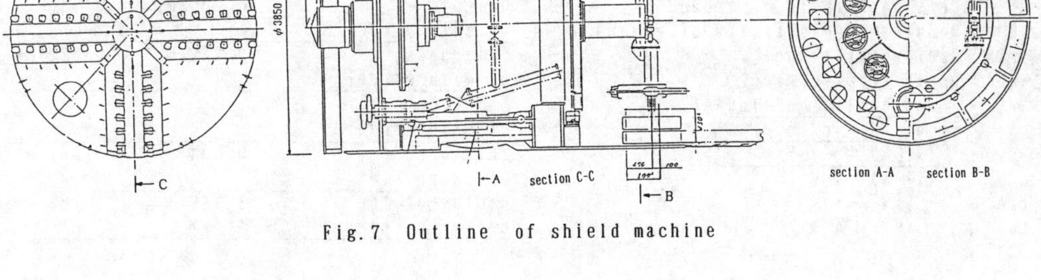

Fig.7 Outline of shield machine

5 EVALUATION OF SUBSIDENCE OF THE LEVEE SUBSEQUENT TO THE DIGGING OF THE TUNNEL

Elastic analysis, elastic-plastic analysis and viscoelastic analysis are in general examined to analyze the effects with tunnel digging. Elastic analysis are used this time.

5.1 The points to be exmined for the subsidence of the levee subsequent to the digging of the tunnel

The examination were carried out at the direct underneath of both banks of the Yodo River.

5.2 Elastic analysis

Ground displacement at any optional points can be computed by Elastic analysis.
All the ground conditions shall be regarded as equally semi-infinity uniform plastic ground as the model sketch is shown in Fig. 8.

• Vertical displacement

$$V = -(1+\nu)\cdot P/E \cdot r_o^2 \ \{ [2(1-\nu)h_o/a-(1-2\nu)] \cos\theta_1/r_1 - [2(1-\nu)h_o/a+(1-2\nu)] \cos\theta_2/r_2 + X/a\,[(h_o-a)\cos 2\theta_1/r_1^2 - (h_o+a)\cos 2\theta_2/r_2^2] \}$$

• Horizontal displacement

$$H = -(1+\nu)\cdot P/E \cdot r_o^2 \ \{ [2(1-\nu)-(1-2\nu)h_o/a]\sin\theta_1/r_1 + [2(1-\nu)+(1-2\nu)h_o/a] \sin\theta_2/r_2 + X/a\,[(h_o-a)\sin 2\theta_1/r_1^2 - (h_o+a)\sin 2\theta_2/r_2^2]$$

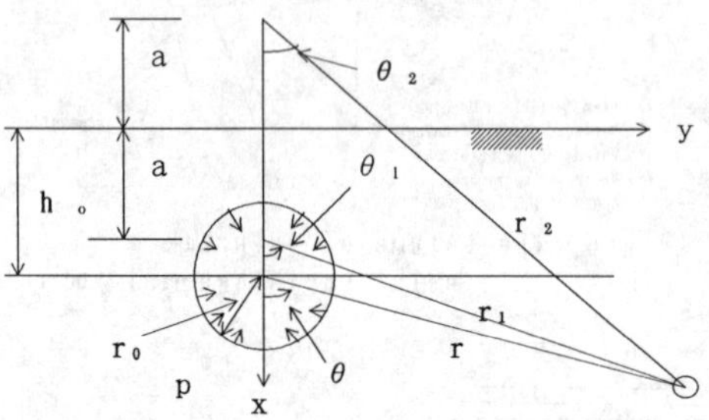

Fig. 8 model sketch of the analysis

<u>Explanatory notes</u>

- P : Uniformly distributed pressure around tunnel
- ν : Poisson's ratio of ground
- E : Modulus of elasticity of ground
- r_o : Radious of tunnel
- h_o : Depth from surface to central of tunnel
- a : $\sqrt{h_o^2 - r_o^2}$
- r_1 : $\sqrt{(x-a)^2 + y^2}$
- r_2 : $\sqrt{(x+a)^2 + y^2}$
- $\tan\theta_1$: $y/x - a$
- $\tan\theta_2$: $y/x + a$

5.3 The analysis results

The figures of subsidence of the levee subsequent to by elastic are shown in table-1 below.

table 1 Analysis result

	Left levee	Right levee
Top of levee	0.7 mm	0.5 mm
Bottom of levee	0.7 mm	0.6 mm
(ν)	0.3	0.4
(E)	555 kgf/cm²	640 kgf/cm²

As a result of this analysis, the maximum amount of the subsidence is only 0.7mm which is considered to be very small to have the levee effected.

6 THE STEEL SEGMENTAL DESIGN

6.1 The sectional designs

The calculation of the sectional steel segmental was made providing that the segment-ring has the same flexural rigidity. The loadings are designed that all the earth pressure from the ground surface shall be affected. The design conditions are shown in table 2.

table 2 Design conditions

Dead load of large scale levee	0.13 MPa
Dead load of building	0.24 MPa
Earth covering	41.83 m
Ground-water level	G.L.-1.5m
Coefficient of earth pressure	0.5
Coefficient of sectional reaction ground	2.4 kgf/cm³

	case.1	case.2
Outside diameter of lining segment	3.7 m	3.85 m
Width of lining segment	0.9 m	0.9 m
Height of main girder	0.225 m	0.300 m
Thickness of main girder	0.022 m	0.022 m
Number of main girder	4	3
Thickness of Skinplate	0.004 m	0.004 m
Radius of the center of figure	1.746 m	1.782 m

Study of segment in Case 1,"4 nos.of main girders" and Case.2, "3 nos.of main girders were carried out as shown in Fig.9.
As a result,Case 1,"4 nos.of main girders" was adopted due to smaller segment radious and economical reason.
No dropping out of the segments can be permited because of the crossing under the bottom of river, wedge type of K-segment was adopted instead of normal K-segment as shown in Fig.10.

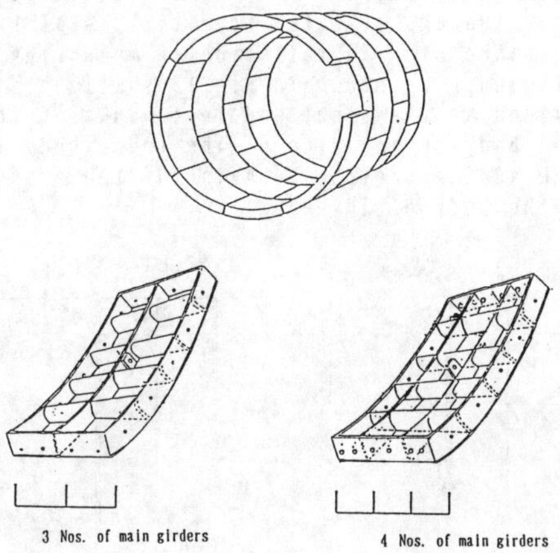

Fig. 9 Segment

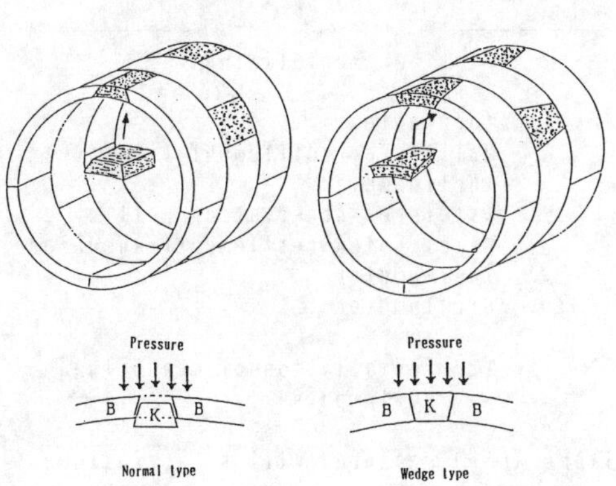

Fig. 1 0 K − Segment

6.2 Calculation of (lining) segment in verticall direction

A model structure applying iron bars for the tunnel and elastic spring for the ground has been made to examine the design of (lining) segment in vertical direction on the condition that the outside force is the amount of consolidation breaking out from the big scale Levee to be constructed in the future.About 5 centimeters of consolidation breakout can be expected at the top of the tunnel in the experiment.It will not be necessary to upgrade the materials according to the analysis result however high tension bolts shall be used as segment bolts for the area of expected big scale Levee.

7 THE FILLING METHOD FOR THE AREA OF OUTSIDE OF THE TUNNEL

After the shield machine goes through,a tail void phenomenon of the earth at the part of outside segments shall be broken out.
Injection of cement mortar must be done immediately after the machine goes through. Otherwise leakage of ground water from the segments and joints in vertical shafts of the tunnels will be occured.The selection of the adequate backfilling method and its material were necessary taking the above fact into consideration.
The selection of criteria for the method are;

1. earlier stabilization of segments can be expected.
2. maltipleport of injection is avilable.

Simultaneous backfilling injection method showen in Fig.11 has been adopted to clear those conditions.
While for the selection criteria for the materials are;

1. it should be gelatinized in a short period of time with great strength.
2. the material dissociation shall be little and proper flow in the pipe can be expected for the length of 2465 meters propulsion.

The material of Air mortar lineage has been adopted to clear the above conditions.

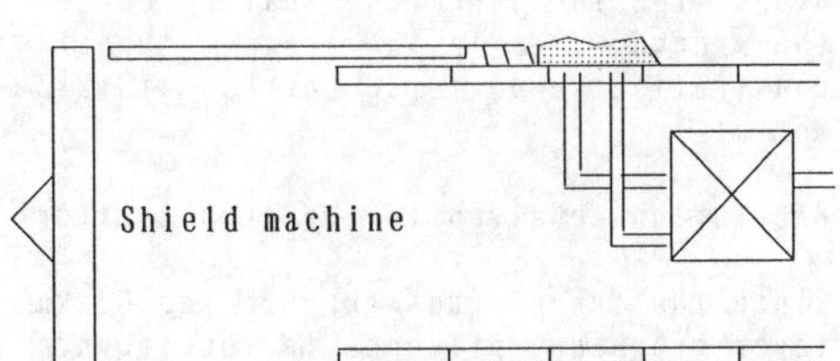

Fig.11 simultaneous backfilling injection method

8 THE INFLUENCE OF GROUND WATER BY THE TUNNEL CONSTRUCTION

8.1 The bed of the section of the Yodo River and its ground water

The formation of the bed of the section of the Yodo River from the top to the bottom, they are, sand layer(2 to 6 meters), holocene clay zone (13 to 16 meters), pleistocene gravel zone (8 to 9 meters), pleistocene clay zone (15 to 16 meters) and pleistocene sand zone.
Accordingly the ground water of the section will not be permeable at the holocene clay zone and pleistocene clay zone and the exsistance of unconfined ground water into the uppermost sand layer and confined ground water into pleistocene deposit can be expected.
The tunnel in question goes thru the Osaka layer group which is composite soil layer of clay and sand zone so that no bad influence on the ground water can be thought.
In this section, there are almost unlimited amount of confined ground water in Pleistocene deposit covering a wide field of area which flow gently and relatively with slow current.
In general many construction works executed in the past in Osaka city verified that there were abundant of ground water in pleistocene sand deposit.

8.2 The utilization of the ground water near the Yodo River

In 1962, two bills about the control of ground water by the government were issued in view of surplus pumping up of ground water which caused soil subside of plains in Osaka. These bills auctully forced us to prohibit from pumping up of the ground water from the depth of within 600 meters in water front area and 500 meters in the other area. The judgement that no one use the ground water in the area of tunnel constrution seems to be correct at this moment.

8.3 Influence given to the ground water

Since the tunnel runs through the Osaka layer group, no influence to the ground water will be given. As described in 8.1 above, there are rich amount of ground water in this area and even if the tunnel cuts off the flow of the ground water partially, enough amount of ground water will be supplied from the other layers.
In the case there arrises the necessity of using ground water, no influence on the supplying can be considered.

9 SAFTY MEASURES

9.1 Safty measures during construction works

Installation of the plenum equipmments (bulkhead, compressor) in the interior of the tunnel shall be done for the accident such as the leakage of water from tails seal by any chance etc. Installation of measuring equipmments as shown in Fig. 12 shall be provided at the right and left banks of the levee body, inland side of the levee body to watch and secure the behavior of the ground and the body of levee.

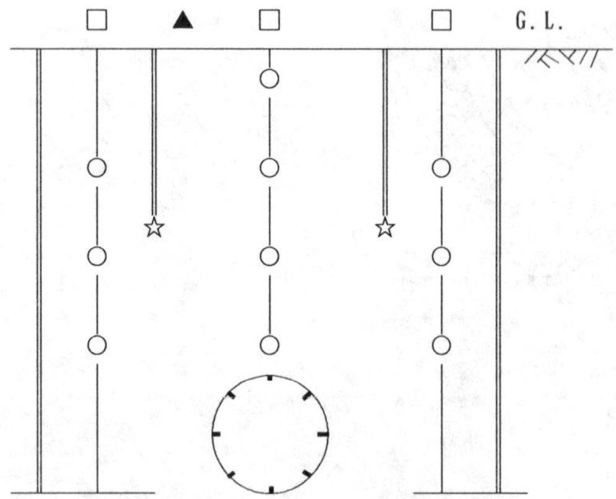

segment lining

explanatory notes
○ :Multilayer settlement transducer
‖ :Inclinometer
☆ :Watelevel observation well
□ :Wateleveled settlementtransducer
▲ :Thermometer
▮ :Strainmeter

Fig. 12 Installation of measuring equipments

Items to be measured will be as follows.

Subsidences of the levee body:
with wateleveled settlement transducer
Incling of the ground:
with Inclinometer
Sinking of the ground:
with Multilayer settlement transducer
Water Pressuring of the ground:
with Pore pressure transducer
Stress of the lining
with Strainmeter

9.2 Safty measures after the completion of works

Installation of watertight wall and water level detector in the tunnel on both ends of the Yodo River and intensive supervision measureing system shall be provided as shown in Fig.13.

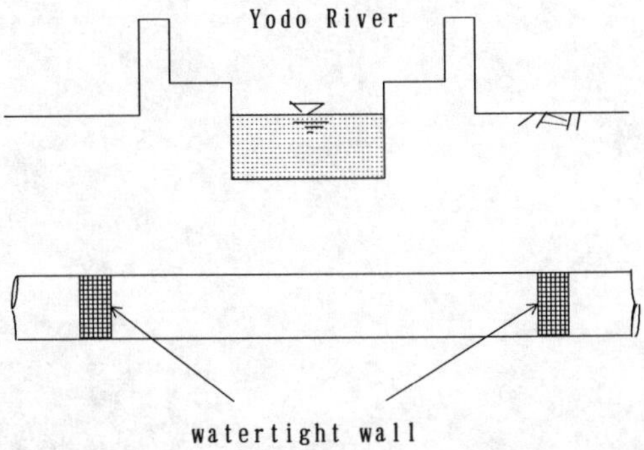

Fig.13 Installation of watertight wall

10 CONCLUSIONS

The construction of this tunnel will be started in December 1993 and scheduled to have it completed in September 1995. The works to pass acrossing the Yodo River is scheduled to complete in June 1994 and deep study and preparation for this project are now on the way to complete it successfully by the shield tunneling method.
We shall be happy if this treatise will be helpful to you in the projects of crossing under the large rivers in the future, if you may have.

Tunnelling and Ground Conditions, Abdel Salam (ed.) © 1994 Balkema, Rotterdam, ISBN 90 5410 363 9

Development and implementation of segment bolt tightening robot

K. Kazama, H. Uyeda & A. Higashide
Obayashi Corporation, Tokyo, Japan
K. Imai, S. Nunomura & T. Morii
Hitachizosen Corporation, Osaka, Japan

ABSTRACT: A multi-purpose manipulator to handle large, heavy load for construction works has been developed, and applied to the segment bolt tightening work for shield tunneling. This is a vertical multi-articulated type, with 6 axis of freedom, and the handling capacity of 500 kgf by a 2.5 m long arm. The system consists of two sets of manipulator with a bolt tightener, a support car and their control system. 120 bolts of one segment ring, of which diameter is 11.3 m, can be full-automatically tightened within one hour. This paper reports on the structure and mechanism of the robot, the details of its development, and the results of performance-validation test.

1 INTRODUCTION

In Japan, the rationalization of works and reduction of labor is an universal subject among industrial circles including construction industry. With the advance of the decrease in the number of the labor population, as well as the shortage and aging of skilled workers, construction works have been expected to be robotized and thus done operatorless. Among numerous construction methods, the research and development of automation and robotization for the shield tunneling have been most actively carried out due to the reasons that many of these works consist of the repetition of simple works. For example, the operation of closed face type shield machines has already been highly automatized.

Developments have been mainly targeted at excavation and segment assembly works that directly affect the construction cycle, namely, excavation supervision, automatic tunnel alignment control, automatic segment transportation systems as well as segment assembly robots, which enabled the automation of a major part of basic shield tunneling works.

Meanwhile, the automation of incidental works such as laying of sleepers and rails, tightening of segment bolts, and wiring (including extension), is far behind as they are considered to be of less importance. The automation of such incidental works, however, becomes essential because the ultimate goal of the development of an automation technology for shield tunneling work is to establish a construction systems which requires only 2 or 3 maintenance personnel.

In order to realize the automation for the incidental works, the authors have developed a manipulator for handling heavy loads, and planned to apply it to one of the such works, segment bolt tightening.

2 OUTLINE AND CHARACTERISTICS OF THE MANIPULATOR

The developed manipulator allows the accurate

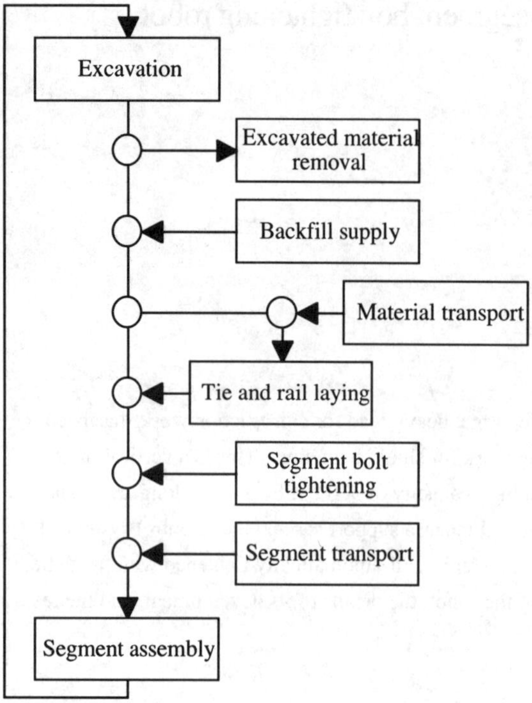

Fig. 1. Shield tunneling work flowchart

Table 1. Current state of automatization

	Implem -ented	Under develop -ment	Not planned
Excavation control	○		
Tunnel alignment control	○		
Simultaneous backfill grouting	○		
Cutting face stability detection	○		
Segment and material transport	○		
Segment assembly	○		
Tie and rail laying			○
Segment bolt tightening		○	
Pipe extension	○		
Wire extension			○

and quick positioning of the heavy load of a rated weight of 500 kg and a maximum weight of 750 kg within an error of 0.1 mm. As conventional articulated manipulators commercially available are intended to handle up to 300 kg, this manipulator has the world's largest handling capacity.

The manipulator mainly consisting of an arm and a replaceable attachment at its tip has the following advantages.

(1) The multi-articulated vertical manipulator consisting of three articulations allows six degrees of freedom, enabling omnidirectional movement of its arm.

(2) The 6 axes of the manipulator consisting of an arm of 2.2 m long and a replaceable attachment at its tip are controlled by alternating current servo motors.

(3) The manipulator uses an image sensor and ultrasonic sensors for the accurate positioning of a load up to 500 kg within an error of 0.1 mm.

(4) A single manipulator can be applicable to multiple works by replacing the attachment installed at the tip of the arm.

3 APPLICATION TO A SEGMENT BOLT TIGHTENING ROBOT

3.1 Significance of bolt tightening

Segment tightening work was chosen as a target of implementation of the newly developed manipulator. Segment bolts are tightened with a required torque at the time of segment erection inside the tail ring of a shield machine. In general, these bolts become loose due to the thrust force of a shield machine and due to the transformation of the shape of a segment ring caused by the unsymmetrical external force such as the earth pressure, ground water pressure, and back-filling grouting pressure on a segment ring. Hence, re-tightening of segment bolts is necessary after these bolts have become free of the influence of such shield thrust force.

As long as torque for segment tightening is concerned, there is a difference of view between Western countries and Japan due to the difference in the concept of segment design. In Western countries where the ground is mostly under stable conditions, it is a prevailing concept that axial force excels bending moment on the cross section of a segment, therefore, bending moment can be transferred to an adjacent segment piece without bolts. This led to the development of boltless segments such as a segment with tenon joint. So, a concept of tightening segment joints does not exist in Western countries.

On the other hand in Japan where the ground is mostly unstable, bending moment often becomes larger than axial force. For this reason, segments are designed so that bending moments at a joint within a ring is to be transferred to a piece of segment of adjoining rings through the tightened bolts. Therefore, the segment bolts should be tightened up with required torque. In addition, sealing tapes are usually placed between segments for water cut-off purposes as the ground water level is mostly high in Japan. To ensure the water cut-off performance, the adequate compressive force should be applied to these sealing tapes. A certain amount of torque is necessary also for this reason.

The reasons why the segment bolt tightening work was firstly chosen to be a target of implementation of the manipulator are; 1) The cross section of tunnels has recently become increasingly large, so bolt tightening works are requested to be done on decks at a considerably high location. Therefore, the automation has been strongly expected to secure safety of the laborers. 2) Particularly in the case of shield tunneling of an ultra-large cross section represented by the Trans-Tokyo Bay Highway Tunnel with a diameter of about 14 m, the required torque is too large, so it is impossible to manually bolt up the segments.

3.2 Structure of the segment bolt tightening robot

The sketch of the segment bolt tightening robot is shown in Fig. 3.

The robot consists of two manipulators mounted

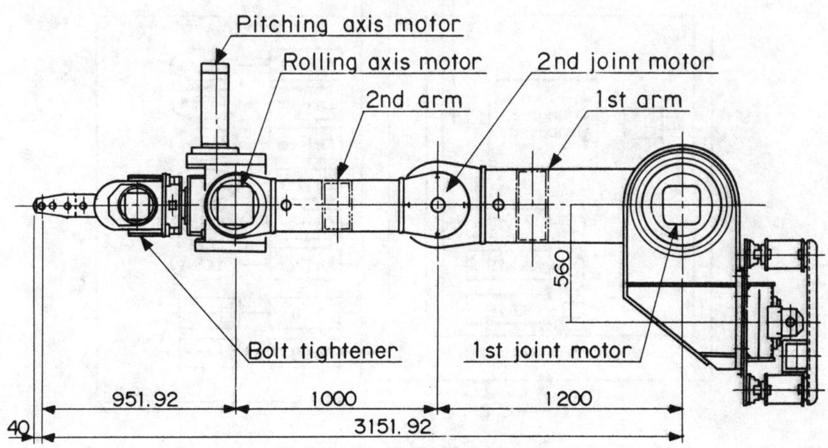

Fig. 2. Manipulator

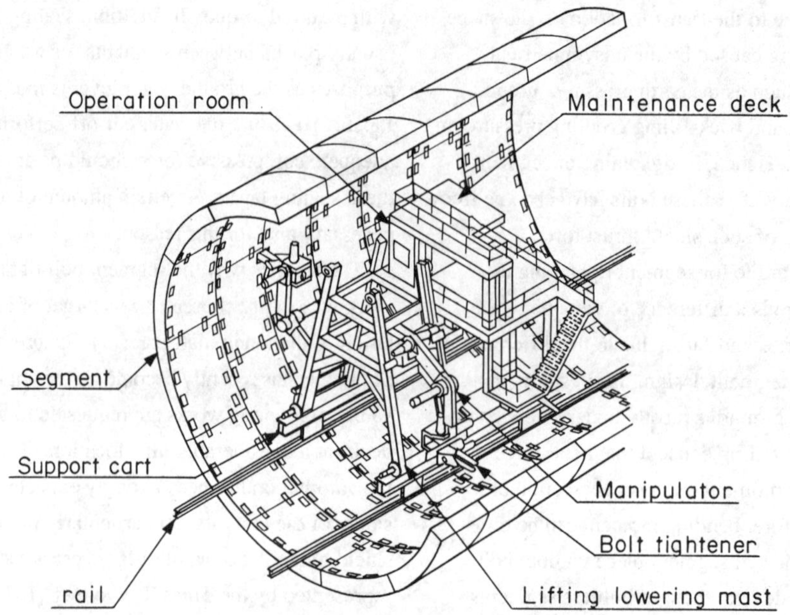

Fig. 3. Segment bolt Tightening Robot

on both sides of a self-advancing support cart that moves on the rail laid in the tunnel. The tips of the manipulators are equipped with bolt tighteners. The robot allows the automation of a series of works includes the detecting of bolt locations, the positioning of the manipulators, and the bolt tightening.

Masts that lift and lower the manipulators are mounted on both sides of the support cart. An operation room is equipped on the upper deck to monitor and control the manipulators.

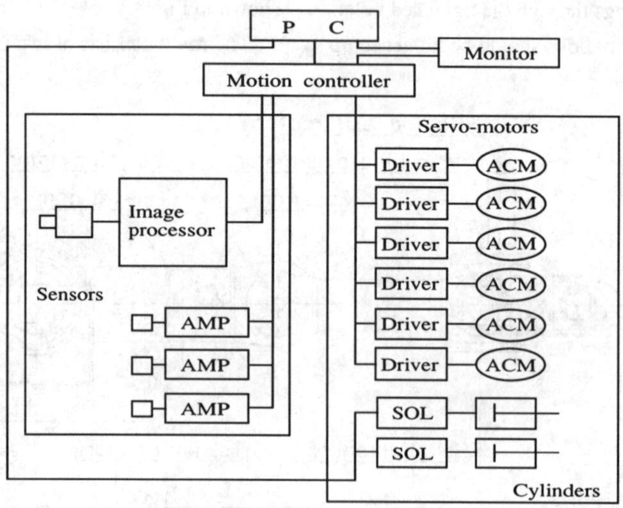

Fig. 4. Control block diagram

3.3 Robot control system

A robot control system consists of sensors mounted in the various parts of the robot, controllers for actuators, and a monitoring system for information control and display. Fig. 4 shows the configuration of the control system. In this system, the programmable controller (PC), the servo controllers for the actuators and sensors are connected by communication lines.

3.4 Bolt tightening procedure

The sequential flow diagram of segment bolt tightening work is presented in Fig. 5. The entire procedure except the transfer of the support cart is full automatically done by the robot.

1) Moving the support cart
The support cart is manually moved to at the location of a target segment ring to be tightened.
2) Rough positioning of the manipulators
The manipulators are moved to the bolt tightening position through sequential numerical control.
3) Bolt box sensing
The bolt tightener is inserted into the center of the bolt box which is previously detected by the image processing function using a television camera.
4) Bolt sensing
After the tightener is inserted into the bolt box, the distance between the center of the bolt box and the inner wall of the box is measured by an ultrasonic sensor attached to the tip of the tightener, so the position of a bolt is relatively detected.
5) Bolt gripping
The gripper of the tighter is passively positioned to maintain co-axial with the bolt when it grips the head of a bolt.
6) Bolt tightening
Bolt tightening is done with the desired torque using a nut runner type bolt tightener.

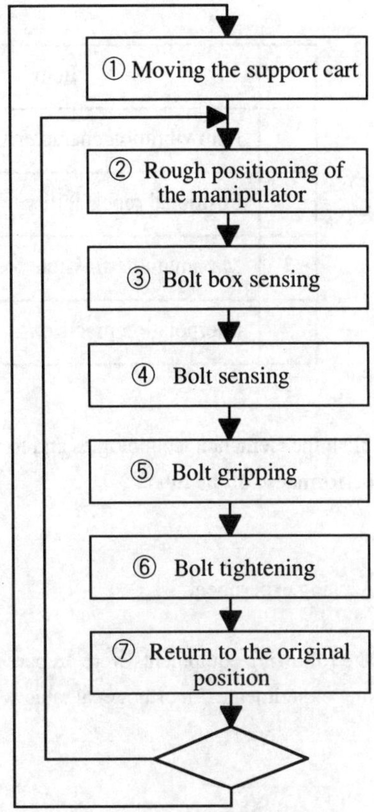

Fig. 5. Operation flowchart

7) After tightening the bolt, the bolt tightener releases its grip of the bolt and returns to the original position.

4 DEVELOPMENT OF ELEMENTAL TECHNOLOGIES AND VALIDATION

The following were the prerequisites for the development of the robot.
1. The establishment of a sensing system that enables the accurate detection of bolts.
2. The development of a manipulator capable of accurate positioning while supporting the weight of the bolt tightener and tightening reactive force.

The following experiments were carried out to

Table 2. Manipulator experimental result

	Item	Main Result
1	Arm vibration characteristic	Natural frequency 4.3 Hz Residual vibration 0.5 mm
2	Positional repeatability	Within +/- 0.1 mm
3	Maximum permissible load capacity	Max. 970 kgf, 18.8 sec.
4	Interpolation precision	Deviation 1.1 mm (Arc interpolation)

establish the elemental technologies and to validate the performance of the robot.

4.1 Sensing experiment

The following equipments were prepared for the sensing experiment. An orthogonal table with 3 degrees of freedom (X, Y and Z directions) is set-up over a model of the bolt box, and a bolt gripper, television camera and ultrasonic sensor are mounted at the end of the Z-axis (vertical axis).

Through the experiment, the sensing system was established to combine an image processing technology and an ultrasonic sensor.

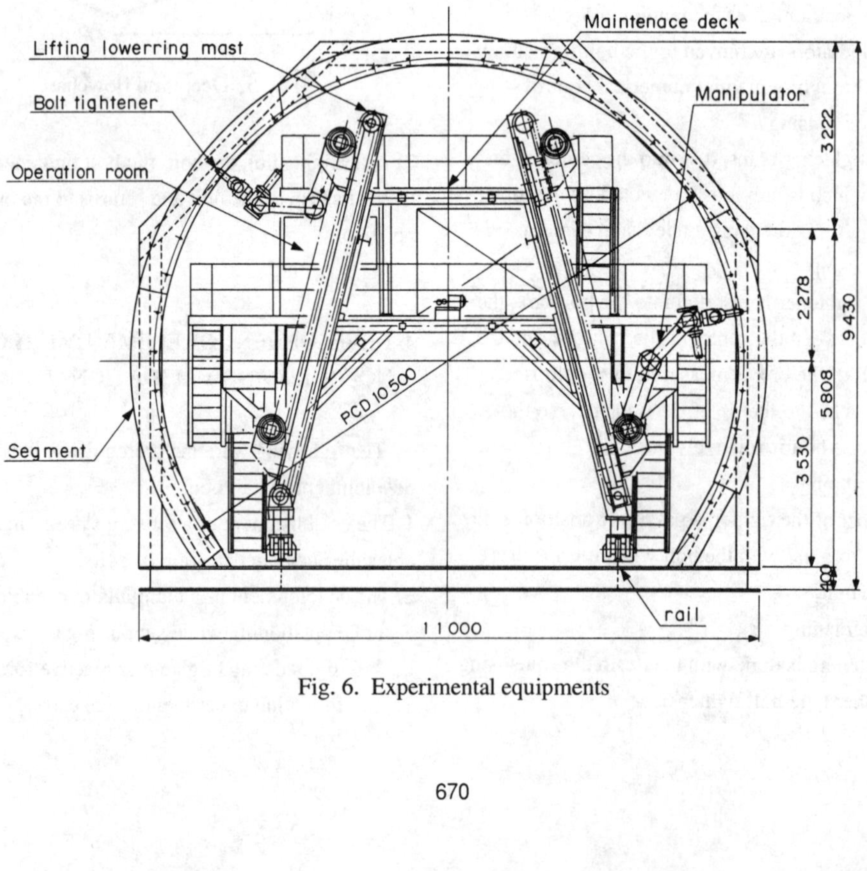

Fig. 6. Experimental equipments

Photo 1. Experimental equipments

4.2 Manipulator experiments (full-scale)

The static and dynamic characteristics of the manipulator were investigated using a full-scale manipulator.

The main items and test results are shown in Table 2. Even though the manipulator is designed to handle heavy load, it was confirmed that the system has high precision with a positioning repeatability of $\pm$ 0.1 mm and sufficient performance as expected.

As for dynamic characteristics, it was evaluated that residual vibration of the arm tip can be suppressed by appropriately controlling the arm speed patterns, and the system has satisfactory characteristics for use in heavy load handling.

It was also proved that the deviation from the track of the arm tip can be controlled within approximately 1.1 mm during the interpolation movement (circular or linear) by employing high-speed motion controller.

Photo 2. Manipulator

4.3 Bolt tightening experiment

Prior to the application to a project, the experiments were carried out at the factory to confirm the functions and performance of the bolt tightening robot, as well as to verify the software by automatically tightening full-scale segment rings. Photo. 1 and 2 show the experiment at the factory.

The experiment results indicated that 60 bolts could be tightened in an hour by a single manipulator.

5 CONCLUSIONS

The newly developed manipulator for handling heavy load is a multi-purpose industrial robot intended for the automation of incidental works involved in shield tunneling. Bolt tightening was chosen as the first target of its application. The segment tightening robot that consists of two manipulators and a self-advancing cart is planned to be implemented in the autumn of 1993 for slurry shield tunneling (length of 1,184 m) with an external diameter of 11.52 m.

It is assumed that the automatization of construction works and the reduction of labor will be increasingly important not only in Japan but in the rest of the world. In view of this situation, the development of such multi-purpose robots will be urgently requested to be developed instead of single-purpose robots.

It is undeniable that the development of robots for constructional purposes has been far behind that of industrial robots. This is due to the fact that robots for constructional purposes must meet the following prerequisites which are not required of industrial robots.

(1) Applicability to a wide range of works
(2) Capability of handling large and heavy load
(3) Capability of self-movement

We would like to continue to strive for the development and implementation of robots that meet various constructional needs to contribute to the automation of construction works.

Tunnelling and Ground Conditions, Abdel Salam (ed.) © 1994 Balkema, Rotterdam, ISBN 90 5410 363 9

Steel fibre reinforced concrete (SFRC) for tunnel linings: A technical approach

D. Moyson
N.V. Bekaert, Belgium

ABSTRACT: Tunnel linings are loaded under the combined action of normal forces and bending moments. In this paper it is shown that the bearing capacity of plain concrete can substantially be enhanced by the addition of steel fibres. The steel fibres provide the concrete an improved bending behaviour resulting in a better bearing capacity, especially for large eccentricities. This improved behaviour together with the practical advantages of the use of steel fibre reinforced concrete (SFRC) makes this composite material particularly interesting for tunnel linings.

1 INTRODUCTION

SFRC is not a new material in the tunnelling world. For several years, steel fibres are commonly used for shotcrete, having several advantages (ref. 1):

1. Technical: the addition of steel fibres provides the shotcrete an important degree of ductility. This ductility results in an improved loading capacity of steel fibre reinforced shotcreted (SFRS) shells compared to plain shotcreted ones. The technical performances of SFRS shells are comparable to conventional reinforced shotcreted shells.

2. Practical: the steel fibres are easy to add to the concrete aggregates. In comparison to conventional mesh a lot of time and labour can be saved.

3. Cost: the shorter execution time and savings in labour costs makes SFRS a competitive alternative.

Since shortly, an increasing growth of SFRC has also been established in the domain of prefabricated tunnel segments and tunnel linings casted in situ. The same practical and technical advantages as for shotcrete are here obtained.

In what follows the bearing capacity of steel fibre reinforced tunnel linings is discussed. In paragraph 2, the toughness ratio of SFRC will be defined. This toughness ratio is the essential property of SFRC and represents the effect of the fibre reinforcement. Based upon the toughness ratio and the stress-strain behaviour of plain concrete, the stress-strain behaviour of SFRC is determined in paragraph 3. By integration of the different stress distribution lines, corresponding to the different considered deformation lines, the bearing capacity in terms of bending moments and normal forces are calculated for a specific cross section. This is done in paragraph 4. In paragraph 5, the design of a SFRC tunnel lining is established for a specific load case according to the method of Bugieva.

2 THE TOUGHNESS OF SFRC

The basic properties of the composite material steel fibre reinforced concrete (SFRC) are described in an increasing number of normative publications. In general they all characterise SFRC by two groups of parameters : flexural strength and flexural toughness.

Flexural strength is expressed at first crack f_o, this is when the pure elastic limit of the material is reached. From this point on, the stress-strain relationship is not linear. A second strength parameter is the modulus of rupture or the ultimate flexural strength f_u. Ultimate flexural strength and flexural strength at first crack are not different for plain concrete nor for SFRC with fibre contents up to 40 kg/m3.

Flexural strength at first crack depends on concrete characteristics and is not influenced by fibre content. Ultimate flexural strength may increase compared to plain concrete at fibre contents ≥ 40 kg/m3, depending mainly on the length/diameter ratio or aspect ratio of the fibre. An increase of ca. 15 % is achieved using 40 kg/m3 of Dramix® ZC 50./50 fibre. The Dramix® ZC 50/.50 fibre is a steel wire drawn fibre with a length of 50 mm and a diameter of .50 mm and is hooked at both ends, see figure 1.

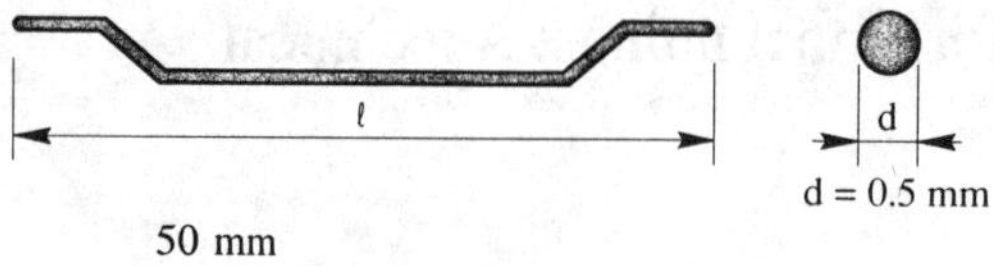

figure 1: Dramix® ZC 50/.50

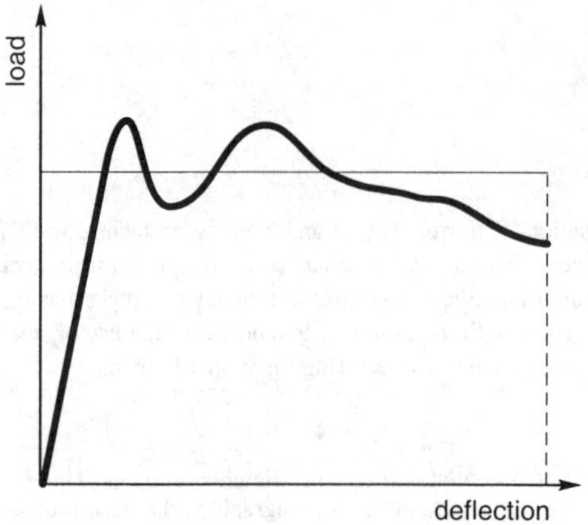

figure 2: Typical load-deflection curve for SFRC

The characteristics of SFRC are measured by means of third-point loading tests, where the beam parameters are fixed in each standard or recommendation. Figure 2 shows a typical load-deflection curve for a SFRC beam reinforced with steel wire drawn fibres with hooked ends.

Flexural toughness is expressed in very different ways. We find definitions of parameters that express absolute toughness, as equivalent flexural strength f_e. This absolute toughness is the average post-crack strength measured on a test beam under bending up to a defined deflection.

Relative toughness R_e can be expressed by relating the average post-crack strength to the flexural strength at first crack, thus creating a toughness rating expressing against what percentage of loading still can be resisted after concrete failure.

In this article we will use the flexural strength and the toughness rating according to the Japanese standards (ref. 2). The average post crack strength or equivalent flexural strength f_e is calculated elastically by dividing the total absorbed energy D during the test (till a deflection of 1/150 of the span length) by the deflection δ, taking into account the dimensions of the beam (1).

$$f_e = \frac{D.l}{\delta.b.h^2} \quad (N/mm^2) \quad (1)$$

where

D = total absorbed energy during the test (Nmm)
l = span length (mm)
δ = maximal deflection (mm)
b = width of the beam (mm)
h = height of the beam (mm)

The equivalent strength f_e is the essential value which characterises the effect of the steel fibre reinforcement. Research has shown that the equivalent flexural strength, measured on beams according to the Japanese standards, varies as a function of the concrete flexural strength. On the other hand, it has been shown that the relative toughness R_e, for concrete qualities up to 50 N/mm2, remains constant and is only dependent on the fibre type and dosage.

The relative toughness R_e can be expressed as a percentage of the first crack flexural strength f_o of the plain concrete (2).

$$R_e = 100 \frac{f_e}{f_o} \quad (\%) \quad (2)$$

R_u is defined as the ratio between the ultimate flexural strength of SFRC and the first crack strength of the plain concrete (3).

$$R_u = \frac{f_u}{f_o} \quad (3)$$

In the figure 3, the toughness ratio R_e as well as the modulus of rupture ratio R_u are expressed as a function of the dosage for the Dramix® ZC 50/.50. The relation is based upon a large amount of tests performed by Bekaert and extern laboratories (ref. 3).

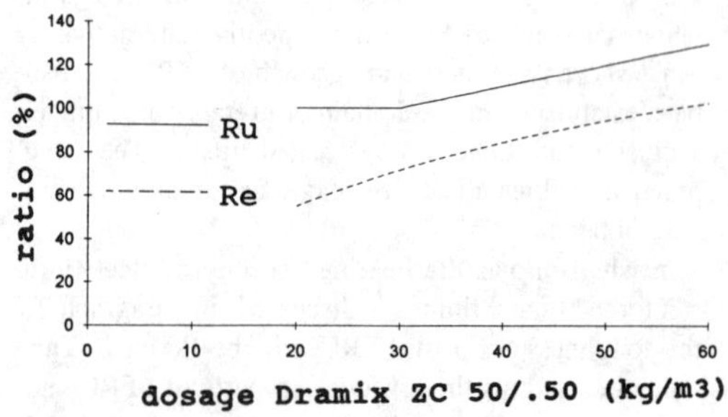

figure 3: R_e and R_u as a function of the dosage of Dramix® ZC 50/.50

From the figure one can see that for a dosage of 50 kg/m3 of Dramix® ZC 50/.50 fibres, the ultimate strength of the concrete is increased by 20 %, while the equivalent strength is 94 % of the first crack strength of the concrete.

Based on these R_e values, the equivalent flexural strength values for all concretes up to 50 N/mm2, and all fibres types and dosages can easily be calculated. Example : Is the concrete giving a 5,0 MPa flexural strength, than the design value of this concrete with steel fibres giving a rating of 90 % will be 4,5 MPa. These 90 % is achieved (according to the diagram) using 45 kg/m3 of Dramix® ZC 50/.50.

3 STRESS-STRAIN RELATIONSHIP FOR SFRC

The equivalent flexural strength f_e, which is the average post crack flexural strength of the SFRC, is deduced from the total absorbed energy (1) assuming a linear variation of the stresses over the height of the element as shown in figure 4a.

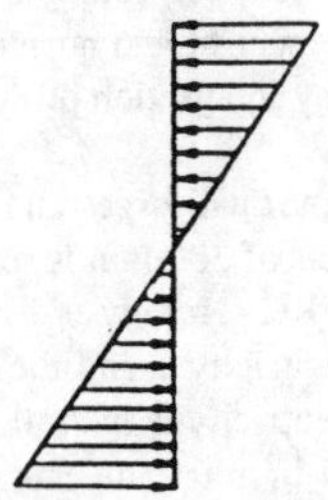

figure 4a: elastic stress distribution

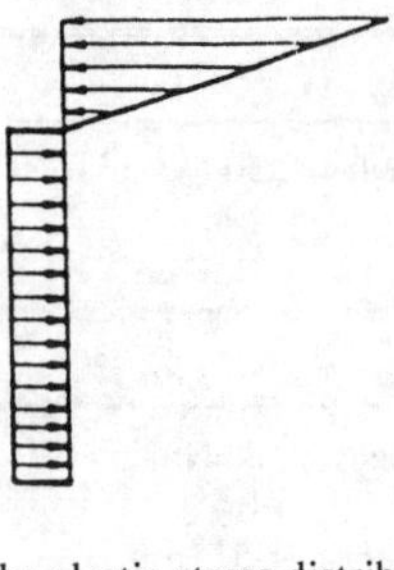

figure 4b: plastic stress distribution

However, in reality the SFRC does not behave in the elastic uncracked state, but the concrete is cracked and the stresses are no more linearly distributed across the cross-section. The plastic stress distribution along the cross-section, corresponding to the average post crack resistance of the test specimen, can simplified be drawn as in figure 4b.

The compression zone can clearly be distinguished from the tensile zone. The width of the compression zone is approximately 0.1 times the width of the tensile zone. The compressive stresses vary linearly, while the tensile stresses are constant. The tensile stress should be considered as a post crack tensile strength or equivalent tensile strength f_{et} provided by the effect of the fibre reinforcement.

Using figures 4a and 4b as a basis, the value of the post-crack tensile strength f_{et} can be determined under the condition that the flexural moment capacity is the same for the plastic as for the elastic stress distribution (4).

$$f_{et} \cdot b \cdot h \cdot 0{,}9 \cdot 0{,}5 \cdot h = f_e \cdot \frac{b \cdot h^2}{6}$$

$$f_{et} = 0{,}37 \cdot f_e \qquad (4)$$

According to (2) the equivalent tensile strength can also be written as a function of the ultimate flexural strength of the plain concrete and the toughness ratio R_e. (5)

$$f_{et} = 0.37 \cdot \frac{R_e}{100} \cdot f_0 \qquad (5)$$

This equivalent tensile strength characterises the effect of the fibre reinforcement. This value is essential, since it represents the tensile strength that can be taken up by the concrete thanks to the addition of steel fibres. Research (ref. 4) has shown that this average strength can effectively be provided by the steel fibres for deformations up to 10 %o.

The addition of steel fibres does not only improve the tensile behaviour of the concrete, the behaviour under compression is also positively modified compared to plain concrete. Although the steel fibres does not enhance the compressive strength of the plain concrete, they do improve the maximal strain under compression. Thanks to the addition of steel fibres, the behaviour of the concrete under compression is modified from a brittle behaviour into a ductile behaviour, providing a larger strain capacity. In (ref. 4) a maximum strain of 5 %o in compression is considered for SFRC, instead of 3.5 %o for plain concrete.

Based upon the above mentioned characteristics for SFRC, the stress-strain behaviour of plain concrete can now be modified. In figure 5, the characteristic stress-strain behaviour for the plain concrete is drawn.

The characteristic stress-strain behaviour is conform to the actual standards.

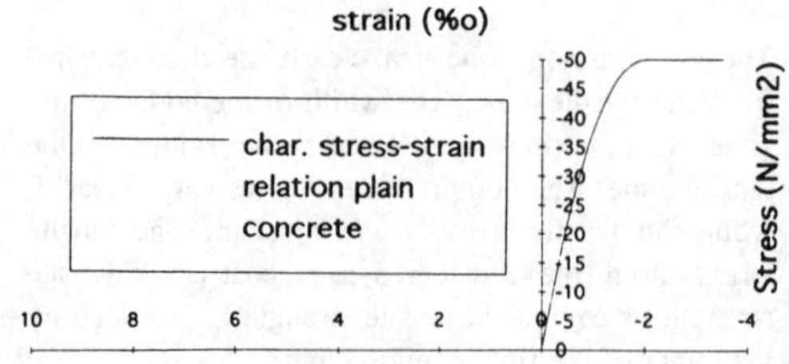

figure 5: Characteristic stress-strain behaviour for plain concrete

The characteristic stress-strain behaviour for SFRC is drawn in figure 6.

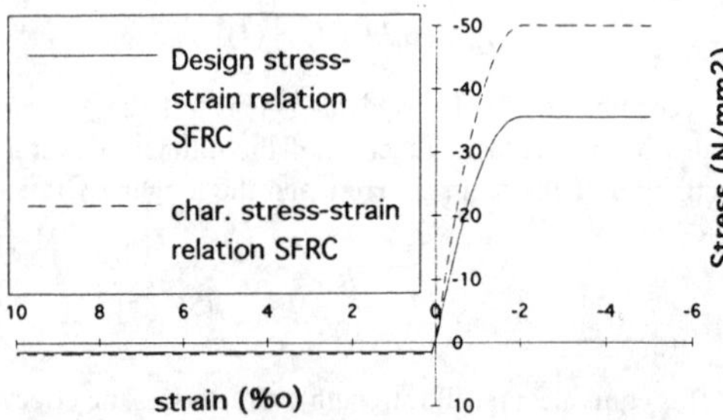

figure 6: Characteristic and design stress-strain behaviour for SFRC

The maximum tensile strength is equal to the equivalent tensile strength f_{et} as defined above and is maintained till a deformation of 10 %o. The stress-strain relation in the elastic range is the same as in compression, apart of the sign. The stress-strain behaviour in compression is equal to the one for plain concrete, except for the maximum strain which is 5 %o instead of 3.5 %o.

In figure 6 the design stress-strain behaviour for SFRC is also drawn. The design stress-strain behaviour can be deduced from the characteristic stress-strain curves by taking into account a safety factor in tension and in compression. In compression a safety factor γ_c of 1.4 is taken. In tension a safety factor γ_t is taken which is a function of the toughness ratio. For a 100 % are more tough material a safety factor of 1.15 is considered, this is the same value as for steel. For a 0 % tough material a safety factor of 1.4 is considered as for plain concrete (6).

$$\gamma_t = \min\left(1 \ ; \ 1{,}4 + (1{,}15 - 1{,}4) \cdot \frac{R_e}{100} \right) \qquad (6)$$

4 THE BEARING CAPACITY OF SFRC

In figure 7 the different deformation conditions for SFRC, which should be considered according to the plastic theory, are drawn.

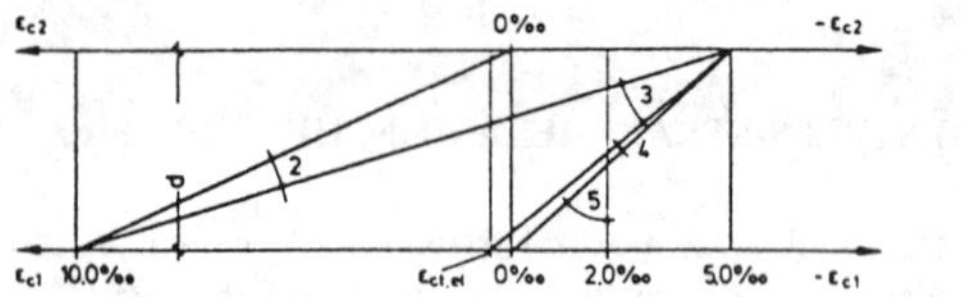

figure 7: Deformation lines for SFRC

To each deformation line corresponds a specific stress distribution. For each deformation line, the corresponding bearing moment M and normal force N can easily be determined by integration of the stresses over the cross-section.

In figure 8, the bearing normal forces and moments of a section with a thickness of 300 mm is represented for plain concrete and SFRC. Hereby e/d represents the ratio between the eccentricity e and the thickness d of the element. The eccentricity is the ratio between M and N and is expressed in mm. The characteristics for both concretes are exposed in table 1.

table 1: Characteristics of plain concrete and SFRC

plain concrete	
char. compressive strength	50 N/mm^2
safety factor	1.4
maximum strain	-3.5 %o
SFRC	
char. compressive strength	50 N/mm^2
safety factor	1.4
maximum strain	-5 %o
char. flexural strength	5 N/mm^2
50 kg/m3 Dramix ZC 50/.50	
toughness ratio R_e	94%
equiv. tensile strength	1.7 N/mm^2
safety factor	1.17
maximum strain	10 %o

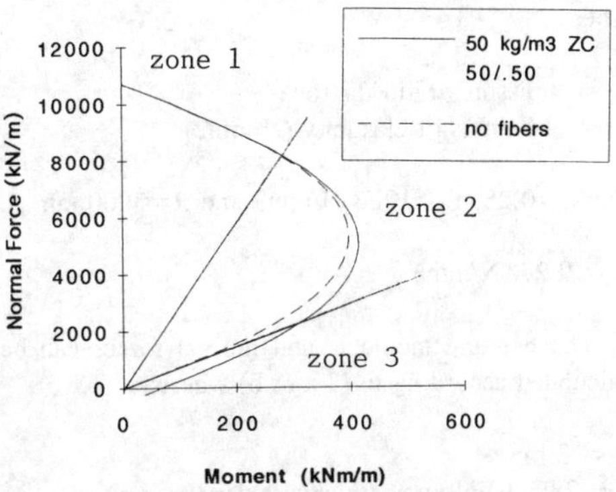

figure 8: Bearing capacity of a plain and SFRC section

From the figure one can distinguish three zones : zone 1, 2 and 3 which corresponds respectively to small, middle and large eccentricities.

For zone 1 the bearing capacity for SFRC is the same as for plain concrete. The whole section is loaded under compression and the maximum compression stress is not influenced by the addition of steelfibres.

In zone 2 an increasement in load bearing capacity can be seen by the addition of steelfibres. The increasement varies from a value of 0 % for the smallest eccentricity till a value of ∞ for the largest eccentricity, since for this eccentricity the plain concrete shows no bearing capacity. In zone 3 the bearing capacity of plain concrete remains 0, while for SFRC important forces can be taken up.

The increasement of loadbearing capacity by the additional steelfibres varies thus as a function of the eccentricity. For large eccentricities the largest enhancements are obtained.

5 DESIGN OF TUNNEL LININGS

The method which will be discussed here to calculate the forces acting on the segmental linings is the method of Bugieva (ref. 5). This method takes into consideration the interaction of the tunnel lining and the surrounding rock. The method developed by Bugieva is based on the analysis of the distribution of bending moments and normal forces in the lining which is supporting a uniform vertical load and is transferring the load onto the surrounding rock in an elastic manner. Figure 9 shows the assumed distribution of elastic reactions around the rock.

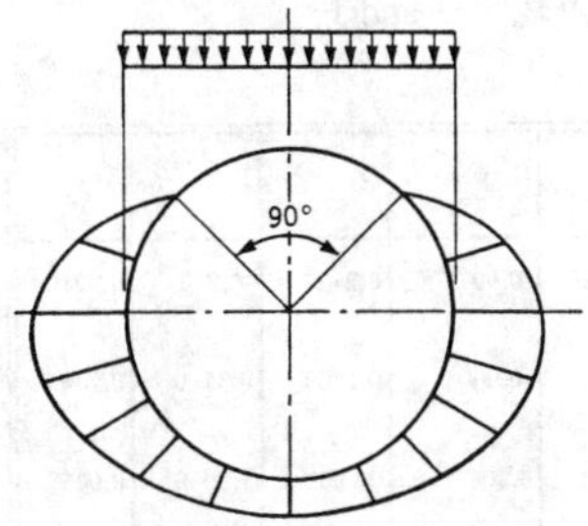

figure 9: Elastic reaction around the rock according to Bugieva

The calculation of the elastic reaction forces in the rock, which are due to the deformation of the lining, makes use of the coefficient of ground reaction. The bending moments and normal forces in a circular lining are the functions of the coefficient of ground reaction and the rigidity of the tunnel lining. An increase in the coefficient of ground reaction or in the flexibility of the tunnel lining will result in a decrease in the acting forces in the cross section of the tunnel.

The bending moments M and normal forces N acting on the cross section of a tunnel lining can be calculated by the following formulas of Bugieva (7 a & b).

$$M = p \cdot R^2 \cdot (A + B + 2 \cdot C \cdot n) \quad (N.mm) \quad (7a)$$

$$N = p \cdot R \cdot (D + F + 2 \cdot G \cdot n) \quad (N) \quad (7b)$$

where

p = uniform vertical load (N/mm2)
R = radius (mm)
I = moment of inertia (mm4)
E = elasticity modulus (N/mm2)
k = coefficient of ground reaction of surrounding rock (N/mm3)

and (8 a & b)

$$n = \frac{1}{m + 0{,}06416} \quad (8a)$$

$$m = \frac{E \cdot I}{R^4 \cdot k} \quad (8b)$$

The values for A,B,C,D and F are exposed in table 2.

table 2: Values for A,B,C,D and F

	A	B	C	D	F	G
a=0	0.01628	0.0872	-0.0070	0.2122	-0.2122	0.02100
Π/4	-0.0250	0.0250	-0.00084	0.1500	0.3500	0.1485
Π/2	-0.1250	-0.1250	0.00825	0.0000	1.0000	0.0057
3Π/4	0.0250	-0.0250	0.00022	-0.1500	0.9000	0.01380
Π	0.0872	0.1628	-0.00837	-0.2122	0.7122	0.02240

Figure 10 shows a typical distribution of the moments along the circumference.

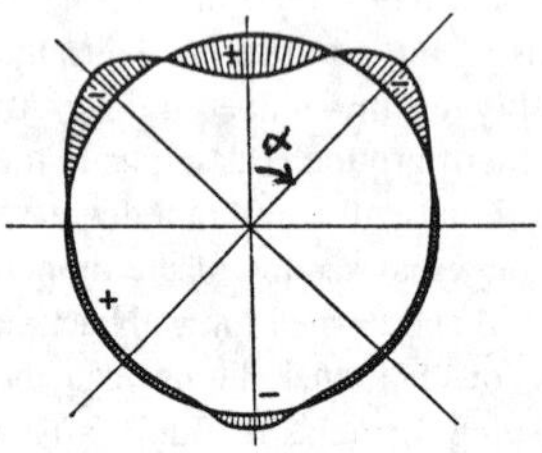

figure 10: Distribution of the moments along the circumference

Example

The maximal uniform vertical load p acting on the crown of the tunnel section for a very blocky and seamy rock is given by Terzaghi (ref. 6) (9) :

$$p = 2 \; . \; 1{,}1 \; . \; R \; . \; \gamma_r \quad (N/m2) \quad (9)$$

where

R = radius of the tunnel (m)
γ_r = unit weight of the rock (N/m3)

For R = 3 m, and for γ_r = 25 000 N/m3:
p = 165 000 N/m2.

The coefficient of ground reaction of the surrounding rock can be written as (10) :

$$k = \frac{E_r}{R\,(\,1 + \nu_r\,)} \quad (N/mm^3) \quad (10)$$

where

ν_r = Poisson's ratio for rock
E_r = Modulus of elasticity (N/mm2)

For ν_r=0.25, E_r=10E3 N/mm2 and R=3000 mm:

k = 0.27 N/mm3

The bending moments and normal forces can be calculated according to (7 a & b) where:

R = 3 m
I = 2.25 10E9 mm4 (thickness d=300 mm)
E = 20 000 N/mm2
k = 0.27 N/mm3

m = 0.002058
n = 15.11

The values for M, N and e/d are represented in table 3.

table 3: Acting moments and normal forces according to Bugieva

α	M (kNm/m)	N (kN)	e/d
α=0	57.2	314	0.61
Π/4	37.7	470	0.27
Π/4	1.2	581	0.01
3Π/4	9.8	578	0.06
Π	4.1	582	0.02

The different values of M and N are also represented in figure 11.
In figure 11, the bearing moments and normal forces for SFRC as well as for plain concrete are also represented.
From the figure one can see that the loads, acting on a section at the crown, exceeds the bearing capacity of the plain elements. The plain elements are thus not appropriate. The steel fibre reinforced elements on the other hand have a bearing capacity which is several times higher than the acting loads. In table 4 the ratio between the bearing capacity and the acting forces are exposed for the plain as well as for the SFRC

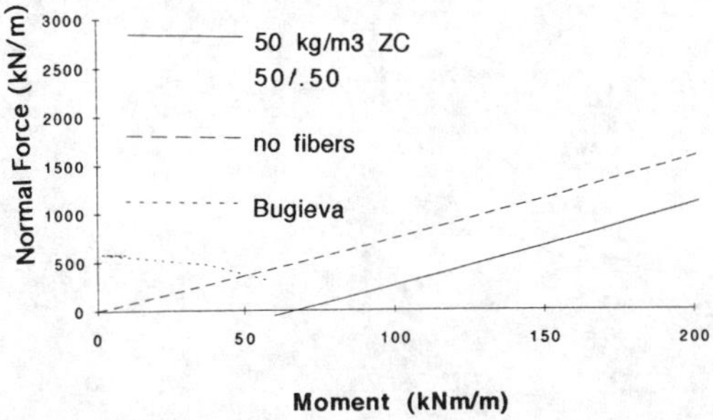

figure 11: Acting and bearing moments and normal forces

elements. From the table one can see that the bearing capacity of the SFRC elements amounts to 3.6 times the acting loads, for plain concrete that value is only 0.7. In other words by addition of 50 kg/m3 of Dramix® ZC 50/.50 steel fibres, the bearing capacity in the most critical section is improved by a factor of 5 (=3.6/0.7).

table 4: The ratio between the bearing capacity and the acting forces

α	plain concrete	SFRC
0	0.7	3.6
$\pi/4$	10.5	11.0
$\pi/2$	18.3	18.3
$3\pi/4$	17.4	17.4
π	14.1	14.1

6 CONCLUSIONS

The bearing capacity of plain concrete can importantly be enhanced by the addition of steel fibres. The steel fibres provide the concrete an improved bending behaviour resulting in a better bearing capacity especially for large eccentricities.

For a specific case it has been shown that the bearing capacity by addition of 50 kg/m3 of Dramix® ZC 50/.50 steel fibres, in the most critical section of the tunnel lining, is enhanced by a factor of 5.

This improved behaviour together with the practical advantages of the use of steel fibre reinforced concrete makes this composite material particularly interesting for tunnel linings.

REFERENCES

1. "Tunnelling the world", M. Vandewalle, 1990.
2. JSLE-SF 1 to 7 : "Method of tests for steel fibre reinforced concrete", Concrete Library of the Japanese Society of Civil Engineers, June 1984.
3. "Steel fibre concrete under flexural load", D. Nemegeer, 1993.
4. "Zur Qualitätsprüfung von Stahlfaserbeton für Tunnelschalen mit Biegezugbeanspruchung", J. Dietrich, Bochum, 1992.
5. "Traité de construction des tunnels", K. Szechy, 1970
6. "Handbuch des Tunnel und Stallenbaus", B. Maidl, 1988

Tunnelling and Ground Conditions, Abdel Salam (ed.) © 1994 Balkema, Rotterdam, ISBN 90 5410 363 9

Behavior and damage of a lined tunnel due to ground surface excavation

T. Yasuoka & T. Kaneda
Obayashi Corporation, Civil Engineering Technical Department, Japan

ABSTRACT: Bench blasting operations were carried out above a lined tunnel in a land reclamation site near Osaka in Japan. The 400 m long, 5.5 m wide tunnel was temporarily constructed for muck transportation by belt conveyors. The ground was excavated by 3 m-step bench blasting and the excavation ended at 1 m above the tunnel crown. Various parameters, such as convergence, induced particle velocities, were monitored during excavation. In this paper, the results of field monitoring as to various geotechnical parameters are presented. It was observed that no cracks appeared in concrete lining at particle velocities of less than 40 kines (cm/sec) and the lining was damaged in the velocity range of 40 to 100 kines. It was confirmed that the tunnel deformed elliptically towards the ground surface under the unloading conditions.

1 INTRODUCTIN

Recent construction works often involve ground surface excavation above a tunnel or blasting in the immediate vicinity of a tunnel. In those construction works, it is important to evaluate the influence of excavation or blasting on a nearby tunnel. A satisfactory theory, however, as to these circumstances is yet to be established, and only a very limited number of tests conducted at construction sites have been reported to date (Sakurai and Shimizu, 1990).

The authors conducted experimental bench blasting at a land reclamation site in Osaka, Japan, in the immediate vicinity of an existing tunnel and monitored the tunnel's behavior to investigate the influence of unloading and blasting on the tunnel. This paper reviews and analyzes the results of the blasting test.

2 OUTLINE OF FIELD MEASUREMENT

2.1 Bench Blasting

The blasting test was conducted at the site of a temporarily constructed mucking tunnel. The ground around the tunnel is a semihard rock formation called Izumi group consisting mainly of weathered sandstone, and has an uniaxial compresson strength of 100-150kgf/cm^2 and a elastic modulus of 150,000–300,000 kgf/cm^2. Fig. 1 shows a longitudinal section of the tunnel, and Fig. 2 illustrates the relationship between bench blasting and zones of excavation. The tunnel support consists mainly of shotcrete and rock bolts, and the lining is made of plain concrete.

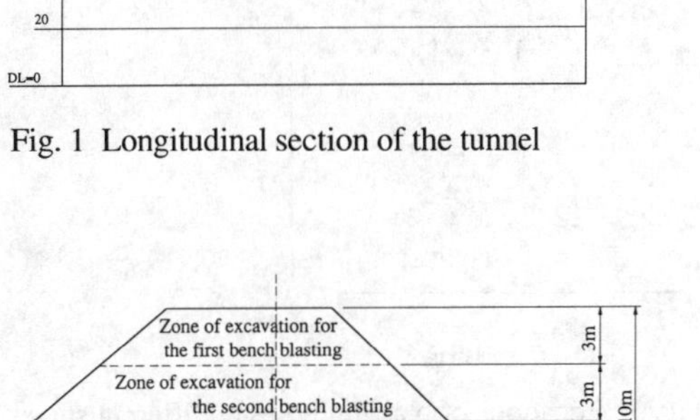

Fig. 1 Longitudinal section of the tunnel

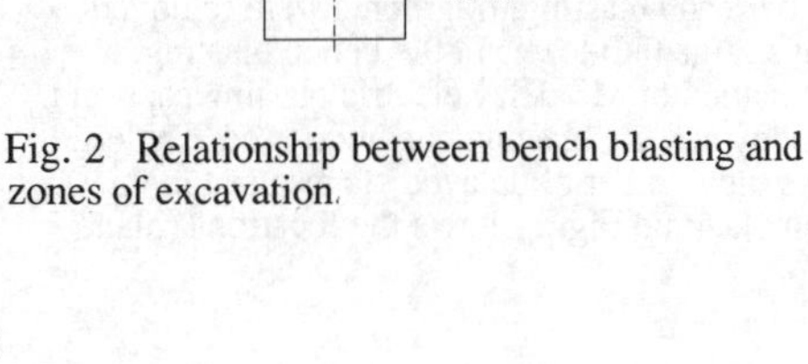

Fig. 2 Relationship between bench blasting and zones of excavation.

Table 1. Delayed stages of bench blasting

1st blasting		2nd blasting		3rd blasting		4th blasting	
Delayed stage	Num. of holes	Delayed stage	Num. of holes	Delayed stage	Num. of holes	Delayed stage	Num. of holes
DS-1	4	MS-8	5	DS-1	4	DS-1	1
DS-5	4	DS-2	5	DS-2	4	DS-5	1
DS-6	4	DS-3	5	DS-3	4	DS-9	1
DS-7	4	DS-4	5	DS-4	4	DS-16	1
DS-8	4	DS-5	5	DS-5	4		
DS-9	4	DS-6	5	DS-6	4		
DS-10	4	DS-7	5	DS-7	4		
DS-11	4	DS-8	5	DS-8	4		
DS-12	4	DS-9	5	DS-9	4		
DS-13	4	DS-10	5	DS-10	4		
DS-14	4	DS-11	5	DS-11	4		
DS-15	3	DS-12	5	DS-12	4		
DS-16	3			DS-13	4		
DS-17	3			DS-14	4		
				DS-15	4		
				DS-16	4		
				DS-17	3		

The weight of explosive per hole in the first to third blasting is 5.2kgf and that in the fourth blasting is 18.7kgf.

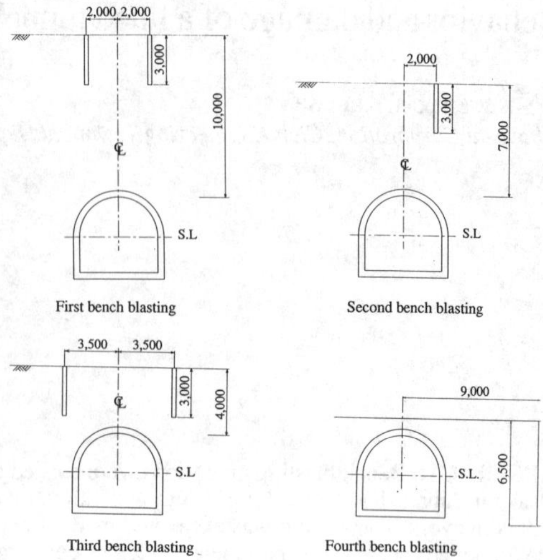

Fig. 4. Locations of instantaneous blasting holes at different stages of bench blasting (cross view)

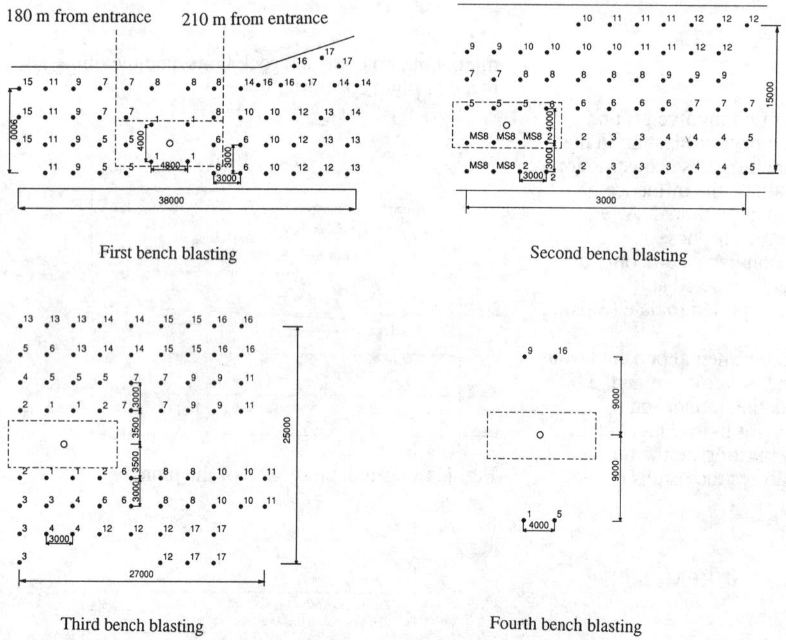

Fig. 3. Locations of blasting holes at different stages of bench blasting (plane view)

Bench blasting was carried out in 3 m steps, so that the overburden became 10 m, 7 m, 4 m, and then 1 m. After each blasting, fragments were removed using a large bulldozer. In the bench blasting, a combination of MS delay electric blasting caps and DS delay electric blasting caps was used. Table 1 shows the number of delayed stages used in the bench blasting. Fig. 3 shows the locations (plane view) of blast holes by delayed stage, and Fig. 4 shows the locations of instantaneous blast holes (cross section). The numbers in Fig. 3 show the number of delays used for the DS electric delay caps. The fourth bench blasting was carried out for the purpose of data acquisition, and unloading was not carried out after this blasting.

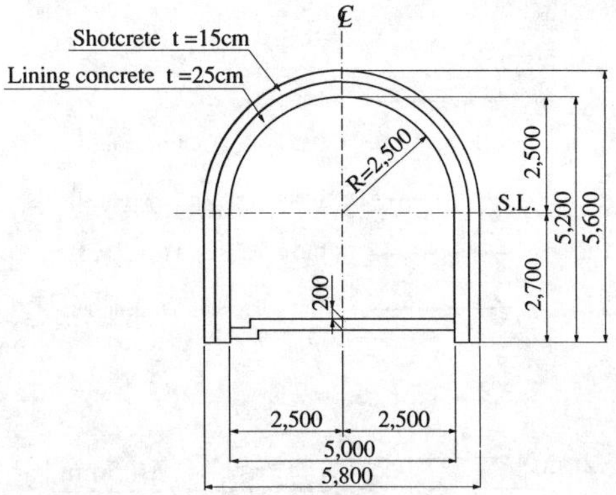

Fig.5 Standard cross section of tunnel support

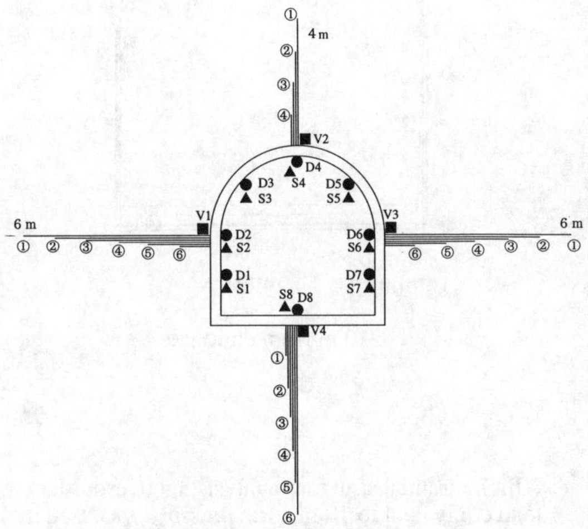

Symbol	Measurement Item	Measurement Point	Number of Points
●	Convergence	D1～D3、D5～D7	6
●	Settlement of crown , Uplift of invert	D4、D8	2
▲	Stress in lining concrete	S1～S8	8
■	Ground displacement	V1～V4	4

Fig. 6. Arrangement of measuring instruments (measurement related to blasting and unloading)

2.2 Measurement Items

The measurement items are divided into items regarding blasting and unloading and those regarding blasting. The former include convergence, settlement of the crown, uplift of the invert, ground displacement, stress in lining concrete, and damage of lining concrete, which were measured before and after each blasting and unloading operation. The latter include particle velocity and stress in lining concrete, which were measured during blasting. Fig. 5 shows a standard cross section of the tunnel support. Fig. 6 shows the arrangement of instruments for the measurements related to blasting and unloading, and Fig. 7 shows the arrangement of instruments for the measurements related to blasting. Table 2 shows the measurement cross section of the tunnel.

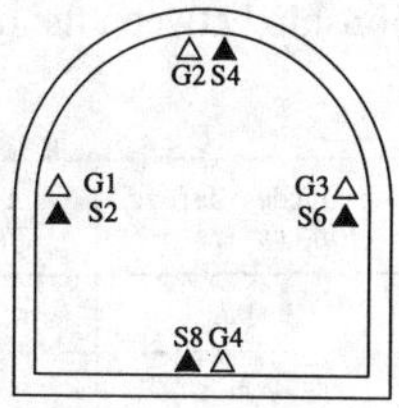

Symbol	Measurement Item	Measurement Point	Number of Points
△	Particle velocity	G1～G4	4
▲	Stress in lining concrete	S2～S8	4

Fig. 7. Arrangement of measuring instruments (measurement related to blasting)

Table 2. Measurement cross section of the tunnel

Measurment item	Measurment cross section
Convergence	180m and 210m from entrance
Settlement of crown	180m and 210m from entrance
Uplift of invert	180m and 210m from entrance
Stress in lining concrete	180m from entrance
Ground displacement	180m from entrance
Particle velocity	180m from entrance
Damage of lining concrete	174-186m from entrance

3 RESULTS

The results of field measurement for unloading and blasting operations are separately shown below.

3.1 Influence of unloading

The influence of unloading is evaluated in terms of convergence, crown settlement, invert uplift, ground displacement, stress in lining concrete, and observed damage of lining concrete. Table 3 shows the relationships between the overburden and the unloading steps.

Table 3. Relationship between the overburden and unloading steps

Unloading step	Ovreburden before unloading	Ovreburden after unloading
Step−1	10m	7m
Step−2	7m	4m
Step−3	4m	1m

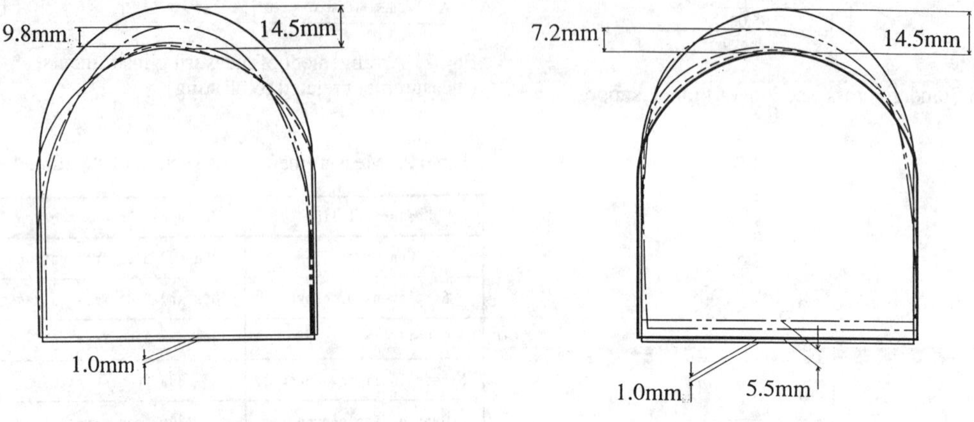

Fig. 8. Tendency of deformation at each unloading step

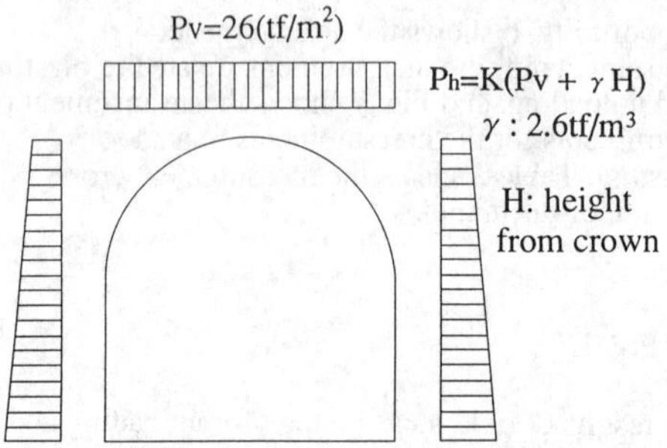

Fig. 9. Analytical model and loads acting on the tunnel

1) Measurement of convergence, crown settlement, and invert uplift

Fig. 8 shows the tendency of deformation at each unloading step. As shown, Fig. 8 compare the measured values with calculated values, which were obtained from numerical analysis using a frame structure model shown in Fig. 9.

In the framed structure analysis, an overburden load equivalent to 10 m earth pressure was used as the initial condition, and deformation at each unloading step was calculated considering unloading as a decrease in load. As for the coefficient of ground reaction, the value of the modulus of deformation obtained from a borehole loading test was used. In the analysis, the combined thickness of shotcrete and lining concrete (40 cm) was used as the support model.

Measured values indicate that most of deformation occurred during unloading steps 2 and 3. Measured values also show the tendency of deformation that the tunnel seems to have expanded upward with the invert remaining almost fixed. The measured values of upward movement at the tunnel crown and at the invert were 8.5 mm (mean of two sections) and 2.8 mm (mean of two sections) respectively, while the calculated values of upward movement at the same locations were 14.5 mm and –1.0 mm (i.e., downward movement).

2) Measurement of ground displacement

We evaluatred ground displacement in terms of displacement per measurement section

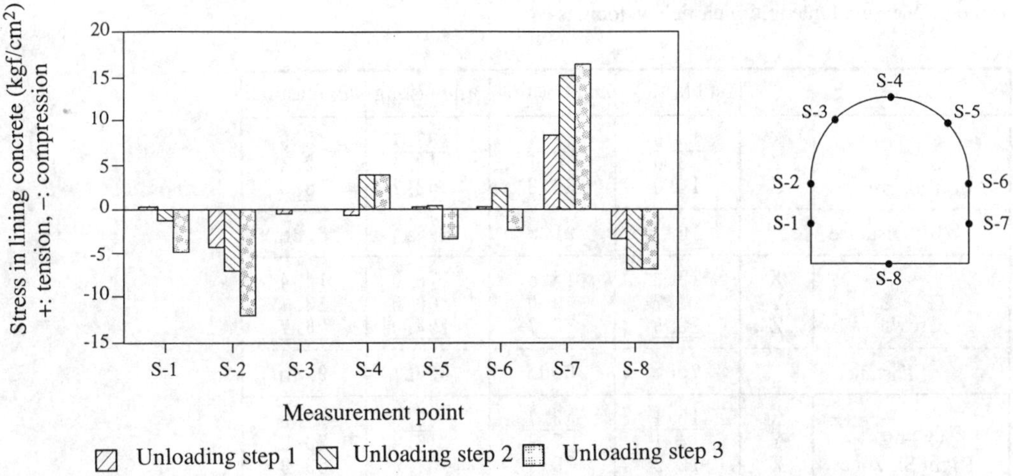

Fig. 10. Measured stresses in lining concrete

(displacement per meter), because it is difficult to determine a fixed point.

On the whole, ground displacement turned out to be very small, and the maximum sectional displacement was 0.6 mm (1–2 m section) in the direction of shrinkage, which occurred during unloading step 3 at the spring line level (right side). This value corresponds to a ground strain of 0.06%.

3) Stress in lining concrete

Stress in lining concrete was obtained using strain gauges installed on the concrete surface. Fig. 10 shows stresses obtained by multiplying measured strains by the modulus of elasticity of the lining concrete. As shown, the maximum stress on both the tension side and the compression side occurred during unloading step 3, and their values were 17 kgf/cm^2 and 13 kgf/cm^2 respectively. The general tendency of stress is such that tensile stress occurred on the right side wall and compressive stress in the left spring line zone.

4) Observed damage of lining concrete

The lining concrete was monitored before and after unloading operations, but the observations revealed little influence of unloading. What was observed here will be reported later in the section regarding the influence of blasting.

Findings from above results are as follows:

1. Measured values of convergence indicate that although the tunnel slightly deformed horizontally, its deformation was mostly upward expansion with the invert remaining motionless. The upward displacement of the crown measured in two cross sections averaged 8.5 mm, which is about 60% of the calculated value. Both the measured and calculated values of displacement at the invert were very small. It can be concluded, from these results, that calculated values agree fairly well with measured values.

2. The maximum tensile stress measured in the lining concrete was about 17 kgf/cm^2. This indicates, if the tensile strength is assumed to be one-tenth of the compressive strength, that a nearly critical level of tensile stress occurred locally. However, the fact that the occurrence or the growth of wide-ranging cracks was hardly observed indicates that even when the overburden decreases from 10 m to 1 m as in the case of this tunnel, the tunnel under the reduced overburden pressure can be deemed fairly stable though some cracks might occur locally.

3. With respect to the influence of unloading on the ground arch of an existing tunnel, it is generally said that arch action around a tunnel acts when the overburden is at least 1D (D: tunnel diameter). In the case of the tunnel considered in this paper, the overburden became less than the critical thickness of 1D during unloading step 2. The results of measurement, however, did not show any noticeable difference between the tendencies observed before and after the critical overburden. This will be partly because the blasting carried out in the immediate vicinity of the tunnel before excavation damaged the ground arch, thereby loosening ground around the tunnel.

Table 4. Measured maximum particle velocities

(unit: kine)

		1st blasting	2nd blasting	3rd blasting	4th blasting
G－1 Left SL zone	X	15.5	43.5	137.7	3.9
	Y	3.0	6.1	12.4	2.4
	Z	14.0	33.8	53.7	5.5
Distance		10.2m	9.2m	5.5m	6.8m
G－2 Crown	X	15.2	15.5	178.0	14.4
	Y	4.7	9.0	60.0	13.7
	Z	23.9	76.7	194.0	26.7
Distance		7.8m	6.0m	3.9m	9.4m
G－3 Right SL zone	X	13.1	43.7	90.0	7.3
	Y	4.1	7.9	16.3	20.5
	Z	12.2	28.1	57.9	59.3
Distance		10.2m	7.3m	5.5m	11.7m
G－4 Invert	X	4.1	5.2	19.3	14.3
	Y	3.5	2.7	9.3	12.0
	Z	14.4	10.4	24.7	29.0
Distance		12.2m	9.9m	6.9m	9.7m

The underlined numbers are the maximum values for the three directions (X, Y, Z).

3.2 Influence of blasting

The influence of blasting is evaluated in terms of particle velocity, stress in lining concrete, and observed damage in lining concrete.

1) Particle velocity

The particle velocity was calculated from data obtained from accelerometers installed at four locations: at the crown, in the right and left spring line zones, and at the invert. Table 4 shows the maximum value of particle velocity during each bench blasting. Most of the maximum values shown in the table resulted from instantaneous blasting. In Table 4, X-, Y- and Z-axis directions are the circumferential direction, the longitudinal direction, and the radial direction respectively. The distance between the blast point and the measurement point was about 12–4m. The frequency of stress-transmitting waves was 50–100 Hz.

The general tendency was that the particle velocity at all points except at the invert peaked during the third blasting. From the viewpoint of locations, particle velocity was maximized at the crown, and from the viewpoint of directions, particle velocity in the vertical direction was greater than that in the other two directions.

2) Stress in lining concrete

As in the case of the measured values of particle velocity, most of the maximum values occurred during instantaneous blasting, and the delayed stages at which the maximum values of particle velocity occurred agree mostly with those at which the maximum values of stress in lining concrete occurred.

Irrespective of the number of bench blasting operations, the maximum value of stress in the tension side remained one-digit numbers. The stress on the compression side showed large values at the second and third bench blastings. In particular, the stress on the compression side during the second bench blasting exceeded 100 kgf/cm^2.

3) Observed damage in lining concrete

Fig. 9 shows observed damage in the lining concrete. Since the observation was made by the eye, mainly 0.5–1 mm or wider cracks were searched.

It was after the third bench blasting that the occurrence and growth of cracks were clearly observed. Crack widths increased at the fourth of bench blasting,and the maximum opening about 3mm was observed.

The Influence of blasting revealed the following:

1. Eye observation of cracks revealed that 1 mm or narrower cracks occurred in the lining concrete during the third bench blasting, and the critical level

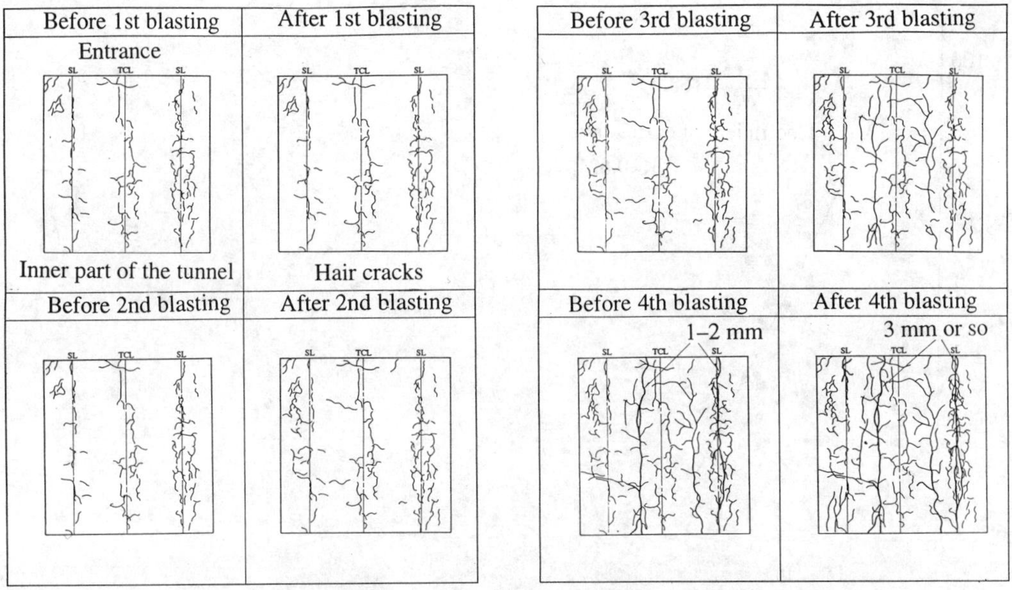

Fig. 11. Observed damage of lining concrete

of particle velocity causing visible cracking is 38–97 kines (the vertical direction at the crown). Although cracks measuring 1–3 mm in width were found after the third and fourth bench blasting, at this stage there occurred no serious failure, such as a toppling of the side wall or a falling of concrete masses. It can be concluded therefore that serious failure such as a toppling of the side wall occurs at 97 kines or more.

2. Occurrence of stress in the lining concrete was studied, assuming the critical tensile stress to be 18 kgf/cm^2 (1/10 of compressive strength). In the left spring line zone, the tensile stress measured around 9 kgf/cm^2 during the third bench blasting. These results and the results of the particle velocity measurement indicate that nearly critical levels of tensile stress occurred even when blasting level was low.

3. For locations considerably removed (usually 200–300 m) from the blast point, the following formula for estimating particle velocity is used

$$V(kine) = K*W^{a}*D^{-b} \quad (1)$$

where

V : particle velocity (kine)

K : constant based on the type of explosive used, ground conditions, and blasting conditions

W : weight of explosive (kgf)

D : distance from blast point to object (measurement point) (m)

a, b : coefficients

Given below is a proposed formula for estimating the particle velocity in the immediate vicinity of a tunnel based on the above formula. Power regression analysis is used for approximation. Since the weight of the explosive used in the test does not vary widely, an empirical value of the constant, a, representing the power of the weight of the explosive is used here (Japan Tunneling Association, 1977). The analysis revealed the relationship between D and ($V/W^{3/4}$) shown in Fig. 12.

4 CONCLUSIONS

The conclusions that can drawn from the test results can be summarized as follows:

1. It was confirmed that when the overburden of a tunnel having a diameter of 5.8 m was reduced to 1 m, the tunnel deformed in such a way that it expanded upward. The uplift of the crown was removed was 8–9 mm. Calculated values obtained including the displacement of the crown agree fairly well with measured values.

2. Unloading caused nearly critical levels of local tensile stress in the lining concrete. Since, however, neither the occurrence of new cracks nor the growth of existing ones was observed, the tunnel can be considered to be generally stable.

3. The critical overburden of the ground arch was studied through the unloading of the tunnel, but the measurement data obtained did not exhibit any pronounced tendency of displacement or stress that can be used to determine the critical overburden. One of the reasons for this is probably that close-range blasting carried out during excavation

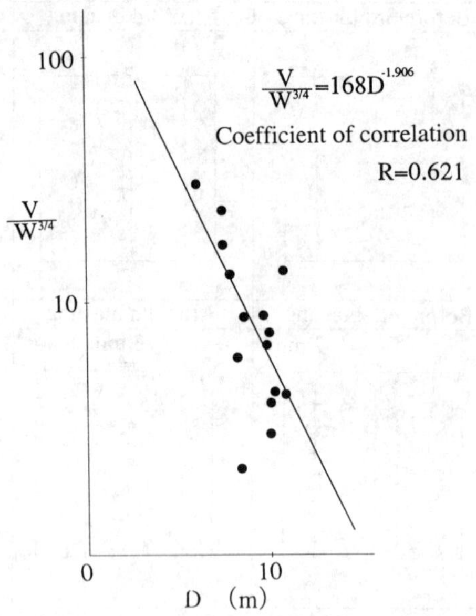

Fig. 12. Approximation of the relationship between D and $(V/W^{3/4})$

damaged the ground arch.

4. With respect to the degree of damage to the lining concrete due to blasting, the lowest particle velocity that causes cracks as large as are visible to the naked eye was 38–97 kines. Serious failure, such as a falling of concrete masses and a toppling of the side wall, occurs at 97 kines or more.

5. A conventional formula for estimating praticle velocity was used to derive a formula applicable to close-range (distance between the blast point and the object: 12–4 m) blasting like the blastings carried out in the test. As a result, $(V/W^{3/4})=168*D^{-1.906}$ (coefficient of correlation R=0.621) was obtained.

5 REFERENCES

Japan Tunneling Association 1977. A Survey Report on Measures to Control Vibration and Noise Caused by Blasting in Tunnel Excavation, p. 20–24.

Sakurai, S. & N. Shimizu 1990. Deformational Behavior of Existing Tunnels Due to Ground Surface Excavations, *Proceedings of 8th Japan Symposium on Rock Mechanics*.

Tunnelling and Ground Conditions, Abdel Salam (ed.) © 1994 Balkema, Rotterdam, ISBN 90 5410 363 9

Author index